Greenland
(Kalatdlit-Nunat)
Alaska
Yukon
N.W.T.
Nunavut
[Labrador]
B.C.
Alta.
Sask.
Man.
Ont.
Que.
Nfld.
St.Pierre
and Miquelon
P.E.I.
N.B.
N.S.
Vt.
Maine
N.H.
Mass.
R.I.
Conn.
N.J.
Del.
Md.
Wash.
Mont.
N.Dak.
Minn.
Oreg.
Idaho
S.Dak.
Wis.
Mich.
N.Y.
Wyo.
Iowa
Pa.
Nev.
Nebr.
Ohio
Utah
Colo.
Ill.
Ind.
W.Va.
Va.
Kans.
Mo.
Ky.
Calif.
N.C.
Tenn.
Okla.
Ark.
S.C.
Ariz.
N.Mex.
Ala.
Ga.
Miss.
Tex.
La.
Fla.
Mexico

Flora of North America

Contributors to Volume 12

William R. Anderson†
W. Scott Armbruster
Peter W. Ball
Paul E. Berry
David E. Boufford
Joshua M. Brokaw
N. Ivalú Cacho
Bijan Dehgan
J. Arturo De-Nova
Craig C. Freeman
Dmitry V. Geltman
Lynn J. Gillespie
W. John Hayden
James Henrickson
Petra Hoffmann
Larry D. Hufford
Michael J. Huft
Donald H. Les
Geoffrey A. Levin
Jin-shuang Ma
Mark Mayfield
Ronald L. McGregor†
Michael O. Moore†
Jeffery J. Morawetz
Nancy R. Morin
Zack E. Murrell
Guy L. Nesom
Daniel L. Nickrent
Christopher F. Nixon†
Tracy J. Park
Jess A. Peirson
Derick B. Poindexter
Duncan M. Porter
Yocupitzia Ramírez-Amezcua
Hamid Razifard
Ricarda Riina
Dayle E. Saar
John O. Sawyer†
John J. Schenk
Clifford L. Schmidt†
Beryl B. Simpson
Victor W. Steinmann
Gordon C. Tucker
Roberto J. Urtecho
Benjamin W. van Ee
Michael A. Vincent
Maria S. Vorontsova
Alan S. Weakley
Jun Wen
R. David Whetstone
Dieter H. Wilken
Kenneth J. Wurdack
Ya Yang

Editors for Volume 12

David E. Boufford
Taxon Editor for Cervantesiaceae, Comandraceae, Santalaceae, Schoepfiaceae, Thesiaceae, Viscaceae, and Ximeniaceae, and Co-Taxon Editor for Celastraceae

Luc Brouillet
Taxon Editor for Chrysobalanaceae

Kanchi Gandhi
Nomenclatural Editor

David E. Giblin
Taxon Editor for Linaceae

Lynn Gillespie
Co-Lead Editor, Taxon Editor for Phyllanthaceae, Picrodendraceae, Putranjivaceae, Rhizophoraceae, and Vitaceae, and Co-Taxon Editor for Euphorbiaceae

Cassandra L. Howard
Assisting Technical Editor

Robert W. Kiger
Bibliographic Editor and Taxon Editor for Elatinaceae

Geoffrey A. Levin
Co-Lead Editor, Taxon Editor for Cornaceae, Eucommiaceae, Garryaceae, Loasaceae, Malpighiaceae, Nyssaceae, Simmondsiaceae, and Zygophyllaceae, and Co-Taxon Editor for Euphorbiaceae

Jackie M. Poole
Taxon Editor for Krameriaceae

Andrew C. Pryor
Technical Editor

Richard K. Rabeler
Taxon Editor for Hydrangeaceae

Heidi H. Schmidt
Managing Editor

John L. Strother
Reviewing Editor

Debra K. Trock
Taxon Editor for Rhamnaceae

Gordon C. Tucker
Taxon Editor for Oxalidaceae

Grady L. Webster†
Preliminary Taxon Editor for Euphorbiaceae, Phyllanthaceae, Picrodendraceae, and Putranjivaceae

Elizabeth F. Wells
Co-Taxon Editor for Celastraceae

James L. Zarucchi
Editorial Director

Volume 12 Composition

Kristin Pierce
Compositor and Editorial Assistant

Heidi H. Schmidt
Production Coordinator and Managing Editor

Euphorbia marginata

Flora of North America

North of Mexico

Edited by FLORA OF NORTH AMERICA EDITORIAL COMMITTEE

VOLUME 12

Magnoliophyta: Vitaceae to Garryaceae

NEW YORK OXFORD • OXFORD UNIVERSITY PRESS • 2016

Oxford University Press is a department of the University of Oxford.
It furthers the University's objective of excellence in research,
scholarship, and education by publishing worldwide.

Oxford New York
Auckland Cape Town Dar es Salaam Hong Kong Karachi Kuala Lumpur
Madrid Melbourne Mexico City Nairobi New Delhi Shanghai Taipei Toronto

With offices in
Argentina Austria Brazil Chile Czech Republic France Greece Guatemala Hungary Italy
Japan Poland Portugal Singapore South Korea Switzerland Thailand Turkey Ukraine Vietnam

Published by Oxford University Press, Inc.
198 Madison Avenue, New York, New York 10016
www.oup.com

Library of Congress Cataloging-in-Publication Data
(Revised for Volume 12)
Flora of North America North of Mexico
edited by Flora of North America Editorial Committee.
Includes bibliographical references and indexes.
Contents: v. 1. Introduction—v. 2. Pteridophytes and gymnosperms—
v. 3. Magnoliophyta: Magnoliidae and Hamamelidae—
v. 22. Magnoliophyta: Alismatidae, Arecidae, Commelinidae (in part), and Zingiberidae—
v. 26. Magnoliophyta: Liliidae: Liliales and Orchidales—
v. 23. Magnoliophyta: Commelinidae (in part): Cyperaceae—
v. 25. Magnoliophyta: Commelinidae (in part): Poaceae, part 2—
v. 4. Magnoliophyta: Caryophyllidae (in part): part 1—
v. 5. Magnoliophyta: Caryophyllidae (in part): part 2—
v. 19, 20, 21. Magnoliophyta: Asteridae (in part): Asteraceae, parts 1–3—
v. 24. Magnoliophyta: Commelinidae (in part): Poaceae, part 1—
v. 27. Bryophyta, part 1—
v. 8. Magnoliophyta: Paeoniaceae to Ericaceae—
v. 7. Magnoliophyta: Salicaceae to Brassicaceae—
v. 28. Bryophyta, part 2—
v. 9. Magnoliophyta: Picramniaceae to Rosaceae—
v. 6. Magnoliophyta: Cucurbitaceae to Droseraceae—
v. 12. Magnoliophyta: Vitaceae to Garryaceae

ISBN: 978-0-19-064372-0 (v. 12)
1. Botany—North America.
2. Botany—United States.
3. Botany—Canada.
I. Flora of North America Editorial Committee.
QK110.F55 2002 581.97 92-30459

1 2 3 4 5 6 7 8 9
Printed in the United States of America on acid-free paper

Contents

Courtesy of Hunt Institute for Botanical Documentation,
Carnegie Mellon University, Pittsburgh, Pennsylvania.
Photograph by Walter Henricks Hodge.

Courtesy of Hunt Institute for Botanical Documentation,
Carnegie Mellon University, Pittsburgh, Pennsylvania.
Photograph by Loraine Yeatts.

This volume is dedicated to the memory of our friend
and esteemed colleague
James Lauritz Reveal
1941 – 2015
eminent student of North American botany
and major contributor to the
Flora of North America project.

FOUNDING MEMBER INSTITUTIONS

Flora of North America Association

Agriculture and Agri-Food Canada
Ottawa, Ontario

Arnold Arboretum
Jamaica Plain, Massachusetts

Canadian Museum of Nature
Ottawa, Ontario

Carnegie Museum of Natural History
Pittsburgh, Pennsylvania

Field Museum of Natural History
Chicago, Illinois

Fish and Wildlife Service
United States Department of the Interior
Washington, D.C.

Harvard University Herbaria
Cambridge, Massachusetts

Hunt Institute for Botanical Documentation
Carnegie Mellon University
Pittsburgh, Pennsylvania

Jacksonville State University
Jacksonville, Alabama

Jardin Botanique de Montréal
Montréal, Québec

Kansas State University
Manhattan, Kansas

Missouri Botanical Garden
St. Louis, Missouri

New Mexico State University
Las Cruces, New Mexico

The New York Botanical Garden
Bronx, New York

New York State Museum
Albany, New York

Northern Kentucky University
Highland Heights, Kentucky

Université de Montréal
Montréal, Québec

University of Alaska
Fairbanks, Alaska

University of Alberta
Edmonton, Alberta

The University of British Columbia
Vancouver, British Columbia

University of California
Berkeley, California

University of California
Davis, California

University of Idaho
Moscow, Idaho

University of Illinois
Urbana-Champaign, Illinois

University of Iowa
Iowa City, Iowa

The University of Kansas
Lawrence, Kansas

University of Michigan
Ann Arbor, Michigan

University of Oklahoma
Norman, Oklahoma

University of Ottawa
Ottawa, Ontario

University of Southwestern Louisiana
Lafayette, Louisiana

The University of Texas
Austin, Texas

University of Western Ontario
London, Ontario

University of Wyoming
Laramie, Wyoming

Utah State University
Logan, Utah

For their support of the preparation of this volume, we gratefully acknowledge and thank:

Franklinia Foundation

The Philecology Foundation

The Andrew W. Mellon Foundation

The David and Lucile Packard Foundation

an anonymous foundation

Chanticleer Foundation

The Stanley Smith Horticultural Trust

Hall Family Charitable Fund

WEM Foundation

William T. Kemper Foundation

For sponsorship of illustrations included in this volume, we express sincere appreciation to:

Erick M. Adams, Olympia, Washington
Cornus canadensis, Cornaceae

Mac Alford, Hattiesburg, Mississippi
Berchemia scandens, Rhamnaceae – in honor of E. Earl Alford and M. Pauline Alford

Arizona Native Plant Society
Jamesia americana, Hydrangeaceae

Richard Carter, Valdosta, Georgia
Lepuropetalon spathulatum, Celastraceae – in memory of Angus Kemp Gholson, Jr.
Licania michauxii, Chrysobalanaceae – in memory of Wayne Reynolds Faircloth
Cornus florida, Cornaceae – in memory of Jim Ella Wells Aden
Stillingia sylvatica, Euphorbiaceae – in memory of Oneita Wells Keith
Hydrangea quercifolia, Hydrangeaceae – in memory of Bessie May Wells

Christopher Davidson and Sharon Christoph, Botanical Research Foundation of Idaho, Boise
Mortonia scabrella, Celastraceae
Parnassia palustris, Celastraceae
Paxistima canbyi, Celastraceae
Pyrularia pubera, Cervantesiaceae
Cornus sericea, Cornaceae
Comandra umbellata, Comandraceae
Eucnide urens, Loasaceae
Petalonyx nitidus, Loasaceae
Simmondsia chinensis, Simmondsiaceae
Buckleya distichophylla, Thesiaceae

David Giblin, Seattle, Washington
Linum lewisii, Linaceae

Huntington Botanical Garden, San Marino, California
Euphorbia, Euphorbiaceae
Volume 12 Frontispiece, *Euphorbia marginata*, Euphorbiaceae

Geoffrey A. Levin, Champaign, Illinois
Acalypha californica, Euphorbiaceae
Acalypha ostryifolia, Euphorbiaceae
Acalypha rhomboidea, Euphorbiaceae
Phyllanthus abnormis, Phyllanthaceae
Phyllanthus caroliniensis, Phyllanthaceae
Phyllanthus polygonoides, Phyllanthaceae

Naples Botanical Garden, Naples, Florida
Hippocratea volubilis, Celastraceae

Nevada Native Plant Society, Reno, Nevada
Mentzelia laevicaulis, Loasaceae

Smithsonian Institution, National Museum of Natural History, Botany Department
Vitaceae

University of Michigan Herbarium, Ann Arbor
Malpighiaceae

Christopher Woods, Benicia, California
Carpenteria californica, Hydrangeaceae

FLORA OF NORTH AMERICA EDITORIAL COMMITTEE (as of 30 April 2016)

Emeritus/a Members of the Editorial Committee

Project Staff — past and present involved with the preparation of Volume 12

Barbara Alongi, *Illustrator*
Burgund Bassüner, *Information Coordinator*
Ariel S. Buback, *Editorial Assistant*
Trisha K. Distler, *GIS Analyst*
Karin Douthit, *Illustrator*
Pat Harris, *Editorial Assistant*
Linny Heagy, *Illustrator*
Cassandra L. Howard, *Assisting Technical Editor*
Ruth T. King, *Editorial Assistant*
Marjorie C. Leggitt, *Illustrator*
John Myers, *Illustrator and Illustration Compositor*
Kristin Pierce, *Editorial Assistant and Compositor*
Andrew C. Pryor, *Technical Editor*
Hong Song, *Programmer*
Alice Tangerini, *Illustrator*
Yevonn Wilson-Ramsey, *Illustrator*

Contributors to Volume 12

William R. Anderson†
University of Michigan
Ann Arbor, Michigan

W. Scott Armbruster
University of Portsmouth
Portsmouth, England

Peter W. Ball
University of Toronto
Mississauga, Ontario

Paul E. Berry
University of Michigan
Ann Arbor, Michigan

David E. Boufford
Harvard University Herbaria
Cambridge, Massachusetts

Joshua M. Brokaw
Abilene Christian University
Abilene, Texas

N. Ivalú Cacho
University of California, Davis
Davis, California

Bijan Dehgan
University of Florida
Gainesville, Florida

J. Arturo De-Nova
Universidad Nacional Autónoma de México
Mexico City, Mexico

Craig C. Freeman
The University of Kansas
Lawrence, Kansas

Dmitry V. Geltman
V. L. Komarov Botanical Institute
St. Petersburg, Russia

Lynn J. Gillespie
Canadian Museum of Nature
Ottawa, Ontario

W. John Hayden
University of Richmond
Richmond, Virginia

James Henrickson
The University of Texas at Austin
Austin, Texas

Petra Hoffmann
Royal Botanic Gardens
Kew, Richmond, Surrey, England

Larry D. Hufford
Washington State University
Pullman, Washington

Michael J. Huft
Valparaiso, Indiana

Donald H. Les
University of Connecticut
Storrs, Connecticut

Geoffrey A. Levin
Illinois Natural History Survey
Champaign, Illinois

Jin-shuang Ma
Shanghai Chenshan Botanical Garden
Shanghai, People's Republic of China

Mark Mayfield
Kansas State University
Manhattan, Kansas

Ronald L. McGregor†
The University of Kansas
Lawrence, Kansas

Michael O. Moore†
University of Georgia
Athens, Georgia

Jeffery J. Morawetz
Rancho Ana Botanic Garden
Claremont, California

Nancy R. Morin
Point Arena, California

Zack E. Murrell
Appalachian State University
Boone, North Carolina

Guy L. Nesom
Fort Worth, Texas

Daniel L. Nickrent
Southern Illinois University-Carbondale
Carbondale, Illinois

Christopher F. Nixon†
Anniston Museum of Natural History
Anniston, Alabama

Tracy J. Park
Eastern Illinois University
Charleston, Illinois

Jess A. Peirson
University of Michigan
Ann Arbor, Michigan

Derick B. Poindexter
The University of North Carolina at Chapel Hill
Chapel Hill, North Carolina

Duncan M. Porter
Virginia Polytechnic Institute and State University
Blacksburg, Virginia

Yocupitzia Ramírez-Amezcua
Instituto de Ecología
Pátzcuaro, Michoacán, Mexico

Hamid Razifard
University of Connecticut
Storrs, Connecticut

Ricarda Riina
Real Jardín Botánico de Madrid
Madrid, Spain

Dayle E. Saar
Murray State University
Murray, Kentucky

John O. Sawyer†
Humboldt State University
Arcata, California

John J. Schenk
Georgia Southern University
Statesboro, Georgia

Clifford L. Schmidt†
Salem, Oregon

Beryl B. Simpson
The University of Texas at Austin
Austin, Texas

Victor W. Steinmann
Instituto de Ecología
Pátzcuaro, Michoacán, Mexico

Gordon C. Tucker
Eastern Illinois University
Charleston, Illinois

Roberto J. Urtecho
College of the Sequoias
Visalia, California

Benjamin W. van Ee
University of Puerto Rico-Mayagüez
Mayagüez, Puerto Rico

Michael A. Vincent
Miami University
Oxford, Ohio

Maria S. Vorontsova
Royal Botanic Gardens
Kew, Richmond, Surrey, England

Alan S. Weakley
The University of North Carolina at Chapel Hill
Chapel Hill, North Carolina

Jun Wen
National Museum of Natural History
Smithsonian Institution
Washington, DC

R. David Whetstone
Jacksonville State University
Jacksonville, Alabama

Dieter H. Wilken
Santa Barbara Botanic Garden
Santa Barbara, California

Kenneth J. Wurdack
National Museum of Natural History
Smithsonian Institution
Washington, DC

Ya Yang
University of Michigan
Ann Arbor, Michigan

Taxonomic Reviewers

W. John Hayden
University of Richmond
Richmond, Virginia

James Henrickson
The University of Texas at Austin
Austin, Texas

Larry D. Hufford
Washington State University
Pullman, Washington

Michael J. Huft
Valparaiso, Indiana

Allison J. Miller
Saint Louis University
St. Louis, Missouri

Amy Pool
Missouri Botanical Garden
St. Louis, Missouri

Kenneth J. Wurdack
National Museum of Natural History
Smithsonian Institution
Washington, DC

Regional Reviewers

ALASKA / YUKON

Bruce Bennett
Yukon Department of Environment
Whitehorse, Yukon

Robert Lipkin
Alaska Natural Heritage Program
University of Alaska
Anchorage, Alaska

David F. Murray
University of Alaska Museum of the North
University of Alaska Fairbanks
Fairbanks, Alaska

Carolyn Parker
University of Alaska Museum of the North
University of Alaska Fairbanks
Fairbanks, Alaska

Mary Stensvold
U.S.D.A. Forest Service
Sitka, Alaska

PACIFIC NORTHWEST

Edward R. Alverson
The Nature Conservancy
Eugene, Oregon

Adolf Ceska
British Columbia Conservation Data Centre
Victoria, British Columbia

Richard R. Halse
Oregon State University
Corvallis, Oregon

Eugene N. Kozloff
Shannon Point Marine Center
Anacortes, Washington

Jim Pojar
British Columbia Forest Service
Smithers, British Columbia

Cindy Roché
Talent, Oregon

Peter F. Zika
University of Washington
Seattle, Washington

SOUTHWESTERN UNITED STATES

Tina J. Ayers
Northern Arizona University
Flagstaff, Arizona

Walter Fertig
Arizona State University
Tempe, Arizona

G. Frederic Hrusa
California Department of Food and Agriculture,
Sacramento, California

Max Licher
Northern Arizona University
Flagstaff, Arizona

James D. Morefield
Nevada Natural Heritage Program
Carson City, Nevada

Donald J. Pinkava
Arizona State University
Tempe, Arizona

Jon P. Rebman
San Diego Natural History Museum
San Diego, California

Glenn Rink
Northern Arizona University
Flagstaff, Arizona

Gary D. Wallace
Rancho Santa Ana Botanic Garden
Claremont, California

WESTERN CANADA

William J. Cody†
Agriculture and Agri-Food Canada
Ottawa, Ontario

Lynn Gillespie
Canadian Museum of Nature
Ottawa, Ontario

A. Joyce Gould
Alberta Tourism, Parks and Recreation
Edmonton, Alberta

Vernon L. Harms
University of Saskatchewan
Saskatoon, Saskatchewan

Elizabeth Punter
University of Manitoba
Winnipeg, Manitoba

ROCKY MOUNTAINS

Curtis R. Björk
University of Idaho
Moscow, Idaho

Bonnie Heidel
University of Wyoming
Laramie, Wyoming

B. E. Nelson
University of Wyoming
Laramie, Wyoming

NORTH CENTRAL UNITED STATES

Anita F. Cholewa
University of Minnesota
St. Paul, Minnesota

Neil A. Harriman
University of Wisconsin Oshkosh
Oshkosh, Wisconsin

Bruce W. Hoagland
University of Oklahoma
Norman, Oklahoma

Robert B. Kaul
University of Nebraska
Lincoln, Nebraska

Deborah Q. Lewis
Iowa State University
Ames, Iowa

Ronald L. McGregor†
The University of Kansas
Lawrence, Kansas

Lawrence R. Stritch
U.S.D.A. Forest Service
Washington, DC

George Yatskievych
The Universtiy of Texas at Austin
Austin, Texas

SOUTH CENTRAL UNITED STATES

David E. Lemke
Southwest Texas State University
San Marcos, Texas

Robert C. Sivinski
University of New Mexico
Albuquerque, New Mexico

EASTERN CANADA

Sean Blaney
Atlantic Canada Conservation Data Centre
Sackville, New Brunswick

Jacques Cayouette
Agriculture and Agri-Food Canada
Ottawa, Ontario

Frédéric Coursol
Mirabel, Québec

William J. Crins
Ontario Ministry of Natural Resources
Peterborough, Ontario

John K. Morton†
University of Waterloo
Waterloo, Ontario

Marian Munro
Nova Scotia Museum of Natural History
Halifax, Nova Scotia

Michael J. Oldham
Natural Heritage Information Centre
Peterborough, Ontario

GREENLAND

Geoffrey Halliday
University of Lancaster
Lancaster, England

NORTHEASTERN UNITED STATES

Ray Angelo
New England Botanical Club
Cambridge, Massachusetts

Tom S. Cooperrider
Kent State University
Kent, Ohio

Arthur Haines
Bowdoin, Maine

Michael A. Homoya
Indiana Department of Natural Resources
Indianapolis, Indiana

Robert F. C. Naczi
The New York Botanical Garden
Bronx, New York

Anton A. Reznicek
University of Michigan
Ann Arbor, Michigan

Edward G. Voss†
University of Michigan
Ann Arbor, Michigan

Kay Yatskievych
Missouri Botanical Garden
St. Louis, Missouri

SOUTHEASTERN UNITED STATES

Mac H. Alford
University of Southern Mississippi
Hattiesburg, Mississippi

J. Richard Carter Jr.
Valdosta State University
Valdosta, Georgia

L. Dwayne Estes
Austin Peay State University
Clarksville, Tennessee

Paul J. Harmon
West Virginia Natural Heritage Program
Charleston, West Virginia

W. John Hayden
University of Richmond
Richmond, Virginia

John B. Nelson
University of South Carolina
Columbia, South Carolina

Bruce A. Sorrie
University of North Carolina
Chapel Hill, North Carolina

R. Dale Thomas
Seymour, Tennessee

Lowell E. Urbatsch
Louisiana State University
Baton Rouge, Louisiana

Thomas F. Wieboldt
Virginia Polytechnic Institute and State University
Blacksburg, Virginia

B. Eugene Wofford
University of Tennessee
Knoxville, Tennessee

FLORIDA

Loran C. Anderson
Florida State University
Tallahassee, Florida

Bruce F. Hansen
University of South Florida
Tampa, Florida

Richard P. Wunderlin
University of South Florida
Tampa, Florida

Preface for Volume 12

Since the publication of *Flora of North America* Volume 6 (the nineteenth volume in the *Flora* series) in mid 2015, the membership of the Flora of North America Association [FNAA] Board of Directors has undergone changes. Ronald L. Hartman, Timothy K. Lowrey, Robert F. C. Naczi, Leila M. Shultz, and Debra K. Trock have retired from the board. New board members include Michael J. Huft, Alexander Krings, Mare Nazaire, and George Yatskievych (Taxon Editors). As a result of a reorganization finalized in 2003, the FNAA Board of Directors succeeded the former Editorial Committee; for the sake of continuity of citation, authorship of *Flora* volumes is still to be cited as "Flora of North America Editorial Committee, eds."

Most of the editorial process for this volume was done at the University of Illinois in Champaign, Illinois, the Canadian Museum of Nature in Ottawa, Ontario, and Missouri Botanical Garden in St. Louis. Final processing and composition took place at the Missouri Botanical Garden; this included pre-press processing, typesetting and layout, plus coordination for all aspects of planning, executing, and scanning the illustrations. Other aspects of production, such as art panel composition plus labeling and occurrence map generation, were carried out in Gaston, Oregon, and Miami, Florida, respectively.

Line drawings published in this volume were executed by seven very talented artists: Barbara Alongi prepared illustrations for Celastraceae, Elatinaceae, Euphorbiaceae in greater part (excluding *Adelia* & *Ricinus*) including the frontispiece depicting *Euphorbia marginata*, Phyllanthaceae, Picodendraceae, Putranjivaceae, Simmondsiaceae, and Zygophyllaceae; Karin Douthit illustrated Malpighiaceae; Linny Heagy illustrated taxa of Krameraceae and Loasaceae; Marjorie C. Leggitt illustrated Chrysobalanaceae; John Myers illustrated *Philadelphus* (Hydrangeaceae), Linaceae, and Rhizophoraceae; Alice Tangerini illustrated Vitaceae; and Yevonn Wilson-Ramsey prepared illustrations for Cervantesiaceae, Comandraceae, Cornaceae, Eucommiaceae, *Adelia* and *Ricinus* (Euphorbiaceae), Garryaceae, Hydrangeaceae (excluding *Philadelphus*), Nyssaceae, Oxalidaceae, Rhamnaceae, Thesiaceae, Viscaceae, and Ximeniaceae. In addition to preparing various illustrations, John Myers composed and labeled all of the line drawings that appear in this volume.

Starting with Volume 8 published in 2009, the circumscription and ordering of some families within the *Flora* have been modified so they mostly reflect that of the Angiosperm Phylogeny Group [APG] rather than the previously followed Cronquist organizational structure. The groups of families found in this and future volumes in the series are mostly ordered following E. M. Haston et al. (2007); since APG views of relationships and circumscriptions have evolved, and will certainly change further through time, some discrepancies in organization will occur. Volume 30 of the *Flora of North America* will contain a comprehensive index to the published volumes.

Support from many institutions and by numerous individuals has enabled the *Flora* to be produced. Members of the Flora of North America Association remain deeply thankful to the many people who continue to help create, encourage, and sustain the *Flora*.

Introduction

Scope of the Work

Flora of North America North of Mexico is a synoptic account of the plants of North America north of Mexico: the continental United States of America (including the Florida Keys and Aleutian Islands), Canada, Greenland (Kalâtdlit-Nunât), and St. Pierre and Miquelon. The *Flora* is intended to serve both as a means of identifying plants within the region and as a systematic conspectus of the North American flora.

The *Flora* will be published in 30 volumes. Volume 1 contains background information that is useful for understanding patterns in the flora. Volume 2 contains treatments of ferns and gymnosperms. Families in volumes 3–26, the angiosperms, were first arranged according to the classification system of A. Cronquist (1981) with some modifications, and starting with Volume 8, the circumscriptions and ordering of families generally follow those of the Angiosperm Phylogeny Group [APG] (see E. Haston et al. 2007). Bryophytes are being covered in volumes 27–29. Volume 30 will contain the cumulative bibliography and index.

The first two volumes were published in 1993, Volume 3 in 1997, and Volumes 22, 23, and 26, the first three of five volumes covering the monocotyledons, appeared in 2000, 2002, and 2002, respectively. Volume 4, the first part of the Caryophyllales, was published in late 2003. Volume 25, the second part of the Poaceae, was published in mid 2003, and Volume 24, the first part, was published in January 2007. Volume 5, completing the Caryophyllales plus Polygonales and Plumbaginales, was published in early 2005. Volumes 19–21, treating Asteraceae, were published in early 2006. Volume 27, the first of two volumes treating mosses in North America, was published in late 2007. Volume 8, Paeoniaceae to Ericaceae, was published in September 2009, and Volume 7, Salicaceae to Brassicaceae, appeared in 2010. In 2014, Volume 28 was published, completing the treatment of mosses for the flora area, and at the end of 2014, Volume 9, Picramniaceae to Rosaceae was published. Volume 6, which covered Cucurbitaceae to Droseraceae, was published in 2015. The correct bibliographic citation for the *Flora* is: Flora of North America Editorial Committee, eds. 1993+. Flora of North America North of Mexico. 20+ vols. New York and Oxford.

Volume 12 treats 765 species in 122 genera contained in 29 families. For additional statistics please refer to Table 1 on p. xx.

Contents · General

The *Flora* includes accepted names, selected synonyms, literature citations, identification keys, descriptions, phenological information, summaries of habitats and geographic ranges, and other biological observations. Each volume contains a bibliography and an index to the taxa included in that volume. The treatments, written and reviewed by experts from throughout the systematic botanical community, are based on original observations of herbarium specimens and, whenever possible, on living plants. These observations are supplemented by critical reviews of the literature.

Table 1. *Statistics for Volume 12 of Flora of North America.*

Family	Total Genera	Endemic Genera	Introduced Genera	Total Species	Endemic Species	Introduced Species	Conservation Taxa
Vitaceae	6	0	1	30	19	5	0
Krameriaceae	1	0	0	4	0	0	0
Zygophyllaceae	6	0	2	15	1	3	2
Rhamnaceae	15	0	2	105	45	13	21
Celastraceae	12	0	0	34	13	5	1
Oxalidaceae	1	0	0	36	11	13	0
Rhizophoraceae	1	0	0	1	0	0	0
Euphorbiaceae	24	1	8	259	79	42	34
Picrodendraceae	1	0	0	3	1	0	1
Phyllanthaceae	7	0	4	23	0	11	4
Elatinaceae	2	0	0	11	5	2	1
Malpighiaceae	8	0	2	9	0	2	0
Chrysobalanaceae	2	0	0	2	1	0	0
Putranjivaceae	1	0	0	2	0	0	0
Linaceae	4	1	1	52	29	7	16
Ximeniaceae	1	0	0	1	0	0	0
Schoepfiaceae	1	0	0	1	0	0	0
Comandraceae	2	1	0	2	1	0	0
Thesiaceae	2	0	1	2	1	1	0
Cervantesiaceae	1	0	0	1	1	0	0
Santalaceae	2	1	1	2	1	1	0
Viscaceae	3	0	1	15	2	1	0
Simmondsiaceae	1	0	0	1	0	0	0
Cornaceae	1	0	0	20	13	3	0
Nyssaceae	1	0	0	5	4	0	1
Hydrangeaceae	9	2	1	25	12	5	9
Loasaceae	4	0	0	94	67	0	24
Eucommiaceae	1	0	1	1	0	1	0
Garryaceae	2	0	1	9	3	1	0
Totals	**122**	**6**	**26**	**765**	**309**	**116**	**114**

Italic = introduced

Basic Concepts

Our goal is to make the *Flora* as clear, concise, and informative as practicable so that it can be an important resource for both botanists and nonbotanists. To this end, we are attempting to be consistent in style and content from the first volume to the last. Readers may assume that a term has the same meaning each time it appears and that, within groups, descriptions may be compared directly with one another. Any departures from consistent usage will be explicitly noted in the treatments (see References).

Treatments are intended to reflect current knowledge of taxa throughout their ranges worldwide, and classifications are therefore based on all available evidence. Where notable differences of opinion about the classification of a group occur, appropriate references are mentioned in the discussion of the group.

Documentation and arguments supporting significantly revised classifications are published separately in botanical journals before publication of the pertinent volume of the *Flora*. Similarly, all new names and new combinations are published elsewhere prior to their use in the *Flora*. No nomenclatural innovations will be published intentionally in the *Flora*.

Taxa treated in full include extant and recently extinct or extirpated native species, named hybrids that are well established (or frequent), introduced plants that are naturalized, and cultivated plants that are found frequently outside cultivation. Taxa mentioned only in discussions include waifs known only from isolated old records and some non-native, economically important or extensively cultivated plants, particularly when they are relatives of native species. Excluded names and taxa are listed at the ends of appropriate sections, for example, species at the end of genus, genera at the end of family.

Treatments are intended to be succinct and diagnostic but adequately descriptive. Characters and character states used in the keys are repeated in the descriptions. Descriptions of related taxa at the same rank are directly comparable.

With few exceptions, taxa are presented in taxonomic sequence. If an author is unable to produce a classification, the taxa are arranged alphabetically and the reasons are given in the discussion.

Treatments of hybrids follow that of one of the putative parents. Hybrid complexes are treated at the ends of their genera, after the descriptions of species.

We have attempted to keep terminology as simple as accuracy permits. Common English equivalents usually have been used in place of Latin or Latinized terms or other specialized terminology, whenever the correct meaning could be conveyed in approximately the same space, for example, "pitted" rather than "foveolate," but "striate" rather than "with fine longitudinal lines." See *Categorical Glossary for the Flora of North America Project* (R. W. Kiger and D. M. Porter 2001; also available online at http://huntbot.andrew.cmu.edu) for standard definitions of generally used terms. Very specialized terms are defined, and sometimes illustrated, in the relevant family or generic treatments.

References

Authoritative general reference works used for style are *The Chicago Manual of Style*, ed. 14 (University of Chicago Press 1993); *Webster's New Geographical Dictionary* (Merriam-Webster 1988); and *The Random House Dictionary of the English Language*, ed. 2, unabridged (S. B. Flexner and L. C. Hauck 1987). *B-P-H/S. Botanico-Periodicum-Huntianum/Supplementum* (G. D. R. Bridson and E. R. Smith 1991), *BPH-2: Periodicals with Botanical Content* (Bridson 2004), and *BPH Online* [http://fmhibd.library.cmu.edu/HIBD-DB/bpho/findrecords.php] (Bridson and D. W. Brown) have been used for abbreviations of serial titles, and *Taxonomic Literature*, ed. 2 (F. A. Stafleu and R. S. Cowan 1976–1988) and its supplements by Stafleu et al. (1992–2009) have been used for abbreviations of book titles.

Graphic Elements

All genera and more than 27 percent of the species in this volume are illustrated. The illustrations may show diagnostic traits or complex structures. Most illustrations have been drawn from herbarium specimens selected by the authors. Data on specimens that were used and parts that were illustrated have been recorded. This information, together with the archivally preserved original drawings, is deposited in the Missouri Botanical Garden Library and is available for scholarly study.

Specific Information in Treatments

Keys

Dichotomous keys are included for all ranks below family if two or more taxa are treated. More than one key may be given to facilitate identification of sterile material or for flowering versus fruiting material.

Nomenclatural Information

Basionyms of accepted names, with author and bibliographic citations, are listed first in synonymy, followed by any other synonyms in common recent use, listed in alphabetical order, without bibliographic citations.

The last names of authors of taxonomic names have been spelled out. The conventions of *Authors of Plant Names* (R. K. Brummitt and C. E. Powell 1992) have been used as a guide for including first initials to discriminate individuals who share surnames.

If only one infraspecific taxon within a species occurs in the flora area, nomenclatural information (literature citation, basionym with literature citation, relevant other synonyms) is given for the species, as is information on the number of infraspecific taxa in the species and their distribution worldwide, if known. A description and detailed distributional information are given only for the infraspecific taxon.

Descriptions

Character states common to all taxa are noted in the description of the taxon at the next higher rank. For example, if sexual condition is dioecious for all species treated within a genus, that character state is given in the generic description. Characters used in keys are repeated in the descriptions. Characteristics are given as they occur in plants from the flora area. Characteristics that occur only in plants from outside the flora area may be given within square brackets, or instead may be noted in the discussion following the description. In families with one genus and one or more species, the family description is given as usual, the genus description is condensed, and the species are described as usual. Any special terms that may be used when describing members of a genus are presented and explained in the genus description or discussion.

In reading descriptions, the reader may assume, unless otherwise noted, that: the plants are green, photosynthetic, and reproductively mature; woody plants are perennial; stems are erect; roots are fibrous; leaves are simple and petiolate; flowers are bisexual, radially symmetric, and pediceled; perianth parts are hypogynous, distinct, and free; and ovaries are superior. Because measurements and elevations are almost always approximate, modifiers such as "about," "circa," or "±" are usually omitted.

Unless otherwise noted, dimensions are length × width. If only one dimension is given, it is length or height. All measurements are given in metric units. Measurements usually are based on dried specimens but these should not differ significantly from the measurements actually found in fresh or living material.

Chromosome numbers generally are given only if published and vouchered counts are available from North American material or from an adjacent region. No new counts are published intentionally in the *Flora*. Chromosome counts from nonsporophyte tissue have been converted to the $2n$ form. The base number ($x =$) is given for each genus. This represents the lowest known haploid count for the genus unless evidence is available that the base number differs.

Flowering time and often fruiting time are given by season, sometimes qualified by early, mid, or late, or by months. Elevations over 200 m generally are rounded to the nearest 100 m; those 100 m and under are rounded to the nearest 10 m. Mean sea level is shown as 0 m, with the understanding that this is approximate. Elevation often is omitted from herbarium specimen labels, particularly for collections made where the topography is not remarkable, and therefore precise elevation is sometimes not known for a given taxon.

The term "introduced" is defined broadly to refer to plants that were released deliberately or accidentally into the flora and that now are naturalized, that is, exist as wild plants in areas in which they were not recorded as native in the past. The distribution of introduced taxa are often poorly documented and changing, so the distribution statements for those taxa may not be fully accurate.

If a taxon is globally rare or if its continued existence is threatened in some way, the words "of conservation concern" appear before the statements of elevation and geographic range.

Criteria for taxa of conservation concern are based on NatureServe's (formerly The Nature Conservancy)—see http://www.natureserve.org—designations of global rank (G-rank) G1 and G2:

G1 Critically imperiled globally because of extreme rarity (5 or fewer occurrences or fewer than 1000 individuals or acres) or because of some factor(s) making it especially vulnerable to extinction.

G2 Imperiled globally because of rarity (5–20 occurrences or fewer than 3000 individuals or acres) or because of some factor(s) making it very vulnerable to extinction throughout its range.

The occurrence of species and infraspecific taxa within political subunits of the *Flora* area is depicted by dots placed on the outline map to indicate occurrence in a state or province. The Nunavut boundary on the maps has been provided by the GeoAccess Division, Canada Centre for Remote Sensing, Earth Science. Authors are expected to have seen at least one specimen documenting each geographic unit record (except in rare cases when undoubted literature reports may be used) and have been urged to examine as many specimens as possible from throughout the range of each taxon. Additional information about taxon distribution may be presented in the discussion.

Distributions are stated in the following order: Greenland; St. Pierre and Miquelon; Canada (provinces and territories in alphabetic order); United States (states in alphabetic order); Mexico (11 northern states may be listed specifically, in alphabetic order); West Indies; Bermuda; Central America (Belize, Costa Rica, El Salvador, Guatemala, Honduras, Nicaragua, Panama); South America; Europe, or Eurasia; Asia (including Indonesia); Africa; Atlantic Islands; Indian Ocean Islands; Pacific Islands; Australia; Antarctica.

Discussion

The discussion section may include information on taxonomic problems, distributional and ecological details, interesting biological phenomena, and economic uses.

Selected References

Major references used in preparation of a treatment or containing critical information about a taxon are cited following the discussion. These, and other works that are referred to in discussion or elsewhere, are included in Literature Cited at the end of the volume.

CAUTION

The Flora of North America Editorial Committee **does not encourage, recommend, promote, or endorse** any of the folk remedies, culinary practices, or various utilizations of any plant described within this volume. Information about medicinal practices and/or ingestion of plants, or of any part or preparation thereof, has been included only for historical background and as a matter of interest. Under no circumstances should the information contained in these volumes be used in connection with medical treatment. Readers are strongly cautioned to remember that many plants in the flora are toxic or can cause unpleasant or adverse reactions if used or encountered carelessly.

Key to boxed codes following accepted names:

- [C] of conservation concern
- [E] endemic to the flora area
- [F] illustrated
- [I] introduced to the flora area
- [W] weedy, based mostly on R. H. Callihan et al. (1995) and/or D. T. Patterson et al. (1989)

Flora of North America

VITACEAE Jussieu

• Grape Family

Michael O. Moore†

Jun Wen

Vines or lianas, occasionally shrubby [trees], synoecious, dioecious, or polygamomonoecious; commonly with multicellular, stalked, caducous, spheric structures (pearl glands); tendrils usually present, rarely absent. **Leaves** alternate, simple or palmately or pinnately compound; stipules present; petiole present; blade often palmately lobed, margins dentate, serrate, or crenate; venation palmate or pinnate. **Inflorescences** bisexual or functionally unisexual, axillary or terminal and appearing leaf-opposed, cymes or thyrses [spikes]. **Flowers** bisexual or unisexual; perianth and androecium hypogynous; hypanthium absent; sepals 4–5(–9), connate most or all of length; petals (3–)4–5(–9), distinct (connate distally, forming calyptra, in *Vitis*) [connate basally], valvate, free; nectary intrastaminal; stamens (3–)4–5(–9), opposite petals, distinct; anthers dehiscing by longitudinal slits; staminodes present in functionally pistillate flowers; pistil 1, 2[–3]-carpellate, ovary superior, 2[–3]-locular, placentation axile, sometimes appearing parietal; ovules 2 per locule, apotropous or anatropous; style 1; stigma 1[4]. **Fruits** berries. **Seeds** 1–4 per fruit.

Genera 15, species ca. 900 (6 genera, 30 species in the flora): North America, Mexico, West Indies, Central America, South America, Eurasia, Africa, Indian Ocean Islands, Pacific Islands, Australia; primarily tropical and subtropical with a few genera in warm temperate to temperate regions.

In the past decade, considerable effort has been made to reconstruct the phylogeny of Vitaceae (M. Rossetto et al. 2001, 2002; M. J. Ingrouille et al. 2002; A. Soejima and Wen J. 2006; Wen et al. 2007; Ren H. et al. 2011; A. Trias-Blasi et al. 2012; Lu L. M. et al. 2013; Wen et al. 2013b; Zhang N. et al. 2015). These analyses have generally supported five major clades within Vitaceae: the *Ampelopsis* clade, the *Ampelocissus-Vitis* clade, the *Parthenocissus-Yua* clade, the core *Cissus* clade, and the *Cayratia-Cyphostemma-Tetrastigma* clade.

Placentation type and ovary locule number in Vitaceae have been interpreted in different ways. Most authors state that the gynoecium is bilocular and the placentation type is axile (for example, J. E. Planchon 1887; P. K. Endress 2010), but others report the gynoecium as unilocular and refer to the placentation type as parietal (V. Puri 1952; N. C. Nair and

K. V. Mani 1960). Recent anatomical studies confirm that the ovary in Vitaceae is usually bilocular [occasionally trilocular in *Cayratia* Jussieu, or unilocular in some species of *Cyphostemma* (Planchon) Alston, such as *Cyphostemma sandersonii* (Harvey) Descoings], and that the placentation is axile, although it may sometimes appear parietal (S. M. Ickert-Bond et al. 2014).

SELECTED REFERENCES Brizicky, G. K. 1965. The genera of Vitaceae in the southeastern United States. J. Arnold Arbor. 46: 48–67. Ren, H. et al. 2011. Phylogenetic analysis of the grape family (Vitaceae) based on the noncoding plastid *trn*C-*pet*N, *trn*H-*psb*A, and *trn*L-F sequences. Taxon 60: 629–637. Soejima, A. and Wen J. 2006. Phylogenetic analysis of the grape family (Vitaceae) based on three chloroplast markers. Amer. J. Bot. 93: 278–287. Wen, J. 2007. Vitaceae. In: K. Kubitzki et al., eds. 1990+. The Families and Genera of Vascular Plants. 10+ vols. Berlin etc. Vol. 9, pp. 467–479.

1. Leaves 2–3-pinnately compound . . . . 4. *Nekemias*, p. 20
1. Leaves simple or palmately compound.
    2. Pith brown; petals connate distally, forming calyptra; bark exfoliating (adherent in *V. rotundifolia*) . . . . 1. *Vitis*, p. 4
    2. Pith white; petals distinct; bark adherent.
        3. Petals 5; stamens 5.
            4. Tendrils 3–12-branched; nectaries annular, adnate to base of ovary, or absent; styles short, conic . . . . 2. *Parthenocissus*, p. 16
            4. Tendrils 2-branched; nectaries cup-shaped, adnate proximally to base of ovary, free distally; styles elongate, cylindric . . . . 3. *Ampelopsis*, p. 18
        3. Petals 4; stamens 4.
            5. Inflorescences leaf-opposed; seeds 1(–4) per fruit . . . . 5. *Cissus*, p. 20
            5. Inflorescences axillary; seeds 2–4 per fruit . . . . 6. *Causonis*, p. 22

1. VITIS Linnaeus, Sp. Pl. 1: 202. 1753; Gen. Pl. ed. 5, 95. 1754 • Grape [Latin, vine]

Lianas, climbing by tendrils, sprawling, or occasionally shrubby, functionally dioecious (synoecious in *V. vinifera*). **Branches:** bark exfoliating (adherent in *V. rotundifolia*); pith brown, interrupted by nodal diaphragms (continuous through nodes in *V. rotundifolia*); tendrils 2–3-branched (unbranched in *V. rotundifolia*), rarely absent, without adhesive discs. **Leaves** simple. **Inflorescences** functionally unisexual (bisexual in *V. vinifera*), leaf-opposed, thyrses. **Flowers** functionally unisexual (bisexual in *V. vinifera*); calyx a minute rim, entire or 5-toothed; petals (3–)5(–9), connate distally, forming calyptra; nectary free, (3–)5(–9) glands alternating with stamens; stamens usually (3–)5(–9), sometimes 0 in pistillate flowers; style conic, short. **Berries** purple or black. **Seeds** 1–4 per fruit. $x = 10$.

Species ca. 70 (19, including 3 hybrids, in the flora): North America, Mexico, West Indies, Central America, n South America, Eurasia; introduced nearly worldwide.

Vitis is nearly restricted to temperate regions of the northern hemisphere, with one extremely variable species (*V. tiliifolia* Humboldt & Bonpland ex Schultes) extending into northern South America.

The North American species of *Vitis* are of considerable economic importance and have played a significant part in commercial viticulture over the last century. The introduction of native North American species into France in the mid 1800s led to the introduction of grape phylloxera, *Daktulosphaira vitifoliae*, an insect to which the grape of commerce (*V. vinifera*) is susceptible. Many European vineyards were virtually destroyed by the 1860s. The reconstruction of the European vineyards was made possible by using native North American species, many of which are resistant to grape phylloxera, as rootstocks for and in hybridizations with *V. vinifera*. Native North American species of *Vitis* have also played a major role in establishing viticulture as an industry in North America, particularly in areas other than California and Oregon, such as Florida, New York, North Carolina, Ontario, Texas, and Virginia.

Two subgenera of *Vitis* commonly have been recognized. Subgenus *Vitis*, which includes the majority of species, is widely distributed in the Northern Hemisphere; subg. *Muscadinia*, with two species, is restricted to the southeastern United States, the West Indies, and Mexico (G. K. Brizicky 1965; Wen J. 2007). J. K. Small (1903) elevated subg. *Muscadinia* to generic rank, a treatment followed by some authors, for example A. S. Weakley et al. (2012). However, a recent phylogenetic study of the *Vitis-Ampelocissus* clade by Liu X. Q. et al. (2015) showed that the Central American *Ampelocissus erdvendbergianus* Planchon is sister to *Vitis* and that subg. *Muscadinia* and subg. *Vitis* are sister groups. To maintain the monophyly of *Vitis* and *Ampelocissus* Planchon, *A. erdvendbergianus* needs to be transferred to *Vitis*. Recognizing *Muscadinia* as a distinct genus would require description of a monospecific new genus for *A. erdvendbergianus*; it makes better sense to maintain a broadly circumscribed *Vitis*.

In the key and descriptions that follow, nodal diaphragm thickness is in the current year's growth, leaf blade pubescence is on fully mature leaves unless otherwise noted, and berry diameter is of 3–4-seeded berries.

SELECTED REFERENCES Bailey, L. H. 1934. The species of grapes peculiar to North America. Gentes Herbarum 3: 154–244. Comeaux, B. L., W. B. Nesbitt, and P. R. Fantz. 1987. Taxonomy of the native grapes of North Carolina. Castanea 52: 197–215. Moore, M. O. 1987. A study of selected *Vitis* (Vitaceae) taxa in the southeastern United States. Rhodora 89: 75–91. Moore, M. O. 1991. Classification and systematics of eastern North American *Vitis* L. (Vitaceae) north of Mexico. Sida 14: 339–367. Munson, T. V. 1909. Foundations of American Grape Culture. New York. Tröndle, D. et al. 2010. Molecular phylogeny of the genus *Vitis* (Vitaceae) based on plastid markers. Amer. J. Bot. 97: 1168–1178. Zecca G. et al. 2012. The timing and the mode of evolution of wild grapes. Molec. Phylogen. Evol. 62: 736–747.

1. Tendrils unbranched; bark adherent; lenticels prominent; pith continuous through nodes [1a. subg. *Muscadinia*]. . . . . 1. *Vitis rotundifolia*
1. Tendrils branched or absent; bark exfoliating; lenticels inconspicuous or absent; pith interrupted by nodal diaphragms [1b. subg. *Vitis*].
    2. Flowers bisexual; berries oblong to ellipsoid, skin adhering to pulp . . . . 2. *Vitis vinifera*
    2. Flowers functionally unisexual; berries globose, skin separating from pulp.
        3. Leaf blade abaxial surface glaucous (sometimes obscured by hairs) . . . . 3. *Vitis aestivalis*
        3. Leaf blade abaxial surface not glaucous (concealed by hairs in *V. labrusca*).
            4. Tendrils or inflorescences present at 3+ consecutive nodes or almost all nodes.
                5. Leaf blade abaxial surface densely and persistently arachnoid, concealed (except sometimes veins) by hairs; nodal diaphragms 0.5–2.5 mm thick; tendrils at almost all nodes . . . . 4. *Vitis labrusca*
                5. Leaf blade abaxial surface ± densely arachnoid when young, sparsely arachnoid when mature, visible through hairs; nodal diaphragms 0.3–1.1 mm thick; tendrils usually not at all nodes . . . . 5. *Vitis* ×*novae-angliae*
            4. Tendrils or inflorescences present at only 2 consecutive nodes.
                6. Leaf blade abaxial surface densely tomentose, concealed (except sometimes veins) by hairs; berries 12+ mm diam.
                    7. Stipules 1.5–4 mm; nodal diaphragms 1.5–3 mm thick; Alabama, Arkansas, Louisiana, Oklahoma, Texas. . . . . 6. *Vitis mustangensis*
                    7. Stipules to 1 mm; nodal diaphragms 2.5–6 mm thick; Florida. . . . .7. *Vitis shuttleworthii*
                6. Leaf blade abaxial surface usually glabrous, moderately arachnoid, or hirtellous, sometimes tomentose (California and s Oregon), visible through hairs; berries 4–12 mm diam. (except 12+ mm in *V.* ×*doaniana* and *V.* ×*champinii*).
                    8. Leaf blades reniform, abaxial surface usually glabrous, sometimes sparsely hirtellous on veins and in vein axils; tendrils absent or only at distalmost nodes . . . . 8. *Vitis rupestris*
                    8. Leaf blades usually cordate to cordate ovate, sometimes orbiculate or nearly reniform, abaxial surface glabrous or hairy; tendrils along length of branchlets.

[9. Shifted to left margin.—Ed.]

9. Nodal diaphragms to 0.5(–1) mm thick; branchlet growing tips enveloped by unfolding leaves.
   10. Plants low to moderately high climbing, much branched; tendrils soon deciduous if not attached to support; branchlets arachnoid or glabrate, growing tips sparsely to densely hairy; inflorescences 3–7(–9) cm . . . . 9. *Vitis acerifolia*
   10. Plants usually moderate to high climbing, sometimes sprawling, sparsely branched; tendrils persistent; branchlets glabrous or sparsely hirtellous, growing tips glabrous or sparsely hairy; inflorescences (4–)9–12 cm . . . . 10. *Vitis riparia*
9. Nodal diaphragms 1–4 mm thick; branchlet growing tips not enveloped by unfolding leaves.
   11. Berries 12+ mm diam.
      12. Leaf blade abaxial surface moderately to densely arachnoid, hirtellous on veins; berries glaucous . . . . 11. *Vitis* ×*doaniana*
      12. Leaf blade abaxial surface sparsely arachnoid to glabrate, not hirtellous; berries usually not, sometimes very slightly, glaucous. . . . . 12. *Vitis* ×*champinii*
   11. Berries 4–12 mm diam.
      13. Leaf blade abaxial surface sparsely to densely tomentose; California, s Oregon.
         14. Berries moderately to heavily glaucous, 8–10 mm diam.; branchlet tomentum thinning in age; nodal diaphragms 3–4 mm thick. . . . . 13. *Vitis californica*
         14. Berries slightly or not glaucous, 4–6 mm diam.; branchlet tomentum usually persistent; nodal diaphragms 1.5–3 mm thick . . . . 14. *Vitis girdiana*
      13. Leaf blade abaxial surface glabrous or sparsely to densely arachnoid or hirtellous; much of United States, but not California or Oregon.
         15. Plants sprawling to low climbing, shrubby, much branched; tendrils soon deciduous if not attached to means of support; Arizona, Nevada, New Mexico, trans-Pecos Texas, Utah . . . . 15. *Vitis arizonica*
         15. Plants usually moderate to high climbing, sometimes ± shrubby and sprawling when without support, sparsely branched; tendrils persistent; e United States, including Texas, not in trans-Pecos region.
            16. Branchlets ± angled, densely hirtellous and/or sparsely to densely arachnoid, to glabrate; berries 4–8 mm diam.; nodes sometimes red-banded. . . . . 16. *Vitis cinerea*
            16. Branchlets terete or subterete, glabrous or sparsely arachnoid; berries 8–12 mm diam.; nodes not red-banded.
               17. Nodal diaphragms 2.5–4 mm thick; leaf blades usually deeply lobed, apices long acuminate; branchlets uniformly red, purplish red, or chestnut. . . . . 17. *Vitis palmata*
               17. Nodal diaphragms 1–2.5 mm thick; leaf blades unlobed or shallowly lobed, sometimes deeply lobed on ground shoots, apices acute to short acuminate; branchlets gray to green or brown, if purplish only on one side.
                  18. Berries usually with lenticels; inflorescences 3–7 cm; branchlet growing tips sparsely to densely hairy; leaf blades 5–8(–10) cm; branchlets sparsely arachnoid or glabrous . . . . 18. *Vitis monticola*
                  18. Berries without lenticels; inflorescences 9–19 cm; branchlet growing tips glabrous to sparsely hairy; leaf blades (5–)9–18 cm; branchlets glabrous. . . . . 19. *Vitis vulpina*

1a. VITIS Linnaeus subg. MUSCADINIA (Planchon) Rehder, Man. Cult. Trees, 601. 1927

Vitis sect. *Muscadinia* Planchon in A. L. P. P. de Candolle and C. de Candolle, Monogr. Phan. 5: 324. 1887; *Muscadinia* (Planchon) Small

Branches: bark adherent; lenticels prominent; pith continuous through nodes; tendrils unbranched. $x = 20$.

Species 2 (1 in the flora): se United States, Mexico, West Indies.

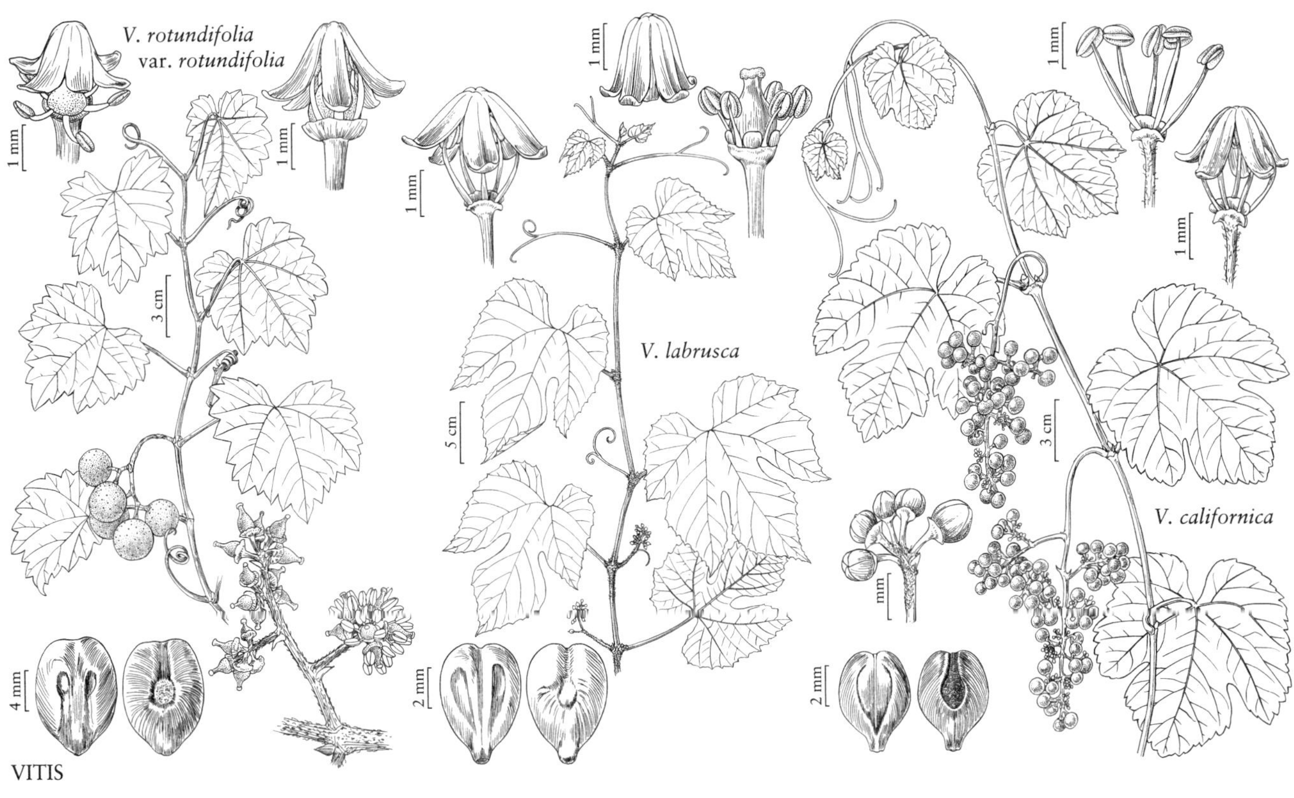

1. Vitis rotundifolia Michaux, Fl. Bor.-Amer. 2: 231. 1803 • Muscadine, scuppernong [E] [F]

Muscadinia rotundifolia (Michaux) Small

Plants usually high climbing or sprawling, sometimes shrubby, usually sparsely branched. **Branches:** branchlets terete to slightly angled, usually sparsely grayish arachnoid, glabrescent, growing tips not enveloped by unfolding leaves; tendrils along length of branchlets, persistent, tendrils (or inflorescences) at only 2 consecutive nodes; nodes sometimes red-banded. **Leaves:** stipules 1–2 mm; petiole mostly ± equaling blade; blade cordate to nearly reniform, 2–12 cm, rarely lobed, apex short acuminate, abaxial surface not glaucous, glabrous, visible through hairs, veins and vein axils sometimes sparsely to densely hirtellous, adaxial surface glabrous. **Inflorescences** 3–8(–10) cm. **Flowers** functionally unisexual. **Berries** usually black or purplish, sometimes bronze when ripe, glaucous, globose, 8–25 mm diam., skin separating from pulp; lenticels present.

Varieties 2 (2 in the flora): c, e United States.

1. Berries 15–25 mm diam., usually 2–8 per infructescence; seeds 9+ mm . . . . 1a. *Vitis rotundifolia* var. *rotundifolia*
1. Berries 8–12 mm diam., 12–30 per infructescence; seeds to 6 mm . . . 1b. *Vitis rotundifolia* var. *munsoniana*

1a. Vitis rotundifolia Michaux var. **rotundifolia** [E] [F]

Plants high climbing. **Leaf blades** (4–)5–9(–12) cm diam. **Berries** usually 2–8 per infructescence, 15–25 mm diam. **Seeds** 9+ mm. **2*n*** = 40.

Flowering late Apr–May; fruiting late Jul–Sep. Upland and lowland forests, swamps, thickets, forest edges, rocky ledges, hammocks, dunes; 0–1900 m; Ala., Ark., Del., Fla., Ga., Ky., La., Md., Miss., Mo., N.C., Okla., S.C., Tenn., Tex., Va., W.Va.

1b. Vitis rotundifolia Michaux var. **munsoniana** (J. H. Simpson ex Planchon) M. O. Moore, Sida 14: 345. 1991 • Everbearing or bullace or pygmy grape, little muscadine [E]

Vitis munsoniana J. H. Simpson ex Planchon in A. L. P. P. de Candolle and C. de Candolle, Monogr. Phan. 5: 615. 1887; *Muscadinia munsoniana* (J. H. Simpson ex Planchon) Small; *V. rotundifolia* var. *pygmaea* McFarlin ex D. B. Ward

Plants high climbing to sometimes shrubby or trailing. **Leaf blades** 3–8 cm diam. **Berries** 12–30 per infructescence, 8–12 mm diam. **Seeds** to 6 mm.

Flowering and fruiting year-round in peninsular Florida, late Apr–May farther north; fruiting late Jul–Sep. Floodplain forests, riverbanks, hammocks, pinelands, sand pine scrub; 0–200 m; Ala., Fla., Ga.

Variety *munsoniana* is distinguished from the more widespread var. *rotundifolia* by its smaller and more numerous berries, and these traits remain consistent under greenhouse conditions. It may well represent a species distinct from *Vitis rotundifolia*.

Variety *pygmaea* was described based on collections from Polk and Highlands counties, Florida, with shrubby or trailing habit, small leaves, and narrow and ridged seed raphe. Field studies by J. Wen suggest that these characteristics intergrade with those of var. *munsoniana* and that it is an extreme xeric ecotype of var. *munsoniana*, which occurs in the same areas.

1b. VITIS Linnaeus subg. VITIS

Branches: bark exfoliating; lenticels inconspicuous or absent; pith interrupted by nodal diaphragms; tendrils branched or absent. $x = 19$.

Species ca. 70 (18, including 3 hybrids, in the flora): North America, Mexico, West Indies, Central America, South America, Eurasia; introduced nearly worldwide.

2. Vitis vinifera Linnaeus, Sp. Pl. 1: 202. 1753 • Wine grape, grape of commerce [I]

Plants sprawling to moderately high climbing, sparsely branched. **Branches:** bark exfoliating in shreds or plates; nodal diaphragms 3–5 mm thick; branchlets terete to slightly angled, pubescent, sometimes glabrescent, growing tips not enveloped by unfolding leaves; tendrils along length of branchlets, persistent, tendrils (or inflorescences) at only 2 consecutive nodes; nodes not red-banded. **Leaves:** stipules usually more than 3.5 mm; petiole ± equaling blade; blade cordate-ovate to cordate-orbiculate, 12–20 cm, usually 3-shouldered to 3–5-lobed, sometimes deeply so, apex acute to short acuminate, abaxial surface not glaucous, sparsely pubescent to glabrate, visible through hairs, adaxial surface usually glabrous. **Inflorescences** 10–20 cm. **Flowers** bisexual. **Berries** usually reddish purple to nearly black, sometimes yellow-green, ± glaucous, oblong to ellipsoid, 8–25 mm diam., skin adhering to pulp; lenticels absent. $2n = 38, 76$.

Flowering Apr–Jun; fruiting Jul–Sep. Riparian areas, disturbed sites; 0–1200 m; introduced; B.C.; Calif., Idaho, Mass., N.H., N.Y., Oreg., Pa., Wash.; Europe; introduced widely.

Vitis vinifera and cultivars formed by hybridization between it and native North American species or through selection are cultivated in Europe, many parts of the United States and southern Canada, and parts of Central and South America, Africa, Asia, and Australia; these have been reported persisting from cultivation (for example, in California, Utah, and Virginia) and occasionally escaping. Some specimens keying here may represent naturally occurring hybrids between native species and *V. vinifera* or its hybrid cultivars.

3. Vitis aestivalis Michaux, Fl. Bor.-Amer. 2: 230. 1803 • Summer or post oak or big summer or silverleaf or blue grape, vigne d'été [E] [W]

Vitis aestivalis var. *argentifolia* Fernald; *V. aestivalis* var. *bicolor* Deam; *V. aestivalis* var. *linsecomii* (Buckley) Munson ex L. H. Bailey; *V. aestivalis* var. *sinuata* Pursh; *V. aestivalis* subsp. *smalliana* (L. H. Bailey) W. M. Rogers; *V. aestivalis* var. *smalliana* (L. H. Bailey) Comeaux; *V. araneosa* J. Le Conte; *V. gigas* J. L. Fennell; *V. labrusca* Linnaeus var. *aestivalis* (Michaux) Regel; *V. linsecomii* Buckley; *V. linsecomii* var. *glauca* Munson; *V. rufotomentosa* Small; *V. sinuata* (Pursh) G. Don; *V. smalliana* L. H. Bailey; *V. vinifera* Linnaeus var. *aestivalis* (Michaux) Kuntze

Plants high climbing, sparsely branched. **Branches:** bark exfoliating in shreds; nodal diaphragms 1–4 mm thick; branchlets terete, tomentose, arachnoid-floccose, or glabrous, growing tips not enveloped by unfolding leaves; tendrils along length of branchlets, persistent, tendrils (or inflorescences) at only 2 consecutive nodes; nodes not red-banded. **Leaves:** stipules 1–4 mm; petiole ± equaling blade; blade cordate to orbiculate, 7–25 cm, unlobed to 3-shouldered or 3–5-lobed, sometimes deeply so, apex acute to short acuminate, abaxial surface glaucous, ± arachnoid or floccose, visible through hairs, hairs usually rusty, sometimes whitish, veins and vein axils sometimes hirtellous, adaxial surface glabrous or puberulent. **Inflorescences** 7–20 cm. **Flowers** functionally unisexual. **Berries** black, glaucous, globose, 8–20 mm diam., skin separating from pulp; lenticels absent. $2n = 38$.

Flowering Apr–Jun; fruiting Jul–Sep. Woodlands, woodland borders, thickets, fence- and hedgerows, scrub, stabilized dunes, stream or riverbanks, floodplain and upland forests, lowland woods; 0–2000 m; Ont.; Ala., Ark., Conn., Del., D.C., Fla., Ga., Ill., Ind., Iowa, Kans., Ky., La., Maine, Md., Mass., Mich., Minn., Miss., Mo., Nebr., N.H., N.J., N.Y., N.C., Ohio, Okla., Pa., R.I., S.C., Tenn., Tex., Vt., Va., W.Va., Wis.

Vitis aestivalis is sometimes confused with *V. cinerea*. However, the glaucous abaxial leaf surfaces, more heavily glaucous and larger berries, terete branchlets that are less evenly pubescent, preference for better drained, drier habitats, and earlier blooming period distinguish *V. aestivalis* from *V. cinerea*.

Several varieties have been recognized based on leaf and young stem pubescence and fruit sizes (B. L. Comeaux et al. 1987; M. O. Moore 1991). The boundaries of the varieties are highly inconsistent and it is often difficult to apply the varietal concepts in the field. Recent molecular evidence supports the monophyly of *Vitis aestivalis*, but none of the varieties were supported as monophyletic (Wan Y. et al. 2013).

4. Vitis labrusca Linnaeus, Sp. Pl. 1: 203. 1753 • Fox grape, vigne lambruche [E] [F] [W]

Vitis labrusca var. *alba* W. R. Prince; *V. labrusca* var. *labruscoides* Eaton; *V. labrusca* var. *rosea* W. R. Prince; *V. labrusca* var. *subedentata* Fernald

Plants high climbing, sparsely branched. **Branches:** bark exfoliating in shreds; nodal diaphragms 0.5–2.5 mm thick; branchlets terete, densely tomentose to arachnoid-floccose or glabrous, sometimes with spinose, gland-tipped hairs, growing tips not enveloped by unfolding leaves; tendrils along length of branchlets, persistent, branched, tendrils (or inflorescences) at almost all nodes; nodes not red-banded. **Leaves:** stipules 2–4 mm; petiole ± equaling blade; blade cordate, usually 3-shouldered, sometimes unlobed or deeply 3(–5)-lobed, 10–20 cm, apex usually acute, abaxial surface not glaucous, densely and persistently arachnoid, concealed (except sometimes veins) by hairs, adaxial surfaces glabrous or sparsely pubescent. **Inflorescences** 6–14 cm. **Flowers** functionally unisexual. **Berries** black, usually not, sometimes slightly, glaucous, globose, 12+ mm diam., skin separating from pulp; lenticels absent. $2n = 38$.

Flowering May–Jun; fruiting Sep–Oct. Upland and lowland woods and forests, intermittently flooded bottomlands, forest edges, thickets, roadsides; 0–2000 m; Ont.; Ala., Conn., Del., D.C., Ga., Ill., Ind., Ky., Maine, Md., Mass., Mich., Miss., N.H., N.J., N.Y., N.C., Ohio, Pa., R.I., S.C., Tenn., Vt., Va., Wis.

Hybrids between *Vitis labrusca* and *V. vinifera*, such as "Concord," are widely cultivated. The Concord grape is the result of crosses with *V. vinifera* as the maternal parent and *V. labrusca* as the paternal parent, with the F_1 backcrossed with *V. labrusca* as the paternal parent (J. Wen, unpubl.). *Vitis* ×*labruscana* L. H. Bailey is the name applied to these hybrids between *V. labrusca* and *V. vinifera*, some of which have escaped from cultivation and become naturalized in New Brunswick, Nova Scotia, Utah, and western British Columbia, Oregon, and Washington.

5. Vitis ×novae-angliae Fernald, Rhodora 19: 146. 1917, as species • New England grape [E]

Plants high climbing, sparsely branched. **Branches:** bark exfoliating in shreds; nodal diaphragms 0.3–1.1 mm thick; branchlets terete, sparsely arachnoid or glabrous, growing tips not enveloped by unfolding leaves; tendrils along length of branchlets, persistent, branched, tendrils (or inflorescences) at 3+ consecutive nodes, usually not at all nodes; nodes not red-banded. **Leaves:** stipules 2.5–6 mm; petiole ½ to ± equaling blade; blade cordate, 10–20 cm, usually 3-shouldered, apex acute to short acuminate, abaxial surface not glaucous, ± densely arachnoid when young, sparsely arachnoid when mature, visible through hairs, adaxial surface usually glabrous. **Inflorescences** 7–13 cm. **Flowers** functionally unisexual. **Berries** black, slightly glaucous, globose, 12+ mm diam., skin separating from pulp; lenticels absent.

Flowering Jun; fruiting Aug–Sep. Thickets, roadsides, pond and stream margins, fence- and hedgerows; 70–1400 m; Ont.; Conn., Maine, Mass., N.H., N.J., N.Y., Pa., R.I., Vt.

Vitis ×*novae-angliae* is a presumed hybrid between *V. labrusca* and *V. riparia* and is common in the New England region (M. O. Moore 1991).

6. Vitis mustangensis Buckley, Proc. Acad. Nat. Sci. Philadelphia 13: 451. 1862 • Mustang grape [E] [W]

Vitis candicans Engelmann ex Durand var. *diversa* L. H. Bailey; *V. mustangensis* var. *diversa* (L. H. Bailey) Shinners

Plants high climbing, sparsely branched. **Branches:** bark exfoliating in shreds; nodal diaphragms 1.5–3 mm thick; branchlets subterete to terete, densely to sparsely tomentose, growing tips not enveloped by unfolding leaves; tendrils along length of branchlets, persistent, branched, tendrils (or inflorescences) at only 2 consecutive nodes; nodes not red-banded. **Leaves:** stipules 1.5–4 mm; petiole ½–¾ blade; blade cordate to nearly reniform, 6–14 cm, usually unlobed, sometimes 3-shouldered or deeply 3–5 lobed, apex acute to obtuse, abaxial surface not glaucous, densely white to rusty tomentose, concealed (except sometimes veins) by hairs, adaxial surface floccose to glabrate. **Inflorescences** 4–10 cm. **Flowers** functionally unisexual. **Berries** usually black, sometimes dark red, slightly or not glaucous, globose, 12+ mm diam., skin separating from pulp; lenticels absent. **2*n*** = 38.

Flowering Apr–early Jun; fruiting Aug–Sep. Woodland edges, fencerows, thickets, lowland woods, disturbed areas; 0–700 m; Ala., La., Miss., Okla., Tex.

In several early publications (for example, T. V. Munson 1909), *Vitis mustangensis* was known as *V. candicans* Engelmann ex A. Gray. M. O. Moore (1991) argued that the name *V. candicans* is ambiguous and not identifiable with any species based on the original description, making the more recent name *V. mustangensis* the valid and legitimate one for this species.

7. Vitis shuttleworthii House, Amer. Midl. Naturalist 7: 129. 1921 • Calloosa or leatherleaf or Florida grape [E]

Vitis coriacea Shuttleworth ex Planchon in A. L. P. P. de Candolle and C. de Candolle, Monogr. Phan. 5: 345. 1887, not Miquel 1863; *V. candicans* Engelmann ex Durand var. *coriacea* L. H. Bailey

Plants moderately high climbing, sparsely branched. **Branches:** bark exfoliating in shreds; nodal diaphragms 2.5–6 mm thick, sometimes continuing halfway into internode; branchlets subterete to terete, densely to sparsely tomentose, growing tips not enveloped by unfolding leaves; tendrils along length of branchlets, persistent, branched, tendrils (or inflorescences) at only 2 consecutive nodes; nodes not red-banded. **Leaves:** stipules less than 1 mm; petiole ½–¾ blade; blade broadly cordate to nearly reniform, 4–12 cm, usually unlobed, sometimes 3-shouldered, infrequently deeply 3–5 lobed, apex acute to obtuse, abaxial surface not glaucous, densely white to rusty tomentose, concealed (except sometimes veins) by hairs, adaxial surface floccose to glabrate. **Inflorescences** 4–10 cm. **Flowers** functionally unisexual. **Berries** dark red to purple-black, slightly or not glaucous, globose, 12+ mm diam., skin separating from pulp; lenticels absent. **2*n*** = 38.

Flowering Apr–early May; fruiting Jun–Aug. Well-drained pinelands, thickets; 0–100 m; Fla.

Vitis shuttleworthii is endemic to peninsular Florida and apparently is the closest relative of *V. mustangensis*.

8. Vitis rupestris Scheele, Linnaea 21: 591. 1848 • Rock or sand grape [E]

Vitis rupestris var. *dissecta* Eggert ex L. H. Bailey

Plants sprawling to low climbing, shrubby, much branched. **Branches:** bark tardily exfoliating in plates; nodal diaphragms to 1 mm thick; branchlets terete, usually glabrous, sometimes sparsely hirtellous, growing tips enveloped by unfolding leaves; tendrils absent or only at distalmost nodes, soon deciduous if not attached to support, branched, tendrils (or inflorescences) at only 2 consecutive nodes; nodes not red-banded. **Leaves:** stipules 3–6.5 mm; petiole ½ blade; blade reniform, conduplicately folded, 5–10 cm, apex acute to short acuminate, usually 3-shouldered, rarely shallowly 3-lobed, abaxial surface not glaucous, usually glabrous, visible through hairs, veins and vein axils sometimes sparsely hirtellous, adaxial surface usually glabrous. **Inflorescences** 4–7 cm. **Flowers** functionally unisexual. **Berries** black, slightly glaucous, globose, 8–12 mm diam., skin separating from pulp; lenticels absent. **2*n*** = 38.

Flowering Apr–May; fruiting Aug–Sep. Gravelly banks, river bottoms, stream beds, washes, often calcareous soils; 70–500 m; Ark., D.C., Ind., Ky., Md., Mo., Okla., Pa., Tenn., Tex., Va., W.Va.

Vitis rupestris once was widely scattered throughout most of its range but now mostly is rare and may have been extirpated in many locations, apparently due to habitat loss. It is most common in the Ozark region of northern Arkansas and the southern half of Missouri, but is imperiled elsewhere (http://explorer.natureserve.org). It is persisting from cultivation in California and some other locations (J. Wen, pers. obs.; E. B. Wada and M. A. Walker, http://ucjeps.berkeley.edu/cgi-bin/get_IJM.pl?tid=48433). Reports from Illinois were based on misidentifications (R. H. Mohlenbrock 2014). The species was used to develop many grape hybrids due to its resistance to disease (J. Gerrath et al. 2015).

9. Vitis acerifolia Rafinesque, Med. Fl. 2: 130, plate 99, fig. C. 1830 • Bush or panhandle or mapleleaf grape [E]

Vitis cordifolia Michaux var. *solonis* Planchon; *V. longii* Prince; *V. longii* var. *microsperma* (Munson) L. H. Bailey; *V. nuevomexicana* Lemmon ex Munson; *V. solonis* (Planchon) Engelmann ex Millardet; *V. solonis* var. *microsperma* Munson

Plants low to moderately high climbing, much branched. **Branches:** bark tardily exfoliating in plates; nodal diaphragms to 0.5(–1) mm thick; branchlets terete, whitish arachnoid or glabrate, growing tips enveloped by unfolding leaves, sparsely to densely hairy; tendrils along length of branchlets, soon deciduous if not attached to support, branched, tendrils (or inflorescences) at only 2 consecutive nodes; nodes not red-banded. **Leaves:** stipules 3–6 mm; petiole ½–⅔ blade; blade broadly cordate, 7–12 cm, usually 3-shouldered to shallowly 3-lobed, apex usually short acuminate, abaxial surface not glaucous, sparsely arachnoid to glabrate, visible through hairs, veins sparsely hirtellous, adaxial surface slightly arachnoid to glabrate. **Inflorescences** 3–7(–9) cm. **Flowers** functionally unisexual. **Berries** black, heavily glaucous, globose, 8–12 mm diam., skin separating from pulp; lenticels absent. $2n = 38$.

Flowering Apr–May; fruiting Jul–Aug. Stream and riverbanks, prairie ravines, alluvial floodplain woodlands, dunes, rocky slopes, fencerows; 200–2300 m; Colo., Kans., N.Mex., Okla., Tex.

10. Vitis riparia Michaux, Fl. Bor.-Amer. 2: 231. 1803 • Riverbank or frost grape, vigne de rivages [E]

Vitis cordifolia Michaux var. *riparia* (Michaux) A. Gray; *V. riparia* var. *syrticola* (Fernald & Wiegand) Fernald; *V. vulpina* Linnaeus subsp. *riparia* (Michaux) R. T. Clausen; *V. vulpina* var. *syrticola* Fernald & Wiegand

Plants usually moderate to high climbing, sometimes sprawling, sparsely branched. **Branches:** bark exfoliating in shreds; nodal diaphragms to 0.5 mm thick; branchlets terete, glabrous or sparsely hirtellous, growing tips enveloped by unfolding leaves, glabrous or sparsely hairy; tendrils along length of branchlets, persistent, branched, tendrils (or inflorescences) at only 2 consecutive nodes; nodes not red-banded. **Leaves:** stipules 3–5 mm; petiole ½ to ± equaling blade; blade cordate, 6–20 cm, 3-shouldered to shallowly 3-lobed, apex short acuminate, abaxial surface not glaucous, glabrate, visible through hairs, veins and vein axils hirtellous, adaxial surface glabrous. **Inflorescences** (4–)9–12 cm. **Flowers** functionally unisexual. **Berries** black, heavily glaucous, globose, 8–12 mm diam., skin separating from pulp; lenticels absent. $2n = 38$.

Flowering Apr–Jun; fruiting Aug–Sep. Stream and riverbanks, pond margins, alluvial woodlands, ravines, thickets, roadsides, fencerows; 0–2200 m; Man., N.B., N.S., Ont., Que., Sask.; Ala., Ark., Colo., Conn., D.C., Ill., Ind., Iowa, Kans., Ky., La., Maine, Md., Mass., Mich., Minn., Miss., Mo., Mont., Nebr., N.H., N.J., N.Mex., N.Y., N.C., N.Dak., Ohio, Okla., Oreg., Pa., R.I., S.Dak., Tenn., Tex., Vt., Va., Wash., W.Va., Wis., Wyo.

Vitis riparia is native throughout much of its range. It has become naturalized in a few locations in Saskatchewan, Oregon, and Washington. Plants on dunes around the Great Lakes with hairier petioles and leaf blades sometimes have been recognized as var. *syrticola*; variation in hairiness is essentially continuous, however, and the form is not worthy of taxonomic recognition (P. M. Catling and G. Mitrow 2005).

11. Vitis ×doaniana Munson ex Viala, Mission Vitic. Amér., 101. 1889 • Panhandle or Doan's grape [E]

Plants usually high climbing, sometimes sprawling and ± shrubby when without support, sparsely branched. **Branches:** bark tardily exfoliating in shreds; nodal diaphragms 1–2 mm thick; branchlets terete, densely tomentose to glabrate, growing tips not enveloped by unfolding leaves; tendrils along length of branchlets, persistent, branched, tendrils (or inflorescences) at only 2 consecutive nodes; nodes not red-banded. **Leaves:** stipules 3–6 mm; petiole ½ blade; blade cordate, 5–15 cm, usually 3-shouldered to shallowly 3-lobed, apex acute to short acuminate, abaxial surface not glaucous, moderately to densely arachnoid, visible through hairs, veins hirtellous, adaxial surface sparsely to moderately arachnoid. **Inflorescences** 4–10 cm. **Flowers** functionally unisexual. **Berries** black, heavily glaucous, globose, 12+ mm diam., skin separating from pulp; lenticels absent. $2n = 38$.

Flowering Apr–May; fruiting Jul–Aug; 200–400 m; Okla., Tex.

Vitis ×doaniana is endemic to well-drained soils of the Rolling Plains and Cross Timbers and Prairies regions in north-central Texas and adjacent Oklahoma.

Vitis ×doaniana is a hybrid between *V. mustangensis* and *V. acerifolia*, and was once more common in nature than it is at present. It was named for Judge J. Doan of Wilbarger County, Texas, who manufactured wine from the berries of this species.

12. Vitis ×champinii Planchon, Vigne Amér. Vitic. Eur. 6: 22. 1882 • Champin's grape [E]

Plants moderate to high climbing, sparsely branched. **Branches:** bark tardily exfoliating in shreds; nodal diaphragms 1.5–2.5 mm thick; branchlets terete, sparsely arachnoid, becoming glabrate, growing tips not enveloped by unfolding leaves; tendrils along length of branchlets, persistent, branched, tendrils (or inflorescences) at only 2 consecutive nodes, nodes not red-banded. **Leaves:** stipules 2–5.5 mm; petiole ½ blade; blade usually cordate, sometimes nearly reniform, 5–15 cm, usually 3-shouldered to very shallowly 3-lobed, apex acute to short acuminate, abaxial surface not glaucous, sparsely arachnoid to glabrate, not hirtellous, visible through hairs, adaxial surface usually glabrous. **Inflorescences** 3–7 cm. **Flowers** functionally unisexual. **Berries** black, usually not, sometimes very slightly, glaucous, globose, 12+ mm diam., skin separating from pulp; lenticels absent. $2n = 38$.

Flowering Apr–May; fruiting Jul–Aug. Well drained, calcareous soils; 200–700 m; Tex.

Vitis ×champinii is found on and adjacent to the Edwards Plateau and is a hybrid between *V. mustangensis* and *V. rupestris*; it is now rare in nature (B. L. Comeaux 1987).

13. Vitis californica Bentham, Bot. Voy. Sulphur, 10. 1844 • Pacific or California wild or northern California grape [E] [F]

Plants high climbing, sparsely branched. **Branches:** bark exfoliating in plates; nodal diaphragms 3–4 mm thick; branchlets terete to slightly angled, tomentose, tomentum thinning in age, sometimes also hirtellous, growing tips not enveloped by unfolding leaves; tendrils along length of branchlets, persistent, branched, tendrils (or inflorescences) at only 2 consecutive nodes; nodes not red-banded. **Leaves:** stipules usually less than 3.5 mm; petiole ± equaling blade; blade cordate to orbiculate or nearly reniform, 7–15 cm, unlobed or shallowly 3–5-lobed, apex acute to short acuminate, abaxial surface not glaucous, moderately to sparsely tomentose, visible through hairs, adaxial surface sparsely tomentose to glabrate. **Inflorescences** 5–10 cm. **Flowers** functionally unisexual. **Berries** purple to black, moderately to heavily glaucous, globose, 8–10 mm diam., skin separating from pulp; lenticels absent. $2n = 38$.

Flowering May–Jun; fruiting Aug–Sep. Stream banks, perennial springs, canyons; 10–1500 m; Calif., Oreg.

Vitis californica occurs in central and northern California (from San Luis Obispo to Inyo counties north) to southern Oregon. It is variable and intergrades with *V. girdiana* in southern California, and hybridizes with both *V. girdiana* and *V. vinifera*, often making identification difficult.

14. Vitis girdiana Munson, Proc. Annual Meetings Soc. Promot. Agric. Sci. 8: 59. 1887 • Southern California or desert wild grape

Plants high climbing, sparsely branched. **Branches:** bark exfoliating in plates; nodal diaphragms 1.5–3 mm thick; branchlets terete to slightly angled, tomentose, tomentum usually persistent, growing tips not enveloped by unfolding leaves; tendrils along length of branchlets, persistent, branched, tendrils (or inflorescences) at only 2 consecutive nodes; nodes not red-banded. **Leaves:** stipules usually 3.5+ mm; petiole ½ to ± equaling blade; blade cordate, 5–10 cm, usually unlobed, sometimes 3-shouldered, rarely shallowly 3-lobed, apex acute to short acuminate, abaxial surface not glaucous, moderately to densely tomentose, visible through hairs, adaxial surface sparsely to moderately tomentose, glabrescent. **Inflorescences** 8–18 cm. **Flowers** functionally unisexual. **Berries** dark purple to black, slightly or not glaucous, globose, 4–6 mm diam., skin separating from pulp; lenticels absent. $2n = 38$.

Flowering Apr–Jun; fruiting Jul–Sep. Stream banks, canyon bottoms; 10–2000 m; Calif.; Mexico (Baja California).

Vitis girdiana is known from Inyo, Kern, and Santa Barbara counties and southward. It is morphologically very similar to and intergrades and hybridizes with *V. californica* in central California. Hybridization with *V. vinifera* is also probable, making identifications of some specimens quite difficult. *Vitis girdiana* intergrades also with *V. arizonica* in eastern Inyo and San Bernardino counties near the California-Nevada border. The *V. californica-girdiana-arizonica* species complex is in need of in-depth field and experimental studies.

15. Vitis arizonica Engelmann, Amer. Naturalist 2: 321. 1868 • Canyon grape

Vitis arizonica var. *galvinii* Munson; *V. arizonica* var. *glabra* Munson; *V. treleasei* Munson ex L. H. Bailey

Plants sprawling to low climbing, shrubby, much branched. **Branches:** bark exfoliating in plates; nodal diaphragms 1.5–3 mm thick; branchlets slightly angled when young, becoming terete, arachnoid or arachnoid-floccose, sometimes glabrescent, growing tips not enveloped by unfolding leaves; tendrils along length of branchlets, soon deciduous if not attached to support, branched, tendrils (or inflorescences) at only 2 consecutive nodes; nodes not red-banded. **Leaves:** stipules 1.5–3 mm; petiole ½ to ± equaling blade; blade cordate to cordate-ovate, 5–12 cm, usually unlobed to 3-shouldered, sometimes shallowly 3-lobed, apex acute to acuminate, abaxial surface not glaucous, moderately to sparsely arachnoid, visible through hairs, veins and vein axils sometimes only hirtellous, adaxial surface sparsely arachnoid or glabrous. **Inflorescences** 4–12 cm. **Flowers** functionally unisexual. **Berries** black, slightly or not glaucous, globose, 6–10 mm diam., skin separating from pulp; lenticels absent. $2n = 38$.

Flowering Apr–Jun; fruiting Jul–Oct. Stream banks, canyon bottoms; 400–3000 m; Ariz., Nev., N.Mex., Tex., Utah; Mexico (Chihuahua, Coahuila, Nuevo León, Sinaloa, Sonora).

Vitis arizonica is variable and intergrades with *V. girdiana* in southern Nevada; it is in need of in-depth field and experimental studies. Some authors have recognized two varieties of this species (vars. *arizonica* and *glabra*), but the characters used to distinguish them intergrade so freely that their recognition does not seem justified.

16. Vitis cinerea (Engelmann) Millardet, Mém. Soc. Sci. Phys. Nat. Bordeaux, sér. 2, 3: 319, 336. 1880 • Downy or sweet winter or graybark grape, parra silvestre

Vitis aestivalis Michaux var. *cinerea* Engelmann in A. Gray, Manual ed. 5, 676. 1867

Plants high climbing, sparsely branched. **Branches:** bark exfoliating in shreds; nodal diaphragms 1.5–3.5 mm thick; branchlets slightly to distinctly angled, densely hirtellous and/or sparsely to densely arachnoid, to glabrate, growing tips not enveloped by unfolding leaves; tendrils along length of branchlets, persistent, branched, tendrils (or inflorescences) at only 2 consecutive nodes; nodes sometimes red-banded. **Leaves:** stipules 1–3 mm; petiole ± equaling blade; blade cordate, 6–20 cm, usually unlobed to 3-shouldered, sometimes 3-lobed, apex acute to acuminate, abaxial surface not glaucous, sparsely to densely arachnoid or glabrous, visible through hairs, veins and vein axils hirtellous, adaxial surface glabrous or hairy. **Inflorescences** 10–25 cm. **Flowers** functionally unisexual. **Berries** black, slightly or not glaucous, globose, 4–8 mm diam., skin separating from pulp, lenticels absent.

Varieties 5 (4 in the flora): c, e United States, ne Mexico.

Vitis cinerea var. *tomentosa* (Planchon) Comeaux is endemic to northeastern Mexico (B. L. Comeaux and J. Lu 2000).

Vitis cinerea is sometimes confused with *V. aestivalis*; see the discussion under that species. *Vitis cinerea* as defined here is highly variable and is in need of field studies and phylogeographic analysis, along with its tropical relatives *V. biforma* Rose and *V. tiliifolia*. Wan Y. et al. (2013) concluded that *V. cinerea* is not monophyletic.

1. Berries moderately to heavily glaucous; leaf blade abaxial surface sparsely hirtellous and arachnoid or glabrate . . . . 16b. *Vitis cinerea* var. *helleri*
1. Berries slightly glaucous; leaf blade abaxial surface sparsely to densely arachnoid, sparsely to moderately hirtellous, or glabrous.
   2. Branchlets densely hirtellous and arachnoid; leaf blade abaxial surface moderately arachnoid and hirtellous . . . . 16a. *Vitis cinerea* var. *cinerea*
   2. Branchlets sparsely to densely arachnoid, not evidently hirtellous; leaf blade abaxial surface sparsely to densely arachnoid, not, or sometimes very sparsely, hirtellous.
      3. Branchlets sparsely to densely arachnoid; nodes usually not red-banded, sometimes so; leaf blade abaxial surface sparsely to densely arachnoid; se Coastal Plain . . . . 16c. *Vitis cinerea* var. *floridana*
      3. Branchlets sparsely arachnoid, becoming glabrate; nodes usually red-banded; leaf blade abaxial surface glabrous or sparsely arachnoid; Piedmont and mountains . . . . 16d. *Vitis cinerea* var. *baileyana*

16a. Vitis cinerea (Engelmann) Millardet var. **cinerea** E

Vitis aestivalis Michaux var. *canescens* Engelmann; *V. cinerea* var. *canescens* (Engelmann) L. H. Bailey

Branchlets distinctly angled, densely hirtellous and arachnoid; nodes sometimes red-banded. **Leaf blade** apex acuminate, abaxial surface moderately arachnoid and hirtellous. **Berries** slightly glaucous. $2n$ = 38.

Flowering late May–Jun; fruiting Jul–Oct. Flood plains, lowland woods, pond and stream margins; 0–500 m; Ala., Ark., Ill., Ind., Iowa, Kans., Ky., La., Miss., Mo., Nebr., Okla., Tenn., Tex.

Variety *cinerea* is most common in rich bottomlands of the Mississippi River basin; in Texas, it is found only in the eastern part of the state, where it intergrades with var. *helleri* southwest of the Brazos River.

16b. Vitis cinerea (Engelmann) Millardet var. **helleri** (L. H. Bailey) M. O. Moore, Sida 14: 352. 1991 • Little mountain or Spanish grape, uva cimarrona

Vitis cordifolia Michaux var. *helleri* L. H. Bailey in A. Gray et al., Syn. Fl. N. Amer. 1(1,2): 424. 1897; *V. berlandieri* Planchon; *V. cinerea* var. *berlandieri* (Planchon) Comeaux; *V. helleri* (L. H. Bailey) Small

Branchlets slightly angled, arachnoid, not hirtellous; nodes usually not red-banded. **Leaf blade** apex acute, abaxial surface sparsely hirtellous and arachnoid to glabrate. **Berries** moderately to heavily glaucous. $2n$ = 38.

Flowering May–Jun; fruiting Aug–Oct. Canyons, limestone slopes, flood plains; 100–700 m; Tex.; Mexico (Coahuila, Nuevo León, Tamaulipas).

Variety *helleri* in the flora area is most common on the Edwards Plateau but found also in the Cross Timbers and Prairies region and the Blackland Prairie. It has more compact and shorter infructescences and more glaucous fruits than the other varieties, and may be more appropriately recognized at the species rank as *Vitis berlandieri*.

16c. Vitis cinerea (Engelmann) Millardet var. **floridana** Munson, Rev. Vitic. 6: 424. 1896 • Simpson's grape E

Vitis aestivalis Michaux subsp. *divergens* W. M. Rogers; *V. aestivalis* subsp. *sola* (L. H. Bailey) W. M. Rogers; *V. austrina* Small; *V. simpsonii* Munson; *V. sola* L. H. Bailey

Branchlets slightly angled, sparsely to densely arachnoid, not evidently hirtellous; nodes usually not red-banded, sometimes so. **Leaf blade** apex acuminate, abaxial surface sparsely to densely arachnoid, usually not, sometimes very sparsely, hirtellous. **Berries** slightly glaucous.

Flowering late May–Jun; fruiting Jul–Oct. Flood plains, lowland woods, stream banks, marsh and pond margins; 0–100 m; Ala., Fla., Ga., La., Md., Miss., N.C., S.C., Va.

Variety *floridana* is found on the coastal plain and intergrades with var. *baileyana* along the fall line at the edge of the Piedmont. Some workers recognized it at the species rank as *Vitis simpsonii* (for example, L. H. Bailey 1934; D. B. Ward 2006b), despite the confusion on the name by Munson; see discussions in B. L. Comeaux and P. R. Fantz (1987). It is sometimes confused with 3. *V. aestivalis*; see discussion under that species.

16d. Vitis cinerea (Engelmann) Millardet var. **baileyana** (Munson) Comeaux, Castanea 52: 212. 1987 • Possum grape E

Vitis baileyana Munson, Vitis baileyana, [1]. 1893

Branchlets slightly angled, sparsely arachnoid, becoming glabrate, not evidently hirtellous; nodes usually red-banded. **Leaf blade** apex acute to acuminate, abaxial surfaces glabrous or sparsely arachnoid, usually not, sometimes very sparsely, hirtellous. **Berries** slightly glaucous.

Flowering mid Jun–Jul; fruiting Jul–Oct. Moist soils, flood plains, lowland woods, stream and pond margins; 100–2000 m; Ala., Ga., Ind., Ky., Miss., N.C., Ohio, Pa., S.C., Tenn., Va., W.Va.

Variety *baileyana* is found in the Piedmont and mountains; it and var. *floridana* intergrade along the fall line between the Piedmont and coastal plain and may not be distinct.

17. Vitis palmata Vahl, Symb. Bot. 3: 42. 1794 • Catbird or red grape [E]

Vitis rubra Michaux ex Planchon

Plants high climbing, sparsely branched. **Branches:** bark exfoliating in shreds; nodal diaphragms 2.5–4 mm thick; branchlets uniformly red, purplish red, or chestnut, subterete, glabrous or very sparsely arachnoid, growing tips not enveloped by unfolding leaves; tendrils red-pigmented when young, along length of branchlets, persistent, branched, tendrils (or inflorescences) at only 2 consecutive nodes; nodes not red-banded. **Leaves:** stipules 1.5–3 mm; petiole somewhat shorter than blade; blade usually cordate, 8–14 cm, usually deeply 3(–5)-lobed, apex long acuminate, abaxial surface not glaucous, glabrous, visible, veins and vein axils sometimes hirtellous, adaxial surface glabrous. **Inflorescences** 6–18 cm. **Flowers** functionally unisexual. **Berries** bluish black to black, slightly or not glaucous, globose, 8–10 mm diam., skin separating from pulp; lenticels absent. $2n = 38$.

Flowering mid Jun–early Jul; fruiting Aug–Oct. Riverbanks, sloughs, alluvial floodplain woodlands; 0–200 m; Ala., Ark., Fla., Ga., Ill., Ind., Ky., La., Miss., Mo., Okla., Tenn., Tex.

Reports of *Vitis palmata* from Virginia (for example, in A. S. Weakley et al. 2012) appear to be based on misidentified material of *V. vulpina* with somewhat lobed leaves.

18. Vitis monticola Buckley, Proc. Acad. Nat. Sci. Philadelphia 13: 450. 1862 • Sweet mountain grape [E]

Vitis aestivalis Michaux var. *monticola* (Buckley) Engelmann

Plants moderate to high climbing, sparsely branched. **Branches:** bark exfoliating in shreds; nodal diaphragms 1–2.5 mm thick; branchlets gray to green or brown, if purplish only on one side, terete, sparsely arachnoid or glabrous, growing tips not enveloped by unfolding leaves, sparsely to densely hairy; tendrils along length of branchlets, persistent, branched, tendrils (or inflorescences) at only 2 consecutive nodes; nodes not red-banded. **Leaves:** stipules 1.5–3 mm; petiole ½ blade; blade cordate, 5–8(–10) cm, unlobed or shallowly 3-lobed, apex acute to short acuminate, abaxial surface not glaucous, glabrous or sparsely hirtellous, visible through hairs, adaxial surface usually glabrous. **Inflorescences** 3–7 cm. **Flowers** functionally unisexual. **Berries** black, usually not, sometimes very slightly, glaucous, globose, 8–10 mm diam., skin separating from pulp; lenticels usually present. $2n = 38$.

Flowering May; fruiting Jul–Aug. Limestone hills and ridges; 300–700 m; Tex.

Vitis monticola is endemic to dry areas on the Edwards Plateau.

19. Vitis vulpina Linnaeus, Sp. Pl. 1: 203. 1753 • Chicken or winter grape, vigne des renards 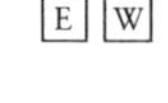

Vitis cordifolia Michaux; *V. cordifolia* var. *sempervirens* Munson; *V. illex* L. H. Bailey; *V. pullaria* J. Le Conte

Plants moderate to high climbing, sparsely branched. **Branches:** bark exfoliating in shreds; nodal diaphragms 1–2.5 mm thick; branchlets gray to green or brown, if purplish only on one side, terete, glabrous, growing tips not enveloped by unfolding leaves, glabrous or sparsely hairy; tendrils along length of branchlets, persistent, branched, tendrils (or inflorescences) at only 2 consecutive nodes; nodes not red-banded. **Leaves:** stipules 1.5–3 mm; petiole ± equaling blade; blade cordate, (5–)9–18 cm, usually unlobed or shallowly 3-lobed, sometimes deeply lobed on ground shoots, apex acute to short acuminate, abaxial surface not glaucous, glabrous, visible, veins and vein axils sometimes hirtellous, adaxial surface usually glabrous, sometimes sparsely hirtellous. **Inflorescences** 9–19 cm. **Flowers** functionally unisexual. **Berries** black, usually not, sometimes very slightly, glaucous, globose, 8–12 mm diam., skin separating from pulp; lenticels absent. $2n = 38$.

Flowering May; fruiting Jul–Aug. Upland forests, floodplain forests, woodland borders, prairies, fencerows, thickets, roadsides; 0–2000 m; Ont.; Ala., Ark., Del., D.C., Fla., Ga., Ill., Ind., Iowa, Kans., Ky., La., Md., Mich., Miss., Mo., Nebr., N.J., N.Y., N.C., Ohio, Okla., Pa., S.C., S.Dak., Tenn., Tex., Va., W.Va.

2. PARTHENOCISSUS Planchon in A. L. P. P. de Candolle and C. de Candolle, Monogr. Phan. 5: 447. 1887, name conserved • [Greek *parthenos*, virgin, and *kissos*, ivy; equivalent of vigne vierge, French name for type species, *P. quinquefolia*]

Landukia Planchon; *Psedera* Necker ex Greene

Lianas, climbing by tendrils or scrambling without support and producing adventitious roots, synoecious. **Branches:** bark adherent; pith white, continuous through nodes; tendrils 3–12-branched, with or without adhesive discs. **Leaves** usually palmately compound, sometimes simple. **Inflorescences** usually bisexual, leaf-opposed or terminal, paniclelike or corymblike cymes, sometimes compound. **Flowers** bisexual; calyx cup-shaped, shallowly 5-lobed; petals 5, distinct; nectary adnate to base of ovary, annular, entire, or absent; stamens 5; style conic, short. **Berries** dark blue or black. **Seeds** 2–4 per fruit. $x = 10$.

Species 15 (4 in the flora): North America, Mexico, West Indies, Central America, Asia.

1. Leaves usually simple, if palmately compound then leaflets 3 .... 1. *Parthenocissus tricuspidata*
1. Leaves palmately compound, leaflets (4–)5–7.
    2. Leaflets thick, fleshy, (5–)7, 3–5 cm; Edwards Plateau and Lampasas Cut Plain, Texas .... 2. *Parthenocissus heptaphylla*
    2. Leaflets thin, herbaceous, (4–)5(–7), 4–12 cm; throughout North America, including Edwards Plateau and Lampasas Cut Plain.
        3. Inflorescences dichotomously branched, without distinct central axes; berries 8–12 mm diam.; tendrils (2–)3-5-branched, usually without, rarely with, adhesive discs; leaflets lustrous adaxially .... 3. *Parthenocissus vitacea*
        3. Inflorescences divergently branched, with distinct central axes; berries 5–8 mm diam.; tendrils 4–12-branched, with adhesive discs; leaflets dull adaxially ....4. *Parthenocissus quinquefolia*

1. Parthenocissus tricuspidata (Siebold & Zuccarini) Planchon in A. L. P. P. de Candolle and C. de Candolle, Monogr. Phan. 5: 452. 1887 • Boston or Japanese ivy, vigne vierge tricuspidée [I]

Ampelopsis tricuspidata Siebold & Zuccarini, Abh. Math.-Phys. Cl. Königl. Wiss. 4(2): 196. 1845

Lianas, high climbing. **Tendrils** 5–10-branched, with adhesive discs. **Leaves** usually simple, sometimes palmately compound on older plants; petiole usually longer than, sometimes ± equaling, blade; blade lustrous adaxially, ovate to cordate-ovate or cordate-orbiculate, 4.5–17 × 4–16 cm, 3-lobed or leaflets 3, thin, herbaceous, base truncate to slightly cordate, margins crenate to crenate-serrate, apex acute to short-acuminate, surfaces glabrous or abaxial veins puberulent. **Inflorescences** ± divergently branching, without distinct central axis. **Flowers** yellowish green. **Berries** globose, 5–8 mm diam. $2n = 40$.

Flowering Jun–Jul; fruiting Sep–Oct. Thickets, forest edges, disturbed places; 50–500 m; introduced; Ont.; Conn., D.C., Ill., Ind., Iowa, Ky., Maine, Mass., Mo., Nebr., N.J., N.Y., N.C., Ohio, Pa., S.C., Tenn.; e Asia.

Parthenocissus tricuspidata has escaped from cultivation and become locally naturalized throughout much of eastern North America.

2. Parthenocissus heptaphylla (Planchon) Britton in J. K. Small, Fl. S.E. U.S., 759. 1903 • Sevenleaf creeper [E]

Parthenocissus quinquefolia (Linnaeus) Planchon var. *heptaphylla* Planchon in A. L. P. P. de Candolle and C. de Candolle, Monogr. Phan. 5: 449. 1887, based on *Ampelopsis heptaphylla* Buckley, Proc. Acad. Nat. Sci. Philadelphia 13: 450. 1862, not (Linnaeus) Roemer & Schules 1819; *Psedera heptaphylla* (Planchon) Rehder; *P. texana* (Buckley ex Durand) Greene

Lianas, scrambling to moderately high climbing. **Tendrils** 3–4(–5)-branched, without adhesive discs. **Leaves** palmately compound; petiole ± equaling blade; leaflets (5–)7, lustrous adaxially, oblanceolate to oblong-oblanceolate, 3–5 × 1–2 cm, thick, fleshy, base cuneate, margins coarsely dentate, apex acuminate, surfaces glabrous or abaxial puberulent. **Inflorescences** dichotomously branched, without distinct central axis. **Flowers** greenish to reddish green. **Berries** subglobose, 6–10 mm diam.

Flowering May–Jun; fruiting Aug–Sep. Woods, riverbanks, rocky or sandy soils; 300–700 m; Tex.

Parthenocissus heptaphylla is endemic to the Edwards Plateau and Lampasas Cut Plain.

3. **Parthenocissus vitacea** (Knerr) Hitchcock, Key Spring Fl. Manhattan, 26. 1894 • Woodbine, thicket creeper, vigne vierge commune [E]

Ampelopsis quinquefolia (Linnaeus) Michaux var. *vitacea* Knerr, Bot. Gaz. 18: [70]. 1893; *Parthenocissus inserta* (A. Kerner) Fritsch var. *laciniata* (Planchon) Rehder; *P. laciniata* (Planchon) Small; *P. quinquefolia* (Linnaeus) Planchon var. *vitacea* (Knerr) L. H. Bailey; *P. vitacea* var. *dubia* Rehder; *P. vitacea* var. *laciniata* (Planchon) Rehder; *P. vitacea* var. *macrophylla* (Lauche) Rehder; *Psedera vitacea* (Knerr) Greene

Lianas, high climbing or scrambling. **Tendrils** (2–)3–5-branched, usually without, rarely with, adhesive discs. **Leaves** palmately compound; petiole ± equaling blade; leaflets 5(–6), lustrous adaxially, oblong-obovate to elliptic, 4–10 × 2–4 cm, thin, herbaceous, base cuneate, margins coarsely serrate, usually distally, apex acuminate, surfaces glabrous or abaxial puberulent. **Inflorescences** dichotomously branched, without distinct central axis. **Flowers** greenish to reddish green. **Berries** subglobose, 6–12 mm diam. $2n = 40$.

Flowering May–Jun; fruiting Aug–Sep. Open woods, hillsides, thickets, ravines, fencerows, roadsides, waste places; 40–2500 m; B.C., Man., N.B., N.S., Ont., Que., Sask.; Ariz., Calif., Colo., Conn., Del., D.C., Idaho, Ill., Ind., Iowa, Kans., Maine, Md., Mass., Mich., Minn., Mo., Mont., Nebr., Nev., N.H., N.J., N.Mex., N.Y., N.Dak., Ohio, Okla., Oreg., Pa., R.I., S.Dak., Tex., Utah, Vt., Wash., Wis., Wyo.

The species treated here as *Parthenocissus vitacea* has sometimes been called *P. inserta* (for example, C. C. Deam 1940; M. L. Fernald 1950; H. J. Scoggan 1978–1979; J. S. Pringle 2010). According to the original description and illustration of *Vitis inserta*, the latter's basionym, that species has much branched tendrils and shows adhesive discs or tendencies of such discs (enlarged tendril tips). These traits are different from those of *P. vitacea*, which bears 2–4(–5) branched tendrils without enlarged tips, and therefore those names are treated here as synonyms of *P. quinquefolia* (but see arguments by Pringle).

4. **Parthenocissus quinquefolia** (Linnaeus) Planchon in A. L. P. P. de Candolle and C. de Candolle, Monogr. Phan. 5: 448. 1887 • Virginia creeper, hiedra, parra, vigne vierge à cinq folioles [F] [W]

Hedera quinquefolia Linnaeus, Sp. Pl. 1: 202. 1753; *Ampelopsis quinquefolia* (Linnaeus) Michaux; *Parthenocissus hirsuta* (Pursh) Graebner; *P. inserta* (A. Kerner) Fritsch; *P. quinquefolia* var. *murorum* (Focke) Rehder; *Psedera quinquefolia* (Linnaeus) Greene; *Vitis inserta* A. Kerner; *V. quinquefolia* (Linnaeus) Lamarck

Lianas, high climbing or scrambling. **Tendrils** 4–12-branched, with adhesive discs. **Leaves** palmately compound; petiole ± equaling blade; leaflets (4–)5(–7), dull adaxially, obovate to elliptic, 6–12 × 2–5 cm, thin, herbaceous, base cuneate, margins coarsely serrate, usually distally, apex acuminate, surfaces glabrous or abaxial puberulent. **Inflorescences** divergently branched, with distinct central axis. **Flowers** greenish to reddish green. **Berries** globose, 4–8 mm diam. $2n = 40$.

Flowering May–Jun; fruiting Aug–Sep. Open woods, prairie ravines, rocky banks and ledges, thickets, fencerows, roadsides, waste places; 0–1500 m; N.B., N.S., Ont., P.E.I., Que., Sask.; Ala., Ark., Colo., Conn., Del., D.C., Fla., Ga., Ill., Ind., Iowa, Kans., Ky., La., Maine, Md., Mass., Mich., Minn., Miss., Mo., Mont., Nebr., N.H., N.J., N.Y., N.C., Ohio, Okla., Pa., R.I., S.C., S.Dak., Tenn., Tex., Utah, Vt., Va., W.Va., Wis.; e Mexico; West Indies (Bahamas, Cuba); Central America (El Salvador, Guatemala).

The leaves of *Parthenocissus quinquefolia* are usually (4–)5(–6)-foliolate. Specimens with 7-foliolate leaves have been collected from sandy areas in Dare County, North Carolina. This species appears to be introduced in most, if not all, of its range in Canada and in Colorado, Montana, and Utah.

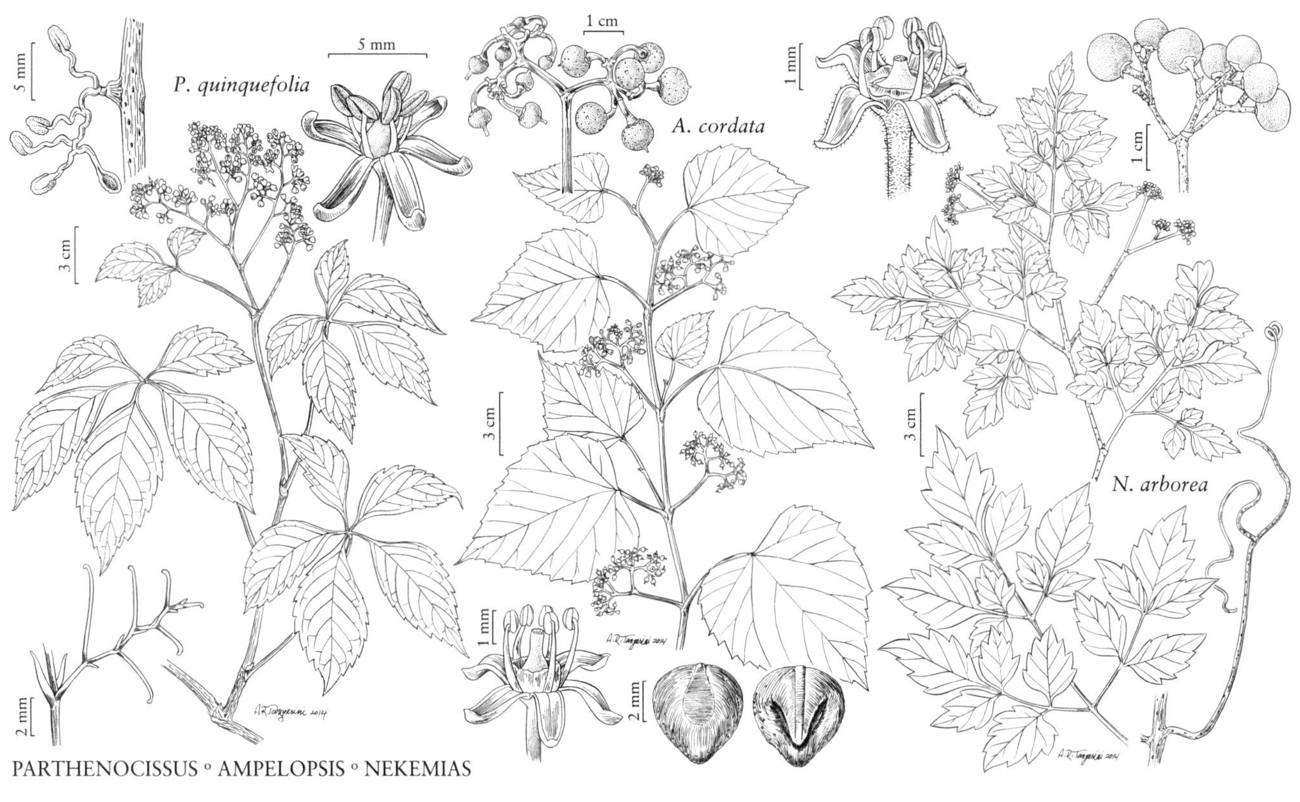

PARTHENOCISSUS ◦ AMPELOPSIS ◦ NEKEMIAS

3. AMPELOPSIS Michaux, Fl. Bor.-Amer. 1: 159. 1803 • [Greek *ampelos*, grapevine, and *-opsis*, similarity]

Lianas, climbing by tendrils, synoecious. **Branches:** bark adherent; pith white, continuous through nodes; tendrils 2-branched, without adhesive discs. **Leaves** simple or palmately compound. **Inflorescences** usually bisexual, leaf-opposed, corymblike cymes, sometimes compound. **Flowers** bisexual; calyx saucer-shaped, indistinctly undulate; petals 5, distinct; nectary adnate proximally to base of ovary, free distally, cup-shaped, slightly lobed; stamens 5; style cylindric, elongate. **Berries** rose violet, purple, blue, black, orange, or yellow. **Seeds** 1–4 per fruit. $x = 10$.

Species ca. 15 (3 in the flora): United States, Mexico, Asia; temperate to tropical regions.

1. Leaves palmately compound, or so deeply lobed as to appear compound, leaflets or leaf lobes (3 or)5 .... 3. *Ampelopsis aconitifolia*
1. Leaves simple.
    2. Branchlets glabrous, green; leaf blades usually unlobed, sometimes shallowly 3-shouldered .... 1. *Ampelopsis cordata*
    2. Branchlets usually sparsely puberulent, sometimes becoming glabrate, purplish green; leaf blades usually shallowly, sometimes deeply, 3(–5)-lobed.................. 2. *Ampelopsis glandulosa*

1. Ampelopsis cordata Michaux, Fl. Bor.-Amer. 1: 159. 1803 • Raccoon or false grape [E] [F]

Vitis indivisa Willdenow

Lianas, moderately high climbing. **Branchlets** green, glabrous. **Leaves** simple; petiole ± equaling blade; blade ovate, 5–12 × 4–9 cm, usually unlobed, sometimes shallowly 3-shouldered, base truncate to subcordate, margins coarsely and irregularly serrate, apex acute to acuminate, surfaces glabrous or abaxial surface sparsely puberulent. **Inflorescences** ± equaling to longer than leaves. **Flowers** yellowish green. **Berries** ripening from green to orange, rose, purple and finally blue, subglobose, 6–10 mm diam. **2*n*** = 40.

Flowering late Apr–Jun; fruiting Jul–Sep. Stream and riverbanks, floodplain forests, wet woodland borders, lowland thickets, fencerows; 0–1000 m; Ala., Ark., Fla., Ga., Ill., Ind., Iowa, Kans., Ky., La., Md., Miss., Mo., Nebr., N.C., Ohio, Okla., S.C., Tenn., Tex., Va., W.Va.

Reports of *Ampelopsis cordata* from Connecticut (for example, G. C. Tucker 1995) are based on misidentified material of *A. aconitifolia.*

2. Ampelopsis glandulosa (Wallich) Momiyama, Bull. Univ. Mus. Univ. Tokyo 2: 78. 1971 • Porcelainberry [I]

Vitis glandulosa Wallich in W. Roxburgh, Fl. Ind. 2: 479. 1824; *Ampelopsis brevipedunculata* (Maximowicz) Trautvetter; *A. brevipedunculata* var. *citrulloides* (Lebas) L. H. Bailey; *A. brevipedunculata* var. *maximowiczii* (Regel) Rehder; *Cissus brevipedunculata* Maximowicz

Lianas, low to moderately high climbing. **Branchlets** purplish green, usually sparsely puberulent, sometimes becoming glabrate. **Leaves** simple; petiole ± equaling blade; blade ovate, 6–14 × 4–11 cm, usually shallowly, sometimes deeply, 3(–5)-lobed (but leaf never appearing compound), base cordate to subcordate, margins crenate-dentate to irregularly serrate, apex short acuminate, surfaces glabrous or abaxial surface puberulent. **Inflorescences** usually shorter than leaves. **Flowers** yellowish green. **Berries** ripening from green to white, rose, blue, or lilac, ± globose, 6–9 mm diam. **2*n*** = 40.

Flowering late May–Jul; fruiting Jul–Oct. Forest edges, pond margins, stream banks, thickets, disturbed areas; 0–800 m; introduced; Ark., Conn., Del., D.C., Ga., Ky., Md., Mass., Mich., N.H., N.J., N.Y., N.C., Ohio, Pa., R.I., S.C., Tenn., Vt., Va., W.Va., Wis.; Asia.

The infraspecific classification of *Ampelopsis glandulosa* is highly controversial. Five varieties were recognized by Chen Z. D. et al. (2007). A broad concept of this taxon is followed here.

Native to eastern Asia, *Ampelopsis glandulosa* has become naturalized and weedy in the eastern United States. Reports from Alabama are based on plants persisting from cultivation; it does not appear to be naturalized there.

This taxon was also known as *Ampelopsis heterophylla* (Thunberg) Siebold & Zuccarini, which is a later homonym (not Blume 1825; see Chen Z. D. et al. 2007).

3. Ampelopsis aconitifolia Bunge, Enum. Pl. China Bor., 12. 1833 [I]

Ampelopsis aconitifolia var. *glabra* Diels & Gilg; *A. aconitifolia* var. *palmiloba* (Carrière) Rehder

Lianas, low to moderately high climbing. **Branchlets** purplish green, usually sparsely pilose, sometimes glabrate or glabrous. **Leaves** palmately compound or blade so deeply lobed that leaf appears compound; petiole ± equaling blade; blade broadly ovate to nearly circular in outline, 4–9 × 2–6 cm, base cuneate, margins deeply serrate, apex acuminate, surfaces glabrous or abaxial puberulent; leaflets or lobes (3 or)5, lanceolate to rhombic-lanceolate, each pinnatifidly lobed. **Inflorescences** usually shorter than leaves. **Flowers** yellowish green. **Berries** orange or yellow at maturity, sometimes bluish prior to maturity, globose, 6–8 mm diam. **2*n*** = 40.

Flowering May–Aug; fruiting Jun–Oct. Thickets, forest edges; 50–400 m; introduced; Conn., Mass., Mich., N.J., N.Y., N.C., Ohio, Pa.; Asia (n, c China).

4. NEKEMIAS Rafinesque, Sylva Tellur., 87. 1838 • [Etymology uncertain; perhaps Latin *nec*, not, and Greek *mya*, unknown plant, alluding to segregation from *Ampelopsis* and *Vitis*]

Lianas, climbing by tendrils, synoecious. **Branches:** bark adherent; pith white, continuous through nodes; tendrils 2-branched, without adhesive discs. **Leaves** [1–]2–3-pinnately compound. **Inflorescences** bisexual, leaf-opposed, corymblike cymes, sometimes compound. **Flowers** bisexual; calyx saucer-shaped, indistinctly undulate; petals 5, distinct; nectary proximally adnate to base of ovary, distally free, cup-shaped; stamens 5; style conic, ± elongate. **Berries** purple, blue, or black. **Seeds** 1–4 per fruit.

Species 9 (1 in the flora): United States, West Indies, e, se Asia.

Ampelopsis has been shown to be paraphyletic by recent phylogenetic analyses (A. Soejima and Wen J. 2006; Ren H. et al. 2011; Nie Z. L. et al. 2012), with molecular data placing the African *Rhoicissus* Planchon and the South American *Cissus striata* Ruiz & Pavón complex within *Ampelopsis*. To maintain the monophyly of genera of Vitaceae, Wen et al. (2014) separated *Ampelopsis* into two genera: *Ampelopsis* and *Nekemias*. The two genera can be distinguished by their leaf and bud morphology. *Nekemias* has pinnately compound leaves (versus simple, trifoliolate, or palmately compound leaves in *Ampelopsis*) and complex axillary buds (versus serial accessory buds).

SELECTED REFERENCE Wen, J., J. K. Boggan, and Nie Z. L. 2014. Synopsis of *Nekemias* Raf., a segregate genus from *Ampelopsis* Michx. (Vitaceae) disjunct between eastern/southeastern Asia and eastern North America, with ten new combinations. PhytoKeys 42: 11–19.

1. Nekemias arborea (Linnaeus) J. Wen & Boggan, PhytoKeys 42: 13. 2014 • Pepper vine [E] [F] [W]

Vitis arborea Linnaeus, Sp. Pl. 1: 203. 1753; *Ampelopsis arborea* (Linnaeus) Koehne

Lianas, moderate to high climbing. **Branchlets** glabrous or glabrate. **Leaves** ternately 2-pinnate or partially 3-pinnate; petiole shorter than blade; blade triangular-ovate in outline, 10–20 cm diam.; leaflets ovate to rhombic-ovate, 2–5 cm, base rounded, truncate, or cuneate, margins coarsely serrate, apex acute to acuminate, abaxial surface sparsely hairy, adaxial surface glabrous. **Inflorescences** usually shorter than leaves. **Flowers** yellowish green. **Berries** oblate to subglobose, 7–15 mm diam. $2n = 40$.

Flowering late May–Jun; fruiting Sep–Oct. Stream banks, edges of swamp forests, floodplain forests, maritime woodlands, moist to wet hammocks, shrubby interdune swales, fence- and hedgerows, thickets, roadsides, waste places; 0–400 m; Ala., Ark., D.C., Fla., Ga., Ill., Ind., Ky., La., Miss., Mo., N.Mex., N.C., Ohio, Okla., S.C., Tenn., Tex., Va., W.Va.

5. CISSUS Linnaeus, Sp. Pl. 1: 117. 1753; Gen. Pl. ed. 5, 53. 1754 • [Greek *kissos*, ivy]

Lianas, climbing by tendrils, synoecious or polygamomonoecious. **Branches:** bark adherent; pith white, continuous through nodes; tendrils unbranched or 2-branched [3–6-branched], without adhesive discs. **Leaves** simple or palmately compound [pinnately compound]. **Inflorescences** bisexual or functionally unisexual, leaf-opposed, corymblike cymes, sometimes compound. **Flowers** bisexual or unisexual; calyx cup-shaped, indistinctly 4-lobed; petals 4, distinct; nectary adnate to base of ovary, cup-shaped, entire or 4-lobed; stamens 4; style conic or cylindric, elongate. **Berries** blue-black to black. **Seeds** 1(–4) per fruit. $x = 12$.

Species ca. 350 (2 in the flora): United States, Mexico, West Indies, Bermuda, Central America, South America, Asia, Africa, Indian Ocean Islands (Madagascar), Pacific Islands, Australia; mostly tropical and subtropical regions.

Cissus antarctica Ventenat, kangaroo vine, has been reported as escaped in California but probably is not naturalized there. Like *C. verticillata*, it has simple leaves, but it can be distinguished from that species by having rusty to brown, appressed (versus grayish white, erect) hairs on branchlets and leaves, prominent domatia in the abaxial secondary vein axils (versus no domatia), and dark blue (versus black) fruits.

1. Leaves simple, blades unlobed, surfaces usually hairy, sometimes glabrous . . . . 1. *Cissus verticillata*
1. Leaves usually 3-foliolate, sometimes simple and blades 3-lobed, rarely unlobed, surfaces glabrous. . . . . 2. *Cissus trifoliata*

1. Cissus verticillata (Linnaeus) Nicolson & C. E. Jarvis, Taxon 33: 727. 1984 • Possum grape, seasonvine, waterwithe treebine, bejuco loco [W]

Viscum verticillatum Linnaeus, Sp. Pl. 2: 1023. 1753; *Cissus argentea* Linden; *C. cordifolia* Linnaeus; *C. sicyoides* Linnaeus

Lianas, low to moderately high climbing, often scrambling over low vegetation. **Branches** usually hairy, sometimes glabrous or glabrate; branchlets succulent to subsucculent when young, becoming woody; growing tips usually hairy; tendrils 2-branched. **Leaves** simple; petiole shorter than blade; blade oblong to ovate, 5–15 × 2–8 cm, unlobed, margins coarsely to finely serrate, surfaces usually hairy, sometimes glabrous. **Flowers** greenish or yellowish green. **Berries** black, 6–10 mm diam. $2n = 48$.

Flowering and fruiting year-round. Coastal hammocks, low ground; 0–20 m; Fla.; Mexico; West Indies; Bermuda; Central America; South America.

Cissus verticillata in the flora area is found in the southern two-thirds of peninsular Florida. The inflorescences of *C. verticillata*, and less often *C. trifoliata*, are sometimes greatly expanded and deformed by the smut *Mycosyrnix cissi* (de Candolle) Beck, with the individual flowers being transformed into subcylindric structures containing the spores of the fungus.

2. Cissus trifoliata (Linnaeus) Linnaeus, Syst. Nat. ed. 10, 2: 897. 1759 (as trifoliat) • Marine vine or ivy, sorrell vine, hierba del buey, ivy treebine [F]

Sicyos trifoliatus Linnaeus, Sp. Pl. 2: 1013. 1753 (as trifoliata); *Cissus incisa* Des Moulins

Lianas, stout, scrambling or sprawling over low vegetation or small trees. **Branches** usually glabrous; branchlets succulent when young, becoming woody, sometimes rooting at nodes; growing tips usually glabrous; tendrils unbranched. **Leaves** usually 3-foliolate, sometimes simple; petiole usually shorter than blade; blade succulent, broadly ovate to ovate-reniform, 2–8 × 2–7 cm, if simple usually 3-lobed, rarely unlobed, margins coarsely and irregularly toothed, surfaces glabrous; leaflets (compound leaves) ovate to oblong. **Flowers** greenish, greenish yellow, whitish, or purplish. **Berries** black to blue-black, 6–12 mm diam.

Flowering late Apr–Jun; fruiting Aug–Sep. Rocky wooded hillsides, stream banks, prairie ravines, glades, bluffs, chaparral, coastal hammocks and dunes, maritime woodlands, shell mounds in salt marshes, roadsides, waste places; 0–2000 m; Ala., Ariz., Ark., Fla., Ga., Kans., La., Miss., Mo., N.Mex., Okla., Tex.; Mexico; West Indies; Central America; n South America.

Many previous authors treated *Cissus incisa* and *C. trifoliata* as distinct species, but the characters used to separate them (size of leaflets, branching patterns of cymes, and berry shape) appear to intergrade abundantly, particularly in Florida, where their geographical ranges overlap. It appears that much of the basis for separating these two species is geographical distribution and habitat, with *C. trifoliata* being chiefly coastal and tropical and *C. incisa* being chiefly subtropical and temperate continental. Some authors (for example, R. P. Wunderlin 1982; R. K. Godfrey 1988; J. A. Lombardi 2000) therefore have treated *C. incisa* as a synonym of *C. trifoliata*, a conclusion that is followed here.

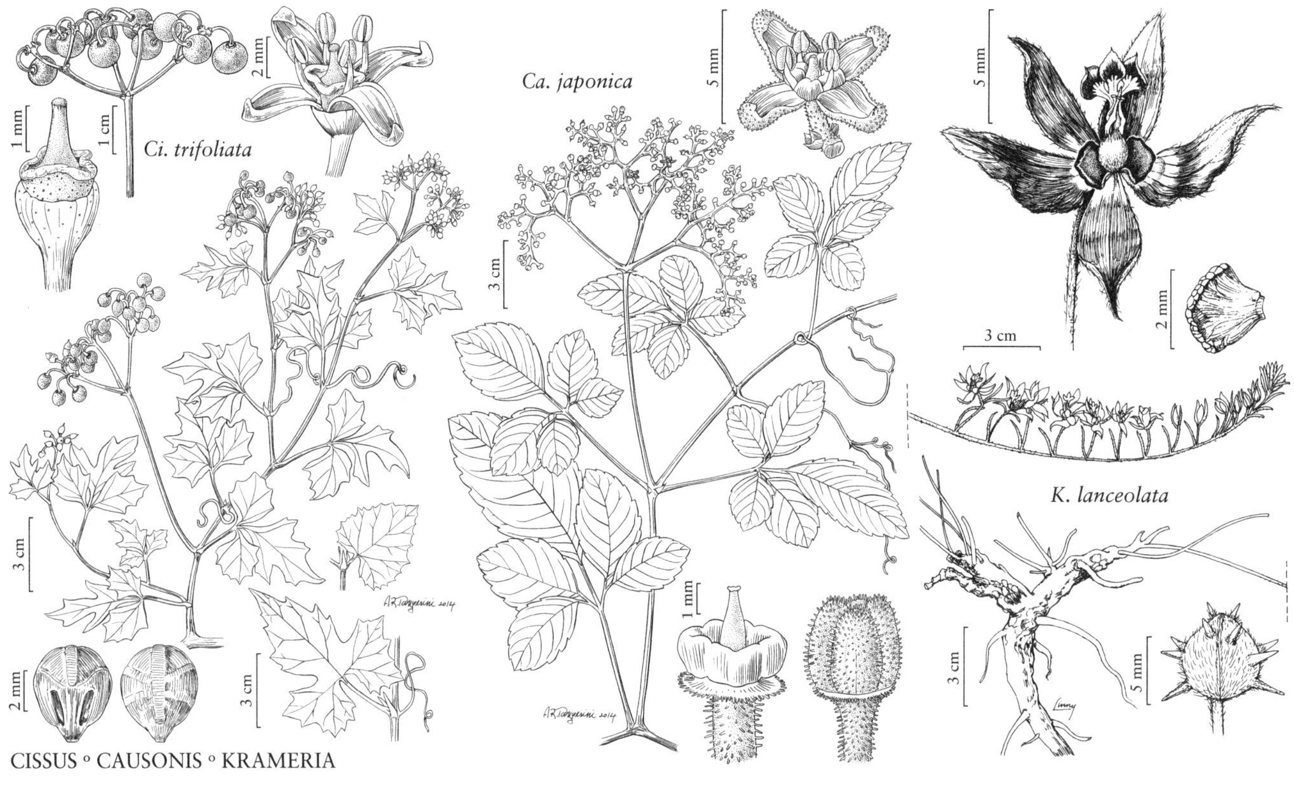

CISSUS ◦ CAUSONIS ◦ KRAMERIA

6. CAUSONIS Rafinesque, Med. Fl. 2: 122. 1830 • [Derivation unknown; perhaps Latin *causa*, reason, and *onus*, necessity, alluding to segregation from *Cissus*] [I]

Vines [lianas], sprawling to moderately high climbing, synoecious or polygamomonoecious. **Branches:** bark adherent; pith white, continuous through nodes; tendrils usually 2–3[–5]-branched, without [with] adhesive discs. **Leaves** palmately (pedately) compound. **Inflorescences** usually bisexual, axillary, corymblike cymes, compound. **Flowers** bisexual or unisexual; calyx cup-shaped, indistinctly 4-lobed; petals 4, distinct; nectary adnate to base of ovary, cup-shaped, 4-lobed; stamens 4; style conic, short. **Berries** dark blue to black. **Seeds** 2–4 per fruit. $x = 10$.

Species ca. 25 (1 in the flora): introduced; e, se Asia, Australia.

Causonis is separated from *Cayratia* (Wen J. et al. 2013) based on phylogenetic evidence (Wen et al. 2007; Lu L. M. et al. 2013). It corresponds to the Asian and Australian *Cayratia* sect. *Discypharia* Suessenguth (K. Suessenguth 1953; A. Latiff 1981). *Causonis trifolia* (Linnaeus) Rafinesque, native to southeast Asia and Australia, is an aggressive weed at Fairchild Tropical Garden in Miami, Florida. No other locations are known for this species in the flora area, but it may become naturalized in southern Florida. It can be distinguished from *C. japonica* by having 3-foliolate leaves and tendrils that are 3–5-branched and usually have adhesive discs at their tips.

1. Causonis japonica (Thunberg) Rafinesque, Sylva Tellur., 87. 1838 (as Causonia) • Bushkiller [F] [I]

Vitis japonica Thunberg in J. A. Murray, Syst. Veg. ed. 14, 244. 1784; Fl. Jap., 104. 1784; *Cayratia japonica* (Thunberg) Gagnepain; *Cissus japonica* (Thunberg) Willdenow; *Columella japonica* (Thunberg) Merrill

Branchlets usually purplish. **Leaves:** petiole slightly shorter to ± equaling blade; blade triangular-ovate in outline; leaflets (3–)5, lateral short-petiolulate, terminal long-petiolulate, leaflets narrowly ovate to oblong, 1–14.5 × 0.5–4.5 cm, terminal larger than lateral, base cuneate to rounded, margins crenate to crenate-serrate, apex usually acute or short-acuminate, sometimes obtuse to rounded, surfaces glabrous or abaxial puberulent. **Flowers** green to yellowish green. **Berries** globose, 8–12 mm diam. $2n$ = 40, 60.

Flowering late May–Jun; fruiting Aug–Sep. Forest edges and openings, pastures, waste areas; 0–300 m; introduced; Ala., La., Miss., N.C., Tex.; e, se Asia; Australia.

Causonis japonica was first reported naturalized in Louisiana in the 1960s (L. H. Shinners 1964). It has since escaped from cultivation elsewhere and become naturalized in scattered places in the southeastern United States.

KRAMERIACEAE Dumortier
• Rhatany Family

Beryl B. Simpson

Herbs or shrubs, perennial, mostly evergreen. **Leaves** alternate, simple [3-foliolate]; stipules absent; petiole absent [present]; blade margins entire; venation pinnate with major veins ± longitudinal. **Inflorescences** terminal, secund racemes [panicles], or axillary, solitary flowers. **Flowers** bisexual, bilaterally symmetric; perianth and androecium hypogynous; hypanthium absent; sepals [4–]5, distinct; petals [4–]5, proximal 2 distinct, modified into fleshy oil-secreting structures flanking ovary, distal [2–]3 distinct or connate basally, petaloid, forming flag adaxial to ovary; nectary absent; stamens [3–]4, usually distinct, sometimes connate basally, free and inserted at base of petaloid petals or adnate to connate bases of petaloid petals; anthers dehiscing by terminal pores; pistil 1, 2-carpellate but appearing 1-carpellate by suppression of second carpel, ovary superior, 1-locular, placentation apical; ovules 2 per locule, anatropous; style 1; stigma 1. **Fruits** capsules, tardily and irregularly dehiscent, spinose. **Seeds** 1 per fruit.

Genus 1, species 18 (4 in the flora): United States, Mexico, West Indies, Central America, South America.

Krameria flowers do not produce nectar, and pollen is rarely collected. The flowers reflect their obligate pollination system with oil-collecting bees (B. B. Simpson 1989). Two petals are modified into fleshy structures that secrete oils, forming blisters under the cuticle (Simpson et al. 1977). Female bees rupture the cuticle (J. L. Neff and Simpson 1981) and collect the oils for nest linings (S. B. Vinson et al. 1996) and larval provisions.

All species of *Krameria* examined are hemiparasitic with roots attaching to a wide range of hosts (W. A. Cannon 1910; J. Kuijt 1969). There is no evidence (B. B. Simpson 1991) that root extracts are medicinally effective in treating bleeding and blood disorders (H. Ruiz López 1799), nor that a tea made from the roots causes esophageal cancer (J. F. Morton 1968).

Because the zygomorphic flowers of *Krameria* superficially resemble those of caesalpinioid legumes or members of the Polygalaceae, *Krameria* has been placed in these families; it was convincingly segregated into its own family based on cytological studies (B. L. Turner 1958) and reinforced by later anatomical evidence (T. H. Milby 1971). Molecular evidence has shown that the family is sister to, but distinct from, Zygophyllaceae (P. A. Gadek et al. 1996). An infrageneric molecular study of *Krameria* (B. B. Simpson et al. 2004) showed that the species cluster into four well-supported clades. A biogeographic analysis based on this phylogeny suggested that there were two independent dispersals or vicariant events between North America and South America (Simpson et al.).

SELECTED REFERENCES Simpson, B. B. 1989. Krameriaceae. In: Organization for Flora Neotropica. 1968+. Flora Neotropica. 110+ nos. New York. No. 49, pp. 1–108. Simpson, B. B. 1991. The past and present uses of rhatany (*Krameria*, Krameriaceae). Econ. Bot. 45: 397–409. Simpson, B. B. 2007. Krameriaceae. In: K. Kubitzki et al., eds. 1990+. The Families and Genera of Vascular Plants. 10+ vols. Berlin, etc. Vol. 9, pp. 207–212. Simpson, B. B. et al. 2004. Species relationships in *Krameria* (Krameriaceae) based on ITS sequences and morphology: Implications for character utility and biogeography. Syst. Bot. 29: 97–108.

1. KRAMERIA Loefling, Iter Hispan., 176, 195, 231. 1758 • Rhatany or ratany [For either Johann Georg Heinrich Kramer, 1684–1744, Austrian Army physician and botanist, or his son William Heinrich Kramer, d. 1765, Austrian physician and naturalist, or both]

Stems consisting of long shoots only or long and short shoots, canescent, strigose, or villous; bark gray or brown. **Leaves:** blade linear to ovate, apex mucronate, acute, or obtuse, surfaces canescent, strigose, or villous, sometimes with stalked glandular hairs. **Inflorescences** with leafy bracts. **Flowers:** sepals purple, pink, or magenta [yellow or brown], lanceolate to ovate or oblong; petals smaller than sepals, secretory petals pink, red, red-brown, green, yellow, purple, or orange, orbiculate to cuneate, petaloid petals green or yellow with expanded distal portions partially purple or pink; stamens pink, green, or white, didynamous or equal; ovary strigose or tomentose; style curved, red, pink, or greenish white, glabrous; stigma recessed. **Capsules** globose or slightly compressed laterally, circular or cordate in outline with longitudinal ridge across 1 or both faces; spines slender to stout, glabrous or hairy proximally, smooth or with retrorse or recurved barbs distally. **Seeds** brown. $x = 6$.

Species 18 (4 in the flora): United States, Mexico, West Indies, Central America, South America.

1. Herbs, decumbent; inflorescences terminal, secund racemes; stamens equal. . . . . 2. *Krameria lanceolata*
1. Shrubs, erect; inflorescences axillary, solitary flowers; stamens didynamous.
   2. Sepals reflexed; secretory petals with oil-filled blisters covering outer surfaces; petaloid petals distinct, narrowly oblanceolate; spines on capsules slender, bearing unicellular hairs basally and amber-colored recurved barbs to 1 mm near tip. . . . . 1. *Krameria bicolor*
   2. Sepals ± cupped inward or around petals and gynoecium; secretory petals with oil-filled blisters mostly on distal portions of outer surfaces; petaloid petals connate basally, free portions oblanceolate to reniform; spines on capsules stout, usually bearing curved multicellular hairs on basal ½ and retrorse barbs near tip, or glabrous, sometimes with minute serrations near tip.
      3. Shrubs 1–2 m; leaf blades linear to linear-lanceolate, sometimes with multicellular, stipitate, glandular hairs; styles pink; spines on capsules 2–5 mm. . . . . 3. *Krameria erecta*
      3. Shrubs 0.3–1 m; leaf blades linear to ovate, lacking glandular hairs; styles greenish white; spines on caspules 0.5–1.5 mm. . . . . 4. *Krameria ramosissima*

1. Krameria bicolor S. Watson, Proc. Amer. Acad. Arts 21: 417. 1886 • Gray's or white ratany

Krameria grayi Rose & J. H. Painter

Shrubs, mound-forming, 0.2–1.5 m. **Stems** erect, long shoots only, young branches green, becoming blue-green with age, canescent, tips thorny. **Leaves:** blade linear or linear-lanceolate, 4–20 × 1–5 mm, apex mucronate, surfaces canescent, lacking glandular hairs. **Inflorescences** axillary, solitary flowers. **Flowers:** sepals reflexed, purple or dark maroon, lanceolate, 7–13 mm; secretory petals dark purple, red-brown, pink, or yellow, 1.5–4.5 mm, with oil-filled blisters covering outer surfaces; petaloid petals 3–6 mm, distinct, green basally, pink or purple distally, narrowly oblanceolate; stamens didynamous; ovary tomentose; style red or pink. **Capsules** cordate to circular in outline, often with conspicuous longitudinal ridge, 5.5–10 mm diam., canescent, sericeous, or tomentose, spines slender, 1.5–5.5 mm, each bearing unicellular hairs basally and amber-colored recurved barbs to 1 mm near tip. $2n = 12$.

Flowering Mar–May. Deserts on limestone, volcanic, or igneous-derived soils; 0–1800 m; Ariz., Calif., Nev., Tex., Utah; Mexico (Baja California, Baja California Sur, Chihuahua, Coahuila, Durango, Hidalgo, Jalisco, Michoacán, Nayarit, Sinaloa, Sonora, Zacatecas).

The name *Krameria grayi* has generally been used for this species in the United States (for example, B. L. Turner et al. 2003) but must be replaced by the older correct name (B. B. Simpson 2013). The species was originally described in 1852 as *K. canescens* by A. Gray, but this name is a later homonym of *K. canescens* Willdenow ex Schultes. Rose and Painter, realizing that the name was illegitimate, renamed the species *K. grayi* in 1906.

Krameria bicolor was reported from New Mexico by W. C. Martin and C. R. Hutchins (1980), but no New Mexico specimens of this species were seen by the author.

2. Krameria lanceolata Torrey, Ann. Lyceum Nat. Hist. New York 2: 168. 1827 • Texas ratany, crameria [F]

Krameria spathulata Small ex Britton

Herbs, spreading, to 2 m diam. **Stems** decumbent, long shoots only, all branches green, densely tomentose to sparsely strigose, tips soft. **Leaves:** blade linear or linear-lanceolate, 5–25 × 0.9–4 mm, apex acute or mucronate, surfaces strigose, lacking glandular hairs. **Inflorescences** terminal, secund racemes. **Flowers:** sepals spreading, purple, lanceolate, 8–16 mm; secretory petals pink, orange, or red, 1.5–3 mm, with oil-filled blisters on distal ½ of outer surfaces and distal margin; petaloid petals 5–7 mm, connate basally, distinct portions green basally, purple or pink distally, reniform, 1–3 mm; stamens equal; ovary strigose; style pink. **Capsules** circular or slightly cordate in outline, with longitudinal ridge on each face, 5–8 mm diam., hairy, spines stout, 1.8–5.3 mm, each bearing conspicuous white hairs proximally and minute, retrorse barbs near tip. $2n = 12$.

Flowering May–Aug. Grasslands, savannas, sandy, calcareous, or clay-based soils; 0–1800 m; Ariz., Colo., Fla., Ga., Kans., N.Mex., Okla., Tex.; Mexico (Chihuahua, Coahuila).

Nuttall reported *Krameria lanceolata* as occurring in Arkansas, and his account was repeated by Delzie Demaree and later workers. As reported by J. H. Peck (2003), the species is not known to occur in the state, and the original claim presumably was based on a specimen collected in the Arkansas Territory at a locality that is now in Oklahoma.

3. Krameria erecta Willdenow in J. A. Schultes & J. H. Schultes, Mant. 3: 303. 1827 • Range or little leaf ratany, heart-nut

Krameria glandulosa Rose & J. H. Painter; *K. palmeri* Rose; *K. parvifolia* Bentham; *K. parvifolia* var. *glandulosa* (Rose & J. H. Painter) J. F. Macbride

Shrubs, mound-forming, 1–2 m. **Stems** erect, with long and short shoots, young branches green, becoming gray with age, strigose, tips often thorny. **Leaves:** blade linear to linear-lanceolate, on both long and short shoots 0.5–15 × 0.5–2 mm, apex acute to mucronate, surfaces strigose, sometimes with multicellular, stipitate, glandular hairs. **Inflorescences** axillary, solitary flowers (on short shoots only, often appearing clustered because of shoot compression). **Flowers:** sepals slightly cupped inward, bright pink or magenta, lanceolate to ovate, 8–10 mm; secretory petals pink, 2–3 mm, with oil-filled blisters mostly on distal portions of outer surfaces; petaloid petals 4–6 mm, connate basally, distinct portions green basally and pink or pink with purple edges distally, oblanceolate to reniform, 2–4 mm; stamens didynamous; ovary strigose; style pink. **Capsules** cordate in outline, with longitudinal ridge more pronounced on 1 face, 6 mm diam., glabrous or densely strigose, spines stout, 2–5 mm, each usually bearing curved, multicellular hairs on basal ½ and retrorse barbs near tip. $2n = 12$.

Flowering Mar–Oct. Sands, gravel ridges, limestone derived soils; 0–1700 m; Ariz., Calif., Nev., N.Mex., Tex., Utah; Mexico (Baja California, Baja California Sur, Chihuahua, Coahuila, Durango, San Luis Potosí, Sinaloa, Zacatecas).

4. Krameria ramosissima (A. Gray) S. Watson, Proc. Amer. Acad. Arts 17: 326. 1882 • Agarito, calderona

Krameria parvifolia Bentham var. *ramosissima* A. Gray, Smithsonian Contr. Knowl. 3(5): 42. 1852

Shrubs, slightly mounding, 0.3–1 m. **Stems** erect, with long and short shoots, young branches strigose, light green, becoming gray with age, tips of long shoots thorny. **Leaves:** blade linear to ovate, on long shoots to 10 × 0.7–1 mm, on short shoots 3–5 × 0.7–2 mm, apex obtuse, surfaces strigose to villous, lacking glandular hairs. **Inflorescences** axillary, solitary flowers (on both long and short shoots, often appearing clustered in short shoots). **Flowers:** sepals ± cupped around petals and gynoecium, pink, ovate to oblong, 7–10 mm; secretory petals dark pink to brick red, 2–3 mm, with oil-filled blisters mostly on distal portions of outer surfaces; petaloid petals 5–6 mm, connate basally, distinct portions yellow or green basally, purple or dark pink tinged with lavender or purple distally, oblanceolate to reniform, 2–3 mm; stamens didynamous; ovary strigose; style greenish white. **Capsules** circular to cordate in outline, with longitudinal ridge on each face, 5–6 mm diam., sparsely to moderately strigose, spines stout, 0.5–1.5 mm, each glabrous, sometimes with minute serrations near tip. **2*n*** = 12.

Flowering Apr–May. Arid areas, sand, limestone, caliche, shale; 0–600 m; Tex.; Mexico (Coahuila, Nuevo León, Tamaulipas).

In Texas, *Krameria ramosissima* is known from counties along the Mexican border from the middle to lower Rio Grande valley.

ZYGOPHYLLACEAE R. Brown

• Creosote Bush Family

Duncan M. Porter

Herbs, subshrubs, shrubs, or trees, annual or perennial, branching usually divaricate, growth sympodial, nodes angled or swollen, evergreen [deciduous], synoecious [dioecious]. **Leaves** opposite or fascicled [alternate or on short lateral branches], palmately or even- [odd-]pinnately compound [simple]; stipules present; petiole present [absent]; blade often fleshy or coriaceous, margins entire; venation pinnate. **Inflorescences** pseudoaxillary [terminal], flowers solitary or in 2-flowered clusters [cymes]. **Flowers** bisexual [unisexual], usually regular, sometimes slightly irregular; perianth and androecium hypogynous; hypanthium absent; sepals 4–5, usually distinct, rarely connate basally; petals 4–5, distinct [rarely connate basally], often clawed, sometimes twisted; nectary usually present, extrastaminal and/or intrastaminal, rarely absent; stamens [5–](8–)10 in 2 whorls, outer usually opposite petals, often alternately unequal in length or sterile, distinct, free or adnate to petal bases, inserted on or proximal to nectary, frequently glandular or appendaged at base; anthers dehiscing by longitudinal slits; pistil 1, (2–)5-carpellate, ovary superior, (2–)5–10-locular; placentation axile [basal]; ovules (1–)2–10 per locule, anatropous; style 1; stigma 1. **Fruits** capsules, dehiscence septicidal or loculicidal, or schizocarps splitting into 5 or 10 mericarps. **Seeds** 1–5(–10) per locule.

Genera 27, species ca. 240 (6 genera, 15 species in the flora): North America, Mexico, West Indies, Central America, South America, Eurasia, Africa, Atlantic Islands, Indian Ocean Islands, Pacific Islands, Australia; mostly tropical or subtropical regions, mainly in arid and semiarid areas.

Zygophyllaceae are most closely related to Krameriaceae, and the two families make up the isolated order Zygophyllales (Angiosperm Phylogeny Group 2003). A number grow in saline soils. Three species are cultivated in milder winter areas of the southeastern United States for their colorful flowers: the South American *Bulnesia arborea* (Jacquin) Engler and *B. sarmientoi* Lorentz ex Grisebach, both verawood, and the Caribbean *Guaiacum officinale* Linnaeus, lignum vitae. *Guaiacum coulteri* A. Gray, guayacán, from western Mexico and Guatemala, is grown in southern Arizona. *Peganum*, often placed in Zygophyllaceae, is now recognized to be a member of the unrelated Nitrariaceae (Angiosperm Phylogeny Group).

SELECTED REFERENCES Beier, B.-A. et al. 2004. Phylogeny and taxonomy of the subfamily Zygophylloideae (Zygophyllaceae) based on molecular and morphological data. Pl. Syst. Evol. 240: 11–40. Porter, D. M. 1972. The genera of Zygophyllaceae in the southeastern United States. J. Arnold Arbor. 53: 531–552. Porter, D. M. 1974b. Disjunct distributions in the New World Zygophyllaceae. Taxon 23: 339–346. Sheahan, M. C. and M. W. Chase. 1996. A phylogenetic analysis of Zygophyllaceae R. Br. based on morphological, anatomical and *rbc*L DNA sequence data. Bot. J. Linn. Soc. 122: 279–300. Sheahan, M. C. and M. W. Chase. 2000. Phylogenetic relationships within Zygophyllaceae based on DNA sequences of three plastid regions, with special emphasis on Zygophylloideae. Syst. Bot. 25: 371–384.

1. Leaflets 2.
   2. Leaflets distinct . . . . 2. *Zygophyllum*, p. 32
   2. Leaflets connate basally, leaves appearing simple and 2-lobed . . . . 3. *Larrea*, p. 33
1. Leaflets (1–)3 or (4–)6–16(–20).
   3. Leaves palmately compound; leaflets (1–)3, apices spinose or spinulose . . . . 1. *Fagonia*, p. 29
   3. Leaves pinnately compound; leaflets (4–)6–16(–20), apices not spinose.
      4. Trees or shrubs; petals usually blue to purple, rarely white; fruits capsules, 2–5-lobed . . . . 4. *Guaiacum*, p. 34
      4. Herbs; petals white or yellow to bright orange, bases sometimes green or red; fruits schizocarps, 5-angled or 10-lobed.
         5. Ovaries 5-lobed, 5-locular; fruits 5-angled, spiny, breaking into 5 mericarps (rarely fewer); petals yellow, bases darker; nectary 10 glands in 2 whorls . . . . . 5. *Tribulus*, p. 36
         5. Ovaries 10-lobed, 10-locular; fruits 10-lobed, not spiny, breaking into 10 mericarps (sometimes fewer); petals white to bright orange, bases white to bright orange or green to red; nectary 5 glands at bases of filaments opposite petals . . . . 6. *Kallstroemia*, p. 38

1. FAGONIA Linnaeus, Sp. Pl. 1: 386. 1753; Gen. Pl. ed. 5, 182. 1754 • [For Guy-Crescent Fagon, 1638–1718, French botanist and chemist, physician to Louis XIV]

Herbs, subshrubs, or shrubs, perennial. **Stems** erect to spreading or ± prostrate, highly branched, angled or ridged [terete], less than 1 m, becoming woody at least at base, hairy and glandular to almost glabrate [glabrous]. **Leaves** opposite, palmately compound [rarely scalelike]; stipules persistent, stiff [herbaceous], spinelike, apex spinose or spinulose; petiolules usually present; leaflets (1–)3[–7], inserted on petiole apex, distinct, [linear] lanceolate to ovate [obovate], terminal largest, base cuneate, apex spinose, surfaces stipitate-glandular, glabrate, or glabrous [hairy]. **Pedicels** in leaf axils, erect, becoming reflexed in fruit. **Flowers** usually solitary, regular to slightly irregular by twisting of petals; sepals deciduous or persistent, 5, distinct, green to purple, equal, margins undifferentiated or sometimes membranous, apex acute-attenuate, hairy or glandular to glabrate [glabrous]; petals soon deciduous, 5, imbricate, spreading, often twisted (propellerlike), purple to pink [rarely white], ± obovate, base clawed, apex rounded, usually apiculate; nectary rudimentary; stamens 10, ± equal; filaments free, filiform, unappendaged; anthers sagittate; ovary sessile, 5-lobed, 5-locular, hairy and usually glandular; ovules (1–)2 per locule; style persistent, forming beak on fruit; stigma minute. **Fruits** capsules, ovoid, deeply 5-lobed, loculicidally dehiscent. **Seeds** usually 1 per locule, brownish to black, flat, ± ovate, seed coat mucilaginous when wet.

Species 35 (3 in the flora): sw United States, nw Mexico, sw South America, w, s Asia, n, sw Africa, Atlantic Islands (Canary Islands, Cape Verde Islands); warm-arid habitats.

Although *Fagonia* species are perennial, they sometimes facultatively form flowers and reproduce during their first season. The seeds of *Fagonia* are sticky when wet, which probably aids in dispersal (D. M. Porter 1974b).

Although *Fagonia laevis* and *F. pachyacantha* do not cluster next to one another in the trees illustrating a phylogenetic study of the genus (B.-A. Beier et al. 2004b), they were placed together in the taxonomic revision (Beier 2005). Beier reported *F. californica* Bentham from Arizona and California, considering *F. laevis* and *F. longipes* to be synonyms; however, *F. californica* is a species of Baja California Sur and Sonora and does not occur in the flora area (D. M. Porter 1963). *Fagonia californica* differs from *F. longipes* in being prostrate-spreading, not erect to spreading; having leaflets that are elliptic to oblong-lanceolate or oblong, not linear to linear-lanceolate; and having pedicels that are stout and 1–6 mm, not slender and 8–20 mm. It differs from *F. laevis*, which has erect to spreading stems and linear-elliptic leaflets and is mostly glabrous, not densely stipitate-glandular.

SELECTED REFERENCES Beier, B.-A. 2005. A revision of *Fagonia* (Zygophyllaceae). Syst. Biodivers. 3: 221–263. Beier, B.-A. et al. 2004b. Phylogenetic relationships and biogeography of the desert plant genus *Fagonia* (Zygophyllaceae), inferred by parsimony and Bayesian model averaging. Molec. Phylogen. Evol. 33: 91–108.

1. Stems densely short-stipitate- or subsessile-glandular, glands golden, to 1 mm diam.; older stems glabrate basally; stipules straight, 3–16 mm, about as long as petioles....... 3. *Fagonia pachyacantha*
1. Stems glabrous or densely to sparsely minutely stipitate-glandular, glands clear or yellow, to 0.1 mm diam.; older stems mostly glabrous or becoming glabrate; stipules curved, 1–6 mm, shorter than petioles.
    2. Pedicels 1.5–7 mm; ovaries and capsules glabrous, puberulent, or minutely strigose, not glandular; plants generally glabrous, but often pedicels and sepals, sometimes stipules and petioles, and rarely ultimate branches sparsely stipitate-glandular; lateral leaflets often narrower than terminal leaflet .... 1. *Fagonia laevis*
    2. Pedicels 8–20 mm; ovaries and capsules puberulent and minutely stipitate-glandular; stems, stipules, petioles, pedicels, and sepals stipitate-glandular; lateral leaflets about same width as terminal leaflet .... 2. *Fagonia longipes*

1. Fagonia laevis Standley, Proc. Biol. Soc. Wash. 24: 249. 1911 • Smooth-stemmed fagonia [F]

Fagonia californica Bentham subsp. *laevis* (Standley) Wiggins; *F. chilensis* Hooker & Arnott var. *laevis* (Standley) I. M. Johnston

Subshrubs or shrubs, less than 1 m, to 1 m diam. **Stems** erect to spreading, intricately branched, dark green, not noticeably slender, mostly glabrous, but ultimate branches rarely sparsely stipitate-glandular, glands clear or yellow, to 0.1 mm diam.; older branches and sometimes younger parts scabrous, older branches becoming stolonlike, bearing many erect smaller branches. **Leaves** (1–)3-foliolate; stipules curved, reflexed to spreading, subulate, 1–6 mm, shorter than petioles, glabrous or sometimes sparsely stipitate-glandular; petiole 2–15 mm, glabrous or sometimes sparsely stipitate-glandular; leaflets linear-elliptic, glabrous, generally longer than petiole, apex spinose, terminal 3–18 × 1–5 mm, laterals to 15 × 3 mm, shorter and narrower than terminal, one or both commonly caducous. **Pedicels** 1.5–7 mm, glabrous or often sparsely stipitate-glandular. **Flowers** 1 cm diam.; sepals green to purple, elliptic to lanceolate, 2–3 × 1 mm, glabrous or often sparsely stipitate-glandular; petals pink to dark red-purple, 4–7 × 1.5–3 mm; stamen filaments 3–4.5 mm; ovary 2–3 mm, glabrous or puberulent, not glandular; style 1–2 mm. **Capsules** 4–5 × 3–6 mm, usually minutely strigose or puberulent, rarely glabrous, not glandular; style 1–2 mm, wider at base.

Flowering Nov–Jun. Rocky desert hillsides to sandy washes; 0–1200 m; Ariz., Calif.; Mexico (Baja California, Baja California Sur, Sonora).

Fagonia laevis is restricted to the Mojave and Sonoran deserts, where it appears to have a more southern distribution than *F. longipes*. There has been some controversy about whether *F. laevis* and *F. longipes* are separate species. They are superficially similar to one another; however, *F. longipes* can be distinguished by its slender branches and longer pedicels and its minute stipitate-glandular hairs on stems, stipules, petioles, leaflets, pedicels, sepals, ovaries, and capsules (D. M. Porter 1963). Although *F. laevis* is usually glabrous, often pedicels and sepals, sometimes stipules and petioles, and rarely ultimate branches have a few small stipitate glands, but these never occur on ovaries or capsules. In addition, although the more southerly *F. laevis* and more northerly *F. longipes* overlap in their distributions in Arizona and California, and both species are found in dry, rocky or sandy, usually hillside habitats, they apparently do not grow together (R. S. Felger, pers. comm.). Also according to Felger, the herbage of the two differs, with *F. laevis* being dark green and *F. longipes* light to grayish green.

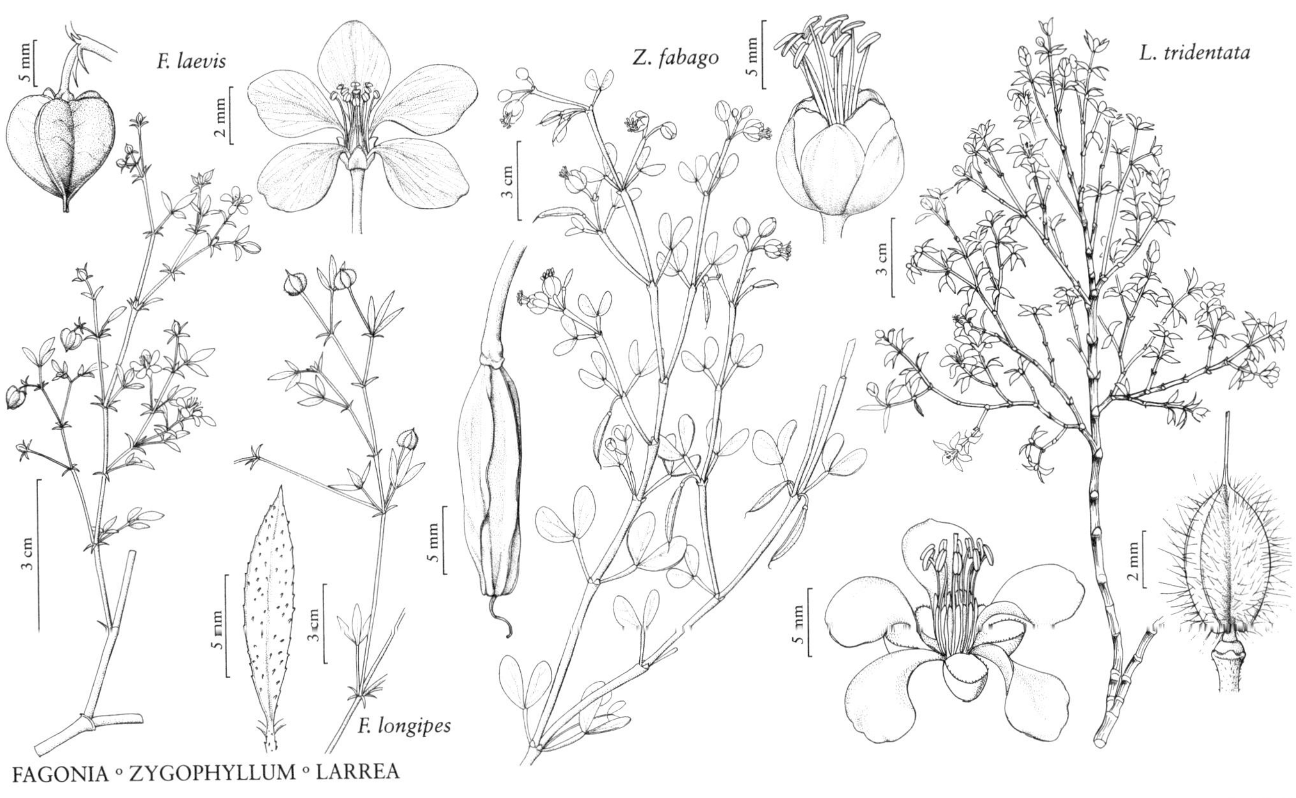

FAGONIA ◦ ZYGOPHYLLUM ◦ LARREA

2. Fagonia longipes Standley, Proc. Biol. Soc. Wash. 24: 250. 1911 • Thin-stemmed fagonia [F]

Fagonia californica Bentham subsp. *longipes* (Standley) Felger & C. H. Lowe

Subshrubs or shrubs, (0.1–)0.3–0.5 m, less than 1 m diam. **Stems** erect to spreading, intricately branched, light green, very slender, sparsely to densely minutely stipitate-glandular, glands clear or yellow, to 0.1 mm diam.; older stems and branches becoming glabrate, scaberulous. **Leaves** 3-foliolate; stipules curved, slightly reflexed to spreading, subulate, 1–2 mm, ½ as long as petioles or shorter, minutely stipitate-glandular to glabrate; petiole 4–10 mm, minutely stipitate-glandular to glabrate; leaflets linear to linear-lanceolate, minutely stipitate-glandular, apex spinose, terminal 3–10 × 1 mm, laterals 2–9 × 1 mm, slightly shorter than but as wide as terminal, sometimes caducous. **Pedicels** 8–20 mm, slender, stipitate-glandular. **Flowers** to 1.5 cm diam.; sepals green, lanceolate to oblong-lanceolate, 2–3 × 1 mm, minutely stipitate-glandular; petals rose-purple, 7–9 × 2–3 mm; stamen filaments 4–5 mm; ovary 2–3 mm, puberulent and minutely stipitate-glandular; style 1–3 mm, wider at base. **Capsules** 4–5 × 4–6 mm, puberulent and minutely stipitate-glandular; style 1.5 mm, wider at base.

Flowering Jan–Aug. Rocky soils of dry hillsides, bajadas and roadsides; 300–900 m; Ariz., Calif., Nev., Utah; Mexico (Sonora).

Fagonia longipes is restricted to the Mojave and Sonoran deserts, where it appears to have a more northern distribution than *F. laevis*. See 1. *F. laevis* for a discussion of the differences between these species.

3. Fagonia pachyacantha Rydberg in N. L. Britton et al., N. Amer. Fl. 25: 105. 1910 • Sticky fagonia

Fagonia californica Bentham var. *glutinosa* Pringle ex Vail

Perennial herbs or subshrubs, to 0.6 m, to 1 m diam. **Stems** ± prostrate, radiating from plant base, sparsely branched, dark green, not noticeably slender, ultimate branches densely short-stipitate- to subsessile-glandular, glands globular (drying to cup-shaped), golden (making branches appear yellowish from a distance), 1 mm diam.; older stems glabrate basally, not scabrous. **Leaves** (1–)3-foliolate; stipules straight, spreading to slightly reflexed, stout, linear-subulate, 3–16 mm, about as long as petioles, glandular to glabrate; petiole 2–16 mm, glandular to glabrate; leaflets ovate to elliptic, slightly obovate, or linear and terete, glandular, becoming glabrate, ± as long as or longer than petiole, apex spinulose, often fleeting, terminal to 26 × 10 mm, laterals to 20 × 7 mm, shorter and narrower than terminal, one or both commonly caducous. **Pedicels** 1–7 mm, glandular. **Flowers** to 1.5 cm diam.; sepals green to purple, ovate-lanceolate, 2–4 × 1–1.5 mm, apiculate, glandular to glabrate; petals light to dark red-purple, 5–8 × 2.5–5.5 mm; stamen filaments 3.5–5 mm; ovary 2 mm, glandular, hairy; style 2–3 mm. **Capsules** 3.5–5.5 × 4–5 mm, puberulent, usually glandular; style 1.5–4 mm, not or barely wider at base.

Flowering Nov–May. Flat, sandy or rocky desert habitats; 0–500 m; Ariz., Calif.; Mexico (Baja California, Baja California Sur, Sonora).

Fagonia pachyacantha is found only in the Sonoran Desert.

2. ZYGOPHYLLUM Linnaeus, Sp. Pl. 1: 385. 1753; Gen. Pl. ed. 5, 182. 1754 • Bean-caper [Greek *zygon*, yoke, and *phyllon*, leaf, alluding to conjugate leaflets as in *Z. fabago*] [I]

Herbs or subshrubs [shrubs], perennial. **Stems** ± erect, highly branched, terete, less than 1 m, fleshy [becoming woody basally], glabrous [hairy]. **Leaves** opposite, 2-foliolate [even-pinnately compound or simple]; stipules deciduous [persistent], ± herbaceous [membranaceous], triangular, apex acute; petiolules absent; leaflets 2[–10], opposite, distinct, obovate, flat [terete], equal, base oblique, apex rounded [acute or obtuse], fleshy, surfaces glabrous [hairy]. **Pedicels** in leaf axils, reflexed downward [erect]. **Flowers** 1–2, slightly irregular by twisting of petals [regular]; sepals deciduous or persistent, [4–]5, distinct, green, often unequal, margins often membranous, apex rounded [obtuse to acute], glabrous; petals deciduous, [0 or 4–]5, imbricate, ± erect, slightly twisted [not twisted to twisted], white or yellow [orange], base red-orange [same as rest of petal], obovate, base clawed, apex rounded; nectary annular; stamens [8–]10, ± equal; filaments free, subulate, each with basal scale; anthers ovate; ovary sessile, [3–]5-lobed, [3–]5-locular, glabrous; ovules [2–]10 per locule; style persistent, not forming beak on fruit; stigma minute. **Fruits** capsules, oblong-cylindric [to globose], [4–]5-angled [or winged], septicidally dehiscent [indehiscent]. **Seeds** [1–]10 per locule, gray-brown, obovoid [ovoid].

Species ca. 70 (1 in the flora): introduced; Eurasia, n, s Africa, Australia.

The common name of *Zygophyllum* results from the pickled flower buds of several species being used as substitutes for capers (*Capparis spinosa* Linnaeus, Capparaceae) in flavoring foods.

1. Zygophyllum fabago Linnaeus, Sp. Pl. 1: 385. 1753 • Syrian bean-caper [F] [I] [W]

Herbs or subshrubs. Stems: branches ± spreading. **Leaves** 2–6 cm; proximal stipules basally connate, distal distinct, green, lanceolate to ovate or elliptic, 4–10 mm; petiole 1–1.5 cm; leaflets 1–4.5 × 0.6–3 cm; awn between leaflets linear or lanceolate, 1 mm. **Pedicels** 4–10 mm. **Flowers** 6–7 mm diam.; sepals ovate to elliptic, 5–7 × 3.5–5.5 mm, margins white; petals obovate, 7–8 mm; stamens exserted [included], 11–12 mm; filaments red-orange, ± linear, basal scales red-orange, apex notched; anthers red-orange. **Capsules** 1–3.5 × 0.4–0.5 cm; style threadlike, to 7 mm. **Seeds** 2–3 mm.

Flowering Apr–Jun. Dry disturbed areas; 0–1000 m; introduced; Calif., Idaho, Mont., Wash.; s Europe; w, c Asia; n Africa.

Zygophyllum fabago has been declared a noxious weed by California, Idaho, Nevada, Oregon, and Washington. Native to the Old World, it has been reported as a waif in Colorado, Kansas, Nevada (Churchill County), New Mexico, New York, Pennsylvania (Philadelphia), and Texas (El Paso County). In the flora area, the species has been referred to as *Z. fabago* var. *brachycarpum* Boissier, an invalid name.

In Spain, *Zygophyllum fabago* has been found to grow in coarse mineral soils contaminated with heavy metals and to accumulate cadmium (I. Lefèvre et al. 2005). Thus, it may potentially cause heavy metal poisoning in grazing stock, as well as alkaloid poisoning.

3. LARREA Cavanilles, Anales Hist. Nat. 2: 119, plates 18, 19. 1800, name conserved • [For Juan Antonio Pérez Hernández de Larrea, 1730–1803, Catholic bishop of Valladolid, Spain]

Shrubs [prostrate subshrubs]. Stems erect [prostrate], branched, swollen at nodes, quadrangular, becoming cylindric, [less than 1] to 4[–7] m, hairy, becoming glabrate. **Leaves** opposite, 2-foliolate [even- or odd-pinnately compound]; stipules persistent, herbaceous, ovate [to triangular], apex obtuse to acuminate [apiculate]; petiolules absent; leaflets [1–]2[–17], opposite [alternate], connate basally 2–4 mm (leaves appearing simple and 2-lobed) [distinct], obliquely lanceolate to falcate [oblong to ovate], equal [unequal], base angular, apex mucronate, surfaces appressed-pubescent, becoming glabrate. **Pedicels** in leaf axils, erect or spreading. **Flowers** solitary, often irregular by twisting of petals; sepals deciduous, 5, distinct, green, ± unequal, margins undifferentiated, apex obtuse, appressed-pubescent; petals deciduous, 5, imbricate, spreading, twisted, yellow, oblong to oblong-lanceolate [ovate], base clawed, apex obtuse; nectary 10-lobed; stamens 10, ± equal; filaments free, subulate, with ± equal basal scales with toothed [entire] margins; anthers sagittate, versatile; ovary short-stalked, 5-lobed, 5-locular, hirsute-pilose; ovules [6–]8 per locule; style persisting on young fruit, later deciduous, not forming beak; stigma minute. **Fruits** schizocarps, globose, 5-lobed, splitting septicidally into 5 mericarps. **Seeds** 1 per locule, brown, boat-shaped. $x = 13$.

Species 5 (1 in the flora): sw United States, Mexico, South America (Argentina, Bolivia, Chile, Peru); warm desert regions.

SELECTED REFERENCES Campos López, E., T. J. Mabry, and S. Fernández Tavizon. 1979. *Larrea.* Mexico City. Laport, R. G., R. L. Minckley, and J. Ramsey. 2012. Phylogeny and cytogeography of the North American creosote bush (*Larrea tridentata*, Zygophyllaceae). Syst. Bot. 37: 153–164. Lia, V. V. et al. 2001. Molecular phylogeny of *Larrea* and its allies (Zygophyllaceae): Reticulate evolution and the probable time of creosote bush arrival to North America. Molec. Phylogen. Evol. 21: 309–320. Mabry, T. J., J. H. Hunziker, and D. R. DiFeo. 1977. Creosote Bush: Biology and Chemistry of *Larrea* in New World Deserts. Stroudsburg, Pa.

1. Larrea tridentata (de Candolle) Coville, Contr. U.S. Natl. Herb. 4: 75. 1893 • Creosote bush, gobernadora, hediondilla [F] [W]

Zygophyllum tridentatum de Candolle in A. P. de Candolle and A. L. P. P. de Candolle, Prodr. 1: 706. 1824; *Larrea divaricata* Cavanilles var. *arenaria* (L. D. Benson) Felger; *L. divaricata* subsp. *tridentata* (de Candolle) Felger & C. H. Lowe; *L. glutinosa* Engelmann; *L. tridentata* var. *arenaria* L. D. Benson; *L. tridentata* var. *glutinosa* Jepson

Shrubs, divaricately branched, multistemmed, strong-scented, resinous. **Stems** reddish when young, becoming gray or black, black-banded, slender. **Leaves:** stipules spreading, not clasping stem, 1–4 mm, fleshy, resinous; petiole to 2 mm; leaflets green to olive brown, 4–18 × 1–8.5 mm, inequilateral, coriaceous, surfaces glutinous; awn between leaflets deciduous, to 2 mm. **Pedicels** 3–12 mm, 4–13 mm in fruit. **Flowers** to 3 cm diam.; sepals ovoid, 5–8 × 3–4.5 mm, appressed-pubescent; petals twisted at claw and appearing propellerlike, 7–11 × 2.5–5.5 mm, claw brownish; stamens 5–9 mm, filaments 4–8 mm, basal scales 2–8 × to 3 mm, ½ to as long as filaments; ovary 2–5 mm, stalk 1 mm, densely hairy; style cylindric, 4–6 mm. **Schizocarps** 4.5 mm diam., pilose-woolly, hairs white, turning reddish brown with age. **Seeds** 4–5 mm. $2n = 26, 52, 78$.

Flowering year-round, following rains. Creosote-bush scrub; -50–1700 m; Ariz., Calif., Nev., N.Mex., Tex., Utah; Mexico.

Larrea tridentata is the dominant shrub in the lower elevations of the Chihuahuan, Mojave, and Sonoran deserts.

There has been much controversy as to whether North American *Larrea* is a separate species from the diploid South American *L. divaricata* ($2n = 26$), and it has been called *L. divaricata* subsp. *tridentata*. However, morphologic (D. M. Porter 1963) and molecular (V. V. Lia et al. 2001) data confirm that they are closely related, but separate, species. The presence of three chromosome races in the three warm desert regions of North America (Chihuahuan, diploid, $2n = 26$; Sonoran, tetraploid, $2n = 52$; Mojave, hexaploid, $2n = 78$) has led to speculation on recognizing three subspecific taxa in *L. tridentata*. However, the overlapping of chromosome races in southeastern California and central and western Arizona (T. W. Yang 1970; R. G. Laport et al. 2012), and the presence of all three races in Baja California (R. S. Felger 2000), plus the lack of morphologic differentiation between them, argue against doing so.

Populations of slender, upright individuals growing on sand dunes in northeastern Baja California, northwestern Sonora, and southeastern California (Imperial County) have been called *Larrea tridentata* var. *arenaria* (*L. divaricata* var. *arenaria*) (R. S. Felger 2000). The California plants are tetraploid and in a few sites are sympatric with hexaploid individuals; they represent a dune-adapted ecotype (R. G. Laport et al. 2012).

Black banding at the nodes of *Larrea tridentata* is caused by resin secreted by glands on the inner surfaces of the stipules. Surfaces of younger parts are resinous and sticky; the odoriferous resins give it the common name, creosote bush. The plant has been widely used medicinally by Native Americans (R. S. Felger 2007). Nordihydroguaiaretic acid is abundant in *L. tridentata* resins and has been used as an antioxidant in foods, pharmaceuticals, and industrial materials (T. J. Mabry et al. 1977).

Clones of *Larrea tridentata* in the Mojave Desert in San Bernardino County, California, have been estimated to be about 11,700 years old based on present growth rates (F. C. Vasek 1980). However, the oldest dated *Larrea* pollen from nearby packrat middens is no more than 5880 years old, so the clones may be younger than thought (R. S. Felger et al. 2012). Fossil *L. tridentata* wood from Yuma County, Arizona, has been radiocarbon dated as 10,850 plus or minus 500 years old (P. V. Wells and J. H. Hunziker 1977). Fossil leaves from Yuma County have been dated to 18,700 plus or minus 1050 years (T. R. Van Devender 1990), while pollen from Inyo County, California lake sediments has been dated to 47,000 (Owens Lake) (W. S. Woolfenden 1996) and 109,000 years old (Badwater) (N. E. Bader 1999). The species, or its progenitor, is estimated to have dispersed from South America between about 0.5 and 8.4 million years ago (V. V. Lia et al. 2001; R. G. Laport et al. 2012).

4. GUAIACUM Linnaeus, Sp. Pl. 1: 381. 1753; Gen. Pl. ed. 5, 179. 1754, name and orthography conserved • Lignum vitae [Spanish mispronunciation of "Huaicum," Bahamas Islands Taino Amerindians' name for the tree and the medicine derived from its resin]

Shrubs or trees; branches spreading or straggling, slightly angled, stout, hairy, becoming glabrate. **Leaves** opposite or fascicled [sometimes crowded on short lateral branchlets], even-pinnately compound; stipules usually deciduous, sometimes persistent, stiff or herbaceous, subulate or ovate, apex acute to acuminate, usually mucronulate, sometimes spinescent; petiolules absent or nearly so; leaflets [2–](4–)6–16, opposite, distinct, [narrowly] elliptic to linear-oblong, linear-spatulate, obliquely oblong, or obovate [ovate], somewhat unequal in size, basal and middle [apical] pairs largest, base oblique, apex obtuse or rounded [acute or retuse] and mucronate, [membranous] subcoriaceous or coriaceous, surfaces glabrous [hairy to glabrate]. **Pedicels** in axils of minute bracts [crowded on short lateral branchlets], erect or spreading. **Flowers** solitary to several [many] together, slightly irregular by twisting of petals; sepals deciduous, 4–5, slightly connate basally, green, unequal, margins undifferentiated, apex obtuse, hairy; petals ± persistent, 4–5, imbricate, spreading, twisted, blue to purple, rarely white [violet], drying yellow, obovate to elliptic, base clawed, apex rounded to lobed or notched; nectary annular; stamens 8–10, ± equal; filaments free, subulate or base slightly winged, sometimes with small basal scale; anthers sagittate or cordate; ovary on short stalk, 2–5-lobed, 2–5-locular, glabrous or hairy; ovules 8–10 per locule; style persisting, forming beak on fruit; stigma minute. **Fruits** capsules, [green] becoming greenish yellow to bright orange[-brown], obovoid to obcordiform, flattened, 2–5-lobed, 2–5-winged, base narrowed into short stalk, broadest apically, coriaceous, smooth or reticulate, septicidally dehiscent. **Seeds** 1 per locule, 1–5 maturing per fruit, yellowish brown, brown, or black, ellipsoid to ovoid, surrounded by thick fleshy red aril.

Species ca. 6 (2 in the flora): sc, se United States, Mexico, West Indies, Central America, n South America.

Guaiacum wood is hard and resinous, long used in turnery and medicine. Several species are grown as ornamentals for their showy flowers in southern Florida, southern Arizona, Mexico, the West Indies, and Central America. *Guaiacum officinale* Linnaeus is known from Florida not only in cultivation, but also as a local escape; however, it has not become naturalized. It can easily be distinguished from *G. sanctum* by its leaves having only (2–)4–6 shiny, green, obovate leaflets that are 15–35(–60) × 25–35 mm. The fruits are 2-lobed and contain only one or two seeds.

1. Leaves 1–3 cm; leaflets 8–16; capsules obcordiform, flattened, 2(–4)-lobed, ± 2(–4)-winged, hairy . . . . 1. *Guaiacum angustifolium*
1. Leaves (4–)6–10 cm; leaflets (4–)6–8(–10); capsules obovoid, 4–5-lobed, 4–5-winged, glabrous. . . . . 2. *Guaiacum sanctum*

1. Guaiacum angustifolium Engelmann in F. A. Wislizenus, Mem. Tour N. Mexico, 113. 1848 • Guayacán, soap-bush [F]

Porlieria angustifolia (Engelmann) A. Gray

Shrubs or trees, to 3(–7) m; trunk well defined, to 0.3 m diam.; bark fissured vertically; branches many, spreading or straggling, knotty; crown diffuse. **Leaves** opposite or fascicled, 1–3 cm, folded at night and often also in heat of day; stipules persistent, subulate, 1 mm, apex acute, somewhat spinescent, glabrous; petiole absent or nearly so; leaflets 8–16, dark green, linear-oblong to linear-spatulate, 5–15 × 2–3 mm, apex obtuse, coriaceous, surfaces reticulate. **Pedicels** hairy. **Flowers** axillary, mostly solitary, sometimes clustered, 1.2–2 cm diam.; sepals (4–)5, obovate, to 5 mm, outer smaller; petals (4–)5, usually blue to purple, rarely white, obovate to elliptic, 6–10 × 2–3 mm, base short-clawed, apex often notched; stamens (8–)10, shorter than petals; filaments each with small crenate scale at base; ovary obcordiform, flattened, 2(–4)-lobed, 2(–4)-locular, hairy; style subulate. **Capsules** becoming orange, obcordiform, flattened, 10–20 mm diam., 2(–4)-lobed, 2(–4)-locular, ± 2(–4)-winged, apex abruptly attenuate-apiculate, reticulate, hairy. **Seeds** yellowish brown, ovoid to reniform, 10–11 mm.

Flowering Mar–Sep. Shrubby vegetation, limestone soils; 0–1200 m; Tex.; Mexico (Chihuahua, Coahuila, Nuevo León, San Luis Potosí, Tamaulipas).

Guaiacum angustifolium occurs in southern, central, and western Texas. The root bark is used as a source of soap, and root extracts are used in folk medicine to treat various diseases. The stems are used for fence posts, tool handles, and firewood.

2. Guaiacum sanctum Linnaeus, Sp. Pl. 1: 382. 1753 • Holywood lignum vitae [C]

Trees, 2.5–10 m; trunk sometimes to 1 m diam.; bark scaly; branches many, spreading, smooth; crown dense, rounded. **Leaves** opposite, (4–)6–10 cm, not folded at night; stipules usually deciduous, ovate, 3 mm, apex acuminate, usually mucronulate, hairy; petiole shorter than leaflets; leaflets (4–)6–8(–10), green, elliptic to obliquely oblong or obovate, 20–35 × 8–15 mm, apex rounded, subcoriaceous. **Pedicels** glabrous. **Flowers** appearing terminal, solitary to several in axils of upper leaves, 2–3 cm diam.; sepals 4–5, obovate, 5–7 mm, outer smaller; petals 4–5, blue, obovate, 7–11 × 5–7 mm, base clawed, apex rounded to lobed; stamens 8–10, ± as long as petals; filaments subulate to base slightly winged; ovary 4–5-lobed, 4–5-locular, obovoid, glabrous; style subulate. **Capsules** becoming greenish yellow to bright orange, obovoid, 14–20 × 12–14 mm, 4–5-lobed, 4–5-winged, apex pointed, smooth, glabrous. **Seeds** brown or black, ellipsoid, 10–11 mm.

Flowering Mar–Jul. Tropical hardwood hammocks on islands; of conservation concern; 0–10 m; Fla.; se Mexico; West Indies; Central America.

Since the early sixteenth century, an extract of the heartwood of *Guaiacum sanctum* has been considered a remedy for venereal disease; today the resin is used medicinally to check for occult blood in human stools. The wood is exceptionally strong and resinous, and hence has been used to make bearings and bushing blocks to line stern tubes of propeller shafts in steamships. Today the wood is used mainly to make such turned objects as mallets, pulley sheaves, caster wheels, bowling balls, stencil and chisel blocks, mortars and pestles, brush backs, and planes (D. M. Porter 1972). Overexploitation of *G. sanctum* has caused it to become extirpated or

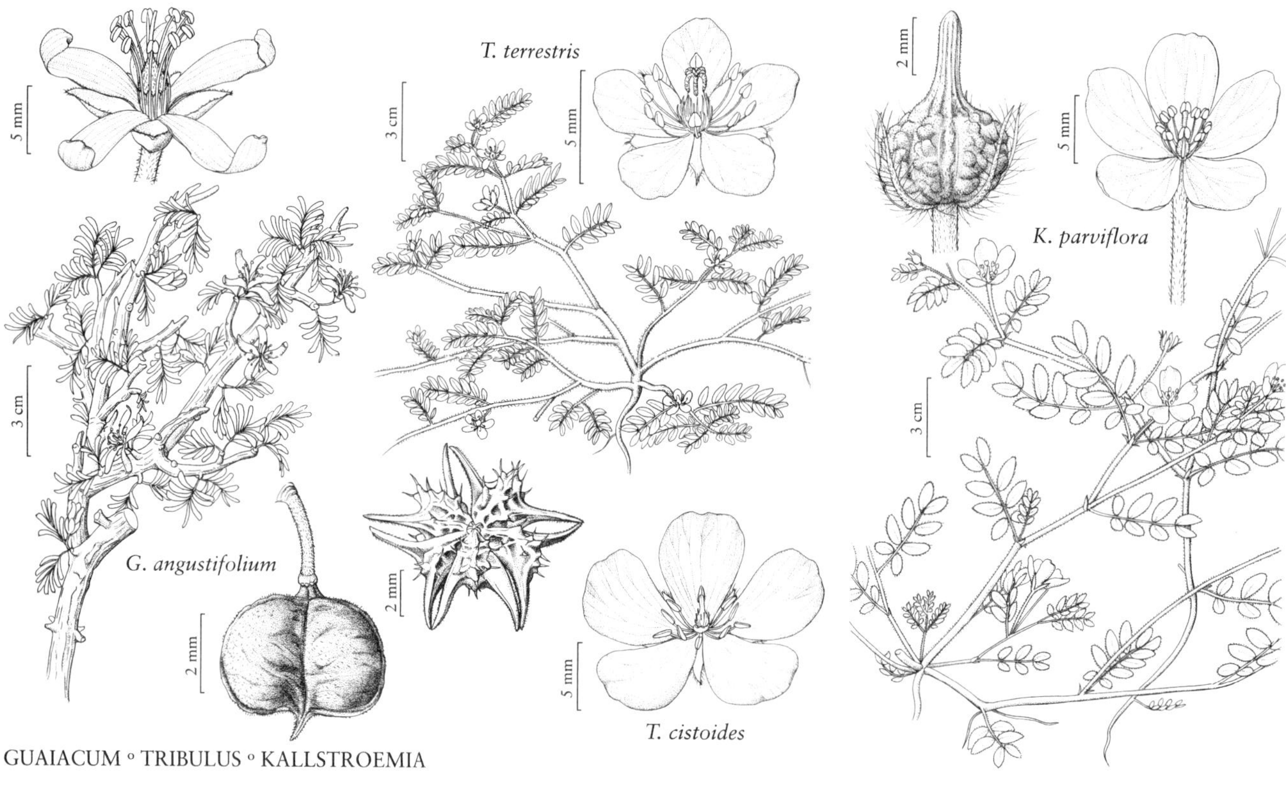

GUAIACUM ◦ TRIBULUS ◦ KALLSTROEMIA

rare in many of the localities where it once occurred in abundance. The species is a state endangered species in Florida, where it is restricted to Miami-Dade and Monroe (Keys only) counties.

Tree ring analyses of *Guaiacum sanctum* individuals in the largest remaining U.S. population, in Lignumvitae Key Botanical State Park, indicate that some may be more than 1000 years old (E. O. Wilson and T. Eisner 1968). This population is under threat from a parasitic scale insect (*Toumeyella lignumvitae*), which is endemic to the Keys and known to feed only on *G. sanctum*. It is a direct threat to the viability and continued existence of this last intact United States population (J. R. Dertien and M. R. Duvall 2009).

SELECTED REFERENCES Dertien, J. R. and M. R. Duvall. 2009. Biogeography and divergence in *Guaiacum sanctum* (Zygophyllaceae) revealed in chloroplast DNA: Implications for conservation in the Florida Keys. Biotropica 41: 120–127. Wilson, E. O. and T. Eisner. 1968. Lignumvitae: Relict island. Nat. Hist. 77: 52–57.

5. TRIBULUS Linnaeus, Sp. Pl. 1: 386. 1753; Gen. Pl. ed. 5, 183. 1754 • Caltrop [Greek *treis*, three, and *bolos*, a point, for a kind of caltrop, alluding to fruits resembling that ancient metal instrument with three or four spines arranged so that one always projects upward, thrown on the ground to stop cavalry and foot soldiers] [I]

Herbs, usually annual, sometimes perennial [rarely subshrubs]. **Stems** prostrate to suberect, diffusely branched, terete, becoming ridged on drying, to 1[–3] m, somewhat succulent, densely hairy to glabrate. **Leaves** opposite, even-pinnately compound, one of each pair alternately smaller or sometimes abortive; stipules persistent, herbaceous, subulate to falcate, apex acute; petiolules very short to absent, less than 1 mm; leaflets 6–16(–20), opposite, distinct, oblong to ovate or elliptic, middle pairs largest, base oblique, apex acute or obtuse, surfaces hairy, terminal pair pointed forward. **Pedicels** emerging from axils of alternately smaller leaves, erect. **Flowers** solitary, regular; sepals caducous, 5, distinct, green, equal, margins scarious, apex acute, hairy;

petals deciduous, 5, imbricate, spreading hemispherically, not twisted, usually bright yellow, rarely white, base often darker, oblong to obovate-cuneate, base not clawed, apex rounded to slightly lobed; nectary 10 glands in 2 whorls, outer whorl 5 ± 2-lobed glands, inner whorls 5 distinct or connate glands; stamens [5–]10, 5 opposite petals somewhat longer; filaments adnate at base to petals, subulate [filiform], unappendaged; anthers cordate to oblong-cordate or sagittate; ovary sessile, 5-lobed, 5-locular, hirsute with stiff, upward-spreading bulbous-based hairs; ovules 3–5 per locule; style deciduous; stigma terminal, slightly asymmetric, 5-lobed, pyramidal or globose, papillose. **Fruits** schizocarps, horizontally depressed, 5-angled, at maturity dividing septicidally into 5 or rarely fewer mericarps and leaving no central axis; mericarps broadly triangular, apex dorsally crested, tuberculate, spiny [or winged or rarely only tuberculate abaxially], each divided internally by oblique transverse septa into 2–5 1-seeded compartments. **Seeds** usually 3–5 per mericarp, white, oblong-ovoid. $x = 6$.

Species ca. 25 (2 in the flora): introduced; Mexico, West Indies, Central America, South America, s Europe, w, s Asia, Africa, Indian Ocean Islands, Pacific Islands, Australia; generally weedy occupants of dry disturbed habitats, several species widely naturalized in temperate and tropical areas of the world.

SELECTED REFERENCES Porter, D. M. 1968. The basic chromosome number in *Tribulus* (Zygophyllaceae). Wasmann J. Biol. 26: 5–6. Porter, D. M. 1971. Notes on the floral glands in *Tribulus* (Zygophyllaceae). Ann. Missouri Bot. Gard. 58: 1–5.

1. Flowers 5(–10) mm diam.; plants annual; leaves 2–4.5 × 1 cm; leaflets 6–12(–16). . . . . . . 1. *Tribulus terrestris*
1. Flowers 15–25 mm diam.; plants perennial; leaves 2.5–8.5 × 1–2.6 cm; leaflets 12–16(–20) . . . . 2. *Tribulus cistoides*

1. Tribulus terrestris Linnaeus, Sp. Pl. 1: 387. 1753 • Puncture vine, goat head F I W

Herbs, annual; herbage hairy (whitish), especially young shoots, becoming glabrate. **Stems** prostrate, green to reddish, to 1 m, ± hirsute, sericeous. **Leaves** 2–4.5 × 1 cm; stipules 1–5 × to 1 mm; leaflets 6–12(–16), ovate to elliptic, largest 4–11 × 2–4 mm, densely sericeous, younger parts silvery, becoming glabrate. **Pedicels** shorter to longer than shorter pair of leaves, in flower 2–7 mm, in fruit 5–15 mm, apex bent. **Flowers** 5(–10) mm diam.; sepals ovate-lanceolate, 2–4 × 1.5–2 mm, minutely ciliate, hirsute; petals oblong, 2.5–5 × 1–3 mm; outer whorl of nectary glands yellowish, inner whorl distinct, yellow, triangular, 0.2 mm; stamen filaments 2–3 mm; anthers yellow, cordate, 1 mm; ovary 1–5 mm diam.; style 5-ridged, cylindric, stout, 1–1.5 mm; stigma globose. **Schizocarps** (7–)10–15 mm diam. excluding 4–12 mm spines; mericarps bearing 2 conic spreading 3–7 mm dorsal spines and sometimes 2 smaller retrorse spines near base, body green to gray or yellowish, hispid, strigose, or glabrate.

Flowering Mar–Oct. Agricultural lands, roadsides, railways, other disturbed areas; 0–2300 m; introduced; B.C., Ont.; Ala., Ariz., Ark., Calif., Colo., Del., Fla., Ga., Idaho, Ill., Ind., Iowa, Kans., Ky., La., Md., Mass., Mich., Minn., Miss., Mo., Mont., Nebr., Nev., N.J., N.Mex., N.Y., N.C., N.Dak., Ohio, Okla., Oreg., Pa., S.C., S.Dak., Tenn., Tex., Utah, Wash., Wis., Wyo.; Europe; n Africa; introduced also in Mexico, South America (Argentina, Ecuador, Peru), s Africa, Pacific Islands, Australia.

Tribulus terrestris, native to the Mediterranean region, is now found throughout drier temperate areas of the world. The species was collected once in Virginia (in 1978) but does not appear to have become established there. *Tribulus terrestris* is often spread by the spiny mericarps sticking to bicycle or automobile tires and to the feet or coats of livestock. The mericarps also wreak havoc with the bare foot. For an extended synonymy, see H. G. Schweickerdt (1937).

Tribulus terrestris officially has been declared an introduced noxious weed by Arizona, California, Colorado, Hawaii, Idaho, Iowa, Michigan, Nebraska, Nevada, North Carolina, Oregon, Washington, and Wyoming. After attempting unsuccessfully for some 50 years to control it by chemical means, the California Department of Agriculture in 1961 imported two species of weevils from India for biological control. The larvae of both feed selectively on *T. terrestris*: *Microlarinus lareynii* on the seeds and *M. lypriformis* on the stems. These weevils were thought to be well on their way to controlling *T. terrestris* in the Southwest by the 1970s (D. M. Porter 1972). Unfortunately, however, control is cyclic and not always effective, and the species remains a pest.

In the United States, Africa, and Australia, ingestion of the plant by sheep leads to so-called geeldikkop or bighead, a fatal disease involving hepatogenic photosensitization. Both nitrate and selenium poisoning may be involved (J. M. Kingsbury 1964). Extracts of *Tribulus terrestris* are now sold on the Internet to treat male infertility and to help build muscle mass.

2. Tribulus cistoides Linnaeus, Sp. Pl. 1: 387. 1753
• Burr-nut, Jamaican feverplant F I W

Herbs, perennial; herbage hairy (often silvery gray), becoming glabrate. **Stems** prostrate to suberect, green to reddish, to 0.8 m, densely sericeous, ± hirsute, especially at nodes. **Leaves** 2.5–8.5 × 1–2.6 cm; stipules 3–9 × 1–4 mm; leaflets 12–16(–20), obliquely oblong to elliptic, largest 6–21 × 2.5–9 mm, densely sericeous when young, whitish abaxially. **Pedicels** longer than shorter pair of leaves, in flower (6–)19–35 mm, in fruit 11–34 mm, apex bent. **Flowers** 15–25 mm diam.; sepals lanceolate, 5–9 × 1.5–3.5 mm, ciliate, densely strigose, silky-pubescent; petals obovate-cuneate, (5.5–)7–17 × (3–)5–11 mm; outer whorl of nectary glands green, inner whorl basally connate into 5-lobed urceolate ring surrounding base of ovary, yellow, broadly triangular, to 1 mm; stamen filaments 2.5–5 mm; anthers yellow, oblong-cordate to narrowly sagittate, 1–3 mm; ovary 1.5–3 mm diam.; style 5-ridged, cylindric, stout, 1–2 mm; stigma globose to pyramidal. **Schizocarps** 8–10 mm diam. excluding 7 mm spines; mericarps bearing 2 conic spreading 5–7 mm dorsal spines and sometimes 2 smaller retrorse spines near base [rarely spines absent and mericarps tuberculate, or spine 1], body green to gray, hispid, densely sericeous to strigose or glabrate.

Flowering year-round. Dry disturbed habitats, especially in maritime areas; 0–10 m; introduced; Fla., Ga., La.; s Africa; introduced also in Mexico, West Indies, n South America, Pacific Islands.

Tribulus cistoides, native to tropical and subtropical southern Africa, is occasionally grown in sandy soils as a garden ornamental or along roads to stabilize shifting soils (D. M. Porter 1972). Although *T. cistoides* has been reported from Texas, no specimen has been seen, and it apparently does not occur there. The species has been introduced to the Galápagos Islands but not to mainland Ecuador.

6. KALLSTROEMIA Scopoli, Intr. Hist. Nat., 212. 1777 • [Derivation obscure, perhaps for Anders Kallström, 1733–1812, a contemporary of Scopoli]

Herbs, usually annual, sometimes perennial. **Stems** prostrate to decumbent or ascending, diffusely branched, terete, becoming ridged on drying, to 1(–1.5) m, somewhat succulent, densely hairy to glabrate. **Leaves** opposite, even-pinnate, one of each pair alternately smaller or sometimes abortive; stipules persistent, herbaceous, falcate, apex acuminate; petiolules very short to absent, less than 1 mm; leaflets 6–16(–20), opposite, distinct, elliptic to broadly oblong, ovate, or obovate, ± unequal in size, those on one side of rachis slightly smaller, base oblique, apex acute to obtuse, mucronate to apiculate, surfaces hairy to glabrate, terminal pair pointed forward. **Pedicels** emerging from axils of alternately smaller leaves, erect, usually reflexed in fruit. **Flowers** solitary, regular; sepals usually persistent, rarely deciduous, 5, distinct, green, equal, margins scarious, apex acute to obtuse, hairy; petals fugacious, usually marcescent, 5, convolute, spreading hemispherically, not twisted, white to bright orange, base white to bright orange or green to red, obovate, base not clawed, apex rounded to truncate, sometimes irregularly notched; nectary 5 2-lobed glands at bases of filaments opposite sepals; stamens 10, 5 opposite petals somewhat longer; filaments adnate at base to petals, filiform, subulate, or rarely basally winged, unappendaged; anthers usually ovoid to oblong [or globose], rarely linear, those opposite sepals rarely aborting; ovary sessile, 10-lobed, 10-locular, glabrous or hairy; ovules 1 per locule, sometimes 1 or more aborting; style persisting, forming beak on fruit; stigma terminal, rarely extending down distal ⅓ [almost to base] of style, 10-ridged or obscurely 10-lobed, capitate, oblong, or clavate, papillose, rarely coarsely canescent. **Fruits** schizocarps, ovoid [pyramidal or conic], 10-lobed, at maturity dividing septicidally and separating from persistent styliferous axis into 10, or sometimes fewer, mericarps; mericarps obliquely triangular, wedge-shaped, tuberculate, rugose, cross-ridged, or ± keeled abaxially, not spiny. **Seeds** 1 per mericarp, white, oblong-ovoid.

Species 17 (6 in the flora): United States, Mexico, West Indies, Central America, South America; open, disturbed, dry habitats.

Kallstroemia is the largest New World genus of the family. *Kallstroemia pubescens* (D. Don) Dandy, native to the Caribbean, was collected in the United States at least six times in four states (Arizona, Florida, Georgia, and Texas) between 1833 and 1898 but apparently has not become naturalized. It most closely resembles *K. hirsutissima* and *K. maxima*, and is to be expected in sandy, coastal areas from Georgia to Florida and along the Gulf Coast to Texas. *Kallstroemia pubescens* can be distinguished from *K. hirsutissima* by having leaflets that are appressed-hairy to glabrate (versus densely hirsute) and sepals that are lanceolate and spreading in fruit with sharply involute margins (versus subulate, clasping the fruit, and with only the scarious margins involute). It can be distinguished from *K. maxima* by having ovaries and fruits that are densely hairy (versus glabrous or sometimes basally strigose) and styles that are densely short-pilose at the base (versus glabrous).

SELECTED REFERENCE Porter, D. M. 1969. The genus *Kallstroemia* (Zygophyllaceae). Contr. Gray Herb. 198: 41–163.

1. Leaves obovate in outline, terminal pairs of leaflets largest.
   2. Sepals usually deciduous; mericarps tuberculate, 4–5 tubercles oblong, 1–1.5 mm, other tubercles rounded, much less than 1 mm . . . . 1. *Kallstroemia californica* (in part)
   2. Sepals persistent; mericarps tuberculate, cross-ridged, or slightly keeled, tubercles if present all rounded, to 1 mm.
      3. Schizocarps strigillose; styles hairy . . . . 3. *Kallstroemia hirsutissima*
      3. Schizocarps usually glabrous, sometimes strigose at base, or rarely to base of beak; styles glabrous . . . . 4. *Kallstroemia maxima*
1. Leaves elliptic in outline, middle pairs of leaflets largest.
   4. Plants perennial; stigmas extending along distal ⅓ of style . . . . 6. *Kallstroemia perennans*
   4. Plants annual; stigmas terminal.
      5. Flowers 20–60 mm diam.; petals 10–34 × 7–22 mm, 2-colored; pedicels in fruit 30–105 mm . . . . .2. *Kallstroemia grandiflora*
      5. Flowers 8–25 mm diam.; petals 3–12 × 2.5–5 mm, 1-colored; pedicels in fruit 8–40 mm.
         6. Pedicels shorter than subtending leaves; sepals usually deciduous; petals yellow, 3–6 × 2.5–3 mm . . . . 1. *Kallstroemia californica* (in part)
         6. Pedicels usually longer than, sometimes equaling, subtending leaves; sepals persistent; petals orange, 5–12 × 3.5–5 mm . . . . .5. *Kallstroemia parviflora*

1. Kallstroemia californica (S. Watson) Vail, Bull. Torrey Bot. Club 22: 230. 1895

Tribulus californicus S. Watson, Proc. Amer. Acad. Arts 11: 125. 1876; *Kallstroemia brachystylis* Vail; *K. californica* var. *brachystylis* (Vail) Kearney & Peebles

Herbs, annual. **Stems** prostrate to decumbent, to 0.7 m, hirsute and strigose with white antrorse hairs, becoming glabrate. **Leaves** usually elliptic in outline, sometimes obovate, 1.5–6 × 1–3 cm; stipules 1.5–5 × 1 mm; leaflets 6–12(–14), elliptic to oblong, 4–17 × 1.5–9 mm, middle pairs usually largest, terminal pairs sometimes largest, surfaces appressed-hirsute, veins and margins sericeous, becoming glabrate. **Pedicels** to 15 mm in flower, 8–33 mm in fruit, shorter than subtending leaves, thickened distally, bent sharply at base and straight distally. **Flowers** 8–10 mm diam.; sepals usually deciduous, if persistent spreading from base of schizocarp, lanceolate, 2–4 × 1–1.5 mm, ⅔ as long as petals, in flower longer than style, in fruit not reaching tops of mature fruit body, margins becoming involute, strigose or hirsutulous and strigillose; petals marcescent, 1-colored, yellow, drying white or orange, obovate, 3–6 × 2.5–3 mm; stamens as long as style; anthers yellow, ovoid, to 1 mm; ovary ovoid, 1 mm diam., hairy; style conic, to 1 mm, shorter than ovary, strigillose; stigma terminal. **Schizocarps** ovoid, to 4 × 3–5 mm (including tubercles), strigillose; beak cylindric, 2–4 mm, shorter than fruit body, base conic, glabrous or base sparingly strigillose; mericarps 3 × 1 mm, abaxially tuberculate, 4–5 tubercles blunt oblong, 1–1.5 mm (tubercles becoming more prominent as fruits mature), other tubercles rounded, much less than 1 mm, sides pitted or smooth, adaxial edge angled.

Flowering Mar–Oct. Flat, sandy, disturbed areas in desert and semiarid grasslands; 0–1600 m; Ariz., Calif., Nev., N.Mex., Tex., Utah; c, n Mexico.

Kallstroemia californica is found throughout the Chihuahuan, Mojave, and Sonoran deserts, mainly at lower elevations.

Kallstroemia californica plants with fewer leaflets and less pronounced fruit tubercles have been called *K. brachystylis* (*K. californica* var. *brachystylis*), but variation in these characters is continuous and two taxa cannot be distinguished. Recognition of two taxa in the past arose primarily from the determination of a number of immature or depauperate specimens of *K. parviflora* as *K. brachystylis* (D. M. Porter 1969).

2. Kallstroemia grandiflora Torrey ex A. Gray, Smithsonian Contr. Knowl. 3(5): 28. 1852 • Arizona- or desert- or Mexican- or summer-poppy [W]

Herbs, annual. **Stems** decumbent to ascending, to 1 m, densely sericeous with white hairs and hispid with white or yellow antrorse hairs, rarely becoming glabrate. **Leaves** elliptic in outline, 1.5–7 × 2–3 cm; stipules 4–10 × 1–2 mm; leaflets 8–16 (–20), elliptic to slightly obovate, 8–25 × 2–3 mm, middle pairs largest, surfaces appressed-hirsute, veins and margins sericeous, becoming glabrate. **Pedicels** 30–105 mm in flower and fruit, longer than subtending leaves (extending flowers well above herbage), slightly thickened distally, sharply bent at base and straight distally. **Flowers** 20–60 mm diam.; sepals persistent, lanceolate, 6–16 × 1.5–2.5 mm, ½ as long as petals, in flower longer than style, in fruit much surpassing mature fruit body but shorter than beak, also shriveling and turning brown, margins becoming strongly involute making sepals appear linear, hispid and strigose; petals marcescent, 2-colored, basally green to red, distally white to yellow or bright orange, fading white to orange, broadly obovate, 10–34 × 7–22 mm; stamens as long as style; anthers red, orange, rarely yellow (same color as petal base), ovoid or oblong, rarely linear, 2–3 mm; ovary ovoid, 2–3 mm diam., hairy; style cylindric but slightly conic basally, 6–8 mm, 2–3 times as long as ovary, strigose at base or to stigma base; stigma terminal. **Schizocarps** ovoid, 4–5 mm diam., strigose; beak cylindric, 6–18 mm, 3 times as long as fruit body, base conic, strigose at base or to stigma base; mericarps 3.5 × 1 mm, abaxially tuberculate, all tubercles rounded, less than 1 mm, sides slightly pitted, adaxial edge angled.

Flowering mainly Jun–Oct. Flat sandy areas throughout Chihuahuan and Sonoran deserts; 0–2000 m; Ariz., N.Mex., Tex.; Mexico.

Kallstroemia grandiflora has been collected twice in California, in San Diego and Riverside counties, and although not yet established in the state, it may be expected to become so. In Mexico, *K. grandiflora* is widespread from the northern border south to Oaxaca but does not occur in the Baja California peninsula.

Although petal color in *Kallstroemia grandiflora* is rather variable, that of individual populations appears to be quite stable. In favorable years, the species sometimes forms large populations many meters in extent, all plants having the same petal color. The relatively large, colorful flowers have resulted in this species being given multiple common names, unlike most of the smaller-flowered species.

3. Kallstroemia hirsutissima Vail ex Small, Fl. S.E. U.S., 670, 1333. 1903 • Carpet weed

Herbs, annual. **Stems** prostrate, to 0.7 m, copiously sericeous and hirsute with white or gray antrorse hairs. **Leaves** obovate in outline, 1–4 × 2–4 cm, copiously and conspicuously hairy; stipules 3–6 × 1 mm; leaflets 6–8, broadly elliptic to oblong-ovate or broadly ovate, 12–19 × 5–11 mm, terminal pair largest, surfaces densely hirsute, veins and margins sericeous. **Pedicels** 5–12 mm in flower and fruit, shorter than subtending leaves, thickened distally, curved or straight. **Flowers** 5–9 mm diam.; sepals persistent, subulate, 2.5–4 × 1 mm, as long as petals, in flower as long as style, in fruit clasping mature fruit body and ½ as long, only scarious margins becoming involute, hirsute, sparingly strigose; petals marcescent, 1-colored, yellow, fading white or orange, obovate, 2–4 × 1.5 mm; stamens as long as style; anthers yellow, ovoid, less than 1 mm; ovary globose, 1 mm diam., hairy; style broadly conic, 0.3–0.5 mm, ⅓–½ as long as ovary, hairy; stigma terminal, appearing almost sessile on ovary. **Schizocarps** broadly ovoid, 4–5 × 6–8 mm, strigillose; beak conic, 1–4 mm, shorter than fruit body, base broadly conic, hirsute; mericarps 4 × 1 mm, abaxially tuberculate, all tubercles rounded, less than 1 mm, sides pitted, adaxial edge angled.

Flowering Jun–Oct, following summer rains. Desert scrub, semiarid grasslands; 0–1700 m; Ariz., N.Mex., Tex.; Mexico (Chihuahua, Coahuila, Nuevo León, San Luis Potosí, Tamaulipas).

Kallstroemia hirsutissima is restricted to the Chihuahuan Desert and adjacent grasslands, mainly at higher elevations. The spreading, copiously hairy stems form a dense carpetlike mat. The species is reported to poison cattle, goats, and sheep (J. M. Kingsbury 1964).

4. **Kallstroemia maxima** (Linnaeus) Hooker & Arnott, Bot. Beechey Voy., 282. 1838

Tribulus maximus Linnaeus, Sp. Pl. 1: 386. 1753

Herbs, annual. **Stems** prostrate to decumbent, to 1(–1.5) m, sericeous and sparingly hirsute with white or yellow antrorse hairs, becoming glabrate. **Leaves** obovate in outline, 1–6 × 1.5–5 cm; stipules 3–5 × 1 mm; leaflets 6–8(–12), broadly oblong to elliptic, 5–29 × 3–14 mm, terminal pair usually largest, surfaces appressed-hirsute to glabrate, veins and margins sericeous. **Pedicels** 10–50 mm in flower and fruit, at first shorter than subtending leaves, equaling them or longer in fruit, little thickened distally, straight or curved. **Flowers** 7–25 mm diam.; sepals persistent, ovate, 3–8 × 2–3 mm, as long as or little shorter than petals, in flower as long as style, in fruit clasping but not entirely covering mature fruit body and shorter than beak, only scarious margins becoming involute, hirsute; petals marcescent, usually ± 2-colored, basally usually white to yellow-green or green, rarely red (often brighter than distal portion), fading white to bright orange, distally white to yellow or pale orange, obovate, 5–12 × 4–10 mm; stamens as long as style; anthers yellow or red-orange, usually ovoid, rarely linear, 1 mm; ovary ovoid, 1 mm, usually glabrous, sometimes strigose at base, rarely to base of style; style cylindric but slightly conic basally, 2–3 mm, 2–3 times as long as ovary, glabrous; stigma terminal. **Schizocarps** ovoid, 5–6 mm diam., usually glabrous, sometimes strigose at base or rarely to base of beak; beak cylindric, 3–7 mm, usually as long as fruit body, base widely conic, glabrous; mericarps 3–4 × 1 mm, abaxially tuberculate, cross-ridged, or slightly keeled, tubercles if present all rounded, less than 1 mm, sides pitted, adaxial edge angled.

Flowering year-round. Weedy habitats, disturbed areas; 0–1400 m; Ala., Fla., Ga., S.C., Tex.; Mexico; West Indies; Central America; n South America.

The most widespread species of the genus, *Kallstroemia maxima* is usually found at lower elevations. The species probably has been taken to many places inadvertently by humans.

5. **Kallstroemia parviflora** Norton, Rep. (Annual) Missouri Bot. Gard. 9: 153, plate 46. 1898 • Carpet weed [F]

Kallstroemia intermedia Rydberg

Herbs, annual. **Stems** prostrate to decumbent or ascending, to 1(–1.5) m, ± coarsely hirsute and sericeous with white or gray antrorse hairs, becoming glabrate. **Leaves** elliptic in outline, 1–6 × to 3 cm; stipules 5–7 × 1–3 mm; leaflets 6–10 (–12), elliptic to oblong, 8–19 × 3.5–9 mm, middle pairs largest, surfaces appressed-hirsute, veins and margins sericeous. **Pedicels** 10–40 mm in flower and fruit, usually longer than, sometimes equaling, subtending leaves, thickened distally, straight or sharply bent at base and straight distally. **Flowers** 10–25 mm diam.; sepals persistent, lanceolate, 4–7 × 1–2 mm, 2/3 as long as petals, in flower longer than style, in fruit shriveling and turning brown, appressed to mericarps and reaching from top of fruit body to tip of style, margins becoming sharply involute, hispid and strigose; petals marcescent, 1-colored, orange, drying white to yellow, narrowly obovate, 5–12 × 3.5–5 mm; stamens as long as style; anthers yellow, usually ovoid, rarely linear, less than 1 mm; ovary ovoid, 1 mm diam., hairy; style cylindric, 1 mm, as long or longer than ovary, strigose or glabrous; stigma terminal. **Schizocarps** ovoid, 3–4 × 4–6 mm, strigose; beak cylindric, 3–9 mm, as long as to 3 times as long as fruit body, base scarcely conic, strigose or glabrous; mericarps 3–4 × 1 mm, abaxially rugose to tuberculate, tubercles if present all rounded, less than 1 mm, sides lightly to strongly pitted, adaxial edge angled.

Flowering Jul–Oct. Disturbed areas, grasslands, dry areas; 100–2600 m; Ariz., Calif., Colo., Ill., Kans., La., Mo., Nev., N.Mex., Okla., Tex., Utah; Mexico; introduced in w South America.

Kallstroemia parviflora has been reported as a waif in the District of Columbia, Maryland, Mississippi, and Pennsylvania. Once considered a waif in southeastern California, it has been collected there regularly since 2003. The species was introduced into western Peru before September/October 1780 and into southern Ecuador before May 1974 (D. M. Porter 2005).

6. Kallstroemia perennans B. L. Turner, Field & Lab. 18: 155. 1950 [C] [E]

Kallstroemia hirsuta L. O. Williams, Ann. Missouri Bot. Gard. 22: 49, plate 4. 1935, not (Bentham) Engler 1890

Herbs, perennial. **Stems** prostrate to ascending, to 0.2 m, densely hispid with white or yellow bulbous-based hairs and strigose with white antrorse hairs. **Leaves** elliptic in outline, 2.5–5 × 2–3 cm; stipules 3–5 × 1–1.5 mm; leaflets 8–10, oblong to ovate, 13–18 × 6–10 mm, middle pairs largest, surfaces densely appressed-hirsute, veins and margins sericeous. **Pedicels** 15–20 mm in flower, 25–36 mm in fruit, shorter than subtending leaves, little thickened distally, bent sharply at base and straight distally. **Flowers** 35–45 mm diam.; sepals persistent, lanceolate, 13–15 × 1.5–2.5 mm, 1/2 as long as petals, in flower longer than style, in fruit spreading from base of mature mericarps and longer than fruit body but shorter than beak, margins becoming sharply involute, densely hispid and strigose; petals not marcescent, 1-colored, usually orange, sometimes pale orange to almost salmon-colored, obovate, 19–26 × 10 mm; stamens 2/3 as long as style; anthers orange, ovoid, 1.5 mm; ovary ovoid, 3 mm diam., hairy; style cylindric, 6 mm, 2 times as long as ovary, hispid; stigma extending along distal 1/3 of style. **Schizocarps** broadly ovoid, 5–6 × 8–10 mm, hispid and strigose; beak cylindric, 6–10 mm, longer than fruit body, base slightly conic, hirsute; mericarps 4 × 2.5 mm, abaxially cross-ridged, ± keeled, sides pitted, adaxial edge straight.

Flowering May–Aug, following spring and summer rains. Somewhat barren limestone or gypseous soils; of conservation concern; 400–900 m; Tex.

Kallstroemia perennans is a rare western Texas endemic, found only in Val Verde County on the Edwards Plateau and in Brewster and Presidio counties in the trans-Pecos. The species is in the Center for Plant Conservation's National Collection of Endangered Plants.

RHAMNACEAE Jussieu

• Buckthorn Family

Guy L. Nesom

Shrubs, trees, or woody vines [**herbs,** annual or perennial], evergreen or deciduous, synoecious, dioecious, or polygamous [monoecious], sometimes armed with thorns or stipular spines. **Leaves** alternate, fascicled, clustered, or opposite, simple; stipules present; petiole present (absent in *Condalia*); blade margins entire, serrate, serrulate, crenate, crenulate, dentate, or denticulate, sometimes spinose or spinulose; venation pinnate (sometimes obscurely, appearing 1-veined) or 3[–5]-veined from base (acrodromous). **Inflorescences** bisexual or unisexual, axillary or terminal, fascicles, umbels, panicles, cymes, or thyrses (these sometimes spikelike, racemelike, or paniclelike), or flowers solitary. **Flowers** bisexual or unisexual; perianth and androecium epigynous or perigynous [hypogynous]; hypanthium free or adnate to ovary proximally and free distally [absent]; sepals [3–]4–5(–8), distinct, valvate; petals 0 or [3–]4–5(–8), distinct; nectary present, intrastaminal, sometimes lining hypanthium; stamens [3–]4–5(–8), opposite petals, distinct, adnate to petal bases; anthers dehiscing by longitudinal slits; pistil 1, 2–4-carpellate, ovary superior to inferior, (1–)2–4-locular, placentation basal; ovules 1 per locule (2 per locule in *Karwinskia*), anatropous; styles 1–4, connate proximally to completely; stigmas 1–4. **Fruits** capsules, dehiscence loculicidal, schizocarps, samaras, or drupes. **Seeds** 1 (sometimes 2 in *Karwinskia*) per locule.

Genera 50–52, species 900–950 (15 genera, 105 species in the flora): North America, Mexico, West Indies, Central America, South America, Europe, Asia, Africa, Atlantic Islands, Indian Ocean Islands, Pacific Islands, Australia; mainly subtropical to tropical areas.

The fused, cuplike basal portion of the flower has been variously called a hypanthium (as here), floral cup, floral tube, and calyx tube.

The bark, leaves, and fruits of *Frangula alnus*, *F. purshiana*, and *Rhamnus cathartica* have been used as laxatives. Old World species of *Rhamnus* provide yellow and green dyes as well as drugs. Wood of some species (*Alphitonia* Reissek ex Endlicher, *Colubrina*, *Hovenia*, and *Ziziphus*) is used for construction, high quality furniture, carving, lathe work, and musical instruments. *Ziziphus mauritiana* (Indian jujube) and *Z. jujuba* (common jujube) are commercially cultivated for edible fruit. *Hovenia dulcis* is grown for its edible, fleshy inflorescence stalks. Some species of *Ceanothus*, *Hovenia, Paliurus,* and *Rhamnus* are cultivated as ornamentals.

SELECTED REFERENCES Aagesen, D. et al. 2005. Phylogeny of the tribe Colletieae (Rhamnaceae)—A sensitivity analysis of the plastid region *trn*L-*trn*F combined with morphology. Pl. Syst. Evol. 250: 197–214. Brizicky, G. K. 1964b. The genera of Rhamnaceae in the southeastern United States. J. Arnold Arbor. 45: 439–463. Medan, D. and C. Schirarend. 2004. Rhamnaceae. In: K. Kubitzki et al., eds. 1990+. The Families and Genera of Vascular Plants. 10+ Vols. Berlin etc. Vol. 6, pp. 320–338. Richardson, J. E. et al. 2000. A phylogenetic analysis of Rhamnaceae using *rbc*L and *trn*L-F plastid DNA sequences. Amer. J. Bot. 87: 1309–1324. Richardson, J. E. et al. 2000b. A revision of the tribal classification of Rhamnaceae. Kew Bull. 55: 311–340.

1. Woody vines.
   2. Tendrils present; fruits schizocarps; ovaries inferior . . . . 15. *Gouania*, p. 109
   2. Tendrils absent; fruits drupes; ovaries superior.
      3. Stems twining, glabrous; leaf blade secondary, and usually tertiary, veins strongly parallel . . . . 3. *Berchemia*, p. 59
      3. Stems not twining, hairy; leaf blade secondary veins not strongly parallel, tertiary veins reticulate . . . . 6. *Sageretia* (in part), p. 64
1. Shrubs (sometimes clambering) or trees.
   4. Sepals usually incurved, sometimes becoming spreading, sepals and petals white, cream, blue, purple, or rarely pink; hypanthium less than 0.5 mm wide . . . . 13. *Ceanothus*, p. 77
   4. Sepals spreading or erect, usually yellowish, yellowish green, green, or greenish white to white, sometimes brown, orange, or purple; petals white, cream, yellow, yellow-green, or greenish, or 0; hypanthium 1–4 mm wide.
      5. Leaf blades 3-veined from near base (venation acrodromous).
         6. Inflorescences compound dichasia, appearing repeatedly dichotomously branched; peduncles and pedicels fleshy in fruit . . . . 9. *Hovenia*, p. 68
         6. Inflorescences fascicles, cymes, or thyrses, or flowers solitary; peduncles and pedicels not fleshy in fruit.
            7. Plants unarmed; fruits capsules . . . . 12. *Colubrina* (in part), p. 73
            7. Plants armed with stipular spines; fruits drupes or samaras.
               8. Fruits samaras; leaf blade secondary veins distal to basal veins well developed; ovaries ½ inferior; bud scales present. . . . . 10. *Paliurus*, p. 68
               8. Fruits drupes; leaf blade secondary veins absent or poorly developed distal to basal veins, except sometimes near apex; ovaries superior; bud scales absent. . . . . 11. *Ziziphus* (in part), p. 70
      5. Leaf blades pinnately veined, sometimes basal pair of secondary veins more prominent or blade obscurely veined, appearing 1-veined or vaguely 3-veined from base.
         9. Leaves alternate, sometimes fascicled or clustered terminally.
            10. Fruits capsules, sometimes tardily dehiscent; ovaries ½ inferior . . . . 12. *Colubrina* (in part), p. 73
            10. Fruits drupes; ovaries superior.
                11. Inflorescences axillary or terminal, spikelike or spicate, paniclelike thyrses, overtopping or extending beyond foliage . . . . 6. *Sageretia* (in part), p. 64
                11. Inflorescences axillary, fascicles, umbels, cymes or thyrses, or flowers solitary, within foliage.
                    12. Drupes with 2–4 stones.
                        13. Leaf blade secondary veins arching, not parallel; bud scales present, buds glabrate; flowers usually unisexual, rarely some bisexual (plants polygamous); sepals thin, spreading, not keeled adaxially; petals 0 or 4, not clawed; styles 2–4, connate proximally; hypanthia usually not circumscissile, rarely so at or just proximal to sepal bases; stones longitudinally dehiscent; seeds not beaked, longitudinally furrowed . . . . 1. *Rhamnus* (in part), p. 45
                        13. Leaf blade secondary veins nearly straight, parallel; bud scales absent, buds hairy; flowers bisexual; sepals fleshy, usually ± erect, sometimes spreading, keeled adaxially; petals (rarely 4–)5, clawed; styles 1; hypanthia usually circumscissile far below sepal bases, rarely not circumscissile; stones indehiscent but open at base; seeds with cartilaginous beak, not furrowed . . . . 2. *Frangula*, p. 52

12. Drupes with 1 stone.
    14. Leaves not borne on short shoots; plants unarmed; sepals crested adaxially . . . . 7. *Krugiodendron* (in part), p. 66
    14. Leaves mostly borne on short shoots, if not, then plants armed with thorns; sepals keeled adaxially.
        15. Petals 0 (5 in *C. ericoides*, with linear, revolute leaves); nectaries absent, thin, or margins thickened, lining hypanthium; styles 1 . . . . 5. *Condalia*, p. 61
        15. Petals 5; nectaries fleshy, filling hypanthium; styles 2(–4) . . . . 11. *Ziziphus* (in part), p. 70

[9. Shifted to left margin.—Ed.]

9. Leaves opposite or subopposite.
   16. Plants armed with thorns.
       17. Leaves early deciduous, usually absent by flowering; fruits capsules . . . . 14. *Adolphia*, p. 108
       17. Leaves persistent or deciduous, present at flowering; fruits drupes.
           18. Inflorescences fascicles or flowers solitary, within foliage; sepals 4, not keeled adaxially; petals 0 or 4; trees or shrubs, erect or spreading . . . . 1. *Rhamnus* (in part), p. 45
           18. Inflorescences spikelike or spicate paniclelike thyrses, overtopping or extending beyond foliage; sepals 5, keeled adaxially; petals 5; shrubs arching, sprawling, drooping, or clambering . . . . 6. *Sageretia* (in part), p. 64
   16. Plants unarmed.
       19. Leaves deciduous; petals 4–5.
           20. Leaf blades not gland-dotted abaxially, secondary veins arching; petals 4; ovules 1 per locule; drupe stones 2–4, dehiscent . . . . 1. *Rhamnus* (in part), p. 45
           20. Leaf blades gland-dotted abaxially (secondary veins appearing light- and dark-banded), secondary veins relatively straight; petals 5; ovules 2 per locule; drupe stones 1, indehiscent . . . . 4. *Karwinskia*, p. 61
       19. Leaves persistent; petals 0.
           21. Leaf blade secondary veins arching near margins, higher order veins not forming adaxially raised reticulum; sepals crested adaxially; drupes 5–8 mm . . . . 7. *Krugiodendron* (in part), p. 66
           21. Leaf blade secondary veins straight nearly to margins, higher order veins forming adaxially raised reticulum enclosing isodiametric areoles; sepals small-keeled, not crested, adaxially; drupes 10–20 mm . . . . 8. *Reynosia*, p. 67

1. RHAMNUS Linnaeus, Sp. Pl. 1: 193. 1753; Gen. Pl. ed. 5, 89. 1754 • Buckthorn, nerprun [Greek *rhamnos*, prickly shrubs, such as buckthorn]

Guy L. Nesom

John O. Sawyer†

Shrubs or trees, erect or spreading, unarmed or armed with thorns; bud scales present, buds glabrate. **Leaves** deciduous or persistent, present at flowering, alternate or opposite to subopposite, sometimes fascicled on short shoots; blade not gland-dotted; pinnately veined, secondary veins arching, not parallel. **Inflorescences** axillary, within foliage, fascicles or cymes, or flowers solitary; peduncles and pedicels not fleshy in fruit. **Pedicels** present. **Flowers** usually unisexual, rarely some bisexual (plants polygamous); hypanthium campanulate to cupulate, 2–3 mm wide, usually not circumscissile, rarely so at or just proximal to sepal bases; sepals 4–5, spreading, yellowish, yellowish green, green, or greenish white to white, ovate-triangular, thin, not keeled or crested adaxially; petals 0 or 4[–5], yellowish to cream, hooded, spatulate, not clawed; nectary thin, lining hypanthium; stamens 4–5 (rudimentary in pistillate flowers); ovary superior, 2–4-locular; styles 2–4, connate proximally. **Fruits** drupes; stones 2–4, longitudinally dehiscent. **Seeds** obovoid or oblong-obovoid, not beaked, longitudinally furrowed. $x = 12$.

Species ca. 150 (14 in the flora): North America, Mexico, Central America, South America, Europe, e Asia, Africa.

Rhamnus in the flora area includes six species (species 9–14) introduced primarily from Europe and Asia. The native species for the most part are clearly delimited, but the *R. crocea* complex needs taxonomic study using modern techniques.

SELECTED REFERENCES Johnston, L. A. 1975. Revision of the *Rhamnus serrata* complex. Sida 6: 67–79. Wolf, C. B. 1938. The North American species of *Rhamnus*. Rancho Santa Ana Bot. Gard. Monogr., Bot. Ser. 1.

1. Leaves persistent, alternate, blades distinctly coriaceous; petals 0 (rarely 4 in *R. pilosa*).
    2. Leaf blade secondary veins with basal pair diverging much more acutely than distal pairs; inflorescences cymes or flowers solitary; sepals 5; drupes dark red, becoming black, stones 3. . . . . .9. *Rhamnus alaternus*
    2. Leaf blade secondary veins all diverging at nearly same angle; inflorescences fascicles or flowers solitary; sepals 4(–5); drupes red, stones 2.
        3. Leaf blade margins entire or coarsely serrate, blunt-serrate, or spinulose-serrate.
            4. Shrubs, 1.2–2(–4) m, usually armed with thorns; leaf blades 0.7–2 cm, elliptic to obovate or ovate; California mainland. . . . . .1. *Rhamnus crocea*
            4. Shrubs or trees, 2.5–6(–10) m, unarmed; leaf blades (1.5–)2–5(–6) cm, broadly elliptic to oblong or ovate-oblong; California Channel Islands . . . . 2. *Rhamnus pirifolia*
        3. Leaf blade margins spinulose to spinose-dentate.
            5. Branchlets stiff, glabrous or densely and softly hirtellous; both leaf blade surfaces glabrous . . . . 3. *Rhamnus ilicifolia*
            5. Branchlets flexible, sparsely to densely softly hirtellous; both leaf blade surfaces sparsely to densely softly hirtellous . . . . 4. *Rhamnus pilosa*
1. Leaves deciduous, alternate, subopposite, or opposite, sometimes fascicled on short shoots, blades herbaceous to subcoriaceous; petals 0 or 4.
    6. Leaves usually alternate, sometimes subopposite, blade secondary veins 4–11 pairs; shrubs or small trees 0.5–3(–4) m, unarmed.
        7. Sepals 5; petals 0; drupe stones 3; shrubs 0.5–1(–1.5) m; wetlands and seeps . . . 8. *Rhamnus alnifolia*
        7. Sepals 4; petals 4; drupe stones 2 (3 in *R. arguta*); shrubs or small trees 0.5–3(–4) m; drier habitats.
            8. Petioles 10–28 mm; pedicels 10–24(–30) mm; drupe stones 3; leaf blades usually broadly ovate, ovate, oblong-ovate, or ovate-cordate, sometimes ovate-orbiculate, margins spinulose-serrate . . . . 14. *Rhamnus arguta* (in part)
            8. Petioles 3–8 mm; pedicels (0–)0.5–4 mm; drupe stones 2; leaf blades lanceolate, elliptic-lanceolate, oblong-lanceolate, elliptic, or elliptic-oblong, margins crenulate, crenulate-serrate, or serrulate.
                9. Branchlets 6+ cm; leaf blade apices acute to acuminate; c, e United States . . . . .5. *Rhamnus lanceolata*
                9. Branchlets 2–5 cm; leaf blade apices acute to obtuse or rounded; w United States.
                    10. Bud scales thin, glossy, pale golden to yellowish; leaf blades usually lanceolate, sometimes elliptic-lanceolate or oblong-lanceolate, 3–8 cm, both surfaces glabrous . . . . 6. *Rhamnus smithii*
                    10. Bud scales coriaceous, dull, dark red to reddish brown; leaf blades usually elliptic to elliptic-lanceolate or elliptic-oblong, 1.5–5(–5.5) cm, both surfaces minutely hirtellous. . . . . 7. *Rhamnus serrata*
    6. Leaves usually opposite to subopposite, sometimes alternate or fascicled on short shoots, blade secondary veins 2–7(–8) pairs; shrubs or trees (1–)2–10 m, armed with thorns.
        11. Leaf blade bases rounded to truncate or cordate; drupe stones 3–4.
            12. Leaf blade margins crenate-serrate, secondary veins 2–4 pairs; pedicels 2–4 mm; drupes 5–6(–8) mm . . . . 10. *Rhamnus cathartica*
            12. Leaf blade margins spinulose-serrate, secondary veins (3–)4–5 pairs; pedicels 10–24(–30) mm; drupes 7–10 mm . . . . 14. *Rhamnus arguta* (in part)

[11. Shifted to left margin.—Ed.]

11. Leaf blade bases cuneate or acute; drupe stones 2.
    13. Leaf blades obovate to elliptic-oblanceolate or broadly oblanceolate, adaxial surfaces usually minutely hirtellous at least along main veins on distal 1/3, rarely glabrous . . . . 13. *Rhamnus japonica*
    13. Leaf blades usually elliptic to oblong, oblong-obovate, obovate-elliptic, or oblong-lanceolate, sometimes ovate to obovate, rarely oblanceolate-elliptic, adaxial surfaces glabrous or sparsely pilose on veins.
        14. Leaf blades green abaxially, abaxial surface with straight, colorless hairs on veins, secondary veins 3–5(–6) pairs; pedicels 10–20 mm . . . . 11. *Rhamnus davurica*
        14. Leaf blades yellowish green abaxially, abaxial surface with wavy or curled, yellow hairs along veins or at least in vein axils, secondary veins (4–)5–7(–8) pairs; pedicels 5–10 mm . . . . 12. *Rhamnus utilis*

1. **Rhamnus crocea** Nuttall in J. Torrey and A. Gray, Fl. N. Amer. 1: 261. 1838 (as croceus) • Spiny redberry [E]

Shrubs, 1.2–2(–4) m, usually armed with thorns. **Branchlets** reddish to red-purple or orange-brown, glabrous. **Leaves** persistent, alternate; petiole 2–6 mm; blade usually dull green abaxially, shiny dark green adaxially, elliptic to obovate or ovate, 0.7–2 cm, distinctly coriaceous, base cuneate to truncate, margins coarsely serrate, spinulose-serrulate, or entire, apex obtuse to emarginate, both surfaces glabrous; secondary veins 5–7 pairs, all diverging at nearly same angle. **Inflorescences** fascicles or flowers solitary. **Pedicels** 2–3 mm. **Sepals** 4(–5). **Petals** 0. **Drupes** red, globose, 5–8 mm; stones 2.

Flowering (Jan–)Mar–May. Dry washes and canyons, coastal and inland dunes, alluvial fans, gravel flood plains, disturbed sandy flats, brushy slopes, steep granitic slopes, woodlands, coastal sage scrub, chaparral; 50–1200 m; Calif.

Rhamnus crocea and closely related species were considered conspecific by C. B. Wolf (1938), who treated those in the flora area as subspp. *crocea*, *ilicifolia*, *pilosa*, and *pirifolia*; he also recognized subsp. *insula* (Kellogg) C. B. Wolf from Mexico (Baja California). The taxa are distinctive, but intermediates exist. Wolf identified *R. crocea/ilicifolia* intermediates from Marin County to the California/Mexico boundary, and *R. ilicifolia/insula*, *R. crocea/pilosa*, and *R. ilicifolia/pilosa* intermediates in southern California, especially in San Diego County.

Arizona plants (Pima County, especially in the Ajo Mountains) identified as *Rhamnus crocea* (for example, K. Christie 2006) appear to be populational variants of the single species present there, *R. ilicifolia* (R. Felger, pers. comm.). In California, where the two are sympatric over a relatively broad region, *R. crocea* is distinct from *R. ilicifolia* in usually bearing thorns and in having smaller leaves with less spinulose margins.

2. **Rhamnus pirifolia** Greene, Pittonia 3: 15. 1896 • Island redberry

Rhamnus crocea Nuttall subsp. *pirifolia* (Greene) C. B. Wolf; *R. crocea* var. *pirifolia* (Greene) Little

Shrubs to small trees, 2.5–6(–10) m, unarmed. **Branchlets** purple to gray, glabrous or puberulent. **Leaves** persistent, alternate; petiole 5–10 mm; blade dull reddish to yellowish brown abaxially, usually shiny green adaxially, broadly elliptic to oblong or ovate-oblong, (1.5–)2–5(–6) cm, distinctly coriaceous, base cuneate to nearly truncate, margins blunt-serrate or entire, apex acute to rounded, apiculate, both surfaces glabrous; secondary veins 5–9 pairs, all diverging at nearly same angle. **Inflorescences** fascicles or flowers solitary. **Pedicels** 3–6 mm. **Sepals** 4. **Petals** 0. **Drupes** red, globose, 6–8 mm; stones 2.

Flowering (Jan–)Feb–Jun. Steep slopes, canyon walls and bottoms, dunes, grasslands, coastal sage scrub, chaparral, pine woodlands; 10–500 m; Calif.; Mexico (Baja California).

Rhamnus pirifolia is known in California on the Channel Islands and in Baja California on Guadalupe Island; it has not been implicated in hybridization with other members of the *R. crocea* complex.

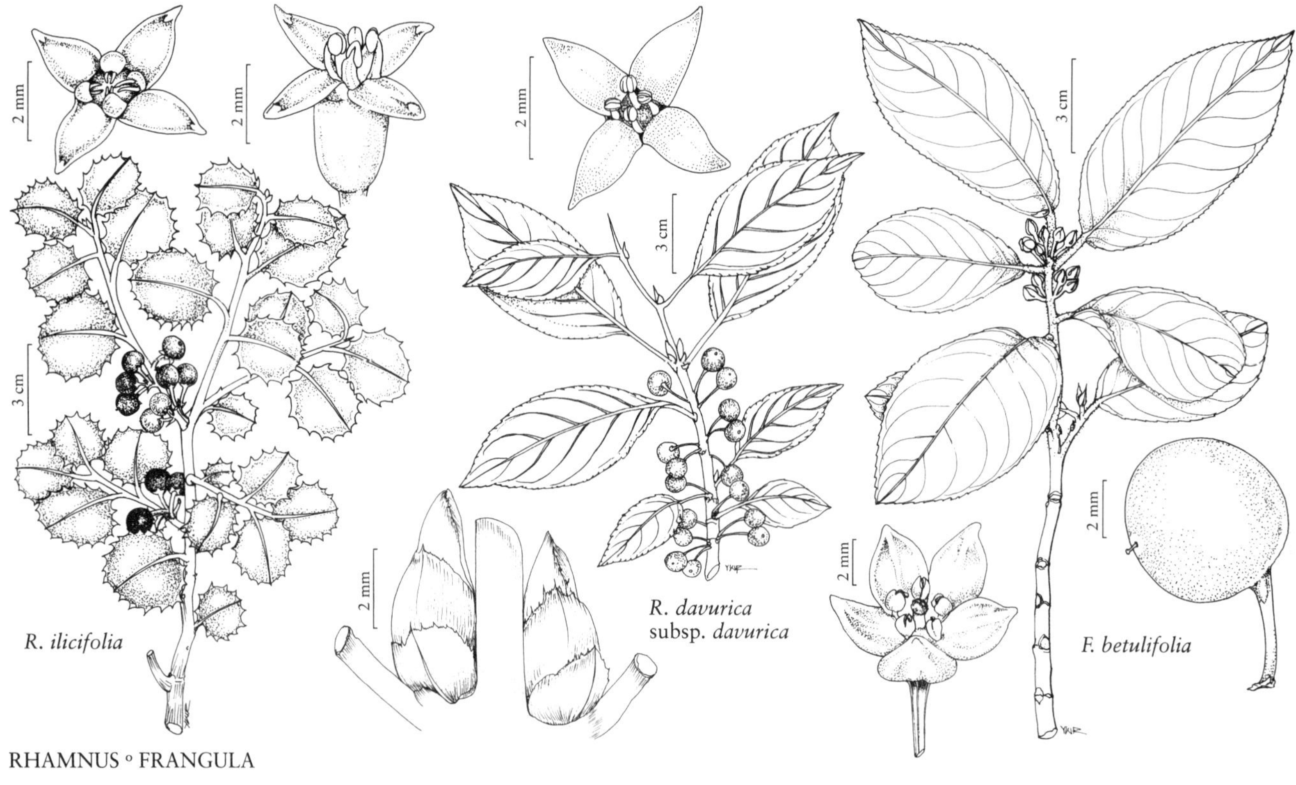

RHAMNUS ◦ FRANGULA

3. **Rhamnus ilicifolia** Kellogg, Proc. Calif. Acad. Sci. 2: 37. 1863 (as ilicifolius) • Hollyleaf redberry [F]

Rhamnus crocea Nuttall subsp. *ilicifolia* (Kellogg) C. B. Wolf; *R. crocea* var. *ilicifolia* (Kellogg) Greene

Shrubs or small trees, 1–4 m, unarmed. **Branchlets** gray, stiff, glabrous or densely and softly hirtellous. **Leaves** persistent, alternate; petiole 2–10 mm; blade usually dull green, sometimes reddish brown, abaxially, glossy to dull green adaxially, broadly elliptic to oblong, orbiculate, or ovate, 2–4.5 cm, distinctly coriaceous, base cuneate to nearly truncate, margins spinulose to spinose-dentate, apex rounded or obtuse to truncate or emarginate, both surfaces glabrous; secondary veins 5–7 pairs, all diverging at nearly same angle. **Inflorescences** fascicles or flowers solitary. **Pedicels** 2–4 mm. **Sepals** 4. **Petals** 0. **Drupes** red, globose, 4–6(–8) mm; stones 2. **2*n*** = 24.

Flowering Jan–Jun. Canyon slopes and bottoms, rock faces, open hillsides, sandstone ridges, serpentine slopes, roadsides, stream benches, riparian areas, meadows, coastal sage scrub, chaparral/desert transition, chaparral, woodlands, montane forests; 100–2200(–2400) m; Ariz., Calif., Nev., Oreg.; Mexico (Baja California).

4. **Rhamnus pilosa** (Trelease ex Curran) Abrams, Bull. Torrey Bot. Club 37: 153. 1910 • Hairyleaf redberry [E]

Rhamnus crocea Nuttall var. *pilosa* Trelease ex Curran, Proc. Calif. Acad. Sci., ser. 2, 1: 251. 1888; *R. crocea* subsp. *pilosa* (Trelease ex Curran) C. B. Wolf

Shrubs, 1–3 m, unarmed. **Branchlets** gray, flexible, sparsely to densely softly hirtellous. **Leaves** persistent, alternate; petiole 2–5 mm; blade dull green abaxially, darker green adaxially, ovate to broadly oblong or orbiculate, 1.5–2 cm, distinctly coriaceous, base cuneate to nearly truncate, margins spinulose, apex rounded, both surfaces sparsely to densely softly hirtellous; secondary veins 5–7 pairs, all diverging at nearly same angle. **Inflorescences** fascicles or flowers solitary. **Pedicels** 2–4 mm. **Sepals** 4. **Petals** usually 0, rarely 4. **Drupes** red, globose, 5–6 mm; stones 2.

Flowering Apr–Jun. Granitic substrates, clay soils, sandstone, woodlands, chaparral, coastal sage scrub; 100–1700 m; Calif.

Rhamnus pilosa is known only from San Diego County and extreme southwestern Riverside County. The species has been reported from Arizona and Mexico (Baja California) (USDA Plants Database, based on L. Abrams and R. S. Ferris 1923–1960, vol. 2), but no specimens have been found to support this claim.

5. **Rhamnus lanceolata** Pursh, Fl. Amer. Sept. 1: 166. 1813 (as lanceolatus) • Eastern lanceleaf buckthorn [E]

Rhamnus lanceolata subsp. *glabrata* (Gleason) Kartesz & Gandhi; *R. lanceolata* var. *glabrata* Gleason

Shrubs, 0.5–2(–4) m, unarmed. **Branchlets** greenish to gray, more than 5 cm, glabrous or sparsely to densely softly hirtellous and glabrescent. **Leaves** deciduous, alternate; petiole 3–8 mm; blade dull green abaxially, shiny darker green adaxially, lanceolate to elliptic, (3–)5–8 cm, herbaceous, base cuneate, margins finely crenulate or crenulate-serrate, apex acute to acuminate, both surfaces glabrous or hairy; secondary veins 4–5 pairs, all diverging at nearly same angle. **Inflorescences** fascicles or flowers solitary. **Pedicels** (1–)2–4 mm. **Sepals** 4. **Petals** 4. **Drupes** black, globose to subglobose, 5–7 mm; stones 2.

Flowering Apr–Jun. Dry to moist thickets over calcareous rocks, seeps, rock outcrops, bottomlands; 50–1200 m; Ala., Ark., Del., Ill., Ind., Iowa, Kans., Ky., La., Md., Miss., Mo., Nebr., N.J., N.C., Ohio., Okla., Pa., S.Dak., Tenn., Tex., Va., W.Va., Wis.

Plants of *Rhamnus lanceolata* with hirtellous-pubescent stems and leaves (subsp. *lanceolata* if given formal recognition; the type from Tennessee) are more restricted in distribution than glabrous plants, but the distinction often seems arbitrary. Densely hairy and glabrous plants both occur in Alabama and Tennessee. Plants with sparsely hairy leaves occur commonly in Kentucky and Illinois, rarely in Missouri and Iowa, and the density of the vestiture is variable.

6. **Rhamnus smithii** Greene, Pittonia 3: 17. 1896 • Smith's buckthorn [E]

Shrubs, 1–3 m, unarmed. **Branchlets** greenish to brown, 5 cm, minutely puberulent, glabrescent. **Bud scales** pale golden to yellowish, thin, glossy. **Leaves** deciduous, usually alternate, rarely subopposite; petiole 3–6(–8) mm; blade green abaxially, glossy green or yellow-green adaxially, usually lanceolate, sometimes elliptic-lanceolate or oblong-lanceolate, 3–8 cm, herbaceous to subcoriaceous, base cuneate, margins crenulate to serrulate, apex acute to obtuse or rounded, both surfaces glabrous; secondary veins 7–11 pairs, all diverging at nearly same angle. **Inflorescences** axillary fascicles or flowers solitary. **Pedicels** (0–)0.5–2.5 mm. **Sepals** 4. **Petals** 4. **Drupes** black, globose or slightly elongate, 4–6(–8) mm; stones 2.

Flowering May–Jun. Dry, grassy hillsides, gravelly terraces, shale knolls, mesic slopes, rocky meadows, sandy alluvium, sagebrush, conifer forests; 2100–2600 (–3000) m; Colo., N.Mex.

7. **Rhamnus serrata** Humboldt & Bonpland ex Willdenow in J. J. Roemer et al., Syst. Veg. 5: 295. 1819 • Sawleaf buckthorn

Rhamnus fasciculata Greene; *R. serrata* var. *guatemalensis* L. A. Johnson; *R. smithii* Greene subsp. *fasciculata* (Greene) C. B. Wolf

Shrubs to small trees, 0.8–2.5 m, unarmed. **Branchlets** gray-brown, 2 cm, puberulent, glabrescent. **Bud scales** dark red to reddish brown, coriaceous, dull. **Leaves** deciduous, alternate to subopposite; petiole 3–4 mm; blade usually yellowish brown, rarely green, abaxially, green adaxially, elliptic to elliptic-lanceolate or elliptic-oblong, 1.5–5(–5.5) cm, herbaceous, base cuneate to rounded or truncate, margins serrulate to crenulate, apex obtuse, both surfaces minutely hirtellous, often more densely so abaxially; secondary veins 5–8(–11) pairs, all diverging at nearly same angle. **Inflorescences** fascicles or flowers solitary. **Pedicels** 1–4 mm. **Sepals** 4. **Petals** 4. **Drupes** black, globose or slightly elongate, 4–6(–8) mm; stones 2.

Flowering Apr–Jun. Canyons, stream banks, open hillsides, ledges, ridges, among boulders, gravelly limestone hills, roadsides, mesic forests, woodlands; 1500–2600 m; Ariz., N.Mex., Tex.; Mexico; Central America (Guatemala).

8. **Rhamnus alnifolia** L'Héritier, Sert. Angl., 3. 1789 (as alnifolius) • Alderleaf buckthorn, American alder-buckthorn, dwarf alder, nerprun á feuilles d'aulne [E]

Shrubs, 0.5–1(–1.5) m, unarmed. **Branchlets** gray to brown, glabrous or pubescent. **Leaves** deciduous, alternate; petiole 5–15 mm; blade dark green to olive green on both surfaces, lanceolate-oblong to elliptic or lanceolate-ovate, 4.5–11 cm, herbaceous, base cuneate to rounded, truncate, or subcordate, margins crenate to crenate-serrate, apex obtuse to acute or acuminate, abaxial surface glabrous or puberulent along veins, adaxial surface glabrous or glabrate; secondary veins (4–)5–7 pairs, all diverging at nearly same angle. **Inflorescences** fascicles or flowers solitary. **Pedicels** 2–10 mm. **Sepals** 5. **Petals** 0. **Drupes** black, globose or slightly elongate, 6–8 mm; stones 3.

Flowering May–Jul. Fens and swamps, generally calcareous, riparian thickets, interdunal swales, shore lines, marshes and mats, wet meadow edges, outcrops, deciduous and coniferous forests; 10–2700 m; Alta., B.C., Man., N.B., Nfld. and Labr. (Nfld.), N.S., Ont., P.E.I., Que., Sask.; Calif., Conn., Idaho, Ill., Ind., Iowa, Maine, Md., Mass., Mich., Minn., Mont., N.H., N.J., N.Y., N.Dak., Ohio, Oreg., Pa., R.I., S.Dak., Tenn., Utah, Vt., Va., Wash., W.Va., Wis., Wyo.

Rhamnus alnifolia is a primary host for the soybean aphid.

9. Rhamnus alaternus Linnaeus, Sp. Pl. 1: 193. 1753 • Italian or evergreen buckthorn [I]

Shrubs, 0.5–6(–10) m, unarmed. **Branchlets** purplish brown, puberulent. **Leaves** persistent, alternate; petiole 4–10 mm; blade dull to glossy yellowish green on both surfaces, elliptic to elliptic-obovate, ovate, or ovate-lanceolate, 2–4(–6) cm, distinctly coriaceous, base acute to obtuse, margins sharply serrate to spinulose-serrate, apex acute to subspinulose, both surfaces glabrous except abaxial vein axils tomentose; secondary veins (3–)4–5 pairs, basal pair diverging much more acutely than distal pairs. **Inflorescences** cymes or flowers solitary. **Pedicels** 2–6 mm. **Sepals** 5. **Petals** 0. **Drupes** dark red, becoming black, globose, 5–7 mm; stones 3.

Flowering (Feb–)Mar–Jun. Woodland edges, fencerows, low areas, stream banks, canyons, arroyos; 0–400 m; introduced; Calif.; Europe; introduced also in Pacific Islands (New Zealand), Australia.

10. Rhamnus cathartica Linnaeus, Sp. Pl. 1: 193. 1753 (as catharticus) • European or common buckthorn, nerprun cathartique [I]

Shrubs or trees, (1–)2–8 m, armed with thorns. **Branchlets** dark to reddish gray or purple, glabrous. **Leaves** deciduous (often present well after frost), usually opposite to subopposite, rarely alternate, sometimes fascicled on short shoots; petiole 10–27 mm; blade dull green abaxially, glossy darker green adaxially, usually ovate to elliptic-ovate, sometimes broadly elliptic or nearly orbiculate, (2–)4–7 cm, usually 1–2 times longer than wide, herbaceous, base rounded to rounded-truncate or slightly subcordate, margins crenate-serrate, apex acute to rounded, often abruptly short-acuminate, both surfaces glabrous; secondary veins 2–4 pairs, all diverging at nearly same angle or proximal diverging more obtusely. **Inflorescences** fascicles or flowers solitary. **Pedicels** 2–4 mm. **Sepals** 4. **Petals** 4. **Drupes** black, globose to depressed-globose, 5–6(–8) mm; stones 3–4. $2n = 24$.

Flowering Apr–Jun. Vacant lots, fields, forest edges, fencerows, roadsides, stream channels, riverbanks, ravines, flood plains, swampy habitats, deciduous forests; 10–1000(–2000) m; introduced; Alta., Man., N.B., N.S., Ont., P.E.I., Que., Sask.; Ariz., Colo., Conn., Del., Idaho, Ill., Ind., Iowa, Kans., Ky., Maine, Md., Mass., Mich., Minn., Mo., Mont., Nebr., N.H., N.J., N.Y., N.C., N.Dak., Ohio, Pa., R.I., S.Dak., Utah, Vt., Va., Wash., W.Va., Wis., Wyo.; Europe; c, sw Asia (China, Russia in w Siberia); nw Africa.

Rhamnus cathartica was introduced to North America as an ornamental shrub in the mid 1800s and was originally used for hedges, farm shelterbelts, and wildlife habitat; it is an aggressive invader of woods and prairies and is able to completely displace native vegetation. W. H. Brewer collected *R. cathartica* (UC 18526) in the 1800s from an unknown location in California. No specimens have been collected in the state since that time, and it apparently is not naturalized there.

Rhamnus cathartica is a primary host for the soybean aphid, *Aphis glycines*, native to eastern Asia. It uses the buckthorn as a winter host and spreads to soybean in the spring. The insect was first discovered in North America in 2000 in Wisconsin and subsequently has spread to at least 20 states in the United States and three provinces in Canada. The orange-colored wood of *R. cathartica* is sometimes used by woodcarvers.

SELECTED REFERENCE Kurylo, J. S. 2007. *Rhamnus cathartica*: Native and naturalized distribution and habitat preferences. J. Torrey Bot. Soc. 134: 420–430.

11. Rhamnus davurica Pallas, Reise Russ. Reich. 3: 721. 1776 (as davuricus) • Dahurian buckthorn [F] [I]

Subspecies 2 (1 in the flora): introduced; Asia.

11a. Rhamnus davurica Pallas subsp. **davurica** [F] [I]

Shrubs or small trees, 2–10 m, armed with thorns. **Branchlets** brown or red-brown, glabrous. **Leaves** deciduous, opposite or sometimes fascicled on short shoots; petiole 5–25 mm; blade green abaxially, darker green adaxially, usually oblong-obovate, sometimes broadly elliptic or ovate to obovate, rarely oblanceolate-elliptic, (4–)5–10(–13) cm, usually 2–3 times longer than wide, herbaceous, base cuneate, margins crenulate, apex usually acute or shortly acuminate to acuminate, rarely obtuse or rounded, abaxial surface sparsely pilose on veins, hairs straight, colorless, adaxial surface glabrous or sparsely pilose on veins; secondary veins 3–5(–6)

pairs, all diverging at nearly same angle. **Inflorescences** fascicles or flowers solitary. **Pedicels** 10–20 mm. **Sepals** 4. **Petals** 4. **Drupes** black, globose, 5–6 mm; stones 2. **2*n*** = 24.

Flowering Apr–Jun. Flood plains, forest edges and openings, stream banks, suburban woodlands; 50–300 m; introduced; Conn., Del., Ill., Iowa, Ky., Md., Mich., Mo., Nebr., N.J., N.Y., N.C., N.Dak., Ohio, Pa., R.I., S.Dak., Tenn., Va.; Asia (China, Korea, Mongolia, Russia).

Plants of *Rhamnus davurica* from the flora area have sometimes been identified as subsp. *nipponica* (Makino) Kartesz & Gandhi [*R. davurica* var. *nipponica* Makino, *R. nipponica* (Makino) Grubov; the type from Japan] (for example, D. W. Magee and H. E. Ahles 1999), but it is not clear whether they are correctly identified or even whether they are outside the range of variability in subsp. *davurica*. They also sometimes have been called *R. citrifolia* (Weston) W. J. Hess & Stern, an illegitimate later homonym of *R. citrifolia* Rusby 1907.

12. Rhamnus utilis Decaisne, Compt. Rend. Hebd. Séances Acad. Sci. 44: 1141. 1857 • Chinese buckthorn I

Varieties 3 (1 in the flora): introduced; Asia.

Infraspecific taxa in *Rhamnus* are usually described at the rank of subspecies, whereas those of *R. utilis* (all Chinese) are described as varieties.

12a. Rhamnus utilis Decaisne var. **utilis** I

Shrubs or small trees, 1–4 m, armed with thorns. **Branchlets** brown to purple-red, glabrate. **Leaves** deciduous, opposite to subopposite, sometimes fascicled on short shoots; petiole 5–15 mm; blade light green abaxially, glossy darker green adaxially, commonly yellowish or pale in age, oblong to elliptic, obovate-elliptic, or oblong-lanceolate, (4–)6–11(–13) cm, usually (2.5–)2.8–3.5 (–3.8) times longer than wide, subcoriaceous, base cuneate, margins serrulate to crenate-serrate, apex acute, abaxial surface puberulent along veins or at least in vein axils, hairs wavy or curled, yellow, adaxial surface glabrous; secondary veins (4–)5–7(–8) pairs, all diverging at nearly same angle. **Inflorescences** fascicles. **Pedicels** 5–10 mm. **Sepals** 4. **Petals** 4. **Drupes** black, globose to subglobose, 5–7 mm; stones 2.

Flowering Apr–Jun. Old fields, thickets, meadows, degraded woods, woods edges, commonly with *Rhamnus cathartica* and *R. davurica*; 200–300 m; introduced; Conn., Ill., Iowa, Mich., Ohio; Asia (China, Japan, Korea).

According to E. G. Voss (1972–1996, vol. 2), some plants of *Rhamnus utilis* in Michigan are similar to forms of *R. davurica* and perhaps represent hybrids between the two. Hybridization between *R. cathartica* and *R. utilis* has been documented (N. L. Gil-Ad and A. A. Reznicek 1997).

13. Rhamnus japonica Maximowicz, Mém. Acad. Imp. Sci. Saint Pétersbourg, Sér. 7, 10(11): 11, figs. 52–64. 1866 • Japanese buckthorn I

Shrubs or small trees, 1–6 m, armed with thorns. **Branchlets** reddish brown to gray, glabrous. **Leaves** deciduous, opposite; petiole 5–16 mm; blade green on both surfaces, obovate to elliptic-oblanceolate or broadly oblanceolate, (2–)3–7 cm, herbaceous, base acute, margins crenate-serrulate, apex acuminate, abaxial surface glabrous or glabrate, adaxial surface minutely hirtellous at least along main veins on distal ⅓, rarely glabrous; secondary veins 4–5 pairs, all diverging at nearly same angle. **Inflorescences** fascicles or flowers solitary. **Pedicels** 3–7 mm. **Sepals** 4. **Petals** 4. **Drupes** black, globose, 5–6(–8) mm; stones 2.

Flowering Apr–May. Second-growth woodlands, disturbed sites; 100–300 m; introduced; Ill., Mo.; e Asia (Japan).

Rhamnus japonica was first reported as naturalized in North America from the Morton Arboretum and its immediate vicinity in DuPage County, Illinois (F. Swink and G. Wilhelm 1994).

14. Rhamnus arguta Maximowicz, Mém. Acad. Imp. Sci. Saint Pétersbourg, Sér. 7, 10(11): 6, figs. 48–51. 1866 • Sawtooth buckthorn I

Shrubs or small trees, 2–3 m, usually unarmed, rarely armed with weak thorns. **Branchlets** dark purple or purple-red, glabrous. **Leaves** deciduous, usually opposite to subopposite, sometimes alternate, usually fascicled on short shoots, separating with age; petiole 10–28 mm; blade green abaxially, darker green adaxially, usually broadly ovate, ovate, oblong-ovate, or ovate-cordate, sometimes ovate-orbiculate, 2.5–6(–7) cm, herbaceous, base truncate to cordate, margins spinulose-serrate, apex obtuse-rounded or acute, both surfaces glabrous or abaxial puberulent along veins; secondary veins (3–)4–5 pairs, all diverging at nearly same angle or proximal diverging more obtusely. **Inflorescences** fascicles or flowers solitary. **Pedicels** 10–24(–30) mm. **Sepals** 4. **Petals** 4. **Drupes** black, globose to obovoid-globose, 7–10 mm; stones 3. **2*n*** = 24.

Flowering May–Jun. Disturbed sites; 200 m; introduced; Ind.; Asia (China).

The only report of *Rhamnus arguta* as naturalized in North America is from Jasper County (F. Swink and G. Wilhelm 1994). It was identified by Swink and Wilhelm as *R. arguta* var. *velutina* Handel-Mazzetti, which is puberulent on the petioles and leaf blades (at least abaxially on veins), but considerable populational variability apparently exists within the species.

2. FRANGULA Miller, Gard. Dict. Abr. ed. 4, vol. 1. 1754 • Buckthorn [Probably from Latin *frango*, to break, and *-ula*, diminutive, alluding to brittleness of twigs]

John O. Sawyer†

Guy L. Nesom

Rhamnus Linnaeus sect. *Frangula* (Miller) de Candolle; *Rhamnus* subg. *Frangula* (Miller) Gray

Shrubs or trees, unarmed; bud scales absent, buds hairy. **Leaves** usually deciduous, sometimes persistent, alternate [rarely opposite], rarely fascicled on short shoots; blade not gland-dotted; pinnately veined, secondary veins nearly straight, parallel. **Inflorescences** axillary, within foliage, umbels or fascicles, or flowers solitary; peduncles and pedicels not fleshy in fruit. **Pedicels** present. **Flowers** bisexual; hypanthium cup-shaped, 1–3 mm wide, usually circumscissile far below sepal bases, rarely not circumscissile; sepals (rarely 4–)5, usually ± erect, sometimes spreading, yellowish to green or white, ovate-triangular, fleshy, keeled adaxially; petals (rarely 4–)5, yellowish, hooded, broadly obovate to obcordate, clawed; nectary thin, lining hypanthium; stamens (rarely 4–)5; ovary superior, (2–)3-locular; style 1. **Fruits** drupes; stones 2–3(–4), indehiscent but open at base. **Seeds** obovoid to lenticular, with cartilaginous beak protruding through opening in stone, not furrowed. x = 20–26.

Species ca. 50 (7 in the flora): North America, Mexico, West Indies, Central America, South America, Europe, e Asia, n Africa.

The difference between *Frangula* and *Rhamnus* has long been recognized, but treatments in taxonomic rank have been inconsistent. *Frangula* was treated within *Rhamnus* by M. C. Johnston and L. A. Johnston (1978), who noted that *Frangula* and *Rhamnus* are more closely related to each other than to other taxa. This observation has been corroborated by molecular studies that show them as sister taxa (J. E. Richardson et al. 2000; K. Bolmgren and B. Oxelman 2004). *Frangula* was also included within *Rhamnus* by D. Medan and C. Schirarend (2004) and by Chen Y. L. and Schirarend (2007), but various other recent treatments, as here, have maintained them as separate genera, emphasizing the differences outlined above in couplet 13 of the key to genera.

The key to species emphasizes geography as a primary character, reflecting the close similarities among the species and allowing morphological contrasts between sympatric taxa.

1. Leaves persistent, semideciduous, or deciduous, blades distinctly coriaceous.
   2. Leaves persistent; drupes 10–15 mm . . . . 1. *Frangula californica*
   2. Leaves semideciduous or deciduous; drupes 5–10 mm.
      3. Leaf blades glaucous adaxially when fresh; Plumas County, California. . . . . 3. *Frangula purshiana* (in part)
      3. Leaf blades not glaucous adaxially; Arizona, Colorado, Nevada, Utah. . . . . 5. *Frangula obovata* (in part)

1. Leaves deciduous, blades herbaceous to subcoriaceous.
    4. Eastern North America, as far west as Manitoba, Nebraska, and central Texas.
        5. Inflorescences umbels, pedunculate; drupes 8–10 mm; leaf blades oblong to elliptic or obovate-elliptic, margins serrulate to crenulate to nearly entire; drupe stones 3 . . . . 6. *Frangula caroliniana*
        5. Inflorescences fascicles, sessile; drupes 6–8 mm; leaf blades broadly elliptic-obovate to broadly elliptic or broadly oblong, margins entire; drupe stones 2(–3) . . . . 7. *Frangula alnus* (in part)
    4. Western North America, as far east as Colorado, New Mexico, trans-Pecos Texas, and Wyoming.
        6. Inflorescences fascicles, sessile. 7. *Frangula alnus* (in part)
        6. Inflorescences umbels, pedunculate.
            7. British Columbia, California, Idaho, Montana, w Nevada, Oregon, Washington.
                8. Leaf blades 1.5–8.5 cm; drupe stones 2(–3); inflorescences (2–)4–15-flowered; California, Nevada, Oregon . . . . 2. *Frangula rubra*
                8. Leaf blades 6–15 cm; drupe stones 3; inflorescences 10–25-flowered; British Columbia, California, Idaho, Montana, Nevada, Oregon, Washington . . . . 3. *Frangula purshiana* (in part)
            7. Arizona, Colorado, s Nevada, New Mexico, Texas, Utah.
                9. Leaf blades elliptic to oblong, elliptic-ovate, or narrowly ovate, 1.6–2.6(–2.9) times longer than wide, ± herbaceous, secondary veins (8–)9–13 pairs; se Arizona, New Mexico, Texas. . . . . 4. *Frangula betulifolia*
                9. Leaf blades obovate to oblong-obovate or oblong, 1.2–1.8(–2.5) times longer than wide, subcoriaceous, secondary veins (5–)6–8(–9) pairs; n Arizona, Colorado, Nevada, Utah . . . . 5. *Frangula obovata* (in part)

1. Frangula californica (Eschscholtz) A. Gray, Gen. Amer. Bor. 2: 178. 1849 • California coffeeberry

Rhamnus californica Eschscholtz, Mém. Acad. Imp. Sci. St. Pétersbourg Hist. Acad. 10: 285. 1826; *R. purshiana* de Candolle var. *californica* (Eschscholtz) Rehder

Shrubs, 0.5–5 m. **Stems** red to gray or brown, glabrous or hairy. **Leaves** usually persistent, rarely deciduous; petiole 3–10 mm; blade bright green to green, gray-green, yellowish green, or yellow abaxially, dark green to yellowish green or greenish white adaxially, not glaucous, ovate, elliptic, or oblong-elliptic, 2–10 cm, distinctly coriaceous, base cuneate to rounded or subcordate, margins entire or serrate, serrulate, or dentate-serrulate, apex acute or acuminate to obtuse, rounded, or truncate, abaxial surface glabrate or densely and closely white stellate-hairy, adaxial surface glabrous, glabrate, or sparsely hirsutulous; secondary veins 7–11(–12) pairs. **Inflorescences** umbels, pedunculate, 5–60-flowered. **Pedicels** 10–20 mm. **Stigmas** 2–3-parted. **Drupes** black, globose or slightly elongate, 10–15 mm; stones 2–3.

Subspecies 6 (6 in the flora): w United States, nw Mexico; introduced in Pacific Islands (Hawaii).

Frangula californica grows throughout most of California, and the subspecies are more or less separated geographically, but intermediates exist among all the subspecies except subsp. *ursina* (C. B. Wolf 1938). In California, subsp. *ursina* occurs only in eastern San Bernardino County. Subspecies *californica* is the most coastal, growing from the western Klamath Mountains south to the Agua Tibia Mountain in southern California. Subspecies *occidentalis* is characteristic of mafic and ultramafic substrates in northwest California; plants on other substrates approach subsp. *californica*, but leaf blades are equally green (not yellow-green) on both surfaces. Wolf reported that a form of *F. californica* found abundantly from the San Francisco Bay region to Santa Barbara County has leaves that are whitened beneath, but the hairs are much shorter than in subsp. *tomentella*. Plants in Los Angeles and Orange counties have leaf blades with sparse (not dense) tomentum and a few long hairs beneath as in subsp. *cuspidata*. Intermediates between subspp. *crassifolia* and *tomentella* are rare in northern California, but many plants in San Diego County that Wolf considered subsp. *tomentella* have narrowly to broadly elliptic leaf blades, thus resembling subsp. *crassifolia* in the Inner Coast Range.

1. Leaf blade surfaces glabrous or abaxial slightly puberulent; drupe stones 2–3.
    2. Leaf blades bright green or yellow abaxially, dark green adaxially; drupe stones 2 . . . . 1a. *Frangula californica* subsp. *californica*
    2. Leaf blades yellowish green on both surfaces; drupe stones 3 . . . . 1b. *Frangula californica* subsp. *occidentalis*
1. Leaf blade abaxial surface white stellate-hairy, sometimes with intermixed simple, erect hairs, adaxial surface glabrous, glabrate, sparsely hirsutulous, or sparsely stellate-hairy; drupe stones 2.
    3. Leaf blade abaxial surfaces without simple, erect hairs.
        4. Leaf blades narrowly elliptic, green and glossy adaxially, abaxial surfaces densely and closely white stellate-hairy, adaxial surfaces glabrous . . . . 1c. *Frangula californica* subsp. *tomentella*
        4. Leaf blades broadly elliptic to oblong-elliptic, green to gray-green adaxially, both surfaces white stellate-hairy, sparsely so adaxially. . . . . .1d. *Frangula californica* subsp. *crassifolia*
    3. Leaf blade abaxial surfaces with simple, erect hairs, at least along veins.
        5. Leaf blade apices abruptly acuminate to acute, margins sharply serrate to dentate-serrulate, sometimes strongly revolute . . . . 1e. *Frangula californica* subsp. *cuspidata*
        5. Leaf blade apices acute to rounded, margins entire or serrulate to serrate, flat to slightly revolute . . . 1f. *Frangula californica* subsp. *ursina*

1a. Frangula californica (Eschscholtz) A. Gray subsp. **californica** [E]

Rhamnus laurifolia Nuttall

Leaf blades bright green or yellow abaxially, dark green adaxially, narrowly to broadly elliptic, 2–8 cm, margins entire or sharply serrate, flat, apex acute, both surfaces glabrous or abaxial slightly puberulent; veins prominent abaxially. **Drupe stones** 2.

Flowering May–Jul. Coastal sage scrub, desert scrub, chaparral, woodlands, forest edges; 0–2800 m; Calif.; introduced in Pacific Islands (Hawaii).

1b. Frangula californica (Eschscholtz) A. Gray subsp. **occidentalis** (Howell ex Greene) Kartesz & Gandhi, Phytologia 76: 449. 1994 [E]

Rhamnus occidentalis Howell ex Greene, Pittonia 2: 15. 1889; *R. californica* Eschscholtz subsp. *occidentalis* (Howell ex Greene) C. B. Wolf; *R. californica* var. *occidentalis* (Howell ex Greene) Jepson

Leaf blades yellowish green on both surfaces, ovate to elliptic, 2–8 cm, margins entire or serrulate, flat, apex acute to obtuse, both surfaces glabrous or abaxial slightly puberulent; veins not prominent abaxially. **Drupe stones** 3.

Flowering Mar–Jun. Chaparral, pine woodlands, serpentine substrates, creek bottoms; 40–2300 m; Calif., Oreg.

1c. Frangula californica (Eschscholtz) A. Gray subsp. **tomentella** (Bentham) Kartesz & Gandhi, Phytologia 76: 449. 1994 • Chaparral coffeeberry

Rhamnus tomentella Bentham, Pl. Hartw., 303. 1849 (as tomentellus); *Frangula californica* var. *tomentella* (Bentham) A. Gray; *F. tomentella* (Bentham) Grubov; *R. californica* Eschscholtz subsp. *tomentella* (Bentham) C. B. Wolf; *R. purshiana* de Candolle var. *tomentella* (Bentham) K. L. Brandegee

Leaf blades green and glossy adaxially, narrowly elliptic, 3–7 cm, margins entire or serrulate, flat, apex acute, abaxial surface densely and closely white stellate-hairy, without intermixed simple, erect hairs, adaxial surface glabrous; veins not prominent abaxially. **Drupe stones** 2.

Flowering Jan–Apr. Chaparral, woodlands; 0–2200 m; Calif.; Mexico (Baja California).

Subspecies *tomentella* frequently intergrades with subsp. *cuspidata*.

1d. Frangula californica (Eschscholtz) A. Gray subsp. **crassifolia** (Jepson) Kartesz & Gandhi, Phytologia 76: 448. 1994 • Thickleaf coffeeberry [E]

Rhamnus californica Eschscholtz var. *crassifolia* Jepson, Man. Fl. Pl. Calif., 615. 1925; *R. californica* subsp. *crassifolia* (Jepson) C. B. Wolf; *R. tomentella* Bentham subsp. *crassifolia* (Jepson) J. O. Sawyer

Leaf blades green to gray-green adaxially, broadly elliptic to oblong-elliptic, 8–10 cm, margins usually entire, sometimes serrulate, flat, apex obtuse to rounded, abaxial surface densely and closely white stellate-hairy, without intermixed simple, erect hairs, adaxial surface moderately to densely hirsutulous and sparsely stellate-hairy; veins prominent abaxially. **Drupe stones** 2.

Flowering Feb–Apr. Chaparral, woodlands; 300–1400 m; Calif.

1e. Frangula californica (Eschscholtz) A. Gray subsp. **cuspidata** (Greene) Kartesz & Gandhi, Phytologia 76: 449. 1994 [E]

Rhamnus cuspidata Greene, Leafl. Bot. Observ. Crit. 1: 64. 1904; *Frangula viridula* (Jepson) Grubov; *R. californica* Eschscholtz subsp. *cuspidata* (Greene) C. B. Wolf; *R. californica* var. *viridula* Jepson; *R. tomentella* Bentham subsp. *cuspidata* (Greene) J. O. Sawyer

Leaf blades green adaxially, elliptic, 2–5(–6) cm, margins sharply serrate to dentate-serrulate with gland-tipped teeth, sometimes strongly revolute, apex abruptly acuminate to acute, abaxial surface sparsely to moderately densely white stellate-hairy, with intermixed longer, simple, erect hairs at least along veins, adaxial surface glabrous or sparsely hirsutulous; veins not prominent abaxially. **Drupe stones** 2.

Flowering Apr–Jun. Chaparral, desert scrub, montane woodlands; 400–2300 m; Calif.

When present, the strongly revolute leaf margins of subsp. *cuspidata* contrast with those of other taxa of *Frangula californica*. Plants identifiable as subsp. *cuspidata* but with flat-margined leaves perhaps show the influence of subsp. *tomentella*.

1f. Frangula californica (Eschscholtz) A. Gray subsp. **ursina** (Greene) Kartesz & Gandhi, Phytologia 76: 449. 1994 • Desert hoary coffeeberry

Rhamnus ursina Greene, Leafl. Bot. Observ. Crit. 1: 63. 1904; *Frangula ursina* (Greene) Grubov; *R. californica* Eschscholtz subsp. *ursina* (Greene) C. B. Wolf; *R. californica* var. *ursina* (Greene) McMinn; *R. castorea* Greene; *R. tomentella* Bentham subsp. *ursina* (Greene) J. O. Sawyer

Leaf blades green adaxially, elliptic to ovate, 3–8.5 cm, margins entire to serrulate or serrate with gland-tipped teeth, flat to slightly revolute, apex acute to rounded, abaxial surface densely and closely white stellate-hairy, usually with simple, erect hairs at least along veins (sometimes relatively inconspicuous), adaxial surface nearly glabrous; veins not prominent abaxially. **Drupe stones** 2.

Flowering May–Sep. Chaparral, desert scrub, woodlands; 700–2600 m; Ariz., Calif., Nev., N.Mex.; Mexico (Sonora).

C. B. Wolf (1938) considered a collection from Cochise County in southeastern Arizona named as *Rhamnus blumeri* Greene [*Frangula* ×*blumeri* (Greene) Kartesz & Gandhi] to be a hybrid between *Frangula californica* var. *ursina* and *F. betulifolia*. Study of an isotype of *R. blumeri*, and specimens of *Wolf 2592* (MO), *2593* (MO), *2595* (MO), and *Wolf & Everett 11384* (TEX), essentially corroborate the observations by Wolf. However, from examination of numerous other collections of both species from the Chiricahua Mountains, it appears that hybridization between *F. betulifolia* and *F. californica* is not common.

2. Frangula rubra (Greene) Grubov, Trudy Bot. Inst. Akad. Nauk S.S.S.R., Ser. 1, Fl. Sist. Vyssh. Rast. 8: 271. 1949 • Sierra coffeeberry [E]

Rhamnus rubra Greene, Pittonia 1: 68. 1887

Shrubs, 0.5–2 m. **Stems** red to gray, glabrous or hairy. **Leaves** deciduous; petiole 2–12 mm; blade light to bright green abaxially, green or dull green adaxially, narrowly elliptic to oblong or obovate, 1.5–8.5 cm, ± herbaceous, base rounded to obtuse or acute, margins entire or serrulate to denticulate, apex acute to obtuse or rounded, surfaces glabrous or short-puberulent, or abaxial puberulent on midrib and veins; secondary veins (7–)8–11 pairs. **Inflorescences** umbels, pedunculate, (2–)4–15-flowered. **Pedicels** 1–12 mm. **Stigmas** 2-lobed. **Drupes** black, globose or pyriform, 8–12 mm; stones 2(–3).

Subspecies 5 (5 in the flora): w United States.

The *Frangula rubra* complex is a group of closely related populations that needs study. In Nevada, they occur only in Douglas and Washoe counties. Descriptions by C. B. Wolf (1938) provide only a single distinct character to separate the subspecies, and he reported much intergradation between subsp. *obtusissima* and all the other subspecies. Field and herbarium studies argue for the recognition of infraspecific taxa despite the intermediates.

Subspecies *yosemitana* and *Frangula californica* subsp. *cuspidata* grow along the east side of the Sierra Nevada and can be easily confused. Plants of both taxa can be deciduous, but Wolf noted differences in pubescence and leaf margin to differentiate the two, although both are variable throughout their ranges.

1. Leaf blade surfaces short-puberulent. . . . . 2d. *Frangula rubra* subsp. *yosemitana*
1. Leaf blade surfaces glabrous or abaxial puberulent on midrib and veins.
    2. Young branches gray; leaves clustered on short-shoots . . . 2c. *Frangula rubra* subsp. *modocensis*
    2. Young branches usually red, rarely gray; leaves scattered along branchlets.
        3. Leaf blade bases and apices rounded . . . . 2b. *Frangula rubra* subsp. *obtusissima*
        3. Leaf blade bases and apices acute to obtuse.
            4. Leaf blades bright green; drupes globose . . . . . 2a. *Frangula rubra* subsp. *rubra*
            4. Leaf blades usually dull green, especially abaxially; drupes pyriform . . . . 2e. *Frangula rubra* subsp. *nevadensis*

2a. Frangula rubra (Greene) Grubov subsp. **rubra** [E]

Young branches red. **Leaves** scattered along branchlets; blade bright green, narrowly elliptic to oblong, 2–6 cm, base acute, margins minutely serrulate, apex acute, abaxial surface glabrous or puberulent on midrib and veins, adaxial glabrous. **Drupes** globose.

Flowering Apr–Jul. Chaparral, woodlands; 1000–2700 m; Calif., Nev., Oreg.

2b. Frangula rubra (Greene) Grubov subsp. **obtusissima** (Greene) Kartesz & Gandhi, Phytologia 76: 449. 1994 [E]

Rhamnus obtusissima Greene, Leafl. Bot. Observ. Crit. 1: 64. 1904; *R. rubra* Greene subsp. *obtusissima* (Greene) C. B. Wolf

Young branches usually red, rarely gray. **Leaves** scattered along branchlets; blade light green abaxially, green or dull green adaxially, oblong to obovate, 2.5–6 cm, base rounded, margins entire or serrulate, apex rounded, abaxial surface glabrous or puberulent on midrib and veins, adaxial glabrous. **Drupes** globose.

Flowering Apr–Jun. Chaparral, montane forests; 800–2100 m; Calif., Nev.

2c. Frangula rubra (Greene) Grubov subsp. **modocensis** (C. B. Wolf) Kartesz & Gandhi, Phytologia 76: 449. 1994 [E]

Rhamnus rubra Greene subsp. *modocensis* C. B. Wolf, Rancho Santa Ana Bot. Gard. Monogr., Bot. Ser. 1: 89, fig. 35. 1938; *R. rubra* var. *modocensis* (C. B. Wolf) McMinn

Young branches gray. **Leaves** clustered on short shoots; blade light green abaxially, dull green adaxially, narrowly elliptic, 1.5–4 cm, base acute, margins serrulate, apex acute, abaxial surface glabrous or puberulent on midrib and veins, adaxial glabrous. **Drupes** globose.

Flowering Apr–Jun. Montane forests, sagebrush steppe; 1000–2200 m; Calif.

2d. Frangula rubra (Greene) Grubov subsp. **yosemitana** (C. B. Wolf) Kartesz & Gandhi, Phytologia 76: 449. 1994 [E]

Rhamnus rubra Greene subsp. *yosemitana* C. B. Wolf, Rancho Santa Ana Bot. Gard. Monogr., Bot. Ser. 1: 90, figs. 36, 37. 1938; *R. rubra* var. *yosemitana* (C. B. Wolf) McMinn

Young branches red to gray. **Leaves** scattered along branchlets; blade light green abaxially, green adaxially, narrowly elliptic to oblong, 3–7 cm, base acute to rounded, margins finely denticulate, apex acute to rounded, both surfaces short puberulent. **Drupes** globose.

Flowering Apr–Jun. Chaparral, montane forests; (400–)1000–2200 m; Calif.

2e. Frangula rubra (Greene) Grubov subsp. **nevadensis** (A. Nelson) Kartesz & Gandhi, Phytologia 76: 449. 1994 • Nevada coffeeberry [E]

Rhamnus nevadensis A. Nelson, Proc. Biol. Soc. Wash. 18: 174. 1905; *R. rubra* Greene subsp. *nevadensis* (A. Nelson) C. B. Wolf

Young branches red. **Leaves** scattered along branchlets; blade usually dull green, especially abaxially, elliptic to oblong, 3–8.5 cm, base acute to obtuse, margins serrulate, apex acute to obtuse, both surfaces glabrous. **Drupes** pyriform.

Flowering Apr–Jun. Pine forests, sagebrush steppe; 1200–1700 m; Nev.

3. Frangula purshiana (de Candolle) A. Gray in War Department [U.S.], Pacif. Railr. Rep. 12(2): 57. 1860 • Cascara [E]

Rhamnus purshiana de Candolle in A. P. de Candolle and A. L. P. P. de Candolle, Prodr. 2: 25. 1825 (as purshianus)

Shrubs or trees, 1–12 m. **Stems** red to brown, gray, or green, glabrous or densely hairy. **Leaves** deciduous or semideciduous, alternate; petiole 6–23 mm; blade usually pale green abaxially, green to bluish or greenish gray adaxially, not glaucous or glaucous when fresh, elliptic to oblong or oblong-obovate, (3.5–)5–15 cm, herbaceous or distinctly coriaceous, base rounded to subcordate, obtuse, or cuneate, margins entire, irregularly toothed, or serrulate, apex obtuse or truncate, both surfaces glabrous or sparsely to densely hairy, or adaxial velvety; secondary veins 9–11 pairs. **Inflorescences** umbels, pedunculate, 10–25-flowered. **Pedicels** 5–15 mm. **Stigmas** 2–3-lobed. **Drupes** black, globose to depressed-globose, 5–10 mm; stones 3.

Subspecies 3 (3 in the flora): w North America.

1. Leaf blades distinctly coriaceous, glaucous adaxially when fresh, surfaces papillate . . . . 3c. *Frangula purshiana* subsp. *ultramafica*
1. Leaf blades herbaceous, not glaucous adaxially when fresh, surfaces not papillate.
    2. Leaf blade bases usually rounded to subcordate . . . . 3a. *Frangula purshiana* subsp. *purshiana*
    2. Leaf blade bases usually cuneate . . . . 3b. *Frangula purshiana* subsp. *anonifolia*

3a. Frangula purshiana (de Candolle) A. Gray subsp. **purshiana** [E]

Plants 1–12 m, usually treelike. **Twigs** red to brown, glabrescent. **Leaves** deciduous (often persistent on seedlings and small saplings); blade widely oblong to widely elliptic, 6–15 cm, herbaceous, base usually rounded to subcordate, margins irregularly toothed or entire, apex obtuse to truncate, surfaces not papillate, sparsely hairy to glabrous, not glaucous adaxially when fresh.

Flowering (Feb–)Mar–Jul. Coniferous forests, forest edges, deciduous woodlands, stream banks, coastal sage scrub, non-serpentine substrates; 0–2000 m; B.C.; Calif., Idaho, Mont., Oreg., Wash.

A. R. Kruckeberg (1996) noted that a population similar to subsp. *purshiana* was discovered on serpentine slopes in the Wenatchee Mountains of Washington. In their shrubby habit, these plants appear to be differentiated from non-serpentine populations elsewhere in the Pacific Northwest, which are treelike in habit. The Wenatchee plants retain their shrubby habit when garden-grown from seeds.

The bark of subsp. *purshiana* is a valuable medicinal crop for its cathartic properties. The tree attracts an abundance of bees and other pollinators during its long flowering season, and the fruits are an important food for songbirds, pileated woodpeckers, band-tailed pigeons, and other wildlife.

3b. Frangula purshiana (de Candolle) A. Gray subsp. **anonifolia** (Greene) J. O. Sawyer & S. W. Edwards, Madroño 54: 173. 2007 (as annonifolia) [E]

Rhamnus anonifolia Greene, Pittonia 3: 16. 1896 (as anonaefolia); *R. purshiana* de Candolle var. *anonifolia* (Greene) Jepson

Plants 1–5 m, usually treelike. **Twigs** red to brown, glabrescent. **Leaves** deciduous; blade widely elliptic or obovate, 6–15 cm, herbaceous, base usually cuneate, margins irregularly toothed or entire, apex obtuse to truncate, surfaces not papillate, sparsely hairy or glabrous, not glaucous adaxially when fresh.

Flowering Mar–Jun. Coniferous forest edges, stream banks, non-serpentine substrates; 1000–2000 m; Calif., Oreg.

3c. Frangula purshiana (de Candolle) A. Gray subsp. **ultramafica** J. O. Sawyer & S. W. Edwards, Madroño 54: 172, fig. 1. 2007 • Caribou coffeeberry [E]

Plants 1–2 m, shrublike. **Twigs** green to gray or dull brown, densely hairy. **Leaves** deciduous or semideciduous; blade broadly oblong or broadly elliptic to ovate or obovate, (3.5–)5–10 cm, distinctly coriaceous, base mostly obtuse or tapered, margins serrulate or entire, often wavy, apex obtuse, often notched, surfaces papillate, sparsely to densely hairy or adaxial velvety, glaucous adaxially when fresh.

Flowering Apr–Jun. Seeps, montane chaparral, open forests over mafic and ultramafic substrates; 800–2000 m; Calif.

Of the three subspecies of *Frangula purshiana*, subsp. *ultramafica* is the most distinctive and might warrant species status. It appears to be restricted to the Feather River complex of serpentinized peridotite and associated mafic and ultramafic substrates near Bucks Lake in Plumas County. Its firm, bluish or greenish gray leaves are suggestive of evergreen *F. californica* subsp. *tomentella*, but they are broader and larger and bear only simple, erect hairs. The leaves are deciduous as in *F. rubra*, but the large, broad leaves and fruits with three stones are like those of *F. purshiana*.

4. Frangula betulifolia (Greene) Grubov, Trudy Bot. Inst. Akad. Nauk S.S.S.R., Ser. 1, Fl. Sist. Vyssh. Rast. 8: 268. 1949 • Birchleaf buckthorn [F]

Rhamnus betulifolia Greene, Pittonia 3: 16. 1896 (as betulaefolia)

Shrubs or small trees, 1–4 m. **Stems** brown to gray-brown, glabrous or pubescent. **Leaves** deciduous; petiole (2–)5–16 mm; blade yellowish green abaxially, green adaxially, elliptic to oblong, elliptic-ovate, or narrowly ovate, (4–)4.5–10 × (2–)2.5–5.5 cm, 1.6–2.6(–2.9) times longer than wide, ± herbaceous, base obtuse to truncate or rounded, margins serrate to subcrenate, apex usually acute to obtuse, sometimes slightly acuminate, both surfaces hirtellous, glabrescent; secondary veins (8–)9–13 pairs. **Inflorescences** umbels, pedunculate, 2–20(–38)-flowered. **Pedicels** 3–7 mm. **Stigmas** 3-lobed. **Drupes** black, globose, 5–10 mm; stones (2–)3(–4).

Flowering Apr–Jun. Moist canyons, stream banks, rocky slopes, cliff bases, ledges, ridges, roadsides, deciduous, coniferous, and mixed woodlands; 900–2800 m; Ariz., N.Mex., Tex.; Mexico (Chihuahua, Coahuila, Durango, Nuevo León, Sonora, Tamaulipas).

In the flora area, *Frangula betulifolia* is found in southeastern Arizona, the southern two-thirds of New Mexico, and trans-Pecos Texas. It and *F. obovata* are allopatric and morphologically distinct. C. B. Wolf (1938) considered a collection from Cochise County in southeastern Arizona, named as *Rhamnus blumeri* (*Frangula* ×*blumeri*), to be a hybrid between *F. betulifolia* and 1f. *F. californica* var. *ursina*; see the discussion of the latter taxon for more information.

5. Frangula obovata (Kearney & Peebles) G. L. Nesom & J. O. Sawyer, Phytologia 91: 302. 2009 • Pearleaf buckthorn [E]

Rhamnus betulifolia Greene var. *obovata* Kearney & Peebles, J. Wash. Acad. Sci. 29: 486. 1939 (as betulaefolia); *Frangula betulifolia* (Greene) Grubov subsp. *obovata* (Kearney & Peebles) Kartesz & Gandhi

Shrubs, 1–2.5 m. **Stems** red to brown or gray-brown, glabrous or pubescent. **Leaves** deciduous; petiole 5–14 mm; blade usually equally green on both surfaces, not glaucous, obovate to oblong-obovate or oblong, (4–)5–9 × 3.2–6 cm, 1.2–1.8(–2.5) times longer than wide, subcoriaceous to distinctly coriaceous, base truncate to subcordate, margins serrulate to nearly entire, apex obtuse to truncate or rounded, both surfaces minutely puberulous to hirtellous, glabrescent; secondary veins (5–)6–8(–9) pairs. **Inflorescences** umbels, pedunculate, 2–12-flowered. **Pedicels** 3–10 mm. **Stigmas** 3-lobed. **Drupes** black, globose, 5–8 mm; stones 3.

Flowering Apr–Jun. Hanging gardens, cliff faces, talus, canyon bottoms, seepage below cliffs, stream and creek banks; 1300–2400 m; Ariz., Colo., Nev., Utah.

Frangula obovata is known from Clark County, Nevada, across northern Arizona and southern Utah to southwestern Colorado; its distribution does not overlap that of *F. betulifolia*.

6. Frangula caroliniana (Walter) A. Gray, Gen. Amer. Bor. 2: 178. 1849 • Carolina buckthorn, Indian cherry [E]

Rhamnus caroliniana Walter, Fl. Carol., 101. 1788 (as carolinianus); *R. caroliniana* var. *mollis* Fernald

Shrubs or small trees, 2–6 (–10) m. **Stems** gray, glabrous or pubescent. **Leaves** deciduous; petiole 8–20 mm; blade dull green abaxially, glossy dark green adaxially, oblong to elliptic or obovate-elliptic, (3–)5–13 cm, herbaceous, base cuneate to rounded, margins serrulate or crenulate to nearly entire, apex acute to acuminate or obtuse, abaxial surface puberulent on veins, adaxial surface glabrous; secondary veins 6–9 (–10) pairs. **Inflorescences** umbels, pedunculate, 1–14-flowered. **Pedicels** 3–6 mm. **Stigmas** 3-lobed. **Drupes** black, globose, 8–10 mm; stones 3.

Flowering Apr–Jun. Dry to moist barrens, sandy and gravelly flats, roadsides, ravines, bluffs, limestone bluffs, shell middens, bottomlands, swamp and pond edges, coastal hammocks, deciduous and coniferous forests; 50–500 m; Ala., Ark., Fla., Ga., Ill., Ind., Ky., La., Md., Miss., Mo., N.C., Ohio, Okla., S.C., Tenn., Tex., Va.

Rhamnus caroliniana var. *mollis* (type from Illinois, with leaves persistently soft-pubescent abaxially) has sometimes been recognized as a western/Ozarkian entity (for example, H. A. Gleason and A. Cronquist 1991), but intermediates and intergrades with the typical form (with leaves glabrescent abaxially) are as numerous as the extremes.

7. **Frangula alnus** Miller, Gard. Dict. ed. 8, Frangula no. 1. 1768 • Glossy buckthorn, European alder-buckthorn [I]

Rhamnus frangula Linnaeus, Sp. Pl. 1: 193. 1753

Shrubs or small trees, 2–5 (–7) m. **Stems** greenish to brown, sparsely puberulent, glabrescent. **Leaves** deciduous; petiole 10–19 mm; blade dull green abaxially, glossy darker green adaxially, broadly elliptic-obovate to broadly elliptic or broadly oblong, 4–7(–11) cm, herbaceous, base cuneate to rounded, margins entire, apex rounded to acute or abruptly short-acuminate, abaxial surface sometimes minutely strigose along midveins, adaxial surface glabrous; secondary veins 6–10 pairs. **Inflorescences** fascicles, sessile, 1–8-flowered. **Pedicels** 3–10 mm. **Stigmas** 2–3-lobed. **Drupes** black, globose or slightly elongate, 6–8 mm; stones 2(–3). $2n$ = 20–26.

Flowering May–Sep. Alkaline and acid tamarack, red maple, and cedar swamps, peatlands, bogs, fens, disturbed areas, riparian thickets, lakeshores, ditches, fencerows, hedgerows, low woods, beaver meadows; 10–400 m; introduced; Man., N.B., N.S., Ont., P.E.I., Que., Sask.; Colo., Conn., Idaho, Ill., Ind., Iowa, Ky., Maine, Mass., Mich., Minn., Nebr., N.H., N.J., N.Y, Ohio, Pa., R.I., Tenn., Vt., W.Va., Wis., Wyo.; Europe; introduced also in Asia (China, Russia), n Africa.

Two horticultural forms of *Frangula alnus* are widely sold and planted in North America. The cultivar 'Asplenifolia' (*F. alnus* var. *asplenifolia* Dippel; fern leaf or cutleaf buckthorn) has linear-oblong leaves with coarsely toothed to incised margins and commonly is grown as a specialty plant. The cultivar 'Columnaris' (tallhedge glossy buckthorn) has a narrow, upright habit and is used in hedging. Both forms were noted by M.H.Brand (http://www.hort.uconn.edu/plants/r/rhafra/rhafra1.html) to spread invasively by seeds. 'Asplenifolia' has been reported as naturalized in Ontario (A. W. Dugal 1989, 1992), Illinois (A. Branhagen, pers. comm.), and Ohio (M. K. Delong et al. 2005).

3. BERCHEMIA Necker ex de Candolle in A. P. de Candolle and A. L. P. P. de Candolle, Prodr. 2: 22. 1825, name conserved • Rattan [For Jacob Pierre Berthoud van Berchem, eighteenth-century Dutch mineralogist and naturalist]

Guy L. Nesom

Woody vines [shrubs, trees], tendrils absent, unarmed; bud scales present. **Stems** twining, glabrous [hairy]. **Leaves** deciduous [persistent], alternate; blade not gland-dotted; pinnately veined, secondary, and usually tertiary, veins strongly parallel. **Inflorescences** axillary or terminal, paniclelike thyrses [corymblike cymes or fascicles]; peduncles and pedicels not fleshy in fruit. **Pedicels** present. **Flowers** functionally unisexual (plants functionally dioecious) [bisexual]; hypanthium patelliform, cupulate, or hemispheric, 2–3 mm wide; sepals 5, staminate spreading, pistillate erect, greenish, triangular [rarely linear or narrowly lanceolate], keeled adaxially; petals 5, cream or yellowish to greenish white, flat, spatulate to lanceolate, short-clawed; nectary fleshy, 10-lobed, filling hypanthium; stamens 5; ovary superior, 2-locular; style 1. **Fruits** drupes; stone 1, indehiscent.

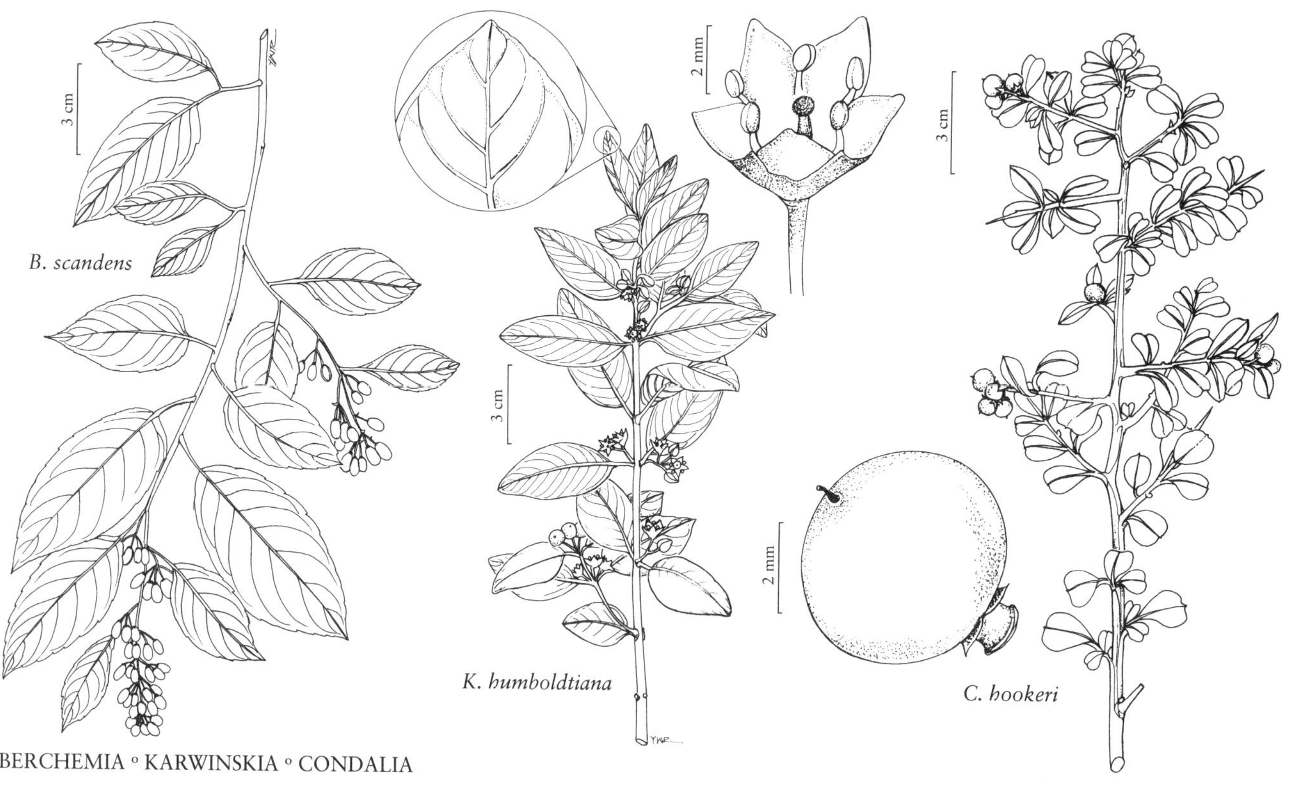

BERCHEMIA ◦ KARWINSKIA ◦ CONDALIA

Species ca. 12 (1 in the flora): c, e United States, Mexico (Chiapas), Central America (Guatemala), Asia, Africa; tropical to warm temperate regions.

Berchemia scandens is the only New World species in the genus. The disjunction of *B. scandens* from the southeastern United States to Chiapas and Guatemala is remarkable, but there seem to be no morphological differences.

1. Berchemia scandens (Hill) K. Koch, Dendrologie 1: 602. 1869 • Alabama supplejack, American rattan [F]

Rhamnus scandens Hill, Hort. Kew., 453, plate 20. 1768

Vines, usually climbing-scandent, twining to some extent, extending into crowns of trees. **Stems** to 10 cm diam., bark smooth, glabrous. **Leaves** glabrate (petioles sometimes sparsely short-pubescent at base); blade ovate to elliptic-ovate or elliptic, 3–6(–8) cm, base truncate to rounded or obtuse, margins entire or shallowly undulate-crenate (teeth at vein endings), apex obtuse to short acuminate; secondary veins 8–12 pairs, strongly parallel. **Inflorescences** 7–20-flowered. **Drupes** mostly blue-black to purple-black or purple-red, cylindric to oblong-ellipsoid, 5–8 mm, often glaucous.

Flowering Mar–May. Riparian areas, ravines, swamps, bottomlands, upland forests; 0–200 m; Ala., Ark., Fla., Ga., Ill., Ky., La., Md., Miss., Mo., N.C., Okla., S.C., Tenn., Tex., Va.; Mexico (Chiapas); Central America (Guatemala).

According to R. K. Godfrey and J. W. Wooten (1981), the flowers of *Berchemia scandens* are functionally unisexual, the plants functionally dioecious. The staminodes remain enclosed by the petals in pistillate flowers.

4. KARWINSKIA Zuccarini, Flora 15(2, Beibl.): 70. 1832; Abh. Math.-Phys. Cl. Königl. Bayer. Akad. Wiss. 1: 349, plate 16. 1832 • [For Baron W. F. von Karvinsky, 1780–1855, botanical collector in Brazil and Mexico]

Guy L. Nesom

Shrubs or trees, unarmed [armed with thorns]; bud scales present. **Leaves** deciduous, opposite; blade gland-dotted abaxially (secondary veins appearing light- and dark-banded); pinnately veined, secondary veins relatively straight. **Inflorescences** axillary, umbel-like cymes or flowers solitary; peduncles and pedicels not fleshy in fruit. **Pedicels** present. **Flowers** bisexual; hypanthium hemispheric to turbinate, 2–3 mm wide, gland-dotted; sepals 5, spreading, greenish, triangular, keeled adaxially, margins often gland-dotted; petals 5, white to yellowish or greenish yellow, hooded, triangular to deltate, not clawed, gland-dotted; nectary thin, lining hypanthium; stamens 5; ovary superior, 2–3-locular; ovules 2 per locule; style 1. **Fruits** drupes; stone 1, indehiscent.

Species ca. 18 (1 in the flora): Texas, Mexico, West Indies (Cuba, Hispaniola), Central America, South American (Colombia).

1. Karwinskia humboldtiana (Schultes) Zuccarini, Abh. Math.-Phys. Cl. Königl. Bayer. Akad. Wiss. 1: 353. 1832 • Coyotillo [F]

Rhamnus humboldtiana Schultes, Syst. Veg. 5: 295. 1819

Shrubs or small trees, 1.5–4(–6) m, glabrous or glabrate. **Leaves:** petiole 2–10 mm; blade oblong to elliptic-oblong, or elliptic-ovate, (1.5–)3–7(–8) cm, base rounded or truncate to acute, margins entire or weakly crenate (teeth at vein tips), apex rounded or truncate to acute, surfaces glabrous or abaxial surface sparsely puberulent along veins. **Inflorescences** 1–3-flowered. **Drupes** black, globose, 9–13 mm.

Flowering (Mar–)Apr–Oct. Limestone ridges and hillsides, roadsides, juniper woodlands, brushlands, mesquite woodlands, sandy clay, sandy loam, deep sand; 10–600(–700) m; Tex.; Mexico.

5. CONDALIA Cavanilles, Anales Hist. Nat. 1: 39, plate 4. 1799, name conserved • Snakewood [For Antonio Condal, 1745–1804, Spanish physician who accompanied Peter Loefling on a journey up the Orinoco River]

Guy L. Nesom

Shrubs or small trees, usually armed with thorns, sometimes unarmed; bud scales absent. **Leaves** deciduous, alternate, mostly borne on short shoots and usually fascicled; blade not gland-dotted; pinnately veined. **Inflorescences** axillary, within foliage, fascicles or flowers solitary; peduncles and pedicels not fleshy in fruit. **Pedicels** present or absent. **Flowers** bisexual; hypanthium hemispheric, 1–1.5 mm wide; sepals 5, spreading, greenish abaxially, yellowish adaxially, deltate, keeled adaxially; petals 0 (5, yellow, hooded, spatulate, short-clawed in *C. ericoides*); nectary absent, thin, or margin thickened, lining hypanthium; stamens 5; ovary superior, 2-locular at least early in development, 1 locule often suppressed; style 1. **Fruits** drupes; stone 1, indehiscent.

Species 18 (7 in the flora): w United States, Mexico, Central America, South America.

A close relationship between *Condalia* and *Ziziphus* might be inferred from numerous synonyms of one genus within the other—drupaceous fruits are produced in both genera. However, the closest relatives of *Condalia* appear to be *Karwinskia* and *Rhamnidium* Reissek (J. E. Richardson et al. 2000) of Central America, South America, and the West Indies, and only slightly more distantly, *Berchemia*, *Rhamnus*, and *Sageretia*.

SELECTED REFERENCE Johnston, M. C. 1962. Revision of *Condalia* including *Microrhamnus* (Rhamnaceae). Brittonia 14: 332–368.

1. Leaf blades linear; petals 5. . . . . 1. *Condalia ericoides*
1. Leaf blades spatulate to obovate or elliptic; petals 0.
    2. Leaf blades spatulate to spatulate-elliptic, venation conspicuous abaxially.
        3. Leaf blade surfaces glabrous, abaxial intervein surfaces with rounded transverse ridges, surface appearing as if molded in wax; branches glabrous. . . . . 2. *Condalia spathulata*
        3. Leaf blade surfaces hispidulous, puberulent, or densely short-villous to velutinous, abaxial intervein surfaces microvesiculate, not appearing waxy; branches hispidulous or densely short-villous to velutinous.
            4. Pedicels 2.5–4.5(–6.5) mm; internodes 2–7 mm; sepals deciduous in fruit; leaf blade apices obtuse to mucronate. . . . . 3. *Condalia globosa*
            4. Pedicels 0.5–3 mm; internodes 0.5–2(–3) mm; sepals persistent in fruit; leaf blade apices acute. . . . . 4. *Condalia warnockii*
    2. Leaf blades obovate to elliptic or elliptic-oblong, venation inconspicuous abaxially.
        5. Leaf blades (10–)15–20(–31) × (5–)9–12(–19) mm, apices on a single plant consistently rounded to truncate-emarginate. . . . . 5. *Condalia hookeri*
        5. Leaf blades 5–11(–21) × 2.5–6(–10) mm, apices on a single plant acute to rounded, truncate, or emarginate.
            6. Secondary branches sparsely hispidulous or glabrate; petioles 3–10 mm; drupes 5–6 mm . . . . 6. *Condalia viridis*
            6. Secondary branches densely hispidulous; petioles 1–3 mm; drupes 8 mm. . . . 7. *Condalia correllii*

1. Condalia ericoides (A. Gray) M. C. Johnston, Brittonia 14: 364. 1962 • Javelina bush

Microrhamnus ericoides A. Gray, Smithsonian Contr. Knowl. 3(5): 34. 1852

Shrubs, 0.3–1(–1.3) m; primary branches thorn-tipped, secondary branches thorn-tipped, with short shoots and some tertiary thorns, glabrous; internodes 1–2(–4) mm. **Leaves:** petiole 0–0.1 mm; blade linear, 2–13 × 1 mm, coriaceous, margins entire, appearing revolute, apex short-acute, surfaces glabrous; venation not evident. **Inflorescences** on short shoots, 1–6-flowered. **Pedicels** 1–2.5 mm. **Flowers:** petals 5, yellow. **Drupes** ellipsoid-globose to ellipsoid-fusiform, 7–12 mm; stones 1(–2)-seeded.

Flowering Mar–Sep. Sand hills, gypsum hills, eroded clay breaks, limestone talus slopes, ridge and mesa tops, rocky flats, dry streambeds, shrublands, grasslands, pastures, roadsides, roadcuts; 400–1600 m; Ariz., N.Mex., Tex.; Mexico (Chihuahua, Coahuila, Durango, San Luis Potosí, Zacatecas).

2. Condalia spathulata A. Gray, Smithsonian Contr. Knowl. 3(5): 32. 1852 • Squaw bush

Shrubs, 0.5–1(–2) m; primary branches usually not thorn-tipped, secondary branches thorn-tipped, with short shoots and often some thorn-tipped tertiary branches, glabrous; internodes 1–3(–5) mm. **Leaves:** petiole 1–2 mm; blade spatulate, 4–12(–14) × 1.6–3 mm, subcoriaceous, margins entire, revolute, apex rounded to emarginate or acute, surfaces glabrous; venation raised, conspicuous abaxially, abaxial intervein surfaces with rounded transverse ridges, surface appearing as if molded in wax. **Inflorescences** on short shoots, 1–5-flowered. **Pedicels** (1.5–)2–3 mm. **Flowers:** petals 0. **Drupes** globose, 3.5–4.5 mm; stones 1-seeded.

Flowering Apr–Jul. Stream banks, low hills, sandstone bluffs, gravelly slopes, disturbed sites, shrublands, shortgrass grasslands; 0–600 m; Tex.; Mexico (Coahuila, Nuevo León, Tamaulipas).

In the flora area, *Condalia spathulata* occurs in central and southern Texas.

3. Condalia globosa I. M. Johnston, Proc. Calif. Acad. Sci., ser. 4, 12: 1086. 1924 • Bitter snakewood

Varieties 2 (1 in the flora): sw United States, nw Mexico.

Variety *globosa* occurs in northwestern Mexico.

3a. Condalia globosa I. M. Johnston var. **pubescens** I. M. Johnston, Proc. Calif. Acad. Sci., ser. 4, 12: 1087. 1924

Shrubs or small trees, 1–4 (–6) m; primary branches usually not thorn-tipped, secondary branches suppressed, hispidulous, thorn-tipped, with short shoots; internodes 2–7 mm. **Leaves:** petiole 1–2 mm; blade dull green, occasionally burnt orange abaxially, spatulate to spatulate-elliptic, 3–13(–22) × 1.5–5 mm, subcoriaceous, margins entire, not revolute, apex obtuse to mucronate, surfaces sparsely to densely hispidulous or puberulent, abaxial intervein surfaces microvesiculate, not appearing waxy; venation thick and raised, conspicuous abaxially. **Inflorescences** on short shoots, 1–7-flowered. **Pedicels** 2.5–4.5(–6.5) mm. **Flowers:** sepals deciduous in fruit; petals 0. **Drupes** usually globose, 3.4–5.1 mm; stones 1-seeded.

Flowering mainly Jan–Apr, sporadically year-round. Dry desert washes, drainages, canyons, open slopes, creosote bush scrub; 100–1500 m; Ariz., Calif.; Mexico (Baja California, Baja California Sur, Sonora).

4. Condalia warnockii M. C. Johnston, Brittonia 14: 352. 1962 • Warnock's snakewood

Shrubs, 0.5–3 m; primary branches thorn-tipped, secondary branches thorn-tipped, with short shoots and occasionally a few tertiary thorns, hispidulous or densely short villous to velutinous; internodes 0.5–2(–3) mm. **Leaves:** petiole 0–2 mm; blade yellowish to brownish olive or gray-green adaxially, spatulate, 3–7(–10) × 0.5–2.5(–5) mm, subcoriaceous, margins entire, not revolute, apex acute, surfaces hispidulous or densely short villous to velutinous, abaxial intervein surfaces microvesiculate, not appearing waxy; venation raised and conspicuous abaxially. **Inflorescences** on short shoots, mostly 1–2-flowered. **Pedicels** 0.5–3 mm. **Flowers:** sepals persistent in fruit; petals 0. **Drupes** globose to fusiform-globose or depressed-globose, 4–6 mm; stones 1–2-seeded.

Varieties 2 (2 in the flora): sw, sc United States, n Mexico.

1. Internodes 1–2(–3) mm; leaf blades 1–2.5(–5) mm wide, length usually 2–2.5 times width . . . . 4a. *Condalia warnockii* var. *warnockii*
1. Internodes 0.5–1(–2.5) mm; leaf blades 0.5–1.5 (–2) mm wide, length usually 2.5–3.5 times width . . . . 4b. *Condalia warnockii* var. *kearneyana*

4a. Condalia warnockii M. C. Johnston var. **warnockii**

Internodes 1–2(–3) mm. **Leaf blades** 1–2.5(–5) mm wide, length usually 2–2.5 times width.

Flowering Jun–Aug. Rocky slopes, canyons, sandhills; 800–1700 m; N.Mex., Tex.; Mexico (Chihuahua, Coahuila, Zacatecas).

4b. Condalia warnockii M. C. Johnston var. **kearneyana** M. C. Johnston, Brittonia 14: 354. 1962

Internodes 0.5–1(–2.5) mm. **Leaf blades** 0.5–1.5(–2) mm wide, length usually 2.5–3.5 times width.

Flowering Jul–Sep(–Oct). Rocky slopes, canyons, sandhills, desert washes; 500–1700 m; Ariz.; Mexico (Sonora).

Variety *kearneyana* is widespread in the southern third of Arizona.

5. Condalia hookeri M. C. Johnston, Brittonia 14: 362. 1962 • Brazilian bluewood [F]

Condalia obovata Hooker, Icon. Pl. 3: plate 287. 1840, not Ruiz & Pavón 1798; *C. hookeri* var. *edwardsiana* (Cory) M. C. Johnston; *C. obovata* var. *edwardsiana* Cory

Shrubs or small trees, (1–) 2–3.5(–6) m; primary branches not thorn-tipped, secondary branches thorn-tipped, with short shoots and few thorn-tipped tertiary branches, glabrous or densely hispidulous; internodes 2–5(–11) mm. **Leaves:** petiole 3–10 mm; blade obovate to elliptic, (10–)15–20(–31) × (5–)9–12(–19) mm, herbaceous, margins entire or distally few-toothed, not revolute, apex on a single plant consistently rounded to truncate-emarginate, surfaces glabrous or occasionally hispidulous; venation flush and inconspicuous abaxially. **Inflorescences** on short shoots, 1–3-flowered. **Pedicels** 0.8 mm. **Flowers:** petals 0. **Drupes** globose, 5–6 mm; stones 1–2-seeded.

Flowering (Mar–)Apr–Jul(–Oct). Limestone slopes, sandstone bluffs, sandy clay, clay dunes, shell ridges, thorn scrub, juniper woodlands, riparian woods; 10–400 m; Tex.; Mexico (Nuevo León, Tamaulipas).

Condalia hookeri var. *edwardsiana* was noted by M. C. Johnston (1962) to be known only from a single thicket at the type locality in Edwards County, Texas; it was described as differing from the typical variety in its leaf blades being 2.5–3 times longer than wide (versus 1–2.5 times longer than wide). This morph has never been relocated despite repeated searching and appears to represent a populational variant. In the flora area, *C. hookeri* is widespread in central and southern Texas.

6. **Condalia viridis** I. M. Johnston, J. Arnold Arbor. 20: 234. 1939 • Green snakewood

Shrubs, 0.7–4 m; primary branches not thorn-tipped, secondary branches thorn-tipped, with short shoots, sparsely hispidulous to nearly glabrous; internodes 1–2(–4) mm. **Leaves:** petiole 3–10 mm; blade pale green, obovate to elliptic-obovate or elliptic-oblong, 5–8(–18) × 2.5–4(–10) mm, coriaceous, margins entire, not revolute, apex acute to rounded, truncate, or emarginate, surfaces glabrous or sparsely and minutely hispidulous; venation closely reticulate, inconspicuous abaxially. **Inflorescences** on short shoots, 1–3-flowered. **Pedicels** 0.4–1 mm. **Flowers:** petals 0. **Drupes** globose to depressed-globose, 5–6 mm; stones 1(–2)-seeded.

Flowering May–Aug. Limestone hills, arroyos, flats, floodplain thickets, mesquite scrub, juniper and oak woodlands; 300–800 m; Tex.; Mexico (Coahuila, Nuevo León, San Luis Potosí).

In the flora area, *Condalia viridis* occurs from central to trans-Pecos Texas.

7. **Condalia correllii** M. C. Johnston, Brittonia 14: 357. 1962 • Correll's snakewood

Shrubs, 1–3 m; primary branches occasionally thorn-tipped, secondary branches thorn-tipped, with short shoots, densely hispidulous; internodes 1–5(–7) mm. **Leaves:** petiole 1–3 mm; blade elliptic-obovate to elliptic, 7–11(–21) × 4–6(–10) mm, subcoriaceous, margins entire, not revolute, apex on a single plant acute to rounded and emarginate, surfaces glabrous; venation closely reticulate, inconspicuous abaxially. **Inflorescences** on season's twigs, 1–4-flowered. **Pedicels** 0.4–1 mm. **Flowers:** petals 0. **Drupes** globose, 8 mm; stones 2-seeded.

Flowering Mar–May. Dry slopes, drainages, canyons; 1000–1500 m; Ariz., N.Mex.; Mexico (Chihuahua, Coahuila, Sonora).

6. SAGERETIA Brongniart, Mém. Fam. Rhamnées, 52, plate 2, fig. 2. 1826 • Mock buckthorn [For Augustin Sageret, 1763–1851, French botanist]

Guy L. Nesom

Shrubs or woody vines [trees], arching, sprawling, drooping, or clambering [erect], tendrils absent, armed with thorns (sometimes not prominent); bud scales present. **Stems** not twining, hairy. **Leaves** persistent or tardily deciduous, present at flowering, usually opposite to subopposite, sometimes alternate distally; blade not gland-dotted; pinnately veined, secondary veins not strongly parallel [± parallel], tertiary veins reticulate. **Inflorescences** terminal and axillary, overtopping or extending beyond foliage, spikelike or spicate, paniclelike thyrses, [5–]30–120[–150]-flowered; peduncles and pedicels not fleshy in fruit. **Pedicels** usually absent, rarely present. **Flowers** bisexual; hypanthium shallowly cupulate to hemispheric, 1–2 mm wide; sepals 5, erect, yellowish green, triangular, ± fleshy, keeled adaxially; petals 5, white to yellow, hooded, spatulate, short-clawed; nectary fleshy, cupulate, distally free from hypanthium; stamens 5; ovary superior, 2–3-locular; style 1. **Fruits** drupes; stones (2–)3, tardily dehiscent.

Species 30–35 (3 in the flora): United States, Mexico, se Asia.

Most species of *Sageretia* are from southeast Asia, with 15 endemic to China (Chen Y. L. and C. Schirarend 2007). Four are native to the Americas (G. L. Nesom 1993h).

1. Shrubs, densely and intricately branched; branches erect or spreading to arching or arcuate-decumbent, densely short-strigillose; leaf blades 0.5–2(–3) × 0.5–1(–2) cm, veins not prominently raised abaxially; inflorescence primary axes 0.5–1 cm, lateral branches 0(–2) . . . . 1. *Sageretia wrightii*
1. Shrubs or vines, loosely to compactly branched; branches often becoming sprawling, trailing, or climbing, tomentose or villous-tomentose, sometimes with understory of minute, erect, glandular hairs; leaf blades 1–4.5(–6) × 0.5–2.5 cm, veins prominently raised abaxially; inflorescence primary axes 2–15 cm, lateral branches 4–8.
    2. Leaf blade apices acute to acuminate; inflorescence primary axes 5–15 cm . . . . . . 2. *Sageretia minutiflora*
    2. Leaf blade apices acute to obtuse or rounded; inflorescence primary axes 2–5 cm. . . . . . . 3. *Sageretia thea*

1. Sageretia wrightii S. Watson, Proc. Amer. Acad. Arts 20: 358. 1885 • Wright's mock buckthorn

Shrubs, densely and intricately branched, usually weak and straggling, sometimes tall and narrow. **Branches** erect or spreading to arching or arcuate-decumbent, densely short-strigillose, hairs arching antrorsely. **Leaves** persistent, opposite to subopposite; blade broadly oblanceolate to oblong-elliptic, 0.5–2(–3) × 0.5–1(–2) cm, base rounded to obtuse, margins entire or mucronate to remotely serrate, apex obtuse, rounded, or retuse, surfaces sparsely tomentose, quickly glabrescent; veins not prominently raised abaxially. **Inflorescences** terminal or in axils of distalmost well-developed leaves, primary axis 0.5–1 cm, lateral branches 0(–2). **Drupes** purple to black, subglobose to depressed-globose, 5 mm; stones (2–)3.

Flowering Mar–Sep. Rocky canyons and hillsides, riparian areas, washes, desert grasslands, scrub, oak and pinyon-juniper woodlands; 900–1500 m; Ariz., N.Mex., Tex.; Mexico (Baja California Sur, Chihuahua, Coahuila, Durango, Sonora).

2. Sageretia minutiflora (Michaux) C. Mohr, Contr. U.S. Natl. Herb. 6: 609. 1901 • Small-flower mock buckthorn [E] [F]

Rhamnus minutiflora Michaux, Fl. Bor.-Amer. 1: 154. 1803 (as minutiflorus)

Shrubs or vines, loosely to compactly branched. **Branches** often sprawling, trailing, or clambering into trees, villous-tomentose, sometimes with understory of minute, erect, glandular hairs. **Leaves** persistent or tardily deciduous, opposite to subopposite; blade ovate to elliptic-ovate or broadly oblong-ovate, 1.5–4(–6) × 1–2 cm, base rounded to truncate-rounded or very slightly subcordate, margins shallowly serrate, apex acute to acuminate, surfaces glabrous; veins prominently raised abaxially. **Inflorescences** terminal and axillary, primary axis 5–15 cm, lateral branches 4–8. **Drupes** purple, subglobose to obovoid, 5–9 mm; stones 3.

Flowering Aug–Sep. Calcareous rocky bluffs, forested shell middens on barrier islands, shell hammocks, evergreen hammocks, beach borders, live oak, palm, and deciduous woods; 0–30 m; Ala., Fla., Ga., Miss., N.C., S.C.

Sageretia minutiflora are apparently most commonly weak-stemmed shrubs clambering over and through other shrubs, but they also may be distinctly viny, reaching to 8 m in trees. Lateral branches may self-prune, resulting in ropelike lianas.

Plants of central Mexico identified by R. Fernández (1996) as *Sageretia minutiflora*, far disjunct from its range in the southeastern United States, instead are *S. mexicana* G. L. Nesom.

Sageretia michauxii Brongniart is a superfluous name that pertains here.

3. Sageretia thea (Osbeck) M. C. Johnston, J. Arnold Arbor. 49. 378. 1968 • Pauper's or Chinese tea [I]

Rhamnus thea Osbeck, Dagb. Ostind. Resa, 232. 1757

Shrubs or vines, loosely to compactly branched. **Branches** usually erect, sometimes sprawling or clambering, closely and thinly brown-tomentose. **Leaves** persistent, opposite to subopposite proximally, becoming alternate distally; blade oblong to elliptic, ovate, or elliptic-ovate [suborbiculate], 1–4.5 × 0.5–2.5 cm, base rounded to subcordate, margins serrulate, apex acute to obtuse or rounded, surfaces glabrous [densely white-tomentose abaxially]; veins prominently raised adaxially, reticulate. **Inflorescences** terminal and axillary, primary axis 2–5 cm, lateral branches 4–8. **Drupes** black to purple-black or dark brown, ovoid or obovoid to subglobose, 4–5 mm; stones 3.

Flowering Aug–Sep. Roadsides, bottomland hardwood-pine forests; 10 m; introduced; Tex.; Asia (Afghanistan, China, India, Iran, Japan, Korea, Nepal, Pakistan, Thailand, Vietnam).

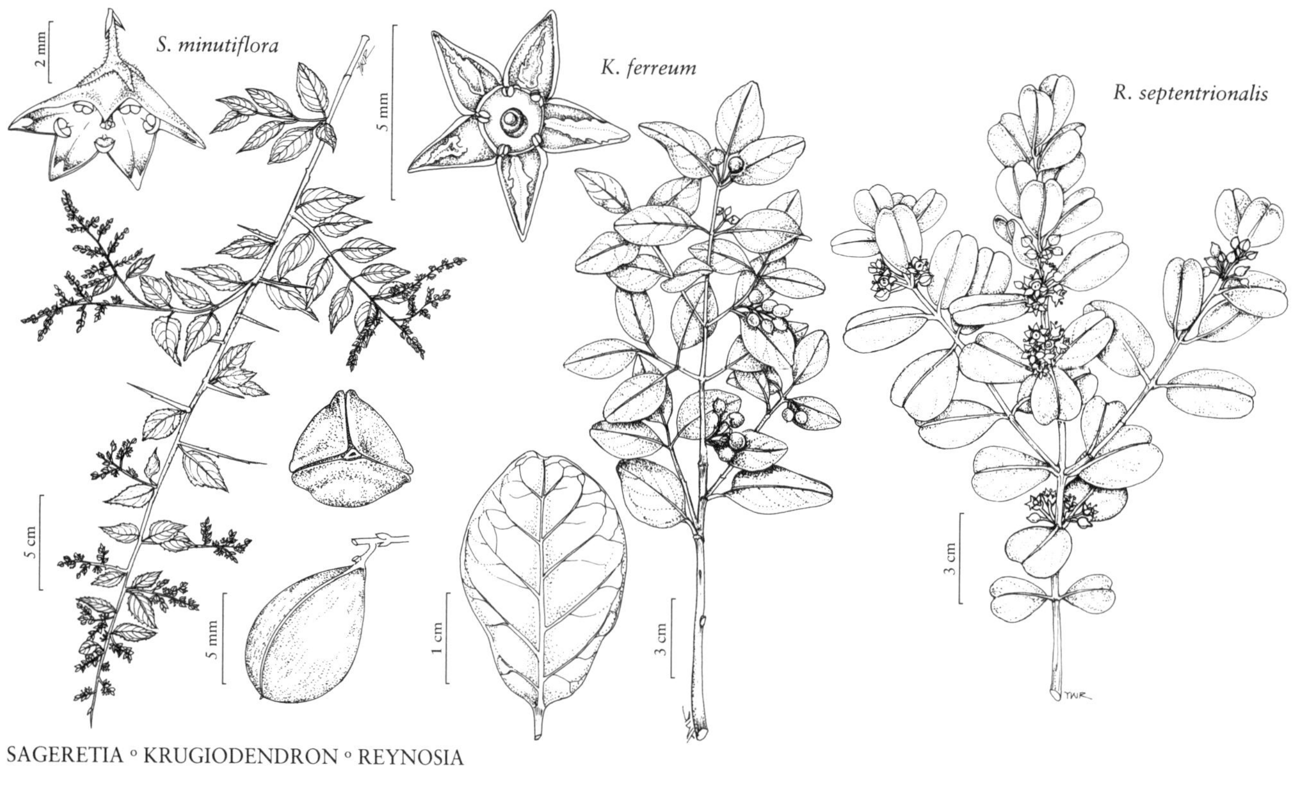

SAGERETIA ∘ KRUGIODENDRON ∘ REYNOSIA

Two non-typical varieties are recognized within *Sageretia thea* (Chen Y. L. and C. Schirarend 2007), each differing from var. *thea* at least in a more prominent leaf vestiture: var. *cordiformis* Y. L. Chen & P. K. Chou and var. *tomentosa* (C. K. Schneider) Y. L. Chen & P. K. Chou (Chen and Schirarend; M. Qaiser and S. Nazimuddin 1981). *Sageretia brandrethiana* Aitchison also has sometimes been treated at varietal rank within *S. thea*. Qaiser and Nazimuddin noted that *S. thea* is a highly polymorphic species, variable particularly in the degree of stems and leaf vestiture and in leaf shape and margin (crenulate, minutely serrate, or entire).

The native range of *Sageretia thea* sometimes is said to extend to northeastern Africa (M. Qaiser and S. Nazimuddin 1981), but this appears not to be clearly established. The type of the species is from China (M. C. Johnston 1968).

Sageretia thea was noted by L. E. Brown and K. N. Gandhi (1989) as occurring beside highways near Alvin, Brazoria County, where it apparently had been planted. Evidence that the species is naturalized, rather than persistent from plantings, is in a collection made in 2007 from within woods near the locality where the plants were earlier observed as common roadside shrubs.

The illegitimate, superfluous names *Rhamnus theezans* Linnaeus and *Sageretia theezans* (Linnaeus) Brongniart pertain here.

7. KRUGIODENDRON Urban, Symb. Antill. 3: 313. 1902 • Leadwood [For Carl Wilhelm Krug, 1833–1898, major collaborator with Urban on the West Indian flora, and Greek *dendron*, tree]

Guy L. Nesom

Shrubs or trees, unarmed; bud scales present. **Leaves** persistent, usually opposite or subopposite, rarely alternate, not borne on short shoots; blade not gland-dotted; pinnately veined, secondary veins arching near margins, higher order veins not forming adaxially raised reticulum. **Inflorescences** axillary, within foliage, umbels; peduncles and pedicels not fleshy in fruit.

Pedicels present. **Flowers** bisexual; hypanthium very shallowly cupulate, 3–4 mm wide; sepals 5, spreading, greenish yellow, triangular to triangular-ovate, crested adaxially; petals 0; nectary fleshy, filling hypanthium, margin pentagonal, decagonal, or weakly subcrenulate; stamens 5; ovary superior, 2-locular; style 1. **Fruits** drupes, 5–8[–12] mm; stone 1, indehiscent.

Species 1: Florida, s Mexico, West Indies, Central America (s to Costa Rica).

1. **Krugiodendron ferreum** (Vahl) Urban, Symb. Antill. 3: 314. 1902 • Black ironwood F

Rhamnus ferrea Vahl, Symb. Bot. 3: 41, plate 58. 1794 (as ferreus); *Krugiodendron acuminatum* J. Á. Gonzales & Poveda; *Rhamnidium ferreum* (Vahl) Sargent

Shrubs or trees, 1–10 m, glabrous or glabrescent; trunks to 5 dm diam. **Leaves:** petiole 3–6 mm; blade ovate to elliptic or broadly elliptic to nearly oval, 2–6 cm, margins undulate, apex rounded to obtuse, usually truncate-emarginate, rarely acuminate. **Inflorescences** 3–5-flowered. **Pedicels** 1–6 mm. **Flowers** almond-fragrant. **Drupes** purplish red to nearly black at maturity.

Flowering sporadically year-round. Hammocks, thickets, mangrove woodlands; 0–10 m; Fla.; s Mexico; West Indies; Central America.

Krugiodendron ferreum in the flora area is found along the Atlantic coast from Brevard to Monroe counties. It is cultivated in gardens and parks as a specimen tree. *Krugiodendron acuminatum* was described from collections far-disjunct in Costa Rica and in Veracruz, Mexico. A. Pool (pers. comm.) considers these collections to fall within the range of variability of *K. ferreum*.

8. REYNOSIA Grisebach, Cat. Pl. Cub., 33. 1866 • Red ironwood [For Alvaro Reynoso, 1829–1888, Cuban chemist and agriculturalist, who revolutionized the sugar industry]

Guy L. Nesom

Shrubs or small trees, unarmed; bud scales present. **Leaves** persistent, opposite; blade not gland-dotted; pinnately veined, secondary veins straight nearly to margins, higher order veins forming adaxially raised reticulum enclosing isodiametric areoles. **Inflorescences** axillary, fascicles or flowers solitary; pedicels not fleshy in fruit. **Pedicels** present. **Flowers** bisexual; hypanthium cupulate, 2–4 mm wide; sepals [4–]5, spreading, yellow-green, triangular-ovate, small-keeled adaxially, not crested; petals 0[4–5]; nectary fleshy, filling hypanthium, margin entire; stamens 5; ovary superior, 2-locular; style 1. **Fruits** drupes, 10–20 mm; stone 1, indehiscent.

Species ca. 15 (1 in the flora): Florida, West Indies, Central America (Guatemala).

SELECTED REFERENCE Schirarend, C. and P. Hoffmann. 1993. Untersuchungen zur Blutenmorphologie der Gattung *Reynosia* Griseb. (Rhamnaceae). Flora, Morphol. Geobot. Ecophysiol. 188: 275–286.

1. **Reynosia septentrionalis** Urban, Symb. Antill. 1: 356. 1899 • Darling-plum F

Shrubs or small trees, 1.5–7 m, glabrous; trunks to 2 dm diam. **Leaves:** petiole 1.5–3 mm; blade elliptic-oblong to oval or obovate, 2–4 cm, coriaceous, base cuneate to truncate, margins entire, revolute, apex usually truncate-emarginate, sometimes rounded. **Inflorescences** 1–4-flowered. **Pedicels** 4–9 mm. **Drupes** dark purple to black, globose to ovoid or ellipsoid.

Flowering sporadically year-round. Coastal hammocks, open woods, thickets, mangrove margins, dunes; 0–10 m; Fla.; West Indies.

Reynosia septentrionalis is found in the flora area in Miami-Dade and Monroe counties.

9. HOVENIA Thunberg, Nov. Gen. Pl. 1: 7. 1781 • Raisin tree [For David Hoven, 1724–1787, Dutch senator and botanical patron] [I]

Guy L. Nesom

Trees [shrubs], unarmed; bud scales present. **Leaves** deciduous, alternate; blade not gland-dotted; pinnately veined, basal pair of secondary veins more prominent. **Inflorescences** terminal or axillary, compound dichasia (appearing repeatedly dichotomously branched); peduncles and pedicels becoming fleshy [not fleshy] in fruit. **Pedicels** present. **Flowers** bisexual; hypanthium hemispheric, 2–3 mm wide; sepals 5, spreading, white, ovate-triangular to triangular, keeled adaxially; petals 5, white or yellow-green, inrolled full length, elliptic to ovate, short-clawed; nectary fleshy, filling hypanthium; stamens 5; ovary ½ inferior, 3-locular; styles 2–3, connate basally. **Fruits** capsules, appearing drupaceous, tardily dehiscent, exocarp sloughing off from leathery mesocarp prior to dehiscence.

Species 7 (1 in the flora): introduced; Asia; introduced also in South America.

SELECTED REFERENCE Koller, G. L. and J. H. Alexander. 1979. The raisin tree—its use, hardiness and size. Arnoldia (Jamaica Plain) 39: 6–15.

1. Hovenia dulcis Thunberg, Nov. Gen. Pl. 1: 8. 1781 • Japanese raisin tree [F] [I]

Hovenia dulcis var. *glabra* Makino; *H. dulcis* var. *latifolia* Nakai ex Y. Kimura

Trees, 5–10 m, branchlets glabrous. **Leaves:** petiole 2–4.5 cm; blade ovate, broadly oblong, or elliptic-ovate, 7–17 × 4–11 cm, herbaceous, base truncate or rarely cordate to subrounded, margins serrate, apex shortly acuminate or acuminate, both surfaces glabrous or abaxial pilose on major veins. **Inflorescences** usually terminal, rarely axillary, 50–80-flowered. **Flowers** 6–8 mm diam.; petals 2.4–2.6 mm. **Capsules** purplish to black at maturity, subglobose, 6.5–7.5 mm, glabrous.

Flowering May–Jun. Cliff bases, suburban woodlots, thickets; 100–300 m; introduced; N.C., Tex.; Asia (China, Japan, Korea, Thailand); introduced also in South America (Argentina, Paraguay).

Hovenia dulcis is known as naturalized in Texas only from several trees about 10–15 m, growing along the base of a northwest-facing limestone cliff in Austin (Travis County). They probably are the progeny of trees originally cultivated in a nearby University of Texas College of Pharmacy Drug Garden, defunct since the mid 1940s. No seedlings have been observed in the area. In North Carolina, the species was collected in 1949 in woods south of the North Carolina State College University campus in Wake County.

The swollen, juicy peduncles and pedicels of the infructescence are sweet and used for making wine and candy. The timber is used for building construction and high-quality furniture.

10. PALIURUS Miller, Gard. Dict. Abr. ed. 4, vol. 3. 1754 • Christ-thorn [Classical Greek name, perhaps derived from *pálin*, again or once more, and *oúron* or *oureó*, urine or to make water, alluding to diuretic properties of roots and leaves of *P. spina-christi*] [I]

Guy L. Nesom

Shrubs [or trees], sometimes clambering, armed with stipular spines; bud scales present. **Leaves** deciduous [persistent], alternate; blade not gland-dotted; 3-veined from base (acrodromous), secondary veins distal to basal veins well developed. **Inflorescences** axillary, cymes; peduncles and pedicels not fleshy in fruit. **Pedicels** present. **Flowers** bisexual; hypanthium hemispheric to patelliform, 2–3 mm wide; sepals 5, spreading, pale yellow, deltate, keeled adaxially;

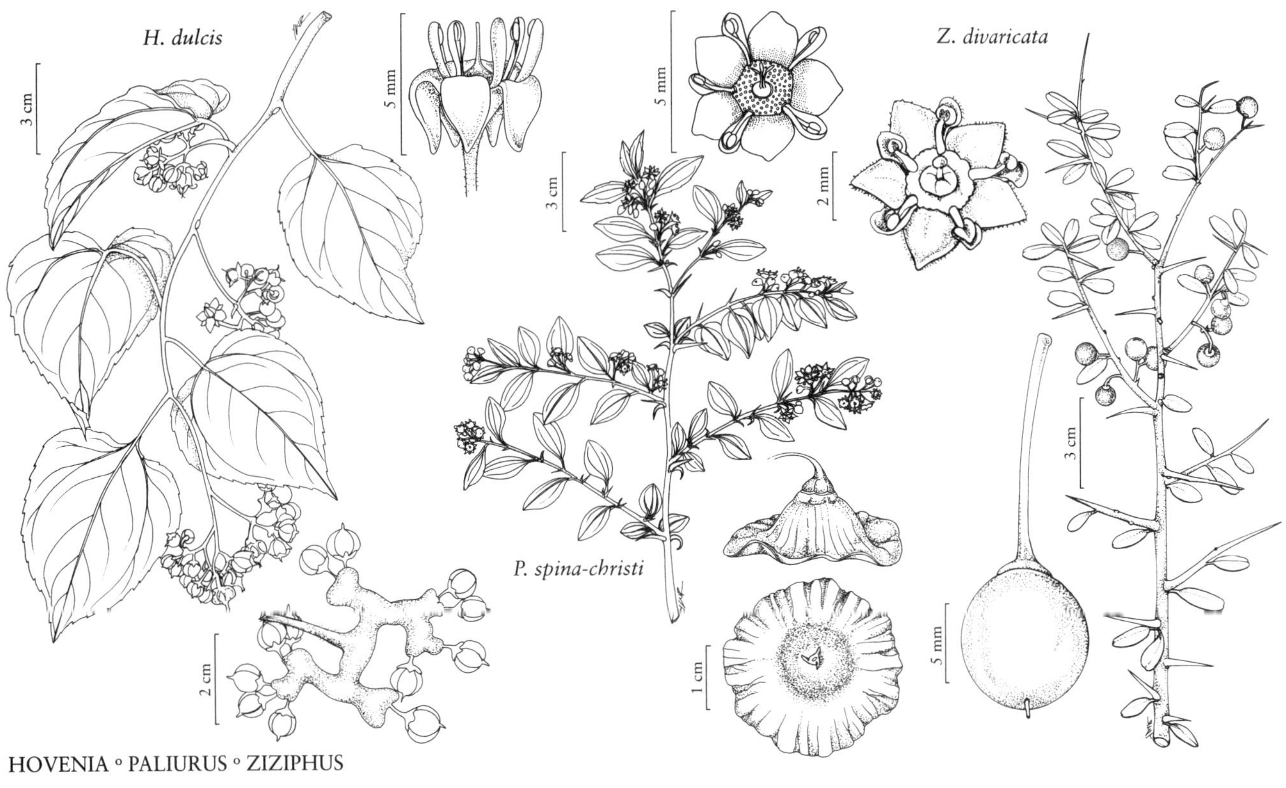

HOVENIA ◦ PALIURUS ◦ ZIZIPHUS

petals 5, white, hooded, obovate, clawed; nectary fleshy, filling hypanthium, margin entire or 5-angled; stamens 5; ovary ½ inferior, 2–3-locular; styles 2–3, connate proximally. **Fruits** samaras. $x = 12$.

Species 5 (1 in the flora): introduced, Texas; Europe, e Asia.

The fruit of *Paliurus* was described by C. Schirarend and M. N. Olabi (1994) as a dry, indehiscent drupe. The gynoecium base in *Paliurus* begins to swell laterally following fertilization, forming a persistent, disciform, cupulate, or hemispheric wing of varying thickness (D. O. Burge and S. R. Manchester 2008). This enlarged and winglike receptacle is diagnostic of the genus.

SELECTED REFERENCES Burge, D. O. and S. R. Manchester. 2008. Fruit morphology, fossil history, and biogeography of *Paliurus* (Rhamnaceae). Int. J. Pl. Sci. 169: 1066–1085. Schirarend, C. and M. N. Olabi. 1994. Revision of the genus *Paliurus* Tourn. ex Mill. (Rhamnaceae). Bot. Jahrb. Syst. 116: 333–359.

1. **Paliurus spina-christi** Miller, Gard. Dict. ed. 8, Paliurus no. 1. 1768 • Jerusalem-thorn, Crown-of-thorns [F] [I]

Rhamnus paliurus Linnaeus, Sp. Pl. 1: 194. 1753

Shrubs, 2–4 m, branchlets brownish pubescent, glabrescent. **Stipular spines** 2 per node, 1 straight, 1–2 cm, 1 shorter and recurved. **Leaves:** blade ovate to oblong-ovate or elliptic, 2–4 cm, herbaceous, base obtuse to cordate, often slightly oblique, margins shallowly and sometimes obscurely serrate or denticulate, apex obtuse, surfaces glabrous or abaxial slightly hairy along veins. **Inflorescences** 5–11-flowered. **Samaras** tan to brownish, glabrous; wing disciform, ± circular, perpendicular to fruit axis, thin, 1.5–3.5 cm diam., with fine radiating, dichotomizing, and anastomosing venation, margin entire, undulate. $2n = 24$.

Flowering spring. Flood plains, riparian scrub, open post oak woodlands, pastures; 150–400 m; introduced; Tex.; s Europe; introduced also in sw, c Asia.

According to R. J. O'Kennon (1991), *Paliurus spina-christi* is naturalized in thicket-forming populations for about 25 km along the flood plains of the Pedernales River and two tributaries in Gillespie County; it is now documented from several locations downstream along the Pedernales in adjacent Blanco County. In this area, it apparently spread from hedgerows planted in the late 1880s along one of the tributaries. It is also documented as naturalized in open woods in Brazos County, perhaps escaping from plantings at Texas A&M University in College Station.

11. ZIZIPHUS Miller, Gard. Dict. Abr. ed. 4, vol. 3. 1754 • Jujube [Latinized Arabic vernacular name *zizouf* for common jujube, *Z. jujuba*]

Guy L. Nesom

Condalia Cavanilles subg. *Condaliopsis* Weberbauer; *Condaliopsis* (Weberbauer) Suessenguth

Shrubs or small trees, armed with thorns or stipular spines [unarmed]; bud scales present or absent. **Leaves** deciduous or persistent, alternate, sometimes fascicled on short shoots; blade not gland-dotted; pinnately veined (obscurely, appearing 1-veined) or 3[–5]-veined from base (acrodromous), secondary veins absent or poorly developed distal to basal veins, except sometimes near apex. **Inflorescences** axillary [terminal], within foliage, corymblike cymes, thyrses, or fascicles, or rarely flowers solitary; peduncles and pedicels not fleshy in fruit. **Pedicels** present. **Flowers** bisexual; hypanthium shallowly cupulate to hemispheric, 1–3 mm wide; sepals 5, spreading, greenish, greenish white, yellowish, yellow-green, orangish, or purplish, ovate-triangular to triangular, keeled adaxially; petals [0–]5, yellow to pale yellow or white, ± flat, obovate or spatulate, clawed; nectary fleshy, filling hypanthium, 5–10-lobed; stamens 5; ovary superior, 2–3(–4)-locular; styles (1–)2–3(–4), connate basally to proximally. **Fruits** drupes; stone 1. $x = 12$.

Species ca. 170 (6 in the flora): United States, Mexico, Central America, South America, Eurasia, Africa.

The molecular and morphological study by M. B. Islam and M. P. Simmons (2006) corroborated earlier observations (for example, M. C. Johnston 1963) that *Ziziphus* comprises two clades, with the Old World species of *Ziziphus* more closely related to *Paliurus* than to the New World species. The New World species have paired or unpaired thorns, tangential to diagonal bands of axial parenchyma, and deciduous, non-spinose stipules. Old World *Ziziphus* and *Paliurus* do not have thorns tangential to diagonal bands of axial parenchyma but do have stipular spines. The Islam and Simmons study found the Florida endemic *Z. celata* to be more closely related to *Berchemia* and *Rhamnus* than to either group of *Ziziphus*. Further study by Islam and R. P. Guralnick (2015) showed *Z. celata* and *Z. parryi* to be sister species that are most closely related to *Condalia*. Maintaining monophyletic genera will require reclassification of *Ziziphus* as treated here.

Ziziphus jujuba and *Z. mauritiana* are grown commercially for fruits in their native or introduced countries; fruits also are harvested in the native or naturalized regions (O. P. Pareek 2001).

SELECTED REFERENCES Islam, M. B. and R. P. Guralnick. 2015. Generic placement of the former *Condaliopsis* (Rhamnaceae) species. Phytotaxa 236: 25–39. Islam, M. B. and M. P. Simmons. 2006. A thorny dilemma: Testing alternative intrageneric classifications within *Ziziphus* (Rhamnaceae). Syst. Bot. 31: 826–842. Johnston, M. C. 1963. The species of *Ziziphus* indigenous to United States and Mexico. Amer. J. Bot. 50: 1020–1027.

1. Leaf blades 3-veined from base; secondary branches not thorn-tipped, axillary thorns absent; stipular spines usually present, sometimes absent.
    2. Secondary branches and leaf blade abaxial surfaces glabrous; stipular spines 15–40 mm. . . . . 5. *Ziziphus jujuba*
    2. Secondary branches and leaf blade abaxial surfaces tomentose; stipular spines 2–3 mm. . . . . 6. *Ziziphus mauritiana*

1. Leaf blades 1-veined from base; secondary branches thorn-tipped and with axillary thorns; stipular spines absent.
   3. Inflorescences usually pedunculate thyrses, rarely flowers solitary; secondary branches gray-green to white, pruinose; drupes 5–10 mm.
      4. Secondary branches usually glabrous, sometimes sparsely pilose; leaf blades subcoriaceous, surfaces usually glabrous; hypanthia moderately to densely strigose, hairs loose, curved; inflorescences (1–)2–6-flowered, peduncles (0.5–)1–2 mm, nearly equaling or shorter than pedicels . . . . 1. *Ziziphus obtusifolia*
      4. Secondary branches minutely hirtellous to short-villous, usually densely so, glabrescent; leaf blades relatively thin-herbaceous, surfaces persistently hirtellous to short-villous; hypanthia densely and persistently hirtellous; inflorescences (5–)10–30-flowered; peduncles 2–4 mm, equaling or longer than pedicels . . . . .2. *Ziziphus divaricata*
   3. Inflorescences usually fascicles, rarely flowers solitary; secondary branches gray or pale greenish yellow to purplish, not pruinose; drupes 10–20(–25) mm.
      5. Leaf blades 1–2.5(–3) cm; hypanthia and sepals purplish to greenish; drupes brownish to orange or purplish brown; California . . . . .3. *Ziziphus parryi*
      5. Leaf blades 0.5–1 cm; hypanthia and sepals greenish; drupes yellow, orange, or brownish; Florida . . . . .4. *Ziziphus celata*

1. Ziziphus obtusifolia (Hooker ex Torrey & A. Gray) A. Gray, Gen. Amer. Bor. 2: 170. 1849 (as Zizyphus) • Graythorn, lotebush, prairie bush, gumdrop-tree, Texas buckthorn

Rhamnus obtusifolia Hooker ex Torrey & A. Gray, Fl. N. Amer. 1: 685. 1840 (as obtusifolius); *Condalia lycioides* (A. Gray) Weberbauer; *C. obtusifolia* (Hooker ex Torrey & A. Gray) Weberbauer; *Condaliopsis lycioides* (A. Gray) Suessenguth; *Ziziphus lycioides* A. Gray

Shrubs, 0.5–1(–2.5) m; secondary branches gray-green to white, pruinose, usually glabrous, sometimes sparsely pilose, thorn-tipped, axillary thorns solitary, with 0–1 nodes, 2–10 mm; stipular spines absent. **Leaves** deciduous, alternate, sometimes fascicled; blade gray-green, glaucous abaxially, dull green adaxially, ovate, oblong, or elliptic to nearly linear, (0.5–)1–2.5(–5) cm, subcoriaceous, base rounded to truncate, margins entire or wider leaves distally toothed, apex rounded to slightly retuse, surfaces usually glabrous, abaxial sometimes sparsely short-villous to short-strigose; 1-veined from base. **Inflorescences** usually pedunculate thyrses, rarely flowers solitary, 5–10 mm, (1–)2–6-flowered, peduncles (0.5–)1–2 mm, nearly equaling or shorter than pedicels. **Flowers:** hypanthium yellow to orangish or slightly purple, moderately to densely strigose, hairs loose, curved; sepals yellow to orangish or slightly purple; petals yellow to pale yellow or nearly white. **Drupes** dark blue to blue-black or purplish, globose to slightly elongate, 5–10 mm.

Flowering (Mar–)Apr–Jun. Silty and sandy flood plains and washes, stream banks, gravelly slopes, brushy hills, sand and clay dunes, plains, gypsum outcrops and roadsides, desert grasslands, scrublands; (10–)100–1600(–1800) m; Ariz., N.Mex., Okla., Tex.; Mexico (Chihuahua, Coahuila, Nuevo León, San Luis Potosí, Sonora, Tamaulipas, Veracruz, Zacatecas).

In Arizona, *Ziziphus obtusifolia* occurs only on gypsum substrates in the southeastern corner of Cochise County.

2. Ziziphus divaricata (A. Nelson) Davidson & Moxley, Fl. S. Calif., 226. 1923 • Graythorn, lotebush, gumdrop-tree F

Condalia divaricata A. Nelson, Bot. Gaz. 47: 427. 1909; *C. lycioides* (A. Gray) Weberbauer var. *canescens* (A. Gray) Trelease; *Condaliopsis lycioides* (A. Gray) Suessenguth var. *canescens* (A. Gray) Suessenguth; *Ziziphus lycioides* A. Gray var. *canescens* A. Gray; *Z. obtusifolia* (Hooker ex Torrey & A. Gray) A. Gray var. *canescens* (A. Gray) M. C. Johnston

Shrubs or small trees, 1–3(–4) m; secondary branches gray-green to white, pruinose, minutely hirtellous to short-villous, usually densely so, glabrescent, thorn-tipped, axillary thorns solitary, with 0–1 nodes, 2–10 mm, stipular spines absent. **Leaves** deciduous, alternate or fascicled; blade gray-green, glaucous abaxially, dull green adaxially, ovate, oblong, or elliptic to nearly linear, 0.3–2(–2.5) cm, relatively thin-herbaceous, base rounded to truncate, margins entire or wider leaves distally toothed, apex rounded to slightly retuse, both surfaces persistently hirtellous to short-villous; 1-veined from base. **Inflorescences** pedunculate thyrses, 10–20 mm, (5–)10–30-flowered, peduncles 2–4 mm, equaling or longer than pedicels; hypanthium yellow to orangish or slightly purple, densely and persistently hirtellous; sepals yellow to orangish or slightly purple; petals yellow to pale yellow or nearly white. **Drupes** dark blue to blue-black or purplish, globose to slightly elongate, 7–10 mm.

Flowering Apr–Jul(–Sep). Washes, basin edges, roadsides, mesquite and tamarisk thickets; (20–) 200–1300 m; Ariz., Calif., Nev., Utah; Mexico (Baja California, Sonora).

Ziziphus divaricata usually has been regarded as *Z. obtusifolia* var. *canescens*, but *Z. divaricata* and *Z. obtusifolia* are discontinuously distinct morphologically and geographically. Although it seems clear that they have a sister relationship, treatment of each at specific rank emphasizes their apparent evolutionary independence.

3. Ziziphus parryi Torrey in W. H. Emory, Rep. U.S. Mex. Bound. 2(1): 46. 1859 (as Zizyphus) • Parry's jujube or abrojo, California crucillo, lotebush

Condalia parryi (Torrey) Weberbauer; *C. parryi* var. *microphylla* I. M. Johnston; *Condaliopsis parryi* (Torrey) Suessenguth; *Ziziphus parryi* var. *microphylla* (I. M. Johnston) M. C. Johnston

Shrubs, 1–4 m; secondary branches pale greenish yellow to purplish, not pruinose, glabrous, thorn-tipped, axillary thorns solitary, with 0–1 nodes, without tertiary thorns, stipular spines absent. **Leaves** deciduous, alternate, sometimes fascicled; blades olive green, elliptic to obovate, 1–2.5(–3) cm, herbaceous, base obtuse to rounded, margins entire, apex rounded to cuneate or shallowly emarginate, surfaces glabrous; 1-veined. **Inflorescences** usually fascicles, 2–5-flowered, rarely flowers solitary. **Flowers:** hypanthium and sepals purplish to greenish; petals white. **Drupes** brownish to orange or purplish brown, ovoid to ellipsoid, 10–20(–25) mm.

Flowering (Feb–)Mar–May. Chaparral, pinyon-juniper woodlands, rocky washes and arroyos, hill slopes; 600–1000(–1600) m; Calif.; Mexico (Baja California).

Ziziphus parryi is known in the flora area only from Imperial, Riverside, San Bernardino, and San Diego counties, but the distribution continues southward in Baja California as far as Cedros Island. Small-leaved plants from southern Baja California have been identified as *Z. parryi* var. *microphylla*, but similar variants occur sporadically into California.

4. Ziziphus celata Judd & D. W. Hall, Rhodora 86: 382, fig. 1. 1984 • Florida jujube [C] [E]

Condalia celata (Judd & D. W. Hall) M. B. Islam

Shrubs or small trees, 1–2 m; secondary branches gray, not pruinose, glabrous, thorn-tipped, axillary thorns solitary or paired and unequal, with 1–3 nodes, bearing several short shoots and often small tertiary thorns; stipular spines absent. **Leaves** deciduous, alternate, sometimes fascicled; blade dark green, oblong-elliptic to elliptic-ovate or elliptic-obovate, 0.5–1 cm, coriaceous, base cuneate to slightly attenuate, margins entire, apex rounded to shallowly emarginate, surfaces glabrous; 1-veined. **Inflorescences** fascicles on short shoots, 2–4-flowered, or flowers solitary. **Flowers:** hypanthium and sepals greenish; petals white. **Drupes** yellow, orange, or brownish, globose to ovoid or oblong, 10 mm.

Flowering Mar–Apr. White sand, pastures on former sand hills, Turkey oak and oak-hickory scrub; of conservation concern; 100 m; Fla.

Ziziphus celata is endemic to the Lake Wales Ridge in Highlands and Polk counties. A molecular phylogenetic study by M. B. Islam and R. P. Guralnick (2015) found that *Z. celata* and *Z. parryi* are sister species and most closely related to *Condalia*, and on that basis, they treated both species within *Condalia*. However, their *Ziziphus*-like morphology is conspicuously out of place within *Condalia*.

Ziziphus celata is ranked as G1 by NatureServe and federally listed as endangered; it is also in the Center for Plant Conservation's National Collection of Endangered Plants.

5. Ziziphus jujuba Miller, Gard. Dict. ed. 8, Ziziphus no. 1. 1768, name conserved • Common jujube [I]

Rhamnus zizyphus Linnaeus, Sp. Pl. 1: 194. 1753

Shrubs or small trees, 2–12 m; secondary branches reddish, glabrescent, not thorn-tipped, axillary thorns absent; stipular spines usually present, straight or curving, 15–40 mm, solitary or paired, sometimes absent. **Leaves** deciduous, alternate; blade green abaxially, darker green and glossy adaxially, ovate to ovate-lanceolate or elliptic-oblong, 3–6 cm, coriaceous, base oblique, margins crenate-serrate, apex usually obtuse to rounded, rarely acute, surfaces glabrous; 3-veined from base. **Inflorescences** cymes, 2–8-flowered, or rarely flowers solitary. **Flowers:** hypanthium and sepals yellow-green, petals pale yellow. **Drupes** ripening through yellow-green to dark red or reddish purple, ellipsoid to narrowly ovoid, 15–20(–30) mm. **2*n*** = 24, 36, 48, 60, 72, 96.

Flowering Mar–May. Old home and ranch sites, fencerows, fields, pastures, roadsides, weedy riparian woods, alluvial slopes; 50–700 m; introduced; Ala., Ariz., Calif., Fla., Ga., La., Tex., Utah; Eurasia.

The fruits of *Ziziphus jujuba* have a datelike taste and are eaten fresh, dried, candied, or preserved. Hundreds of cultivars have been developed in China. The spineless var. *inermis* (Bunge) Rehder sometimes is identified as a cultivar.

The species was first introduced to North America in 1837 and has spread widely. It is not always clear whether it is naturalized or persisting from earlier plantings.

The name *Ziziphus zizyphus* (Linnaeus) H. Karsten, sometimes used for this species, is a tautonym and therefore illegitimate.

6. Ziziphus mauritiana Lamarck in J. Lamarck et al., Encycl. 3: 319. 1789 • Indian jujube [I]

Rhamnus jujuba Linnaeus, Sp. Pl. 1: 194. 1753, not *Ziziphus jujuba* Miller 1768

Shrubs or small trees, 3–10(–15) m; secondary branches white-silvery to grayish, becoming brown, tomentose, glabrescent, not thorn-tipped, axillary thorns absent; stipular spines usually present, straight or recurving, 2–3 mm, solitary or paired, sometimes absent. **Leaves** persistent, alternate; blade whitish to tawny abaxially, dark green adaxially, oblong to elliptic or elliptic-ovate, 2.5–8 cm, base obtuse to rounded, usually oblique, margins serrulate, apex rounded, abaxial surface tomentose, adaxial surface glabrous; 3-veined from base. **Inflorescences** cymes, 2–8-flowered. **Flowers:** hypanthium and sepals greenish to greenish white; petals white. **Drupes** ripening from yellow or orange to red or reddish brown, globose to ovoid or oblong, 20–30 cm. **2*n*** = 24.

Flowering summer. Disturbed sites, thicket edges, rockland hammocks; 0–10 m; introduced; Fla.; c, se Asia; introduced also in West Indies (Jamaica, Lesser Antilles, Puerto Rico), Europe, elsewhere in Asia, Africa, Indian Ocean Islands (Seychelles Islands), Pacific Islands (Fiji, Hawaii, New Caledonia, Papua New Guinea, Philippines), Australia.

Florida records for *Ziziphus mauritiana* are from Broward and Miami-Dade counties. Reports of the species from California (that is, USDA Plants Database) presumably refer to *Z. jujuba*.

Ziziphus mauritiana is a major commercial fruit-producing species in India, with many cultivars varying in fruiting season and in fruit form, size, color, flavor, and keeping quality. The fruit is rich in vitamin C and is eaten raw, pickled, or used in beverages.

The illegitimate name *Ziziphus jujuba* (Linnaeus) Gaertner is sometimes used for this species.

12. COLUBRINA Richard ex Brongniart, Mém. Fam. Rhamnées, 61, plate 4, fig. 3. 1826, name conserved • Nakedwood [Latin *coluber*, racer snake, perhaps alluding to twisting of deep furrows on stems of some species]

Guy L. Nesom

Shrubs or trees, unarmed [armed with thorns]; bud scales present at least on short shoots. **Leaves** persistent or deciduous, alternate [opposite], sometimes fascicled on short shoots; blade not gland-dotted; pinnately veined or 3-veined from base (acrodromous). **Inflorescences** axillary, usually cymes, thyrses, or fascicles, rarely flowers solitary; peduncles and pedicels not fleshy in fruit. **Pedicels** present. **Flowers** bisexual; hypanthium hemispheric, 2–3 mm wide; sepals 5, spreading, brown to greenish, ovate-triangular to triangular, keeled adaxially; petals 5, greenish or yellowish to creamy white, ± hooded, spatulate or obovate, clawed; nectary fleshy, adnate to and sometimes ± filling hypanthium; stamens 5; ovary ½ inferior to inferior, 3(–4)-locular; styles 3, connate proximally. **Fruits** capsules, sometimes tardily dehiscent, 3-locular.

Species ca. 30 (8 in the flora): United States, Mexico, West Indies, Central America, South America, se Asia, Indian Ocean Islands (Madagascar), Pacific Islands (Hawaii); introduced widely.

Branches of some *Colubrina* species are rigid but rarely produce thorns. Those of *C. californica* have been described as spinescent or subspinescent, alluding to short branches that sometimes are attenuate to relatively sharp points, but these are not the same as the clearly defined thorns that occur in other genera of the family.

SELECTED REFERENCES Johnston, M. C. 1971. Revision of *Colubrina* (Rhamnaceae). Brittonia 23: 2–53. Nesom, G. L. 2013. Taxonomic notes on *Colubrina* (Rhamnaceae). Phytoneuron 2013-4: 1–21.

1. Leaf blade margins entire, mucronulate to obscurely serrulate, or crenulate.
   2. Leaf blades 1–2.5(–3.5) cm; Arizona, California, Nevada . . . . 3. *Colubrina californica* (in part)
   2. Leaf blades 4–15 cm; Florida.
      3. Leaf blade abaxial surfaces tawny-tomentose, adaxial surfaces villous-strigose, glabrescent, secondary veins relatively straight, 6–12 pairs; inflorescences 20–50 (–70)-flowered; peduncles 8–15 mm . . . . 6. *Colubrina cubensis*
      3. Leaf blade abaxial surfaces sparsely strigose or red-brown tomentose, glabrescent or persistently tomentose on veins, adaxial surfaces glabrous or tomentose and glabrescent, secondary veins arcuate, (4–)5–9 pairs; inflorescences 8–30-flowered; peduncles 1–10 mm.
         4. Leaves deciduous, blades herbaceous, 4–9 cm, abaxial surfaces sparsely strigose, glabrescent, adaxial surfaces glabrous; peduncles 1–7 mm . . . . 7. *Colubrina elliptica*
         4. Leaves persistent, blades subcoriaceous, 5–15 cm, both surfaces tomentose, abaxial glabrescent except persistently tomentose on veins, adaxial glabrescent; peduncles (3–)5–10 mm . . . . 8. *Colubrina arborescens*
1. Leaf blade margins serrate, serrulate, or crenate-serrate.
   5. Shrubs, erect to sprawling or clambering; leaf blade surfaces glabrous or glabrate; Florida . . . . 5. *Colubrina asiatica*
   5. Shrubs or trees, erect, (rarely clambering in *C. greggii*); leaf blade surfaces (one or both) hairy; sc, sw United States.
      6. Leaf blades 1–3(–4) cm, margins with 2–20 teeth per side.
         7. Leaf blades (1–)3–4 cm, margins with 10–20 teeth per side; inflorescences fascicles, 2–4(–7)-flowered, or flowers solitary, peduncles absent . . . . 1. *Colubrina texensis*
         7. Leaf blades 1–2.5(–3.5) cm, margins with 2–7 teeth per side; inflorescences usually cymes or thyrses, 2–12-flowered, sometimes flowers solitary, peduncles 1–2 mm . . . . 3. *Colubrina californica* (in part)
      6. Leaf blades 3–12 cm, margins with 40–100 teeth per side.
         8. Leaf blades 3–7.5 cm, margins with 40–70 teeth per side; inflorescences 6–15-flowered, peduncles 2–8 mm . . . . 2. *Colubrina stricta*
         8. Leaf blades (3–)4–12 cm, margins with 50–100 teeth per side; inflorescences (10–)20–40-flowered, peduncles 5–20 mm . . . . 4. *Colubrina greggii*

1. Colubrina texensis (Torrey & A. Gray) A. Gray, Boston J. Nat. Hist. 6: 169. 1850 • Texan hog-plum [E] [F]

Rhamnus texensis Torrey & A. Gray, Fl. N. Amer. 1: 263. 1838

Shrubs or small trees, erect, 1–2(–2.8) m. **Stems** zigzag, white tomentose-sericeous, becoming glabrate. **Leaves** deciduous, sometimes fascicled on short shoots; petiole 1–4 mm; blade ovate to elliptic, oblong-obovate, or obovate, 1–3(–4) cm, subcoriaceous, base rounded to subcordate, margins shallowly serrate, teeth 10–20 per side, apex rounded, often apiculate, abaxial surface loosely sericeous, adaxial glabrate; pinnately veined, secondary veins (2–)3–4 pairs, arcuate, basal pair prominent. **Inflorescences** fascicles, 2–4(–7)-flowered, or flowers solitary; peduncles absent; fruiting pedicels 5–13 mm. **Capsules** 6–9 mm.

Flowering Mar–May. Fencerows, roadsides, disturbed sites, clay banks, shell ridges, loose sand, sandy loam, rocky limestone slopes and crevices, gravel hills, stream banks, alluvial terraces, gravelly flood plains, shrub-grasslands, mesquite shrublands, oak-mesquite, oak-juniper, and mesquite-hackberry woodlands; 0–800 m; Tex.

Colubrina texensis is widespread in central and southern Texas.

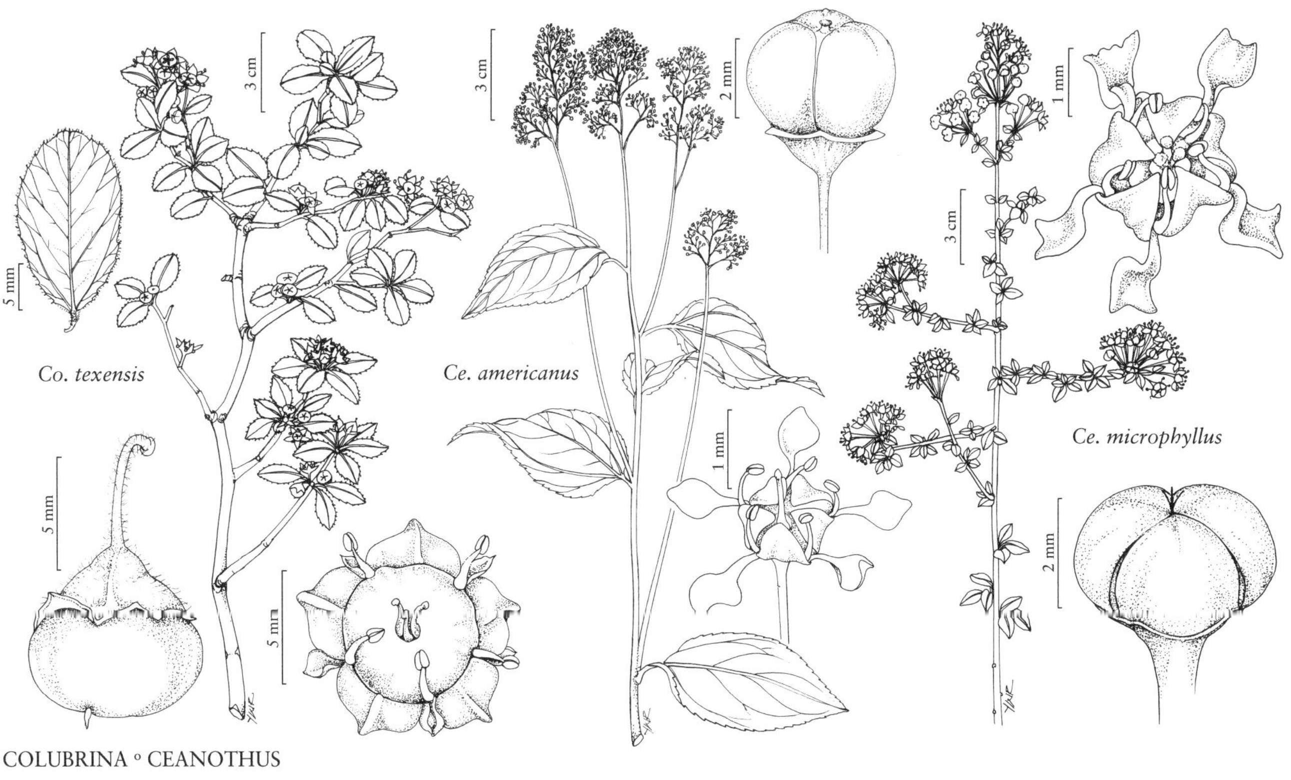

COLUBRINA ◦ CEANOTHUS

2. **Colubrina stricta** Engelmann, Smithsonian Contr. Knowl. 3(5): 33. 1852 • Comal nakedwood [C]

Colubrina texensis (Torrey & A. Gray) A. Gray var. *pedunculata* M. C. Johnston

Shrubs, erect, 1–2 m. **Stems** straight or slightly zigzag, loosely sericeous to glabrate. **Leaves** deciduous; petiole 3–10 mm; blade ovate to ovate-oblong, 3–7.5 cm, herbaceous, base rounded to truncate, margins serrulate, teeth 40–70 per side, villous-tufted, apex rounded to acute, abaxial surface persistently sparsely villous at least among veins, adaxial surface sparsely strigose, becoming glabrate; pinnately veined, secondary veins (3–)4–5 pairs, arcuate, basal pair slightly more prominent. **Inflorescences** thyrses, 6–15-flowered; peduncles 2–8 mm; fruiting pedicels 5–6 mm. **Capsules** 7–8 mm.

Flowering Mar–Jun. Rocky open slopes; of conservation concern; 200–1600 m; Tex.; Mexico (Chihuahua, Coahuila, Durango, Nuevo León).

Colubrina stricta is known in the flora area only from historic collections (during 1850) at the type locality in Comal County and a few more recent collections from one site in Hueco Tanks State Park in El Paso County. It is more continuously distributed and abundant in Mexico (G. L. Nesom 2013). Validation of the epithet by M. C. Johnston (1969) was superfluous.

3. **Colubrina californica** I. M. Johnston, Proc. Calif. Acad. Sci., ser. 4, 12: 1085. 1924 • Las Animas nakedwood, California snakebush

Colubrina texensis (Torrey & A. Gray) A. Gray var. *californica* (I. M. Johnston) L. D. Benson

Shrubs, erect, 1–2(–3, rarely to 8) m. **Stems** ± straight, intricately branched, white-tomentose, glabrescent. **Leaves** deciduous, sometimes fascicled; petiole 1–4 mm; blade elliptic to oblong-elliptic or elliptic-obovate, 1–2.5(–3.5) cm, coriaceous, base rounded to cuneate, margins entire or mucronulate to obscurely serrulate, teeth (0–)2–7 per side, apex rounded to truncate, both surfaces usually minutely hirtellous to sparsely pilose, abaxial sometimes strigose; pinnately veined, secondary veins 3–4(–5) pairs, ± straight. **Inflorescences** cymes or thyrses, 2–12-flowered, or flowers solitary; peduncles 1–2 mm; fruiting pedicels 2–4 mm. **Capsules** 7–9 mm.

Flowering (Dec–)Mar–May. Sandy washes, arroyos, alluvial slopes and fans, granite slopes, creosote bush and desert scrubs; (100–)300–1000 m; Ariz., Calif., Nev.; Mexico (Baja California, Sonora).

In Arizona, *Colubrina californica* occurs in La Paz, Maricopa, Pima, Pinal, and Yuma counties; in California, all known records are from Imperial, Riverside, and San Diego counties; in Nevada, it is known from Clark County.

4. Colubrina greggii S. Watson, Proc. Amer. Acad. Arts 17: 336. 1882 • Sierran nakedwood

Shrubs, usually erect, rarely clambering, 2–3 m. **Stems** zigzag, loosely sericeous to glabrate. **Leaves** deciduous; petiole 4–20 mm; blade ovate to lanceolate-ovate or elliptic-ovate, (3–)4–12[–23] cm, herbaceous, base rounded to shallowly cordate, margins serrulate, teeth 50–100 per side, villous-tufted, apex obtuse to acute or barely acuminate [acuminate], abaxial surface persistently tomentose-sericeous at least among veins, adaxial surface sparsely strigose; pinnately veined, secondary veins 4–5(–6) pairs, arcuate, basal pair usually more prominent. **Inflorescences** thyrses, (10–)20–40 [–80]-flowered; peduncles 5–20 mm; fruiting pedicels 5–10 mm. **Capsules** 8–10 mm.

Flowering Feb–Mar. Remnant Sabal palm groves, shrublands; 0–10 m; Tex.; Mexico (Coahuila, Guanajuato, Hidalgo, Nuevo León, Querétaro, Tamaulipas, Veracruz).

Colubrina greggii is known in the flora area only in Cameron County. These plants and those in northern Tamaulipas have sparser foliar vestiture than those elsewhere in the range. The Texas plants also have leaves with obtuse to acute or barely acuminate apices, in contrast to the more typical acuminate apices over the rest of the range.

Colubrina greggii var. *yucatanensis* M. C. Johnston (southern Mexico and Central America) and *C. greggii* var. *angustior* M. C. Johnston (east-central Mexico) both have been treated at specific rank (G. L. Nesom 2013).

5. Colubrina asiatica (Linnaeus) Brongniart, Mém. Fam. Rhamnées, 62. 1826 • Asian nakedwood, latherleaf [I]

Ceanothus asiaticus Linnaeus, Sp. Pl. 1: 196. 1753; *Rhamnus asiatica* (Linnaeus) Lamarck ex Poiret

Varieties 2 (1 in the flora): introduced, Florida; Asia, Pacific Islands (Philippines); introduced also in many areas worldwide.

Colubrina asiatica is an almost pantropical species reported for many coastal habitats of the New and Old World tropics. The type is from Sri Lanka.

Variety *subpubescens* (Pitard) M. C. Johnston occurs in southeast Asia.

5a. Colubrina asiatica (Linnaeus) Brongniart var. **asiatica** [I]

Shrubs, erect to sprawling or clambering, 1.5–8 m. **Stems** zigzag, glabrous. **Leaves** persistent; petiole 7–17 mm; blade ovate to broadly ovate, 4–8 cm, herbaceous, base obtuse to rounded or subcordate, margins crenate-serrate, teeth 10–30 per side, apex acuminate, glabrous or glabrate on both surfaces; 3-veined from base, distal secondary veins 1–2(–4) pairs, arcuate. **Inflorescences** thyrses, 2–7-flowered, or flowers solitary; peduncles (0–)1 mm; fruiting pedicels 4–6 mm. **Capsules** 7–8 mm.

Flowering year-round. Disturbed fill, roadside thickets, fencerows, hammocks, shell mounds, limestone ridges, mangrove thickets, beach strands; 0–10 m; introduced; Fla.; s, se Asia; Pacific Islands (Philippines); introduced also in West Indies, Africa, elsewhere in Pacific Islands, Australia.

6. Colubrina cubensis (Jacquin) Brongniart, Mém. Fam. Rhamnées, 62. 1826 • Cuban nakedwood

Rhamnus cubensis Jacquin, Enum. Syst. Pl., 16. 1760

Varieties 3 (1 in the flora): Florida, West Indies (Cuba, Hispaniola).

Varieties *cubensis* and *ekmanii* M. C. Johnston occur in the West Indies (Cuba and Hispaniola, respectively).

6a. Colubrina cubensis (Jacquin) Brongniart var. **floridana** M. C. Johnston, Wrightia 3: 96. 1963

Shrubs or trees, erect, 5–8 m. **Stems** straight, tawny puberulent-tomentose. **Leaves** deciduous; petiole 5–15 mm; blade elliptic-oblong to ovate-oblong or obovate-oblong, 4–10 cm, subcoriaceous, base rounded to obtuse, margins entire or obscurely serrulate to crenulate, apex acute to rounded or emarginate, abaxial surface tawny-tomentose, adaxial surface villous-strigose, glabrescent; pinnately veined, secondary veins 6–12 pairs, relatively straight, basal pair prominent. **Inflorescences** corymblike thyrses, 20–50(–70)-flowered; peduncles 8–15 mm; fruiting pedicels 4–9 mm. **Capsules** 7–9 mm.

Flowering Feb–Mar, possibly year-round. Rockland hammocks; 0–10 m; Fla.; West Indies (Cuba, Hispaniola).

Variety *floridana* is known in the flora area only from Miami-Dade County.

7. Colubrina elliptica (Swartz) Brizicky & W. L. Stern, Trop. Woods 109: 95. 1958 • Soldierwood

Rhamnus elliptica Swartz, Prodr., 50. 1788 (as ellipticus); *Colubrina reclinata* (L'Héritier) Brongniart

Shrubs or trees, erect, 2–6 m. **Stems** straight, strigose, glabrescent. **Leaves** deciduous; petiole 5–20 mm; blade broadly elliptic to elliptic-ovate or ovate-lanceolate, 4–9 cm, herbaceous, base cuneate to rounded, 2 prominent glands at petiole/blade junction, margins entire, apex acute to acuminate, abaxial surface sparsely strigose, glabrescent, adaxial surface glabrous; pinnately veined, secondary veins 5–8 pairs, arcuate. **Inflorescences** thyrses, 8–20-flowered; peduncles 1–7 mm; fruiting pedicels 8–15 mm. **Capsules** 6–7 mm.

Flowering Sep–Feb, possibly year-round. Rockland hammocks, hammock margins, thickets, rocky strands; 0–10 m; Fla.; Mexico; West Indies; Central America; South America (Venezuela).

Colubrina elliptica is known in the flora area only from Miami-Dade and Monroe counties.

8. Colubrina arborescens (Miller) Sargent, Trees & Shrubs 2: 167. 1911 • Greenheart, coffee colubrina, wild coffee

Ceanothus arborescens Miller, Gard. Dict. ed. 8, Ceanothus no. 3. 1768; *Colubrina ferruginosa* Brongniart

Shrubs or trees, erect, 1–8 m. **Stems** straight, densely reddish tomentose, glabrescent. **Leaves** persistent; petiole 5–20 mm; blade ovate to ovate-oblong or elliptic or oblong-obovate, 5–15 cm, subcoriaceous, base obtuse to rounded or subcordate, 2 prominent glands at petiole/blade junction, margins entire, sometimes with glands, apex acute to acuminate, both surfaces red-brown tomentose, abaxial glabrescent except persistently tomentose on veins, adaxial glabrescent; pinnately veined, secondary (4–)5–9 pairs, arcuate, abaxial glands at ends of intermediate secondary veins. **Inflorescences** cymes or thyrses, 10–30-flowered; peduncles (3–)5–10 mm; fruiting pedicels 4–10 mm. **Capsules** 6–8 mm.

Flowering Nov–Feb, possibly year-round. Dunes, glade margins, mangrove margins, hammocks, canal edges; 0–10 m; Fla.; Mexico; West Indies; Central America.

Colubrina arborescens is known in the flora area only from Miami-Dade and Monroe counties.

13. CEANOTHUS Linnaeus, Sp. Pl. 1: 195. 1753; Gen. Pl. ed. 5, 90. 1754 • [Greek *keanothus*, name used by Dioscorides for some spiny plant]

Clifford L. Schmidt†

Dieter H. Wilken

Shrubs, rarely arborescent, armed with thorns or unarmed; bud scales present. **Leaves** persistent or deciduous, alternate or opposite, sometimes fascicled on short shoots; blade not gland-dotted; pinnately veined or 3-veined from base (acrodromous). **Inflorescences** terminal or axillary, cymes aggregated into umbel-like clusters, or latter aggregated into racemelike or paniclelike thyrses; peduncles and pedicels not fleshy in fruit. **Pedicels** present. **Flowers** bisexual; hypanthium shallowly cupulate to hemispheric, less than 0.5 mm wide; sepals 5 (or (5–)6(–8) in *C. jepsonii*), usually incurved, sometimes becoming spreading, usually white to cream, blue, or purple, rarely pink, lanceolate to deltate, keeled adaxially; petals 5 (or (5–)6(–8) in *C. jepsonii*), usually white to cream, blue, or purple, rarely pink, hooded, spatulate or obovate, clawed; nectary fleshy, free from hypanthium; stamens 6(–8); ovary ½ inferior, 3-locular; styles 3 (sometimes 4 in *C. jepsonii*), connate basally. **Fruits** capsules, leathery exocarp sloughing off prior to dehiscence. $x = 12$.

Species 58 (51 in the flora): North America, Mexico, Central America (Costa Rica, Guatemala, Panama).

Among the *Ceanothus* species found in the flora area, only three occur entirely east of the Rocky Mountains. Among the remaining species, a few of which are widespread in western North America, 42 are endemic to the California Floristic Province. Four species are entirely restricted to ultramafic (serpentine, gabbro) soils, while others occur on a diversity of substrates. Several widespread species are co-dominant shrubs in chaparral, or are important understory shrubs in woodlands and forests, especially in western North America. In addition to reproduction by seeds, many species of subg. *Ceanothus* respond to fire by developing sprouts from the root crown, whereas all species in subg. *Cerastes* reproduce strictly from seeds (F. I. Pugnaire et al. 2006). Many species form mycorrhizal associations (subterranean coralloid root clusters) with actinomycete symbionts (*Frankia*) and thus are capable of nitrogen fixation (S. L. Rose 1980; S. G. Conard et al. 1985).

Hybridization is widespread in the genus, with at least 44 interspecific combinations reported in the literature (H. McMinn 1944; D. Fross and D. H. Wilken 2006), resulting from a common diploid chromosome number of $2n = 24$ and the absence of strong isolating mechanisms. At least one putative hybrid swarm has been documented to include four species, *Ceanothus cuneatus*, *C. divergens*, *C. gloriosus*, and *C. sonomensis* (J. T. Howell 1940; M. A. Nobs 1963). The widespread occurrence of some hybrids often contributes to difficulty in identifying specimens and, in some cases, may have contributed to complex local and regional patterns of variation in flower color or leaf morphology. Most hybrids are between taxa within the same subgenus. Intersubgeneric hybrids are few and are characterized by high levels of sterility. *Ceanothus* is a popular source of horticultural cultivars, with over 200 named selections (Fross and Wilken). One of the first and most popular hybrids in the 1830s was *C.* ×*delilianus* Spach, which was developed in France from a cross between *C. americanus* (eastern North America) and *C. caeruleus* Lagasca (Mexico).

Some species and varieties of *Ceanothus* are considered difficult to identify. Some identification problems result from both local and geographical variation within species and intergradation following hybridization (M. Van Rensselaer and H. McMinn 1942; M. A. Nobs 1963). The dependence on both flower color and mature fruit morphology for accurate identification is exacerbated by a delay in fruit maturation following a short duration of flowering. Careful attention to life form, flower color, and fruit morphology is critical to identification. Knowledge of geographic distribution and edaphic substrate preference, especially in the California Floristic Province, can be helpful in determining a number of species.

In the keys and descriptions that follow, tooth number is per leaf.

SELECTED REFERENCES Fross, D. and D. H. Wilken. 2006. *Ceanothus*. Portland. Nobs, M. A. 1963. Experimental studies on species relationships in *Ceanothus*. Publ. Carnegie Inst. Wash. 623. Van Rensselaer, M. and H. McMinn. 1942. *Ceanothus*. Santa Barbara.

1. Stipules thin, deciduous; leaves alternate, blades usually herbaceous, sometimes leathery, pinnately veined or 3-veined from base; inflorescences usually racemelike to paniclelike, sometimes umbel-like; capsules not horned, sometimes crested . . . . . . 13a. *Ceanothus* subg. *Ceanothus*, p. 79
1. Stipules thick, persistent, wartlike; leaves opposite (alternate in *C. verrucosus* and *C. megacarpus*), blades leathery, pinnately veined; inflorescences usually umbel-like, rarely racemelike; capsules horned or not, usually not crested (crested in *C. divergens* and *C. gloriosus*) . . . . 13b. *Ceanothus* subg. *Cerastes*, p. 94

13a. Ceanothus Linnaeus subg. Ceanothus

Shrubs, evergreen or deciduous. **Branchlets** thorn-tipped or not. **Leaves** alternate; stipules deciduous, thin; blade usually herbaceous, sometimes leathery, margins entire or teeth gland-tipped, at least when young, stomata on abaxial surface not in crypts; pinnately veined or 3-veined from base. **Inflorescences** usually racemelike or paniclelike, sometimes umbel-like. **Capsules** not horned, crested or not; ridges between valves absent.

Species 32 (28 in the flora): North America, Mexico, Central America (Costa Rica, Guatemala, Panama).

Species of subg. *Ceanothus* not accounted for here are: *Ceanothus buxifolius* Willdenow ex Schultes f., *C. caeruleus*, *C. depressus* Bentham, and *C. ochraceus* Suessenguth; all are native to Mexico and Central America.

1. Branchlets rigid, thorn-tipped; shrubs evergreen.
    2. Leaf blades pinnately veined, adaxial surfaces green, shiny, abaxial surfaces pale green .... 10. *Ceanothus spinosus* (in part)
    2. Leaf blades 3-veined from base (lateral basal veins sometimes obscure in *C. fendleri*), adaxial surfaces pale green to dark green or grayish green, dull, abaxial surfaces pale to grayish green.
        3. Leaf blades 20–60 × 10–30 mm, abaxial surfaces pale green, appressed-puberulent, glabrescent, adaxial surfaces grayish green, glabrate .... 12. *Ceanothus incanus* (in part)
        3. Leaf blades 5–30 × 3–18 mm, abaxial surfaces pale green to grayish green, glabrous, sparsely puberulent, or appressed-villosulous to tomentulose, especially along veins, adaxial surfaces dark to pale green or grayish green, glabrous, glabrate, or appressed-villosulous.
            4. Shrubs 1.5–4 m, stems erect; inflorescences usually paniclelike, sometimes racemelike, (3–)5–15 cm .... 11. *Ceanothus leucodermis*
            4. Shrubs 0.5–1.5 m, stems erect, ascending, or spreading; inflorescences umbel-like or racemelike, 1–3.5(–4) cm.
                5. Stems ascending to spreading, not rooting at nodes; leaf blade adaxial surfaces pale to grayish green, glabrate .... 13. *Ceanothus cordulatus*
                5. Stems erect, ascending, or spreading, rooting at proximal nodes; leaf blade adaxial surfaces dark green, appressed villosulous or glabrous .... 14. *Ceanothus fendleri*
1. Branchlets flexible to rigid, not thorn-tipped (sometimes weakly thorn-tipped in *C. oliganthus*); shrubs deciduous, semideciduous, or evergreen.
    6. Leaf blades pinnately veined (proximal pair of secondary veins rarely more prominent and longer than distal pairs).
        7. Leaf blade margins usually entire, rarely denticulate distally, adaxial surfaces glabrous or weakly puberulent.
            8. Shrubs 0.3–0.8 m; inflorescences usually umbel-like, sometimes racemelike.
                9. Sepals, petals, and nectaries white; petioles 0.5–1 mm... 4. *Ceanothus microphyllus* (in part)
                9. Sepals, petals, and nectaries blue; petioles 1–3 mm .... 26. *Ceanothus foliosus* (in part)
            8. Shrubs 1–3 m; inflorescences racemelike to paniclelike.
                10. Shrubs deciduous; leaf blades flat.
                    11. Leaf blades 6–25 mm, adaxial surfaces ± shiny; capsules usually not lobed, sometimes weakly lobed .... 7. *Ceanothus parvifolius* (in part)
                    11. Leaf blades 20–80 mm, adaxial surfaces dull; capsules lobed .... 8. *Ceanothus integerrimus* (in part)
                10. Shrubs evergreen or semideciduous; leaf blades ± cupped.
                    12. Sepals and petals usually white, rarely pale blue; capsules lobed, valves viscid, crested; leaf blade adaxial surfaces dull to ± shiny ... 9. *Ceanothus palmeri*
                    12. Sepals and petals pale blue to blue; capsules not lobed, valves not conspicuously viscid, not or weakly crested; leaf blade adaxial surfaces shiny .... 10. *Ceanothus spinosus* (in part)

7. Leaf blade margins serrulate to denticulate (sometimes entire in *C. hearstiorum*), adaxial surfaces puberulent, pilosulous, villosulous, strigillose, or glandular-papillate (sometimes glabrous in *C. foliosus*).
 13. Leaf blade margins not revolute.
  14. Shrubs 0.2–0.3(–0.5) m; stems spreading; capsules 4–5 mm wide; leaf blade surfaces pilosulous . . . . 28. *Ceanothus diversifolius*
  14. Shrubs 1–3.5 m; stems erect to ascending; capsules 3–4 mm wide; leaf blade surfaces villosulous to strigillose or glabrate.
   15. Branchlets green; petioles 1–3 mm; leaf blade margins usually wavy, sometimes not . . . . 26. *Ceanothus foliosus* (in part)
   15. Branchlets pale green to grayish green and glaucous; petioles 2–8 mm; leaf blade margins not wavy . . . . 27. *Ceanothus lemmonii* (in part)
 13. Leaf blade margins revolute.
  16. Branchlets lanate to woolly, glabrescent; leaf blade abaxial surfaces cobwebby, soon glabrescent; inflorescences paniclelike, 5–15 cm. . . . . 19. *Ceanothus parryi* (in part)
  16. Branchlets puberulent to tomentulose; leaf blade abaxial surfaces variously hairy, but neither cobwebby nor glabrescent; inflorescences umbel- or racemelike, 1–5(–8) cm.
   17. Leaf blade adaxial surfaces strigillose or sparsely puberulent, not glandular papillate.
    18. Leaf blades oblong, elliptic, or suborbiculate, 6–25 × 4–15 mm, adaxial surfaces sparsely puberulent, veins furrowed . . . . . 22. *Ceanothus impressus*
    18. Leaf blades narrowly elliptic, narrowly oblong, or linear, 5–16 × 2–8 mm, adaxial surfaces strigillose, veins not furrowed . . . 23. *Ceanothus dentatus*
   17. Leaf blade adaxial surfaces glandular-papillate, sometimes also puberulent or villosulous.
    19. Shrubs 1–5 m; stems erect to ascending; leaf blades 12–50 × 6–15 mm; capsules 2–3 mm wide . . . . 24. *Ceanothus papillosus*
    19. Shrubs 0.1–0.3 m; stems prostrate to spreading, some flowering branches ascending; leaf blades 8–20 × 2–10 mm; capsules 4–5 mm wide . . . . 25. *Ceanothus hearstiorum*

[6. Shifted to left margin.—Ed.]

6. Leaf blades 3-veined from base (proximal pair of secondary veins longer than those above, sometimes equaling central vein).
 20. Leaf blade margins usually entire, rarely denticulate or serrulate distally.
  21. Petioles 0.5–1 mm; leaves often fascicled, blades 2–10 × 1–6 mm; se United States . . . . .4. *Ceanothus microphyllus* (in part)
  21. Petioles (1–)1.5–12 mm; leaves not fascicled, blades 6–80 × 3–45 mm; w United States.
   22. Branchlets light gray; capsules prominently rugose. . . . . 12. *Ceanothus incanus* (in part)
   22. Branchlets pale green or green to grayish green or brown; capsules smooth to ± rugulose.
    23. Shrubs 0.5–1 m; leaf blades widely elliptic to suborbiculate; inflorescences umbel-like to racemelike, 1.5–4 cm; sepals and petals white; nectaries yellow to yellow-green. . . . . 6. *Ceanothus martini* (in part)
    23. Shrubs 1–4 m; leaf blades elliptic, oblong-elliptic, lanceolate, or ovate; inflorescences racemelike to paniclelike, 3–25 cm; sepals and petals white or blue; nectaries usually white or blue, rarely pink.
     24. Leaf blades 6–25 mm, adaxial surfaces ± shiny; capsules usually not lobed, sometimes weakly lobed . . . . .7. *Ceanothus parvifolius* (in part)
     24. Leaf blades (10–)20–80 mm, adaxial surfaces dull; capsules lobed . . . . 8. *Ceanothus integerrimus* (in part)

[20. Shifted to left margin.—Ed.]

20. Leaf blade margins serrate, serrulate, or denticulate (rarely entire in *C. cyaneus*).
    25. Leaf blades (20–)25–100(–130) × 10–64 mm, margins serrate to serrulate, teeth (35–)40–150+; sepals and petals white to cream (sometimes pink-tinged in *C. sanguineus*, blue in *C. arboreus*).
        26. Leaf blades leathery, resinous, aromatic, adaxial surfaces shiny . . . . 5. *Ceanothus velutinus*
        26. Leaf blades herbaceous, not resinous, not aromatic, adaxial surfaces dull.
            27. Leaf blades usually elliptic to lanceolate, sometimes ovate or oblanceolate; inflorescences terminal, globose to hemispheric . . . . 2. *Ceanothus herbaceus*
            27. Leaf blades ovate, oblong-ovate, or widely elliptic; inflorescences terminal or axillary, cylindric to conic.
                28. Shrubs evergreen; branchlets brown, tomentulose; sepals, petals, and nectaries blue; California Channel Islands . . . . 15. *Ceanothus arboreus*
                28. Shrubs deciduous; branchlets green to reddish brown, puberulent, glabrescent; sepals, petals, and nectaries usually white to cream, sometimes pink-tinged; mainland United States.
                    29. Leaf blade apices usually acuminate to acute, rarely obtuse; capsules not lobed, valves ± rugulose, crested; e North America . . . . 1. *Ceanothus americanus*
                    29. Leaf blade apices acute to rounded; capsules weakly lobed near apex, valves smooth, not or sometimes weakly crested; w North America (disjunct in Michigan) . . . . 3. *Ceanothus sanguineus*
    25. Leaf blades 5–50 × 3–30 mm, margins usually denticulate, sometimes serrulate (rarely entire in *C. cyaneus*), teeth 19–71; sepals and petals blue (white in *C. martini*, rarely white in *C. oliganthus* and *C. thyrsiflorus*).
        30. Leaf blade margins revolute.
            31. Branchlets usually round, sometimes ± angled, in cross section, lanate to woolly, glabrescent; leaf blade abaxial surfaces cobwebby, soon glabrescent . . . . 19. *Ceanothus parryi* (in part)
            31. Branchlets angled in cross section, sparsely puberulent or glabrous; leaf blade abaxial surfaces puberulent to villosulous . . . . 21. *Ceanothus griseus*
        30. Leaf blade margins not revolute (sometimes incompletely revolute in *C. thyrsiflorus*).
            32. Branchlets angled in cross section.
                33. Branchlets often tuberculate (tubercles minute, brownish); inflorescences 15–30(–40) cm, paniclelike; capsules deeply lobed. . . . . .18. *Ceanothus cyaneus*
                33. Branchlets not tuberculate; inflorescences 2.5–9 cm, usually racemelike, rarely paniclelike; capsules weakly lobed . . . . 20. *Ceanothus thyrsiflorus*
            32. Branchlets round in cross section (rarely angled in cross section in *C. oliganthus*).
                34. Sepals and petals white; nectaries pale yellow to yellow-green; leaf blade margins serrulate distal to middle . . . . 6. *Ceanothus martini* (in part)
                34. Sepals and petals pale blue to blue; nectaries usually pale to dark blue or purplish blue, rarely white; leaf blade margins denticulate or serrulate most of length.
                    35. Branchlets reddish brown to brown; leaf blades ovate to widely elliptic.
                        36. Leaf blade abaxial surfaces glabrate to hirtellous; petioles 3–8 mm; capsules 4–7 mm wide . . . . .16. *Ceanothus oliganthus*
                        36. Leaf blade abaxial surfaces tomentose to tomentulose; petioles 1–4 mm; capsules 3–4 mm wide . . . . 17. *Ceanothus tomentosus*
                    35. Branchlets green, pale green, or grayish green; leaf blades narrowly elliptic, oblong-elliptic, or oblanceolate.
                        37. Branchlets green; leaf blades ± folded lengthwise, margins usually ± wavy, sometimes not; capsule valves not or weakly crested . . . . .26. *Ceanothus foliosus* (in part)
                        37. Branchlets pale green to grayish green and glaucous; leaf blades flat, margins not wavy; capsule valves crested. . . . .27. *Ceanothus lemmonii* (in part)

1. Ceanothus americanus Linnaeus, Sp. Pl. 1: 195. 1753 • New Jersey-tea, Céanothe d'Amérique [E] [F]

Ceanothus americanus var. *intermedius* (Pursh) Torrey & A. Gray; *C. americanus* var. *pitcheri* Torrey & A. Gray; *C. intermedius* Pursh

Shrubs, deciduous, 0.8–1.5 m. **Stems** erect to ascending, not rooting at nodes; branchlets usually green, sometimes reddish brown, not thorn-tipped, round or slightly angled in cross section, flexible, puberulent, glabrescent. **Leaves:** petiole 4–13 mm; blade not aromatic, flat, ovate to ovate-oblong, (20–)30–100 × 15–64 mm, herbaceous, not resinous, base rounded, margins serrate to serrulate, teeth 54–130+, apex usually acuminate to acute, rarely obtuse, abaxial surface pale green, puberulent, especially on veins, adaxial surface dark green, dull, puberulent, especially on major veins; 3-veined from base. **Inflorescences** terminal or axillary, paniclelike, cylindric to conic, 3–14 cm. **Flowers:** sepals, petals, and nectary white. **Capsules** 4–6 mm wide, not lobed; valves ± rugulose, crested. $2n = 24$.

Flowering May–Aug. Open areas in forests and woodlands, abandoned fields, sandhills, prairies; 80–300 m; Ont., Que.; Ala., Ark., Conn., Del., D.C., Fla., Ga., Ill., Ind., Iowa, Kans., Ky., La., Maine, Md., Mass., Mich., Minn., Miss., Mo., Nebr., N.H., N.J., N.Y., N.C., Ohio, Okla., Pa., R.I., S.C., Tenn., Tex., Vt., Va., W.Va., Wis.

Ceanothus americanus is closely related to *C. herbaceus* and *C. sanguineus*. *Ceanothus herbaceus* differs in having narrower leaf blades, short peduncles, and globose to hemispheric inflorescences. *Ceanothus sanguineus*, which occurs in western North America (except for a disjunct population in the Upper Peninsula of Michigan), differs by smooth fruit valves and inflorescences borne on stems older than the current year. Three varieties, vars. *americanus*, *intermedius*, and *pitcheri*, have been recognized within *C. americanus*, based on leaf shape and indumentum (M. L. Fernald 1950; H. A. Gleason and A. Cronquist 1991), but N. C. Coile (1988) provided evidence for clinal intergradation among them.

An infusion of *Ceanothus americanus* leaves or bark was used widely by Native Americans as an anti-inflammatory and to treat gastrointestinal ailments (D. E. Moerman 1998). The dried leaves were used as a tea substitute during the American Revolution (N. L. Britton and A. Brown 1896–1898, vol. 2).

2. Ceanothus herbaceus Rafinesque, Med. Repos., hexade 2, 5: 360. 1808 • Inland Jersey tea, Céanothe á feuilles étroites

Ceanothus herbaceus var. *pubescens* (S. Watson) Shinners; *C. ovatus* Desfontaines; *C. ovatus* var. *pubescens* S. Watson

Shrubs, deciduous, 0.6–1 m. **Stems** erect to ascending, not rooting at nodes; branchlets green, brown, or reddish, not thorn-tipped, round in cross section, flexible, ± appressed-puberulent or villosulous, glabrescent. **Leaves:** petiole 2–6(–10) mm; blade not aromatic, flat, usually elliptic to lanceolate, sometimes ovate or oblanceolate, (20–)25–70 × 10–30 mm, herbaceous, not resinous, base cuneate to rounded, margins serrate to serrulate, teeth (37–)45–71, apex acute to obtuse, abaxial surface pale green, glabrous or puberulent, especially on veins, adaxial surface dark green, dull, villosulous or glabrate; 3-veined from base. **Inflorescences** terminal, umbel-like, globose to hemispheric, 4–8 cm. **Flowers:** sepals, petals, and nectary white. **Capsules** 3–5 mm wide, lobed; valves smooth, usually not crested, sometimes weakly crested near apex.

Flowering Mar–Aug. Open rocky areas or on sandy soils, slopes and bluffs in shrublands, prairies, forests; 10–1800 m; Man., Ont., Que.; Ark., Colo., Ill., Ind., Iowa, Kans., Ky., La., Mass., Mich., Minn., Mo., Mont., Nebr., N.H., N.Mex., N.Y., N.Dak., Ohio, Okla., S.Dak., Tenn., Tex., Vt., Va., Wis., Wyo.; Mexico (Coahuila).

Ceanothus herbaceus is an older name than *C. ovatus*, which has been used widely in botanical and horticultural literature (G. K. Brizicky 1964c). Plants with persistently puberulent leaves occur principally east of the Mississippi River Valley and have been called var. *pubescens*, but the extent of intergradation occurs over a broad geographic area and deserves further study (N. C. Coile 1988). Putative hybrids between *C. herbaceus* and *C. fendleri* in the eastern foothills of the Rocky Mountains were named *C.* ×*subsericeus* Rydberg.

3. Ceanothus sanguineus Pursh, Fl. Amer. Sept. 1: 167. 1813 • Redstem ceanothus [E]

Ceanothus oreganus Nuttall

Shrubs, deciduous, 1–2.5 m. **Stems** erect to ascending, not rooting at nodes; branchlets greenish to reddish brown, not thorn-tipped, round in cross section, flexible to ± rigid, puberulent, glabrescent. **Leaves:** petiole 6–25 mm; blade not aromatic, flat, ovate, ovate-elliptic, or widely elliptic, 25–100 × (17–)20–60 mm, herbaceous, not resinous, base rounded or subcordate, margins serrulate, teeth 50–100+, apex acute to rounded, abaxial surface pale green, glabrous or puberulent, especially on veins, adaxial surface green, dull, glabrate; 3-veined from base. **Inflorescences** axillary, paniclelike, cylindric, 5–12 cm. **Flowers:** sepals and petals usually white to cream, sometimes pink-tinged; nectary cream. **Capsules** 4–5 mm wide, weakly lobed near apex; valves smooth, usually not crested, sometimes weakly crested.

Flowering Apr–Jul. Open areas in forests, clear-cuts, rocky hillsides, slopes, prairies, burns; 0–1400 m; B.C.; Calif., Idaho, Mich., Mont., Oreg., Wash.

Putative hybrids between *Ceanothus sanguineus* and *C. velutinus* have been reported from British Columbia and Oregon (H. McMinn 1944). The occurrence of *C. sanguineus* in the Upper Peninsula of Michigan (Keweenaw County) is a significant disjunction from the nearest locations in western Montana.

4. Ceanothus microphyllus Michaux, Fl. Bor.-Amer. 1: 154. 1803 • Sandflat ceanothus [E] [F]

Ceanothus serpyllifolius Nuttall

Shrubs, deciduous, 0.4–0.7 m. **Stems** erect to ascending, not rooting at nodes; branchlets reddish green or yellow-green, not thorn-tipped, round in cross section, flexible, usually puberulent, sometimes strigillose. **Leaves** often fascicled; petiole 0.5–1 mm; blade flat, elliptic, ovate-elliptic, or narrowly obovate, 2–10 × 1–6 mm, base cuneate, margins entire or weakly denticulate distally, not wavy, teeth 5–9, apex rounded or obtuse, abaxial surface pale green, puberulent on veins, adaxial surface green, glabrous; pinnately veined or 3-veined from base (venation obscure). **Inflorescences** axillary or terminal, umbel-like or ± racemelike, 1–3 cm. **Flowers:** sepals, petals, and nectary white. **Capsules** 3–4.5 mm wide, lobed; valves smooth, not crested.

Flowering Mar–Jun. Sandy flats, shrublands, pine-oak woodlands; 0–200 m; Ala., Fla., Ga.

Short-statured plants of *Ceanothus microphyllus* with ovate-elliptic leaves, evident venation, and racemelike inflorescences have been called *C. serpyllifolius* (M. Van Rensselaer and H. McMinn 1942; W. H. Duncan and J. T. Kartesz 1981). *Ceanothus serpyllifolius* is treated here as part of *C. microphyllus* because of continuous variation in leaf size, shape, and inflorescence architecture. *Ceanothus serpyllifolius* also has been applied to small-leaved, short-statured plants of *C. americanus*, some of which may be hybrids between that species and *C. microphyllus* (N. C. Coile 1988).

5. Ceanothus velutinus Douglas in W. J. Hooker, Fl. Bor.-Amer. 1: 125, plate 45. 1831 • Varnish-leaf ceanothus [E]

Shrubs, sometimes arborescent, evergreen, 1–6 m. **Stems** ascending to erect, not rooting at nodes; branchlets brown, not thorn-tipped, round in cross section, flexible or ± rigid, puberulent, glabrescent. **Leaves:** petiole 9–32 mm; blade aromatic, flat, widely elliptic to ovate-elliptic, (25–)40–80(–130) × (13–)20–55(–60) mm, leathery, resinous, base subcordate to rounded, margins glandular-serrulate, teeth 93–150+, apex obtuse, abaxial surface pale green, velvety puberulent, especially on veins, or glabrous, adaxial surface dark green, shiny, glabrous; 3-veined from base. **Inflorescences** axillary, paniclelike, 5–12 cm. **Flowers:** sepals and petals cream; nectary yellow-tinged. **Capsules** 3–4 mm wide, lobed at apex; valves smooth or ± rugose, sometimes viscid, weakly crested or not crested.

Varieties 2 (2 in the flora): w, c North America.

A common shrub on mountain slopes, *Ceanothus velutinus*, which reproduces by both seeds and layering, often forms large colonies, especially following fires or forest clearing. The leaves of *C. velutinus* are strongly aromatic (often vanilla-scented) when crushed, and the adaxial faces, especially in var. *laevigatus*, appear varnished.

An infusion of leaves of *Ceanothus velutinus* was used by Native Americans in cleansing and to treat skin inflammations (D. E. Moerman 1998).

1. Leaf blade abaxial surfaces velvety puberulent, especially on veins; capsule valves smooth . . . . 5a. *Ceanothus velutinus* var. *velutinus*
1. Leaf blade abaxial surfaces glabrous, veins glabrous or sparsely puberulent; capsule valves ± rugose . . . . 5b. *Ceanothus velutinus* var. *laevigatus*

5a. Ceanothus velutinus Douglas var. **velutinus** [E]

Shrubs, 1–3 m. **Leaf blades:** abaxial surface velvety puberulent, especially on veins. **Capsules:** valves smooth. $2n = 24$.

Flowering May–Aug. Open sites, rocky slopes, montane chaparral, conifer forests; 400–3400 m; Alta., B.C.; Calif., Colo., Idaho, Mont., Nev., Oreg., S.Dak., Wash., Wyo.

Putative hybrids between var. *velutinus* and *Ceanothus cordulatus*, reported from the Klamath Mountains, the southern Cascade Range, and the Sierra Nevada, have been called *C.* ×*lorenzenii* (Jepson) McMinn. Putative hybrids between *C. velutinus* var. *velutinus* and *C. prostratus* (subg. *Cerastes*) and named *C.* ×*rugosus* Greene have been reported from northeastern California (H. McMinn 1944). *Ceanothus* ×*rugosus*, with spreading stems, opposite leaves, and wartlike, semipersistent stipules, is notable in being one of few intersubgeneric hybrids in *Ceanothus*.

5b. Ceanothus velutinus Douglas var. **laevigatus** Torrey & A. Gray, Fl. N. Amer. 1: 686. 1840 [E]

Ceanothus laevigatus Hooker, Fl. Bor.-Amer. 1: 125. 1831, not (Vahl) de Candolle 1825

Shrubs, sometimes arborescent, 2–6 m. **Leaf blades:** abaxial surface glabrous, veins glabrous or sparsely puberulent. **Capsules:** valves ± rugose. $2n = 24$.

Flowering Mar–Jun. Clearings, open sites, mixed evergreen and conifer forests; 0–1400 m; B.C.; Calif., Oreg., Wash.

Variety *laevigatus* is distributed primarily in coastal mountains from California north to British Columbia and on the west slope of the Cascade Range. *Ceanothus velutinus* var. *hookeri* M. C. Johnston, based on *C. laevigatus* Hooker, is an illegitimate name. Putative hybrids between *C. velutinus* var. *laevigatus* and *C. thyrsiflorus* have been named *C.* ×*mendocinensis* McMinn.

6. Ceanothus martini M. E. Jones, Contr. W. Bot. 8: 41. 1898 • Utah mountain lilac [E]

Ceanothus utahensis Eastwood

Shrubs, deciduous, 0.5–1 m. **Stems** erect, ascending, or spreading, not rooting at nodes; branchlets green to grayish green or brown, not thorn-tipped, glaucous, round in cross section, flexible or ± rigid, strigillose. **Leaves** not fascicled; petiole (1–)3–7 mm; blade flat, widely elliptic to suborbiculate, 12–30 × 8–20 mm, base rounded or ± cuneate, margins entire or serrulate distal to middle, not revolute, teeth 23–41, apex obtuse or rounded, abaxial surface pale green, glabrous or veins puberulent, adaxial surface green, slightly shiny, glabrous, veins strigillose; 3-veined from base. **Inflorescences** axillary, umbel-like or racemelike, 1.5–4 cm. **Flowers:** sepals and petals white; nectary pale yellow to yellow-green. **Capsules** 4–5 mm wide, lobed; valves smooth or ± rugulose, crested.

Flowering May–Jul. Rocky soils, shrublands, pine-oak and pinyon pine-juniper woodlands, open sites in conifer forests; 1800–3200 m; Ariz., Colo., Nev., Utah, Wyo.

Leaves of *Ceanothus martini* are similar to those of *C.* ×*lorenzenii* and small-leaved forms of *C. velutinus*, with which it has sometimes been confused.

7. Ceanothus parvifolius Trelease, Proc. Calif. Acad. Sci., ser. 2, 1: 110. 1888 • Little-leaf ceanothus [E]

Ceanothus integerrimus Hooker & Arnott var. *parvifolius* S. Watson in W. H. Brewer et al., Bot. California 1: 102. 1876, [illegitimate name] based on *C. integerrimus* var. *parviflorus* S. Watson, Proc. Amer. Acad. Arts 10: 334. 1875

Shrubs, deciduous, 1–2.5 m. **Stems** ± erect or ascending, not rooting at nodes; branchlets green, not thorn-tipped, round in cross section, flexible, tomentulose, glabrescent. **Leaves** not fascicled; petiole 1.5–5 mm; blade flat, oblong-elliptic to elliptic, 6–25 × 3–13 mm, base cuneate, margins usually entire, sometimes denticulate distally, teeth 3–5, apex acute to obtuse, abaxial surface pale green, usually glabrous, veins sometimes strigillose, adaxial surface green, ± shiny, glabrous; pinnately veined or ± 3-veined from base. **Inflorescences** axillary, racemelike, 3–8 cm. **Flowers:** sepals and petals pale to deep blue; nectary blue. **Capsules** 4–5 mm wide, usually not lobed, sometimes weakly lobed; valves smooth, weakly viscid, usually not crested, sometimes weakly crested. $2n = 24$.

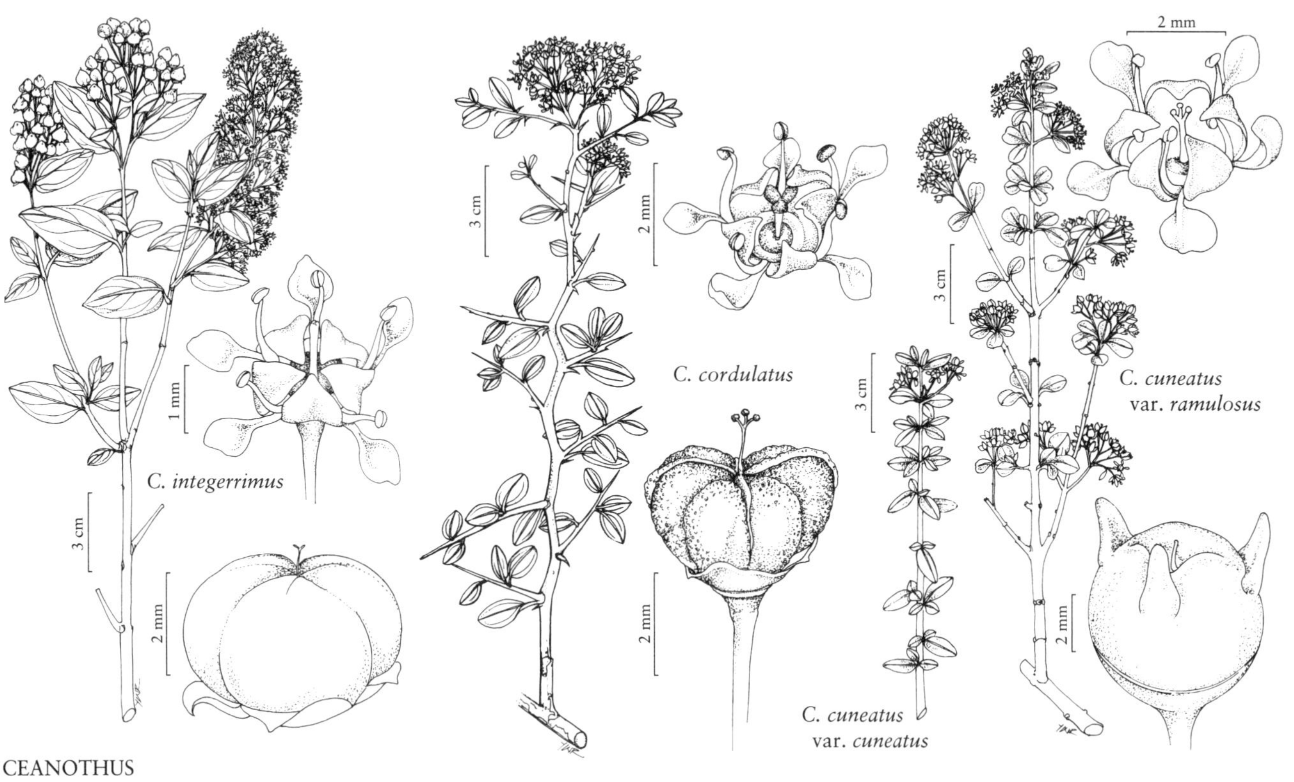

CEANOTHUS

Flowering May–Jul. Open sites and flats, conifer forests; 1300–2100 m; Calif.

Ceanothus parvifolius is restricted to the western slope of the Sierra Nevada from Plumas County south to Tulare County. Putative hybrids with *C. cordulatus* have been reported (H. McMinn 1944).

8. **Ceanothus integerrimus** Hooker & Arnott, Bot. Beechey Voy., 329. 1838 • Deer brush [E] [F]

Ceanothus andersonii Parry; *C. integerrimus* var. *californicus* (Kellogg) G. T. Benson; *C. integerrimus* var. *macrothyrsus* (Torrey) G. T. Benson; *C. integerrimus* var. *puberulus* (Greene) Abrams

Shrubs, deciduous, 1.5–3 m. **Stems** ascending to erect, not rooting at nodes; branchlets pale green, not thorn-tipped, round in cross section, flexible, glabrous or strigillose. **Leaves** not fascicled; petiole 3–12 mm; blade flat, lanceolate, elliptic, oblong-elliptic, or ovate, (10–)20–80 × 10–45 mm, base rounded, margins entire or denticulate distally, teeth 3–5, apex acute to obtuse, abaxial surface pale green, glabrate or puberulent, especially on veins, sometimes glabrescent, adaxial surface green, dull, glabrous or sparsely puberulent; usually 3-veined from base, rarely pinnately veined. **Inflorescences** usually terminal, rarely axillary, racemelike or paniclelike, 5–25 cm. **Flowers:** sepals, petals, and nectary usually white or pale to dark blue, rarely pink. **Capsules** 4–5 mm wide, lobed; valves smooth, viscid, not or weakly crested. **2*n*** = 24.

Flowering May–Jul. Rocky slopes, open sites, chaparral, conifer and mixed evergreen forests; 20–2600 m; Ariz., Calif., N.Mex., Oreg., Wash.

G. T. Benson (1930) recognized four varieties in *Ceanothus integerrimus*, based on flower color, leaf blade shape, venation, and indumentum. With the exception of regional clines in indumentum density, N. C. Coile (1988) was unable to detect consistent patterns of variation correlated with ecological or geographical distribution. The types of *C. integerrimus* (*D. Douglas* in 1831) and *C. andersonii* (*C. C. Parry* in 1888) are notable in that they bear oblong leaf blades with one principal vein from the base; the basal pair of secondary veins are shorter than those distal to it and relatively short compared to those in leaves elsewhere in the species. This venation is restricted to only a few specimens from near Ben Lomond, Santa Cruz Mountains, California, and has not been seen since the collection by Parry. W. L. Jepson (1909–1943, vol. 2) suggested that the venation pattern may be correlated with the oblong leaf shape that is shown by the type specimens.

Young shoots of *Ceanothus integerrimus* were used by Native Americans in basketry and for arrow foreshafts (D. E. Moerman 1998).

9. Ceanothus palmeri Trelease, Proc. Calif. Acad. Sci., ser. 2, 1: 109. 1888 • Palmer's ceanothus

Ceanothus spinosus Nuttall var. *palmeri* (Trelease) K. L. Brandegee

Shrubs, evergreen or semi-deciduous, 1–2 m. **Stems** ascending to erect, not rooting at nodes; branchlets green or gray-green, not thorn-tipped, round in cross section, flexible or rigid, glabrous or glabrate. **Leaves** not fascicled; petiole 3–8 mm; blade ± cupped, elliptic, oblong, or oblong-ovate, 13–30 × 9–15 mm, base cuneate to obtuse, margins entire, apex acute to rounded, abaxial surface pale green, glabrous, adaxial surface green, dull to ± shiny, glabrous; usually pinnately veined, rarely 3-veined. **Inflorescences** terminal or axillary, paniclelike, 7.5–22 cm. **Flowers:** sepals, petals, and nectary usually white, rarely pale blue. **Capsules** 4.5–7 mm wide, lobed; valves smooth, viscid, crested.

Flowering Mar–Jun. Rocky slopes, chaparral, conifer forests; 1100–1800 m; Calif.; Mexico (Baja California).

Ceanothus palmeri occurs in the Transverse and Peninsular ranges of California, with a disjunct distribution in the central Sierra Nevada foothills (Amador and El Dorado counties). *Ceanothus palmeri* appears to be closely related to and intermediate between *C. integerrimus* and *C. spinosus* (W. L. Jepson 1909–1943, vol. 2; M. Van Rensselaer and H. McMinn 1942). We treat this species as distinct from *C. spinosus* on the basis of its white flowers, and viscid, lobed fruit, but note that some populations intergrade, especially in the Transverse Ranges of southern California. Leaf blades of both *C. palmeri* and *C. spinosus* have somewhat shiny adaxial faces, but those of *C. integerrimus* appear dull. Included here are plants with weakly three-veined leaf blades with somewhat shiny adaxial surfaces from the Transverse Ranges, previously treated by Van Rensselaer and McMinn as part of *C. integerrimus*.

10. Ceanothus spinosus Nuttall in J. Torrey and A. Gray, Fl. N. Amer. 1: 267. 1838 • Green-bark ceanothus

Shrubs, sometimes arborescent, evergreen, 2–6 m. **Stems** erect, not rooting at nodes; branchlets green, thorn-tipped or not, round to ± angled in cross section, rigid, glabrous or sparsely puberulent. **Leaves** not fascicled; petiole 4–7 mm; blade ± cupped, elliptic to oblong, 11–35 × 8–29 mm, base cuneate to rounded, margins entire, apex obtuse to weakly retuse, abaxial surface pale green, not glaucous, sometimes puberulent on midribs, adaxial surface green, shiny, glabrous; pinnately veined. **Inflorescences** axillary, paniclelike, 4–17 cm. **Flowers:** sepals, petals, and nectary pale blue to blue. **Capsules** 4–6 mm wide, not lobed; valves smooth, not conspicuously viscid, not or weakly crested. $2n = 24$.

Flowering Jan–May. Rocky slopes, coastal sage scrub, chaparral; 60–900 m; Calif.; Mexico (Baja California).

Like *Ceanothus incanus*, plants of *C. spinosus* are polymorphic for the presence of thorn-tipped branchlets. Putative hybrids with *C. thyrsiflorus* and *C. oliganthus* var. *sorediatus* have been reported (M. Van Rensselaer and H. McMinn 1944).

11. Ceanothus leucodermis Greene, Bull. Misc. Inform. Kew 1895: 15. 1895 • Chaparral whitethorn

Shrubs, evergreen, 1.5–4 m. **Stems** erect, not rooting at nodes; branchlets light green or grayish green and glaucous, thorn-tipped, round in cross section, rigid, glabrous. **Leaves:** petiole 3–7 mm; blade flat, ovate to oblong-elliptic, 5–30 × 3–15 mm, base rounded to subcordate, margins usually entire to minutely glandular-serrulate, rarely serrate, glands 16–20, apex acute to obtuse, abaxial surface grayish green, glaucous, glabrous, adaxial surface green, ± glaucous, dull, usually glabrous, sometimes appressed villosulous; 3-veined from base. **Inflorescences** axillary, usually paniclelike, sometimes racemelike, (3–)5–15 cm. **Flowers:** sepals, petals, and nectary white to blue. **Capsules** 3–5 mm wide, lobed; valves smooth, viscid, not crested. $2n = 24$.

Flowering Apr–Jun. Rocky slopes, chaparral, oak woodlands, conifer forests; 200–1800 m; Calif.; Mexico (Baja California).

Ceanothus leucodermis is a common, often dominant species of chaparral in the mountains of central and southern California. This species is notable in producing serrate to serrulate leaves on stump sprouts following fire (M. Van Rensselaer and H. McMinn 1942). Putative hybrids and advanced generation intermediates with *C. oliganthus* occur throughout the southern Coast and Transverse ranges of California (McMinn 1944).

12. Ceanothus incanus Torrey & A. Gray, Fl. N. Amer. 1: 265. 1838 • Coast whitethorn [E]

Shrubs, evergreen, 1–4 m. **Stems** erect, not rooting at nodes; branchlets light gray, glaucous, thorn-tipped or not, round or slightly angled in cross section, rigid, puberulent, glabrescent. **Leaves** not fascicled; petiole 3–12 mm; blade flat, elliptic, ovate, or suborbiculate, 20–60 × 10–30 mm, base rounded or subcordate, margins entire, sometimes minutely glandular-denticulate above middle, glands 26–52, apex obtuse, abaxial surface pale green, appressed-puberulent, glabrescent, adaxial surface grayish green, dull, glabrate; 3-veined from base. **Inflorescences** axillary, usually paniclelike, sometimes racemelike, 3–6 cm. **Flowers:** sepals, petals, and nectary usually white to cream, sometimes pink. **Capsules** 4–5 mm wide, ± lobed; valves rugose, viscid when young, not or weakly crested. $2n = 24$.

Flowering Apr–Jun. Flats, slopes, chaparral, open sites in conifer and mixed evergreen forests; 60–1000 m; Calif.

Ceanothus incanus is restricted to the Klamath and Santa Cruz mountains and North Coast Ranges of California. Some populations are evidently polymorphic for the presence of thorn-tipped or non-thorn-tipped branchlets (F. K. Klein 1970). Putative hybrids with *C. papillosus* and *C. parryi* have been reported (H. McMinn 1944); hybrids with *C. thyrsiflorus* have been called *C.* ×*vanrensselaeri* Roof.

13. Ceanothus cordulatus Kellogg, Proc. Calif. Acad. Sci. 2: 124, fig. 39. 1863 • Mountain whitethorn [F]

Shrubs, evergreen, 0.5–1.5 m. **Stems** ascending to spreading, not rooting at nodes; branchlets yellowish or grayish green, glaucescent, thorn-tipped, round in cross section, rigid, puberulent, glabrescent. **Leaves:** petiole 2–8 mm; blade flat to cupped, ovate to elliptic, 10–30 × 6–18 mm, base rounded, margins usually entire, sometimes minutely glandular-denticulate distally, glands 18–30, apex obtuse, abaxial surface pale grayish green, sparsely puberulent or glabrous, sometimes villosulous along veins, adaxial surface pale green to grayish green, glaucous, dull, glabrate; 3-veined from base. **Inflorescences** axillary, umbel-like or racemelike, sometimes densely clustered, 1.2–2(–4) cm. **Flowers:** sepals, petals, and nectary usually white, rarely pink. **Capsules** 3.5–5 mm wide, lobed; valves rugose, viscid when young, weakly crested. $2n = 24$.

Flowering May–Jul. Rocky ridges and slopes, chaparral, conifer and mixed evergreen forests; 400–3400 m; Calif., Nev., Oreg.; Mexico (Baja California).

Ceanothus cordulatus is one of the most common shrubs in montane chaparral and forests of the Coast Ranges and Cascades of southern Oregon and northern California, southward through the Sierra Nevada, Transverse and Peninsular ranges of California, to the mountains of northern Baja California, and occurs disjunctly in the Charleston Mountains of Nevada.

Putative hybrids between *Ceanothus cordulatus* and *C. velutinus* var. *velutinus*, reported from the Klamath Mountains, the southern Cascade Range, and the Sierra Nevada, have been called *C.* ×*lorenzenii* (Jepson) McMinn. A rare intersectional hybrid between *C. cordulatus* and *C. prostratus* in the Lake Tahoe basin has been named *C.* ×*serrulatus* McMinn. Putative hybrids of *C. cordulatus* with *C. diversifolius* and *C. integerrimus* also have been reported (H. McMinn 1944).

14. Ceanothus fendleri A. Gray, Mem. Amer. Acad. Arts, n. s. 4: 29. 1849 • Fendler's ceanothus

Ceanothus fendleri var. *venosus* Trelease; *C. fendleri* var. *viridis* A. Gray ex Trelease

Shrubs, evergreen, 0.5–1.5 m. **Stems** erect, ascending, or spreading, rooting at proximal nodes; branchlets green to grayish green, thorn-tipped, round in cross section, rigid, canescent, often glaucous. **Leaves:** petiole 1–4 mm; blade flat, elliptic, ovate, or orbiculate, 8–25(–30) × 3–8(–14) mm, base cuneate to rounded, margins usually entire, rarely serrulate near apex, teeth 3–7, apex obtuse to rounded, abaxial surface pale green or grayish green and glaucous, appressed-villosulous to tomentulose, especially along veins, adaxial surface dark green, dull, appressed-villosulous or glabrous; 3-veined from base (lateral veins sometimes obscure). **Inflorescences** terminal or axillary, usually umbel-like, sometimes racemelike, 1–3.5 cm. **Flowers:** sepals, petals, and nectary white or pinkish. **Capsules** 4–6 mm wide, lobed; valves smooth to rugose, viscid, usually not crested, sometimes weakly crested.

Flowering Jan–Jul. Rocky soils, slopes, open sites, chaparral, oak-pine woodlands, conifer forests; 1400–2700 m; Ariz., Colo., N.Mex., S.Dak., Tex., Utah, Wyo.; Mexico (Chihuahua, Coahuila, Sonora).

Plants of *Ceanothus fendleri* with glabrous leaves found throughout its range have been called var. *viridis*. The name *C. fendleri* var. *venosus* has been applied to plants with widely elliptic, villosulous leaf blades. Such plants are similar to *C. buxifolius* of northern Mexico (Chihuahua and Sonora), which has glabrous or sparsely puberulent leaf blades and more or less persistent glands on denticulate leaf margins. Putative hybrids between *C. fendleri* and *C. herbaceus* in the eastern foothills of the Rocky Mountains were named *C.* ×*subsericeus* Rydberg.

15. Ceanothus arboreus Greene, Bull. Calif. Acad. Sci. 2: 144. 1886 • Island ceanothus

Ceanothus arboreus var. *glaber* Jepson

Shrubs, sometimes arborescent, evergreen, 2–7 m. **Stems** erect, not rooting at nodes; branchlets brown, not thorn-tipped, round or slightly angled in cross section, flexible, tomentulose. **Leaves:** petiole 8–25 mm; blade not aromatic, flat to cupped, widely ovate to elliptic, 25–80 × 20–40 mm, herbaceous, not resinous, base rounded, margins serrulate, teeth (35–)40–65, apex acute to obtuse, abaxial surface pale green, usually densely tomentulose, rarely glabrate, adaxial surface green, dull, glabrous or sparsely puberulent; 3-veined from base. **Inflorescences** axillary or terminal, paniclelike, conic, 5–12 cm. **Flowers:** sepals and petals pale blue to blue; nectary blue to dark blue. **Capsules** 6–8 mm wide, ± lobed; valves rugose, ± viscid, weakly crested. $2n = 24$.

Flowering Feb–May. Rocky slopes, ridges, chaparral, oak woodlands, closed-cone pine woodlands; 60–600 m; Calif.; Mexico (Baja California).

Ceanothus arboreus is restricted to Santa Catalina, Santa Cruz, and Santa Rosa islands, California, and Guadalupe Island, Baja California. Plants with glabrate leaves on Santa Cruz and Santa Rosa islands have been called *C. arboreus* var. *glaber*.

16. Ceanothus oliganthus Nuttall in J. Torrey and A. Gray, Fl. N. Amer. 1: 266. 1838

Shrubs, sometimes arborescent, evergreen, 2–3(–6) m. **Stems** erect, not rooting at nodes; branchlets reddish brown or brown, usually not, sometimes weakly, thorn-tipped, usually round, sometimes ± angled, in cross section, flexible to rigid, glabrous, puberulent, or villosulous. **Leaves:** petiole 3–8 mm; blade flat, ovate to widely elliptic, 8–35 × 4–25 mm, base obtuse to subcordate, margins denticulate most of length, not revolute, teeth 19–71, apex obtuse, rounded, or acute, abaxial surface usually pale green, sometimes gray-green, sometimes glaucous, glabrate to hirtellous, adaxial surface dark green, villosulous, especially on the veins, or glabrate; 3-veined from base. **Inflorescences** axillary, racemelike, 1.5–5.2 cm. **Flowers:** sepals, petals, and nectary usually pale to deep blue or purplish blue, rarely white. **Capsules** 4–7 mm wide, lobed, ± depressed at apex; valves smooth or rugose, viscid, crested or not.

Varieties 3 (3 in the flora): California, nw Mexico.

M. Van Rensselaer and H. McMinn (1942) treated var. *oliganthus* and var. *sorediatus* as species, although they and R. F. Hoover (1970) discussed intergradation between the two forms throughout part of their distribution, especially in the southern Coast Ranges and Transverse Ranges of California.

1. Ovaries and nectaries pilosulous; capsules glabrescent, valves rugose, weakly crested dorsally....... 16b. *Ceanothus oliganthus* var. *orcuttii*
1. Ovaries and nectaries glabrous; capsules glabrous, valves smooth, crested or not.
    2. Branchlets puberulent to villosulous; leaf blade adaxial surfaces usually villosulous, sometimes glabrate; capsule valves usually crested, sometimes not.................. .... 16a. *Ceanothus oliganthus* var. *oliganthus*
    2. Branchlets glabrous to sparsely puberulent; leaf blade adaxial surfaces sparsely villosulous to glabrate; capsule valves usually not, sometimes weakly, crested ....16c. *Ceanothus oliganthus* var. *sorediatus*

16a. Ceanothus oliganthus Nuttall var. **oliganthus** • Hairy ceanothus [E]

Ceanothus divaricatus Nuttall; *C. hirsutus* Nuttall

Stems: branchlets ± flexible, puberulent to villosulous. **Leaf blades** ovate, elliptic, or elliptic-oblong, adaxial surface usually villosulous, sometimes glabrate. **Flowers:** sepals, petals, and nectary blue or purplish blue; nectary and ovary glabrous. **Capsules** glabrous; valves smooth, usually crested, sometimes not.

Flowering Mar–Jun. Open slopes and ridges, chaparral, woodlands; 60–1500 m; Calif.

Variety *oliganthus* occurs primarily in the Coast Ranges south of San Francisco Bay, the Transverse Ranges, and the northern Peninsular Ranges. Putative hybrids and intergradation with var. *orcuttii* and *Ceanothus tomentosus* var. *olivaceus* have been reported from the Santa Ana Mountains, California (M. Van Rensselaer and H. McMinn 1942).

16b. Ceanothus oliganthus Nuttall var. **orcuttii** (Parry) Trelease ex Jepson, Man. Fl. Pl. Calif., 621. 1925 • Orcutt's ceanothus

Ceanothus orcuttii Parry, Proc. Davenport Acad. Nat. Sci. 5: 194. 1889

Stems: branchlets ± flexible, puberulent to villosulous. **Leaf blades** ovate, elliptic, or elliptic-oblong, adaxial surface sparsely villosulous. **Flowers:** sepals, petals, and nectary blue; nectary and ovary pilosulous. **Capsules** glabrescent; valves rugose, weakly crested.

Flowering Feb–Jun. Open slopes and ridges, chaparral, pine forests; 200–1800 m; Calif.; Mexico (Baja California).

Variety *orcuttii* occurs in the Peninsular Ranges of southern California and northern Baja California, Mexico. The pilosulous nectary and ovary are unique in *Ceanothus*.

16c. Ceanothus oliganthus Nuttall var. **sorediatus** (Hooker & Arnott) Hoover, Leafl. W. Bot. 10: 349. 1966 • Jim-brush [E]

Ceanothus sorediatus Hooker & Arnott, Bot. Beechey Voy., 328. 1838

Stems: branchlets sometimes rigid and becoming weakly thorn-tipped, glabrous to sparsely puberulent. **Leaf blades** elliptic or ovate, adaxial surface sparsely villosulous to glabrate. **Flowers:** sepals, petals, and nectary usually pale to deep blue, rarely white; nectary and ovary glabrous. **Capsules** glabrous, valves usually smooth, usually not crested, sometimes weakly crested. 2*n* = 24.

Flowering Feb–Jun. Open slopes and ridges, chaparral, oak woodlands, mixed evergreen forests; 60–1300 m; Calif.

Variety *sorediatus* occurs in the Coast Ranges of California from Humboldt County south to San Luis Obispo County, and in the Transverse and northern Peninsular ranges of southern California. Putative hybrids have been reported with *Ceanothus integerrimus* (Santa Lucia Mountains), *C. leucodermis* (San Gabriel Mountains), and *C. thyrsiflorus* (Santa Cruz Mountains).

17. Ceanothus tomentosus Parry, Proc. Davenport Acad. Nat. Sci. 5: 190. 1889

Shrubs, evergreen, 1.5–2.5 m. **Stems** erect, rarely spreading, not rooting at nodes; branchlets reddish brown, not thorn-tipped, round in cross section, flexible, tomentulose, glabrescent. **Leaves:** petiole 1–4 mm; blade flat, ovate to widely elliptic, 10–25 × 5–12 mm, base obtuse to rounded, margins serrulate to denticulate most of length, not revolute, teeth 39–71, apex obtuse, abaxial surface light or gray-green, tomentose to tomentulose, adaxial surface dark green, villosulous or puberulent, glabrescent; usually 3-veined from base, rarely pinnately veined (veins sometimes obscured by indumentum). **Inflorescences** axillary or terminal, racemelike or paniclelike, 2–6 cm. **Flowers:** sepals, petals, and nectary usually deep blue, sometimes pale blue. **Capsules** 3–4 mm wide, slightly lobed at apex; valves smooth, viscid when young, crested.

Varieties 2 (2 in the flora): California, nw Mexico.

1. Leaf blade abaxial surfaces uniformly and densely tomentulose, veins obscured; c Sierra Nevada, California . . . . 17a. *Ceanothus tomentosus* var. *tomentosus*
1. Leaf blade abaxial surfaces sparsely to moderately tomentulose, veins ± evident; s California . . . . 17b. *Ceanothus tomentosus* var. *olivaceus*

17a. Ceanothus tomentosus Parry var. **tomentosus** • Woolly-leaf ceanothus [E]

Leaf blades: margins serrulate, abaxial surface uniformly and densely tomentulose, veins obscured. 2*n* = 24.

Flowering Mar–May. Open sites, chaparral, oak-pine woodlands; 100–1600 m; Calif.

Variety *tomentosus* is restricted to the western slope of the Sierra Nevada, from Nevada County south to Mariposa County. A rare, low-growing form with spreading stems occurs on nutrient-poor, sandy soils of the Ione Formation in Amador County.

17b. Ceanothus tomentosus Parry var. **olivaceus** Jepson, Man. Fl. Pl. Calif., 621. 1925 • Ramona lilac

Leaf blades: margins denticulate, abaxial surface sparsely to moderately tomentulose, veins ± evident.

Flowering Feb–May. Open sites, chaparral, oak and pine woodlands; 10–1100 m; Calif.; Mexico (Baja California).

Variety *olivaceus* occurs in southern California from the Transverse Ranges south through the Peninsular Ranges into northern Baja California.

18. Ceanothus cyaneus Eastwood, Proc. Calif. Acad. Sci., ser. 4, 16: 361. 1927 • Lakeside ceanothus [C]

Shrubs, evergreen, 2–5 m. **Stems** erect, not rooting at nodes; branchlets light green, not thorn-tipped, angled in cross section, flexible, often tuberculate (tubercles minute, brownish), puberulent, glabrescent. **Leaves:** petiole 2–6 mm; blade flat, ovate-elliptic, 20–50 × 15–20 mm, base rounded, margins usually denticulate to serrulate, rarely entire, not revolute, teeth 23–58, apex acute to obtuse, abaxial surface pale green, veins puberulent, adaxial surface dark green, glabrous or sparsely puberulent; 3-veined from base. **Inflorescences** terminal, paniclelike, 15–30(–40) cm. **Flowers:** sepals, petals, and nectary deep blue. **Capsules** 3–5 mm wide, deeply lobed; valves smooth, weakly crested. **2*n*** = 24.

Flowering Apr–Jun. Rocky or gravelly slopes, chaparral; of conservation concern; 40–600 m; Calif.; Mexico (Baja California).

Ceanothus cyaneus is known in the flora area only from San Diego County and is threatened throughout its range.

19. Ceanothus parryi Trelease, Proc. Calif. Acad. Sci., ser. 2, 1: 109. 1888 • Parry's ceanothus [E]

Shrubs, evergreen, 2–6 m. **Stems** erect or ascending, not rooting at nodes; branchlets grayish green to brown, not thorn-tipped, usually round, sometimes ± angled, in cross section, flexible, lanate to woolly, glabrescent. **Leaves:** petiole 1–8 mm; blade flat to slightly cupped, oblong or ± elliptic, 12–50 × 6–20 mm, base obtuse to rounded, margins entire or obscurely glandular-denticulate, narrowly revolute, glands 21–36, apex obtuse, abaxial surface green, cobwebby, soon glabrescent, adaxial surface dark green, shiny, villosulous, glabrescent; usually 3-veined from base, rarely pinnately veined. **Inflorescences** axillary or terminal, paniclelike, 5–15 cm. **Flowers:** sepals, petals, and nectary deep blue. **Capsules** 2.5–4 mm wide, lobed; valves smooth, not or weakly crested. **2*n*** = 24.

Flowering Apr–May. Rocky soils, open sites, flats, mixed evergreen and redwood forests; 30–800 m; Calif., Oreg.

Ceanothus parryi occurs in the outer coast ranges of Oregon (Benton and Lane counties) and from Humboldt County south to Napa County in California; it is reported to hybridize with *C. foliosus*, *C. incanus*, and *C. thyrsiflorus* (H. McMinn 1944). The deep blue sepals and petals, cobwebby indumentum on young leaves, and narrowly revolute leaf margins are diagnostic.

20. Ceanothus thyrsiflorus Eschscholtz, Mém. Acad. Imp. Sci. St. Pétersbourg Hist. Acad. 10: 285. 1826 (as thyrsiflora) • Blueblossom

Ceanothus thyrsiflorus var. *chandleri* Jepson; *C. thyrsiflorus* var. *repens* McMinn

Shrubs, sometimes arborescent, evergreen, 0.5–6 m. **Stems** erect, usually ascending to arcuate, rarely prostrate, not rooting at nodes; branchlets green, not thorn-tipped, angled in cross section, flexible, not tuberculate, sparsely puberulent or glabrous. **Leaves:** petiole 3–10 mm; blade flat to cupped, elliptic to ovate, 10–40(–50) × 5–15(–20) mm, base obtuse to rounded, margins denticulate to serrulate, usually not revolute, sometimes incompletely revolute, teeth glandular, 23–48, apex obtuse, abaxial surface pale green, sparsely puberulent to villosulous or glabrate, veins prominently raised, puberulent to villosulous, adaxial surface dark green, glabrate; 3-veined from base. **Inflorescences** terminal, usually racemelike, rarely paniclelike, 2.5–9 cm. **Flowers:** sepals, petals, and nectary usually pale to deep blue, rarely white. **Capsules** 3–4 mm wide, weakly lobed; valves smooth, viscid, not crested. **2*n*** = 24.

Flowering Mar–Jun. Sandy or rocky flats and slopes, maritime chaparral, open sites in mixed evergreen and conifer forests; 10–600 m; Calif., Oreg.; Mexico (Baja California).

Ceanothus thyrsiflorus occurs along the coast from Coos County, Oregon, south to Santa Barbara County, California, and disjunctly near Eréndira, Baja California. A wide range of growth forms characterize this species and the closely related *C. griseus*, including plants ranging from almost prostrate to arborescent, sometimes with single trunks. Prostrate plants from several maritime bluffs along the California coast have been called *C. thyrsiflorus* var. *repens* McMinn; they retain their stature under cultivation. Named hybrids include *C. ×regius* (Jepson) McMinn (*C. thyrsiflorus* × *C. papillosus*) and *C. ×vanrensselaeri* Roof (*C. thyrsiflorus* × *C. incanus*). H. McMinn (1944) reported hybrids with *C. foliosus*.

21. Ceanothus griseus (Trelease) McMinn in M. van Rensselaer and H. McMinn, Ceanothus, 210. 1942 • Carmel ceanothus [E]

Ceanothus thyrsiflorus Eschscholtz var. *griseus* Trelease in A. Gray et al., Syn. Fl. N. Amer. 1(1,2): 415. 1897; *C. griseus* var. *horizontalis* McMinn

Shrubs, sometimes arborescent, evergreen, 0.5–4 m. **Stems** erect, ascending to arcuate, rarely prostrate, not rooting at nodes; branchlets green, not thorn-tipped, angled in cross section, flexible, sparsely puberulent or glabrous. **Leaves:** petiole 5–10 mm; blade flat to cupped, ovate to suborbiculate, 10–45 × 10–30 mm, base obtuse to rounded, margins denticulate, ± revolute, teeth 21–45, apex obtuse to rounded, abaxial surface pale green, puberulent to densely villosulous, veins prominently raised, puberulent to villosulous, adaxial surface dark green, glabrate; 3-veined from base. **Inflorescences** axillary, paniclelike, 2–7 cm. **Flowers:** sepals, petals, and nectary blue. **Capsules** 3–4 mm wide, weakly lobed at apex; valves smooth, viscid, not crested.

Flowering Mar–Jun. Sandy or rocky flats and slopes, maritime chaparral, open sites in pine and cypress forests; 10–200 m; Calif.

Ceanothus griseus is distributed along the coast from Mendocino County south to Santa Barbara County. Plants with incompletely revolute leaf margins and abaxial surfaces intermediate or similar to those of *C. thyrsiflorus* are encountered frequently. Whether this pattern is a result of primary or secondary intergradation is not known. Prostrate plants with wide elliptic leaves have been called *C. griseus* var. *horizontalis* McMinn; they retain their stature under cultivation. Putative hybrids with *C. dentatus* have been named *C. ×lobbianus* Hooker (M. Van Rensselaer and H. McMinn 1942). *Ceanothus ×veitchianus* Hooker is a rare intersubgeneric hybrid between *C. griseus* and *C. rigidus*, first collected by William Lobb near Monterey in 1853, that is cultivated in Great Britain as an ornamental.

22. Ceanothus impressus Trelease, Proc. Calif. Acad. Sci., ser. 2, 1: 112. 1888 [E]

Shrubs, evergreen, 0.5–3 m. **Stems** erect, not rooting at nodes; branchlets brown, not thorn-tipped, round in cross section, flexible, puberulent. **Leaves:** petiole 0–4 mm; blade ± flat to strongly cupped, oblong, elliptic, or suborbiculate, 5–20(–25) × 2.5–17(–20) mm, base rounded, margins serrulate, thick to weakly or strongly revolute, teeth 11–29, apex obtuse to rounded, abaxial surface pale green, villosulous, adaxial surface dark to medium green, sparsely puberulent, not glandular papillate; pinnately veined, veins furrowed. **Inflorescences** axillary or terminal, umbel-like to racemelike, 1.2–3.5 cm. **Flowers:** sepals, petals, and nectary blue. **Capsules** 3–4 mm wide, lobed; valves smooth, crested.

Varieties 2 (2 in the flora): California.

1. Leaf blades oblong to elliptic, 5–11(–14) × 2.5–8(–12) mm, strongly cupped, margins strongly revolute. . . . . . 22a. *Ceanothus impressus* var. *impressus*
1. Leaf blades widely elliptic to suborbiculate, 11–20(–25) × 7–17(–20) mm, ± flat to cupped, margins thick to weakly revolute . . . . 22b. *Ceanothus impressus* var. *nipomensis*

22a. Ceanothus impressus Trelease var. **impressus** • Santa Barbara ceanothus [E]

Shrubs, ± compact, intricately branched, 0.5–1.5 m. **Leaf blades** strongly cupped, oblong to elliptic, 5–11(–14) × 2.5–8(–12) mm, margins strongly revolute, teeth not evident adaxially; veins deeply furrowed. $2n = 24$.

Flowering Feb–Apr. Open sites, chaparral, pine forests; 50–300 m; Calif.

Variety *impressus* is restricted to coastal sites in southwestern San Luis Obispo and western Santa Barbara counties.

22b. Ceanothus impressus Trelease var. **nipomensis** McMinn in M. van Rensselaer and H. McMinn, Ceanothus, 219, figs. 12, 13. 1942 • Nipomo ceanothus [C] [E]

Shrubs, open, not intricately branched, 1.5–3 m. **Leaf blades** ± flat to cupped, widely elliptic to suborbiculate, 11–20(–25) × 7–17(–20) mm, margins thick to weakly revolute, teeth evident adaxially; veins moderately furrowed.

Flowering Feb–May. Sandy or gravelly, open sites, chaparral, oak woodlands; of conservation concern; 50–200 m; Calif.

Variety *nipomensis* occurs primarily on Nipomo Mesa and the eastern San Luis Range of southern San Luis Obispo County. R. F. Hoover (1970) regarded it as derived from hybridization between var. *impressus* and *Ceanothus oliganthus*. However, there is little evidence to indicate that populations of these two taxa overlap in geographic and ecological range. We treat var. *nipomensis* as part of *C. impressus*, based on its furrowed veins and fruit morphology. Urbanization has reduced the number of known populations to a relatively few, scattered localities.

Variety *nipomensis* is in the Center for Plant Conservation's National Collection of Endangered Plants.

23. Ceanothus dentatus Torrey & A. Gray, Fl. N. Amer. 1: 268. 1838 • Cropleaf ceanothus [E]

Ceanothus dentatus subsp. *floribundus* (Hooker) Trelease; *C. floribundus* Hooker

Shrubs, evergreen, 0.5–1.5 m. **Stems** erect, ascending or spreading, not rooting at nodes; branchlets brown to reddish brown, not thorn-tipped, round in cross section, rigid, puberulent. **Leaves:** petiole 1–2 mm; blade flat to cupped, narrowly elliptic to narrowly oblong or linear, 5–16 × 2–8 mm, base obtuse, margins obscurely glandular-denticulate, strongly revolute, glands 14–36, apex truncate to retuse, abaxial surface pale green, villosulous to strigillose, especially on veins, adaxial surface dark green, strigillose, not glandular papillate; pinnately veined, veins not furrowed. **Inflorescences** axillary or terminal, racemelike, 1.5–3 cm. **Flowers:** sepals, petals, and nectary deep blue. **Capsules** 2.5–4 mm wide, not lobed to weakly lobed; valves smooth, crested or not.

Flowering Mar–Jun. Sandy soils, slopes, flats, maritime chaparral, cypress and pine forests; 0–50 m; Calif.

Ceanothus dentatus is a narrow endemic, restricted to the Monterey Bay region. Some specimens of this species have been mistaken for *C. papillosus*. Papillose glands are restricted to leaf blade margins and are absent from adaxial leaf surfaces in *C. dentatus*, but are evenly distributed over the adaxial leaf surfaces in *C. papillosus*. Putative hybrids between *C. dentatus* and *C. griseus* have been named *C.* ×*lobbianus* Hooker (M. Van Rensselaer and H. McMinn 1942).

24. Ceanothus papillosus Torrey & A. Gray, Fl. N. Amer. 1: 268. 1838 • Wartleaf ceanothus

Ceanothus papillosus subsp. *roweanus* (McMinn) Munz; *C. papillosus* var. *roweanus* McMinn

Shrubs, evergreen, 1–5 m. **Stems** erect to ascending, not rooting at nodes; branchlets green to reddish brown, not thorn-tipped, round in cross section, ± flexible to rigid, densely tomentulose. **Leaves:** petiole 1–3 mm; blade cupped to flat, linear, narrowly oblong, or oblong-elliptic, 12–50 × 6–15 mm, base obtuse to rounded, margins minutely glandular-denticulate, revolute, glands 17–31, apex obtuse, truncate, or retuse, abaxial surface pale green, densely villosulous to tomentulose, adaxial surface dark green, sparsely puberulent and glandular-papillate; pinnately veined. **Inflorescences** axillary or terminal, racemelike, 2–8 cm. **Flowers:** sepals, petals, and nectary deep blue. **Capsules** 2–3 mm wide, lobed; valves smooth, viscid when young, not or weakly crested. 2*n* = 24.

Flowering Mar–May. Rocky ridges, slopes, and flats, chaparral, mixed evergreen forests; 20–1500 m; Calif.; Mexico (Baja California).

Ceanothus papillosus occurs in the Coast Ranges from San Francisco Bay south to the Santa Ynez Mountains, Ventura County, with disjunct populations in the Santa Ana Mountains, California, and Cerro Bola, in northern Baja California. The name *C. papillosus* var. *roweanus* was originally applied to low-growing plants with ascending to spreading, arcuate branches (H. McMinn 1939). M. Van Rensselaer and McMinn (1942) later emended the circumscription to include plants with linear leaves and retuse to truncate leaf apices, but these are found throughout the range of the species. Leaves with obtuse to truncate or retuse leaf apices also can be found on the same plant. Putative hybrids with *C. integerrimus* and *C. oliganthus* have been documented (McMinn 1944). Hybrids with *C. thyrsiflorus* have been named *C.* ×*regius* (Jepson) McMinn. Some putatively advanced generation hybrids have narrowly elliptic, weakly papillate leaves with obtuse apices, and sometimes have been misinterpreted as belonging to *C. papillosus*.

25. Ceanothus hearstiorum Hoover & Roof, Four Seasons 2(1): 4. 1966 • Hearst ceanothus C E

Shrubs, evergreen, 0.1–0.3 m, matlike or moundlike. **Stems** spreading or prostrate, not rooting at nodes, some flowering branches ascending; branchlets green to reddish brown, not thorn-tipped, round or slightly angled in cross section, flexible, densely puberulent. **Leaves:** petiole 1–2 mm; blade flat to cupped, linear, oblong, or oblong-obovate, 8–20 × 2–10 mm, base cuneate to obtuse, margins entire or obscurely glandular-denticulate, weakly revolute, glands 23–31, apex truncate or retuse, abaxial surface green, densely tomentulose, adaxial surface dark green, glandular-papillate and sometimes villosulous; pinnately veined, veins ± furrowed. **Inflorescences** terminal or axillary, umbel-like or racemelike, 1–5 cm. **Flowers:** sepals, petals, and nectary deep blue. **Capsules** 4–5 mm wide, not lobed to weakly lobed; valves smooth, not crested.

Flowering Mar–Apr. Consolidated alluvial or serpentine soils, maritime chaparral, coastal prairies; of conservation concern; 20–200 m; Calif.

Ceanothus hearstiorum occurs in a small area of coastal bluffs in northern San Luis Obispo County, growing in close proximity to another local endemic, *C. maritimus* (subg. *Cerastes*).

26. Ceanothus foliosus Parry, Proc. Davenport Acad. Nat. Sci. 5: 172. 1889 E

Shrubs, evergreen, 0.3–3.5 m. **Stems** erect, ascending, arcuate, or prostrate, not rooting at nodes; branchlets green, not thorn-tipped, round or slightly angled in cross section, flexible, villosulous. **Leaves:** petiole 1–3 mm; blade flat or ± folded lengthwise, narrowly elliptic to oblong-elliptic or oblanceolate, 5–24 × 3–13 mm, base obtuse to rounded, margins entire or weakly glandular-denticulate most of length, ± thick, not revolute, usually wavy, sometimes not wavy, teeth 31–42, apex obtuse, abaxial surface pale green to grayish green, villosulous or glabrous and sometimes sparsely puberulent on veins, adaxial surface dark green, glabrous or sparsely puberulent; pinnately veined or faintly 3-veined from base. **Inflorescences** axillary, umbel-like to racemelike, 1–3.5(–7) cm. **Flowers:** sepals, petals, and nectary pale to dark blue. **Capsules** 3–4 mm wide, weakly lobed at apex; valves smooth, not or weakly crested.

Varieties 3 (3 in the flora): California.

1. Leaf blade abaxial surfaces moderately to densely villosulous . . . . 26b. *Ceanothus foliosus* var. *medius*
1. Leaf blade abaxial surfaces glabrous or sometimes sparsely puberulent on veins.
   2. Leaf blades ± folded lengthwise, margins wavy, glandular-denticulate . . . . 26a. *Ceanothus foliosus* var. *foliosus*
   2. Leaf blades flat to ± folded lengthwise, margins not to weakly wavy, entire or weakly denticulate near apex . . . . 26c. *Ceanothus foliosus* var. *vineatus*

26a. Ceanothus foliosus Parry var. **foliosus** • Wavyleaf ceanothus E

Ceanothus dentatus Torrey & A. Gray var. *dickeyi* Fosberg

Shrubs, 1.5–3.5 m. **Stems** ascending to erect. **Leaf blades** ± folded lengthwise, margins glandular-denticulate, wavy, abaxial surface pale green to grayish green, glabrous, sometimes sparsely puberulent on veins. $2n = 24$.

Flowering Jan–May. Rocky soils, chaparral, mixed evergreen and conifer forests; 100–1100 m; Calif.

Variety *foliosus* occurs in the outer North Coast Ranges and Santa Cruz Mountains, from Humboldt to Santa Clara counties. H. McMinn (1944) reported putative hybrids with *Ceanothus parryi* and *C. thyrsiflorus*.

26b. Ceanothus foliosus Parry var. **medius** McMinn in M. van Rensselaer and H. McMinn, Ceanothus, 222. 1942 • La Cuesta ceanothus E

Ceanothus austromontanus Abrams

Shrubs, 1–2 m. **Stems** ascending to erect. **Leaf blades** ± folded lengthwise, margins glandular-denticulate, wavy, abaxial surface pale green, moderately to densely villosulous.

Flowering Feb–May. Rocky soils, chaparral, mixed evergreen and conifer forests; 100–1500 m; Calif.

Variety *medius* appears restricted to the Santa Lucia Mountains and La Panza Range of Monterey and San Luis Obispo counties, with disjunct populations in the Cuyamaca Mountains of San Diego County.

26c. Ceanothus foliosus Parry var. **vineatus** McMinn in M. van Rensselaer and H. McMinn, Ceanothus, 221. 1942 • Vine Hill ceanothus [C] [E]

Shrubs, 0.3–0.8 m. **Stems** ascending, arcuate, or prostrate. **Leaf blades** flat to ± folded lengthwise, margins entire or weakly denticulate near apex, not to weakly wavy, abaxial surface pale green, glabrous, sometimes sparsely puberulent on veins.

Flowering Mar–May. Sandy soils, chaparral; of conservation concern; 50–100 m; Calif.

Variety *vineatus* is known only from two localities in Sonoma County.

27. Ceanothus lemmonii Parry, Proc. Davenport Acad. Nat. Sci. 5: 192. 1889 (as lemmoni) • Lemmon's ceanothus [E]

Shrubs, evergreen, 0.5–1 m. **Stems** ascending to spreading, not rooting at nodes; branchlets pale green to grayish green and glaucous, not thorn-tipped, round in cross section, flexible to ± rigid, sparsely villosulous. **Leaves:** petiole 2–6 mm; blade flat, narrowly elliptic to oblong-elliptic, 13–35 × 6–15 mm, base cuneate to rounded, margins serrulate to denticulate most of length, not revolute, not wavy, teeth 34–45, apex acute to obtuse, abaxial surface pale green to grayish green and glaucous, villosulous, especially on veins, adaxial surface green, strigillose; pinnately veined or weakly 3-veined from base. **Inflorescences** axillary or terminal, umbel-like to racemelike, 2–6.5 cm. **Flowers:** sepals, petals, and nectary pale to deep blue. **Capsules** 3–4 mm wide, lobed near apex; valves smooth, crested. $2n = 24$.

Flowering Apr–May. Rocky slopes and flats, open sites, conifer forests, oak and pine woodlands; 200–1300 m; Calif.

Ceanothus lemmonii occurs in the inner North Coast Ranges, Klamath Mountains, and the western slope of the Cascade Range and northern Sierra Nevada. H. McMinn (1944) reported putative hybrids with *C. foliosus*, *C. integerrimus*, and *C. oliganthus* var. *sorediatus*.

28. Ceanothus diversifolius Kellogg, Proc. Calif. Acad. Sci. 1: 58. 1855 • Pine mat [E]

Ceanothus decumbens S. Watson

Shrubs, evergreen, 0.2–0.3 (–0.5) m, matlike. **Stems** spreading, sometimes rooting at nodes; branchlets green, sometimes tinged red, not thorn-tipped, usually round, sometimes angled, in cross section, flexible, puberulent. **Leaves:** petiole 3–11 mm; blade flat, elliptic to widely ovate, 12–45 × 6–20 mm, base obtuse to rounded, margins serrulate to denticulate, not revolute, usually not wavy, sometimes wavy, teeth 27–42, apex ± obtuse to rounded, abaxial surface pale green, pilosulous, adaxial surface green, pilosulous; usually pinnately veined, rarely 3-veined from base. **Inflorescences** axillary, umbel-like to racemelike, 1.3–4 cm. **Flowers:** sepals, petals, and nectary usually blue to pale blue, rarely white. **Capsules** 4–5 mm wide, weakly lobed near apex; valves smooth, crested.

Flowering Apr–Jun. Well-drained slopes and canyons, open to shaded sites, mixed evergreen and conifer forests; 700–2300 m; Calif.

Ceanothus diversifolius occurs in the North Coast Ranges and the western slopes of the Cascade Range and the Sierra Nevada; it often forms mats to 2 m wide. Marginal teeth on young leaves are notable in having more or less persistent, narrowly conic glands, not seen elsewhere in *Ceanothus*.

13b. Ceanothus Linnaeus subg. Cerastes (S. Watson) Weberbauer in H. G. A. Engler and K. Prantl, Nat. Pflanzenfam. 128[III,5]: 414. 1896

Ceanothus sect. *Cerastes* S. Watson, Proc. Amer. Acad. Arts 10: 338. 1875

Shrubs, evergreen. **Branchlets** not thorn-tipped. **Leaves** opposite (alternate in *C. megacarpus* and *C. verrucosus*); stipules persistent, thick, wartlike; blade leathery, margins entire or teeth not gland-tipped, often spinulose, stomata on abaxial surface in crypts (crypts appearing as areolae aligned in rows between secondary veins); pinnately veined. **Inflorescences** umbel-like (rarely racemelike in *C. pauciflorus*). **Capsules** usually horned (horns sometimes minute or weakly developed bulges), sometimes not horned, usually not crested (crested in *C. divergens* and *C. gloriosus*); ridges between valves present or absent.

Species 25 (23 in the flora): w, sc United States, Mexico.

Species of subg. *Cerastes* not accounted for here are: *Ceanothus australis* Rose and *C. bolensis* S. Boyd & J. Keeley, both endemic to Mexico.

In the following key, references to indumentum do not include the hairs associated with the stomatal crypts, which in all species of subg. *Cerastes* appear microscopically tomentulose or densely puberulent; these hairs are much shorter than hairs borne on the abaxial surfaces and veins between the crypts.

1. Leaves alternate.
    2. Leaf blades widely obovate to suborbiculate, 5–14 mm, apices truncate to retuse; capsules 4–6 mm wide, valves smooth, horns minute or absent. . . . . 29. *Ceanothus verrucosus*
    2. Leaf blades elliptic to obovate, 10–25(–33) mm, apices obtuse; capsules 7–12 mm wide, valves rugulose to weakly ridged near apex, horns prominent. . . . . . . 30. *Ceanothus megacarpus* (in part)
1. Leaves opposite.
    3. Shrubs matlike to moundlike, 0.1–1 m; stems spreading to weakly ascending (sometimes erect to ascending in *C. maritimus* and *C. sonomensis*), sometimes rooting at nodes.
        4. Shrubs matlike, 0.1–0.3 m; stems prostrate or spreading.
            5. Leaf blade margins entire or denticulate near apex, teeth 0–3.
                6. Leaf blade adaxial surfaces dark green, shiny, apices rounded to retuse; capsule horns prominent . . . . 36. *Ceanothus fresnensis*
                6. Leaf blade adaxial surfaces green to grayish green, dull, apices usually truncate, sometimes obtuse; capsule horns minute or weakly developed bulges . . . . 44. *Ceanothus pumilus* (in part)
            5. Leaf blade margins sharply dentate to spinose-dentate, teeth 3–9.
                7. Leaves not crowded (shorter than internodes); leaf blade adaxial surfaces green, dull; capsules 4–6 mm wide. 42. *Ceanothus confusus* (in part)
                7. Leaves often crowded (usually longer than internodes); leaf blade adaxial surfaces dark green, shiny; capsules 6–9 mm wide. . . . . . . 43. *Ceanothus prostratus* (in part)
        4. Shrubs moundlike, 0.1–1 m; stems ± prostrate, spreading, or weakly ascending.
            8. Leaf blade margins spinose-dentate.
                9. Leaf blades widely obovate to suborbiculate, 5–12 mm, apices widely notched, marginal teeth 2–4. . . . . 40. *Ceanothus sonomensis*
                9. Leaf blades elliptic, ± oblong, or obovate, 10–20 mm, apices acute or retuse, with an apical tooth, marginal teeth 3–9(–11).
                    10. Shrubs 0.5–1.5 m; stems ascending to erect. . . . . 41. *Ceanothus divergens* (in part)
                    10. Shrubs 0.1–0.6 m; stems spreading to weakly ascending . . . . 42. *Ceanothus confusus* (in part)
            8. Leaf blade margins entire, denticulate, or serrulate.
                11. Leaf blade abaxial surfaces tomentulose . . . . 33. *Ceanothus maritimus*
                11. Leaf blade abaxial surfaces strigillose, sometimes only on or between veins, puberulent, or glabrate.
                    12. Leaf blade margins dentate to denticulate or serrulate most of length, teeth 5–31.
                        13. Leaves crowded, blades elliptic to obovate, marginal teeth 5–9 . . . . 43. *Ceanothus prostratus* (in part)
                        13. Leaves not crowded, blades widely elliptic, obovate, or suborbiculate, marginal teeth 9–31.
                            14. Leaf blades 23–31(–45) mm, marginal teeth 13–31; capsule horns minute, intermediate ridges absent; coastal habitats, 30–200 m. . . . . 38. *Ceanothus gloriosus* (in part)
                            14. Leaf blades 10–20 mm, marginal teeth 9–15; capsule horns prominent, intermediate ridges present; montane habitats, 1600–2600 m. . . . . 45. *Ceanothus pinetorum* (in part)
                    12. Leaf blade margins entire, denticulate near apex, or remotely denticulate, teeth 1–5(–7).

15. Leaf blades narrowly oblanceolate to narrowly oblong-lanceolate, margins thick to ± revolute, apex usually truncate, sometimes obtuse . . . . 44. *Ceanothus pumilus* (in part)
15. Leaf blades elliptic to oblanceolate, margins sometimes thick but not revolute, apex acute, obtuse, retuse, or rounded, or ± truncate.
    16. Leaf blades folded lengthwise; capsule horns absent or weakly developed bulges; axillary short shoots, if present, erect. . . . . 37. *Ceanothus roderickii*
    16. Leaf blades flat; capsule horns prominent to minute or absent; axillary short shoots, if present, ascending to spreading.
        17. Leaf blades apices rounded to obtuse, adaxial surfaces pale green; capsule horns subapical, erect . . . . 35. *Ceanothus arcuatus*
        17. Leaf blades apices acute to ± truncate, adaxial surfaces grayish green; capsule horns minute or absent, or lateral, spreading . . . . 48. *Ceanothus pauciflorus* (in part)

[3. Shifted to left margin.—Ed.]

3. Shrubs not matlike or moundlike, 0.5–6 m; stems usually erect to ascending, sometimes spreading, not rooting at nodes.
    18. Leaf blade margins spinose-dentate.
        19. Stems ± flexible; leaves spreading, abaxial leaf surfaces grayish green. . . . . 41. *Ceanothus divergens* (in part)
        19. Stems rigid; leaves spreading to deflexed, abaxial leaf surfaces pale green or pale yellowish green.
            20. Sepals, petals, and stamens 5; leaf blade adaxial surfaces green to dark green; capsule valves smooth, horns slender, intermediate ridges absent; on volcanic soils . . . . 46. *Ceanothus purpureus*
            20. Sepals, petals, and stamens (5–)6(–8); leaf blade adaxial surfaces pale green; capsule valves rugose, horns thick, intermediate ridges present; on serpentine soils . . . . 47. *Ceanothus jepsonii*
    18. Leaf blade margins entire, dentate, denticulate, or serrulate.
        21. Leaf blade margins entire or remotely denticulate, teeth if present minute, 1–7 (8–19 in *C. crassifolius*).
            22. Leaves both fascicled and not fascicled on same plant.
                23. Petioles 1–3 mm; leaf blades elliptic, oblanceolate, obovate, or orbiculate, 6–22(–30) × 3–12(–22) mm; capsules 4–6 mm wide, horns prominent . . . . 32. *Ceanothus cuneatus* (in part)
                23. Petioles 0–1 mm; leaf blades narrowly oblanceolate to narrowly obovate, 3–7 × 1–3 mm; capsules 3–4 mm wide, horns minute or absent. . . 51. *Ceanothus ophiochilus*
            22. Leaves not fascicled.
                24. Branchlets light gray to ashy gray, puberulent to tomentulose, hairs curly or wavy, glabrescent; leaf blade adaxial surfaces grayish green, puberulent, hairs curly or wavy, glabrescent . . . . 48. *Ceanothus pauciflorus* (in part)
                24. Branchlets grayish brown to brown, puberulent or tomentulose, hairs straight, sometimes glabrate; leaf blade adaxial surfaces green, glabrous or sparsely tomentulose, hairs straight, glabrescent.
                    25. Leaf blades widely oblanceolate to widely obovate, 5–11 × 4–7 mm; sepals, petals, and nectaries usually lavender, sometimes pale blue; capsules 5–6 mm wide. . . . . 32. *Ceanothus cuneatus* (in part)
                    25. Leaf blades elliptic to obovate, 10–40 × 5–15(–20) mm; sepals and petals white; nectaries blue or black; capsules 6–12 mm wide.
                        26. Leaf blade abaxial surfaces glabrous or strigillose on veins; capsules 7–12 mm wide, horns weakly developed or absent . . . . 30. *Ceanothus megacarpus* (in part)
                        26. Leaf blade abaxial surfaces tomentulose; capsules 6–9 mm wide, horns prominent . . . . 31. *Ceanothus crassifolius* (in part)

[21. Shifted to left margin.—Ed.]

21. Leaf blade margins dentate or denticulate, at least distal to middle, teeth 5–35 (3–5 in *C. otayensis*).
    27. Sepals and petals white to cream; nectaries tan to brown, yellow to green, blue, purple, or black.
        28. Leaf blade margins revolute or thick, abaxial surfaces tomentulose.
            29. Leaf blade margins with 8–19 teeth most of length, apices obtuse to rounded; capsules 5–9 mm wide, horns prominent . . . . 31. *Ceanothus crassifolius* (in part)
            29. Leaf blade margins with 3–5 teeth near apex, apices truncate, retuse, or cuspidate; capsules 4–6 mm wide, horns minute or absent .50. *Ceanothus otayensis*
        28. Leaf blade margins thick or not, not revolute, abaxial surfaces sparsely puberulent (hairs curly) or strigillose, sometimes glabrescent.
            30. Leaf blade adaxial surfaces dark green; nectaries dark blue to purple; capsules 7–9 mm wide, horns subapical, prominent; serpentine substrates. . . . . . . .34. *Ceanothus ferrisiae*
            30. Leaf blade adaxial surfaces green to yellowish green; nectaries yellow to green; capsules 4–6 mm wide, horns lateral, usually minute, sometimes absent; granitic or metamorphic substrates . . . . 49. *Ceanothus perplexans*
    27. Sepals, petals, and nectaries pale blue, blue, or purple.
        31. Leaf blades 4–10 × 4–6 mm, margins denticulate distal to middle, teeth 5–9 . . . . 32. *Ceanothus cuneatus* (in part)
        31. Leaf blades 7–40 × 4–22 mm, denticulate most of length, teeth 9–35.
            32. Sepals, petals, and nectaries pale blue to blue; capsule horns prominent, rugose, intermediate ridges present; montane habitats, 1600–2600 m . . . . .45. *Ceanothus pinetorum* (in part)
            32. Sepals, petals, and nectaries deep blue to purple; capsule horns minute, not rugose, intermediate ridges absent; coastal habitats, 30–500 m.
                33. Leaf blades widely elliptic to suborbiculate, 3–40 × 17–22 mm, marginal teeth 13–35 . . . . .38. *Ceanothus gloriosus* (in part)
                33. Leaf blades usually elliptic or oval, sometimes suborbiculate, 7–21 × 4–13 mm, marginal teeth 9–17 . . . . .39. *Ceanothus masonii*

29. Ceanothus verrucosus Nuttall in J. Torrey and A. Gray, Fl. N. Amer. 1: 267. 1838 • Barranca brush

Shrubs, 1–3 m. **Stems** erect to ascending, not rooting at nodes; branchlets grayish brown, rigid, tomentulose. **Leaves** alternate, not fascicled, often crowded; petiole 1–3 mm; blade flat to cupped, widely obovate to suborbiculate, 5–14 × 3–10 mm, base cuneate to rounded, margins not revolute, entire to weakly denticulate distal to middle, teeth 9–16, apex truncate to retuse, abaxial surface pale green, sparsely strigillose, glabrescent, adaxial surface dark green, glabrous. **Inflorescences** axillary, 1–1.5 cm. **Flowers:** sepals and petals white; nectary black. **Capsules** 4–6 mm wide, weakly lobed; valves smooth, horns minute or absent, intermediate ridges absent. $2n = 24$.

Flowering Jan–Apr. Slopes, coastal mesas, chaparral, pine woodlands; 20–800 m; Calif.; Mexico (Baja California).

Ceanothus verrucosus occurs from San Diego County to the foothills of the Sierra San Pedro Mártir and Cedros Island in northern Baja California, Mexico; its habitat is threatened by development and urbanization throughout its range.

30. Ceanothus megacarpus Nuttall, N. Amer. Sylv. 2: 46. 1846 [E]

Ceanothus macrocarpus Nuttall in J. Torrey and A. Gray, Fl. N. Amer. 1: 267. 1838, not Cavanilles 1795

Shrubs, 1.5–6 m. **Stems** erect to ascending, not rooting at nodes; branchlets grayish brown to brown, flexible, appressed puberulent to tomentulose, hairs straight. **Leaves** sometimes alternate, not fascicled, sometimes crowded; petiole 1–5 mm; blade flat to cupped, oval, elliptic, or oblanceolate, 10–25(–33) × 5–19 mm, base cuneate to rounded, margins thick, usually entire, rarely remotely denticulate, teeth 5–7, apex obtuse, abaxial surface pale green, glabrous or sparsely strigillose on veins, adaxial surface green, glabrous. **Inflorescences** axillary, 1–2 cm.

Flowers: sepals and petals white; nectary blue to black. **Capsules** 7–12 mm wide, not lobed; valves smooth or rugulose to weakly ridged near apex; horns subapical, prominent, erect, often rugulose, or weakly developed to absent, intermediate ridges absent. $2n = 24$.

Varieties 2 (2 in the flora): California.

Some populations of *Ceanothus megacarpus* are polymorphic for leaf arrangement. Plants assignable to either var. *insularis* or var. *megacarpus* based on leaf arrangement may have fruits intermediate to both varieties. Putative hybrids between var. *megacarpus* and *C. cuneatus* were reported by H. McMinn (1944), but their variable leaf arrangement (alternate and opposite on the same plant) may also represent intermediates between var. *insularis* and var. *megacarpus*.

1. Leaves mostly alternate; capsule valves rugulose to weakly ridged near apex; horns prominent . . . . 30a. *Ceanothus megacarpus* var. *megacarpus*
1. Leaves mostly opposite; capsule valves smooth; horns weakly developed or absent. . . . . .30b. *Ceanothus megacarpus* var. *insularis*

30a. Ceanothus megacarpus Nuttall var. megacarpus [E]

Leaves mostly alternate; petiole 1–3 mm; blade oval, elliptic, or oblanceolate, 10–25(–33) × 5–19 mm. **Capsules:** valves rugulose to weakly ridged near apex; horns prominent.

Flowering Nov–Apr. Rocky slopes, canyons, chaparral; 10–900 m; Calif.

Variety *megacarpus* occurs on coastal slopes of mountains from Santa Barbara County south to San Diego County and at isolated localities on several California Channel Islands.

30b. Ceanothus megacarpus Nuttall var. insularis (Eastwood) Munz, Bull. S. Calif. Acad. Sci. 31: 68. 1932 • Island big-pod ceanothus [E]

Ceanothus insularis Eastwood, Proc. Calif. Acad. Sci., ser. 4, 16: 362. 1927; *C. megacarpus* subsp. *insularis* (Eastwood) P. H. Raven

Leaves mostly opposite; petiole 2–5 mm; blade oval to elliptic, 15–30 × 10–18 mm. **Capsules:** valves smooth; horns weakly developed or absent.

Flowering Feb–Apr. Rocky slopes, canyons, chaparral; 10–500 m; Calif.

Variety *insularis* occurs on the California Channel islands and at several isolated localities in the Santa Monica and Santa Ynez mountains.

31. Ceanothus crassifolius Torrey in War Department [U.S.], Pacif. Railr. Rep. 4(5): 75. 1857 • Hoary-leaf ceanothus

Shrubs, 1.5–4 m. **Stems** erect, not rooting at nodes; branchlets grayish brown to brown, ± flexible to rigid, tomentulose, hairs straight. **Leaves** not fascicled; petiole 2–6 mm; blade flat to ± cupped, elliptic to widely elliptic, 12–25(–30) × 8–15(–23) mm, base obtuse to rounded, margins thick or revolute, entire or denticulate most of length, teeth 8–19, apex obtuse to rounded, abaxial surface pale green to white, tomentulose to glabrate, adaxial surface green, sparsely tomentulose, hairs straight, glabrescent. **Inflorescences** axillary or terminal, 1–2 cm. **Flowers:** sepals and petals white; nectary blue to black. **Capsules** 5–9 mm wide, sometimes weakly lobed at apex; valves viscid, smooth, horns lateral, prominent, erect, intermediate ridges absent. $2n = 24$.

Varieties 2 (2 in the flora): California, nw Mexico.

1. Leaf blade abaxial surfaces densely tomentulose, veins obscured, margins denticulate and revolute . . . . 31a. *Ceanothus crassifolius* var. *crassifolius*
1. Leaf blade abaxial surfaces sparsely tomentulose to glabrate, veins evident, margins entire or weakly denticulate, thick to weakly revolute. . . . . .31b. *Ceanothus crassifolius* var. *planus*

31a. Ceanothus crassifolius Torrey var. crassifolius

Leaf blades 13–25(–30) × 9–15 mm, margins denticulate, revolute, abaxial surface densely tomentulose, veins obscured. **Capsules** 5–7 mm wide.

Flowering Jan–Apr. Ridges, slopes, chaparral; 200–1300 m; Calif.; Mexico (Baja California).

Variety *crassifolius* occurs on coastal slopes of the Transverse Ranges in Los Angeles and San Bernardino counties, the outer Peninsular Ranges, and northern Baja California. Some specimens in the Transverse Ranges suggest intergradation with var. *planus*. Putative hybrids between var. *crassifolius* and *Ceanothus ophiochilus* have been reported (S. Boyd et al. 1991).

31b. Ceanothus crassifolius Torrey var. **planus** Abrams, Bull. New York Bot. Gard. 6: 415. 1910 E

Leaf blades 12–20(–27) × 8–15(–23) mm, margins entire or weakly denticulate, thick to weakly revolute, abaxial surface sparsely tomentulose to glabrate, veins evident. **Capsules** 6–9 mm wide.

Flowering Jan–Apr. Ridges, slopes, chaparral; 200–1200 m; Calif.

Variety *planus* is restricted to coastal slopes of San Luis Obispo County and the western Transverse Ranges in Santa Barbara, Ventura, and western Los Angeles counties. Reports of var. *planus* from the outer Peninsular Ranges may be based on hybrids between var. *crassifolius* and *Ceanothus cuneatus* (M. Van Rensselaer and H. McMinn 1942).

32. Ceanothus cuneatus (Hooker) Nuttall in J. Torrey and A. Gray, Fl. N. Amer. 1: 267. 1838 F

Rhamnus cuneata Hooker, Fl. Bor.-Amer. 1: 124. 1831 (as cuneatus)

Shrubs, 0.5–3.5 m. **Stems** erect, ascending, or spreading, not rooting at nodes; branchlets grayish brown to brown, rigid or flexible, glabrate, puberulent, or tomentulose, hairs straight. **Leaves** usually both fascicled and not fascicled on same plant, rarely none fascicled; petiole 1–3 mm; blade flat to cupped, elliptic, oblanceolate, obovate, or orbiculate, 4–22(–30) × 3–12(–22) mm, base rounded, margins thick, not revolute, entire or denticulate distal to middle, teeth 0–9, apex obtuse, rounded, truncate, or retuse, abaxial surface pale green, glabrate or glabrous, adaxial surface green, glabrous. **Inflorescences** axillary or terminal, 0.8–2.5 cm. **Flowers:** sepals, petals, and nectary white to lavender or blue. **Capsules** 4–6 mm wide, weakly lobed; valves smooth, horns subapical, prominent, erect, intermediate ridges absent.

Varieties 4 (4 in the flora): w United States, nw Mexico.

1. Leaf blades of fascicled and non-fascicled leaves elliptic to widely oblanceolate, length usually 2+ times width; sepals, petals, and nectaries usually white, sometimes pale blue or pale lavender . . . . 32a. *Ceanothus cuneatus* var. *cuneatus*
1. Leaf blades of fascicled and non-fascicled leaves widely oblanceolate, widely obovate, or orbiculate, length usually less than 2 times width, or of fascicled leaves elliptic to narrowly oblanceolate (in var. *fascicularis*); sepals, petals, and nectaries usually lavender to blue, sometimes pale blue, rarely white.
   [2. Shifted to left margin.—Ed.]
2. Leaf blades of fascicled leaves elliptic to narrowly oblanceolate, 9–15 × 3–6 mm, length usually 2+ times width . . . 32b. *Ceanothus cuneatus* var. *fascicularis*
2. Leaf blades of fascicled leaves widely oblanceolate, widely obovate, or orbiculate, 4–15 × 3–12 mm, length less than 2 times width.
   3. Leaf blade margins usually entire, rarely 1–4-toothed, apices rounded, truncate, or retuse . . . . 32c. *Ceanothus cuneatus* var. *ramulosus*
   3. Leaf blade margins 5–9-toothed, apices rounded to truncate . . . . 32d. *Ceanothus cuneatus* var. *rigidus*

32a. Ceanothus cuneatus (Hooker) Nuttall var. **cuneatus** • Buck brush F

Ceanothus cuneatus var. *submontanus* (Rose) McMinn; *C. oblanceolatus* Davidson

Shrubs, 1.5–3.5 m. **Stems** erect; branchlets grayish brown to light gray, glaucous. **Leaf blades** of fascicled and non-fascicled leaves flat, elliptic to widely oblanceolate, 6–22(–30) × 3–12(–22) mm, length usually 2+ times width, margins entire, apex usually obtuse to rounded, rarely truncate. **Flowers:** sepals, petals, and nectary usually white, sometimes pale blue or pale lavender. **Capsules** 4–6 mm wide. 2*n* = 24.

Flowering Jan–May. Rocky slopes, ridges, sometimes on serpentine, chaparral, oak and oak-pine woodlands, conifer forests, gravelly flood plains; 10–1900 m; Calif., Oreg.; Mexico (Baja California).

Variety *cuneatus* in Oregon and in the Klamath Mountains of northern California is characterized by relatively small, elliptic to oblanceolate leaf blades 6–12 mm. The type specimen, collected by David Douglas in the upper Willamette Valley of Oregon, falls within this range. Low-growing, moundlike plants in the Klamath Mountains, less than eight tenths of a meter, with spreading stems, leaves similar in size and shape, and white to pale blue sepals and petals, are treated here as *Ceanothus arcuatus*. Elsewhere, var. *cuneatus* is characterized by leaf blades 9–30 mm.

Shrubs to 3.5 m with large leaf blades 15–30 × 9–18(–22) mm have been named *Ceanothus cuneatus* var. *dubius* J. T. Howell, and are restricted to sandy soils and open sites in chaparral and mixed evergreen forests of the Santa Cruz Mountains. Plants in the Transverse and Peninsular ranges of southern California, with narrowly oblanceolate leaf blades with sparsely canescent abaxial surfaces, have been named *C. oblanceolatus*. Putative hybrids with *C. pauciflorus* have been reported from several localities in the southern Sierra Nevada (H. McMinn 1944).

Formally named hybrids involving var. *cuneatus* include *C.* ×*connivens* Greene (either with *C. prostratus* or *C. fresnensis*), *C.* ×*flexilis* McMinn (with *C. prostratus*), and *C.* ×*humboldtensis* Roof (with *C. pumilus*).

Wood of var. *cuneatus* was used by Native Americans to make tools and arrow foreshafts (D. E. Moerman 1998).

32b. Ceanothus cuneatus (Hooker) Nuttall var. **fascicularis** (McMinn) Hoover, Leafl. W. Bot. 10: 350. 1966 • Lompoc ceanothus [C] [E]

Ceanothus ramulosus (Greene) McMinn var. *fascicularis* McMinn in M. van Rensselaer and H. McMinn, Ceanothus, 250, figs. 24, 25. 1942

Shrubs, 1.5–2.5 m. **Stems** erect to spreading; branchlets brown to grayish brown. **Leaf blades** of non-fascicled leaves flat or cupped, widely oblanceolate to widely obovate, 5–11 × 4–7 mm, length usually less than 2 times width, margins usually entire, rarely 1–3-toothed, apex truncate or retuse; of fascicled leaves ± flat, elliptic to narrowly oblanceolate, 9–15 × 3–6 mm, length usually 2+ times width, apex obtuse to rounded. **Flowers:** sepals, petals, and nectary usually lavender, sometimes pale blue. **Capsules** 5–6 mm wide. $2n = 24$.

Flowering Feb–Apr. Sandy soils, maritime chaparral; of conservation concern; 10–200 m; Calif.

Variety *fascicularis* is endemic to sandy soils on marine terraces and coastal slopes of western San Luis Obispo and Santa Barbara counties. Plants are occasionally found without fascicled leaves, but can be identified by their lavender sepals and petals and capsules with weakly developed horns. Some specimens from south of Morro Bay, San Luis Obispo County, have pale blue flowers and denticulate nodal leaf blades similar to var. *rigidus*.

32c. Ceanothus cuneatus (Hooker) Nuttall var. **ramulosus** Greene, Fl. Francisc., 86. 1891 • Coast buck brush [E] [F]

Shrubs, (0.5–)1–2.5 m. **Stems** erect, ascending, or spreading; branchlets gray to grayish brown. **Leaf blades** of fascicled and non-fascicled leaves cupped, widely oblanceolate, widely obovate, or orbiculate, 5–15 × 3–12 mm, length usually less than 2 times width, margins usually entire, rarely 1–4-toothed, apex rounded, truncate, or retuse. **Flowers:** sepals, petals, and nectary lavender to blue. **Capsules** 5–6 mm wide. $2n = 24$.

Flowering Feb–May. Rocky slopes, often on serpentine, chaparral, pine woodlands; 10–800 m; Calif.

Variety *ramulosus* occurs disjunctly in the San Francisco Bay area (Alameda, Contra Costa, Lake, Marin, Napa, Santa Cruz, and Sonoma counties) and on coastal slopes in western San Luis Obispo and Santa Barbara counties. In both areas, plants occur primarily but not exclusively on serpentine soils. Some specimens in the San Francisco Bay area are intermediate to var. *cuneatus*, while others in San Luis Obispo and Santa Barbara counties suggest intergradation with var. *fascicularis*. Plants on Point Sal Ridge in western Santa Barbara County are low-growing, less than 1 m, and have ascending to spreading stems, features that are retained in cultivation.

32d. Ceanothus cuneatus (Hooker) Nuttall var. **rigidus** (Nuttall) Hoover, Leafl. W. Bot. 10: 350. 1966 • Monterey buck brush [C] [E]

Ceanothus rigidus Nuttall in J. Torrey and A. Gray, Fl. N. Amer. 1: 268. 1838; *C. rigidus* var. *albus* Roof; *C. rigidus* var. *pallens* Sprague

Shrubs, 0.5–1.5 m. **Stems** erect, ascending, or spreading; branchlets gray to grayish brown. **Leaf blades** of fascicled and non-fascicled leaves flat to ± cupped, widely obovate to orbiculate, 4–10 × 4–6 mm, length usually less than 2 times width, margins 5–9-toothed, apex rounded to truncate. **Flowers:** sepals, petals, and nectary usually pale to deep blue, rarely white. **Capsules** 5–6 mm wide. $2n = 24$.

Flowering Feb–Apr. Sandy soils, flats, dune swales, maritime chaparral, pine forests; of conservation concern; 10–400 m; Calif.

The typical form of var. *rigidus* is restricted almost entirely to sandy soils of coastal hills and mesas in the southern Monterey Bay region, although plants with denticulate leaf blades occur near Morro Bay in San Luis Obispo County. The name *Ceanothus rigidus* var. *albus* was applied to low-growing plants with spreading stems and white flowers. Such plants apparently have been extirpated on the Monterey peninsula and survive only in cultivation.

33. Ceanothus maritimus Hoover, Leafl. W. Bot. 7: 111. 1953 • Maritime ceanothus [C] [E]

Shrubs, 0.3–1 m, moundlike. **Stems** usually prostrate to ascending, rarely erect, sometimes rooting at proximal nodes; branchlets reddish to grayish brown, rigid, smooth to slightly ridged, tomentulose, glabrescent. **Leaves** not fascicled; petiole 1–2 mm; blade flat or cupped, obovate to oblong-obovate, 8–20 × 4–12(–15) mm, base cuneate, margins thick to revolute, usually entire, sometimes denticulate near apex, teeth 3–5, apex acute to rounded, truncate, or retuse, abaxial surface grayish green, tomentulose, adaxial surface green, glabrous. **Inflorescences** axillary, 0.8–1.5 cm. **Flowers:** sepals and petals pale to deep blue, sometimes tinged with lavender; nectary dark purplish green. **Capsules** 5–8 mm wide, not to weakly lobed; valves smooth, horns subapical, minute, erect, intermediate ridges absent.

Flowering Feb–May. Maritime terraces and bluffs, alluvial or serpentine soils, coastal prairies, open sites in maritime chaparral; of conservation concern; 10–60 m; Calif.

Ceanothus maritimus is restricted to a small area of coastal bluffs in northern San Luis Obispo County, growing in close proximity to another local endemic, *C. hearstiorum* (subg. *Ceanothus*).

34. Ceanothus ferrisiae McMinn, Madroño 2: 89. 1933 (as ferrisae) • Coyote ceanothus [C] [E]

Shrubs, 1–2 m. **Stems** erect, not rooting at nodes; branchlets grayish brown, glaucous, rigid, puberulent. **Leaves** not fascicled; petiole 1–3 mm; blade flat or ± cupped, widely elliptic to widely obovate, 11–30 × 7–18 mm, base obtuse to rounded, margins not revolute, usually denticulate, rarely entire, teeth 6–13, apex rounded, abaxial surface pale green, sparsely strigillose between veins, adaxial surface dark green, glabrate. **Inflorescences** terminal, 1.2–1.5(–2) cm. **Flowers:** sepals and petals white; nectary dark blue to purple. **Capsules** 7–9 mm wide, weakly lobed; valves ± smooth, horns subapical, prominent, erect, intermediate ridges absent. $2n = 24$.

Flowering Jan–May. Serpentine soils and outcrops, chaparral, pine and oak woodlands; of conservation concern; 100–500 m; Calif.

Ceanothus ferrisiae, federally listed as endangered, occurs at a few localities in the foothills of the Mount Hamilton Range northeast of Morgan Hill, Santa Clara County.

35. Ceanothus arcuatus McMinn in M. van Rensselaer and H. McMinn, Ceanothus, 247, fig. 82. 1942 • Arching ceanothus [E]

Shrubs, 0.3–0.8 m, moundlike. **Stems** ascending or spreading, not rooting at nodes; branchlets brown to grayish brown, ± rigid, tomentulose, glabrescent. **Leaves** sometimes fascicled, axillary short shoots ascending to spreading; petiole 1–2 mm; blade flat, elliptic to oblanceolate, 4–10 × 2–5 mm, base rounded, margins thick, not revolute, usually entire, rarely denticulate near apex, teeth 1–3, apex rounded to obtuse, abaxial surface green to pale green, glabrate, adaxial surface pale green, glabrous. **Inflorescences** axillary or terminal, 0.8–2.5 cm. **Flowers:** sepals and petals white to pale blue; nectary pale blue or yellow tinged. **Capsules** 4–6 mm wide, not to weakly lobed; valves smooth to rugulose, horns subapical, prominent, erect, intermediate ridges absent.

Flowering May–Jun. Granitic or serpentine soils, conifer forests; 900–2300 m; Calif., Oreg.

Ceanothus arcuatus was explicitly described as a species by McMinn but inexplicably treated as a hybrid between *C. fresnensis* and *C. cuneatus* by P. A. Munz (1959). Munz may have been influenced by McMinn's hypothesis that *C. arcuatus* was derived through hybridization between the two species. Some specimens from the Klamath Mountains have been either interpreted as *C. cuneatus* or misidentified as hybrids between *C. pumilus* and *C. cuneatus*. Populations of *C. arcuatus* are relatively uniform throughout their geographic distribution and often represent the dominant understory shrub in conifer forests of the Klamath Mountains and Sierra Nevada. *Ceanothus arcuatus* occurs primarily on metamorphic substrates in the Klamath Mountains and the northern Sierra Nevada, but in the central Sierra Nevada (Nevada County south to Madera County), the most common substrate is granitic.

36. Ceanothus fresnensis Dudley ex Abrams, Bot. Gaz. 53: 68. 1912 • Fresno mat [C] [E]

Shrubs, 0.1–0.3 m, matlike. **Stems** prostrate or spreading, rooting at proximal nodes; branchlets brown to reddish or grayish brown, rigid, tomentulose. **Leaves** not fascicled, sometimes crowded, not obscuring internodes; petiole 1–2 mm; blade flat or ± cupped, elliptic to oblanceolate, 4–12 × 3–8 mm, base obtuse to rounded, margins ± thick, not revolute, usually entire, sometimes minutely denticulate near apex, teeth 0–3, apex rounded to retuse, abaxial surface pale green, strigillose, adaxial surface dark green, shiny, puberulent, glabrescent. **Inflorescences** axillary, 1–2.3 cm. **Flowers:** sepals, petals, and nectary usually blue, rarely pale blue. **Capsules** 4–6 mm wide, weakly lobed; valves smooth, horns subapical, prominent, erect, intermediate ridges absent.

Flowering May–Jun. Granitic soils and outcrops, semishaded sites, conifer forests; of conservation concern; 400–2200 m; Calif.

Ceanothus fresnensis occurs infrequently along the western slopes of the Sierra Nevada, from Nevada County south to Fresno County. Despite its similarity to *C. arcuatus*, the two species can be separated by differences in life form and intensity of flower color. *Ceanothus fresnensis* tends to occur in relatively dense forests, whereas *C. arcuatus* occurs in open, relatively exposed sites, often at higher elevations.

37. Ceanothus roderickii W. Knight, Four Seasons 2(4): 23. 1968 (as rodericki) • Pine Hill ceanothus [C] [E]

Shrubs, 0.1–0.5 m, moundlike. **Stems** prostrate or spreading, arcuate, often rooting at distal nodes; branchlets brown to grayish brown, rigid, puberulent, glabrescent. **Leaves** both fascicled and not on same plant, axillary short shoots erect; petiole 1–2 mm; blade folded lengthwise abaxially, elliptic to oblanceolate, 4–12 × 2–6 mm, base obtuse to cuneate, margins not revolute, entire or denticulate near apex, teeth 3–5, apex obtuse, abaxial surface pale green, glabrate or sparsely strigillose between the veins, adaxial surface green, glabrate. **Inflorescences** terminal or axillary, 0.6–1.4 cm. **Flowers:** sepals and petals white to pale blue; nectary blue. **Capsules** 4–5 mm wide, usually not, sometimes weakly lobed; valves smooth or slightly rugulose, sometimes ridged, horns absent or weakly developed bulges, intermediate ridges absent.

Flowering Apr–Jun. Rocky soils derived from gabbro, chaparral, pine woodlands; of conservation concern; 200–600 m; Calif.

Ceanothus roderickii is restricted to a few localities in the foothills of the Sierra Nevada (El Dorado County). A close relationship to *C. cuneatus* var. *cuneatus* is supported by molecular data (T. M. Hardig et al. 2000b). The ability to root at remote, distal nodes was shown to enhance density and recovery, long after episodic establishment from seeds following fires (R. S. Boyd 2007).

Ceanothus roderickii is in the Center for Plant Conservation's National Collection of Endangered Plants.

38. Ceanothus gloriosus J. T. Howell, Leafl. W. Bot. 2: 43. 1937 [E]

Shrubs, 0.1–3 m, matlike to moundlike. **Stems** prostrate, spreading, ascending, or erect, sometimes rooting at proximal nodes; branchlets green to brown or reddish brown, flexible to rigid, strigillose or tomentulose. **Leaves** not fascicled, not crowded; petiole 1–4 mm; blade flat to ± cupped or folded lengthwise adaxially, widely elliptic, obovate, or suborbiculate, 10–40(–45) × 5–24 mm, base cuneate to ± rounded, margins not revolute, sometimes slightly thickened, dentate to denticulate most of length, teeth 9–35, apex rounded, truncate, or retuse, abaxial surface pale green, sparsely strigillose or glabrate, adaxial surface dark green, ± shiny, glabrous. **Inflorescences** axillary, 0.9–2.5 cm. **Flowers:** sepals, petals, and nectary deep blue to bluish purple. **Capsules** 4–6 mm wide, lobed; valves usually smooth, sometimes rugulose or crested distal to middle, horns subapical, minute, not rugose, intermediate ridges absent.

Varieties 3 (3 in the flora): California.

Ceanothus gloriosus is composed of three varieties occurring along the northern California coast from Humboldt County to Marin County. Variety *gloriosus* and var. *porrectus* generally differ primarily by leaf shape, length and width, and the number of marginal teeth. Variety *exaltatus* differs from the other two varieties primarily in stature. Complex hybrids with *C. cuneatus* var. *ramulosus*, *C. divergens*, and *C. sonomensis* were studied by J. T. Howell (1940).

1. Shrubs 0.8–3 m; stems erect to ascending . . . . 38c. *Ceanothus gloriosus* var. *exaltatus*
1. Shrubs 0.1–0.5 m; stems prostrate to spreading.
    2. Leaf blades widely obovate to suborbiculate, 23–31(–45) × 17–24 mm, marginal teeth 13–31 . . . . 38a. *Ceanothus gloriosus* var. *gloriosus*
    2. Leaf blades elliptic, obovate, or narrowly obovate, 10–21 × 5–15 mm, marginal teeth 9–19 . . . . 38b. *Ceanothus gloriosus* var. *porrectus*

38a. Ceanothus gloriosus J. T. Howell var. **gloriosus** • Point Reyes ceanothus [E]

Ceanothus prostratus Bentham var. *grandifolius* (Torrey) Jepson

Shrubs, 0.1–0.4 m. **Stems** prostrate or spreading. **Leaf blades** widely obovate to suborbiculate, 23–31(–45) × 17–24 mm, marginal teeth 13–31. **2*n*** = 24.

Flowering Mar–May. Sandy soils, coastal bluffs, maritime chaparral, pine forests; 30–200 m; Calif.

Variety *gloriosus* occurs from Humboldt County south to Marin County and sometimes intergrades with var. *exaltatus*.

38b. Ceanothus gloriosus J. T. Howell var. **porrectus** J. T. Howell, Leafl. W. Bot. 4: 31. 1944 • Mount Vision ceanothus [C] [E]

Shrubs, 0.2–0.5 m. **Stems** spreading. **Leaf blades** elliptic, obovate, or narrowly obovate, 10–21 × 5–15 mm, marginal teeth 9–19. **2*n*** = 24.

Flowering Mar–May. Sandy soils, coastal bluffs, maritime chaparral, pine forests, mixed evergreen forests; of conservation concern; 100–200 m; Calif.

Variety *porrectus* is restricted to a few localities on Inverness Ridge, Marin County.

38c. Ceanothus gloriosus J. T. Howell var. **exaltatus** J. T. Howell, Leafl. W. Bot. 2: 44. 1937 • Navarro ceanothus [E]

Shrubs, 0.8–3 m. **Stems** erect to ascending. **Leaf blades** widely elliptic to suborbiculate, 13–40 × 17–22 mm, marginal teeth 13–35. **2*n*** = 24.

Flowering Mar–Jun. Sandy or rocky slopes, ridges, chaparral, conifer forests; 30–500 m; Calif.

Variety *exaltatus* occurs from Mendocino County south to Marin County, often more inland than var. *gloriosus*.

39. Ceanothus masonii McMinn, Madroño 6: 171. 1942 • Mason's ceanothus [C] [E]

Shrubs, 0.6–2 m. **Stems** erect to ascending, not rooting at nodes; branchlets dark brown, rigid, ± tomentulose. **Leaves** not fascicled; petiole 1–2 mm; blade flat, usually elliptic or oval, sometimes suborbiculate, 7–21 × 4–13 mm, base rounded to ± cuneate, margins not revolute, denticulate most of length, teeth 9–17, apex obtuse, rounded to truncate, abaxial surface pale green to grayish green, strigose on veins, glabrate, adaxial surface dark green, glabrous. **Inflorescences** axillary, 1–2.5 cm. **Flowers:** sepals, petals, and nectary deep blue to purple. **Capsules** 3–4 mm wide, not lobed; valves smooth, horns apical, minute, not rugose, intermediate ridges absent. **2*n*** = 24.

Flowering Feb–Apr. Soils derived from serpentine, chaparral, pine forests; of conservation concern; 100–500 m; Calif.

Ceanothus masonii occurs only at a few localities on Bolinas Ridge, Marin County. With the exception of its leaf morphology, it bears a close resemblance to *C. gloriosus* var. *exaltatus*.

40. Ceanothus sonomensis J. T. Howell, Leafl. W. Bot. 2: 162. 1939 • Sonoma ceanothus [C] [E]

Shrubs, 0.5–1 m, often moundlike. **Stems** erect to ascending, not rooting at nodes; branchlets gray to grayish brown, rigid, strigillose, glabrescent. **Leaves** not fascicled; petiole 0–1 mm; blade cupped, widely obovate to suborbiculate, 5–12 × 2–10 mm, base cuneate, margins not revolute, wavy, spinose-dentate, teeth 2–4, apex widely notched; abaxial surface pale green or grayish green and glaucous, strigillose on veins, adaxial surface shiny green, glabrous. **Inflorescences** axillary or terminal, 0.8–1.5 cm. **Flowers:** sepals, petals, and nectary blue to lavender. **Capsules** 4–5 mm wide, usually not, sometimes weakly, lobed; valves smooth, horns subapical, minute to ± prominent, erect, intermediate ridges absent. **2*n*** = 24.

Flowering Mar–Apr. Sandy to rocky soils derived mostly from volcanic substrates, slopes, ridges, chaparral; of conservation concern; 100–700 m; Calif.

Ceanothus sonomensis is distinctive in having spinose-dentate, few-toothed leaves, and slender fruit horns 2–3 mm; it occurs at a few scattered localities in the mountains of Napa and Sonoma counties.

41. Ceanothus divergens Parry, Proc. Davenport Acad. Nat. Sci. 5: 173. 1889 • Calistoga ceanothus [C] [E]

Shrubs, 0.5–1.5 m, sometimes moundlike. **Stems** erect to ascending, not rooting at nodes; branchlets brown to grayish brown, sometimes glaucous, ± flexible, glabrous or sparsely puberulent. **Leaves** not fascicled, spreading; petiole 0–2 mm; blade flat to ± cupped or weakly folded lengthwise, elliptic to ± oblong or obovate, 10–20 × 5–12 mm, base obtuse to cuneate, margins thick or slightly revolute, slightly wavy, spinose-dentate, teeth 5–9(–11), apex sharply acute or retuse with an apical tooth, abaxial surface grayish green, veins strigillose, adaxial surface green, glabrous. **Inflorescences** axillary, 1.2–2.5 cm. **Flowers:** sepals and petals deep blue to purple; nectary dark blue or purple. **Capsules** 5–6 mm wide, lobed; valves smooth, crested, horns subapical, prominent, erect, intermediate ridges weakly developed. $2n = 24$.

Flowering Feb–Apr. Rocky soils apparently derived from serpentine or volcanic substrates, chaparral, oak and pine woodlands; of conservation concern; 100–1000 m; Calif.

Ceanothus divergens is restricted to a few localities in Napa and Sonoma counties.

42. Ceanothus confusus J. T. Howell, Leafl. W. Bot. 2: 160. 1939 • Rincon Ridge ceanothus [C] [F]

Shrubs, 0.1–0.6 m, matlike to moundlike. **Stems** prostrate, spreading, or weakly ascending, often rooting at proximal nodes; branchlets brown to reddish brown, ± rigid, glabrous or sparsely puberulent. **Leaves** not fascicled, not crowded, shorter than internodes; petiole 0–2 mm; blade flat to ± cupped, elliptic to ± oblong or obovate, 10–20 × 5–14 mm, base obtuse to cuneate, margins thick or slightly revolute, slightly wavy, sharply dentate to spinose-dentate, teeth 3–9, apex acute or retuse, with an apical tooth, abaxial surface grayish green, strigillose on veins, adaxial surface green, dull, glabrous. **Inflorescences** axillary, 1.5–3 cm. **Flowers:** sepals, petals, and nectary blue, lavender, or purple. **Capsules** 4–6 mm wide, lobed; valves smooth, crested, horns subapical, prominent, erect, intermediate ridges weakly developed. $2n = 24$.

Flowering Feb–May. Rocky soils apparently derived from serpentine or volcanic substrates, chaparral, oak and pine woodlands, conifer forests; of conservation concern; 70–1000 m; Calif.

Ceanothus confusus is weakly defined and perhaps best treated as a part of *C. divergens* (L. Abrams and R. S. Ferris 1923–1960, vol. 3). At least some populations in the Hood Mountains (Napa and Sonoma counties) include plants with the habit and leaf morphology of both species, while other, more uniform populations appear intermediate; it remains to be determined whether this pattern is a product of primary or secondary intergradation.

43. Ceanothus prostratus Bentham, Pl. Hartw., 302. 1849 [E] [F]

Shrubs, 0.1–0.3 m, matlike to moundlike. **Stems** prostrate, spreading, or ascending, rooting at distal nodes; branchlets reddish brown, ± flexible, puberulent, glabrescent. **Leaves** not fascicled, crowded, usually longer than internodes and obscuring them; petiole 1–3 mm; blade flat to ± cupped, elliptic to obovate, 6–30 × 4–16 mm, base cuneate, margins sometimes thick, not revolute, sometimes wavy, sharply dentate to spinose-dentate, teeth 3–9, apex rounded, abaxial surface pale green, glabrous except sparsely strigillose on veins, adaxial surface dark green, shiny, glabrate. **Inflorescences** axillary, 0.9–2 cm. **Flowers:** sepals, petals, and nectary pale to deep blue or purplish blue. **Capsules** 6–9 mm wide, lobed; valves smooth to rugulose, horns subapical, prominent, erect or spreading, rugose or not, intermediate ridges present. $2n = 24$.

Varieties 2 (2 in the flora): w United States.

Putative hybrids between *Ceanothus prostratus* and *C. velutinus* var. *velutinus*, named *C.* ×*rugosus*, have been reported from northeastern California (H. McMinn 1944). A rare putative hybrid between *C. prostratus* and *C. cordulatus* in the Lake Tahoe basin has been named *C.* ×*serrulatus*. Both *C.* ×*rugosus* and *C.* ×*serrulatus* are intersubgeneric hybrids. Formally named hybrids between *C. prostratus* and *C. cuneatus* var. *cuneatus* include *C.* ×*flexilis* and possibly *C.* ×*connivens*, but the latter could have *C. fresnensis* as one of the parents rather than *C. prostratus*.

1. Leaf blades flat, margins not wavy, teeth usually 3–5(–7); capsule horns erect . . . . 43a. *Ceanothus prostratus* var. *prostratus*
1. Leaf blades slightly folded lengthwise adaxially, margins ± wavy, teeth 5–9; capsule horns spreading . . . . 43b. *Ceanothus prostratus* var. *occidentalis*

43a. Ceanothus prostratus Bentham var. **prostratus** • Mahala mat [E] [F]

Ceanothus prostratus var. *laxus* Jepson

Shrubs, 0.1–0.2 m, matlike. **Stems** prostrate, with short erect shoots. **Leaf blades** flat, margins not wavy, teeth usually 3–5(–7). **Capsule horns** erect, rugose.

Flowering Apr–Jun. Sandy or gravelly soils, open flats, gentle slopes, conifer forests; 400–2700 m; Calif., Idaho, Nev., Oreg., Wash.

Variety *prostratus* often forms mats 1–2.5 m wide.

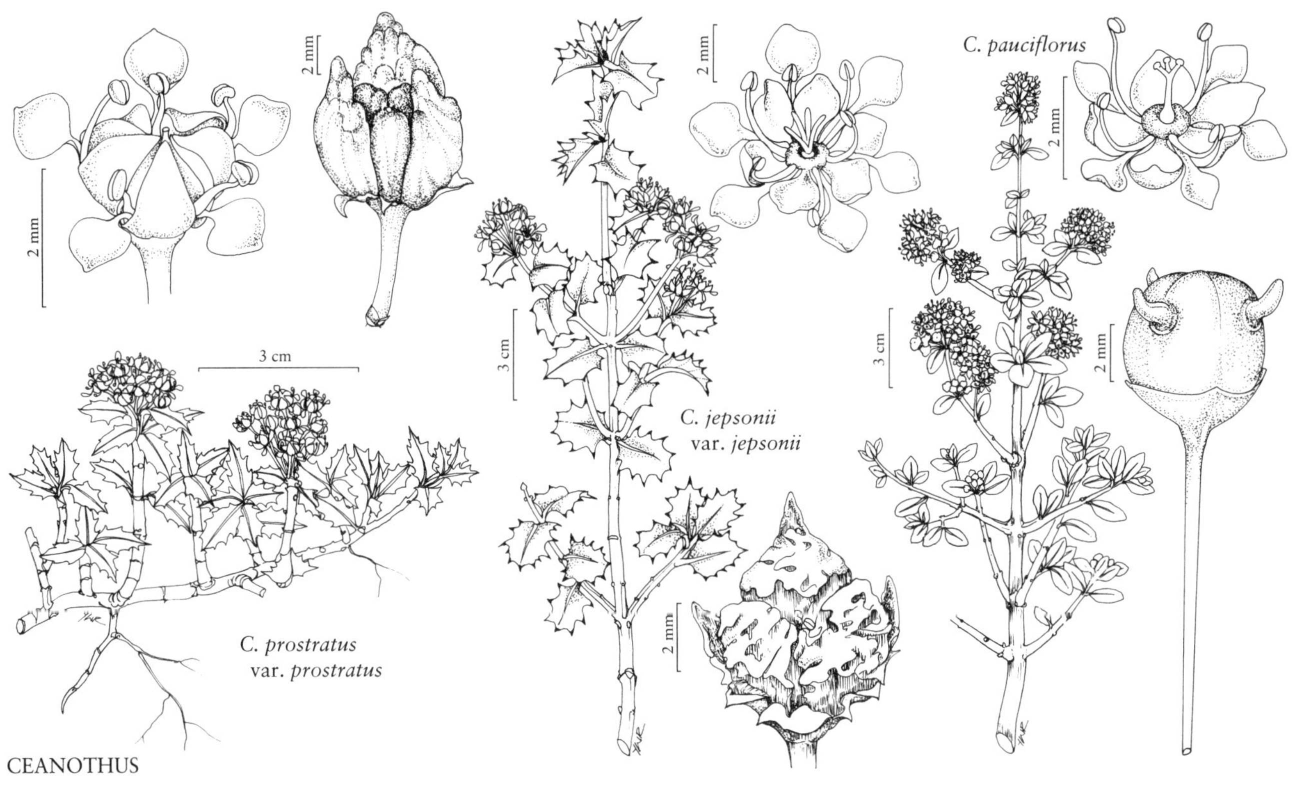

CEANOTHUS

43b. Ceanothus prostratus Bentham var. **occidentalis** McMinn in M. van Rensselaer and H. McMinn, Ceanothus, 262. 1942 • Cobb Mountain ceanothus [C] [E]

Ceanothus divergens Parry subsp. *occidentalis* (McMinn) Abrams

Shrubs, 0.1–0.3 m, matlike to moundlike. **Stems** prostrate, spreading, or ascending. **Leaf blades** slightly folded lengthwise adaxially, margins ± wavy, teeth 5–9. **Capsule horns** spreading, not conspicuously rugose. **2*n*** = 24.

Flowering Apr–May. Gravelly or rocky soils derived from volcanic substrates, open flats and ridges, conifer forests; of conservation concern; 800–1400 m; Calif.

Variety *occidentalis* is known from only a few localities in the mountains of Lake, Mendocino, Napa, and Sonoma counties. L. Abrams and R. S. Ferris (1923–1960, vol. 3) treated it as part of *Ceanothus divergens*, based partly on leaf morphology, but its life form and fruit morphology suggest a closer relationship to *C. prostratus*.

44. Ceanothus pumilus Greene, Erythea 1: 149. 1893 • Siskiyou mat [E]

Ceanothus prostratus Bentham var. *profugus* Jepson

Shrubs, 0.1–0.4 m, matlike to moundlike. **Stems** prostrate to spreading, sometimes rooting at proximal nodes; branchlets reddish brown, flexible to ± rigid, tomentulose. **Leaves** not fascicled; petiole 1–2 mm; blade flat to ± cupped, slightly folded lengthwise adaxially, narrowly oblanceolate to narrowly oblong-oblanceolate, 5–15 × 3–6 mm, base cuneate, margins thick to ± revolute, usually denticulate near apex, sometimes entire, teeth (0 or) 2–3, apex usually truncate, sometimes obtuse, abaxial surface pale green, sparsely strigillose to glabrous, adaxial surface green to grayish green, dull, glabrous, sometimes glaucous. **Inflorescences** axillary, 1–1.7 cm. **Flowers:** sepals, petals, and nectary pale blue to lavender. **Capsules** 4–6 mm wide, lobed; valves smooth, horns subapical, minute or weakly developed bulges, intermediate ridges absent. **2*n*** = 24.

Flowering Apr–Jun. Rocky soils derived from serpentine, open flats and slopes, chaparral, conifer forests; 100–2200 m; Calif., Oreg.

Ceanothus pumilus is endemic to the Klamath Mountains, where it occurs strictly on serpentine soils; it sometimes has been confused with *C. arcuatus* and *C. prostratus*, from which it differs principally by its oblanceolate to oblong-lanceolate leaf blades with a truncate, 3-toothed apex.

Hybrids between *Ceanothus pumilus* and *C. cuneatus* have been called *C.* ×*humboldtensis* Roof.

45. Ceanothus pinetorum Coville, Contr. U.S. Natl. Herb. 4: 80, plate 6. 1893 • Coville's ceanothus [E]

Shrubs, 0.5–1.5 m, sometimes moundlike. **Stems** erect, spreading, or weakly ascending, sometimes arcuate, rooting at proximal nodes; branchlets reddish to grayish brown, sometimes glaucous, rigid, glabrous or sparsely puberulent. **Leaves** sometimes fascicled, not crowded; petiole 1–3 mm; blade flat to slightly cupped, slightly folded lengthwise adaxially, widely elliptic to suborbiculate, 10–20 × 8–19 mm, base rounded, margins thick to slightly revolute, dentate to denticulate most of length, teeth 9–15, apex rounded, abaxial surface pale green, glabrous except on veins, adaxial surface dark green, glabrous. **Inflorescences** axillary, 1.2–2.1 cm. **Flowers:** sepals, petals, and nectary pale blue to blue. **Capsules** 6–9 mm wide, weakly lobed; valves smooth to rugulose, horns subapical, prominent, erect, rugose, intermediate ridges present.

Flowering May–Jun. Rocky granitic or metamorphic slopes and ridges, open pine forests; 1600–2600 m; Calif.

Ceanothus pinetorum occurs disjunctly in the southern Trinity Mountains (Shasta and Trinity counties) and in the southern Sierra Nevada (Kern and Tulare counties).

46. Ceanothus purpureus Jepson, Fl. W. Calif., 258. 1901 (as purpurea) • Hollyleaf ceanothus [C] [E]

Shrubs, 1–2 m. **Stems** erect to ascending, not rooting at nodes; branchlets reddish brown, sometimes glaucescent, rigid, glabrate. **Leaves** not fascicled, spreading to deflexed; petiole 0–2 mm; blade ± cupped, folded lengthwise adaxially, widely elliptic to widely obovate, 12–25 × 7–20 mm, base obtuse to cuneate, margins not revolute, spinose-dentate, teeth 7–15, apex rounded to sharply acute, abaxial surface pale green, sparsely strigillose, especially on veins, adaxial surface green to dark green, glabrous. **Inflorescences** axillary, 1.2–2.5 cm. **Flowers:** sepals, petals, and nectary deep blue to purple. **Capsules** 4–5 mm wide, lobed; valves smooth, horns subapical, prominent, slender, erect, intermediate ridges absent. $2n = 24$.

Flowering Feb–Apr. Rocky slopes and outcrops derived from volcanic substrates, chaparral, oak woodlands; of conservation concern; 100–400 m; Calif.

Ceanothus purpureus is endemic to the Vaca Mountains (Napa and Solano counties); it has been confused with *C. jepsonii*, which differs by its 6–8-merous flowers, rugose capsule horns, prominent ridges between the horns, and distribution on serpentine soils.

47. Ceanothus jepsonii Greene, Man. Bot. San Francisco, 78. 1894 • Jepson's ceanothus [E] [F]

Shrubs, 0.5–1.5 m. **Stems** erect to ascending, not rooting at nodes; branchlets reddish to grayish brown, rigid, puberulent, glabrescent. **Leaves** not fascicled, deflexed; petiole 0–2 mm; blade ± cupped, slightly folded lengthwise adaxially, elliptic to ± oblong, 10–20 × 5–13 mm, base rounded, margins thick or slightly revolute, spinose-dentate, teeth 7–11, apex rounded or sharply acute, abaxial surface pale yellowish green, glabrous, adaxial surface pale green, glabrous. **Inflorescences** axillary or terminal, 1–2 cm. **Flowers:** sepals and petals (5–)6(–8), usually blue to lavender or white, rarely pink; nectary blue; stamens (5–)6(–8). **Capsules** 5–7 mm wide, lobed; valves rugose, horns subapical, prominent, thick, erect, rugose, intermediate ridges present.

Varieties 2 (2 in the flora): California.

Ceanothus jepsonii, composed of two allopatric varieties, is the only species with mostly six (rarely eight) sepals and petals, and cymules reduced to solitary flowers (M. A. Nobs 1963). T. M. Hardig et al. (2000) provided evidence showing that the two varieties may not form a monophyletic group. H. McMinn (1942) and Nobs reported putative hybrids with *C. cuneatus* and *C. prostratus*.

1. Sepals and petals usually pale blue to lavender, rarely pink or white; capsules globose . . . . 47a. *Ceanothus jepsonii* var. *jepsonii*
1. Sepals and petals white; capsules ± oblong . . . . 47b. *Ceanothus jepsonii* var. *albiflorus*

47a. Ceanothus jepsonii Greene var. **jepsonii** E F

Stems erect to ascending. **Flowers:** sepals and petals usually pale blue to lavender, rarely pink or white. **Capsules** globose. **2*n*** = 24.

Flowering Feb–Apr. Rocky serpentine slopes and ridges, chaparral; 200–800 m; Calif.

Variety *jepsonii* occurs in the outer North Coast Ranges of California, from Mendocino County south to Marin County. Plants in one population from Marin County with five sepals and petals have been called *Ceanothus decornutus* V. T. Parker.

47b. Ceanothus jepsonii Greene var. **albiflorus** J. T. Howell, Leafl. W. Bot. 3: 231. 1943 E

Stems erect. **Flowers:** sepals and petals white. **Capsules** ± oblong. **2*n*** = 24.

Flowering Feb–Apr. Rocky serpentine slopes and ridges, chaparral; 300–1000 m; Calif.

Variety *albiflorus* occurs in the inner North Coast Ranges from Glenn County south to Napa County.

48. Ceanothus pauciflorus de Candolle in A. P. de Candolle and A. L. P. P. de Candolle, Prodr. 2: 33. 1825 F

Ceanothus greggii A. Gray; *C. greggii* var. *franklinii* S. L. Welsh; *C. vestitus* Greene

Shrubs, 0.2–2 m, sometimes moundlike. **Stems** erect or weakly ascending to spreading, not rooting at nodes; branchlets light gray to ashy gray, rigid, puberulent to tomentulose, hairs curly or wavy, glabrescent. **Leaves** not fascicled; petiole (0–)1–3 mm; blade flat to ± cupped, elliptic, oblong-elliptic, obovate, or suborbiculate, 5–15(–20) × 3–14 (–15) mm, base cuneate to rounded, margins thick, not revolute, entire or remotely denticulate, teeth 1–5(–7), apex acute to ± truncate, abaxial surface pale green to grayish green, glabrate or puberulent, hairs curly or wavy, glabrescent, adaxial surface grayish green, puberulent, hairs curly or wavy, glabrescent. **Inflorescences** axillary, rarely racemelike, 0.7–3 cm. **Flowers:** sepals and petals white to cream, sometimes pale blue or lavender; nectary yellowish green, brown, or blue. **Capsules** 3.5–6 mm wide, usually not, sometimes weakly, lobed; valves smooth, horns lateral, prominent to minute or absent, spreading, intermediate ridges absent.

Flowering Feb–May. Rocky slopes, ridges, alluvial fans, sagebrush and montane shrublands, pinyon and/or juniper and montane conifer woodlands; 900–2900; Ariz., Calif., Nev., N.Mex., Tex., Utah; n, c Mexico.

Ceanothus pauciflorus as circumscribed here includes plants that have flat to cupped leaf blades with a sparse to dense but not intertwined indumentum composed of short curly or wavy hairs, at least when young; this indumentum also occurs on the petiole and ultimate branchlets. Such plants in the United States have been treated either as *C. greggii* or *C. vestitus* (M. Van Rensselaer and H. McMinn 1942). However, R. McVaugh (1998) and D. O. Burge and K. Zhukovsky (2013) provided evidence that they should be treated as *C. pauciflorus*. Specimens from the desert slopes of the Sierra Nevada and San Bernardino Mountains, California, and a few scattered localities in western Arizona, have leaves similar in shape and dentation to those of *C. perplexans*, suggesting local hybridization.

49. Ceanothus perplexans Trelease in A. Gray et al., Syn. Fl. N. Amer. 1(1,2): 417. 1897

Ceanothus greggii A. Gray var. *perplexans* (Trelease) Jepson

Shrubs, 1–2 m. **Stems** erect, not rooting at nodes; branchlets brown to grayish brown, rigid, glabrate or tomentulose, glabrescent. **Leaves** not fascicled; petiole 1–3 mm; blade flat to ± cupped, elliptic, widely obovate, or suborbiculate, 10–20 × 7–18 mm, base rounded, margins thick, not revolute, usually sharply denticulate, sometimes weakly denticulate to almost entire, teeth 7–15, apex rounded to ± truncate, abaxial surface pale green to yellowish green, puberulent, hairs curly, glabrescent, adaxial surface green to yellowish green, sparsely puberulent, hairs curly, glabrescent. **Inflorescences** axillary, 0.7–2 cm. **Flowers:** sepals and petals white to cream; nectary yellow to green. **Capsules** 4–6 mm wide, usually not, sometimes weakly, lobed; valves smooth, horns lateral, usually minute, sometimes absent, spreading, intermediate ridges absent. **2*n*** = 24.

Flowering Jan–Apr. Granitic or metamorphic substrates, rocky slopes, ridges, alluvial fans, chaparral, montane shrublands, pinyon and/or juniper and montane conifer woodlands; 500–1900 m; Ariz., Calif.; Mexico (Baja California).

Ceanothus perplexans occurs in southwestern Arizona, on the desert slopes of the San Bernardino Mountains and Peninsular Ranges of southern California, and in Baja California. *Ceanothus* specimens from Guadalupe Island, Baja California, with entire or weakly denticulate leaf margins have been referred to either *C. crassifolius* or *C. cuneatus*, but in their leaf shape and indumentum they more closely resemble *C. perplexans*.

50. Ceanothus otayensis McMinn in M. van Rensselaer and H. McMinn, Ceanothus, 273, fig. 102. 1942, as hybrid • Otay Mountain ceanothus [C]

Shrubs, 1–3.5 m. **Stems** erect to ascending, not rooting at nodes; branchlets grayish brown to brown, ± flexible to rigid, tomentulose. **Leaves** not fascicled; petiole 0–2 mm; blade flat to ± cupped, widely elliptic to obovate, 5–13 × 4–10 mm, base cuneate, margins revolute, sometimes wavy, coarsely denticulate near apex, teeth 3–5, apex truncate, retuse, or cuspidate, abaxial surface green, tomentulose, adaxial surface green, glabrous or sparsely puberulent. **Inflorescences** axillary or terminal, 0.6–1.8 cm. **Flowers:** sepals and petals white; nectary tan to brown. **Capsules** 4–6 mm wide, weakly lobed; valves smooth, horns minute or absent, intermediate ridges absent.

Flowering Jan–Apr. Rocky slopes, chaparral; of conservation concern; 500–1100 m; Calif.; Mexico (Baja California).

Ceanothus otayensis is known from the Otay and San Miguel Mountains, southern San Diego County, with at least one locality in northern Baja California. McMinn described *C. otayensis* as a hybrid between *C. crassifolius* and *C. perplexans*. However, neither of the putative parents occurs sympatrically with *C. otayensis*, and its populations do not display the increased variation expected from hybridization.

51. Ceanothus ophiochilus S. Boyd, T. S. Ross & Arnseth, Phytologia 70: 29, figs. 1–4. 1991 • Vail Lake ceanothus [C] [E]

Shrubs, 1–2 m. **Stems** erect to ascending, not rooting at nodes; branchlets reddish brown to gray, terete, ± flexible to rigid, glabrate. **Leaves** both fascicled and not fascicled on same plant; petiole 0–1 mm; blade folded lengthwise abaxially, narrowly oblanceolate to narrowly obovate, 3–7 × 1–3 mm, base cuneate, margins not revolute, usually entire, rarely denticulate near apex, teeth 1–4, apex obtuse, rounded, or cuspidate, abaxial surface pale green, glabrate, adaxial surface pale to yellowish green, glabrate. **Inflorescences** axillary or terminal, 0.7–2 cm. **Flowers:** sepals, petals, and nectary pale blue or pink-tinged. **Capsules** 3–4 mm wide, not lobed; valves smooth, horns minute or absent, intermediate ridges absent.

Flowering Mar–Apr. Rocky soils from deeply weathered gabbro or pyroxene substrates, slopes and ridges, chaparral; of conservation concern; 600–700 m; Calif.

Ceanothus ophiochilus, known only from the northern part of the Palomar Mountains in southwestern Riverside County, is distinctive by having small, more or less terete, fascicled leaves; it is known to hybridize with *C. crassifolius* at one locality.

Ceanothus ophiochilus is in the Center for Plant Conservation's National Collection of Endangered Plants.

14. ADOLPHIA Meisner, Pl. Vasc. Gen. 1: 70; 2: 50. 1837 • Prickbush, spinebrush [For Adolphe Brongniart, 1801–1876, French botanist and student of Rhamnaceae]

Guy L. Nesom

Shrubs, armed with thorns, secondary branches and branchlets green; bud scales absent. **Leaves** early deciduous, usually absent by flowering, opposite or subopposite; blade not gland-dotted; pinnately veined (obscurely, appearing 1-veined or vaguely 3-veined from base). **Inflorescences** axillary, cymes or flowers solitary; peduncles and pedicels not fleshy in fruit. **Pedicels** present. **Flowers** bisexual; hypanthium broadly obconic to hemispheric-campanulate, 1.5–3 mm wide; sepals (4–)5, spreading, whitish to greenish white, ovate-triangular or triangular-deltate, keeled adaxially; petals (4–)5, white, sometimes yellow-tipped, hooded, spatulate, clawed; nectary fleshy, 5-angled, lining hypanthium; stamens (4–)5; ovary superior, 3-locular; style 1. **Fruits** capsules.

Species 2 (2 in the flora): sw United States, Mexico.

SELECTED REFERENCES Mantese, A. and D. Medan. 1993. Anatomía y arquitectura foliares de *Colletia* y *Adolphia* (Rhamnaceae). Darwiniana 32: 91–97. Tortosa, R. D. 1993. Revisión del género *Adolphia* (Rhamnaceae: Colletieae). Darwiniana 32: 185–189.

1. Secondary branches and branchlets glabrous or glabrate, branchlets 1–1.5 mm diam.; leaf blades elliptic-oblong to obovate, 1- or vaguely 3-veined from base; hypanthia broadly obconic, 2.5–3 mm wide, sides nearly straight; sepals triangular-deltate, length equaling width, margins straight; petals 2–2.5 mm, deeply hooded . . . . 1. *Adolphia californica*
1. Secondary branches and branchlets minutely and persistently short-hispid, slowly glabrescent, branchlets usually 0.5–1 mm diam.; leaf blades narrowly oblanceolate to linear, 1-veined from base; hypanthia hemispheric-campanulate, 1.5 mm wide, sides convex; sepals ovate-triangular, length nearly 2 times width, margins curved; petals 1.2–1.5 mm, shallowly hooded . . . . 2. *Adolphia infesta*

1. **Adolphia californica** S. Watson, Proc. Amer. Acad. Arts 11: 126. 1876 • California prickbush or adolphia, spineshrub

Shrubs, 0.5–1.5 m, secondary branches and branchlets glabrous or glabrate, branchlets 1–1.5 mm diam. **Leaves:** blade elliptic-oblong to obovate, 3–12 mm, margins entire, surfaces puberulent; 1- or vaguely 3-veined from base. **Pedicels** 3–6 mm. **Flowers:** hypanthium broadly obconic, 2.5–3 mm wide, sides nearly straight; sepals triangular-deltate, length equaling width, margins straight; petals 2–2.5 mm, deeply hooded.

Flowering Dec–Apr(–May). Rocky slopes, dry canyons and washes, fields, roadsides, arroyos and hillsides near seeps, streambeds, chaparral, coastal sage scrub; 10–200 m; Calif.; Mexico (Baja California).

R. D. Tortosa (1993) treated *Adolphia californica* as a synonym of *A. infesta*, although within the single broadly conceived species he recognized a Texas form and California form, these differing in shape and size of leaves, insertion of staminal filaments, size of the hypanthia, and features of leaf anatomy (according to A. Mantese and D. Medan 1993). He did not find a geographic zone of intergradation but noted that some plants from Zacatecas appear intermediate in floral features and that one collection from southern Mexico was the California type in leaf anatomy and one from Baja California was the Texas type.

F. Shreve and I. L. Wiggins (1964) noted contrasts (as emphasized in the key here) between the two taxa, which they treated as separate species. In addition to the morphological differences, the two are allopatric and appear to be ecologically differentiated. *Adolphia californica* is known in the flora area only from San Diego County.

2. **Adolphia infesta** (Kunth) Meisner, Pl. Vasc. Gen. 2: 50. 1837 • Junco, Texas adolphia [F]

Ceanothus infestus Kunth in A. von Humboldt et al., Nov. Gen. Sp. 7(fol.): 47; 7(qto.): 61; plate 614. 1824 (as infesta); *Colubrina infesta* (Kunth) Schlechtendal

Shrubs, 0.3–1(–2) m, secondary branches and branchlets minutely and persistently short-hispid, slowly glabrescent, branchlets usually 0.5–1 mm diam. **Leaves:** blade narrowly oblanceolate to linear, 3–10 mm, margins entire [rarely dentate], surfaces glabrous; 1-veined from base. **Pedicels** 1–2.5 mm. **Flowers:** hypanthium hemispheric-campanulate, 1.5 mm wide, sides convex; sepals ovate-triangular, length nearly 2 times width, margins curved; petals 1.2–1.5 mm, shallowly hooded.

Flowering Mar–Aug(–Sep). Open slopes and washes, rocky outcrops, limestone and igneous substrates, roadsides, chaparral, oak-juniper and mixed conifer-oak woodlands; 1200–1800 m; Ariz., N.Mex., Tex.; c, n Mexico.

15. GOUANIA Jacquin, Select. Stirp. Amer. Hist., 263, plate 179, fig. 40. 1763 • Chewstick, jaboncillo [For Antoine Gouan, 1733–1821, French botanist and ichthyologist at Montpellier, director of botanical garden in 1767, later professor of botany and medicine]

Guy L. Nesom

Woody vines [climbing shrubs], tendrils present [rarely absent], unarmed; bud scales present. **Leaves** deciduous [persistent], alternate; blade not gland-dotted; pinnately veined [3-veined

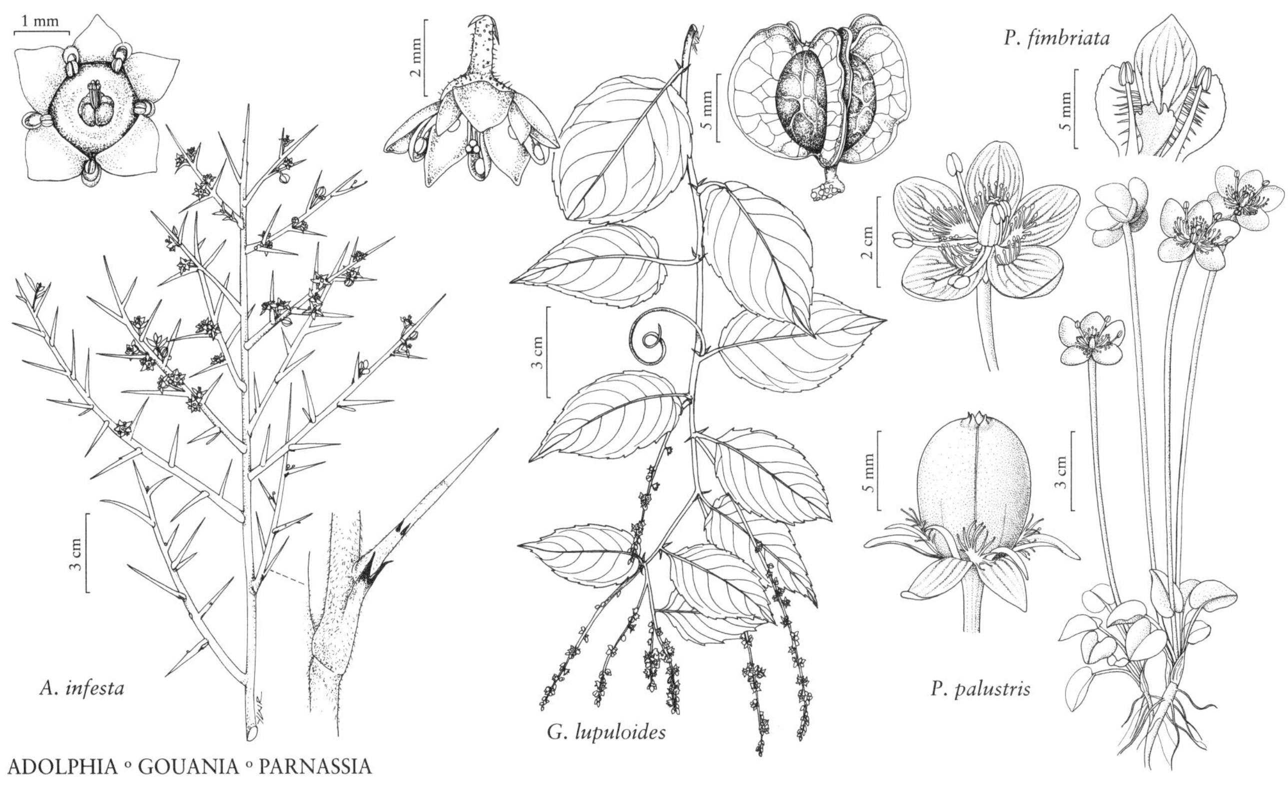

ADOLPHIA ◦ GOUANIA ◦ PARNASSIA

from base]. **Inflorescences:** axillary or terminal, racemelike or paniclelike thyrses; peduncles and pedicels not fleshy in fruit. **Pedicels** present or absent. **Flowers** bisexual [unisexual, plants polygamous]; hypanthium cupulate to campanulate, 1.5–3 mm wide; sepals 5, spreading, pale green, ovate-triangular, keeled adaxially; petals 5 [rarely 0], white to yellowish or greenish, hooded, spatulate, clawed; nectary fleshy, margins 5-lobed [unlobed], lining hypanthium but distally free; stamens 5, enfolded by petals; ovary inferior, 3-locular; styles 3, connate basally. **Fruits** schizocarps, breaking into 3 2-winged samaras [rarely nutlets].

Species 50–70 (1 in the flora): Florida, Mexico, West Indies, Central America, South America, Asia, Africa, Indian Ocean Islands, Pacific Islands, Australia.

Stems of at least some species of *Gouania* are chewed to clean the teeth and harden the gums, hence the common names.

1. Gouania lupuloides (Linnaeus) Urban, Symb. Antill. 4: 378. 1910 • Whiteroot, toothbrush tree [F]

Banisteria lupuloides Linnaeus, Sp. Pl. 1: 427. 1753; *Gouania domingensis* Linnaeus; *G. glabra* Jacquin

Woody vines scrambling and climbing to 7–12 m. **Leaves:** petiole hairy; blade elliptic to ovate or lanceolate, 4–10 cm, base rounded to subcordate, margins serrate to crenate-serrulate, apex acute-acuminate, abaxial surface glabrous [densely hairy] except veins sparsely puberulent-pubescent, adaxial surface glabrous [densely hairy, glabrescent]. **Inflorescences:** racemelike portions 5–20 cm. **Pedicels** (0.5–)1–3 mm, densely hairy. **Flowers:** hypanthium densely [white to] brown-hairy externally, glabrous internally except at orifice; nectary lobes chartaceous. **Schizocarps** 6–13 mm, glabrous; samaras butterfly-shaped, wings reniform, ½ elliptic, 7–14 × 2–6 mm.

Flowering Aug–Mar. Mangroves, coastal hammocks; 0–10 m; Fla.; Mexico; West Indies; Central America.

Some treatments of *Gouania* treated *G. polygama* (Jacquin) Urban as a synonym of *G. lupuloides*, but A. Pool (2014) considered them to be distinct species and identified all *Gouania* in the flora area as *G. lupuloides*. In the flora area, *G. lupuloides* is known from Brevard, Indian River, Manatee, Martin, Miami-Dade, and Monroe counties. *Gouania polygama* occurs in Mexico, the West Indies, Central America, and South America.

CELASTRACEAE R. Brown

• Staff-tree or Bittersweet Family

Jinshuang Ma
Peter W. Ball
Geoffrey A. Levin

Herbs, shrubs, trees, or vines, annual or perennial, deciduous or evergreen, synoecious, dioecious, or polygamomonoecious. **Leaves** alternate, subopposite, opposite, whorled, or fascicled, simple; stipules absent or present; petiole present or absent; blade margins serrate, dentate, spiny, or entire; venation pinnate, palmate, or 1-veined, sometimes obscure. **Inflorescences** unisexual or bisexual, terminal or axillary, cymes, racemes, panicles, thyrses, or fascicles, or flowers solitary. **Flowers** bisexual or unisexual, radially symmetric or weakly asymmetric; perianth and androecium hypogynous or perigynous; hypanthium free, completely adnate to ovary, or absent; sepals (3–)4–5[–7], distinct or connate proximally; petals 0 or (3–)4–5[–7], distinct; nectary present, rudimentary, or absent; stamens 3–5[–10], distinct, free or adnate to nectary; anthers dehiscing by longitudinal slits; staminodes 0 or [4–]5[–7]; pistil 1, 1–5-carpellate, ovary superior, often embedded in nectary, to ½ inferior, 1–5-locular, placentation axile or parietal; ovules 1–2[–4] or 100–2000+ per locule, anatropous; styles 0, 1, or 3, connate proximally; stigmas 2–5. **Fruits** capsules, dehiscence loculicidal, drupes, or nutlike (small, hard-walled, indehiscent, 1-locular, and 1-seeded) [berries or samaras]. **Seeds** 1, 2, 40–70, or 100–2000+ per locule, often winged or covered by brightly colored, pulpy aril.

Genera ca. 100, species ca. 1400 (12 genera, 34 species in the flora): North America, Mexico, West Indies, Central America, South America, Eurasia, Africa, Atlantic Islands, Indian Ocean Islands, Pacific Islands, Australia.

As treated here, Celastraceae include Hippocrateaceae Jussieu and Parnassiaceae Martinov. Both DNA sequence data and morphology place *Hippocratea* and its relatives nested within Celastraceae as subfam. Hippocrateoideae Lindley (M. P. Simmons et al. 2001; Simmons 2004, 2004b). Placement of Parnassiaceae (*Lepuropetalon* and *Parnassia*) is less certain. The group has long been associated with Saxifragaceae (such as by J. D. Hooker 1865b; A. Cronquist 1981), but broad-scale phylogenetic analyses utilizing DNA sequences have aligned Parnassiaceae with Celastraceae, either as a sister family or as a basal member of Celastraceae (M. W. Chase et al. 1993; Simmons et al. 2001b; Simmons 2004; Zhang L. B. and Simmons 2006). Including Parnassiaceae within Celastraceae follows APGIII (Angiosperm Phylogeny Group 2009).

Glossopetalon (sometimes under the illegitimate name *Forsellesia* Greene), included within Celastraceae in many local floras in North America, belongs in Crossosomataceae (R. F. Thorne and R. Scogin 1978; V. Sosa and M. W. Chase 2003); see *Flora of North America North of Mexico*, volume 9, page 9.

Some members of Celastraceae are of economic importance. Species of *Celastrus*, *Euonymus*, *Maytenus*, and *Paxistima* are grown as ornamentals, and *Euonymus, Hippocratea,* and *Maytenus* have medicinal uses.

SELECTED REFERENCE Brizicky, G. K. 1964. The genera of Celastraceae in the southeastern United States. J. Arnold Arbor. 45: 206–234.

1. Herbs; staminodes opposite petals.
   2. Perennials; petals 3–22 mm, usually longer than sepals; stigmas 4; capsules 4-valved . . . . 1. *Parnassia*, p. 113
   2. Annuals; petals 0–0.4 mm, much shorter than sepals; stigmas 3; capsules 3-valved. . . . . 2. *Lepuropetalon*, p. 117
1. Shrubs, trees, or vines; staminodes alternate with petals or 0.
   3. Vines.
      4. Plants climbing by adventitious roots; sepals and petals 4; ovaries and capsules 4-locular . . . . 7. *Euonymus* (in part), p. 122
      4. Plants twining or clambering; sepals and petals 5; ovaries and capsules 3-locular.
         5. Leaves opposite, persistent; seeds 5–6 per locule, winged, arils absent . . . . 3. *Hippocratea*, p. 118
         5. Leaves alternate, deciduous; seeds 2 per locule, not winged, arils red. . . . . . . 5. *Celastrus*, p. 120
   3. Shrubs or trees.
      6. Leaves caducous, plants appearing leafless, blades 1–2 mm; fruits capsules, apices beaked . . . . 4. *Canotia*, p. 119
      6. Leaves deciduous or persistent, blades 3–160 mm; fruits capsules, drupes, or nutlike, apices not beaked except when nutlike.
         7. Leaves alternate or fascicled.
            8. Sepals, petals, and stamens 4; fruits drupes . . . . 12. *Schaefferia*, p. 131
            8. Sepals, petals, and stamens 5; fruits capsules or nutlike.
               9. Fruits capsules; arils red; inflorescences axillary; leaf blade venation pinnate . . . . 9. *Maytenus*, p. 126
               9. Fruits nutlike; arils absent; inflorescences terminal; leaf blades 1-veined . . . . 10. *Mortonia*, p. 128
         7. Leaves opposite or whorled.
            10. Leaves deciduous; arils yellow, orange, or red . . . . 7. *Euonymus* (in part), p. 122
            10. Leaves persistent; arils white, yellow, or absent.
               11. Branchlets terete; ovaries 4-locular; stigmas 4; fruits red drupes . . . . 6. *Crossopetalum*, p. 121
               11. Branchlets 4-angled; ovaries 2-locular; stigmas 2; fruits capsules or bluish black drupes.
                  12. Shrubs or trees, to 8 m; flowers unisexual (plants dioecious); fruits drupes . . . . 8. *Gyminda*, p. 125
                  12. Shrubs, 0.1–1 m; flowers bisexual; fruits capsules. . . . . 11. *Paxistima*, p. 129

1. PARNASSIA Linnaeus, Sp. Pl. 1: 273. 1753; Gen. Pl. ed. 5, 133. 1754 • Grass of Parnassus, bog star, parnassie [Greek *Parnassos*, alluding to fabled origin on slopes of Mount Parnassus]

Peter W. Ball

Herbs, perennial with caudices or rarely rhizomes. **Stems** erect, unbranched, scapelike. **Leaves:** basal in rosettes or 1–2 per node on rhizomes (*P. caroliniana*), cauline (0–)1[–8], alternate; stipules absent; petiole present in basal leaves, usually absent in cauline leaf; blade margins entire; venation palmate. **Inflorescences** terminal, flowers solitary. **Flowers** bisexual, radially symmetric or ± asymmetric; perianth and androecium hypogynous or perigynous; hypanthium absent or completely adnate to ovary; sepals 5, connate proximally; petals 5, white [yellowish] with distinct yellowish or greenish or gray-brown veins, 3–22 mm, usually longer than sepals; nectary absent; stamens 5; staminodes 5, opposite petals, usually deeply divided, sometimes undivided, gland-tipped or glandular at apex [without glands]; pistil [3–]4[–5]-carpellate; ovary superior to ½ inferior, 1-locular, placentation parietal; style absent or essentially so; stigmas [3–]4[–5]; ovules 100–2000+. **Fruits** capsules, 1-locular, [3–]4[–5]-valved, ellipsoid to globose, apex not beaked. **Seeds** 100–2000+ per fruit, oblong, winged; aril absent. $x = 9$.

Species ca. 70 (9 in the flora): North America, Mexico, Europe, Asia, nw Africa.

The North American species of *Parnassia* usually occur in moist to wet sites on neutral to base-rich substrates, but *P. asarifolia* often occurs on acidic substrates.

The treatment of *Parnassia cirrata* and *P. fimbriata* follows that proposed by R. B. Phillips (1980).

SELECTED REFERENCE Phillips, R. B. 1980. Systematics of *Parnassia* L. (Parnassiaceae): Generic Overview and Revision of North American Taxa. Ph.D. dissertation. University of California, Berkeley.

1. Petal margins fimbriate proximally; cauline leaf usually on middle to distal ½ of stem, rarely absent.
   2. Staminodes irregularly divided into oblong, obtuse lobes, usually glandular at apex but tip not differentiated into distinct gland; sepal margins denticulate, erose, or short-fimbriate distally . . . . . 4. *Parnassia fimbriata*
   2. Staminodes scalelike proximally, distally divided into gland-tipped filaments; sepal margins usually entire, rarely minutely denticulate distally . . . . . 5. *Parnassia cirrata*
1. Petal margins entire or undulate; cauline leaf usually on proximal ½ to middle of stem or absent, rarely on distal ½.
   3. Sepal margins not hyaline; staminodes unlobed or divided distally into 3–27 filaments.
      4. Petal lengths 1.5–2 times sepals; anthers 1.5–2.8 mm . . . . . 1. *Parnassia palustris*
      4. Petal lengths 0.8–1.5 times sepals; anthers 0.7–1.6 mm.
         5. Petals 5–13-veined; staminodes with 5–7(–9) filaments; anthers 1–1.6 mm . . . . . 2. *Parnassia parviflora*
         5. Petals usually 3-veined; staminodes unlobed or with 3–5 filaments; anthers 0.7–1 mm . . . . . 3. *Parnassia kotzebuei*
   3. Sepal margins hyaline, mostly 0.2–0.5 mm wide; staminodes 3-fid almost to base.
      6. Leaf blades mostly wider than long; petal bases abruptly contracted to claw . . . . . 6. *Parnassia asarifolia*
      6. Leaf blades longer than to ca. as long as wide; petal bases cuneate to rounded.
         7. Staminodes 4–7 mm, shorter than or ca. equaling stamens . . . . . 7. *Parnassia glauca*
         7. Staminodes 9–16 mm, longer than stamens.

[8. Shifted to left margin.—Ed.]

8. Staminode glands elliptic to subglobose, 0.4–0.6 mm; plants with caudices; basal leaves in rosettes; ovaries green, sometimes whitish at base . . . . 8. *Parnassia grandifolia*
8. Staminode glands lanceolate, 1–1.7 mm; plants with creeping rhizomes; basal leaves 1–2 per node on rhizomes; ovaries white . . . . 9. *Parnassia caroliniana*

1. **Parnassia palustris** Linnaeus, Sp. Pl. 1: 273. 1753 • Parnassie des marais [F]

Parnassia californica (A. Gray) Greene; *P. montanensis* Fernald & Rydberg; *P. palustris* var. *californica* A. Gray; *P. palustris* var. *montanensis* (Fernald & Rydberg) C. L. Hitchcock; *P. palustris* subsp. *neogaea* (Fernald) Hultén; *P. palustris* var. *neogaea* Fernald

Herbs with caudices. **Stems** 8–35(–50) cm. **Leaves:** basal in rosettes; petiole 2–10 cm; blade (of larger leaves) ovate to suborbiculate, 6–40 × 4–30 mm, base rounded to cordate, apex rounded; cauline usually on proximal ½ of stem, rarely on distal ½. **Flowers:** sepals spreading or ± reflexed in fruit, linear to narrowly lanceolate, 4–11 mm, margins not hyaline, entire, apex subacute; petals 5–11-veined, obovate, (7–)8–17(–20) × 5–12 mm, length 1.5–2 times sepals, base rounded, margins entire; stamens 6–8 mm; anthers 1.5–2.8 mm; staminodes obovate, divided distally into (7–)9–27 gland-tipped filaments, 5–9 mm, ca. as long as stamens, apical glands suborbicular to ovoid, 0.2–0.5 mm; ovary green. **Capsules** (6–)8–10 mm. **2*n*** = 18, 36.

Flowering summer. Seasonally wet meadows, fens, thickets, ditches, shores, rocky or gravelly seashores, wet rocks; 0–3800 m; Alta., B.C., Man., Nfld. and Labr., N.W.T., Nunavut, Ont., Que., Sask., Yukon; Alaska, Ariz., Calif., Colo., Idaho, Mich., Minn., Mont., Nev., N.Mex., N.Dak., Oreg., S.Dak., Utah, Wash., Wis., Wyo.; Eurasia; nw Africa; temperate and subarctic regions.

The North American populations of *Parnassia palustris* have been treated as subsp. *neogaea*, which has petals usually with five to nine veins and staminodes tapering proximally. Most European populations of var. *palustris* have petals with 10 to 17 veins and staminodes abruptly clawed proximally. Populations from northern Europe and from some of the mountains of central Europe often have petals and staminodes similar to those of var. *neogaea*. European authors have sometimes treated those populations as subsp. *obtusiflora* (Ruprecht) D. A. Webb, but there is no general agreement on how the variation in this species should be treated.

Chromosome numbers in Eurasia have been reported as 2*n* = 17, 18, 27, 32–37, 43–45, 54. Attempts to correlate morphology with differences in chromosome number have been unsuccessful except for slight differences in pollen diameter and seed length (U.-M. Hultgård 1987; R. J. Gornall and J. E. Wentworth 1993). As noted by Gu C. Z. and Hultgård (2001), *Parnassia palustris* shows much variation in stem height; basal leaf shape and size; cauline leaf position, shape, and size; sepal and petal shape, size, and venation; and number and length of staminode filaments through much of Eurasia, making it very difficult to subdivide it into consistently distinct taxa over the total range. This diversity in Eurasia is paralleled in North America.

Within North America several variants have sometimes been recognized and even treated as distinct species. Variety *californica* (*Parnassia californica*) has the cauline leaf in the distal half of the stem, and most populations in California and southern Oregon display this feature, but such plants rarely occur elsewhere in North America and some populations in eastern California have the cauline leaf in the proximal half of the stem. Populations in the Rocky Mountains and some adjacent ranges are sometimes recognized as *P. montanensis*, which has petals 7–9 × 5–7 mm and staminodes with seven to ten filaments. These are somewhat intermediate between *P. palustris* and *P. parviflora*, and it has been suggested that they may be of hybrid origin, although a similar variant occurs in northern Asia in the absence of *P. parviflora*. It is unclear whether all *Parnassia* from Arizona are *P. parviflora* or whether some are this small-flowered variant of *P. palustris*.

2. **Parnassia parviflora** de Candolle in A. P. de Candolle and A. L. P. P. de Candolle, Prodr. 1: 320. 1824 [E]

Parnassia palustris Linnaeus var. *parviflora* (de Candolle) B. Boivin

Herbs with caudices. **Stems** 2–35 cm. **Leaves:** basal in rosettes; petiole 0.4–2 cm; blade (of larger leaves) ovate to oblong, 6–35 × 5–25 mm, base cuneate to subcordate, apex rounded to subacute; cauline on proximal ½ to middle of stem. **Flowers:** sepals spreading in fruit, linear-lanceolate to oblong or elliptic-oblong, 3–6 mm, margins not hyaline, entire, apex obtuse; petals 5–13-veined, oblong to elliptic, 3.5–10 × 4–6 mm, length 1.1–1.5 times sepals, base rounded to cuneate, margins entire; stamens (2–)4–7 mm; anthers 1–1.6 mm; staminodes obovate, divided distally into 5–7(–9) gland-tipped filaments, (2–)3.5–5 mm, shorter than stamens, apical glands suborbicular, 0.2–0.3 mm; ovary green. **Capsules** 7–10 mm. **2*n*** = 36.

Flowering summer. Wet, calcareous shores, meadows, fens, seepy scree slopes; 10–2900 m; Alta., B.C., Man., Nfld. and Labr., N.S., Nunavut, Ont., P.E.I., Que., Sask.; Alaska, Ariz., Calif., Colo., Idaho, Mich., Mont., Nev., N.Dak., S.Dak., Utah, Wash., Wis., Wyo.

Parnassia parviflora has been included in *P. palustris* by some authors. Small-flowered plants of *P. palustris* usually have the staminodes divided into about nine filaments distally and the anthers exceed 1.5 mm, but rarely some plants cannot be clearly assigned to one or other of these species. In Nunavut, *P. parviflora* is known only from Akimiski Island in James Bay.

3. Parnassia kotzebuei Chamisso ex Sprengel, Syst. Veg. 1: 951. 1824

Parnassia kotzebuei var. *pumila* C. L. Hitchcock & Ownbey

Herbs with caudices. **Stems** 2–15(–25) cm. **Leaves:** basal in rosettes; petiole 0.2–1(–2) cm; blade (of larger leaves) deltate-ovate to rhombic-ovate, 3.5–12(–30) × 4–10(–25) mm, base cordate to cuneate, apex acute to obtuse; cauline on proximal ½ of stem or absent. **Flowers:** sepals spreading in fruit, oblong-lanceolate to oblanceolate, 4–8 mm, margins not hyaline, entire, apex obtuse; petals usually 3-veined, oblong to elliptic, 3–7 × 2–3 mm, length 0.8–1.3 times sepals, base rounded or cuneate, margins entire; stamens 3–4.5 mm; anthers 0.7–1 mm; staminodes obovate, unlobed or divided distally into 3–5 gland-tipped filaments, 1.5–3 mm, shorter than stamens, apical glands suborbicular, 0.1–0.2 mm; ovary green. **Capsules** 6–12 mm. **2*n*** = 18, 36.

Flowering summer. Moist or seasonally dry shores, stream banks, riverbanks, meadows, tundra, seepage areas, talus, snowbeds, wet calcareous rocky places, open conifer forests; 0–3800 m; Greenland; Alta., B.C., Man., Nfld. and Labr., N.W.T., Nunavut, Ont., Que., Sask., Yukon; Alaska, Colo., Idaho, Mont., Nev., Wash., Wyo.; ne Asia.

Variety *pumila* was described as endemic to the western mountains, but identical plants occur throughout much of the range of the species.

4. Parnassia fimbriata K. D. König, Ann. Bot. (König & Sims) 1: 391. 1804 E F

Herbs with caudices. **Stems** 10–35 cm. **Leaves:** basal in rosettes; petiole 2–16 cm; blade (of larger leaves) orbiculate-reniform to reniform, 10–50 × 15–60 mm, base cordate, apex rounded; cauline usually on middle to distal ½ of stem. **Flowers:** sepals reflexed in fruit, oblong-elliptic to ovate, 4–6 mm, margins hyaline, 0.2–0.5 mm wide, denticulate, erose, or short-fimbriate distally, apex rounded; petals 5–7-veined, oblanceolate to obovate, 8–14 × 4–8 mm, length 2 times sepals, base cuneate, margins fimbriate proximally; stamens 6–9 mm; anthers 1.5–2.5 mm; staminodes oblong-obovate, irregularly divided into 5–10+ oblong, obtuse lobes, usually glandular at apex but tip not differentiated into distinct gland, 3–5 mm, shorter than stamens; ovary green. **Capsules** 8–12 mm. **2*n*** = 36.

Flowering summer. Stream banks, wet meadows, fens, seeps in forest glades, alpine ravines; 10–3200 m; Alta., B.C., N.W.T., Yukon; Alaska, Calif., Colo., Idaho, Mont., Nev., N.Mex., Oreg., Utah, Wash., Wyo.

5. Parnassia cirrata Piper, Erythea 7: 128. 1899

Herbs with caudices. **Stems** 15–40 cm. **Leaves:** basal in rosettes; petiole 1–12 cm; blade (of larger leaves) ovate-orbiculate to elliptic-ovate, 15–60 × 7–50 mm, base cuneate, rounded, or weakly cordate, apex rounded; cauline usually on middle to distal ½ of stem, rarely absent. **Flowers:** sepals reflexed in fruit, elliptic or oblong-lanceolate to ovate, 4–7 mm, margins sometimes scarious, to 0.1 mm wide, usually entire, rarely minutely denticulate distally, apex rounded; petals 5–7-veined, oblanceolate to obovate or elliptic, 8–15 × 3.3–9.8 mm, length 2 times sepals, base cuneate, margins fimbriate proximally; stamens 6–9 mm; anthers 1.5–2.2 mm; staminodes scalelike proximally, distally divided into 5–15 gland-tipped filaments, 3.5–6 mm, shorter than stamens, apical glands globose, 0.2–0.4 mm; ovary green. **Capsules** 10 mm.

Varieties 2 (2 in the flora): w North America, nw Mexico.

1. Larger petals 3.3–5.2(–7) mm wide, longer fimbriae (3.3–)3.5–6.5 mm; s California . . . . 5a. *Parnassia cirrata* var. *cirrata*
1. Larger petals (4–)5–10 mm wide, longer fimbriae 1–3(–3.5) mm; British Columbia, n California, Idaho, Nevada, Oregon, Washington . . . . 5b. *Parnassia cirrata* var. *intermedia*

5a. Parnassia cirrata Piper var. **cirrata**

Leaves: blade (of larger leaves) 7–25(–35) mm wide. **Flowers:** sepals 1.8–2.6(–4) mm wide; larger petals 3.3–5.2(–7) mm wide, longer fimbriae (3.3–) 3.5–6.5 mm.

Flowering summer. Marshes, wet meadows, streamsides; 1000–3000 m; Calif.; Mexico (Durango).

Variety *cirrata* occurs in the flora area only in the San Bernardino and San Gabriel mountains.

5b. Parnassia cirrata Piper var. **intermedia** (Rydberg) P. K. Holmgren & N. H. Holmgren in A. Cronquist et al., Intermount. Fl. 3(A): 61. 1997 [E]

Parnassia intermedia Rydberg in N. L. Britton et al., N. Amer. Fl. 22: 78. 1905; *P. fimbriata* Koenig var. *hoodiana* C. L. Hitchcock; *P. fimbriata* var. *intermedia* (Rydberg) C. L. Hitchcock

Leaves: blade (of larger leaves) (10–)13–50 mm wide. **Flowers:** sepals (2–)2.5–3.5 mm wide; larger petals (4–)5–10 mm wide, longer fimbriae 1–3 (–3.5) mm.

Flowering summer. Marshes, fens, wet meadows, streamsides; 1000–3000 m; B.C.; Calif., Idaho, Nev., Oreg., Wash.

In California, var. *intermedia* is known only from Shasta, Siskiyou, and Trinity counties.

6. Parnassia asarifolia Ventenat, Jard. Malmaison 1: plate 39. 1804 [E]

Herbs with caudices. **Stems** 18–50 cm. **Leaves:** basal in rosettes; petiole 6–17 cm; blade (of larger leaves) reniform to reniform-orbiculate, 20–60 × 25–100 mm, mostly wider than long, base cordate, apex rounded; cauline on proximal ½ to middle of stem or absent. **Flowers:** sepals reflexed in fruit, oblong to obovate, 2.5–6.5 mm, margins hyaline, 0.2–0.4 mm wide, entire, apex rounded; petals 11–18-veined, ovate-elliptic, 10–18 × 7–11 mm, length 2–3 times sepals, base abruptly contracted to 2–3.5 mm claw, margins entire or undulate; stamens 8.5–11.5 mm; anthers 2.2–3.2 mm; staminodes 3-fid almost to base, gland-tipped, 5–9 mm, shorter than stamens, apical glands ovoid-conical, 0.5–0.9 mm; ovary green. **Capsules** 12 mm. **2*n*** = 32.

Flowering summer–early fall. Fens, wet woods, rocky banks, often on acidic soils; 200–1500 m; Ala., Ark., Ga., Ky., Md., N.C., S.C., Tenn., Tex., Va., W.Va.

Parnassia asarifolia is considered endangered in Kentucky and Maryland. It is uncommon throughout its range.

7. Parnassia glauca Rafinesque, Autik. Bot., 42. 1840 [E]

Herbs with caudices. **Stems** 10–60 cm. **Leaves:** basal in rosettes; petiole 1.5–17 cm; blade (of larger leaves) oblong to orbiculate-ovate, 20–70 × 10–50 mm, longer than to ca. as long as wide, base cuneate to subcordate, apex obtuse to subacute; cauline on proximal ½ of stem or absent. **Flowers:** sepals reflexed in fruit, oblong, elliptic, or ovate, 2.5–5 mm, margins hyaline, 0.2–0.5 mm wide, entire, apex rounded; petals 5–12-veined, oblong to ovate, 9–18 × 6–10 mm, length 2–4 times sepals, base cuneate to rounded, margins entire or undulate; stamens 7–10 mm; anthers 1.2–2.8 mm; staminodes 3-fid almost to base, gland-tipped, 4–7 mm, shorter than to ca. equaling stamens, apical glands elliptic-oblong to subreniform, 0.3–0.7 mm; ovary green. **Capsules** 12–14 mm. **2*n*** = 32.

Flowering summer–early fall. Seasonally wet meadows, shores, fens, ditches, seeps, wet, calcareous soils; 50–700 m; Man., N.B., Nfld. and Labr. (Nfld.), Ont., Que., Sask.; Conn., Ill., Ind., Iowa, Maine, Mass., Mich., Minn., N.H., N.J., N.Y., N.Dak., Ohio, Pa., R.I., S.Dak., Vt., Wis.

The name *Parnassia caroliniana* has been misapplied to this species (for example, N. L. Britton and A. Brown 1913).

8. Parnassia grandifolia de Candolle in A. P. de Candolle and A. L. P. P. de Candolle, Prodr. 1: 320. 1824 [E]

Herbs with caudices. **Stems** 12–70 cm. **Leaves:** basal in rosettes; petiole 3–15 cm; blade (of larger leaves) oblong-ovate to suborbiculate, 25–80 × 20–70 mm, longer than to ca. as long as wide, base rounded to subcordate, apex obtuse; cauline on proximal ½ of stem or absent. **Flowers:** sepals reflexed in fruit, oblong to ovate, 2.5–5 mm, margins hyaline, 0.2–0.5 mm wide, entire, apex obtuse; petals 7–11-veined, oblong to ovate, 15–22 × 6–10 mm, length 3–4 times sepals, base cuneate to rounded, margins entire or undulate; stamens 7–9 mm; anthers 2–3 mm; staminodes 3-fid almost to base, gland-tipped, 10–16 mm, longer than stamens, apical glands elliptic to subglobose, 0.4–0.6 mm; ovary green, sometimes whitish at base. **Capsules** 10–15 mm. **2*n*** = 32.

Flowering mid summer–fall. Wet, calcareous rocks, shores, meadows, fens; 200–1400 m; Ala., Ark., Fla., Ga., Ky., La., Miss., Mo., N.C., Okla., S.C., Tenn., Tex., Va., W.Va.

Parnassia grandifolia is uncommon throughout most of its range; it is listed as endangered in Florida and Kentucky, threatened in North Carolina, and of special concern in Tennessee.

9. **Parnassia caroliniana** Michaux, Fl. Bor.-Amer. 1: 184. 1803 [E]

Herbs with horizontal creeping rhizomes. **Stems** 20–60 cm. **Leaves:** basal 1–2 per node on rhizomes; petiole 8–22 cm; blade (of larger leaves) ovate to suborbiculate, 20–75 × 15–70 mm, longer than to ca. as long as wide, base rounded to subcordate, apex obtuse; cauline on proximal ½ of stem or absent. **Flowers:** sepals reflexed in fruit, oblong to oblong-elliptic, 3.5–5 mm, margins hyaline, 0.2 mm wide, entire, apex obtuse; petals 7–12-veined, broadly ovate, 14–20 × 9–12 mm, length 3–4 times sepals, base rounded, margins entire or undulate; stamens 7–11 mm; anthers 1.8–3 mm; staminodes 3-fid almost to base, gland-tipped, 9–14 mm, longer than stamens, apical glands lanceolate, 1–1.7 mm; ovary white. **Capsules** 10–15 mm.

Flowering late summer–fall. Wet pine savannas, seepage slopes, streamhead ecotones, all subject to recurring fires; 10–300 m; Fla., N.C., S.C.

Parnassia caroliniana is rare throughout its range; it is listed as endangered in Florida and North Carolina.

2. LEPUROPETALON Elliott, Sketch Bot. S. Carolina 1: 370. 1817 • [Greek *lepyron*, scale, and *petalon*, petals, alluding to scalelike petals inserted into calyx]

Dayle E. Saar

Herbs, winter annual, often hemispheric tufts. **Stems** erect, branched or unbranched. **Leaves** basal and cauline, alternate or subopposite; stipules absent; petiole present; blade margins entire; venation palmate. **Inflorescences** terminal, cymules or flowers solitary. **Flowers** bisexual, radially symmetric or weakly asymmetric; perianth and androecium perigynous; hypanthium completely adnate to ovary; sepals 5, connate proximally; petals 0 or (4–)5, white, 0–0.4 mm, much shorter than sepals; nectary absent; stamens 5; staminodes 5, opposite petals, undivided, not gland-tipped; pistil 3-carpellate; ovary inferior, 1-locular, placentation parietal; styles 3, weakly connate proximally; stigmas 3; ovules 40–70. **Fruits** capsules, 1-locular, 3-valved, obpyramidal, apex not beaked. **Seeds** 40–70 per fruit, oblong, not winged; aril absent. ***x*** = 23.

Species 1: sc, se United States, Mexico, South America (Chile, Uruguay).

SELECTED REFERENCE Ward, D. B. and A. K. Gholson. 1987. The hidden abundance of *Lepuropetalon spathulatum* (Saxifragaceae) and its first reported occurrence in Florida. Castanea 52: 59–66.

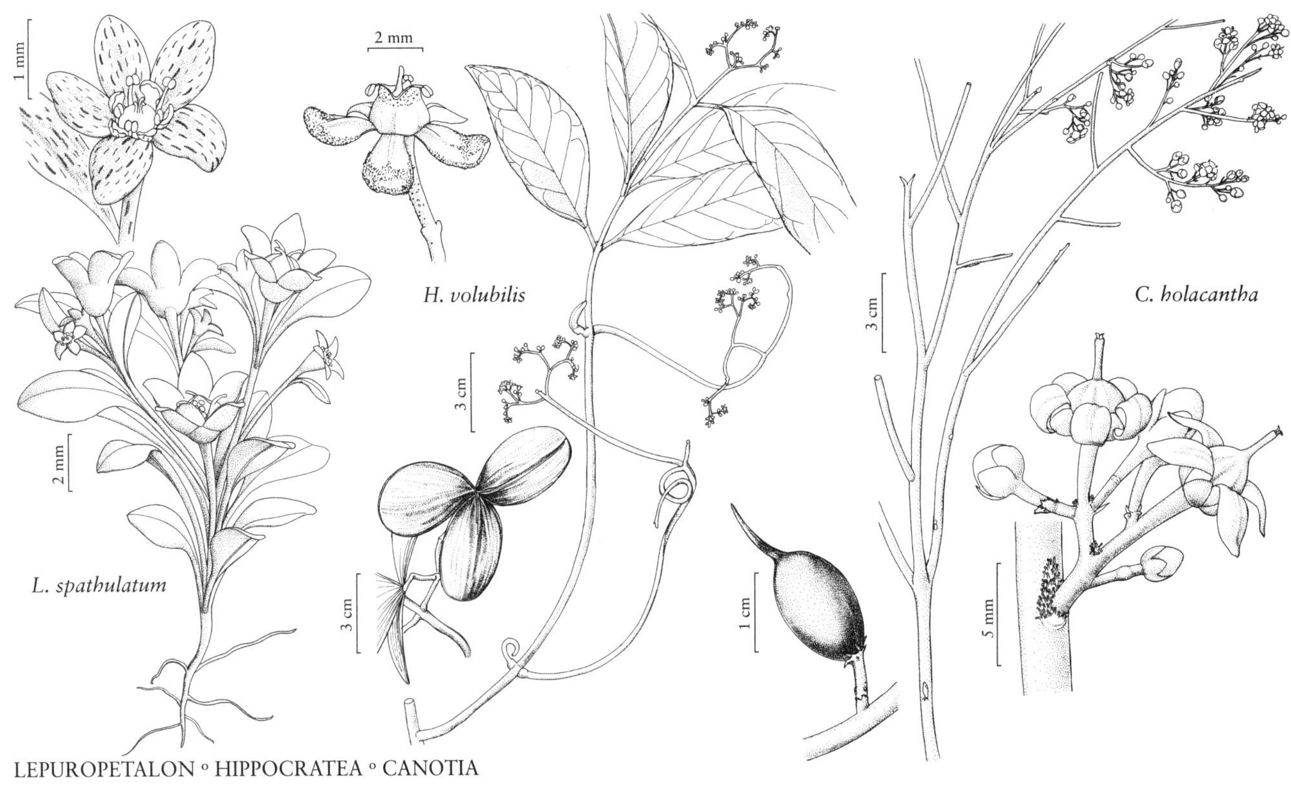

LEPUROPETALON ◦ HIPPOCRATEA ◦ CANOTIA

1. Lepuropetalon spathulatum Elliott, Sketch Bot. S. Carolina 1: 370. 1817 • Petiteplant, little people [F]

Herbs to 3 cm. **Stems:** internodes congested. **Leaves:** rosettes 1–3 cm, usually with 2 minute leaves proximally; blade spatulate, 3–10 × 1–2 mm, usually red-punctate in rows, surfaces glabrous. **Pedicels** 0.5–3 mm. **Flowers:** calyx 2–3 mm diam., sepals 1–1.5 mm; petals 0–0.4 mm. **Capsules** 1.3–1.6 mm. **Seeds** 0.2–0.3 mm, reticulate; endosperm absent. **2*n*** = 46.

Flowering spring. Damp mineral soils, sandy ditches, granitic outcroppings, sandy or gravelly substrates, disturbed areas; 0–300 m; Ala., Ark., Fla., Ga., La., Miss., N.C., Okla., S.C., Tex.; e Mexico; South America (Chile, Uruguay).

Lepuropetalon spathulatum is among the smallest terrestrial angiosperms (D. R. Morgan and D. E. Soltis 1993).

3. HIPPOCRATEA Linnaeus, Sp. Pl. 2: 1191. 1753; Gen. Pl. ed. 5, 498. 1754 • [For Hippocrates, ca. 460–370 BCE, Greek physician]

Jinshuang Ma

Vines, clambering. **Branchlets** terete to 4-angled. **Leaves** persistent, opposite; stipules present; petiole present; blade margins crenate-serrate; venation pinnate. **Inflorescences** axillary, cymes. **Flowers** bisexual, radially symmetric; perianth and androecium hypogynous; hypanthium absent; sepals 5, distinct; petals 5, pale yellow to white; nectary extrastaminal, annular, fleshy; stamens 3, adnate to nectary margin; staminodes 0; pistil 3-carpellate; ovary superior, immersed in and adnate to nectary, 3-locular, placentation axile; style 1; stigmas 3; ovules 6–8 per locule. **Fruits** capsules, 3-locular, deeply 3-parted longitudinally, segments obovate-elliptic or narrowly elliptic, apex not beaked. **Seeds** 5–6 per locule, ellipsoid, winged in proximal ½; aril absent.

Species 3 (1 in the flora): Florida, Mexico, West Indies, Central America, South America, Africa.

SELECTED REFERENCE Smith, A. C. 1940. The American species of Hippocrateaceae. Brittonia 3: 356–367.

1. **Hippocratea volubilis** Linnaeus, Sp. Pl. 2: 1191. 1753 • Medicine vine [F]

Vines 20–25 m. **Stems** much-branched, ascending, green. **Leaves:** blade elliptic to ovate or obovate, 5–14 × 2–8 cm. **Inflorescences** 4–12 cm, brownish tomentulose or puberulent. **Flowers** 4–8 mm diam.; sepals 0.5–1.2 mm; petals 2.5–4 mm. **Capsules** 4–8 × 1.5–5 cm. **Seeds** 1.3–2.5 × 4–7 mm, wing obovate-oblong, 2–4 cm.

Flowering spring–summer; fruiting fall–winter. Hammocks, stream banks, shores, mangrove and hardwood swamps; 0–10 m; Fla.; s Mexico; West Indies; Central America; South America.

Hippocratea volubilis often climbs to the tops of trees, binding itself and the supporting vegetation into an impenetrable network.

4. CANOTIA Torrey ex A. Gray in J. C. Ives, Rep. Colorado R. 4: 15. 1861 • [Spanish name in Mexico]

Jinshuang Ma

Geoffrey A. Levin

Shrubs or trees. Branchlets terete. **Leaves** caducous, scalelike (plants appearing leafless), alternate; stipules absent; petiole absent; blade margins entire; venation obscure. **Inflorescences** axillary, cymes or thyrses. **Flowers** bisexual, radially symmetric; perianth and androecium hypogynous; hypanthium absent; sepals 5, distinct; petals 5, greenish white to yellowish white; nectary absent; stamens 5; staminodes 0; pistil 5-carpellate; ovary superior, 5-locular, placentation axile; style 1; stigmas 5; ovules 2 per locule. **Fruits** capsules, 5-locular, oblong, apex beaked. **Seeds** 1–2 per locule, oval to oblong, winged in proximal ½; aril absent. $x = 15$.

Species 2 (1 in the flora): Arizona, n Mexico.

The systematic position of *Canotia* has been much disputed. Morphologic (M. C. Johnston 1975; H. Tobe and P. H. Raven 1993) and DNA-sequence data (M. P. Simmons et al. 2008) support placement in Celastraceae near *Euonymus* and *Paxistima*.

The other species of *Canotia*, *C. wendtii* M. C. Johnston, is restricted to Chihuahua, Mexico.

1. **Canotia holacantha** Torrey ex A. Gray in J. C. Ives, Rep. Colorado R. 4: 15. 1861 • Crucifixion thorn [F]

Shrubs or trees 2–5(–10) m. **Stems** many-branched, virgate, principal branches ascending, pale green. **Flowers:** sepals 2–4 mm; petals oblong, 4–5 × 2 mm; stamens 5, filaments as long as petals; styles united into beak. **Capsules** narrowly obovoid, 21–27 × 6–10 mm (including beak), opening from apex to middle. **Seeds** basally compressed. $2n = 30$.

Flowering spring–fall; fruiting late summer–winter, fruit persisting into spring. Open roadsides, deserts; 0–2000 m; Ariz.; Mexico (Sonora).

Canotia holacantha has been reported from Utah (J. R. Hastings et al. 1972; M. C. Johnston 1975) but was not included in *A Utah Flora* (S. L. Welsh et al. 1993, 2003); no corroborating specimens have been located.

5. CELASTRUS Linnaeus, Sp. Pl. 1: 196. 1753; Gen. Pl. ed. 5, 91. 1754 • Bittersweet [Greek *kelastros,* ancient name for holly, *Ilex aquifolium*]

Jinshuang Ma

Geoffrey A. Levin

Vines, twining, polygamodioecious. **Branchlets** terete. **Leaves** deciduous [persistent], alternate; stipules present; petiole present; blade margins denticulate; venation pinnate. **Inflorescences** terminal or axillary, panicles or cymes [racemes]. **Flowers** bisexual and unisexual, radially symmetric; perianth and androecium hypogynous; hypanthium absent; sepals 5, distinct; petals 5, white or greenish white; nectary intrastaminal, fleshy. **Bisexual flowers:** stamens 5, free from and inserted under nectary; staminodes 0; pistil 3-carpellate; ovary superior, 3-locular; placentation axile; style 1; stigmas 3; ovules 2 per locule. **Staminate flowers:** stamens 5; free from and inserted under nectary; staminodes 0; pistillode present. **Pistillate flowers:** staminodes 5, alternate with petals, undivided, not gland-tipped, minute; pistil 3-carpellate; ovary superior, 3-locular, placentation axile; style 1; stigmas 3; ovules 2 per locule. **Fruits** capsules, 3-locular, globose or subglobose, 3-lobed distally, apex not beaked. **Seeds** [1–]2 per locule, ellipsoid, not winged; aril red, completely surrounding seed. $x = 23$.

Species ca. 30 (2 in the flora): North America, e Asia; primarily tropics and subtropics, some temperate regions.

SELECTED REFERENCES Hou, D. 1955. A revision of the genus *Celastrus.* Ann. Missouri Bot. Gard. 42: 215–302. Leicht-Young, S. A. et al. 2007. Distinguishing native (*Celastrus scandens* L.) and invasive (*C. orbiculatus* Thunb.) bittersweet species using morphological characteristics. J. Torrey Bot. Soc. 134: 441–450. Pooler, M. R., R. L. Dix, and J. Feely. 2002. Interspecific hybridizations between the native bittersweet, *Celastrus scandens*, and the introduced invasive species, *C. orbiculatus*. SouthE. Naturalist 1: 69–76.

1. Inflorescences terminal, panicles; capsules orange; pollen yellow; leaf blades oblong, aestivation involute . . . . 1. *Celastrus scandens*
1. Inflorescences axillary, cymes; capsules yellow; pollen white; leaf blades suborbiculate to broadly oblong-obovate or ovate-orbiculate, aestivation conduplicate. . . . . 2. *Celastrus orbiculatus*

1. Celastrus scandens Linnaeus, Sp. Pl. 1: 196. 1753 • American bittersweet [E] [F]

Vines to 15 m. **Leaves:** blade oblong, 5–10 × 2–4 cm, aestivation involute. **Inflorescences** terminal, panicles, 3–8 cm. **Bisexual and staminate flowers:** pollen yellow. **Capsules** orange, subglobose, 4–8 × 5–9 mm, glabrous. **Seeds** dark brown, 6 mm. $2n = 46$.

Flowering spring–summer. Thickets, woodland margins, roadsides, usually rich soils; 0–1400 m; Man., Ont., Que., Sask.; Ala., Ark., Conn., Del., D.C., Ga., Ill., Ind., Iowa, Kans., Ky., La., Maine, Md., Mass., Mich., Minn., Miss., Mo., Mont., Nebr., N.H., N.J., N.Y., N.C., N.Dak., Ohio, Okla., Pa., R.I., S.C., S.Dak., Tenn., Tex., Vt., Va., W.Va., Wis., Wyo.

2. Celastrus orbiculatus Thunberg in J. A. Murray, Syst. Veg. ed. 14, 237. 1784 • Oriental bittersweet [I] [W]

Vines 40+ m. **Leaves:** blade suborbiculate to broadly oblong-obovate or ovate-orbiculate, 4–6 × 3–5 cm, aestivation conduplicate. **Inflorescences** axillary, cymes, 1–2 cm. **Bisexual and staminate flowers:** pollen white. **Capsules** yellow when mature, globose, 7–10 mm diam., glabrous. **Seeds** orange, 6 mm. $2n = 46$.

Flowering spring–summer. Thickets, woodland margins, open woods, roadsides, usually rich soils; 0–900 m; introduced; N.B., Ont., P.E.I., Que.; Ala., Ark., Conn., Del., D.C., Ga., Ill., Ind., Iowa, Kans., Ky., Maine, Md., Mass., Mich., Minn., Mo., N.H., N.J., N.Y., N.C., Ohio, Pa., R.I., S.C., Tenn., Vt., Va., Wash., W.Va., Wis.; Asia (China, Japan, Korea).

Celastrus orbiculatus has become a seriously invasive plant in much of eastern North America.

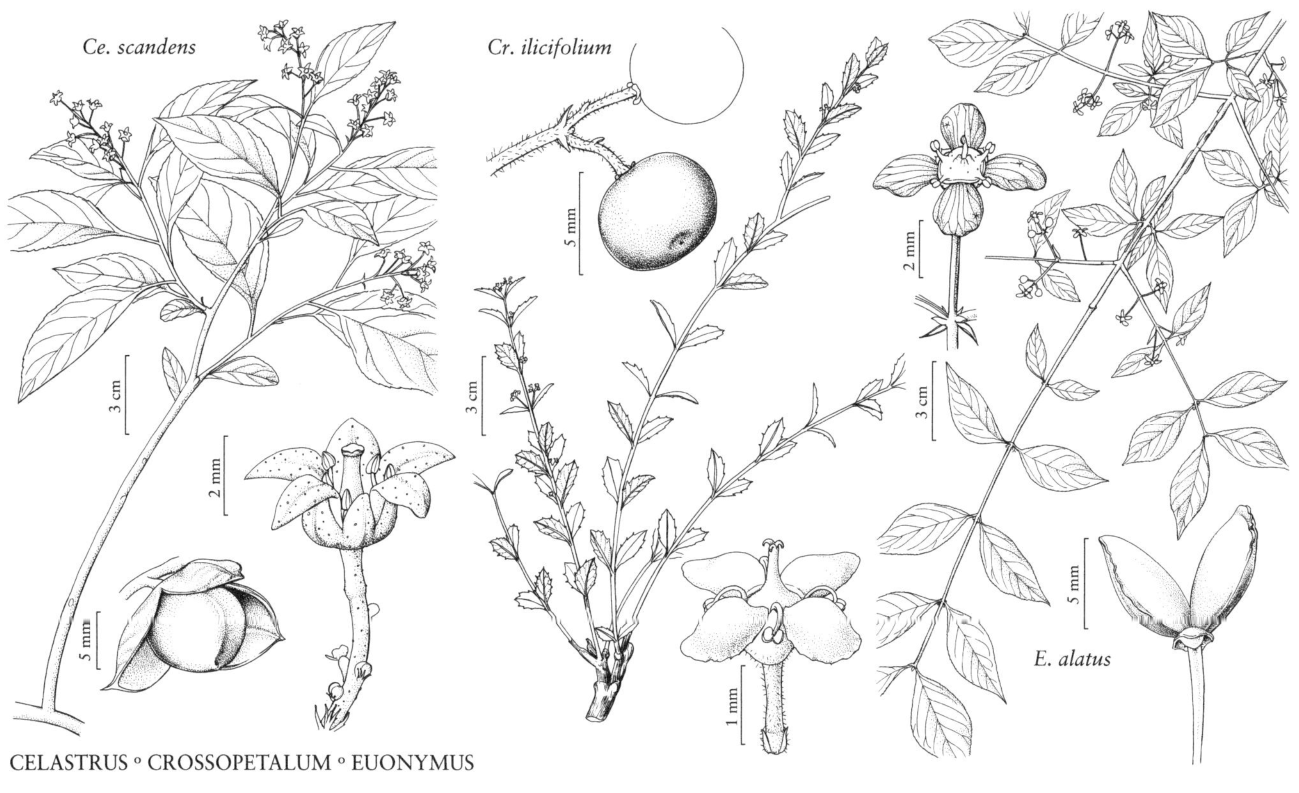

CELASTRUS ◦ CROSSOPETALUM ◦ EUONYMUS

6. CROSSOPETALUM P. Browne, Civ. Nat. Hist. Jamaica, 145, plate 16, fig. 1. 1756

• [Greek *krossos*, fringe, and *petalon*, petal, alluding to fimbriate petals of the type species]

Jinshuang Ma

Shrubs or trees. Branchlets terete. **Leaves** persistent, opposite or whorled; stipules present; petiole present; blade margins spiny-toothed, crenate, or entire; venation pinnate. **Inflorescences** axillary, cymes. **Flowers** bisexual, radially symmetric; perianth and androecium hypogynous; hypanthium absent; sepals 4, distinct; petals 4, white, pale green, reddish, or purplish; nectary intrastaminal, annular, fleshy; stamens 4, free from and inserted under nectary; staminodes 0; pistil 4-carpellate; ovary superior, immersed in and adnate to nectary, 4-locular, placentation axile; style 1; stigmas 4; ovule 1 per locule. **Fruits** drupes, red, 1[–2]-locular by abortion, obovoid or subglobose, apex not beaked. **Seeds** 1[–2] per fruit, obovoid, not winged; aril absent.

Species ca. 35 (2 in the flora): Florida, Mexico, West Indies, Central America, South America.

1. Shrubs, to 0.5 m; stems spreading or prostrate; leaf margins spiny-toothed; pedicels 3–7 mm . . . . 1. *Crossopetalum ilicifolium*
1. Shrubs or trees, 3–4 m; stems erect; leaf margins crenate or entire; pedicels 7–12 mm. . . . . .2. *Crossopetalum rhacoma*

1. Crossopetalum ilicifolium (Poiret) Kuntze, Revis. Gen. Pl. 1: 116. 1891 • Christmas berry [F]

Myginda ilicifolia Poiret in J. Lamarck et al., Encycl. 4: 396. 1797; *Rhacoma ilicifolia* (Poiret) Trelease

Shrubs, to 0.5 m. **Stems** spreading or prostrate, much-branched from base; branchlets pubescent. **Leaf blades** elliptic to ovate, 10–15 × 5–7 mm, margins coarsely spiny-toothed. **Inflorescences:** peduncle 3–7 mm. **Flowers:** petals white or pale green, orbiculate, 1 mm. **Drupes** subglobose, 3–4 × 1–2 mm.

Flowering year-round; fruiting year-round. Pinelands, rocky pine woods, roadsides; 0–10 m; Fla.; West Indies.

In the flora area, *Crossopetalum ilicifolium* occurs in Collier, Miami-Dade, and Monroe (Keys only) counties.

2. Crossopetalum rhacoma Crantz, Inst. Rei Herb. 2: 321. 1766 • Rhacoma

Rhacoma crossopetalum Linnaeus, Syst. Nat. ed. 10, 2: 896, 1361. 1759; *Myginda rhacoma* (Crantz) Swartz

Shrubs or trees, 3–4 m. **Stems** erect, branched; branchlets glabrous. **Leaf blades** ovate or sometimes elliptic, 10–40 × 5–7 mm, margins shallowly crenate or entire. **Inflorescences:** peduncle 7–12 mm. **Flowers:** petals reddish or purplish, ovate to suborbiculate, 1 mm. **Drupes** obovoid, 5–6 × 2–3 mm.

Flowering year-round; fruiting year-round. Pinelands, hammocks, sand dunes; 0–10 m; Fla.; s Mexico; West Indies.

In the flora area, *Crossopetalum rhacoma* occurs in Miami-Dade, Monroe (Keys only), and Sarasota counties.

7. EUONYMUS Linnaeus, Sp. Pl. 1: 197. 1753 (as Evonymus); Gen. Pl. ed. 5, 91. 1754, name and orthography conserved • [Greek *eu*-, good, and *onyma*, name, apparently applied ironically, the genus having had the bad reputation of poisoning cattle]

Jinshuang Ma

Geoffrey A. Levin

Shrubs, trees, or vines, climbing by adventitious roots. **Branchlets** terete or 4-angled. **Leaves** deciduous or persistent, opposite; stipules present; petiole present; blade margins entire or toothed; venation pinnate. **Inflorescences** terminal or axillary, cymes. **Flowers** bisexual, radially symmetric; perianth and androecium hypogynous; hypanthium absent; sepals 4–5, distinct; petals 4–5, white, green, yellow, red, or purple; nectary intrastaminal, annular, fleshy; stamens 4–5, adnate to nectary margin; staminodes 0; pistil 4–5-carpellate; ovary superior, 4–5-locular, placentation axile; style 1; stigma 1; ovules 2 per locule. **Fruits** capsules, (1–)2–5-locular, globose, subglobose, or obovoid, unlobed or 2–5-lobed, apex not beaked. **Seeds** 2 per locule, ellipsoid, ovoid, or subglobose, not winged; aril yellow, orange, or red, completely surrounding seed. $x = 16$.

Species ca. 140 (7 in the flora): North America, Mexico, West Indies, Central America, Europe, e, se Asia, Africa; tropics and subtropics.

Several cultivated species of *Euonymus*, all native to eastern Asia, are established locally in the flora area and, although apparently not naturalized, should be watched for invasive tendencies. Two of these, *E. hamiltonianus* Wallich, reported from Illinois, Indiana, and Michigan, and *E. maackii* Ruprecht (= *E. bungeanus* Maximowicz), reported from Colorado, Florida, Georgia, Illinois, and South Carolina, are similar to each other and to *E. europaeus*, from which they both differ by having purple (versus white) anthers. In addition, *E. hamiltonianus* can be distinguished by its red (versus orange or yellow) arils, and *E. maackii* by its leaf blades with smooth (versus rough) surfaces. *Euonymus japonicus* Thunberg, reported from Indiana, Louisiana, Mississippi, North Carolina, and Virginia, is similar to *E. fortunei* but is a shrub with erect stems that never produce adventitious roots. *Euonymus phellomanus* Loesener, reported from Connecticut and Massachusetts, has corky winged branches like *E. alatus*, but has larger leaves (6–10 × 2–3 cm) and yellow-brown to red-brown, 4-angled capsules.

SELECTED REFERENCES Blakelock, R. A. 1951. A synopsis of the genus *Euonymus* L. Kew Bull. 6: 210–290. Ma, J. S. 2001. A revision of *Euonymus* (Celastraceae). Thaiszia 11: 1–264.

1. Vines, climbing by adventitious roots; leaves persistent . . . . 1. *Euonymus fortunei*
1. Shrubs or trees, creeping, arching, or erect; leaves deciduous.
   2. Young branches becoming corky winged; capsules deeply lobed to base or only 1 lobe developing, lobes nearly distinct . . . . 6. *Euonymus alatus*
   2. Young branches not corky winged; capsules unlobed to deeply lobed, rarely only 1 lobe developing, lobes clearly connate.
      3. Sepals, petals, and stamens 4; young branches terete.
         4. Petals dark purple, 1.5–2 mm; inflorescences 7–20-flowered; capsules 11–13 × 15–17 mm; seeds 5–7 mm, arils red . . . . 4. *Euonymus atropurpureus*
         4. Petals yellow or white, 3–4 mm; inflorescences 1–7(–15)-flowered; capsules 8–10 × 12–15 mm; seeds 7–8 mm, arils yellow or orange . . . . 7. *Euonymus europaeus*
      3. Sepals, petals, and stamens 5; young branches 4-angled.
         5. Capsules smooth, moderately 3(–5) lobed; petals brownish purple, 3–6.5 mm; seeds 6–8 mm. . . . . 5. *Euonymus occidentalis*
         5. Capsules spiny, unlobed or very shallowly 4–5-lobed; petals pale green, often suffused with purple, 2–3 mm; seeds 4.5–5.5 mm.
            6. Stems erect or arching, not rooting at nodes; petioles 1–3 mm; leaf blades oval to lanceolate. . . . . 2. *Euonymus americanus*
            6. Stems creeping, rooting at nodes; petioles 3–5 mm; leaf blades obovate . . . . 3. *Euonymus obovatus*

1. Euonymus fortunei (Turczaninow) Handel-Mazzetti, Symb. Sin. 7: 660. 1933 (as Evonymus) • Wintercreeper, climbing euonymus [I]

Elaeodendron fortunei Turczaninow, Bull. Soc. Imp. Naturistes Moscou 36: 603. 1863 (as Elaeodendrum), name conserved; *Euonymus hederaceus* Champion ex Bentham, name rejected; *E. kiautschovicus* Loesener

Vines, to 20 m. **Stems** prostrate to erect, climbing by adventitious roots; young branches terete, not corky winged. **Leaves** persistent; petiole 5–10 mm; blade lanceolate, ovate, elliptic, to broadly obovate-elliptic, 1–9 × 0.5–5 cm, base cuneate, acute, obtuse, to rounded, margins crenate-serrate, apex obtuse, acute, or acuminate. **Inflorescences** axillary, 5–15-flowered. **Flowers:** sepals 4; petals 4, white to pale green, oblong, 3–4 × 2–3 mm; stamens 4; ovary smooth. **Capsules** straw colored to orange, globose, 6–8 mm diam., unlobed or very shallowly 4-lobed, lobes clearly connate, surface smooth. **Seeds** ellipsoid, 4–6 mm; aril orange. **2*n*** = 32.

Flowering summer; fruiting summer–fall. Moist woods, stream banks, riverbanks, disturbed areas; 0–300 m; introduced; Ont.; Ala., Ark., Conn., Del., D.C., Ga., Ill., Ind., Kans., Ky., Md., Mass., Mich., Miss., Mo., N.J., N.C., Ohio, Pa., R.I., S.C., Tenn., Tex., Va., Wis.; e Asia.

Euonymus fortunei is widely planted as an ornamental and has escaped widely. Plants grow horizontally until they encounter a vertical surface like a rock, wall, or tree, which they then climb using adventitious roots. They sometimes form dense mats over other vegetation, excluding other plants.

2. Euonymus americanus Linnaeus, Sp. Pl. 1: 197. 1753 • Strawberry-bush [E]

Shrubs, 2–3.5 m. **Stems** erect or arching, not rooting at nodes; young branches 4-angled, not corky winged. **Leaves** deciduous; petiole 1–3 mm; blade oval to lanceolate, 3–10 × 0.8–3 cm, base cuneate, obtuse, to rounded, margins crenate-serrate, apex acute or acuminate. **Inflorescences** terminal or axillary, 1–3-flowered. **Flowers:** sepals 5; petals 5, pale green, often suffused with purple, round, 2–3 mm diam.; stamens 5; ovary spiny. **Capsules** red or pink to purple, subglobose, 11–13 mm diam., unlobed or very shallowly 4–5-lobed, lobes clearly connate, surface spiny. **Seeds** subglobose, 4.5–5.5 mm; aril red. **2*n*** = 64.

Flowering spring–summer; fruiting late summer–early winter. Rich woods, bluffs, flood plains, along streams, hammocks, sandy banks; 0–400 m; Ala., Ark., Del., D.C., Fla., Ga., Ill., Ind., Ky., La., Md., Miss., Mo., N.J., N.Y., N.C., Ohio, Okla., Pa., S.C., Tenn., Tex., Va., W.Va.

3. Euonymus obovatus Nuttall, Gen. N. Amer. Pl. 1: 155. 1818 • Running strawberry-bush [E]

Euonymus americanus Linnaeus var. *obovatus* (Nuttall) Dippel

Shrubs, 0.1–0.5 m. **Stems** creeping, rooting at nodes; young branches 4-angled, not corky winged. **Leaves** deciduous; petiole 3–5 mm; blade obovate, 2–8 × 1.5–4.5 cm, base attenuate, margins crenate-serrate, apex acute or acuminate. **Inflorescences** terminal or axillary, 1–5-flowered. **Flowers:** sepals 5; petals 5, pale green, often suffused with purple, round, 2–3 mm diam.; stamens 5; ovary spiny. **Capsules** pink, subglobose, 10–12 mm diam., unlobed or very shallowly 4–5-lobed, lobes clearly connate, surface spiny. **Seeds** subglobose, 4.5–5.5 mm; aril bright orange.

Flowering spring; fruiting summer–fall. Rich, dry, or moist woods; 100–1000 m; Ont.; Ark., Ga., Ill., Ind., Ky., Mich., Mo., N.Y., N.C., Ohio, Pa., S.C., Tenn., W.Va.

Euonymus obovatus can be difficult to separate from *E. americanus*, but the rooting stems of *E. obovatus* clearly differentiate them.

4. Euonymus atropurpureus Jacquin, Hort. Bot. Vindob. 2: 55, plate 120. 1772/1773 • Burning-bush, wahoo, spindle-tree [E] [W]

Euonymus atropurpureus var. *cheatumii* Lundell

Shrubs or trees, to 8 m. **Stems** erect; young braches terete, not corky winged. **Leaves** deciduous; petiole 6–20 mm; blade elliptic, oval, ovate, or obovate, 5–16 × 1–3 cm, base broadly cuneate to rounded, margins serrate, apex acuminate. **Inflorescences** terminal or axillary, 7–20-flowered. **Flowers:** sepals 4; petals 4, dark purple, nearly triangular, obovate, or oblong, 1.5–2 × 1.2–1.5 mm; stamens 4; ovary smooth. **Capsules** pinkish purple, obovoid, 11–13 × 15–17 mm, deeply 4-lobed, lobes clearly connate, surface smooth. **Seeds** ellipsoid, 5–7 × 4–5 mm; aril red. $2n = 32$.

Flowering spring–summer; fruiting late summer–fall. Rich moist woods and thickets, hillsides; 0–400 m; Ont.; Ala., Ark., Conn., Del., D.C., Fla., Ga., Ill., Ind., Iowa, Kans., Ky., La., Maine, Md., Mass., Mich., Minn., Miss., Mo., Nebr., N.H., N.J., N.Y., N.C., N.Dak., Ohio, Okla., Pa., R.I., S.C., S.Dak., Tenn., Tex., Va., W.Va., Wis.

Euonymus atropurpureus is widely cultivated and has become naturalized in New England (Connecticut, Maine, Massachusetts, New Hampshire, and Rhode Island). The root bark is used medicinally.

5. Euonymus occidentalis Nuttall ex Torrey in War Department [U.S.], Pacif. Railr. Rep. 4(5): 74. 1857 • Western spindle-tree [E]

Shrubs, 2–6 m. **Stems** erect or scandent; young branches 4-angled, not corky winged. **Leaves** deciduous; petiole 3–15 mm; blade ovate to obovate, 4–10 × 2–7 cm, base rounded to narrowly cuneate, margins serrulate, apex obtuse, rounded, or acuminate. **Inflorescences** axillary, 1–5 flowered. **Flowers:** sepals 5; petals 5, brownish purple, sometimes spotted, round or oblong, 3–6.5 × 2–4 mm; stamens 5; ovary smooth. **Capsules** pink to purple, subglobose to obovoid, 11–12 × 12–15 mm, moderately 3(–5)-lobed, lobes clearly connate, surface smooth. **Seeds** obovoid, 6–8 × 4–5 mm; aril orange to red.

Varieties 2 (2 in the flora): w North America.

1. Leaf blade apices acuminate; twigs greenish; inflorescences 1–5-flowered. . . . . 5a. *Euonymus occidentalis* var. *occidentalis*
1. Leaf blade apices obtuse or rounded; twigs whitish; inflorescences 3–5-flowered . . . . 5b. *Euonymus occidentalis* var. *parishii*

5a. Euonymus occidentalis Nuttall ex Torrey var. **occidentalis** [E]

Twigs greenish. **Leaf blade apices** acuminate. **Inflorescences** 1–5-flowered.

Flowering spring–summer; fruiting summer–fall. Deep moist woods, shaded stream banks; 0–1600 m; B.C.; Calif., Oreg., Wash.

Variety *occidentalis* occurs from southwestern British Columbia south to central California.

5b. Euonymus occidentalis Nuttall ex Torrey var. **parishii** (Trelease) Jepson, Man. Fl. Pl. Calif., 610. 1925 E

Euonymus parishii Trelease, Trans. Acad. Sci. St. Louis 5: 354. 1889

Twigs whitish. **Leaf blade apices** obtuse or rounded. **Inflorescences** 3–5-flowered.

Flowering spring; fruiting summer. Shaded canyons; 1300–2000 m; Calif.

Variety *parishii* is restricted to the mountains of southern California in western Riverside and San Diego counties.

6. Euonymus alatus (Thunberg) Siebold, Verh. Batav. Genootsch. Kunst. 12: 49. 1830 (as Evonimus) • Oriental spindle-tree F I

Celastrus alatus Thunberg in J. A. Murray, Syst. Veg. ed. 14, 237. 1784; *Euonymus alatus* var. *apterus* Loesener

Shrubs, 1–4(–7) m. **Stems** erect; young branches 4-angled, becoming corky winged. **Leaves** deciduous; petiole 0.5–4 mm; blade narrowly elliptic, 2.5–6 × 0.5–2.5 cm, base attenuate or cuneate, margins denticulate, apex acuminate. **Inflorescences** axillary, (1–)3(–7)-flowered. **Flowers:** sepals 4; petals 4, yellowish green or white, oblong, 2–3 × 1.5–2 mm; stamens 4; ovary smooth. **Capsules** purple-brown, obovoid, 8–10 × 5–15 mm, deeply 2–4-lobed to base or only 1 lobe developing, lobes nearly distinct, surface smooth. **Seeds** ellipsoid, 7–8 × 4–5 mm; aril yellow or orange. $2n = 64$.

Flowering spring–fall; fruiting summer–fall. Roadsides, old fields, thickets, woodlands; 0–400 m; introduced; Ont.; Conn., Del., D.C., Ga., Ill., Ind., Iowa, Kans., Ky., Maine, Md., Mass., Mich., Minn., Mo., Mont., N.H., N.J., N.Y., N.C., Ohio, Pa., R.I., S.C., Utah, Vt., Va., W.Va., Wis.; e Asia.

The leaves of *Euonymus alatus* become bright red or reddish purple in fall, one of the reasons for its widespread use as an ornamental.

7. Euonymus europaeus Linnaeus, Sp. Pl. 1: 197. 1753 (as Evonymus) • European spindle-tree I W

Shrubs or trees, 2–10 m. **Stems** erect; young branches terete, not corky winged. **Leaves** deciduous; petiole 4–12 mm; blade ovate-elliptic, 2.5–10 × 1.5–3.5 cm, base attenuate to broadly cuneate, margins minutely denticulate, apex acuminate. **Inflorescences** axillary, 1–7 (–15)-flowered. **Flowers:** sepals 4; petals 4, yellow or white, oblong, 3–4 × 1–2 mm; stamens 4; ovary smooth. **Capsules** pink, obovoid, 8–10 × 12–15 mm, deeply (2–)4-lobed, rarely only 1 lobe developing, lobes clearly connate, surface smooth. **Seeds** obovoid, 7–8 × 4–5 mm; aril orange or yellow.

Flowering spring–summer; fruiting summer–fall. Roadsides, thickets, woodlands; 0–300 m; introduced; N.B., Ont., P.E.I., Que.; Ala., Ark., Conn., Ill., Ind., Ky., Maine, Mass., Mich., Miss., Mo., Nebr., N.H., N.J., N.Y., Ohio, Pa., R.I., Tenn., Utah, Vt., Va., Wis.; Europe.

The leaves of *Euonymus europaeus* vary greatly in shape and size, especially in cultivated plants. The species was introduced to the flora area as a garden ornamental.

8. GYMINDA Sargent, Gard. & Forest 4: 4. 1891 • [Anagram of *Myginda*, to which these species had been referred]

Jinshuang Ma

Shrubs or trees, dioecious. **Branchlets** 4-angled. **Leaves** persistent, opposite; stipules present; petiole present; blade margins entire or toothed; venation pinnate. **Inflorescences** axillary, cymes. **Flowers** unisexual, radially symmetric; perianth and androecium hypogynous; hypanthium absent; sepals (3–)4, distinct; petals (3–)4, white; nectary intrastaminal, 4-lobed, fleshy. **Staminate flowers:** stamens 4, free from and inserted between lobes of nectary; staminodes 0; pistillode present. **Pistillate flowers:** staminodes 0; pistil 2-carpellate; ovary superior, adnate to nectary at base, 2-locular, placentation axile; style absent; stigmas 2; ovule 1 per locule. **Fruits** drupes, bluish black, 2-locular, ellipsoid to ovoid, apex not beaked. **Seeds** 1 per locule, sometimes 1 per fruit by abortion, ovoid, not winged; aril absent.

Species 2 (1 in the flora): Florida, s Mexico, West Indies, Central America.

Gyminda orbicularis Borhidi & O. Muñiz is endemic to Cuba.

1. Gyminda latifolia (Swartz) Urban, Symb. Antill. 5: 80. 1904 • False boxwood [F]

Myginda latifolia Swartz, Prodr., 39. 1788

Shrubs or trees to 8 m. **Stems** much-branched; branches 4-angled. **Leaves:** petiole 0–2 mm; blade bright green, obovate to elliptic-obovate, 2–4 × 1–3 cm, base acuminate to acute, margins entire or obscurely crenulate-serrate, especially distally, reflexed, apex round or emarginate. **Staminate flowers:** sepals 1 mm; petals elliptic to obovate-elliptic, 1.5–2 mm. **Pistillate flowers:** sepals 1 mm; petals elliptic to obovate-elliptic, 1.5–2 mm. **Drupes** 7–8 × 3–5 mm; peduncle to 8 mm. **Seeds** black or dark blue, ovoid.

Flowering summer; fruiting late summer–fall. Hammocks; 0–10 m; Fla.; s Mexico; West Indies.

The dark heartwood of *Gyminda latifolia* is heavy and hard. In the flora area, *G. latifolia* occurs in Miami-Dade and Monroe (Keys only) counties.

9. MAYTENUS Molina, Sag. Stor. Nat. Chili, 177, 349. 1782 • [Vernacular Chilean *mayten*, name for type species]

Jinshuang Ma

Geoffrey A. Levin

Tricerma Liebmann

Shrubs or trees, polygamomonoecious. **Branchlets** terete. **Leaves** persistent, alternate; stipules present; petiole present; blade margins entire or serrate; venation pinnate. **Inflorescences** axillary, cymes or flowers solitary. **Flowers** bisexual and unisexual, radially symmetric; perianth and androecium hypogynous; hypanthium absent; sepals 5, distinct; petals 5, white or greenish white; nectary intrastaminal, annular, fleshy. **Bisexual flowers:** stamens 5, free and inserted under nectary; staminodes 0; pistil 2–4-carpellate; ovary superior, immersed in and adnate to nectary, 2–4-locular, placentation axile; style 1; stigmas 2–4; ovules 1–2 per locule. **Staminate flowers:** stamens 5, free from and inserted under nectary; staminodes 0; pistillode present. **Pistillate flowers:** staminodes 5, alternate with petals, undivided, not gland-tipped, minute; pistil 2–4-carpellate; ovary superior, immersed in and adnate to nectary, 2–4-locular, placentation axile; style 1; stigmas 2–4; ovules 1–2 per locule. **Fruits** capsules, 2–4-locular, obovoid, 2–4-angled, apex not beaked. **Seeds** 1 per locule, ellipsoid, not winged; aril bright red, completely surrounding seed.

Species ca. 200 (2 in the flora): United States, Mexico, West Indies, Central America, South America, Eurasia, Africa, Atlantic Islands, Indian Ocean Islands, Pacific Islands, Australia.

Tricerma, sometimes separated from *Maytenus* (for example, by C. L. Lundell 1971), is embedded within *Maytenus* (M. J. McKenna et al. 2011) and not accepted here.

1. Leaf blade margins entire or remotely serrulate, apices obtuse to rounded; branchlets erect or spreading; capsules bright red, 8–12 mm; Florida, Texas. . . . . 1. *Maytenus phyllanthoides*
1. Leaf blade margins regularly serrulate, apices acute; branchlets pendent; capsules yellow or straw colored, 5–9 mm; California . . . . 2. *Maytenus boaria*

G. latifolia

Ma. phyllanthoides var. *phyllanthoides*

GYMINDA ◦ MAYTENUS ◦ MORTONIA

1. Maytenus phyllanthoides Bentham, Bot. Voy. Sulphur, 54. 1844 • Gutta-percha, leather leaf, mangle dulce [F]

Tricerma phyllanthoides (Bentham) Lundell

Shrubs or trees, 0.2–7 m. **Stems** erect to spreading, or decumbent to prostrate; branchlets erect or spreading, puberulent. **Leaves:** petiole 1–6 mm; blade oblong-elliptic to obovate, 1.5–5(–6) × 1–2.5(–3.5) cm, base cuneate or rounded, margins entire or remotely serrulate, sometimes wavy, apex obtuse to rounded, sometimes mucronate. **Flowers:** sepals reddish, rounded; petals 2–3 mm. **Capsules** bright red, 8–12 × 4–6 mm.

Varieties 2 (2 in the flora): sc, se United States, Mexico, West Indies (Bahamas, Cuba).

Recognition of two varieties follows G. L. Nesom (2009e).

1. Leaf blade bases cuneate; stems erect to spreading . . . . 1a. *Maytenus phyllanthoides* var. *phyllanthoides*
1. Leaf blade bases rounded; stems decumbent to prostrate . . . 1b. *Maytenus phyllanthoides* var. *ovalifolia*

1a. Maytenus phyllanthoides Bentham var. **phyllanthoides** • Florida mayten [F]

Shrubs or trees, 1–7 m. **Stems** erect to spreading. **Leaves:** petiole 2–6 mm; blade obovate, 1.5–5(–6) × 1–2.5(–3.5) cm, base cuneate, margins entire, sometimes wavy, apex rounded.

Flowering early spring–summer; fruiting summer–winter. Hammocks, dunes, edges of mangrove forests; 0–10 m; Fla.; Mexico; West Indies (Bahamas, Cuba).

In the flora area, var. *phyllanthoides* occurs along the Gulf coast of peninsular Florida from Levy County south and on the Atlantic coast in Miami-Dade County and the Keys. The leaves yield a gum that has been used as a substitute for gutta-percha, a rubberlike substance derived from *Palaquium* Blanco spp. (Sapotaceae) of southeast Asia and used in dentistry and historically for electrical insulation and golf balls.

1b. Maytenus phyllanthoides Bentham var. **ovalifolia** Loesener, Repert. Spec. Nov. Regni Veg. 8: 291. 1910

Maytenus texana Lundell; *Tricerma texanum* (Lundell) Lundell

Shrubs, 0.2–2 m. **Stems** decumbent to prostrate. **Leaves:** petiole 1–3 mm; blade oblong-elliptic to elliptic-obovate, 1.5–4(–6) × 1–2(–3.5) cm, base rounded, margins entire or remotely serrulate, apex obtuse to rounded, sometimes mucronate.

Flowering early spring–summer; fruiting summer–winter. Coastal prairies, marshes, clay or sand-clay mounds, often saline sites; 0–10 m; Tex.; Mexico (Tamaulipas).

In the flora area, var. *ovalifolia* occurs along the Gulf coast of Texas from Cameron to Nueces counties.

2. Maytenus boaria Molina, Sag. Stor. Nat. Chili, 177, 349. 1782 • Mayten tree [I]

Trees, to 15 m. **Stems** erect to spreading; branchlets gray-brown, pendent, glabrous. **Leaves:** petiole 2–5 mm; blade elliptic to lanceolate, 1–6 × 0.4–2 cm, base cuneate, margins regularly serrulate, apex acute. **Flowers:** sepals green, rounded; petals 1.5–3 m. **Capsules** yellow or straw colored, 5–9 × 5–6.5 mm.

Flowering spring; fruiting summer. Disturbed scrub and oak woodlands; 100–400 m; introduced; Calif.; South America (Argentina, Chile).

Maytenus boaria is naturalized in the San Francisco Bay area, where it spreads after fires.

10. MORTONIA A. Gray, Smithsonian Contr. Knowl. 3(5): 34, plate 4. 1852 • [For Samuel George Morton, 1799–1851, North American naturalist]

Jinshuang Ma

Geoffrey A. Levin

Shrubs. Branchlets terete, scabrous. **Leaves** persistent, alternate, crowded; stipules present; petiole essentially absent; blade margins entire; 1-veined. **Inflorescences** terminal, cymose panicles or racemes. **Flowers** bisexual, radially symmetric; perianth and androecium perigynous; hypanthium free from ovary; sepals 5, distinct; petals 5, white; nectary intrastaminal, usually lining hypanthium, fleshy; stamens 5, adnate to nectary margin; staminodes 0; pistil 5-carpellate; ovary superior, 5-locular, placentation axile; style 1; stigmas 5; ovules 2 per locule. **Fruits** nutlike (small, hard-walled, indehiscent), 1-locular, oblong to cylindric, apex beaked. **Seeds** 1 per fruit, oblong, not winged; aril absent.

Species 8–10 (4 in the flora): sw United States, Mexico, Central America.

Mortonia is poorly understood taxonomically.

1. Leaf blades linear-oblanceolate or oblanceolate, 15–25 mm, length 5 times width, glabrous, pliable, margins often thickened, not revolute . . . . 1. *Mortonia greggii*
1. Leaf blades ovate, elliptic, oblong-elliptic, suborbiculate, or obovate, 3–16 mm, length 1.2–2.5 times length, usually scabridulous, rarely glabrous, rigid, margins not thickened, revolute.
    2. Leaf blades oblong-elliptic to narrowly obovate, 3–5 × 1.5–2 mm, length 2–2.5 times width. . . . . 3. *Mortonia sempervirens*
    2. Leaf blades ovate or broadly elliptic to suborbiculate, 5–16 × 4–12 mm, length 1.2–1.4 times width.
        3. Leaves 5–6 × 4–5 mm; fruits 3.5–4.5 mm . . . . 2. *Mortonia scabrella*
        3. Leaves 7–16 × 5–12 mm; fruits 5–7 mm . . . . 4. *Mortonia utahensis*

1. **Mortonia greggii** A. Gray, Smithsonian Contr. Knowl. 3(5): 35. 1852 • Afinador

Shrubs to 2 m. **Leaves:** blade linear-oblanceolate or oblanceolate, 15–25 × 3–5 mm, length 5 times width, pliable, base tapering to petiole, margins often thickened, not revolute, apex obtuse, mucronate, surfaces glabrous. **Inflorescences** cymose panicles. **Flowers:** sepals narrowly triangular, 1 mm; petals oblong, 2 mm. **Fruits** 4–5 mm.

Flowering winter–spring; fruiting summer–fall. Thickets, gravelly hills; 0–100 m; Tex.; Mexico (Chihuahua, Coahuila, Nuevo León, Sonora).

In the flora area, *Mortonia greggii* occurs in the lower Rio Grande valley in Hidalgo and Starr counties.

2. **Mortonia scabrella** A. Gray, Smithsonian Contr. Knowl. 5(6): 28. 1853 [F]

Shrubs to 2.5 m. **Leaves:** blade broadly elliptic to suborbiculate, 5–6 × 4–5 mm, length 1.2–1.3 times width, rigid, base round, margins not thickened, revolute, apex round to broadly acute, surfaces scabridulous. **Inflorescences** cymose panicles. **Flowers:** sepals rounded-deltate, 1 mm; petals obovate-oblong, 2–2.5 mm. **Fruits** 3.5–4.5 mm.

Flowering spring and fall; fruiting summer–fall. Rocky slopes, hillsides, dry plains, ridges, ledges, rimrock; 300–1700 m; Ariz., N.Mex., Tex.; Mexico (Chihuahua, Coahuila).

In the flora area, *Mortonia scabrella* occurs from trans-Pecos Texas to southeastern Arizona.

3. **Mortonia sempervirens** A. Gray, Smithsonian Contr. Knowl. 3(5): 35, plate 4. 1852 • Texas mortonia

Shrubs 0.6–1.5 m. **Leaves:** blade oblong-elliptic to narrowly obovate, 3–5 × 1.5–2 mm, length 2–2.5 times width, rigid, base rounded, margins not thickened, revolute, apex narrowly obtuse to acute, mucronate, surfaces usually scabridulous, rarely glabrous. **Inflorescences** cymose racemes. **Flowers:** sepals deltate, 1 mm; petals oblong, 2 mm. **Fruits** 5–6 mm.

Flowering spring; fruiting summer. Rocky limestone hills, ledges, stony prairies; 500–1800 m; Tex.; Mexico (Chihuahua, Durango, Sonora).

In the flora area, *Mortonia sempervirens* occurs in trans-Pecos Texas.

4. **Mortonia utahensis** (Coville ex Trelease) A. Nelson, Bot. Gaz. 47: 427. 1909 • Utah mortonia [E]

Mortonia scabrella A. Gray var. *utahensis* Coville ex Trelease in A. Gray et al., Syn. Fl. N. Amer. 1(1,2): 400. 1897

Shrubs 0.3–1.2 m. **Leaves:** blade ovate or broadly elliptic to suborbiculate, 7–16 × 5–12 mm, length 1.3–1.4 times width, rigid, base rounded to broadly cuneate, margins not thickened, revolute, apex rounded to acute, sometimes mucronulate, surfaces scabridulous. **Inflorescences** cymose panicles. **Flowers:** sepals triangular, 1–2.3 mm; petals ovate, 2.2–3 mm. **Fruits** 5–7 mm.

Flowering spring; fruiting late spring–summer. Limestone, dry desert slopes, canyon bottoms; 500–2100 m; Ariz., Calif., Nev., Utah.

Mortonia utahensis is restricted to the Mojave Desert.

11. PAXISTIMA Rafinesque, Sylva Tell., 42. 1838 • [Greek *pachos*, thick, and *stigma*, alluding to slightly enlarged stigma]

Jinshuang Ma

Shrubs. Branchlets 4-angled, with corky ridges. **Leaves** persistent, opposite; stipules present; petiole present or absent; blade margins finely toothed; venation pinnate. **Inflorescences** axillary, cymes or flowers solitary. **Flowers** bisexual, radially symmetric; perianth and androecium perigynous; hypanthium free from ovary; sepals 4, distinct; petals 4, reddish brown to yellowish green; nectary intrastaminal, lining hypanthium, fleshy; stamens 4, adnate to nectary margin; staminodes 0; pistil 2-carpellate; ovary superior, immersed in and adnate to nectary, 2-locular, placentation axile; style 1; stigmas 2; ovules 2 per locule. **Fruits** capsules, 2-locular, ellipsoid or obovoid, apex not beaked. **Seeds** 1 per locule, ellipsoid or obovoid, not winged; aril yellow or white, surrounding base and 1 side of seed, lacerate or fringed. $x = 16$.

Species 2 (2 in the flora): North America, n Mexico.

Pachistima and *Pachystima* are orthographic variants used frequently in the flora area.

SELECTED REFERENCES Navaro, A. M. and W. H. Blackwell. 1990. A revision of *Paxistima* (Celastraceae). Sida 14: 231–249. Wheeler, L. C. 1943. History and orthography of the celastraceous genus "*Pachystima*" Rafinesque. Amer. Midl. Naturalist 29: 792–795.

1. Shrubs 1–4 dm; leaves 5–20 × 2–4 mm, usually linear to narrowly elliptic, rarely oblanceolate; capsules 3–4 mm . . . . 1. *Paxistima canbyi*
1. Shrubs 5–10 dm; leaves 8–34 × 3–12 mm, ovate, elliptic, or oblanceolate; capsules 4–7 mm . . . . 2. *Paxistima myrsinites*

1. Paxistima canbyi A. Gray, Proc. Amer. Acad. Arts 8: 623. 1873 (as Pachystima) • Canby's mountain-lover, rat-stripper, cliff green [C] [E] [F]

Shrubs 1–4 dm. **Stems** diffuse, creeping. **Leaves:** petiole absent; blade usually linear to narrowly elliptic, rarely oblanceolate, 5–20 × 2–4 mm, base obtuse, margins serrulate, apex obtuse. **Inflorescences** 1–5-flowered. **Flowers:** sepals obtuse-deltate, 0.8–1 mm; petals obovate, 1.5 mm; filaments shorter than anthers. **Capsules** ellipsoid, 3–4 × 1.5–2 mm. **Seeds** black, ellipsoid; aril yellow or white.

Flowering spring; fruiting summer. Limestone cliffs, shaded banks, dry gravelly soils; of conservation concern; 0–600 m; Ky., Md., N.C., Ohio, Pa., Tenn., Va., W.Va.

There are only about 50 to 60 extant populations of *Paxistima canbyi*.

2. Paxistima myrsinites (Pursh) Rafinesque, Sylva Tellur., 42. 1838 • Myrtle box-leaf, Oregon boxwood, mountain-lover

Ilex myrsinites Pursh, Fl. Amer. Sept. 1: 119. 1813; *Paxistima myrsinites* subsp. *mexicana* Navaro & W. H. Blackwell

Shrubs 5–10 dm. **Stems** prostrate to spreading. **Leaves:** petiole 1 mm; blade ovate, elliptic, or oblanceolate, 8–34 × 3–12 mm, base cuneate, margins serrate, apex rounded to acute. **Inflorescences** 1–3-flowered. **Flowers:** sepals obtuse-deltate, 0.8–1 mm; petals ovate, 1–1.2 mm; filaments 2 times longer than anthers. **Capsules** obovoid, 4–7 × 2–4 mm. **Seeds** dark brown to black, obovoid; aril white.

Flowering spring–summer; fruiting summer. Shaded forests in mountains; 600–3400 m; Alta., B.C.; Ariz., Calif., Colo., Idaho, Mont., N.Mex., Oreg., Utah, Wash., Wyo.; Mexico (Coahuila, Nuevo León, Tamaulipas).

The leaves of *Paxistima myrsinites* vary in size and shape with elevation and latitude, but the variation is continuous and no differences to separate infraspecific taxa have been found.

Paxistima myrsinites was reported from Bexar County, Texas, based on a specimen collected by E. H. Wilkinson in 1900 (C. L. Lundell 1942–1969, vol. 2; D. S. Correll and M. C. Johnston 1970). However, it was not reported from Texas by B. L. Turner et al. (2003), and no appropriate habitat occurs in that county.

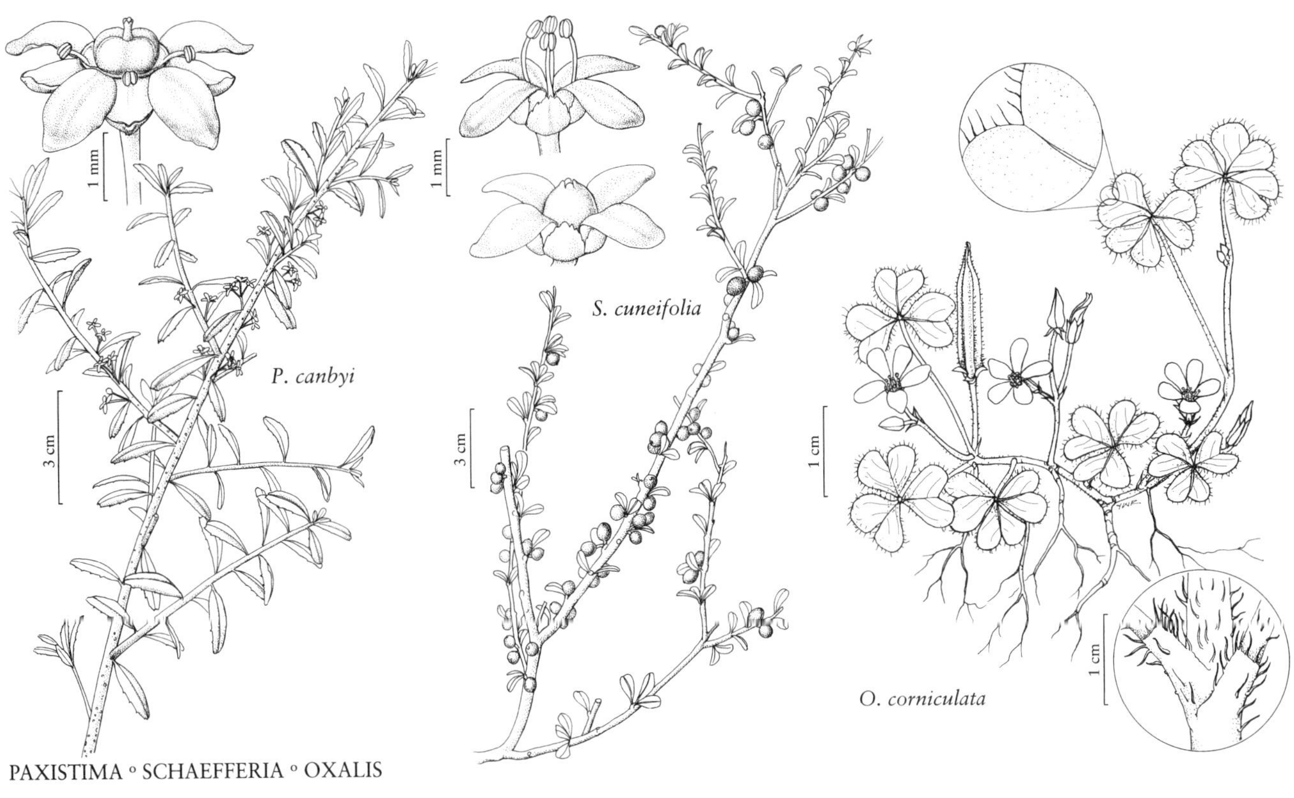

PAXISTIMA ◦ SCHAEFFERIA ◦ OXALIS

12. SCHAEFFERIA Jacquin, Enum. Syst. Pl., 10, 33. 1760 • [For Jacob Christian Schaeffer, 1718–1790, German botanist, zoologist, theologian, and clergyman]

Jinshuang Ma

Shrubs or trees, dioecious. **Branchlets** terete. **Leaves** persistent, alternate or fascicled; stipules absent; petiole present or absent; blade margins entire; venation pinnate. **Inflorescences** axillary, fascicles or flowers solitary. **Flowers** unisexual, radially symmetric; perianth and androecium hypogynous; hypanthium absent; sepals 4, distinct; petals 4, greenish white to white; nectary rudimentary or absent. **Staminate flowers:** stamens 4; staminodes 0; pistillode present. **Pistillate flowers:** staminodes 0; pistil 2-carpellate; ovary superior, 2-locular, placentation axile; style 1; stigmas 2; ovule 1 per locule. **Fruits** drupes, 2-locular, subglobose, apex not beaked. **Seeds** 1 per locule, oblong, not winged; aril absent.

Species 5–15 (2 in the flora): sc, se United States, Mexico, West Indies, Central America, South America.

1. Shrubs, 1–2 m; twigs ± thorn-tipped; short shoots present; leaves 5–23 × 2–15 mm; fruiting pedicels 0–2 mm . . . . 1. *Schaefferia cuneifolia*
1. Trees, to 12 m; twigs not thorn-tipped; short shoots absent; leaves 40–60 × 15–30 mm; fruiting pedicels 4–6 mm . . . . 2. *Schaefferia frutescens*

1. Schaefferia cuneifolia A. Gray, Smithsonian Contr. Knowl. 3(5): 35. 1852 • Desert yaupon, capul, panalero [F]

Schaefferia cuneifolia var. *pedicellata* Lundell

Shrubs 1–2 m. **Stems:** twigs ± thorn-tipped; short shoots present. **Leaves** alternate or fascicled; petiole absent; blade cuneate-obovate to oblanceolate, 5–23 × 2–15 mm, base cuneate, apex broadly obtuse to shallowly emarginate. **Pedicels** (fruiting) 0–2 mm. **Flowers:** sepals 0.5 mm; petals oblong, 3 mm. **Drupes** orange to bright red, 3–5 mm. **Seeds** yellow.

Flowering Feb–Jul. Rocky hillsides, breaks; 100–600 m; Tex.; Mexico (Baja California, Baja California Sur, Coahuila, Nuevo León, San Luis Potosí, Sonora, Tamaulipas).

In the flora area, *Schaefferia cuneifolia* occurs in southern and western Texas.

2. Schaefferia frutescens Jacquin, Enum. Syst. Pl., 10, 33. 1760 • Florida boxwood, yellow wood

Shrubs or trees to 12 m. **Stems:** twigs not thorn-tipped; short shoots absent. **Leaves** alternate; petiole present; blade elliptic, ovate, or oblanceolate, 40–60 × 15–30 mm, base attenuate, apex round or broadly obtuse. **Pedicels** (fruiting) 4–6 mm. **Flowers:** sepals 0.7 mm; petals oblanceolate, 2.5 mm. **Drupes** bright red or scarlet, 4–6 mm. **Seeds** yellow.

Flowering spring–fall. Hammocks; 0–10 m; Fla.; e Mexico; West Indies; Central America; ne South America.

In the flora area, *Schaefferia frutescens* occurs in Miami-Dade and Monroe (Keys only) counties. The bright yellow heartwood is close grained and heavy; it is sometimes used as a substitute for boxwood in engraving.

OXALIDACEAE R. Brown

• Wood-sorrel Family

Guy L. Nesom

Herbs [subshrubs, shrubs, vines or trees], annual or perennial. **Leaves** alternate or whorled, usually palmately or pinnately compound, sometimes 1-foliolate; stipules usually present, rarely apparently absent; petiole present; blade margins entire; venation pinnate or subpalmate. **Inflorescences** terminal or axillary, cymes or racemes, or flowers solitary. **Flowers** bisexual, perianth and androecium hypogynous; hypanthium absent; sepals 5, distinct or slightly connate basally; petals 5, distinct or slightly connate basally; nectary present; stamens 10 in 2 whorls, connate basally (monadelphous), free; anthers dehiscing by longitudinal slits; pistil 1, 5-carpellate, ovary superior, 5-locular; placentation axile; ovules (1–)3–8(–10) per locule, anatropous; styles 5, distinct; stigmas 5. **Fruits** capsules [berries], dehiscence loculicidal, often elastically. **Seeds** 1–10 per locule.

Genera 5, species ca. 800 (1 genus, 36 species in the flora): North America, Mexico, West Indies, Central America, South America, Eurasia, Africa, Atlantic Islands, Indian Ocean Islands, Pacific Islands, Australia; introduced in Bermuda.

Oxalidaceae occurs mostly in the tropics and subtropics of both hemispheres but extends into temperate regions. Species of *Averrhoa* Linnaeus and *Sarcotheca* Blume are trees or shrubs, those of *Dapania* Korthals lianas; all have fleshy, baccate fruits. Species of *Biophytum* de Candolle and *Oxalis* are herbs, subshrubs, shrubs, or rarely vines with capsular fruits; *Biophytum* has even-pinnate leaves fascicled at the stem tips. Seeds of Oxalidaceae develop an elastic, translucent, arilliform epidermis that turns inside out, explosively ejecting them from the capsule (K. R. Robertson 1975).

Oxalis tuberosa Molina (oca or New Zealand yam), of Andean South America, is cultivated for its edible tubers. *Averrhoa* is widely cultivated in the tropics for its fruits, which are eaten fresh, used in drinks, or made into jelly and jam. Both species of *Averrhoa* [*A. bilimbi* Linnaeus (bilimbi or cucumber tree), *A. carambola* Linnaeus (starfruit or carambola)] are known only in cultivation or as escapes from cultivation.

1. OXALIS Linnaeus, Sp. Pl. 1: 433. 1753; Gen. Pl. ed. 5, 198. 1754 • Wood-sorrel, sourgrass, false shamrock [Greek *oxys*, acid, and *-alis*, with the nature of, alluding to supposed medicinal use]

Bolboxalis Small; *Hesperoxalis* Small; *Ionoxalis* Small; *Lotoxalis* Small; *Otoxalis* Small; *Pseudoxalis* Rose; *Sassia* Molina; *Xanthoxalis* Small

Herbs [subshrubs, shrubs, or vines], caudex absent (present in *O. dichondrifolia*, *O. frutescens*, *O. montana*, and *O. texana*), usually with tubers, bulbs, or rhizomes, sometimes fibrous-rooted or taprooted. **Aerial stems** erect to decumbent, prostrate, or creeping, or absent. **Leaves** basal or cauline, palmately compound (pinnately compound in *O. frutescens*, 1-foliolate in *O. dichondrifolia*); leaflets pulvinate, usually 2-lobed, sometimes not lobed, often deflexed and folded together at night, laminae and margins often with black or orange oxalate dots or stripes. **Flowers** tristylous, distylous, semihomostylous (one whorl of anthers at level of stigmas, other proximal or distal), or homostylous (all anthers at level of stigmas); petals yellow, white, pink, salmon, red, lavender, purple, or violet; stamens: filaments 5 long and 5 short, alternating with one another, or equal length. x = (5–)7(–12).

Species ca. 700 (36 in the flora): North America, Mexico, West Indies, Central America, South America, Eurasia, Africa, Atlantic Islands, Indian Ocean Islands, Pacific Islands, Australia; introduced in Bermuda.

Species of *Oxalis* occur in the tropics and subtropics, mostly of North America, South America, and South Africa; they extend as well into temperate regions of North America, Europe, and Asia. The largest numbers of species are in South America and South Africa. Many species are cultivated as ornamentals because of their showy flowers and leaf diversity, and at least 100 species are offered for sale via the Internet. They are usually grown as container plants and need sun for flowering.

Oxalate deposits, usually as orange or blackish dots or stripes, sometimes are produced on the leaf lamina and along the margins and on the sepal apices. Their arrangement, especially on the leaves, usually is characteristic of a species.

The stipules of some caulescent species are characterized here as rudimentary; they are completely or nearly completely adnate to the petiole base and appear only as a thickened, more or less broadly oblong region, perhaps suggestive of a pulvinus.

Canadian distributions below are based on VASCAN (http://data.canadensys.net/vascan); not all have been verified for this treatment.

SELECTED REFERENCES Denton, M. F. 1973. A monograph of *Oxalis*, section *Ionoxalis* (Oxalidaceae) in North America. Publ. Mus. Michigan State Univ., Biol. Ser. 4: 455–615. Eiten, G. 1963. Taxonomy and regional variation of *Oxalis* section *Corniculatae*. I. Introduction, keys and synopsis of the species. Amer. Midl. Naturalist 69: 257–309. Gardner, A. G. et al. 2012. Diversification of the American bulb-bearing *Oxalis* (Oxalidaceae): Dispersal to North America and modification of the tristylous breeding system. Amer. J. Bot. 99: 152–164. Lourteig, A. 1975. Oxalidaceae extra-Austroamericanae: 1. *Oxalis* L. sectio *Thamnoxys* Planchon. Phytologia 29: 449–471. Lourteig, A. 1979. Oxalidaceae extra-Austroamericanae: 2. *Oxalis* L. sectio *Corniculatae* DC. Phytologia 42: 57–198. Lourteig, A. 1982. Oxalidaceae extra-Austroamericanae: 4. *Oxalis* L. sectio *Articulatae* Knuth. Phytologia 50: 130–142. Lourteig, A. 1994. *Oxalis* L. subgénero *Thamnoxys* (Endl.) Reiche emend. Lourt. Bradea 7: 1–199. Lourteig, A. 2000. *Oxalis* L. subgeneros *Monoxalis* (Small) Lourt., *Oxalis* y *Trifidus* Lourt. Bradea 7: 201–629. Nesom, G. L. 2009b. Again: Taxonomy of yellow-flowered caulescent *Oxalis* (Oxalidaceae) in eastern North America. J. Bot. Res. Inst. Texas 3: 727–738. Nesom, G. L. 2009c. Notes on *Oxalis* sect. *Corniculatae* (Oxalidaceae) in the southwestern United States. Phytologia 91: 527–533. Nesom, G. L. 2009d. Taxonomic notes on acaulescent *Oxalis* (Oxalidaceae) in the United States. Phytologia 91: 501–526. Nesom, G. L., D. D. Spaulding, and H. E. Horne. 2014. Further observations on the *Oxalis dillenii* group (Oxalidaceae). Phytoneuron 2014-12: 1–10. Ornduff, R. 1972. The breakdown of trimorphic incompatibility in *Oxalis* section *Corniculatae*. Evolution 26: 52–65. Robertson, K. R. 1975. The Oxalidaceae in the southeastern United States. J. Arnold Arb. 56: 223–239. Rosenfeldt, S. and B. G. Galati. 2009. The structure of the stigma and the style of *Oxalis* spp. (Oxalidaceae). J. Torrey Bot. Soc. 136: 33–45. Weller, S. G. and M. F. Denton. 1976. Cytogeographic evidence for the evolution of distyly from tristyly in the North American species of *Oxalis* section *Ionoxalis*. Amer. J. Bot. 63: 120–125. Wiegand, K. M. 1925. *Oxalis corniculata* and its relatives in North America. Rhodora 27: 113–124, 133–139.

1. Leaflets 1 .... 1. *Oxalis dichondrifolia*
1. Leaflets 3–11.
   2. Leaves basal; plants acaulous.
      3. Petals deep golden yellow .... 17. *Oxalis pes-caprae*
      3. Petals white to pink, salmon, red, lavender, purple, or violet, sometimes yellow or greenish yellow basally or proximally.
         4. Rhizomes present; bulbs or bulblets present or absent.
            5. Leaflets dark purple, often with lighter violet splotches radiating from midvein adaxially, lobes apically truncate to slightly convex; rhizomes short, branching, covered with thickened, overlapping scales .... 32. *Oxalis triangularis*
            5. Leaflets usually green, sometimes purplish to deep purple abaxially, lobes apically convex or leaflets not lobed; rhizomes woody and irregularly nodulate-segmented, or fleshy-thickened.
               6. Sepal apices with 2 orange tubercles; rhizomes thick, woody, irregularly nodulate-segmented, often with persistent, thickened, woody petiole bases .... 36. *Oxalis articulata*
               6. Sepal apices without tubercles; rhizomes slender or fleshy-thickened, sparsely or densely scaly or nonscaly.
                  7. Bulbs or bulblets present; rhizomes slender or thickened, sparsely scaly.
                     8. Inflorescences 1-flowered; scapes 1.5–6(–8) cm, sparsely to moderately villous (hairs eglandular); bulbs 1–2.5 cm or plants with clustered bulblets; leaflets not lobed.......... 23. *Oxalis purpurea*
                     8. Inflorescences 4–12-flowered; scapes 15–20 cm, densely glandular-puberulent; bulbs 2–4 cm; leaflets lobed.......... 24. *Oxalis bowiei*
                  7. Bulbs or bulblets absent; rhizomes fleshy-thickened, densely scaly.
                     9. Inflorescences 2–9(–15)-flowered; petals white to pinkish, sometimes greenish proximally, without prominent veins; capsules narrowly fusiform, 15–25(–30) mm .... 20. *Oxalis trilliifolia*
                     9. Inflorescences 1-flowered; petals white to deep pink, usually with yellow spot sub-basally and rose or purple veins; capsules subglobose to ovoid, 2–8(–12) mm.
                        10. Scapes 4–15 cm; petals 10–15 mm; capsules 2–4 mm; e North America .... 21. *Oxalis montana*
                        10. Scapes (6–)11–25 cm; petals (8–)15–25 mm; capsules 6–8(–12) mm; w North America .... 22. *Oxalis oregana*
         4. Rhizomes absent (rarely present in *O. violacea*, where slender, scaly, bearing bulblets at tips); bulbs or bulblets present.
            11. Leaflets (3–)4–11.
                12. Leaflets (3–)4(–5), obtriangular to obcordate, 5–22 mm, lobed $\frac{1}{5}$–$\frac{1}{2}$ length; bulb scales (3–)5–7-nerved.......... 25. *Oxalis caerulea*
                12. Leaflets (3–)5–11, narrowly oblong-oblanceolate to narrowly oblong or linear, (10–)12–38(–72) mm, lobed ($\frac{1}{6}$–)$\frac{1}{2}$–$\frac{2}{3}$(–$\frac{9}{10}$) length; bulb scales 9–15+-nerved.......... 26. *Oxalis decaphylla*
            11. Leaflets 3.
                13. Leaflets: oxalate deposits in dots at least around distal margins, often evenly over surface, abaxial surfaces hirsute .... 33. *Oxalis debilis*
                13. Leaflets: oxalate deposits absent or in band or lines along margins at base of notch, abaxial surfaces glabrous, strigose, or hirsute-strigose, sometimes densely hirsute at very base.
                    14. Leaflets obtriangular to broadly obtriangular, lobes apically truncate; stolons often present, with bulblets at tips; bulbs usually clustered, sometimes solitary.......... 31. *Oxalis intermedia*
                    14. Leaflets obtriangular to obcordate, rounded-obcordate, obreniform, or obdeltate, lobes apically rounded to convex or nearly truncate; stolons absent; bulbs solitary or clustered.

15. Petals with dark purple veins proximally; leaflet abaxial surfaces strigose to hirsute-strigose, sometimes densely hirsute at very base.
    16. Leaflet abaxial surfaces strigose to hirsute-strigose, densely hirsute at very base, margins prominently ciliate with stiff, sharp-pointed hairs; outer bulb scales 3[–5]-nerved; sepal apices with 2 elongate, orange tubercles . . . . 34. *Oxalis hispidula*
    16. Leaflet abaxial surfaces sparsely but evenly strigose with fine hairs, margins glabrous to sparsely irregularly ciliate with loose, fine hairs; outer bulb scales 5–8[–13]-nerved; sepal apices without tubercles . . . . 35. *Oxalis brasiliensis*
15. Petals with green veins proximally; leaflet abaxial surfaces glabrous.
    17. Bulbs usually surrounded by cluster of bulblets; leaflets: oxalate deposits usually in narrow band 0.5–1.5 mm along margins at base of notch, sometimes evident on one surface but not other, rarely apparently absent. . . . . 27. *Oxalis metcalfei*
    17. Bulbs solitary, bulblets absent or rarely at tips of rhizomes; leaflets: oxalate deposits absent or in lines along margins at base of notch.
        18. Bulb scales (5–)7–9(–11)-nerved; sepal apices with tubercles not confluent; sw United States. . . . . 30. *Oxalis latifolia*
        18. Bulb scales 3-nerved; sepal apices with tubercles apically confluent; primarily e and s central United States.
            19. Leaflets obtriangular to obcordate, (6–)14–34 mm, oxalate deposits absent . . . . 28. *Oxalis drummondii*
            19. Leaflets rounded-obcordate to obreniform, (5–)8–15(–20) mm, oxalate deposits in lines along margins at base of notch . . . . .29. *Oxalis violacea*

[2. Shifted to left margin.—Ed.]

2. Leaves cauline or basal and cauline; plants caulescent.
    20. Petals white to pink, rosy purple, or pinkish purple.
        21. Petioles 0–0.2 cm; stems villous; leaflets linear to narrowly oblanceolate or oblong-cuneate, not lobed, abaxial surfaces villous; axillary bulblets absent; petals rosy purple to pink or white. . . . . 18. *Oxalis hirta*
        21. Petioles 2–5(–7) cm; stems glabrous; leaflets obcordate, lobed ¼ length, abaxial surfaces glabrous; axillary bulblets often present; petals white to pale pinkish purple. . . . . .19. *Oxalis incarnata*
    20. Petals yellow, sometimes with red lines.
        22. Leaves pinnately compound (terminal leaflets on extended petiolules); leaflets cuneate to obovate or oblong-obovate, not lobed . . . . 2. *Oxalis frutescens*
        22. Leaves palmately compound; leaflets obcordate, lobed ⅕–⅓ length.
            23. Inflorescences racemes; plants annual; capsules 3–5 mm .16. *Oxalis laxa*
            23. Inflorescences cymes; plants perennial (annual in *O. corniculata*, sometimes annual in *O. stricta*); capsules 6–20(–25) mm.
                24. West of the Mississippi River.
                    25. Stipule margins with wide, free flanges, apical auricles free; stems prostrate or decumbent, often rooting at nodes; rhizomes absent . . . . 3. *Oxalis corniculata* (in part)
                    25. Stipules rudimentary or margins narrowly to very narrowly flanged or without free portions, apical auricles slightly free or absent; stems erect, ascending, decumbent, or prostrate, rooting at nodes or not; rhizomes present or absent.

26. Stems villous; petioles, and usually stems, with septate and nonseptate hairs; rhizomes present; stems 1(–3) from base.
    27. Petals (6–)8–11 mm; inflorescences (1–)5–7(–15)-flowered, regular or irregular cymes; stems 20–60(–90) cm; rhizomes short . . . . 12. *Oxalis stricta* (in part)
    27. Petals 12–20 mm; inflorescences 1–3-flowered, umbelliform cymes; stems 10–25 cm; rhizomes long . . . . 15. *Oxalis suksdorfii*
26. Stems usually strigose, strigillose, puberulent, hirtellous-puberulent, pilose, rarely villous proximally (*O. florida*, *O. texana*), or glabrous; petioles and stems glabrous or with nonseptate hairs; rhizomes present or absent; stems (1–)2–8 from base (usually 1 in *O. florida*).
    28. Stems usually strigillose or strigose, hairs straight, antrorsely appressed to closely ascending, Louisiana plants rarely villous proximally, hairs spreading.
        29. Inflorescences 1–3(–8)-flowered; flowers mostly homostylous; petals without red lines, (2.5–)4–8 mm . . . . 4. *Oxalis dillenii* (in part)
        29. Inflorescences (2–)3–5(–8)-flowered; flowers distylous; petals with prominent red lines proximally, (10–)12–16 (–17) mm . . . . 5. *Oxalis texana* (in part)
    28. Stems glabrous or puberulent, hirtellous-puberulent, pilose, or sparsely to moderately strigose, hairs curved or crisped, or, if straight, spreading.
        30. Sepals glabrous.
            31. Stems usually glabrate to sparsely or moderately strigose, sometimes sparsely villous proximally; stipules rudimentary; inflorescences 1–2(–3)-flowered; Arkansas, Louisiana, Missouri, Texas . . . . 8. *Oxalis florida* (in part)
            31. Stems glabrous or very sparsely short-puberulent; stipules oblong; inflorescences 1(–3)-flowered; Arizona, California . . . . 11. *Oxalis californica*
        30. Sepals strigose to hirsute-strigose.
            32. Stems puberulent to hirtellous-puberulent, hairs usually antrorsely curved or crisped, sometimes ± straight, longer hairs 0.2–0.3(–0.8) mm . . . . 9. *Oxalis albicans*
            32. Stems sparsely to densely pilose, hairs spreading irregularly to ± deflexed, longer hairs 0.6–1.2 mm . . . 10. *Oxalis pilosa*

[24. Shifted to left margin.—Ed.]

24. East of the Mississippi River.
    33. Stems prostrate or decumbent, rooting at nodes; rhizomes and stolons absent; stipules oblong, margins with wide, free flanges, apical auricles free; seeds brown, transverse ridges brown . . . . 3. *Oxalis corniculata* (in part)
    33. Stems erect to ascending, sometimes decumbent, rarely rooting at nodes; rhizomes or stolons usually present; stipules rudimentary or apparently absent, or, if oblong, margins narrowly to very narrowly flanged or without free portions, apical auricles slightly free or absent; seeds brown to blackish brown, transverse ridges brown, white, or with grayish or white lines.
        34. Stems strigose or strigillose, hairs straight, antrorsely appressed to closely ascending (central Louisiana populations of *O. texana* rarely irregularly villous proximally).
            35. Inflorescences 1–3(–8)-flowered; flowers mostly homostylous; petals (2.5–) 4–8 mm, without red lines . . . . 4. *Oxalis dillenii* (in part)
            35. Inflorescences (2–)3–5(–8)-flowered; flowers distylous; petals 11–15 mm (Arkansas, Louisiana, Texas) or 6–12 mm (Alabama), with red lines proximally . . . . 5. *Oxalis texana* (in part)
        34. Stems glabrate or villous-hirsute, hirsute-pilose, pilose, villous, or strigose, hairs ± straight or slightly curved (if stems strigose), spreading, deflexed, or antrorse.

[36. Shifted to left margin.—Ed.]

36. Stems usually sparsely to moderately strigose or glabrate distally, sometimes sparsely villous proximally, hairs slightly curved, antrorse; petals 4–8 mm. . . . . 8. *Oxalis florida* (in part)
36. Stems villous, villous-hirsute, hirsute-pilose, or pilose, sometimes glabrate on at least proximal ⅔, hairs curved or ± straight, spreading or deflexed; petals (6–)8–20(–23) mm.
    37. Petioles and stems with nonseptate hairs; inflorescences umbelliform cymes; flowers well above level of leaves.
        38. Petals (13–)15–20(–23) mm; pedicels villous, hairs long, spreading; capsules sparsely to densely hirsute-pilose, hairs long; plants strongly colonial; limestone and chalk habitats. . . . . 6. *Oxalis macrantha*
        38. Petals 9–15 mm; pedicels strigose, hairs short, curved antrorsely; capsules glabrate to puberulent or sparsely hirsute-strigose, hairs short; plants cespitose or weakly colonial; wide variety of habitats, including limestone . . . . 7. *Oxalis colorea*
    37. Petioles, and usually stems, with septate and nonseptate hairs; inflorescences regular, irregular, or umbelliform cymes; flowers above or at level of leaves.
        39. Petals (6–)8–11 mm; inflorescences (1–)5–7(–15)-flowered . . . . 12. *Oxalis stricta* (in part)
        39. Petals 10–18 mm; inflorescences 1–4(–8)-flowered.
            40. Petals 10–14 mm, with or without faint red lines proximally; rhizomes becoming woody, without tubers; leaflet lobes apically usually rounded, rarely ± truncate, margins green or brownish purple; flowers above level of leaves . . . . 13. *Oxalis grandis*
            40. Petals 12–18 mm, with prominent red lines proximally; rhizomes herbaceous, with tubers or tuberlike thickenings; leaflet lobes apically truncate, margins green; flowers mostly at level of leaves . . . . 14. *Oxalis illinoensis*

1. Oxalis dichondrifolia A. Gray, Smithsonian Contr. Knowl. 3(5): 27. 1852 (as dichondraefolia) • Peony-leaf wood-sorrel, agrito

Monoxalis dichondrifolia (A. Gray) Small

Herbs perennial, caulescent, caudex present, rhizomes and stolons absent, bulbs absent, taproot sometimes with tuberlike portions. **Aerial stems** mostly 1–3 from base, erect, 5–20(–30) cm, becoming woody proximally, finely hirtellous-villous. **Leaves** cauline; stipules brownish, linear-setiform, margins without flanges, apical auricles absent; petiole (0.5–)1–3 cm; leaflet 1, dull gray-green, suborbiculate to oblong-obovate or ovate, 5–37 mm, not lobed, apex concave or truncate to retuse and apiculate, surfaces strigose-hirsute, oxalate deposits absent. **Inflorescences** 1-flowered, axillary at distal nodes; peduncles 15–25(–50) cm. **Flowers** heterostylous; sepal apices without tubercles; petals yellow to orange-yellow, 11–13 mm. **Capsules** broadly cylindric, 5–8(–10) mm, densely pilose.

Flowering Feb–Jun, sporadically year-round. Gravelly hills, clay dunes, limestone slopes, calcareous marl, sand, sandy loam, sandy silt, alluvial soils, brushlands, mesquite thickets, chaparral, roadsides, fields, ditch and river margins; 0–500 m; Tex.; Mexico (Coahuila, Nuevo León, San Luis Potosí, Tamaulipas, Veracruz).

Oxalis dichondrifolia is fairly widespread in southern Texas.

2. Oxalis frutescens Linnaeus, Sp. Pl. 1: 435. 1753 • Shrubby wood-sorrel

Subspecies 4 (1 in the flora): sw, sc United States, Mexico, West Indies, Central America, South America.

2a. Oxalis frutescens Linnaeus subsp. **angustifolia** (Kunth) Lourteig, Phytologia 29: 463. 1975

Oxalis angustifolia Kunth in A. von Humboldt et al., Nov. Gen. Sp. 5(fol.): 193; 5(qto.): 249. 1822; *O. berlandieri* Torrey; *O. lindheimeri* Torrey ex R. Knuth; *O. neaei* de Candolle; *O. yucatanensis* (Rose) R. Knuth

Herbs perennial, caulescent, caudex present, rhizomes and stolons absent, bulbs absent. **Aerial stems** usually 1 from base, erect, 8–35 cm, becoming woody proximally, loosely villous to puberulent, hairs long, curved, spreading to deflexed, orientation often mixed. **Leaves** cauline, pinnately compound, terminal leaflet on extended petiolule; stipules rudimentary; petiole 10–15 cm; leaflets 3, green, sometimes purplish abaxially, cuneate to obovate or oblong-obovate, 6–17(–25) mm, not lobed, apex of lateral leaflets singly or doubly notched, of terminal leaflet emarginate or notched, surfaces sparsely strigose-villous to glabrate, oxalate deposits absent. **Inflorescences** umbelliform cymes, 1–5-flowered; peduncles 10–25 cm. **Flowers** heterostylous;

sepal apices without tubercles; petals yellow, 10–16 mm. **Capsules** broadly cylindric, 6–8 mm, sparsely villous on angles, glabrous between.

Flowering Mar–May, Jul–Oct. Pastures, roadsides, grasslands, live oak woods, sand, sandy loam; 0–200 m; N.Mex., Tex.; Mexico; Central America; South America.

Subspecies *angustifolia* (type from central Mexico) has the widest distribution of the four subspecies recognized by A. Lourteig (1975). Subspecies *frutescens* (type from Martinique) occurs in the West Indies and into the South American Andes; subsp. *borjensis* (Kunth) Lourteig (type from Colombia) and subsp. *pentantha* (Jacquin) Lourteig (type from Venezuela) are both restricted to the Andes.

3. Oxalis corniculata Linnaeus, Sp. Pl. 1: 435. 1753 • Creeping yellow wood-sorrel [F] [I] [W]

Oxalis corniculata var. *atropurpurea* Planchon; *O. corniculata* var. *domingensis* (Urban) Moscoso; *O. corniculata* var. *langloisii* (Small) Wiegand; *O. corniculata* var. *lupulina* (Kunth) Zuccarini; *O. corniculata* var. *macrophylla* Arsène ex R. Knuth; *O. corniculata* var. *repens* (Thunberg) Zuccarini; *O. corniculata* var. *villosa* (M. Bieberstein) Hohenacker; *O. corniculata* var. *viscidula* Wiegand; *O. langloisii* (Small) Fedde; *O. repens* Thunberg ; *Xanthoxalis corniculata* (Linnaeus) Small; *X. langloisii* Small

Herbs annual, caulescent, rhizomes and stolons absent, bulbs absent. **Aerial stems** commonly 2–8 from base, prostrate or decumbent, stolonlike, rooting at nodes, 4–10(–30) cm, herbaceous, sparsely and loosely strigose to strigose-villous or villous, hairs nonseptate. **Leaves** basal and cauline; stipules oblong, membranous, margins with wide, free flanges, apical auricles free; petiole 1–5 cm; leaflets 3, green or bronze-purple to maroon, obcordate, (4–)6–12 mm, lobed 1/5–1/3 length, margins often prominently villous-ciliate, surfaces glabrous, oxalate deposits absent. **Inflorescences** irregular or umbelliform cymes, 1–3(–6)-flowered; peduncles (1–)2–4 (–8) cm. **Flowers** mostly homostylous; sepal apices without tubercles; petals yellow, 4–8 mm. **Capsules** angular-cylindric, gradually or abruptly tapering to apex, 8–17(–20) mm, sparsely puberulent to glabrate or glabrous. **Seeds** brown, transverse ridges brown. $2n$ = 24, 36, 42, 44, 48.

Flowering Mar–Aug(–Oct), sporadically year-round. Disturbed areas, gardens, greenhouses, lawns, fields, roadsides, hammocks, beach margins, open pine woods, grasslands; 10–500(–2500) m; introduced; B.C., Man., N.S., Ont., P.E.I., Que., Sask.; Ala., Ariz., Ark., Calif., Colo., Conn., Del., D.C., Fla., Ga., Idaho, Ill., Ind., Ky., La., Maine, Md., Mass., Mich., Miss., Mo., Mont., Nebr., N.J., N.C., Ohio, Okla., Oreg., Pa., S.C., S.Dak., Tex., Vt., Va., Wash., W.Va.; Mexico; West Indies; Central America; South America; introduced also in Europe, Asia (China, India, Japan), Africa, Pacific Islands, Australia.

Oxalis corniculata in the flora area is recognized by a combination of its small flowers (petals yellow, 4–8 mm); sparsely hairy, herbaceous stems creeping and rooting at nodes; and stipules with free flanges and apical auricles. Peduncles and leaves (one to three) are produced at the nodes, short erect stems less commonly so. Specimens have been seen documenting its distribution in the United States as listed above; it may also occur in intervening areas.

The typical form of *Oxalis corniculata* is strictly annual with consistently herbaceous, prostrate stems. At least some populations in western Oregon are distinctly more erect, with decumbent-ascending stems, than those of the eastern United States. In contrast, stems of *O. dillenii* characteristically are initially erect but may become decumbent to prostrate, occasionally rooting at the nodes; they almost always become more or less woody. Stems arising from nodes of laterally oriented stems characteristically are erect. In most of the United States, *O. corniculata* usually occurs in urban and highly disturbed habitats, but along the Gulf Coast it occasionally grows in less obviously disturbed sites and might be native there. However, assignment of nativity awaits a clearer understanding of patterns of variation within what is recognized as a highly variable species.

Variants of *Oxalis corniculata* and closely similar forms occur in Mexico, the West Indies, Central America, and South America, as well as in other parts of the world, including the flora area. Plants with bronze-purple to maroon leaves and hairy capsules have been recognized as var. *atropurpurea* (for example, in Florida, D. B. Ward 2004; in California, L. Abrams and R. S. Ferris 1923–1960, vol. 3). Variety *atropurpurea* in Malaysia has features of a distinct species, differing from typical *O. corniculata* in karyotype as well as in floral and vegetative morphology and is isolated by post-pollination reproductive barriers (B. R. Nair and P. Kuriachan 2004). Australasian variants sometimes identified as *O. corniculata* recently have been treated at specific rank (for example, P. J. de Lange et al. 2005). In view of the significant variation in ploidy level reported for the species, formal recognition of these and probably still other segregates may be justified.

SELECTED REFERENCE Turner, B. L. 1994. Regional variation in the North American elements of *Oxalis corniculata* (Oxalidaceae). Phytologia 77: 1–7.

4. Oxalis dillenii Jacquin, Oxalis, 28. 1794 • Common yellow wood-sorrel [E] [W]

Oxalis corniculata Linnaeus var. *dillenii* (Jacquin) Trelease; *O. dillenii* var. *radicans* Shinners; *O. florida* Salisbury subsp. *prostrata* (Haworth) Lourteig; *O. lyonii* Pursh; *O. prostrata* Haworth; *Xanthoxalis dillenii* (Jacquin) Holub

Herbs perennial, caulescent, rhizomes present (sometimes appearing taprootlike), stolons absent, bulbs absent. **Aerial stems** (1–)2–8 from base, erect initially, often becoming decumbent or prostrate and stolonlike, rarely rooting at nodes, 10–25 cm, usually herbaceous, sometimes becoming woody proximally, densely and evenly strigillose to strigose from base to peduncles and pedicels, hairs straight, antrorsely appressed, nonseptate, sharp-pointed. **Leaves** basal and cauline; stipules oblong, margins narrowly flanged or without free portions, apical auricles absent; petiole 1–4 cm, hairs nonseptate; leaflets 3, green, obcordate, (4–)6–15(–21) mm, lobed 1/5–1/3 length, abaxial surface sparsely strigillose, adaxial surface glabrous, oxalate deposits absent. **Inflorescences** usually umbelliform cymes, rarely irregular cymes, 1–3(–8)-flowered; peduncles 1–6(–10) cm. **Flowers** mostly homostylous; sepal apices without tubercles; petals yellow, without red lines, (2.5–)4–8 mm. **Capsules** angular-cylindric, abruptly tapering to apex, 12–20(–25) mm, densely strigose-pilose, hairs both appressed and spreading, with puberulent understory. **Seeds** brown, transverse ridges with strong grayish or white lines. $2n$ = 18, 20, 22, 24.

Flowering Feb–May(–Oct). Pastures, roadsides, lawns, river bottoms, sandy, rocky, or gravelly soils; 0–300 m; B.C., Man., N.B., N.S., Ont., P.E.I., Que., Sask.; Ala., Ariz., Ark., Colo., Conn., Del., D.C., Fla., Ga., Idaho, Ill., Ind., Iowa, Kans., Ky., La., Maine, Md., Mass., Mich., Minn., Miss., Mo., Mont., Nebr., N.H., N.J., N.Mex., N.Y., N.C., N.Dak., Ohio, Okla., Oreg., Pa., R.I., S.C., S.Dak., Tenn., Tex., Utah, Vt., Va., Wash., W.Va., Wis., Wyo.; introduced in Bermuda, Europe.

Decumbent stems of *Oxalis dillenii* often appear stolonlike, producing erect branches and leaves at the nodes, rarely producing a few adventitious roots. Such plants sometimes are misidentified as *O. corniculata*, but they differ in their overall habit, stems and rhizomes that become woody, reduced stipules, strigillose cauline vestiture, denser fruit vestiture, and seed color. Plants of *O. dillenii* in Canada and the western United States appear to be adventive.

Plants of *Oxalis dillenii* flowering into November and December in Texas and North Carolina, and probably elsewhere, characteristically are depressed in habit, with creeping stems and forming matlike colonies. The corollas are small, with petals 2.5–3 mm. Although similar in habit to typical *O. corniculata*, these plants do not have stems that root at nodes, and the strigillose vestiture, though reduced in density, and the stipule morphology are like *O. dillenii*. Whether these are the same genotype as spring-flowering plants or a different entity remains to be investigated. At least some of the variability in *O. dillenii*, particularly in habit, may be genetically partitioned, as dysploid chromosome races apparently exist.

5. Oxalis texana (Small) Fedde, Just's Bot. Jahresber. 32(1): 410. 1905 • Texas wood-sorrel [E]

Xanthoxalis texana Small, Fl. S.E. U.S., 667, 1332. 1903; *Oxalis priceae* Small subsp. *texana* (Small) G. Eiten; *O. recurva* Elliot var. *texana* (Small) Wiegand

Herbs perennial, caulescent, cespitose, caudex present, rhizomes or stolons present, bulbs absent. **Aerial stems** usually 2–6 from base, erect to ascending, 5–15 cm, becoming woody proximally, usually evenly strigose to strigillose from base to peduncles and pedicels, hairs straight, antrorsely appressed to closely ascending, (rarely Louisiana plants proximally villous, hair spreading), nonseptate. **Leaves** basal and cauline; stipules oblong, margins usually very narrowly flanged, apical auricles usually slightly free; petiole 2–6 cm, hairs nonseptate; leaflets 3, green to purple, cordate, (4–)6–12(–18) mm, lobed 1/5–1/3 length, abaxial surface sparsely strigose, adaxial surface glabrous or sparsely strigose, oxalate deposits absent. **Inflorescences** umbelliform cymes, very rarely irregular cymes, (2–)3–5(–8)-flowered; peduncles 4–10 cm. **Flowers** distylous; sepal apices without tubercles; petals yellow, with prominent red lines proximally, (10–)12–16(–17) mm (Arkansas, Louisiana, Texas) or 6–12 mm (Alabama). **Capsules** angular-cylindric, abruptly tapering to apex, 8–15 mm, moderately to densely puberulent to puberulent-villous. **Seeds** brown, transverse ridges white.

Flowering Mar–May(–Jun). Commonly in undisturbed habitats and usually in deep, loose sand, but also fields, roadsides, edges and openings in pine, pine-oak, and mixed hardwood woods; 10–200 m; Ala., Ark., La., Tex.

Oxalis texana is similar to *O. dillenii* in its evenly strigose to strigillose stems but differs primarily in its more numerous flowers per inflorescence and larger, distylous flowers with red-lined corolla throats. Plants of *O. dillenii* with larger flowers on elevated peduncles might be mistaken for *O. texana*, yet the two taxa exist sympatrically in the range of *O. texana*, and it is clear that they are separate species.

All Alabama plants identified as *Oxalis texana* (weighting orientation of cauline vestiture in the identifications) are from Dauphin Island and localities in and around Mobile. Compared to those in Louisiana and Texas, the Alabama plants have shorter petals [6–12 versus (10–)12–16(–17) mm], a more colonial habit (usually with long, lateral, stolonlike branches versus commonly with short basal offsets), and they grow in disturbed sites (versus mostly undisturbed sites, usually within woods). It is plausible that they may prove to be more closely related to *O. colorea* than to the western *O. texana*.

6. Oxalis macrantha (Trelease) Small, Bull. Torrey Bot. Club 23: 268. 1896 • Price's wood-sorrel

Oxalis corniculata Linnaeus var. *macrantha* Trelease, Mem. Boston Soc. Nat. Hist. 4: 88, plate 11, fig. 5. 1888, *O. hirsuticaulis* Small, *O. priceae* Small; *O. recurva* Elliott var. *macrantha* (Trelease) Wiegard; *Xanthoxalis hirsuticaulis* (Small) Small; *X. macrantha* (Trelease) Small; *X. priceae* (Small) Small

Herbs perennial, caulescent, strongly colonial, rhizomes or stolons usually present, bulbs absent. **Aerial stems** usually 2–8 from base, erect initially, usually becoming decumbent, 5–20(–40) cm, becoming woody proximally, hirsute-pilose on at least proximal ⅔, hairs curved, loosely and irregularly spreading, nonseptate. **Leaves** basal and cauline; stipules oblong, margins narrowly flanged or without free portions, apical auricles absent; petiole 2–7 cm; leaflets 3, green, obcordate, 3.5–12 mm, lobed ⅕–⅓ length, surfaces usually strigose-hirsute, sometimes glabrate, oxalate deposits absent. **Inflorescences** umbelliform cymes, less commonly irregular cymes, (1–)3–8-flowered; peduncles (3–)5–10 (–15) cm. **Pedicels** villous, hairs long, spreading. **Flowers** distylous, well above level of leaves; sepal apices without tubercles; petals yellow to yellow-orange, with prominent red lines proximally, (13–)15–20(–23) mm. **Capsules** angular-cylindric, abruptly tapering to apex, 10–15 mm, sparsely to densely hirsute-pilose, hairs long, sometimes mostly along angles. **Seeds** brown, transverse ridges usually white.

Flowering Mar–May. Dry limestone glades, cedar barrens, chalk prairies, limestone bluffs and outcrops; 100–300 m; Ala., Ky., Tenn.; Mexico (Nuevo León).

Oxalis macrantha is restricted mostly to limestone glades in Alabama, Kentucky, and Tennessee. It is recognized by its villous to villous-hirsute stems, flowers in umbelliform cymes, and large yellow to yellow-orange corollas with red lines proximally. The lines in the throat remain visible after drying and usually can be seen on herbarium specimens even from the outside of the flower. A similar pattern also occurs in other species, especially *O. grandis*, *O. illinoensis*, and *O. texana*.

Seemingly disjunct plants of native habitats in Nuevo León, Mexico, identified as *Oxalis macrantha* apparently are more common than reported by G. L. Nesom (2009b). Whether these are actually disjunct or a parallel morphological expression derived from some Mexican species needs to be investigated.

7. Oxalis colorea (Small) Fedde, Just's Bot. Jahresber. 32(1): 410. 1905 • Small's wood-sorrel [E]

Xanthoxalis colorea Small, Fl. S.E. U.S. 668, 1333. 1903; *Oxalis priceae* Small subsp. *colorea* (Small) G. Eiten

Herbs perennial, caulescent, cespitose or weakly colonial, rhizomes or stolons sometimes present, short, bulbs absent. **Aerial stems** usually 1–4 from base, mostly erect, 5–15(–25) cm, often becoming woody proximally, hirsute-pilose proximally, hairs curved, irregularly spreading or slightly deflexed, sometimes ± antrorse on peduncles and pedicels, nonseptate. **Leaves** basal and cauline; stipules oblong, margins narrowly flanged or without free portions, apical auricles absent; petiole 2–7 cm; leaflets 3, green, obcordate, 3–8 mm, lobed ⅕–⅓ length, surfaces usually sparsely strigose-hirsute, sometimes glabrate, oxalate deposits absent. **Inflorescences** usually umbelliform cymes, sometimes irregular cymes, (1–)2–4(–5)-flowered; peduncles (3–)5–10(–15) cm. **Pedicels** strigose, hairs short, curved antrorsely. **Flowers** tristylous, well above level of leaves; sepals 3.5–5 mm, apices without tubercles; petals yellow to orange-yellow, usually with prominent, rarely faint, red lines proximally, very rarely lines absent, 9–15 mm. **Capsules** angular-cylindric, abruptly tapering to apex, 6–16 mm, glabrate to puberulent or sparsely hirsute-strigose, hairs short, sometimes only along angles or at apex. **Seeds** brown, transverse ridges usually white.

Flowering (Mar–)Apr–May(–Oct). Longleaf pine, longleaf pine-scrub oak, pine-mixed hardwood, hardwood, beech-magnolia, and alluvial woods, shale slopes, sandstone outcrops, granite outcrops, limestone, river and stream banks, hillsides and ridges, bluffs, ravines, clearings, roadsides; 30–300 m; Ala., Fla., Ga., La., Miss., Mo., N.J., N.C., S.C., Tenn., Va., W.Va.

Oxalis colorea is common in southeastern Alabama, Mississippi, and probably the northern half of Georgia; outliers apparently occur in a wider area toward the north and northeast. Scattered variants from widely scattered localities in Alabama, Mississippi, and Tennessee, from within the geographic range of *O. colorea*, have the habit and prominently red-lined petals of *O. colorea*; stem hairs are relatively long but are antrorsely oriented, as in *O. florida*.

8. Oxalis florida Salisbury, Prodr. Stirp. Chap. Allerton, 322. 1796 • Slender eastern wood-sorrel [E] [W]

Oxalis brittoniae Small; *O. dillenii* Jacquin subsp. *filipes* (Small) G. Eiten; *O. dillenii* subsp. *recurva* (Elliott) C. F. Reed; *O. filipes* Small; *O. florida* var. *filipes* (Small) H. E. Ahles; *O. florida* var. *recurva* (Elliott) H. E. Ahles; *O. recurva* Elliott; *Xanthoxalis brittoniae* Small; *X. filipes* (Small) Small; *X. recurva* (Elliott) Small

Herbs perennial, caulescent, rhizomes or stolons usually present, bulb absent. **Aerial stems** usually 1 from base, usually erect, rarely leaning and decumbent, not rooting at nodes, (5–)8–30(–35) cm, herbaceous, glabrous, glabrate, or sparsely to moderately strigose, sometimes sparsely villous proximally, hairs slightly curved, antrorse, nonseptate. **Leaves** basal and cauline; stipules rudimentary; petiole 2–5 cm, hairs nonseptate; leaflets 3, green, obcordate, 4–11 mm, lobed 1/5–1/3 length, abaxial surface sparsely strigose, adaxial surface glabrous, oxalate deposits absent. **Inflorescences** umbelliform cymes, 1–2(–3)-flowered; peduncles (2–)3–8 cm. **Flowers** tristylous, at or slightly above level of leaves; sepal apices without tubercles, surfaces glabrous; petals yellow, sometimes with faint red lines proximally, 4–8 mm. **Capsules** angular-cylindric, abruptly tapering to apex, 7–10 mm, glabrous or glabrate to sparsely puberulent, hairs short, sometimes only along angles. **Seeds** brown, transverse ridges brown. $2n = 16$.

Flowering Mar–May(–Aug). Low woods, swamp forests, rich woods, pine woods, sandy sites, burned-over woods, ditches, roadside banks, flood plains, low fields, lake edges, stream banks, pastures, disturbed sites, bluffs, rocky slopes; 10–400 m; Ark., Conn., D.C., Fla., Ga., Ind., Ky., La., Maine, Md., Mass., Miss., Mo., N.H., N.J., N.Y., N.C., Pa., S.C., Tex., Vt., Va., W.Va.

Oxalis florida is recognized by its mostly erect stems, thin stems and peduncles (compared to other species), sparsely strigose cauline vestiture of relatively short, slightly curved hairs, rudimentary stipules, and relatively small flowers with petals that usually lack red lines proximally. It is a species primarily of the Atlantic states and Gulf coast, but it also occurs in Arkansas, Louisiana, southwestern Mississippi, Missouri, and Texas, apparently disjunct westward from its main range.

Intermediates between *Oxalis florida* and *O. colorea* apparently occur where their ranges come into contact. According to G. Eiten (1963), *O. florida* intergrades with *O. dillenii* in forming intermediate homogenous populations as well as hybrid swarms. Eiten treated *O. florida* as *O. dillenii* subsp. *filipes*, but D. B. Ward (2004) noted that the differences between *O. florida* and *O. dillenii* are appreciable and intermediates seem few. K. M. Wiegand (1925) observed that *O. florida* and *O. filipes* have the appearance of hybrids between *O. stricta* and either *O. dillenii* or *O. corniculata* but that their absence in much of the region where the possible parents both occur argues against this hypothesis.

9. Oxalis albicans Kunth in A. von Humboldt et al., Nov. Gen. Sp. 5(fol.): 189; 5(qto.): 244. 1822 • Western yellow wood-sorrel

Oxalis californica (Abrams) R. Knuth var. *subglabra* Wiegand; *O. corniculata* Linnaeus subsp. *albicans* (Kunth) Lourteig; *O. corniculata* var. *wrightii* (A. Gray) B. L. Turner; *O. pilosa* Nuttall var. *wrightii* (A. Gray) Wiegand; *O. wrightii* A. Gray; *Xanthoxalis albicans* (Kunth) Small; *X. wrightii* (A. Gray) Abrams

Herbs perennial, caulescent, rhizomes and stolons absent, bulbs absent. **Aerial stems** usually 2–8 from base, usually decumbent to prostrate, less commonly ascending, sporadically rooting at nodes, 10–40 cm, becoming woody proximally, puberulent to hirtellous-puberulent, hairs usually antrorsely curved or crisped, sometimes ± straight, nonseptate, longer hairs 0.2–0.3 (–0.8) mm. **Leaves** basal and cauline; stipules rudimentary or oblong, margins narrowly flanged, apical auricles absent; petiole (1–)3–7(–10) cm, hairs nonseptate; leaflets 3, glaucous and gray-green to yellowish green, obcordate, 5–10(–15) mm, lobed 1/5–1/3 length, surfaces glabrous to loosely strigose, oxalate deposits absent. **Inflorescences** umbelliform cymes, 1–2(–3)-flowered; peduncles 1.5–5 cm. **Flowers** semihomostylous, within level of leaves; sepal apices without tubercles, surfaces strigose to hirsute-strigose; petals yellow, rarely with red lines proximally, 6–10(–12) mm. **Capsules** angular-cylindric, abruptly tapering to apex, 14–20 mm, strigose-hirsute. **Seeds** brown to blackish brown, transverse ridges rarely with whitish lines or spots.

Flowering Apr–Aug(–Oct). Desert scrub, grasslands, mesquite-acacia, pinyon-juniper, oak-pine-juniper, oak-buckthorn, riparian woodlands (sycamore-hackberry-walnut-ash-willow), stream banks, meadows, washes, hillsides, ravines, canyons, disturbed sites; (600–) 700–1900(–2100) m; Ariz., N.Mex., Tex.; Mexico; Central America (Guatemala).

10. Oxalis pilosa Nuttall in J. Torrey and A. Gray, Fl. N. Amer. 1: 212. 1838 • Hairy western wood-sorrel

Oxalis albicans Kunth subsp. *pilosa* (Nuttall) G. Eiten; *O. corniculata* Linnaeus subsp. *pilosa* (Nuttall) Lourteig; *O. corniculata* var. *pilosa* (Nuttall) B. L. Turner; *O. wrightii* A. Gray var. *pilosa* (Nuttall) Wiegand; *Xanthoxalis pilosa* (Nuttall) Small

Herbs perennial, caulescent, rhizomes and stolons absent, bulbs absent. **Aerial stems** usually 2–8 from base, decumbent to ascending, 10–40 cm, becoming woody proximally, sparsely to densely pilose, hairs spreading irregularly to ± deflexed, nonseptate, longer hairs 0.6–1.2 mm. **Leaves** basal and cauline; stipules oblong, margins narrowly flanged, apical auricles absent; petiole (1–)2–6 cm, hairs nonseptate; leaflets 3, glaucous and gray-green to yellowish green, obcordate, 5–12 mm, lobed ⅕–⅓ length, surfaces glabrous to loosely strigose to hirsute-villous, oxalate deposits absent. **Inflorescences** umbelliform cymes, 1–2(–3)-flowered; peduncles 1.5–5 cm. **Flowers** semihomostylous or distylous, within level of leaves; sepal apices without tubercles, surfaces strigose to hirsute-strigose; petals yellow, rarely with red lines proximally, 8–12 mm. **Capsules** angular-cylindric, abruptly tapering to apex, 12–17(–20) mm, strigose-hirsute. **Seeds** brown to blackish brown, transverse ridges rarely with whitish lines or spots.

Flowering (Feb–)Mar–Jun(–Oct). Juniper-grasslands, pinyon-juniper, oak-juniper, oak, oak-pine, rocky and grassy hillsides, riparian woods (sycamore-walnut, cottonwood-willow), canyons, stream banks, washes, gravel bars; (700–)900–1900(–2000) m; Ariz., Calif., Nev., N.Mex., Oreg., Utah; Mexico (Baja California, Chihuahua, Coahuila, Durango, Nuevo León, Sonora).

Oxalis pilosa has been treated as an infraspecific entity within *O. albicans*, but the two are sympatric in the southwestern United States and although each is variable, there appear to be relatively few unequivocal intermediates. Reports of *O. pilosa* from Texas apparently were based on misidentifications of *O. albicans* and perhaps also of *O. dillenii*.

11. Oxalis californica (Abrams) R. Knuth, Notizbl. Bot. Gart. Berlin-Dahlem 7: 300. 1919 • California wood-sorrel

Xanthoxalis californica Abrams, Bull. Torrey Bot. Club 34: 264. 1907; *Oxalis albicans* Kunth subsp. *californica* (Abrams) G. Eiten

Herbs perennial, caulescent, rhizomes and stolons absent, bulbs absent. **Aerial stems** usually 2–8 from base, erect to ascending, 10–40 cm, becoming woody proximally, glabrous or very sparsely short-puberulent, hairs curved-ascending, nonseptate. **Leaves** basal and cauline; stipules oblong, margins narrowly flanged, apical auricles absent; petiole 3–8 cm, hairs nonseptate; leaflets 3, gray-green, obcordate, 8–15 mm, lobed ⅕–⅓ length, surfaces sparsely strigose, oxalate deposits absent. **Inflorescences** umbelliform cymes, 1(–3)-flowered; peduncles 2–9 cm. **Flowers** distylous; sepal apices without tubercles, surfaces glabrous; petals yellow, with red lines proximally, 7–11(–13) mm. **Capsules** angular-cylindric, abruptly tapering to apex, 10–15 mm, strigillose, hairs very short, nonseptate.

Flowering (Dec–)Feb–Apr(–Jun). Slopes and flats, brushy ridges, roadside banks, canyon bottoms, rock outcrops, grasslands, oak chaparral, coastal sage scrub; (0–)30–800 m; Ariz., Calif.; Mexico (Baja California).

Oxalis californica is recognized by its caulescent habit, stems with reduced vestiture (glabrous or very sparsely short-puberulent), one (to three) flowers on long peduncles and pedicels, the yellow corollas often drying with a blue or purplish tinge, and relatively wide, glabrous, and usually purplish or pinkish tinged sepals.

12. Oxalis stricta Linnaeus, Sp. Pl. 1: 435. 1753 • Upright yellow wood-sorrel [E] [F] [W]

Oxalis bushii Small; *O. coloradensis* Rydberg; *O. cymosa* Small; *O. europaea* Jordan; *O. europaea* var. *bushii* (Small) Wiegand; *O. fontana* Bunge; *O. fontana* var. *bushii* (Small) H. Hara; *O. interior* (Small) Fedde; *O. rufa* Small; *O. stricta* var. *decumbens* Bitter; *O. stricta* var. *piletocarpa* Wiegand; *O. stricta* var. *rufa* (Small) Farwell; *O. stricta* var. *villicaulis* (Wiegand) Farwell; *Xanthoxalis bushii* (Small) Small; *X. cymosa* (Small) Small; *X. rufa* (Small) Small; *X. stricta* (Linnaeus) Small

Herbs annual or short-lived perennial, caulescent, rhizomes present, short, stolons absent, bulbs absent. **Aerial stems** 1(–3) from base, erect or later leaning or falling over and decumbent, not rooting at nodes, 20–60

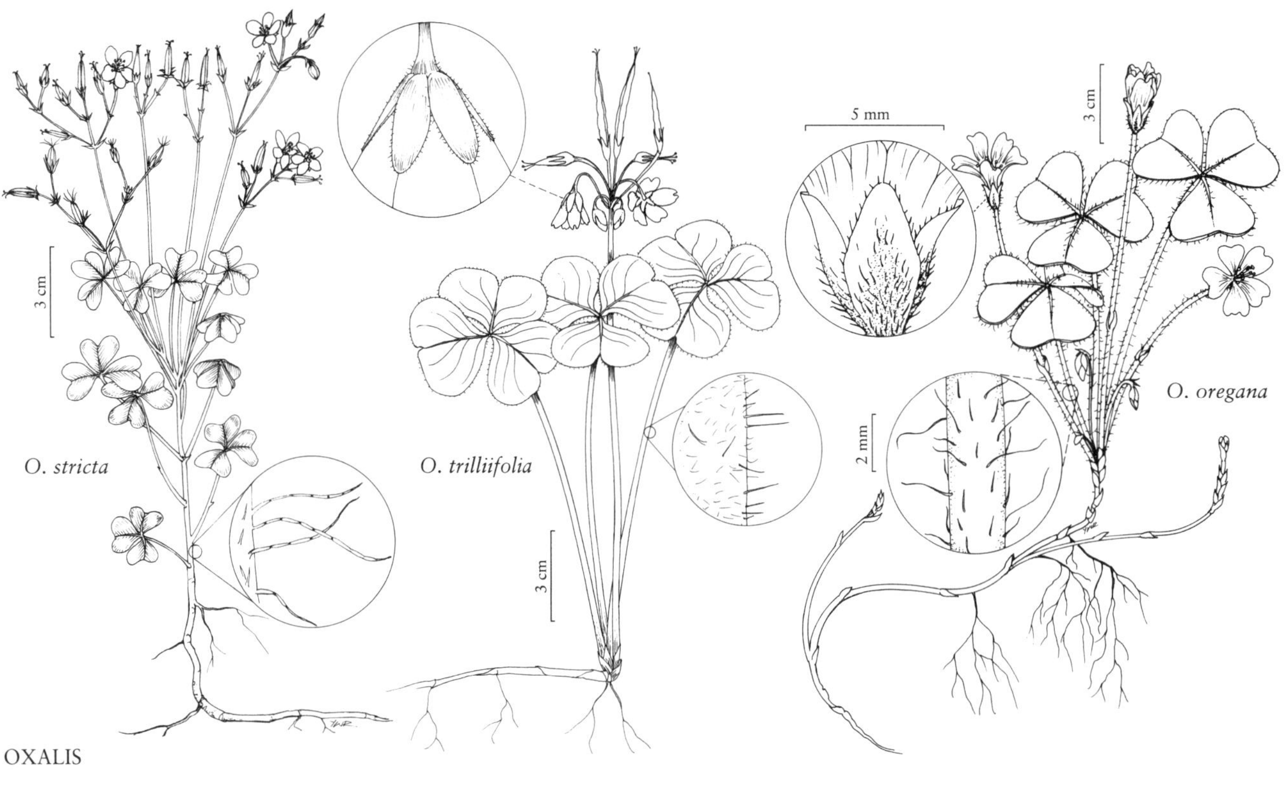

OXALIS

(–90) cm, herbaceous, villous, hairs ± straight, spreading, usually nonseptate and septate, septate hairs commonly concentrated at nodes, very rarely only nonseptate. **Leaves** basal and cauline; stipules rudimentary; petiole 2–8 cm, hairs septate and nonseptate; leaflets 3, light green to yellowish green, obcordate, (8–)10–20(–30) mm, lobed 1/5–1/3 length, surfaces glabrous, oxalate deposits absent. **Inflorescences** usually regular cymes, rarely irregular cymes, (1–)5–7(–15)-flowered; peduncles 3–9(–11) cm. **Flowers** homostylous or slightly to strongly heterostylous, usually within or slightly above level of leaves; sepal apices without tubercles; petals yellow, without red lines, (6–)8–11 mm. **Capsules** ± cylindric, nearly terete, abruptly tapering to apex, 8–15 mm, villous to puberulent or glabrate, hairs septate. **Seeds** brown, transverse ridges rarely white. 2*n* = 18, 24.

Flowering (Apr–)Jul–Oct. Prairie ravines, riverbanks, sandbars, low woods, mesic forests, flood plains, roadsides, fields, lawns, gardens; 20–1200 m; B.C., Man., N.B., Nfld. and Labr. (Nfld.), N.S., Ont., P.E.I., Que., Sask.; Ala., Ariz., Ark., Colo., Conn., Del., D.C., Ga., Idaho, Ill., Ind., Iowa, Kans., Ky., La., Maine, Md., Mass., Mich., Minn., Miss., Mo., Mont., Nebr., N.H., N.J., N.Mex., N.Y., N.C., N.Dak., Ohio, Okla., Pa., R.I., S.C., S.Dak., Tenn., Vt., Va., Wash., W.Va., Wis., Wyo.; introduced in Europe, Asia, Africa, Pacific Islands (New Zealand), Australia.

Oxalis stricta is uncommon and probably adventive in Canada and the western United States. The species is recognized by the combination of its tall (rarely up to nearly a meter), erect stems from a short, simple rhizome; septate hairs; cymose inflorescence; and small flowers. Septate hairs on the stems and petioles are easily recognized (lens) by their colored cross-walls, but they vary greatly in density. In *villicaulis* forms, the hairs are dense and evenly distributed, but in most plants over the range they are localized around the nodes and intermixed with nonseptate hairs. Often they are sparse; rarely they appear to be absent on plants with greatly reduced vestiture overall.

A. Lourteig (1979) identified this species as *Oxalis fontana*, typified by a plant from northern China, and applied the name O. *stricta* to the species identified by G. Eiten (1955, 1963) and here as O. *dillenii*. The basis for the difference lies in selection of lectotypes. Choice of *stricta* as the correct epithet acknowledges that the description of O. *stricta* by Linnaeus best matches these plants and that selection of a Morison illustration as lectotype characterizes the species long-naturalized and weedy in Europe and probably known first-hand by Linnaeus. C. E. Jarvis (2007) has confirmed this choice of lectotype. *Oxalis dillenii* is appropriately lectotypified by a John Clayton collection from Virginia. The situation has been summarized well by D. B. Ward (2004).

13. Oxalis grandis Small, Bull. Torrey Bot. Club 21: 475. 1894 • Great yellow wood-sorrel [E]

Oxalis recurva Trelease, Mem. Boston Soc. Nat. Hist. 4: 89. 1888, not Elliott 1821; *Xanthoxalis grandis* (Small) Small

Herbs perennial, caulescent, rhizomes present, without tubers or tuberlike thickenings, stolons absent, bulbs absent. **Aerial stems** 1(–3) from base, erect, (10–)25–60(–100) cm, herbaceous, glabrate to sparsely or densely pilose or villous, hairs ± straight, spreading, septate and nonseptate. **Leaves** cauline, mostly on distal ½ of stem; stipules apparently absent; petiole 5–7.5 cm, hairs septate and nonseptate; leaflets 3, green, obcordate, 5–25(–30) mm, lobed ⅕ length, lobes apically usually rounded, rarely ± truncate, margins green or brownish purple, ciliate, surfaces glabrous, oxalate deposits absent. **Inflorescences** regular, irregular, or umbelliform cymes, 1–4(–8)-flowered; peduncles 7–12 cm. **Flowers** tristylous, above level of leaves; sepal apices without tubercles; petals yellow, with or without faint red lines proximally, 10–14 mm. **Capsules** ovoid to ovoid-oblong, 6–10 mm, sparsely puberulent. **Seeds** brown, transverse ridges brown. 2*n* = 28.

Flowering May–Aug. Sandy woods, alluvial soils; 100–1100 m; Ala., D.C., Ga., Ind., Ky., Md., N.C., Ohio, Pa., S.C., Tenn., Va., W.Va.

14. Oxalis illinoensis Schwegman, Phytologia 50: 467. 1982 • Illinois wood-sorrel [E]

Herbs perennial, caulescent, rhizomes present, with horizontal, white, fusiform tubers or tuberlike thickenings, stolons absent, bulbs absent. **Aerial stems** 1(–3) from base, erect, 15–40 cm, herbaceous, glabrate to sparsely to densely villous, hairs ± straight, spreading, septate and nonseptate. **Leaves** cauline; stipules rudimentary; petiole 4–7.5 cm, hairs septate and nonseptate; leaflets 3, green, obcordate, (12–)20–30(–35) mm, lobed ⅕ length, lobes apically truncate, margins green, ciliate, surfaces glabrous, oxalate deposits absent. **Inflorescences** regular or irregular cymes, 1–3(–6)-flowered; peduncles 3–10 cm. **Flowers** tristylous, mostly at level of leaves; sepal apices without tubercles; petals yellow, with prominent red lines proximally, 12–18 mm. **Capsules** oblong-ovoid, 7–10 mm, sparsely puberulent to villous. **Seeds** brown, transverse ridges brown.

Flowering Apr–Sep. Slopes, bluffs, ravines, flood plains, mesic forests, sometimes forming dominant ground cover, commonly on limestone, shale, or calcareous loess; 200–500 m; Ill., Ind., Ky., Tenn.

Differences between *Oxalis illinoensis* and *O. grandis* are subtle, but they appear to be correlated with geography. *Oxalis illinoensis* occupies the western part of the range of *O. grandis* in the wide sense. The tuberous portions of the rhizomes of *O. illinoensis* are diagnostic, but they are commonly broken off during collection. M. E. Medley (1993) observed that the two taxa intergrade, and a hybrid population has been identified in Indiana (A. L. Heikens 2003).

Oxalis illinoensis is listed as threatened in Illinois and rare in Indiana.

15. Oxalis suksdorfii Trelease, Mem. Boston Soc. Nat. Hist. 4: 89. 1888 • Western yellow wood-sorrel [E]

Xanthoxalis suksdorfii (Trelease) Small

Herbs perennial, caulescent, rhizomes present, long, stolons absent, bulbs absent. **Aerial stems** 1(–3) from base, mostly erect, 10–25 cm, herbaceous, sparsely to moderately villous, hairs nonseptate and usually septate. **Leaves** cauline; stipules rudimentary; petiole (2–)3–5(–6) cm, hairs septate and nonseptate; leaflets 3, green, obcordate, (8–)10–16(–20) mm, lobed ⅕–⅓ length, margins green, surfaces glabrous to sparsely strigose, oxalate deposits absent. **Inflorescences** umbelliform cymes, 1–3-flowered; peduncles (2–)4–8(–10) cm. **Flowers** tristylous, above level of leaves; sepal apices without tubercles; petals yellow, 12–20 mm. **Capsules** oblong-cylindric, 10–15 mm, densely puberulent. 2*n* = 24.

Flowering May–Aug. Open woods, fir, Douglas fir-oak woodlands, dry shrublands, roadsides, disturbed areas; 0–700 m; B.C.; Calif., Oreg., Wash.

16. Oxalis laxa Hooker & Arnott, Bot. Beechey Voy., 13. 1830 • Dwarf wood-sorrel [I]

Oxalis corniculata Linnaeus var. *sericea* R. Knuth; *O. micrantha* Bertero ex Savi; *O. radicosa* A. Richard; *O. simulans* Baker

Herbs annual, caulescent, sometimes densely cespitose, rhizomes and stolons absent, bulbs absent. **Aerial stems** 1–5 from base, erect, 0.5–7 cm, usually herbaceous, sometimes becoming ± woody proximally, hirtellous to villous-hirtellous. **Leaves** cauline; stipules rudimentary; petiole 1.5–6 cm; leaflets 3, green, obcordate, 5–12 mm, lobed ⅕ length, lobes apically convex, surfaces glabrous, oxalate deposits absent. **Inflorescences** racemes, 6–14-flowered; peduncles 3–15 cm. **Flowers** heterostylous; sepal apices without tubercles; petals yellow, 6–12 mm. **Capsules** ovoid to spheric, 3–5 mm, puberulent.

Flowering Apr–Jun. Disturbed sites, riparian woodlands, riverbanks, gravelly beaches, rock crevices, foothill woodlands; 10–800 m; introduced; Calif.; South America (Chile).

Oxalis laxa is widespread in California in the eastern part of the Central Valley and along the central coast.

17. Oxalis pes-caprae Linnaeus, Sp. Pl. 1: 434. 1753 • African wood-sorrel, Bermuda buttercup, soursob [I] [W]

Oxalis cernua Thunberg

Herbs perennial, acaulous, rhizomes present, vertical, white, rootlike, stolons absent, bulb usually solitary, sometimes with bulblets at base; bulb scales not observed. **Leaves** basal, rarely absent at flowering; petiole 3–12 cm; leaflets 3, green, rarely mottled with purplish red spots, angular-obcordate, (5–)7–20 mm, lobed 1/4–2/5 length, lobes apically convex, margins and abaxial surface villous, adaxial surface glabrous, oxalate deposits absent. **Inflorescences** umbelliform cymes, 2–12(–20)-flowered; scapes often becoming fistulose proximally, 15–30 cm, sparsely villous to pilose. **Flowers** tristylous in diploids and tetraploids, consistently short-styled in pentaploids; sepal apices with 2 orange tubercles; petals deep golden yellow, 15–20 mm. **Capsules** not seen. **$2n$** = 14, 28, 35.

Flowering Nov–Apr. Disturbed areas, orchards, fields, grasslands, oak woodlands, coastal sage, dunes; 10–500 m; introduced; Ariz., Calif.; s Africa; introduced also in West Indies, Bermuda, South America, Europe, Asia (China, Iran, Turkey), n Africa, Australia.

Outside its native range, *Oxalis pes-caprae* is mostly represented by a sterile pentaploid morph, although tetraploids also are known. The occurrence of both pentaploid and tetraploid individuals in the exotic range may be the result of independent introductions (P. Michael 1964; R. Ornduff 1986). Fruit production has not been observed in North America, and the plants are assumed to be seed-sterile (Ornduff 1987). Bulbs of *O. pes-caprae* are rarely collected, as they detach easily from the vertical, rootlike stems. Each bulb may produce over 20, small, whitish bulblets each year. Bulblets may also be formed at the soil surface crown.

Oxalis pes-caprae was reported by J. K. Small (1933) to occur in waste places and cultivated grounds in northern Florida, but as noted by D. B. Ward (2004), no Florida specimens are known.

SELECTED REFERENCES Galil, J. 1968. Vegetative dispersal in *Oxalis cernua*. Amer. J. Bot. 55: 787–792. Ornduff, R. 1986. The origin of weediness in *Oxalis pes-caprae*. Amer. J. Bot. 73: 779–780. Ornduff, R. 1987. Reproductive systems and chromosome races of *Oxalis pes-caprae* L. and their bearing on the genesis of a noxious weed. Ann. Missouri Bot. Gard. 74: 79–84. Putz, N. 1994. Vegetative spreading of *Oxalis pes-caprae* (Oxalidaceae). Pl. Syst. Evol. 191: 57–67.

18. Oxalis hirta Linnaeus, Sp. Pl. 1: 434. 1753 • Tropical wood-sorrel [I]

Varieties 5–7 (1 in the flora): introduced, California; Africa (South Africa); introduced also in Europe, Australia.

Oxalis hirta is recognized by its rhizomatous habit, sessile to subsessile leaves with unlobed leaflets, and large, solitary, axillary flowers. Plants apparently do not fruit in California. T. M. Salter (1944) referred to *O. hirta* as a polymorphous group-species and recognized seven varieties in South Africa, primarily based on variation in habit and corolla color (white to pink, purplish, or yellowish) and shape (funnelform to cylindric).

18a. Oxalis hirta Linnaeus var. **hirta** [I]

Herbs perennial, caulescent, rhizomes absent, stolons present, bulbs absent. **Aerial stems** (1–)2–5 from base, erect to ascending, 10–30 cm, herbaceous, villous. **Leaves** cauline; stipules rudimentary; petiole 0–0.2 cm; leaflets 3, green, linear to narrowly oblanceolate or oblong-cuneate, (3–)5–15 mm, not lobed, abaxial surface villous, adaxial surface glabrous, oxalate deposits absent. **Inflorescences** 1-flowered, axillary, produced evenly along stems; peduncles 1.5–2.5(–7) cm. **Flowers** heterostylous; sepal apices without tubercles; petals rosy purple to pink or white, 15–20(–25) mm. **Capsules** not seen.

Flowering Mar–Nov. Disturbed sites, especially near gardens, cemeteries, sidewalks; 0–100 m; introduced; Calif.; Africa (South Africa); introduced also in Europe.

19. Oxalis incarnata Linnaeus, Sp. Pl. 1: 433. 1753 • Crimson wood-sorrel [I]

Herbs perennial, caulescent, rhizomes present, 3–8 cm, slender, sometimes producing small tubers, stolons absent, bulblets often present on rhizomes and in leaf axils. **Aerial stems** mostly 1–4 from base, mostly erect, 5–25 cm, herbaceous, glabrous. **Leaves** cauline, usually in pseudowhorls of 4–8, sometimes opposite proximally; stipules rudimentary; petiole 2–5(–7) cm; leaflets 3, green, sometimes purplish abaxially, obcordate, 6–10(–15) mm, lobed ¼ length, lobes apically convex, surfaces glabrous, oxalate deposits absent. **Inflorescences** 1-flowered; peduncles 5–7 cm. **Flowers:** stamen/style arrangement not seen; sepal apices with 2 orange tubercles; petals white to pale pinkish purple with darker veins, 10–20 mm. **Capsules** not seen.

Flowering Jan–May. Shady, disturbed, generally urban sites, greenhouses, roadsides, yards; 0–200 m; introduced; Calif.; Africa (South Africa); introduced also in Europe, Pacific Islands (New Zealand), Australia.

Oxalis incarnata is recognized by its rhizomatous habit, small leaves in pseudowhorls, and large, solitary, flowers with white to pink or purple petals. Plants apparently are seed-sterile in California.

SELECTED REFERENCE Thoday, D. 1926. The contractile roots of *Oxalis incarnata*. Ann. Bot. (Oxford) 40: 571–583.

20. Oxalis trilliifolia Hooker, Fl. Bor.-Amer. 1: 118. 1831 (as trilliifolium) • Trillium-leaf wood-sorrel [E] [F]

Hesperoxalis trilliifolia (Hooker) Small

Herbs perennial, acaulous, rhizomes present, fleshy-thickened, densely scaly, stolons absent, bulbs absent. **Leaves** basal, clustered at rhizome tips; petiole 15–30 cm; leaflets 3, green, broadly obcordate, 20–40(–60) mm, lobed ⅙–¼ length, lobes apically convex, surfaces sparsely villous, oxalate deposits absent. **Inflorescences** umbelliform cymes, 2–9(–15)-flowered; scapes 15–25 cm, glabrous or sparsely villous. **Flowers** heterostylous; sepal apices without tubercles; petals white to pinkish, sometimes greenish proximally, without prominent veins, 8–14 mm. **Capsules** narrowly fusiform, 15–25(–30) mm, glabrous.

Flowering May–Sep. Redwood, spruce-fir, Douglas fir, hemlock, hemlock-cedar, hemlock-alder woodlands, stream margins, swamps; 20–1800 m; Calif., Idaho, Oreg., Wash.

21. Oxalis montana Rafinesque, Amer. Monthly Mag. & Crit. Rev. 2: 266. 1818 • Sleeping-beauty, American or white wood-sorrel [E]

Oxalis acetosella Linnaeus subsp. *montana* (Rafinesque) Hultén; *O. acetosella* var. *rhodantha* (Fernald) R. Knuth

Herbs perennial, acaulous, caudex present, branched, scaly, rhizomes present, fleshy-thickened, densely scaly, stolons absent, bulbs absent. **Leaves** basal; petiole (2.5–)3–9 cm, villous, hairs reddish; leaflets 3, green, broadly obcordate, 10–16(–20) mm, lobed ⅕ length, lobes apically convex, surfaces glabrous, oxalate deposits absent. **Inflorescences** 1-flowered; scapes 4–15 cm, glabrous or sparsely villous, hairs reddish. **Flowers** heterostylous; sepal apices without tubercles; petals white with orange-yellow spot sub-basally, rose colored band proximally, and prominent rose colored veins, 10–15 mm. **Capsules** subglobose, 2–4 mm, glabrous. $2n = 22$.

Flowering May–Aug. Spruce-fir, spruce-hemlock, spruce-cedar, spruce-birch, mixed conifer-hardwoods, beech-maple, damp and swampy woods; 100–2200 m; St. Pierre and Miquelon; N.B., Nfld. and Labr., N.S., Ont., P.E.I., Que.; Conn., Ga., Ky., Maine, Md., Mass., Mich., Minn., N.H., N.J., N.Y., N.C., Ohio, Pa., Tenn., Vt., Va., W.Va., Wis.

Oxalis montana of eastern North America and *O. oregana* of the Pacific region have sometimes been treated as disjunct geographical taxa of the European (or Eurasian, depending on taxonomic interpretation) *O. acetosella* Linnaeus. The three are very similar and surely are closely related. Section *Acetosellae* Reiche also includes the Asian *O. griffithii* Edgeworth & Hooker f., *O. leucolepsis* Diels, and *O. obtriangulata* Maximowicz, as well as *O. magellanica* G. Forster (South America, New Zealand, Tasmania); the first two of these also have been treated as taxa within *O. acetosella* (see synonyms in Liu Q. R. and M. F. Watson 2008). Among all these, typical *O. acetosella*, the two American taxa, and *O. magellanica* have erect flowers, while those of the strictly Asian taxa are distinctly nodding, perhaps suggesting that the latter are monophyletic.

Oxalis oregana stands apart from *O. montana* and typical *O. acetosella* in its larger leaves and flowers and its strigose-villous (versus glabrous) sepals. Each of the three is distinct in petal coloration, and this difference between *O. acetosella* and *O. montana* is perhaps the only one between them besides the geographical disjunction. In *O. montana*, the orange-yellow region near the petal base is constricted to a spot (versus a lateral band in typical *O. acetosella*) and a light and diffuse but distinct rose colored band lies immediately distal to the spot, connecting among the petals to form a circle. *Oxalis acetosella* in the strict sense occurs from Iceland to southern Europe (and possibly northern Africa), reportedly stretching as a broad band completely across Eurasia to Japan to Korea (J. F. Veldkamp 1971; Liu Q. R. and M. F. Watson 2008). Attributions of *O. acetosella* to Pakistan and other Himalayan localities apparently are based on plants of *O. griffithii*.

SELECTED REFERENCE Jasieniuk, M. and M. J. Lechowicz. 1987. Spatial and temporal variation in chasmogamy and cleistogamy in *Oxalis montana* (Oxalidaceae). Amer. J. Bot. 74: 1672–1680.

22. Oxalis oregana Nuttall in J. Torrey and A. Gray, Fl. N. Amer. 1: 211. 1838 • Redwood-sorrel [E] [F]

Oxalis acetosella Linnaeus subsp. *oregana* (Nuttall) D. Löve; O. *oregana* var. *smalliana* (R. Knuth) M. Peck; O. *oregana* var. *tracyi* Jepson; O. *smalliana* R. Knuth

Herbs perennial, acaulous, rhizomes present, fleshy-thickened, densely scaly, stolons absent, bulbs absent. **Leaves** basal, clustered at rhizome tips; petiole 5–15(–21) cm, sparsely to densely villous, hairs rusty; leaflets 3, green, broadly obcordate, 10–30(–40) mm, lobed 1/5–1/4 length, lobes apically convex, surfaces sparsely villous, oxalate deposits absent. **Inflorescences** 1-flowered; scapes (6–)11–25 cm, glabrous or sparsely villous. **Flowers** heterostylous; sepal apices without tubercles; petals white to deep pink, usually with yellow spot sub-basally and prominent purple veins, (8–)15–25 mm. **Capsules** ovoid, 6–8(–12) mm, glabrate.

Flowering Feb–Sep. Douglas fir, mixed fir, cedar-spruce, mixed conifer, and hemlock-maple forests, maple woodlands, alder glens, *Gaultheria* thickets, stream banks; 10–800(–1000) m; B.C.; Calif., Oreg., Wash.

Oxalis oregana has sometimes been treated as a disjunct geographical taxon of the European O. *acetosella* Linnaeus (see comments under 21. O. *montana*).

23. Oxalis purpurea Linnaeus, Sp. Pl. 1: 433. 1753 • Purple wood-sorrel [I]

Oxalis variabilis Jacquin

Herbs perennial, acaulous, rhizomes present, slender, sparsely scaly, stolons absent, bulb solitary, 1–2.5 cm, or with clustered bulblets; bulb scales black, thickened, not prominently nerved. **Leaves** basal, rarely absent at flowering; petiole (1.5–)3–5 cm; leaflets 3, green to deep purple abaxially, green adaxially, broadly obovate to obtriangular or broadly rounded-rhombic, 10–20 mm, not lobed, apex truncate to rounded or obtuse, rarely slightly emarginate, margins and abaxial surface hairy, adaxial surface glabrous, oxalate deposits absent. **Inflorescences** 1-flowered; scapes 1.5–6(–8) cm, sparsely to moderately villous, hairs eglandular. **Flowers** tristylous; sepal apices without tubercles; petals yellow basally, usually purple to red, pink, salmon, or white, rarely yellow, distally, 25–35 mm. **Capsules** not seen.

Flowering Feb–Apr. Waste places, especially near gardens; 20–100 m; introduced; Calif.; s Africa; introduced also in Europe, Australia.

Oxalis purpurea is widely cultivated as an ornamental because of its large, solitary flowers in many color forms, borne on short scapes barely higher than the level of the leaves. Plants of O. *purpurea* apparently do not produce fertile fruit in California, where it is naturalized in scattered central and southern coastal counties.

24. Oxalis bowiei Aiton ex G. Don, Gen. Hist. 1: 761. 1831 (as bowii) • Red-flower wood-sorrel [I]

Oxalis purpurata Jacquin var. *bowiei* (Aiton ex G. Don) Sonder

Herbs perennial, acaulous, rhizomes present, vertical, slender or thickened, sparsely scaly, stolons absent, bulbs solitary, ovate, 2–4 cm; bulb scales 5-nerved. **Leaves** basal, rarely absent at flowering; petiole (4–)6–16 cm, densely glandular-puberulent; leaflets 3, green to purplish abaxially, green adaxially, obcordate, (12–)30–60 mm, lobed 1/6–1/3 length, lobes apically convex, often fleshy, surfaces densely glandular-puberulent, oxalate deposits absent. **Inflorescences** umbelliform cymes, 4–12-flowered; scapes 15–20 cm, densely glandular-puberulent. **Flowers** heterostylous; sepal apices without tubercles; petals greenish yellow basally, pink to deep rose pink or red distally, 15–20 mm. **Capsules** not seen.

Flowering Oct–Dec, Apr–Jun. Disturbed areas; 300 m; introduced; Calif.; Africa (South Africa); introduced also in Europe, Asia (China), Australia.

Oxalis bowiei is a naturalized garden escape in Oroville (Butte County; V. H. Oswald and L. Ahart 1994). *Oxalis bowiei* Aiton ex G. Don, from the Cape of Good Hope, was described as hoary-pubescent with peduncles about equal in length to the leaves and with red flowers. It perhaps is not the same species as O. *bowiei* Herbert (1833), provenance unspecified, but the color illustration clearly shows the commonly cultivated plant of contemporary commerce. Apparently neither name has been typified.

25. Oxalis caerulea (Small) R. Knuth, Notizbl. Bot. Gart. Berlin-Dahlem 7: 316. 1919 • Blue wood-sorrel

Ionoxalis caerulea Small in N. L. Britton et al., N. Amer. Fl. 25: 33. 1907

Herbs perennial, acaulous, rhizomes and stolons absent, bulb solitary; bulb scales (3–)5–7-nerved. **Leaves** basal; petiole 3–10(–13) cm; leaflets (3–)4(–5), green to purplish abaxially, green adaxially, obtriangular to obcordate, 5–22 mm, lobed 1/5–1/2 length, lobes apically rounded, surfaces glabrous, oxalate deposits absent or as few punctate tubercles near lobe apices. **Inflorescences** umbelliform cymes, (1–)2–7-flowered; scapes 6–12(–15) cm, glabrous. **Flowers** distylous; sepal apices with 2 orange, linear tubercles; petals red to pinkish lavender, 8–10 mm. **Capsules** ellipsoid, 3.5–6 mm, glabrous.

Flowering Jun–Sep. Streambeds, stream banks, meadows, pinyon-juniper, pine-oak-juniper, or pine-aspen woodlands; (1800–)2000–2600 m; Ariz., Colo., N.Mex.; Mexico (Chihuahua, Durango, Sonora).

26. Oxalis decaphylla Kunth in A. von Humboldt et al., Nov. Gen. Sp. 5(fol. & qto.): plate 468. 1822 • Ten-leaf wood-sorrel

Oxalis grayi (Rose) R. Knuth

Herbs perennial, acaulous, rhizomes and stolons absent, bulbs solitary or clustered; bulb scales 9–15+-nerved. **Leaves** basal, rarely absent at flowering; petiole 7–32(–46) cm; leaflets (3–)5–11, green to purplish abaxially, green adaxially, sometimes with purplish transverse medial band, narrowly oblong-oblanceolate to narrowly oblong or linear, (10–)12–38(–72) mm, lobed (1/6–)1/2–2/3(–9/10) length, lobes apically subacute, surfaces glabrous, oxalate deposits absent. **Inflorescences** umbelliform cymes, (2–)6–11(–15)-flowered; scapes 7–35 cm, glabrous. **Flowers** distylous; sepal apices with 2 orange, linear, thick tubercles; petals green proximally, rose purple or lavender to pink, rarely white, distally, with green veins, (7–)9–17(–22) mm. **Capsules** ellipsoid, 3–11 mm, glabrous. **2*n*** = 28, 56.

Flowering Jun–Aug. Sycamore-walnut, oak, pine-oak, ponderosa pine, pine-spruce-aspen, or spruce-fir woodlands, canyons, meadows, seeps, streamsides; (1700–)2200–3000(–3200) m; Ariz., N.Mex.; c, n Mexico.

27. Oxalis metcalfei (Small) R. Knuth, Notizbl. Bot. Gart. Berlin-Dahlem 7: 314. 1919 • Metcalfe's wood-sorrel

Ionoxalis metcalfei Small in N. L. Britton et al., N. Amer. Fl. 25: 39. 1907; *I. monticola* Small; *Oxalis bulbosa* A. Nelson; *O. neomexicana* R. Knuth

Herbs perennial, acaulous, rhizomes and stolons absent, bulb 5–10 mm, usually surrounded by dense cluster of bulblets, 3–4 mm (sometimes obscuring bulb); bulb scales 3-nerved. **Leaves** basal; petiole 7–15 cm; leaflets 3, green, obtriangular-obcordate, 11–25 mm, lobed 1/6–1/3 length, lobes apically rounded to shallowly convex, surfaces glabrous, oxalate deposits usually in narrow band 0.5–1.5 mm along margins at base of notch, sometimes evident on one surface but not other, rarely absent. **Inflorescences** umbelliform cymes, 3–7-flowered; scapes 7–22 cm, glabrous. **Flowers** tristylous and distylous; sepal apices with 2 orange, narrow-elongate, nonconfluent tubercles; petals white to pale green proximally with green veins, purplish to lavender or pink distally, (9–)12–16 mm. **Capsules** cylindric, 6 mm, glabrous. **2*n*** = 28, 42.

Flowering (Jun–)Jul–Sep(–Oct). Stream banks, wet meadows, canyon bottoms, talus, rocky banks, crevices, juniper-chaparral, *Cercocarpus*, pine, yellow pine-Douglas fir-oak, Douglas fir-aspen, pine-white fir-Douglas fir, spruce-fir, or spruce woodlands; 1800–3100(–3400) m; Ariz., Colo., N.Mex., Tex.; Mexico (Chihuahua, Durango, Sonora, Zacatecas).

Oxalis metcalfei has mostly been identified as *O. alpina* (Rose) Rose ex R. Knuth, but the latter is a species of south-central Mexico, far from the populations in northwestern Mexico and the southwestern United States. *Oxalis alpina* has leaflets with dotlike oxalate deposits scattered throughout the lamina, concentrated near margins, or as continuous, filiform marginal bands around the lobe apices; the corollas usually are white. *Oxalis metcalfei* is consistently different in the nature of its foliar oxalate deposits and the corollas usually are purplish to lavender or pink.

Plants with chromosome numbers of 2*n* = 28 are found in both Arizona and New Mexico; those with 2*n* = 42 are found only in New Mexico (S. C. Weller and M. F. Denton 1976).

28. Oxalis drummondii A. Gray, Smithsonian Contr. Knowl. 5(6): 25. 1853 • Drummond's wood-sorrel [F]

Oxalis vespertilionis Torrey & A. Gray, Fl. N. Amer. 1: 679. 1840, not Zuccarini 1834; *O. amplifolia* (Trelease) R. Knuth

Herbs perennial, acaulous, rhizomes and stolons absent, bulb solitary; bulb scales 3-nerved, margins villous-ciliate on distal 1/3–1/2. **Leaves** basal; petiole 5–16 cm; leaflets 3, green, sometimes with red splotches in irregular medial band adaxially, obtriangular to obcordate, (6–)14–34 mm, lobed 1/4–4/5 length, lobes apically convex to nearly truncate, surfaces glabrous, oxalate deposits absent. **Inflorescences** umbelliform cymes, 3–10-flowered; scapes (7–)11–23 cm, glabrous. **Flowers** distylous or rarely homostylous; sepal apices with 2(–6) orange, linear, thickened, apically confluent tubercles; petals white to pale green proximally with

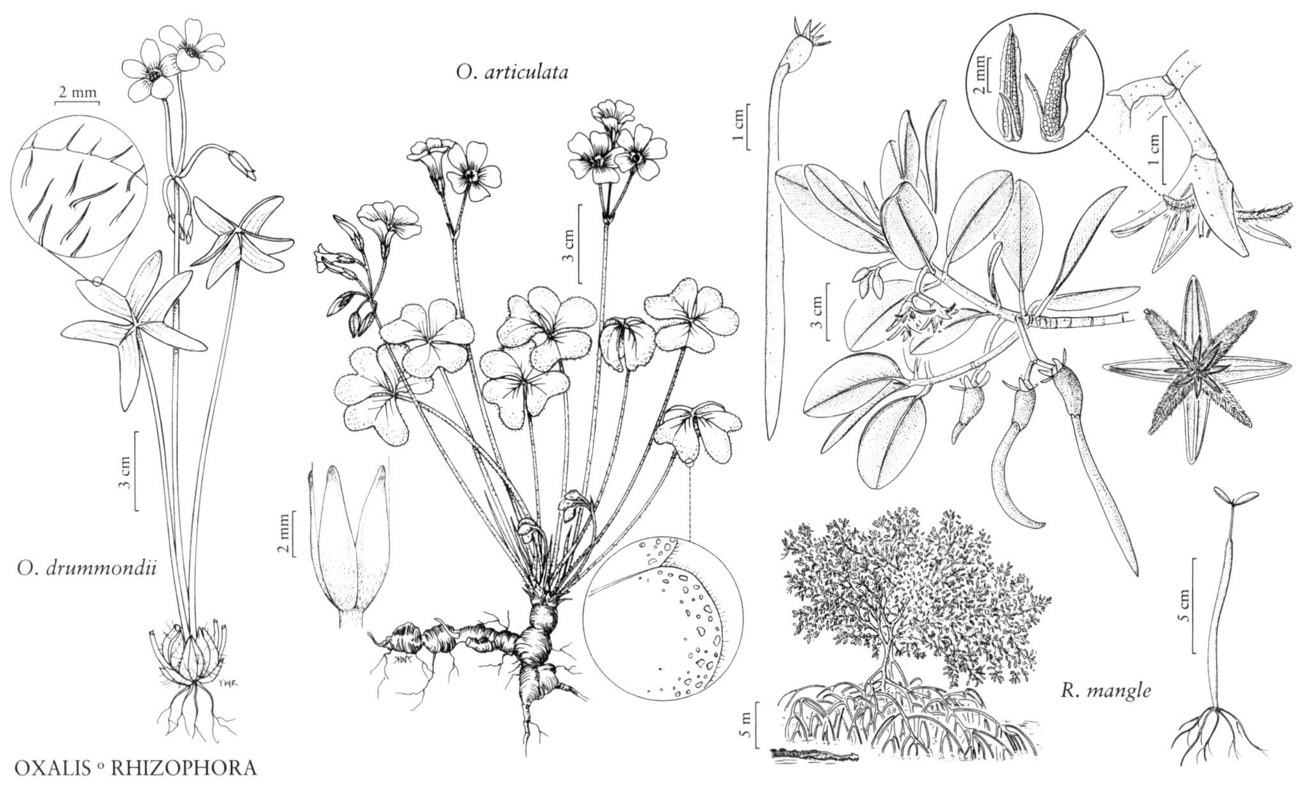

OXALIS ◦ RHIZOPHORA

green veins, pink to violet or purple-violet distally, (8–)15–23 mm. **Capsules** cylindric, 4–12 mm, hairy. **2*n*** = 14.

Flowering Mar–Nov. Sandy-gravelly soils, limestone soils, disturbed areas, prairies, limestone hills, open woodlands, chaparral; 20–300 m; Tex.; Mexico (Coahuila, Nuevo León, San Luis Potosí, Tamaulipas).

Oxalis drummondii is found in the flora area in central and southern Texas. Reports of this species from Arizona, New Mexico, and trans-Pecos Texas are based on misidentifications of *O. latifolia* and *O. metcalfei.*

29. Oxalis violacea Linnaeus, Sp. Pl. 1: 434. 1753 • Violet wood-sorrel [W]

Ionoxalis violacea (Linnaeus) Small; *Oxalis violacea* var. *trichophora* Fassett

Herbs perennial, acaulous, rhizomes usually absent, rarely present, slender, scaly, stolons absent, bulb solitary, bulblets at rhizome tips; bulb scales 3-nerved, margins villous-ciliate on distal ⅓–½. **Leaves** basal, rarely absent at flowering; petiole (4–)7–13(–24) cm; leaflets 3, green to purple abaxially, green adaxially, often with purplish, lateral band across lobes of each leaflet, rounded-obcordate to obreniform, (5–)8–15(–20) mm, lobed ¼–⅓ length, lobes apically convex, surfaces glabrous, oxalate deposits in lines along margins at base of notch. **Inflorescences** umbelliform cymes, (1–)2–8(–19)-flowered; scapes (6–)9–23(–31) cm, glabrous. **Flowers** distylous; sepal apices with 2 orange, linear, apically confluent tubercles; petals white to pale green proximally with green veins, rose purple or lavender to pink or white distally, 10–18 mm. **Capsules** ovoid, 4–5 mm, glabrous. **2*n*** = 28.

Flowering Apr–May(–Jul) (with leaves) and Aug–Oct (usually without leaves, following rains). Sandy soils, gravelly soils, prairies, limestone glades, hills of granite, limestone, and rocky clay, rock outcrops, bluffs, bottomland, oak-pine/heath, oak-hickory, live oak, or juniper woodlands, cutover pine forests, roadsides, disturbed sites, abandoned fields; 50–400(–1000) m; Ala., Ark., Conn., Del., D.C., Fla., Ga., Ill., Ind., Iowa, Kans., Ky., La., Md., Mass., Mich., Minn., Miss., Mo., Nebr., N.J., N.Y., N.C., N.Dak., Ohio, Okla., Oreg., Pa., R.I., S.C., S.Dak., Tenn., Tex., Vt., Va., W.Va., Wis., Wyo.; Mexico (Coahuila).

Oxalis violacea in the flora area is native to the eastern United States, reaching westward as far as the line of states from North Dakota to Texas; it is apparently non-native in Oregon and Wyoming, if those plants are correctly identified. Plants identified by M. F. Denton (1973) as *O. violacea* and those considered to have affinity to that species from Arizona, Colorado, and New Mexico are identified here as *O. latifolia* and *O. metcalfei.*

30. Oxalis latifolia Kunth in A. von Humboldt et al., Nov. Gen. Sp. 5(fol. & qto.): plate 467. 1822 • Broad-leaf wood-sorrel

Ionoxalis latifolia (Kunth) Rose

Herbs perennial, acaulous, rhizomes and stolons absent, bulb solitary; bulb scales (5–)7–9 (–11)-nerved, inner thickened, white, margins hyaline. **Leaves** basal; petiole (6–)10–25 cm; leaflets 3, green, obtriangular to obcordate, (15–)25–40 mm, lobed 1/5–1/2 length, lobes apically rounded, surfaces glabrous, oxalate deposits absent. **Inflorescences** umbelliform cymes, (3–)6–10(–12)-flowered; scapes (7–)10–20(–30) cm, glabrous. **Flowers** usually tristylous, rarely homostylous; sepal apices with 2 orange, short to elongate, nonconfluent tubercles; petals white to pale green proximally with green veins, purple to lavender or pink distally, 9–12 mm. **Capsules** cylindric, 4–6 mm, glabrous. $2n$ = 14, 28, 42.

Flowering Jun–Sep(–Oct). Rocky slopes, ledges and crevices, ridge tops, canyons, sandy washes, flood plains, mesquite-baccharis, mesquite-acacia, hackberry-willow, oak-juniper-pinyon, oak-maple, or pine-oak woodlands; (1100–)1300–2200(–2800) m; Ariz., Calif., N.Mex., Tex.; Mexico; introduced in West Indies, Europe, Asia (Pakistan), Africa (South Africa), Atlantic Islands (Canary Islands), Pacific Islands (New Zealand), Australia.

Collections of *Oxalis latifolia* from trans-Pecos Texas and eastern New Mexico were mostly identified by M. F. Denton (1973) as *O. alpina* and *O. drummondii*. *Oxalis latifolia* in the narrow sense (type from Campeche, Mexico), with somewhat angular leaflets, occurs in Mexico as far north as Veracruz, San Luis Potosí, and Tamaulipas. The northern populations tend to have more rounded leaflets and perhaps may be better identified by a different name, but they are neither *O. alpina* (with bulblet clusters) nor *O. drummondii* (with three-nerved bulb scales).

M. F. Denton (1973) observed that *Oxalis latifolia* probably consists of several races and perhaps hybrids with other species. A. Lourteig (2000) recognized four subspecies of *O. latifolia*. Her concept of subsp. *latifolia* restricted it to South America and the West Indies (even though the type was collected in Mexico); subsp. *schraderiana* (Kunth) Lourteig is entirely South American; and subsp. *vespertilionis* (Zuccarini) Lourteig occurs in montane Mexico south to Guatemala. She recognized subsp. *galeottii* (Turczaninow) Lourteig as occurring from Mexico into the United States. In contrast, most others have treated *O. galeottii* Turczaninow as a distinct species of central and southern Mexico, broadly sympatric with *O. latifolia*.

31. Oxalis intermedia A. Richard, Hist. Phys. Cuba, Pl. Vasc., 315. 1841 • West Indian wood-sorrel [I]

Ionoxalis intermedia A. St.-Hilaire

Herbs perennial, acaulous, rhizomes absent, stolons often present, numerous, slender, with bulblets at tips, bulbs usually clustered, sometimes solitary; bulb scales (3–)5–7-nerved. **Leaves** basal, rarely absent at flowering; petiole 10–22 cm; leaflets 3, green, obtriangular to broadly obtriangular, 20–50 mm, lobed 1/5–1/3 length, lobes apically truncate, surfaces glabrous, oxalate deposits absent. **Inflorescences** umbelliform cymes, 3–12(–18)-flowered; scapes 7–30 cm, glabrous or sparsely hairy. **Flowers** semihomostylous; sepal apices with 2 orange tubercles; petals usually lavender to purple, less commonly pink or white, 8–12 mm. **Capsules** ellipsoid, 3–8 mm, glabrous.

Flowering Apr–Sep. Gardens, fields, orchards, roadsides, moist waste areas, fencerows; 0–100 m; introduced; Fla., La., Tex.; West Indies; introduced also in Mexico (Chiapas, San Luis Potosí, Veracruz).

Oxalis intermedia is recognized by a combination of its large, obtriangular leaflets; numerous, small flowers; and usually clustered bulbs. It was collected in California in 1934 and Massachusetts in 1940 but does not appear to have become naturalized in either state. Plants in the flora area are usually without fertile fruit.

32. Oxalis triangularis A. St.-Hilaire, Fl. Bras. Merid. 1(qto.): 128. 1825 • Purple shamrock, scurvy grass [I]

Oxalis palustris A. St.-Hilaire; *O. papilionacea* Hoffmannsegg ex Zuccarini; *O. regnellii* Miquel; *O. triangularis* subsp. *papilionacea* (Hoffmannsegg ex Zuccarini) Lourteig

Herbs perennial, acaulous, rhizomes present, branching, short, 1 cm diam., densely scaly, stolons absent, bulblets sometimes present, clustered; bulb scales (1–)3-nerved, margins glandular. **Leaves** basal, rarely absent at flowering; petiole 12–20 cm; leaflets 3, dark purple, commonly with lighter violet splotches radiating from midvein adaxially, obtriangular to obovate-triangular, (20–)30–50(–60) mm, lobed 1/10 length or apex merely notched, lobes apically truncate to slightly convex, surfaces glabrous, oxalate deposits absent or in short, marginal lines on both sides of notch. **Inflorescences** umbelliform cymes, (1–)2–5(–9)-flowered; scapes 15–35 cm, glabrous. **Flowers** heterostylous; sepal apices with 2 orange tubercles; petals white to pinkish or pale purple, 15–22 mm. **Capsules** ovoid-ellipsoid, 12–18 mm, glabrous.

Flowering Apr–May. Disturbed sites, near gardens; 0–100 m; introduced; Fla., La.; South America (Argentina, Bolivia, Brazil, Paraguay, Peru).

In East Feliciana Parish, Louisiana, north of Baton Rouge, *Oxalis triangularis* has spread from a planter pot into adjacent woods (G. F. Guala, pers. comm.). In Leon County, Florida, a population is growing and slowly spreading at the edge of a woodland remnant within the city of Tallahassee (L. C. Anderson, pers. comm.).

A. Lourteig (2000) recognized two subspecies sympatric over much of their native ranges (as cited, subsp. *triangularis* in Argentina, Bolivia, Brazil, and Paraguay, subsp. *papilionacea* in Bolivia, Brazil, Paraguay, and Peru) and differing by the following contrasts: subsp. *triangularis* has sepals acute, oxalate tubercles small or absent, and petals white to purplish, three to four times longer than the sepals; subsp. *papilionacea* has sepals obtuse to subacute with oxalate tubercles thickened and petals pink to purplish, rarely white, about two to two and a half times longer than the sepals. With these subtle differences and broad sympatry, it seems probable that only a single evolutionary entity exists. In any case, the few North American collections studied here would be *Oxalis triangularis* in the strict sense. Forms of *O. triangularis* are sometimes recognized as "*atropurpurea*" but apparently this is a horticultural name.

33. Oxalis debilis Kunth in A. von Humboldt et al., Nov. Gen. Sp. 5(fol.): 183; 5(qto.): 236. 1822 • World-wide wood-sorrel [I]

Ionoxalis martiana (Zuccarini) Small; *Oxalis corymbosa* de Candolle; *O. debilis* subsp. *corymbosa* (de Candolle) O. Bolòs & Vigo; *O. debilis* var. *corymbosa* (de Candolle) Lourteig; *O. martiana* Zuccarini

Herbs perennial, acaulous, rhizomes and stolons absent, bulblets clustered; bulb scales 3-nerved. **Leaves** basal; petiole 10–25 cm; leaflets 3, green to yellowish green, rounded-obcordate, 17–40(–50) mm, lobed 1/6–1/5 length, lobes apically convex, abaxial surface hirsute, adaxial surface glabrous, oxalate deposits in dots at least around distal margins, often evenly over surface. **Inflorescences** irregular cymes, (3–)8–14(–28)-flowered; scapes 15–28 cm, moderately villous to glabrate. **Flowers** mostly homostylous, infrequently tristylous; sepal apices with 2 orange tubercles; petals violet to lavender or rose purple, 10–16(–20) mm. **Capsules** not observed. 2*n* = 14, 28, rarely 35.

Flowering Dec–May, rarely again in summer. Fencerows, yards, flower beds, roadsides, disturbed areas, hammock margins, sandy live oak woods, mesic woods, stream and river terraces; 0–100 m; introduced; Ala., Calif., Fla., Ga., La., Miss., Oreg., S.C., Tex., Wash.; South America; introduced also in Mexico, West Indies, Central America, Europe, se Asia (Malesia), Pacific Islands, Australia.

Oxalis debilis appears to be spreading rapidly in the United States. The species produces numerous bulblets in a basal cluster and apparently also can spread laterally by production of bulblets at the tips of filiform roots or rhizomes; it can form large, dense colonies. A. Lourteig (1980) noted that plants of this species occasionally fruit but consistently reproduce through abundant bulblets. They apparently are seed-sterile in North America.

Oxalis corymbosa and *O. debilis* were differentiated by A. Lourteig (2000) primarily by the distribution of oxalate deposits in the leaf lamina. In *O. debilis*, the dotlike deposits are crowded along the margins and absent to distinctly less abundant elsewhere. In *O. corymbosa*, the deposits are evenly distributed over the whole lamina. In their native range in South America, the two expressions are broadly sympatric and intermediates are common, as they are in the flora area. Intermediates have the oxalate dots along the margins as well as over the whole surface or sometimes mostly on the outer third of the blades, near the margins. There is no justification for formal recognition of two entities.

34. Oxalis hispidula Zuccarini, Denkschr. Königl. Akad. Wiss. München 9: 143. 1825 [I]

Herbs perennial, acaulous, rhizomes and stolons absent, bulbs solitary or clustered, mostly 8–15 mm diam.; outer bulb scales 3[–5]-nerved, inner scales thick, reddish brown, rugose. **Leaves** basal; petiole 1.5–15 cm, sparsely villous or glabrous; leaflets 3, green, rounded-obcordate, 4–18 mm, lobed 1/6–1/5 length, lobes apically convex to nearly truncate, margins prominently ciliate, hairs stiff, sharp-pointed, abaxial surface strigose to hirsute-strigose, densely hirsute at very base, adaxial surface glabrous, oxalate deposits absent. **Inflorescences** umbelliform cymes, 1(–2)[–4]-flowered; scapes 3–27 cm, glabrous or sparsely hirsute-villous proximally. **Flowers** apparently tristylous (mid-styled flowers observed); sepals yellowish green, apices with 2 orange, elongate tubercles; petals yellow basally, otherwise deep rose to purple or violet, with dark purple veins proximally, 11–20 mm. **Capsules** fusiform, mature size not observed, indumentum not seen.

Flowering Oct–Nov. Wet ditches, disturbed roadsides; 10–90 m; introduced; Ala.; South America (Argentina, Brazil, Paraguay, Uruguay).

Oxalis hispidula is naturalized in Baldwin County (H. E. Horne et al. 2013). The species is recognized by its leaves without oxalate deposits, outer bulb scales with mostly three nerves, flowers one (or two) per scape, and corollas violet-purple with dark veins. It was noted by S. Rosenfeldt and B. G. Galati (2009) to be tristylous.

35. Oxalis brasiliensis G. Loddiges et Drapiez, Encyclogr. Règne Vég. 1: unnumb. 1833–1834 • Brazilian wood-sorrel [I]

Herbs perennial, acaulous, rhizomes and stolons absent, bulb solitary or clustered, 5–20 × 5–17 mm; outer bulb scales 5–8[–13]-nerved, margins ciliate, inner scales thick, orangish. **Leaves** basal; petiole often purplish proximally, 3–13[–20] cm, glabrous [sparsely and finely strigose]; leaflets 3, light green, obdeltate with rounded angles, [2–]10–21[–32] mm, lobed 1/20–1/10 length, lobes apically shallowly convex to nearly truncate, margins glabrous or sparsely irregularly ciliate, hairs loose, fine, abaxial surface sparsely but evenly strigose, adaxial surface glabrous, oxalate deposits absent. **Inflorescences** umbelliform cymes, 1(–2)[–5]-flowered; scapes 14–17[–30] cm, glabrous. **Flowers** apparently tristylous (mid-styled flowers observed); sepals purplish, apices without tubercles, surfaces glabrous; petals violet-purple, with dark purple veins proximally, 18–20 mm. **Capsules** narrowly cylindric, 15–22 mm, indumentum not seen.

Flowering Mar–Jul. Disturbed roadsides; 30–50 m; introduced; Ala.; South America (Argentina, Brazil, Uruguay); introduced also in e Asia (Japan), Australia.

Oxalis brasiliensis is naturalized in Dallas County (H. E. Horne et al. 2013). The species is recognized by its leaves without oxalate deposits and sepals without tubercles; outer bulb scales with five to eight (to 13) nerves; one or two (or three) flowers per scape; and violet-purple, dark purple-veined corollas. The large, showy flowers make this species popular in the horticultural trade (see http://www.pacificbulbsociety.org/pbswiki/index.php/SouthAmericanOxalis#brasiliensis for additional horticultural information on the species). Growth habit in the Dallas County population ranged from small clumps to dense mats along the roadside, extending to the margin of the woodland.

A low percentage of the Dallas County *Oxalis brasiliensis* population was reproducing by tiny propagules produced at the bract region of the scape. These propagules apparently are highly foreshortened stems, as they produce whorls of small leaves; they do not produce scales and thus are not the so-called aerial bulbils, as in the miniature bulbs described in some South African species (see http://www.pacificbulbsociety.org/pbswiki/index.php/SouthAfricanOxalis).

36. Oxalis articulata Savigny in J. Lamarck et al., Encycl. 4: 686. 1798 • Windowbox wood-sorrel [F] [I]

Oxalis articulata subsp. *rubra* (A. St.-Hilaire) Lourteig; *O. rubra* A. St.-Hilaire

Herbs perennial, acaulous, rhizomes present, thick, woody, irregularly nodulate-segmented, often covered with persistent petiole bases, stolons absent, bulbs absent. **Leaves** basal; petiole 11–30 cm; leaflets 3, green to purplish abaxially, green adaxially, rounded-obcordate, 18–20 mm, margins densely loosely ciliate, lobed 1/5–1/3 length, lobes apically convex, surfaces evenly strigose-villous to strigose-hirsute, oxalate deposits in dots concentrated mostly toward margins or over whole surface. **Inflorescences** usually umbelliform cymes, less commonly in irregular cymes, 3–12-flowered; scapes 12–28 cm, sparsely strigose. **Flowers** heterostylous; sepal apices with 2 orange tubercles; petals usually purplish rose to red, rarely white, 10–14 mm. **Capsules** ovoid, 4–8 mm, sparsely strigose. 2*n* = 42.

Flowering Mar–Jul. Disturbed places, especially near gardens, lawns, fields, roadsides; 0–250 m; introduced; Ala., Ark., Calif., Fla., Ga., La., Miss., N.C., Okla., Oreg., S.C., Tex., Va.; South America (Argentina, Brazil, Uruguay); introduced also in Europe, Pacific Islands (New Zealand), Australia.

Oxalis articulata in the United States commonly has been identified as *O. rubra*. *Oxalis rubra* was treated at subspecific rank by A. Lourteig (1982), but subsp. *articulata* and subsp. *rubra* have essentially the same native range and occur in similar habitats. Lourteig identified both subspecies in the United States, noting in her key that vestiture is reduced and the sepals are broader in subsp. *rubra*. Evidence is weak for recognizing more than a single entity. In the Flora of Panama (Lourteig 1980), she recognized only *O. articulata*, noting that it is naturalized in other parts of America and in the Old World.

RHIZOPHORACEAE Persoon

David E. Boufford
Lynn J. Gillespie

Shrubs or trees, evergreen, synoecious [dioecious]. **Leaves** opposite, simple; stipules present, interpetiolar; petiole present; blade margins entire [serrulate near apex]; venation pinnate. **Inflorescences** axillary, cymes [fascicles or flowers solitary]. **Flowers** bisexual [rarely unisexual]; perianth and androecium perigynous [hypogynous or epigynous]; hypanthium completely adnate to ovary [adnate to ovary proximally and free distally, or absent]; sepals 4[–16], distinct or connate basally, valvate; petals 4[–16], distinct; nectary present [absent]; stamens [4 or]8[–32], distinct [connate basally], free; anthers dehiscing by adaxial valve [longitudinal slits]; pistil 1, 2[–5(–20)]-carpellate; ovary ½ inferior [superior to inferior], [1–]2[–5(–10)]-locular, placentation apical-axile; ovules 2[–8] per locule, anatropous [hemitropous]; style [0–]1; stigmas [1–]2[–4]. **Fruits** berries [capsules or drupes]. **Seeds** 1 per fruit [1 per locule].

Genera 17, species ca. 120 (1 in the flora): nearly worldwide; tropics and subtropics.

Bruguiera gymnorrhiza (Linnaeus) Savigny (large-leafed orange mangrove), native to the Indian and western Pacific Islands, has been found as an escape in Miami-Dade County, Florida. It is considered invasive in Florida and efforts are underway to eradicate it. *Bruguiera* may be distinguished from *Rhizophora* by its 8–16 sepals and petals and lack of aerial stilt roots.

SELECTED REFERENCES Graham, S. A. 1964. The genera of Rhizophoraceae and Combretaceae in the southeastern United States. J. Arnold Arbor. 45: 285–301. Tomlinson, P. B. 1986. The Botany of Mangroves. Cambridge.

1. RHIZOPHORA Linnaeus, Sp. Pl. 1: 443. 1753; Gen. Pl. ed. 5, 202. 1754

• Mangrove [Greek *rhiza*, root, and *phoros*, bearing, alluding to conspicuous prop roots]

Shrubs or trees, with aerial prop roots and swollen stem nodes. **Leaves:** stipules sheathing terminal bud, caducous; blade surfaces glabrous. **Inflorescences** dense, dichotomously branched cymes; bracteoles forming cup just below flower. **Flowers:** sepals persistent in fruit; filaments absent [much shorter than anthers]; ovary apically partly surrounded by nectary, free part elongating after anthesis. **Berries** brown, ovoid-conic [ovoid or pyriform], leathery. **Seeds:** germination viviparous; hypocotyl protruding and elongating before seedling falls from fruit.

Species 8 or 9 (1 in the flora): seacoasts nearly throughout tropics and subtropics.

1. Rhizophora mangle Linnaeus, Sp. Pl. 1: 443. 1753 • Red mangrove [F] [W]

Rhizophora americana Nuttall

Shrubs or trees, 6–7[–25] m. **Prop roots** numerous, arching, 2–4.5 m. **Leaves:** stipules lanceolate, leaving ring-shaped scar; petioles 1.5–2 cm; blade elliptic, 6–12 × 2.2–6 cm, thick, leathery, base rounded-cuneate, apex acute to obtuse, midvein extended into caducous point, abaxial surface pale green, black punctate, adaxial surface dark green, shiny; midvein conspicuous, higher order veins obscure. **Inflorescences** mostly 2–4-flowered, 4–7 cm. **Flowers** 2 cm diam.; hypanthium campanulate-funnelform, 5 mm; sepals widely spreading, pale yellow, lanceolate or narrowly triangular, 12 mm, leathery; petals recurved, creamy white, turning brown in age, narrowly lanceolate to linear, 6–10 × 1.5–2 mm, adaxially wooly, soon deciduous; anthers sessile; ovary conic; style slender; stigma receptive after stamens and petals fall. **Berries** persistent on tree until after seed germination, rusty or dark brown, 3 cm; sepals spreading to reflexed. **Seed** with hypocotyl becoming elongate, cylindric, to 12–25 × 1.2 cm, apex sharply pointed. $2n = 36$.

Flowering spring; fruiting late summer–fall. Shallow, brackish to saline water in sand, silt, mud, or clay of coastal and estuarine sand flats and swamps; 0 m; Fla., Tex.; Mexico; West Indies; Central America; South America; Africa; Pacific Islands.

Rhizophora mangle is native to the Americas and west Africa; it was introduced to Hawaii early in the 20th century for erosion control and has become invasive there. The species is particularly susceptible to freezing temperatures, which limit the northern (and southern) extent of its range. In Florida, it is found on the immediate Atlantic and Gulf seacoasts north to Saint Johns County at almost 30° north latitude (W. B. Zomlefer et al. 2006). K. C. Cavanaugh et al. (2014) determined that its abundance has increased significantly at the northern edge of the range, likely due to a decline in the frequency of severe cold events. *Rhizophora mangle* has recently become established along the Texas coast, presumably also as a result of a warming climate (P. A. Montagna et al. 2011).

Seedling propagules may float in the ocean and remain viable for up to a year before reaching a suitable substrate where they resume growth. Propagules occasionally wash ashore and become temporarily established north of its range (to North Carolina), but plants do not currently survive long-term (A. S. Weakley 2012). Records from the Florida panhandle are also assumed to be temporarily established plants (B. Hansen, pers. comm.).

EUPHORBIACEAE Jussieu
• Spurge Family

Geoffrey A. Levin
Lynn J. Gillespie

Herbs, subshrubs, shrubs, trees, or vines [lianas], annual, biennial, or perennial, deciduous or evergreen, monoecious or dioecious; latex present or absent. **Leaves** alternate, opposite, whorled, or fascicled on short shoots, simple (3-foliolate in *Tragia laciniata*) [palmately compound]; stipules present or absent; petiole present or absent; blade sometimes palmately lobed, margins entire, subentire, repand, crenate, serrate, or dentate; venation pinnate, palmate, or palmate at base and pinnate distally. **Inflorescences** unisexual or bisexual, axillary, terminal, or leaf-opposed [cauliflorous], racemes, panicles, spikes, thyrses, cymes, fascicles, or pseudanthia, or flowers solitary. **Flowers** unisexual; perianth hypogynous; hypanthium absent; sepals 0 or 2–12, distinct or connate basally to most of length; petals 0 or (3–)5(–6), distinct or connate; nectary present or absent; stamens 1–35(–1000), distinct or connate, free; anthers dehiscing by longitudinal slits; pistil 1, (1–)3–5(–20)-carpellate, ovary superior, (1–)3–5(–20)-locular, placentation axile; ovules 1 per locule, anatropous; styles 1–5(–9), distinct or connate, unbranched, 2-fid, or multifid; stigmas 1–32+. **Fruits** usually capsules, dehiscence septicidal, (usually schizocarpic with cocci separating from persistent columnella, coccus usually dehiscent loculicidally), sometimes schizocarps, drupes, or achenes [berries]. **Seeds** 1 per locule.

Genera ca. 220, species ca. 6500 (24 genera, 259 species in the flora): North America, Mexico, West Indies, Bermuda, Central America, South America, Eurasia, Africa, Atlantic Islands, Indian Ocean Islands, Pacific Islands, Australia; mostly tropical to warm temperate regions.

Molecular phylogenetic studies have shown Euphorbiaceae, as traditionally treated (G. L. Webster 1994b, 2014; A. Radcliffe-Smith 2001), to be polyphyletic, forming several groups in Malpighiales (C. C. Davis et al. 2005; T. Tokuoka and H. Tobe 2006; K. Wurdack and Davis 2009; Z. Xi et al. 2012). The treatment here reflects those findings. Genera formerly placed in subfamilies Oldfieldioideae Eg. Köhler & G. L. Webster and Phyllanthoideae Beilschmied, characterized by two ovules per locule, are treated as Picrodendraceae and Phyllanthaceae respectively, with two genera segregated from the latter as Putranjivaceae; the first two families appear to be sister taxa, possibly near much of the remaining traditional Euphorbiaceae, whereas the third is placed elsewhere in Malpighiales (Xi et al.). The remaining genera, characterized by one ovule per locule, are treated as Euphorbiaceae in the strict sense and Peraceae Klotzsch (not represented in the flora area).

Four subfamilies currently are recognized in the narrowly defined Euphorbiaceae (K. Wurdack et al. 2005; G. L. Webster 2014). Traditionally, three subfamilies, Acalyphoideae Beilschmied, Crotonoideae Beilschmied, and Euphorbioideae Beilschmied, were recognized based on laticifer presence or absence and pollen morphology (Webster 1994b; A. Radcliffe-Smith 2001). Phylogenetic analyses of molecular data (Wurdack et al.; T. Tokuoka 2007) support the monophyly of Euphorbioideae, represented in the flora area by *Ditrysinia*, *Euphorbia*, *Gymnanthes*, *Hippomane*, *Hura*, *Microstachys*, *Pleradenophora*, *Stillingia*, and *Triadica*. Most of the traditional Crotonoideae is moderately supported as monophyletic, including *Astraea*, *Cnidoscolus*, *Croton*, *Jatropha*, *Manihot*, and *Vernicia* in the flora area. These studies also strongly support the monophyly of most of Acalyphoideae (including *Acalypha*, *Adelia*, *Argythamnia*, *Bernardia*, *Caperonia*, *Dalechampia*, *Mercurialis*, *Ricinus*, and *Tragia* in the flora area), and segregation of a fourth, small subfamily, Cheilosoideae K. Wurdack & Petra Hoffmann (not represented in the flora area). The remaining genera of Acalyphoideae and Crotonoideae, none of which are in the flora area, were placed in several small clades whose relationships to each other and to the four subfamilies were not well supported.

Euphorbiaceae in the strict sense are diverse morphologically, but most are characterized by schizocarpic capsules in which the cocci separate from the persistent columella, often explosively; the seeds are dispersed when the cocci split septicidally and usually also loculicidally. Similar capsules also are found in many Phyllanthaceae and Picrodendraceae but, as noted above, they have two ovules per locule, versus one in Euphorbiaceae. White or whitish latex is found in Euphorbioideae and colored latex in many Crotonoideae, whereas Acalyphoideae lack latex. Pseudanthia evolved independently and are structurally different in *Dalechampia* and *Euphorbia*.

Euphorbiaceae are most species-rich and ecologically important in tropical and subtropical regions. The same is true of genera found in the flora area, many of which are represented there only by a small proportion of their global diversity.

Notable economically important Euphorbiaceae are *Hevea* Aublet, a major source of rubber; *Manihot*, cassava or manioc, the starchy tubers of which are a major food source in much of the tropics; *Ricinus*, the source of castor oil; and *Vernicia* (and its close relative *Aleurites* J. R. Forster & G. Forster), sources of tung and other finishing oils. Some species of Euphorbiaceae are used horticulturally, especially members of *Euphorbia*; probably the best known of these is the poinsettia, *E. pulcherrima* Willdenow ex Klotzsch. Also well known is *Codiaeum variegatum* (Linnaeus) A. Jussieu, the horticultural croton. Some introduced Euphorbiaceae have become problematic invasives in the flora area, particularly in areas with mild climates. Most notable among these invasive species are leafy spurge (*E. virgata*, commonly incorrectly called *E. esula*), Chinese tallowtree (*Triadica sebifera*), and tung-oil tree (*V. fordii*).

Mallotus japonicus (Linnaeus f.) Müller Arg., food wrapper plant, has escaped locally in Durham and Orange counties, North Carolina. The Orange County population may have been eradicated, and the status of the Durham County population is not known. This dioecious shrub or small tree, native to eastern Asia, has stellate hairs and in the key below would come under the first lead of couplet 5 with *Astraea*, *Bernardia*, and *Croton*. It can be distinguished from those three genera by its pale yellow to white glandular scales on the leaves and stems, staminate flowers with 70–100 stamens, unbranched styles, and ovaries and capsules covered with soft spines and reddish orange glandular scales.

Sapium haematospermum Müller Arg. from South America was collected on ballast in Pensacola, Florida, in 1901; this collection generally has been incorrectly reported as *S. glandulosum* (Linnaeus) Morong. Although the species does not appear to have become naturalized in the flora area, it could become adventive in subtropical areas. In the key below, *S. haematospermum* would come under the second lead of couplet 12 and can be distinguished from the five genera there by its combination of petioles with apical glands, inflorescences with two glands subtending each bract, staminate flowers with two to three sepals that are connate basally, pistillate flowers with three-carpellate pistils bearing three styles, and seeds covered in a red aril.

In the key and descriptions below, laminar glands are those borne on the surface of the leaf blade, not those extending from the margins or teeth or borne at the leaf base at the junction with the petiole.

SELECTED REFERENCES Punt, W. 1962. Pollen morphology of the Euphorbiaceae with special reference to taxonomy. Wentia 7: 1–116. Radcliffe-Smith, A. 2001. Genera Euphorbiacearum. Kew. Tokuoka, T. 2007. Molecular phylogenetic analysis of Euphorbiaceae sensu stricto based on plastid and nuclear DNA sequences and ovule and seed character evolution. J. Pl. Res. 120: 511–522. Webster, G. L. 1967. The genera of Euphorbiaceae in the southeastern United States. J. Arnold Arbor. 48: 303–361, 363–430. Webster, G. L. 1994. Synopsis of the genera and suprageneric taxa of Euphorbiaceae. Ann. Missouri Bot. Gard. 81: 33–144. Webster, G. L. 1994b. Classification of the Euphorbiaceae. Ann. Missouri Bot. Gard. 81: 3–32. Webster, G. L. 2014. Euphorbiaceae. In: K. Kubitzki et al., eds. 1990+. The Families and Genera of Vascular Plants. 10+ vols. Berlin etc. Vol. 11, pp. 51–216. Wurdack, K., P. Hoffmann, and M. W. Chase. 2005. Molecular phylogenetic analysis of uniovulate Euphorbiaceae (Euphorbiaceae sensu stricto) using plastid *rbc*L and *trn*L-F DNA sequences. Amer. J. Bot. 92: 1397–1420.

1. Inflorescences pseudanthia, each consisting of an involucre enclosing 1 or 3 pistillate flowers and (0–)1–80 staminate flowers.
    2. Vines; pseudanthia with involucre of 2 showy bracts and 3 pistillate flowers; latex absent . . . . 9. *Dalechampia*, p. 191
    2. Herbs, subshrubs, or shrubs; pseudanthia with cuplike involucre and 1 pistillate flower; latex white. . . . . 24. *Euphorbia*, p. 237
1. Inflorescences thyrses, cymes, racemes, panicles, spikes, or fascicles, or flowers solitary.
    3. Stinging hairs present (sometimes inconspicuous in *Tragia* except on ovaries and capsules).
        4. Sepals usually green, sometimes reddish green, not petaloid; inflorescences racemes; petioles without glands at apices; latex absent . . . . 8. *Tragia*, p. 184
        4. Sepals white, petaloid; inflorescences dichasial cymes; petioles with glands at apices; latex white. . . . . 11. *Cnidoscolus*, p. 196
    3. Stinging hairs absent.
        5. Hairs stellate or scalelike, sometimes also unbranched.
            6. Stamens ± straight in bud; laminar glands abaxial, crateriform; staminate petals 0, nectaries intrastaminal; latex absent; caruncle absent. . . . . 7. *Bernardia*, p. 182
            6. Stamens inflexed in bud; laminar glands absent; staminate petals 0 or (3–)5(–6), nectaries extrastaminal; latex usually present, colorless to reddish; caruncle present.
                7. Leaf blades usually palmately lobed, sometimes unlobed; hairs unbranched and stellate; pistillate sepals usually not touching in bud; seeds rectangular-oblong. . . . . 13. *Astraea*, p. 205
                7. Leaf blades unlobed; hairs stellate or scalelike; pistillate sepals imbricate or valvate; seeds ellipsoid, oblong, ovoid, globose, or lenticular . . . . 14. *Croton*, p. 206
        5. Hairs unbranched, malpighiaceous, 2-fid, or absent.
            8. Leaves opposite or subopposite.
                9. Latex absent; plants dioecious; inflorescences axillary; capsules hispid . . . . 2. *Mercurialis*, p. 161
                9. Latex white; plants monoecious; inflorescences terminal; capsules glabrous. . . . . 23. *Stillingia* (in part), p. 233
            8. Leaves alternate or fascicled.

[10. Shifted to left margin.—Ed.]

10. Glands subtending each inflorescence bract 2 or 10–14.
11. Leaf blades palmately lobed; stamens to 1000; pistillate sepals 5; styles 2-fid . . . . 1. *Ricinus*, p. 160
11. Leaf blades unlobed; stamens 2–3; pistillate sepals 0 or 2–3; styles unbranched.
12. Pistils 6–9-carpellate; fruits drupes . . . . 21. *Hippomane*, p. 232
12. Pistils 2–3-carpellate; fruits capsules.
13. Glands subtending inflorescence bracts 10–14; pistillate sepals 2; pistils 2-carpellate; styles 2 . . . . 22. *Pleradenophora*, p. 232
13. Glands subtending inflorescence bracts 2; pistillate sepals 0 or 3; pistils 3-carpellate; styles 3.
14. Petioles with glands at apices; staminate sepals connate most of length; outer seed coat fleshy . . . . 16. *Triadica*, p. 226
14. Petioles without glands at apices; staminate sepals distinct or connate basally; outer seed coat dry.
15. Plants without hairs; staminate sepals 2; stamens 2; capsule base persisting as 3-lobed gynobase. . . . . 23. *Stillingia* (in part), p. 233
15. Plants with hairs; staminate sepals 3; stamens 3; capsule base not persisting.
16. Herbs, annual; latex white; leaf blade margins serrulate . . . 18. *Microstachys*, p. 229
16. Shrubs; latex absent; leaf blade margins entire . . . . 19. *Ditrysinia*, p. 230
10. Glands subtending each inflorescence bract 0.
17. Petals 5(–6) (pistillate sometimes rudimentary or 0 in *Argythamnia*).
18. Sepals 2(–3); petioles with glands at apices; petals longer than 24 mm; inflorescences paniclelike thyrses. . . . . 15. *Vernicia*, p. 225
18. Sepals 5–8(–10); petioles without glands at apices (sometimes stipitate-glandular or with tack-shaped glands along length); petals less than 18 mm or absent; inflorescences cymes, fascicles, racemes, or spikes, or flowers solitary.
19. Latex colorless, cloudy-whitish, yellow, or red; inflorescences cymes or fascicles, or flowers solitary; staminate sepals imbricate; leaf blades palmately lobed or unlobed . . . . 12. *Jatropha*, p. 198
19. Latex absent; inflorescences spikes or racemes; staminate sepals valvate; leaf blades unlobed.
20. Leaf blade secondary veins arcuate, not closely spaced; styles 2-fid, branches 6 per flower; capsules not muricate; flower nectaries 5 glands; hairs usually malpighiaceous, sometimes unbranched, rarely absent. . . . . . 4. *Argythamnia*, p. 172
20. Leaf blade secondary veins straight, closely spaced; styles deeply multifid, branches 12–21 per flower; capsules muricate; flower nectaries absent; hairs unbranched . . . . 5. *Caperonia*, p. 179
17. Petals 0 (sepals petaloid in *Manihot*).
21. Petioles with conspicuous lateral glands at apex; pistils 5–20-carpellate; style 1; stamens 10–80, connate entire length forming thick column; trees, trunks with broad-based conic thorns . . . . 20. *Hura*, p. 230
21. Petioles with inconspicuous adaxial glands at apex or glands absent; pistils (1–)3(–4)-carpellate; styles (1–)3(–4), equal to carpel number; stamens (2–)4–17, distinct or connate basally; herbs, subshrubs, shrubs, or trees, unarmed (branchlets sometimes stiff and thorn-tipped in *Adelia*).
22. Inflorescences racemes or panicles; staminate sepals 5, petaloid, 7–20 mm, connate ½ length; latex white; leaf blades usually deeply palmately lobed, rarely unlobed . . . . 10. *Manihot*, p. 192
22. Inflorescences fascicles or racemelike or spikelike thyrses, or flowers solitary; staminate sepals 0 or 4–5, not petaloid, 1–2 mm, distinct; latex colorless or absent; leaf blades unlobed.
23. Sepals 0; latex colorless; styles unbranched; stamens (2–)4(–5) . . . . .17. *Gymnanthes*, p. 227
23. Staminate sepals 4–5, pistillate sepals 3 or 5(–6); latex absent; styles usually multifid or laciniate, rarely 2-fid or unbranched; stamens 4–8 or 14–17.

[24. Shifted to left margin.—Ed.]

24. Leaves alternate, blade margins serrate or crenate to subentire; inflorescences spikelike thyrses; bracts subtending pistillate flowers enlarging in fruit; stamens 4–8; pistillate sepals 3 . . . . 3. *Acalypha*, p. 162
24. Leaves fascicled on short shoots, blade margins entire; inflorescences fascicles or flowers solitary; bracts subtending pistillate flowers minute, not enlarging in fruit; stamens 14–17; pistillate sepals 5(–6) . . . . 6. *Adelia*, p. 181

1. RICINUS Linnaeus, Sp. Pl. 2: 1007. 1753; Gen. Pl. ed. 5, 437. 1754 • Castor bean, castor oil plant [Latin, tick or louse, alluding to appearance of seeds] [I]

Lynn J. Gillespie

Herbs [shrubs], annual or perennial, monoecious; hairs absent; latex absent. **Leaves** deciduous, alternate, simple; stipules present, caducous; petiole present, glands present at apex and usually proximally; blade palmately lobed, margins serrate, laminar glands present; venation palmate. **Inflorescences** bisexual (staminate flowers proximal, pistillate distal) or pistillate, terminal or leaf-opposed, racemelike thyrses; glands subtending each bract 2. **Pedicels** present. **Staminate flowers:** sepals 3–5, valvate, connate basally; petals 0; nectary absent; stamens to 1000, connate proximally in numerous slender, irregularly branched columns; pistillode absent. **Pistillate flowers:** sepals 5, distinct or connate basally; petals 0; nectary absent; pistil 3-carpellate; styles 3, distinct or slightly connate basally, 2-fid. **Fruits** capsules. **Seeds** ovoid or ovoid-ellipsoid; caruncle present. $x = 10$.

Species 1: introduced; ne Africa; widely cultivated and often naturalized in tropical, subtropical, and warm temperate regions worldwide.

1. Ricinus communis Linnaeus, Sp. Pl. 2: 1007. 1753 [F] [I] [W]

Plants 1–4[–5+] m. **Stems** erect. **Leaves:** stipules 2–3 cm, connate, leaving conspicuous scar around stem; petiole 10–55 cm, with (0–)1–3 cuplike glands on proximal adaxial surface, 2 glands at apex adaxially; blade circular in outline, 15–50 cm diam., base peltate, marginal teeth gland-tipped, laminar glands scattered on adaxial surface, lobes 7–12, lanceolate or oblanceolate, increasing in size apically, apex narrowly acute to acuminate. **Inflorescences** 6–30 cm, to 45 cm in fruit; bracts caducous except for 2 persistent glands. **Pedicels:** staminate 5–15 mm; pistillate 0.5–5 mm, elongating to 40 mm in fruit. **Staminate flowers:** calyx lobes ovate, 7–8 mm; stamen cluster ± spheric, 10–12 mm diam. **Pistillate flowers:** sepals ovate, 4–5 mm; ovary densely covered in slender-conic, bristle-tipped outgrowths; styles red or orange-red, 4–5 mm; stigmas distinctly spreading, papillose. **Capsules** dark red, echinate, subglobose, 1.5–2 cm diam. **Seeds** mottled brown, 8–11 mm, shiny. $2n = 20$.

Flowering and fruiting summer–late fall. Waste ground, riverbanks, sand and gravel bars, ravines, margins of cultivated fields, roadsides, along railways; 0–700 m; introduced; Ala., Ariz., Ark., Calif., Fla., Ga., La., Miss., Mo., Tex.; Africa; introduced also in Mexico, West Indies, Bermuda, Central America, South America, Eurasia, Atlantic Islands, Indian Ocean Islands, Pacific Islands, Australia.

Ricinus communis, native to northeastern Africa, is cultivated as an ornamental throughout subtropical and temperate North America and is naturalized in the southern United States. It has been reported throughout the eastern United States as far north as Michigan and New Hampshire, and north to Utah and Kansas farther west, but appears to be a nonpersisting garden escape that is only adventive and not naturalized in these areas. In Missouri, it is considered to be naturalized (and uncommon) only in the extreme southeastern corner. Plants are shrubs in tropical and subtropical regions and very large annual herbs in cooler regions. Numerous horticultural cultivars exist, including some with dark red stems and leaves that are commonly planted in North America.

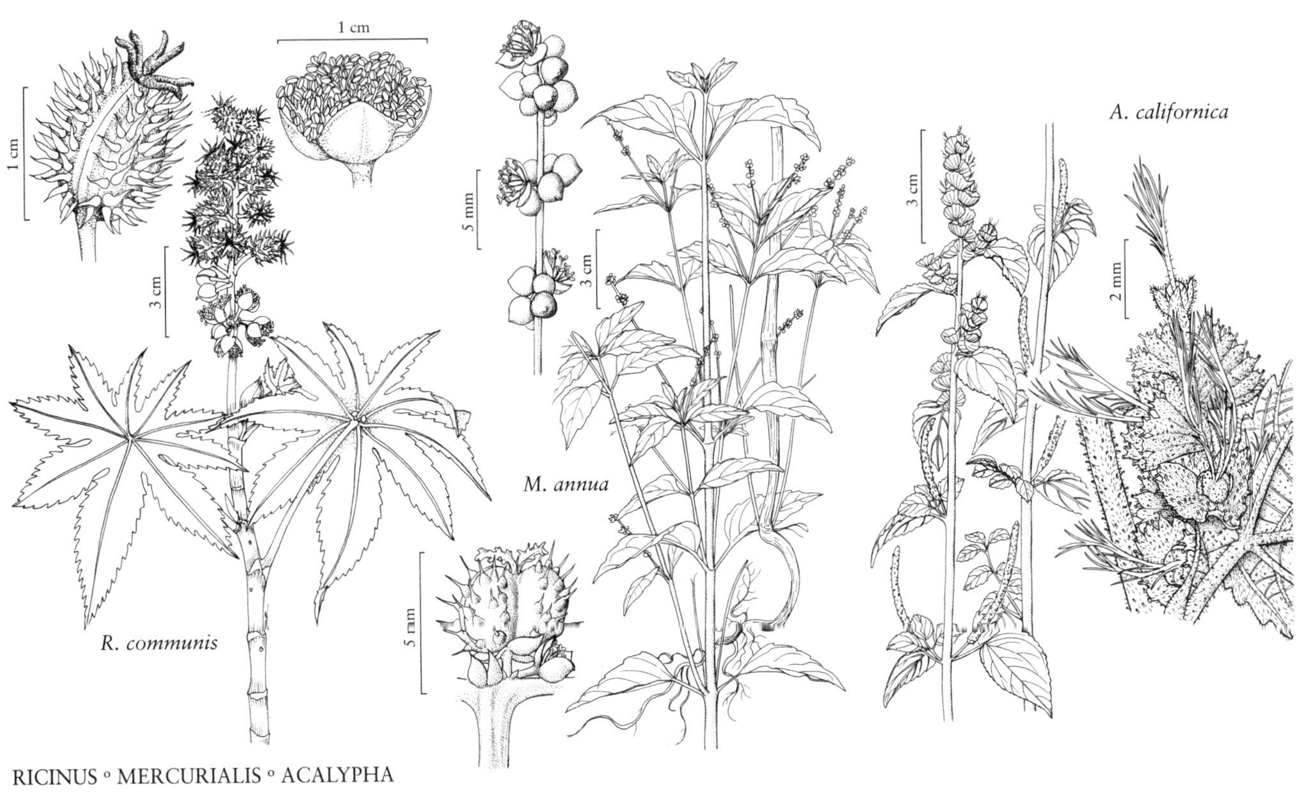

RICINUS ◦ MERCURIALIS ◦ ACALYPHA

All plant parts are poisonous due to the water soluble protein ricin. In tropical regions (mainly Brazil and India), the species is widely cultivated for its seed oil. Castor oil, derived from the seeds (highly refined oil does not contain ricin), is used in cosmetics, medicines, paints, plastics, and lubricants. The use of castor oil in traditional medicine dates at least from ancient Egyptian times. Castor oil is used worldwide for a variety of medicinal purposes, most commonly as a laxative and for skin ailments. In North America, it was formerly commonly given to children as a general cure-all for internal ailments.

2. MERCURIALIS Linnaeus, Sp. Pl. 2: 1035. 1753; Gen. Pl. ed. 5, 457. 1754 • Mercury [Latin *Mercurius*, Roman mythological deity, and *-alis*, belonging to, alluding to belief that it was discovered by him] [I]

Lynn J. Gillespie

Herbs, annual [perennial], dioecious or monoecious; hairs unbranched; latex absent. **Leaves** opposite, simple; stipules present, persistent; petiole present [absent], glands present at apex; blade unlobed, margins serrate or crenate, laminar glands absent; venation pinnate. **Inflorescences** usually unisexual, rarely bisexual, axillary, staminate elongate spikelike thyrses, pistillate and bisexual fascicles or cymules [short spikelike thyrses]; glands subtending each bract 0. **Pedicels:** staminate rudimentary or absent, pistillate present. **Staminate flowers**: sepals 3, valvate, distinct; petals 0; nectary absent; stamens 8–12(–20), distinct; pistillode absent. **Pistillate flowers**: sepals 3, distinct; petals 0; nectary 2 glands; pistil 2-carpellate; styles 2, distinct or connate basally, unbranched. **Fruits** capsules, hispid [glabrous]. **Seeds** ovoid; caruncle present. $x = 8$.

Species 10 (1 in the flora): introduced, California; Eurasia, n Africa, Atlantic Islands (Macaronesia); introduced also in s South America, s Africa, Pacific Islands (New Zealand).

1. Mercurialis annua Linnaeus, Sp. Pl. 2: 1035. 1753 • Annual mercury [F] [I] [W]

Herbs 1–6 dm. **Stems** erect, glabrous or glabrescent. **Leaves:** petiole 0.2–2 cm, with pair of small, usually stalked, rarely sessile, glands at apex; blade ovate to rhombic-ovate, 2.5–5.5 × 1–2.5 cm, base rounded to acute, apex acute to ± acuminate, both surfaces glabrate or glabrescent, margins ciliate. **Staminate inflorescences:** peduncle 2–5 cm, fertile portion 1–5 cm, 2–12 glomerate clusters of flowers separated by internodes to 1 cm. **Pedicels:** pistillate 0.5–1 mm, elongating to 0.5–7 mm in fruit. **Staminate flowers:** sepals broadly ovate, 1.5–1.8 mm; stamens 1 mm; filaments slender; anther locules subglobose. **Pistillate flowers:** sepals ovate, 1–1.5 mm; ovary densely hispid, hairs usually with bulbous bases. **Capsules** 2.5–3 × 3–4 mm, 3-lobed, hispid; sepals and columella persistent. **Seeds** gray or grayish, 2 × 1.5 mm, surface tuberculate. $2n$ = 16, 48, 64, 80, 96, 112 (Europe).

Flowering and fruiting Feb–Nov. Disturbed areas, waste places, cultivated fields; 0–300 m; introduced; Calif.; Europe; w Asia; n Africa; Atlantic Islands (Macaronesia); introduced also in South America (Argentina), s Africa, Pacific Islands (New Zealand).

Mercurialis annua is one of two annual species in the genus; it is complex and morphologically variable, comprising diploids and polyploids, and with a variety of breeding systems including dioecious, androdioecious, and monoecious (M. Krähenbühl et al. 2002). Although several species and infraspecific taxa (for example, *M. ambigua* Linnaeus) have been recognized based on ploidy and sexual expression, these characters are not always correlated, and there are no good morphological characters to distinguish taxa. The species is treated here in the broad sense, following T. G. Tutin (1968) and R. Govaerts et al. (2000). Both dioecious and monoecious populations appear to be naturalized in California. The majority are dioecious and belong to diploid *M. annua* in the strict sense, while several recent collections are from monoecious populations and likely correspond to the polyploid *M. ambigua*.

Mercurialis annua is naturalized in the San Francisco Bay area (northern Bay area to Monterey County) and has been recorded there at least since the 1930s; it is becoming increasingly common as a weed in cultivated fields and waste areas. It has been collected recently in Riverside and Yolo counties, but it is unknown whether it is established in either county. There are old collections, mostly from the 1870s to 1920s, from eastern Canada (New Brunswick, Nova Scotia, Ontario, Quebec) and the eastern United States (for example, Alabama, Delaware, Florida, Maryland, Massachusetts, New Jersey, New York, Ohio, Pennsylvania, South Carolina), but it has not been collected recently in eastern North America and appears not to be established. These old collections appear to be ballast waifs—the majority are from coastal areas or shores along inland shipping routes with a habit recorded as on ballast or in waste areas in the vicinity of ports. There are several old collections (early 1900s) and one recent collection (Lane County, weed in garden) from Oregon, but there is no evidence that it has become naturalized there.

3. ACALYPHA Linnaeus, Sp. Pl. 2: 1003. 1753; Gen. Pl. ed. 5, 436. 1754 • Threeseeded mercury, copperleaf [Greek *akalephe*, stinging nettle, from *a*-, without, *kalos*, good, and *haphe*, touch, alluding to some species resembling *Urtica* (though not stinging)]

Geoffrey A. Levin

Herb or shrubs [trees], annual or perennial, unarmed, monoecious or dioecious; hairs unbranched [stellate], sometimes glandular, or absent; latex absent. **Leaves** persistent or drought-deciduous, alternate, simple; stipules present, persistent or deciduous; petiole present, glands absent or present at apex, adaxial, inconspicuous [conspicuous]; blade unlobed, margins deeply serrate or crenate to subentire, laminar glands absent; venation palmate at base, pinnate distally [pinnate]. **Inflorescences** unisexual or bisexual (pistillate flowers proximal, staminate distal [staminate proximal, pistillate distal]), axillary or terminal, spikelike [paniclelike] thyrses; allomorphic pistillate flowers sometimes present; bracts subtending pistillate flowers enlarging in fruit [remaining minute]; glands subtending bracts 0. **Pedicels:** staminate present, pistillate absent [present], allomorphic present or absent. **Staminate flowers:** sepals 4, not petaloid, 1–2[–3] mm, valvate, distinct [connate]; petals 0; nectary absent; stamens 4–8, distinct; anthers elongated and twisted at maturity; pistillode absent. **Pistillate flowers:** sepals 3 [or 5], distinct [connate]; petals 0; nectary absent; pistil (1–)3-carpellate; styles (1–)3, distinct or connate basally, usually multifid or laciniate, rarely 2-fid or unbranched, branches threadlike. **Fruits** capsules, allomorphic fruits achenes or schizocarps. **Seeds** ellipsoid to subglobose; caruncle present, sometimes rudimentary. $x = 10$.

Species ca. 450 (18 in the flora): North America, Mexico, West Indies, Central America, South America, Asia, Africa, Indian Ocean Islands, Pacific Islands, Australia; introduced in Europe; primarily tropical and subtropical regions, reaching temperate regions in eastern North America and eastern Asia.

Some species of *Acalypha* are cultivated as ornamentals, notably *A. herzogiana* Pax & K. Hoffmann, *A. hispida* Burman f., and *A. wilkesiana*; the last has become naturalized in Florida. *Acalypha herzogiana* has escaped locally on Dauphin Island, Alabama (H. Horne, pers. comm.). It may escape locally elsewhere but might not become naturalized in the flora area, because the cultivated form is sterile, although pieces of the plant root readily if they are spread. It is an herbaceous perennial readily recognized by its erect, feathery, red spikes of sterile pistillate flowers (V. W. Steinmann and G. A. Levin 2011).

Acalypha mexicana Müller Arg. [*A. indica* Linnaeus var. *mexicana* (Müller Arg.) Pax & K. Hoffmann], native from central Mexico to Guatemala, was collected twice early in the twentieth century in southeastern Arizona but has not been collected there since and presumably did not become established. It will key here to *A. australis*; *A. mexicana* differs in having pistillate bracts that are 10 mm and eglandular (versus 10–15 mm and glandular) and allomorphic flowers that are common, long-pedicelled, and 1-carpellate (versus rare, sessile, and 2-carpellate).

Some *Acalypha* species, including about half of those in the flora area, produce allomorphic pistillate flowers (A. Radcliffe-Smith 1973). These flowers may be mixed with normal pistillate or staminate flowers or be terminal on the inflorescence. Their pistils generally have fewer carpels than normal pistillate flowers of the same species, bear sub-basal rather than terminal styles, and develop into nutlets or schizocarps, frequently bearing bristles or variously ornamented outgrowths that presumably facilitate dispersal. Unlike normal pistillate flowers, allomorphic flowers frequently lack bracts and are borne on elongate pedicels. Characters of these flowers generally are species-specific and useful for identification (Radcliffe-Smith).

The bracts subtending normal pistillate flowers of most *Acalypha* species, including all in the flora area, enlarge as the fruits develop. Measurements for these bracts, referred to here as pistillate bracts, and the pistillate portion of the inflorescences given in the key and descriptions are post-anthesis, after the bracts have completed most or all of their growth.

Seed descriptions pertain to those from normal pistillate flowers.

The sequence of species below starts with shrubs (species 1 and 2), followed by subshrubs and perennial herbs (species 3 to 6), and concludes with annual herbs (species 7 to 18). Within each growth form, similar species are grouped together.

SELECTED REFERENCES Levin, G. A. 1995. Systematics of the *Acalypha californica* complex (Euphorbiaceae). Madroño 41: 254–265. Levin, G. A. 1999. Evolution in the *Acalypha gracilens/monococca* complex (Euphorbiaceae): Morphological analysis. Syst. Bot. 23: 269–287. Levin, G. A. 1999b. Notes on *Acalypha* (Euphorbiaceae) in North America. Rhodora 101: 217–233. Miller, L. W. 1964. A Taxonomic Study of the Species of *Acalypha* in the United States. Ph.D. dissertation. Purdue University. Radcliffe-Smith, A. 1973. Allomorphic female flowers in the genus *Acalypha* (Euphorbiaceae). Kew Bull. 28: 525–529.

1. Shrubs.
   2. Leaf blades 1–5 cm; stems stipitate-glandular; plants 5–10 dm; Arizona, California . . . . 1. *Acalypha californica*
   2. Leaf blades 9–20 cm; stems not glandular; plants 20–50 dm; Florida . . . . 2. *Acalypha wilkesiana*

1. Herbs or subshrubs.
    3. Pistillate bract lobes linear or proximally deltate with linear tips.
        4. Pistillate bracts densely crowded (inflorescence axis not or sparingly visible between bracts), abaxial surfaces long-hirsute (nonglandular hairs to 2 mm), lobes proximally deltate with linear tips, smooth.
            5. Stems, petioles, and peduncles stipitate-glandular; pistillate inflorescences terminal; styles unbranched or rarely 2-fid . . . . 7. *Acalypha alopecuroidea*
            5. Stems, petioles, and peduncles not stipitate-glandular; pistillate (and bisexual) inflorescences axillary; styles multifid or laciniate. . . . . .8. *Acalypha arvensis*
        4. Pistillate bracts loosely arranged (inflorescence axis visible between bracts), abaxial surfaces glabrous or pubescent (nonglandular hairs to 0.3 mm, glandular hairs may be longer), lobes linear, muricate.
            6. Leaf bases cordate; pistillate bracts pubescent and stipitate-glandular; capsules spiny; seeds tuberculate . . . . 9. *Acalypha ostryifolia*
            6. Leaf bases obtuse to rounded or truncate; pistillate bracts glabrous; capsules smooth; seeds minutely pitted . . . . 10. *Acalypha setosa*
    3. Pistillate bract lobes deltate (without linear tips), triangular, attenuate, narrowly oblong, lanceolate, spatulate, or rounded.
        7. Inflorescences (all or some) terminal.
            8. Stems erect; leaf blades 2–6 cm.
                9. Inflorescences usually bisexual (rarely staminate portion replaced by allomorphic pistillate flowers), all terminal; petioles 0.2–1 cm; allomorphic pistillate flowers rare, pedicels 3–5 mm, without bracts. . . . . 3. *Acalypha phleoides*
                9. Inflorescences unisexual, pistillate terminal (sometimes on short lateral branches, appearing axillary), staminate axillary; petioles 1–4 cm; allomorphic pistillate flowers common, pedicels rudimentary, bracts like those of normal pistillate flowers . . . . .11. *Acalypha neomexicana*
            8. Stems prostrate to ascending; leaf blades 0.3–2.5 cm.
                10. Petioles 0.1–0.5 cm, less than ⅓ leaf blade length; inflorescences all bisexual and terminal; Florida . . . . 4. *Acalypha chamaedrifolia*
                10. Petioles 0.4–2.5 cm, more than ⅔ leaf blade length; inflorescences unisexual or bisexual, terminal and axillary; Texas.
                    11. Leaf blade margins shallowly crenate; pistillate bract lobes ¼ bract length . . . . 5. *Acalypha monostachya*
                    11. Leaf blade margins deeply crenate; pistillate bract lobes ½ bract length . . . . 6. *Acalypha radians*
        7. Inflorescences all axillary.
            12. Pistillate bracts 4–5 mm; styles unbranched . . . . 12. *Acalypha poiretii*
            12. Pistillate bracts 6–15(–20) mm; styles multifid or laciniate.
                13. Pistillate bract lobes rounded, 1/20 bract length . . . . 13. *Acalypha australis*
                13. Pistillate bract lobes deltate, triangular, lanceolate, or narrowly oblong, 1/10–¾ bract length.
                    14. Leaf blades linear, linear-lanceolate, or oblong-lanceolate; pistillate bracts sessile-glandular, lobes deltate, 1/10–¼ bract length.
                        15. Pistils 3-carpellate. . . . . 17. *Acalypha gracilens*
                        15. Pistils 1-carpellate. . . . . 18. *Acalypha monococca*
                    14. Leaf blades broadly lanceolate to ovate or rhombic; pistillate bracts not sessile-glandular (sometimes stipitate-glandular), lobes triangular to lanceolate or narrowly oblong, ¼–¾ bract length.
                        16. Pistillate bract abaxial surfaces hirsute (sometimes also stipitate-glandular), lobes (9–)10–14(–16), ¼–½ bract length; stems usually hirsute. . . . . 16. *Acalypha virginica*
                        16. Pistillate bract abaxial surfaces sparsely pubescent (usually also stipitate-glandular), lobes (5–)7–9(–11), ⅓–¾ bract length; stems usually pubescent or glabrate, rarely hirsute.
                            17. Pistils (normal flowers) 2-carpellate; seeds 2.4–3.2 mm . . . .14. *Acalypha deamii*
                            17. Pistils (normal flowers) 3-carpellate; seeds (1.2–)1.5–1.7(–2) mm . . . . 15. *Acalypha rhomboidea*

1. Acalypha californica Bentham, Bot. Voy. Sulphur, 51. 1844 • California copperleaf [F]

Acalypha pringlei S. Watson

Shrubs, 5–10 dm, monoecious. **Stems** erect, hirsute and stipitate-glandular, becoming glabrate. **Leaves** persistent or drought-deciduous; petiole 0.5–2.5 cm; blade ovate to cordate, 1–5 × 0.5–4 cm, base truncate to rounded or cordate, margins serrate-crenate, apex acute or obtuse. **Inflorescences** unisexual and bisexual, axillary and terminal; staminate peduncle 0.3–2.5 cm, fertile portion 1–4 cm; pistillate peduncle 0.4–3 cm, fertile portion 1–3 × 0.8–1.2 cm; bisexual similar to staminate, with 1–3 pistillate bracts near base; allomorphic pistillate flowers absent. **Pistillate bracts** loosely arranged (inflorescence axis visible between bracts), 3–6 × 5.5–11 mm, abaxial surface pubescent, sessile- and stipitate-glandular; lobes (8–)10–18, rounded, 1/5 bract length. **Pistillate flowers:** pistil 3-carpellate; styles multifid or laciniate. **Capsules** smooth, pubescent and stipitate-glandular. **Seeds** 1.5–2 mm, minutely pitted. **2*n*** = 20.

Flowering and fruiting year-round, especially spring and fall. Arid rocky slopes, desert washes; 10–1400 m; Ariz., Calif.; Mexico (Baja California, Baja California Sur, Sinaloa, Sonora).

Plants in Arizona and Sonora have been segregated as *Acalypha pringlei* based on having long nonglandular hairs mixed with shorter hairs on the stem (versus hairs all of one length). This trait appears throughout the range of *A. californica* and cannot be used to distinguish two species (G. A. Levin 1995).

2. Acalypha wilkesiana Müller Arg. in A. P. de Candolle and A. L. P. P. de Candolle, Prodr. 15(2): 817. 1866 • Painted copperleaf, beefsteak plant, match-me-if-you-can [I]

Acalypha amentacea Roxburgh subsp. *wilkesiana* (Müller Arg.) Fosberg

Shrubs, 20–50 dm, monoecious. **Stems** erect, sparsely to densely pubescent, not glandular. **Leaves** persistent; petiole 1–6 cm; blade ovate to broadly ovate or suborbiculate, 9–20 × 4–15 cm, base obtuse to rounded or subcordate, margins serrate-crenate, apex acuminate. **Inflorescences** usually unisexual, rarely bisexual, axillary; staminate peduncle 0.1–1.5 cm, fertile portion 10–20 cm; pistillate peduncle 1–2 cm, fertile portion 4–15 × 0.5–0.8 cm; bisexual similar to staminate, with 1–2 pistillate bracts near base; allomorphic pistillate flowers absent. **Pistillate bracts** loosely arranged (inflorescence axis visible between bracts), 2–4 × 3–5 mm, abaxial surface sparsely to moderately pubescent; lobes 7–9, ovate to lanceolate, 1/4 bract length, except terminal lobe to 1/2 bract length. **Pistillate flowers:** pistil 3-carpellate; styles multifid or laciniate. **Capsules** unknown. **Seeds** unknown.

Flowering spring–fall. Old home sites, disturbed areas; 0–10 m; introduced; Fla.; Pacific Islands; introduced also in Mexico, West Indies, Central America, n South America, se Asia, Africa.

Acalypha wilkesiana is not known in the wild but presumably originated in the southwestern Pacific Islands (Bismarck Archipelago east to Fiji). The species is commonly cultivated as an ornamental for its leaves that may be various shades of green, purple, red, orange, and yellow (sometimes variegated), and sometimes contorted into unusual shapes. Despite low seed set, it occasionally becomes naturalized in tropical and subtropical areas. Naturalized plants often lack the distinctive leaf coloration found in cultivated plants. Although sometimes treated as *A. amentacea* subsp. *wilkesiana*, DNA sequence data show that *A. wilkesiana* and *A. amentacea* are distinct species (V. G. Sagun et al. 2010).

3. Acalypha phleoides Cavanilles, Anales Hist. Nat. 2: 139. 1800 • Shrubby copperleaf

Acalypha lindheimeri Müller Arg.; *A. lindheimeri* var. *major* Pax & K. Hoffmann

Herbs, perennial, 2–5 dm, monoecious. **Stems** erect, short-pubescent and hirsute. **Leaves:** petiole 0.2–1 cm; blade rhombic-ovate to ovate, or proximal suborbiculate, 2–6 × 1–3 cm, base acute to rounded, margins serrate to crenate-serrate, apex acute to acuminate. **Inflorescences** bisexual, terminal; peduncle 0.3–0.5(–1) cm, pistillate portion 4–7 × 1.6–2 cm, staminate portion 0.5–3.5 cm; allomorphic pistillate flowers rarely present, replacing all or part of staminate portion of inflorescence. **Pistillate bracts** (normal flowers) loosely arranged (inflorescence axis visible between bracts), 8–12 × 7–11 mm, abaxial surface sparsely pubescent and stipitate-glandular; lobes (3–)5–7(–8), triangular to attenuate, 1/5–1/3 bract length or terminal lobe longer; of allomorphic flowers absent. **Pedicels** of allomorphic flowers 3–5 mm. **Pistillate flowers:** pistil 3-carpellate (normal flowers), 2-carpellate (allomorphic flowers); styles multifid or laciniate. **Capsules** muricate, pubescent; allomorphic fruits obovoid, 2 × 1.5 mm, muricate, pubescent. **Seeds** 1.5–2 mm, minutely pitted. **2*n*** = 40 (Mexico).

Flowering and fruiting spring–fall. Rocky areas, grasslands, oak, pine, or juniper woodlands; 100–2600 m; Ariz., N.Mex., Tex.; Mexico; Central America (Guatemala).

Plants from the United States have nearly always been called *Acalypha lindheimeri*, distinguished from *A. phleoides* on the basis of leaf shape and bract lobing. Although plants from Texas generally can be distinguished from plants from central Mexico southward, plants from intervening regions in the United States and Mexico include a full range of intermediates (G. A. Levin 1999b).

4. **Acalypha chamaedrifolia** (Lamarck) Müller Arg. in A. P. de Candolle and A. L. P. P. de Candolle, Prodr. 15(2): 879. 1866 • Everglades or bastard copperleaf

Croton chamaedryfolius Lamarck in J. Lamarck et al., Encycl. 2: 215. 1786 (as chamaedrifolium)

Herbs, perennial, 1–2.5 dm, monoecious. **Stems** prostrate to ascending, pubescent. **Leaves:** petiole 0.1–0.5 cm; blade ovate to orbiculate, 0.3–2.1 × 0.3–1.2 cm, base cordate or rounded, margins serrate-crenate, apex obtuse or acute. **Inflorescences** bisexual, terminal; peduncle 0.2–1 cm, pistillate portion 1.5–3 × 1–1.5 cm, staminate portion 0.8–2.5 cm; allomorphic pistillate flowers absent. **Pistillate bracts** crowded (inflorescence axis not or sparingly visible between bracts), 4–6 × 7–10 mm, abaxial surface pubescent and sessile-glandular; lobes (7–)10–13, deltate to triangular, ⅕ bract length. **Pistillate flowers:** pistil 3-carpellate; styles multifid or laciniate. **Capsules** smooth, pubescent. **Seeds** 1.2–1.4 mm, minutely pitted.

Flowering and fruiting year-round, mainly spring–fall. Rocky pine woods, disturbed areas; 0–10 m; Fla.; West Indies.

In the flora area, *Acalypa chamaedrifolia* is native to Miami-Dade and Monroe counties but has been sparingly, and apparently accidentally, introduced farther north.

5. **Acalypha monostachya** Cavanilles, Anales Hist. Nat. 2: 138, plate 21, fig. 3. 1800 • Round copperleaf

Acalypha hederacea Torrey

Herbs or subshrubs, perennial, 1.5–4 dm, monoecious or dioecious (staminate plants rare). **Stems** prostrate to ascending, short-pubescent and hirsute. **Leaves:** petiole 0.5–2.5 cm; blade orbiculate or reniform, 0.7–2.5 × 0.8–2.5 cm, base cordate or rounded, margins shallowly crenate, apex rounded. **Inflorescences** unisexual or bisexual, terminal (staminate, pistillate, and bisexual) and axillary (pistillate); staminate peduncle 0.8–3 cm, fertile portion 1–4 cm; pistillate peduncle 0.4–1.5 cm, fertile portion 1–2 × 0.8–1.2 cm; bisexual similar to staminate, with 1–3 pistillate bracts near base; allomorphic pistillate flowers absent. **Pistillate bracts** crowded (inflorescence axis not visible between bracts), 6–8.5 × 8–12 mm, abaxial surface hirsute, sessile- and stipitate-glandular; lobes (8–)10–12(–14), rounded, ¼ bract length. **Pistillate flowers:** pistil 3-carpellate; styles multifid or laciniate. **Capsules** smooth, pubescent. **Seeds** 1.5–1.8 mm, minutely pitted.

Flowering and fruiting spring–fall. Dry, open, rocky, gravelly, or sandy areas; 0–900 m; Tex.; Mexico.

Acalypha hederacea, the name most frequently used for these plants in the United States, and *A. monostachya*, commonly used for Mexican plants, were thought to differ in sexuality and staminate inflorescence length, but plants throughout Mexico and Texas show no consistent differences among populations and should be treated as a single species (G. A. Levin 1999b).

In the flora area, *Acalypha monostachya* is widespread in central and southern Texas.

6. **Acalypha radians** Torrey in W. H. Emory, Rep. U.S. Mex. Bound. 2(1): 200. 1859 • Cardinal feather, palmate copperleaf

Herbs or subshrubs, perennial, 1.5–4 dm, dioecious. **Stems** prostrate to ascending, short-pubescent and hirsute. **Leaves:** petiole 0.4–1.6 cm; blade reniform or suborbiculate, 0.5–1.5 × 0.8–2 cm, base cordate or rounded, margins deeply crenate, apex rounded. **Inflorescences** unisexual, terminal (staminate and pistillate) and axillary (pistillate); staminate peduncle 0.5–3 cm, fertile portion 1–4(–5) cm; pistillate peduncle 0.1–0.5 cm, fertile portion 1–2.5 × 0.8–1.2 cm; allomorphic pistillate flowers absent. **Pistillate bracts** crowded (inflorescence axis not visible between bracts), 7–10 × 12–16 mm, abaxial surface hirsute, sessile- and stipitate-glandular; lobes (7–)8–10(–13), spatulate, ½ bract length. **Pistillate flowers:** pistil 3-carpellate; styles multifid or laciniate. **Capsules** smooth, pubescent and hirsute. **Seeds** 1.8–2 mm, minutely pitted.

Flowering and fruiting spring–fall. Grassy openings, dunes, and oak or mesquite woodlands, usually on deep sand; 0–200 m; Tex.; Mexico (Tamaulipas).

Acalypha radians is found in the flora area from the Edwards Plateau south to the Mexican border.

7. Acalypha alopecuroidea Jacquin, Collectanea 3: 196. 1791 • Foxtail copperleaf [I]

Acalypha aristata Kunth

Herbs, annual, 2–6 dm, monoecious. **Stems** erect, short-pubescent and stipitate-glandular. **Leaves:** petiole 0.5–7 cm, stipitate-glandular; blade ovate to broadly ovate, 2–8 × 1.5–5 cm, base rounded or subcordate, margins serrate, apex acuminate. **Inflorescences** unisexual, axillary (staminate) and terminal (pistillate); staminate peduncle 0.1–0.6 cm, stipitate-glandular, fertile portion 0.2–0.8 cm; pistillate peduncle 0.2–1 cm, stipitate-glandular, fertile portion 2–6 × 0.8–1.5 cm; allomorphic pistillate flowers common, terminal on pistillate or, rarely, staminate inflorescences. **Pistillate bracts** (normal flowers) crowded (inflorescence axis not visible between bracts), 8–12 × 3–4 mm, abaxial surface long-hirsute (hairs to 2 mm) and stipitate-glandular; lobes 3–5, proximally deltate with linear tips, ¾ bract length, smooth; of allomorphic flowers absent. **Pedicels** of allomorphic flowers 5–15 mm. **Pistillate flowers:** pistil 3-carpellate (normal flowers), 1(–2)-carpellate (allomorphic flowers); styles unbranched or rarely 2-fid. **Capsules** smooth, pubescent and stipitate-glandular or glabrate; allomorphic fruits obovoid, 1–1.5 × 0.9–1.2 mm, muricate, hirsute. **Seeds** 1–1.1 mm, minutely pitted.

Flowering and fruiting late summer–fall. Disturbed areas; 0–40 m; introduced; Ala., Fla., La.; Mexico; West Indies; Central America; n South America.

Acalypha alopecuroidea has been established in the United States since at least the 1950s.

8. Acalypha arvensis Poeppig in E. F. Poeppig and S. L. Endlicher, Nov. Gen. Sp. Pl. 3: 21. 1841 • Field copperleaf [I]

Herbs, annual, 2–8 dm, monoecious. **Stems** erect to ascending, short-pubescent and densely [sparsely] hirsute. **Leaves:** petiole 0.5–4 cm, not stipitate-glandular; blade rhombic-ovate to rhombic-lanceolate, 2–9(–12) × 1.2–5 (–6.5) cm, base cuneate to rounded, margins serrate, apex obtuse to acute. **Inflorescences** unisexual or bisexual, axillary; staminate peduncle 0.3–2.5 cm, not stipitate-glandular, fertile portion 1.5–3[–6] cm; pistillate peduncle 0.4–3 cm, not stipitate-glandular, fertile portion 2.5–4[–8] × 1–2 cm; bisexual similar to pistillate, with staminate portion 0.4–0.7 cm; allomorphic pistillate flowers common, terminal on pistillate and bisexual inflorescences. **Pistillate bracts** (normal flowers) very densely crowded (inflorescence axis not visible between bracts), 6–12 × 4–6 mm, abaxial surface long-hirsute (hairs to 2 mm) and stipitate-glandular; lobes 3–7, proximally deltate with linear tips, ½–⅔ bract length, smooth; of allomorphic flowers absent. **Pedicels** of allomorphic flowers rudimentary if borne above staminate flowers or 10–18 mm if borne above normal pistillate flowers. **Pistillate flowers:** pistil 3-carpellate (normal flowers), (1–)2-carpellate (allomorphic flowers); styles multifid or laciniate. **Capsules** smooth, hispidulous; allomorphic fruits ovoid, 1.5–1.6 × 1.2–1.3 mm, muricate, sparsely to densely puberulent. **Seeds** 1.1–1.5 mm, minutely pitted.

Flowering and fruiting summer–fall. Disturbed areas; 10–50 m; introduced; Fla.; Mexico; West Indies; Central America; n, c South America.

Acalypha arvensis became established in the United States in the 1980s and is now known from scattered localities throughout peninsular Florida. Some recent literature misapplies *A. aristata* to this species, but that name is a synonym of *A. alopecuroidea*.

9. Acalypha ostryifolia Riddell ex J. M. Coulter, Mem. Torrey Bot. Club 5: 213. 1894 (as ostryaefolia) • Hophornbeam copperleaf, pineland threeseeded mercury [F] [W]

Herbs, annual, 3.5–7 dm, monoecious. **Stems** erect, pubescent, sparsely hirsute, and stipitate-glandular. **Leaves:** petiole 1.5–6.5 cm; blade ovate, 3–8 × 1.5–5 cm, base cordate, margins serrate, apex acute to short acuminate. **Inflorescences** unisexual, axillary (staminate) and terminal (pistillate; sometimes on short lateral branches, appearing axillary); staminate peduncle 0.5–1.5 cm, fertile portion 0.5–3.5 cm; pistillate peduncle 0.1–1 cm, fertile portion 3–7 × 0.7–1 cm; allomorphic pistillate flowers common, near apices of pistillate inflorescences. **Pistillate bracts** (normal and allomorphic flowers) loosely arranged (inflorescence axis visible between bracts), 3–6 × 6–8 mm, abaxial surface pubescent (hairs to 0.3 mm) and sparsely stipitate-glandular; lobes (9–)13–17, linear, ⅔ bract length, muricate. **Pedicels** of allomorphic flowers rudimentary. **Pistillate flowers:** pistil 3-carpellate (normal flowers), 1(–3)-carpellate (allomorphic flowers); styles multifid or laciniate. **Capsules** spiny, pubescent; allomorphic fruits obovoid, 2 irregular flanges near apex, 2–2.2 × 1.6–1.8 mm, smooth, pubescent. **Seeds** 1.6–2 mm, tuberculate.

Flowering and fruiting summer–fall. Stream banks, edges of woods, disturbed areas, agricultural fields; 0–1700 m; Ala., Ariz., Ark., Fla., Ga., Ill., Ind., Iowa, Kans., Ky., La., Miss., Mo., Nebr., N.C., Ohio, Okla., S.C., Tenn., Tex., Va., W.Va.; Mexico; West Indies; introduced to Central America.

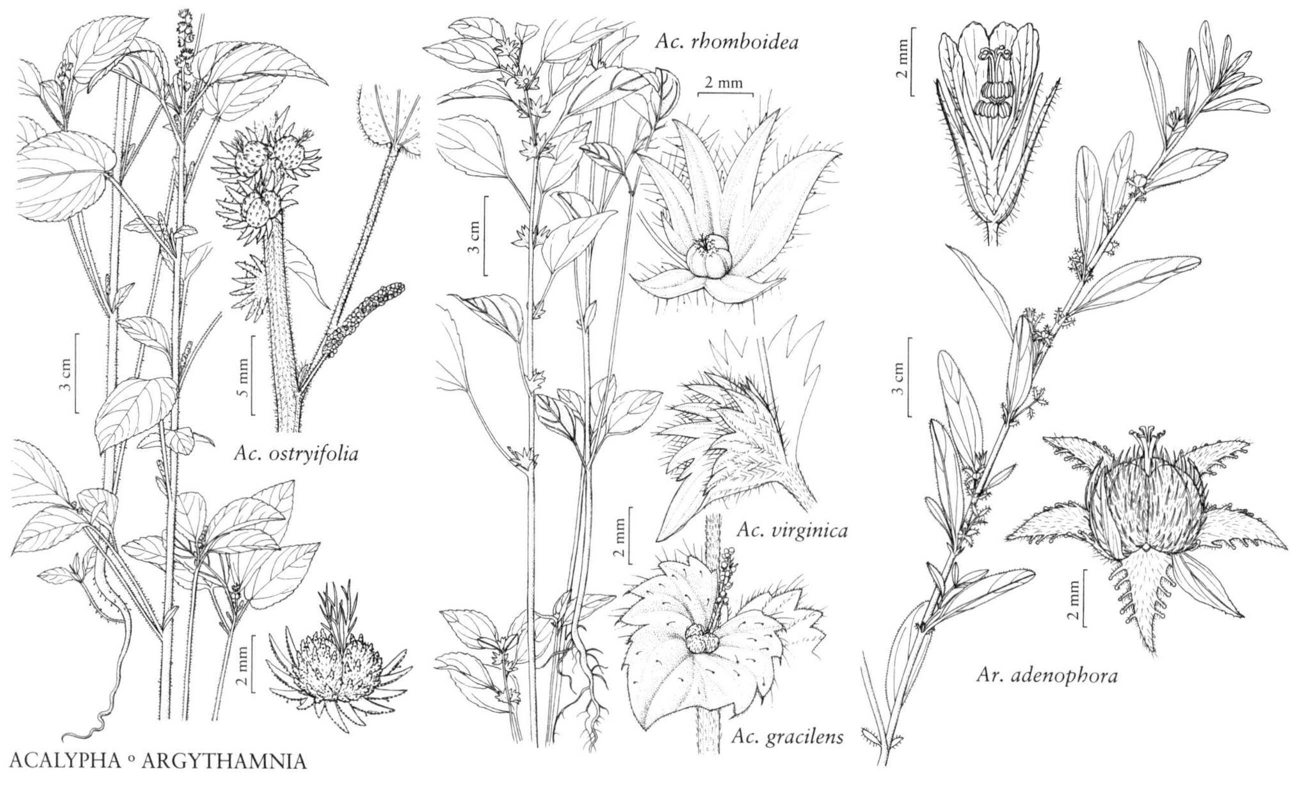

ACALYPHA ◦ ARGYTHAMNIA

J. Torrey [in W. H. Emory 1857–1859, vol. 2(1)] reported *Acalypha ostryifolia* (as *A. caroliniana* Elliott) from New Mexico on the basis of *Bigelow s.n.*, collected near "the Copper Mines" in Grant County. Although no specimen has been located by the author, and this is the only report of this species from New Mexico, it is to be expected in the southwestern part of the state. It was collected in the late nineteenth century in New Jersey and Pennsylvania, but has not been collected there since.

Acalypha ostryifolia may not be native to the northern part of its range. Specimen collection dates suggest that the species is spreading northward, and in much of its range, it is found primarily in areas with human disturbance.

10. Acalypha setosa A. Richard in R. Sagra, Hist. Fis. Cuba 11: 204. 1850 • Cuban copperleaf [I]

Herbs, annual, 3–5 dm, monoecious. **Stems** erect, sparsely pubescent. **Leaves:** petiole 1–7 cm; blade ovate, 2.5–10 × 1.5–8 cm, base broadly obtuse to rounded or truncate, margins serrate, apex acuminate. **Inflorescences** unisexual, axillary (staminate) and terminal (pistillate; sometimes on short lateral branches, appearing axillary); staminate peduncle 0.5–0.7 cm, fertile portion 0.5–1.5 cm; pistillate peduncle 0.1–0.5 cm, fertile portion 3–12 × 0.7–1 cm; allomorphic pistillate flowers common, near apices of pistillate inflorescences and mixed with flowers in staminate inflorescences. **Pistillate bracts** (normal flowers) loosely arranged (inflorescence axis visible between bracts), 5–7 × 3–4 mm, abaxial surface glabrous; lobes 7–9(–13), linear, nearly bract length, muricate; of allomorphic flowers on pistillate inflorescences like those of normal flowers, on staminate inflorescences absent. **Pedicels** of allomorphic flowers rudimentary. **Pistillate flowers:** pistil 3-carpellate (normal flowers), 1(–3)-carpellate (allomorphic flowers); styles multifid or laciniate. **Capsules** smooth, sparsely pubescent; allomorphic fruits oblate ellipsoidal, longitudinally ridged, an irregular flange on each side, 1.4–1.7 × 1.5–2 mm, smooth, puberulent. **Seeds** 1.2–1.4 mm, minutely pitted.

Flowering and fruiting summer–fall. Disturbed areas; 0–100 m; introduced; Ala., Fla., Ga., La., Miss., S.C.; c, s Mexico; West Indies; Central America; n South America.

Acalypha setosa, indigenous from Mexico to northern South America and the West Indies, was first introduced to the United States in the late 1800s and now occurs sporadically through the southeastern states.

11. Acalypha neomexicana Müller Arg., Linnaea 34: 19. 1865 • New Mexico copperleaf [W]

Herbs, annual, 1–3.5 dm, monoecious. **Stems** erect, pubescent. **Leaves:** petiole 1–4 cm; blade ovate-lanceolate, 2–5 × 0.7–2.7 cm, base rounded (sometimes asymmetric), margins crenate to serrate, apex acute. **Inflorescences** unisexual, axillary (staminate) and terminal (pistillate; sometimes on short lateral branches, appearing axillary); staminate peduncle 0.1–0.3 cm, fertile portion 0.1–0.6 cm; pistillate peduncle 0.1–0.2 cm, fertile portion 2–5(–7) × 1.5–2.5 cm; allomorphic pistillate flowers common, near apices of pistillate inflorescences. **Pistillate bracts** (normal and allomorphic flowers) crowded (inflorescence axis not or sparingly visible between bracts), 12–15 × 7–10 mm, abaxial surface pubescent and stipitate-glandular; lobes (5–) 9–13(–17), deltate to triangular, 1/8–1/4 bract length, except terminal lobe to 1/3 bract length. **Pedicels** of allomorphic flowers rudimentary. **Pistillate flowers:** pistil 3-carpellate (normal flowers), 2-carpellate (allomorphic flowers); styles multifid or laciniate. **Capsules** smooth, pubescent; allomorphic fruits obovoid, longitudinally 2-ridged near apex, 1.4–1.6 × 1–1.2 mm, muricate, pubescent. **Seeds** 1.2–1.5 mm, coarsely pitted.

Flowering and fruiting late summer–fall. Moist or shaded areas, oak or pine woodlands, desert grasslands; 600–2500 m; Ariz., N.Mex., Tex.; n, c Mexico.

Acalypha neomexicana is known in Texas only from the trans-Pecos region.

12. Acalypha poiretii Sprengel, Syst. Veg. 3: 879. 1826 (as poireti) • Poiret's copperleaf

Acalypha macrostachyos Poiret in J. Lamarck et al., Encycl. 6: 208. 1804, not *A. macrostachya* Jacquin 1797

Herbs, annual, 1–4 dm, monoecious. **Stems** erect, pubescent and sparsely hirsute. **Leaves:** petiole 1–4.5 cm; blade ovate to elliptic, 2–5 × 1–3.5 cm, base obtuse to rounded, margins serrate-crenate, apex acute. **Inflorescences** bisexual, axillary; peduncle 0.1–0.5 cm, pistillate portion 2–4 × 0.8–1.2 cm (shorter on proximal inflorescences), staminate portion 0.3–1 cm; allomorphic pistillate flowers common, terminal on staminate portion of inflorescences. **Pistillate bracts** (normal flowers) crowded (inflorescence axis not visible between bracts), 4–5 × 6–8 mm, abaxial surface pubescent and sparsely stipitate-glandular; lobes 7–9, triangular, 1/5 bract length; of allomorphic flowers absent. **Pedicels** of allomorphic flowers rudimentary. **Pistillate flowers:** pistil 3-carpellate (normal flowers), 1-carpellate (allomorphic flowers); styles unbranched. **Capsules** smooth, pubescent; allomorphic fruits obovoid, 1.2–1.5 × 1–1.2 mm, muricate, pubescent. **Seeds** 1.2–1.5 mm, minutely pitted.

Flowering and fruiting late summer–fall. Disturbed areas; 10–100 m; Tex.; c, e Mexico; Central America (Guatemala); introduced in West Indies, South America, Africa.

Acalypha poiretii is known in the flora area from the lower Rio Grande valley (Cameron, Hidalgo, and Starr counties). It was collected in the late nineteenth century on ballast dumps in Alabama, Florida, and New Jersey, but has not been reported again from any of these states.

Some authors (for example, R. Govaerts et al. 2000) have treated *Acalypha poiretii* and *A. alnifolia* Poiret as synonyms, in which case the latter would be the correct name for this species; however, the types of the two names clearly belong to different species.

13. Acalypha australis Linnaeus, Sp. Pl. 2: 1004. 1753

• Asian copperleaf [I]

Herbs, annual, 3–6 dm, monoecious. **Stems** erect, densely to sparsely pilose. **Leaves:** petiole 0.5–4 cm; blade ovate to broadly lanceolate, 2–8 × 1.5–4 cm, base cuneate to obtuse, margins serrate, apex acute to short-acuminate. **Inflorescences** bisexual, axillary; peduncle 0.5–4(–6) cm, pistillate portion 1–2 × 1.5–2.5 cm or pistillate bract solitary, staminate portion 0.5–3 cm; allomorphic pistillate flowers rare, when present replacing staminate part of inflorescence. **Pistillate bracts** (normal flowers) loosely arranged (inflorescence axis visible between bracts) or solitary, 10–15(–20) × 8–12 mm, abaxial surface sparsely pubescent; lobes 12–15, rounded, 1/20 bract length; of allomorphic flowers absent. **Pedicels** of allomorphic flowers rudimentary. **Pistillate flowers:** pistil 3-carpellate (normal flowers), 2-carpellate (allomorphic flowers); styles multifid or laciniate. **Capsules** muricate, pubescent; allomorphic fruits obovoid, 2 × 1.2 mm, muricate, pubescent. **Seeds** 1.5–1.8 mm, minutely pitted.

Flowering and fruiting late summer–fall. Disturbed areas; 0–20 m; introduced; N.J., N.Y.; Asia (China, Japan, Korea, Taiwan); Pacific Islands (Philippines).

Acalypha australis, native to eastern Asia, became established in metropolitan New York City in the 1980s. It was also collected once in 1900 in Oregon (*Suksdorf 2892*, GH), and apparently has not persisted there.

14. Acalypha deamii (Weatherby) H. E. Ahles in G. N. Jones and G. D. Fuller, Vasc. Pl. Illinois, 301. 1955 • Largeseeded mercury, Deam's threeseeded mercury [E]

Acalypha virginica Linnaeus var. *deamii* Weatherby, Rhodora 29: 197. 1927; *A. rhomboidea* Rafinesque var. *deamii* (Weatherby) Weatherby

Herbs, annual, 3–7 dm, monoecious. **Stems** erect, glabrate. **Leaves:** petiole 2.5–7 cm; blade ovate to broadly rhombic, 4–12 × 2–7 cm, base obtuse, margins serrate, apex acute to acuminate. **Inflorescences** bisexual, axillary; peduncle 0.1–0.6 cm, pistillate portion 0.8–1.5 × 1–2 cm or pistillate bract solitary, staminate portion 0.3–0.7 cm; allomorphic pistillate flowers common, solitary in axils near base of stem. **Pistillate bracts** (normal flowers) loosely arranged to crowded (inflorescence axis clearly to scarcely visible between bracts) or solitary, 8–16 × 11–20 mm, abaxial surface sparsely pubescent and usually stipitate-glandular; lobes (5–)7–9, lanceolate to narrowly oblong, ½–¾ bract length; of allomorphic flowers absent. **Pedicels** of allomorphic flowers rudimentary. **Pistillate flowers:** pistil 2-carpellate (normal flowers), 1-carpellate (allomorphic flowers); styles multifid or laciniate. **Capsules** smooth, pubescent; allomorphic fruits obovoid, 2.5–3 × 1.8–2 mm, spiny, pubescent. **Seeds** 2.4–3.2 mm, shallowly pitted.

Flowering and fruiting late summer–fall. Moist bottomland woods, near streams or rivers, rarely in moist upland forests; 100–400 m; Ala., Ark., Ill., Ind., Iowa, Kans., Ky., Md., Mo., Ohio, Pa., Tenn., Va., W.Va.

Acalypha deamii strongly resembles robust individuals of *A. rhomboidea* but is generally restricted to moist bottomland woods. Some accounts state that the leaves of *A. deamii* droop; this characteristic is not consistent and cannot be used for identification.

The distribution of *Acalypha deamii* is probably not as patchy as collections suggest. It is frequently overlooked because it looks so much like the widespread and abundant *A. rhomboidea* and is difficult to recognize until fruits mature. Collections may also be limited because *A. deamii* shares its habitat with stinging nettles.

15. Acalypha rhomboidea Rafinesque, New Fl. 1: 45. 1836 • Rhombic or common threeseeded mercury, ricinelle rhomboïde [E] [F] [W]

Acalypha virginica Linnaeus var. *rhomboidea* (Rafinesque) Cooperrider

Herbs, annual, 1.5–6 dm, monoecious. **Stems** usually erect, sometimes ascending, usually sparsely pubescent, rarely sparsely hirsute. **Leaves:** petiole 0.4–7 cm; blade ovate to broadly rhombic, 2–9 × 0.8–5 cm, base obtuse, margins crenate to serrate, apex acute to acuminate. **Inflorescences** bisexual, axillary; peduncle 0.1–0.6 cm, pistillate portion 0.7–1.5(–2) × 1–2(–2.5) cm or pistillate bract solitary, staminate portion 0.3–1 cm; allomorphic pistillate flowers rare, solitary in axils near base of stem. **Pistillate bracts** (normal flowers) loosely arranged to crowded (inflorescence axis clearly to scarcely visible between bracts) or solitary, 6–15 × 9–23 mm, abaxial surface sparsely pubescent and stipitate-glandular; lobes (5–)7–9(–11), lanceolate to triangular, ⅓–⅔ bract length; bracts of allomorphic flowers absent. **Pedicels** of allomorphic flowers 1–2 mm. **Pistillate flowers:** pistil 3-carpellate (normal flowers), 1-carpellate (allomorphic flowers); styles multifid or laciniate. **Capsules** smooth, pubescent; allomorphic fruits obovoid, 2 × 1 mm, spiny, pubescent. **Seeds** (1.2–)1.5–1.7(–2) mm, minutely pitted.

Flowering and fruiting summer–fall. Deciduous and evergreen woods, moist depressions, swampy areas, riverbanks, agricultural fields, disturbed areas; 0–1500 m; N.B., N.S., Ont., Que.; Ala., Ark., Conn., Del., D.C., Fla., Ga., Ill., Ind., Iowa, Kans., Ky., La., Maine, Md., Mass., Mich., Minn., Miss., Mo., Nebr., N.H., N.J., N.Y., N.C., N.Dak., Ohio, Okla., Pa., R.I., S.C., S.Dak., Tenn., Tex., Vt., Va., W.Va., Wis.

Acalypha rhomboidea often has been called *A. virginica* due to controversy about the typification of that name. Conservation of the type of *A. virginica* resolved the issue. This nomenclatural problem, combined with use of inappropriate characters to distinguish *A. rhomboidea* and *A. virginica*, has resulted in considerable confusion between these amply distinct species. The two can be distinguished most readily by the pistillate bracts, which are clearly hirsute abaxially in *A. virginica* but sparsely pubescent abaxially in *A. rhomboidea*. In addition, the bracts of *A. virginica* have (8–)10–14(–16) triangular lobes one fourth to one half the bract length, whereas those of *A. rhomboidea* have (5–)7–9(–11) lanceolate or triangular lobes one third to two thirds the bract length, and the stems of *A. virginica* usually are hirsute whereas the stems of *A. rhomboidea* are rarely so.

In the southern part of its range, many *Acalypha rhomboidea* plants have been confused with *A. gracilens*. These plants are more delicate than *A. rhomboidea* from farther north, with notably smaller pistillate bracts, and they usually produce allomorphic flowers, whereas more robust or northerly plants rarely do. They can be distinguished from *A. gracilens* by having relatively wider leaves and pistillate bracts with fewer lobes and no red sessile glands.

16. Acalypha virginica Linnaeus, Sp. Pl. 2: 1003. 1753, name conserved • Virginia threeseeded mercury E F W

Acalypha digyneia Rafinesque

Herbs, annual, 1–5 dm, monoecious. **Stems** usually erect, sometimes ascending, pubescent and usually hirsute. **Leaves:** petiole 0.3–7 cm; blade narrowly rhombic to broadly lanceolate, 1–8(–11) × 0.5–3 (–4) cm, base acute, margins serrate, apex acute to acuminate. **Inflorescences** bisexual, axillary; peduncle 0.1–0.6 cm, pistillate portion 0.7–1.5 × 1.3–1.7 cm or pistillate bract solitary, staminate portion 0.3–1(–1.8) cm; allomorphic pistillate flowers absent. **Pistillate bracts** loosely arranged to crowded (inflorescence axis clearly to scarcely visible between bracts) or solitary, 6–13 × 9–20 mm, abaxial surface hirsute and sometimes stipitate-glandular; lobes (9–)10–14(–16), triangular, 1/4–1/2 bract length. **Pistillate flowers:** pistil 3-carpellate; styles multifid or laciniate. **Capsules** smooth, pubescent. **Seeds** (1.2–)1.5–1.6(–1.8) mm, minutely pitted.

Flowering and fruiting summer–fall. Deciduous and evergreen woods, riverbanks, agricultural fields, disturbed areas; 30–1200 m; Ala., Ark., Conn., Del., D.C., Ga., Ill., Ind., Iowa, Kans., Ky., La., Md., Mass., Mich., Miss., Mo., Nebr., N.H., N.J., N.Y., N.C., Ohio, Okla., Pa., R.I., S.C., Tenn., Tex., Va., W.Va.

Acalypha virginica overlaps geographically with *A. rhomboidea* but is more frequently found in grassy or prairielike habitats. See the discussion of 15. *A. rhomboidea* for notes on the nomenclatural and taxonomic confusion associated with these species.

Acalypha virginica and *A. gracilens* can sometimes be difficult to distinguish, especially when young. They can generally be distinguished by the pistillate bracts, which in *A. virginica* are hirsute and lack red sessile glands abaxially and in *A. gracilens* are sparsely pubescent and bear some red sessile glands abaxially.

Reports of *Acalypha virginica* from states other than those listed here are based on misidentifications.

17. Acalypha gracilens A. Gray, Manual, 408. 1848 • Slender threeseeded mercury E F W

Acalypha gracilens var. *delzii* Lillian W. Miller; *A. gracilens* var. *fraseri* (Müller Arg.) Weatherby

Herbs, annual, 1–6 dm, monoecious. **Stems** erect, pubescent. **Leaves:** petiole 0.2–1.2(–1.8) cm; blade oblong-lanceolate to linear-lanceolate, 1.7–6 × 0.4–2 cm, base cuneate, margins serrate to crenate to subentire, apex obtuse to acute. **Inflorescences** bisexual, axillary; peduncle 0.1–0.6 cm, pistillate portion 0.7–1.3 × 0.8–1.5 cm or pistillate bract solitary, staminate portion 0.2–2.6 cm; allomorphic pistillate flowers absent. **Pistillate bracts** loosely arranged to crowded (inflorescence axis clearly to scarcely visible between bracts) or solitary, 8–14 × 11–17 mm, abaxial surface sparsely pubescent, red sessile-glandular, and sometimes stipitate-glandular; lobes (7–)9–13(–15), deltate, 1/10–1/4 bract length. **Pistillate flowers:** pistil 3-carpellate; styles multifid or laciniate. **Capsules** smooth, pubescent. **Seeds** 1.1–1.9 mm, minutely pitted.

Flowering and fruiting mostly summer–fall. Pine and pine-oak woods, dry hardwood forests, glades, prairies, disturbed areas, usually on sand or shallow rocky soils; 0–1100 m; Ala., Ark., Conn., Del., D.C., Fla., Ga., Ill., Ind., Iowa, Ky., La., Maine, Md., Mass., Miss., Mo., N.H., N.J., N.Y., N.C., Ohio, Okla., Pa., R.I., S.C., Tenn., Tex., Vt., Va., W.Va., Wis.

Acalypha gracilens varies considerably throughout its range; some of the extremes have been named. The variation shows no discrete breaks, and no infraspecific taxa warrant recognition (G. A. Levin 1999). Populations in central Illinois, Indiana, and Iowa have been alleged to be introduced, but they show slight morphological differences from other populations and appear to be native. *Acalypha gracilens* is introduced in Wisconsin. See 16. *A. virginica* for a discussion of the differences between *A. gracilens* and that species.

18. Acalypha monococca (Engelmann ex A. Gray) Lillian W. Miller & Gandhi, Sida 13: 123. 1988 • Oneseeded mercury E

Acalypha gracilens A. Gray var. *monococca* Engelmann ex A. Gray, Manual ed. 2, 390. 1856; *A. gracilens* subsp. *monococca* (Engelmann ex A. Gray) G. L. Webster

Herbs, annual, 1–4 dm, monoecious. **Stems** erect, pubescent. **Leaves:** petiole 0.2–1(–1.2) cm; blade linear-lanceolate to linear, 1.7–6(–7) × 0.3–1.2 cm, base cuneate, margins usually subentire, sometimes shallowly serrate, apex acute. **Inflorescences** bisexual, axillary; peduncle 0.1–0.6 cm, pistillate portion 0.6–1.2 × 0.8–1.5 cm or pistillate bract solitary, staminate portion 0.1–2.5 cm; allomorphic pistillate flowers absent. **Pistillate bracts** loosely arranged to crowded (inflorescence axis clearly to scarcely visible between bracts) or solitary, 8–13 × 11–16 mm, abaxial surface sparsely to densely pubescent, red sessile-glandular, and rarely stipitate-glandular; lobes (7–)9–13(–17), deltate, 1/10–1/4 bract length. **Pistillate flowers:** pistil 1-carpellate; styles multifid or laciniate. **Capsules** smooth, pubescent. **Seeds** 1.6–2.4 mm, shallowly pitted.

Flowering and fruiting summer–fall. Pine and oak woods, prairies, barrens, on sandy or shallow rocky soils; 80–600 m; Ark., Ill., Kans., Ky., La., Mo., Okla., Tex.

Even when fruits have dehisced, *Acalypha monococca* is easily distinguished from *A. gracilens* and all other species in the genus by its curved, needlelike columellae.

4. ARGYTHAMNIA P. Browne, Civ. Nat. Hist. Jamaica, 338. 1756 • Silverbush, wild-mercury [Greek *argyros*, silver-white, and *thamnos*, shrub, alluding to trunk and branches covered with whitish bark]

Yocupitzia Ramírez-Amezcua

Aphora Nuttall; *Ditaxis* Vahl ex A. Jussieu; *Serophyton* Bentham

Herbs, subshrubs, or shrubs [trees], annual or perennial, monoecious or dioecious; hairs usually malpighiaceous (appressed and attached by the middle), sometimes unbranched [stellate], rarely absent; latex absent. **Leaves** drought deciduous or persistent, alternate, simple; stipules present, persistent or deciduous; petiole absent or present, glands usually absent (tack-shaped glands along length in *A. adenophora*); blade unlobed, margins entire or serrate-dentate, laminar glands absent; venation palmate (3- or 5-veined), secondary veins arcuate, not closely spaced. **Inflorescences** unisexual or bisexual (pistillate flowers proximal, staminate distal), axillary, racemes; glands subtending each bract 0. **Pedicels** present. **Staminate flowers:** sepals [4–]5, valvate, distinct; petals [4–]5, distinct, free or adnate to androphore, white, sometimes pale yellow-green or pale purple proximally; nectary extrastaminal, [4–]5 glands; stamens [4–](7–)10[–12] in [1–]2 whorls, connate proximally forming androphore; staminodes 0–5, at apex of androphore; pistillode absent. **Pistillate flowers:** sepals 5, distinct; petals usually 5, sometimes rudimentary or 0, distinct, white, sometimes pale yellow-green or pale purple proximally; nectary 5 glands; pistil 3(–4)-carpellate; styles 3, distinct or connate proximally, 2-fid, branches 6 per flower, [2 times 2-fid]. **Fruits** capsules, not muricate. **Seeds** globose to ovoid; caruncle absent [present].

Species ca. 80 (12 in the flora): United States, Mexico, West Indies, Central America, South America; tropical and subtropical regions.

There has been controversy surrounding the taxonomic status of *Argythamnia*. Some authors have recognized *Ditaxis*, which includes all of the species in the flora area, at the generic level (G. L. Webster 1994b; A. Radcliffe-Smith 2001), whereas others have treated it as a subgenus of *Argythamnia* (J. W. Ingram 1980; R. McVaugh 1995). There are several morphological characters that distinguish these taxa, and pollen morphology supports their generic recognition (W. Punt 1962). However, recent molecular phylogenetic studies demonstrate that recognizing *Ditaxis* makes *Argythamnia* paraphyletic (Y. Ramírez-Amezcua 2011), so they are treated here as a single genus.

Argythamnia heterantha (Zuccarini) Müller Arg., from Mexico, is cultivated; the seeds are used as a saffron substitute and represent a potential resource for dye, oil, and protein (M. D. Méndez-Robles et al. 2004).

M. C. Johnston (1990) reported *Argythamnia astroplethos* J. W. Ingram from the Chinati Mountains, Presidio County, Texas, but no specimens were cited and none have been located. This species grows nearby in Chihuahua, Mexico, and may eventually be documented from Texas. It belongs to subg. *Chiropetalum* (A. Jussieu) J. W. Ingram and can be distinguished from other *Argythamnia* species in the flora area by its indumentum of stellate hairs in addition to malpighiaceous hairs, tetramerous staminate flowers, and styles that are twice 2-fid.

SELECTED REFERENCES Ingram, J. W. 1980. The generic limits of *Argythamnia* (Euphorbiaceae) defined. Gentes Herbarum 11: 427–436. Ramírez-Amezcua, Y. 2011. Relaciones Filogenéticas en *Argythamnia* (Euphorbiaceae) Sensu Lato. Tesis de Maestría. Universidad Michoacana de San Nicolás de Hidalgo.

1. Glands present on margins of stipules, leaf blades, bracts, and pistillate sepals.
   2. Stems and leaf blades with simple hairs; margins of stipules, leaf blades, bracts, and pistillate sepals with tack-shaped glands . . . . 1. *Argythamnia adenophora*
   2. Stems and leaf blades with simple and malpighiaceous hairs; margins of stipules, leaf blades, bracts, and pistillate sepals with conic glands . . . . 6. *Argythamnia claryana*
1. Glands absent on margins of stipules, leaf blades, bracts, and pistillate sepals.
   3. Petioles usually absent, rarely to 4 mm on proximal leaves.
      4. Pistillate petals 5, 2.8–3.5 mm, obovate to spatulate; stipules punctiform. . . . . 7. *Argythamnia cyanophylla*
      4. Pistillate petals 0 or 5, 0–1.7 mm, elliptic, lanceolate, linear, or punctiform; stipules elliptic, ovate, subulate, or linear-lanceolate.
         5. Nectary glands of pistillate flowers linear, of staminate flowers linear to obovate . . . . 10. *Argythamnia mercurialina*
         5. Nectary glands of pistillate and staminate flowers ovate or oblong.
            6. Stamens 10; staminate petals cuneate-elliptic to cuneate-obovate; ovary tomentose to lanulose . . . . 2. *Argythamnia aphoroides*
            6. Stamens 7–8; staminate petals elliptic; ovary strigose to hispidulous . . . . 12. *Argythamnia simulans*
   3. Petioles present, 1–18 mm.
      7. Plants dioecious.
         8. Flowers usually releasing pink dye when wetted; styles pilose; staminate nectary glands 0.1–0.2 × 0.1 mm; stems and leaf blades densely hairy, hairs silvery . . . . 4. *Argythamnia argyraea* (in part)
         8. Flowers without pink dye when wetted; styles glabrous; staminate nectary glands 0.3–1 × 0.2–0.4 mm; stems and leaf blades hairy or glabrous, hairs silvery or not.
            9. Staminate petals free from androphore; pistillate petals elliptic to filiform, 0.3–1.7 × 0.3–0.4 mm . . . . 8. *Argythamnia humilis* (in part)
            9. Staminate petals adnate to androphore; pistillate petals elliptic, 1.8–3 × 0.7–1.4 mm . . . . 9. *Argythamnia lanceolata* (in part)
      7. Plants monoecious.
         10. Staminate petals adnate to androphore.
            11. Inflorescences 1.5–5.5 cm; staminate sepals 4–6 × 0.8–1 mm; staminate petals 5–7 × 1.2–2.5 mm; pistillate petals 5–6 × 1.5–1.6 mm; petioles 7–18 mm . . . . 5. *Argythamnia brandegeei*
            11. Inflorescences 0.4–1.3 cm; staminate sepals 2–3 × 0.5–0.8 mm; staminate petals 2.1–3.4 × 0.7–1.2 mm; pistillate petals 1.5–3 × 0.6–1.4 mm; petioles 1–5 mm.
               12. Stigmas terete or slightly flattened; pistillate petals 1.5–1.8 × 0.6–1 mm; Florida. . . . . 3. *Argythamnia argothamnoides*
               12. Stigmas flattened; pistillate petals 1.8–3 × 0.7–1.4 mm; Arizona, California . . . . 9. *Argythamnia lanceolata* (in part)

[10. Shifted to left margin.—Ed.]

10. Staminate petals free from androphore.
    13. Stigmas terete; staminate sepals 0.9–2.5 mm . . . . 11. *Argythamnia serrata*
    13. Stigmas flattened; staminate sepals 2–5 mm.
        14. Flowers usually releasing pink dye when wetted; staminate nectary glands ovate, 0.1–0.2 × 0.1 mm . . . . 4. *Argythamnia argyraea* (in part)
        14. Flowers not releasing pink dye when wetted; staminate nectary glands ovate to linear, 0.4–0.8 × 0.2–0.3 mm . . . . 8. *Argythamnia humilis* (in part)

1. Argythamnia adenophora A. Gray, Proc. Amer. Acad. Arts 8: 294. 1870 (as Argyrothamnia) • Glandular silverbush [F]

Ditaxis adenophora (A. Gray) Pax & K. Hoffmann

Herbs or subshrubs, perennial, monoecious, to 10 dm. **Stems** erect, usually hairy, hairs simple. **Leaves:** stipules usually persistent, elliptic to ovate, frequently divided in 2 unequal segments, 1–8 mm, margins with tack-shaped glands; petiole 4–15 mm, with tack-shaped glands; blade elliptic, 1.5–6.5 × 0.7–1.8 cm, margins glandular-serrulate, with tack-shaped glands, surfaces usually hairy, hairs simple. **Inflorescences** bisexual, 0.9–1.8 cm; bracts elliptic to lanceolate, to 7 mm, margins with tack-shaped glands. **Flowers** without pink dye when wetted. **Staminate flowers:** sepals lanceolate, 3–4 × 0.4–1.2 mm; petals elliptic to obovate, 4–7 × 1.4–1.8 mm, adnate to androphore; nectary glands ovate, 0.5–0.7 × 0.3 mm, adnate to androphore, glabrous; stamens 10, staminodes 3–5, glabrous. **Pistillate flowers:** sepals ovate, 4–6.5 × 1.4–2.1 mm, with tack-shaped glands; petals 5, elliptic, 4–6 × 1.3–2.1 mm; nectary glands ovate, 0.5–1 × 0.3–0.6 mm, glabrous; ovary hispidulous, hairs malpighiaceous; styles 1.5–2.5 mm, pilose; stigmas flattened. **Capsules** 4–4.7 mm, hispidulous, long and short malpighiaceous hairs mixed. **Seeds** 3–3.3 mm, with shallow depressions and sometimes also striate.

Flowering Jan–Mar and Jul–Oct. Desert scrub, rocky soils; 0–700 m; Ariz.; Mexico (Sonora).

Argythamnia adenophora is closely related to *A. claryana*, and they sometimes have been treated as a single species (F. Shreve and I. L. Wiggins 1964, vol. 1). *Argythamnia claryana* has malpighiaceous hairs throughout the plant and non-stipitate, conic glands along the margins of the leaves, stipules, and pistillate sepals. In contrast, *A. adenophora* has malpighiaceous hairs exclusively on the ovaries and capsules and simple hairs elsewhere, and the marginal glands on the leaves, stipules, and pistillate sepals are clearly stipitate and tack-shaped. In the flora area, *A. adenophora* is found in southwestern Arizona.

2. Argythamnia aphoroides Müller Arg., Linnaea 34: 146. 1865 (as Argyrothamnia) • Hill country silverbush [C] [E]

Ditaxis aphoroides (Müller Arg.) Pax

Herbs, perennial, monoecious or dioecious, to 5 dm. **Stems** erect to ascending, hairy, hairs malpighiaceous. **Leaves:** stipules deciduous, linear-lanceolate, to 1 mm, margins not glandular; petiole absent; blade elliptic, 1.5–4 × 0.6–2 cm, margins entire, without glands, surfaces hairy, hairs malpighiaceous. **Inflorescences** unisexual, 3–8 cm; bracts linear-lanceolate, 2–4.5 mm, margins without glands. **Flowers** without pink dye when wetted. **Staminate flowers:** sepals lanceolate to narrowly elliptic, 4–5.5 × 1.1–1.4 mm; petals cuneate-elliptic to cuneate-obovate, 3.3–4 × 1.6–3 mm, free from androphore; nectary glands oblong, 0.4–0.5 × 0.3–0.4 mm, adnate to androphore, glabrous; stamens 10, staminodes 0. **Pistillate flowers:** sepals lanceolate to elliptic, 6–7 × 1.7–3.2 mm, without glands; petals 0 or 5, linear, 0–1.7 mm; nectary glands oblong, 0.5 × 0.7 mm, glabrous; ovary tomentose to lanulose; styles 1.5–3 mm, tomentose; stigmas flattened. **Capsules** 7–9 mm, tomentose to lanulose. **Seeds** 3.7–5 mm, smooth, lightly tuberculate, or striate.

Flowering Jan–Jul. Bluestem-grama grasslands, oak woodlands, calcareous, often rocky, clay or loam soils; of conservation concern; 300–600 m; Tex.

Argythamnia aphoroides is restricted to the Edwards Plateau.

3. Argythamnia argothamnoides (Bertero ex Sprengel) J. W. Ingram, Bull. Torrey Bot. Club 80: 423. 1953 • Blodgett's silverbush

Croton argothamnoides Bertero ex Sprengel, Syst. Veg. 3: 872. 1826; *Aphora blodgettii* Torrey ex Chapman; *Argythamnia blodgettii* (Torrey ex Chapman) Chapman; *A. fendleri* Müller Arg.; *A. savanillensis* Kuntze; *Ditaxis argothamnoides* (Bertero ex Sprengel) Radcliffe-Smith & Govaerts; *D. blodgettii* (Torrey ex Chapman) Pax; *D. rubricaulis* Pax & K. Hoffmann

Herbs or shrubs, perennial, monoecious, to 5 dm. **Stems** erect, densely to sparsely hairy, hairs malpighiaceous. **Leaves:** stipules usually persistent, subulate, 0.6–0.8 mm, margins not glandular; petiole 1–5 mm; blade elliptic, ovate to obovate, 1.5–6 × 0.6–3.3 cm, margins minutely serrulate to serrate, without glands, surfaces densely to sparsely hairy, hairs malpighiaceous. **Inflorescences** bisexual, 0.8–1 cm; bracts ovate-elliptic, 1–1.7 mm, margins without glands. **Flowers** without pink dye when wetted. **Staminate flowers:** sepals lanceolate, 2.5–3 × 0.5–0.7 mm; petals elliptic, 2.1–2.6 × 0.9 mm, adnate to androphore; nectary glands ovate, 0.3–0.4 × 0.2–0.4 mm, adnate to androphore, usually glabrous, rarely pubescent; stamens 10, staminodes 3–5, sometimes minute, usually pubescent, rarely glabrous. **Pistillate flowers:** sepals lanceolate, 3.3–4 × 1–1.9 mm, without glands; petals 5, elliptic, 1.5–1.8 × 0.6–1 mm; nectary glands oblate, 0.2–0.3 × 0.3–0.4 mm, glabrous; ovary hispidulous; styles 1–1.4 mm, hispidulous to pilose; stigmas terete or slightly flattened. **Capsules** 3–5 mm, hispidulous. **Seeds** 1.5–3 mm, with shallow depressions, striate or reticulate.

Flowering year-round. Open cactus hammocks, pinelands, mixed forests, rocky and sandy soils, limestone; 0–10 m; Fla.; West Indies; n South America (Colombia, Venezuela).

Most floras have treated *Argythamnia argothamnoides* and *A. blodgettii* as separate species, with the former found in northern South America and adjacent Caribbean islands and the latter in Florida. However, the two are morphologically indistinguishable and treated here as a single species.

In the flora area, *Argythamnia argothamnoides* is found only in Miami-Dade and Monroe counties.

4. Argythamnia argyraea Cory, Madroño 8: 92. 1945
• Silky silverbush, silvery wild-mercury [C] [E]

Herbs, perennial, dioecious, rarely monoecious, to 3.5 dm. **Stems** erect, densely hairy, hairs silvery, malpighiaceous. **Leaves:** stipules deciduous, oblanceolate or obovate to ovate, 0.2–0.3 mm, margins not glandular; petiole 1–3 mm; blade elliptic to ovate, 0.9–4 × 0.4–0.8 cm, margins entire, without glands, surfaces densely hairy, hairs silvery, malpighiaceous. **Inflorescences** unisexual, pistillate 0.5–1 cm, flowers solitary, staminate 1–2 cm; bracts ovate to elliptic, 0.4–0.5 mm, margins without glands. **Flowers** usually releasing pink dye when wetted. **Staminate flowers:** sepals linear to lanceolate, 2–5 × 0.3–2 mm; petals linear-lanceolate, 3–5 × 0.5–0.6 mm, free from androphore; nectary glands ovate, 0.1–0.2 × 0.1 mm, free from androphore, glabrous; stamens 10, staminodes 5, punctiform, glabrous. **Pistillate flowers:** sepals elliptic, 3–4.3 × 0.8–2 mm, without glands; petals 5, lanceolate to linear, 1.5–2 × 0.3–0.5 mm; nectary glands oblong to linear, 0.8 × 0.5 mm, glabrous; ovary hispidulous; styles 1–1.7 mm, pilose; stigmas flattened. **Capsules** 4–5 mm, hairs adpressed, short, silvery. **Seeds** 2.3–2.5 mm, finely papillate.

Flowering Apr–Jul. Shortgrass grasslands or open shrublands on whitish, clay soils; of conservation concern; 100–400 m; Tex.

There are very few collections of *Argythamnia argyraea*, which is endemic to south Texas in Kinney, La Salle, and Maverick counties. Plants of *A. argyraea* are usually dioecious, but William Mahler found that larger plants are occasionally monoecious (J. Poole, pers. comm.).

5. Argythamnia brandegeei Millspaugh, Proc. Calif. Acad. Sci., ser. 2, 2: 220. 1889 (as brandegei)
• Sonoran or Brandegee's silverbush

Ditaxis brandegeei (Millspaugh) Rose & Standley

Varieties 2 (1 in the flora): Arizona, nw Mexico.

5a. Argythamnia brandegeei Millspaugh var. **intonsa** (I. M. Johnston) J. W. Ingram ex L. C. Wheeler in L. D. Benson and R. A. Darrow, Trees & Shrubs Southw. Deserts, 394. 1954 (as brandegei)

Ditaxis brandegeei (Millspaugh) Rose & Standley var. *intonsa* I. M. Johnston, Proc. Calif. Acad. Sci., ser. 4, 12: 1062. 1924 (as brandegei)

Herbs or shrubs, perennial, monoecious, to 15 dm. **Stems** erect, sparsely hairy, sometimes glabrescent, hairs malpighiaceous. **Leaves:** stipules deciduous or sometimes persistent, subulate, 0.4–0.6 mm, margins not glandular; petiole 7–18 mm; blade elliptic to oblong, 4.5–9.6 × 0.9–2.9 cm, margins serrulate to serrate, without glands, surfaces sparsely hairy, sometimes glabrescent, hairs malpighiaceous. **Inflorescences** bisexual, 1.5–5.5 cm; bracts elliptic to ovate, 1.5–2.2 mm, margins without glands. **Flowers** without pink dye when wetted. **Staminate flowers:** sepals lanceolate, 4–6 × 0.8–1 mm; petals elliptic, 5–7 × 1.2–2.5 mm, adnate to androphore; nectary glands ovate to elliptic, 0.5–0.7 × 0.3–0.5 mm, adnate to or free from androphore, glabrous; stamens 10, staminodes 0–3, glabrous. **Pistillate flowers:** sepals lanceolate, 5.5–9 × 1.5–2.5 mm, without glands; petals 5, elliptic, lanceolate, or rhombic, 5–6 × 1.5–1.6 mm; nectary glands oblate, 0.7–1 × 0.9–1.3 mm, glabrous; ovary hispidulous; styles 2.4–4 mm, glabrous; stigmas flattened. **Capsules** 4.5–7 mm, hispidulous. **Seeds** 3–4 mm, smooth or rugose.

Flowering year-round. Desert scrub, rocky soils; 100–400 m; Ariz.; Mexico (Baja California Sur, Sonora).

Variety *brandegeei* is endemic to Mexico and usually is glabrous, although a few inconspicuous hairs are present on a fruit of the isotype: *T. S. Brandegee s.n.* (UC). In contrast, var. *intonsa* has hispidulous ovaries and capsules, and usually pubescent stems and leaves.

6. Argythamnia claryana Jepson, Fl. Calif. 2: 419. 1936 (as clariana) • Desert silverbush [E]

Ditaxis claryana (Jepson) G. L. Webster

Herbs or subshrubs, perennial, monoecious, to 10 dm. **Stems** erect, hairy, hairs simple and malpighiaceous. **Leaves:** stipules persistent, elliptic, lanceolate, or subulate, sometimes divided in 2 unequal segments, 1.7–2.5 mm, margins with conic glands; petiole 2–6 mm; blade elliptic, 0.8–1.9 × 0.4–1.2 cm, margins serrate, with conic glands, surfaces hairy, hairs simple and malpighiaceous. **Inflorescences** bisexual, to 0.5 cm; bracts ovate, 2–3 mm, margins with conic glands. **Flowers** without pink dye when wetted. **Staminate flowers:** sepals lanceolate, 2.5–3.5 × 0.4–0.5 mm; petals elliptic, 3–4 × 0.8–1.2 mm, adnate to androphore; nectary glands narrowly elliptic, oblong, or ovate, 0.3–0.7 × 0.2–0.3 mm, free from androphore, glabrous; stamens 10, staminodes 3–5, glabrous. **Pistillate flowers:** sepals lanceolate to ovate, 2.5–4.5 × 0.8–1.3 mm, margins with conic glands; petals 5, elliptic, 2–3 × 0.7–1.2 mm, margins with glandular teeth; petals elliptic, 2–3 × 0.7–1.2 mm; nectary glands ovate to linear, 0.3–0.5 × 0.2–0.5 mm, glabrous; ovary hispidulous; styles 1–2 mm, glabrous; stigmas flattened. **Capsules** 3–4.5 mm, hispidulous. **Seeds** 2.5–3 mm, with shallow depressions, lightly striate.

Flowering Feb–Apr and Oct–Nov. Desert scrub, rocky soils, basaltic hills; 60–900 m; Ariz., Calif.

Argythamnia claryana is known with certainty only from very arid regions of southeastern California and southwestern Arizona but also is expected in Baja California. I. L. Wiggins (1980) reported *A. adenophora* from northeastern Baja California, but considering that these two species are very similar and that Wiggins (in F. Shreve and Wiggins 1964) previously treated *A. claryana* as a synonym of *A. adenophora*, it is quite possible that this report actually refers to *A. claryana*. Unfortunately, voucher specimens that confirm the occurrence of either of these species in Baja California have not been found. Reports from Nevada appear to be based on misidentified specimens.

7. Argythamnia cyanophylla (Wooton & Standley) J. W. Ingram, Bull. Torrey Bot. Club 80: 423. 1953 • Charleston Mountain silverbush, bruise leaf [E] [F]

Ditaxis cyanophylla Wooton & Standley, Bull. Torrey Bot. Club 36: 106. 1909; *D. diversiflora* Clokey

Herbs, perennial, monoecious, to 2.5 dm. **Stems** erect, sparsely hairy, hairs malpighiaceous. **Leaves:** stipules apparently persistent, punctiform, minute, margins glandular; petiole absent; blade elliptic, obovate, or linear, 1–5 × 0.5–2.2 cm, margins entire, without glands, surfaces sparsely hairy, hairs malpighiaceous. **Inflorescences** unisexual or bisexual, to 4 cm; bracts ovate to lanceolate, to 1.5 mm, margins without glands. **Flowers** without pink dye when wetted. **Staminate flowers:** sepals lanceolate, 3.5–7 × 1–1.3 mm; petals spatulate to obovate, 3–5 × 1.5 mm, adnate to androphore; nectary glands elliptic, 0.3 × 0.2 mm, adnate to androphore, glabrous; stamens (7–)10, staminodes 0 or rudimentary. **Pistillate flowers:** sepals lanceolate to ovate, 3.5–7 × 1.5–2.2 mm, without glands; petals 5, obovate to spatulate, 2.8–3.5 × 0.7–2.8 mm; nectary glands obovate, 0.5 × 0.7 mm, glabrous; ovary sericeous; styles 1.6–2 mm, glabrous; stigmas flattened. **Capsules** 3–5.5 mm, with scattered adpressed malpighiaceous hairs. **Seeds** 3.5–4 mm, smooth, maculate.

Flowering Mar–Jun and Aug–Oct. Scrub, pinyon-juniper woodlands, limestone-derived, rocky soils, damp zones near lakes; 2000–2600 m; Ariz., Nev., N.Mex.

Argythamnia cyanophylla is found from southern Nevada across northern Arizona to central New Mexico.

8. Argythamnia humilis (Engelmann & A. Gray) Müller Arg., Linnaea 34: 147. 1865 (as Argyrothamnia) • Low silverbush [F]

Aphora humilis Engelmann & A. Gray, Boston J. Nat. Hist. 5: 262. 1845; *A. laevis* A. Gray ex Torrey; *A. mercurialina* Nuttall var. *pumila* Torrey; *Argythamnia humilis* var. *laevis* (A. Gray ex Torrey) Shinners; *A. humilis* var. *leiosperma* Waterfall; *A. laevis* (A. Gray ex Torrey) Müller Arg.; *Ditaxis humilis* (Engelmann & A. Gray) Pax; *D. humilis* var. *leiosperma* (Waterfall) Radcliffe-Smith & Govaerts; *D. laevis* (A. Gray ex Torrey) A. Heller

Herbs, perennial, usually monoecious, rarely dioecious, to 3 dm. **Stems** erect or prostrate, hairy or glabrous, hairs not silvery, malpighiaceous. **Leaves:** stipules persistent, ovate to lanceolate, 0.3–1.4 mm, margins not

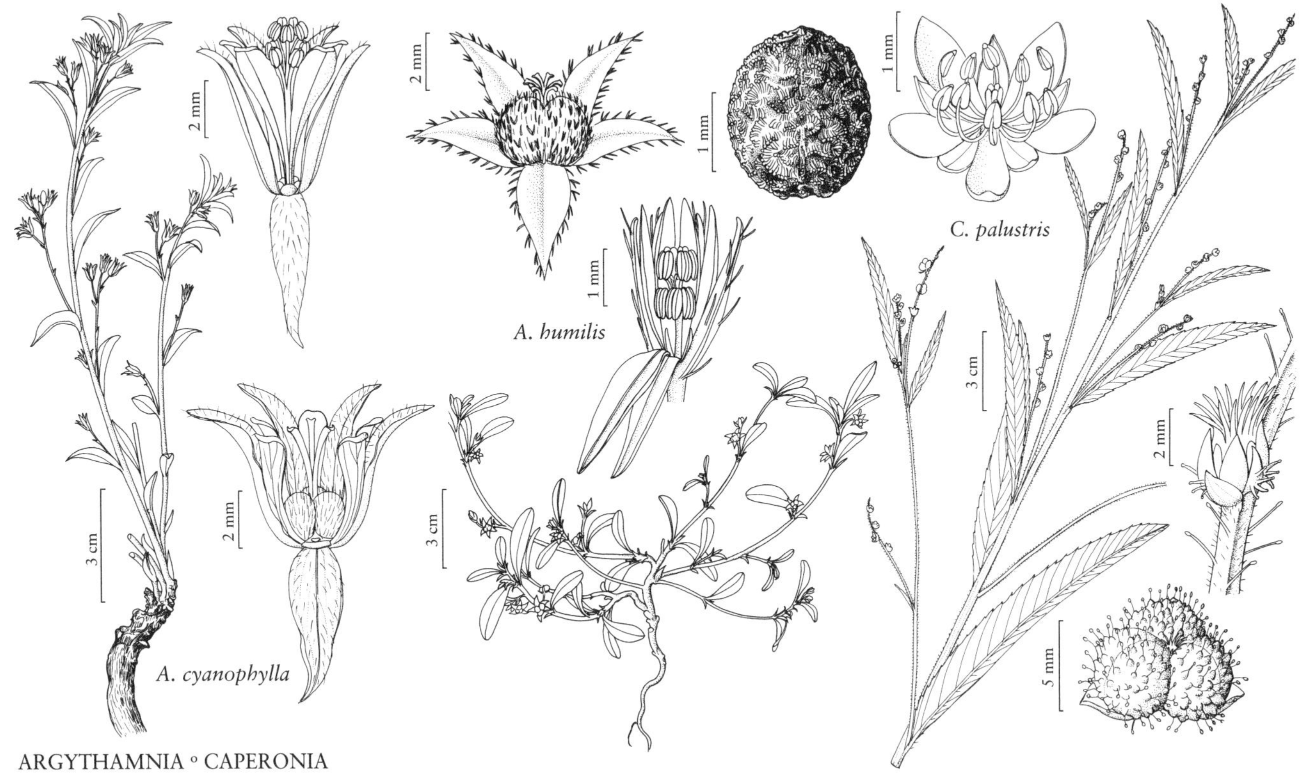

ARGYTHAMNIA ◦ CAPERONIA

glandular; petiole 2–10 mm; blade obovate to linear, 0.5–7 × 0.2–2.2 cm, margins entire, sometimes slightly revolute, without glands, surfaces hairy or glabrous, hairs not silvery, malpighiaceous. **Inflorescences** bisexual, rarely unisexual (staminate), 0.7–0.9 cm; bracts ovate to lanceolate, 0.8–2.5 mm, margins without glands. **Flowers** without pink dye when wetted. **Staminate flowers:** sepals lanceolate, 2.4–3.3 × 0.3–0.8 mm; petals elliptic, 1.5–2.6 × 0.4–0.8 mm, free from androphore; nectary glands ovate to linear, 0.4–0.8 × 0.2–0.3 mm, adnate to or free from androphore, glabrous, rarely pubescent; stamens (7–)10, staminodes 2–3, glabrous. **Pistillate flowers:** sepals elliptic, 4–5.2 × 0.9–2.5 mm, without glands, petals 5, elliptic to filiform, 0.3–1.7 × 0.3–0.4 mm; nectary glands linear to oblong, 0.3–1.5 × 0.1–0.4 mm, glabrous; ovary hispidulous, strigillose, or glabrous; styles 0.6–1.3 mm, glabrous; stigmas flattened. **Capsules** 1.6–4.2 mm, hispidulous, strigillose, or glabrous. **Seeds** 1.5–2.5 mm, smooth or rugose-roughened to papillate.

Flowering year-round. Pine-oak forests, desert scrub, grasslands, limestone, sandstone, and clay soils; 0–1600 m; Colo., Kans., N.Mex., Okla., Tex.; Mexico (Chihuahua, Coahuila, Nuevo León, Tamaulipas).

Glabrous plants have been recognized as *Argythamnia laevis* (and other combinations of *Aphora laevis* at different taxonomic ranks). B. L. Turner (2001b) studied this variation and concluded that the glabrous plants do not form distinct populations but instead are individuals in otherwise pubescent populations. *Argythamnia humilis* var. *leiosperma* was distinguished by possessing smooth seeds, in contrast to the markedly rugose-roughened seeds of var. *humilis*. When he described the variety, Waterfall mentioned that var. *leiosperma* is found in populations of western Oklahoma (Cimarron County) and central Texas (Brown and Hamilton counties). However, this taxon has not been recognized in recent treatments of the plants of Texas (G. M. Diggs et al. 1999; Turner et al. 2003) and Oklahoma (R. J. Tyrl et al. 2010) and is not accepted here.

9. Argythamnia lanceolata (Bentham) Müller Arg., Linnaea 34: 148. 1865 (as Argyrothamnia) • Narrowleaf silverbush

Serophyton lanceolatum Bentham, Bot. Voy. Sulphur, 52. 1844; *Aphora lanceolata* (Bentham) Engelmann & A. Gray; *Argythamnia palmeri* S. Watson; *A. sericophylla* A. Gray ex S. Watson; *A. sericophylla* var. *verrucosemina* Millspaugh; *Ditaxis lanceolata* (Bentham) Pax & K. Hoffmann; *D. palmeri* (S. Watson) Pax & K. Hoffmann; *D. sericophylla* (A. Gray ex S. Watson) A. Heller

Herbs or subshrubs, perennial, monoecious or dioecious, to 10 dm. **Stems** erect, hairy, hairs silvery or not, malpighiaceous. **Leaves:** stipules persistent, subulate, 0.5–1.3 mm, margins not glandular; petiole 1–3 mm;

blade lanceolate to ovate, 1–4.5 × 0.4–1.4 cm, margins usually entire, rarely serrulate, without glands, surfaces hairy, hairs silvery or not, malpighiaceous. **Inflorescences** bisexual or unisexual, 0.4–1.3 cm; bracts ovate to elliptic, 1–2 mm, margins without glands. **Flowers** without pink dye when wetted. **Staminate flowers:** sepals lanceolate, 2–3 × 0.5–0.8 mm; petals elliptic, 2.5–3.4 × 0.7–1.2 mm, adnate to androphore; nectary glands ovate to subulate, 0.7–1 × 0.3–0.4 mm, free from androphore, glabrous; stamens 10, staminodes 3–5, glabrous. **Pistillate flowers:** sepals lanceolate, 3–6.5 × 0.7–1.8 mm, without glands; petals 5, elliptic, 1.8–3 × 0.7–1.4 mm; nectary glands oblate, elliptic to ovate, 0.4–1 × 0.5–0.7 mm, glabrous, rarely pubescent; ovary strigose to hispidulous; styles 1.5–2 mm, glabrous; stigmas flattened. **Capsules** 3–5 mm, strigose to hispidulous. **Seeds** 1.8–2.5 mm, foveolate, striate.

Flowering year-round. Desert scrub, rocky soils; 0–1000 m; Ariz., Calif.; Mexico (Baja California, Baja California Sur, Sonora).

Argythamnia lanceolata is a common shrub of the Sonoran Desert and is easily recognized by its densely strigose, lanceolate to ovate, usually entire leaves, strongly flattened stigmas, and staminate flower petals adnate to the androphore. Plants from shady canyons tend to have broader and less hairy leaves.

10. Argythamnia mercurialina (Nuttall) Müller Arg., Linnaea 34: 148. 1865 (as Argyrothamnia) • Tall silverbush [E]

Aphora mercurialina Nuttall, Trans. Amer. Philos. Soc., n. s. 5: 174. 1835; *Ditaxis mercurialina* (Nuttall) J. M. Coulter

Herbs, perennial, monoecious or dioecious, to 8 dm. **Stems** usually erect, rarely prostrate (var. *pilosissima*), hairy, hairs malpighiaceous. **Leaves:** stipules deciduous, subulate to ovate, 0.3–0.5 mm, margins not glandular; petiole usually absent, rarely to 4 mm on proximal leaves; blade lanceolate, elliptic, ovate to obovate, 1.5–7.5 × 0.6–3.5 cm, margins entire or serrulate, without glands, surfaces hairy, hairs malpighiaceous. **Inflorescences** bisexual or unisexual, to 12 cm; bracts ovate to lanceolate, 1.5–2.5 mm, margins without glands. **Flowers** without pink dye when wetted. **Staminate flowers:** sepals lanceolate to narrowly elliptic, 2.3–3 × 0.7–0.8 mm; petals elliptic to spatulate, 2–2.5 × 0.8–1 mm, free from androphore; nectary glands linear to obovate, 0.3–0.5 × 0.1–0.2 mm, free from androphore, glabrous or pubescent; stamens 8, staminodes 0. **Pistillate flowers:** sepals elliptic, 1–7 × 0.3–3 cm, without glands; petals 0 or 5, elliptic, linear, or punctiform, less than 0.5 mm; nectary glands linear, 0.7 × 1.5 mm, glabrous or pubescent; ovary strigose; styles 0.8–1.5 mm, pubescent; stigmas flattened. **Capsules** 6–14 mm, hispidulous to strigose. **Seeds** 3.2–4.5 mm, smooth to slightly papillate.

Varieties 2 (2 in the flora): sc United States.

1. Nectary glands of staminate and pistillate flowers glabrous . . . . 10a. *Argythamnia mercurialina* var. *mercurialina*
1. Nectary glands of staminate and pistillate flowers hairy . . . 10b. *Argythamnia mercurialina* var. *pilosissima*

10a. Argythamnia mercurialina (Nuttall) Müller Arg. var. **mercurialina** [E]

Aphora drummondii (Bentham) Engelmann & A. Gray; *Serophyton drummondii* Bentham

Nectary glands of staminate and pistillate flowers glabrous.

Flowering Mar–Aug. Flats, swamps, prairies, mesquite-yucca short grass communities, rocky hills, sandy soils, limestone; 0–1800 m; Ark., Colo., Kans., N.Mex., Okla., Tex.

10b. Argythamnia mercurialina (Nuttall) Müller Arg. var. **pilosissima** (Bentham) Shinners, Field & Lab. 24: 38. 1956 [E]

Serophyton pilosissimum Bentham, Bot. Voy. Sulphur, 53. 1844; *Aphora pilosissima* (Bentham) Torrey; *Argythamnia pilosissima* (Bentham) Müller Arg.; *Ditaxis pilosissima* (Bentham) A. Heller

Nectary glands of staminate and pistillate flowers hairy.

Flowering from Mar–Jul. Oak woodlands and forests, grasslands, sandy soils; 10–150 m; Tex.

Variety *pilosissima* occurs from east-central to south Texas, whereas var. *mercurialina* is widely distributed. Both varieties appear to prefer sandy soils.

11. Argythamnia serrata (Torrey) Müller Arg., Linnaea 34: 147. 1865 (as Argyrothamnia) • Yuma or New Mexico silverbush

Aphora serrata Torrey in W. H. Emory, Rep. U.S. Mex. Bound. 2(1): 197. 1859; *Argythamnia californica* Brandegee; *A. dressleriana* J. W. Ingram; *A. gracilis* Brandegee; *A. micrandra* Croizat; *A. neomexicana* Müller Arg.; *A. serrata* var. *magdalenae* Millspaugh; *Ditaxis californica* (Brandegee) A. Heller; *D. dressleriana* (J. W. Ingram) Radcliffe-Smith & Govaerts;

D. gracilis Rose & Standley; *D. micrandra* (Croizat) Radcliffe-Smith & Govaerts; *D. neomexicana* (Müller Arg.) A. Heller; *D. odontophylla* Rose & Standley; *D. serrata* (Torrey) A. Heller; *D. serrata* var. *californica* (Brandegee) V. W. Steinmann & Felger; *D. serrata* var. *magdalenae* (Millspaugh) Eastwood

Herbs, annual or perennial, monoecious, to 5 dm. **Stems** erect or prostrate, usually densely to sparsely hairy, rarely glabrous, hairs malpighiaceous. **Leaves:** stipules deciduous, ovate, lanceolate to subulate, 0.5–2.3 mm, margins not glandular; petiole 1–10 mm; blade elliptic to ovate, obovate, or narrowly lanceolate, 1–10.5 × 0.2–3 cm, margins serrate, serrulate, or entire, without glands, surfaces usually densely to sparsely hairy, rarely glabrous, hairs malpighiaceous. **Inflorescences** bisexual, 0.4–1.5 cm; bracts ovate to elliptic, 0.8–2 mm, margins without glands. **Flowers** without pink dye when wetted. **Staminate flowers:** sepals lanceolate, 0.9–2.5 × 0.4–1.4 mm; petals elliptic to obovate, 1.5–3.5 × 0.6–2 mm, free from androphore; nectary glands elliptic to oblate, 0.3–0.5 × 0.1–0.2 mm, free from androphore, glabrous; stamens 10, staminodes 0 or 3–5, glabrous. **Pistillate flowers:** sepals lanceolate to elliptic, 2.3–6.5 × 0.6–1.6 mm, without glands; petals 5, elliptic, 0.7–3 × 0.4–1.6 mm; nectary glands elliptic, 0.4–0.6 × 0.3–0.4 mm, glabrous; ovary sericeous, strigose, or hispidulous; styles 1–1.8 mm, glabrous; stigmas terete. **Capsules** 1.8–4.5 mm, strigose-hispidulous. **Seeds** 1.5–2.4 mm, smooth, punctate, striate, or reticulate. 2*n* = 26 (Baja California Sur, Mexico).

Flowering year-round. Desert scrub, sandy, rocky, calcareous soils; 0–1600 m; Ariz., Calif., Nev., N.Mex., Tex.; Mexico; Central America (Guatemala, Nicaragua).

Argythamnia serrata is widely distributed and variable; six species have been described to recognize this variation. However, these groups overlap morphologically and geographically, and many specimens cannot be determined with confidence. A broadly defined *A. serrata* is therefore recognized here.

12. Argythamnia simulans J. W. Ingram, Bull. Torrey Bot. Club 84: 421. 1958 • Plateau silverbush

Ditaxis simulans (J. W. Ingram) Radcliffe-Smith & Govaerts

Herbs or subshrubs, perennial, usually monoecious, rarely dioecious, to 6 dm. **Stems** erect, hairy, hairs malpighiaceous. **Leaves:** stipules deciduous, elliptic, 0.3–0.4 mm, margins not glandular; petiole absent; blade ovate, elliptic, or obovate, 1.6–7.5 × 0.4–3.8 cm, margins entire or undulate, without glands, surfaces hairy, hairs malpighiaceous. **Inflorescences** bisexual, 3.5–12 cm; bracts elliptic, 1–1.5 mm, margins without glands. **Flowers** without pink dye when wetted. **Staminate flowers:** sepals narrowly elliptic, 4–5.3 × 0.9–1.6 mm; petals elliptic, 2–4 × 1.2–1.5 mm, adnate to or free from androphore; nectary glands ovate, 0.5–0.8 × 0.3–0.6 mm, adnate to androphore, glabrous; stamens 7–8, staminodes 0. **Pistillate flowers:** sepals narrowly elliptic, 6–9 × 1.7–2.7 mm, without glands; petals 0 or 5, lanceolate, less than 0.5 mm; nectary glands ovate to oblong, 0.3–0.4 × 0.5–0.6 mm, glabrous; ovary strigose to hispidulous; styles 1.5–2.1 mm, glabrous; stigmas flattened. **Capsules** 3.5–9 mm, hispidulous. **Seeds** 3.5–4.2 mm, faintly and irregularly alveolate.

Flowering Mar–Nov. Grasslands, oak woodlands, loamy to clayey, calcareous, often rocky, soils; 100–500 m; Tex.; Mexico (Coahuila).

Argythamnia simulans was described by Ingram as having staminate petals free from the androphore (*Clemans 433*, RSA), but in some specimens the staminate petals are adnate to the androphore (*Riskind 2370*, MEXU, TEX, DAV). This character is usually constant within a species, although it is in some instances difficult to discern. Further study is needed to determine the significance of this variation within *A. simulans*.

In the flora area, *Argythamnia simulans* is found primarily on the Edwards Plateau.

5. CAPERONIA A. St.-Hilaire, Hist. Pl. Remarq. Brésil 3/4: 244. 1825 • False croton [For Natalis (Noël) Caperon or Capperon, d. 1572, apothecary of Orleans] ☐

Lynn J. Gillespie

Herbs [subshrubs], annual [perennial], monoecious [rarely dioecious]; hairs unbranched, sometimes glandular; latex absent. **Leaves** alternate, simple; stipules present, persistent; petiole present, glands absent; blade unlobed, margins serrate, laminar glands absent; venation pinnate or weakly palmate at base, pinnate distally, secondary veins straight, closely spaced, and parallel [arched, moderately spaced]. **Inflorescences** bisexual (pistillate flowers proximal, staminate distal) [unisexual], axillary, spikes or racemes; glands subtending each bract 0. **Pedicels** present or absent. **Staminate flowers:** sepals 5, valvate, connate basally; petals 5, distinct, adnate to base of staminal column, white; nectary absent; stamens 10, in 2 whorls, connate basally; pistillode present, at top of staminal column. **Pistillate flowers:** sepals persistent, often enlarging in fruit, 5–8(–10), connate basally, unequal, small outer lobes often present alternating with larger lobes; petals 5(–6) [often rudimentary], distinct, white; nectary absent; pistil 3-carpellate; styles 3, connate basally [distinct], deeply multifid, branches [9–]12–21 per flower. **Fruits** capsules, densely muricate. **Seeds** subglobose; caruncle absent. $x = 11$.

Species 34 (2 in the flora): introduced; Mexico, West Indies, Central America, South America, Africa; introduced also in Pacific Islands (Guam).

1. Plants without glandular hairs . . . . 1. *Caperonia castaneifolia*
1. Plants with coarse gland-tipped hairs (especially abundant on stems and petioles). . . . . 2. *Caperonia palustris*

1. Caperonia castaneifolia (Linnaeus) A. St.-Hilaire, Hist. Pl. Remarq. Brésil 3/4: 245. 1825 (as castanefolia) • Chestnutleaf false croton [I]

Croton castaneifolius Linnaeus, Sp. Pl. 2: 1004. 1753 (as castaneifolium)

Herbs, 30–140 cm; indumentum of nonglandular hairs. **Stems** erect, glabrescent or sparsely hairy, hairs fine, usually appressed; older stems 6–13 mm diam. **Leaves:** petiole 0.3–1.5 cm; blade linear, linear-lanceolate, lanceolate, narrowly ovate, or elliptic, 5–15 × 0.4–3 cm (L/W = (2–)4–24), base narrowly acute, apex narrowly acute, acute, or obtuse, surfaces glabrescent or sparsely appressed-hairy. **Inflorescences** 1.5–7 cm, peduncle 0.5–3 cm, fertile portion 0.7–5 cm, with 1–2 pistillate flowers. **Staminate flowers:** petals narrowly obovate, 2–2.3 mm, often unequal, exserted well beyond calyx. **Pistillate flowers:** sepals ovate or elliptic, longest 2–2.4 mm, becoming 3–4.5 mm in fruit; petals 1.9–2.9 mm; ovary densely covered in bulbous-based, nonglandular trichomes. **Capsules** 5–6 mm wide, trichomes conic proximally, hairlike distally, nonglandular. **Seeds** pale brown, 3 mm diam.

Flowering and fruiting Aug–Nov. Shallow water or wet margins and banks of ditches, canals, marshy areas, lakes; 0–30 m; introduced; Fla.; Mexico; West Indies; Central America; South America.

Native to tropical America from Mexico and the West Indies to Brazil, *Caperonia castaneifolia* is known in the flora area from southern Florida and from one isolated population in Sumter County in central Florida. Collections in the flora area date from the late 1960s; older records attributed to this species appear to have been based on misidentified specimens of *C. palustris*. A report from Mississippi (C. T. Bryson and D. A. Skojac 2011) also was based on a misidentified specimen of *C. palustris*.

Caperonia castaneifolia is easily distinguished from *C. palustris* by the lack of glandular hairs, and also differs in its tendency to have thicker stems, narrower leaves, shorter inflorescences with fewer pistillate flowers, and larger staminate petals. Both species exhibit considerable inter-individual (and to a lesser extent within-individual) variation in leaf blade shape, but in the flora area, *C. castaneifolia* more frequently has narrower, linear or linear-lanceolate leaves.

2. Caperonia palustris (Linnaeus) A. St.-Hilaire, Hist. Pl. Remarq. Brésil 3/4: 245. 1825 • Sacatrapo, Texasweed [F] [I] [W]

Croton palustris Linnaeus, Sp. Pl. 2: 1004. 1753 (as palustre)

Herbs, 25–100 cm; indumentum of glandular and nonglandular hairs, glandular hairs coarse, erect, thick-based, and gland-tipped (especially abundant on stems and petioles). **Stems** erect, moderately to densely hairy, with gland-tipped hairs and finer, usually appressed, nonglandular hairs; older stems 4–7 mm diam. **Leaves:** petiole 0.3–2.5(–3.5) cm; blade narrowly ovate, lanceolate, or linear-lanceolate, 6–15 × (0.6–)1–6 cm (L/W = 2–7(–11)), base usually rounded or obtuse, rarely acute, apex acute or narrowly acute, surfaces glabrescent or sparsely, mostly appressed-hairy. **Inflorescences** 2–14 cm, peduncle 1–7 cm, fertile portion 1–9 cm, with (1–)2–4 pistillate flowers. **Staminate flowers:** petals narrowly obovate, 1.4 mm, ± equal, not or somewhat exserted beyond calyx. **Pistillate flowers:** sepals ovate or elliptic, longest 2–3.2 mm, becoming 3.5–5.5 mm in fruit; petals 1–2(–2.4) mm; ovary densely covered in bulbous-based, gland-tipped trichomes. **Capsules** 5–7 mm wide, trichomes conic proximally, hairlike distally, gland-tipped. **Seeds** brown, 2.5–3 mm diam. $2n = 22$.

Flowering and fruiting Jul–Nov. Disturbed wet areas, ditches, swampy areas, rice fields; 0–100 m; introduced; Ala., Ark., Fla., La., Miss., Tex.; West Indies; Central America; South America.

Caperonia palustris is a major weed in rice fields in parts of the southern United States (R. K. Godara et al. 2011). Known in the flora area at least since the 1920s from Texas, this species was first collected in Arkansas in 1971 and Mississippi in 1982.

6. ADELIA Linnaeus, Syst. Nat. ed. 10, 2: 1285, 1298. 1759, name conserved • Wild lime [Greek *a*-, not, and *delos*, evident, alluding to small, obscure flowers]

J. Arturo De-Nova

Shrubs [trees], unarmed or branchlets sometimes stiff and thorn-tipped, dioecious [monoecious], hairs unbranched; latex absent. **Leaves** deciduous, fascicled on short shoots [alternate], simple; stipules present, deciduous; petiole present, glands absent; blade unlobed, margins entire [crenate], laminar glands absent; venation palmate at base and pinnate distally [pinnate]. **Inflorescences** unisexual [bisexual], axillary, fascicles [racemes] or flowers solitary; bracts subtending pistillate flowers minute, not enlarging in fruit; glands subtending each bract 0. **Pedicels** present. **Staminate flowers:** sepals [4–]5, not petaloid, 2[–5] mm, valvate, distinct; petals 0; nectary extrastaminal, annular [5 glands], adnate to calyx; stamens [6–]14–17[–30], connate basally [distinct]; pistillode present [absent]. **Pistillate flowers:** sepals 5(–6)[–7], distinct; petals 0; nectary annular; pistil (2–)3(–4)-carpellate; styles (2–)3(–4), distinct [connate basally], deeply multifid. **Fruits** capsules. **Seeds** subglobose; caruncle absent.

Species 9 (1 in the flora): Texas, Mexico, West Indies, Central America, South America; tropical and subtropical regions.

Phylogenetic analyses of morphological and molecular data support *Adelia* as a monophyletic group sister to the Caribbean genera *Lasiocroton* Grisebach and *Leucocroton* Grisebach (J. A. De-Nova and V. Sosa 2007). Three principal lineages were recognized in *Adelia*: the first includes only *A. cinerea* (Wiggins & Rollins) A. Cervantes, V. W. Steinmann & Flores Olvera, a species endemic to Mexico; the second includes four Mexican species (including *A. vaseyi*); and the third comprises the remaining four species from the West Indies, Central America, and South America.

SELECTED REFERENCES De-Nova, J. A. and V. Sosa. 2007. Phylogenetic relationships and generic delimitation in *Adelia* (Euphorbiaceae s.s.) inferred from nuclear, chloroplast, and morphological data. Taxon 56: 1027–1036. De-Nova, J. A., V. Sosa, and V. W. Steinmann. 2007. A synopsis of *Adelia* (Euphorbiaceae s.s.). Syst. Bot. 32: 583–595.

1. Adelia vaseyi (J. M. Coulter) Pax & K. Hoffmann in H. G. A. Engler, Pflanzenr. 63[IV,147]: 69. 1914 • Vasey's wild lime [F]

Euphorbia vaseyi J. M. Coulter, Contr. U.S. Natl. Herb. 1: 48. 1890

Shrubs, 1–2(–3) m; bark gray to brown, branches usually ascending. **Leaves:** petiole 0.2 cm; blade pale green, spatulate to obovate, 2–3(–4) × 0.5–1 cm, base cuneate, apex usually rounded, rarely emarginate, surfaces glabrous; 3-veined at base, secondary veins 4–7 pairs. **Inflorescences** 1–3-flowered. **Pedicels:** staminate 2 mm, pistillate 3–5 mm in flower to 20 mm in fruit. **Staminate flowers** greenish white; sepals lanceolate, acute, 2 mm, hairy; stamens whitish; filaments 1.5 mm; pistillode 3-parted. **Pistillate flowers** yellowish green; sepals reflexed at anthesis, ovate to lanceolate, acute, 2–3 mm, puberulent; ovary 1.5–2.5 mm diam.; styles 2 mm. **Capsules** subglobose, 1–1.3 cm diam., hairy. **Seeds** 4–5 mm.

Flowering Jan–Jun, fruiting Jun. Subtropical semideciduous woodlands on loamy soils, occasionally shrublands on sandy to gravelly soils; 10–100 m; Tex.; Mexico (Tamaulipas).

In the flora area, *Adelia vaseyi* is restricted to the lower Rio Grande Valley (Cameron, Hidalgo, Starr, and Willacy counties), where much of its habitat has been destroyed.

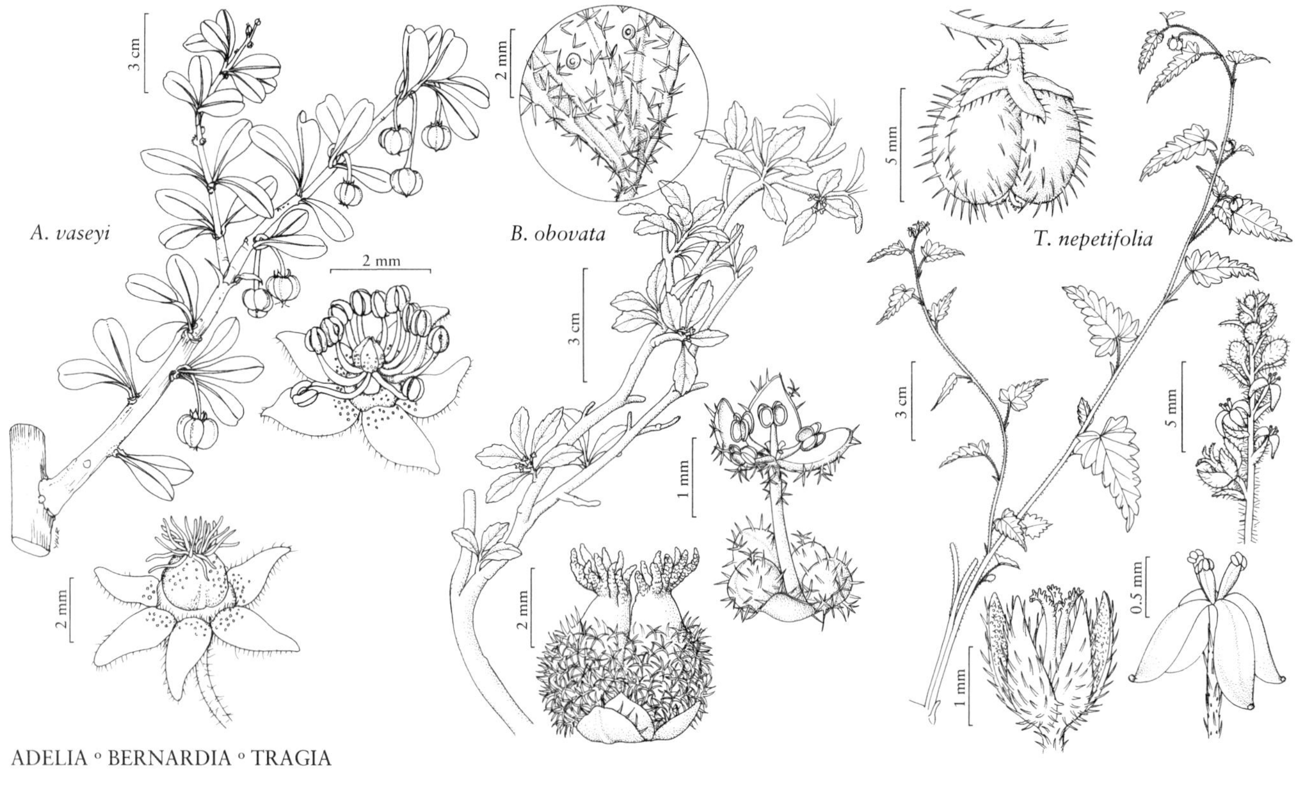

ADELIA ◦ BERNARDIA ◦ TRAGIA

7. BERNARDIA Houstoun ex Miller, Gard. Dict. Abr. ed. 4, vol. 1. 1754 • Myrtlecroton, oreja de raton [Probably for Bernard de Jussieu, 1699–1777, French botanist]

Mark H. Mayfield

Shrubs [**herbs or subshrubs**], dioecious [monoecious]; hairs stellate [unbranched or absent]; latex absent. **Leaves** persistent, alternate, simple; stipules present, persistent or caducous; petiole present [absent], glands absent; blade unlobed, margins coarsely crenate to crenate-serrate [serrate or entire], laminar glands usually abaxial, proximal, crateriform, occasionally absent on some leaves [absent]; venation pinnate (with strong secondary veins ascending from base). **Inflorescences** unisexual, axillary, often on short, lateral shoots; staminate spicate thyrses, pistillate flowers solitary [terminal spikes]; glands subtending bracts 0. **Pedicels:** staminate present, pistillate absent [present]. **Staminate flowers**: sepals 3(–4), valvate, distinct; petals 0; nectary intrastaminal, 1 to several glands; stamens 3–15(–20)[–50], ± straight in bud, distinct; pistillode absent. **Pistillate flowers**: sepals 3–5, distinct; petals 0; nectary absent [present]; pistil 2–3-carpellate; styles 2–3, distinct, 2-fid, branches flattened, adaxial surface stigmatic. **Fruits** capsules. **Seeds** subglobose; caruncle absent. $x = 13$.

Species ca. 70 (3 in the flora): sw, sc United States, Mexico, West Indies, Central America, South America.

The species that occur in the flora area are distinct from most species of the genus in being shrubs with relatively small leaves and stellate vestiture; most *Bernardia* species are perennial herbs or subshrubs. The rounded shrubs native to the flora area grow well in cultivation and would make attractive native borders within their range.

1. Leaf blade abaxial surfaces green, sparsely spreading stellate-pubescent, veins not prominent abaxially; pistils 2-carpellate. . . . . 3. *Bernardia obovata*
1. Leaf blade abaxial surfaces grayish white, densely appressed and/or spreading stellate-pubescent, veins prominent abaxially; pistils 3-carpellate.
    2. Stipules persistent, yellowish brown to black, bases thickened, with dark resinous exudate; stamens (3–)5–7 . . . . 1. *Bernardia incana*
    2. Stipules caducous, green to straw colored, bases not thickened, without dark resinous exudate; stamens (10–)12–15(–20) . . . . 2. *Bernardia myricifolia*

1. Bernardia incana C. V. Morton, J. Wash. Acad. Sci. 29: 376. 1939

Shrubs to 1.5 m. **Leaves:** stipules persistent, yellowish brown to black, base thickened, with dark resinous exudate; petiole 1.5–2.5(–3.4) mm; blade usually broadly elliptic to suborbiculate, rarely cuneate, 0.8–2.5 × 0.3–1.1 cm, margins revolute, crenate, laminar glands (0–)2–4(–6), abaxial surface grayish white, densely appressed stellate-pubescent, adaxial surface green, glabrate; veins prominent abaxially. **Inflorescences:** staminate thyrses 5–20 mm. **Staminate flowers:** stamens (3–)5–7, nectary glands peltiform. **Pistillate flowers:** pistil 3-carpellate; styles 3, lobulate adaxially. **Capsules** 6 mm, 3-lobed.

Flowering spring–summer; fruiting summer–fall. Protected slopes in desert canyon washes; 700–1600 m; Ariz., Calif.; Mexico (Baja California, Chihuahua, Sonora).

As indicated by V. W. Steinmann and R. S. Felger (1997), *Bernardia incana* is readily distinguished from *B. myricifolia*, with which it has often been merged (for example, G. L. Webster 1993b), by its dark persistent stipules that are thickened by a dried, resinous exudate. Within the flora area, *B. incana* occupies a mostly montane Sonoran Desert range in southern Arizona and California (extending just into the southern Mojave Desert), with outlying populations in the Grand Canyon. No specimens have been located to document a reported distribution in Nevada, but it is present to within a few miles of the border along the Grand Wash Cliffs in Arizona.

2. Bernardia myricifolia (Scheele) S. Watson in W. H. Brewer et al., Bot. California 2: 70. 1880 (as myricaefolia)

Tyria myricifolia Scheele, Linnaea 25: 581. 1853 (as myricaefolia)

Shrubs to 3 m. **Leaves:** stipules caducous, green to straw colored, base not thickened, without dark resinous exudate; petiole 0.5–2(–2.2) mm; blade broadly elliptic to orbiculate, 1.5–4.5 × 1–2.5 cm, margins revolute, crenate, laminar glands (0–)2–4(–6), abaxial surface grayish white, densely spreading and/or appressed stellate-pubescent, adaxial surface green, glabrate; veins prominent abaxially. **Inflorescences:** staminate thyrses 5–15 mm. **Staminate flowers:** stamens (10–)12–15(–20), nectary glands claviform. **Pistillate flowers:** pistil 3-carpellate; styles 3, densely penicillate adaxially. **Capsules** 7–8 mm, 3-lobed.

Flowering spring–summer; fruiting summer–fall. Shrub communities on rocky limestone hills in oak-juniper woodlands, thornscrub; 100–1300 m; Tex.; Mexico (Chihuahua, Coahuila, Durango, Nuevo León, San Luis Potosí, Sonora, Tamaulipas).

In the flora area, *Bernardia myricifolia* is known from southern Texas, with a single outlying collection from Brewster County (*H. B. Parks 1724*, MO; B. L. Turner et al. 2003). Reports of *B. myricifolia* from Arizona and California are based on plants identified here as *B. incana*. Reports of *B. myricifolia* from New Mexico are based on specimens of *B. obovata*. In Mexico, *B. myricifolia* is most abundant in thornscrub in the northeast to montane areas of the Chihuahuan Desert.

3. Bernardia obovata I. M. Johnston, J. Arnold Arbor. 21: 261. 1940 [F]

Shrubs to 0.8 m. **Leaves:** stipules caducous, green to straw colored, not thickened, without dark resinous exudate; petiole 0.5–3(–4) mm; blade usually obovate to cuneate, rarely broadly elliptic to suborbiculate, 0.6–3 × 0.5–2.5 cm, margins flat, crenate-serrate, laminar glands 1(–2), abaxial surface green, sparsely spreading stellate-pubescent, adaxial surface green, sparsely stellate-pubescent to glabrate; veins not prominent abaxially, ± level with surface. **Inflorescences:** staminate thyrses 5–10 mm. **Staminate flowers:** stamens 3–4(–6), nectary glands claviform. **Pistillate flowers:** pistil 2-carpellate; styles 2, irregularly dissected adaxially. **Capsules** 5 mm, 2-lobed.

Flowering spring; fruiting summer–fall. Canyon washes in high desert scrub; 800–1900 m; N.Mex., Tex.; Mexico (Chihuahua, Coahuila, Sonora).

Bernardia obovata is a Chihuahuan Desert endemic, occurring in the flora area only from the trans-Pecos of west Texas to the San Andres Mountains of southern New Mexico (K. W. Allred 2002b). The bicarpellate pistil is apparently unique within *Bernardia*.

8. TRAGIA Linnaeus, Sp. Pl. 2: 980. 1753; Gen. Pl. ed. 5, 421. 1754 • Noseburn [For Hieronymus Bock, 1498–1553, German botanist; from Greek *tragos*, goat, bock being the German equivalent]

Roberto J. Urtecho

Herbs, subshrubs, or vines, perennial, monoecious [dioecious]; hairy, hairs unbranched, always some stinging (sometimes inconspicuous except on ovaries and capsules), sometimes glandular; latex absent. **Leaves** deciduous, alternate, simple (usually 3-foliolate in *T. laciniata*); stipules present, persistent; petiole present, glands absent; blade usually unlobed, sometimes lobed basally (sometimes deeply 3-lobed in *T. laciniata*) [palmately lobed], margins serrate, crenate, dentate, or entire, laminar glands absent; venation pinnate or palmate at base, pinnate distally [palmate]. **Inflorescences** bisexual (pistillate flowers proximal, staminate distal) [unisexual], axillary, terminal, or leaf-opposed, racemes [rarely with single pistillate branch]; glands subtending each bract 0. **Pedicels** present, staminate with persistent base, pistillate elongated in fruit. **Staminate flowers:** sepals 3–5, usually green, sometimes reddish green, not petaloid, valvate, distinct; petals 0; nectary absent [present]; stamens 2–6(–10)[–25], distinct or connate basally (connate ½ length in *T. nigricans*); pistillode present [absent]. **Pistillate flowers:** sepals 6, usually green, sometimes reddish green, not petaloid, connate basally; petals 0; nectary absent; pistil 3-carpellate; styles 3, connate basally to ½ [most of] length, unbranched. **Fruits** capsules, usually 3 carpels maturing, except often 1 maturing in *T. brevispica*. **Seeds** globose to ovoid; caruncle absent.

Species ca. 175 (15 in the flora): United States, Mexico, West Indies, Central America, South America, Asia, Africa, Australia; primarily tropical and subtropical regions.

Tragia is a taxonomically difficult genus that is characterized by stinging hairs. Although many species of *Tragia* are twining vines, most species in the flora area are subshrubs or herbs. Some species are used medicinally for their anti-inflammatory, analgesic, vermifugic, and antihyperglycemic properties. Two sections are represented in the flora area: *Tragia* and *Leptobotrys* (Baillon) Müller Arg. Molecular phylogenetic analysis (W. M. Cardinal-McTeague and L. J. Gillespie, unpubl.) suggests that *Tragia* is polyphyletic and that sect. *Leptobotrys* (*T. smallii*, *T. urens*) should be segregated as a distinct genus; these results are supported by pollen morphology (L. J. Gillespie 1994). *Tragia volubilis* Linnaeus was collected from Florida once (*Rugel s.n.*, 1842–1848, US), but has not been collected there since and is presumed extirpated in the flora area. This species is widespread in the Caribbean and Latin America.

SELECTED REFERENCES Miller, K. I. 1964. A Taxonomic Study of the Species of *Tragia* in the United States. Ph.D. thesis. Purdue University. Miller, K. I. and G. L. Webster. 1967. A preliminary revision of *Tragia* (Euphorbiaceae) in the United States. Rhodora 69: 241–305. Urtecho, R. J. 1996. A Taxonomic Study of the Mexican Species of *Tragia* (Euphorbiaceae). Ph.D. dissertation. University of California, Davis.

1. Inflorescence glands stipitate.
   2. Fruiting pedicels 3–7 mm; leaf blades narrowly ovate to lanceolate, bases shallowly cordate to truncate; persistent base of staminate pedicels 0.3–0.7 mm; stigmas smooth to undulate; Texas . . . . . 5. *Tragia glanduligera*
   2. Fruiting pedicels 7–11 mm; leaf blades ovate to triangular-ovate, bases deeply cordate; persistent base of staminate pedicels 1.8–2 mm; stigmas undulate to subpapillate; Arizona . . . . . 6. *Tragia jonesii*
1. Inflorescence glands absent or sessile (*T. nepetifolia* and *T. ramosa*).
   3. Capsules 11–13 mm wide; leaf blades 4.5–10(–13) cm, bases cordate; petioles 15–85 mm; stamens 3 . . . . . 4. *Tragia cordata*
   3. Capsules 4–11 mm wide (9–13 mm in *T. smallii*); leaf blades 1–8(–10) cm, bases acute, obtuse, subcuneate, cuneate, truncate, subcordate, cordate, subhastate, or hastate; petioles 0–38(–41) mm; stamens 2–6(–10).
      4. Stamens 2.
         5. Leaf blades orbiculate to elliptic, margins serrate to crenate . . . . . 13. *Tragia smallii*
         5. Leaf blades usually oblanceolate to linear, sometimes elliptic, margins entire or irregularly sinuate . . . . . 14. *Tragia urens*
      4. Stamens 3–6(–10).
         6. Leaves usually 3-foliolate, sometimes 3-lobed nearly to base . . . . . 7. *Tragia laciniata*
         6. Leaves simple, usually unlobed, sometimes lobed basally.
            7. Stems purple-green to reddish black or brownish red to maroon-green; staminate flowers 2–5 per raceme.
               8. Leaf blades acicular to narrowly oblong, 1–6 cm, margins usually entire, sometimes serrulate, petioles 0.5–2 mm; stamens connate basally; capsules 4–5 mm wide . . . . . 8. *Tragia leptophylla*
               8. Leaf blades oblong to oblanceolate, 3–7 cm, margins coarsely serrate, petioles 1–5 mm; stamens connate ½ length; capsules 6–7 mm wide . . . . . 10. *Tragia nigricans*
            7. Stems green, whitish green, reddish green, dark green, or gray-green; staminate flowers 2–80 per raceme.
               9. Stigmas papillate.
                  10. Leaf blade margins coarsely dentate to coarsely serrate; staminate sepals reddish green; Arizona, Colorado, New Mexico . . . . . 9. *Tragia nepetifolia*
                  10. Leaf blade margins serrate; staminate sepals green, sometimes red-tinged; c, e United States.
                      11. Staminate flowers 15–80 per raceme, distally clustered; persistent bases of staminate pedicels 0.3–0.6 mm, shorter than subtending bract; pistillate sepals 1.8–5 mm . . . . . 2. *Tragia betonicifolia*
                      11. Staminate flowers 11–40 per raceme, evenly distributed; persistent bases of staminate pedicels 1–1.8 mm, longer than subtending bract; pistillate sepals 1.3–2.3 mm . . . . . 15. *Tragia urticifolia*
               9. Stigmas smooth, undulate, or subpapillate.
                  12. Leaf blades suborbiculate to ovate; Florida . . . . . 12. *Tragia saxicola*
                  12. Leaf blades linear-lanceolate, lanceolate, ovate, triangular, subhastate or cordate; sc, sw United States.
                      13. Staminate flowers 2–8 per raceme; capsules with often 1 carpel maturing; stems decumbent, twining, or erect, apices usually flexuous . . . . . 3. *Tragia brevispica*
                      13. Staminate flowers 2–20 per raceme; capsules with usually 3 carpels maturing; stems erect to trailing, apices flexuous or not.

[14. Shifted to left margin.—Ed.]

14. Leaf blades usually triangular to subhastate, sometimes ovate, bases cordate, hastate, or truncate; stems gray-green, apices often flexuous; stigmas undulate to subpapillate; styles connate to ⅓ length, short-exserted; stamens 3–4 . . . . 1. *Tragia amblyodonta*
14. Leaf blades linear-lanceolate to narrowly ovate, bases truncate to weakly cordate; stems dark green to light green, apices rarely flexuous; stigmas smooth to undulate; styles connate ⅓–½ length, long-exserted; stamens 3–6(–10) . . . . 11. *Tragia ramosa*

1. **Tragia amblyodonta** (Müller Arg.) Pax & K. Hoffmann in H. G. A. Engler, Pflanzenr. 68[IV,147]: 51. 1919 • Dog-tooth or blunt-toothed noseburn

Tragia nepetifolia Cavanilles var. *amblyodonta* Müller Arg. in A. P. de Candolle and A. L. P. P. de Candolle, Prodr. 15(2): 934. 1866 (as nepetaefolia)

Subshrubs, 1.2–5 dm. **Stems** erect to trailing, gray-green, apex often flexuous. **Leaves:** petiole 4–20(–30) mm; blade usually triangular to subhastate, sometimes ovate, 1–4.5 × 0.8–3 cm, base cordate, hastate, or truncate, margins crenate to serrate, apex acute to obtuse. **Inflorescences** terminal or axillary, glands absent, staminate flowers 5–16 per raceme; staminate bracts 0.9–2 mm. **Pedicels:** staminate 0.7–1.2 mm, persistent base 0.2–0.8 mm; pistillate 1.5–4 mm in fruit. **Staminate flowers:** sepals 3–4, green, 0.9–1.2 mm; stamens 3–4, filaments 0.2–0.7 mm. **Pistillate flowers:** sepals lanceolate, 1–2.5 mm; styles connate to ⅓ length, short-exserted; stigmas undulate to subpapillate. **Capsules** 7–8 mm wide. **Seeds** brown with tan mottling, 2.5–3.5 mm. $2n = 110$.

Flowering spring–fall; fruiting summer–late fall. Dry, rocky, exposed slopes in xerophytic scrub; 10–1400 m; Ariz., N.Mex., Tex.; Mexico (Chihuahua, Coahuila, Durango, Nuevo León, Tamaulipas).

Tragia amblyodonta is easily distinguished from other members of *Tragia* by the combination of usually triangular to subhastate leaf blades, gray-green coloration, and painfully stinging hairs. Both stomata diameter and pollen grain size of *T. amblyodonta* are larger than in any other North American species of *Tragia* (K. I. Miller and G. L. Webster 1967).

2. **Tragia betonicifolia** Nuttall, Trans. Amer. Philos. Soc., n. s. 5: 173. 1835 (as betonicaefolia) • Betony-leaf noseburn [E] [W]

Tragia urticifolia Michaux var. *texana* Shinners

Herbs or subshrubs, 2–5 dm. **Stems** erect to trailing, green to whitish green, apex never flexuous. **Leaves:** petiole 10–40 mm; blade triangular-lanceolate to triangular-ovate, 1.5–6 × 1–3.5 cm, base cordate to truncate, margins serrate, apex acute. **Inflorescences** terminal (often appearing leaf-opposed), glands absent, staminate flowers 15–80 per raceme, distally clustered; staminate bracts 1–2 mm. **Pedicels:** staminate 0.7–1 mm, persistent base 0.3–0.6 mm, shorter than subtending bract; pistillate 3–4 mm in fruit. **Staminate flowers:** sepals 3–4, green, sometimes red-tinged, 1.2–2.3 mm; stamens 3(–4), filaments 0.4–1 mm. **Pistillate flowers:** sepals lanceolate, 1.8–5 mm; styles connate ⅓ length; stigmas papillate. **Capsules** 7–9 mm wide. **Seeds** dark brown with light brown streaks, 3–4 mm.

Flowering late spring–summer; fruiting summer–fall. Dry, sandy soils, disturbed fields, prairies, open woods; 0–400 m; Ala., Ark., Kans., La., Miss., Mo., Okla., Tenn., Tex., Va.

Plants of *Tragia betonicifolia* resemble those of *T. urticifolia* but differ in the greater number of branches from the root crowns, the shorter length of the persistent staminate flower pedicel bases, the longer, narrower pistillate sepals, and the distally clustered arrangement of the staminate flowers.

3. **Tragia brevispica** Engelmann & A. Gray, Boston J. Nat. Hist. 5: 262. 1845 • Short-spike noseburn

Tragia nepetifolia Cavanilles var. *scutellariifolia* (Scheele) Müller Arg.; *T. nepetifolia* var. *teucriifolia* (Scheele) Müller Arg.; *T. scutellariifolia* Scheele; *T. teucriifolia* Scheele

Herbs or vines, 2–12 dm. **Stems** decumbent, twining, or erect, light green, apex usually flexuous. **Leaves:** petiole 6–38 mm; blade triangular to cordate, 1.9–6 × 1.5–3.5 cm, base truncate to cordate, margins serrate to crenate, apex acute. **Inflorescences** terminal (often appearing leaf-opposed), glands absent, staminate flowers 2–8[–10] per raceme; staminate bracts 1–1.8 mm. **Pedicels:** staminate 0.7–2 mm, persistent base 0.4–1.5 mm; pistillate 2–4 mm in fruit. **Staminate flowers:** sepals 3–4[–5], green, 1–1.5 mm; stamens 3–4(–5), filaments 0.3–0.6 mm. **Pistillate flowers:** sepals ovate, 1.3–3.5 mm; styles connate ⅓ length; stigmas subpapillate to undulate. **Capsules** 6.5–7 mm wide, often 1 carpel maturing. **Seeds** dark brown, 2.5–3.8 mm. $2n = 44$.

Flowering spring–fall; fruiting late summer–fall. Open forests, scrublands, disturbed roadsides, open fields, often on loam and clay soils; 10–500 m; La., Okla., Tex.; Mexico (Nuevo León).

The leaves of *Tragia brevispica* are highly variable and frequently resemble those of *T. ramosa*, which differs in having smooth stigmas and leaf blades much longer than wide. The presence of 1-carpellate fruit in *T. brevispica* is unique in *Tragia* in the flora area.

4. Tragia cordata Michaux, Fl. Bor.-Amer. 2: 176. 1803 • Heart-leaf noseburn [E]

Vines, 15–20 dm. **Stems** usually decumbent or twining, rarely erect, gray-green to light green, apex flexuous. **Leaves:** petiole 15–85 mm; blade ovate to broadly cordate, 4.5–10(–13) × 3.5–10 cm, base cordate, margins serrate, apex acuminate. **Inflorescences** terminal (often appearing leaf-opposed), glands absent, staminate flowers 20–60 per raceme; staminate bracts 1.5–2 mm. **Pedicels:** staminate 1.5–2.2 mm, persistent base 0.7–1 mm; pistillate 2.5–3 mm in fruit. **Staminate flowers:** sepals 3, green, 0.7–1 mm; stamens 3, filaments 0.2–0.5 mm. **Pistillate flowers:** sepals elliptic to ovate, 1.5–2 mm; styles connate ¼–⅓ length; stigmas papillate. **Capsules** 11–13 mm wide. **Seeds** dark brown, 4.3–5.3 mm.

Flowering spring–fall; fruiting summer–late fall. Rich deciduous forests, riverbanks, rocky thickets; 50–500 m; Ala., Ark., Fla., Ga., Ill., Ind., Ky., La., Miss., Mo., Okla., Tenn., Tex.

Both the morphology and ecology of *Tragia cordata* make it unique among American members of *Tragia*. The relatively large, heart-shaped leaves separate it from the other *Tragia* in the flora area; it is the only twining species of *Tragia* found in the deciduous forest of the Midwest.

5. Tragia glanduligera Pax & K. Hoffmann in H. G. A. Engler, Pflanzenr. 68[IV,147]: 55. 1919 • Brush or sticky noseburn

Subshrubs or vines, 3–10 dm. **Stems** trailing or twining, dark green, apex flexuous. **Leaves:** petiole 6–22 mm; blade narrowly ovate to lanceolate, 2.5–4 × 1.5–2 cm, base shallowly cordate to truncate, margins serrate to crenate, apex acute to acuminate. **Inflorescences** terminal (often appearing leaf-opposed), glands stipitate, prominent throughout, staminate flowers 10–30 per raceme; staminate bracts 0.5–1.5 mm. **Pedicels:** staminate 1–2 mm, persistent base 0.3–0.7 mm; pistillate 3–7 mm in fruit. **Staminate flowers:** sepals 3, green, 0.7–1.2 mm; stamens 3, filaments 0.2–0.4 mm. **Pistillate flowers:** sepals lanceolate, 0.7–1.5 mm; styles connate ⅓ length; stigmas smooth to undulate. **Capsules** 4–5 mm wide. **Seeds** dark brown to black, 1.9–2.2 mm.

Flowering late spring; fruiting late summer–fall. Dry, sandy limestone soils, abandoned home sites, mesquite scrub; 10–80 m; Tex.; s, e Mexico; Central America (Guatemala).

Southern Texas is the northernmost distribution of *Tragia glanduligera*. In Mexico, it is found in tropical deciduous forests in Campeche, Nuevo León, Tabasco, Veracruz, and Yucatan. This species and *T. jonesii* are the only species in the flora area with stipitate glands on the inflorescence. *Tragia glanduligera* differs from *T. jonesii* by its leaf blade margins with 10–15 smaller teeth per side, shorter staminate pedicels, and truncate to weakly cordate leaf blade bases.

6. Tragia jonesii Radcliffe-Smith & Govaerts, Kew Bull. 52: 480. 1997 • Jones's noseburn

Tragia scandens M. E. Jones, Contr. W. Bot. 18: 49. 1933, not Linnaeus 1754

Subshrubs, 4–5 dm. **Stems** decumbent, trailing, or erect, green to gray-green, apex flexuous. **Leaves:** petiole 3–10(–15) mm; blade ovate to triangular-ovate, 0.9–2(–3) × 0.5–1.5(–2) cm, base deeply cordate, margins serrate, apex acute. **Inflorescences** terminal (often appearing leaf-opposed), glands stipitate, prominent throughout, staminate flowers 10–30 per raceme; staminate bracts 0.8–1 mm. **Pedicels:** staminate 2.2–2.4 mm, persistent base 1.8–2 mm; pistillate 7–11 mm in fruit. **Staminate flowers:** sepals 3–4, green, 0.9–1.1 mm; stamens 2–3, filaments 0.2–0.3 mm. **Pistillate flowers:** sepals ovate,

1.5 mm; styles connate ⅓–½ length; stigmas undulate to subpapillate. **Capsules** 5 mm wide. **Seeds** mottled brown-purple, 2.5–3 mm.

Flowering spring–summer; fruiting summer–fall. Sonoran desert scrub; 10–900 m; Ariz.; Mexico (Baja California Sur, Sonora).

In the flora area, *Tragia jonesii* is confined to Pima County in southern Arizona. Identified as *T. amblyodonta* in several floras, it differs from that species by its stipitate glands and twining habit. *Tragia jonesii* resembles *T. glanduligera* from southern Texas and eastern Mexico in the presence of stipitate glands, but differs in leaf blade shape and base, the number of teeth on the leaf blade margin (4–9 teeth per side in *T. jonesii*, 10–15 teeth per side in *T. glanduligera*), and the longer fruiting pedicel.

7. **Tragia laciniata** (Torrey) Müller Arg., Linnaea 34: 182. 1865 • Sonoita or Sonoran noseburn

Tragia urticifolia Michaux var. *laciniata* Torrey in W. H. Emory, Bot. U.S. Mex. Bound. 2(1): 200. 1859 (as urticaefolia)

Subshrubs, 2.5–5 dm. **Stems** erect to decumbent, dark green, apex never flexuous. **Leaves** usually 3-foliolate, sometimes 3-lobed nearly to base; petiole 7–18 mm; leaflets: blade lanceolate, base acute, margins deeply and coarsely serrate, sinuses often extending ½+ to midvein, apex acute, central one 2.5–4 × 1–2 cm, lateral ones often with basal lobe. **Inflorescences** terminal (often appearing leaf opposed), glands absent, staminate flowers 10–20 per raceme; staminate bracts 0.5–1.5 mm. **Pedicels:** staminate 0.8–1.6 mm, persistent base 0.3–0.7 mm; pistillate 2.5–3 mm in fruit. **Staminate flowers:** sepals 3–4, green, 1–1.4 mm; stamens 3, filaments 3.5–4 mm. **Pistillate flowers:** sepals lanceolate, 2–3 mm; styles connate ¼–½ length; stigmas undulate to subpapillate. **Capsules** 6–7 mm wide. **Seeds** dark brown, 3–3.2 mm.

Flowering summer–fall; fruiting late summer–fall. Oak woodlands, ravines, stream banks; 1200–1700 m; Ariz.; Mexico (Chihuahua, Sonora).

Tragia laciniata is the only *Tragia* in the flora area with compound leaves. Some plants from Sonora, Mexico, are intermediate between *T. laciniata* and *T. nepetifolia* var. *dissecta* Müller Arg. In the flora area, *T. laciniata* is known from southern Arizona.

8. **Tragia leptophylla** (Torrey) I. M. Johnston, Contr. Gray Herb. 68: 91. 1923 • Fine-leaf noseburn [E]

Tragia ramosa Torrey var. *leptophylla* Torrey in W. H. Emory, Bot. U.S. Mex. Bound. 2(1): 201. 1859; *T. stylaris* Müller Arg. var. *leptophylla* (Torrey) Müller Arg.

Herbs, 1–4.5 dm. **Stems** erect, brownish red to maroon-green, apex never flexuous. **Leaves:** petiole 0.5–2 mm; blade acicular to narrowly oblong, 1–6 × 0.2–0.6 cm, base acute to subcuneate, margins usually entire, sometimes serrulate, apex acute to obtuse. **Inflorescences** terminal (appearing leaf opposed) or axillary, glands absent, staminate flowers 2–3(–5) per raceme; staminate bracts 2–2.5 mm. **Pedicels:** staminate 1–2 mm, persistent base 0.5–0.8 mm, pistillate 2–3 mm in fruit. **Staminate flowers:** sepals 3–4(–5), green, 1–2.5 mm; stamens 3–4, filaments 1.2–1.4 mm, connate basally. **Pistillate flowers:** sepals lanceolate, 1.5–2 mm; styles connate ⅕ length; stigmas undulate to slightly papillate. **Capsules** 4–5 mm wide. **Seeds** mottled dark olive brown, 2.5–3 mm.

Flowering spring–fall; fruiting late spring and fall. Dry streams and river margins with limestone cobble substrates; 400–700 m; Tex.

Tragia leptophylla is known from the western part of the Edwards Plateau region in west-central Texas. K. I. Miller and G. L. Webster (1967) did not recognize *T. leptophylla* as a distinct species; they treated it as a synonym of *T. ramosa*. *Tragia leptophylla* differs from *T. ramosa* in its less branching habit, dark reddish stems, usually entire leaf blade margins, fewer staminate flowers per inflorescence, and riparian limestone cobble habitat.

9. **Tragia nepetifolia** Cavanilles, Icon. 6: 37, plate 557, fig. 1. 1800 (as nepetaefolia) • Catnip noseburn [F]

Subshrubs, 1.5–5 dm. **Stems** erect to trailing, green to reddish green, apex never flexuous. **Leaves:** petiole 3–25(–41) mm; blade triangular to ovate [linear], proximal broadly ovate to sometimes suborbiculate, 1.8–5 × 0.9–3.6 cm, often red-green, base truncate to cordate, margins coarsely dentate to coarsely serrate, apex acute. **Inflorescences** terminal (often appearing leaf opposed), glands sessile or absent, staminate flowers 8–40 per raceme, distally clustered [evenly distributed]; staminate bracts 1.3–1.6 mm. **Pedicels:** staminate 1.4–1.7 mm, persistent base 0.5–0.7 mm; pistillate 2.9–3.3 mm in fruit. **Staminate flowers:** sepals 3–4, reddish green,

1–2 mm; stamens 3–4, filaments 0.3–0.6 mm. **Pistillate flowers:** sepals lanceolate [ovate], 1.4–2.3 mm; styles connate ¼–⅓ length; stigmas papillate. **Capsules** 6–8 mm wide. **Seeds** brownish black, 3–4 mm.

Flowering late spring; fruiting late summer–fall. Pine-oak woodlands; 1500–2500 m; Ariz., Colo., N.Mex.; Mexico; Central America.

Tragia nepetifolia is typically found at high elevations in Mexico and the southwestern United States. Since it was described more than 200 years ago, many collections of *Tragia* in Mexico and the United States have been identified mistakenly as this species.

Tragia nepetifolia includes four varieties in Mexico, but none match plants occurring in the United States. These most closely resemble var. *dissecta* of western Mexico, sharing inflorescences with distally clustered staminate flowers and a tendency toward reddish coloration, but differing in that their leaf blades are not as deeply toothed.

10. Tragia nigricans Bush ex Small, Fl. S.E. U.S., 702. 1903 • Dark noseburn [E]

Herbs, 1.5–5.5 dm. **Stems** erect, purple-green to reddish black, apex never flexuous. **Leaves:** petiole 1–5 mm; blade oblong to oblanceolate, 3–7 × 1–2.8 cm, base acute to obtuse, margins coarsely serrate, tooth apices often somewhat recurved, apex acute. **Inflorescences** terminal (appearing leaf opposed), glands absent, staminate flowers 2–5 per raceme; staminate bracts 1–2 mm. **Pedicels:** staminate 1.3–1.6 mm, persistent base 0.2–0.4 mm; pistillate 2–3 mm in fruit. **Staminate flowers:** sepals 3–4, green, 1.5–2.5 mm; stamens 4–5, filaments 0.7–1.3 mm, connate ½ length. **Pistillate flowers:** sepals rhombic-lanceolate, 1–4 mm; styles connate ¼ length; stigmas undulate. **Capsules** 6–7 mm wide. **Seeds** dark brown, 2.5–3.2 mm.

Flowering spring–summer; fruiting midsummer–fall. Open oak woodlands; 100–700 m; Tex.

The combination of relatively large, coarsely serrate leaf blades, dark stems, and filaments connate to half of length make *Tragia nigricans* unique within the genus in North America. It appears to be most closely related to *T. leptophylla*, which also has dark stems and few staminate flowers per inflorescence. Like *T. leptophylla*, it is found only in the Edwards Plateau, but is restricted to the eastern part; they overlap only in Uvalde County. They also differ in habitat preference.

11. Tragia ramosa Torrey, Ann. Lyceum Nat. Hist. New York 2: 245. 1827 • Branched or desert or common noseburn [F] [W]

Tragia angustifolia Nuttall; *T. nepetifolia* Cavanilles var. *angustifolia* (Müller Arg.) Müller Arg.; *T. nepetifolia* var. *ramosa* (Torrey) Müller Arg.; *T. ramosa* var. *latifolia* (Müller Arg.) Pax & K. Hoffmann; *T. stylaris* Müller Arg.; *T. stylaris* var. *angustifolia* Müller Arg.; *T. stylaris* var. *latifolia* Müller Arg.

Subshrubs, 1.2–5 dm. **Stems** erect to trailing, dark green to light green, apex rarely flexuous. **Leaves:** petiole 2–20 mm; blade linear-lanceolate to narrowly ovate, 1–4 × 0.5–2 cm, base truncate to weakly cordate, margins serrate, apex acute. **Inflorescences** terminal (often appearing leaf opposed), glands few, sessile, staminate flowers 2–20 per raceme; staminate bracts 1.5–2 mm. **Pedicels:** staminate 0.7–2 mm, persistent base 0.4–1.5 mm; pistillate 2–2.5 mm in fruit. **Staminate flowers:** sepals 3–4, green, 1–2.2 mm; stamens 3–6(–10), filaments 0.3–1 mm. **Pistillate flowers:** sepals lanceolate, 0.8–2.5 mm; styles connate ⅓–½ length, long-exserted; stigmas smooth to undulate. **Capsules** 6–8 mm wide. **Seeds** dark brown, 2.5–3.5 mm. **2*n*** = 44.

Flowering spring–fall; fruiting late spring–fall. Mesquite, desert scrub, pine-juniper, oak woodlands; 200–2800 m; Ariz., Ark., Calif., Colo., Kans., Mo., Nebr., Nev., N.Mex., Tex., Utah; Mexico (Baja California, Chihuahua, Coahuila, Nuevo León, Sonora, Tamaulipas).

Tragia ramosa is a variable species showing much environmental plasticity. Collections from the western United States and western Mexico have much broader leaves than those from Texas and Nuevo León and were previously referred to as *T. stylaris*. Smooth stigmatic surfaces, three to six (rarely to ten) stamens, and narrow apical leaves are characters consistent with *T. ramosa*.

12. Tragia saxicola Small, Fl. S.E. U.S. 702, 1333. 1903 • Florida Keys noseburn [E]

Herbs or subshrubs, 1.2–3.5 dm. **Stems** erect, green, apex flexuous. **Leaves:** petiole 5–13 mm; blade suborbiculate to ovate, 1.2–3 × 1–2.3 cm, base subcordate, margins dentate to serrate, apex acute. **Inflorescences** terminal (often appearing leaf opposed), glands absent, staminate flowers 12–20 per raceme; staminate bracts 0.8–1.2 mm. **Pedicels:** staminate 1.5–1.9 mm, persistent base 0.5–0.7 mm; pistillate 3.2–3.7 mm in fruit.

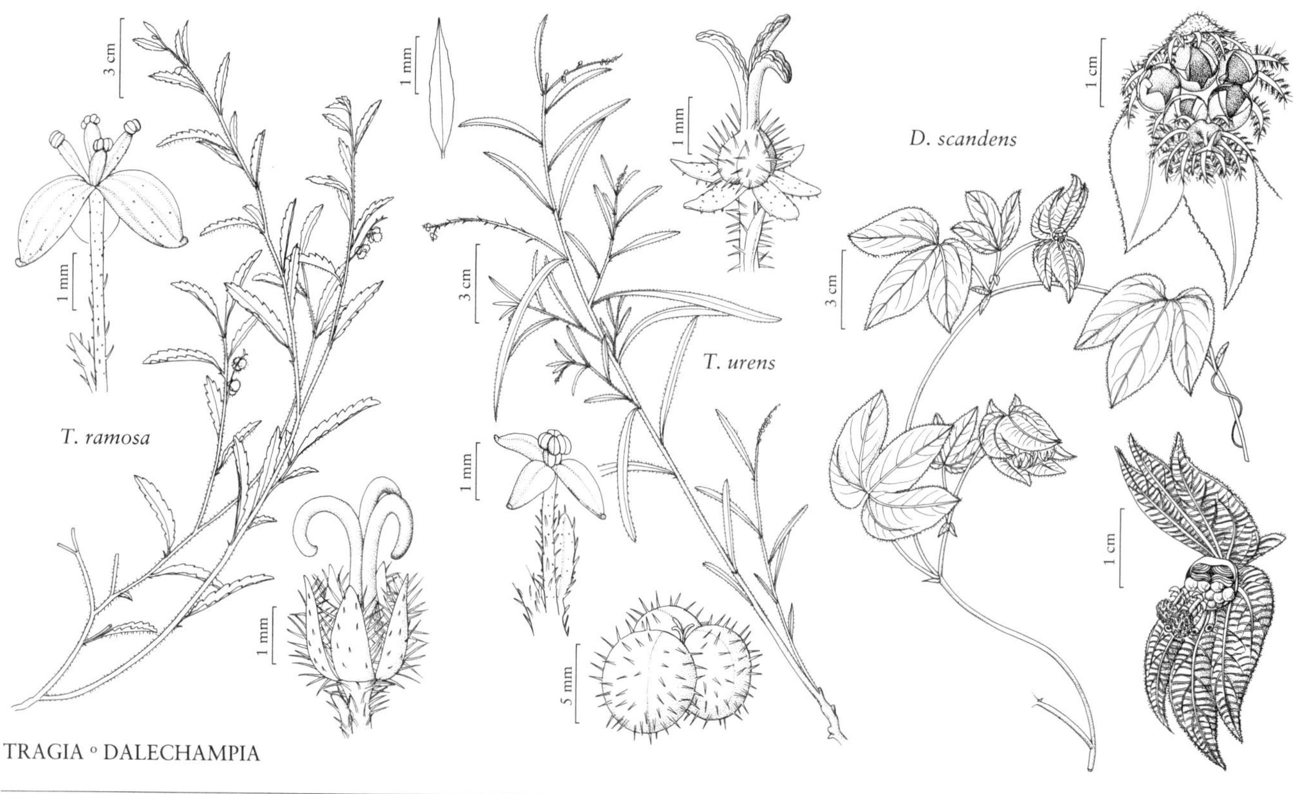

TRAGIA ◦ DALECHAMPIA

Staminate flowers: sepals 3–4, green, 1–1.5 mm; stamens 3–4, filaments 0.4–0.6 mm. **Pistillate flowers:** sepals lanceolate, 1.5–3 mm; styles connate ¼–⅓ length; stigmas undulate. **Capsules** 6–7 mm wide. **Seeds** dark brown with light brown streaks, 2.4–3 mm.

Flowering late winter–fall; fruiting spring–early winter. Dry pinelands and hammocks on limestone substrates; 0–10 m; Fla.

Tragia saxicola occurs in south Florida and the Florida Keys. Although similar to *T. smallii* in its suborbiculate leaf blades, it differs in having longer petioles and smaller seeds.

13. Tragia smallii Shinners, Field & Lab. 24: 37. 1956
• Small's noseburn [E]

Herbs or subshrubs, 1.2–2.5 dm. **Stems** erect, purple-green, apex never flexuous. **Leaves:** petiole 1–4 mm; blade orbiculate to elliptic, 2–5 × 0.8–3 cm, base acute, margins serrate to crenate, apex acute to blunt. **Inflorescences** terminal (often appearing leaf opposed), glands absent, staminate flowers 4–11 per raceme; staminate bracts 0.8–1.2 mm. **Pedicels:** staminate 1.5–1.9 mm, persistent base 0.4–0.6 mm; pistillate 2.8–3.4 mm in fruit. **Staminate flowers:** sepals 4–5, green, 0.9–1.5 mm; stamens 2, filaments 0.2–0.5 mm. **Pistillate flowers:** sepals lanceolate, 1.3–2.3 mm; styles connate ¼ length; stigmas undulate. **Capsules** 9–13 mm wide. **Seeds** dark brown with light brown streaks, 4–4.5 mm. **2*n*** = 44.

Flowering late spring–summer; fruiting summer–fall. Dry, sandy oak-pine forests, prairies, disturbed grasslands; 0–100 m; Ala., Fla., Ga., La., Miss., Tex.

Plants of *Tragia smallii* are easily distinguished from those of most *Tragia* species by the combination of orbiculate to elliptic leaf blades, two stamens, and relatively large seeds. K. I. Miller and G. L. Webster (1967) placed it and *T. urens*, which also has two stamens, in sect. *Leptobotrys*.

Tragia smallii is found on the Gulf Coastal Plain from easternmost Texas to peninsular Florida.

14. Tragia urens Linnaeus, Sp. Pl. ed. 2, 2: 1391. 1763
• Wavyleaf noseburn [E] [F]

Leptobotrys discolor Baillon; *Tragia discolor* (Baillon) Müller Arg.; *T. discolor* var. *linearis* (Michaux) Müller Arg.; *T. discolor* var. *subovalis* (Michaux) Müller Arg.; *T. linearifolia* Elliott; *T. urens* var. *innocua* (Walter) Pax & K. Hoffmann; *T. urens* var. *lanceolata* Michaux; *T. urens* var. *linearis* Michaux; *T. urens* var. *subovalis* Michaux

Herbs or subshrubs, 2–5 dm. **Stems** erect, green to purple-green, apex never flexuous. **Leaves:** petiole

0–2 mm; blade usually oblanceolate to linear, sometimes elliptic, 2–8(–10) × 0.2–1.4 cm, base acute, margins entire or irregularly and shallowly sinuate, apex acute. **Inflorescences** terminal (often appearing leaf opposed), glands absent, staminate flowers 3–45 per raceme; staminate bracts 1–1.5 mm. **Pedicels:** staminate 1.3–2 mm, persistent base 0.3–0.6 mm; pistillate 3.5–4 mm in fruit. **Staminate flowers:** sepals 4–5, green, 1–1.5 mm; stamens 2, filaments 0.2–0.4 mm. **Pistillate flowers:** sepals linear, 1–1.8 mm; styles connate ¼–⅓ length; stigmas undulate. **Capsules** 7–8 mm wide. **Seeds** brown with tan streaks, 3–4 mm. **2*n*** = 44.

Flowering late spring; fruiting summer–fall. Dry, sandy pinelands, oak barrens, disturbed fields; 0–300 m; Ala., Fla., Ga., La., Miss., N.C., S.C., Tex., Va.

Tragia urens is widespread from the Gulf Coast to the mid-Atlantic states and displays considerable foliar variation. Leaf blades that usually are oblanceolate to linear with entire to sinuate margins, two stamens, and sparse, stinging hairs are diagnostic traits for this species.

15. Tragia urticifolia Michaux, Fl. Bor.-Amer. 2: 176. 1803 • Nettle-leaf noseburn [E]

Herbs or subshrubs, 2–7 dm. **Stems** erect, green, apex never flexuous. **Leaves:** petiole 3–15 mm; blade triangular-lanceolate, 2.7–6.7 × 1–3 cm, base truncate to cordate, margins serrate, apex acute. **Inflorescences** terminal (appearing leaf opposed), glands absent, staminate flowers 11–40 per raceme, evenly distributed; staminate bracts 1–1.5 mm. **Pedicels:** staminate 1.5–2 mm, persistent base 1–1.8 mm, longer than subtending bract; pistillate 3–4 mm in fruit. **Staminate flowers:** sepals 3, green, 1.2–2.1 mm; stamens 3, filaments 0.3–0.8 mm. **Pistillate flowers:** sepals lanceolate-ovate, 1.3–2.3 mm; styles connate ⅓ length; stigmas papillate. **Capsules** 7–8 mm wide. **Seeds** dark brown, 3–4 mm. **2*n*** = 44.

Flowering late spring–summer; fruiting summer–fall. Dry, sandy soils, open pine-oak woods, hardwoods, rock ledges, fields; 10–600 m; Ala., Ark., Fla., Ga., La., Miss., N.C., S.C., Tenn., Tex., Va.

Plants of *Tragia urticifolia* are easily distinguished from those of other non-glandular North American members of *Tragia* by the persistent bases of their staminate pedicels, which are long-extended beyond their subtending bracts. Although similar to *T. betonicifolia*, *T. urticifolia* has fewer stems, shorter pistillate sepals, inflorescences with fewer staminate flowers that are not distally clustered, and leaf blades with very light green abaxial surfaces.

9. DALECHAMPIA Linnaeus, Sp. Pl. 2: 1054. 1753; Gen. Pl. ed. 5, 473. 1754 • [For Jacques Daléchamps (or D'Aléchamps), 1513–1588, French surgeon and botanist] [I]

W. Scott Armbruster

Vines [lianas or shrubs], twining [trailing], monoecious [gynodioecious]; hairs unbranched, always some stinging [none stinging, sometimes glandular]; latex absent. **Leaves** deciduous [persistent], alternate, simple [palmately compound]; stipules present, persistent; petiole present, glands absent; stipels present, often with basal glands; blade palmately lobed [unlobed], margins serrulate-crenulate [subentire], laminar glands absent; venation palmate [pinnate]. **Inflorescences** bisexual, apparently axillary (technically terminal on axillary short shoots), pseudanthia (consisting of 2, often showy, involucral bracts subtending 2 cymules, one with [1–]3 pistillate flowers and 1–3 involucellar bractlets, and one with [4–]10[–40] staminate flowers, 1–5 involucellar bractlets, and a series of resiniferous [non-resiniferous] bractlets forming condensed gland; involucral bracts open when flowers receptive, enveloping fruits [deciduous] as seeds develop); glands subtending each bract 0. **Pedicels** present. **Staminate flowers:** sepals 3–6, valvate, distinct; petals 0; nectary absent; stamens [5–]25–35[–90], distinct, borne on elongated [flat, dome-shaped] receptacle, thus appearing connate proximally; pistillode absent. **Pistillate flowers:** sepals [5–]8–12, distinct; petals 0; nectary absent; pistil 3-carpellate; style 1, unbranched. **Fruits** capsules. **Seeds** globose to subglobose; caruncle absent. x = 11, 13.

Species ca. 130 (1 in the flora): introduced, Florida; Mexico, West Indies, Central America, South America, s, se Asia, Africa, Indian Ocean Islands (Madagascar), Pacific Islands (Java, Sumatra).

Some species of *Dalechampia* are cultivated as ornamentals in North America, notably *D. aristolochiifolia* Kunth and *D. spathulata* (Scheidweiler) Baillon. The former is grown outdoors in the southern United States and may naturalize in the future, especially in peninsular Florida, where its specialized pollinator *Euglossa viridissima* has become naturalized (R. W. Pemberton and H. Liu 2008). *Dalechampia scandens* can be distinguished from *D. aristolochiifolia* by the latter's pink bracts and from *D. spathulata* by the latter's monopodial-shrubby habit.

SELECTED REFERENCES Armbruster, W. S. 1985. Patterns of character divergence and the evolution of reproductive ecotypes of *Dalechampia scandens* (Euphorbiaceae). Evolution 39: 733–752. Pemberton, R. W. and H. Liu. 2008b. Naturalization of *Dalechampia scandens* in southern Florida. Caribbean J. Sci. 44: 417–419. Webster, G. L. and W. S. Armbruster. 1991. A synopsis of the neotropical species of *Dalechampia*. Bot. J. Linn. Soc. 105: 137–177.

1. **Dalechampia scandens** Linnaeus, Sp. Pl. 2: 1054. 1753 • Spurgecreeper [F] [I]

Stems 1–5 m, hirsute, with stinging hairs. **Leaves:** stipules lanceolate to ovate, 3–8 × 2 mm; petiole 2–8 cm; blade ovate, 3–10 × 3–13 cm, usually deeply 3-lobed, base shallowly to deeply cordate, margins weakly serrulate-crenulate, apex acute or acuminate. **Inflorescences:** peduncle of inflorescence 2–6 cm; involucral bracts greenish white at anthesis, broadly ovate, 1.5–3 × 1.5–3 cm, deeply 3-lobed, with 5 major veins; peduncle of staminate cymule 1–10 mm; pistillate cymule nearly sessile. **Pistillate flowers:** sepals pinnatifid, with stiff, detaching hairs and, often, stalked, resiniferous glands; styles 4–8 mm; stigma 0.5 mm diam. **Capsules** glabrescent to finely pubescent. **Seeds** 3.5–4 mm, smooth.

Flowering and fruiting year-round, especially at end of summer/rainy season. Dry subtropical forests and scrub; 0–10 m; introduced; Fla.; Mexico; West Indies; Central America; South America.

Dalechampia scandens is widespread, variable (W. S. Armbruster 1985), and probably a complex of cryptic species (C. Pélabon et al. 2005). The species is naturalized in Broward and Hillsborough counties; the geographic origin of these populations is uncertain (see R. W. Pemberton and H. Liu 2008b).

10. MANIHOT Miller, Gard. Dict. Abr. ed. 4, vol. 2. 1754 • Cassava, yuca, manioc [From Brazilian vernacular name *mani oca*, wood spirit root, alluding to use]

W. John Hayden

Herbs, subshrubs, shrubs, or trees, perennial, unarmed, usually monoecious, rarely dioecious; hairs unbranched or absent; latex white. **Leaves** persistent or deciduous, alternate, simple [palmately compound]; stipules present, deciduous; petiole present [rudimentary], glands absent; stipels present; blade usually palmately lobed, rarely unlobed, lobes undivided or secondarily lobed, margins entire, repand, or serrate, laminar glands absent; venation palmate (pinnate in lobes). **Inflorescences** bisexual (pistillate flowers proximal, staminate distal), terminal or axillary, racemes or panicles; glands subtending each bract 0. **Pedicels** present, pistillate often elongating in fruit. **Staminate flowers:** sepals 5, petaloid, 7–20 mm, valvate, connate ½ length; petals 0; nectary intrastaminal, cushion-shaped, lobed; stamens (6–8)–10, in 2 whorls, distinct;

pistillode absent. **Pistillate flowers:** sepals 5, petaloid, distinct; petals 0; nectary annular, lobed or unlobed; pistil 3-carpellate; styles 3, connate basally, unbranched, flabellate, prominently papillate. **Fruits** capsules. **Seeds** globose to oblong; caruncle present. $x = 9$.

Species ca. 100 (6 in the flora): s United States, Mexico, Central America, South America.

Manihot is one of the most economically important members of Euphorbiaceae, primarily because of the starchy food-bearing roots of *M. esculenta*, now cultivated throughout the tropics. Also, *M. glaziovii* Müller Arg., from northeastern Brazil, was once an important source of Ceará rubber. *Manihot* appears to be most closely related to *Cnidoscolus*, a conclusion supported by morphological (G. L. Webster 1994) and DNA sequence data (K. Wurdack et al. 2005). Four species of sect. *Parvibracteatae* Pax, as defined by D. J. Rogers and S. G. Appan (1973), barely extend across the borders of Arizona and Texas from Mexico. In addition, two species are naturalized in the southeastern United States.

Leaf blade lobe characters (length, outline) are best developed in the median and immediately adjacent lobes; lateral lobes are progressively smaller and tend to have simpler outlines with distance from the median lobe.

SELECTED REFERENCES Croizat, L. 1942. A study of *Manihot* in North America. J. Arnold Arbor. 23: 216–225. Rogers, D. J. 1963. Studies of *Manihot esculenta* Crantz and related species. Bull. Torrey Bot. Club 90: 43–54. Rogers, D. J. and S. G. Appan. 1973. *Manihot, Manihotoides* (Euphorbiaceae). In: Organization for Flora Neotropica. 1968+. Flora neotropica. 110+ nos. New York. No. 13. Webster, G. L. 1994. Synopsis of the genera and suprageneric taxa of Euphorbiaceae. Ann. Missouri Bot. Gard. 81: 33–144. Wurdack, K., P. Hoffman, and M. W. Chase. 2005. Molecular phylogenetic analysis of uniovulate Euphorbiaceae (Euphorbiaceae sensu stricto) using plastid *rbc*L and *trn*L-F DNA sequences. Amer. J. Bot. 92: 1397–1420.

1. Shrubs or trees; inflorescences panicles; Gulf and Atlantic coastal plains.
   2. Leaf blade lobes without secondary lobes; stipules lanceolate, entire; stem nodes conspicuously swollen; leaf and stipule scars elevated, especially on older stems; capsules usually winged. . . . . 3. *Manihot esculenta*
   2. Median and adjacent leaf blade lobes with secondary lobes, lateral lobes without secondary lobes; stipules linear, remotely serrate; stem nodes not swollen; leaf and stipule scars not elevated; capsules not winged . . . . 4. *Manihot grahamii*
1. Herbs or subshrubs; inflorescences racemes; Texas and Arizona.
   3. Leaf blade secondary lobes acute, proximal; leaf blade margins remotely serrate.
      4. Leaf blade margins neither thickened nor revolute; inflorescences axillary; capsules finely tuberculate; s Arizona and adjacent Mexico . . . . 1. *Manihot angustiloba*
      4. Leaf blade margins thickened and revolute; inflorescences terminal; capsules smooth; s Texas and adjacent Mexico. . . . . 5. *Manihot subspicata*
   3. Leaf blade secondary lobes rounded, distal or distal and proximal to middle; leaf blade margins entire.
      5. Leaf blades basally attached; staminate calyces campanulate; capsules nearly smooth; s Arizona and adjacent Mexico . . . . 2. *Manihot davisiae*
      5. Leaf blades peltate; staminate calyces tubular, midsection constricted; capsules verrucose-rugose; s Texas . . . . 6. *Manihot walkerae*

1. Manihot angustiloba (Torrey) Müller Arg. in A. P. de Candolle and A. L. P. P. de Candolle, Prodr. 15(2): 1073. 1866 • Desert mountain manihot, narrow-leaved cassava, pata de gallo

Janipha manihot (Linnaeus) Kunth var. *angustiloba* Torrey in W. H. Emory, Rep. U.S. Mex. Bound. 2(1): 199. 1859

Herbs or subshrubs, 1–3 m. **Roots** thickened. **Stems** erect, terete when young; nodes not swollen; leaf and stipule scars not elevated. **Leaves:** stipules lanceolate, entire; petiole 3–12 cm; blade basally attached, 5–7-lobed, lobes with acute secondary lobes proximally, median lobe 5–15 cm, margins neither thickened nor revolute, remotely serrate, apex acute, surfaces glabrous, abaxial smooth. **Inflorescences** axillary, racemes, to 12 cm. **Pedicels:** staminate 3–8 mm; pistillate 10–25 mm in fruit, downcurved. **Staminate flowers:** calyx campanulate, 10–18 mm, lobes erect or spreading; stamens 10. **Capsules** 1.5 cm, finely tuberculate, not winged. **Seeds** globose, 12 mm.

Flowering Jul–Oct. Desert scrub, thorn scrub, oak woodlands, oak grasslands; 30–2000 m; Ariz.; Mexico (Baja California Sur, Chihuahua, Sinaloa, Sonora).

D. J. Rogers and S. G. Appan (1973) noted the overall similarity and nearly identical geographic ranges of *Manihot angustiloba* and *M. davisiae*, yet they maintained these taxa as separate species, presumably because of their (nearly) constant and consistent differences in leaf lobe outline. *Manihot angustiloba* has generally narrow, nearly linear, primary lobes with a pair of serrate secondary lobes forming the widest portion of the lobe proximal to the middle; *M. davisiae* has generally broader leaf lobes with one pair of rounded secondary lobes that form the widest portion of the lobe distal to the middle, or two pairs of nearly equal, rounded, secondary lobes proximal and distal to the middle. Some specimens exhibit an intermediate condition: primary lobes that are narrow distally but also bear a pair of rounded secondary lobes proximal to the middle.

In the flora area, *Manihot angustiloba* is restricted to Cochise, Pima, and Santa Cruz counties.

2. Manihot davisiae Croizat, J. Arnold Arbor. 23: 224. 1942 • Arizona manihot, pata de gallo

Herbs or subshrubs, 1–3 m. **Roots** thickened. **Stems** erect, terete when young; nodes not swollen; leaf and stipule scars not elevated. **Leaves:** stipules lanceolate, entire; petiole 2–9 cm; blade basally attached, 5–7-lobed, lobes with rounded secondary lobes distal to middle or with 2 pairs of rounded secondary lobes of nearly equal width proximal and distal to middle, median lobe 4–11 cm, margins neither thickened nor revolute, entire, apex acuminate, surfaces glabrous, abaxial smooth. **Inflorescences** axillary, racemes, to 8 cm. **Pedicels:** staminate 4–9 mm; pistillate to 15 mm in fruit, downcurved. **Staminate flowers:** calyx campanulate, 7–15 mm, lobes erect or spreading; stamens 10. **Capsules** 1.5 cm, nearly smooth, not winged. **Seeds** globose, 12 mm.

Flowering Jul–Oct. Desert scrub, thorn scrub, oak woodlands, oak grasslands; 100–2000 m; Ariz.; Mexico (Chihuahua, Sinaloa, Sonora).

In the flora area, *Manihot davisiae* is limited to Pima and Santa Cruz counties.

3. Manihot esculenta Crantz, Inst. Rei Herb. 1: 167. 1766 [I]

Shrubs, 1–4 m. **Roots** thickened. **Stems** erect, terete when young; nodes conspicuously swollen; leaf and stipule scars elevated, especially on older stems. **Leaves** persistent; stipules lanceolate, entire; petiole 3–20 cm; blade basally attached, usually 3–10-lobed, sometimes unlobed, lobes without secondary lobes, median lobe 5–18 cm, margins neither thickened nor revolute, entire to ± repand, apex acuminate, surfaces glabrous or hairy, abaxial finely reticulate. **Inflorescences** axillary, panicles, 2–10 cm. **Pedicels:** staminate 2–4 mm; pistillate 20 mm in fruit, straight. **Staminate flowers:** calyx campanulate, 10–15 mm, lobes erect or spreading; stamens 10. **Capsules** 1.5 cm, usually winged. **Seeds** subglobose to oblong, 12 mm. $2n = 36$.

Flowering year-round, mostly fall and winter. Disturbed areas, spreading from cultivation; 0–200 m; introduced; Ala., Fla., Tex.; South America (Brazil); introduced widely in tropical and subtropical regions worldwide.

The enlarged storage roots of *Manihot esculenta* yield a starchy staple, now much consumed in tropical regions around the world. Tapioca, a pelletized and

MANIHOT ◦ CNIDOSCOLUS ◦ JATROPHA

partially hydrolyzed form of cassava starch, is the chief form of consumption in temperate regions. Multiple cultivars are known. These are generally characterized as bitter (containing cyanogenic glycosides, which must be removed before consumption) or sweet (cyanogenic glycosides absent or at low levels). A form with variegated leaves is sometimes grown for ornament. Cassava was cultivated throughout the Neotropics in pre-Columbian times. Because cassava is a root crop with poor storage qualities adapted to humid regions, archeological remains are few, leading to much speculation in the literature about the origin of this important crop. Molecular data reported by K. Olsen and B. A. Schaal (1999, 2001) indicate that cultivated cassava constitutes subsp. *esculenta*, derived by artificial selection from its sole wild ancestor, *M. esculenta* subsp. *flabellifolia* (Pohl) Ciferri from the southern border of the Amazon basin. Under this classification, all North American plants belong to subsp. *esculenta*.

SELECTED REFERENCES Allem, A. C. 1994. The origin of *Manihot esculenta* Crantz (Euphorbiaceae). Genet. Resources Crop Evol. 41: 133–150. Olsen, K. and B. A. Schaal. 1999. Evidence on the origin of cassava: Phylogeography of *Manihot esculenta*. Proc. Natl. Acad. Sci. U.S.A. 96: 5586–5591. Olsen, K. and B. A. Schaal. 2001. Microsatellite variation in cassava (*Manihot esculenta*, Euphorbiaceae) and its wild relatives: Further evidence for a southern Amazonian origin of domestication. Amer. J. Bot. 88: 131–142. Rogers, D. J. 1965. Some botanical and ethnological considerations of *Manihot esculenta*. Econ. Bot. 19: 369–377.

4. **Manihot grahamii** Hooker, Icon. Pl. 6: plate 530. 1843 (as grahami) • Hardy tapioca, Graham's manihot or cassava [F] [I]

Shrubs or trees, 2–6[–7] m. **Roots** not thickened. **Stems** erect, angled when young; nodes not swollen; leaf and stipule scars not elevated. **Leaves** deciduous; stipules linear, remotely serrate; petiole 5–33 cm; blade basally attached, 5–13-lobed, median and adjacent lobes with pair of weakly defined rounded secondary lobes distal to middle, lateral lobes without secondary lobes, median lobe 5–24 cm, margins neither thickened nor revolute, entire, apex acuminate, surfaces glabrous, abaxial smooth. **Inflorescences** axillary, panicles, to 30 cm. **Pedicels:** staminate 4–10 mm; pistillate 10–40 mm in fruit, straight. **Staminate flowers:** calyx campanulate, 10–15 mm, lobes erect or spreading; stamens 10. **Capsules** 1.8 cm, smooth, not winged. **Seeds** oblong, 10–12 mm.

Flowering Apr–Aug; fruiting Jun–Sep. Disturbed areas, spreading from cultivation; 0–600 m; introduced; Ala., Ark., Fla., Ga., La., Miss., Tex.; South America.

Manihot grahamii is native to northern Argentina, southeastern Brazil, Paraguay, and Uruguay, and is sometimes cultivated for its distinctive, attractive foliage. The flowers are relatively inconspicuous, but are much-visited by bees. This is the most cold-tolerant *Manihot* species; above-ground stems survive light frosts, and if severe cold kills the aerial shoot system outright, new stems can regenerate from underground parts. It survives well and self-sows in garden settings as far north as tidewater Virginia; northern limits for the persistence of plants escaping from cultivation have yet to be established. In addition to characteristics noted in the key, herbarium specimens frequently exhibit contracted petiole bases.

5. **Manihot subspicata** D. J. Rogers & Appan in Organization for Flora Neotropica, Fl. Neotrop. 13: 62, figs. 19D, 20A–C. 1973 • Spiked manihot, palo mulato

Herbs or subshrubs, to 1 m. **Roots** thickened. **Stems** lax (often leaning on other vegetation), terete when young; nodes not swollen; leaf and stipule scars not elevated. **Leaves:** stipules lanceolate, entire; petiole 2–10 cm; blade basally attached to subpeltate, 5-lobed, lobes with acute secondary lobes near base, median lobe 2–10 cm, margins thickened and revolute, remotely serrate, apex acute to acuminate (bristle-tipped), surfaces glabrous, abaxial smooth. **Inflorescences** terminal, racemes, 25 cm. **Pedicels:** staminate 1–5 mm; pistillate 10–20 mm in fruit, downcurved. **Staminate flowers:** calyx campanulate to conic, 8–13 mm, lobes reflexed; stamens 10. **Capsules** 1.5 cm, smooth, not winged. **Seeds** oblong, 10 mm.

Flowering Jun–Aug. Savannas and grasslands with scattered shrubs and trees; 30–60 m; Tex.; Mexico (Coahuila, Nuevo León, Tamaulipas).

Although D. J. Rogers and S. G. Appan (1973) characterized leaf blades of *Manihot subspicata* as peltate, seldom are leaves unambiguously so. Typically, just the thickened margins of lateral lobes are confluent across the distal end of the petiole. Though relatively common in northern Mexico, *M. subspicata* is known in Texas only from the vicinity of Lake Corpus Christi (Jim Wells and Live Oak counties); whether it is native or introduced there is unresolved. In Mexico, *M. subspicata* appears to be tolerant of disturbance, frequently colonizing roadsides and similar habitats.

6. **Manihot walkerae** Croizat, Bull. Torrey Bot. Club 69: 452. 1942 • Walker's manihot, Texas tapioca [C]

Herbs or subshrubs, to 1.5 m. **Roots** thickened. **Stems** decumbent to ascending (often growing through associated vegetation), terete when young; nodes not swollen; leaf and stipule scars not elevated. **Leaves:** stipules lanceolate, entire; petiole 1–7 cm; blade peltate, 3–5-lobed, lobes with rounded secondary lobes distal to middle or with 2 pairs of rounded secondary lobes of nearly equal width proximal and distal to middle, median lobe 2–7 cm, margins neither thickened nor revolute, entire, apex cuspidate, surfaces glabrous, abaxial smooth. **Inflorescences** axillary, subspicate racemes, 5–10 cm. **Pedicels:** staminate 1–3 mm; pistillate 15 mm in fruit, downcurved. **Staminate flowers:** calyx tubular, base gibbous, midsection constricted, 10–20 mm, lobes erect or spreading; stamens 6–8. **Capsules** 1 cm, verrucose-rugose, not winged. **Seeds** globose, 8–9 mm.

Flowering Apr–Sep, following rains. Shrublands and grasslands; of conservation concern; 20–200 m; Tex.; Mexico (Tamaulipas).

Manihot walkerae is a globally endangered species known from Duval, Hidalgo, and Starr counties and nearby Tamaulipas, Mexico. It is restricted to areas of sandy or gravelly calcareous soils overlying caliche or limestone bedrock. An estimated 95 percent of its habitat in the United States portion of the lower Rio Grande Valley has been converted to largely agricultural uses (www.natureserve.org).

The stamen number (six to eight) of *Manihot walkerae* is notable relative to that of other species in the genus, which typically have ten stamens per staminate flower.

Manihot walkerae is in the Center for Plant Conservation's National Collection of Endangered Plants.

11. CNIDOSCOLUS Pohl, Pl. Bras. Icon. Descr. 1: 56, plates 49–52. 1827 • Bull-nettle, mala mujer [Greek *cnide*, nettle, and *skolos*, thorn, alluding to stinging hairs]

Geoffrey A. Levin

Herbs [shrubs or trees], perennial, monoecious; hairs unbranched, always some stinging; latex white. **Leaves** [persistent or deciduous], alternate, simple; stipules present, deciduous [persistent]; petiole present, glands present at apex; blade palmately lobed [unlobed], margins entire or dentate, laminar glands absent; venation palmate (pinnate in lobes) [pinnate].

Inflorescences bisexual (pistillate flowers central, staminate lateral), terminal, dichasial cymes; glands subtending each bract 0. **Pedicels** present. **Staminate flowers:** sepals 5, white, petaloid, imbricate, connate ½ length; petals 0; nectary intrastaminal, annular; stamens [8–]10[–25] in 2[–6] whorls, outer whorl distinct or connate proximally, inner whorl connate proximally; staminodes absent or present at apex of staminal column; pistillode absent. **Pistillate flowers:** sepals 5, white, petaloid, distinct [connate proximally]; petals 0; nectary annular; staminodes often present; pistil 3-carpellate; styles 3, connate basally, [1–]2–3 times 2-fid. **Fruits** capsules. **Seeds** ovoid; caruncle present. $x = 9$.

Species ca. 50 (3 in the flora): s United States, Mexico, West Indies, Central America, South America.

The characteristic stinging hairs of *Cnidoscolus* immediately distinguish it from *Manihot*, which appears to be its closest relative, as shown by morphological (G. L. Webster 1994) and DNA sequence data (K. Wurdack et al. 2005). Although the chemistry of the stinging hairs has been little studied, those of *C. texanus* have been shown to contain serotonin but not histamine and acetylcholine as in the stinging hairs of some true nettles in Urticaceae (S. E. Lookadoo and A. J. Pollard 1991). The roots of *Cnidoscolus*, like those of *Manihot*, are rich in edible starch; the seeds also are edible. The leaves of two tropical species, *C. aconitifolius* (Miller) I. M. Johnston and *C. chayamansa* McVaugh, are eaten as a vegetable.

1. Leaf blades moderately lobed, lobes ⅓–½ blade length, lobe apices and marginal teeth acuminate, aristate; staminate flower calyx tubes 4–6 mm; stamens subequal, filaments all connate most of length . . . . 1. *Cnidoscolus angustidens*
1. Leaf blades deeply lobed, lobes (½–)⅗–⁹⁄₁₀ blade length, lobe apices and marginal teeth acute to obtuse, not aristate; staminate flower calyx tubes 8–17 mm; stamens of outer whorl shorter than inner whorl, filaments of outer whorl distinct or connate basally, of inner whorl connate most of length.
   2. Staminate flower calyx tubes 8–11 mm, distally straight or constricted, without stinging hairs, lobes 7–10 mm; capsules 10–12 mm; seeds 8–9 mm. . . . . 2. *Cnidoscolus stimulosus*
   2. Staminate flower calyx tubes 12–17 mm, distally flaring, with stinging hairs, lobes 10–17 mm; capsules 15–20 mm; seeds 14–18 mm . . . . .3. *Cnidoscolus texanus*

1. Cnidoscolus angustidens Torrey in W. H. Emory, Rep. U.S. Mex. Bound. 2(1): 198. 1859

Plants 15–100 cm. **Leaves:** stipules 5–6 mm, margins deeply toothed; petiole 2–10 cm; blade ± round in outline, 8–15 cm diam., moderately lobed, lobes ⅓–½ blade length, base broadly cordate, margins dentate, teeth and lobe apices acuminate, aristate. **Staminate flowers:** calyx funnel-shaped, tube 4–6 mm, distally flaring, stinging hairs absent, lobes 4–8 mm; stamens subequal, filaments all connate most of length; staminodes 3, threadlike. **Pistillate flowers:** sepals 10–12 mm; stigmas 12. **Capsules** 10–12 mm. **Seeds** brown, sometimes mottled, 9–11 mm.

Flowering May–Sep. Grasslands, desert scrub, oak woodlands; 1100–1600 m; Ariz.; Mexico (Baja California Sur, Sinaloa, Sonora).

In the flora area, *Cnidoscolus angustidens* is restricted to southern Arizona.

2. Cnidoscolus stimulosus (Michaux) Engelmann & A. Gray, Boston J. Nat. Hist. 5: 234. 1845 • Tread softly, finger rot [E] [F] [W]

Jatropha stimulosa Michaux, Fl. Bor.-Amer. 2: 216. 1803; *Bivonea stimulosa* (Michaux) Rafinesque; *Cnidoscolus urens* (Linnaeus) Arthur var. *stimulosus* (Michaux) Govaerts

Plants 10–120 cm. **Leaves:** stipules 2–3.5 mm, margins entire; petiole 3–8 cm; blade ovate to round in outline, 5–17 × 4–12 cm, deeply lobed, lobes (½–)¾–⁹⁄₁₀ blade length, base broadly cordate to truncate, margins usually dentate, rarely entire, teeth and lobe apices acute to obtuse, not aristate. **Staminate flowers:** calyx salverform, tube 8–11 mm, distally straight or constricted, stinging hairs absent, lobes 7–10 mm; stamens of outer whorl shorter than inner, filaments of outer whorl distinct, of inner whorl connate most of length; staminodes 0. **Pistillate flowers:** sepals 10–15 mm; stigmas 12–24. **Capsules** 10–12 mm. **Seeds** brown, sometimes mottled, 8–9 mm. $2n = 36$.

Flowering Mar–Aug. Sandhills, dry sandy woods, sandy old fields; 0–600 m; Ala., Fla., Ga., Ky., La., Miss., N.C., S.C., Va.

In Kentucky, *Cnidoscolus stimulosus* is occasionally naturalized along railroads. Although closely related to *C. urens* of Mexico, Central America, and South America, *C. stimulosus* differs in habit, leaf pubescence, and seed shape, and the two are treated here as distinct species.

3. **Cnidoscolus texanus** (Müller Arg.) Small, Fl. S.E. U.S., 706. 1903 • Texas bull-nettle [W]

Jatropha texana Müller Arg., Linnaea 34: 211. 1865

Plants (30–)40–50(–100) cm. **Leaves:** stipules 3–4 mm, margins usually deeply toothed, rarely entire; petiole 5–18 cm; blade ± round in outline, 6–15 cm diam., deeply lobed, lobes 3/5–4/5 blade length, base broadly cordate, margins dentate, teeth and lobe apices acute, not aristate. **Staminate flowers:** calyx funnel-shaped, tube 12–17 mm, distally flaring, stinging hairs present, lobes 10–17 mm; stamens of outer whorl shorter than inner, filaments of outer whorl distinct or connate basally, of inner whorl connate most of length; staminodes 0. **Pistillate flowers:** sepals 15–25 mm; stigmas 12–24. **Capsules** 15–20 mm. **Seeds** brown, sometimes mottled, 14–18 mm. $2n = 36$.

Flowering mainly Apr–Jul(–Nov). Sandy open woods, fields, disturbed areas; 0–900 m; Ark., La., Okla., Tex.; Mexico (Tamaulipas).

Both *Cnidoscolus texanus* and *C. stimulosus* are found in Louisiana, but their distributions are separated by about 250 km, with *C. texanus* found in the western half of the state and *C. stimulosus* restricted to St. Tammany and Washington parishes in the east.

12. JATROPHA Linnaeus, Sp. Pl. 2: 1006. 1753; Gen. Pl. ed. 5, 437. 1754 • [Greek *iatros*, physician, and *trophe*, food, alluding to use of *J. curcas* as purgative]

Bijan Dehgan

Herbs, subshrubs, shrubs, or trees, perennial, monoecious or dioecious [gynodioecious]; hairs unbranched, sometimes glandular, or absent; latex colorless, cloudy-whitish, yellow, or red. **Leaves** deciduous or persistent, alternate but sometimes appearing fascicled, simple; stipules absent or present, persistent or deciduous; petiole absent or present, glands absent at apex, sometimes stipitate-glandular along length; blade unlobed or palmately lobed, margins entire, serrate, or dentate, laminar glands absent; venation pinnate or palmate. **Inflorescences** unisexual or bisexual (pistillate flowers central, staminate lateral), axillary or terminal, cymes or fascicles, or flowers solitary; glands subtending each bract 0. **Pedicels** present. **Staminate flowers:** sepals 5, imbricate, distinct or connate to 1/2 length; petals 5, distinct or connate basally to most of length, white, greenish yellow, pink, red, or purple [yellow, yellow-brown, orange, or 2-colored]; nectary extrastaminal, annular and 5-lobed or of 5 glands; stamens [6–]8 or 10 in 1–2 whorls, distinct or connate basally to most of length; pistillode absent. **Pistillate flowers:** sepals 5, imbricate, distinct or connate to 1/2 length; petals 5, distinct or connate basally to most of length, white, greenish yellow, pink, red, or purple [yellow, yellow-brown, orange, or 2-colored]; nectary annular and 5-lobed or 5 glands; staminodes sometimes present; pistil 1–3-carpellate; styles (1–)3, distinct or connate basally to most of length [absent], 2-fid. **Fruits** capsules, ± fleshy, sometimes tardily dehiscent. **Seeds** ellipsoid to globose; caruncle present (sometimes rudimentary) or absent. $x = 11$.

Species ca. 190 (10 in the flora): s United States, Mexico, West Indies, Central America, South America, s Asia (India), Africa; introduced elsewhere in Asia, Pacific Islands, Australia; tropical and subtropical regions.

Some species of *Jatropha* are cultivated as ornamentals throughout the tropical and subtropical regions of the world, notably *J. integerrima*, *J. multifida*, and *J. podagrica* Hooker. These and *J. curcas* and *J. gossypiifolia* have escaped from cultivation in subtropical regions. *Jatropha curcas* (physic nut), which probably originated in Central America, is now pantropical and is extensively cultivated for production of biodiesel from its seeds, which are also eaten as roasted nuts and used as a purgative and for other medicinal purposes. More than 50 New World species are known from cultivation in the United States, either as ornamentals or for medicinal purposes, many of which are being studied. Some African species are in cultivation, primarily by collectors of succulent plants.

SELECTED REFERENCES Dehgan, B. 2012. *Jatropha.* In: Organization for Flora Neotropica. 1968+. Flora Neotropica. 110+ nos. New York. No. 110. Dehgan, B. and B. Schutzman. 1994. Contributions toward a monograph of neotropical *Jatropha*: Phenetic and phylogenetic analysis. Ann. Missouri Bot. Gard. 81: 349–367. Dehgan, B. and G. L. Webster. 1979. Morphology and infrageneric relationships of the genus *Jatropha* (Euphorbiaceae). Univ. Calif. Publ. Bot. 74: 1–73. McVaugh, R. 1945. The genus *Jatropha* in America: Principal intergeneric groups. Bull. Torrey Bot. Club 72: 271–294.

1. Perennial herbs or rhizomatous subshrubs, to 1 m; stems herbaceous or rubbery-succulent.
   2. Perennial herbs with subterranean caudices, to 0.5 m; stems green; plants monoecious; carpels 3.
      3. Caudices woody, stem scars crescent-shaped; leaf blades lobed nearly to base; corollas deep red; stamens distinct at maturity; Texas. 2. *Jatropha cathartica*
      3. Caudices ± fleshy, stem scars round; leaf blades lobed to middle; corollas light pink; stamens: outer 5 distinct, inner 3 connate to ½ length; Arizona, New Mexico, Texas . . . . 5. *Jatropha macrorhiza*
   2. Rhizomatous subshrubs, 0.5–1 m; stems reddish brown; plants dioecious; carpel 1.
      4. Petioles 1–2.5 cm; leaf blades widely ovate-deltate, 1.5–2.6 cm wide, unlobed, margins sinuate to weakly serrate-crenate; Arizona . . . . 9. *Jatropha cardiophylla*
      4. Petioles 0–0.2 cm; leaf blades linear-spatulate to narrowly obovate, 0.2–0.7 cm wide, sometimes 3-lobed, margins entire; Texas . . . . 10. *Jatropha dioica*
1. Shrubs or trees, 1–10 m; stems woody or woody-succulent.
   5. Short shoots present; plants dioecious; corollas usually white, sometimes pinkish, petals connate ½+ length; Arizona.
      6. Latex cloudy-whitish; leaves mostly ± evenly distributed on long shoots, rarely on short shoots, petioles 1.3–3 cm, blades cordate to broadly ovate, 2.2–3.5 × 1.6–3 cm, canescent abaxially, sparingly hairy adaxially; corollas subglobose-urceolate; carpels 2 . . . . 7. *Jatropha canescens*
      6. Latex yellow in young shoots, red in older shoots; leaves usually fascicled on short shoots, petioles 0(–0.2) cm, blades obovate-spatulate, 0.7–1.9 × 0.3–0.9 cm, glabrous; corollas tubular-urceolate; carpel 1 . . . . 8. *Jatropha cuneata*
   5. Short shoots absent; plants monoecious; corollas greenish yellow, pink, red, orange, or purple, petals distinct or connate to ½ length; Florida.
      7. Stipules persistent, filiform-divided; stamens 8; styles distinct or connate to ¼ length.
         8. Stipules, petioles, and/or leaf margins glandular; leaf blades 3–5-lobed; petals connate ¼–½ length . . . . 1. *Jatropha gossypiifolia*
         8. Stipules, petioles, and leaf margins not glandular; leaf blades 9–11-lobed; petals distinct . . . . 3. *Jatropha multifida*
      7. Stipules caducous (narrowly lanceolate) or absent; stamens 10; styles connate ½+ length.
         9. Shrubs, 2.5–5 m; corollas rotate, bright red to scarlet or pink; capsules explosively dehiscent . . . . 4. *Jatropha integerrima*
         9. Trees, to 10 m; corollas campanulate, greenish yellow; capsules drupaceous. . . 6. *Jatropha curcas*

1. Jatropha gossypiifolia Linnaeus, Sp. Pl. 2: 1006. 1753 (as gossypifolia) • Bellyache bush [I] [W]

Shrubs, to 3 m, monoecious. **Stems** erect, brown, sparsely to much-branched, woody-succulent, hirsute, glandular when young; short shoots absent; latex viscous, colorless. **Leaves** persistent or drought-deciduous, ± evenly distributed on long shoots; stipules persistent, 2.5–12 mm, filiform-divided, each segment ending in stipitate gland; petiole 3–14.5 cm, stipitate-glandular; blade cordate to ovate in outline, 4–18.2 × 4.2–13.4 cm, 3–5-lobed, base cordate, margins usually serrulate-denticulate or glandular-ciliate, rarely entire, apex acuminate, membranous, surfaces glabrous or sparsely hairy especially on veins; venation palmate. **Inflorescences** bisexual, terminal and subterminal, cymes; peduncle 2.5–10.5 cm; bracts 6–16 mm, margins entire, glandular-ciliate. **Pedicels** 1–2 mm. **Staminate flowers:** sepals distinct or connate to 1/4 length, lanceolate to ovate-lanceolate, 1.2–2.5 × 2.5–4 mm, margins entire, apex round, surfaces glabrous or sparsely hairy, glandular-ciliate; corolla orange-red to purple, sometimes with lighter center, campanulate, petals connate 1/4–1/2 length, 3.5–5.5 × 1.8–3 mm, glabrous or sparsely hairy on 1 or both surfaces; stamens 8 in 2 whorls (5 + 3); filaments of each whorl connate 1/4–1/2+ length, outer whorl 1.4–3 mm, inner whorl 1.8–4 mm. **Pistillate flowers** resembling staminate, but sepals 2.5–4 × 1–1.7 mm; petals 4–6.5 × 2–3.5 cm; staminodes sometimes present; carpels 3; styles connate to 1/4 length, 1–2 mm. **Capsules** ellipsoidal, 1–1.2 × 0.8–1 cm, explosively dehiscent. **Seeds** gray-brown mottled with dark brown spots, ovoid, 6.5–7 × 3.8–4.5 mm; caruncle prominent. $2n$ = 22 (Mexico).

Flowering and fruiting year-round. Disturbed sites; 0–50 m; introduced; Fla.; Mexico; West Indies; Central America; South America; introduced also in Asia, Africa, Pacific Islands, Australia.

Jatropha gossypiifolia is native to tropical America and has been introduced throughout the tropics, including southern Florida, and in some regions it is invasive; it is widely cultivated for medicinal and landscape purposes. It is a complex species with more than 40 described varieties, subspecies, and forms, some of which are sometimes considered distinct species.

2. Jatropha cathartica Terán & Berlandier in J. L. Berlandier, Mem. Comis. Limites, 9. 1832 • Berlandier's nettlespurge

Jatropha berlandieri Torrey

Herbs, perennial, to 0.3 m, monoecious, with woody subterranean caudex to 13.5 cm diam., stem and root scars crescent-shaped. **Stems** erect, green, sparsely branched, herbaceous, somewhat succulent, glabrous; short shoots absent; latex watery, cloudy-whitish. **Leaves** ± evenly distributed on long shoots; stipules persistent, 2–3 mm, deeply divided into linear-lanceolate segments; petiole 6–10 cm, not stipitate-glandular; blade cordate in outline, 6–11 × 3.5–6 cm, deeply 5–7-lobed nearly to base, segments deeply lobed, base cordate, margins coarsely dentate, apex acuminate, membranous, surfaces glabrous; venation palmate. **Inflorescences** bisexual, terminal and subterminal, cymes; peduncle 7.5–11 cm; bracts 3–10 mm, margins entire, glabrous. **Pedicels** 9–13 mm. **Staminate flowers:** sepals connate 1/2 length, ovate, 2–2.5 × 1–1.2 mm, margins usually entire, rarely 1–2-lobed, apex acute, surfaces glabrous; corolla deep red, rotate-campanulate, petals distinct, 7–10 × 2–3 mm, surfaces glabrous; stamens 8 in 2 whorls (5 + 3); filaments of each whorl appearing connate 1/2 length at anthesis, distinct at maturity, both whorls 2.5–3 mm. **Pistillate flowers** resembling staminate, but sepals distinct, lanceolate, 2.5–3.5 × 0.8–1 mm; corolla rotate, petals 7–9 × 2.5–2.8 mm; carpels 3; styles connate to 3/4 length, 0.5–1.5 mm. **Capsules** spheric, 1–1.5 × 1–1.5 cm, explosively dehiscent. **Seeds** brown with darker markings, ellipsoidal, 9–13 × 5–7 mm; caruncle prominent. $2n$ = 22.

Flowering and fruiting spring–summer. Grassy clay-rocky and saline flats; 1000–2500 m; Tex.; Mexico (Coahuila, Nuevo León, Tamaulipas).

Jatropha cathartica is attractive and is threatened by extensive collecting by growers and collectors of succulent plants. In the flora area, the species is known from south Texas.

3. Jatropha multifida Linnaeus, Sp. Pl. 2: 1006. 1753 • Coral plant, French physic or physic nut, yucca [I]

Shrubs or trees, to 7 m, monoecious. **Stems** erect, yellow-brown, sparsely branched, woody-succulent, glabrous; short shoots absent; latex viscous, cloudy-whitish. **Leaves** persistent or drought-deciduous, mostly borne on or near branch tips; stipules persistent, (3–)6–15(–25) mm, filiform-divided; petiole 11–29 cm, not stipitate-glandular; blade ovate-cordate in outline, 16–30 × 10–22 cm, deeply 9–11-lobed nearly to base, base rounded, margins incised, apex acuminate, membranous, surfaces glabrous; venation palmate. **Inflorescences** bisexual, terminal and subterminal, cymes; peduncle 12–30 cm; bracts 2–4.5 mm, margins entire, glabrous. **Pedicels** 3.5–6 mm. **Staminate flowers:** sepals connate to ½ length, ovate, 1.5–3 × 1–1.2 mm, margins entire, apex round, surfaces glabrous; corolla orange-red, campanulate, petals distinct, 5–7 × 2.5–3 mm, surfaces glabrous; stamens 8 in 1 whorl; filaments distinct, 2.2–2.5 mm. **Pistillate flowers** resembling staminate, but sepals 2–3.5 × 1.1–1.3 mm; petals 6–8.2 × 2–3.2 mm; carpels (1–)3; styles distinct, 1–1.2 mm. **Capsules** ellipsoidal, winged, 3–3.4 × 2.7–2.9 cm, tardily dehiscent. **Seeds** yellow to light brown, mottled with dark brown spots or stripes, spheric, 14–18 mm; caruncle rudimentary. $2n$ = 22 (Puerto Rico).

Flowering and fruiting year-round. Disturbed sites; 0–50 m; introduced; Fla.; West Indies; introduced also in Mexico, Central America, South America, Asia, Africa, Pacific Islands, Australia.

The geographical origin of *Jatropha multifida* cannot be determined with certainty; it is probably native to the West Indies. The species is widely cultivated throughout the tropics as an ornamental and has escaped and naturalized in many areas, including central and southern Florida.

4. Jatropha integerrima Jacquin, Enum. Syst. Pl., 32. 1760 • Peregrina [I]

Shrubs, to 2.5–5 m, monoecious. **Stems** erect, dark brown, striate, much-branched, woody, glabrous; short shoots absent; latex watery, colorless in younger shoots, cloudy-whitish in older branches. **Leaves** persistent, ± evenly distributed on long shoots; stipules absent; petiole 1–5.5 cm, not stipitate-glandular; blade elliptic-ovate, obovate, lyrate, or panduriform, 7.5–15.3 × 2.9–12.5 cm, unlobed or shallowly 3-lobed, base rounded, cordate, or cuneate, margins entire (sometimes with 2–4 glands or hairs at base), apex acuminate, membranous to ± coriaceous, surfaces glabrous; venation pinnate (palmate if lobed). **Inflorescences** bisexual, terminal and subterminal, cymes; peduncle 5.2–21 cm; bracts 1–12 mm, margins entire, glabrous. **Pedicels** 2–8 mm. **Staminate flowers:** sepals distinct, ovate, 2.5–3(–4) × 1–1.7 mm, margins entire, apex obtuse, surfaces glabrous; corolla bright red to scarlet or pink, rotate, petals distinct, 8.4–12.1 × 2.5–4.3 mm, abaxial surface glabrous, adaxial with tufts of hairs near base; stamens 10 in 2 whorls (5 + 5); filaments of each whorl connate ½–¾ length, outer whorl 4–9 mm, inner whorl 5–12 mm. **Pistillate flowers** resembling staminate, but sepals 3.1–3.8 × 1.2–2.2 mm; petals 9–17 × 5–10 mm; carpels 3; styles connate ½ length, 3–4 mm. **Capsules** ovoid, 1–1.3 × 0.7–1.1 cm, explosively dehiscent. **Seeds** cream, mottled with red and black spots, ellipsoidal, 7–10 × 4–6.5 mm; caruncle relatively small, conspicuous. $2n$ = 22 (cult. Fla.).

Flowering and fruiting year-round. Disturbed sites; 0–50 m; introduced; Fla.; West Indies; introduced also in Central America, South America, Asia, Pacific Islands, Australia.

Jatropha integerrima, native to the West Indies, is one of the more common landscape plants in subtropical and tropical regions and has become naturalized in many areas; it is part of a complex hybrid group involving three or four species that grow sympatrically in western Cuba. There are many cultivars in the trade.

5. Jatropha macrorhiza Bentham, Pl. Hartw., 8. 1839 • Jirawilla, jicamilla, bahada, ragged nettlespurge [F]

Jatropha arizonica I. M. Johnston; *J. macrorhiza* var. *septemfida* Englemann

Herbs, perennial, to 0.5 m, monoecious, with ± fleshy subterranean caudex to 11 cm diam., stem and root scars round. **Stems** erect, green, usually sparsely branched, herbaceous, somewhat succulent, glabrous; short shoots absent; latex watery, colorless. **Leaves** ± evenly distributed on long shoots; stipules persistent, 4–10 mm, filiform-divided; petiole 4.8–10.5(–13.5) cm, not stipitate-glandular; blade cordate in outline, 11–16 × 9.3–11.2 cm, (3–)5–7(–9)-lobed to middle, base cordate, margins coarsely dentate, apex acuminate, membranous, abaxial surface glabrous, adaxial glabrous except puberulent on veins, margins sometimes puberulent or with setae; venation palmate. **Inflorescences** bisexual, terminal and subterminal, cymes; peduncle 1.5–4.5 cm; bracts (6–)8–16 mm, margins serrate, glabrous. **Pedicels** 2.5–4 mm. **Staminate flowers:** sepals distinct or connate to ½ length, lanceolate, 5–7 × 1–2 mm, margins deeply divided, apex acuminate, surfaces glabrous; corolla light pink, often with white striations, rotate, petals distinct or connate 1/4 length, 8–11.5 × 2.5–4.5 mm, surfaces glabrous; stamens 8(–9) in 2 whorls (5 + 3); filaments of outer whorl distinct, of inner whorl connate to 1/2 length, outer whorl 3.5–6 mm, inner whorl 4.5–9 mm. **Pistillate flowers** resembling staminate, but slightly larger; carpels 3; styles distinct or connate only at base, 3.5–4 mm. **Capsules** ± spheric, 1.2–1.3 × 1.2–1.3 cm, distinctly 3-lobed, explosively dehiscent. **Seeds** pale gray, ellipsoidal, 8–9 × 6–6.5 mm; caruncle prominent. **2*n*** = 22.

Flowering and fruiting spring–summer. Hillsides, mesas, sandy washes; 1000–2600 m; Ariz., N.Mex., Tex.; Mexico (Chihuahua, Sonora).

Jatropha macrorhiza is known in Texas only from a single collection made in Presidio County in 1938; it is common in nearby Chihuahua and appears to be native to trans-Pecos Texas, but may now be extirpated there.

Plants from the United States often have been called var. *septemfida*, named in reference to having seven (as opposed to five) leaf lobes or segments. However, leaf segment number varies, even on the same plant, from four to nine (usually five to seven). Hence, a distinct variety cannot be recognized. R. McVaugh (1945) reported this species to have ten stamens; the author has not seen any live or herbarium specimens with ten; two collections have recorded nine.

6. **Jatropha curcas** Linnaeus, Sp. Pl. 2: 1006. 1753 • Physic or purging nut, piñón [I]

Trees, to 10[–15] m, monoecious. **Stems** erect, gray-green, much-branched, woody-succulent, glabrous; short shoots absent; latex watery, colorless in younger branches, cloudy-whitish in older shoots. **Leaves** persistent, ± evenly distributed on long shoots; stipules caducous, narrowly lanceolate, 5 mm, undivided; petiole 9–19 cm, not stipitate-glandular; blade round in outline, 9–15 × 9–15 cm, usually shallowly 3–5-lobed, rarely unlobed, base cordate, margins entire or glandular (young leaves), apex acuminate, membranous, surfaces glabrous; venation palmate. **Inflorescences** bisexual, terminal and subterminal, cymes; peduncle 5–10 cm; bracts 3–10 mm, margins entire, glabrous. **Pedicels** 1–3 mm. **Staminate flowers:** sepals distinct, ovate-elliptic, 4–6 × 2–3 mm, margins entire, apex acute, surfaces glabrous; corolla greenish yellow, campanulate, petals distinct or connate 1/4 length, 6–8 × 2–3.5 mm, glabrous abaxially, tomentose adaxially; stamens 10, ± in 2 whorls (5 + 5); filaments of both whorls connate to top or nearly so, outer whorl 3–4.5 mm, inner whorl 3–5 mm. **Pistillate flowers** resembling staminate, but sepals connate to 1/2 length, 5–7.5 × 2–5 mm; petals 4–5 × 2–2.5(–3) mm; staminodes infrequent; carpels 3; styles connate most of length, 0.5–1.5 mm. **Capsules** ellipsoidal, 2.6–3 × 2.2–2.8 cm, drupaceous. **Seeds** black or black mottled with white spots, ellipsoidal, 18–20 × 11–13 mm; caruncle rudimentary. **2*n*** = 22 (Puerto Rico).

Flowering and fruiting spring (late summer–early fall). Disturbed sites; 0–50 m; introduced; Fla.; Mexico; Central America; introduced also in West Indies, South America, Asia, Africa, Pacific Islands, Australia.

Jatropha curcas now has a circumtropical distribution but probably originated in Central America; it is naturalized in southern Florida. The latex of *J. curcas* is used for soap making and for medicinal purposes; the seeds are used for biofuel production.

7. **Jatropha canescens** (Bentham) Müller Arg. in A. P. de Candolle and A. L. P. P. de Candolle, Prodr. 15(2): 1079. 1866 • Sangre de drago, Arizona nettlespurge

Mozinna canescens Bentham, Bot. Voy. Sulphur, 52, plate 25. 1844

Shrubs, to 1.2–2.5 m, dioecious. **Stems** erect, grayish white, branched from base, woody-succulent, canescent; short shoots common; latex watery, cloudy-whitish. **Leaves** deciduous, mostly ± evenly distributed on long shoots, rarely on short shoots; stipules absent; petiole 1.3–3 cm, not stipitate-glandular; blade cordate to broadly ovate, 2.2–3.5 × 1.6–3 cm, unlobed or shallowly 3-lobed, base truncate-cordate, margins entire, apex rounded, ± coriaceous, abaxial surface canescent, adaxial surface sparingly hairy; venation pinnate (palmate if lobed). **Inflorescences** terminal on branches or on short shoots, staminate cymes, pistillate fascicles, or flowers solitary; peduncle 1–2.6 cm; bracts 1–2.5 mm, margins entire, sparsely hairy. **Pedicels** 2–3 mm. **Staminate flowers:** sepals connate to 1/4 length, lanceolate, 2–2.5 × 0.5–0.8 mm, margins entire, apex acute, canescent abaxially, glabrous or sparingly hairy adaxially; corolla grayish white, sometimes pinkish abaxially, subglobose-urceolate, petals connate 3/4–4/5 length, 5–8 × 1–2 mm, surfaces sparingly hairy; stamens 10 in 2 whorls (5 + 5); filaments of outer whorl connate 1/2 length, of inner whorl connate 1/4 length, outer whorl 2–3 mm, inner whorl 4–5 mm. **Pistillate flowers** resembling staminate, but sepals connate only at base, 3–3.5 × 1.5–2 mm; petals connate 1/2–3/4 length, 8–11 × 3–5 mm; carpels 2[–3]; styles connate 3/4 their lengths, 2–5 mm. **Capsules** compressed ellipsoidal, 1.2–1.5 × 2–2.5 cm, 2-lobed [ellipsoidal, 3-lobed], tardily dehiscent. **Seeds** solid brown, subspheric, 9–12 mm; caruncle absent. **2*n*** = 22 (Mexico).

Flowering and fruiting late spring(–summer). Sandy washes, sand dunes; 0–500 m; Ariz.; Mexico (Baja California, Baja California Sur, Sinaloa, Sonora).

In Arizona, *Jatropha canescens* is found only in Pima County, primarily in Organ Pipe Cactus National Monument.

Jatropha canescens is part of a hybrid complex that includes the Mexican species *J. cinerea* (Ortega) Müller Arg. and, probably, *J. giffordiana* Dehgan & G. L. Webster. These can be difficult to distinguish. R. McVaugh (1945) suggested considering *J. canescens* as a synonym of *J. cinerea*, and F. Shreve and I. L. Wiggins (1964), as well as others, have done so. *Jatropha canescens* may be distinguished from *J. cinerea* and *J. giffordiana* most reliably by its crowded inflorescences of staminate flowers with smaller subglobose (as opposed to urceolate) whitish gray corollas (sometimes with some red or pink on the adaxial surface) as opposed to larger darker red corollas of *J. cinerea* and *J. giffordiana*. In addition, *J. canescens* generally has more numerous, longer and darker colored short shoots, and smaller and less often shallowly 3-lobed leaves. Although *J. cinerea* (in the strict sense) does not occur in northern Sonora, it is one of the more common plants in Baja California, Baja California Sur, and western mainland Mexico.

8. **Jatropha cuneata** Wiggins & Rollins, Contr. Dudley Herb. 3: 272, plate 62, fig. 1. 1943 • Leatherplant, limberbush, sangre de drago [F]

Shrubs, to 2 m, dioecious. **Stems** spreading, yellow to yellow-brown [gray], much-branched, woody-succulent, glabrous; short shoots common; latex watery, yellow in young shoots, red in older shoots. **Leaves** deciduous, usually fascicled on short shoots; stipules absent; petiole 0(–0.2) cm, not stipitate-glandular; blade obovate-spatulate, 0.7–1.9 × 0.3–0.9 cm, unlobed (shallowly 3-lobed on active shoots), base attenuate, margins entire, apex usually rounded, sometimes emarginate, coriaceous, surfaces glabrous; venation pinnate (palmate if lobed). **Inflorescences** on short shoots, cymes or flowers solitary; peduncle 0.4–0.6 cm; bracts 0.4–1 mm, margins entire, glabrous. **Pedicels** 1–4(–6.5) mm. **Staminate flowers:** sepals connate basally, ovate-lanceolate, 2.5–4 × 0.6–1 mm, margins entire, apex acute, surfaces glabrous; corolla white, tubular-urceolate, petals connate most of length, 4–6 × 1.3–2 mm, surfaces glabrous; stamens 10 in 2 whorls (5 + 5); filaments of both whorls connate almost to top, outer whorl 1–2.5 mm, inner whorl 2.5–4 mm. **Pistillate flowers** resembling staminate, but sepals distinct, 1.5–3 × 0.5–1 mm; petals 4–4.5 × 1–1.5 mm; carpel 1; styles 0.5–1 mm. **Capsules** spheric, 0.9–1.1 × 0.9–1.1 cm, tardily dehiscent. **Seeds** solid dark to golden brown, spheric, 10 × 10 mm; caruncle absent. $2n = 44$ (Mexico).

Flowering and fruiting late spring–summer. Dry rocky limestone mesas, sandy areas, bajadas; 0–800 m; Ariz.; Mexico (Baja California, Baja California Sur, Sonora).

In Arizona, *Jatropha cuneata* is found only in Pima County, primarily in Organ Pipe Cactus National Monument and Cabeza Prieta National Wildlife Refuge. Its stems have been used in basket making.

9. **Jatropha cardiophylla** (Torrey) Müller Arg. in A. P. de Candolle and A. L. P. P. de Candolle, Prodr. 15(2): 1079. 1866 • Sangre de Cristo, heartleaf dragon's blood or limberbush

Mozinna cardiophylla Torrey in W. H. Emory, Rep. U.S. Mex. Bound. 2(1): 198. 1859

Subshrubs, to 0.5–1 m, dioecious, rhizomatous, often forming large clumps. **Stems** erect or ascending, reddish brown, much-branched, rubbery-succulent, glabrous; short shoots common; latex watery, colorless to cloudy-whitish in young shoots, blood red in basal portions of older shoots and rhizomes. **Leaves** deciduous, borne profusely on long and short shoots in rainy season; stipules absent; petiole 1–2.5 cm, not stipitate-glandular; blade widely ovate-deltate, 1.8–4.6 × 1.5–2.6 cm, unlobed, base truncate, margins sinuate to weakly serrate-crenate with glands on apices of crenations in younger leaves, apex acuminate, membranous, surfaces glabrous; venation palmate. **Inflorescences** axillary, cymes; staminate with peduncle 0.8–3 cm, pistillate with peduncle absent; bracts 1–1.5 mm, margins entire, or sometimes with glands, glabrous. **Pedicels** 2–3 mm. **Staminate flowers:** sepals distinct, ovate to obovate, 2–2.5 × 0.8–1 mm, margins entire, apex acute, surfaces glabrous; corolla light pink to white, tubular, petals connate to ¾ length, 5–7.5 × 1.8–2.2 mm, surfaces glabrous; stamens 10 in 2 whorls (5 + 5); filaments of both whorls distinct, outer whorl 1 mm, inner whorl 2 mm. **Pistillate flowers** resembling staminate, but sepals obovate, 1–2.5 × 0.8–2 mm, apex rounded; petals connate to ½ length, 6.5–9 × 1.8–2.7 mm; carpel 1; style 1 mm. **Capsules** spheric, apiculate, 1.1–1.8 × 1.1–1.8 cm, tardily dehiscent. **Seeds** mottled gray-brown, spheric, 12 mm diam.; caruncle rudimentary. $2n = 22$.

Flowering and fruiting spring–summer. Gravelly desert washes and volcanic hillsides; 600–1500 m; Ariz.; Mexico (Sonora).

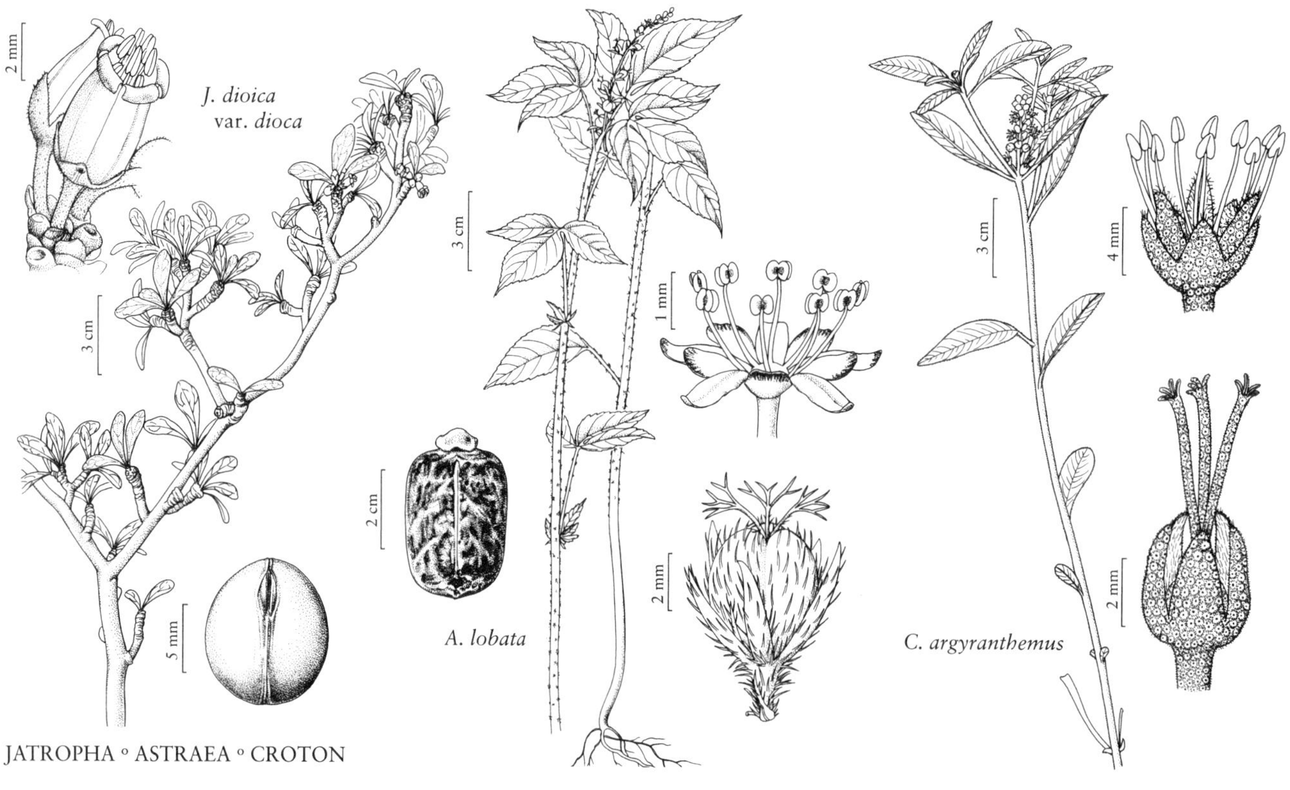

JATROPHA ◦ ASTRAEA ◦ CROTON

Jatropha cardiophylla and *J. dioica* often have been confused because of the similarity of their spreading rhizomatous habits and red, rubbery branches during the dormant season. The two species are easily distinguished by their leaves. Furthermore, *J. cardiophylla* is restricted to the Sonoran Desert; *J. dioica* is found only in the Chihuahuan Desert and areas to the east.

10. Jatropha dioica Sessé, Gaz. Lit. México 3(suppl.): 4. 1794 • Sangre de drago, sangregado, leatherstem [F]

Subshrubs, to 1 m, dioecious, rhizomatous, often forming colonies. **Stems** spreading, reddish brown (when actively growing) or red to dark red or grayish (when dormant), much-branched, rubbery-succulent, glabrous; short shoots common; latex watery, colorless to cloudy-whitish in fast growing shoots, blood red in basal portion of older shoots and rhizomes. **Leaves** deciduous, fascicled on short shoots; stipules deciduous, linear, 1–2 mm, undivided; petiole 0–0.2 cm, not stipitate-glandular; blade linear-spatulate to narrowly obovate, 1.5–5(–7) × 0.2–0.7 cm, usually unlobed, sometimes shallowly, rarely deeply, 3-lobed, base attenuate, margins entire, apex rounded, membranous, surfaces glabrous; venation pinnate (palmate if lobed). **Inflorescences** terminal or axillary, fascicles; peduncle absent; bracts absent. **Pedicels** 1–2.5 mm. **Staminate flowers:** sepals distinct or connate basally, ovate to elliptic-lanceolate, 2.5–3 × 1.2–1.4 mm, margins entire, apex acute, abaxial surface hairy, adaxial glabrous; corolla white to pinkish white, urceolate-tubular, petals connate most of length, 5–6 × 1.4–1.5 mm, abaxial surface sparsely puberulent, adaxial glabrous; stamens 10 in 2 whorls (5 + 5); filaments of both whorls connate from $^{1}/_{5}$–$^{3}/_{4}$+ length, outer whorl 2–3 mm, inner whorl 3–5 mm. **Pistillate flowers** resembling staminate; carpel 1; style 2.5–3 mm. **Capsules** ellipsoidal, 1.2–1.4 × 1.1–1.3 cm, tardily dehiscent. **Seeds** solid gray-brown, spheric, 12 mm diam.; caruncle rudimentary. **2*n*** = 44.

Varieties 2 (2 in the flora): Texas, n Mexico.

R. McVaugh (1945b) recognized two varieties in *Jatropha dioica* based primarily on leaf shape; the two are not readily recognizable, and probably hybridization between them has resulted in intermediate populations, as McVaugh noted. The varieties are treated here with some reservation.

1. Leaf blades oblanceolate to narrowly obovate, 3.5–5(–7) cm; sepals distinct; stamen filaments connate $^{3}/_{4}$+ length; petals pinkish white . . . . 10a. *Jatropha dioica* var. *dioica*
1. Leaf blades linear-spatulate, 1.5–2.5 cm; sepals connate basally; stamen filaments connate to $^{1}/_{4}$ length; petals white . . . . 10b. *Jatropha dioica* var. *graminea*

10a. Jatropha dioica Sessé var. **dioica** F

Jatropha dioica var. *sessiflora* (Hooker) McVaugh

Plants to 1 m. **Leaves:** blade oblanceolate to narrowly obovate, 3.5–5(–7) × 0.3–0.7 cm, usually unlobed, sometimes deeply 3-lobed (in moister sites); secondary veins visible. **Flowers:** sepals distinct; petals pinkish white; stamen filaments connate ¾+ length.

Flowering and fruiting late spring–early summer. Dry sandy or rocky soils; 400–2500 m; Tex.; Mexico (Chihuahua, Coahuila, Durango, Nuevo León, San Luis Potosí, Tamaulipas, Zacatecas).

Variety *dioica* is widespread in eastern Mexico south to Zacatecas. R. McVaugh (1945b) reported the range as south to Oaxaca, but most of the specimens annotated by him and others as this taxon are currently recognized as two related but morphologically distinct species (*Jatropha elbae* J. Jiménez Ramírez and *J. oaxacana* J. Jiménez Ramírez & R. Torres). In Texas, it is found mainly south and east of the Pecos River.

10b. Jatropha dioica Sessé var. **graminea** McVaugh, Bull. Torrey Bot. Club 72: 39. 1944

Plants to 0.5 m. **Leaves:** blade linear-spatulate, 1.5–2.5 × 0.2–0.4 cm, unlobed; only midvein visible. **Flowers:** sepals connate basally; petals white; stamen filaments connate to ¼ length.

Flowering and fruiting late spring–early summer. Dry sandy soils; 100–900 m; Tex.; Mexico (Chihuahua, Coahuila, San Luis Potosí, Zacatecas).

The distribution of var. *graminea* lies west of that of var. *dioica* and does not extend as far south. In Texas, it is restricted to the trans-Pecos region. Some Mexican specimens erroneously annotated as var. *graminea* are actually *Jatropha rzedowskii* J. Jiménez Ramírez.

13. ASTRAEA Klotzsch, Arch. Naturgesch. (Berlin) 7: 194. 1841 • [For Greek mythological Astraea (star maiden), daughter of Zeus and Themis] I

Paul E. Berry

Benjamin W. van Ee

Herbs [subshrubs, shrubs], annual [perennial], monoecious; hairs unbranched and stellate; latex colorless. **Leaves** alternate, simple [palmately compound]; stipules present, persistent; petiole present, glands present at apex; blade unlobed or palmately lobed, margins serrate [entire], laminar glands absent; venation palmate at base, pinnate distally. **Inflorescences** bisexual (pistillate flowers proximal, staminate distal), terminal, racemes [thyrses]; glands subtending each bract 0. **Pedicels** present. **Staminate flowers:** sepals 5, imbricate, distinct; petals 5, distinct, white [to pink]; nectary extrastaminal, 5 glands; stamens 8–15, inflexed in bud, distinct; pistillode absent. **Pistillate flowers:** sepals 5(–7), usually not touching in bud, connate basally; petals 0; nectary 5 glands; pistil 3-carpellate; styles 3, connate basally [distinct], multifid. **Fruits** capsules. **Seeds** oblong-rectangular; caruncle present. $x = 9$.

Species ca. 12 (1 in the flora): introduced, Florida; Mexico, West Indies, Central America, South America; introduced also in Asia (Arabian Peninsula, India), Africa; tropical and subtropical areas.

Astraea was treated as a section of *Croton* by G. L. Webster (1993). However, the molecular phylogeny of P. E. Berry et al. (2005) showed that it represents a lineage distinct from *Croton*. Morphological characters that support this separation include the markedly rectangular seeds, the often deeply lobed leaves, and the mixture of simple and stellate hairs. *Astraea* is most diverse in southeastern Brazil.

1. **Astraea lobata** (Linnaeus) Klotzsch, Arch. Naturgesch. (Berlin) 7: 194. 1841 • Lobed croton F I W

Croton lobatus Linnaeus, Sp. Pl. 2: 1005. 1753 (as lobatum)

Herbs, 2–10 dm. **Stems** erect. **Leaves:** stipules linear-lanceolate, 2–6 mm, sometimes split at base; petiole 2–10 cm, glands 2, papillate, lobed or digitate; blade usually deeply 3(–5)-lobed, sometimes unlobed, central lobe elliptic to oblanceolate, 3–9 × 2–10 cm, larger than lateral ones, base truncate to rounded, apex acuminate, abaxial surface hairy, adaxial surface glabrate or hairy. **Inflorescences** 2–6(–15) cm; staminate flowers 3–15(–20), pistillate flowers 1–5(–7). **Pedicels:** staminate 2–3 mm, pistillate 1–2 mm (to 3 mm in fruit). **Staminate flowers:** sepals usually purple-tinged, ± unequal, narrowly ovate-oblong, 1–1.3 mm, enveloping bud before anthesis, apex blunt and ± hooded; petals hyaline, oblong, 1.2–1.8 mm. **Pistillate flowers:** sepals oblanceolate to spatulate, 4–7 × 1–2 mm, apex acute, semiappressed-hairy; ovary usually hairy, rarely glabrous; styles 2–4.5 mm, 3-fid, each branch then 2-fid, terminal segments 18. **Capsules** 5–8 × 5–7 mm, sparsely hairy. **Seeds** 4.5–5.5 × 2.7–3.5 mm, 2.5–3 mm thick, rugulose; caruncle shortly stipitate, reniform-peltate, 0.5–0.8 × 1.2–2 mm. $2n = 18$.

Flowering Jun–Nov. Open areas, disturbed habitats; 0–50 m; introduced; Fla.; Mexico; West Indies; Central America; South America; introduced also in Asia (Arabian Peninsula, India), tropical Africa.

Astraea lobata is a weedy annual widely distributed throughout the New World tropics and subtropics. It is not clear where in this area the species is native and where it has been introduced. The records from the flora area, which are restricted to Manatee, Miami-Dade, and Pinellas counties, probably represent introductions rather than natural occurrences.

Astraea lobata is highly variable, and it has been divided into numerous subspecies or distinct species. It is treated here as a single species; it is not clear if any of the subspecies warrant recognition. The Florida material agrees well with the type material of the species from Veracruz, Mexico.

14. CROTON Linnaeus, Sp. Pl. 2: 1004. 1753; Gen. Pl. ed. 5, 436. 1754 • [Greek *kroton*, tick, alluding to resemblance of seeds]

Benjamin W. van Ee

Paul E. Berry

Herbs, subshrubs, or shrubs [trees], annual or perennial, monoecious or dioecious; hairs stellate or scalelike; latex colorless to reddish [absent]. **Leaves** persistent, semideciduous, or drought deciduous, alternate, simple, often with lemony, pungent, or acrid odor when crushed, older leaves often turning orange before falling; stipules absent or present, persistent, deciduous, or caducous; petiole present [absent], glands present at apex or absent; blade unlobed [palmately lobed], margins entire, crenate, denticulate, serrulate, or serrate-dentate, laminar glands absent [at base, on margins or abaxial surface]; venation pinnate or palmate at base, pinnate distally. **Inflorescences** unisexual or bisexual (pistillate flowers proximal, staminate distal), terminal or axillary, spikes, racemes, or thyrses; glands subtending each bract 0. **Pedicels** present or absent. **Staminate flowers:** sepals (3–)5(–6), valvate or slightly imbricate, distinct or connate basally; petals (3–)5(–6) or 0, distinct, white; nectary extrastaminal, usually 5 glands; stamens 3–35 [–50], inflexed in bud, distinct; pistillode absent. **Pistillate flowers:** sepals [3–](4–)5(–9)[–10] or 0, imbricate or valvate, distinct (connate for ½+ length in *C. argyranthemus*); petals 5 (sometimes rudimentary) or 0, distinct or connate basally, white or pale green; nectary annular, 5 glands, or absent; pistil (1–)3-carpellate; styles (1–)3, distinct or connate basally, unbranched, 2-fid, or multifid. **Fruits** usually capsules (achenes in *C. michauxii*). **Seeds** ellipsoid, oblong, ovoid, globose, or lenticular; caruncle present. $x = 8, 9, 10, 14$.

Species ca. 1250 (31 in the flora): North America, Mexico, West Indies, Bermuda, Central America, South America, Asia, Africa, Indian Ocean Islands, Pacific Islands, Australia; most diverse in dry tropical and warm-temperate regions.

Croton species are the main larval food of the leafwing butterfly (*Anaea* spp.) and some related genera. Their seeds are also a favorite food for doves and other birds. The circumscription of *Croton* was consolidated by the molecular phylogeny of P. E. Berry et al. (2005), which separated *C. lobatus* into the genus *Astraea* and confirmed the inclusion of the genera *Crotonopsis* Michaux, *Eremocarpus* Bentham, and *Julocroton* Martius as sections of *Croton*. The molecular phylogeny and sectional treatment of B. W. van Ee et al. (2011) further updated the sectional taxonomy of G. L. Webster (1993) for the New World species.

Croton trinitatis Millspaugh, native to Central and South America, was collected once (in 1886) on ballast piles in Pensacola, Florida, but has not been reported in the flora area since. It keys with *C. glandulosus*, from which it differs in its larger, more deltate leaves with deeper, sharper marginal teeth. *Croton bonplandianus* Baillon, which is native to Argentina and Paraguay and naturalized elsewhere, was collected several times in the 1950s on ballast ore piles in the ports of Baltimore, Maryland, and Newport News, Virginia (C. F. Reed 1964), but does not appear to have persisted there. *Croton bonplandianus* would also key with *C. glandulosus* but differs in its larger, shallowly serrate-margined, narrowly ovate leaves; inflorescences 5–14 cm; and sessile glands at the base of the leaf blade.

SELECTED REFERENCES Berry, P. E. et al. 2005. Molecular phylogenetics of the giant genus *Croton* (Euphorbiaceae sensu stricto) using ITS and *trn*L–F DNA sequence data. Amer. J. Bot. 92: 1520–1534. Ferguson, A. M. 1901. Crotons of the United States. Rep. (Annual) Missouri Bot. Gard. 1901: 33–73. van Ee, B. W. and P. E. Berry. 2009. The circumscription of *Croton* section *Crotonopsis* (Euphorbiaceae), a North American endemic. Harvard Pap. Bot. 14: 61–70. van Ee, B. W. and P. E. Berry. 2010. Taxonomy and phylogeny of *Croton* section *Heptallon* (Euphorbiaceae). Syst. Bot. 35: 151–167. van Ee, B. W., R. Riina, and P. E. Berry. 2011. A revised infrageneric classification and molecular phylogeny of New World *Croton* (Euphorbiaceae). Taxon 60: 791–823. Webster, G. L. 1993. A provisional synopsis of the sections of the genus *Croton* (Euphorbiaceae). Taxon 42: 793–823.

1. Leaf blade margins coarsely crenate to serrate-dentate; petioles with 2 cuplike glands at apex . . . . 12. *Croton glandulosus*
1. Leaf blade margins entire, denticulate, or serrulate; petioles without cuplike glands at apex (sessile or stipitate glands often present at apex in *C. linearis*, stipitate-glandular in *C. ciliatoglandulifer*).
    2. Shrubs; leaf blade abaxial surfaces densely lepidote; pistillate petals 5, ovate, pale green; styles 3, usually unbranched, rarely 2-fid . . . . 1. *Croton alabamensis*
    2. Herbs, subshrubs, or shrubs; if shrubs, leaf blade abaxial surfaces mainly stellate-hairy or stellate-tomentose, seldom stellate-lepidote; pistillate petals 0, rudimentary, or, if 5 (sometimes in *C. humilis* and *C. soliman*), linear or subulate, white; styles 2–3, 2-fid to multifid, or 1, unbranched.
        3. Ovaries 1-locular; styles unbranched; leaves mostly clustered near inflorescences; annual herbs . . . . 26. *Croton setigerus*
        3. Ovaries 2–3-locular; styles 2-fid or multifid; leaves clustered near inflorescences or not; annual or perennial herbs, subshrubs, or shrubs.
            4. Leaf blade abaxial surfaces densely lepidote or stellate-lepidote, often silvery.
                5. Inflorescences bisexual; staminate petals 5; columellae 3-angled (not markedly 3-winged) or absent (when fruit an achene); annual or perennial herbs or subshrubs.
                    6. Leaf blades on proximal parts of stems oval to narrowly obovate, on distal parts oblong to lanceolate-oblong or broadly elliptic, apices obtuse to rounded, adaxial surfaces sparsely lepidote or glabrate; inflorescences racemes; ovaries 3-locular; fruits capsules; staminate sepals 5 mm, petals 5 mm, pedicels 1–5 mm; pistillate sepals connate for ½+ length . . . . 3. *Croton argyranthemus*
                    6. Leaf blades lanceolate-linear to ovate-lanceolate, apices acute, adaxial surfaces sparsely stellate-hairy; inflorescences spikes; ovaries 1-locular; fruits achenes; staminate sepals 1 mm, petals 0.6–1 mm, pedicels 0–1 mm; pistillate sepals distinct . . . . 20. *Croton michauxii*
                5. Inflorescences usually unisexual (sometimes bisexual in *C. punctatus*);

staminate petals 0; columellae 3-winged; subshrubs or shrubs.

7. Plants monoecious (sometimes appearing dioecious); leaf blades more than ½ as wide as long, petioles ½ to as long as leaf blade; Atlantic and Gulf coasts . . . . 24. *Croton punctatus* (in part)
7. Plants dioecious; leaf blades usually less than ½ as wide as long, petioles usually less than ½ as long as leaf blade; c Texas to California.
    8. Subshrubs, 2–5(–9) dm; petioles 0.2–0.8(–2) cm; New Mexico, Texas . . . . 9. *Croton dioicus*
    8. Shrubs, 4–11 dm; petioles 1–4(–4.5) cm; Arizona, California, Nevada, Utah.
        9. Pedicels of pistillate flowers less than or equal to 1 mm (1–3 mm in fruit); seeds 4–5.5 × 3.5–5.5 mm; Arizona, California, Nevada, Utah . . . . 4. *Croton californicus* (in part)
        9. Pedicels of pistillate flowers 1–2 mm (4–7 mm in fruit); seeds 6.5–7 × 2–3 mm; sw Arizona, se California . . . . . . 31. *Croton wigginsii* (in part)

[4. Shifted to left margin.—Ed.]

4. Leaf blade abaxial surfaces usually stellate-hairy, if stellate-lepidote not markedly silvery.
    10. Annual herbs.
        11. Plants dioecious; inflorescences unisexual; staminate petals 0; capsules verrucose, columellae 3-winged.
            12. Plants 5–15 dm; leaf blade abaxial surfaces densely stellate-tomentose; capsules 8–9 mm; s Texas. . . . . 22. *Croton parksii*
            12. Plants 2–7(–9) dm; leaf blade abaxial surfaces densely appressed stellate-hairy; capsules 5–8 mm; widespread in s and c United States, mostly not s Texas . . . . 30. *Croton texensis*
        11. Plants monoecious; inflorescences bisexual or unisexual; staminate petals 3–5; capsules smooth, columellae not markedly 3-winged (usually 3-angled or apex with 3 lobes).
            13. Leaf blade abaxial surfaces appearing brown-dotted, some stellate hairs with dark brown centers; stamens 3–5; styles 2, 2-fid, terminal segments 4; ovaries 2-locular, only 1 fertile . . . . 21. *Croton monanthogynus*
            13. Leaf blade abaxial surfaces not appearing brown-dotted, no stellate hairs with brown centers; stamens 7–16; styles 3, 2-fid or multifid, terminal segments 6–18(–24); ovaries 3-locular.
                14. Pistillate sepal margins laciniate. . . . . 2. *Croton argenteus*
                14. Pistillate sepal margins entire.
                    15. Inflorescences 4–7 cm; staminate flowers 15–25, sepals 3 mm, petals 3–3.5 mm, stamens 14–16; pistillate flowers 8–15. . . . . 8. *Croton coryi*
                    15. Inflorescences 0.5–4 cm; staminate flowers 1–15, sepals 0.8–2 mm, petals 0.8–1.5 mm, stamens 7–13; pistillate flowers 1–8.
                        16. Pistillate sepals 5–6; styles 2-fid, terminal segments 6; herbs 1–5 dm.
                            17. Pistillate sepals unequal, 3 outer 5–7 mm, 2 inner 1–2 mm . . . . 16. *Croton leucophyllus*
                            17. Pistillate sepals equal, 3 mm . . . . 18. *Croton lindheimerianus*
                        16. Pistillate sepals 6–9; styles multifid, terminal segments 12–18 (–24); herbs 3–20 dm.
                            18. Leaf blades 0.2–0.8 cm wide, if wider then pistillate sepal apices recurved; staminate sepals and petals 0.8–1 mm; styles 2–3 mm.
                                19. Leaf blades ovate to lanceolate-elliptic, 3–8(–15) × 1–4 cm; pistillate sepals 7–10(–15 in fruit) mm, apices recurved. . . . . 5. *Croton capitatus*
                                19. Leaf blades lanceolate to oblong, 2–5.5 × 0.2–0.8 cm; pistillate sepals 5–6 mm, apices incurved . . . . 10. *Croton elliottii*
                            18. Leaf blades 1–5 cm wide, pistillate sepal apices straight or slightly incurved; staminate sepals 1–2 mm, petals 1–1.5 mm;

styles 3–4 mm.

20. Pistillate sepals whitish appressed-tomentose; styles each appearing 4-fid. . . . . .13. *Croton heptalon*

20. Pistillate sepals yellowish woolly-tomentose; styles each 2 times 2-fid. . . . . .17. *Croton lindheimeri*

[10. Shifted to left margin.—Ed.]

10. Shrubs or perennial herbs.

21. Staminate petals 0; columellae 3-winged.

22. Plants monoecious or sometimes apparently dioecious; leaf blades much less than 2 times as long as wide, petioles 1/2 to equal blade length; beaches and coastal dunes, Atlantic and Gulf coasts. . . . . 24. *Croton punctatus* (in part)

22. Plants strictly dioecious; leaf blades more than 2 times as long as wide, petioles usually less than 1/2 blade length; dunes and sandy areas, inland as well as coastal, Arizona, California, Nevada, Utah.

23. Pedicels of pistillate flowers to 1 mm, 1–3 mm in fruit; seeds 4–5.5 × 3.5–5.5 mm; various habitats, Arizona, California, Nevada, Utah. . . . . .4. *Croton californicus* (in part)

23. Pedicels of pistillate flowers 1–2 mm, 4–7 mm in fruit; seeds 6.5–7 × 2–3 mm; dunes of se-most California and sw Arizona . . . . .31. *Croton wigginsii* (in part)

21. Staminate petals (4–)5(–6); columella apices with 3 rounded, inflated lobes or 3 sharp projections.

24. Perennial herbs; columella apices with 3 sharp projections. . . . . .23. *Croton pottsii*

24. Shrubs; columella apices with 3 rounded, inflated lobes.

25. Petioles mostly 1/4–7/10+ leaf blade length, if shorter (*C. fruticulosus*) then leaf blade margins finely serrulate and staminate sepals 0.8–1.2 mm; leaf blades usually ovate to lanceolate or oblong, margins entire, minutely glandular-denticulate, or serrulate, abaxial surfaces stellate-hairy or glabrous; staminate petal abaxial surfaces glabrous except margins often villous or ciliate; seeds shiny; styles 2- or 4-fid, terminal segments 6 or 12.

26. Styles 2-fid, terminal segments 6; petioles mostly 1/8–1/2 leaf blade length; stamens 9–16; stipules not glandular-ciliate; petioles not stipitate-glandular at apices (*C. sonorae* glandular-ciliate on leaf blade margins).

27. Staminate pedicels 2.5–4 mm, pistillate 0–0.5 mm; seeds 4–5 mm; leaf blades 2–8 cm, adaxial surfaces puberulent, margins serrulate, often slightly undulate . . . . 11. *Croton fruticulosus*

27. Staminate pedicels 1–1.5 mm, pistillate 1–4 mm; seeds 6–7 mm; leaf blades 0.8–4 cm, adaxial surfaces glabrate, margins entire, not undulate . . . . 28. *Croton sonorae*

26. Styles 4-fid, terminal segments 12; petioles 3/8–7/10+ leaf blade length; stamens 15–35; stipules glandular-ciliate or not; petioles stipitate-glandular or without glands at apices.

28. Stipules and leaf blade margins glandular-ciliate; petioles stipitate-glandular at apices. . . . . 6. *Croton ciliatoglandulifer*

28. Stipules not glandular-ciliate (may be small clusters of stipitate glands or have short-stipitate glandular processes); leaf blade margins not glandular-ciliate (sometimes glandular-denticulate or with glandular-capitate processes); petioles without glands at apices.

29. Leaf blade abaxial surfaces densely stellate-hairy; staminate sepals 3–4 mm, petals 3–4 mm; pistillate sepals 4 mm, petals 1 mm; seeds 3–4 × 2.5–3 mm. . . . . .14. *Croton humilis*

29. Leaf blade abaxial surfaces glabrous or sparsely stellate-hairy (usually along margins); staminate sepals 0.8–1 mm, petals 1–1.3 mm; pistillate sepals 5–6 mm, petals 0 or 2.5–3.5 mm; seeds

4–5 × 3–4 mm. . . . . 27. *Croton soliman*

[25. Shifted to left margin.—Ed.]

25. Petioles mostly 1/10–1/4 leaf blade length, if longer (*C. incanus*) then leaf blade surfaces stellate-tomentose, staminate petal abaxial surfaces villous, and seeds dull; leaf blades usually obovate, elliptic, oblong, or linear, sometimes lanceolate or ovate, margins entire or minutely denticulate, abaxial surfaces stellate-hispid, stellate-tomentose, or stellate-hairy; staminate sepals 1.5–2.5 mm; staminate petal abaxial surfaces villous, glabrate, or glabrous except base sometimes villous and margins ciliate or tomentose; seeds shiny or dull; styles 2-fid, terminal segments 6.
 30. Leaf blade adaxial surfaces glabrous or minutely stellate-puberulent; plants dioecious.
  31. Leaf blades narrowly obovate, oblong, elliptic, or lanceolate, mostly less than 3.5 times as long as wide, petioles 1–1.5 cm; Texas . . . . 7. *Croton cortesianus*
  31. Leaf blades linear to narrowly oblong, mostly more than 4 times as long as wide, petioles 0.3–1 cm; se Florida . . . . 19. *Croton linearis*
 30. Leaf blade adaxial surfaces stellate-tomentose or stellate-hairy; plants monoecious or dioecious.
  32. Shrubs 10–20 dm, much-branched distally; stipules linear-subulate, 2–3 mm; stems stellate-velutinous . . . . 15. *Croton incanus*
  32. Shrubs 1–5 dm, much-branched proximally; stipules papilliform, 0.1–0.5 mm; stems coarsely stellate-tomentose.
   33. Plants dioecious; stipules each 1 glandular papilla, 0.1 mm; capsules 5.5–6 mm wide; seeds 3.6–4.7 mm . . . . 25. *Croton sancti-lazari*
   33. Plants monoecious; stipules each 5–10 glandular papillae, 0.2–0.5 mm; capsules 6–8 mm wide; seeds 5.5–7 mm. . . . . 29. *Croton suaveolens*

1. Croton alabamensis E. A. Smith ex Chapman, Fl. South. U.S. ed. 2, 648. 1883 E

Shrubs, 5–35 dm, monoecious. **Stems** usually well branched distally, lepidote. **Leaves** clustered at branch tips; stipules absent; petiole 0.6–2 cm, glands absent at apex; blade elliptic, ovate, or oblong, 3–10 × 1.5–5 cm, base rounded to obtuse, margins entire, sometimes ± undulate, apex acute, rounded, or emarginate, abaxial surface silvery or coppery, densely lepidote, adaxial surface green, sparsely lepidote. **Inflorescences** bisexual or unisexual, racemes, 2–4.5 cm, staminate flowers 0–15, pistillate flowers 0–10. **Pedicels:** staminate 2.2–4 mm, pistillate 2.2–7.5 mm (7–11 mm in fruit). **Staminate flowers:** sepals 5, 1.1–2.9 mm, abaxial surface lepidote; petals 5, oblong-ovate, 2–3.1 mm, abaxial surface glabrous except margins stellate-ciliate; stamens 10–22. **Pistillate flowers:** sepals 5, equal, 2–4.5 mm, margins entire, apex incurved, abaxial surface lepidote; petals 5, pale green, ovate, 2–3.5 mm; ovary 3-locular; styles 3, 2–5 mm, usually unbranched, rarely 2-fid, terminal segments 3 (or 6). **Capsules** 1.6–2.5 × 2–3 mm, smooth; columella 3-angled. **Seeds** 6.7–8 × 5.2–6 mm, shiny.

Varieties 2 (2 in the flora): sc United States.

B. W. van Ee et al. (2006) examined populations of *Croton alabamensis* using DNA sequence and AFLP data, and their results supported the recognition of two varieties and an isolated position in the genus. Buds develop in the summer and fall for the following spring's flowers.

1. Leaf blade abaxial surfaces silvery, scales mostly unpigmented or light amber, some dark blackish brown scales sometimes present; inflorescences 9–18-flowered, producing 0–11 fruits; Alabama . . . . 1a. *Croton alabamensis* var. *alabamensis*
1. Leaf blade abaxial surfaces coppery, some scales unpigmented, others with dark reddish brown center and reddish amber rays; inflorescences 6–14-flowered, producing 0–6 fruits; Texas . . . . 1b. *Croton alabamensis* var. *texensis*

1a. Croton alabamensis E. A. Smith ex Chapman var. **alabamensis** • Alabama croton E

Leaf blades: abaxial surface silvery, scales mostly unpigmented or light amber, some dark blackish brown scales sometimes present. **Inflorescences** 9–18-flowered, producing 0–11 fruits. **2*n*** = 16, 32.

Flowering mostly late Feb–early Apr; fruiting mostly May–Jun. Limestone glades, forest understories, shale or sandstone outcrops, often on steep slopes; 50–150 m; Ala.

Variety *alabamensis* is restricted to Bibb and Tuscaloosa counties. A report of this variety from Tennessee is almost certainly erroneous (K. Wurdack 2006).

1b. Croton alabamensis E. A. Smith ex Chapman var. **texensis** Ginzbarg, Sida 15: 42, fig. 1. 1992 • Texabama croton [C] [E]

Leaf blades: abaxial surface coppery, some scales unpigmented, others with dark reddish brown center and reddish amber rays. **Inflorescences** 6–14-flowered, producing 0–6 fruits.

Flowering mostly late Feb–early Apr; fruiting mostly May–Jun. Mesic hardwood forest understories, soils overlying limestone, canyon slopes, flat terraces; of conservation concern; 200–400 m; Tex.

Variety *texensis* is restricted to Bell, Coryell, and Travis counties.

2. Croton argenteus Linnaeus, Sp. Pl. 2: 1004. 1753 (as argenteum) • Silver July croton [I]

Julocroton argenteus (Linnaeus) Didrichsen

Herbs, annual, 2–12 dm, monoecious. **Stems** branching once into 2–3 branches, tomentose. **Leaves** sometimes clustered near inflorescences; stipules linear-subulate, 4–11 mm, unlobed or deeply divided; petiole 0.3–8 cm, glands absent at apex; blade ovate to ovate-oblong, 2–15 × 1.5–8 cm, base obtuse, cuneate, or subtruncate, margins denticulate, apex obtuse to rounded, abaxial surface pale green, not appearing brown-dotted, no stellate hairs with brown centers, densely stellate-hairy, adaxial surface green, less densely stellate-hairy. **Inflorescences** bisexual, congested racemes, 1–4 cm, staminate flowers 4–10, pistillate flowers 3–6. **Pedicels:** staminate 2–3 mm, pistillate 1–4 mm (3–5 mm in fruit). **Staminate flowers:** sepals 5, 1.5–2 mm, abaxial surface stellate-hairy; petals 5, oblong, 2–3 mm, abaxial surface glabrous except margins ciliate; stamens 10–13. **Pistillate flowers:** sepals 5, unequal, 4–8 mm, margins deeply laciniate, apex incurved, abaxial surface stellate-hairy; petals 0; ovary 3-locular; styles 3, 2–3 mm, 4-fid, terminal segments 12. **Capsules** 5 × 7 mm, smooth; columella 3-angled. **Seeds** 3.2–3.8 × 2.4–3 mm, dull.

Flowering Jun–Dec. Disturbed sites, waste areas, levees; 0–50 m; introduced; Tex.; e Mexico; West Indies; Central America; South America.

Croton argenteus, which in the flora area is known only from Cameron and Hidalgo counties, may be a fairly recent introduction into the United States; the earliest known collections date from 1923.

3. Croton argyranthemus Michaux, Fl. Bor.-Amer. 2: 215. 1803 (as argyranthemum) • Healing or silver croton [F]

Herbs or subshrubs, annual or short-lived perennial, 1–6 dm, monoecious. **Stems** several from base, branching distally into 2–4 branches, lepidote. **Leaves** not clustered; stipules subulate, 0.2 mm; petiole 0.2–1(–2) cm, glands absent at apex; blade on proximal parts of stems oval to narrowly obovate, on distal parts oblong to lanceolate-oblong or broadly elliptic, 1–5 × 0.5–2(–3) cm, base rounded to narrowed, margins entire, apex obtuse to rounded, abaxial surface silvery, densely lepidote, adaxial surface darker green, sparsely lepidote or glabrate. **Inflorescences** bisexual, racemes, 2–5 cm, staminate flowers 15–35, pistillate flowers 2–8. **Pedicels:** staminate 1–5 mm, pistillate 0–2 mm. **Staminate flowers:** sepals 5, 5 mm, abaxial surface lepidote; petals 5, oblong-spatulate, 5 mm, abaxial surface densely lepidote, scales translucent, petals appearing hyaline; stamens 10–15. **Pistillate flowers:** sepals 5–7, connate for ½+ length, equal, 3–4 mm, margins entire, apex incurved, abaxial surface lepidote; petals 0; ovary 3-locular; styles 3, 2–4 mm, 1–3 times 2-fid at apex, terminal segments 6–24. **Capsules** 5–6 × 3–4 mm, smooth; columella 3-angled. **Seeds** 4–5 × 2.5–3 mm, dull.

Flowering Mar–Sep. Deep sandy soils in pinelands, pine-oak scrub, sandhills; 0–200 m; Ala., Fla., Ga., La., Miss., Tex.; Mexico (Nuevo León, Tamaulipas).

Croton argyranthemus is closely related to *C. coryi*. Although they have very different kinds of vegetative indumentum, the staminate petals are characteristically silvery-lepidote in both species.

4. Croton californicus Müller Arg. in A. P. de Candolle and A. L. P. P. de Candolle, Prodr. 15(2): 691. 1866 • California croton

Hendecandra procumbens Eschscholtz, Mém. Acad. Imp. Sci. St. Pétersbourg Hist. Acad. 10: 287. 1826, not *Croton procumbens* Jacquin 1760; *C. californicus* var. *longipes* (M. E. Jones) A. M. Ferguson; *C. californicus* var. *mohavensis* A. M. Ferguson; *C. californicus* var. *tenuis* (S. Watson) A. M. Ferguson; *C. longipes* M. E. Jones; *C. mohavensis* (A. M. Ferguson) Tidestrom; *C. tenuis* S. Watson

Subshrubs or shrubs, 4–11 dm, dioecious. **Stems** loosely branched, stellate-lepidote. **Leaves** not clustered; stipules absent; petiole 1–3.5(–4.5) cm, usually less than ½ blade

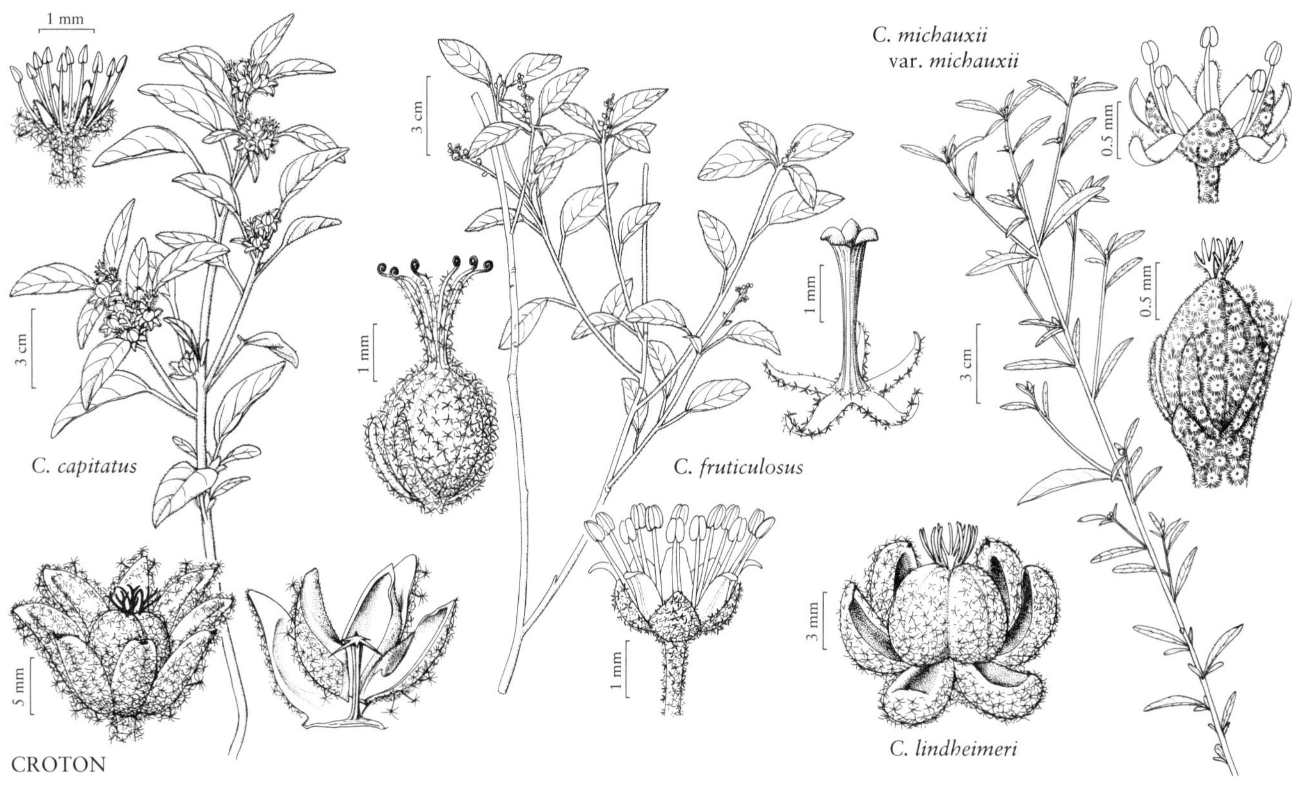

length, glands absent at apex; blade elliptic to narrowly oblong, 2–5.5(–7) × 0.8–2(–2.5) cm, usually more than 2 times as long as wide, base obtuse, margins entire, apex obtuse to rounded, abaxial surface light green, adaxial surface darker green, both stellate-lepidote. **Inflorescences** unisexual, racemes or thyrses; staminate 0.5–3.5(–10) cm, flowers 3–8(–20); pistillate 0.5–1 cm, flowers 1–6. **Pedicels:** staminate 1–5.5(–7) mm, pistillate to 1 mm (1–3 mm in fruit). **Staminate flowers:** sepals 5, 1 mm, abaxial surface stellate-lepidote; petals 0; stamens 10–15. **Pistillate flowers:** sepals 5, equal, 2 mm, margins entire, apex incurved, abaxial surface stellate-lepidote; petals 0; ovary 3-locular; styles 3, 2–2.5 mm, 4-fid, terminal segments usually 12. **Capsules** 6–8 × 5–7 mm, smooth; columella 3-winged. **Seeds** 4–5.5 × 3.5–5.5 mm, dull. **2*n*** = 28.

Flowering Feb–Oct. Sandy soils, sage scrub, dunes, washes; 0–900 m; Ariz., Calif., Nev., Utah; Mexico (Baja California, Sinaloa).

5. Croton capitatus Michaux, Fl. Bor.-Amer. 2: 214. 1803 (as capitatum) • Capitate or woolly croton, hogwort [E] [F] [W]

Heptallon graveolens Rafinesque; *Oxydectes capitata* (Michaux) Kuntze; *Pilinophytum capitatum* (Michaux) Klotzsch

Herbs, annual, 3–8 dm, monoecious. **Stems** well branched distally, stellate-hairy, hairs whitish to pale yellow, glabrescent. **Leaves** not clustered; stipules filiform, 2–5 mm; petiole 0.5–6 cm, glands absent at apex; blade ovate to lanceolate-elliptic, 3–8(–15) × 1–4 cm, base rounded to cuneate, margins entire, apex obtuse to acute or rounded, abaxial surface light green, not appearing brown-dotted, no stellate hairs with brown centers, adaxial surface darker green, both appressed stellate-hairy. **Inflorescences** bisexual, racemes or capitate, 1.5–3 cm, staminate flowers 5–15, pistillate flowers 3–6. **Pedicels:** staminate 0.5–1.5 mm, pistillate 0–1.5 mm. **Staminate flowers:** sepals 5, 0.8–1 mm, abaxial surface stellate-hairy; petals 5, linear-oblong-lanceolate, 0.8–1 mm, abaxial surface stellate-hairy; stamens 7–12. **Pistillate flowers:** sepals 6–9, equal, 7–10 mm (–15 mm in fruit), opening broadly and persistent after capsule dehiscence, margins entire, apex recurved, abaxial surface stellate-hairy; petals 0; ovary

3-locular; styles 3, 2–3 mm, 2–3 times 2-fid, terminal segments 12–18(–24). **Capsules** 7–9 × 5–6 mm, smooth; columella tipped with 3-pronged grappling hooklike appendage. **Seeds** 5 × 4.5–5 mm, shiny. **2*n*** = 20.

Flowering Jun–Oct. Pastures, old fields, cultivated land, prairies, flood plains, longleaf pinelands, sandy to loamy soils; 0–300 m; Ala., Ark., Del., D.C., Ill., Ind., Iowa, Kans., Ky., Md., Mass., Miss., Mo., Nebr., N.J., N.C., Ohio, Okla., Pa., S.C., Tenn., Tex., Va., W.Va.

Croton capitatus is frequently confused with *C. heptalon* and *C. lindheimeri*. *Croton capitatus* usually has some leaves that are blunt-tipped and mucronate (versus all or most acute in *C. lindheimeri*), and the pistillate sepals are longer than the ovary, with tips that flare outward at anthesis. Also, *C. capitatus* has petioles roughly equal in length from the middle to the tip of the stem, whereas *C. lindheimeri* has petioles that decrease more markedly in length from the middle to the tip of the stem. Occurrences of *C. capitatus* east of the Appalachian Mountains are likely adventive.

6. **Croton ciliatoglandulifer** Ortega, Nov. Pl. Descr. Dec., 51. 1797 (as cilato-glanduliferum) • Mexican croton

Shrubs, 2–10 dm, monoecious. **Stems** ± dichotomously branched, stellate-hairy. **Leaves** not clustered; stipules glandular-ciliate, 2–3 mm; petiole 1–3.5 cm, ½–7/10+ leaf blade length, stipitate-glandular at apex; blade ovate, 2–5 × 1–3 cm, base rounded to subcordate, margins entire, glandular-ciliate, apex acute to acuminate, abaxial surface pale green, densely stellate-hairy, adaxial surface darker green, sparsely stellate-hairy. **Inflorescences** bisexual, racemes, 1–2.5 cm, staminate flowers 3–10, pistillate flowers 1–8. **Pedicels:** staminate 0.5–1.5 mm, pistillate 2–3 mm. **Staminate flowers:** sepals 5, 3.5–4 mm, abaxial surface stellate-hairy; petals 5, spatulate, 2.5–3 mm, abaxial surface glabrous; stamens 20–30. **Pistillate flowers:** sepals 5, equal, 3–5 mm, margins entire, glandular-ciliate, apex straight, abaxial surface stellate-hairy; petals 0; ovary 3-locular; styles 3, 4–5 mm, 4-fid, terminal segments 12. **Capsules** 4–5 × 5–7 mm, smooth; columella apex with 3 rounded, inflated lobes. **Seeds** 4–5.5 × 2.5–3 mm, shiny. **2*n*** = 20.

Flowering Oct–Dec. Rocky forested slopes, sandy flats; 0–2200 m; Ariz., Tex.; Mexico; West Indies.

Croton ciliatoglandulifer is characterized by the glandular projections on the leaf blade margins, stipules, and pistillate sepals. The species is closely related to *C. humilis* and *C. soliman*, which may intergrade with it. *Croton ciliatoglandulifer* presents an interesting disjunction in the United States from lowland southern Texas (0–150 m elevation) to upland southern Arizona (to 2200 m elevation).

7. **Croton cortesianus** Kunth in A. von Humboldt et al., Nov. Gen. Sp. 2(fol.): 66; 2(qto.): 83. 1817 • Cortez's croton, palillo

Croton chichenensis Lundell; *C. trichocarpus* Torrey

Shrubs, 10–30 dm, dioecious. **Stems** di- or trichotomously branched, stellate-hairy, young stems often blackened by fungus. **Leaves** not clustered; stipules rudimentary; petiole 1–1.5 cm, 1/6–1/3 leaf blade length, glands absent at apex; blade narrowly obovate, oblong, elliptic, or lanceolate, 3–10 × 1–5 cm, mostly less than 3.5 times as long as wide, base subacute to rounded, margins minutely denticulate, apex acute or acuminate, abaxial surface light green, densely ochraceous stellate-hispid when young, pale stellate-hairy when mature, adaxial surface darker green, glabrous. **Inflorescences** unisexual, racemes; staminate 7–20 cm, flowers 10–30; pistillate 1–5 cm, flowers 8–20(–30). **Pedicels:** staminate 0–1 mm, pistillate 0–0.5 mm. **Staminate flowers:** sepals 5, 2.5 mm, abaxial surface stellate-hairy; petals 5, oblong, 3 mm, abaxial surface glabrous except margins ciliate basally; stamens 12–16. **Pistillate flowers:** sepals 5, equal, 1–1.5 mm, margins entire, apex straight to slightly incurved, abaxial surface stellate-hispid; petals rudimentary; ovary 3-locular; styles 3, 4–5 mm, 2-fid, terminal segments 6. **Capsules** 5–6 × 4.5–5.5 mm, smooth; columella apex with 3 rounded, inflated lobes. **Seeds** 4–5 × 3–3.5 mm, shiny (with silvery sheen). **2*n*** = 20.

Flowering Feb–Oct, possibly year-round. Scrub forests; 0–50 m; Tex.; e, se Mexico; Central America.

Croton cortesianus is known in the flora area only from Cameron, Hidalgo, and Starr counties.

8. **Croton coryi** Croizat, Bull. Torrey Bot. Club 69: 446, 457. 1942 • Cory's croton [E]

Herbs, annual, 5–10 dm, monoecious. **Stems** trichotomously branched, mostly densely, grayish white, long stellate-hairy. **Leaves** sometimes clustered near inflorescences; stipules absent; petiole 1.5–4 cm, glands absent at apex; blade narrowly ovate to elliptic, 3.5–7 × 1.5–3.5 cm, base rounded to subtruncate or subcordate, margins entire, apex acute, both surfaces light olive green, abaxial surface not appearing brown-dotted, no stellate hairs with brown centers, whitish stellate-hispid. **Inflorescences** bisexual, subspicate racemes, 4–7 cm, staminate flowers 15–25, pistillate flowers 8–15. **Pedicels:** staminate 0.5–2.5 mm, pistillate 0–3 mm.

Staminate flowers: sepals 5, 3 mm, abaxial surface stellate-hairy; petals 5, ligulate, 3–3.5 mm, abaxial surface densely lepidote, scales translucent, petals appearing hyaline; stamens 14–16. **Pistillate flowers:** sepals 6–8, equal, 7 mm, margins entire, apex straight to slightly incurved, abaxial surface shaggy-tomentose; petals 0 or rudimentary; ovary 3-locular; styles 3, 3–4 mm, 2 times 2-fid, terminal segments 12. **Capsules** 4.5–5.5 × 4 mm, smooth; columella 3-angled. **Seeds** 3 × 2 mm, dull.

Flowering May–Nov. Grasslands, woodland openings, deep well-drained sands, barrier island dunes, similar habitats inland; 0–50 m; Tex.

Croton coryi is endemic to the South Texas Sand Sheet of coastal southern Texas (Brooks, Hidalgo, Jim Hogg, Kenedy, Kleberg, Nueces, Starr, and Willacy counties). Despite the marked difference in vegetative indumenta, *C. coryi* is closely related to *C. argyranthemus*. The most obvious feature linking them is the silvery lepidote staminate petals.

9. **Croton dioicus** Cavanilles, Icon. 1: 4, plate 6. 1791 (as dioicum) • Grassland croton, hierba del gato, rosval, rubaldo

Croton elaeagnifolius Vahl; *C. gracilis* Kunth; *C. neomexicanus* Müller Arg.; *C. vulpinus* Sessé & Mociño

Subshrubs, 2–5(–9) dm, dioecious. **Stems** well branched from base, stellate-lepidote. **Leaves** not clustered; stipules rudimentary or absent; petiole 0.2–0.8(–2) cm, usually less than ½ blade length, glands absent at apex; blade narrowly elliptic-ovate to lanceolate, 1–6.5 × 0.6–2.2 cm, usually less than ½ as wide as long, base rounded, margins entire, apex acute to rounded, abaxial surface pale green, densely silvery lepidote or stellate-lepidote, adaxial surface darker green, less densely lepidote. **Inflorescences** unisexual, racemes; staminate 2–8 cm, flowers 4–16; pistillate 0.5–1 cm, flowers 2–5. **Pedicels:** staminate 1–4 mm, pistillate 2–5 mm. **Staminate flowers:** sepals 5, 1 mm, abaxial surface lepidote; petals 0; stamens 10–12. **Pistillate flowers:** sepals 5, equal, 1.5–2 mm, margins entire, apex straight to slightly incurved, abaxial surface stellate-lepidote; petals 0; ovary 3-locular; styles 3, 0.5–1.5 mm, 2–3 times 2-fid, terminal segments 12–24. **Capsules** 5–6 mm diam., smooth; columella 3-winged. **Seeds** 3.5–5 × 3–4 mm, shiny or dull. $2n$ = 28, 56.

Flowering Mar–Nov. Limestone and igneous mountains, canyons, mesas, flats, disturbed areas; 30–2000 m; N.Mex., Tex.; Mexico.

10. **Croton elliottii** Chapman, Fl. South. U.S., 407. 1860 • Elliott's or pondshore croton [E]

Herbs, annual, 3–8 dm, monoecious. **Stems** usually single from base, then well branched from first reproductive node, appressed stellate-hairy. **Leaves** sometimes clustered near inflorescences; stipules absent; petiole 0.5–2 cm, glands absent at apex; blade lanceolate to oblong, 2–5.5 × 0.2–0.8 cm, base rounded, margins entire, apex subacute, abaxial surface pale green, not appearing brown-dotted, no stellate hairs with brown centers, densely long stellate-hairy, adaxial surface darker green, less densely short stellate-hairy. **Inflorescences** bisexual, congested racemes, 0.8–1.5 cm, staminate flowers 5–15, pistillate flowers 3–6. **Pedicels:** staminate 0.4–0.9 mm, pistillate 0–0.8 mm. **Staminate flowers:** sepals 5, 0.8–1 mm, abaxial surface stellate-hairy; petals 5, linear-oblong to lanceolate, 0.8–1 mm, abaxial surface glabrous; stamens 7–10. **Pistillate flowers:** sepals 6–7, equal, 5–6 mm, margins entire, apex incurved, abaxial surface stellate-hairy; petals 0; ovary 3-locular; styles 3, 2–3 mm, 2 times 2-fid, terminal segments 12. **Capsules** 4–5 mm diam., smooth; columella slightly 3-winged distally. **Seeds** 4–4.5 × 3–4 mm, shiny.

Flowering Jul–Sep. Depression ponds, depression meadows, clay-based Carolina bays, usually on exposed pond edges or bottoms; 0–100 m; Ala., Fla., Ga., S.C.

Croton elliottii is most closely related to *C. capitatus*.

11. **Croton fruticulosus** Engelmann ex Torrey in W. H. Emory, Rep. U.S. Mex. Bound. 2(1): 194. 1859 (as fruticulosum) • Bush croton, encinilla, hierba loca [F]

Shrubs, 2–10 dm, monoecious. **Stems** much branched distally, stellate-hairy. **Leaves** not clustered; stipules rudimentary or absent; petiole 0.4–1(–1.5) cm, ⅛–⅕ blade length, glands absent at apex; blade ovate to ovate-lanceolate, 2–8 × 2–4 cm, base truncate to cordate, margins serrulate, often slightly undulate, apex attenuate, acute, abaxial surface pale green, stellate-hairy, adaxial surface darker green, puberulent. **Inflorescences** bisexual, racemes, 3–12 cm, staminate flowers 10–20, pistillate flowers 2–5. **Pedicels:** staminate 2.5–4 mm, pistillate 0–0.5 mm. **Staminate flowers:** sepals 5, 0.8–1.2 mm, abaxial surface stellate-tomentose; petals 5, oblanceolate to spatulate, 2 mm, abaxial surface glabrous except margins densely fimbrillate-villous; stamens 9–16. **Pistillate flowers:** sepals 5, equal, 2.2 mm, margins entire, apex incurved, abaxial surface stellate-hairy; petals 0; ovary 3-locular; styles 3, 3–4.5 mm, 2-fid to base, terminal segments 6. **Capsules** 5–6 mm diam., smooth; columella apex with 3 rounded, inflated lobes. **Seeds** 4–5 × 3.2–3.8 mm, shiny.

Flowering May–Dec. Limestone or basalt hills; 100–1700 m; Ariz., N.Mex., Tex.; Mexico (Chihuahua, Coahuila, Nuevo León, San Luis Potosí, Tamaulipas).

Croton fruticulosus is known in the flora area from southeastern Arizona through southern New Mexico and trans-Pecos Texas to the Edwards Plateau.

12. Croton glandulosus Linnaeus, Syst. Nat. ed. 10, 2: 1275. 1759 (as glandulosum) • Sand or tooth-leaved or tropic croton, vente conmigo

Decarinium glandulosum (Linnaeus) Rafinesque; *Geiseleria glandulosa* (Linnaeus) Klotzsch; *Oxydectes glandulosa* (Linnaeus) Kuntze

Herbs, annual, 1–12 dm, monoecious. **Stems** much branched distally, usually coarsely stellate-hairy, rarely glabrescent. **Leaves** not clustered; stipules linear-subulate, glandular or not, to 0.5 mm or absent; petiole 0.2–1 cm, glands at apex 2, yellow, sessile to shortly stipitate, cuplike; blade ovate proximally, oblong-lanceolate distally, 0.6–3.5(–7) × 0.3–1.5(–3) cm, base obtuse to truncate, margins coarsely crenate to serrate-dentate, apex obtuse to rounded, both surfaces green, stellate-hairy, glabrate, or rarely glabrous. **Inflorescences** bisexual, racemes, 1–3 cm, staminate flowers 10–20, pistillate flowers 1–4. **Pedicels:** staminate 0.8–2 mm, pistillate 0–5 mm. **Staminate flowers:** sepals 5, 0.8–1.2 mm, abaxial surface stellate-hairy; petals 5, oblanceolate, 1–1.3 mm, abaxial surface glabrous except margins ciliate; stamens 7–13. **Pistillate flowers:** sepals 5, subequal, 6–7.5 mm, margins entire, apex straight to slightly incurved, abaxial surface glabrous except stellate-hairy apically; petals 0 or 5, rudimentary; ovary 3-locular; styles 3, 1–2.5 mm, deeply 2-fid, terminal segments 6. **Capsules** 3.5–6 × 4–5 mm, smooth; columella 3-angled. **Seeds** 3–4 × 2–2.5 mm, shiny. $2n = 16$.

Varieties ca. 20 (5 in the flora): United States, Mexico, West Indies, Central America, South America.

Croton glandulosus is widespread in the New World, with a complex pattern of variation. The classification here follows B. W. van Ee et al. (2009).

1. Leaf blades glabrate or sparsely to moderately stellate-hairy.
   2. Plants 2–12 dm; leaf blades 2–7 cm . . . . 12e. *Croton glandulosus* var. *septentrionalis* (in part)
   2. Plants 1–2 dm; leaf blades 0.6–2(–3) cm.
      3. Leaf blades glabrate, bases markedly 3-veined; petiole apical glands sessile; Florida . . . . 12b. *Croton glandulosus* var. *floridanus*
      3. Leaf blades sparsely stellate-hairy, bases obscurely 3-veined; petiole apical glands stipitate; Kansas, New Mexico, Oklahoma, Texas . . . . . . 12c. *Croton glandulosus* var. *lindheimeri* (in part)
1. Leaf blades moderately to densely stellate-hairy.
   4. Leaf blades firm-thick, length mostly 2 times width or less, marginal teeth rounded; petiole apical glands sessile . . . . 12a. *Croton glandulosus* var. *arenicola*
   4. Leaf blades membranous, length mostly more than 2 times width, marginal teeth pointed; petiole apical glands sessile or stipitate.
      5. Leaf blades 1–2(–3) × 0.3–0.8(–1.3) cm; petiole apical glands stipitate, circular when dry, 0.1–0.4 mm diam. . . . . 12c. *Croton glandulosus* var. *lindheimeri* (in part)
      5. Leaf blades 2–7 × 0.7–3 cm; petiole apical glands sessile, wavy-wrinkled when dry, 0.5–0.8 mm diam.
         6. Stems densely stellate-hairy, hairs spreading, radii unequal; leaf blades densely stellate-villous . . . 12d. *Croton glandulosus* var. *pubentissimus*
         6. Stems moderately stellate-hairy, hairs appressed, radii equal; leaf blades moderately stellate-hairy . . . . 12e. *Croton glandulosus* var. *septentrionalis* (in part)

12a. Croton glandulosus Linnaeus var. **arenicola** (Small) B. W. van Ee, P. E. Berry & Ginzbarg, Harvard Pap. Bot. 14: 49. 2009 [E]

Croton arenicola Small, Bull. New York Bot. Gard. 3: 428. 1905

Plants 1–2 dm. **Stems** densely stellate-hairy, hairs appressed, radii equal. **Leaves:** petiole apical glands sessile, circular when dry, 0.2 mm diam.; blade 0.7–3.5 × 0.5–1.5 cm, length mostly 2 times width or less, firm-thick, marginal teeth rounded, both surfaces densely stellate-hairy; base obscurely 3-veined.

Flowering year-round. Beaches, old fields, disturbed sites, sandy waste places; 0–50 m; Fla.

Variety *arenicola* is restricted to Miami-Dade, Monroe, and Palm Beach counties. A. M. Ferguson (1901) treated a collection of this variety as *Croton betulinus* Vahl, but that is a distinct, perennial species from the West Indies.

12b. Croton glandulosus Linnaeus var. **floridanus** (A. M. Ferguson) R. W. Long, Rhodora 72: 22. 1970 [E]

Croton floridanus A. M. Ferguson, Rep. (Annual) Missouri Bot. Gard. 12: 50, plate 15. 1901

Plants 1–2 dm. **Stems** sparsely stellate-hairy, hairs appressed, radii equal, central radius lacking. **Leaves:** petiole apical glands sessile, circular when dry, 0.2–0.3 mm diam.; blade 0.6–1.5(–3) × 0.5–1.2 cm, length mostly 2 times width or less, membranous, marginal teeth rounded, both surfaces glabrate; base markedly 3-veined.

Flowering year-round. Beaches, sand dunes, old fields, disturbed sites, waste places; 0–50 m; Fla.

Variety *floridanus* grows in Broward, Collier, Lee, Manatee, Martin, and Pinellas counties.

12c. Croton glandulosus Linnaeus var. **lindheimeri** Müller Arg. in A. P. de Candolle and A. L. P. P. de Candolle, Prodr. 15(2): 685. 1866

Croton glandulosus var. *parviseminus* Croizat

Plants 1–2 dm. **Stems** stellate-hairy, hairs appressed, radii equal. **Leaves:** petiole apical glands stipitate, circular when dry, 0.1–0.4 mm diam.; blade 1–2(–3) × 0.3–0.8(–1.3) cm, length mostly more than 2 times width, membranous, marginal teeth pointed, both surfaces sparsely to moderately stellate-hairy; base obscurely 3-veined.

Flowering May–Dec. Old fields, roadsides, waste places; 0–900 m; Kans., N.Mex., Okla., Tex.; e Mexico.

12d. Croton glandulosus Linnaeus var. **pubentissimus** Croizat, J. Arnold Arbor. 26: 188. 1945 [E]

Croton glandulosus var. *hirsutus* Shinners

Plants 1–2 dm. **Stems** densely stellate-hairy, hairs spreading, radii unequal, central radius prominent. **Leaves:** petiole apical glands sessile, wavy-wrinkled when dry, 05–0.8 mm diam.; blade 2–7 × 0.7–3 cm, length mostly more than 2 times width, membranous, marginal teeth pointed, both surfaces densely stellate-villous; base obscurely 3-veined.

Flowering May–Nov. Beaches, sand dunes, roadsides; 0–50 m; Tex.

Variety *pubentissimus* grows primarily in the Texas Gulf Coast counties of Aransas, Brazoria, Cameron, Kenedy, Kleberg, Nueces, and San Patricio, but a few collections are known from inland Colorado and Victoria counties.

12e. Croton glandulosus Linnaeus var. **septentrionalis** Müller Arg. in A. P. de Candolle and A. L. P. P. de Candolle, Prodr. 15(2): 686. 1866 [E] [W]

Croton glandulosus var. *angustifolius* Müller Arg.; *C. glandulosus* var. *crenatifolius* A. M. Ferguson; *C. glandulosus* var. *shortii* A. M. Ferguson; *C. glandulosus* var. *simpsonii* A. M. Ferguson

Plants 2–12 dm. **Stems** moderately stellate-hairy, hairs appressed, radii equal. **Leaves:** petiole apical glands sessile, wavy-wrinkled when dry, 0.5–0.8 mm diam.; blade 2–7 × 0.7–3 cm, length mostly more than 2 times width, membranous, marginal teeth pointed, both surfaces moderately stellate-hairy; base 3-veined.

Flowering May–Nov. Sand dunes, old fields, roadsides, waste places, cultivated land; 0–900 m; Ala., Ark., Del., Fla., Ga., Ill., Ind., Iowa, Kans., Ky., La., Md., Mich., Minn., Miss., Mo., Nebr., N.J., N.C., Ohio, Okla., Pa., S.C., Tenn., Tex., Va., W.Va., Wis.

Variety *septentrionalis* is by far the most widespread variety of *Croton glandulosus* in North America, and the only one present north of Florida in the east and north of Oklahoma and Kansas in the Midwest.

13. Croton heptalon (Kuntze) B. W. van Ee & P. E. Berry, Syst. Bot. 35: 159. 2010 • Woolly croton

Oxydectes heptalon Kuntze, Revis. Gen. Pl. 2: 610. 1891, based on *Croton berlandieri* Müller Arg., Linnaea 34: 141. 1865, not Torrey 1859; *C. albinoides* (A. M. Ferguson) Croizat; *C. capitatus* Michaux var. *albinoides* (A. M. Ferguson) Shinners; *C. engelmannii* A. M. Ferguson var. *albinoides* A. M. Ferguson; *C. muelleri* J. M. Coulter var. *albinoides* (A. M. Ferguson) Croizat; *Heptallon aromaticum* Rafinesque

Herbs, annual, 5–15 dm, monoecious; stems, leaves, and buds whitish-hairy when young, becoming glabrate. **Stems** well branched distally, stellate-hairy. **Leaves** not clustered; stipules linear, 2–7 mm; petiole 0.5–5 cm, glands absent at apex; blade ovate-lanceolate, 3–10 × 1–5 cm, base cordate to rounded, margins entire, apex acute, abaxial surface pale green, not appearing brown-dotted, no stellate hairs with brown centers, adaxial surface darker green, both stellate-hairy. **Inflorescences** bisexual, racemes, 2–4 cm, staminate flowers 3–10, pistillate flowers 4–8. **Pedicels:** staminate 2–4 mm, pistillate 1–2 mm. **Staminate flowers:** sepals (4–)5, 1–2 mm, abaxial surface stellate-hairy; petals 5, linear-oblong-lanceolate, 1–1.5 mm, abaxial surface stellate-hairy; stamens 9–12. **Pistillate flowers:** sepals 7–8, subequal, 3–6 mm, margins entire, apex straight to slightly incurved, abaxial surface whitish appressed-tomentose; petals 0; ovary 3-locular; styles 3, 3–4 mm, 4-fid, terminal segments 12. **Capsules** 6–8 × 6–7 mm, smooth; columella tipped with 3-pronged grappling hooklike appendage. **Seeds** 4–5 × 2–2.5 mm, shiny.

Flowering May–Dec. Beaches, coastal dunes, roadsides; 0–50 m; Tex.; e Mexico.

Morphological differences among *Croton heptalon* and its multifid-styled relatives in sect. *Heptallon*, especially *C. capitatus* and *C. lindheimeri*, can be quite subtle. In general, *C. heptalon* can be distinguished from *C. capitatus* by its more elongate pistillate part of the inflorescence, non-recurving sepal tips in the pistillate flowers, and more cordate leaf bases on larger basal leaves. Whitish pubescence on its young growth and styles that branch once into four terminal segments distinguish *C. heptalon* from *C. lindheimeri*.

Croton muelleri J. M. Coulter, which is an illegitimate name, pertains here.

14. Croton humilis Linnaeus, Syst. Nat. ed. 10, 2: 1276. 1759 (as humile) • Low croton, pepperbush, salvia

Croton berlandieri Torrey

Shrubs, 3–8 dm, monoecious. **Stems** much branched, stellate-hairy, viscid. **Leaves** not clustered; stipules 2–5 stipitate glands, to 0.5 mm; petiole 0.7–3(–5) cm, 3/8–5/8 leaf blade length, glands absent at apex; blade ovate to oblong, 1.5–8 × 1–2(–5) cm, base rounded to subcordate, margins entire to minutely glandular-denticulate, apex abruptly acute to acuminate, both surfaces pale green, abaxial densely stellate-hairy, adaxial tomentose, glabrescent. **Inflorescences** bisexual or unisexual, racemes, 3–7 cm, staminate flowers 20–35, pistillate flowers 2–6. **Pedicels:** staminate 3–4 mm, pistillate 1–2(–3) mm. **Staminate flowers:** sepals 5, 3–4 mm, abaxial surface stellate-hairy; petals 5, spatulate, 3–4 mm, abaxial surface glabrous except margins ciliate basally; stamens 15–35. **Pistillate flowers:** sepals 5, equal, 4 mm, margins entire, sessile- or shortly stipitate-glandular, apex incurved, abaxial surface stellate-hairy; petals 0 or 5, white, subulate, 1 mm; ovary 3-locular; styles 3, 3–5 mm, 4-fid, terminal segments 12. **Capsules** 4–5 × 4 mm, smooth; columella apex with 3 rounded, inflated lobes. **Seeds** 3–4 × 2.5–3 mm, shiny. $2n = 20$.

Flowering year-round. Hammocks, thickets, disturbed areas; 0–50 m; Fla., Tex.; e, se Mexico; West Indies.

Croton humilis is a mainly West Indian species extending from southernmost Florida (Collier and Monroe counties) to the Bahamas, Cuba, Hispaniola, Jamaica, and Puerto Rico, and up the Caribbean coast of Mexico to southernmost Texas (Cameron, Hidalgo, Starr, Willacy, and Zapata counties). Texas plants have more stamens (30–35) than Florida plants (15–20).

15. Croton incanus Kunth in A. von Humboldt et al., Nov. Gen. Sp. 2(fol.): 58; 2(qto.): 73. 1817 • Salvia, Torrey's croton, vara blanca

Croton suaveolens Torrey var. *oblongifolius* Torrey; *C. torreyanus* Müller Arg.

Shrubs, 10–20 dm, monoecious. **Stems** much branched distally, stellate-velutinous. **Leaves** sometimes clustered near inflorescences; stipules linear-subulate, 2–3 mm; petiole 0.7–1.5 cm, (1/4–)3/8–1/2 leaf blade length, glands absent at apex; blade oblong, ovate-oblong, or elliptic-oblong, 1.5–4(–6) × 1.5–3 cm, base rounded to obtuse, margins entire, apex acute to obtuse, abaxial surface whitish, densely stellate-tomentose, adaxial surface darker green, stellate-tomentose. **Inflorescences** bisexual, racemes, 2–5 cm, staminate flowers 10–25, pistillate flowers 2–4. **Pedicels:** staminate 1.7–3 mm, pistillate 1–2 mm. **Staminate flowers:** sepals 5, 1.5–2 mm, abaxial surface densely tomentose; petals 5, oblanceolate, 1.5–2 mm, abaxial surface villous; stamens 10–16. **Pistillate flowers:** sepals 5, equal, 1.2–3.5 mm, margins entire, apex straight to slightly incurved, abaxial surface tomentose; petals 0; ovary 3-locular; styles 3, 1.8–3.5 mm, 2-fid to base, terminal segments 6. **Capsules** 6–8 × 4–5 mm, smooth; columella apex with 3 rounded, inflated lobes. **Seeds** 4.5–7 × 3–4 mm, dull. **2*n*** = 20.

Flowering Mar–Nov. Calcareous loams, xeric rocky limestone slopes and canyons; 0–700 m; Tex.; n, c Mexico.

Croton incanus is known in the flora area from the trans-Pecos region to south Texas.

16. Croton leucophyllus Müller Arg., Linnaea 34: 139. 1865 • Two-color croton

Oxydectes leucophylla (Müller Arg.) Kuntze

Varieties 2 (1 in the flora): Texas, ne Mexico.

16a. Croton leucophyllus Müller Arg. var. **leucophyllus**

Herbs, annual, 1–3(–5) dm, monoecious. **Stems** dichotomously branched throughout, stellate-hairy. **Leaves** not clustered; stipules subulate, 1–2.5 mm; petiole 0.5–3 cm, glands absent at apex; blade ovate to ovate-oblong, 1–2.5(–5) × 0.5–1.5(–3) cm, base rounded to cuneate, margins entire, apex rounded to subacute, abaxial surface pale green, not appearing brown-dotted, no stellate hairs with brown centers, densely stellate-hairy, adaxial surface slightly darker green, sparsely stellate-hairy. **Inflorescences** bisexual, racemes, 0.5–2 cm, staminate flowers 1–15, pistillate flowers 3–6. **Pedicels:** staminate 1.5–2 mm, pistillate 2–2.5 mm. **Staminate flowers:** sepals 5, 0.8–1 mm, abaxial surface stellate-hairy; petals 5, linear-oblong-lanceolate, 0.8–1 mm, abaxial surface glabrous except margins villous; stamens 7–10. **Pistillate flowers:** sepals 5, unequal, 3 outer 5–7 mm, 2 inner 1–2 mm, margins entire, apex incurved, abaxial surface stellate; petals 0; ovary 3-locular; styles 3, 1–2 mm, 2-fid to base, terminal segments 6. **Capsules** 4–5 mm diam., smooth; columella apex with 3 short recurved lobes. **Seeds** 3.5–4 × 1.8–2 mm, shiny.

Flowering Jun–Nov. Bottomlands, flood plains; 0–100 m; Tex.; Mexico (Coahuila, Nuevo León, Sonora, Tamaulipas).

Variety *leucophyllus* is known in the flora area from at least Brewster County southeast to Cameron and Nueces counties. *Croton leucophyllus* var. *trisepalis* A. M. Ferguson (*C. palmeri* var. *ovalis* Fernald) differs from the typical variety in having the two inner sepals abortive in the pistillate flowers so that they appear to have only three sepals. Variety *trisepalis* is found on shale plains in the Mexican states of Nuevo León and Tamaulipas, and may be expected in nearby Texas.

17. Croton lindheimeri (Engelmann & A. Gray) Alph. Wood, Class-book Bot. ed. s.n.(b), 631. 1861 • Goatweed, woolly croton [E] [F]

Pilinophytum lindheimeri Engelmann & A. Gray, Boston J. Nat. Hist. 5: 232. 1845; *Croton capitatus* Michaux var. *lindheimeri* (Engelmann & A. Gray) Müller Arg.

Herbs, annual, 5–20 dm, monoecious, stems, leaves, and buds with yellow-brown (ochraceous) pubescence when young, becoming glabrate. **Stems** branching distally, stellate-hairy. **Leaves** not clustered; stipules linear, 0–5 mm; petiole 1.5–7 cm, glands absent at apex; blade ovate-lanceolate, 3–7 × 1–3 cm, base cordate to rounded or subcordate, margins entire, apex acute, abaxial surface pale green, not appearing brown-dotted, no stellate hairs with brown centers, densely stellate-hairy, adaxial surface greener, more sparsely hairy. **Inflorescences** bisexual, racemes, 1.5–3 cm, staminate flowers 8–15, pistillate flowers 2–7. **Pedicels:** staminate 0.5–3 mm, pistillate 0–1 mm. **Staminate flowers:** sepals (4–)5, 1.5–2 mm, abaxial surface stellate-hairy; petals 5, linear-oblong, 1–1.5 mm, abaxial surface stellate-hairy; stamens 9–13. **Pistillate flowers:** sepals 7–8, equal, 5–7 mm, margins entire, apex straight to slightly incurved, abaxial surface yellowish woolly-tomentose; petals 0; ovary 3-locular; styles 3, 3–4 mm, 2 times 2-fid, terminal segments 12. **Capsules** 6–8 × 8–9 mm, smooth; columella tipped with 3-pronged grappling hooklike appendage. **Seeds** 4–5 × 4–4.5 mm, shiny. **2*n*** = 20.

Flowering May–Dec. Old pastures, forest openings, bottomlands, fence rows, disturbed areas; 0–300 m; Ala., Ark., Fla., Ga., La., Miss., Mo., S.C., Tenn., Tex.

Croton lindheimeri is very similar to *C. capitatus*, and the two can sometimes be found growing together in northeastern Texas. *Croton lindheimeri* can be distinguished by its more ochraceous pubescence on young growth, consistently acute leaf tips, somewhat more elongated pistillate part of the raceme, and pistillate sepal tips that do not recurve after anthesis. *Croton lindheimeri* was reported from Indiana, adventive on ballast in 1898, and from Kansas based on a single 1883 collection from Miami County, but apparently did not become established in either state.

18. Croton lindheimerianus Scheele, Linnaea 25: 580. 1852 • Three-seed croton

Herbs, annual, 1–5 dm, monoecious. **Stems** dichotomously branched, densely stellate-tomentose. **Leaves** sometimes clustered near inflorescences; stipules rudimentary; petiole 0.5–3 cm, glands absent at apex; blade lanceolate to suborbiculate, 1–5(–8) × 0.8–2.8(–3.5) cm, base rounded, margins entire, apex rounded to acute, abaxial surface pale green, stellate-hairy, adaxial surface darker green, sparsely stellate-hairy. **Inflorescences** bisexual, racemes, 1–3 cm, not appearing brown-dotted, no stellate hairs with brown centers, staminate flowers 3–8, pistillate flowers 1–3. **Pedicels:** staminate 2–3 mm, pistillate 1–5(–15) mm. **Staminate flowers:** sepals (4–)5, 1.5–2 mm, abaxial surface stellate-hairy; petals 5, linear-oblong-lanceolate, 1–1.5 mm, abaxial surface glabrous except margins villous; stamens 7–9(–12). **Pistillate flowers:** sepals 5–6, equal, 3 mm, margins entire, apex incurved, abaxial surface stellate-hairy; petals 0; ovary 3-locular; styles 3, 2–3 mm, 2-fid to base, terminal segments 6. **Capsules** 4–5 × 4–4.5 mm, smooth; columella apex with 3 sharp projections. **Seeds** 3.2–3.6 × 1.5 mm, shiny.

Varieties 2 (2 in the flora): United States, n Mexico.

Croton lindheimerianus is one of the bifid-styled species in sect. *Heptallon.*

1. Leaf blades suborbiculate, length mostly less than 2 times width, apex rounded to broadly acute; plants densely velvety appressed-tomentose; pedicels, at least some, recurved in fruit . . . . 18a. *Croton lindheimerianus* var. *lindheimerianus*
1. Leaf blades lanceolate, length mostly more than 2 times width, apex acute; plants roughly tomentose; pedicels erect in fruit. . . . . 18b. *Croton lindheimerianus* var. *tharpii*

18a. Croton lindheimerianus Scheele var. **lindheimerianus**

Croton eutrigynus A. Gray

Plants densely velvety appressed-tomentose. **Leaf blades** suborbiculate, length mostly less than 2 times width, apex rounded to broadly acute. **Pedicels,** at least some, recurved in fruit.

Flowering May–Nov. Floodplains, old pastures, loamy soils, roadsides; 50–600 m; Ariz., Ark., Ill., Ind., Kans., La., Mo., N.Mex., N.C., Okla., S.C., Tex.; Mexico (Coahuila, Nuevo León, Tamaulipas).

Variety *lindheimerianus* is somewhat weedy and is adventive to central North Carolina, where it has persisted at the same site for over 50 years; it presumably is introduced also in South Carolina.

18b. Croton lindheimerianus Scheele var. **tharpii** M. C. Johnston, SouthW. Naturalist 3: 188. 1959 • Tharp's three-seed croton

Plants roughly tomentose. **Leaf blades** lanceolate, length mostly more than 2 times width, apex acute. **Pedicels** erect in fruit.

Flowering Jul–Oct. Loamy or rocky flats, limestone washes; 1000–1600 m; Ariz., Tex.; Mexico (Coahuila).

Variety *tharpii* is found in the flora area only in southern Arizona and western Texas.

19. Croton linearis Jacquin, Enum. Syst. Pl., 32. 1760 (as lineare) • Grannybush, pineland croton

Croton fergusonii Small

Shrubs, 10–20 dm, dioecious. **Stems** well branched from base, stellate-hairy. **Leaves** not clustered; stipules rudimentary; petiole 0.3–1 cm, $^1/_{10}$–$^1/_7$ leaf blade length, often with 2 sessile or stipitate glands at apex; blade linear to narrowly oblong, 3–7 × 0.3–1.5 cm, mostly more than 4 times as long as wide, base cuneate to obtuse, margins entire, apex obtuse to acute, abaxial surface whitish yellow, densely appressed stellate-hairy, adaxial surface green, glabrous or minutely stellate-puberulent. **Inflorescences** unisexual, racemes; staminate 4–10 cm, flowers 10–30; pistillate 3–5 cm, flowers 5–12. **Pedicels:** staminate 1–2 mm, pistillate 2–3 mm. **Staminate flowers:** sepals 5(–6), 2–2.5 mm, abaxial surface stellate-hairy; petals 5–6, spatulate, obtuse, 1.5–2.5 mm, abaxial surface glabrous except margins ciliate; stamens 12–17. **Pistillate flowers:** sepals 5, equal, 2.5–3.5 mm, margins entire, apex incurved, abaxial surface canescent; petals rudimentary or 0; ovary 3-locular; styles 3, 2–3 mm, 2-fid, terminal segments 6. **Capsules** 5–7 × 4–5 mm, smooth; columella with 3 rounded, inflated lobes. **Seeds** 3–4 × 1.5–2 mm, dull.

Flowering year-round. Rocky limestone pinelands, coastal areas; 0–30 m; Fla.; West Indies.

Croton linearis in the flora area is found only in Martin, Miami-Dade, Monroe (Keys only), Palm Beach, and St. Lucie counties.

There has been some confusion in the past with the application of the name *Croton cascarilla* (Linnaeus) Linnaeus to this taxon, but *C. cascarilla* is now considered to be a synonym of the West Indian *C. eluteria* (Linnaeus) W. Wright (B. W. van Ee and P. E. Berry 2010b).

20. Croton michauxii G. L. Webster, Novon 2: 270. 1992 • Rushfoil [E] [F]

Crotonopsis linearis Michaux, Fl. Bor.-Amer. 2: 186, plate 46. 1803

Herbs, annual, 1–9 dm, monoecious. **Stems** usually single from base, much branched from near middle, densely stellate-lepidote, appearing silvery. **Leaves** not clustered; stipules rudimentary; petiole 0.1–0.5 cm, glands absent at apex; blade lanceolate-linear to ovate-lanceolate, 1.5–4 × 0.1–0.4(–1.5) cm, base acute to cuneate, margins entire, apex acute, abaxial surface silvery, stellate-lepidote, adaxial surface green, sparsely stellate-hairy, hairs with few radii. **Inflorescences** bisexual, spikes, 0.3–3 cm, staminate flowers 1–12, pistillate flowers 1–6. **Pedicels:** staminate 0–1 mm, pistillate absent. **Staminate flowers:** sepals 5, 1 mm, abaxial surface lepidote; petals 5, spatulate, 0.6–1 mm, abaxial surface glabrous except margins villous; stamens 5–7. **Pistillate flowers:** sepals (4–)5, equal, 1–1.5 mm, margins entire, apex straight, abaxial surface lepidote; petals 0; ovary 1-locular; styles 3, 0.5–0.8 mm, minutely 2-fid, terminal segments 3 or 6. **Achenes** 2.5–3 × 1–1.2 mm, smooth; columella absent. **Seeds** 2–2.5 × 1 mm, surface not visible because seed coat adherent to fruit wall.

Varieties 2 (2 in the flora): United States.

1. Inflorescences 1–3 cm; achenes usually 3–6, densely stellate-lepidote, hairs ± spreading, radii separate most of length, umbos usually sharp, sometimes rounded; leaf blade adaxial surface hairs not overlapping . . . . 20a. *Croton michauxii* var. *michauxii*
1. Inflorescences usually less than 1 cm; achenes 1–2, sparsely stellate-lepidote, hairs appressed, radii joined most or all of length, umbos usually blunt, rarely sharp; leaf blade adaxial surface hairs overlapping near midvein . . . . 20b. *Croton michauxii* var. *elliptica*

20a. Croton michauxii G. L. Webster var. **michauxii** • Michaux's croton, narrow-leaf rushfoil 

Crotonopsis abnormis Baillon; *C. spinosa* Nash

Leaf blades lanceolate-linear, 1.8–3.5 × 0.1–0.3 cm, abaxial surface (midstem leaves) hairs with radii separate ½+ length, adaxial surface hairs appressed, not overlapping, radii usually 5–8. **Inflorescences** 1–3 cm, achenes usually 3–6, ± laxly spaced (not appearing sessile in axil of subtending leaf). **Staminate flowers** usually greater than 1 mm diam. **Achenes** densely stellate-lepidote, hairs ± spreading, radii separate most of length, umbos usually sharp, sometimes rounded.

Flowering Jun–Oct. Deep sandy soils, forest openings, disturbed soils, coasts, riverine beaches; 0–100 m; Ala., Ark., Fla., Ga., Ill., Iowa, La., Miss., Mo., S.C., Tenn., Tex.

20b. Croton michauxii G. L. Webster var. **elliptica** (Willdenow) B. W. van Ee & P. E. Berry, Harvard Pap. Bot. 14: 68. 2009 • Elliptic-leaf rushfoil, Willdenow's croton [E]

Crotonopsis elliptica Willdenow, Sp. Pl. 4: 380. 1805; *Croton willdenowii* G. L. Webster; *Leptemon ellipticum* (Wildenow) Rafinesque

Leaf blades lanceolate to ovate-lanceolate, 1.5–4 × 0.1–0.4(–1.5) cm; abaxial surface (midstem leaves) hairs with radii separate less than ½ length, adaxial surface hairs often slightly raised, overlapping near midvein, radii usually 1–3. **Inflorescences** usually less than 1 cm, achenes 1–2, crowded at base (appearing sessile in axil of subtending leaf). **Staminate flowers** usually less than 1 mm diam. **Achenes** sparsely stellate-lepidote, hairs appressed, radii joined most or all of length, umbos usually blunt, rarely sharp.

Flowering Jun–Oct. Sandstone and granitic outcrops (glades), disturbed sandy soils, forest openings, eroded slopes; 0–300 m; Ala., Ark., Conn., Del., Fla., Ga., Ill., Ind., Iowa, Kans., Ky., La., Md., Miss., Mo., N.J., N.C., Okla., Pa., S.C., Tenn., Tex., Va.

21. Croton monanthogynus Michaux, Fl. Bor.-Amer. 2: 215. 1803 (as monanthogynum) • One-seed croton, prairie tea

Engelmannia nuttalliana Klotzsch; *Gynamblosis monanthogyna* (Michaux) Torrey; *Heptallon ellipticum* Rafinesque; *Oxydectes monanthogyna* (Michaux) Kuntze

Herbs, annual, 2–5 dm, monoecious. **Stems** dichotomously branched from near base, stellate-hairy, some stellate hairs with dark brown centers. **Leaves** sometimes clustered near inflorescences; stipules glandlike, 0.1–0.3 mm; petiole 0.3–1.5 cm, glands absent at apex; blade ovate-oblong to nearly round (proximal) to narrowly elliptic (distal), 1–3.5 × 0.5–3 cm, base obtuse to rounded or truncate, margins entire, apex rounded to acute, abaxial surface pale green, sparsely whitish stellate-hairy and appearing brown-dotted, some hairs with dark brown centers, adaxial surface darker green, densely stellate-hairy, hairs without brown centers. **Inflorescences** bisexual or sometimes pistillate, congested racemes, 0.3–1 cm, staminate flowers 3–10, pistillate flowers 1–2(–5). **Pedicels:** staminate 0.5–2 mm, pistillate 1–2.5 mm (2–3 mm and recurved in fruit). **Staminate flowers:** sepals 3–5, 0.7–1 mm, abaxial surface stellate-hairy; petals 3–5, narrowly elliptic-oblanceolate, 0.7–1 mm, abaxial surface glabrous except margins villous; stamens 3–5. **Pistillate flowers:** sepals 5, subequal, 1.5–2 mm, margins entire, apex straight to slightly incurved, abaxial surface stellate-hairy, some hairs with dark brown centers; petals 0; ovary 2-locular, 1 fertile; styles 2, 0.8–1.2(–1.5) mm, 2-fid to base, terminal segments 4. **Capsules** appearing follicular (1-seeded), 3.5–4.5 × 1.8–2.2 mm, smooth; columella ± curved, apparently deciduous upon dehiscence of capsule. **Seeds** 2.5–3.3 × 2–2.5 mm, shiny. **2*n*** = 20.

Flowering May–Nov. Prairies, sandstone and limestone glades, thinly wooded bluffs, fallow fields, other disturbed habitats; 0–1000 m; Ala., Ariz., Ark., Ga., Ill., Ind., Iowa, Kans., Ky., La., Md., Mich., Miss., Mo., Nebr., N.Mex., N.C., Ohio, Okla., Pa., S.C., Tenn., Tex., Va., W.Va., Wis.; Mexico (Coahuila, Nuevo León, San Luis Potosí, Tamaulipas).

Croton monanthogynus may be adventive in the northernmost states.

22. Croton parksii Croizat, Bull. Torrey Bot. Club 69: 445, 457. 1942 • Parks's croton [E]

Herbs, annual, 5–15 dm, dioecious. **Stems** much branched distally, stellate-hairy. **Leaves** not clustered; stipules rudimentary; petiole 0.5–1.5 cm, glands absent at apex; blade elliptic, 2.5–6 × 1–2 cm, base rounded, margins entire, apex short-rounded, abaxial surface grayish green, densely stellate-tomentose, adaxial surface olive green, stellate-tomentose. **Inflorescences** unisexual; staminate subspicate to racemose or paniculate, 3–6 cm, flowers 3–20; pistillate capitate, 0.1–0.7 cm, flowers 1–3. **Pedicels:** staminate 0.5–2 mm, pistillate 0.5–1.5 mm (1–3.5 mm in fruit). **Staminate flowers:** sepals (5–)6, 3 mm, abaxial surface lanose, margins ciliate; petals 0; stamens 10–15. **Pistillate flowers:** sepals 5, equal, 2 mm, margins entire, apex incurved and ± hooded, abaxial surface densely stellate-hairy; petals 0; ovary 3-locular; styles 3, 2 mm, irregularly 2–3 times 2-fid, terminal segments 24, forming matted cluster. **Capsules** 8–9 × 9–10 mm, verrucose; columella 3-winged. **Seeds** 6–7 mm diam., shiny.

Flowering Jun–Dec. Deep sandy soils; 0–50 m; Tex.

Croton parksii, found on the southern Gulf Coastal Plain, is closely related to *C. texensis*, with which it shares capsules with conspicuous scurfy bumps covered by stellate hairs. Although they appear to intergrade where their ranges overlap around Wilson County, they can generally be distinguished by *C. parksii* being more robust, with larger capsules and seeds, and more densely tomentose leaves.

23. Croton pottsii (Klotzsch) Müller Arg. in A. P. de Candolle and A. L. P. P. de Candolle, Prodr. 15(2): 561. 1866 • Leatherweed

Lasiogyne pottsii Klotzsch in B. Seemann, Bot. Voy. Herald, 278. 1853

Herbs, perennial, 1–6 dm, monoecious or dioecious. **Stems** several from base, sparsely branched distally, stellate-hairy. **Leaves** sometimes clustered near inflorescences; stipules rudimentary; petiole 0.5–2 cm, glands absent at apex; blade ovate-oblong, proximal rounder, distal more ovate-elliptic, 1–6 × 1–3 cm, base rounded, margins entire, apex rounded to acute, both surfaces pale to ashy green, softly stellate-hairy. **Inflorescences** bisexual or unisexual, racemes, 0.5–2.5 cm, staminate and pistillate flowers each (12–)20–25. **Pedicels:** staminate 1.5–6 mm, pistillate 1–3.5 mm, ± recurved in fruit. **Staminate flowers:** sepals (4–)5, 1–1.5 mm, abaxial surface densely stellate-hairy; petals (4–)5, oblong-lanceolate, 1–1.3 mm, abaxial surface glabrous except margins villous; stamens 10–15. **Pistillate flowers:** sepals 5, equal, 1.5–3 mm, margins entire, apex straight to slightly incurved, abaxial surface glabrous except stellate-hairy at apex; petals 0; ovary 3-locular; styles 3, 0.5–4.5 mm, 2-fid to base, terminal segments 6. **Capsules** 4–6(–7) × 4–5 mm, smooth; columella apex with 3 sharp projections. **Seeds** 3.5–4 × 2–2.5 mm, shiny. **2*n*** = 20.

Varieties 2 (2 in the flora): sw, sc United States, n, c Mexico.

The affinities of *Croton pottsii* lie with *C. lindheimerianus* and *C. monanthogynus*.

1. Stems straight; erect radii of stellate hairs shorter than lateral radii, thus pubescence appearing appressed and smooth; leaf blades 1.6–6 cm; racemes mostly 1.1–2.5 cm; styles 1.1–4.5 mm; capsules 4–6(–7) mm. . . . 23a. *Croton pottsii* var. *pottsii*
1. Stems somewhat zigzag; erect radii of stellate hairs longer than lateral radii, thus pubescence appearing ± shaggy; leaf blades usually 1–1.4 cm; racemes mostly 0.5–0.9 cm; styles usually 0.5–1 mm; capsules 4 mm. . . . . 23b. *Croton pottsii* var. *thermophilus*

23a. Croton pottsii (Klotzsch) Müller Arg. var. **pottsii** • Encinilla

Croton corymbulosoides Radcliffe-Smith & Govaerts; *C. corymbulosus* Engelmann

Stems straight; erect radii of stellate hairs shorter than lateral radii, thus pubescence appearing appressed and smooth. **Leaf blades** 1.6–6 cm. **Racemes** mostly 1.1–2.5 cm. **Styles** 1.1–4.5 mm. **Capsules** 4–6(–7) mm.

Flowering May–Oct. Desert scrub, arid plains, rocky slopes; 700–2000 m; Ariz., N.Mex., Tex.; n, c Mexico.

23b. Croton pottsii (Klotzsch) Müller Arg. var. **thermophilus** (M. C. Johnston) M. C. Johnston, SouthW. Naturalist 5: 171. 1960 • Desert leatherweed [C]

Croton corymbulosus Engelmann var. *thermophilus* M. C. Johnston, SouthW. Naturalist 3: 187. 1959; *C. thermophilus* (M. C. Johnston) B. L. Turner

Stems somewhat zigzag; erect radii of stellate hairs longer than lateral radii, thus pubescence appearing ± shaggy. **Leaf blades** usually 1–1.4 cm. **Racemes** mostly 0.5–0.9 cm. **Styles** usually 0.5–1 mm. **Capsules** 4 mm.

Flowering Jul–Oct. Basalt and limestone hills; of conservation concern; 700–1800 m; Tex.; Mexico (Coahuila).

Variety *thermophilus* is restricted to Brewster County and nearby Mexico.

24. Croton punctatus Jacquin, Collectanea 1: 166. 1787 (as puntatum) • Beach tea, gulf croton, hierba de jabalí

Croton disjunctiflorus Michaux; *C. maritimus* Walter; *C. plukenetii* Geiseler

Shrubs, 3–10 dm, monoecious or sometimes appearing dioecious. **Stems** trichotomously branching, stellate-hairy to stellate-lepidote. **Leaves** not clustered; stipules absent; petiole 1–4 cm, ½ to equal blade length, glands absent at apex; blade broadly elliptic to suborbiculate, 2–5 × 1.5–4 cm, much less than 2 times as long as wide, base rounded to truncate, margins entire, apex obtuse to rounded, abaxial surface pale grayish green, stellate-lepidote, adaxial surface slightly darker green, stellate-lepidote. **Inflorescences** unisexual or bisexual, racemes, 1–4 cm, staminate flowers 3–7, pistillate flowers 1–3. **Pedicels:** staminate 2–4 mm, pistillate 0–1 mm. **Staminate flowers:** sepals 5–6, 2.5 mm, abaxial surface stellate-lepidote; petals 0; stamens 10–13. **Pistillate flowers:** sepals 5, equal, 3–3.5 mm, margins entire, apex incurved, abaxial surface stellate-lepidote; petals 0; ovary 3-locular; styles 3, 1–2 mm, multifid, terminal segments 12–24. **Capsules** 5–8 × 7–9 mm, smooth; columella 3-winged. **Seeds** 4.5–6 × 3.7–4.5 mm, dull. $2n = 28$.

Flowering year-round. Beaches, dunes; 0–20 m; Ala., Fla., Ga., La., Miss., N.C., S.C., Tex.; e Mexico; West Indies; Central America; n South America.

A report of *Croton punctatus* from Pennsylvania apparently was based on a transient appearance on ballast (E. T. Wherry et al. 1979).

25. Croton sancti-lazari Croizat, J. Arnold Arbor. 26: 185. 1945 • Trans-Pecos croton

Croton abruptus M. C. Johnston

Shrubs, 1–4 dm, dioecious. **Stems** much branched proximally, coarsely stellate-tomentose. **Leaves** not clustered; stipules each 1 glandular papilla, 0.1 mm; petiole 0.1–0.4(–0.6) cm, ⅒–⅕ blade length, glands absent at apex; blade ovate to elliptic-ovate, 1–3(–4.5) × 0.5–1.5(–2) cm, base usually rounded to obtuse, rarely acute, margins entire, apex acute or rounded, abaxial surface pale green to pale yellow, stellate-tomentose, adaxial surface darker green, stellate-tomentose. **Inflorescences** unisexual; staminate racemes, 1–3.3 cm, flowers 1–8; pistillate congested racemes, 0.1–0.3 cm, flowers 1–3. **Pedicels:** staminate 1 mm, pistillate 0–0.5 mm. **Staminate flowers:** sepals (4–)5, 2 mm, abaxial surface stellate-hairy; petals (4–)5, narrowly oblanceolate, 1.8–2 mm, abaxial surface nearly glabrous; stamens 9–12. **Pistillate flowers:** sepals 5, equal, 1 mm, margins entire, apex straight to slightly incurved, abaxial surface stellate-hairy; petals 0 or rudimentary; ovary 3-locular; styles 3, 1.5–2 mm, 2-fid to base, terminal segments 6. **Capsules** 4–5 × 5.5–6 mm, smooth; columella apex with 3 rounded, inflated lobes. **Seeds** 3.6–4.7 × 2.8–3.4 mm, shiny.

Flowering Oct–Nov. Rocky hillsides; 900–1200 m; Tex.; Mexico (Chihuahua, Coahuila).

Croton sancti-lazari in the flora area grows only in Brewster and Presidio counties.

26. Croton setigerus Hooker, Fl. Bor.-Amer. 2: 141. 1838 (as setigerum) • Dove weed, turkey mullein [W]

Eremocarpus setigerus (Hooker) Bentham

Herbs, annual, 0.5–5 dm, monoecious. **Stems** densely and dichotomously shortly branched, forming loose, prostrate circular mats 5–80 cm across, proximally bristly stellate-hairy, central radii spreading, 2–3 mm. **Leaves** mostly clustered near inflorescences; stipules rudimentary; petiole 0.3–5 cm, glands absent at apex; blade ovate to rhombic, 0.8–6.5 × 0.8–4 cm, base cuneate, margins entire, apex obtuse, abaxial surface pale grayish green, adaxial surface grayish green, both densely stellate-hairy. **Inflorescences** unisexual, staminate dense capitate clusters, 1–2 cm, flowers 2–10; pistillate clusters, 1–2 cm, flowers 1–3. **Pedicels:** staminate 0–1 mm, pistillate absent. **Staminate flowers:** sepals 5–6, 2–2.5 mm, abaxial surface densely stellate-hairy; petals 0; stamens 5–9. **Pistillate flowers:** sepals 0; petals 0; ovary 1-locular; style 1, 2–3 mm, unbranched. **Capsules** follicular (1-seeded), 3–6 × 2–3 mm, smooth; columella absent. **Seeds** 3–5 × 2–3 mm, shiny. $2n = 20$.

Flowering Jun–Oct. Coastal sage scrub, foothill woodlands, valley grasslands, oak woodlands, edges of fields, dry stream beds, disturbed areas, roadsides; 0–1900 m; Ariz., Calif., Idaho, Nev., Oreg., Utah, Wash.; Mexico (Baja California); introduced in s South America (Chile), Australia.

Croton setigerus is nearly unique in the genus with its one-locular fruit, single unbranched style, and pistillate flowers devoid of any perianth. The foliage is toxic to animals, and the crushed plants were used by Native Americans to stupefy fish. The seeds are palatable to birds, giving rise to the common names cited above. Individual plants produce either mottled, striped, or solid gray or black seeds. Gray seeds are produced by desiccating plants and appear to be much less palatable to doves than the other color morphs (A. D. Cook et al. 1971).

27. Croton soliman Chamisso & Schlechtendal, Linnaea 6: 361. 1831 • Soliman's croton

Shrubs, 5–8 dm, monoecious. **Stems** sparsely branched distally, sparsely stellate-hairy. **Leaves** not clustered; stipules linear, (0–)1–2 mm, sometimes short stipitate-glandular; petiole 1–2 cm, $\frac{2}{5}$–$\frac{1}{2}$ leaf blade length, glands absent at apex; blade ovate, 2–5 × 1.5–2.5 cm, base rounded-cuneate, margins entire, with scattered glandular-capitate processes 1 mm, usually denser at base, apex acute to acuminate, abaxial surface green, glabrous or sparsely stellate-hairy, usually along margin, adaxial surface slightly darker green, glabrous. **Inflorescences** bisexual, racemes, 2–4 cm, staminate flowers 10–30, pistillate flowers 3–6. **Pedicels:** staminate 1–5 mm, pistillate 2–5 mm. **Staminate flowers:** sepals 5, 0.8–1 mm, abaxial surface sparsely stellate-hairy; petals 5, spatulate, 1–1.3 mm, abaxial surface glabrous except margins ciliate basally; stamens 15–20. **Pistillate flowers:** sepals 5, unequal, 5–6 mm, margins entire, apex straight to slightly incurved, abaxial surface glabrous or with a few stellate hairs; petals 0 or 5, white, linear, 2.5–3.5 mm; ovary 3-locular; styles 3, 3–4 mm, 4-fid, terminal segments 12. **Capsules** 6 × 5–6 mm, smooth; columella apex with 3 rounded, inflated lobes. **Seeds** 4–5 × 3–4 mm, shiny.

Flowering May–Jul. Thickets, low ridges; 0–50 m; Tex.; e, s Mexico.

Croton soliman is most similar to *C. humilis*. In the flora area, *C. soliman* is found only in Cameron County.

28. Croton sonorae Torrey in W. H. Emory, Rep. U.S. Mex. Bound. 2(1): 194. 1859 • Sonora croton

Croton attenuatus M. E. Jones; *C. pringlei* S. Watson

Shrubs, 8–16 dm, monoecious. **Stems** much branched distally, canescent when young, becoming glabrate. **Leaves** sometimes clustered near inflorescences; stipules subulate, to 1 mm; petiole 0.4–1 cm, $\frac{1}{4}$–$\frac{1}{2}$ leaf blade length, glands absent at apex; blade ovate to broadly lanceolate, 0.8–4 × 0.6–2 cm, base cuneate to rounded or cordate, margins entire, not undulate, ciliate-glandular, apex acute to acuminate, abaxial surface pale green, sparsely whitish stellate-hairy, adaxial surface darker green, glabrate. **Inflorescences** bisexual or staminate, racemes, 2–14 cm, staminate flowers 10–25, pistillate flowers 3–5(–8). **Pedicels:** staminate 1–1.5 mm, pistillate 1–4 mm. **Staminate flowers:** sepals 5, 1–1.5 mm, abaxial surface sparsely stellate-hairy; petals 5, lanceolate, 1.5–2 mm, abaxial surface glabrous except margins villous; stamens 13–16. **Pistillate flowers:** sepals 5, equal, 1.5–2 mm, margins entire, apex incurved, abaxial surface sparsely stellate-hairy; petals papillae, 0.5 mm; ovary 3-locular; styles 3, 2–3 mm, 2-fid, terminal segments 6. **Capsules** 7–8 × 5–6 mm, smooth; columella apex with 3 rounded, inflated lobes. **Seeds** 6–7 × 3–4 mm, shiny.

Flowering Jul–Sep. Rocky slopes, desert scrub; 500–1100 m; Ariz.; Mexico (Baja California, Sonora).

Croton sonorae is widely distributed in Mexico and extends into southern Arizona in Maricopa, Pima, and Pinal counties.

29. Croton suaveolens Torrey in W. H. Emory, Rep. U.S. Mex. Bound. 2(1): 194. 1859 • Scented croton

Shrubs, 2–5 dm, monoecious. **Stems** dichotomously much branched proximally, coarsely stellate-tomentose. **Leaves** often clustered near inflorescences; stipules each 5–10 glandular papillae, 0.2–0.5 mm; petiole 0.5–1.5 cm, ¼ leaf blade length, glands absent at apex; blade obovate or ovate to broadly elliptic, 2–5.4 × 1–3.6 cm, base obtuse, margins entire, apex obtuse to acute, abaxial surface pale green to cream, adaxial surface darker green, both densely stellate-hairy. **Inflorescences** bisexual or staminate, racemes, 1–3 cm, staminate flowers 6–10, pistillate flowers 1–3. **Pedicels:** staminate 3–5 mm, pistillate 1–2 mm. **Staminate flowers:** sepals 5, 2 mm, abaxial surface stellate-hairy; petals 5, spatulate or obovate, 0.4–0.6 mm, abaxial surface densely villous at base and along margins, otherwise glabrous; stamens 12–16. **Pistillate flowers:** sepals 5, equal, 2.5 mm, margins entire, apex straight to slightly incurved, abaxial surface stellate-hairy; petals glandular papillae; ovary 3-locular; styles 3, 4–6 mm, 2-fid to base, terminal segments 6. **Capsules** 7–10 × 6–8 mm, smooth; columella apex with 3 rounded, inflated lobes. **Seeds** 5.5–7 × 4–5 mm, shiny.

Flowering Apr–Nov. Rocky slopes, foothills; 1500–2000 m; Tex.; Mexico (Coahuila, Chihuahua, Nuevo León).

Croton suaveolens is found in the flora area only in trans-Pecos Texas.

30. Croton texensis (Klotzsch) Müller Arg. in A. P. de Candolle and A. L. P. P. de Candolle, Prodr. 15(2): 692. 1866 • Skunkweed, Texas croton [F] [W]

Hendecandra texensis Klotzsch, Arch. Naturgesch. (Berlin) 7: 252. 1841; *Croton luteovirens* Wooton & Standley; *C. texensis* var. *utahensis* Cronquist; *C. virens* Müller Arg.

Herbs, annual, 2–7(–9) dm, dioecious. **Stems** loosely branched distally, stellate-hairy. **Leaves** not clustered; stipules absent; petiole 0.3–2 cm, glands absent at apex; blade narrowly ovate-oblong to linear-lanceolate, 1–5 × 0.5–2 cm, base truncate to rounded or subcordate, margins entire, apex rounded to acute, abaxial surface pale green, densely whitish appressed stellate-hairy, adaxial surface darker green, less hairy. **Inflorescences** unisexual; staminate racemes or irregularly branched panicles, 2–8 cm, flowers 10–30; pistillate racemes, 1–2 cm, flowers 1–6. **Pedicels:** staminate 2–3 mm, pistillate 1–3 mm. **Staminate flowers:** sepals 5, 1–2 mm, abaxial surface densely whitish appressed stellate-hairy; petals 0; stamens 8–12. **Pistillate flowers:** sepals 5, equal, 1–1.5 mm, margins entire, apex incurved, abaxial surface densely stipitate-stellate-hairy; petals 0; ovary 3-locular; styles 3, 1–2 mm, multifid, terminal segments 12–32+. **Capsules** 5–8 × 4–5.5 mm, verrucose; columella 3-winged. **Seeds** 3.5–4 × 2.5–3 mm, shiny. $2n = 28$.

Flowering Jun–Nov. Prairies, sandy creek beds, old fields, canyons, disturbed areas; 50–2000 m; Ala., Ariz., Colo., Del., Fla., Ill., Iowa, Kans., Md., Mo., Nebr., N.Mex., Okla., S.Dak., Tex., Utah, W.Va., Wis., Wyo.; Mexico (Chihuahua, Sonora).

Croton texensis, despite being annual, grows larger than the related perennial *C. dioicus*. *Croton texensis* has verrucose fruits similar to those of the closely related *C. parksii*.

There is a single specimen of *Croton texensis* from Massachusetts, collected at a dump in Boston in 1890, but the species did not become established there.

31. Croton wigginsii L. C. Wheeler, Contr. Gray Herb. 124: 37. 1939 • Wiggins's croton [C]

Croton arenicola Rose & Standley, Contr. U.S. Natl. Herb. 16: 12. 1912, not Small 1905

Shrubs, 2–10 dm, dioecious. **Stems** densely branched, appressed-lepidote. **Leaves** not clustered; stipules absent; petiole 1–4 cm, usually less than ½ blade length, glands absent at apex; blade narrowly elliptic to linear-oblong, 2–8.5 × 0.6–1.5 cm, more than 2 times as long as wide, base obtuse, margins entire, apex obtuse to rounded, abaxial surface pale green, adaxial surface darker green, both densely pale stellate-lepidote. **Inflorescences** unisexual, racemes or thyrses; staminate 1–3.5(–10) cm, flowers 3–8(–15); pistillate 0.5–1 cm, flowers 1–6. **Pedicels:** staminate 1–7 mm, pistillate 1–2 mm (4–7 mm in fruit). **Staminate flowers:** sepals 5, 1 mm, abaxial surface stellate-hairy; petals 0; stamens 10–15. **Pistillate flowers:** sepals 5, equal, 2 mm, margins entire, apex incurved, abaxial surface stellate-lepidote; petals 0; ovary 3-locular; styles 3, 1.5–2.5 mm, 2–3 times 2-fid, terminal segments 12–24. **Capsules** 7–10 × 6–8 mm, smooth; columella 3-winged. **Seeds** 6.5–7 × 2–3 mm, dull. $2n = 28$.

Flowering Feb–May. Sand dunes; 10–100 m; of conservation concern; Ariz., Calif.; Mexico (Baja California, Sonora).

Croton wigginsii is closely related to *C. californicus* but more robust in its habit and floral features, and is restricted to sand dunes in a limited area of the Sonoran Desert. In the flora area, *C. wigginsii* is known only from Yuma County, Arizona, and Imperial County, California.

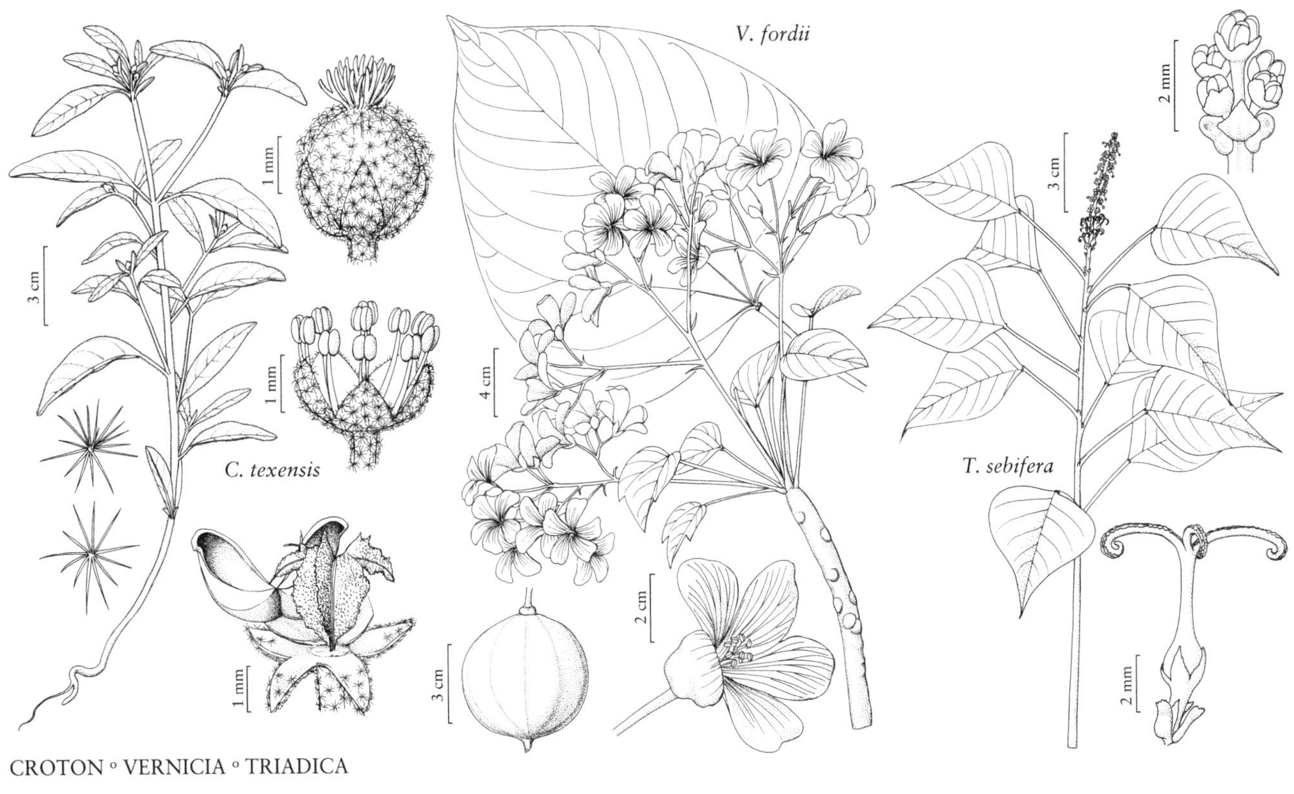

CROTON ◦ VERNICIA ◦ TRIADICA

15. VERNICIA Loureiro, Fl. Cochinch. 2: 541, 586. 1790 • [Latin *vernix*, varnish, alluding to use of seed oil in finishes] [I]

Lynn J. Gillespie

Trees, monoecious [dioecious]; hairs unbranched or 2-fid; latex whitish or reddish (often not apparent). **Leaves** deciduous [persistent], alternate, simple; stipules present, caducous; petiole present, glands present at apex; blade unlobed or palmately lobed, margins entire, laminar glands absent or present in sinuses of lobes; venation palmate. **Inflorescences** bisexual (cymules staminate or bisexual, pistillate flower central, staminate lateral) [unisexual], terminal, paniclelike thyrses; glands subtending each bract 0. **Pedicels** present. **Staminate flowers:** sepals 2(–3), valvate, connate basally; petals 5(–6), distinct, white or pink; nectary extrastaminal, 5(–6) glands; stamens (7–)8–12(–14), in 2 whorls, connate into androphore, outer whorl connate basally, inner whorl longer and connate proximally to much of length; pistillode absent. **Pistillate flowers:** sepals 2(–3), connate basally; petals 5(–6), distinct, white or pink; nectary 5(–6) glands (often inconspicuous); pistil [3–]4(–5)-carpellate; styles [3–]4(–5), distinct or connate basally, 2-fid. **Fruits** capsules, tardily dehiscent. **Seeds** obovoid [subglobose]; caruncle absent. $x = 11$.

Species 3 (1 in the flora): introduced; e Asia; introduced also in Africa, Australia.

The seeds of all three species of *Vernicia* are pressed for oil, which is used in the production of varnish and high quality paints; of these the tung-oil tree (*V. fordii*) is the most important commercially. *Vernicia montana* Loureiro is sometimes cultivated in the southeastern United States but is not known to be naturalized there. It may be distinguished from *V. fordii* by its persistent leaves with stalked, cup-shaped glands at the petiole apex and blades mostly 3-lobed; fruits with distinct grooves and ridges; and inflorescences mostly unisexual.

Species of *Vernicia* previously have been included within *Aleurites* J. R. Forster & G. Forster, but the two genera are now considered distinct and closely related (H. K. Airy Shaw 1967;

W. Stuppy et al. 1999). *Aleurites* may be distinguished by its stellate hairs, 2(–3)-locular ovary, fleshy indehiscent fruit, and 17–32 stamens (versus simple or 2-fid hairs, 3–5-locular ovary, dehiscent fruits, and 7–14 stamens in *Vernicia*). *Aleurites moluccanus* (Linnaeus) Willdenow (candlenut or Indian walnut) is occasionally cultivated in Florida and rarely escapes locally but does not appear to be naturalized there.

SELECTED REFERENCES Airy Shaw, H. K. 1967. Notes on Malaysian and other Asiatic Euphorbiaceae LXXII. Generic segregation in the affinity of *Aleurites* J. R. et G. Forster. Kew Bull. 20: 393–395. Stuppy, W. et al. 1999. A revision of the genera *Aleurites* J. R. Forst. & G. Forst., *Reutealis* Airy Shaw and *Vernicia* Lour. (Euphorbiaceae). Blumea 44: 73–98.

1. **Vernicia fordii** (Hemsley) Airy Shaw, Kew Bull. 20: 394. 1967 • Tung-oil tree [F] [I] [W]

Aleurites fordii Hemsley, Bull. Misc. Inform. Kew 1906: 120. 1906

Trees, to 10[–20] m. **Leaves:** stipules 4–12 mm; petiole 6–22 cm, with pair of round, sessile, cushion-shaped glands at apex; blade broadly ovate or triangular-ovate, 10–25 × 8–20 cm, usually unlobed, sometimes shallowly 3-lobed, base cordate, truncate, or rounded, apex acuminate, both surfaces moderately to sparsely hairy, hairs appressed. **Inflorescences** 6–15 × 6–20 cm, often branching from near base, branches to 15 cm. **Pedicels** 1–2 cm. **Staminate flowers:** sepals green to purplish, 10–12 mm; petals white or pale pink with dark pink to red veins proximally, sometimes yellow basally, obovate, 25–35(–40) × 15–20 mm, narrowed at base; nectary glands awl-shaped to strap-shaped; stamens in outer whorl 8 mm, in inner whorl 13 mm, connate ½–⅔ length. **Pistillate flowers:** sepals and petals as in staminate flowers; ovary hairy. **Capsules** subglobose, 4–6 cm diam., smooth, glabrous or glabrate, short stipitate, apex apiculate. **Seeds** 2.5–3 × 2 cm, surface warty, ridged. $2n = 22$ (China).

Flowering Mar–Apr; fruiting Apr–Aug. Wood and field margins, abandoned fields, roadsides, disturbed woods; 0–150 m; introduced; Ala., Ark., Fla., Ga., La., Miss., N.C., S.C., Tex.; se Asia; introduced also in Australia.

Vernicia fordii was cultivated for its seed oil in plantations along the Gulf coast from Florida to Texas from the 1920s to the 1960s. Although no longer commercially cultivated in the southeastern United States, it is naturalized there and is now listed as an invasive weed in Florida. All parts of the plant are poisonous; seeds have strong purgative properties and may cause poisoning if eaten.

16. TRIADICA Loureiro, Fl. Cochinch. 2: 598, 610. 1790 • [Greek and Latin *triadis*, a group of three, alluding to 3-merous flowers] [I]

Kenneth J. Wurdack

Trees, monoecious; hairs absent; latex white. **Leaves** deciduous, alternate, simple; stipules present, persistent; petiole present, glands present at apex; blade unlobed, margins entire, laminar glands abaxial, submarginal, occasionally absent on some leaves; venation pinnate. **Inflorescences** bisexual (pistillate flowers proximal, staminate distal) or staminate, terminal and subterminal (in axils of branch-tip leaves), racemelike thyrses; glands subtending each bract 2. **Pedicels** present. **Staminate flowers:** sepals 3, apparently imbricate, connate most of length; petals 0; nectary absent; stamens 2–3, distinct; pistillode absent. **Pistillate flowers:** sepals 3, distinct or connate basally; petals 0; nectary absent; pistil 3-carpellate; styles 3, connate ½–⅔ length, unbranched. **Fruits** capsules. **Seeds** ovoid with ventral face distinctly flattened or angled; outer seed coat fleshy; caruncle absent. $x = 11$.

Species 3 (1 in the flora): introduced; Asia.

Although classified historically in *Sapium* sect. *Triadica*, *Triadica* is distinct based on morphology and molecular phylogenetic evidence (K. Wurdack et al. 2005).

SELECTED REFERENCES DeWalt, S. J., E. Siemann, and W. E. Rogers. 2011. Geographic distribution of genetic variation among native and introduced populations of Chinese tallow tree, *Triadica sebifera* (Euphorbiaceae). Amer. J. Bot. 98: 1128–1138. Esser, H.-J. 2002. A revision of *Triadica* Lour. (Euphorbiaceae). Harvard Pap. Bot. 7: 17–21.

1. Triadica sebifera (Linnaeus) Small, Florida Trees, 59. 1913 • Chinese tallowtree, candleberry or popcorn tree [F] [I] [W]

Croton sebifer Linnaeus, Sp. Pl. 2: 1004. 1753 (as sebiferum); *Excoecaria sebifera* (Linnaeus) Müller Arg.; *Sapium sebiferum* (Linnaeus) Dumont de Courset; *Stillingia sebifera* (Linnaeus) Michaux

Trees, to 13 m (fertile from 1 m). **Leaves:** stipules persistent, elliptic, 0.7–1 × 0.5–0.7 mm; petiole 2–7 cm, glands 2, discoid, adaxial; blade ovate to broadly elliptic or rhomboid, 3.5–10 × 3–9.5 cm, base broadly cuneate to nearly truncate, apex acuminate; laminar glands 0–10, elliptic, 0.3 × 0.2 mm, usually on distal ½ of leaf. **Inflorescences** to 20 cm; staminate cymules numerous, 10–20 flowered, bracts ovate, 1.5 mm, subtended by 2[–4] ellipsoid glands; pistillate flowers 0–6 per inflorescence, 1 per bract (often in bisexual cymules with 0–5 staminate flowers), bracts of basal flowers usually not subtended by ellipsoid glands. **Pedicels:** staminate 1.5–3 mm; pistillate 1–2 mm, to 12 mm in fruit. **Staminate flowers** yellow; sepals 0.5–1 mm, shallowly 3-lobed, margins erose; filaments to 0.2–0.3 mm; anthers 0.5 mm. **Pistillate flowers** yellowish green; sepals 2–3 × 1 mm, apex acuminate; styles 4–8 mm, coiled distally. **Capsules** 1–1.3 cm diam., subglobose, trigonous; columella 1 cm. **Seeds** 6–9 × 4–7 mm, usually remaining attached to columella; outer seed coat white, waxy; inner coat woody, brown, smooth. $2n$ = 88.

Flowering Apr–Jun; fruiting Aug–Nov. Low swampy places to uplands; 0–200 m; introduced; Ala., Ark., Calif., Fla., Ga., La., Miss., N.C., S.C., Tex.; e Asia (s China, Taiwan, Vietnam).

Triadica sebifera is economically important in Asia for the waxes and oils in its sarcotesta and seeds and as a source of dye and timber. The species is cultivated in the United States mainly as an ornamental, with introductions dating to the late 1700s; it readily becomes naturalized and is a stand-replacing invasive of native forests, riparian areas, and prairies. The range of *T. sebifera* is predicted to continue expanding from the Sacramento River delta region in California, and also north and west from southeastern United States coastal regions, until it reaches limits for drought, salinity, and cold tolerance (R. Pattison and R. Mack 2007). An isolated population in Kentucky was extirpated to prevent naturalization (E. Comley 2008).

17. GYMNANTHES Swartz, Prodr., 6, 95. 1788 • [Greek *gymnos*, naked, and *anthos*, flower, alluding to highly reduced or absent perianth]

Kenneth J. Wurdack

Shrubs or trees, unarmed, monoecious [dioecious]; hairs absent [unbranched]; latex colorless [white]. **Leaves** persistent [deciduous], alternate, simple; stipules present, persistent; petiole present, glands absent; blade unlobed, margins subentire to serrulate [entire], laminar glands abaxial, submarginal [absent]; venation pinnate. **Inflorescences** bisexual (pistillate flowers proximal, staminate distal) [unisexual], terminal or axillary, racemelike thyrses; glands subtending each bract 0 [or 2]. **Pedicels** present. **Staminate flowers:** sepals 0[–3]; petals 0; nectary absent; stamens (2–)4(–5)[–100], distinct; pistillode absent. **Pistillate flowers:** sepals 0[–3]; petals 0; nectary absent; pistil 3-carpellate; styles 3, connate basally, unbranched. **Fruits** capsules. **Seeds** subglobose; caruncle present [absent].

Species ca. 45 (1 in the flora): Florida, Mexico, West Indies, Central America, South America, Asia, Africa; tropical and subtropical regions.

Members of *Gymnanthes* are found primarily in the New World; some are in Africa and Asia.

SELECTED REFERENCE Webster, G. L. 1983. A botanical Gordian knot: The case of *Ateramnus* and *Gymnanthes* (Euphorbiaceae). Taxon 32: 304–305.

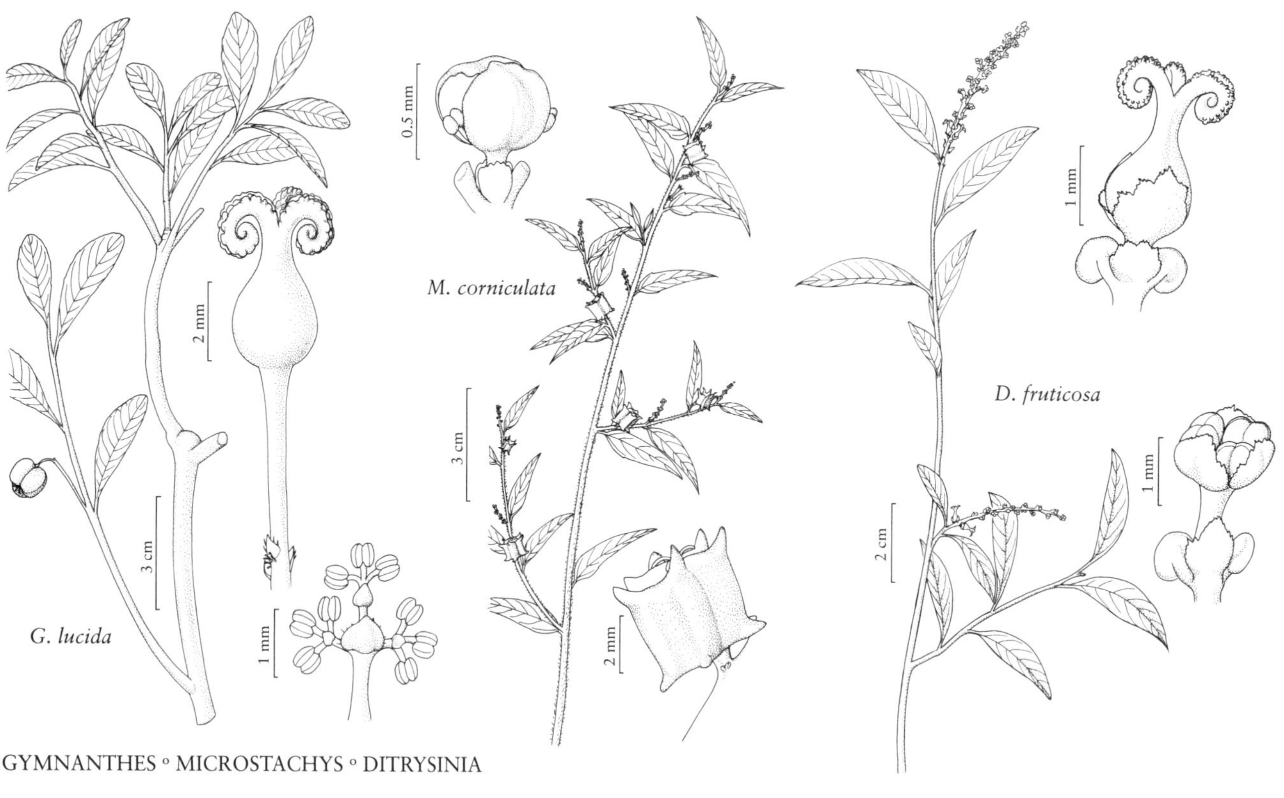

GYMNANTHES ∘ MICROSTACHYS ∘ DITRYSINIA

1. Gymnanthes lucida Swartz, Prodr., 96. 1788 • Crabwood, oysterwood [F]

Ateramnus lucidus (Swartz) Rothmaler; *Excoecaria lucida* (Swartz) Swartz; *Sebastiania lucida* (Swartz) Müller Arg.

Shrubs or trees, 1–10 m, latex watery. **Leaves:** stipules persistent, lanceolate, 1 mm, glands present; petiole 0.2–1.5 cm; blade oblanceolate, 3.5–10 × 1.5–4 cm, base cuneate, margins obscurely gland-toothed, apex acute to obtuse; laminar glands usually 2 near base, elliptic; venation prominent on both surfaces. **Inflorescences** 3–3.5 cm, peduncle 0–0.5 cm; staminate cymules numerous, each usually with 3 flowers per bract, cymule axis 0.2–1.3 mm, bracts reniform, cucullate, 1 mm wide; pistillate flowers usually 1 per inflorescence. **Pedicels:** staminate 0.3–2 mm, usually bearing scalelike bracteole at summit; pistillate to 10 mm, to 30–40 mm in fruit, bearing 1–5 scalelike bracteoles. **Staminate flowers** yellowish green; filaments connate 0.3 mm, distally distinct 0.2–0.4 mm; anthers 0.4–0.5 mm. **Pistillate flowers** yellowish green; styles 2–3 mm, connate ½ length, distally recurved. **Capsules** subglobose, trigonous, 0.8–1 cm diam.; styles persistent. **Seeds** 4–4.5 × 4 mm; seed coat brown, smooth; caruncle discoid, 1.5–2 mm diam.

Flowering spring–summer (Apr–Jun); fruiting summer–fall (Jul–Oct). Subtropical moist hardwood hammocks; 0–10 m; Fla.; Mexico; West Indies; Central America.

The young inflorescences of *Gymnanthes lucida* are formed during the prior growing season and resemble small cones due to the tightly appressed bracts; they expand rapidly at the end of the dry season. The scalelike structures subtending individual staminate and pistillate flowers are here referred to as bracteoles (P. B. Tomlinson 1980) but have also been interpreted as vestiges of sepals (G. L. Webster 1967).

Gymnanthes lucida is known in the flora area from Miami-Dade and Monroe counties.

18. MICROSTACHYS A. Jussieu, Euphorb. Gen., 48. 1824 • [Greek *mikros*, small, and *stachys*, spike, alluding to inflorescence] [I]

Kenneth J. Wurdack

Herbs [shrubs], annual [perennial], monoecious; hairs unbranched [branched]; latex white. **Leaves** alternate, simple; stipules present, persistent; petiole present, glands absent; blade unlobed, margins serrulate [entire], laminar glands abaxial, at base [absent]; venation pinnate. **Inflorescences** appearing unisexual (pistillate and staminate portions usually shortly separated on stem), terminal, leaf-opposed, or axillary, racemelike thyrses; glands subtending each bract [0] 2. **Pedicels** present, often rudimentary. **Staminate flowers:** sepals 3, imbricate, distinct [connate basally]; petals 0; nectary absent; stamens 3, distinct; pistillode absent. **Pistillate flowers:** sepals 3, distinct; petals 0; nectary absent; pistil 3-carpellate; styles [0 or]3, connate basally, unbranched. **Fruits** capsules, base not persisting. **Seeds** oblong, ends truncate [elliptic]; outer seed coat dry; caruncle present [absent].

Species ca. 15 (1 in the flora): introduced, Florida; Mexico, West Indies, Central America, South America, Asia, Africa, Australia; tropical and subtropical regions.

Microstachys is distinct based on morphology and molecular phylogenetic evidence (H.-J. Esser 1998; K. Wurdack et al. 2005), although species have been included historically in *Sebastiania* Sprengel. The inflorescence architecture characteristic of *Microstachys* is unusual: the pistillate and staminate parts are usually shortly separated along the main stem, appearing as two separate partial inflorescences at consecutive nodes, the pistillate one proximal and supra-axillary, and the staminate distal and leaf opposed. The genus is known primarily from the New World with a few species found in Africa, Asia, and Australia.

1. Microstachys corniculata (Vahl) A. Jussieu ex Grisebach, Fl. Brit. W. I., 49. 1859 • Hato tejas [F] [I]

Tragia corniculata Vahl, Eclog. Amer. 2: 55, plate 19. 1798; *Sebastiania corniculata* (Vahl) Müller Arg.; *Stillingia corniculata* (Vahl) Baillon

Herbs, to 0.5 m, sparsely to moderately hirsute. **Stems** erect. **Leaves:** stipules triangular to rounded, 0.2–0.4 mm; petiole 0.1–1.5 cm; blade 2–4.5 × 0.3–2 cm, base cuneate to cordate, margins serrulate with proximal teeth occasionally replaced by orbicular glands, apex acute to acuminate. **Inflorescences:** staminate portion on peduncle 1–2 mm, fertile part 0.5–1.5 cm, to 15 bracts and cymules, flowers 1–3 per bract; pistillate portion with 1 flower; bracts triangular, 0.2 mm, subtended by 2 stipitate glands to 0.2 mm diam. **Pedicels** 0–0.2 mm, pistillate to 1 mm in fruit. **Staminate flowers:** sepals red-purple to greenish yellow, ovate, 0.3–0.5 mm; stamens yellow; filaments 0.2–0.3; anthers 0.2 mm. **Pistillate flowers** green; sepals ovate, 0.2–0.5 mm; ovary bearing short horned appendages; styles 0.3–0.6 mm. **Capsules** subglobose, 3–4 mm diam.; horned appendages 2–3 per valve, to 1 mm. **Seeds** 3 × 2 mm; seed coat brown, warty; caruncle discoid, 0.5–0.8 mm diam., stipitate on projection of seed coat.

Flowering and fruiting summer–fall. Dry, disturbed sandhills; 0–40 m; introduced; Fla.; Mexico; West Indies; Central America; South America.

Microstachys corniculata is a weedy species that is adventive and becoming established in Hillsborough County. It has the potential to become broadly naturalized across subtropical parts of the United States. Because this species is widespread in the West Indies, including Cuba, a 1906 collection from Key West may represent a native occurrence.

19. DITRYSINIA Rafinesque, Neogenyton, 2. 1825 • [Greek *ditry*, two or three, and *syn*, together, alluding to number and union of stamens] E

Kenneth J. Wurdack

Shrubs, monoecious; hairs unbranched; latex absent. **Leaves** semipersistent, alternate, simple; stipules present, persistent; petiole present, glands absent; blade unlobed, margins entire, laminar glands usually abaxial, scattered, sometimes absent; venation pinnate. **Inflorescences** bisexual (pistillate flowers proximal, staminate distal) or staminate, terminal, racemes; glands subtending each bract 2. **Pedicels** present. **Staminate flowers:** sepals 3, imbricate, connate basally; petals 0; nectary absent; stamens 3, distinct; pistillode absent. **Pistillate flowers:** sepals 3, connate basally; petals 0; nectary absent; pistil 3-carpellate; styles 3, connate basally, unbranched. **Fruits** capsules, base not persisting. **Seeds** ovoid-oblong; outer seed coat dry; caruncle present.

Species 1: se United States.

Molecular phylogenetic analyses indicate a close relationship of *Ditrysinia* with *Microstachys*, but taxon sampling is too sparse to determine if *Ditrysinia* should contain other species that are now classified in polyphyletic *Sebastiania* (K. Wurdack et al. 2005).

1. **Ditrysinia fruticosa** (W. Bartram) Govaerts & Frodin in R. Govaerts et al., World Checklist Bibliogr. Euphorbiaceae, 586. 2000 • Gulf sebastian-bush E F

Stillingia fruticosa W. Bartram, Travels Carolina, 476. 1791 (as fructicosa); *Gymnanthes ligustrina* (Michaux) Müller Arg.; *Sebastiania fruticosa* (W. Bartram) Fernald; *S. ligustrina* (Michaux) Müller Arg.; *Stillingia ligustrina* Michaux

Shrubs, 1–3 m. **Leaves:** stipules subulate to triangular, 1–2 × 0.5 mm; petiole 0.2–1 cm, minutely puberulent on adaxial ridges confluent with margins; blade ovate to lanceolate, 2–8 × 1–3 cm, base cuneate, margins minutely puberulent, apex acute; laminar glands 0–5, elliptic, 0.3 × 0.2 mm. **Inflorescences:** peduncle 0.3–2.5 cm; fertile part 1–4 cm; flowers 1 per bract; pistillate flowers 0–4; bracts ovate, 0.7–1 mm, subtended by 2, reniform, stipitate glands 0.6–1.3 × 0.5 mm. **Pedicels:** staminate 0.5–2 mm; pistillate 0.3–1.5 mm, to 6 mm in fruit. **Staminate flowers** yellowish green; sepals ovate, 0.5–0.7 mm, margin erose, ciliate; stamens yellow; filaments 0.2–0.4 mm; anthers 0.5 mm. **Pistillate flowers** yellowish green; sepals ovate, 1–1.5 mm, margin erose, ciliate; ovary 1.5–2 mm diam.; styles 2–2.7 mm, connate 0.5–0.7 mm, distally recurved. **Capsules** subglobose, 0.5–1 cm diam. **Seeds** 4–5 mm; seed coat brown with silver mottling, smooth; caruncle discoid, 2 mm diam.

Flowering Apr–Jun; fruiting Jul–Oct. Hardwood forests, especially wet or moist, shaded stream banks, flood plains, adjacent wooded slopes; 0–100 m; Ala., Ark., Fla., Ga., La., Miss., N.C., S.C., Tex.

20. HURA Linnaeus, Sp. Pl. 2: 1008. 1753; Gen. Pl. ed. 5, 439. 1754 • Sandbox tree [Native American word for poisonous sap, alluding to caustic latex] I

Michael J. Huft

Trees, monoecious; trunk with broad-based, conic thorns; hairs unbranched; latex white or colorless. **Leaves** deciduous, alternate, simple; stipules present, caducous; petiole present, glands present at apex, lateral, conspicuous; blade unlobed, margins serrate or crenate-serrulate, laminar glands absent; venation pinnate. **Inflorescences** usually unisexual, rarely bisexual (pistillate flowers proximal, staminate distal); staminate terminal, spikelike thyrses, cymules densely crowded in conelike structure; pistillate axillary, solitary flowers; bisexual as in staminate with solitary pistillate flower at base; glands subtending each bract 0. **Pedicels** present. **Staminate**

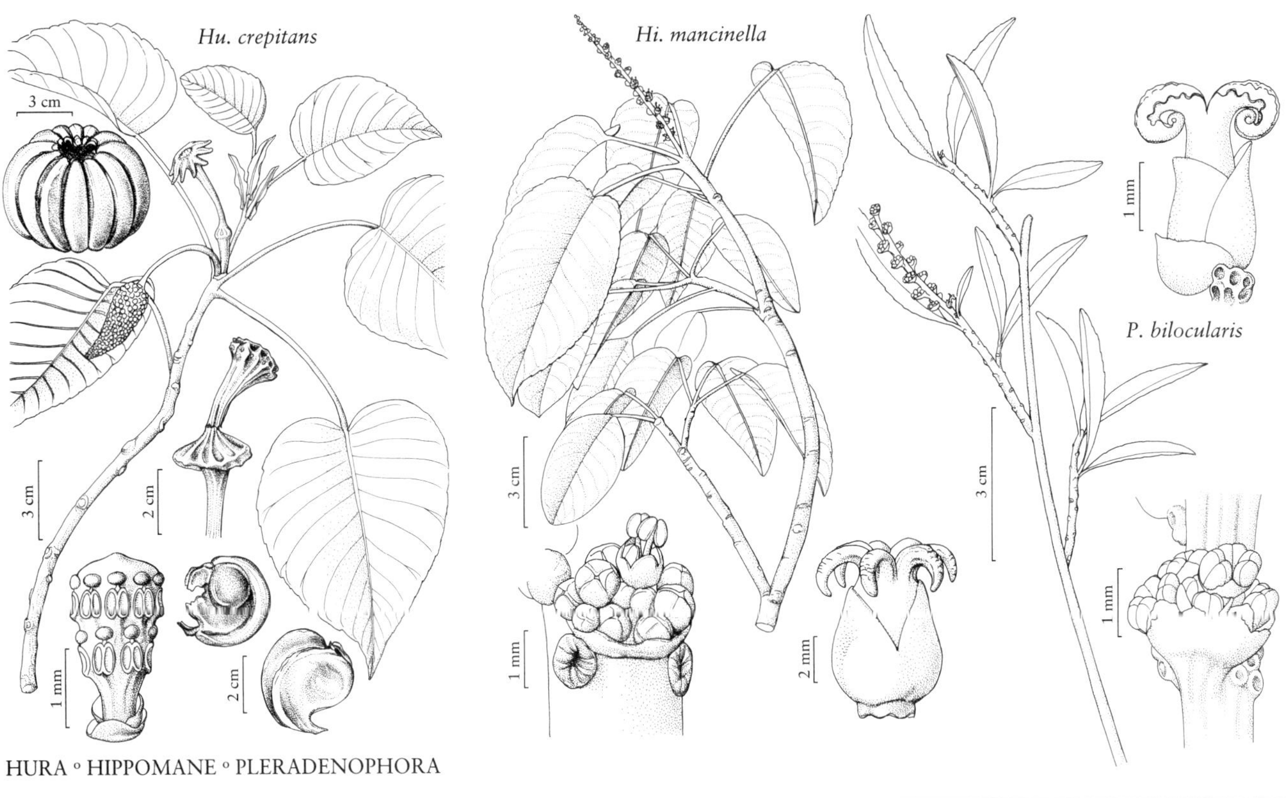

HURA ◦ HIPPOMANE ◦ PLERADENOPHORA

flowers: sepals 5, imbricate, connate most of length; petals 0; nectary absent; stamens 10–80, connate entire length forming thick column; pistillode absent. **Pistillate flowers:** sepals 5, connate entire length; petals 0; nectary absent; pistil 5–20-carpellate; style 1, unbranched, terminating in lobed stigmatic disc. **Fruits** capsules, woody. **Seeds** lenticular; caruncle absent. $x = 11$.

Species 2 (1 in the flora): introduced, Florida; Mexico, West Indies, Central America, South America; introduced also in tropical areas worldwide.

Hura polyandra Baillon is distributed from Mexico to Nicaragua.

1. Hura crepitans Linnaeus, Sp. Pl. 2: 1008. 1753 [F] [I]

Trees, to 30 m; latex copious, caustic. **Leaves:** stipules leaving scar around stem, 6–10 × 2–4 mm; petiole 10–20 cm, glands 2 adaxially at apex; blade broadly ovate to orbiculate, 10–25 × 8–15 cm, base cordate, margins crenate-serrulate with 10–20 teeth on each side, apex acuminate; midvein and secondary veins prominent. **Staminate inflorescences:** peduncle (2–)5–10(–15) cm, fertile portion 2–5 cm, 60–80-flowered; staminate bracts membranaceous, enclosing flowers prior to anthesis. **Pedicels:** pistillate 1–5 cm. **Staminate flowers:** calyx cup-shaped, 1 mm, 3–5-lobed; staminal column to 2.5 mm, anthers usually in 2, rarely in 3, whorls. **Pistillate flowers** red; calyx cup-shaped, 3–5(–8) mm, 5-lobed; style 30–50 mm, terminating in thick apical disc 1 cm diam. with 5–20 radiating lobes 5–10 mm. **Capsules** depressed-globose, 3–5 × 6–10 cm. **Seeds** brownish, 15–20 mm diam., 5–8 mm thick, smooth.

Flowering Jan–May; fruiting (Mar–)May–Sep. Hammocks and disturbed ground; 0–10 m; introduced; Fla.; West Indies; Central America; South America; widely cultivated and often escaped throughout tropical regions of the world.

In the flora area, *Hura crepitans* is found only in Miami-Dade and Monroe (Keys only) counties. The capsules dehisce explosively, producing a loud sound and throwing the seeds up to 45 m (M. D. Swain and T. Beer 1977).

21. HIPPOMANE Linnaeus, Sp. Pl. 2: 1191. 1753; Gen. Pl. ed. 5, 499. 1754 • Manchineel [Greek *hippos*, horse, and *mania*, fury, alluding to effect of the caustic latex on horses]

Michael J. Huft

Trees, monoecious; hairs absent; latex white. **Leaves** persistent, alternate, simple; stipules present, caducous; petiole present, glands at apex; blade unlobed, margins remotely serrate or crenate, laminar glands absent; venation pinnate. **Inflorescences** bisexual (pistillate flowers proximal, staminate distal), terminal, spikelike thyrses; glands subtending each bract 2. **Pedicels:** staminate present, pistillate rudimentary. **Staminate flowers:** sepals 2, imbricate, connate proximally; petals 0; nectary absent; stamens 2, connate basally; pistillode absent. **Pistillate flowers:** sepals 3[–4], connate proximally; petals 0; nectary absent; pistil 6–9-carpellate; styles 6–9, connate basally, unbranched. **Fruits** drupes. **Seeds** elliptic-compressed; caruncle absent. $x = 11$.

Species 2 or 3 (1 in the flora): Florida, Mexico, West Indies, Central America, n South America, Pacific Islands (Galápagos Islands).

Hippomane horrida Urban & Ekman and *H. spinosa* Linnaeus are endemic to Hispaniola.

1. Hippomane mancinella Linnaeus, Sp. Pl. 2: 1191. 1753 [F]

Trees, to 7 m with much-branched spreading crown; bark light gray or light brown, ± smooth, slightly warty and fissured; latex copious, caustic. **Leaves:** stipules caducous, ± narrowly deltate, 1 mm, margins entire; petiole slender, 1–6 cm, varying in length on single branchlets, with 1 disc-shaped sessile gland abaxially at apex; blade broadly ovate-elliptic, 2–9 × 1–7 cm, chartaceous, shiny adaxially, base rounded or slightly cordate, margins remotely crenate, apex obtuse or short-acuminate; midvein prominent, secondary veins less prominent. **Inflorescences** 6–8 cm; staminate cymules subsessile, clearly separated, 10–30-flowered; pistillate flowers 2–3; bracts flabelliform, 1–2 mm. **Staminate pedicels** 1 mm. **Staminate flowers:** calyx 1 mm, lobes broad, shallow. **Pistillate flowers:** styles thick, to 10 mm, strongly recurved. **Drupes** depressed-globose, 1.5–2.5 cm diam. **Seeds** smooth.

Flowering Jan–May; fruiting (Mar–)May–Sep. Hammocks and low ground behind mangrove zone along coasts; 0–10 m; Fla.; Mexico; West Indies; Central America; n South America; Pacific Islands (Galápagos Islands).

In the flora area, *Hippomane mancinella* is found only in Miami-Dade and Monroe counties. The caustic latex can cause severe skin irritation on contact and, if taken internally, can cause severe poisoning, although individual reactions can vary dramatically (R. A. Howard 1981). The applelike appearance of the fruit no doubt has been the cause of more frequent poisonings than might otherwise be the case (G. L. Webster 1967). Smoke from burning wood can be harmful to the eyes (R. W. Long and O. Lakela 1971; W. C. Burger and M. J. Huft 1995).

22. PLERADENOPHORA Esser in A. Radcliffe-Smith, Gen. Euphorb., 377. 2001 • [Greek *pleros*, very many, *aden*-, gland, and *-phoros*, bearing, alluding to many glands on leaves and subtending floral bracts]

Kenneth J. Wurdack

Shrubs or trees, monoecious; hairs absent [unbranched]; latex white. **Leaves** semipersistent, alternate, simple; stipules present, persistent [deciduous]; petiole present, glands present at apex or absent; blade unlobed, margins serrulate [entire], laminar glands absent; venation pinnate. **Inflorescences** bisexual (pistillate flowers proximal, staminate distal) or staminate, terminal, racemelike thyrses; glands subtending each bract 10–14. **Pedicels** present or absent. **Staminate**

flowers: sepals 3, imbricate, connate basally; petals 0; nectary absent; stamens 2[–5], distinct; pistillode absent. **Pistillate flowers:** sepals 2[–3], distinct; petals 0; nectary absent; pistil 2[–3]-carpellate; styles 2[–3], connate proximally, unbranched. **Fruits** capsules. **Seeds** broadly ovoid-oblong; outer seed coat dry; caruncle absent.

Species 5 (1 in the flora): Arizona, Mexico, Central America, South America.

Species of *Pleradenophora* were historically mostly classified within *Sebastiania*. Molecular phylogenetic analyses show that *Sebastiania* in the broad sense is polyphyletic and indicate a close relationship of *Pleradenophora* with a West Indian clade containing *Bonania* A. Richard, *Grimmeodendron* Urban, and *Hippomane* (K. Wurdack et al. 2005).

1. **Pleradenophora bilocularis** (S. Watson) Esser & A. L. Melo, Phytotaxa 81: 34. 2013 • Arrow poison plant, hierba de la fleche [F]

Sebastiania bilocularis S. Watson, Proc. Amer. Acad. Arts 20: 374. 1885; *Sapium biloculare* (S. Watson) Pax; *S. biloculare* var. *amplum* I. M. Johnston

Shrubs or small trees, 1–8 m. **Leaves** distichous, clustered at branch tips; stipules triangular, 0.8–1 × 1.5–2.5 mm, erose, hyaline; petiole 1.2–4 mm, glands 0–2, orbicular, axil with tuft of coarse glandular hairs; blade lanceolate, 2–7 × 0.5–1.8[–3] cm, base acute, margins with fine teeth bearing setae, teeth sometimes replaced by orbicular glands, apex acute. **Inflorescences:** peduncle 0.2–0.6 cm; fertile portion 3–6 cm; staminate cymules numerous, condensed, each with 3–10 flowers per bract; pistillate flowers 0–4 per inflorescence, 1 flower per bract; bracts ovate, 1.5–2 mm, subtended on each side by row or cluster of 5–7 elliptic glands, 0.1–0.3 mm diam. **Pedicels** 0–2 mm. **Staminate flowers** yellowish green; sepals 0.5 mm; filaments 0.3–0.8 mm, anthers 0.2–0.3 mm. **Pistillate flowers** yellowish green; sepals ovate, 1–2 mm; styles 2.5–3.5 mm, connate ⅓ length, free portion coiled. **Capsules** subglobose, 0.5–1 cm diam. **Seeds** 4–5 mm, wider than long; seed coat silver and brown mottled, smooth with verrucose patches. $2n = 22$.

Flowering and fruiting year-round. Sonoran desert scrub, thorn scrub; 0–1000 m; Ariz.; Mexico (Baja California Sur, Sonora).

Pleradenophora bilocularis reaches its northern limit of distribution in southwestern Arizona and is widespread in adjacent Sonora, Mexico. Across its range, there is a six-fold variation in leaf width with collections from Baja California being much wider (to 3 cm) and formerly recognized as a distinct variety (V. W. Steinmann and R. S. Felger 1997). The species is a minor source for "Mexican jumping beans" from seeds parasitized by moth larvae. Indigenous populations used the latex as arrow and fish poison (C. E. Bradley 1956).

23. STILLINGIA Garden in C. Linnaeus, Mant. Pl. 1: 19, 126. 1767; Syst. Nat. ed. 12, 2: 611, 637. 1767 • [For Benjamin Stillingfleet, 1702–1771, British botanist]

Michael J. Huft

Herbs, subshrubs, or shrubs [trees], annual or perennial, monoecious; hairs absent [rarely glandular]; latex white. **Leaves** deciduous, alternate, opposite, or subopposite [whorled], simple; stipules absent or present, persistent; petiole absent or present, glands absent [small sessile gland at apex]; blade unlobed, margins entire, dentate, crenate, serrulate, or spinulose-dentate, laminar glands absent; venation pinnate. **Inflorescences** bisexual (pistillate flowers proximal, staminate distal), terminal, spikes or spikelike thyrses; glands subtending each bract 2. **Pedicels** absent. **Staminate flowers:** sepals 2, imbricate, connate basally; petals 0; nectary absent; stamens 2, connate basally; pistillode absent. **Pistillate flowers:** sepals 0 or [2–]3, distinct; petals 0; nectary absent; pistil [2–]3-carpellate; styles 3 [rarely 2], connate proximally, unbranched. **Fruits** capsules, base persisting as [2–]3-lobed gynobase, glabrous. **Seeds** globose, ovoid, ellipsoid, or cylindric, ± flattened or depressed at hilar end; outer seed coat dry; caruncle absent or present. $x = 11$.

Species ca. 33 (7 in the flora): s United States, Mexico, West Indies, Central America, South America, Indian Ocean Islands (Madagascar), Pacific Islands (Fiji Islands).

Stillingia is distributed primarily in the warmer regions of the western hemisphere, with a major center of diversity extending from the southwestern United States through Mexico to northern Central America and another occupying the region of southern Brazil, northern Argentina, and Paraguay. Other New World species occur in Peru, southern Central America, and the southeastern United States. Outside of the western hemisphere, there are three species in Madagascar and one in Fiji. Among species in the flora area, only *S. sylvatica* is widespread, ranging throughout much of the southern United States from Virginia to New Mexico.

Stillingia is one of the more distinctive genera in the tribe Hippomaneae A. Jussieu ex Spach, which are generally characterized by the presence of white latex and by terminal or axillary spikelike inflorescences with one or more solitary basal pistillate flowers. Among these genera, *Stillingia* is distinguished by the presence of a gynobase, the hardened proximal portion of the ovary that remains as a 3-parted (or 2-parted in a few species outside the flora area) persistent base attached to the pedicel after dehiscence of the fruit. The circumscription of *Stillingia* has remained essentially unchanged since 1880, when Bentham first recognized the importance of the gynobase as the most important distinguishing character (D. J. Rogers 1951).

SELECTED REFERENCE Rogers, D. J. 1951. A revision of *Stillingia* in the New World. Ann. Missouri Bot. Gard. 38: 207–259.

1. Staminate flowers in 3–15-flowered cymules; sepals of pistillate flowers 3, well developed, persistent; seeds with caruncles.
    2. Shrubs with taproots . . . . 1. *Stillingia aquatica*
    2. Herbs or subshrubs with woody caudices or rhizomes.
        3. Leaf blades ovate, elliptic, or lanceolate, to obovate or oblanceolate, teeth without prominent blackened tips, incurved; capsules 6–12 mm diam. . . . . 5. *Stillingia sylvatica*
        3. Leaf blades linear to linear-lanceolate, teeth with prominent blackened tips, not incurved; capsules 6–8 mm diam. . . . . 6. *Stillingia texana*
1. Staminate flowers 1 per node; sepals of pistillate flowers 0 or 3, minute and fugacious; seeds with or without caruncles.
    4. Leaf blades linear, margins entire or remotely minutely denticulate.
        5. Pistillate flowers widely spaced; glands of pistillate bracts long-stalked; seeds without caruncles . . . . 2. *Stillingia linearifolia*
        5. Pistillate flowers crowded; glands of pistillate bracts sessile; seeds with minute caruncles. . . . . 3. *Stillingia paucidentata*
    4. Leaf blades elliptic, elliptic-spatulate, or obovate-spatulate, margins spinulose-dentate.
        6. Leaves opposite or subopposite, stipules absent, blade apices acuminate; inflorescences sessile, 1–1.2(–2) cm; glands of pistillate bracts long-stalked; seeds without caruncles . . . . 4. *Stillingia spinulosa*
        6. Leaves alternate, stipules present, blade apices rounded to obtuse; inflorescences pedunculate, 2.5–5 cm; glands of pistillate bracts ± sessile; seeds with minute caruncles. . . . . 7. *Stillingia treculiana*

1. Stillingia aquatica Chapman, Fl. South. U.S., 405. 1860 • Water toothleaf, corkwood [E]

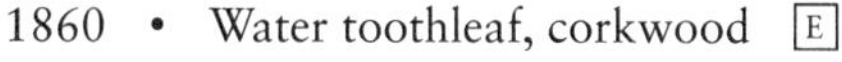

Shrubs with taproot. **Stems** solitary, erect, branching near apex, 4–12 dm. **Leaves** alternate; stipules linear, 1.5–2 mm; petiole 0.1–0.5(–1) cm; blade linear, lanceolate, or narrowly elliptic, 3.5–10 × (0.2–)0.5–2.5 cm, base acute to narrowly obtuse, margins minutely crenulate, teeth without prominent blackened tips, not incurved, apex acute to acuminate; midvein prominent, secondary veins ± obscure. **Inflorescences** sessile, 3–5 cm; staminate cymules ± crowded, 5–15-flowered; pistillate flowers 1–2(–4), crowded; bracts narrowly ovate to oblong, 2 mm, apex acute, glands crateriform, short-stalked, 1–1.2 mm diam. **Staminate flowers:** calyx 1–1.3 mm. **Pistillate flowers:** sepals persistent, 3, well developed, ± orbiculate; styles connate ¾ length, to 4 mm. **Capsules** globose, 4–5 × 6–7 mm, shallowly 3-lobed; lobes of gynobase 3–3.5 mm; columella absent or only short basal part persistent. **Seeds** gray, globose, 4.5 × 4 mm, rugulose; caruncle minute.

Flowering Feb–May; fruiting (Feb–)May–Oct. Floodplain swamps, wet pinelands, shallow standing water, pond edges, maritime interdunal swales, generally in moist soils; 0–200 m; Ala., Fla., Ga., S.C.

2. Stillingia linearifolia S. Watson, Proc. Amer. Acad. Arts 14: 297. 1879 • Queen's-root

Herbs, perennial, with woody taproot. **Stems** fascicled, erect, branching throughout, 1.5–6 (–9) dm. **Leaves** alternate; stipules linear-lanceolate, 0.5–1 mm; petiole absent; blade linear, (1–)2–3(–3.5) × 0.1–0.2(–0.3) cm, base acute, margins entire or remotely minutely denticulate, teeth without prominent blackened tips, not incurved, apex acute; venation not prominent. **Inflorescences** sessile or short-pedunculate, (1.5–)2–4.5(–7) cm, staminate flowers widely spaced, 1 per node; pistillate flowers 3–7, widely spaced; bracts deltate to broadly ovate, to 2 mm, apex acuminate, glands cyathiform, long-stalked, to 1 mm diam. **Staminate flowers:** calyx to 1 mm. **Pistillate flowers:** sepals 0; styles connate only at base, 1.5–2 mm. **Capsules** globose-ovoid, 2–2.5 × 3–3.5 mm, deeply 3-lobed; lobes of gynobase 1–1.5 mm; columella sometimes persistent. **Seeds** gray or gray-brown, globose, 2–2.5 × 1.5 mm, smooth; caruncle absent. $2n = 22$.

Flowering Jan–Apr; fruiting Feb–Jun. Washes, deserts, dry soils; 0–1500 m; Ariz., Calif., Nev.; Mexico (Baja California, Sonora).

3. Stillingia paucidentata S. Watson, Proc. Amer. Acad. Arts 14: 298. 1879 • Mojave toothleaf [E]

Herbs, perennial, with thick taproot. **Stems** fascicled, erect, branching scattered proximally and crowded distally, (1.4–) 2–3.5(–4) dm. **Leaves** alternate; stipules absent; petiole absent; blade linear, 2–4(–6) × 0.1–0.3 (–0.4) cm, base acute, margins entire or remotely denticulate, teeth without prominent blackened tips, not incurved, apex usually acute, rarely acuminate; venation not prominent. **Inflorescences** sessile or rarely short-pedunculate, 6–7 cm; staminate flowers crowded distally, 1 per node; pistillate flowers 3–5, crowded; bracts broadly ovate, to 1.5 mm, apex mucronulate or acuminate, glands patelliform, sessile, to 1.3 mm diam. **Staminate flowers:** calyx to 1 mm. **Pistillate flowers:** sepals 0; styles connate only at base, to 4 mm. **Capsules** oblate, 3 × 4 mm, deeply 3-lobed; lobes of gynobase 1.5–2 mm; columella persistent. **Seeds** brown, often mottled, ovoid, 2.3 × 1.3 mm, smooth; caruncle minute.

Flowering Mar–May; fruiting May–Jun. Sandy flats, dry slopes, 0–1500 m; Calif.

Stillingia paucidentata is widespread in the Mojave Desert and extends into the Sonoran Desert in central Riverside County. It was reported from Arizona by T. H. Kearney and R. H. Peebles (1942, 1960) solely on the basis of the type (*Palmer 517*), purportedly collected in 1876 in the "Colorado Valley, near mouth of Williams River." R. McVaugh (1943b) and McVaugh and Kearney (1943) have cast doubt on whether a number of Palmer collections with labels indicating 1876 were actually made in Arizona; they did not discuss *Palmer 517* specifically. There appear to be no other specimens or reports of this species from Arizona. Because *S. paucidentata* is known from numerous collections in eastern San Bernardino County, California, its presence in bordering areas of Arizona cannot be completely discounted.

4. Stillingia spinulosa Torrey in W. H. Emory, Not. Milit. Reconn., 151. 1848 • Annual toothleaf, broad-leaved stillingia

Herbs, annual or perennial, with taproot. **Stems** fascicled, decumbent to erect, branching distally, 0.5–3 dm. **Leaves** opposite or subopposite; stipules absent; petiole absent; blade elliptic to elliptic-spatulate, 1.5–4(–4.5) × 0.5–1.4(–1.8) cm, base narrowly cuneate, margins prominently spinulose-dentate, teeth without prominent blackened tips, not incurved, apex acuminate; midvein and secondary veins prominent. **Inflorescences** sessile, 1–1.2(–2) cm; staminate flowers crowded, 1 per node; pistillate flowers 1–3, crowded; bracts linear-lanceolate, to 2 mm, apex acute, glands patelliform, long-stalked, 1.5–2 mm diam. **Staminate flowers:** calyx to 1 mm. **Pistillate flowers:** sepals 0; styles connate only at base, to 3 mm. **Capsules** ovoid, 3.5 × 4–4.5 mm, deeply 3-lobed; lobes of gynobase 2 mm; columella sometimes persistent. **Seeds** mottled light gray, cylindric-ovoid, 3.5 × 1.8 mm, smooth; caruncle absent.

Flowering Dec–Mar(–Apr); fruiting Mar–Jun. Dry sandy desert soils; 0–1000 m; Ariz., Calif., Nev.; Mexico (Baja California, Sonora).

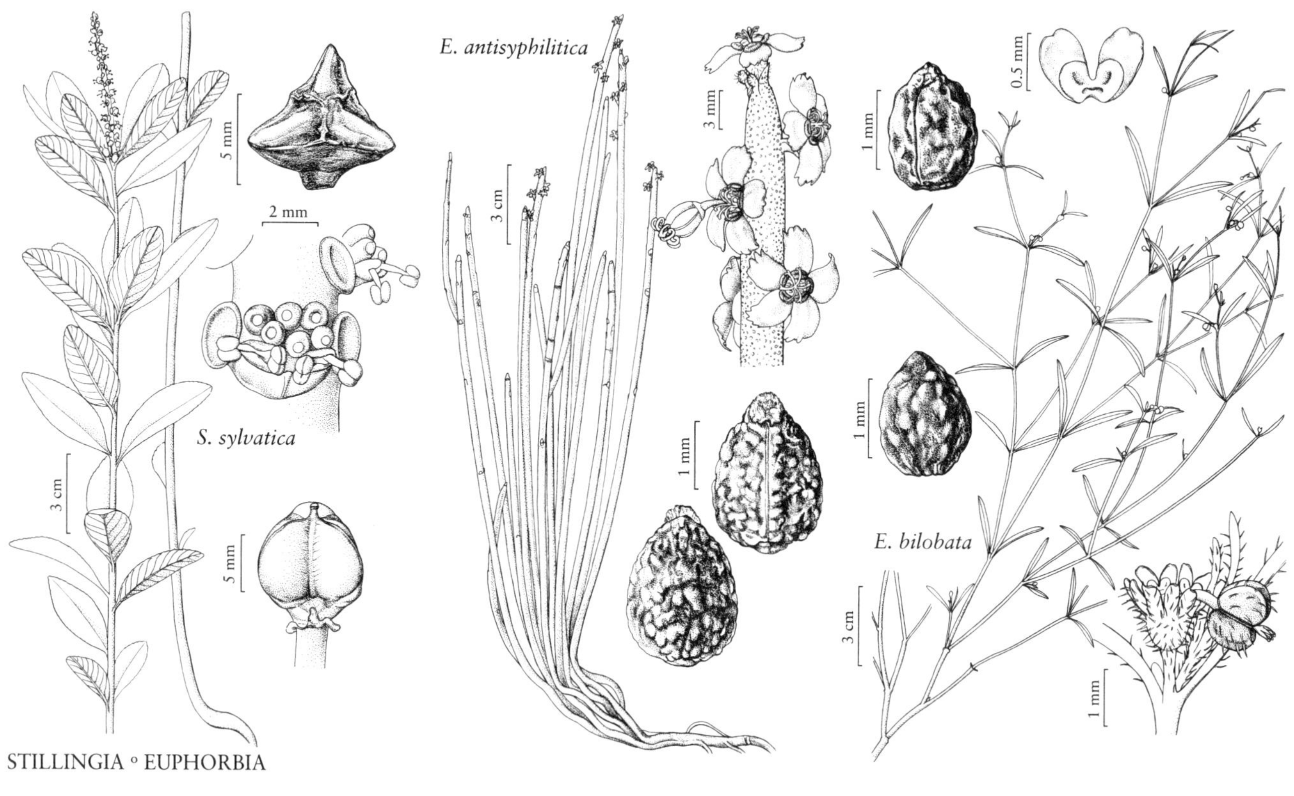

STILLINGIA ◦ EUPHORBIA

5. **Stillingia sylvatica** Linnaeus, Syst. Nat. ed. 12, 2: 637. 1767; Mant. Pl. 1: 126. 1767 • Queen's delight [E] [F]

Stillingia angustifolia (Müller Arg.) Engelmann ex S. Watson; *S. sylvatica* var. *salicifolia* Torrey; *S. sylvatica* subsp. *tenuis* (Small) D. J. Rogers; *S. tenuis* Small

Herbs or subshrubs, perennial, with woody rhizome. **Stems** solitary or fascicled, erect or ascending, mostly unbranched, (1–)2.5–7(–12) dm. **Leaves** alternate; stipules absent; petiole 0–0.4(–0.8) cm; blade ovate, elliptic, lanceolate, obovate, or oblanceolate, 1–10 × 0.5–3 cm, base acute or broadly cuneate, margins serrulate to crenulate, teeth without prominent blackened tips, incurved, apex acute, obtuse, or rounded; midvein prominent, secondary veins ± obscure. **Inflorescences** sessile or short-pedunculate, 3–9 cm; staminate cymules ± crowded, 4–7-flowered; pistillate flowers 3–4, crowded; bracts broadly ovate, 1.5 × 2 mm, apex rounded or obtuse, glands patelliform, sessile, 1.5–2 mm diam. **Staminate flowers:** calyx 1 mm. **Pistillate flowers:** sepals persistent, 3, well developed, elliptic; styles connate ⅓ length, to 5 mm. **Capsules** globose, 6–12 mm diam., shallowly 3-lobed; lobes of gynobase 4–6 mm; columella not persistent. **Seeds** light gray, short cylindric, 4.5 × 3 mm, rugose; caruncle white, broadly crescent-shaped, 1–1.3 mm.

Flowering Mar–Jun; fruiting Apr–Sep. Well-drained sandy soils, sandhills, pine flatwoods; 0–1400 m; Ala., Ark., Colo., Fla., Ga., Kans., La., Miss., N.Mex., N.C., Okla., S.C., Tenn., Tex., Va.

Throughout the range of *Stillingia sylvatica*, leaves vary widely in length/width ratio, though seldom on the same plant.

Some populations in southern Florida have been recognized as *Stillingia tenuis* or *S. sylvatica* subsp. *tenuis* on the basis of linear or narrowly linear-elliptic leaves and slender, reddish inflorescences. Leaf blades of the southern Florida populations vary from linear to broadly elliptic, and the reddish cast of the inflorescence is characteristic of only some of these populations and also occurs in *S. aquatica*. D. J. Rogers (1951) and G. L. Webster (1967) suggested that the characteristics used to define *S. tenuis* may be the result of introgression of *S. aquatica* into *S. sylvatica*, and that the putative subsp. *tenuis* might be an ecotype of predominantly calcareous soils of southern Florida.

6. **Stillingia texana** I. M. Johnston, Contr. Gray Herb. 68: 91. 1923 • Texas toothleaf

Sapium sylvaticum (Linnaeus) Torrey var. *linearifolium* Torrey in W. H. Emory, Rep. U. S. Mex. Bound. 2(1): 201. 1859 (as linearifolia), not *Stillingia linearifolia* S. Watson 1879; *S. sylvatica* Linnaeus var. *linearifolia* (Torrey) Müller Arg.

Herbs or subshrubs, perennial, with woody caudex. **Stems** solitary or fascicled, erect, mostly unbranched, 1.5–4.5(–6) dm. **Leaves** alternate; stipules linear-lanceolate, to 1 mm; petiole absent; blade linear to linear-lanceolate, (1–)3–6(–7) × 0.3–0.6(–1) cm, base cuneate to obtuse, margins crenate-dentate, teeth with prominent blackened tips, not incurved, apex acute; midvein prominent, especially proximally, secondary veins obscure. **Inflorescences** sessile, 3–9 cm; staminate cymules crowded, 3–5-flowered; pistillate flowers 3–4, crowded; bracts broadly deltate, to 1 mm, apex acute, glands patelliform, sessile, 2 mm diam. **Staminate flowers:** calyx 1 mm. **Pistillate flowers:** sepals persistent, 3, well developed, elliptic; styles connate ¾ length, to 4 mm. **Capsules** globose, 6–8 mm diam., shallowly 3-lobed; lobes of gynobase 2–3 mm; columella often persistent. **Seeds** gray, ellipsoid, 5 × 5 mm, smooth; caruncle white, broadly crescent-shaped, 1 × 1.3–1.5 mm.

Flowering late Apr–May(–Jul); fruiting Jun–Sep. Calcareous prairies, open uplands; 200–1500 m; Okla., Tex.; Mexico (Coahuila).

Stillingia texana is widespread in central Texas, extending north to scattered locations in central Oklahoma.

7. **Stillingia treculiana** (Müller Arg.) I. M. Johnston, Contr. Gray Herb. 68: 91. 1923 • Trecul's toothleaf

Gymnanthes treculiana Müller Arg., Linnaea 34: 216. 1865

Herbs, perennial, with woody taproot. **Stems** fascicled, spreading, branching throughout, 1–4.5 dm. **Leaves** alternate; stipules linear, to 1 mm; petiole absent; blade usually obovate-spatulate, rarely narrowly elliptic, 1–4 × 0.5–1.5 cm, base cuneate, margins coarsely spinulose-dentate, teeth without prominent blackened tips, not incurved, apex rounded to obtuse; midrib prominent, secondary veins obscure. **Inflorescences** pedunculate, 2.5–5 cm; staminate flowers ± widely spaced, 1 per node; pistillate flowers 3–4, widely spaced; bracts ovate, 1 mm, apex mucronate, glands patelliform, ± sessile, 0.7 mm diam. **Staminate flowers:** calyx 1 mm. **Pistillate flowers:** sepals fugacious, 3, minute; styles connate only at base, 1.5 mm. **Capsules** ovoid-pyriform, 4 × 5 mm, deeply 3-lobed; lobes of gynobase to 1 mm; columella persistent. **Seeds** gray to black, ellipsoid, 2.5 × 2 mm, smooth; caruncle minute.

Flowering Mar–May; fruiting (Mar–)Apr–Jul(–Dec). Calcareous sandy soils, gravelly soils, and uplands; 0–1000 m; Tex.; Mexico (Coahuila, Nuevo León, Tamaulipas).

Stillingia treculiana is known in the flora area from the western Edwards Plateau south to the lower Rio Grande valley.

24. EUPHORBIA Linnaeus, Sp. Pl. 1: 450. 1753; Gen. Pl. ed. 5, 208. 1754 • Spurge [For Euphorbus, first-century Greek physician]

Paul E. Berry Ricarda Riina
Jess A. Peirson Ya Yang
Victor W. Steinmann Dmitry V. Geltman
Jeffery J. Morawetz N. Ivalú Cacho

Herbs, subshrubs, or shrubs [trees, cactoid succulents, geophytes, vines], annual, biennial, or perennial, monoecious [dioecious]; hairs unbranched or absent; latex white. **Leaves** persistent, deciduous, or small and caducous proximally, alternate, opposite, or whorled, sometimes bractlike and subtending floral structures, simple; stipules absent or present, persistent or deciduous; petiole absent or present, glands absent; blade unlobed, margins entire, crenulate, crenate-dentate, or serrulate, laminar glands absent; venation palmate, palmate at base and pinnate distally, or pinnate, often only midvein prominent. **Inflorescences** bisexual [unisexual], terminal or axillary, pseudanthia (each consisting of cuplike involucre bearing glands on rim, these sometimes with petaloid appendages, enclosing solitary pistillate flower surrounded by (0–)1–80 staminate flowers, entire structure termed the cyathium), in monochasia, dichasia, pleiochasia, cymose clusters, capitate glomerules, or solitary; glands subtending each bract 0. **Pedicels** present. **Staminate flowers**: sepals 0; petals 0; nectary absent; stamen 1; pistillode absent. **Pistillate flowers**: sepals 0 (ovary subtended by a calyxlike structure in *E. floridana*, *E. inundata*, *E. mesembrianthemifolia*, *E. porteriana*, *E. rosescens*, and *E. telephioides*); petals 0; nectary absent; pistil 3-carpellate; styles 3, distinct or connate basally to most of length, unbranched or 2-fid. **Fruits** capsules (tardily dehiscent and with spongy mesocarp in *E. lathyris*) [drupes]. **Seeds** globose to ovoid, oblong, cylindric, deltoid, pyramidal, or bottle-shaped; caruncle present or absent. x = 6, 7, 8, 9, 10.

Species ca. 2000 (139 in the flora): North America, Mexico, West Indies, Bermuda, Central America, South America, Eurasia, Africa, Atlantic Islands, Indian Ocean Islands, Pacific Islands, Australia.

Euphorbia is one of the two or three most species-rich angiosperm genera worldwide. Members of the genus occur in almost all habitat types, and many species prefer disturbed areas. Species in the genus are vegetatively highly diverse; growth forms include diminutive ephemerals, tuberous geophytes, taprooted perennial herbs, vines, various types of shrubs, trees to 25 m tall, and many xerophytic stem-succulents. Although succulents are primarily restricted to the Old World, a handful of independently derived succulents are native to the New World (B. L. Dorsey et al. 2013). Within the flora area these include *E. antisyphilitica* and *E. tithymaloides*. The striking vegetative similarity between the Old World succulent *Euphorbia* and New World cacti is one of the most commonly cited examples of convergent evolution.

The most distinctive feature of *Euphorbia* is its unique pseudanthial inflorescence, the cyathium (G. Prenner and P. J. Rudall 2007). This structure is so similar in appearance to a bisexual flower that many early botanists, including Linnaeus, actually mistook it for one. As with all Euphorbiaceae, the flowers of *Euphorbia* are unisexual, but in contrast to most other members of the family, they are extremely reduced. The pistillate flower comprises a single, perianth-less pistil, and the staminate flower comprises a single, perianth-less stamen. Most species are likely insect-pollinated, but a few species are hummingbird-pollinated (R. L. Dressler 1957).

The latex of all of the species is abundant and in some instances highly caustic, and care should be taken to avoid exposure to it. Nevertheless, some species in the flora area have been used for medicinal purposes (for example, *Euphorbia corollata* and *E. ipecacuanhae*, C. F. Millspaugh 1892).

Euphorbia is best known for its ornamental taxa, in particular *E. pulcherrima* Willdenow ex Klotzsch (the Christmas poinsettia), a native of Mexico. This species is widely grown throughout the flora area but has not become naturalized. Other commonly cultivated species in the flora area include two native species, *E. antisyphilitica* (candelilla) and *E. marginata* (snow-on-the-mountain), and the non-natives *E. characias* (Mediterranean spurge), *E. milii* Des Moulins (crown-of-thorns), and *E. rigida* M. Bieberstein (silver or upright myrtle spurge). *Euphorbia* "Diamond Frost" has in recent years become a popular cultivar to grow in pots or flowerbeds. Its progenitor, *E. graminea*, is introduced in the flora area and appears mainly associated with plantings in several southern states. In addition to several herbaceous European species that have become naturalized in North America, several others that have been recorded in the flora area in the past appear not to be persistent. These include species such as *E. amygdaloides* Linnaeus, *E. epithymoides* Linnaeus (sometimes treated as *E. polychroma* Kerner), *E. lucida* Waldstein & Kitaibel, *E. paralias* Linnaeus, and *E. segetalis* Linnaeus. In addition to these leafy taxa, numerous succulent species are commonly cultivated in botanical gardens and by private growers. Among the most popular are *E. lactea* Haworth, *E. neriifolia* Linnaeus, *E. obesa* Hooker, *E. tirucalli* Linnaeus, and *E. trigona* Miller. Although some of these species may persist around areas where they were previously cultivated, there is no evidence that they are actually naturalized in the region.

One of the most troublesome noxious weeds in the northern part of the flora area is leafy spurge, which was introduced from Eurasia. This species has been widely treated in North America as *Euphorbia esula* Linnaeus, but it turns out to be a misapplication of that name. The true leafy spurge in North America is more appropriately treated as *E. virgata*, a weedy species that is broadly distributed throughout temperate Europe and Asia (D. V. Geltman 1998). The actual *E. esula* is a related species of more restricted distribution in Europe that lacks the weedy tendencies of *E. virgata* (see discussion under 124. *E. virgata* for characters that distinguish the two). As with some of the herbaceous European waif species mentioned above, the real *E. esula* has been recorded historically in different parts of the flora area, but it does not appear to have persisted. It is therefore excluded here, and this should help to dispel the incorrect application of that name to leafy spurge in North America.

Historically, distinctive clades within *Euphorbia* were segregated into a number of satellite genera. In the flora area, these include *Chamaesyce*, *Pedilanthus*, and *Poinsettia*. Although these segregate genera are morphologically well-defined and monophyletic assemblages, recent molecular phylogenetic research has demonstrated that they are all nested within a broadly defined *Euphorbia* (V. W. Steinmann and J. M. Porter 2002; J. W. Horn et al. 2012). These molecular analyses both show *Euphorbia* as comprising four distinct clades, each of which is treated as a subgenus. Three of these subgenera, subg. *Chamaesyce* (Gray) Caesalpinius ex Reichenbach, subg. *Esula*, and subg. *Euphorbia*, are represented in the flora area. Subgenus *Esula* is treated here as a morphologically cohesive unit, whereas subg. *Chamaesyce* and subg. *Euphorbia* are divided into well-defined sections, which are keyed out and treated below.

The pleiochasia in *Euphorbia* are determinate. Each bears a whorl of pleiochasial bracts, which subtends multiple dichasial cymes, termed pleiochasial branches, that arise from a common point. The pleiochasium usually is terminated by a cyathium, but that sometimes aborts.

SELECTED REFERENCES Dorsey, B. L. et al. 2013. Phylogenetics, morphological evolution, and classification of *Euphorbia* subgenus *Euphorbia*. Taxon 62: 291–315. Horn, J. W. et al. 2012. Phylogenetics and the evolution of major structural characters in the giant genus *Euphorbia* L. (Euphorbiaceae). Molec. Phylogen. Evol. 63: 305–326. Prenner, G. and P. J. Rudall. 2007. Comparative ontogeny of the cyathium in *Euphorbia* (Euphorbiaceae) and its allies: Exploring the organ-flower-inflorescence boundary. Amer. J. Bot. 94: 1612–1629.

1. Stems usually prostrate, sometimes erect, ascending, reclining, or decumbent; leaves opposite (rarely whorled in *E. fendleri*), blade bases usually asymmetric; stipules interpetiolar (except in *E. acuta* where at base of petiole, deciduous, sometimes appearing absent) .... 24b. *Euphorbia* [subg. *Chamaesyce*] sect. *Anisophyllum*, p. 251
1. Stems usually erect or ascending, rarely decumbent or prostrate; leaves alternate, opposite, or whorled, blade bases symmetric; stipules at base of petiole or absent.
   2. Stems semisucculent to succulent, zigzag; involucres strongly zygomorphic, spurred and forming tube that encloses glands....... 24c. *Euphorbia* [subg. *Euphorbia*] sect. *Crepidaria*, p. 293
   2. Stems not both succulent and zigzag; involucres ± actinomorphic, not spurred.
      3. Involucral gland appendages usually petaloid, occasionally rudimentary; leaf margins entire.................24a. *Euphorbia* [subg. *Chamaesyce*] sect. *Alectoroctonum*, p. 240
      3. Involucral gland appendages not petaloid (except in *E. bifurcata*, *E. eriantha*, and *E. exstipulata* in sect. *Poinsettia*, but then leaf margins usually toothed); leaf margins entire or toothed (teeth sometimes inconspicuous in *E. eriantha* with linear leaves).

[4. Shifted to left margin.—Ed.]

4. Cyathia in terminal monochasia, dichasia, or condensed pleiochasia; involucral glands shallowly cupped to deeply concave, 1–3 per cyathium (if 4–5 in *E. exstipulata* then involucral gland appendages present; if 4–5 in *E. radians* then terminal clusters of cyathia subtended by white to pinkish, leafy bracts; if 4–5 in *E. eriantha*, then involucral gland appendages fringed, canescent, and folded over glands); involucral gland appendages petaloid, fringed, or absent . . . . 24f. *Euphorbia* [subg. *Chamaesyce*] sect. *Poinsettia*, p. 317
4. Cyathia in pleiochasia; involucral glands slightly concave, flat, or slightly convex, 4–5 per cyathium (2–3 in *E. oblongata*), terminal clusters of cyathia never subtended by white to pinkish bracts; involucral gland appendages hornlike or absent.
   5. Ovary and capsule not subtended by calyxlike structure; seeds with caruncle; involucral gland appendages hornlike or absent . . . . 24d. *Euphorbia* subg. *Esula*, p. 294
   5. Ovary and capsule subtended by calyxlike structure; seeds without caruncle; involucral gland appendages absent . . . .24e. *Euphorbia* [subg. *Euphorbia*] sect. *Nummulariopsis*, p. 313

24a. Euphorbia Linnaeus sect. Alectoroctonum (Schlechtendal) Baillon, Étude Euphorb., 284. 1858

Jess A. Peirson

Victor W. Steinmann

Jeffery J. Morawetz

Alectoroctonum Schlechtendal, Linnaea 19: 252. 1846; *Agaloma* Rafinesque; *Tithymalopsis* Klotzsch & Garcke

Herbs or shrubs [trees, rarely lianas], annual or perennial, with taproot or thickened or tuberous rootstock. **Stems** erect, ascending, decumbent, or prostrate, branched or unbranched, terete, glabrous or variously hairy (covered by exfoliating waxy coat in *E. antisyphilitica*). **Leaves** alternate or opposite; stipules present (sometimes rudimentary in *E. graminea* and *E. hexagona*), at base of petiole; petiole usually present, rarely absent or rudimentary, glabrous or hairy; blade monomorphic (dimorphic in *E. curtisii*, *E. exserta*, *E. ipecacuanhae*, and *E. mercurialina*), base symmetric, margins entire [rarely toothed], surfaces glabrous or hairy; venation pinnate, occasionally inconspicuous. **Cyathial arrangement:** solitary or in terminal monochasia, dichasia, or pleiochasia; individual dichasial or pleiochasial branches unbranched or few-branched at one or more successive nodes; bracts subtending dichasia and pleiochasia (pleiochasial bracts) opposite or whorled, green or with white margins, similar in shape and size to distal stem leaves, those on branches (dichasial or subcyathial bracts) opposite (rarely whorled or alternate in *E. corollata*), distinct; additional cymose branches occasionally present in distal axils, but not subtended by opposite or whorled bracts. **Involucre** ± actinomorphic, not spurred; glands [0–](2–)5, slightly concave, flat, or slightly convex; appendages usually petaloid, occasionally rudimentary. **Staminate flowers** (5–)20–25(–70). **Pistillate flowers:** ovary glabrous or hairy; styles connate basally, 2-fid. **Seeds:** caruncle present or absent.

Species ca. 120 (21 in the flora): North America, Mexico, West Indies, Central America, South America; introduced in Asia, Pacific Islands.

SELECTED REFERENCES Huft, M. J. 1979. A Monograph of *Euphorbia* Section *Tithymalopsis*. Ph.D. dissertation. University of Michigan. Park, K. R. 1998. Monograph of *Euphorbia* sect. *Tithymalopsis* (Euphorbiaceae). Edinburgh J. Bot. 55: 161–208.

1. Shrubs.
   2. Stems pencil-like, covered with flaky, exfoliating layer of wax; cyathia in axillary congested cymes near branch tips or solitary at distal nodes; Arizona, New Mexico, Texas. . . . . 2. *Euphorbia antisyphilitica*
   2. Stems gnarled, not waxy; cyathia solitary on short shoots; s California . . . . 16. *Euphorbia misera*

1. Herbs.
   3. Annual herbs with taproots (*E. graminea* rarely perennial).
      4. Leaves opposite.
         5. Stems 30–70(–100) cm; leaf blades linear-filiform, linear, or elliptic; cyathia solitary in leaf axils or in terminal cymes or dichasia; involucral gland appendages 0.7–1.7 mm; seeds 3.4 × 2.7 mm; c United States, mostly e of Rocky Mountains . . . . 10. *Euphorbia hexagona*
         5. Stems 4–25 cm; leaf blades usually linear- to narrowly-elliptic, occasionally ovate to obovate; cyathia solitary at distal bifurcations of stems; involucral gland appendages 0.2–0.5 mm; seeds 2.3–2.6 × 1.3–1.5 mm; Colorado Plateau of Utah, sw Colorado . . . . 17. *Euphorbia nephradenia*
      4. Leaves mostly alternate (opposite at proximal nodes in *E. bilobata*; some opposite in *E. graminea*).
         6. Dichasial bracts with conspicuous white margins.
            7. Dichasial bracts linear to narrowly oblanceolate; leaf blades pilose . . . . 3. *Euphorbia bicolor*
            7. Dichasial bracts narrowly elliptic to oblanceolate; leaf blades glabrous . . . . 14. *Euphorbia marginata*
         6. Dichasial bracts wholly green or distal ones white.
            8. Stems 10–35 cm; leaves opposite proximally, alternate distally; dichasial bracts wholly green; involucral glands 5; involucral gland appendages usually 2-fid . . . . 4. *Euphorbia bilobata*
            8. Stems 30–80(–110) cm; leaves usually alternate, sometimes some opposite; distal dichasial bracts often white; involucral glands (1–)2–4; involucral gland appendages undivided . . . . 9. *Euphorbia graminea*
   3. Perennial herbs with rootstocks, tubers, or taproots.
      9. Stem leaves usually opposite, occasionally whorled distally, rarely with 1 or 2 alternate leaves; plants with thick, globose to elongated tubers; se Arizona, primarily from Huachuca Mountains . . . . 13. *Euphorbia macropus*
      9. Stem leaves alternate; plants usually with rootstocks or taproots, rarely with elongated tubers; Arizona to e North America.
         10. Involucral glands 4, appendages green; leaf blade adaxial surfaces densely pilose, bases cordate; s coastal Texas . . . . 11. *Euphorbia innocua*
         10. Involucral glands 5, appendages usually white to pink, if greenish then minute and forming rim around gland; leaf blade adaxial surfaces glabrous, rarely villous or strigose (or pilose when young in *E. aaron-rossii*), bases cuneate to rounded; not s coastal Texas.
             11. Stems usually densely clumped, previous year's dead stems often persistent; leaf blades filiform to linear or narrowly ovate to lanceolate or oblanceolate, 0.5–6.5 mm wide; Arizona, New Mexico to wc Texas.
                 12. Cyathia in terminal monochasia; petiole 0.2–2.2 mm; stem leaves usually reflexed, occasionally spreading; endemic to banks of Colorado River in n Arizona . . . . 1. *Euphorbia aaron-rossii*
                 12. Cyathia in terminal dichasia (rarely in pleiochasia in *E. wrightii*); petiole absent; stem leaves spreading or ascending; New Mexico, Texas.
                     13. Leaf blades (2–)4–5 mm wide; involucral gland appendages 0.2 mm, forming narrow rim around distal margin of gland; capsules 3.2–4.5 × 4–6.5 mm, all 3 locules fertile; seeds 3.8 mm; mid and proximal cyathia early deciduous; Texas Panhandle, adjacent New Mexico . . . . 20. *Euphorbia strictior*
                     13. Leaf blades 1–2.5 mm wide; involucral gland appendages 0.5–1 mm, orbiculate; capsules 2.5(–3) × 2.7–3.3(–5) mm, 1 locule usually aborting; seeds 2.2–2.9 mm; cyathia persistent; wc Texas . . . . 21. *Euphorbia wrightii*

[11. Shifted to left margin.—Ed.]

11. Stems usually solitary or few, if densely clumped then previous year's dead stems not persistent; leaf blades filiform, linear or elliptic to lanceolate, ovate, oblanceolate, obovate or orbiculate, 0.8–26 mm wide (often greater than 5 mm wide); e Texas and Oklahoma to e North America.
   14. Involucral gland appendages 0–0.2 mm; peduncles 10–50(–70) mm.
      15. Involucres and glands typically dark red; plants with spreading rootstocks; stems erect or ascending . . . . 8. *Euphorbia exserta*
      15. Involucres and glands yellow or yellow-green; plants with deep, stout taproots; stems decumbent or slightly ascending . . . . 12. *Euphorbia ipecacuanhae*
   14. Involucral gland appendages 0.3–3.5(–4.5) mm; peduncles 1–17 mm (occasionally peduncle of central cyathium greater than 30 mm; occasionally to 40 mm in early May–Jun flowering *E. pubentissima*).
      16. Involucral glands red; leaf blades linear to filiform, 10–20 × 0.8–1.5(–4) mm; c, s peninsular Florida . . . . 18. *Euphorbia polyphylla*
      16. Involucral glands green; leaf blades not linear to filiform, or if linear then 10–55 × 1.5–6 mm; not peninsular Florida.
         17. Involucral gland appendages 0.3–0.6 mm; proximal leaves greatly reduced and often scalelike and appressed.
            18. Leaf blades usually linear, occasionally elliptic, rarely ovate, 1.5–6 mm wide, margins occasionally sparsely ciliate; seeds smooth . . . . 6. *Euphorbia curtisii*
            18. Leaf blades elliptic to ovate-deltate, 20–26 mm wide, margins densely ciliate; seeds with shallow and coarse depressions . . . . 15. *Euphorbia mercurialina*
         17. Involucral gland appendages (0.5–)1–3.5 mm; proximal leaves not reduced, neither scalelike nor appressed.
            19. Leaf blades usually linear, rarely ovate, 1.5–4 mm wide, margins revolute; stems usually densely puberulent to sericeous, rarely glabrous; seeds 2 × 1.2–1.3 mm . . . . 7. *Euphorbia discoidalis*
            19. Leaf blades oblanceolate, obovate, lanceolate, lance-ovate, or elliptic, 5–18 mm wide, margins not revolute or occasionally slightly revolute (*E. corollata*); stems glabrous, slightly pilose, or rarely villous; seeds 2.2–2.8 × 1.6–2.2 mm.
               20. Involucral gland appendages 2.5–3.5(–4.5) × 2.5–3.2 mm; peduncles (1.5–)5–11(–13) mm (proximal to 70 mm); seeds 2.5–2.8 mm . . . 5. *Euphorbia corollata*
               20. Involucral gland appendages 1–2.2 × 1.5 mm; peduncles 1–5 mm (or 15–40 mm in early flowering plants); seeds 2.2–2.4 mm . . . . 19. *Euphorbia pubentissima*

1. Euphorbia aaron-rossii A. H. Holmgren & N. H. Holmgren, Brittonia 40: 357, figs. 1, 2. 1988 • Ross's or Marble Canyon spurge [C] [E]

Herbs, perennial, with deep stout rootstock. **Stems** erect, branched, densely clumped, previous year's dead stems persistent, 25–45(–60) cm, glabrous, striate. **Leaves** alternate, persisting, usually reflexed, occasionally spreading; stipules 0.1–0.3 mm; petiole 0.2–2.2 mm, glabrous; blade narrowly ovate to lanceolate proximally, narrowly lanceolate, linear, or filiform distally, 10–32 × 0.5–6.5 mm, base cuneate, margins entire, apex acute, surfaces pilose when young, sparsely strigose or glabrous with age; venation obscure, only midvein conspicuous. **Cyathia** in terminal monochasia (thus appearing solitary at alternate nodes); peduncle 0.5–2.5(or 10–25) mm, glabrous or sparsely strigose. **Involucre** turbinate to campanulate, 2.2–3.7 × 1.5–2.5 mm, moderately strigose; glands 5, dark green, reniform, 0.7–1.1 × 1–1.6 mm; appendages white to pink, flabellate, 0.5–1.5 × 0.8–2.2 mm, dentate or erose. **Staminate flowers** 20–25. **Pistillate flowers:** ovary strigose; styles 1–1.3 mm, 2-fid at apex. **Capsules** subglobose, 2–3 × 4 mm, sparsely strigose; columella 2–3 mm. **Seeds** gray-green to gray-brown, globose-ovoid, 1.8–2.2 × 1.2–1.6 mm, longitudinally pitted; caruncle absent.

Flowering and fruiting spring–fall. Sandy soils and dunes, occasionally rocky slopes, riparian areas; of conservation concern; 600–1300 m; Ariz.

Euphorbia aaron-rossii is restricted to the banks of the Colorado River in several small areas of the Grand and Marble canyons. The species is most closely related to *E. strictior* and *E. wrightii*, but due to its rarity, it has not been extensively studied.

2. Euphorbia antisyphilitica Zuccarini, Flora 15(2, Beibl.): 58. 1832 • Candelilla, wax plant [F]

Shrubs, with much-branched, fleshy rootstock. **Stems** erect, few branched, 25–50(–100) cm, glabrous or puberulent, pencil-like, in age covered with flaky, exfoliating layer of wax. **Leaves** alternate, usually caducous, sometimes persisting; stipules 0.4–0.5 mm; petiole absent; blade ovate to deltate-subulate, 2.5–4 × 1 mm, thick, fleshy, base usually rounded and swollen, rarely cuneate, margins entire, apex acute, surfaces puberulent, adaxial sometimes canescent; venation inconspicuous. **Cyathia** in axillary congested cymes, near branch tips or solitary at distal nodes; peduncle 0–1 mm, lanulose. **Involucre** campanulate, 1.6–2.2 × 1.6–1.9 mm, puberulent to canescent; glands 5, pinkish, narrowly oblong to reniform, 0.3–0.4 × 0.8–1 mm; appendages white to pink, ovate, oblong, or transversely oblong, 1.3–2.5 × 1.4–2.5 mm, usually erose, rarely entire. **Staminate flowers** 50–70. **Pistillate flowers:** ovary glabrous; styles 0.9–1.1 mm, 2-fid nearly entire length. **Capsules** oblong to ovoid, 3.9–4.2 × 3.6–3.9 mm, glabrous; columella 3.1–3.3 mm. **Seeds** whitish gray, narrowly ovoid, 2.4–3.1 × 1.4–1.6 mm, irregularly rugose-tuberculate; caruncle crescent-shaped, 0.3–0.6 × 0.6–0.8 mm.

Flowering and fruiting year-round in response to sufficient rainfall. Desert scrub, frequently on limestone substrates; 100–1200 m; N.Mex., Tex.; Mexico.

Euphorbia antisyphilitica is the only pencil-stemmed species of *Euphorbia* occurring in the flora area. The species is characteristic of the Chihuahuan Desert scrub of Mexico from Chihuahua and Coahuila south to Hidalgo and Querétaro, and barely enters into the United States in southern New Mexico (Doña Ana and Lincoln counties) and southwest (Brewster, Hudspeth, Presidio, and Terrell counties) and south (Starr and Webb counties) Texas. The stems are covered in a conspicuous coat of exfoliating wax, and the plants historically have been harvested for this product, although the practice is much less prevalent now. The specific epithet refers to its traditional medicinal use in treating sexually transmitted infections.

3. Euphorbia bicolor Engelmann & A. Gray, Boston J. Nat. Hist. 5: 233. 1845 • Snow-on-the-prairie [E]

Herbs, annual, with taproot. **Stems** erect, unbranched or branched, 40–100 cm, pilose. **Leaves** alternate; stipules 0.3–0.4 mm; petiole 0.3–1 mm, pilose; blade narrowly elliptic to lanceolate, 37–54 × 7–17 mm, base cuneate to slightly rounded, margins entire, apex aristate or acute, surfaces pilose; venation obscure, only midvein conspicuous. **Cyathia** in terminal pleiochasia, dichasial and pleiochasial bracts linear to narrowly oblanceolate, with conspicuous white margins; peduncle 1.2–3 mm, densely pilose. **Involucre** campanulate, 2.7–3.5 × 2.2–3 mm, densely pilose; glands 4–5, green to pale greenish yellow, reniform, 0.6–0.7 × 1.4–1.6 mm; appendages white, obdeltate to orbiculate, 1.4–2.5 × 1.7–3 mm, dentate to erose. **Staminate flowers** 30–70. **Pistillate flowers:** ovary pilose; styles 0.7–1.2 mm, 2-fid ½ length. **Capsules** depressed-ovoid, 3.5–7.5 × 6–8.7 mm, densely pilose; columella 4.5–5.5 mm. **Seeds** tan to brown, ovoid, 4.3–4.5 × 3.7–3.9 mm, alveolate; caruncle absent.

Flowering and fruiting summer–fall. Prairies, blackland (calcareous) prairies, pastures and clearings in former blackland prairie areas, roadside clearings; 100–200 m; Ark., La., Okla., Tex.

Euphorbia bicolor is similar in appearance to *E. marginata* but can be distinguished by its linear to narrowly oblanceolate bracts and the presence of hairs on all parts of the plant.

4. Euphorbia bilobata Engelmann in W. H. Emory, Rep. U.S. Mex. Bound. 2(1): 190. 1859 • Blackseed spurge [F] [W]

Herbs, annual, with slender taproot. **Stems** erect, branched, 10–35 cm, glabrous or strigillose (especially when young and around nodes). **Leaves** opposite proximally, alternate distally; stipules 0.1–0.2 mm; petiole 1–4(–6) mm, glabrous, sericeous or strigillose; blade linear to narrowly elliptic, 8–52 × 2–7 mm, base attenuate, margins entire, ciliate-strigose, apex acute, abaxial surface sparsely strigillose to sericeous, adaxial surface usually glabrous; venation obscure, only midvein conspicuous. **Cyathia** solitary at distal nodes or in weakly defined cymes or dichasia, dichasial bracts and distal stem leaves wholly green; peduncle 0.5–3.6 mm, strigillose. **Involucre** obconic, 0.9–1.5 × 0.9–1.3 mm, strigillose to pilose; glands 5, yellow or pink, U-shaped,

EUPHORBIA

0.2–0.3 × 0.4–0.5 mm; appendages greenish, white, or pink, forming narrow rim around gland, or ovate, oblong, or obovate and usually 2-fid, rarely rudimentary, 0.2–1.4 × 0.2–0.6 mm, entire. **Staminate flowers** 20–25. **Pistillate flowers:** ovary glabrous, puberulent, strigillose, or pilose; styles 0.5–0.8 mm, 2-fid ⅓–½ length. **Capsules** oblate, 1.5–2.6 × 2.1–3.3 mm, glabrous or puberulent, strigillose, or pilose; columella 1.2–2.1 mm. **Seeds** brown to grayish black, narrowly ovoid, 3- or 4-angled in cross section, sometimes obscurely so, 1.3–1.9 × 1–1.4 mm, tuberculate, often with shallow depressions; caruncle absent. **2*n*** = 32.

Flowering and fruiting spring–fall. Sandy and rocky soils on slopes and canyon bottoms in pine-juniper woodlands, oak woodlands, grasslands; 1400–2600 m; Ariz., N.Mex., Tex.; Mexico (Chihuahua, Durango, Sonora).

In Texas, *Euphorbia bilobata* is known only from Jeff Davis County.

5. Euphorbia corollata Linnaeus, Sp. Pl. 1: 459. 1753 • Eastern flowering spurge [E] [F] [W]

Euphorbia corollata var. *molle* Millspaugh; *E. corollata* var. *viridiflora* Farwell; *E. marilandica* Greene; *E. olivacea* Small; *Tithymalopsis corollata* (Linnaeus) Klotzsch & Garcke; *T. olivacea* (Small) Small

Herbs, perennial, with deep, spreading rootstock. **Stems** erect or ascending, usually unbranched, occasionally few branched, solitary or few, previous year's dead stems not persistent, 20–100 cm, glabrous or slightly pilose to villous. **Leaves** alternate, ascending; stipules 0.1–0.2 mm; petiole minute or absent; blade oblanceolate, obovate, or elliptic, 25–55 × 5–12 mm, base cuneate to rounded, margins entire, occasionally slightly revolute, apex rounded to subacute, abaxial surface glabrous or pilose to villous, adaxial surface usually glabrous, rarely villous; venation occasionally obscure on small leaves, midvein conspicuous. **Cyathia** in terminal pleiochasia, dichasial bracts occasionally whorled or rarely alternate; peduncle (1.5–)5–11(–13) mm (proximal to 70 mm), glabrous. **Involucre** campanulate, 1.2–1.5 × 1.2–1.5(–2) mm, glabrous or moderately puberulent (especially near glands); glands 5, green, reniform, 0.5 × 0.8–1 mm; appendages white, flabellate, 2.5–3.5(–4.5) ×

2.5–3.2 mm, entire. **Staminate flowers** 20–25. **Pistillate flowers:** ovary glabrous; styles 0.8–1.4 mm, 2-fid at apex to ½ length. **Capsules** globose, 2.3–3 × 3.5–4.2 mm, glabrous; columella 2–2.5 mm. **Seeds** white or light gray, ovoid, 2.5–2.8 × 2.2 mm, with shallow and coarse depressions; caruncle absent.

Flowering and fruiting early summer–fall. Prairies, open fields, upland woods, glades, barrens, borders of swamps, roadsides, disturbed sites; 0–1300 m; Ont.; Ala., Ark., Conn., Del., Ga., Ill., Ind., Iowa, Kans., Ky., La., Maine, Md., Mass., Mich., Minn., Miss., Mo., Nebr., N.H., N.J., N.Y., N.C., Ohio, Okla., Pa., R.I., S.C., S.Dak., Tenn., Tex., Vt., Va., W.Va., Wis.

Euphorbia corollata is morphologically variable and widely distributed across a large part of eastern North America. The species appears to be expanding its range, as adventive populations have been reported from disturbed habitats at the northern edge of its range in Maine, Massachusetts, Michigan, Minnesota, New Hampshire, South Dakota, and Vermont.

6. **Euphorbia curtisii** Engelmann ex Chapman, Fl. South. U.S., 401. 1860 • Curtis's or sandhills spurge [E]

Euphorbia eriogonoides Small; *Tithymalopsis curtisii* (Engelmann ex Chapman) Small; *T. eriogonoides* (Small) Small

Herbs, perennial, with spreading rootstock. **Stems** erect or ascending, branched, solitary or few, previous year's dead stems not persistent, 20–40 cm, usually glabrous, rarely strigose to sericeous at nodes. **Leaves** alternate; stipules to 0.1 mm; petiole to (0–)1–2 mm, glabrous or strigose to sericeous; blade usually linear, occasionally elliptic, rarely ovate, proximal often greatly reduced and often scalelike, 10–30 × 1.5–6 mm, base cuneate, margins entire, occasionally sparsely ciliate, apex rounded or broadly acute, abaxial surface glabrous or sparsely strigose to sericeous, adaxial surface glabrous; venation obscure, only midvein conspicuous. **Cyathia** in terminal pleiochasia (fertile axillary branches occasionally present); peduncle 6.5–17 mm, filiform, glabrous. **Involucre** campanulate, 1–1.2 × 1.3–1.5(–1.7) mm, glabrous or strigose to sericeous on distal ½; glands 5, green, reniform, 0.3 × 0.6 mm; appendages white, semicircular, 0.3–0.4 × 0.6–0.8 mm, entire. **Staminate flowers** 20–25. **Pistillate flowers:** ovary glabrous or sparsely strigose to sericeous; styles 0.6–1.1 mm, 2-fid at apex to ½ length. **Capsules** globose, 2.5–3.2 × 4.3–5.1 mm, glabrous or sparsely strigose to sericeous; columella 2.4–3.1 mm. **Seeds** usually gray to black, occasionally brown, ovoid-globose, 2.2 × 1.8 mm, smooth; caruncle absent.

Flowering and fruiting early spring–summer. Xeric to dry oak or oak-pine scrub of sand hills, pine-oak woodlands, pine-oak savannas; 0–200 m; Fla., Ga., N.C., S.C.

Euphorbia curtisii is found on the Gulf and Atlantic coastal plains.

7. **Euphorbia discoidalis** Chapman, Fl. South. U.S., 401. 1860 • Summer spurge [E]

Tithymalopsis discoidalis (Chapman) Small

Herbs, perennial, with spreading rootstock. **Stems** erect or ascending, unbranched, solitary or few, previous year's dead stems not persistent, 45–70 cm, usually densely puberulent to sericeous, rarely glabrous. **Leaves** alternate; stipules to 0.1 mm; petiole (0–)1–2 mm, densely puberulent; blade usually linear, rarely ovate, 25–55 × 1.5–4 mm, base cuneate, margins entire, revolute, apex rounded, abaxial surface glabrous or puberulent to sericeous, adaxial surface glabrous; venation often obscure on smaller leaves, midvein conspicuous. **Cyathia** in terminal pleiochasia; peduncle 5–15 mm, filiform, glabrous or very sparsely puberulent to sericeous. **Involucre** campanulate, 1.2–1.4 × 1.2–2 mm, sparsely to densely puberulent; glands 5, green, reniform, 0.2–0.3 × 0.5–0.6 mm; appendages white, orbiculate to oblong, (0.5–)1–1.7 × 1–1.5 mm, entire. **Staminate flowers** 20–25. **Pistillate flowers:** ovary glabrous or sparsely strigose; styles 0.5–1.1 mm, 2-fid at apex to ½ length. **Capsules** globose, 1.8–3 × 2.5–4.8 mm, glabrous or sparsely strigose; columella 2.3–2.5 mm. **Seeds** light gray, ovoid, 2 × 1.2–1.3 mm, smooth or with few, very shallow depressions; caruncle absent.

Flowering and fruiting late spring–fall. Sand hills, pine savannas, woodland borders, open fields with sandy soils; 0–150 m; Ala., Fla., Ga., La., Miss., Tex.

M. J. Huft (1979) remarked that *Euphorbia discoidalis* is uncommon west of Alabama and referred many narrow-leaved specimens from Louisiana and Texas to *E. corollata*. K. R. Park (1998) included them in an expanded *E. discoidalis*, and that is followed here. The western populations can be distinguished from *E. corollata* by their shorter involucral gland appendages and revolute leaf margins. Further study of these western populations is warranted.

8. Euphorbia exserta (Small) Coker, Pl. Life Hartsville, 88. 1912 • Maroon or purple sand spurge [E]

Tithymalopsis exserta Small, Fl. S.E. U.S., 717, 1334. 1903; *Euphorbia gracilior* Cronquist; *T. gracilis* Small

Herbs, perennial, with spreading rootstock. **Stems** erect or ascending, unbranched or branched, solitary, few, or occasionally densely clumped, previous year's dead stems not persistent, 20–33 cm, usually glabrous, rarely sparsely villous, glaucescent. **Leaves** alternate; stipules less than 0.1 mm; petiole (0–)1–3 mm, glabrous; blade linear or linear-elliptic to obovate or orbiculate, proximal greatly reduced, scalelike, 15–30 × 1–20 mm, base cuneate to rounded, margins entire, apex rounded, broadly acute, or emarginate, surfaces glabrous, glaucescent; venation often obscure on narrow leaves, midvein conspicuous. **Cyathia** usually in terminal dichasia, sometimes pleiochasia; peduncle 6–33 mm, filiform, glabrous. **Involucre** usually dark red, campanulate, 1.3–1.6 × 1.4–2.1 mm, glabrous; glands 5, usually dark red, rarely greenish red, elliptic reniform, thickened, 0.3–0.5 × 0.8 mm; appendages white or green, often forming narrow rim around distal margin of gland, 0–0.2 mm, entire. **Staminate flowers** 20–25. **Pistillate flowers:** ovary glabrous; styles 0.4–0.8 mm, 2-fid at apex. **Capsules** depressed-globose, 1.8–2.5 × 3.6–4.4 mm, glabrous; columella 1.9–2.4 mm. **Seeds** ashy white, ovoid, 2.1 × 1.3 mm, angled with 5 blunt longitudinal ridges, with shallow and irregular pits; caruncle absent.

Flowering and fruiting spring–summer. Xeric to dry pine-oak scrub of sand hills, pine-oak woodlands, pine-oak savannas; 0–150 m; Fla., Ga., N.C., S.C., Va.

K. R. Park (1998) recognized both *Euphorbia exserta* and *E. gracilior* as distinct species, with the former known only from the holotype. However, this treatment follows M. J. Huft (1997) and treats *E. gracilior* as a synonym of *E. exserta*. Although the type of *E. exserta* is unusual in having greenish red (versus dark red) cyathia and glands, as well as small gland appendages, it is otherwise typical of the species as a whole, including features such as the upright habit, reddish coloration, scalelike proximal leaves, and filiform peduncles to 30 mm that are also common to plants formerly treated as *E. gracilior*. *Euphorbia gracilis* Elliott, which has sometimes been applied to *E. exserta*, is an illegitimate name (a later homonym of *E. gracilis* Loiseleur-Deslongchamps) and pertains here.

The upright habit and usually dark red cyathia and glands distinguish *Euphorbia exserta* from the otherwise similar *E. ipecacuanhae*, while the glaucescent vegetative parts and smaller gland appendages readily separate *E. exserta* from the similar *E. curtisii*. *Euphorbia exserta* is found on the Gulf and Atlantic coastal plains.

9. Euphorbia graminea Jacquin, Select. Stirp. Amer. Hist., 151. 1763 • Grassleaf spurge [I]

Herbs, usually annual, rarely perennial, with slender, rarely tuberous, taproot. **Stems** erect or ascending, branched, 30–80(–110) cm, strigillose or glabrescent, sharply angled. **Leaves** usually alternate, sometimes some opposite; stipules usually 0.2–0.5 mm, rarely rudimentary; petiole 0.4–5.9 mm, strigillose; blade ovate, elliptic, linear-elliptic, or oblong, 10–83 × 3–39 mm, base attenuate, rounded, or cuneate, margins entire, apex acute or obtuse, surfaces strigillose; venation occasionally obscure on narrow leaves, midvein conspicuous. **Cyathia** in usually terminal, rarely axillary, dichasia, distal dichasial bracts often white; peduncle 0.4–4.5 mm (to 15 mm at first node of inflorescence), glabrous. **Involucre** campanulate or obconic, 1–1.8 × 0.8–1.7 mm, glabrous or strigillose toward rim; glands (1–)2–4, yellow to greenish, elliptic or oblong, 0.1–0.3 × 0.2–0.4 mm; appendages white to tinged purple, ovate and often hoodlike or forming narrow rim around distal margin of gland, 0.3–1.6 × 0.4–0.9 mm, entire. **Staminate flowers** 30–40. **Pistillate flowers:** ovary glabrous; styles 0.7–1 mm, 2-fid from 1/2 to nearly entire length. **Capsules** ovoid-oblate, 2.5–3 × 3–3.5 mm, glabrous; columella 1.6–1.9 mm. **Seeds** gray, brown, or nearly black, ovoid, circular or weakly angled in cross section, 1.5–1.7 × 1.3–1.5 mm, coarsely tuberculate with longitudinal rows of shallow pits; caruncle absent or punctiform, 0.1–0.2 mm.

Flowering and fruiting year-round. Disturbed, weedy, or urban areas; 0–500 m; introduced; Ark., Calif., Fla., La., Tex.; Mexico; Central America; South America; introduced also in West Indies, Asia, Pacific Islands.

Euphorbia graminea occurs natively from northern South America to northern Mexico. The species is a variable and taxonomically complex entity whose boundaries are not well defined and are in need of further study. *Euphorbia graminea* is often weedy and has recently become established in warmer areas of the southern United States, where it will likely become more common in the future. In recent years, a cultivar of *E. graminea* has found considerable horticultural success and is marketed under the trade name "Diamond Frost."

10. Euphorbia hexagona Nuttall ex Sprengel, Syst. Veg. 3: 791. 1826 • Six-angle spurge [E] [F]

Herbs, annual, with taproot. **Stems** erect, unbranched or branched, 30–70(–100) cm, sparsely hispid, occasionally densely so at distal nodes. **Leaves** opposite; stipules (0–)0.1 mm; petiole 1–4 mm, pilose; blade linear-filiform, linear, or elliptic, 21–40 × 0.9–7.5 mm, base attenuate, margins entire, apex acute, abaxial surface sparsely hispidulous to strigillose, adaxial surface glabrous; venation obscure, only midvein conspicuous. **Cyathia** solitary in leaf axils or in terminal cymes or dichasia; peduncle 1–2.1 mm, strigillose. **Involucre** campanulate, 1–1.5 × (1–)1.5–1.8 mm, hispid; glands 5, green to deep red, elliptic to reniform, 0.5 × 0.8–1 mm; appendages white to green, tinged red, deltate to ovate, 0.7–1.7 × (0.9–)1.3–1.5 mm, entire. **Staminate flowers** 15–30(–40). **Pistillate flowers:** ovary glabrous; styles 0.7–1.1 mm, 2-fid nearly entire length. **Capsules** subglobose to broadly ovoid, 4.7–6.5 × 4.9–6.5(–7.1) mm, glabrous; columella 3.5–4.5 mm. **Seeds** dark brown or dark gray, ovoid, 3.4 × 2.7 mm, rugose, whitish glaucous; caruncle absent.

Flowering and fruiting late summer–fall. Sand prairies, other sandy soil habitats, stream banks, sand bars, damp places; 200–1300 m; Ark., Colo., Ill., Iowa, Kans., Minn., Mo., Mont., Nebr., N.Mex., Okla., S.Dak., Tex., Wis., Wyo.

Euphorbia hexagona is native to the central United States and is most common from southern South Dakota to Oklahoma and northern Texas.

11. Euphorbia innocua L. C. Wheeler, Contr. Gray Herb. 127: 62, plate 3, fig. D. 1939 • Velvet spurge [E]

Herbs, perennial, with moderately to strongly thickened rootstock. **Stems** prostrate to decumbent or ascending, branched (often near base), 7–45 cm, densely pilose. **Leaves** alternate; stipules to 0.1 mm; petiole (0.7–)1.1–3.5 mm, pilose; blade ovate to orbiculate, 4.6–17(–25) × 4.5–15(–19) mm, base cordate, margins entire, apex rounded to obtuse, surfaces densely pilose; venation obscure, usually only midvein conspicuous. **Cyathia** in terminal dichasia (often weakly defined); peduncle 1–2.7 mm, densely pilose. **Involucre** campanulate, 1–1.3 × 1.2–1.4 mm, pilose; glands 4, yellow to green, elliptic, 0.2–0.3 × 0.5–0.6 mm; appendages green, elliptic, 0.4–0.5 × 0.5–0.9 mm, entire or crenulate, ciliate. **Staminate flowers** 5–10. **Pistillate flowers:** ovary pilose; styles 0.4–0.7 mm, 2-fid ½ length. **Capsules** depressed-ovoid, 2–2.5 × 2.7–3.3 mm, pilose; columella 1.6–2.1 mm. **Seeds** gray to brown, ovoid, 1.5–1.7 × 1.2–1.3 mm, rugose with whitish ridges; caruncle absent.

Flowering and fruiting early winter–late spring. Sandy soils or dunes, grasslands, pastures; 0–20 m; Tex.

Euphorbia innocua is restricted to south coastal Texas in Aransas, Calhoun, Kenedy, Kleberg, Nueces, Refugio, San Patricio, and Willacy counties.

12. Euphorbia ipecacuanhae Linnaeus, Sp. Pl. 1: 455. 1753 • American or Carolina ipecac [E]

Euphorbia arundelana Bartlett; *Tithymalopsis ipecacuanhae* (Linnaeus) Small

Herbs, perennial, with deep, stout rootstock. **Stems** decumbent or slightly ascending, branched, often densely clumped, previous year's dead stems not persistent, 17–27 cm, usually glabrous, rarely sparsely villous. **Leaves** alternate; stipules 0.1–0.2 mm; petiole (0–)1–2 mm, glabrous; blade usually linear, obovate, or oblanceolate to orbiculate, rarely filiform, proximal greatly reduced, scalelike, 15–70 × 1.5–13 mm, gradually smaller proximally, base cuneate, margins entire, apex rounded, broadly acute, or emarginate, surfaces glabrous; venation occasionally obscure on smaller leaves, midvein conspicuous. **Cyathia** usually in terminal dichasia, sometimes pleiochasia; peduncle 10–50(–70) mm, glabrous. **Involucre** yellow or yellow-green, hemispheric, 1–1.2 × 2–2.4 mm, glabrous; glands 5, yellow or yellow-green, obovate or elliptic, 0.7–0.8 × 1–1.2 mm; appendages white or green, often forming narrow rim around distal margin of gland, 0–0.2 mm, entire. **Staminate flowers** 10–20. **Pistillate flowers:** ovary glabrous; styles 0.4–0.8 mm, 2-fid at apex. **Capsules** globose, 2.3–3.4 × 3.5–4.2 mm, glabrous; columella 3–3.1 mm. **Seeds** white or brown, ovoid, 2.3–2.5 × 1.5 mm, angular, with 5 longitudinal ridges, shallowly and irregularly pitted; caruncle absent.

Flowering and fruiting early spring–early summer. Pine and pine-oak savannas, pine-oak sand hills, turkey oak scrub, open sand habitats; 0–150 m; Conn., Del., D.C., Ga., Md., N.J., N.Y., N.C., Pa., S.C., Va.

The vegetative stems of *Euphorbia ipecacuanhae* are often quite short in proportion to the dichasial or pleiochasial branches, thus superficially plants often appear to have mostly opposite leaves. However, careful examination of the base of the plant will reveal alternate leaves. The leaves are extremely variable in both shape and coloration, and the variation can be pronounced within a population or even on a single plant. M. J. Huft (1979) did not recognize infraspecific taxa within *E. ipecacuanhae*, and his treatment is followed here. This species is found on the Atlantic coastal plain.

13. Euphorbia macropus (Klotzsch & Garcke) Boissier in A. P. de Candolle and A. L. P. P. de Candolle, Prodr. 15(2): 52. 1862 • Huachuca mountain spurge

Anisophyllum macropus Klotzsch & Garcke, Abh. Königl. Akad. Wiss. Berlin 1859: 33. 1860; *Euphorbia biformis* S. Watson; *E. plummerae* S. Watson

Herbs, perennial, with thick, globose to elongated tubers, 2–8 cm. **Stems** erect to ascending, branched, 10–45 (–60) cm, glabrous, puberulent, or densely hirsute to setose, often with 2-layered indumentum of long hairs intermixed with short hairs. **Leaves** usually opposite, occasionally whorled distally, or rarely with 1–2 alternate leaves; stipules 0.1–0.2 mm; petiole 0–18 mm, hirsute, sericeous, or strigose; blade linear to ovate or almost orbiculate, 6–54 × 2–19 mm, base rounded to attenuate, margins entire, occasionally ciliate with stiff recurved hairs, apex acute to obtuse, surfaces usually hirsute, sericeous, or strigose, occasionally glabrous adaxially; venation conspicuous. **Cyathia** in weakly-defined terminal dichasia; peduncle 1.4–5.8 mm, glabrous. **Involucre** obconic to campanulate, 1.1–1.4 × 0.5–1.5 mm, glabrous or strigillose; glands 4–5, greenish, oblong, 0.2 × 0.4–0.5 mm; appendages usually yellowish or green, rarely dark purple, ovate, flabellate, semiorbiculate, or oblong, 0.3–0.9 × 0.4–1.1 mm, usually entire. **Staminate flowers** 10–15. **Pistillate flowers:** ovary glabrous, sericeous, or strigillose; styles 0.4–0.6 mm, 2-fid ½ length. **Capsules** oblate, 2.3–3 × 3.1–4.2 mm, glabrous, sericeous, or strigillose; columella 1.6–2.1 mm. **Seeds** black to light brown, broadly ovoid to subglobose, rounded in cross section, 1.5–2.3 × 1.4–1.8 mm, smooth or with low rounded tubercles; caruncle absent.

Flowering and fruiting summer–fall. Stream banks and rocky slopes in pine-oak woodlands, sometimes with juniper, Douglas fir-pine forests; 1500–2200 m; Ariz.; Mexico; Central America (Guatemala, Honduras).

Euphorbia macropus is a widespread and common Mexican species just barely entering the flora area in southeastern Arizona, where most of the collections are from the Huachuca Mountains.

14. Euphorbia marginata Pursh, Fl. Amer. Sept. 2: 607. 1813 • Snow-on-the-mountain, euphorbe marginée [F] [W]

Euphorbia bonplandii Sweet; *Lepadena marginata* (Pursh) Nieuwland

Herbs, annual, with taproot. **Stems** erect, unbranched or branched, 30–85(–150) cm, pilose or glabrous. **Leaves** alternate; stipules 0.1–0.3 mm; petiole 0.2–3 mm, glabrous or minutely pilose; blade broadly ovate to elliptic, 32–62 (–82) × 18–28(–52) mm, base rounded to cuneate, margins entire, often white on distal leaves, apex acute, rarely mucronate, surfaces glabrous; venation obscure, only midvein conspicuous. **Cyathia** in terminal pleiochasia, dichasial bracts narrowly elliptic to oblanceolate, with conspicuous white margins; peduncle 1.8–2.7(–22) mm, densely pilose. **Involucre** campanulate, 2.2–3.5 × 1.3–2.3 mm, margin between glands deeply divided into fringe of fimbriate lobes, pilose; glands 4–5, green to greenish yellow, reniform to subcircular, 0.7–1.1 × 1–1.6 mm; appendages white, orbiculate, 1.5–2.7 × 1.9–2.9 (–3.6) mm, entire. **Staminate flowers** 30–70. **Pistillate flowers:** ovary pilose; styles 1–2.5 mm, 2-fid ½–⅔ length. **Capsules** oblate, 3–5 × 3.5–7.5 mm, moderately to densely pilose; columella 3–4.1 mm. **Seeds** orange-tan to gray, ovoid, 3.7–3.9 × 3–3.3 mm, rugose, with 2 transverse ridges (one dark orange to brown, other inconspicuous); caruncle absent. $2n = 56$.

Flowering and fruiting summer–fall. Disturbed areas and grasslands; 0–1700 m; Man., Ont., Que., Sask.; Ark., Calif., Colo., Conn., Ill., Ind., Iowa, Kans., La., Md., Mass., Mich., Minn., Mo., Mont., Nebr., N.H., N.Mex., N.Y., N.C., Ohio, Okla., Pa., R.I., S.C., S.Dak., Tenn., Tex., Utah, Va., W.Va., Wis., Wyo.; c, s Mexico.

Euphorbia marginata is native to the central United States. The type specimen was collected by Meriwether Lewis along the Yellowstone River in southern Montana in 1806, and it has been reported to be native as far south as Arizona, New Mexico, and Texas, and as far east as southern Minnesota, western Iowa, and Missouri (G. Yatskievych 1999–2013, vol. 2). It is presumably naturalized outside of this area. *Euphorbia marginata* is widely cultivated as an ornamental for its showy, white-margined distal leaves, and it can escape locally.

15. Euphorbia mercurialina Michaux, Fl. Bor.-Amer. 2: 212. 1803 • Mercury spurge [E]

Tithymalopsis mercurialina (Michaux) Small

Herbs, perennial, with thickened, spreading rootstock. **Stems** erect, unbranched or branched, solitary or few, previous year's dead stems not persistent, 20–33 cm, glabrous or villous to lanate. **Leaves** alternate; stipules 0.1–0.2 mm; petiole (1–)2.5–5(–6) mm, ciliate to lanate; blade elliptic to ovate-deltate, proximal greatly reduced, scalelike, 34–55 × 20–26 mm, base rounded or cuneate, margins entire, densely ciliate, apex rounded to acute, abaxial surface sparsely pilose to villous (to lanate on midrib), adaxial surface glabrous; venation obscure, only midvein conspicuous. **Cyathia** usually in terminal pleiochasia, rarely dichasia; peduncle 1.3–2.7 mm (to 40–70 mm for central cyathium), filiform, glabrous. **Involucre** campanulate or hemispheric, 1.5–2.5 × 2–3 mm, glabrous; glands 5, green, elliptic-reniform, 0.5 × 2 mm; appendages white, narrowly transversely-oblong to lunate, 0.6 × 2.5 mm, slightly erose. **Staminate flowers** 10–15. **Pistillate flowers:** ovary glabrous; styles 0.7–1.5 mm, 2-fid at apex. **Capsules** depressed-globose, 2.3–3.3 × 4.4–5 mm, glabrous; columella 2.7–3 mm. **Seeds** tan to dark brown, ovoid, 2.2 × 1.6 mm, with shallow and coarse depressions; caruncle absent.

Flowering and fruiting spring. Dry to mesic wooded slopes and ravines; 100–600 m; Ala., Ga., Ky., N.C., Tenn.

Euphorbia mercurialina is restricted primarily to the Cumberland Plateau and southern Appalachians, with disjunct occurrences in south-central North Carolina in the lower Piedmont. The North Carolina plants are markedly hairier than plants elsewhere, with villous or lanate stems, petioles, and abaxial leaf midribs. *Euphorbia mercurialina* has been reported from Florida and Virginia in the past. The Virginia plants were apparently planted (A. S. Weakley 2010), and the Florida reports are most certainly in error.

16. Euphorbia misera Bentham, Bot. Voy. Sulphur, 51. 1844 • Cliff spurge

Shrubs, soft wooded, with woody rootstock. **Stems** erect to ascending, often gnarled and scraggly, branched, with conspicuous knobby short shoots, 70–150 cm, puberulent-tomentose, bark grayish red to light gray. **Leaves** alternate, well spaced on long shoots or fasciculate on short shoots; stipules 0.6–1.1 mm; petiole 4–12(–19) mm, slender, puberulent to shortly pilose; blade oblong, ovate, orbiculate, elliptic, or obovate, 6–24 × 5–21 mm, base rounded to cuneate, margins entire, apex acute to obtuse, surfaces puberulent-tomentose; venation conspicuous. **Cyathia** usually solitary on short shoots, peduncle 1.8–10.5 mm, puberulent-tomentose. **Involucre** campanulate, 1.4–3.8 × 2.1–4.4 mm, puberulent-tomentose; glands 5, yellow to reddish, oblong to reniform, 0.7–1.3 × 1.1–2.6 mm; appendages green-yellow to yellowish or whitish, oblong to transversely oblong, 0.6–1.9 × 1.3–3.8 mm, crenulate to erose. **Staminate flowers** 40–50. **Pistillate flowers:** ovary glabrous or puberulent; styles 1.6–2.7 mm, 2-fid at apex. **Capsules** oblate, 4.6–5.1 × 6.1–6.7 mm, usually glabrous or glabrescent, occasionally puberulent; columella 2.8–3.6 mm. **Seeds** grayish, subglobose to ovoid, rounded in cross section, 2.7–3.3 × 2.5–2.8 mm, foveolate; caruncle absent.

Flowering and fruiting year-round (but most prolific after winter rains). Rocky soils, sometimes in crevices of vertical cliff faces, coastal scrub, maritime desert scrub, arid desert scrub; 0–400 m; Calif.; Mexico (Baja California, Baja California Sur, Sonora).

Euphorbia misera is relatively infrequent within the flora area, known primarily from coastal southern California and the Channel Islands (although a relictual inland population occurs in the Little San Bernardino Mountains). The species has been considered worthy of conservation, but appears to be under little threat, especially in Mexico where it is frequent and often locally abundant.

17. Euphorbia nephradenia Barneby, Leafl. W. Bot. 10: 314. 1966 • Paria or Utah spurge [C] [E]

Herbs, annual, with slender, little-branched taproot. **Stems** erect to ascending, branched, dichotomous distally and slightly angled, 4–25 cm, glabrous or sparsely strigillose. **Leaves** opposite; stipules 0.1–0.2 mm; petiole 2–6 mm, glabrous or sparsely strigillose; blade usually linear- to narrowly elliptic, occasionally ovate to obovate, 14–42 × 3–10 mm, progressively narrower distally, base attenuate, margins entire, apex usually acute, rarely obtuse, surfaces glabrous or sparsely strigillose; venation inconspicuous. **Cyathia** solitary at distal bifurcations of stems; peduncle 0.6–2.4 mm, glabrous or strigillose. **Involucre** campanulate, 1–1.1 × 1.2–1.4 mm, strigillose at least toward apex; glands 5, green-yellow, oblong, 0.4–0.6 × 0.7–1 mm; appendages whitish to yellow-green, lunate to broadly ovate, 0.2–0.5 × 0.7–1.1 mm, entire or slightly crenulate. **Staminate flowers** 25–30. **Pistillate flowers:** ovary glabrous; styles 0.7–1 mm, 2-fid at apex. **Capsules** oblate to subglobose, 2.9–3.2 × 3.2–3.4 mm, glabrous;

columella 2.8–3.1 mm. **Seeds** light gray to whitish, oblong-ovoid, rounded in cross section, 2.3–2.6 × 1.3–1.5 mm, dimpled and rugulose; caruncle absent.

Flowering and fruiting spring–summer. Saltbush, blackbrush, *Ephedra*-dominated scrub and desert communities; of conservation concern; 1100–1500 m; Colo., Utah.

Euphorbia nephradenia is the only species of the genus endemic to the Colorado Plateau of Utah and adjacent Colorado.

18. Euphorbia polyphylla Engelmann ex Chapman, Fl. South. U.S. ed. 2, repr. 2, 694. 1892 • Lesser Florida spurge [E]

Tithymalopsis polyphylla (Engelmann ex Chapman) Small

Herbs, perennial, with spreading rootstock. **Stems** erect or ascending, branched, solitary, few, or occasionally densely clumped, previous year's dead stems not persistent, 18–33 cm, glabrous. **Leaves** alternate; stipules to 0.1 mm; petiole minute or absent; blade linear to filiform, 10–20 × 0.8–1.5(–4) mm, base cuneate, margins entire, often involute, apex rounded, surfaces glabrous; venation usually obscure, midvein visible at base of wider leaves. **Cyathia** in terminal dichasia or pleiochasia; peduncle 2–6 mm (to 20 mm for central cyathium), glabrous. **Involucre** campanulate, 1.2–2 × 1.3–1.8 mm, glabrous or strigose on distal extreme; glands 5, red, elliptic, 0.5 × 1 mm; appendages white, orbiculate, 0.5–0.8 × 1.3–1.5 mm, erose. **Staminate flowers** 20–25. **Pistillate flowers:** ovary glabrous; styles 0.6–1 mm, 2-fid at apex. **Capsules** globose, 2.3–2.8 × 4.3–5.1 mm, glabrous; columella 2.5–3.2 mm. **Seeds** ashy gray, ovoid, 2.9 × 2 mm, with obscure shallow depressions; caruncle absent.

Flowering and fruiting late spring–late fall. Open sand and pine savannas; 0–10 m; Fla.

Euphorbia polyphylla is endemic to sandy habitats in the southern half of peninsular Florida. The species has been reported from coastal Louisiana, but whether those plants represent native occurrences or plantings is unclear (R. D. Thomas and C. M. Allen 1993–1998, vol. 2). The Florida populations are here recognized as the only native occurrences.

19. Euphorbia pubentissima Michaux, Fl. Bor.-Amer. 2: 212. 1803 • Southeastern flowering spurge [E]

Euphorbia apocynifolia Small; *E. corollata* Linnaeus var. *paniculata* Boissier; *E. corollata* var. *zinniiflora* (Small) H. E. Ahles; *E. zinniiflora* Small; *Tithymalopsis apocynifolia* (Small) Small; *T. paniculata* (Boissier) Small; *T. zinniiflora* (Small) Small

Herbs, perennial, with spreading rootstock. **Stems** erect, usually unbranched, occasionally few branched distally, solitary or few, previous year's dead stems not persistent, 30–65 cm, usually glabrous, rarely villous. **Leaves** alternate; stipules 1 mm; petiole (0–) 1–2(–10) mm, glabrous or densely villous; blade lanceolate, lance-ovate, elliptic, or obovate, 40–68 × 6–18 mm, often reflexed, base cuneate to rounded, margins entire, apex usually rounded, sometimes broadly acute, abaxial surface glabrous or villous, adaxial surface usually glabrous, rarely sparsely villous; venation occasionally obscure on small leaves, midvein conspicuous. **Cyathia** in terminal pleiochasia; peduncle 1–5 mm (or 15–40 mm on early flowering individuals), usually glabrous, occasionally sparsely villous. **Involucre** campanulate or hemispheric, 1.3–1.7 × 1.8–2.2 mm, glabrous or villous; glands 5 (7–10 on central cyathium), green, reniform or broadly elliptic, 0.3–0.5 × 0.5–0.8 mm; appendages white, orbiculate or narrowly flabellate, 1–2.2 × 1.5 mm, entire. **Staminate flowers** 20–25. **Pistillate flowers:** ovary glabrous or sparsely villous; styles 0.6–1.1 mm, 2-fid ½ length. **Capsules** globose, 2–2.4 × 3.3–4.8 mm, glabrous or sparsely villous; columella 1.8–2.1 mm. **Seeds** light gray, ovoid, 2.2–2.4 × 1.6–1.8 mm, with shallow depressions; caruncle absent.

Flowering and fruiting spring–fall. Open fields, cliffs, woods, flood plains; 0–900 m; Ala., Fla., Ga., La., Md., Miss., N.C., S.C., Tenn., Va., W.Va.

The taxonomic history of *Euphorbia pubentissima* is complex. The species has been included within a very broadly defined *E. corollata* in the past. Therefore, its geographic distribution appears significantly more wide-ranging in some treatments. The framework established by M. J. Huft (1979) and later by K. R. Park (1998) is followed here. *Euphorbia pubentissima* is recognized as a variable species restricted to the southeastern United States. The species can be distinguished from *E. corollata* by its shorter involucral gland appendages and smaller seeds. *Euphorbia paniculata* Elliott, which sometimes is applied to *E. pubentissima*, is an illegitimate name (a later homonym of *E. paniculata* Desfontaines).

20. Euphorbia strictior Holzinger, Contr. U.S. Natl. Herb. 1: 214, plate 18. 1892 • Panhandle spurge [E]

Herbs, perennial, with cylindric rootstock. **Stems** erect, branched, densely clumped, previous year's dead stems often persistent, 30–70 cm, glabrous. **Leaves** alternate, persisting, spreading or ascending; stipules to 0.1 mm; petiole absent; blade linear to narrowly oblanceolate, (20–)40–70 × (2–)4–5 mm, base narrowly cuneate, margins entire, apex broadly acute, surfaces glabrous; venation obscure, only midvein conspicuous on wider leaves. **Cyathia** in terminal dichasia; peduncle (2–)4–12(–18) mm, proximal and mid peduncles and cyathia abscising early, sparsely to moderately strigose to sericeous. **Involucre** campanulate or hemispheric, 1.2–2.4 × 2.2–3.2 mm, pilose; glands 5, green, broadly elliptic, 0.7–0.8 × 1.3–1.6 mm; appendages white, forming narrow rim around distal margin of gland, 0.2 × 1.5–1.8 mm, entire or erose. **Staminate flowers** 20–25. **Pistillate flowers:** ovary strigillose to tomentulose; styles 0.4–0.6 mm, 2-fid at apex. **Capsules** globose, all 3 locules fertile, 3.2–4.5 × 4–6.5 mm, sparsely strigillose; columella 2.5–3.9 mm. **Seeds** gray-green to gray-brown, ovoid, 3.8 × 3 mm, shallowly and obscurely pitted; caruncle absent.

Flowering and fruiting spring–fall. Open grasslands and uplands; 900–1400 m; N.Mex., Tex.

Euphorbia strictior is confined to a small area of the Texas Panhandle and adjacent New Mexico. The species is closely related to *E. aaron-rossii* and *E. wrightii*; it can be distinguished from *E. wrightii* by its larger stature, shorter and less petaloid involucral gland appendages, and early abscising proximal and mid cyathia and peduncles. *Euphorbia strictior* also tends to develop all three seeds in the capsule, whereas *E. wrightii* tends to develop only two.

21. Euphorbia wrightii Torrey & A. Gray in War Department [U.S.], Pacif. Railr. Rep. 2(4): 174. 1857 • Wright's spurge

Herbs, perennial with cylindric rootstock or elongated tubers. **Stems** erect, branched, densely clumped, previous year's dead stems often persistent, 20–50 cm, glabrous. **Leaves** alternate, persisting, spreading or ascending; stipules 0.1–0.3 mm; petiole absent; blade linear to linear-filiform, (17–)20–40 × 1–2.5 mm, base cuneate, slightly sheathing stem, margins entire, apex broadly acute to rounded, abaxial surface glabrous or sparsely villous, adaxial surface glabrous; venation obscure, only midvein conspicuous on wider leaves. **Cyathia** usually in terminal dichasia, rarely pleiochasia; peduncle (3–)5–15 mm, all peduncles and cyathia persistent on plant, glabrous. **Involucre** campanulate, 1.5–2 × 1.8–2.5 mm, pilose; glands 5, green, broadly elliptic, 0.7–0.8 × 1 mm; appendages white to pink, orbiculate, 0.5–1 × 1.3–1.8 mm, coarsely erose. **Staminate flowers** 20–25. **Pistillate flowers:** ovary strigose to tomentose; styles 1 mm, 2-fid at apex. **Capsules** depressed-globose, 1 locule usually aborting, 2.5(–3) × 2.7–3.3(–5) mm, sparsely tomentose; columella 2.6–3 mm. **Seeds** gray-green to gray-brown, globose-ovoid, 2.2–2.9 × 1.8–2 mm, shallowly and obscurely pitted; caruncle absent.

Flowering and fruiting spring–fall. Open grasslands and uplands, often on limestone outcrops; 500–1000 m; Tex.; Mexico (Coahuila).

Within the flora area, *Euphorbia wrightii* is endemic to the western Edwards Plateau and adjacent rolling plains in western Texas.

24b. EUPHORBIA Linnaeus sect. ANISOPHYLLUM Roeper in J. É. Duby, Bot. Gall. 1: 412. 1828

Victor W. Steinmann Jeffery J. Morawetz
Paul E. Berry Jess A. Peirson
Ya Yang

Anisophyllum Haworth, Syn. Pl. Succ., 159. 1812, not Jacquin 1763; *Chamaesyce* Gray

Herbs, rarely subshrubs or shrubs, annual or perennial, with taproot or thickened rootstock. **Stems** usually prostrate, sometimes erect, ascending, reclining, or decumbent, branched [unbranched], terete or flattened, glabrous or hairy. **Leaves** opposite (rarely whorled in *E. fendleri*); stipules present, deciduous, (sometimes appearing absent in *E. acuta*), interpetiolar (at base of petiole in *E. acuta*); petiole present [absent], glabrous or hairy; blade monomorphic, base usually asymmetric, rarely symmetric, margins entire or variously toothed, surfaces glabrous or hairy; venation usually palmate or palmate at base and pinnate distally, sometimes pinnate, often only midvein conspicuous. **Cyathial arrangement:** terminal or axillary, solitary or in cymose clusters or capitate glomerules; bracts absent (except for small, bractlike leaves in capitate glomerules). **Involucre** ± actinomorphic, not spurred; glands (2–)4(–5), slightly concave, flat, or slightly convex; appendages petaloid or absent. **Staminate flowers** (0–)1–80. **Pistillate flowers:** ovary glabrous or hairy; styles distinct or connate basally, unbranched or 2-fid. **Seeds:** caruncle absent (except for a carunclelike structure in *E. caruunculata*).

Species ca. 300 (69 in the flora): North America, Mexico, West Indies, Bermuda, Central America, South America, Eurasia, Africa, Atlantic Islands, Indian Ocean Islands, Pacific Islands, Australia.

In the key and descriptions that follow, for species with mostly prostrate stems, the side of the stem toward the ground is called the lower side and the opposite side is call the upper side.

SELECTED REFERENCES Herndon, A. 1993. A revision of the *Chamaesyce deltoidea* (Euphorbiaceae) complex of southern Florida. Rhodora 95: 38–51. Herndon, A. 1993b. Notes on *Chamaesyce* (Euphorbiaceae) in Florida. Rhodora 95: 352–368. Wheeler, L. C. 1941. *Euphorbia* subgenus *Chamaesyce* in Canada and the United States exclusive of southern Florida. Rhodora 43: 97–154, 168–205, 223–286. Yang, Y. and P. E. Berry. 2011. Phylogenetics of the Chamaesyce clade (*Euphorbia*, Euphorbiaceae): Reticulate evolution and long-distance dispersal in a prominent C_4 lineage. Amer. J. Bot. 98: 1486–1503.

1. Ovaries and capsules ± hairy.
    2. Cyathia in capitate glomerules (with reduced, bractlike leaves subtending cyathia).
        3. Stipules distinct when young, connate into deltate scales when older, often with dark glands along margins or at base; styles 0.6–0.9 mm; seeds plumply ovoid, 0.7–0.8 mm wide, dark reddish brown to almost black; s Florida. . . . . 54. *Euphorbia lasiocarpa*
        3. Stipules distinct or connate only at base, deltate, subulate, or linear- or filiform-subulate, without dark glands along margins or at base; styles 0.1–0.6 mm; seeds narrowly ovoid or ovoid-oblong, 0.5–0.7 mm wide, not dark reddish brown to almost black; widespread, including s Florida.
            4. Glomerules of cyathia terminal and axillary, axillary glomerules sessile or at tips of elongated, leafless stalks . . . . 45. *Euphorbia hirta*
            4. Glomerules of cyathia terminal, on main stems or short, leafy, axillary branches.
                5. Stems glabrous, strigillose, or pilose; involucre 0.8–1.6 × 0.7–1.3 mm, appendages 0.2–1.1 × 0.5–1.7 mm; styles 0.4–0.6 mm; capsules pilose; seeds 0.9–1.5 mm . . . . 30. *Euphorbia capitellata* (in part)
                5. Stems usually both strigillose and hirsute; involucre 0.5–0.7 × 0.4–0.6 mm, appendages 0.1–0.2 × 0.1–0.3 mm; styles 0.1–0.3 mm; capsules strigillose; seeds 0.7–0.9(–1.1) mm . . . . 65. *Euphorbia ophthalmica*
    2. Cyathia solitary or in small, cymose clusters at distal nodes, on congested, axillary branches, or at branch tips.
        6. Involucral gland appendages unequal.
            7. Seeds 3-angled to almost round in cross section, with 4–5 rounded, transverse ridges encircling seed; Arizona, California . . . . 68. *Euphorbia pediculifera* (in part)
            7. Seeds conspicuously 4-angled in cross section, variously grooved or ridged, but not with 4–5 rounded, transverse ridges encircling seed; Florida, Louisiana (*E. conferta*, *E. pergamena*, and *E. thymifolia*) or sw United States (*E. indivisa*).
                8. Styles 0.8–1.3 mm, usually unbranched, rarely 2-fid at apex; sw United States from Arizona to Texas. . . . . 50. *Euphorbia indivisa*
                8. Styles 0.4–1 mm, 2-fid; Florida, Louisiana.
                    9. Stems pilose. . . . . 34. *Euphorbia conferta*
                    9. Stems strigose to sericeous or strigose-tomentulose.

10. Capsules well exserted from involucre at maturity; seeds with 3 or 4 transverse sulci alternating with low transverse ridges . . . . 70. *Euphorbia pergamena*

10. Capsules scarcely exserted from involucre, base often remaining inside the involucre and splitting one side of it during maturity; seeds with 4 low transverse ridges, but not sulcate . . . . 85. *Euphorbia thymifolia* (in part)

[6. Shifted to left margin.—Ed.]

6. Involucral gland appendages ± equal in size or rudimentary to absent.

11. Leaf blade margins toothed, at least toward apex.

12. Styles unbranched.

13. Stems strigillose; seeds broadly ovoid, 1.2–1.4 × 1–1.1 mm, with 2 well-defined transverse ridges. . . . . 76. *Euphorbia rayturneri*

13. Stems pilose to lanate; seeds narrowly oblong-ovoid to ellipsoid, 1–1.5 × 0.5–0.6 mm, almost smooth, rugulose, dimpled, or with short and irregularly interrupted furrows (seeds appearing partially and irregularly few-ridged) . . . . 83. *Euphorbia stictospora*

12. Styles 2-fid.

14. Capsules with pubescence concentrated along keels or toward base, often glabrous between keels.

15. Petioles, leaf blade abaxial surfaces, ovaries, and capsules glabrous or sparsely sericeous, pilose, or villous; seeds reddish brown to orange or gray-pink, almost smooth or with faint transverse ridges; s Florida . . . . 59. *Euphorbia mendezii*

15. Petioles, leaf blade abaxial surfaces, ovaries, and capsules crisped-villous to glabrate; seeds white but with barely concealed brown surface beneath, with sharp transverse ridges; widespread, including s Florida . . . . . . 75. *Euphorbia prostrata*

14. Capsules ± evenly hairy or pubescence at least not concentrated only along keels and base, not glabrous between keels.

16. Capsules pilose to villous.

17. Involucral gland appendages not puberulent-ciliate along margins; capsules 1.4–1.9 × 1.5–2 mm; seeds ovoid to narrowly ovoid, 1–1.4 × 0.6–0.9 mm; California . . . . 79. *Euphorbia serpillifolia* (in part)

17. Involucral gland appendages often puberulent-ciliate along margins; capsules 1.5–1.9 × 1.3–1.7 mm; seeds narrowly pyramidal-ovoid, 1.1–1.3 × 0.4–0.6 mm; Texas . . . . 88. *Euphorbia velleriflora*

16. Capsules strigose to sericeous.

18. Capsules scarcely exserted from involucre, base often remaining inside involucre and splitting one side of it during maturity . . . . 85. *Euphorbia thymifolia* (in part)

18. Capsules well exserted from involucre at maturity.

19. Stems rooting at nodes; styles 0.5–0.8 mm; seeds bluntly angled, smooth or papillate. . . . . 47. *Euphorbia humistrata*

19. Stems not rooting at nodes; styles 0.3–0.4 mm; seeds sharply angled, with 3–4 low transverse ridges . . . . 56. *Euphorbia maculata*

11. Leaf blade margins entire (rarely sparsely serrulate in *E. garberi* and *E. laredana*).

20. Capsules with pubescence concentrated along keels or toward base, often glabrous or less hairy between keels . . . . 53. *Euphorbia laredana*

20. Capsules ± evenly hairy or pubescence at least not concentrated only along keels and base, not glabrous between keels.

21. Involucral gland appendages divided into 3–8 triangular to subulate segments.

22. Shrubs; stems ascending, puberulent to shortly hirsute; stipules 0.3–0.5 mm; seeds 1.4–1.5 mm . . . . 51. *Euphorbia jaegeri*

22. Annual herbs; stems prostrate, villous, hairs glandular-glistening; stipules rudimentary to 0.2 mm; seeds 0.8–1 mm . . . . 81. *Euphorbia setiloba*

[21. Shifted to left margin.—Ed.]

21. Involucral gland appendages entire, toothed, or absent.
    23. Proximal leaf blades ovate to ovate-elliptic, distal ones linear to elliptic-linear, more than 6 times as long as wide . . . . 25. *Euphorbia angusta*
    23. Leaf blades not linear (rarely linear in *E. pediculifera*), less than 3 times as long as wide.
        24. Stems glabrous or sparsely hairy . . . . 37. *Euphorbia deltoidea* (in part)
        24. Stems conspicuously and densely hairy.
            25. Stems with appressed hairs.
                26. Leaf blade apices long-acuminate and spinulose; seeds 2.2–2.6 mm. . . . . 23. *Euphorbia acuta*
                26. Leaf blade apices acute to obtuse or rounded; seeds 0.8–1.8(–2) mm.
                    27. Largest leaf blades usually more than 15 mm; seeds 3-angled to almost round in cross section, with 4–5 rounded, transverse ridges encircling seed . . . . 68. *Euphorbia pediculifera* (in part)
                    27. Largest leaf blades less than 15 mm; seeds 3- or 4-angled in cross section, without 4–5 rounded, transverse ridges encircling seed.
                        28. Involucral glands deep red, purple, or purple-black.
                            29. Involucral gland appendages absent or forming narrow rim around distal margin of gland, 0–0.1 × 0–0.6 mm; seeds ovoid, 1–1.4 × 0.6–0.8 mm; Texas . . . . 33. *Euphorbia cinerascens*
                            29. Involucral gland appendages oblong to flabellate, 0.4–0.7(–1) × 0.7–1.2 mm; seeds narrowly oblong, 1–1.2 × 0.4–0.6 mm; Arizona, California. . . . . 58. *Euphorbia melanadenia*
                        28. Involucral glands green to yellow-green or brown.
                            30. Stipules filiform, 0.8–1.3 mm; involucres 2–2.5 × 2.2–2.6 mm; styles 0.8–1.2 mm; capsules 1.9–2.3 × 2–2.4 mm; seeds whitish, 1.5–1.8(–2) × 0.6–0.9 mm; c, s United States but not Florida . . . . 55. *Euphorbia lata*
                            30. Stipules triangular or triangular-subulate, 0.2–0.7 mm; involucres 0.6–1 × 0.5–1.3 mm; styles 0.3–0.7 mm; capsules 1.1–1.6 × 1.3–2.2 mm; seeds gray to reddish brown, 0.8–1.2 × 0.5–0.8 mm; s Florida.
                                31. Stems wiry, less than 1 mm diam., 5–20 cm; leaf blades deltate, 2–5(–7) × 1–4.5 mm . . . . 37. *Euphorbia deltoidea* (in part)
                                31. Stems not wiry, 1–3 mm diam., 15–50 cm; leaf blades ovate to oblong-elliptic, 4–9(–15) × 3–6 mm . . . . 40. *Euphorbia garberi* (in part)
            25. Stems with spreading to erect hairs.
                32. Stems and leaves with glistening hairs; stipules (0–)0.1 mm. . . . . 26. *Euphorbia arizonica*
                32. Stems and leaves without glistening hairs; stipules 0.2–1.6 mm.
                    33. Involucres 0.4–0.6 × 0.5–0.9 mm, glands 0.1 × 0.1–0.2 mm, appendages absent. . . . . 61. *Euphorbia micromera* (in part)
                    33. Involucres 0.6–2.5 × 0.5–2.4 mm, glands 0.1–0.6 × 0.2–0.9 mm, appendages usually present (absent or forming narrow rim in some specimens of *E. deltoidea*; sometimes forming narrow rim in *E. polycarpa*).
                        34. Involucral gland appendages ciliate-puberulent adaxially; perennial herbs . . . . 87. *Euphorbia vallis-mortae*
                        34. Involucral gland appendages not ciliate-puberulent adaxially; annual or perennial herbs.
                            35. Seeds 1.1–1.7 × 0.8–1.3 mm. . . . . 64. *Euphorbia ocellata* (in part)
                            35. Seeds 0.8–1.2 × 0.5–0.8 mm.
                                36. Stems prostrate to ascending, mat-forming; sw United States . . . . 72. *Euphorbia polycarpa* (in part)
                                36. Stems ascending to erect, not mat-forming; s Florida.

37. Stems and leaf blades villous or villous-hirsute; stems ascending to erect, wiry, less than 1 mm diam., 5–20 cm; leaf blades deltate, 2–4.5 × 2–4.5 mm . . . . 37. *Euphorbia deltoidea* (in part)
37. Stems and leaf blades canescent; stems ascending, not wiry, 1–3 mm diam., 15–50 cm; leaf blades ovate to oblong-elliptic, 4–9(–15) × 3–6 mm. . . . 40. *Euphorbia garberi* (in part)

1. Ovaries and capsules glabrous.
38. Leaf blades linear, 5 times or more as long as wide, bases symmetric or subsymmetric (sometimes slightly asymmetric in *E. parryi*).
39. Leaf blade margins serrulate . . . . 39. *Euphorbia florida*
39. Leaf blade margins entire.
40. Stipules entire; seeds 0.9–1.4 mm, 4-angled in cross section.
41. Stipules 0.3–0.5 mm; involucres 0.4–0.5 mm, glands yellow to pink; styles 2-fid; capsules 1.1–1.4 × 1.1–1.4 mm; s Arizona . . . . 44. *Euphorbia gracillima*
41. Stipules 0.5–0.9 mm; involucres 0.7–0.9 mm, glands pink to dark purple; styles unbranched; capsules 1.5–1.8 × 1.6–1.8 mm; sw United States, including s Arizona . . . . 77. *Euphorbia revoluta*
40. Stipules usually deeply and irregularly fringed, lobed, or lacerate and divided into slender segments, rarely entire; seeds 1.4–2 mm, bluntly 3-angled or rounded-angular in cross section.
42. Stems erect or ascending; involucral gland appendages 0.4–2.5 × 1.1–1.7 mm . . . . 62. *Euphorbia missurica* (in part)
42. Stems usually prostrate, rarely ascending-erect; involucral gland appendages 0.2–0.6 × 0.3–0.7(–1.1) mm . . . . 67. *Euphorbia parryi*
38. Leaf blades not linear, 4 times or less as long as wide, bases usually asymmetric, rarely subsymmetric to symmetric.
43. Styles unbranched.
44. Annual herbs with slender taproots; leaf blade margins serrulate to denticulate; styles filiform, 1.8–2.6 mm; California . . . . 46. *Euphorbia hooveri*
44. Perennial herbs with thickened taproots; leaf blade margins entire; styles thickened-clavate, 0.3–0.5 mm; Texas.
45. Petioles 0–0.2(–0.3) mm; leaf blade bases cordate to auriculate; involucral gland appendages 0.1–0.2(–0.5) × 0.4–0.8 mm, entire or dentate-crenate . . . . 27. *Euphorbia astyla*
45. Petioles 0.3–0.9 mm; leaf blade bases rounded to truncate; involucral gland appendages 0.3–0.6 × 0.8–1.2 mm, usually deeply dissected into 4–5 acuminate lobes, rarely entire or crenate . . . . 52. *Euphorbia jejuna*
43. Styles 2-fid at least toward apex.
46. Cyathia in capitate glomerules (with reduced, bractlike leaves subtending cyathia).
47. Stipules distinct, filiform or divided into subulate-filiform segments; leaf blades ovate to narrowly ovate, often with red spot in center, 4–19 × 2–8 mm; stems and leaves glabrous, strigillose, or pilose . . . . 30. *Euphorbia capitellata* (in part)
47. Stipules connate, deltate, entire; leaf blades obliquely oblong-oblanceolate, 10–35 × 7–15 mm; stems and leaves glabrous . . . . 48. *Euphorbia hypericifolia*
46. Cyathia solitary or in small cymose clusters at distal nodes or on congested, axillary branches.
48. Leaf blade margins toothed (at least toward apex or on majority of leaves).
49. Seeds with prominent transverse ridges that interrupt abaxial keel.
50. Stems shortly pilose or puberulent proximally (often glabrous distally); capsules 1.3–1.5 × 1.1–1.5 mm . . . . 22. *Euphorbia abramsiana*
50. Stems glabrous; capsules 1.3–1.9 × 1.6–2 mm . . . . 42. *Euphorbia glyptosperma*
49. Seeds without prominent transverse ridges, or if present, not interrupting abaxial keel.

[51. Shifted to left margin.—Ed.]

51. Largest leaf blades more than 20 mm.
    52. Seeds 1.9–2.3 × 1.3–1.4 mm; stems and leaves glabrous . . . . 86. *Euphorbia trachysperma*
    52. Seeds 1–1.6 × 0.5–1.1 mm; stems and leaves glabrous or hairy.
        53. Stems sparsely to densely pilose or pilose-crinkled proximally, usually glabrous distally; leaf blades glabrous or sparsely pilose toward base (abaxially), glabrous (adaxially); seeds with prominent transverse ridges or coarsely and inconspicuously pitted-reticulate. . . . . 49. *Euphorbia hyssopifolia*
        53. Stems sparsely to moderately pilose to villous or with short, incurved hairs, pubescence often concentrated at nodes and distally (hairs occasionally in 2 bands along opposite sides of stem); leaf blades usually sparsely to moderately pilose, especially toward base, sometimes glabrous; seeds finely and irregularly wrinkled or with indistinct shallow, rounded cross ridges. . . . . .63. *Euphorbia nutans*
51. Largest leaf blades less than 20 mm.
    54. Leaf blade surfaces papillate; cocci of capsule often elongated and terminating in empty portion . . . . 90. *Euphorbia villifera* (in part)
    54. Leaf blade surfaces not papillate; cocci of capsule not elongated or terminating in empty portion.
        55. Stems and leaf blades glabrous.
            56. Seeds pyramidal to oblong-ovoid, weakly 4-angled in cross section, 0.9–1 × 0.7 mm, reddish brown to brown, minutely beaded, with broad, rounded, transverse ridges; s coastal Louisiana and adjacent Texas. . . . . .57. *Euphorbia meganaesos*
            56. Seeds ovoid to narrowly ovoid, 4-angled in cross section, 1–1.4 × 0.6–0.9 mm, pink, light brown or grayish, smooth to dimpled or rugose, or with faint transverse ridges; widespread, but not coastal Louisiana and adjacent Texas . . . . 79. *Euphorbia serpillifolia* (in part)
        55. Stems and leaf blades usually hairy, rarely glabrate.
            57. Leaf blade margins serrate to serrulate, usually with conspicuous teeth at base; capsules 2–2.6 × 3.2–3.7 mm; seeds 1.5–1.8 × 1.1–1.3(–1.5) mm . . . 80. *Euphorbia serrula*
            57. Leaf blade margins usually entire in proximal ½ and serrulate in distal ½ (rarely some leaves with margins nearly entire in *E. vermiculata*; rarely some leaves serrulate nearly to base in *E. serpillifolia*); capsules 1.4–1.9 × 1.5–2.1 mm; seeds 1–1.4 × 0.6–0.9 mm.
                58. Stems prostrate to ascending, often mat-forming; cyathial glands yellow to pink; seeds smooth to dimpled or rugose, or with faint transverse ridges; California . . . . 79. *Euphorbia serpillifolia* (in part)
                58. Stems prostrate to ascending or erect, not mat-forming; cyathial glands red to reddish green; seeds rugulose and sometime also with low transverse ridges; Arizona, New Mexico, ne United States, Canada . . . . .89. *Euphorbia vermiculata*

[48. Shifted to left margin.—Ed.]

48. Leaf blade margins entire (occasionally toothed in *E. blodgettii*; rarely sparsely serrate in *E. porteriana*).
    59. Stipules (at least those of upper side of stem) connate, forming deltate, ligulate, or ovate scale.
        60. Subshrubs or shrubs; stems erect to ascending; leaf blade bases slightly asymmetric . . . . .60. *Euphorbia mesembrianthemifolia*
        60. Herbs; stems prostrate to decumbent and often rooting at nodes; leaf blade bases asymmetric.
            61. Perennials; involucral glands 0.2–0.5 × (0.2–)0.3–0.8 mm, appendages 0.3–1 × 0.6–1.3 mm; leaf blades often with red blotch in center . . . . 24. *Euphorbia albomarginata*
            61. Annuals, rarely short-lived perennials; involucral glands 0.1 × 0.1–0.3 mm, appendages 0–0.2 × 0.1–0.3 mm; leaf blades without red blotch.

62. Stems prostrate to decumbent; involucral gland appendages unequal (pair near sinus lunate to oblong, 0.1–0.2 × 0.1–0.3 mm, distal margins entire, crenulate, or irregularly sinuate, other pair rudimentary, 0–0.1 × 0.1–0.3 mm, distal margins crenulate or entire); peninsular Florida . . . . 28. *Euphorbia blodgettii*
62. Stems prostrate; involucral gland appendages equal, forming narrow rim at edge of gland, 0.1–0.2 × 0.2–0.3 mm, distal margins entire or crenulate; widespread, including peninsular Florida. . . . . 78. *Euphorbia serpens*

[59. Shifted to left margin.—Ed.]

59. Stipules usually distinct, occasionally connate basally (rarely connate to middle in *E. platysperma*), not forming conspicuous, deltate, ligulate, or ovate scale.
63. Perennial herbs or subshrubs with thickened and often woody rootstocks.
64. Seeds 1.6–2.4 mm.
65. Capsules 2.8–3.3 × 2.8–3.4 mm; involucres broadly campanulate to hemispheric, 1.7–2.2 × 1.5–2.7 mm; styles 0.7–0.9 mm; leaf blades ovate or orbiculate-deltate to reniform-deltate, 5–17 × 4–16 mm. . . . . .69. *Euphorbia perennans*
65. Capsules 1.7–2.4 × 1.6–2.5 mm; involucres campanulate to turbinate or broadly cupulate, 0.8–1.7 × 0.8–1.8 mm; styles 0.3–0.4 mm; leaf blades lanceolate, linear- or oblong-lanceolate, ovate, or orbiculate, 3–11 × 0.8–7 mm.
66. Stems usually erect, rarely slightly decumbent; leaf blades ovate to lanceolate or oblong- or linear-lanceolate, 3–11 × 0.8–3(–5) mm, apices acute to short acuminate, bases short-tapered, occasionally one side rounded. . . . . .32. *Euphorbia chaetocalyx*
66. Stems usually prostrate, decumbent, or ascending, very rarely erect; leaf blades usually orbiculate to ovate, rarely almost lanceolate, 3–8 × 2.5–7 mm, apices rounded to obtuse, bases slightly cordate to rounded or obtuse . . . . 38. *Euphorbia fendleri*
64. Seeds 0.8–1.4 mm.
67. Stipules usually glabrous (occasionally sparsely hairy with appressed uncinate hairs in *E. deltoidea*).
68. Stems prostrate; stems and leaf blades glabrous or sparsely hairy with appressed uncinate hairs; s Florida. .37. *Euphorbia deltoidea* (in part)
68. Stems erect to ascending; stems and leaf blades usually villous, rarely glabrous; Texas . . . . 90. *Euphorbia villifera* (in part)
67. Stipules pilose to hirsute.
69. Capsules 1.6–1.7 × 1.6–1.9 mm; seeds 1.2–1.4 mm; involucral gland appendages absent; involucres 1–1.4 mm. . . . . 66. *Euphorbia parishii*
69. Capsules 1.1–1.4 × 1.1–1.4 mm; seeds 0.8–1.1 mm; involucral gland appendages present, occasionally forming rim at margin of gland; involucres 0.6–1.1 mm. . . . . 72. *Euphorbia polycarpa* (in part)
63. Annual, rarely short-lived perennial, herbs with taproots or spreading rootstocks.
70. Capsules 4.7–5.5(–6) mm; seeds (2.8–)4.1–5.2 mm, bottle-shaped, strongly dorsiventrally compressed and weakly 3-angled in cross section, with linear carunclelike structures . . . . .31. *Euphorbia carunculata*
70. Capsules 1–3.5(–4) mm; seeds 0.7–2.8 mm, not bottle-shaped, weakly dorsiventrally compressed or terete to sub-, 3-, or 4-angled in cross section, without carunclelike structures.
71. Seeds terete or bluntly subangled in cross section (weakly dorsiventrally compressed in *E. platysperma* and *E. polygonifolia*, thus appearing semielliptic), usually smooth, occasionally rugose, minutely pitted, or obscurely wrinkled (*E. platysperma* with sharp, linelike, longitudinal ridges on adaxial side).
72. Seeds (2–)2.2–2.8 mm, weakly dorsiventrally compressed in cross section.
73. Involucral gland appendages present; leaf blade apices usually acute to mucronulate, rarely obtuse; capsules 2.7–3.2 × 2.2–2.9 mm; Arizona, California . . . . 71. *Euphorbia platysperma*
73. Involucral gland appendages absent; leaf blade apices obtuse, often mucronulate; capsules 3–3.5(–4) × (2–)2.4–3 mm; Atlantic and Great Lakes shores . . . . 73. *Euphorbia polygonifolia*

72. Seeds 0.7–2.1 mm, terete to bluntly sub- or 3-angled in cross section.
   74. Stems erect or ascending; leaf blades narrowly oblong to narrowly lanceolate-oblong; styles 0.5–1.4 mm; involucral gland appendages present, 0.4–2.5 mm; seeds bluntly 3-angled ....62. *Euphorbia missurica* (in part)
   74. Stems prostrate to slightly ascending; leaf blades oblong, oblong-obovate, narrowly elliptic, or lanceolate, ovate, deltate, or falcate; styles 0.2–0.6 mm; involucral gland appendages present or absent, (0–)0.1–1 mm; seeds terete to bluntly subangled.
      75. Leaf blades ovate to deltate or falcate; w United States........ ....64. *Euphorbia ocellata* (in part)
      75. Leaf blades elliptic, oblong, or lanceolate; c United States to Gulf and Atlantic coasts.
         76. Leaf blades narrowly elliptic to lanceolate, 4–8 × 1–2 mm; seeds 1–1.3 mm; involucral gland appendages 0.1–0.2 mm, often rudimentary; s Florida.................... 36. *Euphorbia cumulicola*
         76. Leaf blades oblong to oblong-elliptic or oblong-obovate, 4–15 × 2–6 mm; seeds 1.1–1.9 mm; involucral gland appendages absent or 0.1–1 mm; widespread, including s Florida.
            77. Leaf blades oblong or elliptic-oblong, 2–3 mm wide; involucral gland appendages (0–)0.1–0.5(–0.7) mm; styles 0.2–0.3 mm; coastal, Texas to Virginia.....29. *Euphorbia bombensis*
            77. Leaf blades oblong, oblong-elliptic, or oblong-obovate, 2–6 mm wide; involucral gland appendages (0.1–)0.5–1 mm; styles 0.2–0.6 mm; not coastal, Texas north through c United States.............................. 41. *Euphorbia geyeri*

[71. Shifted to left margin.—Ed.]

71. Seeds usually 3–4-angled in cross section (± weakly angled in *E. villifera*), smooth to rugose or wrinkled, or with transverse ridges.
   78. Stems erect to ascending.
      79. Stems glabrous; stipules triangular; involucral gland appendages (0–)0.1–0.2 mm; seeds 0.7–1 mm; Florida..74. *Euphorbia porteriana*
      79. Stems usually villous, rarely glabrous; stipules filiform; involucral gland appendages 0.2–0.4 mm; seeds 1–1.4 mm; Texas .... 90. *Euphorbia villifera* (in part)
   78. Stems prostrate or reclining.
      80. Stipules filiform; leaf blade bases cordate to rounded; seeds 1.8–2.1 mm, smooth to rugose.. 35. *Euphorbia cordifolia*
      80. Stipules subulate or scalelike; leaf blade bases rounded, oblique, or cuneate; seeds 0.8–2 mm, sculpture various (if 1.5–2 mm then with 5–7 faint transverse ridges).
         81. Involucres 0.4–0.6 mm..............................61. *Euphorbia micromera* (in part)
         81. Involucres 0.6–1.8 mm (usually more than 0.8 mm).
            82. Seeds 1.3–2 mm.
               83. Involucral glands subcircular; involucral gland appendages present; seeds 1.5–2 mm; leaf blades oblong, ovate-oblong, or narrowly elliptic-oblong; vicinity of Rio Grande, w Texas .... 43. *Euphorbia golondrina*
               83. Involucral glands elliptic; involucral gland appendages absent; seeds 1.3–1.5 mm; leaf blades orbiculate, oval, or shortly oblong; widespread in Texas.. 82. *Euphorbia simulans*
            82. Seeds 0.8–1.2 mm.
               84. Seeds smooth or faintly rippled; involucral glands oblong, appendages 0.1–0.2 mm; leaf blade apices obtuse to acute .... 72. *Euphorbia polycarpa* (in part)
               84. Seeds with deep transverse ridges; involucral glands subcircular to slightly elliptic, appendages (0–)0.1–0.4 mm; leaf blade apices rounded, occasionally slightly emarginate...................84. *Euphorbia theriaca*

22. Euphorbia abramsiana L. C. Wheeler, Bull. S. Calif. Acad. Sci. 33: 109. 1934 • Abrams's sandmat

Chamaesyce abramsiana (L. C. Wheeler) Koutnik

Herbs, annual, with slender taproot. **Stems** prostrate, mat-forming, 10–35(–50) cm, shortly pilose or puberulent at least proximally, often glabrous distally. **Leaves** opposite; stipules distinct, divided into 5–7 subulate-filiform segments, 0.6–1.1 mm, usually glabrous, rarely pilose; petiole 0.5–1 mm, glabrous; blade ovate, elliptic-oblong, or slightly ovate-cordate, 3–11 × 2–5 mm, base asymmetric, truncate to hemicordate, margins serrulate at least toward apex, often entire toward base, apex acute to obtuse, surfaces sometimes with red spot in center, glabrous; usually only the midvein conspicuous. **Cyathia** solitary at distal nodes of primary stems or at nodes of short congested axillary branchlets; peduncle 0.2–0.5 mm. **Involucre** obconic, 0.5–0.6 × 0.4–0.5 mm, glabrous; glands 4, yellowish to pink, circular to oblong, 0.1 × 0.1–0.2 mm; appendages absent, or white to pink, semicircular to broadly ovate, to 0.1 × 0.2 mm, distal margin entire or shallowly lobed. **Staminate flowers** 3–5. **Pistillate flowers:** ovary glabrous; styles 0.1–0.3 mm, 2-fid nearly entire length. **Capsules** ellipsoid to ovoid, 1.3–1.5 × 1.1–1.5 mm, glabrous; columella 1–1.3 mm. **Seeds** light gray to light brown, narrowly ovoid to ovoid, 4-angled in cross section, 1–1.2 × 0.6–0.7 mm, with 3–5 prominent transverse ridges that often interrupt abaxial keel.

Flowering and fruiting summer–fall. Desert scrub and desert grasslands; -40–1400 m; Ariz., Calif., N.Mex., Tex.; Mexico (Baja California, Baja California Sur, Sonora).

23. Euphorbia acuta Engelmann in W. H. Emory, Rep. U.S. Mex. Bound. 2(1): 189. 1859, name proposed for conservation • Pointed sandmat

Chamaesyce acuta Millspaugh; *Euphorbia acuta* var. *stenophylla* Boissier; *E. georgei* Oudejans

Herbs, perennial, with strongly thickened rootstock. **Stems** ascending to erect, 5–30 cm, uniformly and densely canescent or sericeous. **Leaves** opposite; stipules deciduous, sometimes appearing absent, distinct, brown, linear-subulate, thin, 0.3–0.8 mm, canescent; petiole 0.4–1.2 mm, moderately to densely canescent; blade ovate to lanceolate, 6–20 × 3–8 mm, base subsymmetric, rounded to cuneate, margins entire, strongly involute, apex long-acuminate, spinulose, abaxial surface canescent to densely sericeous, adaxial surface glabrous or sparsely canescent; 3-veined from base but only midvein conspicuous. **Cyathia** solitary at distal nodes; peduncle 1.3–3.2 mm. **Involucre** turbinate to urceolate, 2–2.6 × 1.7–2.5 mm, villous to lanate; glands 4, yellow-green to orange or red, slightly concave, oblong-elliptic, 0.2–0.4 × 0.6–1.5 mm; appendages white, flabellate, 1.1–2.1 × 0.2–0.6 mm, distal margin shallowly and irregularly toothed. **Staminate flowers** 20–25. **Pistillate flowers:** ovary strigose, pubescent to villous; styles 0.6–0.9 mm, 2-fid ½ length. **Capsules** subglobose to broadly ovoid, 2.8–3.7 mm diam., strigose, pubescent to villous; columella 2.3–3 mm. **Seeds** white, ovoid, 4-angled in cross section, 2.2–2.6 × 1.1–1.4 mm, smooth to finely reticulate. $2n$ = 28, 48, 56.

Flowering and fruiting spring–fall. Desert scrub, grasslands, oak-juniper savannas, limestone, rocky, sandy, or clay soils; 400–1900 m; N.Mex., Tex.; Mexico (Chihuahua, Coahuila).

Euphorbia acuta is easily distinguished in the field by its relatively large, strongly involute, hairy, and acutely pointed leaves. The name *E. acuta* Engelmann has been proposed for conservation against the earlier name *E. acuta* Bellardi ex Colla (P. E. Berry et al. 2011).

24. Euphorbia albomarginata Torrey & A. Gray in War Department [U.S.], Pacif. Railr. Rep. 2(4): 174. 1857 • Rattlesnake weed, white-margin sandmat or sandwort [W]

Chamaesyce albomarginata (Torrey & A. Gray) Small

Herbs, perennial, with moderately to strongly thickened rootstock. **Stems** prostrate, occasionally mat-forming, frequently rooting at nodes, 10–80 cm, glabrous. **Leaves** opposite; stipules connate into conspicuous, deltate or ovate scale, white, 0.4–1(–2) mm, glabrous; petiole less than 1 mm, glabrous; blade ovate, oblong, or orbiculate, 3–8(–15) × 3–7 mm, base strongly asymmetric, obtuse to hemicordate, margins whitish, entire, apex obtuse, rarely mucronulate, surfaces often with red blotch in center, glabrous; 3-veined from base but usually only midvein conspicuous. **Cyathia** solitary at distal nodes; peduncle 1–4 mm. **Involucre** campanulate, 0.8–1.1 × 0.9–2 mm, glabrous; glands 4, greenish yellow to red, usually oblong to reniform, rarely subcircular, 0.2–0.5 × (0.2–)0.3–0.8 mm; appendages white to pink, flabellate to oblong, 0.3–1 × 0.6–1.3 mm, distal margin entire or crenulate to erose. **Staminate flowers** 15–30. **Pistillate flowers:** ovary glabrous; styles 0.3–0.7 mm, 2-fid nearly entire length. **Capsules** broadly ovoid, 1.1–2.3 × 1.2–2 mm, glabrous; columella 1.1–1.6 mm. **Seeds** white to gray or brownish red, oblong-ovoid, 4-angled in cross section, 1–1.7 × 0.5–0.8 mm, smooth.

Flowering and fruiting year-round. Disturbed areas in desert scrub, grasslands, mesquite woodlands, chaparral; 0–2300 m; Ariz., Calif., Colo., Nev., N.Mex., Okla., Tex., Utah; Mexico; introduced in Pacific Islands (Hawaii).

Euphorbia albomarginata is native to northern and central Mexico and the southwestern and south-central United States. The species occurs in a variety of habitats in western North America and in some areas is quite weedy. It has been recorded as a waif in Louisiana.

25. Euphorbia angusta Engelmann in W. H. Emory, Rep. U.S. Mex. Bound. 2(1): 189. 1859 • Blackfoot sandmat, narrow-leaf spurge

Chamaesyce angusta (Engelmann) Small

Herbs, perennial, with moderately to strongly thickened rootstock. **Stems** erect, 12–43 cm, uniformly strigose. **Leaves** opposite; stipules distinct, linear-subulate or nodiform to papilliform (then often reddish brown), 0.1–0.7 mm, strigose; petiole 0.3–1.2 mm, strigose; blade: proximal ovate to ovate-elliptic, distal linear to elliptic-linear, 7–41 × 2–5 mm, distal leaf blades more than 6 times as long as wide, base asymmetric, cuneate to rounded, margins entire, often involute on drying, apex acute, surfaces usually short strigose, occasionally glabrous adaxially; venation pinnate, only midvein conspicuous. **Cyathia** solitary at distal nodes; peduncle 1.1–2.4 mm. **Involucre** turbinate to campanulate-turbinate, 1–1.5 × 1–1.4 mm, strigose; glands 4, green to yellow-green, concave, narrowly oblong, 0.2–0.4 × 0.4–0.7 mm; appendages white, flabellate, 0.5–1.1 × 0.3–0.5 mm, distal margin shallowly and irregularly toothed. **Staminate flowers** 16–26. **Pistillate flowers:** ovary strigose; styles 0.3–0.5 mm, 2-fid at apex to almost ½ length. **Capsules** broadly ovoid, 2–2.6 × 2.5–3 mm, strigose; columella 1.6–2.2 mm. **Seeds** white, ovoid, 4-angled in cross section, 1.7–2.2 × 1.1–1.2 mm, transversely low-ridged or wrinkled. $2n = 28$.

Flowering and fruiting early spring–fall. Rocky limestone soils; 400–1200 m; Tex.; Mexico (Coahuila).

Euphorbia angusta, which in the flora area is known from the trans-Pecos region to the Edwards Plateau, is easily recognized by its erect habit, linear leaves, and relatively showy involucral gland appendages with toothed margins. The species is closely related to *E. acuta* and the Mexican endemic *E. johnstonii* Mayfield (M. H. Mayfield 1991); it is not only morphologically distinctive but is also the only species in sect. *Anisophyllum* with C_3 photosynthesis (G. L. Webster 1975; R. F. Sage et al. 2011; T. L. Sage et al. 2011).

26. Euphorbia arizonica Engelmann in W. H. Emory, Rep. U.S. Mex. Bound. 2(1): 186. 1859 • Arizona sandmat

Chamaesyce arizonica (Engelmann) Arthur

Herbs, annual or short-lived perennial, with slender to slightly thickened taproot. **Stems** erect to ascending, 10–30 cm, uniformly pilose with glistening hairs. **Leaves** opposite; stipules distinct, deltate, (0–)0.1 mm, glabrous or with few scattered hairs; petiole 0.4–1.5 mm, pilose with glistening hairs; blade usually ovate, rarely elliptic, 3–11 × 2–7 mm, base asymmetric, one side cuneate to rounded, other side rounded to strongly cordate, margins entire, apex obtuse or acute, surfaces pilose with glistening hairs; 3-veined from base but usually only midvein conspicuous. **Cyathia** solitary at distal nodes; peduncle 0.7–3.6 mm. **Involucre** urceolate, 1.1–1.3 × 0.5–0.9 mm, glabrous or pilose with glistening hairs; glands 4, dark maroon, usually oblong to reniform, rarely almost circular, 0.2 × 0.2–0.4 mm; appendages white to pink, flabellate, oblong, or elliptic, 0.3–0.6 × 0.6–0.9 mm, distal margin entire or crenulate. **Staminate flowers** 5–12. **Pistillate flowers:** ovary pilose; styles 0.5–0.6 mm, 2-fid ½ length. **Capsules** broadly ovoid to subglobose, 1.4–1.8 mm diam., pilose; columella 1.1–1.4 mm. **Seeds** gray to light brown, ovoid, 4-angled in cross section, 0.9–1.1 × 0.5–0.6 mm, rugose with 2–5 irregular transverse ridges that sometimes pass through abaxial keel.

Flowering and fruiting year-round in response to sufficient moisture. Washes and rocky slopes, sometimes on limestone, desert scrub communities often with creosote-bush, riparian forests, mesquite woodlands, oak chaparral; 100–1400 m; Ariz., Calif., N.Mex., Tex.; Mexico (Baja California, Baja California Sur, Chihuahua, Coahuila, Durango, Sonora).

Euphorbia arizonica is distinctive and easily recognized by its glistening, translucent hairs that appear somewhat glutinous and are most apparent on the stems.

27. Euphorbia astyla Engelmann ex Boissier in A. P. de Candolle and A. L. P. P. de Candolle, Prodr. 15(2): 40. 1862 • Alkali or Pecos spurge [C]

Chamaesyce astyla (Engelmann ex Boissier) Millspaugh

Herbs, perennial, with woody or fibrous-fleshy taproot, 5–12 mm thick. **Stems** decumbent, ascending, or erect, few to many emerging from woody crown, 5–25(–50) cm, glabrous. **Leaves** opposite; stipules connate into deltate scale, 0.2–0.5 mm, minutely lacerate at apex, glabrous; petiole 0–0.2(–0.3) mm, glabrous; blade orbiculate-reniform to acute-cordate, 2–5(–8) × 2–5(–6) mm, base ± asymmetric, cordate to auriculate, sometimes clasping stem, margins entire, apex narrowly acute, surfaces glabrous; 2- or 3-veined from base, but usually only midvein conspicuous. **Cyathia** solitary at distal nodes; peduncle 0.3–1(–1.5) mm. **Involucre** broadly campanulate, 0.8–1.4 × 0.9–1.4 mm, glabrous; glands 4, yellow-green to brownish, oblong, 0.2–0.3 × 0.5–0.7 mm; appendages white, flabellate to oblong, 0.1–0.2(–0.5) × 0.4–0.8 mm, distal margin entire or dentate-crenate. **Staminate flowers** 22–26. **Pistillate flowers:** ovary glabrous; styles 0.3–0.4 mm, unbranched, thickened-clavate. **Capsules** ovoid and broadly triangular, 1.5–1.9(–2.5) × 1.4–1.6(–2.2) mm, glabrous; columella 1.2–1.8 mm. **Seeds** white, oblong, 4-angled in cross section, adaxial faces slightly concave, with long raphe between, 1.5–1.8 × 0.7–1 mm, markedly foveolate, with irregular to ± parallel or anastomosing ridges. $2n = 28$.

Flowering and fruiting late spring–early fall. Desert, grasslands, limestone substrates, usually on very saline or alkaline soils; of conservation concern; 700–1100 m; Tex.; Mexico (Coahuila, Nuevo León).

Euphorbia astyla is a specialist on halophytic, alkaline soils and is known in the flora area only in part of Pecos County. The species is closely related to *E. jejuna* but differs in its sessile or sub-sessile leaves with a cordate-auriculate base and involucral gland appendages that are not deeply lobed or cleft.

28. Euphorbia blodgettii Engelmann ex Hitchcock, Rep. (Annual) Missouri Bot. Gard. 4: 126, plate 13. 1893 • Limestone sandmat [E]

Chamaesyce blodgettii (Engelmann ex Hitchcock) Small; *C. nashii* Small

Herbs, usually annual, occasionally perennial, with slender to slightly thickened taproot, 3.5 mm diam. **Stems** prostrate to decumbent, loosely mat-forming, often rooting at nodes, 10–45 cm, glabrous. **Leaves** opposite; stipules distinct, subulate filiform segments (lower side), or connate forming conspicuous, broad deltate scale (upper side), toothed, 0.5–1 mm, glabrous; petiole 0.7–1 mm, glabrous; blade ovate to oblong-elliptic, 4–10 × 2–5 mm, base asymmetric, subcordate to rounded, margins usually entire, occasionally toothed, apex usually obtuse to rounded, occasionally acute to apiculate, surfaces without red blotch, glabrous; 3-veined from base, only midvein conspicuous. **Cyathia** solitary or in small, cymose clusters at distal nodes of stem or on congested, axillary branches; peduncle 0.4–0.6 mm. **Involucre** turbinate to campanulate, 0.5–0.6 × 0.4–0.6 mm, glabrous; glands 4, red, slightly concave, elliptic-oblong, 0.1 × 0.1–0.3 mm; appendages white to pink, unequal, pair near sinus lunate to oblong, 0.1–0.2 × 0.1–0.3 mm, distal margin entire, crenulate, or irregularly sinuate, other pair sometimes rudimentary, 0–0.1 × 0.1–0.3 mm, distal margin crenulate or entire. **Staminate flowers** 8–12. **Pistillate flowers:** ovary glabrous; styles 0.3–0.4 mm, 2-fid ½ length. **Capsules** broadly ovoid, 1.2–1.7 × 1.4–1.9 mm, glabrous; columella 1.2–1.6 mm. **Seeds** gray to reddish brown, oblong-ovoid, 4-angled in cross section, 0.9–1.1 × 0.5–0.6 mm, flat or obscurely wrinkled.

Flowering and fruiting year-round. Coastal sand dunes and disturbed upland sandy areas; 0–20 m; Fla.

Euphorbia blodgettii is found only in peninsular Florida. It is closely related to *E. garberi*, *E. porteriana*, and *E. serpens* (Y. Yang and P. E. Berry 2011).

29. Euphorbia bombensis Jacquin, Enum. Syst. Pl., 22. 1760 • Dixie sandmat

Chamaesyce ammannioides (Kunth) Small; *C. bombensis* (Jacquin) Dugand; *C. ingallsii* Small; *Euphorbia ammannioides* Kunth

Herbs, usually annual, rarely perennial, with taproot. **Stems** prostrate or slightly ascending, 10–40 cm, glabrous. **Leaves** opposite; stipules distinct, linear-subulate, usually divided into 3 linear segments, 1–2 mm, glabrous; petiole 1–2 mm, glabrous; blade oblong or elliptic-oblong, 4–15 × 2–3 mm, base asymmetric to nearly symmetric, obtuse, margins entire, apex acute to mucronate, surfaces green to reddish flushed, glabrous; only midvein conspicuous. **Cyathia** solitary or in small, cymose clusters at distal nodes; peduncle 0.5–3 mm. **Involucre** obconic-campanulate, 1.2–1.6 × 1.5–1.7 mm, glabrous; glands 4, green to red, slightly concave, elliptic, oblong, or subcircular, 0.3–0.5 × 0.4–0.6 mm; appendages white or pink, semilunate, fringing edge of gland, sometimes rudimentary, (0–)0.1–0.5(–0.7) × 0.5–0.8 mm, distal margin crenate to entire. **Staminate flowers** 5–16. **Pistillate flowers:** ovary glabrous; styles 0.2–0.3 mm, 2-fid ½ length. **Capsules** broadly ovoid, 2–2.1 × 2.3–2.5 mm, glabrous; columella 1.5–2 mm. **Seeds** ashy white, plumply ovoid, terete to bluntly subangled in cross section, 1.5–1.9 × 1–1.2 mm, smooth or minutely pitted, with smooth brown line from top to bottom on adaxial side.

Flowering and fruiting spring–fall. Coastal dunes and sandy habitats; 0–30 m; Ala., Fla., Ga., La., Miss., N.C., S.C., Tex., Va.; Mexico; Central America; n South America (Venezuela).

Euphorbia bombensis is similar and closely related to *E. cumulicola, E. geyeri,* and *E. polygonifolia*. It differs notably from *E. polygonifolia* in its smaller, plumply ovoid seeds. L. C. Wheeler (1941) suggested that where *E. bombensis* and *E. polygonifolia* are sympatric, *E. bombensis* grows farther away from the shore. *Euphorbia bombensis* is usually distinguished from *E. cumulicola* by the latter's smaller, isomorphic leaves that lack any fleshiness, smaller seeds, and diffuse growth habit. *Euphorbia bombensis* differs from *E. geyeri* in its usually shorter, less conspicuous involucral gland appendages and its geographic restriction to the coastal plain. However, Wheeler pointed out that plants of *E. bombensis* from Texas have more or less conspicuous involucral gland appendages. Examination of specimens confirmed that *E. bombensis* occasionally has conspicuous involucral gland appendages, and because of this, *E. bombensis* and *E. geyeri* are difficult to distinguish in Texas. This clade of closely related sand and dune specialists requires further study.

30. Euphorbia capitellata Engelmann in W. H. Emory, Rep. U.S. Mex. Bound. 2(1): 188. 1859 • Head spurge, capitate sandmat

Chamaesyce capitellata (Engelmann) Millspaugh; *C. pycnanthema* (Engelmann) Millspaugh; *Euphorbia pycnanthema* Engelmann

Herbs, annual or perennial, with slender to thick and woody rootstock. **Stems** usually ascending (but ranging from decumbent to erect), 15–50 cm, glabrous, strigillose, or pilose. **Leaves** opposite; stipules distinct, filiform or divided into 2–3 subulate-filiform segments, without dark, circular glands at base, 0.6–1.5 mm, pilose; petioles 0.6–1.3 mm, glabrous, pilose, or strigillose; blade ovate to narrowly ovate, 4–19 × 2–8 mm, base asymmetric, one side strongly cordate, other side rounded to slightly cordate, margins entire or serrulate (commonly nearly entire with few scattered teeth, often slightly thickened), apex acute, surfaces often with red spot in center, glabrous, pilose, or strigillose; weakly 3-veined from base, usually only midvein conspicuous. **Cyathia** in dense, terminal, capitate glomerules, with reduced, bractlike leaves subtending cyathia, at tips of main stems and short, leafy, axillary branches; peduncle 0.1–1.2 mm. **Involucre** narrowly obconic to narrowly campanulate, 0.8–1.6 × 0.7–1.3 mm, glabrous or pilose; glands 4, yellow-green to pink or maroon, circular to oblong, 0.2–0.4 × 0.2–0.5 mm; appendages white to light pink, oblong to reniform or flabellate, 0.2–1.1 × 0.5–1.7 mm, surfaces glabrous, distal margin entire. **Staminate flowers** 25–40. **Pistillate flowers:** ovary glabrous or pilose; styles 0.4–0.6 mm, 2-fid entire length. **Capsules** ovoid to oblate, 1.3–1.9 × 1.4–2.1 mm, glabrous or pilose; columella 1.1–1.7 mm. **Seeds** pink to pinkish gray, narrowly ovoid to narrowly ovoid-oblong, 4-angled or weakly 3-angled in cross section, 0.9–1.5 × 0.5–0.7 mm, irregularly dimpled, sometimes also with faint transverse ridges that do not pass through abaxial keel. $2n = 14$.

Flowering and fruiting year-round in response to sufficient moisture. Gravelly washes, rocky slopes, basaltic talus, disturbed roadsides, primarily desert scrub, desert grasslands, riparian forests, rarely oak-juniper woodlands; 600–1600 m; Ariz., N.Mex., Tex.; Mexico (Baja California, Baja California Sur, Chihuahua, Coahuila, Sinaloa, Sonora).

Euphorbia capitellata is a characteristic herb in the Sonoran Desert of southern Arizona, ranging east to extreme southwestern Texas. During peak flowering, plants are attractive due to the dense clusters of cyathia with well-developed involucral gland appendages.

31. Euphorbia carunculata Waterfall, Rhodora 50: 63. 1948 • Sand-dune sandmat or spurge

Chamaesyce carunculata (Waterfall) Shinners

Herbs, annual, with taproot. **Stems** prostrate, spreading and lanky or occasionally mat-forming, ± succulent, 70–150 cm, glabrous. **Leaves** opposite; stipules usually distinct, occasionally connate basally (primarily at distal nodes), usually divided into 2–5 subulate to subulate-filiform segments, occasionally forming narrow deltate segments (primarily at distal nodes), 0.8–1.8 mm, glabrous; petiole 3.1–6.3 mm, glabrous; blade ovate to elliptic-oblong, 5–26 × 4–12 mm, base subsymmetric to symmetric, rounded to cuneate, margins entire, apex usually mucronate, rarely acute or obtuse, surfaces glabrous; pinnately veined. **Cyathia** solitary at distal nodes; peduncle 1.4–3.6 mm. **Involucre** campanulate, 1.1–1.8 × 1.5–2.5 mm, glabrous; glands 4, yellowish, sessile or short stipitate, circular to oblong, 0.5–0.7 × 0.5–0.8 mm; appendages white to yellowish, ovate to oblong, occasionally rudimentary, (0–)0.8–1.2 × 0.8–1.5 mm, distal margin entire. **Staminate flowers** 15–25. **Pistillate flowers:** ovary glabrous; styles 0.7–1 mm, 2-fid ½ length. **Capsules** ovoid, 4.7–5.5(–6) × 3.6–5.1 mm, glabrous; columella 4.3–5.1 mm. **Seeds** grayish white to reddish brown mottled, bottle-shaped, strongly dorsiventrally compressed and weakly 3-angled in cross section, (2.8–)4.1–5.2 × 1.2–2 (–3.4) mm, smooth; carunclelike structure linear, 0.4–0.5 × 0.1–0.2 mm.

Flowering and fruiting summer–fall. Sand dunes; 400–1300 m; Kans., N.Mex., Okla., Tex.; Mexico (Chihuahua).

Euphorbia carunculata has a highly localized and scattered distribution. The species is restricted to sand dunes and known from only a handful of localities throughout its relatively wide range. The seeds are unique in being bottle-shaped and strongly laterally compressed, and unlike other members of sect. *Anisophyllum*, there is a minute, linear, carunclelike protuberance at the hilum.

32. Euphorbia chaetocalyx (Boissier) Tidestrom, Proc. Biol. Soc. Wash. 48: 40. 1935 • Bristlecup sandmat

Euphorbia fendleri Torrey & A. Gray var. *chaetocalyx* Boissier in A. P. de Candolle and A. L. P. P. de Candolle, Prodr. 15(2): 39. 1862; *Chamaesyce chaetocalyx* (Boissier) Wooton & Standley

Herbs, perennial, with woody, thickened taproot. **Stems** usually erect, rarely slightly decumbent, often densely clustered from top of woody crown, 3–15 cm, glabrous. **Leaves** opposite; stipules distinct, narrowly linear, usually entire, 0.5–1 mm, glabrous; petiole 0.5–1 mm, glabrous; blade ovate to lanceolate or oblong- or linear-lanceolate, 3–11 × 0.8–3(–5) mm, base slightly asymmetric, short-tapered, occasionally one side slightly rounded, margins entire, apex acute or short-acuminate, surfaces glabrous; only midvein conspicuous. **Cyathia** solitary at distal nodes; peduncle 0.8–1.3 mm. **Involucre** campanulate to turbinate, 0.8–1.4 × 0.8–1 mm, glabrous; glands 4, yellow-brown to reddish, concave or convex, elliptic or oval, 0.2–0.4 × 0.4–0.6 mm; appendages absent or white, lanceolate-deltate to straplike, 0.2–1.1 × 0.2–0.9 mm, distal margin entire, crenate, or deeply cleft or divided. **Staminate flowers** 25–35. **Pistillate flowers:** ovary glabrous; styles 0.3–0.4 mm, 2-fid ½ length. **Capsules** depressed-ovoid to depressed-globose, 1.7–2.1 × 1.6–2.4 mm, glabrous; columella 1.2–1.8 mm. **Seeds** white, ovoid-pyramidal, prominently 4-angled in cross section, 1.6–2 × 1–1.2 mm, smooth to slightly wrinkled.

Varieties 2 (2 in the flora): sw, sc United States, n Mexico.

Euphorbia chaetocalyx is similar to *E. fendleri* but can generally be distinguished from that species by its narrow, acute leaves and usually erect stems. Some authors have used the presence or absence and shape of the involucral gland appendages to help separate *E. chaetocalyx* from *E. fendleri*, but those characters appear highly variable and of little taxonomic utility. Some individuals from western Texas (Culberson and El Paso counties) and southern New Mexico appear intermediate with *E. fendleri*. The specific epithet of *E. chaetocalyx* refers to the bristly perianthlike segments that subtend the ovary, but these structures are found intermittently in both *E. chaetocalyx* and *E. fendleri*.

1. Leaf blades ovate to lanceolate or oblong- or linear-lanceolate; involucral gland appendages absent or lanceolate-deltate, entire, crenate, or deeply cleft; widely distributed in Arizona, New Mexico, w Texas . . . . 32a. *Euphorbia chaetocalyx* var. *chaetocalyx*
1. Leaf blades narrowly ovate to linear-lanceolate; involucral gland appendages straplike, divided into 3–5 linear segments; cliffs and rocks of Boquillas Canyon, sw Texas . . . . .32b. *Euphorbia chaetocalyx* var. *triligulata*

32a. Euphorbia chaetocalyx (Boissier) Tidestrom var. **chaetocalyx**

Leaf blades ovate to lanceolate or oblong- or linear-lanceolate, 3–11 × 0.8–3(–5) mm. **Involucral gland appendages** absent or lanceolate-deltate, entire, crenate, or deeply cleft, 0–1 × 0–0.9 mm. **2*n*** = 52.

Flowering and fruiting spring–late fall. Mesa slopes, ridges, rock crevices, cliff faces, mostly rocky limestone soils, sometimes gypsum, rarely igneous-derived soils; 600–2500 m; Ariz., N.Mex., Tex.; Mexico (Coahuila).

32b. Euphorbia chaetocalyx (Boissier) Tidestrom var. **triligulata** (L. C. Wheeler) M. C. Johnston, Wrightia 5: 139. 1975 • Triligulate bristlecup sandmat [C]

Euphorbia fendleri Torrey & A. Gray var. *triligulata* L. C. Wheeler, Bull. Torrey Bot. Club 63: 445. 1936; *Chamaesyce chaetocalyx* (Boissier) Wooton & Standley var. *triligulata* (L. C. Wheeler) Mayfield

Leaf blades narrowly ovate to linear-lanceolate, 3–11 × 0.8–3 mm. **Involucral gland appendages** straplike, divided into 3–5 linear segments, each 0.8–1.1 × 0.1 mm.

Flowering and fruiting early summer–fall. Canyon walls, crevices in boulders; of conservation concern; 600–900 m; Tex.; Mexico (Coahuila).

Variety *triligulata* was described from the cliffs of Boquillas Canyon, on both sides of the Rio Grande in Texas and adjacent Coahuila, Mexico. The variety has elongate involucral gland appendages that are divided into three to five linear, straplike segments; it appears to represent a distinctive, localized population within the overall distribution of *Euphorbia chaetocalyx*. More generally, populations of *E. chaetocalyx* from elsewhere in northern Coahuila, Mexico, have similarly shaped involucral gland appendages and may represent additional populations of var. *triligulata*.

33. Euphorbia cinerascens Engelmann in W. H. Emory, Rep. U.S. Mex. Bound. 2(1): 186. 1859 • Ashy sandmat

Chamaesyce cinerascens (Engelmann) Small

Herbs, perennial, with moderately to strongly thickened rootstock. **Stems** prostrate to decumbent, mat-forming, 5–30 cm, appressed woolly, strigillose, or short-sericeous. **Leaves** opposite; stipules distinct, subulate, 0.2–0.5 mm, appressed woolly to sericeous; petiole 0.3–0.8 mm, appressed woolly to sericeous; blade ovate to elliptic, 1.5–5.5 × 1.3–4 mm, base asymmetric, obtuse to hemicordate, margins entire, often reddish, apex usually obtuse, occasionally acute (young leaves), surfaces sericeous to strigillose or slightly pilose, adaxial surface often glabrous; 3-veined from base but only midvein conspicuous. **Cyathia** solitary at distal nodes; peduncle 0.2–0.3 mm. **Involucre** turbinate, 0.8–1.3 × 1.2–2 mm, appressed woolly, sericeous, or strigillose; glands 4, purple-black, elliptic to oblong, 0.2–0.3 × 0.4–0.6 mm; appendages absent or reddish pink, forming narrow rim around distal margin of gland, 0–0.1 × 0–0.6 mm, distal margin entire, crenulate, or erose. **Staminate flowers** 15–20. **Pistillate flowers:** ovary canescent; styles 0.3–0.5 mm, 2-fid at apex. **Capsules** subglobose to broadly ovoid, 1.3–1.7 × 1.5–1.8 mm, canescent; columella 1–1.3 mm. **Seeds** white to pinkish or light brown, ovoid, 4-angled in cross section, 1–1.4 × 0.6–0.8 mm, smooth to rugulose or rarely with 1–2 inconspicuous transverse ridges. $2n = 32$.

Flowering and fruiting year-round (mostly spring–fall). Desert scrub, oak and juniper woodlands, thorn scrub, shrublands, grasslands, frequently on limestone substrates; 70–1400 m; Tex.; Mexico.

In the flora area, *Euphorbia cinerascens* is found only in southern and western Texas. In Mexico, it is found from Chihuahua east to Tamaulipas, south to Guanajuato.

34. Euphorbia conferta (Small) B. E. Smith, J. Elisha Mitchell Sci. Soc. 62: 82. 1946 • Everglade Key sandmat [E]

Chamaesyce conferta Small, Fl. S.E. U.S., 713, 1333. 1903

Herbs, annual, with taproot. **Stems** prostrate, ascending, to erect, 10–45 cm, pilose. **Leaves** opposite; stipules distinct, subulate, 1–2 mm, pilose; petiole 0.9–1.5 mm, pilose; blade oblanceolate to elliptic, 3–10 × 1–4 mm, base asymmetric, obtuse to hemicordate, margins coarsely serrate, sometimes revolute, apex rounded to acute, abaxial surface sparsely hispidulous to strigillose (densely so on young leaves), adaxial surface glabrous; 3-veined from base but only midvein conspicuous. **Cyathia** solitary at distal nodes (appearing clustered at points of new growth); peduncle 0.4–0.6 mm. **Involucre** campanulate, 1.1–1.4 × 0.5–0.7 mm, pilose; glands 4, red, narrowly reniform, 0.1 × 0.2–0.4 mm; appendages pink to red, larger 2 petal-like, smaller 2 elliptic, 0.4–1 × 0.3–0.5 mm, distal margin erose. **Staminate flowers** 5–8. **Pistillate flowers:** ovary pilose; styles 0.4–0.8 mm, 2-fid at apex. **Capsules** conic or truncate-ovoid, 1–1.2 mm diam., pilose; columella 0.9–1.1 mm. **Seeds** orange-brown, glaucous, oblong, 4-angled in cross section, 0.7–0.8 × 0.4–0.5 mm, slightly rugose and with 3–4 transverse ridges.

Flowering and fruiting year-round. Sandy, disturbed, wet areas, often roadsides; 0–10 m; Fla.

Euphorbia conferta is known only from Broward, Miami-Dade, and Monroe counties in southern Florida.

35. Euphorbia cordifolia Elliott, Sketch. Bot. S. Carolina 2: 656. 1824 • Heartleaf sandmat [E]

Chamaesyce cordifolia (Elliott) Small

Herbs, annual, with taproot. **Stems** prostrate, occasionally mat-forming, 10–43 cm, glabrous. **Leaves** opposite; stipules usually distinct, occasionally connate at base, filiform, 1–1.2(–2.8) mm, usually glabrous, rarely pilose; petiole 0.4–1 mm, usually glabrous; blade ovate to oblong, 4.4–12 × 2.6–7.6 mm, base asymmetric, cordate to rounded, margins entire, apex rounded to mucronulate, surfaces glabrous; only midvein conspicuous. **Cyathia** solitary at distal nodes; peduncle 0.9–3 mm. **Involucre** campanulate, 1–1.3 × 1–1.3 mm, glabrous; glands 4, yellowish to pink, elliptic, 0.3–0.5 × 0.5–1 mm; appendages whitish to pink, sometimes drying red, elliptic to ovoid, 1.1–1.5 × 1.2–1.9 mm, distal margin entire, retuse, or erose. **Staminate flowers** 5–40. **Pistillate flowers:** ovary glabrous; styles 0.5–0.8 mm, 2-fid nearly entire length. **Capsules** ovoid, 2–3 mm diam., glabrous; columella 1.2–2.7 mm. **Seeds** gray or tan with dark brown mottling, ovoid, bluntly 3–4-angled in cross section, 1.8–2.1 × 1.2–1.4 mm, smooth to rugose.

Flowering and fruiting summer–fall. Xeric oak-pine scrub, pine barrens, sand barrens, sandy stream banks; 0–200 m; Ala., Ark., Fla., Ga., La., Miss., N.C., Okla., S.C., Tex.

Euphorbia cordifolia is easily identified by its cordate to rounded leaf base and distinctive filiform stipules.

36. Euphorbia cumulicola (Small) Oudejans, Phytologia 67: 45. 1989 • Coastal dune sandmat [C] [E]

Chamaesyce cumulicola Small, Man. S.E. Fl. 794, 1505. 1933

Herbs, usually annual, rarely perennial, with taproot. **Stems** prostrate, 10–20 cm, glabrous. **Leaves** opposite; stipules distinct, linear-subulate, usually divided into 3–7 linear segments, 0.5–1 mm, glabrous; petiole 0.5–1 mm, glabrous; blade narrowly elliptic to lanceolate, 4–8 × 1–2 mm, uniform in size, base asymmetric, obtuse to rounded, margins entire, apex obtuse to acute, surfaces green to reddish flushed, glabrous; only midvein conspicuous. **Cyathia** solitary at distal nodes; peduncle 1 mm. **Involucre** campanulate, 0.9–1.1 × 1–1.3 mm, glabrous; glands 4, green to red, slightly stipitate, subcircular, 0.1 × 0.1 mm; appendages white or pink, fringing edge of gland, short-flabellate, often rudimentary, 0.1–0.2 × 0.1 mm, distal margin crenate or entire. **Staminate flowers** 5–8. **Pistillate flowers:** ovary glabrous; styles 0.2–0.3 mm, 2-fid at apex to nearly ½ length. **Capsules** ovoid or subglobose, 1.8–2 × 2–2.3 mm, glabrous; columella 1.2–1.5 mm. **Seeds** white to gray-brown, ovoid, terete to bluntly subangled in cross section, 1–1.3 × 1 mm, smooth or minutely pitted, with smooth brown line from top to bottom on adaxial side.

Flowering and fruiting year-round. Sandy oak hammocks, open sandy areas behind mangroves, disturbed sandy sites; of conservation concern; 0–10 m; Fla.

Euphorbia cumulicola could be confused with *E. bombensis*, both of which are widespread in Florida, but there the latter occurs on beaches close to the ocean, whereas *E. cumulicola* is not a beach-inhabiting species but occurs in more protected sandy habitats such as hammocks or stabilized dunes behind mangroves. Also, *E. bombensis* has fleshier leaves and larger seeds, and is more compact in habit than *E. cumulicola*, which is a more sprawling and densely branched plant with leaves that are usually purplish tinged along the margins.

37. Euphorbia deltoidea Engelmann ex Chapman, Fl. South. U.S. ed. 2, 647. 1883 • Wedge sandmat [C] [E]

Chamaesyce deltoidea (Engelmann ex Chapman) Small

Herbs, perennial, delicate, with woody, thickened taproot, 15 mm diam. **Stems** prostrate, ascending, or erect, often numerous and wiry, less than 0.1 mm diam., 5–20 cm, glabrous, puberulent, canescent, villous, or hirsute, shorter hairs often uncinate and longer hairs straight or irregularly twisted. **Leaves** opposite; stipules distinct, triangular, sometimes lacerate or ciliate, 0.2–0.3 mm, glabrous or hairy; petiole 0.3–1 mm, glabrous or hairy; blade narrowly to broadly deltate, cordate, or reniform, 2–5(–7) × 1–4.5(–5) mm, base asymmetric, cordate to rounded, margins entire, ± revolute, apex obtuse or rounded, surfaces glabrous or hairy; only midvein conspicuous. **Cyathia** solitary at distal nodes; peduncle 0.7–1.5 mm. **Involucre** turbinate to campanulate, 0.8–1 × 1.1–1.3 mm, glabrous or hairy; glands 4, green to yellow-green, oblong to subcircular, 0.2–0.4 × 0.4–0.6 mm; appendages absent or white, forming narrow rim at edge of gland, rarely slightly wider than gland, (0–)0.1(–0.3) × 0.4–0.6 mm, distal margin entire. **Staminate flowers** 8–14. **Pistillate flowers:** ovary glabrous or hairy, subtended by triangular pad of tissue; styles spreading, 0.3–0.4 mm, 2-fid ½ to nearly entire length. **Capsules** broadly deltoid, 1.2–1.5 × 2–2.2 mm, glabrous or hairy; columella 0.9–1.3 mm. **Seeds** reddish brown, ovoid, 4-angled in cross section, 0.8–1.2 × 0.5–0.6 mm, obscurely wrinkled.

Subspecies 4 (4 in the flora): Florida.

Euphorbia deltoidea comprises four narrowly endemic subspecies, all of which are endangered due to habitat loss and fragmentation. The subspecies occur in pine rockland habitat that is free of shrubby undergrowth. Periodic fires are required to keep the rockland habitat open. Subspecies *serpyllum* is restricted to Big Pine Key, Monroe County, whereas the other subspecies are found only in Miami-Dade County.

1. Stems erect or ascending.
   2. Leaves and stems villous, hairs uncinate or irregularly twisted, 0.2–0.5 mm; leaf blades green . . . . 37b. *Euphorbia deltoidea* subsp. *adhaerens* (in part)
   2. Leaves and stems villous-hirsute, hairs straight and spreading, 0.6–0.7 mm; leaf blades silver-green . . . 37c. *Euphorbia deltoidea* subsp. *pinetorum*
1. Stems prostrate.
   3. Leaves and stems glabrous or very sparsely hairy, hairs 0.1–0.2 mm, appressed, uncinate; leaf blades abaxially reddish, adaxially bright green . . . . 37a. *Euphorbia deltoidea* subsp. *deltoidea*
   3. Leaves and stems canescent or villous, hairs either less than 0.1 mm or 0.2–0.5 mm, uncinate or irregularly twisted; leaf blades green or silver-green.
      4. Leaves and stems villous, hairs uncinate or irregularly twisted, 0.2–0.5 mm; leaf blades as long as wide, green . . . . 37b. *Euphorbia deltoidea* subsp. *adhaerens* (in part)
      4. Leaves and stems canescent, hairs less than 0.1 mm; leaf blades 2 times longer than wide, silver-green . . . . . 37d. *Euphorbia deltoidea* subsp. *serpyllum*

37a. Euphorbia deltoidea Engelmann ex Chapman subsp. **deltoidea** C E

Stems prostrate, densely mat-forming, glabrous or very sparsely hairy, hairs appressed, uncinate, 0.1–0.2 mm. **Leaves:** stipules glabrous or sparsely hairy with appressed uncinate hairs; petiole glabrous or sparsely hairy with appressed uncinate hairs; blade 2–4.5 × 2–4.5 mm, as long as wide, abaxial surface reddish, adaxial surface bright green, surfaces glabrous or sparsely hairy with appressed uncinate hairs. **Involucre** glabrous or sparsely hairy with appressed uncinate hairs. **Pistillate flowers:** ovary glabrous or sparsely hairy with appressed uncinate hairs. **Capsules** glabrous or sparsely hairy with appressed uncinate hairs.

Flowering and fruiting year-round. Open pine rocklands; of conservation concern; 0–10 m; Fla.

Subspecies *deltoidea* is the northernmost of the four subspecies. It grows in a thin layer of white sand over limestone bedrock.

37b. Euphorbia deltoidea Engelmann ex Chapman subsp. **adhaerens** (Small) Oudejans, World Cat. Euphorb. Cum. Suppl. I, 11. 1993 C E

Chamaesyce adhaerens Small, Torreya 27: 104. 1928

Stems prostrate to ascending, mat-forming, sparsely to densely villous, hairs uncinate or irregularly twisted, 0.2–0.5 mm. **Leaves:** stipules villous with uncinate or irregularly twisted hairs; petiole villous with uncinate or irregularly twisted hairs; blade 2–4.5 × 2–4.5 mm, as long as wide, surfaces green, villous with uncinate or irregularly twisted hairs. **Involucre** villous with uncinate or irregularly twisted hairs. **Pistillate flowers:** ovary villous with uncinate or irregularly twisted hairs. **Capsules** villous with uncinate or irregularly twisted hairs.

Flowering and fruiting year-round. Open pine rocklands; of conservation concern; 0–10 m; Fla.

Subspecies *adhaerens* is intermediate in both morphology and distribution between subspp. *deltoidea* and *pinetorum*.

37c. Euphorbia deltoidea Engelmann ex Chapman subsp. **pinetorum** (Small) Oudejans, World Cat. Euphorb. Cum. Suppl. I, 11. 1993 C E

Chamaesyce pinetorum Small, Bull. New York Bot. Gard. 3: 429. 1905

Stems ascending to erect, not mat-forming, usually red, villous-hirsute, hairs straight and spreading, 0.6–0.7 mm. **Leaves:** stipules villous-hirsute with straight, spreading hairs; petiole villous-hirsute with straight, spreading hairs; blade 2–4.5 × 2–4.5 mm, as long as wide, surfaces silver-green, villous-hirsute with straight, spreading hairs. **Involucre** villous-hirsute with straight, spreading hairs. **Pistillate flowers:** ovary villous-hirsute with straight, spreading hairs. **Capsules** villous-hirsute with straight, spreading hairs.

Flowering and fruiting year-round. Open pine rocklands; of conservation concern; 0–10 m; Fla.

Subspecies *pinetorum* is the southernmost of the three mainland subspecies.

37d. Euphorbia deltoidea Engelmann ex Chapman subsp. **serpyllum** (Small) Ya Yang, Taxon 61: 783. 2012 [C] [E]

Chamaesyce serpyllum Small, Fl. Florida Keys, 81, 155. 1913; *C. deltoidea* subsp. *serpyllum* (Small) D. G. Burch

Stems prostrate, loosely mat-forming, canescent, hairs appressed, less than 0.1 mm. **Leaves:** stipules canescent with appressed hairs; petiole canescent with appressed hairs; blade 2–5(–7) × 1–2.5(–5) mm, 2 times longer than wide, surfaces silver-green, canescent with appressed hairs. **Involucre** canescent with appressed hairs. **Pistillate flowers:** ovary canescent with appressed hairs. **Capsules** canescent with appressed hairs.

Flowering and fruiting year-round. Open pine rocklands; of conservation concern; 0–10 m; Fla.

Subspecies *serpyllum* is restricted to pine rockland on Big Pine Key, Monroe County.

38. Euphorbia fendleri Torrey & A. Gray in War Department [U.S.], Pacif. Railr. Rep. 2(4): 175. 1857 • Fendler's sandmat

Chamaesyce fendleri (Torrey & A. Gray) Small

Herbs, perennial, with woody, thickened taproot. **Stems** usually prostrate, decumbent, or ascending, very rarely erect, often densely clustered from top of woody crown, 5–12 cm, glabrous. **Leaves** usually opposite, rarely whorled; stipules distinct, narrowly linear, 0.5–1 mm, glabrous; petiole 0.5–1 mm, glabrous; blade usually orbiculate to ovate, rarely almost lanceolate, 3–8 × 2.5–7 mm, base slightly asymmetric, slightly cordate to rounded or obtuse, margins entire, apex rounded to obtuse, surfaces glabrous; obscurely 3–5-veined from base, only midvein conspicuous. **Cyathia** solitary at distal nodes; peduncle 0.7–1.2 mm. **Involucre** campanulate to turbinate or broadly cupulate, 1.1–1.7 × 1.2–1.8 mm, glabrous; glands 4, yellow-green to reddish, elliptic to oblong, 0.2–0.5 × 0.4–0.9 mm; appendages absent or white, rarely pink, often unequal, lunate to flabellate or sometimes forming crenate margin along gland, (0–)0.1–0.6 × (0–)0.5–1.5 mm, distal margin entire or toothed. **Staminate flowers** 25–35. **Pistillate flowers:** ovary glabrous; styles 0.3–0.4 mm, 2-fid ½ length. **Capsules** depressed-globose, 2–2.4 × 2.2–2.5 mm, glabrous; columella 1.7–2.1 mm. **Seeds** white, ovoid-pyramidal, prominently 4-angled in cross section, 1.7–2 × 1–1.2 mm, smooth to slightly wrinkled. $2n = 28$.

Flowering and fruiting spring–fall. Mountain slopes, desert scrub, pinyon-juniper woodlands, hills, canyons, grasslands, washes, roadsides, dry crevices in limestone, often in gravel and sand; 500–2600 m; Ariz., Calif., Colo., Kans., Nebr., Nev., N.Mex., Okla., S.Dak., Tex., Utah, Wyo.; Mexico (Coahuila).

Euphorbia fendleri is similar to *E. chaetocalyx* and may sometimes be confused with that species. Its prostrate to decumbent or ascending stems and small, ovate to orbiculate leaves distinguish it from *E. chaetocalyx*. Some authors have used the presence or absence and shape of the involucral gland appendages to help separate *E. fendleri* from *E. chaetocalyx*, but those characters appear highly variable and of little taxonomic utility; somewhat intermediate individuals occur in western Arizona, New Mexico, and Texas.

39. Euphorbia florida Engelmann in W. H. Emory, Rep. U.S. Mex. Bound. 2(1): 189. 1859 • Chiricahua Mountain sandmat

Chamaesyce florida (Engelmann) Millspaugh

Herbs, annual, with slender taproot. **Stems** erect, 15–60 cm, usually glabrous, rarely puberulent. **Leaves** opposite; stipules distinct, divided into 3–4 subulate-filiform divisions, 0.4–1.6 mm, usually glabrous, rarely puberulent; petiole 0.5–2.5 mm, glabrous; blade usually linear, rarely to narrowly elliptic, 10–40(–60) × 0.5–2.5 mm, base symmetric, attenuate, margins serrulate, often revolute, apex acute, surfaces usually glabrous, rarely puberulent; obscurely pinnately veined. **Cyathia** solitary at nodes or in small, cymose clusters at branch tips; peduncle 1.2–8.1 mm. **Involucre** obconic, 1.7–2.4 × 1.5–2.1 mm, glabrous; glands 4, greenish yellow to slightly pink, circular to oblong, 0.4–0.5 × 0.4–0.6 mm; appendages white to pink, obovoid, circular, flabellate, or oblong, 0.8–2.9 × 1–2.8 mm, distal margin entire. **Staminate flowers** 25–35. **Pistillate flowers:** ovary glabrous; styles 0.8–1.4 mm, 2-fid entire length. **Capsules** oblate, 2.2–2.5 × 2.7–3.1 mm, glabrous; columella 1.8–2.1 mm. **Seeds** light gray to light brown, ovoid, slightly 4-angled in cross section, 1.6–2 × 1.3–1.7 mm, with 2 or 3 well-developed transverse ridges.

Flowering and fruiting summer–late fall. Sandy flats, gravelly washes, rocky hillsides, talus slopes, desert scrub, desert grasslands, mesquite woodlands, rarely oak woodlands; 600–1300 m; Ariz.; Mexico (Sinaloa, Sonora).

Euphorbia florida is known in the flora area from Coconino County south to the Mexican border.

40. Euphorbia garberi Engelmann ex Chapman, Fl. South. U.S. ed. 2, 646. 1883 • Garber's sandmat [C] [E]

Chamaesyce adicioides Small; *C. brachypoda* Small; *C. garberi* (Engelmann ex Chapman) Small; *C. keyensis* Small; *C. mosieri* Small; *C. porteriana* Small var. *keyensis* (Small) D. G. Burch; *Euphorbia porteriana* (Small) Oudejans var. *keyensis* (Small) Oudejans

Herbs, perennial, with moderately thickened rootstock. **Stems** ascending, sometimes slightly woody at base, not wiry, 1–3 mm diam., 15–50 cm, canescent. **Leaves** opposite; stipules distinct or connate at base, triangular-subulate, apex divided into 2–5 subulate to subulate-filiform segments, 0.5–0.7 mm, pilose; petiole 0.5–1.1 mm, pilose; blade ovate to oblong-elliptic, 4–9(–15) × 3–6 mm, base asymmetric, obtuse to rounded, margins usually entire, rarely very sparsely serrulate, apex usually obtuse to rounded, occasionally acute to apiculate, surfaces canescent; 3-veined from base, only midvein conspicuous. **Cyathia** solitary at nodes of short, axillary branches; peduncle 0.2–0.4 mm. **Involucre** turbinate to campanulate, 0.6–0.9 × 0.5–0.9 mm, pilose; glands 4, brown, slightly concave, elliptic to subcircular, 0.1–0.3 × 0.2–0.4 mm; appendages white to pink, as narrow rim along gland, (0–)0.1–0.2 × 0.3–0.5 mm, surfaces pilose, distal margin entire or crenulate. **Staminate flowers** 8–20. **Pistillate flowers:** ovary pilose; styles 0.6–0.7 mm, 2-fid ½ length. **Capsules** subglobose to broadly ovoid, 1.1–1.6 × 1.3–2.1 mm, pilose; columella 1.1–1.6 mm. **Seeds** gray to reddish brown, oblong-ovoid, 4-angled in cross section, 1–1.2 × 0.6–0.8 mm, flat or obscurely wrinkled.

Flowering and fruiting year-round. Beach dunes, coastal rock barrens, disturbed uplands and pine rocklands; of conservation concern; 0–10 m; Fla.

Euphorbia garberi appears to be an interspecific hybrid, but its parentage is not entirely clear (Y. Yang and P. E. Berry 2011). The taxon is closely related to *E. blodgettii*, *E. porteriana*, and *E. serpens*. It is restricted to Collier, Miami-Dade, and Monroe counties in southern Florida and is federally listed as threatened due to impacts from habitat loss, fire suppression, and invasive species. A. Herndon (1993) synonymized *E. porteriana* var. *keyensis* under *E. garberi* because of its uniformly pilose capsules, and that treatment is followed here.

41. Euphorbia geyeri Engelmann, Boston J. Nat. Hist. 5: 260. 1845 • Geyer's sandmat

Chamaesyce geyeri (Engelmann) Small

Herbs, annual, with taproot. **Stems** prostrate or slightly ascending, loosely mat-forming, 4–25(–45) cm, glabrous. **Leaves** opposite; stipules usually distinct, occasionally connate basally on lower side, usually deeply parted into 3 or more filiform segments, 0.7–1.5 mm, glabrous; petiole 1–2 mm, glabrous; blade oblong to oblong-obovate or oblong-elliptic, 4–12 × 2–6 mm, base slightly asymmetric, angled or rounded, with one side usually expanded into small, rounded auricle, margins entire, apex usually truncate, occasionally emarginate, abaxial surface pale grayish green, both surfaces glabrous; only midvein conspicuous or venation obscurely pinnate (larger leaves). **Cyathia** solitary or in small, cymose clusters at distal nodes; peduncle 1–2 mm. **Involucre** broadly campanulate, 1–1.5 × 0.7–0.9 mm, glabrous; glands 4, green to reddish, slightly cupped to folded, elliptic-oblong to nearly circular, 0.2–0.4 × 0.2–0.6 mm; appendages rudimentary to absent or white to reddish-tinged, usually rounded, sometimes pointed, (0–)0.5–1 × (0–)0.1–1.2 mm, distal margin entire or slightly toothed. **Staminate flowers** 5–20. **Pistillate flowers:** ovary glabrous; styles 0.2–0.6 mm, 2-fid nearly ½ length. **Capsules** globose-ovoid, 1.5–2 × 1.5–3 mm, glabrous; columella 1.5–1.9 mm. **Seeds** ashy white, ovoid, terete to bluntly subangled in cross section, 1.1–1.7 × 0.9–1.2 mm, smooth, with smooth brown line from top to bottom on adaxial side.

Varieties 2 (2 in the flora): North America, n Mexico.

The two varieties of *Euphorbia geyeri* have been distinguished in large part by the presence of conspicuous involucral gland appendages in var. *geyeri* and the lack of appendages in var. *wheeleriana*. The two varieties are recognized here, but the variation in the size and presence of involucral gland appendages in the closely related *E. bombensis* suggests that this might be a somewhat variable character in this group of species.

Euphorbia geyeri is widespread throughout the central United States in sandy soils. Populations at the eastern edge of the range are often considered adventive (for example, sandy soils along railroad grades in Michigan). *Euphorbia geyeri* resembles *E. glyptosperma* (both being entirely glabrous), but that species has serrulate leaves (near the apex) and strongly angled, transverse-ridged seeds whereas *E. geyeri* has entire leaves and smooth, rounded seeds.

1. Involucral gland appendages present; staminate flowers 5–9; seeds 1.1–1.4(–1.6) mm . . . . 41a. *Euphorbia geyeri* var. *geyeri*
1. Involucral gland appendages absent or rudimentary; staminate flowers 10–20; seeds 1.6–1.7 mm. . . . 41b. *Euphorbia geyeri* var. *wheeleriana*

41a. Euphorbia geyeri Engelmann var. **geyeri** E

Involucral gland appendages present. **Staminate flowers** 5–9. **Capsules** 1.5–2 × 1.5–2.5 mm. **Seeds** 1.1–1.4(–1.6) × 0.9–1.2 mm.

Flowering and fruiting midsummer–early fall. Sand barrens, riverbanks, disturbed sandy or gravelly areas; 0–1500 m; Man.; Colo., Ill., Ind., Iowa, Kans., La., Mich., Minn., Miss., Mo., Mont., Nebr., N.Mex., N.Dak., Okla., S.Dak., Tex., Utah, Wis., Wyo.

41b. Euphorbia geyeri Engelmann var. **wheeleriana** Warnock & M. C. Johnston, SouthW. Naturalist 14: 128. 1969 C

Chamaesyce geyeri (Engelmann & A. Gray) Small var. *wheeleriana* (Warnock & M. C. Johnston) Mayfield

Involucral gland appendages absent or rudimentary. **Staminate flowers** 10–20. **Capsules** 1.5–2 × 2–3 mm. **Seeds** 1.6–1.7 × 0.9–1.2 mm. **2*n*** = 12.

Flowering and fruiting midsummer–early fall. Sand barrens, riverbanks, disturbed sandy or gravelly areas; of conservation concern; 500–1500 m; N.Mex., Tex.; Mexico (Chihuahua).

Variety *wheeleriana* is known in the flora area from south-central New Mexico and western trans-Pecos Texas.

42. Euphorbia glyptosperma Engelmann in W. H. Emory, Rep. U.S. Mex. Bound. 2(1): 187. 1859 • Ribseed sandmat F W

Chamaesyce glyptosperma (Engelmann) Small

Herbs, annual, with taproot. **Stems** prostrate, loosely mat-forming, 5–40 cm, glabrous, sometimes slightly glaucous. **Leaves** opposite; stipules usually distinct, linear-subulate, usually irregularly fringed or lobed, rarely laciniate, 0.4–2 mm, glabrous; petiole 0.2–2 mm, glabrous; blade narrowly oblong to oblong-obovate or oblong-ovate, 3–15 × 2–7 mm, base asymmetric, one side angled and other rounded to shallowly cordate, margins minutely sparsely serrulate, especially near apex and on longer side, apex rounded to obtuse, abaxial surface usually pale grayish green, adaxial surface sometimes reddish tinged or with reddish blotch, both surfaces glabrous; palmately veined at base, pinnate distally. **Cyathia** solitary or in small, cymose clusters at distal nodes; peduncle 0.5–1.6 mm. **Involucre** obconic, 0.6–0.9 × 0.3–0.6 mm, glabrous; glands 4, red to purple, narrowly oblong-elliptic, 0.1–0.2 × 0.1–0.5 mm; appendages white or pinkish tinged, semilunate and fringing distal margin of gland, 0.1–0.3 × 0.1–0.3 mm, distal margin usually irregularly crenulate or lobed. **Staminate flowers** 1–5. **Pistillate flowers:** ovary glabrous; styles 0.1–0.3 mm, 2-fid ½ length. **Capsules** broadly ovoid, 1.3–1.9 × 1.6–2 mm, glabrous; columella 1.3–1.5 mm. **Seeds** with thin white coat that readily falls off, surface below tan brown, oblong-ovoid, sharply angular in cross section, 1–1.4 × 0.6–0.9 mm, with 3–4(–6) prominent transverse ridges that usually interrupt abaxial keel. **2*n*** = 22.

Flowering and fruiting early summer–fall. Stream and river banks, sand prairies, loess hill prairies, meadows, ballast, open disturbed areas, roadsides; 0–1800 m; Alta., B.C., Man., N.B., N.S., Ont., P.E.I., Que., Sask.; Ariz., Calif., Colo., Conn., Idaho, Ill., Ind., Iowa, Kans., La., Maine, Mass., Mich., Minn., Mo., Mont., Nebr., Nev., N.H., N.Mex., N.Y., N.Dak., Okla., Oreg., S.Dak., Tenn., Tex., Utah, Vt., Wash., Wis., Wyo.; Mexico (Chihuahua, Coahuila, Tamaulipas, Zacatecas).

Euphorbia glyptosperma is one of the most widespread species of the genus in North America, and it is often quite weedy. It is likely native to much of the central United States, but in areas like eastern Canada, it occurs in highly disturbed habitats, and it may be adventive there. *Euphorbia glyptosperma* is characterized by its prominently ridged seeds, laciniate stipules, complete lack of pubescence, and somewhat obscure toothing near the tips of the leaves.

43. Euphorbia golondrina L. C. Wheeler, Proc. Biol. Soc. Wash. 53: 8. 1940 • Canyon or swallow spurge, Boquillas sandmat C

Chamaesyce golondrina (L. C. Wheeler) Shinners

Herbs, annual, with slender taproot. **Stems** prostrate, 5–35 cm, glabrous. **Leaves** opposite; stipules distinct, subulate, 0.5–0.8 mm, glabrous; petiole 0.8–1 mm, glabrous; blade oblong, ovate-oblong, to narrowly elliptic-oblong, 5–11.5 × 1–4 mm, base asymmetric, cuneate to rounded, margins entire, thickened and often revolute on drying, apex obtuse, surfaces glabrous; only midvein conspicuous. **Cyathia** solitary at distal

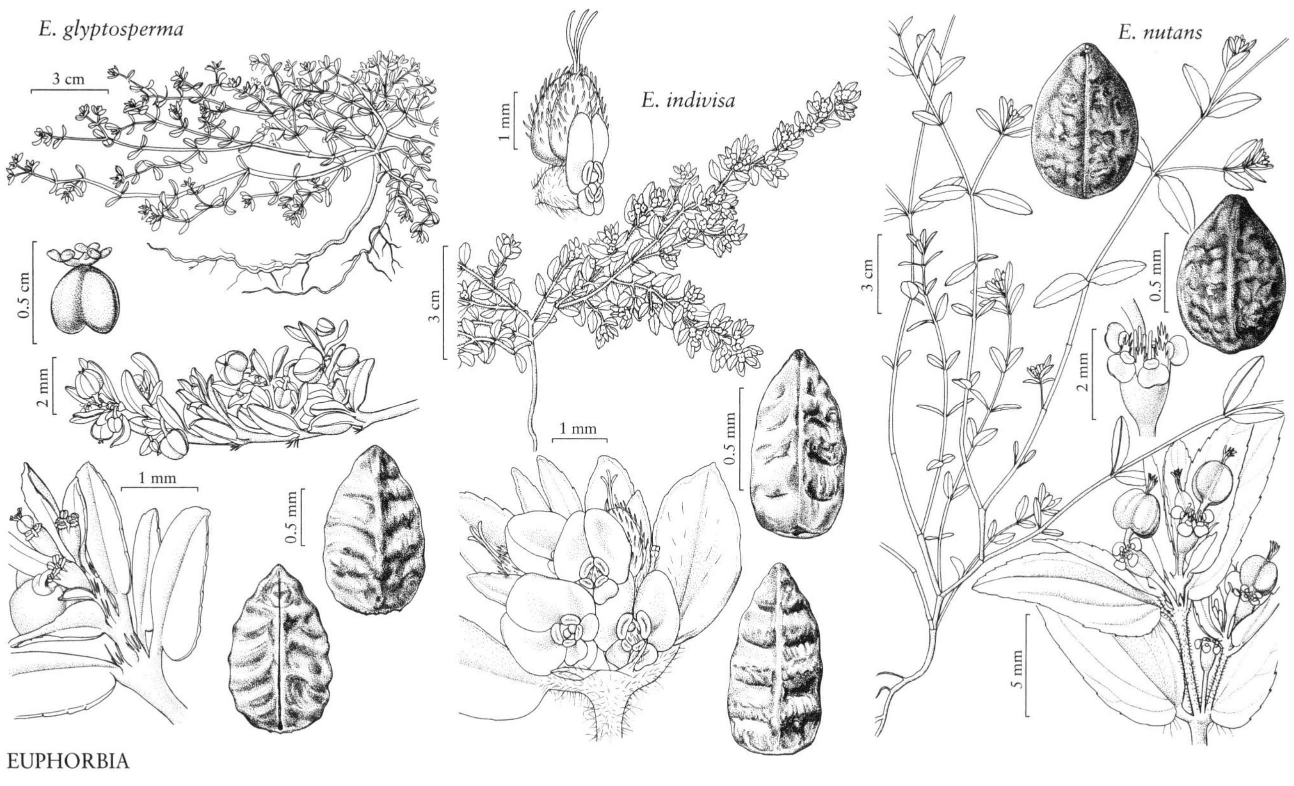

EUPHORBIA

nodes; peduncle 0.9–1.5 mm. **Involucre** turbinate, 0.9–1.5 × 0.8–1.3 mm, glabrous; glands 4, occasionally rudimentary, red to purple, deeply concave, subcircular, 0.3–0.4 × 0.3–0.4 mm; appendages white, semilunate to slightly flabellate, 0.1–0.3 × 0.5–0.8 mm, distal margin entire. **Staminate flowers** 28–40. **Pistillate flowers:** ovary glabrous; styles 0.3–0.4 mm, 2-fid nearly entire length. **Capsules** broadly ovoid, 1.7–2 × 1.5–1.6 mm, glabrous; columella 1.4–1.7 mm. **Seeds** narrowly pyramidal-ovoid, 4-angled in cross section, 1.5–2 × 1.3–1.8 mm, with very faint transverse ridges or wrinkles.

Flowering and fruiting late spring–fall. Deep, sandy riverbanks; of conservation concern; 600 m; Tex.; Mexico (Chihuahua, Coahuila).

Euphorbia golondrina was first collected on a sandy riverbank at the entrance of Boquillas Canyon on the Texas/Mexico border, and the species appears to be restricted to the vicinity of the Rio Grande in western Texas and northern Mexico. Aside from the type locality at Boquillas Canyon, *E. golondrina* has been documented from additional locations along the Rio Grande in Brewster, Hudspeth, and Presidio counties in Texas and from northern Coahuila and Chihuahua, Mexico. Phylogenetic data place *E. golondrina* in a clade of primarily Chihuahuan desert annual and perennial herbaceous species (for example, *E. chaetocalyx*, *E. fendleri*, *E. perennans*, *E. simulans,* and *E theriaca*; Y. Yang and P. E. Berry 2011). *Euphorbia golondrina* is superficially similar to other glabrous species in western Texas (for example, *E. micromera* and *E. theriaca*), but *E. micromera* and *E. theriaca* either lack involucral gland appendages or have shorter, triangular appendages compared to the typical semilunate appendages in *E. golondrina*.

44. Euphorbia gracillima S. Watson, Proc. Amer. Acad. Arts 21: 438. 1886 • Mexican sandmat

Chamaesyce gracillima (S. Watson) Millspaugh

Herbs, annual, with slender taproot. **Stems** erect to ascending, 5–25 cm, glabrous. **Leaves** opposite; stipules distinct, subulate-filiform, entire, 0.3–0.5 mm, glabrous; petiole 0.4–0.9 mm, glabrous; blade narrowly oblong to linear, often slightly falcate, 2–15 × 0.3–0.8 mm, base symmetric to subsymmetric, rounded to attenuate, margins entire, thickened and often revolute, apex acute to obtuse, surfaces glabrous; only midvein conspicuous. **Cyathia** solitary at distal nodes; peduncle 0.1–0.9 mm. **Involucre** turbinate, 0.4–0.5 × 0.4–0.6 mm, glabrous; glands 4, yellow to pink, oblong to slightly reniform, (0–)0.1–1 × 0.1–0.2 mm; appendages white to pink, ovate to oblong, 0.2–0.4 × 0.1–0.3 mm, distal margin usually entire, rarely emarginate. **Staminate flowers** 5–12. **Pistillate flowers:** ovary glabrous; styles 0.1–0.2 mm, 2-fid ½ length. **Capsules** broadly ovoid, 1.1–1.4 mm diam., glabrous; columella 0.8–1.1 mm. **Seeds** orange to tan or reddish brown, narrowly ovoid, 4-angled in cross section, 0.9–1.1 × 0.5–0.6 mm, smooth.

Flowering and fruiting late summer–fall. Rocky slopes and dry washes in desert scrub; 600–900 m; Ariz.; Mexico (Chihuahua, Jalisco, Sinaloa, Sonora).

Euphorbia gracillima occurs from south-central Arizona (Pima and Pinal counties) south through northwestern Mexico.

45. Euphorbia hirta Linnaeus, Sp. Pl. 1: 454. 1753 • Pillpod spurge [I] [W]

Chamaesyce gemella (Lagasca) Small; *C. hirta* (Linnaeus) Millspaugh; *Euphorbia gemella* Lagasca

Herbs, annual or perennial, with slender to thickened taproot. **Stems** usually erect to ascending, rarely prostrate or decumbent, 10–50(–75) cm, usually both strigillose and hirsute. **Leaves** opposite; stipules usually distinct, rarely connate at base, undivided or divided into 2–4 narrowly deltate to linear-subulate segments, 0.5–1.8(–2.9) mm, pilose, often with light-colored, minute circular glands at base; petiole 1–3 mm, usually both strigillose and hirsute; blade ovate to rhombic, 7–43 × 3–18 mm, base strongly asymmetric, one side rounded or slightly cordate to truncate, the other cuneate to attenuate, margins serrulate to double serrulate, apex acute, surfaces often with red spot in center, usually strigose to hirtellous, rarely glabrescent; 3–5-veined from base. **Cyathia** in dense, axillary and terminal, capitate glomerules, with reduced, bractlike leaves subtending cyathia, axillary glomerules either sessile or at tips of elongated, leafless stalks; peduncle 0.4–2.1 mm. **Involucre** obconic, 0.6–1.2 × 0.4–0.9 mm, strigillose; glands 4, greenish to pink, circular, oblong, or reniform, 0.1–0.2 × 0.1–0.2 mm; appendages white to pink, flabellate, subcircular, or transversely oblong, rarely absent, (0–)0.1–0.6 × (0–)0.1–0.7 mm, distal margin usually entire, rarely slightly lobed. **Staminate flowers** 2–8. **Pistillate flowers:** ovary strigillose, often canescent when young; styles 0.2–0.6 mm, 2-fid ½ length. **Capsules** subglobose to slightly oblate, 1–1.3 × 1.1–1.6 mm, strigillose; columella 0.7–1 mm. **Seeds** brownish red to orange or pink, narrowly ovoid, 4-angled in cross section, 0.7–0.9 × 0.5–0.7 mm, usually rugulose or with 3–6 low transverse ridges, rarely nearly smooth.

Flowering and fruiting year-round. Disturbed areas, roadsides, vacant lots, desert grasslands, mesquite woodlands, riparian forests with cottonwoods and willows, floodplain forests, pinelands, deciduous forests; 0–1500 m; introduced; Ala., Ariz., Ark., Calif., Fla., Ga., La., Miss., N.Mex., N.Y., N.C., S.C., Tex.; Mexico; West Indies; Bermuda; Central America; South America; introduced also in Europe, Asia, Africa, Australia.

Euphorbia hirta is a widespread weed that is distributed throughout tropical and subtropical regions of the world. It is probably native to at least central Mexico, and its native range possibly extends from South America to the southern United States. In many places within the flora area the species is certainly introduced, and it has been recorded as a waif from several states (for example, Maryland, Michigan, and Virginia). *Euphorbia pilulifera* Linnaeus, a rejected name that is a heterotypic synonym of the Asian *E. parviflora* Linnaeus, has been misapplied to North American material of *E. hirta* in the past.

46. Euphorbia hooveri L. C. Wheeler, Proc. Biol. Soc. Wash. 53: 9. 1940 • Hoover's sandmat [C] [E]

Chamaesyce hooveri (L. C. Wheeler) Koutnik

Herbs, annual, with slender to slightly thickened taproot. **Stems** prostrate, occasionally mat-forming, 4–10 cm, glabrous. **Leaves** opposite; stipules distinct or connate at base, deeply laciniate into numerous subulate to filiform segments, 0.3–1.3 mm, glabrous; petiole 0–0.5 mm, glabrous; blade broadly ovate, broadly oblong or suborbiculate, 3–7 × 2–5.5 mm, base asymmetric, hemiamplexicaulous, margin sharply serrulate to denticulate, teeth whitish and occasionally setaelike at apex, apex obtuse, surfaces papillate, glabrous; 3-veined from base, usually only midvein conspicuous. **Cyathia** solitary at distal nodes; peduncle 0.5 mm, stout. **Involucre** campanulate, 1.2–1.5 × 1.5–2.2 mm, papillate; glands 4, yellow to reddish, subcircular to oblong, 0.4–0.5 × 0.5–0.7 mm; appendages white to pink, 0.3–0.9 × 0.7–1.6 mm, divided into 4–6 narrowly triangular, acute segments, these occasionally 2-fid, margins entire. **Staminate flowers** 25–35. **Pistillate flowers:** ovary glabrous; styles 1.8–2.6 mm, unbranched, filiform. **Capsules** oblate, 1.5–1.7 × 1.8–2.2 mm, glabrous; columella 1.2–1.4 mm. **Seeds** gray to light brown, ovoid, inconspicuously 4-angled in cross section, 1.2–1.6 × 1–1.1 mm, rugose.

Flowering and fruiting summer–fall. Drying mudflats of vernal pools in grasslands and woodlands; of conservation concern; 20–200 m; Calif.

Euphorbia hooveri is federally listed as threatened; its populations are being affected severely by habitat loss and the invasion of exotic species. The species is endemic to vernal pools in six counties in the Central Valley of California. Molecular data show that *E. hooveri* is a hybrid species, closely related to *E. albomarginata* and *E. serpens* (Y. Yang and P. E. Berry 2011).

47. Euphorbia humistrata Engelmann in A. Gray, Manual ed. 2, 386. 1856 • Spreading sandmat [E] [W]

Chamaesyce humistrata (Engelmann ex A. Gray) Small

Herbs, annual, with taproot. **Stems** prostrate to ascending, usually mat-forming and rooting at nodes, 5–45 cm, sparsely to moderately villous to pilose (densely on young growth). **Leaves** opposite; stipules distinct, linear-subulate, often irregularly 2- or 3-lobed, 1–1.3 mm, sparsely villous to pilose; petiole 0.5–1.5 mm, sparsely to moderately villous to pilose; blade oblong-ovate to ovate-elliptic or oblong-elliptic, 4–18 × 2.5–8 mm, base strongly asymmetric, one side angled and other rounded to auriculate, margin on longer side serrulate, on shorter side subentire, apex rounded or broadly acute, abaxial surface pale grayish green, sparsely lanulose, adaxial surface usually with irregular reddish streak along midvein, usually glabrate, rarely sparsely lanulose; palmately veined at base, pinnate distally. **Cyathia** solitary at distal nodes; peduncle 0.1–0.6(–2) mm. **Involucre** obconic, 0.8–1 × 0.6–0.8 mm, sparsely villous to pilose; glands 4, green to yellow-green (turning pink with age), usually ± unequal, narrowly oblong, 0.1–0.2 × 0.2–0.5 mm; appendages white to reddish tinged, lunate, ± irregular and variable in shape, 0.1–0.3 × 0.2–1.5 mm, distal margin crenulate. **Staminate flowers** 2–5. **Pistillate flowers:** ovary short-sericeous; styles 0.5–0.8 mm, 2-fid ½ length. **Capsules** ovoid, well exserted from involucre at maturity, 1.3–1.5 × 1.2–1.6 mm, sparsely to moderately short-sericeous; columella 0.9–1.2 mm. **Seeds** white to light brown, oblong-ovoid, bluntly angular in cross section, 0.8–1.2 × 0.5–0.9 mm, smooth or papillate.

Flowering and fruiting spring–late summer. Stream and river banks, gravel bars, flood plains, pond edges, disturbed fields, railroads, roadsides; 0–300 m; Ala., Ark., Fla., Ga., Ill., Ind., Kans., Ky., La., Miss., Mo., Ohio, Okla., Tenn., Tex., Va., W.Va.

Euphorbia humistrata is distributed throughout the Mississippi River valley and along other major river systems in the central and eastern United States. There are scattered reports of this species as a waif or as introduced farther north and/or east (for example, Minnesota, New York, North Carolina, Ontario, South Dakota, and Wisconsin), but the authors have not been able to verify these occurrences. *Euphorbia humistrata* is similar to *E. maculata* and is often confused with that species in herbaria. It can be distinguished from *E. maculata* by its tendency to root at the stem nodes, its longer styles, and its seeds that lack low transverse ridges and that are more bluntly angled. When growing side-by-side, *E. humistrata* has an overall less congested appearance and its cyathia are not as numerous or crowded as those of *E. maculata*.

48. Euphorbia hypericifolia Linnaeus, Sp. Pl. 1: 454. 1753 • Graceful sandmat

Chamaesyce glomerifera Millspaugh; *C. hypericifolia* (Linnaeus) Millspaugh; *Euphorbia glomerifera* (Millspaugh) L. C. Wheeler

Herbs, annual, with taproot. **Stems** erect to ascending, 15–50 cm, glabrous. **Leaves** opposite; stipules connate, deltate, usually entire, sometimes laciniate-fringed at tip, 1.5–2.2 mm, glabrous; petiole 1–3 mm, glabrous; blade obliquely oblong-oblanceolate, 10–35 × 7–15 mm, base asymmetric, oblique, margins serrate or serrulate, especially toward apex, apex broadly acute, surfaces glabrous; palmately veined at base, pinnate distally. **Cyathia** in dense, axillary and terminal, capitate glomerules with reduced, bractlike leaves subtending cyathia; peduncle 0.5–1.8 mm. **Involucre** obconic, 0.9–1.1 × 0.4–0.9 mm, glabrous; glands 4, yellow-green to brown, stipitate, subcircular, 0.2 × 0.2 mm, occasionally nearly rudimentary; appendages absent on smaller glands or white to pink, shape highly variable, usually round to ± elliptic, 0.3–0.4 × 0.5–0.7 mm, distal margin entire. **Staminate flowers** (0–)2–20. **Pistillate flowers:** ovary glabrous; styles 0.4 mm, 2-fid ½ length. **Capsules** depressed-globoid, 1.3–1.4 × 1.1–1.5 mm, glabrous; columella 1–1.1 mm. **Seeds** with very thin whitish mucilaginous coat over light brown testa below, ovoid-triangular, bluntly 4-angled in cross section, 0.9–1.1 × 0.5 mm, with shallow irregular depressions alternating with low, smooth ridges.

Flowering and fruiting early spring–late fall. Open, disturbed areas, nurseries; 0–200 m; Ala., Ark., Fla., Ga., La., Okla., S.C., Tex.; West Indies; Central America; South America; introduced in Asia, Pacific Islands.

Euphorbia hypericifolia is native to the New World tropics, and it is most likely adventive in the flora area (where it is most widely distributed in Florida and Texas). Reports from Arizona, California, and Maryland likely represent waifs or misidentifications.

49. Euphorbia hyssopifolia Linnaeus, Syst. Nat. ed. 10, 2: 1048. 1759 • Hyssopleaf sandmat [W]

Chamaesyce hyssopifolia (Linnaeus) Small; *Euphorbia jonesii* Millspaugh; *E. stenomeres* S. F. Blake

Herbs, annual, with taproot. **Stems** erect to ascending, 80 cm, sparsely to densely pilose or pilose-crinkled proximally, usually glabrous distally. **Leaves** opposite; stipules usually connate, irregularly lacerate, 0.5–1 mm, usually glabrous, occasionally with few marginal hairs; petiole 1–2 mm, glabrous; blade lanceolate to oblong or falcate, 8–35 × 7–15 mm, base asymmetric, rounded, margins serrulate, apex broadly acute, abaxial surface glabrous or sparsely pilose toward base, adaxial surface glabrous; palmately veined at base, pinnate distally. **Cyathia** solitary or in small, cymose clusters, occasionally with bractlike leaves, at distal nodes or on congested, axillary branches; peduncle 0.5–2.5 mm. **Involucre** obconic, 0.9–1.1 × 0.7–0.9 mm, glabrous; glands 4(–5) (5th gland without appendage), yellow-green to maroon, elliptic to circular, 0.1–0.2 × 0.1–0.3 mm; appendages spreading, usually white or turning reddish with age, short reniform or semilunate, 0.1–0.3 × 0.2–0.6 mm, distal margin entire or slightly undulate to crenate. **Staminate flowers** 4–15. **Pistillate flowers:** ovary glabrous; styles 0.5–0.9 mm, 2-fid ½ length. **Capsules** depressed-ovoid, 1.5–1.6 × 1.7–1.8 mm, glabrous; columella 1.5–2 mm. **Seeds** brown to grayish white, ovoid, slightly 4-angled in cross section, abaxial faces convex, adaxial faces slightly concave to slightly convex, 1–1.4 × 0.7–1.1 mm, with 2–3 prominent transverse ridges that do not interrupt adaxial keel, or coarsely and inconspicuously pitted-reticulate. $2n$ = 12, 14.

Flowering and fruiting late spring–early fall. Disturbed areas, ditches, gardens; 0–1500 m; Ala., Ariz., Fla., Ga., La., Miss., N.Mex., S.C., Tex., Utah; Mexico; West Indies; Central America; South America; introduced in tropical Asia, Africa, Australia.

Euphorbia hyssopifolia is native to the New World tropics and is probably also native to parts of the southern United States. However, at least some of the records from the flora area appear to be from adventive plants.

50. Euphorbia indivisa (Engelmann) Tidestrom, Proc. Biol. Soc. Wash. 48: 40. 1935 • Royal sandmat [F]

Euphorbia dioeca Kunth var. *indivisa* Engelmann in W. H. Emory, Rep. U.S. Mex. Bound. 2(1): 187. 1859; *Chamaesyce indivisa* (Engelmann) Millspaugh

Herbs, annual or short-lived perennial, with slender taproot to thickened and woody rootstock. **Stems** prostrate, usually mat-forming, terete to slightly flattened, 40 cm, lower surface glabrous, upper surface strigillose, pilose or villous. **Leaves** opposite; stipules distinct, entire or divided into 3–4 subulate to filiform segments, 0.8–2 mm, usually pilose, rarely glabrous; petiole 0.5–1 mm, pilose to villous; blade oblong, ovate or narrowly obovate, 3–10(–12) × 2–6 mm, base strongly asymmetric, hemicordate, margins serrulate, apex obtuse to subacute, surfaces glabrous or slightly pilose; 3-veined from base, often only midvein conspicuous. **Cyathia** usually in small cymose clusters on congested, axillary branches; peduncle rudimentary or to 0.2 mm. **Involucre** narrowly turbinate, 1–1.2 × 0.4–0.7 mm, pilose; glands 4, yellow to pink, unequal, proximal pair oblong or linear, 0.1 × 0.3–0.4(–0.6) mm, distal pair oblong or subcircular, 0.1 × 0.1–0.2 mm; appendages pink to reddish, unequal, on proximal glands oblique, 0.4–0.8(–1) × 0.8–1.4(–2) mm, on distal glands symmetric, 0.2–0.5 × 0.2–0.5 mm, slightly undulate to slightly crenate. **Staminate flowers** 5–15. **Pistillate flowers:** ovary pilose to strigillose in parts, glabrous in other parts; styles 0.8–1.3 mm, usually unbranched, rarely 2-fid at apex, filiform. **Capsules** ovoid-triangular, 1.2–1.5 × 1–1.4 mm, pilose to strigillose in parts, glabrous in other parts; columella 1–1.3 mm. **Seeds** brown to light gray, ovoid, 4-angled in cross section, 0.8–1 × 0.4–0.5 mm, with 4 or 5 deep transverse sulci alternating with low transverse ridges.

Flowering and fruiting summer–fall. Grasslands, oak forests, oak-mesquite woodlands, oak-juniper communities, rarely entering desert scrub; 1000–2000 m; Ariz., N.Mex., Tex.; Mexico.

Euphorbia indivisa is characteristic of grasslands and oak woodlands from extreme western Texas to southeastern Arizona. The species is often treated as a synonym of *E. dioeca* Kunth, but the two species are readily separable on the basis of their seeds. The seeds of *E. indivisa* possess deep transverse sulci, whereas those of *E. dioeca* are merely rippled or with low transverse ridges. *Euphorbia dioeca* is a weedy species that occurs widely throughout tropical America but has yet to be encountered within the flora area.

51. Euphorbia jaegeri V. W. Steinmann & J. M. André, Aliso 30: 1, figs. 1–4. 2012 • Orocopia Mountains spurge [C] [E]

Shrubs, with woody rootstock. **Stems** ascending, diffusely and intricately branched, 15–25 cm, usually puberulent to shortly hirsute, sometimes glabrate, bark grayish. **Leaves** opposite; stipules distinct or connate, subulate, 0.3–0.5 mm, puberulent; petiole 0.7–1.1 mm, puberulent to shortly hirsute; blade ovate or elliptic, 3–9 × 1.5–5 mm, base symmetric to slightly asymmetric, rounded to cuneate, margins entire, apex usually obtuse, sometimes acute, surfaces puberulent to shortly hirsute; 3-veined from base, often only midvein conspicuous. **Cyathia** solitary at distal nodes; peduncle 0.5–1.7 mm. **Involucre** obconic to campanulate, 1.2–1.8 × 1.1–1.4 mm, puberulent to shortly hirsute; glands 4, yellow to pinkish, elliptic to oblong, 0.3 × 0.4–0.5 mm; appendages white to pink, 0.2–0.7 × 0.6–1.2 mm, irregularly divided from halfway to nearly base into 4–8 triangular to subulate segments, segments entire. **Staminate flowers** 25–30. **Pistillate flowers:** ovary canescent; styles 0.3–0.4 mm, 2-fid entire length. **Capsules** oblate, 1.7–2.3 × 1.8–2.7 mm, puberulent; columella 1.4–2 mm. **Seeds** tan to grayish, narrowly oblong-ovoid, ± 3–4-angled in cross section, 1.4–1.5 × 0.7–0.9 mm, irregularly dimpled or with faint transverse ridges that do not interrupt abaxial keel.

Flowering and fruiting fall–spring. Desert scrub, hillsides, arroyos, primarily in rock crevices; of conservation concern; 600–900 m; Calif.

Euphorbia jaegeri is known only from the Orocopia Mountains of Riverside County and the Bristol and Marble Mountains of San Bernardino County. The species is one of few shrubby species of sect. *Anisophyllum* in the flora area.

52. Euphorbia jejuna M. C. Johnston & Warnock, SouthW. Naturalist 5: 97, fig. [p. 98]. 1960 • Dwarf broomspurge [C]

Chamaesyce jejuna (M. C. Johnston & Warnock) Shinners

Herbs, perennial, with woody or fibrous-fleshy, napiform, branched or tuberous taproot, 3–15 mm thick. **Stems** ascending to erect, densely emerging from woody crown, 5–15 cm, glabrous. **Leaves** opposite; stipules connate into lacerate or 2-fid, lanceolate or deltate scale, 0.5–0.9 mm, glabrous; petiole 0.3–0.9 mm, glabrous; blade orbiculate-obovate, ovate, or elliptic, 3–6(–8) × 1.8–5 mm, base moderately asymmetric, rounded to truncate, margins entire, apex blunt to acute, surfaces glabrous; 2–3-veined from base but usually only midvein conspicuous. **Cyathia** solitary at distal nodes; peduncle 0.8–1.5 mm. **Involucre** broadly campanulate, 1.2–1.5 × 0.8–1.2 mm, glabrous; glands 4, yellowish to green or purplish, oblong, 0.2–0.4 × 0.5–1 mm; appendages erect or spreading, white, 0.3–0.6 × 0.8–1.2 mm, usually deeply dissected into 4–5 acuminate lobes, rarely undivided, when divided distal margin rarely entire or crenate. **Staminate flowers** 12–35. **Pistillate flowers:** ovary glabrous; styles 0.4–0.5 mm, unbranched thickened-clavate. **Capsules** ovoid and broadly triangular, 1.8–2.2(–2.7) × 1.5–2.1 mm, glabrous; columella 1.5–2.2 mm. **Seeds** whitish, oblong, 4-angled in cross section, adaxial faces concave, with long raphe between, 1.5–2(–2.3) × 0.6–0.8 mm, dimpled with faint irregular transverse wrinkles or with up to 10 low, rounded transverse ridges.

Flowering and fruiting early spring–summer. Thin calcareous soils (caliche) on limestone hills; of conservation concern; 500–900 m; Tex.; Mexico (Chihuahua, Coahuila).

Euphorbia jejuna is known in the flora area from only a few collections in Mitchell, Nolan, Terrell, and Val Verde counties. The species is very similar to *E. astyla*, but differs in its more deeply divided involucral gland appendages and more definitely petiolate, rounder leaves.

53. Euphorbia laredana Millspaugh, Pittonia 2: 88. 1890 • Laredo sandmat

Chamaesyce laredana (Millspaugh) Small

Herbs, annual, with taproot. **Stems** prostrate, ± mat-forming, 10–20 cm, densely ashy pilose-tomentose. **Leaves** opposite; stipules distinct, filiform, 0.5–1 mm, pilose-tomentose; petiole 0.5–1 mm, pilose-tomentose; blade ovate to elliptic-oblong, 3–6 × 3–5 mm, base markedly asymmetric, rounded to slightly auriculate, margins usually entire, rarely largest leaves sparsely serrulate, apex acute to obtuse, surfaces moderately to densely strigose; 3-veined from base. **Cyathia** solitary or in small, cymose clusters at distal nodes or on congested, axillary branches; peduncle 0.5–1.5 mm. **Involucre** obconic, 0.6–1 × 0.5–1 mm, densely strigose; glands 4, yellowish to reddish, oval to oblong, 0.1 × 0.2–0.3 mm; appendages white to pink, rudimentary or minute, (0–)0.1–0.2 × (0–)0.1–0.3 mm, distal margin crenulate. **Staminate flowers** 3–5. **Pistillate flowers:** ovary densely white villous; styles 0.1–0.2 mm, 2-fid ½ length. **Capsules** broadly ovoid, 1.3–1.5 × 1.4–1.5 mm, villous on keels, often glabrous or less hairy between keels; columella 1.1–1.3 mm. **Seeds** white, barely concealing brown undercoat, 4-angled, sharply angled in cross section, abaxial faces plane to convex, adaxial faces concave, 1.1–1.2 × 0.5–0.7 mm, with several rounded, irregular, transverse ridges.

Flowering and fruiting almost year-round. Open sandy, loamy, or gravelly sites, old dunes, pastures; 0–200 m; Tex.; Mexico (Tamaulipas).

Euphorbia laredana is similar to *E. prostrata* but differs from that species in its more densely tomentose indumentum, leaves with usually entire rather than serrulate margins, and slightly longer seeds with rounded rather than sharp ridges. The species occurs primarily in southern Texas.

54. Euphorbia lasiocarpa Klotzsch, Nov. Actorum Acad. Caes. Leop.-Carol. Nat. Cur. 19(suppl. 1): 414. 1843 • Roadside sandmat [I] [W]

Chamaesyce lasiocarpa (Klotzsch) Arthur

Herbs, annual or perennial, often robust, with slender to moderately thickened taproot. **Stems** erect to ascending, 30–100 cm, pilose to tomentose. **Leaves** opposite; stipules distinct when young, connate into deltate scale when older, erose to laciniate, with dark glands at margin or base, 0.5–1.3 mm, pilose; petiole 0.5–2 mm, glabrescent or pilose; blade ovate or oblong, sometimes slightly falcate, 8–46 × 3–21 mm, base asymmetric, obtuse to hemicordate, margins serrulate, apex obtuse or acute, surfaces often with red spot in center, pilose to sericeous; palmately 3–5(–7)-veined from base. **Cyathia** in dense, usually terminal, capitate glomerules, with reduced, bractlike leaves subtending cyathia; peduncle 0.8–2.3 mm. **Involucre** obconic, 0.9–1.3 × 0.8–1.2 mm, pilose to sericeous; glands 4, yellow or pink, circular or oblong, 0.1–0.3 mm diam.; appendages white or pink, oblong, flabellate, or suborbiculate, 0.1–0.6 × 0.2–1 mm, entire or crenate. **Staminate flowers** 15–25. **Pistillate flowers:** ovary densely pilose to sericeous with yellowish hairs; styles 0.6–0.9 mm, 2-fid ½ length, filiform. **Capsules** subglobose to broadly ovoid, 1.7–2 mm diam., pilose to sericeous with yellowish hairs; columella 1.5–1.7 mm. **Seeds** dark reddish brown to almost black, plumply ovoid, 4-angled in cross section, 1.2–1.3 × 0.7–0.8 mm, with 2 inconspicuous rows of 3–5 shallow depressions separated by low ridges.

Flowering and fruiting year-round. Open disturbed areas, mostly along roadsides and railroad tracks; 0–10 m; introduced; Fla.; Mexico; West Indies; Central America; South America.

Euphorbia lasiocarpa is similar to *E. hypericifolia* but is much more hairy on its stems and leaves. *Euphorbia lasiocarpa* is widespread throughout tropical America, but its precise native range in the New World is not clear. In the flora area, *E. lasiocarpa* is found in southern Florida, where it is likely introduced.

55. Euphorbia lata Engelmann in W. H. Emory, Rep. U.S. Mex. Bound. 2(1): 188. 1859 • Broadleaf spurge

Euphorbia dilatata Torrey & A. Gray in War Department [U.S.], Pacif. Railr. Rep. 2(4): 175. 1857, not Hochstetter ex A. Richard 1850; *Chamaesyce lata* (Engelmann) Small; *E. rinconis* M. E. Jones

Herbs, perennial, with moderately thickened to robust rootstock. **Stems** ascending to erect, or prostrate, 10–25 cm, strigose to short-sericeous or ± villous. **Leaves** opposite; stipules distinct, filiform, 0.8–1.3 mm, strigose to short-sericeous or ± villous; petiole 0.5–2 mm, densely strigose to short-sericeous or ± villous; blade narrowly to broadly ovate-deltate, older ones often falcate, 4–12 × 3–7 mm, base asymmetric, obliquely rounded to obtuse, noticeably wider on one side, margins entire, often ± revolute, apex broadly acute, surfaces strigose to short-sericeous or ± villous; obscurely 3–5-veined from base, midvein prominent abaxially. **Cyathia** solitary at distal nodes; peduncle 1–3 mm. **Involucre** broadly campanulate, 2–2.5 × 2.2–2.6 mm, strigose; glands 4, greenish, oblong to semilunate, 0.2–0.7 × 0.6–1 mm; appendages rudimentary or white, forming narrow band, (0–)0.1–0.2 × (0–)0.6–1 mm, distal margin entire or crenate. **Staminate flowers** 25–35. **Pistillate flowers:** ovary densely strigose to short-sericeous or ± villous; styles dark purplish, 0.8–1.2 mm, 2-fid ½ to nearly entire length. **Capsules** ovoid, 1.9–2.3 × 2–2.4 mm, strigose to short-sericeous or ± villous; columella 1.7–2.2 mm. **Seeds** whitish, oblong, 4-angled in cross section, faces concave, 1.5–1.8(–2) × 0.6–0.9 mm, smooth. $2n = 28, 56$.

Flowering and fruiting spring–fall. Mountain slopes, canyons, basins, rocky prairies, roadsides, disturbed sites, usually in calcareous soils, sometimes in igneous-derived, sandy or rocky soils; 600–2200 m; Colo., Kans., N.Mex., Okla., Tex.; Mexico (Chihuahua, Coahuila).

56. Euphorbia maculata Linnaeus, Sp. Pl. 1: 455. 1753 • Milk purslane, prostrate spurge, spotted sandmat or spurge, euphorbe maculée [W]

Chamaesyce maculata (Linnaeus) Small; *C. mathewsii* Small; *C. supina* (Rafinesque) Moldenke; *C. tracyi* Small; *Euphorbia supina* Rafinesque

Herbs, annual, with taproot. **Stems** usually prostrate, occasionally with ascending tips, often mat-forming, not rooting at nodes, 5–45 cm, densely and evenly short-sericeous to sericeous or villous. **Leaves** opposite; stipules distinct, linear-subulate, sometimes irregularly

2–3-lobed, 1–1.3 mm, sparsely short-sericeous to sericeous or villous; petiole 0.5–1.5 mm, moderately short-sericeous to sericeous or villous; blade oblong-ovate to ovate-elliptic or oblong-elliptic, 4–18 × 2.5–8 mm, base strongly asymmetric, one side usually angled and other ± truncate and expanded into small, rounded auricle, margins serrulate (longer side) or subentire (shorter side), apex rounded or broadly acute, abaxial surface pale grayish green, moderately to densely lanulose to villous, adaxial surface usually with irregular reddish streak along midvein, glabrate or with sparse, long, slender hairs; palmately veined at base, pinnate distally. **Cyathia** solitary or in small, cymose clusters at distal nodes or on congested, axillary branches; peduncle 0.1–0.6 mm. **Involucre** obconic, 0.8–1 × 0.6–0.8 mm, sparsely strigose to short-sericeous; glands 4, green to yellow-green, turning pink with age, usually ± unequal, narrowly oblong to nearly linear, 0.1–0.2 × 0.2–0.5 mm; appendages white to reddish tinged, lunate to oblong, 0.1–0.3 × 0.2–1.5 mm, distal margin crenulate. **Staminate flowers** 2–5. **Pistillate flowers:** ovary sericeous; styles 0.3–0.4 mm, 2-fid at apex. **Capsules** ovoid, well exserted from involucre at maturity, 1.3–1.5 × 1.2–1.4 mm, sparsely to moderately and evenly sericeous; columella 1–1.2 mm. **Seeds** white to light brown, oblong-ovoid, sharply angular in cross section, 1–1.2 × 0.6–0.9 mm, with 3–4 low, transverse ridges that cross angles. $2n = 28$.

Flowering and fruiting spring–fall. Disturbed areas, fallow fields, gardens, sidewalk cracks, railroads, roadsides; 0–1500 m; B.C., N.B., N.S., Ont., P.E.I., Que.; Ala., Ariz., Ark., Calif., Colo., Conn., Del., D.C., Fla., Ga., Idaho, Ill., Ind., Iowa, Kans., Ky., La., Maine, Md., Mass., Mich., Minn., Miss., Mo., Mont., Nebr., Nev., N.H., N.J., N.Mex., N.Y., N.C., N.Dak., Ohio, Okla., Oreg., Pa., R.I., S.C., S.Dak., Tenn., Tex., Utah, Vt., Va., Wash., W.Va., Wis., Wyo.; Mexico; West Indies; Central America; South America; Eurasia; Africa; Pacific Islands (New Zealand); Australia.

Euphorbia maculata is a widespread weed in temperate latitudes, and it also occurs in cool climates at higher elevations in the tropics. It is presumed to be native to eastern and central North America, but given its extremely weedy tendencies, it is difficult to know for sure. It spreads readily in association with greenhouse plants and earth-moving activities, and it is notorious for its ability to colonize sidewalk cracks in the summer, even in congested cities. The name *E. maculata* was misapplied by most earlier botanists (for example, L. C. Wheeler 1941) to plants with ascending stems that are treated here as *E. nutans*. D. G. Burch (1966) reviewed the sources of data used by Linnaeus in his original description and concluded that the name *E. maculata* applies to this prostrate-stemmed taxon. For further discussion of the distinctions between *E. maculata* and the similar 47. *E. humistrata*, see the treatment of that species.

57. **Euphorbia meganaesos** Featherman, Rep. (Annual) Board Supervisors Louisiana State Seminary Learning Military Acad. 1870: 71, 105. 1871 [E]

Herbs, annual, with taproot. **Stems** prostrate to ascending, drooping at tips, 15–30 cm, glabrous. **Leaves** opposite; stipules distinct, divided nearly to base into linear-filiform segments, 1–2 mm, glabrous; petiole 0.5–1.3 mm, glabrous; blade narrowly oblong to oblong-obovate, often ± falcate, 5–16 × 1–4.5 mm, base subsymmetric to strongly oblique, margins sparsely spinulose-serrulate, apex rounded or broadly acute, abaxial surface pale grayish green, adaxial surface sometimes with reddish streak along midvein, both surfaces not papillate, glabrous; 3–5-veined at base. **Cyathia** solitary at distal nodes or on congested, axillary branches; peduncle 0.1–0.4 mm. **Involucre** obconic, 0.7–0.9 × 0.4–0.6 mm, glabrous; glands 4, green to yellow-green, subequal, oblong, 0.1–0.2 × 0.2–0.3 mm; appendages white to reddish tinged, lunate to oblong, 0.1–0.3 × 0.3–0.5 mm (2 ± 2 times longer than other 2), distal margin entire or coarsely toothed. **Staminate flowers** 2–5. **Pistillate flowers:** ovary glabrous; styles 0.2–0.3 mm, 2-fid at apex to nearly ½ length. **Capsules** ovoid, cocci not elongated nor terminating in empty portion, 1.5 × 1.7 mm, glabrous; columella 1.3 mm. **Seeds** reddish brown to brown, pyramidal to oblong-ovoid, weakly 4-angled in cross section, 0.9–1 × 0.7 mm, minutely beaded, with 3–4 broad, rounded, transverse ridges that do not interrupt abaxial keel.

Flowering and fruiting late spring–late summer. Sandy beaches, edges of marshes, coastal prairies, roadsides; 0–10 m; La., Tex.

Euphorbia meganaesos is known only from coastal areas of southern Louisiana and adjacent Texas. This species was often considered conspecific with *E. maculata* in the past, but it differs from that species in being entirely glabrous.

58. **Euphorbia melanadenia** Torrey in War Department [U.S.], Pacif. Railr. Rep. 4(5): 135. 1857 • Squaw sandmat, red-gland spurge

Chamaesyce melanadenia (Torrey) Millspaugh

Herbs, perennial, with moderately to strongly thickened rootstock. **Stems** ascending to erect, 5–20 cm, sericeous to appressed-villous. **Leaves** opposite; stipules distinct (lower side) and connate (upper side), linear, 0.5–1 mm, densely pilose; petiole 0.8–1.5 mm, tomentose; blade ovate, 1.2–5 × 0.8–2.9 mm, base asymmetric, hemicordate, margins entire, apex rounded to acute, surfaces tomentose; venation inconspicuous. **Cyathia** solitary at distal nodes; peduncle (0.6–)1.4–1.9 mm. **Involucre** campanulate, 0.6–1.1 × 0.7–1 mm, tomentose; glands 4, deep red to purple, elliptic, 0.3–0.4 × 0.4–0.7 mm; appendages white or becoming pink with age, oblong to flabellate, 0.4–0.7(–1) × 0.7–1.2 mm, distal margin entire or erose. **Staminate flowers** 45–80. **Pistillate flowers:** ovary tomentose; styles 0.5–0.8 mm, 2-fid nearly entire length. **Capsules** ovoid, 1.4–1.8 × 1.4–1.7 mm, tomentose; columella 1.2–1.5 mm. **Seeds** gray to tan, oblong, 4-angled in cross section, 1–1.2 × 0.4–0.6 mm, smooth to wrinkled or alveolate.

Flowering and fruiting year-round. Rocky slopes, river washes, dry to wet soils; 400–1400 m; Ariz., Calif.; Mexico (Baja California, Sonora).

Euphorbia melanadenia is similar in appearance to *E. cinerascens*, but *E. melanadenia* has conspicuous involucral gland appendages whereas *E. cinerascens* has inconspicuous appendages or lacks them entirely. *Euphorbia melanadenia* occurs in Arizona and southern California, whereas *E. cinerascens* is found only in southern and western Texas.

59. Euphorbia mendezii Boissier, Cent. Euphorb., 15. 1860 • Mendez's sandmat [I] [W]

Chamaesyce mendezii (Boissier) Millspaugh

Herbs, annual or perennial, with slender taproot. **Stems** prostrate, often mat-forming, 8–35 cm, usually villous along margins, lower surface glabrous, upper surface usually strigillose to puberulent, rarely glabrous or glabrate. **Leaves** opposite; stipules usually distinct or connate basally, rarely completely connate, deltate, laciniate, glabrous or pilose (lower side), forming narrow deltate scale, sometimes apically 2-fid or laciniate, glabrous (upper side), 0.4–1.9 mm; petiole 0.3–1.2 mm, glabrous, pilose or villous; blade oblong to obovate, 4–12 × 2–7 mm, base asymmetric, one side attenuate, cuneate or rounded, other rounded or cordate, margins serrulate at least distally, apex obtuse, surfaces glabrous or sparsely sericeous, pilose or villous; 3-veined at base. **Cyathia** solitary at nodes or on short, congested axillary branches; peduncle 0.9–2.5 mm. **Involucre** campanulate or obconic, 0.8–1 × 0.7–0.8 mm, usually glabrous, rarely sparsely pilose toward apex; glands 4, pink, reniform, oblong or elliptic, 0.1 × 0.2–0.3 mm; appendages absent or white to pink, oblong, flabellate or forming narrow rim around edge of gland, 0.1–0.3 × 0.3–0.6 mm, distal margin usually entire, sometimes lobed. **Staminate flowers** 6–15. **Pistillate flowers:** ovary usually pilose or villous with hairs concentrated along keels, rarely glabrous; styles 0.2–0.3 mm, 2-fid ½ length. **Capsules** ovoid, 1.2–1.6 × 1.2–1.4 mm, usually pilose or villous with hairs concentrated along keels, often glabrous in between, very rarely completely glabrous; columella 1–1.4 mm. **Seeds** reddish brown to orange or gray-pink, narrowly ovoid, 4-angled in cross section, 0.9–1.2 × 0.5–0.6 mm, almost smooth or with 5–7 faint transverse ridges that do not pass through abaxial keel.

Flowering and fruiting year-round. Disturbed areas; 0–10 m; introduced; Fla.; Mexico; West Indies; Central America; n South America.

Euphorbia mendezii is a common weed distributed widely throughout Mexico and Central America. Within the flora area the species is known only from southern Florida.

60. Euphorbia mesembrianthemifolia Jacquin, Enum. Syst. Pl., 22. 1760 • Coastal beach sandmat

Chamaesyce buxifolia (Lamarck) Small; *C. mesembrianthemifolia* (Jacquin) Dugand; *Euphorbia buxifolia* Lamarck

Subshrubs or shrubs, perennial, with thickened and often woody rootstock. **Stems** erect to ascending, or nearly decumbent in shifting sand, 25–60 cm, glabrous. **Leaves** opposite; stipules connate, forming conspicuous, ligulate or deltate scale, short cleft or fringed, 1–1.8 mm, glabrous; petiole 0.5–1 mm, glabrous; blade ovate to elliptic, often folded along midrib, 5–12 × 3–8 mm, ± fleshy, base slightly asymmetric, truncate to cordate, partially obscuring stem, margins entire, apex usually obtuse, rarely acute, surfaces yellowish to dark green, glabrous, glaucous; obscurely 3–5-veined at base, pinnate distally, only midvein conspicuous. **Cyathia** solitary at distal nodes; peduncle 0.5–1 mm. **Involucre** campanulate, 1–1.6 × 0.7–1.2 mm, glabrous; glands 4, brown, usually elliptic, occasionally almost round, 0.2–0.4 × 0.5–0.7 mm, fleshy; appendages white, oblong, rarely rudimentary, 0.2–0.4 × 0.5–0.9 mm, distal margin entire or undulate. **Staminate flowers** 12–20. **Pistillate flowers:** ovary glabrous; styles 0.3–0.4 mm, 2-fid ½ length. **Capsules** subglobose, subtended by calyxlike structure, 1.5–2 × 2.2–2.8 mm, glabrous; columella 1–1.5 mm. **Seeds** ashen, broadly ovoid, angled in cross section, faces plump, convex, 1.2–1.3 × 0.9–1.2 mm, obscurely pitted.

Flowering and fruiting year-round. Sandy and rocky shores, associated beach scrub; 0–10 m; Fla.; Mexico; West Indies; Bermuda; South America (Colombia, Venezuela).

Euphorbia mesembrianthemifolia is found in the flora area along the sandy and rocky shores of southern Florida from Pinellas and Volusia counties southward. It is one of the few members of subg. *Chamaesyce* in the flora area that is a shrub or subshrub.

61. Euphorbia micromera Boissier, Proc. Amer. Acad. Arts 5: 171. 1861 • Desert spurge, Sonoran or tiny sandmat [W]

Chamaesyce micromera (Boissier) Wooton & Standley

Herbs, annual, with slender taproot. **Stems** prostrate, mat-forming, 5–35 cm, glabrous or shortly pilose. **Leaves** opposite; stipules distinct, subulate, 0.2–0.4 mm, pilose; petiole 0.5–1.2 mm, glabrous or pilose; blade ovate to elliptic, 6–15 × 2–4 mm, base asymmetric, one side cuneate to rounded, other side rounded, margins entire, apex obtuse, surfaces glabrous or pilose; venation obscure or only midvein conspicuous. **Cyathia** solitary at distal nodes; peduncle 0.4–1.4 mm. **Involucre** campanulate, 0.4–0.6 × 0.5–0.9 mm, glabrous or pilose; glands 4, red, circular to oblong, 0.1 × 0.1–0.2 mm; appendages absent. **Staminate flowers** 2–5. **Pistillate flowers:** ovary usually glabrous, rarely pilose; styles 0.1–0.2 mm, 2-fid at apex. **Capsules** oblong, 1.3–1.5 × 1.1–1.3 mm, usually glabrous, rarely pilose; columella 1–1.2 mm. **Seeds** light gray, narrowly ovoid, 4-angled in cross section, 0.9–1 × 0.5–0.6 mm, smooth to slightly rugose or with 1–4 faint transverse ridges that do not pass through abaxial keel.

Flowering nearly year-round in response to sufficient rainfall. Desert scrub, riparian woods with ash and willow, saltbush scrub, Joshua tree woodlands and grasslands, often in sandy or gravelly areas; -20–1800 m; Ariz., Calif., Nev., N.Mex., Tex., Utah; Mexico (Baja California, Baja California Sur, Chihuahua, Coahuila, Durango, Sonora).

62. Euphorbia missurica Rafinesque, Atlantic J. 1: 146. 1832 • Prairie or Missouri spurge, prairie sandmat [E]

Chamaesyce missurica (Rafinesque) Shinners; *Euphorbia missurica* var. *intermedia* (Engelmann) L. C. Wheeler; *E. petaloidea* Engelmann var. *intermedia* Engelmann

Herbs, annual, with taproot. **Stems** erect or ascending, 10–60(–100) cm, glabrous, sometimes ± glaucous. **Leaves** opposite; stipules usually distinct, occasionally connate basally on one or both sides of stem, linear to triangular-subulate, usually deeply and irregularly fringed or lobed, rarely entire, 0.7–1.5 mm, glabrous; petiole 1–3 mm, glabrous; blade linear to narrowly oblong or narrowly lanceolate-oblong, (4–)8–30 × 3–7 mm, base symmetric or subsymmetric (usually narrower leaves), or slightly asymmetric and angled or short-tapered (wider leaves), margins entire, occasionally ± revolute, apex rounded to truncate, occasionally emarginate or mucronulate, abaxial surface pale green, adaxial surface light to bright green, both surfaces glabrous; venation obscure. **Cyathia** solitary or in small, cymose clusters, these occasionally subtended by reduced, bractlike leaves, at distal nodes or on congested, axillary branches; peduncle 1–5(–11) mm. **Involucre** broadly campanulate, 1.2–1.8 × 1.7–1.9 mm, glabrous; glands 4, yellowish green, broadly oblong to nearly circular, cupped or folded, 0.3–0.6 × 0.3–0.7 mm; appendages white or ± pinkish tinged, ovate to oblong-ovate, 0.4–2.5 × 1.1–1.7 mm, distal margin entire or slightly crenate or emarginate at tip. **Staminate flowers** 24–60. **Pistillate flowers:** ovary glabrous; styles 0.5–1.4 mm, 2-fid ½ length. **Capsules** broadly ovoid-globose, 1.9–2.5 × 2–2.5(–3) mm, glabrous; columella 1.8–2.1 mm. **Seeds** mottled whitish to brown, ovoid to broadly ovoid-triangular, bluntly 3-angled in cross section, 1.5–2 × 1.1–1.4 mm, smooth or slightly wrinkled.

Flowering and fruiting late spring–late summer. Glades, ledges, bluff tops (usually calcareous), dry upland forest margins, sandy or disturbed areas; 50–1500 m; Ark., Colo., Iowa, Kans., Mo., Mont., Nebr., N.Mex., N.Dak., Okla., S.Dak., Tex., Wyo.

Euphorbia missurica is similar to the western *E. parryi* but has a more upright growth habit and more conspicuous involucral gland appendages. Native occurrences have been documented from Minnesota (last collected in Ottertail County in 1936), but it appears to have been extirpated from that state due to habitat loss to agriculture.

63. Euphorbia nutans Lagasca, Gen. Sp. Pl., 17. 1816 • Nodding or upright spotted spurge, eyebane [F] [W]

Chamaesyce lansingii Millspaugh; *C. nutans* (Lagasca) Small; *Euphorbia lansingii* (Millspaugh) Brühl; *E. preslii* Gussone

Herbs, annual, with taproot. **Stems** usually ascending, occasionally erect, often arched at tips, 20–80 cm, sparsely to moderately pilose to villous or with short, incurved hairs, hairs often concentrated at nodes and distally, occasionally in 2 bands along opposite sides of stem. **Leaves** opposite; stipules usually distinct, sometimes connate basally on one side of stem, small scales, irregularly toothed, fringed, or divided, 1–1.5 mm, sparsely to moderately villous distally; petiole 0.3–1.6 mm, moderately pilose to villous; blade oblong to oblong-lanceolate, 8–40 × 3–12 mm, base asymmetric, one side usually angled or rounded, other side ± truncate to cordate-auriculate, margins serrulate, apex angled with blunt tip, abaxial surface pale green or faintly to strongly reddish tinged, adaxial surface usually reddish mottled or with conspicuous reddish spot, both surfaces usually sparsely to moderately pilose, especially toward base, sometimes glabrous; 3–5-veined from base, pinnate distally, veins faint. **Cyathia** solitary at distal nodes or in small, cymose clusters at branch tips; peduncle 0.5–2.5 mm. **Involucre** narrowly obconic, 0.5–1 × 0.3–0.7 mm, glabrous; glands 4, usually green, sometimes reddish purple, oblong to nearly circular, 0.2–0.4 × 0.3–0.5 mm; appendages white or pinkish, ovate to broadly elliptic, 0.2–1 × 0.2–1.5 mm, distal margin entire. **Staminate flowers** 5–28. **Pistillate flowers:** ovary glabrous; styles 0.6–2.5 mm, 2-fid ½ length. **Capsules** ovoid, 1.6–2.3 × 1.5–2.4 mm, glabrous; columella 1.4–1.6 mm. **Seeds** dark brown, sometimes with thin, white coating (often more persistent along angles than faces), elliptic-ovoid to ovoid, rounded-angular in cross section, 1–1.6 × 0.5–0.8 mm, surface finely and irregularly wrinkled, sometimes faintly so, or with indistinct, shallow, rounded cross ridges. $2n = 12, 14, 22$.

Flowering and fruiting spring–early fall. Stream banks, pond edges, disturbed portions of upland prairies, mesic to dry upland forest openings, pastures, fallow fields, railroads, roadsides, gardens, disturbed areas; 0–1600 m; Ont., Que.; Ala., Ark., Calif., Conn., Del., Fla., Ga., Ill., Ind., Iowa, Kans., Ky., La., Md., Mass., Mich., Minn., Miss., Mo., Nebr., N.H., N.J., N.Mex., N.Y., N.C., N.Dak., Ohio, Okla., Pa., R.I., S.C., S.Dak., Tenn., Tex., Vt., Va., W.Va., Wis., Wyo.; Mexico; West Indies; Central America; South America; introduced in Eurasia.

Euphorbia nutans is probably native to at least central and eastern North America, but given its strongly weedy tendencies, it is difficult to know where it may be adventive in parts of the flora area. It is certainly introduced where it occurs in the Old World and probably in South America as well. D. G. Burch (1966) discussed the problems of assigning names to the four main entities in this nomenclatural complex (*E. hypericifolia*, *E. hyssopifolia*, *E. lasiocarpa*, and *E. nutans*) and determined that the oldest valid name for the relatively robust, temperate North American plants with ascending stems is *E. nutans*. See the treatment of 56. *E. maculata* for a discussion of the misapplication of that name to *E. nutans*.

64. Euphorbia ocellata Durand & Hilgard, Pl. Heermann., 46. 1854 • Contura Creek sandmat [E]

Chamaesyce ocellata (Durand & Hilgard) Millspaugh

Herbs, annual, with taproot. **Stems** prostrate, 10–35 cm, glabrous or pilose. **Leaves** opposite; stipules distinct, subulate, 0.5–1.6 mm, glabrous or pilose; petiole 0.3–2 mm, glabrous or pilose; blade ovate to deltate or falcate, 2.3–13 × 1.5–6 mm, base asymmetric, usually cordate, rarely rounded, margins occasionally reddish, entire, often revolute, apex acute to obtuse, occasionally mucronate, surfaces glabrous or pilose; midvein conspicuous, lateral veins frequently visible abaxially. **Cyathia** solitary at distal nodes; peduncle 0.9–2.2 mm. **Involucre** campanulate, 1–2.5 × 1.3–2.4 mm, glabrous or pilose; glands 4, yellow becoming deep red, elliptic or oblong to orbiculate, 0.4–0.6 × 0.5–0.7 mm; appendages absent or whitish, orbiculate, 0.1–0.2 × 0.3–0.5 mm, distal margin entire. **Staminate flowers** 30–70. **Pistillate flowers:** ovary glabrous or pilose; styles 0.4–0.5 mm, 2-fid at apex. **Capsules** subglobose to broadly ovoid, 1.4–2.7 × 1.9–3.1 mm, glabrous or pilose; columella 1.4–2 mm. **Seeds** whitish gray to black, ovoid to oblong, terete to bluntly subangled in cross section, 1.1–1.7 × 0.8–1.3 mm, rugose or smooth.

Subspecies 2 (2 in the flora): w United States.

1. Stems, leaves, ovaries, and capsules glabrous; involucral gland appendages absent . . . . 64a. *Euphorbia ocellata* subsp. *ocellata*
1. Stems, leaves, ovaries, and capsules pilose; involucral gland appendages present. . . . . 64b. *Euphorbia ocellata* subsp. *rattanii*

64a. Euphorbia ocellata Durand & Hilgard subsp. ocellata [E]

Chamaesyce arenicola (Parish) Millspaugh; *C. ocelleta* (Durand & Hilgard) Millspaugh subsp. *arenicola* (Parish) Thorne; *Euphorbia arenicola* Parish; *E. ocellata* subsp. *arenicola* (Parish) Oudejans; *E. ocelleta* var. *arenicola* (Parish) Jepson; *E. ocelleta* var. *kirbyi* J. T. Howell

Stems 10–35 cm, glabrous. **Leaves:** stipules 0.5–1.6 mm, glabrous; petiole 0.3–1.2 mm, glabrous; blade 2.3–13 × 1.5–6 mm, apex acute, occasionally mucronate, surfaces glabrous; lateral veins frequently visible abaxially. **Peduncle** 0.9–2 mm. **Involucre** 1–1.8 × 1.3–2.4 mm, glabrous; gland appendages absent. **Pistillate flowers:** ovary glabrous. **Capsules** 1.4–2.7 × 1.9–2.6 mm, glabrous; columella 1.4–2 mm. **Seeds** whitish gray, ovoid, 1.1–1.6 × 0.9–1.3 mm, rugose or smooth.

Flowering and fruiting early spring–fall. Sandy soils, dunes, river washes, hard clay soils, roadsides; 80–1000 m; Ariz., Calif., Idaho, Nev., Utah.

Plants with smooth, round seeds and larger, ovate to lanceolate (and not usually falcate) leaves have been segregated as *Euphorbia ocellata* subsp. *arenicola*. Leaf size and shape and seed surface sculpturing, however, vary considerably across the species range, and most individuals appear intermediate. Therefore a variable and more broadly defined subsp. *ocellata* is recognized here.

64b. Euphorbia ocellata Durand & Hilgard subsp. **rattanii** (S. Watson) Oudejans, Phytologia 67: 47. 1989 • Rattan's sandmat [E]

Euphorbia rattanii S. Watson, Proc. Amer. Acad. Arts 20: 372. 1885 (as rattani); *Chamaesyce ocellata* (Durand & Hilgard) Millspaugh subsp. *rattanii* (S. Watson) Koutnik; *C. rattanii* (S. Watson) Millspaugh; *E. ocellata* var. *rattanii* (S. Watson) L. C. Wheeler

Stems 10–22 cm, pilose. **Leaves:** stipules 1–1.5 mm, pilose; petiole 0.9–2 mm, pilose; blade 4.2–9.5 × 2.7–5 mm, apex acute to obtuse, surfaces pilose; lateral veins visible abaxially on larger leaves. **Peduncle** 1–2.2 mm. **Involucre** 2–2.5 × 1.7–2.4 mm, pilose; gland appendages present. **Pistillate flowers:** ovary pilose. **Capsules** 1.9–2.4 × 2–3.1 mm, pilose; columella 1.6–2 mm. **Seeds** pale gray to black, oblong, 1.6–1.7 × 0.8–1.2 mm, rugose.

Flowering and fruiting summer–fall, occasionally spring. Roadsides, gravelly or sandy dry stream beds; 80–500 m; Calif.

Subspecies *rattanii* occurs in the Sacramento Valley in Colusa, Glenn, and Tehama counties.

65. Euphorbia ophthalmica Persoon, Syn. Pl. 2: 13. 1806 • Florida hammock sandmat

Chamaesyce ophthalmica (Persoon) D. G. Burch; *Euphorbia hirta* Linnaeus var. *procumbens* (Boissier) N. E. Brown; *E. pilulifera* Linnaeus var. *procumbens* Boissier

Herbs, usually annual, rarely short-lived perennial, with slender to slightly thickened taproot. **Stems** usually prostrate, rarely ascending, 6–22 cm, usually both strigillose and hirsute. **Leaves** opposite; stipules distinct, subulate-filiform, undivided or divided into 2–4 narrowly triangular to linear-subulate segments, no dark, circular glands at base of stipules, 0.9–1.5 mm, pilose or strigillose; petiole 0.3–1.2 mm, glabrescent, strigillose, or sericeous; blade usually ovate or oblong, rarely subrhombic, 4–13 × 3–7 mm, base asymmetric, one side usually angled and other side rounded, margins coarsely serrulate, apex acute, surfaces often with red spot in center, strigillose or sericeous, or adaxial surface glabrescent; 3-veined from base. **Cyathia** in dense, terminal, capitate glomerules, with reduced, bractlike leaves subtending cyathia; peduncles 0–0.8 mm. **Involucre** obconic, 0.5–0.7 × 0.4–0.6 mm, strigillose; glands 4, yellow-green to pink, circular to slightly oblong, 0.1–0.2 × 0.1–0.2 mm; appendages absent or white to pink, forming thin rim around edge of gland or oblong, 0.1–0.2 × 0.1–0.3 mm, distal margin entire or shallowly lobed. **Staminate flowers** 2–8. **Pistillate flowers:** ovary strigillose, often canescent when young; styles 0.1–0.3 mm, 2-fid ½ to nearly entire length. **Capsules** ovoid, 1–1.2 × 1–1.3 mm, strigillose; columella 0.7–1.1 mm. **Seeds** orange-brown to pinkish, narrowly ovoid, 4-angled in cross section, 0.7–0.9(–1.1) × 0.5 mm, usually rugulose, with 3–6 faint, low, transverse ridges, rarely almost smooth.

Flowering and fruiting year-round. Hammock forests, disturbed areas in lawns, roadsides; 0–200 m; Ark., Calif., Fla., Ga., La., Mo., Pa.; Mexico; West Indies; Central America; South America; introduced in Europe.

Euphorbia ophthalmica is a weedy species distributed throughout the Neotropics. It is also adventive in the Old World. Whether it is indeed native to the southeastern United States is questionable; it is introduced in Arkansas, California, Missouri, and Pennsylvania and likely occurs also in other states. Although sometimes treated as *E. hirta* var. *procumbens, E. ophthalmica* appears sufficiently distinct to justify recognition at the rank of species, differing primarily by its mostly prostrate growth form, smaller leaves, and strictly terminal clusters of cyathia.

66. Euphorbia parishii Greene, Bull. Calif. Acad. Sci. 2: 56. 1886 • Parish's sandmat [E]

Chamaesyce parishii (Greene) Millspaugh; *Euphorbia polycarpa* Bentham var. *parishii* (Greene) Jepson

Herbs or subshrubs, perennial, with thickened and often woody rootstock. **Stems** prostrate, sometimes forming dense mounds, 10–50 cm, glabrous. **Leaves** opposite; stipules distinct, subulate-filiform, 0.3–0.9 mm, pilose; petiole 0.3–1.2 mm, glabrous; blade usually ovate, rarely oblong, 2–7 × 1–5 mm, base usually asymmetric, rounded to hemicordate, margins entire, apex usually obtuse, rarely acute, surfaces glabrous; only midvein conspicuous. **Cyathia** solitary at distal nodes; peduncle 0.1–0.6(–2.2) mm. **Involucre** obconic to campanulate, 1–1.4 × 0.9–1.3 mm, glabrous except for pilose lobes; glands 4, pink to maroon, circular, 0.3–0.4 × 0.3–0.4 mm; appendages absent. **Staminate flowers** 40–50. **Pistillate flowers:** ovary glabrous; styles 0.3–0.6 mm, 2-fid ½ length. **Capsules** ovoid to oblate-ovoid, 1.6–1.7 × 1.6–1.9 mm, glabrous; columella 1.2–1.5 mm. **Seeds** whitish to light brown, ovoid, 4-angled in cross section, 1.2–1.4 × 0.6–0.8 mm, rugose or with indistinct, irregular, low transverse ridges.

Flowering and fruiting fall–summer. Desert scrub, often with creosote bush, disturbed roadsides, rocky soils; -90–600 m; Calif., Nev.

Euphorbia parishii is common in the Death Valley region of southern California, where it is often encountered well below sea level. The species is frequently confused with *E. micromera* and *E. polycarpa* but differs from the former in being a more robust plant with larger cyathia and from the latter in lacking involucral gland appendages.

67. Euphorbia parryi Engelmann, Amer. Naturalist 9: 350. 1875 • Dune spurge

Chamaesyce longeramosa (S. Watson) Millspaugh; *C. parryi* (Engelmann) Rydberg; *Euphorbia longeramosa* S. Watson

Herbs, annual, with taproot. **Stems** usually prostrate, rarely ascending-erect, 5–70(–85) cm, glabrous. **Leaves** opposite; stipules distinct, linear-subulate, usually lacerate and divided into 2 or more slender segments, rarely entire, 0.6–1.4 mm, glabrous; petiole 1–2.5 mm, glabrous; blade linear to narrowly oblong, (5–)10–25(–30) × 2–5 mm, base usually symmetric, sometimes slightly asymmetric, attenuate, margins entire, occasionally ± revolute, apex acute to obtuse, mucronulate, surfaces glabrous; only midvein conspicuous. **Cyathia** solitary or in small clusters on short axillary branches at distal nodes; peduncle 1–5 mm. **Involucre** broadly cupuliform-campanulate, 1.2–1.7 × 1.4–1.8 mm, glabrous; glands 4, reddish pink to greenish yellow, deeply concave, elliptic to oblong, 0.2–0.3 × 0.3–0.5 mm; appendages white, elliptic to oblong, usually forming narrow margin around gland, sometimes rudimentary, 0.2–0.6 × 0.3–0.7(–1.1) mm, distal margin entire. **Staminate flowers** 40–55. **Pistillate flowers:** ovary glabrous; styles 0.5–0.7 mm, 2-fid nearly entire length. **Capsules** ovoid-globose, 2–2.3 × 1.5–2.5 mm, glabrous; columella 1.4–2 mm. **Seeds** mottled brown and white because of irregularly loose and tight outer covering, broadly ovoid, rounded-angular in cross section, 1.4–1.8 × 0.8–1 mm, smooth or only inconspicuously roughened.

Flowering and fruiting spring–fall. Sand dunes, other sandy habitats; 200–2200 m; Ariz., Calif., Colo., Nev., N.Mex., Tex., Utah; Mexico (Chihuahua).

Euphorbia parryi is similar to *E. missurica*, differing only by the generally narrow involucral gland appendages and prostrate habit in *E. parryi* as opposed to the conspicuous involucral gland appendages and ascending-erect habit in *E. missurica*. *Euphorbia parryi* has sometimes been considered the western race of *E. missurica* (D. S. Correll and M. C. Johnston 1970).

68. Euphorbia pediculifera Engelmann in W. H. Emory, Rep. U.S. Mex. Bound. 2(1): 186. 1859 • Carrizo Mountain sandmat

Chamaesyce pediculifera (Engelmann) Rose & Standley

Varieties 2 (1 in the flora): sw United States, nw Mexico.

Euphorbia pediculifera can be variable in terms of leaf shape and the extent of pubescence, but the species is readily identified by the short, appressed hairs on the leaves and the distinctive seed that is encircled by a series of round, transverse ridges.

Euphorbia pediculifera var. *linearifolia* S. Watson is found only near Guaymas in southwestern Sonora, Mexico.

68a. Euphorbia pediculifera Engelmann var. **pediculifera**

Herbs, annual or perennial, with taproot. **Stems** prostrate to ascending, 15–50 cm, strigose to short-sericeous when young, glabrate with age. **Leaves** opposite; stipules distinct (lower side) or connate (upper side), usually inconspicuous, filiform to triangular-subulate, 0.5–1 mm, strigose to short-sericeous; petiole 1–2 mm, strigose to short-sericeous; blade usually ovate to oblong, rarely linear or spatulate, 2–37 × 1–10 mm, base oblique to subsymmetric, cuneate to slightly rounded or cordate, margins entire, apex acute, adaxial surface sometimes with reddish blotch in center, both surfaces strigose, short-sericeous, or tomentose to glabrate; venation palmate at base, pinnate distally, only midvein conspicuous. **Cyathia** solitary at distal nodes; peduncle 0.5–1.5 mm. **Involucre** campanulate, 1.5–2 × 1.2–1.6 mm, closely strigose to short-sericeous to glabrate; glands 4, usually dark purple, rarely green-brown, oblong, 0.5 × 0.7–1.3 mm; appendages absent or white, ± unequal, irregular or rudimentary, 0.6–1 × 1–1.5 mm, distal margin entire or crenately lobed. **Staminate flowers** 22–25. **Pistillate flowers:** ovary densely strigose to short-sericeous; styles slender, 1–1.2 mm, 2-fid nearly entire length. **Capsules** broadly ovoid, 2 mm diam., strigose to short-sericeous; columella (1.4–)1.8–2 mm. **Seeds** white, slender ovoid, 3-angled to almost round in cross section, 1.2–1.6 × 0.6–0.7 mm, with 4–5 rounded, transverse ridges encircling seed.

Flowering and fruiting year-round. Desert scrub, thorn scrub, rocks and sandy washes; 0–1200 m; Ariz., Calif.; Mexico (Baja California, Baja California Sur, Sinaloa, Sonora).

69. Euphorbia perennans (Shinners) Warnock & M. C. Johnston, SouthW. Naturalist 5: 170. 1960 • Terlingua spurge, perennial sandmat

Chamaesyce perennans Shinners, Field & Lab. 24: 38. 1956

Herbs, perennial, with strongly thickened, woody rootstock. **Stems** erect, 7–45 cm, glabrous. **Leaves** opposite; stipules distinct, linear-filiform in (1–)2(–3) segments, 0.3–0.4 mm, glabrous; petiole 0.8–2 mm, glabrous; blade ovate or orbiculate-deltate to reniform-deltate, 5–17 × 4–16 mm midstem leaves largest, base symmetric, cuneate, rounded to cordate, margins entire, apex acute to rounded, surfaces glabrous, often glaucous; 3-veined from base, only midvein conspicuous. **Cyathia** solitary at distal nodes or at nodes of short, axillary branches; peduncle 1.8–3 mm. **Involucre** broadly campanulate to hemispheric, 1.7–2.2 × 1.5–2.7 mm, glabrous; glands 4, green to yellow-green, elliptic to oblong, folded longitudinally, 0.3–0.5 × 0.7–1.4 mm; appendages absent. **Staminate flowers** 35–45. **Pistillate flowers:** ovary glabrous; styles 0.7–0.9 mm, 2-fid nearly entire length. **Capsules** subglobose to broadly ovoid, 2.8–3.3 × 2.8–3.4 mm, glabrous; columella 2.2–2.7 mm. **Seeds** white to light brown, ovoid, 3–4-angled in cross section, 2–2.4 × 1–1.2 mm, smooth to faintly transverse-wrinkled.

Flowering and fruiting spring–fall. Desert scrub, on Cretaceous gypseous clay, limestone hills, and flats; 900–1200 m; Tex.; Mexico (Chihuahua).

Euphorbia perennans is a distinctive species with an erect habit and relatively large, firm, deltate midstem leaves. Phylogenetic data place *E. perennans* in a clade of primarily Chihuahuan Desert annual and perennial species (for example, *E. chaetocalyx*, *E. fendleri*, *E. golondrina*, *E. simulans*, and *E. theriaca*; Y. Yang and P. E. Berry 2011).

Euphorbia perennans is known in the flora area only from Brewster County.

70. Euphorbia pergamena Small, Bull. Torrey Bot. Club 25: 615. 1898 • Southern Florida sandmat, rocklands spurge [C]

Chamaesyce adenoptera (Bertoloni) Small subsp. *pergamena* (Small) D. G. Burch; *C. pergamena* (Small) Small; *Euphorbia adenoptera* Bertoloni subsp. *pergamena* (Small) Oudejans

Herbs, perennial, with moderately to strongly thickened rootstock. **Stems** prostrate to ascending, occasionally mat-forming, terete to slightly flattened, 5–18 cm, glabrous on lower surface, strigillose to sericeous on upper surface. **Leaves** opposite; stipules distinct, subulate, better developed on lower side of stem, 0.4–1.3 mm, glabrous or pilose; petiole 0.3–0.5 mm, puberulent, sericeous, or strigillose; blade oblong to ovate, 4–7 × 2–4 mm, base asymmetric, hemicordate, larger side sometimes amplexicaulous, margins entire or serrulate, apex obtuse, surfaces puberulent, sericeous, or strigillose; 3-veined from base, lateral veins inconspicuous. **Cyathia** solitary at distal nodes or at nodes of short, congested, axillary shoots; peduncle to 0.5 mm. **Involucre** campanulate, 1–1.3 × 1–1.1 mm, canescent; glands 4, pinkish, oblong or reniform, 0.2 × 0.4–0.6 mm; appendages white to pink, lunate, unequal, those of proximal glands oblique, 0.6–1 × 1.4–1.6 mm, those of distal glands symmetric, 0.3 × 0.8–0.9 mm, irregularly undulate to incised. **Staminate flowers** 15. **Pistillate flowers:** ovary puberulent, sericeous or strigillose; styles 0.5–1 mm, 2-fid at apex. **Capsules** oblate, well exserted from involucre at maturity, 1–1.3 × 1.3–1.8 mm, puberulent, sericeous or strigillose; columella 1 mm. **Seeds** whitish to gray, ovoid, 4-angled in cross section, 0.7–0.8 × 0.5–0.6 mm, with 3–4 transverse sulci alternating with low transverse ridges.

Flowering and fruiting year-round. Crevices of limestone outcrops in pinelands and pine-palm woods; of conservation concern; 0–10 m; Fla.; West Indies (Cuba, Hispaniola).

Euphorbia pergamena is a Florida-listed threatened species known within the flora area from only Miami-Dade and Monroe counties in extreme southern Florida. The species is often included as a subspecies of *E. adenoptera* (for example, D. G. Burch 1965), but here A. Herndon (1993b) is followed and it is treated as a distinct species.

71. Euphorbia platysperma Engelmann in W. H. Brewer et al., Bot. California 2: 482. 1880 • Dune or flat-seeded spurge, flatseed sandmat [C]

Chamaesyce platysperma (Engelmann) Shinners

Herbs, annual or short-lived perennial, with slender taproot. **Stems** prostrate, spreading and often mat-forming, 10–100 cm, glabrous. **Leaves** opposite; stipules usually distinct, occasionally connate basally, rarely to middle, subulate, narrowly triangular, or divided into 2–4 subulate segments, 0.5–1.1 mm, glabrous; petiole 1–3.6 mm, glabrous; blade oblong to obovate, 5–12 × 3–5 mm, base subsymmetric, cuneate to attenuate, margins entire, apex usually acute to mucronulate, rarely obtuse, surfaces glabrous; only midvein conspicuous. **Cyathia** solitary at distal nodes; peduncle 1.6–4.1 mm. **Involucre** campanulate

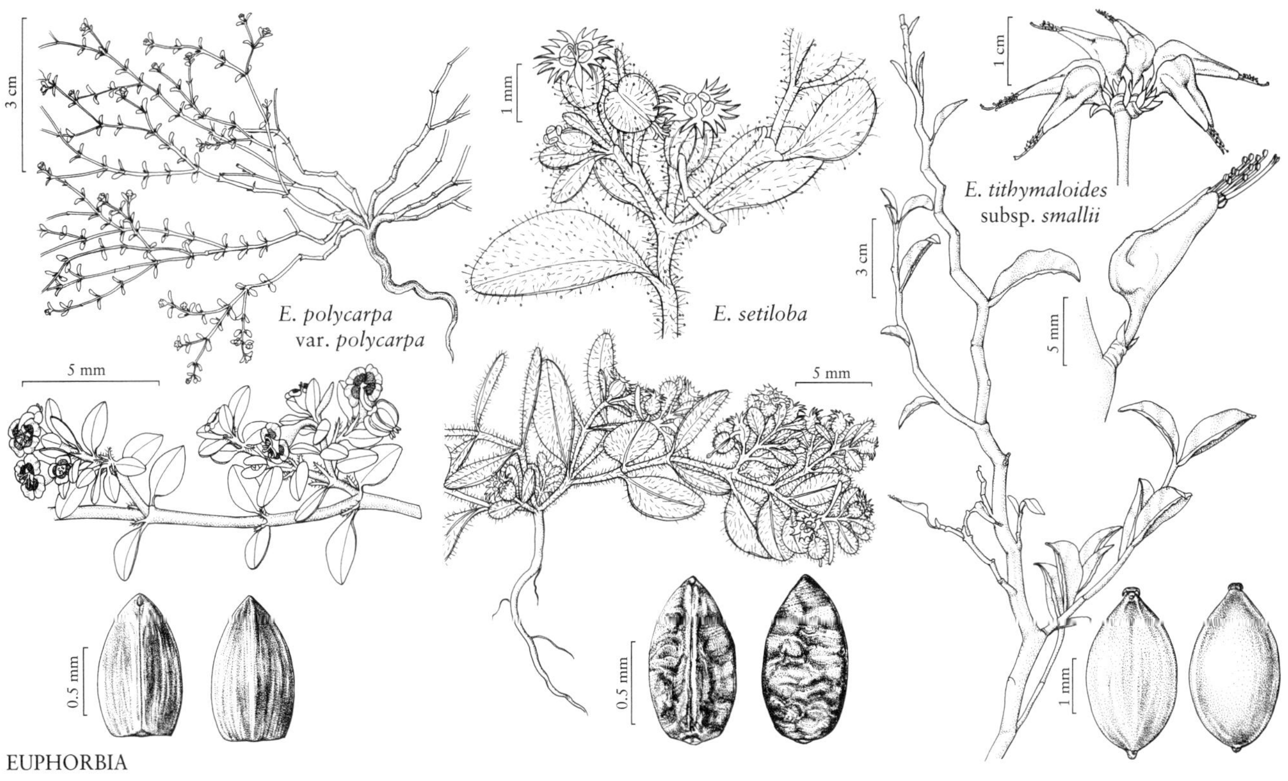

EUPHORBIA

to obconic, 1.5–2 × 1.3–2.5 mm, glabrous; glands 4, yellowish, subcircular to oblong, 0.5–0.6 × 0.5–0.6 mm; appendages white, ovate to oblong or almost triangular, 0.3–0.6 × 0.3–0.8 mm, distal margin entire or shallowly 2–3-lobed. **Staminate flowers** 45–50. **Pistillate flowers:** ovary glabrous; styles 0.4–0.6 mm, 2-fid ½ length. **Capsules** ovoid to ellipsoid, 2.7–3.2 × 2.2–2.9 mm, glabrous; columella 2.6–2.8 mm. **Seeds** whitish, pinkish, or light brown, ellipsoid-oblong, weakly dorsiventrally compressed and semielliptic in cross section, 2.2–2.5 × 1.3–1.6 mm, abaxial side smooth and rounded, adaxial side with sharp linelike longitudinal ridge.

Flowering and fruiting year-round in response to sufficient moisture. Sand dunes in Sonoran Desert scrub; of conservation concern; 60–200 m; Ariz., Calif.; Mexico (Baja California, Sonora).

72. Euphorbia polycarpa Bentham, Bot. Voy. Sulphur, 50. 1844 • Smallseed sandmat [F]

Chamaesyce polycarpa (Bentham) Millspaugh

Varieties 5 (1 in the flora): sw United States, nw Mexico.

Euphorbia polycarpa is a highly variable species distributed throughout the Baja California peninsula, eastern Sonora, and the arid southwestern United States from southwestern Arizona to southern Nevada and central California. L. C. Wheeler (1941) divided the species into seven varieties, the majority of which occur in Baja California Sur. He reported two of these for the United States: vars. *hirtella* and *polycarpa*. Variety *hirtella*, as the name suggests, was applied to hairy plants. However, it is here treated as a synonym of var. *polycarpa*, because variation in pubescence shows no geographic segregation and ranges along a continuum from glabrous or sparsely to densely hairy, and because glabrous and hairy branches can occur on the same individual.

Some of the varieties occurring in Mexico are highly divergent and appear to be sufficiently distinct to merit recognition as species. However, within the flora area, *Euphorbia polycarpa* is relatively uniform, and the only noteworthy variation involves the size of the involucral gland appendages. In the portion of the California Floristic Province occupied by *E. polycarpa* (Los Angeles, Orange, western Riverside, western San Diego, and Ventura counties), the appendages are conspicuously larger than those of plants throughout the remainder of its range in the southwestern United States (0.4–1.2 versus 0.1–0.3 mm). Large-appendaged plants are also common in Baja California and correspond well with the type collection. The taxonomic significance of this variation is not clear at this time, and the small-appendaged plants may merit segregation as an infraspecific taxon. However, no formal changes are proposed, awaiting a comprehensive review of the species throughout its range.

72a. Euphorbia polycarpa Bentham var. **polycarpa** [F]

Chamaesyce polycarpa (Bentham) Millspaugh var. *hirtella* (Boissier) Millspaugh; *Euphorbia polycarpa* var. *hirtella* Boissier

Herbs, annual or perennial, with slender taproot to thickened rootstock. **Stems** prostrate to reclining, usually mound- or mat-forming, 7–25 cm, glabrous, puberulent, shortly pilose, or hirsute. **Leaves** opposite; stipules usually distinct, occasionally connate basally on lower side of stem, subulate, 0.3–0.6 mm, shortly pilose or hirsute; petiole 0.4–0.9 mm, glabrous, puberulent, shortly pilose, or hirsute; blade ovate, elliptic, or oblong, 1.5–5 × 1–3 mm, base asymmetric, rounded to slightly oblique, margins entire, apex obtuse or acute, surfaces glabrous, puberulent, shortly pilose, or hirsute; venation obscure, only midvein sometimes conspicuous. **Cyathia** solitary at distal nodes, sometimes appearing clustered when distal nodes congested; peduncles 0–1.6 mm. **Involucre** campanulate, 0.6–1.1 × 0.5–1.3 mm, glabrous, shortly pilose, or hirsute; glands 4, dark pink to black-purple, oblong, 0.1–0.3 × 0.2–0.5 mm; appendages white to pink, usually ovate or flabellate, sometimes forming narrow rim around distal margin of gland, 0.1–0.2 × 0.3–0.8 mm, distal margin entire or shallowly lobed. **Staminate flowers** 15–30. **Pistillate flowers:** ovary glabrous, puberulent, or shortly pilose; styles 0.2–0.4 mm, 2-fid ½ length. **Capsules** ovoid to subglobose, 1.1–1.4 mm diam., glabrous, puberulent, or shortly pilose; columella 0.9–1.2 mm. **Seeds** pink-brown to light brown, narrowly oblong-ovoid, (3–)4-angled in cross section, 0.8–1.1 × 0.5–0.6 mm, smooth or faintly rippled.

Flowering and fruiting year-round in response to sufficient moisture. Desert scrub, chaparral, coastal sage scrub, disturbed areas, especially roadsides; -20–1200 m; Ariz., Calif., Nev.; Mexico (Baja California, Baja California Sur, Sonora).

73. Euphorbia polygonifolia Linnaeus, Sp. Pl. 1: 455. 1753 • Dune or seaside spurge, euphorbe à feuilles de renouée [E]

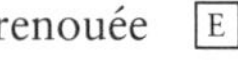

Chamaesyce polygonifolia (Linnaeus) Small

Herbs, annual, with taproot. **Stems** usually prostrate, occasionally ascending, 5–30 cm, glabrous. **Leaves** opposite; stipules usually distinct, occasionally connate basally (distal portion of stem), triangular-subulate, entire or divided, 0.8–1.5 mm, glabrous; petiole 1–3 mm, glabrous; blade oblong, linear-oblong, or linear-lanceolate, 5–16 × 2–4 mm, base slightly asymmetric, obtuse or subcordate, margins entire, apex obtuse, often mucronulate, surfaces uniformly green or reddish tinged, glabrous; venation obscure. **Cyathia** solitary or in small, cymose clusters at distal nodes; peduncle 0.5–5 mm. **Involucre** obconic-campanulate, 1.2–1.7 × 1–1.4 mm, glabrous; glands 4, occasionally rudimentary, green-yellow to tan or orange-tinged, occasionally shortly stipitate, usually broadly oval to subcircular, sometimes figure eight-shaped, shallowly cupped, 0.1–0.3 × 0.2–0.4 mm; appendages absent or rudimentary. **Staminate flowers** 5–14. **Pistillate flowers:** ovary glabrous; styles 0.7–1 mm, 2-fid ½ length. **Capsules** broadly ovoid, 3–3.5(–4) × (2–)2.4–3 mm, glabrous; columella 2–3 mm. **Seeds** ashy white, wedge-shaped to slightly ovoid, weakly dorsiventrally compressed and elliptic-terete to bluntly subangled in cross section, abaxial side strongly rounded, adaxial side slightly rounded, (2–)2.2–2.8 × 1.6–1.9 mm, smooth or minutely pitted, with smooth brown line from top to bottom on adaxial side.

Flowering and fruiting early summer–fall. Sandy maritime and freshwater beaches and foredunes; 0–200 m; N.B., N.S., Ont., P.E.I., Que.; Conn., Del., Fla., Ga., Ill., Ind., Maine, Md., Mass., Mich., N.H., N.J., N.Y., N.C., Ohio, Pa., R.I., S.C., Va., Wis.

Euphorbia polygonifolia is native to coastal beaches and dunes along the Atlantic Ocean from the maritime provinces of Canada south to northern Florida. The species also occurs disjunctly along the shores of the North American Great Lakes. *Euphorbia polygonifolia* was native to Quebec but is now considered extirpated from the province. It has apparently been introduced in Europe, but it is unclear if it has persisted there (L. C. Wheeler 1941). Immature individuals of this species can be somewhat difficult to distinguish from *E. bombensis* where their ranges overlap (Virginia to northern Florida). Where they occur together, *E. polygonifolia* tends to be a pioneer species on the upper beach and foredune front, whereas *E. bombensis* tends to inhabit areas behind the foredune (R. D. Porcher and D. A. Rayner 2002). *Euphorbia polygonifolia* can be distinguished also by its larger capsules and larger, wedge-shaped to slightly ovoid seeds.

74. Euphorbia porteriana (Small) Oudejans, Phytologia 67: 48. 1989 • Porter's sandmat [C] [E]

Chamaesyce porteriana Small, Fl. S.E. U.S. 711, 1333. 1903; *C. porteriana* var. *scoparia* (Small) D. G. Burch; *C. scoparia* Small

Herbs, usually annual, sometimes perennial, with slender to occasionally thickened taproot, 5 mm diam. **Stems** erect to ascending, sometimes slightly woody at base, not mat-forming, 15–60 cm, usually glabrous, young branches rarely very sparsely short pubescent. **Leaves** opposite; stipules usually distinct, occasionally connate basally, triangular, entire or 2–4 parted, apically ciliate, 0.3–0.4 mm, usually glabrous, rarely very sparsely short pubescent; petiole 0.5–1.5 mm, usually glabrous, rarely very sparsely short pubescent; blade ovate, oval, elliptic, oblong-elliptic, or linear-elliptic, 4–12 × 1–7 mm, base asymmetric, rounded or subcordate, margins usually entire, rarely sparsely serrate, apex obtuse to acute, sometimes mucronate, surfaces abaxially often purple or red, adaxially green, usually glabrous, rarely very sparsely short pubescent; 3-veined from base, only midvein conspicuous. **Cyathia** solitary at distal nodes; peduncle 0.5–1.1 mm. **Involucre** turbinate to campanulate, 0.7–1 × 0.8–1 mm, usually glabrous, rarely very sparsely short pubescent; glands 4, brown, slightly concave, elliptic-oblong, 0.1–0.2 × 0.3–0.6 mm; appendages white or pink to dark red, usually oblong or flabellate, occasionally rudimentary and forming narrow rim at edge of gland, (0–)0.1–0.2 × 0.4–1 mm, distal margin entire or crenulate. **Staminate flowers** 8–14. **Pistillate flowers:** ovary glabrous, subtended by triangular calyxlike structure; styles 0.4–0.6 mm, 2-fid ½ length. **Capsules** broadly ovoid, 1–1.5 × 1.7–2.1 mm, glabrous; columella 1–1.4 mm. **Seeds** reddish brown, ovoid, 4-angled in cross section, 0.7–1 × 0.6–0.7 mm, smooth or obscurely wrinkled.

Flowering and fruiting year-round. Pine rocklands, coastal scrub, open hammocks; of conservation concern; 0–10 m; Fla.

Euphorbia porteriana is found in Miami-Dade County and on Big Pine Key, Monroe County. This species is closely related to *E. blodgettii*, *E. garberi*, and *E. serpens*. The capsules of *E. porteriana* are almost always completely glabrous, but the authors have seen three specimens with just a few scattered hairs on the keels. A. Herndon (1993) synonymized *E. porteriana* var. *keyensis* under *E. garberi* because of its uniformly hairy capsules, and that treatment is followed here.

75. Euphorbia prostrata Aiton, Hort. Kew. 2: 139. 1789 • Prostrate spurge or sandmat [W]

Chamaesyce prostrata (Aiton) Small

Herbs, annual, with taproot. **Stems** prostrate to decumbent, usually not mat-forming, 10–30 cm, crisped-villous to glabrate (proximally). **Leaves** opposite; stipules connate (lower side) or distinct (upper side), triangular-subulate, often lacerate distally, 0.5–1 mm, short crisped-villous to glabrate; petiole 0.5–1.5 mm, usually short crisped-villous, sometimes woolly at distal nodes; blade broadly elliptic to elliptic-oblong, ovate-spatulate, or ovate, 3–11(–15) × 3–6(–8) mm, base slightly asymmetric, rounded to slightly cordate and oblique, margins serrulate at least in distal ½, sometimes obscurely so, apex obtuse, abaxial surface finely crisped-villous, adaxial surface usually glabrous or glabrate, sometimes sparsely crisped-villous; 3-veined from base. **Cyathia** solitary or in small, cymose clusters at distal nodes or on congested, axillary branches; peduncle 1–2 mm. **Involucre** obconic, 0.6–0.9 × 0.5 mm, crisped-villous or glabrous; glands 4, reddish, oval to oblong, 0.1 × 0.1–0.2 mm; appendages white to pink, rudimentary, 0–0.2 mm, distal margin entire or irregularly scalloped. **Staminate flowers** 3–6. **Pistillate flowers:** ovary densely crisped-villous; styles 0.1 mm, 2-fid nearly entire length. **Capsules** broadly ovoid, 1.2–2 × 1.4–1.5 mm, crisped-villous along keels and toward base, often glabrous between keels; columella 1–1.2 mm. **Seeds** white but with barely concealed brown surface beneath, ovoid, sharply 4-angled in cross section, abaxial faces plane to convex, adaxial faces concave, 0.8–1.1 × 0.5–0.7 mm, with several narrow, sharp, slightly irregular, transverse ridges. $2n = 18$.

Flowering and fruiting early spring–fall. Disturbed areas, fields, gardens, sidewalks, sandy places, ballast piles; 0–1400 m; Ala., Ariz., Ark., Calif., Colo., Fla., Ga., Ill., Ind., Iowa, Kans., Ky., La., Md., Mass., Mich., Miss., Mo., Nebr., N.Mex., N.C., Ohio, Okla., Pa., S.C., S.Dak., Tenn., Tex., Utah, Va., W.Va., Wyo.; Mexico; West Indies; Central America; South America; introduced in Eurasia, Africa, Pacific Islands, Australia.

Euphorbia prostrata is native to tropical America and possibly into the southern part of the flora area. It is likely adventive throughout most of the northern part of its range. It is widely naturalized throughout much of the rest of the tropics.

76. Euphorbia rayturneri V. W. Steinmann & Jercinovic, Novon 22: 482, figs. 1, 2. 2013 [C] [E]

Herbs, annual, with slender taproot. **Stems** prostrate, 4–8 cm, uniformly strigillose. **Leaves** opposite; stipules distinct, filiform, 0.6–0.9 mm, pilose; petiole 0.5–0.9 mm, strigillose; blade ovate to elliptic, often slightly falcate, 5–11 × 2–5 mm, base asymmetric, one side cordate, other round to attenuate, margins sharply serrulate, apex acute, surfaces often with red spot toward middle, abaxial surface strigillose, adaxial surface glabrescent; only midvein conspicuous. **Cyathia** solitary at distal nodes; peduncle 0.5–1.6 mm. **Involucre** obconic, 0.9–1.2 × 0.8–1.2 mm, strigillose; glands 4, green, yellow, or light pink, circular to oblong, 0.2 × 0.2–0.3 mm; appendages absent or green, yellow, or light pink, forming narrow margin on distal portion of gland, 0–0.1 × 0.2–0.3 mm, distal margin entire. **Staminate flowers** 5–8. **Pistillate flowers:** ovary strigillose-canescent; styles 0.3–0.4 mm, unbranched. **Capsules** oblate, 1.7–2 × 2.2–2.7 mm, strigillose; columella 1.5–1.9 mm. **Seeds** blackish brown, broadly ovoid, 3-angled in cross section, 1.2–1.4 × 1–1.1 mm, with 2 well-defined transverse ridges that do not pass through abaxial keel. $2n$ = 14.

Flowering and fruiting late summer–early fall. Desert grasslands; of conservation concern; 1400–1700 m; N.Mex.

Euphorbia rayturneri is known from only three collections in extreme southwestern New Mexico. Given its close proximity to the Mexican border, the species may also occur in the adjacent states of Chihuahua or Sonora.

77. Euphorbia revoluta Engelmann in W. H. Emory, Rep. U.S. Mex. Bound. 2(1): 186. 1859 • Threadstem spurge

Chamaesyce revoluta (Engelmann) Small

Herbs, annual, with slender taproot. **Stems** erect, 5–25 cm, glabrous. **Leaves** opposite; stipules distinct, subulate-filiform, entire, 0.5–0.9 mm, glabrous; petiole 0.6–1.1 mm, glabrous; blade linear, 6–27 × 0.6–1.2 mm, base symmetric, attenuate, margins entire, revolute, apex acute, sometimes mucronate, surfaces glabrous; only midvein conspicuous. **Cyathia** solitary at distal nodes; peduncles absent or to 1.5 mm. **Involucre** obconic to campanulate, 0.7–0.9 × 0.5–0.7 mm, glabrous; glands 4, pink to dark purple, nearly circular to oblong or reniform, 0.1 × 0.1–0.2 mm; appendages white, oblong, nearly circular, ovate, deltate, or forming thin margin around gland, rarely absent, (0–)0.1–0.2 × (0–)0.1–0.3 mm, distal margin entire. **Staminate flowers** 5–10. **Pistillate flowers:** ovary glabrous; styles 0.2–0.3 mm, unbranched. **Capsules** ovoid to subglobose, 1.5–1.8 × 1.6–1.8 mm, glabrous; columella 1.1–1.5 mm. **Seeds** whitish, brick red, light gray, or light brown, narrowly to broadly ovoid, 4-angled in cross section, 0.9–1.4 × 0.7–1 mm, nearly smooth, rugulose, with faint transverse ridges, or with 2–3 well-defined transverse ridges separated by shallow depressions.

Flowering and fruiting summer–late fall. Desert scrub, sagebrush scrub, juniper woodlands, Joshua tree-pinyon pine woodlands, oak woodlands, grasslands, chaparral, pine-oak forests; 600–2500 m; Ariz., Calif., Colo., Nev., N.Mex., Tex., Utah; Mexico (Baja California, Chihuahua, Coahuila, Sonora, Zacatecas).

Euphorbia revoluta is a distinctive, easily recognizable species by virtue of the combination of unbranched styles and linear leaves with symmetric bases. The species ranges through northern Mexico into the southwestern United States and is composed of three well-marked, geographically distinct races that can be distinguished by their seeds. The first of these races occurs primarily in pine-oak forest of the Sierra Madre Occidental of Mexico (Chihuahua and Sonora) and barely enters the flora area in the Huachuca Mountains of southeastern Arizona; the seeds are brick red and nearly smooth. The second race is widespread in the Chihuahuan Desert from northern Mexico to southeastern New Mexico and southwestern Texas; its seeds are whitish to light gray and possess two or three well-defined transverse ridges separated by shallow depressions. The third race corresponds to the type collection and is characterized by grayish white to light brown seeds that are rugulose or with faint transverse ridges; it occurs in northwestern Mexico (Baja California and Sonora) and throughout the arid southwestern United States. It is probable that further study will justify the taxonomic segregation of these races as either distinct species or subspecies.

78. Euphorbia serpens Kunth in A. von Humboldt et al., Nov. Gen. Sp. 2(fol.): 41; 2(qto.): 52. 1817 • Creeping or round-leafed spurge, matted sandmat [W]

Chamaesyce serpens (Kunth) Small

Herbs, annual, with slender taproot. **Stems** prostrate, frequently mat-forming and rooting at nodes, 15–50 cm, glabrous. **Leaves** opposite; stipules connate into conspicuous, deltate scale, white to pink, membranaceous, 0.5–1(–1.2) mm, glabrous; petiole less than 1 mm, glabrous; blade ovate, oblong, or orbiculate, 2–7(–9) × 2–6 mm, base asymmetric, rounded to subcordate, margins entire, apex rounded, surfaces without red blotch, glabrous; usually only midvein conspicuous. **Cyathia** solitary at distal nodes; peduncle 0.5–1(–2.5) mm. **Involucre** campanulate to turbinate, 0.3–0.7(–1) × 0.4–0.6 mm, glabrous; glands 4, yellow, oblong, 0.1 × 0.2 mm; appendages white to pinkish, equal, forming narrow rim at edge of gland, 0.1–0.2 × 0.2–0.3 mm, distal margin entire or crenulate. **Staminate flowers** 5–10. **Pistillate flowers:** ovary glabrous; styles 0.2 mm, 2-fid ½ to nearly entire length. **Capsules** broadly ovoid, oblate, or subglobose, 1.3–1.4 × 1.3–1.7 mm, glabrous; columella (0.9–)1–1.2 mm. **Seeds** white to gray or light pink, ovoid, bluntly 3–4-angled in cross section, 0.7–1.1 × 0.4–0.7 mm, smooth.

Flowering and fruiting year-round (in warmer areas) or summer (in temperate regions). Mostly sandy or well-drained soils, desert scrub, coastal scrub, chaparral, oak and juniper woodlands, sand dunes, riparian forests, mesquite grasslands, prairies, coniferous and deciduous hardwood forests, disturbed areas; 0–2000 m; Ont., Sask.; Ala., Ariz., Ark., Calif., Colo., Fla., Ga., Ill., Ind., Iowa, Kans., Ky., La., Mich., Miss., Mo., Mont., Nebr., N.J., N.Mex., N.Dak., Ohio, Okla., Pa., S.Dak., Tenn., Tex., Utah, Vt., Wyo.; Mexico; West Indies; Bermuda; Central America; South America; introduced in Europe, Asia, Africa, Australia.

Euphorbia serpens is one of the most widespread species of the genus in the New World. While it may be indigenous to a portion of the flora area, probably in the warmer, southern part of its range, it is weedy and has likely been introduced in many parts of the flora area, such as Canada and the eastern United States. It is also widely distributed in the Old World, where it is certainly introduced. The strictly prostrate habit with stems rooting at the nodes is characteristic. *Euphorbia serpens* is often confused with *E. albomarginata*, a species distributed in the southwestern United States; in addition to the features mentioned in the key, an easy and reliable way to distinguish between them is by the size of the involucral gland appendages: those of *E. albomarginata* are conspicuous to the naked eye, whereas those of *E. serpens* are inconspicuous.

79. Euphorbia serpillifolia Persoon, Syn. Pl. 2: 14. 1806 • Thymeleaf sandmat

Chamaesyce serpillifolia (Persoon) Small

Herbs, annual, with slender taproot. **Stems** prostrate to ascending, often mat-forming, 7–35 cm, glabrous, pilose, or villous. **Leaves** opposite; stipules distinct, divided nearly to base into 3–5 subulate to filiform segments, these sometimes 2-fid toward apex or laciniate, 0.7–2.1 mm, glabrous; petiole 0.5–1 mm, glabrous, villous, or pilose; blade ovate, oblong, elliptic, or obovate, 3–13 × 2–7 mm, base asymmetric, rounded to oblique, margins usually entire in proximal ½ and serrulate in distal ½, rarely serrulate nearly to base, apex obtuse or truncate, surfaces often with red spot in center, not papillate, glabrous, villous, or pilose; weakly 3-veined from base, usually only midvein conspicuous. **Cyathia** solitary or in small, cymose clusters at distal nodes or on congested, axillary branches; peduncle 0.4–1.7 mm. **Involucre** obconic, 0.6–1.1 × 0.6–1 mm, glabrous, villous, or pilose; glands 4, yellow to pink, usually oblong to reniform, 0.1 × 0.2–0.3 mm; appendages white to pink, oblong or flabellate, rarely absent, (0–)0.1–0.2 × (0–)0.3–0.4 mm, distal margin entire or shallowly lobed. **Staminate flowers** 5–20. **Pistillate flowers:** ovary glabrous, villous, or pilose; styles 0.3–0.4 mm, 2-fid ½ length. **Capsules** broadly ovoid to oblate, cocci not elongated nor terminating in empty portion, 1.4–1.9 × 1.5–2 mm, glabrous, pilose, or villous; columella 1.2–1.6 mm. **Seeds** pink, light brown, or grayish, ovoid to narrowly ovoid, 4-angled in cross section, 1–1.4 × 0.6–0.9 mm, smooth to dimpled or rugose, or with faint transverse ridges that do not interrupt abaxial keel.

Subspecies 2 (2 in the flora): North America, Mexico, South America.

Euphorbia serpillifolia is variable, especially in regard to seed sculpturing. L. C. Wheeler (1941) documented and discussed this variation, suggesting that within the United States and Canada the taxon could be further divided into various taxa. However, Wheeler refrained from actually proposing names and commented that further study was needed. The authors concur with Wheeler and maintain a broad delimitation of the species, pending a detailed study of variation throughout its range. Otherwise indistinguishable hairy plants are treated as subsp. *hirtula*. In contrast to the widespread typical subspecies, subsp. *hirtula* ranges from northern Baja California, Mexico, to central California. In this region the two subspecies sometimes grow together.

The spelling of the specific epithet follows the original publication of the name and contrasts with the often-used variant *serpyllifolia*.

1. Stems and leaves glabrous . . . . 79a. *Euphorbia serpillifolia* subsp. *serpillifolia*
1. Stems and leaves pilose or villous . . . . 79b. *Euphorbia serpillifolia* subsp. *hirtula*

79a. Euphorbia serpillifolia Persoon subsp. **serpillifolia**

Stems glabrous. **Leaves:** petiole and blade glabrous. **Involucre** glabrous. **Pistillate flowers:** ovary glabrous. **Capsules** glabrous.

Flowering and fruiting year-round in response to sufficient moisture (but mostly summer–fall). Pine forests, cottonwood-willow riparian forests, temperate deciduous forests, chaparral, grasslands, Joshua tree woodlands, desert scrub, juniper-sagebrush scrub, disturbed areas; 0–2600 m; Alta., B.C., Man., N.B., Ont., Que., Sask.; Ariz., Calif., Colo., Fla., Ga., Idaho, Ill., Iowa, Mich., Minn., Mo., Mont., Nebr., Nev., N.H., N.Mex., N.Y., N.Dak., Okla., Oreg., Pa., S.Dak., Tex., Utah, Wash., Wis., Wyo.; Mexico; South America.

Subspecies *serpillifolia* is likely native to the central-western United States and Canada, and it may be disjunct in northern Argentina. It appears to be adventive in eastern Canada and possibly other northern and eastern parts of its range. A report of the species from Cameron Parish, Louisiana, was based on a misidentified specimen of *Euphorbia glyptosperma*.

79b. Euphorbia serpillifolia Persoon subsp. **hirtula** (Engelmann ex S. Watson) Oudejans, Phytologia 67: 48. 1989 (as serpyllifolia)

Euphorbia hirtula Engelmann ex S. Watson in W. H. Brewer et al., Bot. California 2: 74. 1880; *Chamaesyce hirtula* (Engelmann ex S. Watson) Millspaugh; *C. serpillifolia* (Persoon) Small subsp. *hirtula* (Engelmann ex S. Watson) Koutnik; *E. serpillifolia* var. *hirtula* (Engelmann ex S. Watson) L. C. Wheeler

Stems pilose or villous. **Leaves:** petiole and blade pilose or villous. **Involucre** pilose or villous. **Pistillate flowers:** ovary pilose, villous, or glabrous. **Capsules** pilose, villous, or glabrous.

Flowering and fruiting summer–fall. Chaparral, oak woodlands, oak-pine forests, disturbed roadsides; 700–1900 m; Calif.; Mexico (Baja California).

Subspecies *hirtula* has a scattered and disjunct distribution primarily in montane areas of central and southern California and northern Baja California. In the flora area, it is found in the central Sierra Nevada and the Santa Lucia, San Bernardino, San Jacinto, Cuyamaca, and Laguna mountains.

80. Euphorbia serrula Engelmann in W. H. Emory, Rep. U.S. Mex. Bound. 2(1): 188. 1859 • Sawtooth sandmat

Chamaesyce serrula (Engelmann) Wooton & Standley

Herbs, annual, with slender taproot. **Stems** prostrate or ascending, 5–20 cm, usually pilose to villous, rarely glabrate. **Leaves** opposite; stipules usually distinct, rarely connate at base, triangular or laciniate into subulate segments, 1–1.8 mm, glabrous; petiole 0.3–0.8 mm, glabrous or villous; blade oblong, ovate, or elliptic, sometimes falcate, 3–11 × 2–5 mm, base asymmetric, rounded to hemicordate, margins sharply serrate to serrulate, usually with conspicuous teeth at base of leaf, apex usually obtuse, rarely acute, surfaces frequently with red blotch in center, not papillate, sparsely pilose to glabrate; only midvein conspicuous. **Cyathia** solitary at distal nodes; peduncle 0.4–1.8(–2.3) mm. **Involucre** obconic, 0.8–1.1 × 0.8–1 mm, glabrous; glands 4, greenish yellow, usually reniform to elliptic, rarely circular, 0.1 × 0.1–0.2 mm; appendages usually white, rarely light pink, orbiculate, 0.1–0.3 × 0.2–0.4 mm, distal margin entire or crenulate. **Staminate flowers** 7–15. **Pistillate flowers:** ovary glabrous; styles 0.3–0.5 mm, 2-fid ½ length. **Capsules** oblate, cocci not elongated nor terminating in empty portion, 2–2.6 × 3.2–3.7 mm, glabrous; columella 1.7–2.1 mm. **Seeds** white to light brown, broadly ellipsoid to ovoid, 3–4-angled in cross section, 1.5–1.8 × 1.1–1.3(–1.5) mm, smooth to minutely rugulose or with scattered small depressions.

Flowering and fruiting spring–fall. Desert scrub with creosote bush, grasslands with mesquite and yucca, rarely in ponderosa pine woodlands, often sandy substrates; 300–1900 m; Ariz., N.Mex., Tex.; Mexico.

In Mexico, *Euphorbia serrula* is found from Chihuahua and Coahuila south to Puebla.

81. Euphorbia setiloba Engelmann in War Department [U.S.], Pacif. Railr. Rep. 5(2): 364. 1857 • Fringed or shaggy spurge, Yuma sandmat [F]

Chamaesyce setiloba (Engelmann) Millspaugh

Herbs, annual, with slender taproot. **Stems** prostrate, mat-forming, 5–50 cm, villous with glistening glandular hairs. **Leaves** opposite; stipules distinct, filiform, rudimentary to 0.2 mm, glabrous or sparsely villous with glistening glandular hairs; petiole 0.5–1.5 mm, villous; blade oblong, ovate, or elliptic, 3–7

× 2–4 mm, base asymmetric, rounded, margins entire, apex obtuse, surfaces villous; weakly 3-veined from base, commonly only midvein conspicuous. **Cyathia** solitary at distal nodes, nodes often congested toward tips of branches; peduncle 0.2–1.6 mm. **Involucre** campanulate or urceolate, 0.7–1 × 0.5–0.8 mm, villous; glands 4, red to pink, oblong to slightly reniform, 0.1–0.2 × 0.2–0.3 mm; appendages white to pink, deeply incised into 3–6 triangular to subulate, attenuate, acute segments, 0.3–0.6 × 0.6–1 mm, segments entire. **Staminate flowers** 3–7. **Pistillate flowers:** ovary villous; styles 0.3–0.4 mm, 2-fid ½ length. **Capsules** subglobose to ovoid, 1–1.2 mm diam., villous; columella 0.9–1.1 mm. **Seeds** pink to light gray, narrowly ovoid, 4-angled in cross section, 0.8–1 × 0.5–0.6 mm, dimpled or with faint transverse ridges that do not pass through abaxial keel.

Flowering nearly year-round in response to sufficient moisture. Desert scrub, blackbrush scrub, Joshua tree woodlands, grasslands, often in sandy areas; 20–1600 m; Ariz., Calif., Nev., N.Mex., Tex., Utah; Mexico (Baja California, Baja California Sur, Chihuahua, Durango, Sinaloa, Sonora).

82. Euphorbia simulans (L. C. Wheeler) Warnock & M. C. Johnston, SouthW. Naturalist 5: 170. 1960 • Similar spurge, mimicking sandmat

Euphorbia polycarpa Bentham var. *simulans* L. C. Wheeler, Rhodora 43: 192. 1941; *Chamaesyce simulans* (L. C. Wheeler) Mayfield

Herbs, annual or short-lived perennial, with usually slender, occasionally slightly thickened, rootstock. **Stems** prostrate to reclining, 5–40 cm, glabrous. **Leaves** opposite; stipules distinct, subulate, 0.5–0.7 mm, glabrous; petiole 0.7–1.3 mm, glabrous; blade orbiculate, oval, to shortly oblong, 1–3.2 × 1.5–5 mm, base subsymmetric, rounded, margins entire, apex usually rounded, occasionally emarginate, surfaces glabrous; venation obscure, only midvein conspicuous. **Cyathia** solitary at distal nodes; peduncle 0.3–0.7 mm. **Involucre** turbinate to campanulate, 0.8–1.2 × 0.7–1 mm, glabrous; glands 4, red to purple, slightly concave, elliptic, 0.2–0.3 × 0.4–0.5 mm; appendages absent. **Staminate flowers** 15–36. **Pistillate flowers:** ovary glabrous; styles 0.2–0.3 mm, 2-fid nearly entire length. **Capsules** broadly ovoid, 1.3–1.8 × 1.5–2 mm, glabrous; columella 1.1–1.4 mm. **Seeds** whitish, reddish brown beneath coat, oblong, 4-angled in cross section, 1.3–1.5 × 0.6–0.7 mm, with 5–7 faint transverse ridges or wrinkles. $2n = 28$.

Flowering and fruiting year-round. Desert scrub, mountains, hills, canyons, arroyos, flats, roadsides, clay, sandy, gravelly, and rocky soils; 600–1300 m; Tex.; Mexico (Chihuahua, Coahuila).

Euphorbia simulans, which in the flora area is known only from Brewster, Hudspeth, and Presidio counties, is difficult to distinguish in the field from the sympatric *E. theriaca* var. *theriaca*, because they are mainly distinguished by seed morphology. The latter has smaller seeds with (two or) three (or four) prominent transverse ridges, whereas *E. simulans* has larger seeds that are slightly wrinkled.

83. Euphorbia stictospora Engelmann in W. H. Emory, Rep. U.S. Mex. Bound. 2(1): 187. 1859 • Mat or narrow-seeded spurge, slimseed sandmat

Chamaesyce stictospora (Engelmann) Small

Herbs, annual, with taproot. **Stems** prostrate, often mat-forming, occasionally with ascending tips, 5–45 cm, densely and evenly pilose to lanate. **Leaves** opposite; stipules distinct or connate basally on one side of stem, entire or irregularly toothed or fringed, 0.5–1.2 mm, pilose to lanate; petiole 0.3–1.5 mm, pilose to lanate; blade usually oblong to oblong-obovate, occasionally nearly circular, 3–10(–15) × 2–5(–10) mm, base asymmetric, one side usually angled or rounded and other truncate-auriculate, margins minutely or conspicuously serrulate at least toward apex, apex usually broadly rounded to broadly acute, occasionally emarginate, abaxial surface often ± lighter green and without reddish spot, both surfaces sparsely to moderately pilose to lanate; 3-veined from base or venation obscure. **Cyathia** solitary at leaf nodes or in small, cymose clusters on congested, axillary branches; peduncle 0.7–2.5 mm. **Involucre** obconic, 0.7–1 × 0.4–0.6 mm, moderately to densely pilose to lanate; glands 4, reddish, ± unequal, oblong, 0.1 × 0.1–0.3 mm; appendages white to strongly pinkish or reddish tinged, often unequal, sometimes 1 to all absent, 0–0.3 × 0–0.4 mm, 3-lobed or rudimentarily 1-lobed, distal margin crenate. **Staminate flowers** 3–9. **Pistillate flowers:** ovary pilose to villous, hairs occasionally slightly appressed; styles 0.2–0.5 mm, unbranched or inconspicuously notched at tip. **Capsules** ovoid, 1.6–2.3 × 1.4–1.5 mm, moderately to densely villous with hairs usually slightly appressed, pubescence often concentrated on proximal ½ or along lobes; columella 1.5–2 mm. **Seeds** light to dark brown, usually mottled, sometimes with thin, white coating, often wearing away irregularly, narrowly oblong-ovoid to ellipsoid, 3–4-angled in cross section, 1–1.5 × 0.5–0.6 mm, with short, irregularly interrupted furrows, appearing partially and irregularly few-ridged.

Flowering and fruiting midsummer–early fall. Open disturbed areas, rocky slopes; 100–2100 m; Ariz., Colo., Iowa, Kans., Mo., Nebr., N.Mex., N.Dak., Okla., S.Dak., Tex., Wyo.; Mexico (Chihuahua, Durango, San Luis Potosí).

Euphorbia stictospora has been recorded from New York, but this disjunct occurrence likely represents a waif or misidentification.

84. Euphorbia theriaca L. C. Wheeler, Rhodora 43: 242, plate 660, fig. A. 1941 • Terlingua sandmat

Chamaesyce theriaca (L. C. Wheeler) Shinners

Herbs, annual, with slender taproot. **Stems** prostrate to reclining, not mat-forming, 5–30 cm, glabrous. **Leaves** opposite; stipules usually distinct, occasionally connate basally on lower side of stem, subulate or scalelike, usually entire, occasionally 2-fid or margin sparsely ciliate, 0.4–1 mm, glabrous; petiole 0.7–1.2(–1.5) mm, glabrous; blade ovate, oblong, orbiculate, or obovate, 2–7.1 × 1–3.5 mm, base slightly asymmetric, rounded, margins entire, often revolute on drying, apex usually rounded, occasionally slightly emarginate, surfaces glabrous; venation usually obscure, only midvein conspicuous. **Cyathia** usually solitary at distal nodes, rarely clustered on short, axillary branches; peduncle 0.3–1.3 mm. **Involucre** usually turbinate-campanulate to hemispheric, occasionally suburceolate, 1–1.8 × 0.9–1.4 mm, glabrous; glands 4, yellow-green to red-purple, sessile or short-stipitate, subcircular to slightly elliptic, 0.2–0.5 × 0.2–0.7 mm; appendages absent or white to pink, semilunate or forming rim at edge of gland, (0–)0.1–0.4 × (0–)0.3–0.9 mm, entire or slightly crenate. **Staminate flowers** 15–36. **Pistillate flowers:** ovary glabrous; styles 0.3–0.5 mm, 2-fid ½ length. **Capsules** broadly ovoid, 1.1–1.6 × 1.5–1.8 mm, glabrous; columella 1.2–1.5 mm. **Seeds** whitish, reddish brown beneath coat, ovate, 4-angled in cross section, 0.8–1.2 × 0.5–0.8 mm, with (2–)3(–5) deep transverse ridges. $2n = 28$.

Varieties 2 (2 in the flora): sc United States, n Mexico.

1. Involucral glands sessile, appendages absent . . . . 84a. *Euphorbia theriaca* var. *theriaca*
1. Involucral glands short-stipitate, appendages present, occasionally rudimentary . . . . 84b. *Euphorbia theriaca* var. *spurca*

84a. Euphorbia theriaca L. C. Wheeler var. **theriaca**

Stems 5–30 cm. **Leaf blades** 2–5.5 × 1–3.5 mm. **Cyathia** solitary at distal nodes of stem; peduncle 0.3–0.7 mm. **Involucre** usually turbinate-campanulate, occasionally suburceolate, 1.1–1.5 × 0.9–1.3 mm; glands sessile, subcircular to slightly elliptic, 0.2–0.5 × 0.3–0.7 mm; appendages absent. **Staminate flowers** 15–36. **Capsules** 1.1–1.6 × 1.5–1.8 mm. **Seeds** with (2–)3(–4) prominent transverse ridges.

Flowering and fruiting summer–late fall. Desert scrub, mostly in calcareous soils; 600–1300 m; Tex.; Mexico (Chihuahua, Coahuila).

Variety *theriaca* occurs in Brewster and Presidio counties in trans-Pecos Texas.

84b. Euphorbia theriaca L. C. Wheeler var. **spurca** M. C. Johnston, Wrightia 5: 138. 1975

Chamaesyce spurca (M. C. Johnston) B. L. Turner; *C. theriaca* (L. C. Wheeler) Shinners var. *spurca* (M. C. Johnston) Mayfield

Stems 5–25 cm. **Leaf blades** 3–7.1 × 1–3.5 mm. **Cyathia** usually solitary at distal nodes, rarely clustered on short, axillary branches; peduncle 0.6–1.3 mm. **Involucre** turbinate-campanulate to hemispheric, 1–1.8 × 0.9–1.4 mm; glands short-stipitate, subcircular, 0.2–0.5 × 0.2–0.5 mm; appendages present, occasionally rudimentary. **Staminate flowers** 20–36. **Capsules** 1.4–1.6 × 1.5–1.7 mm. **Seeds** with 3–4(–5) prominent transverse ridges.

Flowering and fruiting late winter–late fall. Desert scrub, limestone or igneous soils; 700–1500 m; N.Mex., Tex.; Mexico (Chihuahua, Coahuila).

Variety *spurca* is partly sympatric with var. *theriaca* in Brewster and Presidio counties but extends farther west into Hudspeth and El Paso counties as well as into adjacent New Mexico.

85. Euphorbia thymifolia Linnaeus, Sp. Pl. 1: 454. 1753 • Gulf sandmat

Chamaesyce thymifolia (Linnaeus) Millspaugh

Herbs, annual or perennial, with taproot. **Stems** prostrate, mat-forming, 15–30 cm, strigose to strigose-tomentulose. **Leaves** opposite; stipules distinct or slightly connate at base, linear-subulate, entire or slightly parted, 0.9–1.2 mm, strigose-tomentulose; petiole 0.5–1 mm, sparsely strigose-tomentulose; blade broadly elliptic to narrowly oblong or ovate-lanceolate, 3–10 × 1.8–5 mm, base asymmetric, one side usually angled or rounded and the other truncate and expanded into small, rounded auricle, margins serrate (larger leaves) to serrulate (smaller leaves), apex blunt to acute, abaxial surface sparsely tomentulose to glabrate, adaxial surface glabrate; palmately veined at base, pinnate distally. **Cyathia** usually in small, cymose clusters on congested, axillary branches; peduncle (0–)0.1–0.3 mm. **Involucre** broadly obconic, becoming distended and distorted by base of partially included capsule, 0.4–0.8 × 0.3–0.5 mm, strigose; glands 4, red, slightly concave, ± unequal, subcircular to broadly oval, 0.1–0.2 × 0.2–0.3 mm; appendages white to pink, usually unequal, occasionally ± equal at distal nodes, elongated toward sinus, sometimes rudimentary, 0.1–0.4 × 0.1–0.3 mm, distal margin entire or crenulate. **Staminate flowers** 3–5. **Pistillate flowers:** ovary densely strigose; styles 0.4–0.5 mm, 2-fid ½ length. **Capsules** conic to truncate-ovoid, scarcely exserted from involucre, base often remaining inside involucre and splitting one side of it during maturation, 0.9–1.2 × 1–1.2 mm, sparsely to moderately strigose; columella 0.6–1 mm. **Seeds** white, tan underneath coat, ovoid to narrowly ovoid, sharply 4-angled in cross section, 0.8–0.9 × 0.4–0.6 mm, with 4 low transverse ridges often slightly extending into angles, not sulcate.

Flowering and fruiting midsummer–early fall. Disturbed areas, often near salt water; 0–20 m; Fla., La.; Mexico; West Indies; South America; introduced in Asia, tropical Africa, Australia.

Euphorbia thymifolia is a widespread tropical and subtropical weed. It is not certain where the species is native, but most likely it originated in the New World and then became widespread in the rest of the tropics. *Euphorbia thymifolia* is present in the flora area in southern Florida and coastal Louisiana, where it is likely adventive. *Euphorbia thymifolia* is generally similar to *E. maculata* but is characterized by its short pistillate pedicels and non-exserted capsules that remain largely enclosed by the involucre and by its unequal involucral gland appendages.

86. Euphorbia trachysperma Engelmann in W. H. Emory, Rep. U.S. Mex. Bound. 2(1): 189. 1859 • San Pedro River sandmat

Chamaesyce trachysperma (Engelmann) Millspaugh

Herbs, annual, with slender taproot. **Stems** erect to ascending, 10–55 cm, glabrous. **Leaves** opposite; stipules distinct, narrowly triangular and often divided into 3–5 subulate to filiform segments, 0.8–1.9 mm, glabrous; petiole 1.2–3.1 mm, glabrous; blade oblong, elliptic, or ovate, 12–43 × 3–14 mm, base subsymmetric to asymmetric, rounded, attenuate, or with one side hemicordate and other side rounded, margins usually serrulate, rarely entire on some leaves or portion of blade, apex obtuse to acute, surfaces often with red blotch in center, glabrous; pinnately veined, often only midvein conspicuous. **Cyathia** solitary at distal nodes; peduncle 1.4–2.4 mm. **Involucre** oblong, campanulate, or funnel-shaped, 1.5–2.6 × 1.1–2.3 mm, glabrous; glands 4, green to yellowish or tinged with red, subcircular to oblong, 0.4–0.7 × 0.5–1 mm; appendages absent or white, often forming narrow rim or oblong to ovate, 0–0.8 × 1.2 mm, distal margin entire. **Staminate flowers** 50–60. **Pistillate flowers:** ovary glabrous; styles 0.5–0.6 mm, 2-fid ½ length. **Capsules** oblate, 2.4–3.2 × 3–3.5 mm, glabrous; columella 2.3–2.5 mm. **Seeds** light brown, ovoid, 4-angled in cross section, abaxial keel well developed, 1.9–2.3 × 1.3–1.4 mm, finely dimpled and papillate.

Flowering and fruiting late summer–early winter. Desert scrub, desert grasslands, mesquite woodlands; 200–1200 m; Ariz.; Mexico (Baja California Sur, Sonora).

Euphorbia trachysperma may be expected in extreme southwestern New Mexico given the close proximity of some collections in southeastern Arizona.

87. Euphorbia vallis-mortae (Millspaugh) J. T. Howell, Madroño 2: 19. 1931 • Death Valley sandmat [E]

Chamaesyce vallis-mortae Millspaugh, Publ. Field Mus. Nat. Hist., Bot. Ser. 2: 403. 1916

Herbs, perennial, with thickened, woody taproot. **Stems** prostrate to ascending, often mat-forming, 10–45 cm, pilose to villous. **Leaves** opposite; stipules distinct or connate, subulate to filiform, 0.4–1.1 mm, densely tomentose; petiole 0.4–1 mm, pilose to villous; blade suborbiculate to oblong-ovate, 3–8 × 2–6 mm, base slightly asymmetric, rounded, margins entire, apex rounded to obtuse, surfaces pilose to villous; 3-veined at base, midvein conspicuous, venation often obscured by pubescence. **Cyathia** solitary at distal nodes; peduncle 0.5–1.8 mm. **Involucre** obconic-campanulate, 1.2–2.3 × 1–1.8 mm, densely pilose to villous; glands 4, yellow to red, subcircular to oblong, 0.2–0.5 × 0.4–0.9 mm; appendages white, flabellate to oblong, 0.1–0.7 × 0.5–1.9 mm, distal margin entire or crenulate, adaxial surface ciliate-puberulent. **Staminate flowers** 15–22(–50). **Pistillate flowers:** ovary densely pilose; styles 0.4–0.8 mm, 2-fid ½ length. **Capsules** ovoid, 1.5–2.2 × 1.8–2.2 mm, tomentose; columella 1.2–1.6 mm. **Seeds** white, gray, or light brown, ovoid, sharply 4-angled in cross section, abaxial faces slightly convex, adaxial faces concave, 1.2–1.7 × 0.6–0.9 mm, smooth.

Flowering and fruiting late spring–fall. Roadsides, desert scrub, streamsides, sandy washes; 700–2000 m; Calif.

The specific epithet of *Euphorbia vallis-mortae* is a misnomer because the species does not occur in Death Valley; instead, it is found at the transition of the northern edge of the Mojave Desert and the foothills of the southern Sierra Nevada in Inyo, Kern, and San Bernardino counties.

88. Euphorbia velleriflora (Klotzsch & Garcke) Boissier in A. P. de Candolle and A. L. P. P. de Candolle, Prodr. 15(2): 40. 1862 • Caliche sandmat [I] [W]

Anisophyllum velleriflorum Klotzsch & Garcke, Abh. Königl. Akad. Wiss. Berlin 1859: 28. 1860; *Chamaesyce velleriflora* (Klotzsch & Garcke) Millspaugh

Herbs, perennial, with strongly thickened and lignified rootstock. **Stems** prostrate, usually mat-forming, terete to slightly flattened and winged, to 30 cm, villous to pilose. **Leaves** opposite; stipules distinct or connate at base, 2-fid to laciniate into 2–5 linear to subulate divisions, 0.5–1(–2) mm, glabrous or villous; petiole 0.5–2 mm, villous; blade usually ovate to oblong, rarely suborbiculate, 5–13 × 4–8 mm, base asymmetric, hemicordate, margins serrulate, apex obtuse, surfaces villous; 3-veined from base, usually only midvein conspicuous. **Cyathia** solitary at distal nodes or at nodes of leafy, congested, axillary branches; peduncle 1–1.5(–2.5) mm. **Involucre** turbinate, 0.9–1.4 × 0.7–1.1 mm, villous; glands 4, pink, oblong to reniform, 0.1–0.2 × 0.3–0.4 mm; appendages white to pink, flabellate, ovate, or oblong, 0.3–0.5 × 0.4–0.7 mm, distal margin entire, crenulate, or erose, often puberulent-ciliate. **Staminate flowers** 8–12. **Pistillate flowers:** ovary villous; styles 0.2–0.4(–0.5) mm, 2-fid ½ length. **Capsules** broadly ovoid, 1.5–1.9 × 1.3–1.7 mm, villous (uniformly so, but most pronounced toward base and along keels); columella 1.2–1.4 mm. **Seeds** white to gray or pink, narrowly pyramidal-ovoid, 4-angled in cross section, 1.1–1.3 × 0.4–0.6 mm, rugulose with shallow depressions separated by inconspicuous transverse ridges.

Flowering and fruiting summer. Disturbed habitats, particularly roadsides, grasslands, live oak-thorn scrub; 50–80 m; introduced; Tex.; Mexico; Central America (Guatemala).

Euphorbia velleriflora is native and widespread from northern Mexico to Guatemala. The species is apparently a recent introduction in southern Texas (W. R. Carr and M. H. Mayfield 1993), as the first records are from the 1990s. It is also expected in southern Arizona due to the presence of collections made a few kilometers south of the border in the Mexican state of Sonora. *Euphorbia velleriflora* is very similar to *E. stictospora*, and the two are sometimes confused. The two can be readily distinguished on the basis of their styles: unbranched in *E. stictospora* and 2-fid in *E. velleriflora*. Also, the involucral gland appendages of *E. velleriflora* are ciliate with short hairs, whereas those of *E. stictospora* are glabrous.

89. Euphorbia vermiculata Rafinesque, Amer. Monthly Mag. & Crit. Rev. 2: 206. 1818 • Wormseed spurge [W]

Chamaesyce rothrockii Millspaugh; *C. vermiculata* (Rafinesque) House; *Euphorbia rothrockii* (Millspaugh) Oudejans

Herbs, annual or short-lived perennial, with slender, fibrous taproot. **Stems** prostrate to ascending or erect, not mat forming, 10–35 cm, usually sparsely to moderately strigillose, pilose, or villous, rarely glabrate, hairs sometimes in longitudinal lines. **Leaves** opposite; stipules distinct or connate, triangular to narrowly triangular or laciniate into subulate to filiform divisions, 0.6–1.3 mm, glabrous or sparsely pilose; petiole 0.2–0.9 mm, glabrous, villous, or strigillose; blade ovate, oblong, or elliptic, often falcate, 5–18 × 3–9 mm, base asymmetric, one side rounded and other cordate, margins usually serrulate especially in distal ½, rarely nearly entire, apex acute or obtuse, surfaces not papillate, sparsely pilose, villous, or sericeous (especially near base), often glabrate (especially older leaves); 3–5-veined from base. **Cyathia** solitary at distal nodes or in small, cymose clusters at branch tips; peduncle 0.2–2.5 mm. **Involucre** obconic, 0.7–1.1 × 0.5–0.8 mm, glabrous; glands (2–3)–4, red to reddish green, circular to oblong, 0.1 × 0.1–0.2 mm; appendages absent or white, turning pink with age, flabellate, oblong, circular, or forming narrow lunate border around margin of gland, 0.1–0.3 × 0.2–0.4 mm, distal margin entire or slightly lobed. **Staminate flowers** 5–15. **Pistillate flowers:** ovary glabrous; styles 0.3–0.5 mm, 2-fid ½ length. **Capsules** oblate to subglobose, cocci not elongated nor terminating in empty portion, 1.4–1.8 × 1.7–2.1 mm, glabrous; columella 1.1–1.5 mm. **Seeds** brown, gray, or almost black, ovoid to oblong, 3–4-angled in cross section, 1.1–1.4 × 0.7–0.8 mm, rugulose and sometimes also with low transverse ridges that do not interrupt abaxial keel.

Flowering and fruiting spring–fall. Juniper-oak woodlands, temperate deciduous forests, grasslands, pine forests, oak forests with sycamores, walnuts and alders, often in disturbed areas; 0–2600 m; N.B., N.S., Ont., Que.; Ariz., Conn., Ill., Ind., Maine, Md., Mass., Mich., Minn., N.H., N.J., N.Mex., N.Y., Ohio, Pa., R.I., Vt., Va., W.Va., Wis.; Mexico.

Euphorbia vermiculata has an interesting disjunct distribution; it ranges from central Mexico to Arizona and New Mexico, and is also present in the northeastern United States and southeastern Canada.

90. Euphorbia villifera Scheele, Linnaea 22: 153. 1849 • Hairy spurge

Chamaesyce stanfieldii Small; *C. villifera* (Scheele) Small; *Euphorbia stanfieldii* (Small) Cory; *E. villifera* var. *nuda* Engelmann ex Boissier

Herbs, annual or perennial, with slender taproot or thickened, woody rootstock. **Stems** usually erect to ascending, rarely prostrate to decumbent, 10–30 cm, papillate, usually villous, sometimes glabrous. **Leaves** opposite; stipules distinct, filiform, usually undivided, rarely divided into 2–3 segments, 0.3–0.7 mm, glabrous, papillate; petiole 0.6–1.8 mm, usually villous, rarely glabrous; blade ovate, 3–12 × 2–10 mm, base asymmetric, rounded to slightly cordate, margins entire or serrulate, apex acute to obtuse, surfaces usually villous, rarely glabrous; only midvein conspicuous. **Cyathia** solitary at distal nodes; peduncles 0–1.8 mm. **Involucre** campanulate, 0.7–0.9 × 0.6–1 mm, glabrous or pilose; glands 4, pink, oval, oblong, or trapezoidal, 0.1–0.2 × 0.2 mm; appendages white to pink, flabellate, oblong, ovate, or nearly rectangular, 0.2–0.4 × 0.2–0.6 mm, distal margin entire. **Staminate flowers** 10–25. **Pistillate flowers:** ovary glabrous; styles 0.3–0.5 mm, 2-fid ½ to nearly entire length. **Capsules** oblate-deltoid, cocci often elongated and terminating in an empty portion, 1.5–2 × 2.1–3.1 mm, glabrous; columella 0.9–1.5 mm. **Seeds** gray-brown to red-brown, ovoid-oblong, weakly 4-angled in cross section, 1–1.4 × 0.6–0.8 mm, smooth, faintly rugose, or with inconspicuous transverse ridges.

Flowering and fruiting early spring–early winter. Riparian forests with walnuts and sycamores, juniper woodlands, pine-oak woodlands, mostly on limestone substrates; 100–1400 m; Tex.; Mexico; Central America.

Although *Euphorbia villifera* has been reported from New Mexico (W. C. Martin and C. R. Hutchins 1980), no vouchers to verify its presence there were located. In Texas, *E. villifera* is known from the Edwards Plateau westward into the trans-Pecos region.

24c. Euphorbia Linnaeus sect. Crepidaria Baillon, Étude Euphorb., 284. 1858

N. Ivalú Cacho

Pedilanthus Necker ex Poiteau

Shrubs [perennial herbs or small trees], with thickened, lignified rootstock. **Stems** semisucculent to succulent [woody], ascending [decumbent], branched [unbranched], zigzag at nodes, terete to flattened, puberulent when young, becoming glabrate [glabrate]. **Leaves** alternate; stipules present, at base of petiole; petiole present, puberulent [glabrous]; blade monomorphic, base ± symmetric, margins entire, surfaces glabrate to sparsely hairy; venation pinnate. **Cyathial arrangement:** terminal or axillary dichasia, branches unbranched or branched at 1 or more successive nodes; bracts on branches and subtending cyathia (dichasial and subcyathial bracts) opposite, distinct, much smaller than distal leaves. **Involucre** strongly zygomorphic, spurred and forming chamber that encloses glands; glands 4 [2 or 6], flat; appendages petaloid, cucullate, partly connate and forming nectar spur that conceals glands. **Staminate flowers** 20–30[–55]. **Pistillate flowers:** ovary glabrous [hairy]; styles connate most of length, 2-fid distally. **Seeds:** caruncle absent.

Species ca. 15 (1 in the flora): Florida, West Indies, Central America, South America.

Section *Crepidaria* is distinguished from other sections of *Euphorbia* by its spurred, zygomorphic involucre that may have evolved as an adaptation to hummingbird pollination. The section was long segregated as the genus *Pedilanthus*, but molecular phylogenetic studies have shown it to be nested within *Euphorbia* subg. *Euphorbia* (B. L. Dorsey et al. 2013). It is native to the New World, with its center of diversity in Mexico (R. L. Dressler 1957; N. I. Cacho et al. 2010). *Euphorbia tithymaloides* is widely cultivated in tropical areas of the world and in greenhouses.

SELECTED REFERENCES Cacho, N. I. et al. 2010. Are spurred cyathia a key innovation? Molecular systematics and trait evolution in the slipper-spurges (Pedilanthus clade – *Euphorbia*, Euphorbiaceae). Amer. J. Bot. 97: 493–510. Dressler, R. L. 1957. The genus *Pedilanthus* (Euphorbiaceae). Contr. Gray Herb. 182: 1–188. Steinmann, V. W. 2003. The submersion of *Pedilanthus* into *Euphorbia* (Euphorbiaceae). Acta Bot. Mex. 65: 44–50.

91. Euphorbia tithymaloides Linnaeus, Sp. Pl. 1: 453. 1753 [F]

Subspecies 8 (1 in the flora): Florida, Mexico, West Indies, Central America, n South America.

91a. Euphorbia tithymaloides Linnaeus subsp. **smallii** (Millspaugh) V. W. Steinmann, Acta Bot. Mex. 65: 50. 2003 • Jacob's ladder, devil's backbone, redbird flower [F]

Pedilanthus smallii Millspaugh, Publ. Field Mus. Nat. Hist., Bot. Ser. 2: 358. 1913; *P. tithymaloides* (Linnaeus) Poiteau subsp. *smallii* (Millspaugh) Dressler; *Tithymalus smallii* (Millspaugh) Small

Shrubs. Stems 0.7–2 m. **Leaves:** stipules dark brown, 0.5–0.6 × 0.5–0.6 mm; petiole 1.5–5 mm; blade usually elliptic- to lance-ovate, 25–70 × 13–32 mm, occasionally bractlike, 8–15 × 3.5–5 mm, base cuneate, apex acuminate, acute, or narrowly obtuse, abaxial surface sparsely pilose-puberulent, adaxial surface glabrate or sparsely puberulent; midvein prominent, keeled or winged adaxially. **Cyathia:** peduncle 6–9 mm; bracts red or pink, oblong-lanceolate to oblong-ovate, 5–7 × 2–3 mm, apex acute or acuminate. **Involucre** red or pink, 8.5–12 mm; glands: medial slightly reniform, subquadrate to oblong, 0.8–1.2 × 0.7–0.8 mm, lateral irregularly ovate or oblong, 1–1.2 × 0.6–1 mm; appendages pink or red. **Pistillate flowers:** styles 5–6 mm. **Seeds** 3.3 × 2.7 mm, faintly keeled dorsally, base rounded, apex apiculate.

Flowering and fruiting winter–early spring. Pinelands, hammocks; 0–10 m; Fla.; West Indies (Cuba).

Subspecies *smallii* is widely cultivated and was likely transported extensively by Spanish explorers very early but appears to be native to coastal areas of southern Florida and Matanzas in northern Cuba. Fieldwork suggests that several if not all Florida populations have apparently been extirpated due to development and human activity on Chokoloskee Island in the Florida Keys.

24d. EUPHORBIA Linnaeus subg. ESULA Persoon, Syn. Pl. 2: 14. 1806

Ricarda Riina Dmitry V. Geltman

Jess A. Peirson Paul E. Berry

Galarhoeus Haworth; *Tithymalus* Gaertner

Herbs or shrubs, annual, biennial, or perennial, with taproot or thickened rootstock. **Stems** woody or herbaceous (succulent in *E. myrsinites*), erect or ascending, branched or unbranched, terete, glabrous or hairy. **Leaves** alternate (opposite in *E. lathyris*); stipules absent; petiole present or absent, glabrous or hairy; blade monomorphic, herbaceous (fleshy in *E. myrsinites*), base symmetric, margins entire or toothed, surfaces glabrous or hairy; venation pinnate, sometimes obscure, midvein often prominent. **Cyathial arrangement:** terminal pleiochasia with (1–)2–17 primary branches; individual pleiochasial branches unbranched or

2–4-branched at 1 or more successive nodes; bracts subtending pleiochasia (pleiochasial bracts) whorled, green, similar in shape and size to distal stem leaves or distinctly different, those on branches and subtending cyathia (dichasial and subcyathial bracts) opposite (alternate in *E. trichotoma*), distinct or connate; additional cymose branches often present in axils of distal leaves, but alternately arranged and without whorled bracts. **Involucre** ± actinomorphic, not spurred; glands 4–5 (2–3 in *E. oblongata*), flat or slightly convex; appendages absent or hornlike, 2, slender with attenuate or rounded tips (thick with rounded and dilated tips in *E. myrsinites*). **Staminate flowers** 5–40. **Pistillate flowers:** ovary glabrous or hairy; styles distinct or partly connate, usually 2-fid, sometimes unbranched. **Seeds:** caruncle present.

Species ca. 480 (34 in the flora): nearly worldwide.

The European natives *Euphorbia paralias* and *E. segetalis* were recorded from Pennsylvania in the early twentieth century. Neither species has been collected more recently, and those early occurrences probably represent waifs that never became established. Other European species here considered waifs in the flora area include *E. amygdaloides*, *E. epithymoides* (sometimes treated as *E. polychroma*), and *E. lucida*. The widespread, introduced leafy spurge is treated here as *E. virgata*, which in North America previously has been mostly misidentified as *E. esula* Linnaeus. *Euphorbia esula* is a related, but more restricted and less weedy, European species that has been recorded historically in the flora area as a waif but is excluded because it is not persistent. Typifications and synonymy of native North American species of subg. *Esula* were published by D. V. Geltman et al. (2011).

SELECTED REFERENCES Geltman, D. V. et al. 2011. Typification and synonymy of the species of *Euphorbia* subgenus *Esula* (Euphorbiaceae) native to the United States and Canada. J. Bot. Res. Inst. Texas 5: 143–151. Norton, J. B. S. 1900. A revision of the American species of *Euphorbia* of the section *Tithymalus* occurring north of Mexico. Rep. (Annual) Missouri Bot. Gard. 11: 85–144. Peirson, J. A. et al. 2014. Phylogeny and taxonomy of the New World leafy spurges, *Euphorbia* section *Tithymalus* (Euphorbiaceae). Bot. J. Linn. Soc. 175: 191–228. Riina, R. et al. 2013. A worldwide molecular phylogeny and classification of the leafy spurges, *Euphorbia* subgenus *Esula* (Euphorbiaceae). Taxon 62: 316–342.

1. Leaves opposite; capsules tardily dehiscent and appearing indehiscent, mesocarp spongy .... 106. *Euphorbia lathyris*
1. Leaves alternate; capsules dehiscent, mesocarp not spongy.
    2. Shrubs .... 100. *Euphorbia dendroides*
    2. Herbs.
        3. Leaves, pleiochasial bracts, and dichasial bracts similar; dichasial bracts alternate; pleiochasial branches 3-branched at nodes; coastal areas of s Florida.... 123. *Euphorbia trichotoma*
        3. Leaves, pleiochasial bracts, and dichasial bracts usually dissimilar; dichasial bracts opposite; pleiochasial branches unbranched or 2–(3–4)-branched at each node; not coastal areas of s Florida.
            4. Cocci verrucose-tuberculate, verrucose, or papillate; involucral gland horns absent.
                5. Perennial herbs with thick rootstock or woody taproot.
                    6. Leaf blade margins finely serrulate; stems often densely villous; involucral glands 2–3; capsules 3–4.5 mm; w United States... 110. *Euphorbia oblongata*
                    6. Leaf blade margins entire; stems glabrous; involucral glands 5; capsules 4.5–5.2 mm; e United States .... 115. *Euphorbia purpurea*
                5. Annual or biennial herbs with taproot.
                    7. Leaf blades sparsely pilose, apices usually acute; terminal pleiochasial branches usually 5, 3-branched; involucre sparsely pilose... 114. *Euphorbia platyphyllos*
                    7. Leaf blades glabrous, apices usually obtuse to rounded; terminal pleiochasial branches usually 3, rarely 5, 2-branched; involucre glabrous.
                        8. Cocci papillate, papillae raised, 0.2–0.5 mm; montane areas of Arizona, New Mexico .... 93. *Euphorbia alta*
                        8. Cocci verrucose, verrucae low and round, 0.1–0.2 mm; widespread, including Arizona, New Mexico .... 119. *Euphorbia spathulata*

[4. Shifted to left margin.—Ed.]

4. Cocci smooth (granulate in *E. cyparissias*, granulate toward abaxial line in *E. agraria* and *E. virgata*, or puncticulate toward abaxial line in *E. exigua*, but then involucral gland horns always present); involucral gland horns present or absent.
   9. Perennial or biennial herbs with rootstocks (taproot in *E. myrsinites*).
      10. Stems succulent; leaf blades fleshy, midvein not prominent; involucral gland horns thick, tips rounded, dilated; capsules 5–7 mm. . . . . 109. *Euphorbia myrsinites*
      10. Stems not succulent; leaf blades not fleshy, midvein prominent; involucral gland horns absent or slender, tips attenuate or rounded; capsules 2–5 mm (4.5–6 mm in *E. serrata*).
         11. Leaf blade and bract margins serrate, serrulate, or irregularly dentate.
            12. Leaf blade margins irregularly serrate; involucres 2–4 mm, gland horns 0–0.6 mm; capsules 4.5–6 mm. . . . . 118. *Euphorbia serrata*
            12. Leaf blade margins finely serrulate; involucres 1.1–2 mm, gland horns 1–2 mm; capsules 2.5–3 mm . . . . 120. *Euphorbia terracina*
         11. Leaf blade and bract margins entire (bract margins occasionally slightly crenate in *E. brachycera* or slightly crenulate in *E. chamaesula*).
            13. Plants with slender, spreading rootstocks; seeds smooth.
               14. Leaf blades oblong-elliptic, 9–20 mm wide, bases truncate or auriculate, apices obtuse to rounded . . . . 92. *Euphorbia agraria*
               14. Leaf blades linear, linear-oblong, linear-ovate, or narrowly oblanceolate, 0.5–12 mm wide, bases attenuate, cuneate, rounded, or truncate, apices acute to rounded (sometimes mucronulate in *E. virgata*).
                  15. Leaf blades 0.5–3 mm wide; stems 10–50 cm; cocci granulate . . . . 99. *Euphorbia cyparissias*
                  15. Leaf blades 3–12 mm wide; stems 20–90 cm; cocci smooth except finely granulate toward abaxial line. . . . . 124. *Euphorbia virgata*
            13. Plants from thick rootstocks; seeds shallowly pitted to almost smooth.
               16. Peduncles 1–3 mm; involucral gland horns usually convergent; capsules 4.3–5 mm . . . . .96. *Euphorbia chamaesula*
               16. Peduncles 0.3–1 mm; involucral gland horns absent or usually divergent; capsules 2–4 mm.
                  17. Involucral gland margins usually entire, occasionally slightly crenate to dentate, horns longer than teeth on gland margins . . . . 95. *Euphorbia brachycera*
                  17. Involucral gland margins irregularly to strongly crenate or dentate, horns absent or equaling to slightly longer than teeth on gland margins.
                     18. Leaf blade apices obtuse to rounded; involucral glands oblong to broadly ovate, 0.5–0.8 × 1–1.6 mm . . . . 108. *Euphorbia lurida*
                     18. Leaf blade apices acute to acuminate; involucral glands semicircular to trapezoidal, 0.8–1.5 × 1–2.2 mm.
                        19. Stems and leaf blades usually glabrous and glaucous . . . . .117. *Euphorbia schizoloba*
                        19. Stems and leaf blades puberulent to lanulose . . . . .125. *Euphorbia yaquiana*
   9. Annual or biennial herbs with taproots.
      20. Leaf blade margins serrulate to crenulate, sometimes only obscurely so; involucral glands without horns; seeds reticulate or areolate.
         21. Terminal pleiochasial branches usually 5; involucres 1.5–2 mm; capsules 2.5–4 mm; seeds 1.6–2.2 mm. . . . . 104. *Euphorbia helioscopia*
         21. Terminal pleiochasial branches 3; involucres 0.5–0.9 mm; capsules 1.6–2 mm; seeds 1.4–1.5 mm. . . . . .122. *Euphorbia texana*
      20. Leaf blade margins entire; involucral glands usually with, occasionally without, horns; seeds smooth, pitted, sulcate, or tuberculate.
         22. Seeds smooth . . . . 105. *Euphorbia helleri*
         22. Seeds pitted, sulcate, or tuberculate.

[23. Shifted to left margin.—Ed.]

23. Seeds sulcate at least on adaxial faces (large-pitted on abaxial faces in *E. peplidion* and *E. peplus*).
    24. Ovaries pilose at base; cocci glabrous or slightly pilose; capsules 2–3 mm; seeds transversely sulcate; involucral glands with or without horns . . . . 102. *Euphorbia falcata*
    24. Ovaries glabrous; cocci glabrous; capsules 1.3–2.3 mm; seeds longitudinally sulcate on adaxial faces, large-pitted on abaxial faces; involucral glands with horns.
        25. Leaf blades linear-oblanceolate to narrowly cuneate-spatulate, 1–4 mm wide; cocci without wings . . . . 112. *Euphorbia peplidion*
        25. Leaf blades obovate, oblong, or suborbiculate, 4–15 mm wide; cocci with low longitudinal wings . . . . 113. *Euphorbia peplus*
23. Seeds pitted or tuberculate.
    26. Seeds tuberculate. . . . . 101. *Euphorbia exigua*
    26. Seeds pitted.
        27. Dichasial bracts connate ¼–½ length (often only on one side or rarely only basally in *E. crenulata*).
            28. Biennial or occasionally annual herbs; dichasial bract margins erose-denticulate to subentire; involucral glands 1.5–2.3 mm wide, horns 0.4–0.6 mm; capsules 2.5–3 mm; seeds 2–2.5 mm; California, sw Colorado, Oregon. . . . . 98. *Euphorbia crenulata*
            28. Annual herbs; dichasial bract margins entire; involucral glands 0.7–1.2 mm wide, horns 0.1–0.4 mm; capsules 2–2.7 mm; seeds 1.6–2 mm; e of Rocky Mountains.
                29. Petioles 0–3 mm; capsules ovoid-globose, 2.6–2.7 mm; seeds with deep, rounded pits in 3–4 regular, vertical rows; c United States but not Texas . . . . 111. *Euphorbia ouachitana*
                29. Petioles 2–5 mm; capsules subglobose, 2–2.5 mm; seeds with scattered, deep and broad pits; Texas . . . . 116. *Euphorbia roemeriana*
        27. Dichasial bracts usually distinct, occasionally subconnate basally.
            30. Dichasial bracts strongly imbricate; seeds gray to purple-gray or sometimes nearly black . . . . 107. *Euphorbia longicruris*
            30. Dichasial bracts not imbricate; seeds white to gray or red-brown to brown.
                31. Biennial or occasionally annual herbs; petioles of proximal leaves 5–10 mm; leaf blades usually oblanceolate to obovate, rarely ovate, 3–10 mm wide (at least some blades more than 5 mm wide); capsules 2.5–3.2 mm . . . . 97. *Euphorbia commutata*
                31. Annual herbs; petioles of proximal leaves 0–2 mm; leaf blades linear to oblanceolate, spatulate-cuneate, or slightly lanceolate, 0.5–5 mm wide; capsules 1.8–2.4 mm.
                    32. Seeds reddish brown to brown, 1.3–1.4 × 0.8–0.9 mm, with 4–6 shallow pits or irregular oblong grooves on adaxial faces, small-pitted or nearly smooth on abaxial faces . . . . 121. *Euphorbia tetrapora*
                    32. Seeds white to gray, 1.4–1.7 × 1–1.6 mm, with pits scattered over entire surface.
                        33. Leaf blades linear to slightly lanceolate or linear-oblanceolate, 0.5–2.5 mm wide; seeds ellipsoid, 1.4–1.7 × 1–1.3 mm; sandy-soiled habitats in Texas. . . . . 94. *Euphorbia austrotexana*
                        33. Leaf blades oblanceolate, 3–5 mm wide; seeds ovoid, 1.6–1.7 × 1.4–1.6 mm; granitic outcrops in Georgia . . . . 103. *Euphorbia georgiana*

92. Euphorbia agraria M. Bieberstein, Fl. Taur.-Caucas. 1: 375. 1808 [I]

Tithymalus agrarius (M. Bieberstein) Klotzsch & Garcke

Herbs, perennial, with slender, spreading rootstock. **Stems** erect or ascending, unbranched or branched, 30–90 cm, glabrous. **Leaves:** petiole absent; blade oblong-elliptic, 20–65 × 9–20 mm, base truncate to auriculate, margins entire, apex obtuse to rounded, surfaces glabrous; venation conspicuously pinnate, midvein prominent. **Cyathial arrangement:** terminal pleiochasial branches 8–15, 1–2 times 2-branched; pleiochasial bracts similar in shape but shorter and narrower than distal leaves; dichasial bracts distinct, rhombic to reniform, base obtuse, margins entire, apex obtuse, mucronate; axillary cymose branches 12–23. **Cyathia:** peduncle 0–2 mm. **Involucre** campanulate, 2.2–3 × 1.8–2 mm, glabrous; glands 4, crescent-shaped, 0.6–1 × 1–2 mm; horns slightly divergent to convergent, 0.1–0.2 mm. **Staminate flowers** 15–20. **Pistillate flowers:** ovary glabrous; styles 1.2–2 mm, 2-fid. **Capsules** globose, 2–2.8 × 2.2–2.7 mm, 3-lobed; cocci rounded, smooth except finely granulate toward abaxial line, glabrous; columella 2.1–2.7 mm. **Seeds** gray or whitish, ovoid-oblong, 2–2.1 × 1.2–1.3 mm, smooth; caruncle ± rounded and flattened, 0.8 × 0.6 mm.

Flowering and fruiting spring–summer. Grasslands, roadside banks, pastures; 200–1600 m; introduced; Alta., Sask.; Kans., Mont., Nebr., N.Y., Pa., Wash., Wyo.; Europe.

93. Euphorbia alta Norton, N. Amer. Euphorbia, 24, plate 24. 1899 • Giant spurge

Tithymalus altus (Norton) Wooton & Standley

Herbs, annual or biennial, with taproot. **Stems** erect, branched, 20–60 cm, glabrous. **Leaves:** petiole 0–1 mm; blade oblong-spatulate, 20–50 × 7–18 mm, base broadly attenuate, margins serrulate, apex rounded to obtuse, surfaces glabrous, ± glaucous; venation pinnate, midvein prominent. **Cyathial arrangement:** terminal pleiochasial branches 3, 2–3 times 2-branched; pleiochasial bracts elliptic-oblanceolate to oblong, similar in size to distal leaves; dichasial bracts distinct, broadly ovate to orbiculate/reniform, base obtuse, margins serrulate, apex obtuse to rounded and often mucronulate; axillary cymose branches 6–20(–25). **Cyathia:** peduncle 0.5–1 mm. **Involucre** narrowly campanulate, 0.8–1.1 × 1.1–1.3 mm, glabrous; glands 4, elliptic, 0.3–0.5 × 0.5–0.7 mm; horns absent. **Staminate flowers** 5–10. **Pistillate flowers:** ovary glabrous; styles 0.5–0.9 mm, 2-fid. **Capsules** depressed-globose, 2–3 × 2.5–3.5 mm, 3-lobed; cocci rounded, papillate, papillae 0.2–0.5 mm, glabrous; columella 1.5–1.9 mm. **Seeds** purple-black, ovoid, 1.6–2 × 1.3–1.7 mm, reticulate and areolate; caruncle reniform, flat, 0.5 × 0.7 mm.

Flowering and fruiting summer. Montane pine-oak and mixed conifer forests, disturbed roadsides, logged areas; 1500–3000 m; Ariz., N.Mex.; Mexico.

Euphorbia alta is a montane species from southern Arizona, New Mexico, and northern and central Mexico that is very similar to and sometimes difficult to distinguish from *E. spathulata*. *Euphorbia alta* tends to be a robust biennial, whereas *E. spathulata* is strictly annual. The most consistent characteristic to separate these two species is that the ovaries and capsules of *E. alta* are distinctly papillate, with the papillae rising sharply above the surface, whereas the ovaries and capsules of *E. spathulata* are merely verrucose, with the protuberances lower and rounded.

94. Euphorbia austrotexana Mayfield, J. Bot. Res. Inst. Texas 7: 634, figs. 1, 2[row 3, left & center]. 2013 [C] [E] [F]

Herbs, annual, with taproot. **Stems** erect, usually branched near base, 6–22 cm, glabrous. **Leaves:** petiole absent; blade linear to slightly lanceolate or linear-oblanceolate, 5–18 × 0.5–2.5 mm, base linear attenuate, margins entire, apex rounded to obtuse or acute, surfaces glabrous; venation pinnate, midvein prominent. **Cyathial arrangement:** terminal pleiochasial branches 3, 1–3 times 2-branched; pleiochasial bracts similar in shape to but slightly shorter and wider than distal leaves; dichasial bracts distinct, not imbricate, reniform-ovate to subdeltate-ovate or broadly ovate-lanceolate, base obliquely truncate to rounded, margins entire, apex obtuse to broadly acuminate; axillary cymose branches 0–3. **Cyathia:** peduncle 0–0.5 mm. **Involucre** infundibular, 0.8–1.1 × 0.6–0.9 mm, glabrous; glands 4, crescent-shaped, 0.2–0.4 × 0.5–0.6 mm; horns divergent, 0.5–0.7 mm. **Staminate flowers** 5–10. **Pistillate flowers:** ovary glabrous; styles 0.3–0.5 mm, 2-fid. **Capsules** ovoid-globose, 1.8–2.2 × 3–3.2 mm, slightly 3-lobed; cocci rounded, smooth, glabrous; columella 1.5–2 mm. **Seeds** white to gray, ellipsoid, 1.4–1.7 × 1–1.3 mm, with deep, irregular to rounded, shallow to concave depressions over entire surface; caruncle reniform-ovate, depressed-conic, 0.5–0.7 × 0.7–1 mm.

Varieties 2 (2 in the flora): Texas.

Euphorbia austrotexana occurs in stabilized sandy soil in the south Texas plains (M. H. Mayfield 2013). It is similar to *E. longicruris* but differs from that species

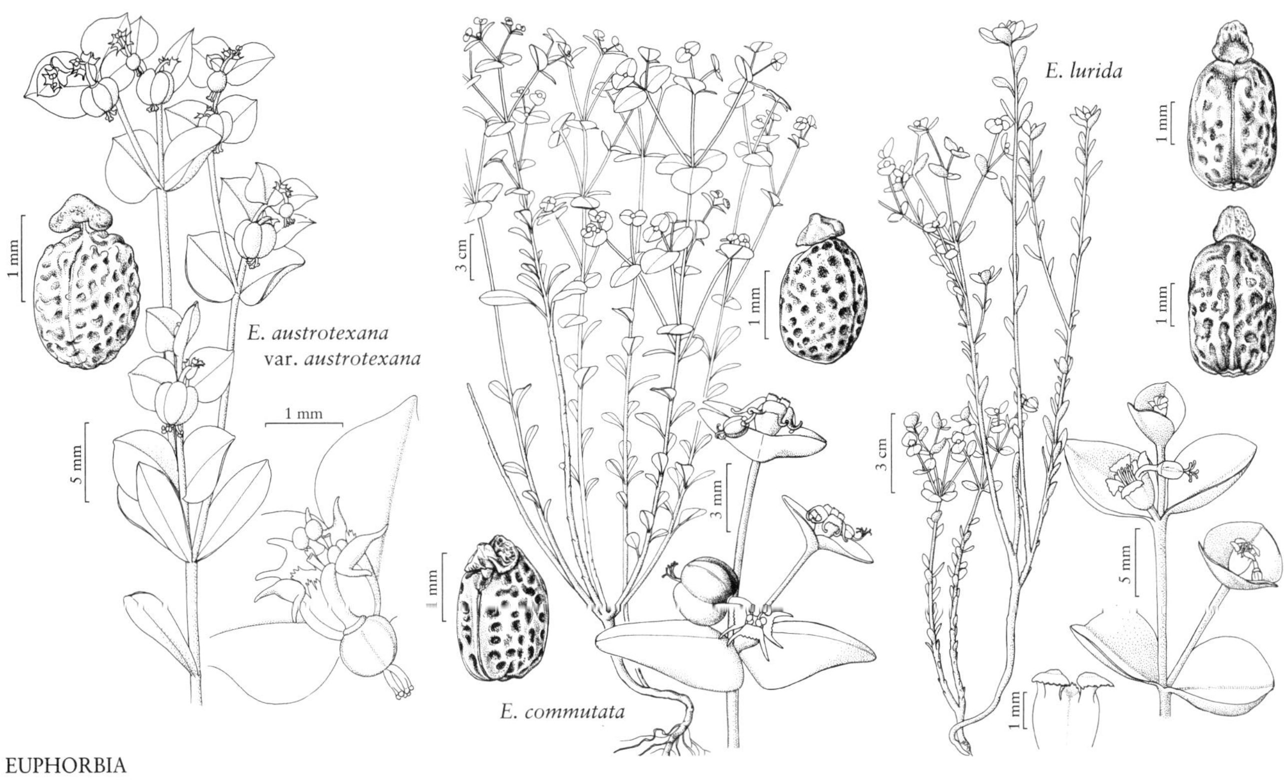

EUPHORBIA

in its often narrowly oblanceolate to linear leaves and its white to gray, ellipsoid seeds that are covered with minute, concave depressions.

1. Stems 10–22 cm; leaf blades linear-oblanceolate, apices rounded; dichasial bracts reniform-ovate to subdeltate-ovate, bases truncate; seeds 1.6–1.7 × 1.1–1.3 mm, surfaces with regular concave depressions; Atascosa, Bexar, Wilson counties .... 94a. *Euphorbia austrotexana* var. *austrotexana*
1. Stems 6–13 cm; leaf blades linear to slightly lanceolate, apices acute; dichasial bracts broadly ovate-lanceolate, bases rounded; seeds 1.4–1.5 × 1–1.1 mm, surfaces with irregular, not obviously concave depressions; Jim Hogg, Kenedy counties .... 94b. *Euphorbia austrotexana* var. *carrii*

94a. Euphorbia austrotexana Mayfield var. **austrotexana** C E F

Stems 10–22 cm. **Leaves:** blade linear-oblanceolate, apex rounded. **Dichasial bracts** reniform-ovate to subdeltate ovate, base truncate. **Seeds** 1.6–1.7 × 1.1–1.3 mm, surface with regular concave depressions. **2*n*** = 26.

Flowering and fruiting spring. Sandy soils; of conservation concern; 200–300 m; Tex.

Variety *austrotexana* is restricted to Atascosa, Bexar, and Wilson counties. According to M. H. Mayfield (2013), only five documented localities are known.

94b. Euphorbia austrotexana Mayfield var. **carrii** Mayfield, J. Bot. Res. Inst. Texas 7: 636, fig. 2[row 3, center]. 2013 C E

Stems 6–13 cm. **Leaves:** blade linear to slightly lanceolate, apex acute. **Dichasial bracts** broadly ovate-lanceolate, base rounded. **Seeds** 1.4–1.5 × 1–1.1 mm, surface with irregular and not obviously concave depressions.

Flowering and fruiting spring. Sandy soils; of conservation concern; 0–200; Tex.

Variety *carrii* is restricted to Jim Hogg and Kenedy counties, Texas; M. H. Mayfield (2013) indicated that populations are confined to the South Texas Sand Sheet and apparently occur in the deepest sand areas of the region. B. L. Turner (2011) mistook collections of this taxon for the introduced, Old World *Euphorbia exigua*. The two species are not closely related and can be readily distinguished based on seed morphology.

95. Euphorbia brachycera Engelmann in W. H. Emory, Rep. U.S. Mex. Bound. 2(1): 192. 1859 • Horned or shorthorn spurge

Euphorbia brachycera var. *robusta* (Engelmann) Dorn; *E. montana* Engelmann [not Rafinesque] var. *robusta* Engelmann; *E. odontadenia* Boissier; *E. robusta* (Engelmann) Small; *Tithymalus brachycerus* (Engelmann) Small; *T. robustus* (Engelmann) Small

Herbs, perennial, with thick rootstock. **Stems** erect or ascending, branched, 10–60 cm, usually glabrous, sometimes puberulent. **Leaves:** petiole 0–0.5 mm; blade oblong-elliptic, lanceolate, or oblanceolate to broadly ovate, 5–25 × 2–7 mm, base truncate, rounded, acute, or attenuate, margins entire, apex acute or obtuse, sometimes mucronate, surfaces usually glabrous, sometimes puberulent; venation inconspicuous, only midvein prominent. **Cyathial arrangement:** terminal pleiochasial branches 3–5(–8), 1–4+ times 2-branched; pleiochasial bracts lanceolate or ovate to broadly ovate, wider than distal leaves; dichasial bracts distinct, broadly ovate or rhombic to triangular-ovate, base obtuse, margins entire or slightly crenate, apex obtuse, mucronate; axillary cymose branches 0–8. **Cyathia:** peduncle 0.4–1 mm. **Involucre** turbinate, campanulate, or cupulate, 1.5–2.6 × 1.4–1.7 mm, usually glabrous, sometimes puberulent; glands 4, crescent-shaped to triangular-ovate, 0.5–0.8 × 0.7–1.7 mm, margins usually entire, occasionally slightly crenate to dentate; horns divergent, 0.2–0.4 mm, longer than teeth on gland margin. **Staminate flowers** 10–15. **Pistillate flower:** ovary usually glabrous, sometimes puberulent; styles 0.5–0.9 mm, 2-fid. **Capsules** depressed-ovoid, 2.8–4 × 3.5–4.5 mm, 3-lobed; cocci rounded, smooth, usually glabrous, sometimes sparsely puberulent; columella 2.4–3.3 mm. **Seeds** light gray, cylindric ovoid-oblong, 2–2.8(–3) × 1.4–2.2 mm, irregularly shallowly pitted; caruncle sessile to shortly stipitate, conic, 0.4–0.5 × 0.8–1.1 mm. **2*n*** = 28.

Flowering and fruiting spring–summer. Montane areas, canyons, rock crevices, sandy or gravelly slopes, pine-oak woodlands, ponderosa pine and mixed coniferous forests; 1200–3200 m; Ariz., Colo., Idaho, Mont., Nebr., N.Mex., N.Dak., S.Dak., Tex., Utah, Wyo.; Mexico (Chihuahua, Coahuila, Sonora).

Euphorbia brachycera is morphologically extremely variable, and plants from the northern part of the species range have sometimes been distinguished as *E. robusta*. *Euphorbia brachycera* is most closely related to the other native, perennial species of subg. *Esula* in western North America, namely *E. chamaesula*, *E. lurida*, *E. schizoloba*, and *E. yaquiana*. The name *E. montana* Engelmann, which has sometimes been applied to *E. brachycera*, is illegitimate because it is a later homonym of *E. montana* Rafinesque.

96. Euphorbia chamaesula Boissier, Cent. Euphorb., 38. 1860 • Mountain spurge

Tithymalus chamaesula (Boissier) Wooton & Standley

Herbs, perennial, with thick rootstock. **Stems** erect, branched, 40–90 cm, glabrous. **Leaves:** petiole 0.5–1 mm; blade elliptic to oblong, 8–20(–40) × 3–6 mm, base truncate, rounded, or attenuate, margins entire, apex obtuse or acute, sometimes slightly mucronate, surfaces glabrous; venation inconspicuous, only midvein prominent. **Cyathial arrangement:** terminal pleiochasial branches 3–5(–6), each 3–4 times 2-branched; pleiochasial bracts ovate-lanceolate to slightly subpandurate, similar in size or wider than distal leaves; dichasial bracts distinct, lanceolate to broadly ovate, base usually truncate to rounded or subcordate, sometimes attenuate, margins entire or slightly crenulate, apex usually obtuse to acute, occasionally acuminate; axillary cymose branches 2–8. **Cyathia:** peduncle 1–3 mm. **Involucre** turbinate or campanulate, 1.5–2.5 × 1.1–1.9 mm, glabrous; glands 4, crescent-shaped to semicircular, 0.5–0.8 × 1–1.8 mm; horns usually convergent, 0.2–0.8 mm. **Staminate flowers** 8–12. **Pistillate flowers:** ovary glabrous; styles 1–1.8 mm, 2-fid. **Capsules** depressed-ovoid, 4.3–5 × 5–6 mm, 3-lobed; cocci rounded, smooth, glabrous; columella 3.5–4 mm. **Seeds** gray to dark brown, ovoid-oblong, truncate at both ends, 2.6–3.4 × 2–2.6 mm, shallowly pitted to almost smooth; caruncle conic, 1 × 0.8 mm. **2*n*** = 26.

Flowering and fruiting spring–summer. Clearings in ponderosa pine forests, montane roadsides, dry streambeds, creek banks, sandy and gravelly soils; 1700–2700 m; Ariz., N.Mex.; Mexico (Chihuahua, Sonora).

Euphorbia chamaesula is easily distinguished from other perennial members of subg. *Esula* in western North America by its larger capsules and the vegetative shoots (without cyathia) that arise from the distal nodes of the stem proximal to the pleiochasia.

97. Euphorbia commutata Engelmann in A. Gray, Manual ed. 2, 389. 1856 • Tinted woodland spurge [E] [F]

Galarhoeus austrinus (Small) Small; *G. commutatus* (Engelmann) Small ex Rydberg; *Tithymalus commutatus* (Engelmann) Klotzsch & Garcke

Herbs, usually biennial, occasionally annual, with taproot. **Stems** erect or ascending, decumbent and often branched near base, 10–40 cm, glabrous. **Leaves:** petiole usually 5–10 mm, 0–1 mm distally; blade usually oblanceolate to obovate, rarely ovate, 5–30 × 3–10 mm, base broadly attenuate, margins entire, apex usually obtuse to rounded, occasionally slightly retuse, surfaces glabrous; venation pinnate, midvein prominent. **Cyathial arrangement:** terminal pleiochasial branches (2–)3(–4), 1–3+ times 2-branched; pleiochasial bracts similar in shape and size to distal leaves; dichasial bracts distinct to basally subconnate, not imbricate, widely ovate, rhombic, or reniform, base cordate, rarely slightly perfoliate, margins entire, apex obtuse to rounded, mucronulate; axillary cymose branches 0–5. **Cyathia:** peduncle 0.5–1 mm. **Involucre** campanulate, 1.7–2.5 × 1.5–2.5 mm, glabrous; glands 4, crescent-shaped, 0.6–1 × 0.8–1.5 mm; horns divergent, 0.5–1.1 mm. **Staminate flowers** 9–15. **Pistillate flowers:** ovary glabrous; styles 0.9–1.3 mm, 2-fid. **Capsules** ovoid-globose, 2.5–3.2 × 3 mm, slightly lobed; cocci rounded to ± flattened, smooth, glabrous; columella 1.5–2 mm. **Seeds** white to gray, broadly oblong-elliptic to ovoid or nearly globose, 1.5–2 × 1.3–1.6 mm, strongly small-pitted; caruncle irregularly winglike, conic, 0.6–1 × 0.7–0.9 mm.

Flowering and fruiting spring–summer. Bottomland and upland forests, bluffs and ledges, stream banks, glades, rarely fen margins; 50–1000 m; Ont.; Ala., Ark., Fla., Ga., Ill., Ind., Iowa, Ky., La., Md., Mich., Miss., Mo., N.C., Ohio, Okla., Pa., S.C., Tenn., Va., W.Va., Wis.

98. Euphorbia crenulata Engelmann in W. H. Emory, Rep. U.S. Mex. Bound. 2(1): 192. 1859 • Chinese caps E

Tithymalus crenulatus (Engelmann) A. Heller

Herbs, usually biennial, occasionally annual, with taproot. **Stems** erect, sometimes decumbent at base, unbranched or branched, 12–40 cm, glabrous. **Leaves:** petiole 0–2 mm; blade obovate-spatulate to oblanceolate, 8–22 × 3–10 mm, base broadly attenuate, margins entire or slightly crisped, apex obtuse to ± rounded, minutely apiculate, surfaces glabrous; venation pinnate, midvein prominent. **Cyathial arrangement:** terminal pleiochasial branches 3, each 2-branched; pleiochasial bracts obovate to orbiculate-reniform, wider than distal leaves; dichasial bracts usually connate 1/3–1/2 length (often only on one side), rarely only connate basally, triangular ovate to reniform, base truncate to perfoliate, margins erose-denticulate to subentire, apex rounded to obtuse, rarely apiculate; axillary cymose branches 0–5. **Cyathia:** peduncle 0–0.5 mm. **Involucre** campanulate, 1.8–2.1 × 1.6–1.8 mm, glabrous; glands 4, crescent-shaped, 0.6–1.2 × 1.5–2.3 mm; horns slightly divergent to slightly convergent, 0.4–0.6 mm. **Staminate flowers** 11–18. **Pistillate flowers:** ovary glabrous; styles 0.9–1.4 mm, 2-fid. **Capsules** subovoid, 2.5–3 × 3.5–4 mm, 3-lobed; cocci rounded, smooth or puncticulate, glabrous; columella 1.9–2.3 mm. **Seeds** cream and brown mottled, oblong-ovoid to nearly globose, 2–2.5 × 1.4–1.7 mm, usually irregularly vermiculate-ridged and large-pitted, occasionally tuberculate or nearly smooth; caruncle reniform, conic, 0.5–0.6 × 0.5–0.7 mm.

Flowering and fruiting spring–summer. Conifer, oak, and mixed forests, coastal scrub, grasslands, barrens and outcrops, roadsides; 30–1800 m; Calif., Colo., N.Mex., Oreg.

Euphorbia crenulata is most common in the central valleys of California and southern Oregon; it occurs disjunctly in southwestern Colorado and northwestern New Mexico. Previous reports from Arizona are based on misidentified specimens. *Euphorbia crenulata* is closely related to *E. commutata.*

99. Euphorbia cyparissias Linnaeus, Sp. Pl. 1: 461. 1753 • Cypress or graveyard spurge, euphorbe cypress I W

Galarhoeus cyparissias (Linnaeus) Small ex Rydberg; *Tithymalus cyparissias* (Linnaeus) Hill

Herbs, perennial, with slender, spreading rootstock. **Stems** erect or ascending, unbranched or branched, often with short axillary vegetative shoots with very narrow leaves, 10–50 cm, glabrous. **Leaves:** petiole absent; blade linear to linear-ovate or linear-oblanceolate, 5–30 × 0.5–3 mm, base rounded to cuneate, margins entire, apex rounded to acute, surfaces glabrous; venation inconspicuous, only midvein prominent. **Cyathial arrangement:** terminal pleiochasial branches 6–25, each 1–2-times 2-branched; pleiochasial bracts similar in shape to distal leaves except shorter and wider; dichasial bracts distinct, widely ovate, rhombic, or reniform, base obtuse to truncate, margins entire, apex acute or obtuse, sometimes mucronulate; axillary cymose branches 0–15. **Cyathia:** peduncle 0–0.5 mm. **Involucre** campanulate to slightly urceolate, 1.5–2 × 0.9–1.1 mm, glabrous; glands 4, crescent-shaped, 0.4–0.6 × 0.7–1.3 mm; horns convergent, 0.1–0.5 mm. **Staminate flowers** 15–25. **Pistillate flowers:** ovary glabrous; styles 1–1.2 mm, 2-fid. **Capsules** subglobose, 2.5–3 × 3–4 mm, slightly lobed; cocci rounded, granulate, glabrous; columella 1.9–2.1 mm. **Seeds** blackish, ovoid-oblong, 1.8–2.5 × 1.4–1.7 mm, smooth; caruncle nipple-shaped or subreniform, 0.2–0.4 × 0.7–1.1 mm.

Flowering and fruiting spring–fall. Fields, roadsides, waste places; 0–1500 m; introduced; B.C., Man., N.B., Nfld. and Labr. (Nfld.), N.S., Ont., P.E.I., Que., Sask.; Ark., Calif., Colo., Conn., Del., D.C., Ga., Idaho, Ill., Ind., Iowa, Ky., Maine, Md., Mass., Mich., Minn., Mo., Mont., Nebr., N.H., N.J., N.Y., N.C., Ohio, Oreg., Pa., R.I., S.C., Tenn., Utah, Vt., Va., Wash., W.Va., Wis., Wyo.; Europe.

100. Euphorbia dendroides Linnaeus, Sp. Pl. 1: 462. 1753 • Tree spurge [I]

Tithymalus dendroides (Linnaeus) Hill

Shrubs, dendroid, with large rootstock. **Stems** erect, stout, densely branched, to 200 cm, glabrous, bark usually reddish and glossy toward tip of branches. **Leaves:** petiole absent; blade linear-lanceolate to oblong-lanceolate, 20–65 × 3–8 mm, base attenuate, margins entire, apex obtuse to acute, sometimes mucronulate, surfaces glabrous; venation pinnate, midvein prominent, base attenuate. **Cyathial arrangement:** terminal pleiochasial branches 4–8, unbranched or 1–2 times 2-branched; pleiochasial bracts similar in shape to but usually shorter and wider than distal leaves; dichasial bracts distinct, yellowish, broadly ovate, orbiculate, rhombic, or reniform, base obtuse, margins entire, apex obtuse, sometimes mucronulate; axillary cymose branches absent. **Cyathia:** peduncle absent. **Involucre** broadly turbinate to hemispheric, 3–4 × 3.5–3.8 mm, glabrous; glands 4, suborbiculate to subtrapezoidal, 1–1.5 × 1.2–1.8 mm, sometimes irregularly lobed; horns absent. **Staminate flowers** 20–25. **Pistillate flowers:** ovary glabrous; styles 2.4–3 mm, 2-fid. **Capsules** subglobose, 4–5.5 × 4–6.5 mm, deeply lobed; cocci laterally compressed, smooth, glabrous; columella 3–4 mm. **Seeds** grayish or blackish, ovoid and strongly laterally compressed, 3–3.5 × 1.5–2 mm, smooth; caruncle semirounded and laterally compressed, 1.2–1.4 × 0.8–1.8 mm.

Flowering and fruiting fall–spring. Coastal plains and basins, hillsides; 0–500 m; introduced; Calif.; Europe; n Africa.

Euphorbia dendroides is native to the Mediterranean region. In the flora area, the species is known from Los Angeles, Santa Barbara, and Ventura counties.

101. Euphorbia exigua Linnaeus, Sp. Pl. 1: 456. 1753 • Dwarf or small spurge [I]

Tithymalus exiguus (Linnaeus) Hill

Herbs, annual, with taproot. **Stems** erect, unbranched or branched, 3–30 cm, glabrous. **Leaves:** petiole 0–0.5 mm; blade linear, linear-oblong, or linear-spatulate, 2–30 × 1–5 mm, base cuneate or truncate, margins entire, apex acute, obtuse, or emarginate, surfaces glabrous; venation inconspicuous, only midvein prominent. **Cyathial arrangement:** terminal pleiochasial branches 3–5, each 1–3 times 2-branched; pleiochasial bracts similar in shape and size to or sometimes slightly longer and wider than distal leaves; dichasial bracts distinct, linear, linear-lanceolate, or lanceolate-ovate, base rounded to subcordate, margins entire, apex acute; axillary cymose branches 0–5. **Cyathia:** peduncle 0–2 mm. **Involucre** cupulate, 0.3–0.5 × 0.6–0.8 mm, glabrous; glands 4, elliptic to crescent-shaped, 0.2–0.4 × 0.3–0.6 mm; horns divergent, 0.2–0.5 mm. **Staminate flowers** 5–8. **Pistillate flowers:** ovary glabrous; styles 0.5–0.7 mm, 2-fid. **Capsules** subglobose, 1–1.8 × 1.3–2 mm, slightly lobed; cocci rounded, smooth, puncticulate toward abaxial line, glabrous; columella 1.1–1.4 mm. **Seeds** blackish to grayish, 4-angled-ovoid, 1–1.5 × 0.5–0.7 mm, white, tuberculate; caruncle conic or subconic, 0.1–0.3 × 0.1–0.3 mm.

Flowering and fruiting summer–fall. Edges of gardens, roadsides, waste places; 0–1500 m; introduced; B.C., N.S., Ont.; Calif., N.Y., W.Va.; Europe; w Asia; n Africa; Atlantic Islands (Macaronesia).

Euphorbia exigua can be easily distinguished from other annual species of the genus in the flora area by its tuberculate seeds.

102. Euphorbia falcata Linnaeus, Sp. Pl. 1: 456. 1753 • Sickle spurge [I]

Tithymalus falcatus (Linnaeus) Klotzsch & Garcke

Herbs, annual, with taproot. **Stems** erect, unbranched or branched, 5–20 cm, glabrous. **Leaves:** petiole absent; blade obovate, linear-oblong, or spatulate, 2–20 × 2–10 mm, base cuneate or attenuate, margins entire, apex acute, obtuse, emarginate, or mucronate, surfaces glabrous; venation usually inconspicuous, sometimes 3-nerved from base, midvein prominent. **Cyathial arrangement:** terminal pleiochasial branches 2–5, each 2–6 times 2-branched; pleiochasial bracts similar in shape to but usually shorter and wider than distal leaves; dichasial bracts distinct, widely ovate, rhombic, or suborbiculate, imbricate, base cordate, truncate, or cuneate, margins finely denticulate, apex acute or obtuse, strongly mucronate; axillary cymose branches 0–10. **Cyathia:** peduncle 0–2 mm. **Involucre** cupulate, 0.5–1.2 × 0.6–1.3 mm, glabrous; glands 4, elliptic to orbiculate, 0.2–0.3 × 0.3–0.8 mm; horns usually absent, occasionally divergent, 0.5–1.2 mm. **Staminate flowers** 6–10. **Pistillate flowers:** ovary pilose only at base; styles 0.9–1.1 mm, 2-fid. **Capsules** subovoid, 2–3 × 1.8–3 mm, slightly lobed; cocci rounded, smooth, glabrous or slightly pilose along abaxial region; columella 1.1–1.8 mm. **Seeds** grayish, whitish, or light brownish, ovoid, 1.2–1.8 × 0.7–1.1 mm, transversely sulcate; caruncle subglobose to subconic, 0.2–0.5 × 0.2–0.5 mm.

Flowering and fruiting spring–summer. Waste places, roadsides; 0–1600 m; introduced; Ky., Md., Ohio, Pa., Tenn., Va., W.Va.; s, c Europe; w, c, s Asia; n Africa; introduced also in South America (Chile).

103. Euphorbia georgiana Mayfield, J. Bot. Res. Inst. Texas 7: 639, fig. 2[row 2, left]. 2013 [C] [E]

Herbs, annual, with taproot. **Stems** erect, often branched near base, 10–18 cm, glabrous. **Leaves:** petiole absent; blade oblanceolate, 5–12 × 3–5 mm, base attenuate, margins entire, apex rounded, surfaces glabrous; venation pinnate, midvein prominent. **Cyathial arrangement:** terminal pleiochasial branches 3, 1–3 times 2-branched; pleiochasial bracts rotund-obovate, wider than distal leaves; dichasial bracts distinct, not imbricate, broadly deltate to subreniform, base truncate to emarginate, margins entire, apex rounded to bluntly acuminate; axillary cymose branches 0–1. **Cyathia:** peduncle 0.3–0.5 mm. **Involucre** infundibular, 1–1.1 × 0.4–0.5 mm, glabrous; glands 4, crescent-shaped, 0.3–0.4 × 0.4–0.5 mm; horns divergent, 0.3–0.5 mm. **Staminate flowers** 5–10. **Pistillate flowers:** ovary glabrous; styles 0.5 mm, 2-fid. **Capsules** depressed-ovoid, 2.2–2.4 × 3.2–3.4 mm, 3-lobed; cocci rounded, smooth, glabrous; columella 2–2.1 mm. **Seeds** gray, ovoid, 1.6–1.7 × 1.4–1.6 mm, with deep, rounded pits irregularly scattered over entire surface; caruncle reniform, subconic, 0.5–0.6 × 0.6–0.8 mm.

Flowering and fruiting spring. Granite outcrops; of conservation concern; 100–200 m; Ga.

Euphorbia georgiana is restricted to granitic outcrops; it is known from Oglethorpe and Wilkes counties (M. H. Mayfield 2013). It is similar to *E. austrotexana* but has larger seeds that are much more deeply pitted and leaves that are oblanceolate instead of linear-oblanceolate to linear. It is also quite distinct from the more robust, biennial or occasionally annual *E. commutata*, the only other closely related species that occurs in the area.

104. Euphorbia helioscopia Linnaeus, Sp. Pl. 1: 459. 1753 • Sun or summer spurge, wartweed, euphorbe réveille-matin [I] [W]

Galarhoeus helioscopius (Linnaeus) Haworth; *Tithymalus helioscopius* (Linnaeus) Hill

Herbs, annual, with taproot. **Stems** erect, unbranched or branched, 5–45 cm, usually glabrous or sparsely pilose. **Leaves:** petiole absent or to 0.5 mm; blade obovate-spatulate, 4–40 × 2–25 mm, base cuneate, attenuate, or auriculate, margins serrulate, apex rounded, surfaces glabrous; venation pinnate, midvein prominent. **Cyathial arrangement:** terminal pleiochasial branches (3–)5, each 1–2 times 2-branched; pleiochasial bracts obovate, wider than distal leaves; dichasial bracts distinct, obovate or rhombic, ± oblique, base rounded, truncate, or attenuate, margins serrulate, apex rounded; axillary cymose branches 0. **Cyathia:** peduncle 0.2–1 mm. **Involucre** cupulate, 1.5–2 × 0.7–1.1 mm, glabrous; glands 4, elliptic, 0.2–0.5 × 0.5–1 mm; horns absent. **Staminate flowers** 10–15. **Pistillate flowers:** ovary glabrous; styles 0.7–1 mm, 2-fid. **Capsules** depressed-globose, 2.5–4 × 3.2–4.2 mm, clearly 3-lobed; cocci rounded, smooth, glabrous; columella 0.9–1.1 mm. **Seeds** dark brown to blackish, subovoid, 1.6–2.2 × 1.5–1.9 mm, reticulate; caruncle elliptic, 0.9–1.1 × 0.4–0.5 mm.

Flowering and fruiting spring–fall. Roadsides, waste places; 0–1400 m; introduced; St. Pierre and Miquelon; Alta., B.C., N.B., Nfld. and Labr. (Nfld.), N.S., Ont., P.E.I., Que., Sask.; Calif., Conn., Del., D.C., Ga., Idaho, Ill., La., Maine, Md., Mass., Mich., Mont., N.H., N.J., N.Y., N.C., Ohio, Oreg., Pa., S.C., Tenn., Tex., Vt., Va., Wash., Wis., Wyo.; Europe; Asia; n Africa; introduced also in South America (Argentina, Chile).

Euphorbia helioscopia was collected once in Minnesota in the late 1800s but apparently did not become established there.

105. Euphorbia helleri Millspaugh, Bot. Gaz. 26: 268, fig. [p. 270]. 1898 • Heller's spurge

Tithymalus helleri (Millspaugh) Small

Herbs, annual, with taproot. **Stems** ascending, branched proximally, 15–30 cm, glabrous. **Leaves:** petiole usually 1–3 mm, absent distally; blade spatulate, 6–15 × 3–5 mm, base broadly attenuate, margins entire, apex usually obtuse to rounded, sometimes retuse, surfaces glabrous; venation pinnate, midvein prominent.

Cyathial arrangement: terminal pleiochasial branches 3, each 1–5 times 2-branched; pleiochasial bracts oblong, similar in size to distal leaves; dichasial bracts distinct, orbiculate-ovate to nearly reniform or subpandurate, base broadly cuneate to truncate, margins entire, apex obtuse, mucronate; axillary cymose branches 0–5. **Cyathia:** peduncle 0.2–0.4 mm. **Involucre** campanulate, 0.9–1.2 × 0.7–1 mm, glabrous; glands 4, elliptic, 0.1–0.2 × 0.3–0.4 mm; horns slightly divergent, 0.1–0.2 mm. **Staminate flowers** 8–10. **Pistillate flowers:** ovary glabrous; styles 0.4–0.5 mm, 2-fid. **Capsules** depressed-globose, 2.1–2.5 × 2.5–3 mm, 3-lobed; cocci flattened, smooth, glabrous; columella 1.2–1.7 mm. **Seeds** white to light gray, ovoid, 1.4–1.6 × 1–1.2 mm, smooth; caruncle 2-lobed, thin, 0.4 × 0.7 mm.

Flowering and fruiting late winter–spring. Forests, stream banks, roadsides, shaded areas with sandy, calcareous soils; 0–50 m; La., Tex.; Mexico (Nuevo León).

The smooth, white to light gray seeds of *Euphorbia helleri* easily distinguish it from other annual members of subg. *Esula* in North America. Collections of *E. helleri* have been made near Brownsville, Texas, and thus it is possible that the species occurs in northern Tamaulipas, Mexico. The Louisiana record (Webster Parish) likely represents introduced plants.

106. Euphorbia lathyris Linnaeus, Sp. Pl. 1: 457. 1753 (as lathyrus) • Mole or gopher plant, caper spurge, euphorbe épurge [I] [W]

Galarhoeus lathyris (Linnaeus) Haworth; *Tithymalus lathyris* (Linnaeus) Hill

Herbs, annual or biennial, with taproot. **Stems** erect, unbranched or branched, to 200 cm, glabrous, glaucous. **Leaves** opposite, decussate; petiole absent; blade linear to oblong-lanceolate, 30–120 × 3–25 mm, base acute, rounded, cordate or clasping, margins entire, apex acute or subobtuse, sometimes mucronate, surfaces glabrous, abaxial ± glaucous; venation pinnate, midvein prominent. **Cyathial arrangement:** terminal pleiochasial branches 2–4, each 1–2 times 2-branched; pleiochasial bracts cordate-lanceolate, shorter and wider than distal leaves; dichasial bracts distinct, ovate-oblong to lanceolate, base subcordate, margins entire, apex acute; axillary cymose branches 0–10. **Cyathia:** peduncle 0–0.5 mm. **Involucre** campanulate, 1.2–2.3 × 1.4–2.5 mm, glabrous; glands 4, elliptic, 0.3–0.6 × 1–1.3 mm; horns divergent, thick, tips rounded, dilated, 0.5–1.4 mm. **Staminate flowers** 25–30. **Pistillate flowers:** ovary glabrous; styles 0.7–2.2 mm, 2-fid. **Capsules** depressed-globose, 9–12 × 12–16 mm, deeply 3-lobed, tardily dehiscent and appearing indehiscent, mesocarp spongy; cocci rounded, smooth, glabrous; columella 4–5.2 mm. **Seeds** brownish or blackish, oblong, 4.5–6 × 3–4.2 mm, rugose, irregularly reticulate; caruncle substipitate, hat-shaped, 1.6–2 × 1.2–1.5 mm.

Flowering and fruiting winter–fall. Roadsides, cultivated fields, stream banks, waste places; 0–1800 m; introduced; B.C., Ont., Que.; Ariz., Calif., Conn., Idaho, Ill., Md., Mass., Mont., N.C., Ohio, Oreg., Pa., S.C., Tenn., Tex., Va., Wash., W.Va.; Europe; Asia; Africa; introduced also in Mexico, South America, Australia.

Euphorbia lathyris is most likely native to the central and eastern Mediterranean region, but it is widely cultivated and often locally escaped in temperate regions worldwide, as in the flora area.

107. Euphorbia longicruris Scheele, Linnaea 22: 152. 1849 • Wedge-leaf spurge [E]

Tithymalus longicruris (Scheele) Small

Herbs, annual, with taproot. **Stems** erect, usually unbranched, occasionally branched later in season, 5–25 cm, glabrous. **Leaves:** petiole 0–0.5 mm; blade cuneate-spatulate to obovate, 5–15 × 2–6 mm, base broadly attenuate, margins entire, apex rounded to obtuse, mucronate, surfaces glabrous; venation pinnate, midvein prominent. **Cyathial arrangement:** terminal pleiochasial branches 3, each many times 2-branched; pleiochasial bracts obovate, similar in size to distal leaves; dichasial bracts basally subconnate, strongly imbricate and often obscuring internodes, reniform to semiorbiculate, base cordate, margins entire, apex rounded; axillary cymose branches 0–5. **Cyathia:** peduncle 0.3–0.5 mm. **Involucre** campanulate, 1.5–2 × 1–1.5 mm, glabrous; glands 4, crescent-shaped to elliptic, 0.4–0.8 × 0.8–1.1 mm; horns divergent, 0.5–0.8 mm. **Staminate flowers** 10–15. **Pistillate flowers:** ovary glabrous; styles 0.5–0.6 mm, 2-fid. **Capsules** ovoid-globose, 2–2.8 × 2.5–3 mm, 3-lobed; cocci rounded, smooth, glabrous; columella 1.6–2.1 mm. **Seeds** gray to purple-gray or sometimes nearly black, oblong, 1.3–1.6 × 0.9–1.2 mm, strongly small-pitted; caruncle umbonate, depressed-conic, 0.5 × 0.7 mm.

Flowering and fruiting spring. Grasslands, open prairies, sites with rocky, usually calcareous soils; 300–800 m; Ark., Kans., Okla., Tex.

Euphorbia longicruris is quite similar to the other small, annual members of subg. *Esula* in the south-central United States and can best be distinguished from those species by its imbricate dichasial bracts that form little tufts of overlapping leaves at the ends of the pleiochasial branches.

108. Euphorbia lurida Engelmann in J. C. Ives, Rep. Colorado R. 4: 26. 1861 • Woodland spurge [F]

Euphorbia palmeri Engelmann ex S. Watson; *E. palmeri* var. *subpubens* (Engelmann ex S. Watson) L. C. Wheeler; *E. subpubens* Engelmann ex S. Watson; *Tithymalus luridus* (Engelmann) Wooton & Standley; *T. palmeri* (Engelmann ex S. Watson) Dayton; *T. subpubens* (Engelmann ex S. Watson) Norton

Herbs, perennial, with thick rootstock. **Stems** erect or ascending, unbranched, sometimes sinuous, 5–30 cm, glabrous or sparsely to densely puberulent. **Leaves:** petiole 0–1 mm; blade oblanceolate to obovate, 8–20 × 3–7 mm, base truncate or cuneate, margins entire, apex obtuse to rounded, minutely mucronate, surfaces puberulent or glabrous; venation inconspicuous, only midvein prominent. **Cyathial arrangement:** terminal pleiochasial branches 3–5, each 1–4 times 2-branched; pleiochasial bracts ovate to broadly ovate or oblanceolate, wider than distal leaves; dichasial bracts distinct, rounded, oblanceolate, or subreniform, base cuneate or obtuse, margins entire, apex obtuse, slightly mucronate; axillary cymose branches 0–4. **Cyathia:** peduncle 0.3–0.9 mm. **Involucre** cupulate, 2–2.2 × 1.3–1.8 mm, glabrous; glands 4, oblong to broadly ovate, usually truncate, 0.5–0.8 × 1–1.6 mm, margins irregularly crenate to strongly dentate; horns absent or usually divergent or straight, 0.1–0.3 mm, usually slightly longer than, occasionally equaling, teeth on gland margin. **Staminate flowers** 10–20. **Pistillate flowers:** ovary glabrous or puberulent; styles 0.7–1 mm, 2-fid. **Capsules** ovoid, 3.5–4 × 4–4.5 mm, 3-lobed; cocci rounded, smooth to slightly rugose, glabrous; columella 3.2–3.5 mm. **Seeds** gray to dark gray, truncate-oblong to truncate-ovoid, 2.8–3 × 1.7–2 mm, irregularly pitted; caruncle conic, 0.6 × 0.7 mm.

Flowering and fruiting spring–summer. Open pine-oak forests, dry slopes and canyons; 1300–2800 m; Ariz., Calif., Nev., N.Mex., Utah; Mexico (Baja California).

Euphorbia lurida has been treated as a complex of several taxa in the past, but only a single, broadly defined species is recognized here. This species is variable in both the pubescence and shape of the bracts subtending the cyathia and also in the degree of crenation of the gland margin. In the northern part of its range, *E. lurida* appears to intergrade with *E. brachycera*, and it can be difficult to distinguish these two species in northern Arizona and New Mexico. A report of the species from Sonora, Mexico, based on a single immature collection (V. W. Steinmann and R. S. Felger 1997) has not been verified.

109. Euphorbia myrsinites Linnaeus, Sp. Pl. 1: 461. 1753 • Myrtle or creeping or blue spurge, donkey tail [I] [W]

Tithymalus myrsinites (Linnaeus) Hill

Herbs, usually perennial, occasionally biennial, with taproot. **Stems** erect or semiprostrate, unbranched or branched, 15–40 cm, succulent, glabrous. **Leaves:** petiole 0–2 mm; blade obovate, obovate-oblong, lanceolate, orbiculate, or suborbiculate, 2–30 × 3–17 mm, fleshy, base truncate or attenuate, margins entire or finely denticulate, apex acute to obtuse, cuspidate or strongly mucronate, surfaces glabrous; venation and midvein inconspicuous. **Cyathial arrangement:** terminal pleiochasial branches 2–12, each 1–2 times 2-branched; pleiochasial bracts similar in shape and size to distal leaves; dichasial bracts distinct, suborbiculate or reniform, base truncate, margins entire or minutely denticulate, apex obtuse, mucronulate; axillary cymose branches 0–4. **Cyathia:** peduncle 0.5–1 mm. **Involucre** campanulate, 2.4–2.6 × 2.3–2.5 mm, glabrous; glands 4, trapezoidal, 1–1.5 × 1.5–2.5 mm; horns divergent, thick, tips rounded, dilated, 0.5–0.9 mm. **Staminate flowers** 6–12. **Pistillate flowers:** ovary glabrous; styles 2.5–2.8 mm, usually unbranched. **Capsules** subglobose, 5–7 × 5–6 mm, unlobed; cocci rounded to subangular, smooth, glabrous; columella 4.5–5 mm. **Seeds** brownish to grayish, oblong, 2.8–4.5 × 2–3.2 mm, vermiculate-rugose; caruncle substipitate, trapezoidal or mushroom-shaped, 1.3–1.5 × 0.6–0.8 mm.

Flowering and fruiting spring–summer. Scrub oak communities, open ground near forests, shrub-steppes; 0–2400 m; introduced; B.C.; Calif., Colo., Idaho, Mont., N.Mex., Oreg., Utah, Wash., Wyo.; s Europe; w Asia.

Euphorbia myrsinites is cultivated in much of the flora area, where it can tolerate cold winters. In some areas, it can locally escape from cultivation.

110. Euphorbia oblongata Grisebach, Spic. Fl. Rumel. 1: 136. 1843 • Egg-leaf or oblong spurge [I] [W]

Tithymalus oblongatus (Grisebach) Soják

Herbs, perennial, with woody taproot. **Stems** erect, unbranched or densely branching, 80 cm, often densely villous (especially young stems and pleiochasial branches). **Leaves:** petiole absent; blade oblong to narrowly obovate or lanceolate, 15–70 × 6–25 mm, base rounded or truncate, margins finely serrulate, apex obtuse, mucronulate, surfaces glabrous;

venation inconspicuously pinnate, midvein prominent. **Cyathial arrangement:** terminal pleiochasial branches 3–5, each 2–3 times 2–4-branched; pleiochasial bracts ovate, similar in size to distal leaves; dichasial bracts distinct, ovate to suborbiculate, base truncate or rounded, margins entire or finely denticulate, apex obtuse, sometimes mucronulate; axillary cymose branches 0–4. **Cyathia:** peduncle 1–5 mm. **Involucre** cupulate to slightly turbinate, 1.5–2.5 × 1.3–1.5 mm, glabrous; glands 2–3, elliptic, 0.6–0.8 × 0.8–1.2 mm; horns absent. **Staminate flowers** 15–40. **Pistillate flowers:** ovary glabrous; styles 1.5–2 mm, 2-fid. **Capsules** globose, 3–4.5 × 3–4.5 mm, slightly 3-lobed; cocci rounded, verrucose-tuberculate, glabrous; columella 2.5–3.3 mm. **Seeds** brown, ovoid, 2.4–2.6 × 1.3–2 mm, smooth; caruncle reniform, 0.2–0.3 × 0.8–0.9 mm.

Flowering and fruiting spring–fall. Waste areas, disturbed sites, roadsides, fields, pastures; 30–900 m; introduced; Calif., Oreg., Wash.; s Europe.

Euphorbia oblongata is listed as a noxious weed by the states of California, Oregon, and Washington.

111. Euphorbia ouachitana Mayfield, J. Bot. Res. Inst. Texas 7: 642, figs. 2[row 2, right], 6. 2013 [E]

Herbs, annual, with taproot. **Stems** erect-ascending, often basally decumbent, often branched near base, 12–28 cm, glabrous. **Leaves:** petiole 0–3 mm; blade broadly oblanceolate to subspatulate, or proximalmost often orbiculate, 3–20 × 3–9 mm, base attenuate, margins entire, apex rounded to emarginate, surfaces glabrous; venation pinnate, midvein prominent. **Cyathial arrangement:** terminal pleiochasial branches 3, 1–3 times 2-branched; pleiochasial bracts ovate-deltate to subrhombic-ovate, shorter and wider than distal leaves; dichasial bracts connate 1/4 length, not imbricate, broadly deltate to subreniform, base truncate to broadly obtuse, margins entire, apex obtuse to bluntly acuminate; axillary cymose branches 1–5. **Cyathia:** peduncle 0.4–0.7 mm. **Involucre** infundibular, 1.3–1.6 × 1–1.2 mm, glabrous; glands 4, crescent-shaped, 0.4–0.5 × 0.8–1 mm; horns divergent, 0.2–0.4 mm. **Staminate flowers** 15–20. **Pistillate flowers:** ovary glabrous; styles 0.8 mm, 2-fid. **Capsules** ovoid-globose, 2.6–2.7 × 2.5–2.7 mm, slightly lobed; cocci rounded to ± flattened, smooth, glabrous; columella 2–2.1 mm. **Seeds** dark brown, oblong-ovoid, 1.5–2 × 1–1.6 mm, with deep, rounded pits in 3–4 regular vertical rows; caruncle reniform-ovate, conic, 0.5–0.7 × 0.5–0.8 mm. 2*n* = 26.

Flowering and fruiting spring. Semiopen forests, bluffs and ledges, stream banks, glades; 100–400 m; Ark., Mo., Okla., Tenn., Tex.

Euphorbia ouachitana is restricted to semiopen forests and woodlands in the south-central United States. It is similar to *E. commutata* and has been most commonly identified as that species in the past. However, it differs in its consistently brown seeds that have pits in regular, vertical rows. The two species also differ in that the proximal leaves of *E. commutata* are elliptic with long petioles, whereas those of *E. ouachitana* are spatulate with orbiculate blades and petiolelike bases. The proximal leaves of are. Aside from the restricted distribution of *E. ouachitana* in Missouri and Tennessee, the ranges of *E. ouachitana* and *E. commutata* do not overlap (see M. H. Mayfield 2013 for a detailed discussion of the distribution). *Euphorbia ouachitana* is most common in the Ouachita Mountains from southeastern Oklahoma to Hot Springs County, Arkansas.

112. Euphorbia peplidion Engelmann in W. H. Emory, Rep. U.S. Mex. Bound. 2(1): 191. 1859 • Low spurge [E]

Tithymalus peplidion (Engelmann) Small

Herbs, annual, with taproot. **Stems** erect, unbranched or branched, 5–20 cm, glabrous. **Leaves:** petiole 0–0.2 mm; blade linear-oblanceolate to cuneate-spatulate, 5–20 × 1–4 mm, base attenuate, margins entire, apex rounded to obtuse, surfaces glabrous; venation inconspicuously pinnate, midvein prominent. **Cyathial arrangement:** terminal pleiochasial branches 3(–5), 1–3 times 2-branched; pleiochasial bracts linear, lanceolate, or spatulate, similar in size to distal leaves; dichasial bracts distinct, rhomboid-lanceolate to ovate, falcate, base rounded to obtuse, margins entire or slightly erose, apex acute; axillary cymose branches 0–3. **Cyathia:** peduncle 0.3–0.5 mm. **Involucre** broadly campanulate-turbinate, 0.8–1 × 0.6–0.9 mm, glabrous; glands 4, elliptic, 0.2–0.3 × 0.5–0.6 mm; horns divergent, 0.4–0.5 mm. **Staminate flowers** 5–10. **Pistillate flowers:** ovary glabrous; styles 0.5–0.7 mm, 2-fid. **Capsules** depressed-globose, 1.8–2.3 × 2.6–3 mm, 3-lobed; cocci slightly flattened, without wings, smooth, glabrous; columella 1.3–1.5 mm. **Seeds** gray to brown, oblong, 1.3–1.6 × 0.9–1.1 mm, abaxial faces irregularly large-pitted, adaxial faces longitudinally sulcate; caruncle flat, umbonate, 2-lobed, 0.4 × 0.7 mm.

Flowering and fruiting late winter–spring. Dry, sandy areas, open areas with poor soils, roadsides, stream banks; 100–300 m; Tex.

Euphorbia peplidion is native to central and south-central Texas.

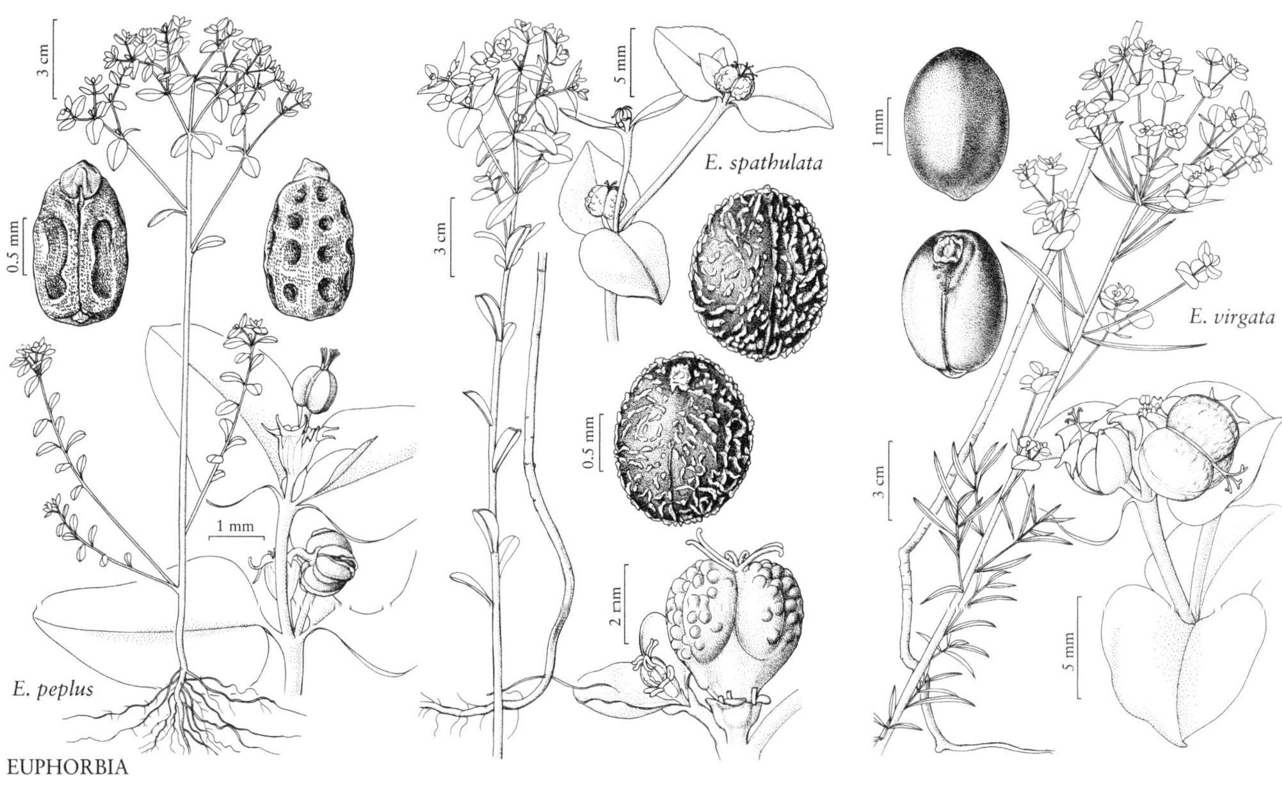

EUPHORBIA

113. Euphorbia peplus Linnaeus, Sp. Pl. 1: 456. 1753 • Petty spurge [F] [I] [W]

Esula peplus (Linnaeus) Haworth; *Galarhoeus peplus* (Linnaeus) Haworth ex Rydberg; *Tithymalus peplus* (Linnaeus) Hill

Herbs, annual, with taproot. **Stems** erect, unbranched or branched, 5–30 cm, glabrous. **Leaves:** petiole 1–10 mm; blade obovate, oblong, or suborbiculate, 5–25 × 4–15 mm, base attenuate or cuneate, margins entire, apex obtuse, surfaces glabrous; venation pinnate, midvein prominent. **Cyathial arrangement:** terminal pleiochasial branches 3–5, usually 2–8 times 2-branched, 1st branching level rarely 3-furcate; pleiochasial bracts similar in shape to and slightly larger than distal leaves; dichasial bracts distinct, ovate to obovate, base obtuse, margins entire, apex obtuse, usually mucronate; axillary cymose branches 0–3. **Cyathia:** peduncle 0.4–1.1 mm. **Involucre** cupulate to slightly turbinate, 0.6–1.1 × 0.7–0.9 mm, glabrous; glands 4, elliptic, 0.2–0.3 × 0.5–0.6 mm; horns slightly convergent to divergent, 0.4–0.6 mm. **Staminate flowers** 10–15. **Pistillate flowers:** ovary glabrous; styles 0.1–0.2 mm, 2-fid. **Capsules** subglobose, 1.3–2 × 1.5–2.2 mm, slightly 3-lobed; cocci rounded, with 2 low longitudinal wings, smooth, glabrous; columella 1.1–1.3 mm. **Seeds** whitish or grayish, subovoid, 1–1.6 × 0.6–1 mm, abaxial faces regularly large-pitted (appearing almost alveolate), adaxial faces longitudinally sulcate; caruncle deciduous, conic, 0.2–0.5 × 0.2–0.7 mm.

Flowering and fruiting year-round. Edges of gardens, weedy flower beds, roadsides, waste places, open ground near forests; 0–1500 m; introduced; Alta., B.C., Man., N.B., Nfld. and Labr. (Nfld.), N.S., Ont., P.E.I., Que., Sask.; Ala., Ariz., Calif., Colo., Conn., Del., Idaho, Ill., Ind., Iowa, Ky., La., Maine, Md., Mass., Mich., Mont., Nev., N.H., N.J., N.Mex., N.Y., N.C., N.Dak., Ohio, Oreg., Pa., R.I., Tenn., Tex., Utah, Vt., Va., Wash., W.Va., Wis.; Europe; w Asia; introduced also in Mexico, West Indies, Bermuda, Central America, South America, Africa, Atlantic Islands, Indian Ocean Islands, Pacific Islands, Australia.

114. Euphorbia platyphyllos Linnaeus, Sp. Pl. 1: 460. 1753 • Broadleaf spurge [I] [W]

Galarhoeus platyphyllos (Linnaeus) Haworth; *Tithymalus platyphyllos* (Linnaeus) Hill

Herbs, annual, with taproot. **Stems** erect, unbranched or branched, 15–80 cm, usually glabrous, rarely pilose. **Leaves:** petiole absent; blade oblanceolate or obovate, 20–50 × 5–10 mm, base subcordate or cuneate, margins finely serrulate, apex usually acute, occasionally obtuse, occasionally mucronulate, surfaces usually sparsely pilose, occasionally glabrate; venation pinnate, midvein prominent. **Cyathial arrangement:** terminal pleiochasial branches (3–)5, proximalmost node 3-branched, more distal ones 1–3 times 2-branched; pleiochasial bracts similar in shape and size to distal leaves; dichasial bracts distinct, ovate-triangular, base obtuse, margins finely serrulate, apex obtuse, mucronulate; axillary cymose branches 0–7(–14). **Cyathia:** peduncle absent. **Involucre** cupulate, 0.9–1.2 × 1.3–1.6 mm, sparsely pilose; glands 4, elliptic to ovate, 0.5–0.7 × 0.7–1 mm; horns absent. **Staminate flowers** 10–12. **Pistillate flowers:** ovary glabrous, styles 1.6–1.8 mm, 2-fid. **Capsules** globose or subglobose, 2.5–3 × 2.5–3 mm, slightly 3-lobed; cocci rounded, sparsely verrucose, glabrous; columella 1.9–2.1 mm. **Seeds** dark brown, ovoid, dorsiventrally compressed, 2–2.2 × 1.5–1.7 mm, smooth; caruncle ± reniform, 0.2–0.5 × 0.2–0.5 mm.

Flowering and fruiting summer–fall. Lake shores, roadsides, waste places; 0–500 m; introduced; Ont.; Mass., Mich., N.Y., N.C., Ohio, Pa., R.I., Tenn., Vt., Wash.; s Europe; introduced also in South America (Argentina, Chile, Paraguay).

115. Euphorbia purpurea (Rafinesque) Fernald, Rhodora 34: 25. 1932 • Glade or Darlington's glade spurge [E]

Agaloma purpurea Rafinesque, Autik. Bot., 94. 1840; *Galarhoeus darlingtonii* (A. Gray) Small

Herbs, perennial, with thick rootstock. **Stems** erect, unbranched, 70–100(–130) cm, glabrous. **Leaves:** petiole 0–2 mm; blade lance-oblong to oblanceolate-oblong, 50–100 × 13–30 mm, base attenuate to cuneate, margins entire, apex usually acute, sometimes obtuse to rounded, minutely apiculate, abaxial surface glabrate to sparsely pilose, adaxial surface glabrous; venation pinnate, midvein prominent. **Cyathial arrangement:** terminal pleiochasial branches 3–6, each unbranched or 1–2 times 2-branched, occasionally appearing pendent; pleiochasial bracts lance-ovate, shorter than distal leaves; dichasial bracts distinct, cordate-deltate to reniform, base subcordate, margins entire, apex rounded; axillary cymose branches 0–10. **Cyathia:** peduncle 0–1 mm. **Involucre** narrowly campanulate, 2.1–3 × 3–4.2 mm, glabrous; glands 5, elliptic to slightly reniform, 1–1.2 × 1.5–2.1 mm; horns absent. **Staminate flowers** 10–15. **Pistillate flowers:** ovary glabrous; styles 3–3.5 mm, 2-fid. **Capsules** globose, 4.5–5.2 × 6–6.8 mm, 3-lobed; cocci rounded, verrucose, sometimes minutely so, glabrous; columella 4–4.8 mm. **Seeds** mottled silver-brown, ovoid-globose, 3–4 × 2.5–3.5 mm, smooth; caruncle subconic, reniform, 0.8–1.1 × 1.4–1.6 mm.

Flowering and fruiting spring–fall. Dry to moist forests and slopes, rock outcrops, swamps or seeps, especially over calcareous rocks; 50–1100 m; Del., Md., N.J., N.C., Ohio, Pa., Va., W.Va.

Euphorbia purpurea is primarily an eastern Appalachian forest species, but it also occurs in Adams, Highland, and Pike counties in southern Ohio. It is listed as endangered by Maryland, New Jersey, Ohio, and Pennsylvania, and is in the Center for Plant Conservation's National Collection of Endangered Plants.

116. Euphorbia roemeriana Scheele, Linnaea 22: 151. 1849 • Roemer's spurge [E]

Tithymalus roemerianus (Scheele) Small

Herbs, annual, with taproot. **Stems** erect, occasionally decumbent at base, branched, 15–30 cm, glabrous. **Leaves:** petiole 2–5 mm; blade oblanceolate to obovate, 5–20 × 5–10 mm (larger leaves on distal portion of stem), base cuneate to attenuate, margins entire, apex rounded, surfaces glabrous; venation pinnate. **Cyathial arrangement:** terminal pleiochasial branches 3, 1–4 times 2-branched; pleiochasial bracts ovate to oblong, similar in size to distal leaves; dichasial bracts connate ½ length, reniform to semicircular, base truncate to perfoliate, margins entire, apex rounded to obtuse; axillary cymose branches 0–3. **Cyathia:** peduncle 0–0.9 mm. **Involucre** campanulate, 1.5–2 × 1–1.6 mm, glabrous; glands 4, elliptic to trapezoidal, 0.4–0.6 × 0.7–1.2 mm; horns slightly convergent, 0.1–0.3 mm. **Staminate flowers** 10–12. **Pistillate flowers:** ovary glabrous; styles 0.6–1 mm, 2-fid. **Capsules** subglobose, 2–2.5 × 2.4–3 mm, slightly lobed; cocci rounded, smooth, glabrous; columella 1.5–1.8 mm. **Seeds** brown, oblong-ovoid, 1.6–1.8 × 1.4–1.5 mm, with scattered, deep and broad pits; caruncle conic, reniform, 0.3–0.4 × 0.4–0.6 mm.

Flowering and fruiting spring. Rich calcareous soils, creek canyons; 100–300 m; Tex.

D. S. Correll and M. C. Johnston (1970) suggested that *Euphorbia roemeriana*, which is restricted to the eastern part of the Edwards Plateau, is the southern counterpart of *E. commutata*, to which it is morphologically very similar. However, molecular phylogenetic analyses indicate that it is most closely related to *E. austrotexana* and *E. longicruris* (J. A. Peirson et al. 2014). *Euphorbia roemeriana* can be distinguished from *E. commutata* and several of the other small, annual species of subg. *Esula* by its consistently connate dichasial bracts.

117. Euphorbia schizoloba Engelmann, Proc. Amer. Acad. Arts 5: 173. 1861 • Mojave spurge [E]

Euphorbia incisa Engelmann; *Tithymalus incisus* (Engelmann) W. A. Weber; *T. schizolobus* (Engelmann) Norton

Herbs, perennial, with thick rootstock. **Stems** slender, ascending, often sinuous, many, densely branched near base, 10–50 cm, usually glabrous, usually glaucous. **Leaves:** petiole 0–1 mm; blade broadly oblanceolate to obovate, 10–20 × 3–9 mm, base usually acute, occasionally short-attenuate, rarely obtuse, margins entire, apex usually acute, occasionally obtuse, acuminate to cuspidate, surfaces usually glabrous, usually glaucous; venation pinnate, sometimes obscure, midvein prominent. **Cyathial arrangement:** terminal pleiochasial branches (1–)3–4(–5), each 1–3 times 2-branched; pleiochasial bracts broadly ovate to subcordate, usually similar in size to, occasionally wider than, distal leaves; dichasial bracts distinct, broadly ovate to almost reniform, base obtuse, margins entire, apex obtuse, acuminate to cuspidate; axillary cymose branches 0–2(–4). **Cyathia:** peduncle 0.3–1 mm. **Involucre** campanulate to broadly turbinate, 2.2–3 × 2–2.5 mm, glabrous; glands 4, irregularly semicircular to trapezoidal or elliptic-truncate, 0.8–1.5 × 1–2.2 mm, margins strongly crenate or dentate; horns usually absent, if present then straight, 0.1–0.2 mm, generally equaling teeth on gland margin. **Staminate flowers** 12–20. **Pistillate flowers:** ovary glabrous; styles 1–1.2 mm, 2-fid. **Capsules** oblong-ovoid, 3.5–4 × 3.8–5 mm, 3-lobed; cocci rounded, smooth, glabrous; columella 3.3–3.8 mm. **Seeds** gray to whitish, oblong cylindric, 2–3 × 1.5 mm, irregularly shallowly pitted to almost smooth; caruncle conic, 0.6 × 0.6 mm.

Flowering and fruiting spring–summer. Desert mountains and canyon slopes, rocky and gravelly soils; 500–1800 m; Ariz., Calif., Nev.

Euphorbia schizoloba is a desert perennial that occurs on bluffs and ledges in the Mojave and Sonoran Deserts. Sparsely pubescent plants of *E. schizoloba* are known from Arizona and are best represented by several collections from the Mazatzal and Sierra Ancha mountains in Gila County from between 1000 to 1800 m. George Engelmann published two names for this species almost simultaneously in 1861 (*E. schizoloba* and *E. incisa*). Although the authors have not been able to determine which publication has priority, Engelmann himself cited *E. incisa* as a synonym of *E. schizoloba* (in W. H. Brewer et al. 1876–1880, vol. 2), as did J. B. S. Norton (1899). Also, the type specimen at MO was annotated by Engelmann as *E. schizoloba*, and there is no mention of the name *E. incisa* on the sheet.

118. Euphorbia serrata Linnaeus, Sp. Pl. 1: 459. 1753 • Sawtoothed or toothed or serrate spurge [I] [W]

Galarhoeus serratus (Linnaeus) Haworth; *Tithymalus serratus* (Linnaeus) Hill

Herbs, perennial, with thick rootstock. **Stems** erect, branched, 10–70 cm, glabrous. **Leaves:** petiole absent; blade lanceolate, ovate-lanceolate, linear, or linear-lanceolate, 10–70 × 2–20 mm, base acute or obtuse, margins irregularly serrate, apex acute to obtuse, surfaces glabrous; venation inconspicuous, only midvein prominent. **Cyathial arrangement:** terminal pleiochasial branches 3–5, each 1–3 times 2-branched; pleiochasial bracts ovate-lanceolate, usually shorter and wider than distal leaves; dichasial bracts distinct, ovate or deltate, base obtuse to cordate, margins irregularly dentate, apex acute or obtuse, mucronate; axillary cymose branches 0–3. **Cyathia:** peduncle 1–5 mm. **Involucre** campanulate, 2–4 × 1.2–3 mm, glabrous; glands 4–5, elliptic, ovate, or suborbiculate, 1.2–1.8 × 1.5–2.7 mm; horns absent or slightly divergent, 0–0.6 mm. **Staminate flowers** 20–40. **Pistillate flowers:** ovary glabrous; styles 1–2 mm, 2-fid. **Capsules** subovoid, 4.5–6 × 4–5 mm, 3-lobed; cocci rounded, smooth, occasionally slightly puncticulate, glabrous; columella 4–4.5 mm. **Seeds** grayish, cylindric, 2.5–3.1 × 1.7–2 mm, smooth or slightly dotted; caruncle subconic, lobed, 1–1.5 × 0.5–1 mm.

Flowering and fruiting spring–fall. Waste places, disturbed sites, roadsides, fields, pastures; 0–300 m; introduced; Calif.; Europe; Atlantic Islands (Macaronesia).

Euphorbia serrata, native to the western Mediterranean region of Europe and Macaronesia, is listed as a noxious weed by the state of California. In the flora area, it has been found in coastal counties from Sonoma to Monterey counties; attempts to eradicate it may have been successful.

119. Euphorbia spathulata Lamarck in J. Lamarck et al., Encycl. 2: 428. 1788 • Warty spurge [F] [W]

Euphorbia arkansana Engelmann & A. Gray; *E. dictyosperma* Fischer & C. A. Meyer; *E. obtusata* Pursh; *Galarhoeus arkansanus* (Engelmann & A. Gray) Small ex Rydberg; *G. obtusatus* (Pursh) Small ex Rydberg; *Tithymalus arkansanus* (Engelmann & A. Gray) Klotzsch & Garcke; *T. dictyospermus* (Fischer & C. A. Meyer) A. Heller; *T. obtusatus* (Pursh) Klotzsch & Garcke; *T. spathulatus* (Lamarck) W. A. Weber

Herbs, usually annual, rarely biennial, with taproot. **Stems** erect or ascending, unbranched or branched, 10–70 cm, glabrous. **Leaves:** petiole absent or to 0.2 mm; blade oblanceolate, oblong-oblanceolate, spatulate, or cuneate, 10–50 × 6–11 mm, base broadly attenuate to rounded or shallowly cordate-clasping, margins finely serrulate (usually distally), apex usually rounded to obtuse, occasionally slightly retuse or obcordate proximally, bluntly mucronate, surfaces glabrous; venation pinnate, midvein prominent. **Cyathial arrangement:** terminal pleiochasial branches 3(–5), each 1–3 times 2-branched; pleiochasial bracts broadly ovate to ovate-oblong, shorter and wider than distal leaves; dichasial bracts distinct, broadly ovate, ovate-triangular, or ovate-elliptic, base cordate-clasping or subcordate to rounded, margins serrulate, apex rounded to obtuse or acute; axillary cymose branches (0–)5–12. **Cyathia:** peduncle 0.3–1(–1.5) mm. **Involucre** campanulate to cupulate, 0.6–1(–1.5) × 0.8–1.2 mm, glabrous; glands 4–5, elliptic, oblong, to slightly reniform, 0.2–0.6 × 0.4–1 mm; horns absent. **Staminate flowers** 3–10. **Pistillate flowers:** ovary glabrous; styles 0.8–1.5 mm, 2-fid. **Capsules** depressed-globose, 2–3.5 × 4 mm, 3-lobed; cocci rounded, verrucose, verrucae 0.1–0.2 mm, glabrous; columella 1.4–2.2 mm. **Seeds** red-brown to dark purple, occasionally ± glaucous, broadly ellipsoid-ovoid to nearly globose, 1.3–2.5 × 1.5–1.8 mm, smooth, reticulate, or finely low-ridged; caruncle irregularly reniform to round, subconic to lenticular, 0.3–0.4 × 0.5–0.6 mm.

Flowering and fruiting spring–summer. Forests, fallow fields, prairies, pastures, glades, stream banks, waste places, roadsides; 0–3500 m; Ont.; Ala., Ariz., Ark., Calif., Colo., D.C., Fla., Ga., Idaho, Ill., Ind., Iowa, Kans., Ky., La., Md., Mich., Minn., Miss., Mo., Mont., Nebr., N.Mex., N.C., N.Dak., Ohio, Okla., Oreg., Pa., S.C., S.Dak., Tenn., Tex., Utah, Va., Wash., W.Va., Wis., Wyo.; Mexico (Chihuahua, Sonora); s South America.

As treated here, *Euphorbia spathulata* is a wide-ranging and variable species. J. B. S. Norton (1900) recognized a number of segregates (for example, *E. arkansana*, *E. dictyosperma*, and *E. obtusata*), all of which are included here in a broadly defined *E. spathulata*. The only segregate species that has been widely recognized in regional floras is *E. obtusata* (for example, M. L. Fernald 1950; T. S. Cooperrider 1995; G. Yatskievych 1999–2013, vol. 2). Authors have generally distinguished the eastern North American *E. obtusata* from the western *E. spathulata* by the former's larger seeds (1.7–2.3 versus 1.5–1.7 mm) with smooth (versus reticulate) surfaces, larger involucres, red (versus yellow) involucral glands, and cordate-clasping (versus rounded to subcordate) dichasial bracts. Examination of specimens of *E. spathulata* in the broad sense from throughout North America showed that there is some geographic patterning to seed size and surface sculpturing, but the variation does not segregate cleanly into two discrete taxa. Plants from western North America typically have small seeds (1.5–1.7 mm) with reticulate surfaces, although some western individuals have seeds 1.8–1.9 mm with reticulate surfaces. Plants from Texas generally have small seeds (1.5–1.6 mm) but with the surfaces either reticulate or completely smooth. Plants from adjacent Louisiana have small seeds with faintly reticulate to almost bumpy surfaces. Plants from eastern North America have larger seeds (2–2.3 mm) with usually smooth surfaces, although individuals from Tennessee and the Carolinas have faintly reticulate surfaces. Involucre height, gland color, and the shape of the dichasial bracts do not segregate with seed size as previous treatments have suggested.

120. Euphorbia terracina Linnaeus, Sp. Pl. ed. 2, 1: 654. 1762 • Terracina or carnation spurge [I] [W]

Tithymalus terracinus (Linnaeus) Klotzsch & Garcke

Herbs, perennial or biennial, with taproot. **Stems** erect, unbranched or branched, 10–100 cm, glabrous. **Leaves:** petiole absent; blade linear, linear-lanceolate, oblong-elliptic, or obovate, 4–50 × 2–10 mm, base obtuse or truncate, margins finely serrulate, apex acute, obtuse, or truncate, sometimes mucronulate, surfaces glabrous; venation inconspicuous, only midvein prominent. **Cyathial arrangement:** terminal pleiochasial branches 2–5, each 1–5 times 2-branched; pleiochasial bracts lanceolate, elliptic, or ovate, similar in size to distal leaves; dichasial bracts distinct, ovate to subreniform, base cuneate to cordate, margins finely serrulate, apex acute, obtuse, or rounded, sometimes mucronulate or cuspidate; axillary cymose branches 0–7. **Cyathia:** peduncle 1–3 mm. **Involucre** cupulate to slightly turbinate, 1.1–2 × 1.3–1.5 mm, glabrous or puberulent; glands 4, elliptic to trapezoidal, 0.6–0.8 × 1–2 mm; horns slightly convergent to divergent, 1–2 mm. **Staminate flowers** 15–20. **Pistillate flowers:** ovary glabrous; styles 1–1.8 mm, 2-fid. **Capsules** depressed-globose, 2.5–3 × 3–4.5 mm, deeply 3-lobed; cocci rounded to subangular, smooth, glabrous; columella 1.9–2.3 mm. **Seeds** pale gray, subovoid, 1.6–2.4 × 1.3–1.8 mm, smooth; caruncle boat-shaped, 0.4–0.6 × 0.4–0.6 mm.

Flowering and fruiting spring–summer. Edges of cultivated fields and woodlands, roadsides, waste areas, pastures, coastal bluffs, dunes, riparian areas; 0–300 m; introduced; Calif.; Europe; introduced also in Mexico, s Africa, Australia.

Euphorbia terracina is native to the Mediterranean region of Europe. This species is invasive and spreading rapidly, displacing native coastal scrub in southern California, and has been listed as a noxious weed by that state.

121. Euphorbia tetrapora Engelmann in W. H. Emory, Rep. U.S. Mex. Bound. 2(1): 191. 1859 • Weak spurge [E]

Tithymalus tetraporus (Engelmann) Small

Herbs, annual, with taproot. **Stems** erect, unbranched, 7–20 cm, glabrous. **Leaves:** petiole 1–2 mm, reduced distally; blade spatulate-cuneate, 8–10 × 4–5 mm (greatly reduced in size proximally), base cuneate, margins entire, apex rounded to emarginate or obcordate, surfaces glabrous; venation pinnate. **Cyathial arrangement:** terminal pleiochasial branches 3, each 1–3(–4) times 2-branched; pleiochasial bracts obovate, similar in size to distal leaves; dichasial bracts distinct or basally subconnate, not imbricate, triangular-ovate, base truncate or cordate, margins entire, apex mucronate; axillary cymose branches 1–4. **Cyathia:** peduncle 0.2–0.6 mm. **Involucre** campanulate, 0.8–1.1 × 0.7–1 mm, glabrous; glands 4, elliptic to trapezoidal, 0.3–0.6 × 0.6–1.2 mm; horns divergent, 0.5–1 mm. **Staminate flowers** 10–15. **Pistillate flowers:** ovary smooth, glabrous; styles 0.6–1 mm, 2-fid. **Capsules** depressed-globose, 1.8–2.2 × 2.2–2.9 mm, slightly lobed; cocci rounded to slightly flattened, smooth, glabrous; columella 1.5–1.8 mm. **Seeds** reddish brown to brown, often glaucous, oblong, 1.3–1.4 × 0.8–0.9 mm, abaxial faces with 15–20 shallow pits or almost smooth, adaxial faces with 4–6 large shallow pits or irregular oblong grooves; caruncle conic, hat-shaped, 0.3–0.4 × 0.4–0.6 mm.

Flowering and fruiting spring. Sandy soils, dry open woods; 0–300 m; La., Okla., Tex.

Euphorbia tetrapora is endemic to a portion of the western Gulf coastal plain. D. S. Correll and M. C. Johnston (1970) included Alabama and Georgia in the distribution of this species as well, probably due to Engelmann's citation of a Georgia specimen from the herbarium of Samuel Boykin. Whether the Boykin specimen came from Georgia, where Boykin was based, is unclear. Because no records to support its occurrence in the eastern Gulf coastal plain (Alabama or Georgia) have been found, those states are here excluded from the distribution of *E. tetrapora*.

122. Euphorbia texana Boissier, Cent. Euphorb., 30. 1860 • Texas spurge [E]

Euphorbia dictyosperma Fischer & C. A. Meyer var. *leiococca* Engelmann; *E. leiococca* (Engelmann) Norton

Herbs, annual, with taproot. **Stems** erect or ascending, unbranched or branched, sometimes extensively at crown, 7–20 cm, glabrous. **Leaves:** petiole 0–0.2 mm; blade oblanceolate to oblong-oblanceolate, 8–15 × 2–5 mm, base usually attenuate, occasionally cuneate, margins crenulate distally, apex rounded to obtuse, bluntly mucronate, surfaces glabrous; venation pinnate, midvein prominent. **Cyathial arrangement:** terminal pleiochasial branches 3, each 1–3 times 2-branched; pleiochasial bracts similar in shape and size to distal leaves; dichasial bracts distinct, elliptic to ovate, base acute to ± truncate, margins crenate distally, apex obtuse to rounded and mucronate; axillary cymose branches 1–3. **Cyathia:** peduncle 0.3–0.6 mm. **Involucre** hemispheric, 0.5–0.9 × 0.6–1 mm, glabrous; glands (4–)5, elongate reniform, 0.1–0.3 × 0.3–0.5 mm; horns absent. **Staminate flowers** 5. **Pistillate flowers:** ovary glabrous; styles 0.8–1 mm, 2-fid. **Capsules** depressed-globose, 1.6–2 × 2.5–3 mm, 3-lobed; cocci rounded, smooth, glabrous; columella 1.2–1.5 mm. **Seeds** brownish black, globose-lenticular, 1.4–1.5 × 1.3–1.4 mm, finely reticulate or areolate with distinct line on back; caruncle low conic, 0.4 × 0.5 mm.

Flowering and fruiting spring. Open ground, prairies; 0–200 m; La., Tex.

Euphorbia texana is related to *E. alta* and *E. spathulata*, but differs from both of those species in its smooth capsules. It is endemic to southeastern Texas and adjacent Louisiana.

123. Euphorbia trichotoma Kunth in A. von Humboldt et al., Nov. Gen. Sp. 2(fol.): 48; 2(qto.): 60. 1817 • Sand-dune spurge [C]

Galarhoeus trichotomus (Kunth) Small; *Tithymalus trichotomus* (Kunth) Klotzsch & Garcke

Herbs, perennial, with thick rootstock. **Stems** erect, branched, 15–40 cm, glabrous. **Leaves:** petiole absent; blade obovate to oblanceolate, 5–20 × 2–7 mm, base broadly attenuate, margins entire, apex acute to obtuse, surfaces glabrous; venation pinnate, very obscure, midvein prominent. **Cyathial arrangement:** terminal pleiochasial branches 3, each usually 1–2 times 3-branched, sometimes unbranched; pleiochasial bracts similar in shape and size to distal leaves, alternate; dichasial bracts distinct, obovate to oblanceolate, base broadly attenuate, margins entire, apex acute to obtuse; axillary cymose branches 0. **Cyathia:** peduncle 0.7–1 mm. **Involucre** hemispheric, 1–1.5 × 1.4–2 mm, glabrous; glands 5, elliptic to slightly crescent-shaped, 0.4–0.6 × 0.9–1.1 mm; horns divergent, 0.1–0.2 mm. **Staminate flowers** 8–10. **Pistillate flowers:** ovary glabrous; styles 0.2–0.4 mm, 2-fid. **Capsules** depressed-ovoid, 1.1–1.5 × 2–3 mm, strongly 3-lobed; cocci rounded, minutely papillate, sometimes appearing smooth, glabrous; columella 1.6–1.9 mm. **Seeds** white, ovoid-globose, 1.8–2 × 1.6–1.8 mm, smooth; caruncle ± rounded and flattened, 0.3 × 0.3 mm.

Flowering and fruiting year-round (primarily spring–summer). Coastal beaches, sand dunes, thickets; 0–10 m; of conservation concern; Fla.; Mexico; West Indies; Central America.

Euphorbia trichotoma is found in coastal peninsular Florida from Hillsborough County south to Key West.

124. Euphorbia virgata Waldstein & Kitaibel, Descr. Icon. Pl. Hung. 2: 176, plate 162. 1803 • Leafy spurge, wolf's milk [F] [I] [W]

Herbs, perennial, with slender, spreading rootstock. **Stems** erect, unbranched or branched, 20–90 cm, glabrous. **Leaves:** petiole 0–1 mm; blade linear to linear-oblanceolate or linear-oblong (margins parallel or almost parallel at midleaf), 40–90 × 3–12 mm, base truncate or abruptly attenuate, margins entire, apex acute or rounded, sometimes mucronulate, surfaces glabrous; venation inconspicuous, only midvein prominent. **Cyathial arrangement:** terminal pleiochasial branches 5–17, each 1–2 times 2-branched; pleiochasial bracts similar in shape to but shorter and wider than distal leaves; dichasial bracts distinct, broadly ovate, rhombic, or reniform, base cordate or cuneate, margins entire, apex obtuse to rounded, mucronulate; axillary cymose branches 0–18. **Cyathia:** peduncle 0–1 mm. **Involucre** campanulate, 1.5–3.5 × 1.7–3 mm, glabrous; glands 4, crescent-shaped, 0.6–1.5 × 1.3–2.5 mm; horns divergent to convergent, 0.2–0.8 mm. **Staminate flowers** 10–25. **Pistillate flowers:** ovary glabrous; styles 1.7–2.5 mm, 2-fid. **Capsules** subglobose, 2.5–3.5 × 3–4.5 mm, slightly lobed; cocci rounded, smooth except finely granulate toward abaxial line, glabrous; columella 2–3.3 mm. **Seeds** yellow-brown to gray or mottled, oblong-ellipsoid to oblong-ovoid, 2.2–2.6 × 1.3–1.6 mm, smooth; caruncle subconic, 0.6–1 × 0.7–0.9 mm. $2n = 60$.

Flowering and fruiting spring–fall. Pastures, fields, waste places, shorelines, railroads, open disturbed areas; 0–2600 m; introduced; Alta., B.C., Man., N.B., N.S., Ont., P.E.I., Que., Sask., Yukon; Alaska, Ariz., Calif., Colo., Conn., Idaho, Ill., Ind., Iowa, Kans., Maine, Md., Mass., Mich., Minn., Mo., Mont., Nebr., Nev., N.H., N.J., N.Mex., N.Y., N.Dak., Ohio, Oreg., Pa., S.Dak., Utah, Vt., Wash., W.Va., Wis., Wyo.; Europe; Asia.

Euphorbia virgata has caused significant economic and ecological impact over large portions of the United States and Canada. It is part of a taxonomically complex group of species native to Europe and Asia, and there has been much confusion over the naming of the species that has become widely established in the New World (A. Radcliffe-Smith 1985; P. M. Catling and G. Mitrow 2012). There has been speculation that hybridization and polyploidy have played a role in the weediness of leafy spurge, and it is possible that the widespread occurrence of leafy spurge in North America is at least partly due to multiple introductions in grain imported from Eurasia (Ma J. S. 2010). Nonetheless, a re-evaluation of the leafy spurge complex by Berry et al. (unpubl.) revealed that *E. esula* Linnaeus and *E. virgata* are two distinct, albeit related species. The true *E. esula* is restricted in range to certain parts of Europe and shows little tendency toward weediness where it occurs. In contrast, *E. virgata* is much more widespread across Europe and temperate Asia, where it shows the same weedy characteristics as leafy spurge in the New World. More importantly, it is morphologically consistent with the North American material of leafy spurge.

According to D. V. Geltman (1998), the best way to distinguish morphologically between *Euphorbia virgata* and *E. esula* is by differences in their leaf shape. In *E. virgata*, the leaf blades are linear to linear-oblanceolate or linear-oblong, 6–15 times longer than wide, with margins that are parallel or almost parallel at the middle of the blade; the apex is usually acute; and the base is truncate or abruptly attenuate. In *E. esula*, the leaf blades are oblanceolate to obovate-elliptic (distinctly wider toward apex), 3–8(–10) times longer than wide, with margins not parallel at the middle of the leaf; the apex is rounded to subacute; and the base is gradually attenuate to cuneate.

There are some herbarium specimens of *Euphorbia esula* from North America that indicate it probably occurred sporadically in certain states in the late 1800s and early 1900s, but the authors have no evidence that it has persisted in any of those places. Therefore in this treatment, *E. esula* is considered to be a waif in the North American flora and, by excluding it here, the authors hope to avoid confusion between it and the widespread *E. virgata*.

125. Euphorbia yaquiana Tidestrom, Proc. Biol. Soc. Wash. 48: 41. 1935 • Hairy Mojave spurge [E]

Euphorbia schizoloba Engelmann var. *mollis* Norton, N. Amer. Euphorbia, 43, plate 43. 1899, not *E. mollis* C. C. Gmelin 1806; *E. incisa* Engelmann var. *mollis* (Norton) L. C. Wheeler

Herbs, perennial, with thick rootstock. **Stems** slender, erect or ascending, sometimes sinuous, densely branched near base, 10–50 cm, moderately to densely puberulent to lanulose. **Leaves:** petiole 0–1 mm; blade usually lanceolate or elliptic-lanceolate, sometimes slightly oblanceolate, 8–30 × 6–14 mm, base usually acute, occasionally short-attenuate, rarely obtuse, margins entire, apex usually acute, occasionally obtuse, acuminate, or cuspidate, surfaces sparsely to moderately puberulent to lanulose; venation pinnate, sometimes obscure, midvein prominent. **Cyathial arrangement:** terminal pleiochasial branches 3–5, each 1–2 times 2-branched; pleiochasial bracts broadly ovate to subcordate, usually similar in size to, occasionally wider than, distal leaves; dichasial bracts distinct, broadly ovate to almost reniform, base obtuse, margins entire, apex obtuse, acuminate to cuspidate; axillary cymose branches 0–5. **Cyathia:** peduncle 0.3–0.8 mm. **Involucre** campanulate to broadly turbinate, 2.2–3 × 2–2.5 mm, puberulent to lanulose; glands 4, semicircular, trapezoidal, or elliptic-truncate, 0.8–1.5 × 1–2.2 mm, margins strongly crenate or dentate; horns usually absent, if present then straight, 0.1–0.2 mm, generally equaling teeth on gland margin. **Staminate flowers** 12–20. **Pistillate flowers:** ovary usually puberulent, occasionally lanulose; styles 1–1.2 mm, 2-fid. **Capsules** oblong-ovoid, 3.5–4 × 3–4 mm, 3-lobed; cocci rounded, smooth, usually puberulent, occasionally lanulose; columella 2.5–3 mm. **Seeds** gray to whitish, oblong cylindric, 2–3 × 1.5–1.8 mm, irregularly shallowly pitted to almost smooth; caruncle conic, 0.6 × 0.6 mm.

Flowering and fruiting spring–summer. Ponderosa pine forests, oak-pine mixed forests, dry stream banks and beds, open scrub areas, roadsides; 1000–2200 m; Ariz.

Euphorbia yaquiana is endemic to Pima and Graham counties in southern Arizona and is known only from the Santa Catalina and Pinaleño mountains. Records of *E. yaquiana* from southwestern Colorado (as *E. incisa* var. *mollis*) likely represent misidentifications of *E. brachycera*; therefore, those disjunct occurrences have been excluded here from the distribution of *E. yaquiana*. *Euphorbia yaquiana* has often been treated as a synonym of *E. schizoloba* var. *mollis*, but molecular phylogenetic data show that it is more closely related to *E. brachycera* and *E. chamaesula* (J. A. Peirson et al. 2014).

24e. EUPHORBIA Linnaeus sect. NUMMULARIOPSIS Boissier in A. P. de Candolle and A. L. P. P. de Candolle, Prodr. 15(2): 71. 1862

Victor W. Steinmann

Jess A. Peirson

Herbs, perennial [rarely annual], with slender to thickened, usually erect, occasionally spreading, rootstock. **Stems** erect or ascending [prostrate], branched or unbranched, terete, glabrous [hairy]. **Leaves** alternate [opposite]; stipules absent [present, at base of petiole]; petiole absent or indistinct; blade dimorphic (at proximal 2–4 nodes triangular and much smaller than at distal nodes) [monomorphic], base symmetric, margins entire, surfaces glabrous; venation pinnate, midvein often prominent. **Cyathial arrangement:** terminal pleiochasia with 2–5 primary branches; individual pleiochasial branches 2-branched at each node; bracts subtending pleiochasia (pleiochasial bracts) whorled, green, similar in shape but slightly smaller than distal stem leaves, those on branches and subtending cyathia (dichasial and subcyathial bracts) opposite, distinct; additional cymose branches occasionally present in axils of distal leaves, but alternately arranged and without whorled bracts. **Involucre** ± actinomorphic, not spurred; glands 5, flat or slightly convex; appendages absent. **Staminate flowers** [10–]20–30. **Pistillate flowers:** ovary glabrous, subtended by calyxlike structure; styles connate ⅛–½ length, 2-fid apically. **Seeds:** caruncle absent.

Species ca. 30 (4 in the flora): se United States, South America.

Section *Nummulariopsis* belongs to subg. *Euphorbia* and has a disjunct distribution, with species native to the southeastern United States and to temperate and subtropical areas of southern South America. In the flora area, the species are restricted to sandy habitats from Florida and southern Georgia to southern Mississippi. All species are adapted to sandy pinelands or scrub vegetation characterized by wide fluctuations in soil moisture and periodic, recurrent natural fires (E. L. Bridges and S. L. Orzell 2002).

In the key and descriptions that follow, leaf refers to the leaves at mid and distal nodes.

SELECTED REFERENCE Bridges, E. L. and S. L. Orzell. 2002. *Euphorbia* (Euphorbiaceae) section *Tithymalus* subsection *Inundatae* in the southeastern United States. Lundellia 5: 59–78.

1. Leaf blades linear, linear-elliptic, linear-lanceolate, or lanceolate, length 7+ times width, largest 15 mm wide or less (usually less than 10 mm wide).
   2. Peduncles (except for that of 1st cyathium at base of pleiochasia) 2–5 mm, not exceeding dichasial bracts; seeds depressed-globose, wider than long . . . . 126. *Euphorbia floridana*
   2. Peduncles 6–24 mm, often exceeding dichasial bracts; seeds ovoid-globose, longer than wide . . . . 127. *Euphorbia inundata*
1. Leaf blades elliptic, narrowly elliptic, narrowly oblong, or obovate, length less than 6 times width, largest usually 16+ mm wide.
   3. Involucres 2.4–3.6 × 2.5–3.6 mm; styles 3.3–3.6 mm, connate ½ length; Highlands County, c peninsular Florida . . . . 128. *Euphorbia rosescens*
   3. Involucres 1.3–2.3 × 1.6–2.5 mm; styles (1–)1.3–1.5 mm, connate ¼ length; Bay, Franklin, and Gulf counties, ec Florida panhandle. . . . . 129. *Euphorbia telephioides*

126. Euphorbia floridana Chapman, Fl. South. U.S., 401. 1860 • Greater Florida spurge [E]

Galarhoeus floridanus (Chapman) Small

Herbs, perennial, with slender to moderately thickened rootstock. **Stems** erect, 20–65 cm. **Leaves:** petiole absent; blade linear to linear-elliptic or linear-lanceolate, 30–105 × 2–4(–7) mm, chartaceous, base attenuate, rounded, or nearly truncate, apex acute; only midvein evident. **Cyathial arrangement:** terminal pleiochasial branches 3–5, 3–16 cm, 3–5(–7) times 2-branched; pleiochasial bracts linear-lanceolate or narrowly ovate, 19–55 × 4–12 mm, margins entire, apex acute or acuminate; dichasial bracts usually ovate, lanceolate, or oblong, rarely deltate, 8–21 × 4–14 mm, margins entire, apex acute, acuminate, or rounded with mucronate tip; axillary cymose branches 1–3(–6). **Cyathia:** peduncle (except for that of 1st cyathium at base of pleiochasia) 2–5 mm (not exceeding dichasial bracts). **Involucre** campanulate or obconic, 1.6–3.3 × 1.8–3.1 mm, lobes ovate, 0.5–0.7 mm, ciliate, exceeded by glands; glands yellow-green, oblong to trapezoidal, 0.5–0.8 × 0.8–1.2 mm, distal margins deeply erose. **Staminate flowers** 20. **Pistillate flowers:** gynophore exserted 2.9–5.3 mm, calyxlike lobes triangular, 0.3–0.7 mm; styles connate ¼–⅓ length, 1.1–1.7 mm. **Capsules** oblate-ovoid, 4.6–5.5 × 8.9–9 mm, strongly 3-lobed; columella 3.4–4.2 mm. **Seeds** brown to blackish, depressed-globose, circular in cross section, 2.8–3 × 3.2–3.6 mm, smooth, base flattened, with punctiform depressions, apex flattened.

Flowering and fruiting spring–fall. Xeric pine-oak sandhills, pine scrub, sandy soils; 20–80 m; Ala., Fla., Ga.

127. Euphorbia inundata Torrey ex Chapman, Fl. South. U.S., 402. 1860 • Florida pineland spurge [E] [F]

Galarhoeus inundatus (Torrey ex Chapman) Small

Herbs, perennial, with thickened rootstock. **Stems** erect or ascending, 15–40 cm. **Leaves:** petiole absent or indistinct, blade linear to linear-elliptic or lanceolate, (25–)30–60(–115) × 1.5–14(–15) mm, chartaceous, base attenuate, apex acuminate, acute, or rounded and mucronate; only midvein evident. **Cyathial arrangement:** terminal pleiochasial branches 2–3, 5–18 cm, 3–8 times 2-branched; pleiochasial bracts linear-lanceolate or narrowly ovate, 25–49 × 5–9 mm, margins entire, apex acute or acuminate; dichasial bracts ovate or lanceolate, 6–21 × 2–6 mm, margins entire, apex acute or acuminate; axillary cymose branches 1–5. **Cyathia:** peduncle 6–24 mm (often exceeding dichasial bracts). **Involucre** obconic or campanulate, 1.3–3.2 × 1.6–3.6 mm,

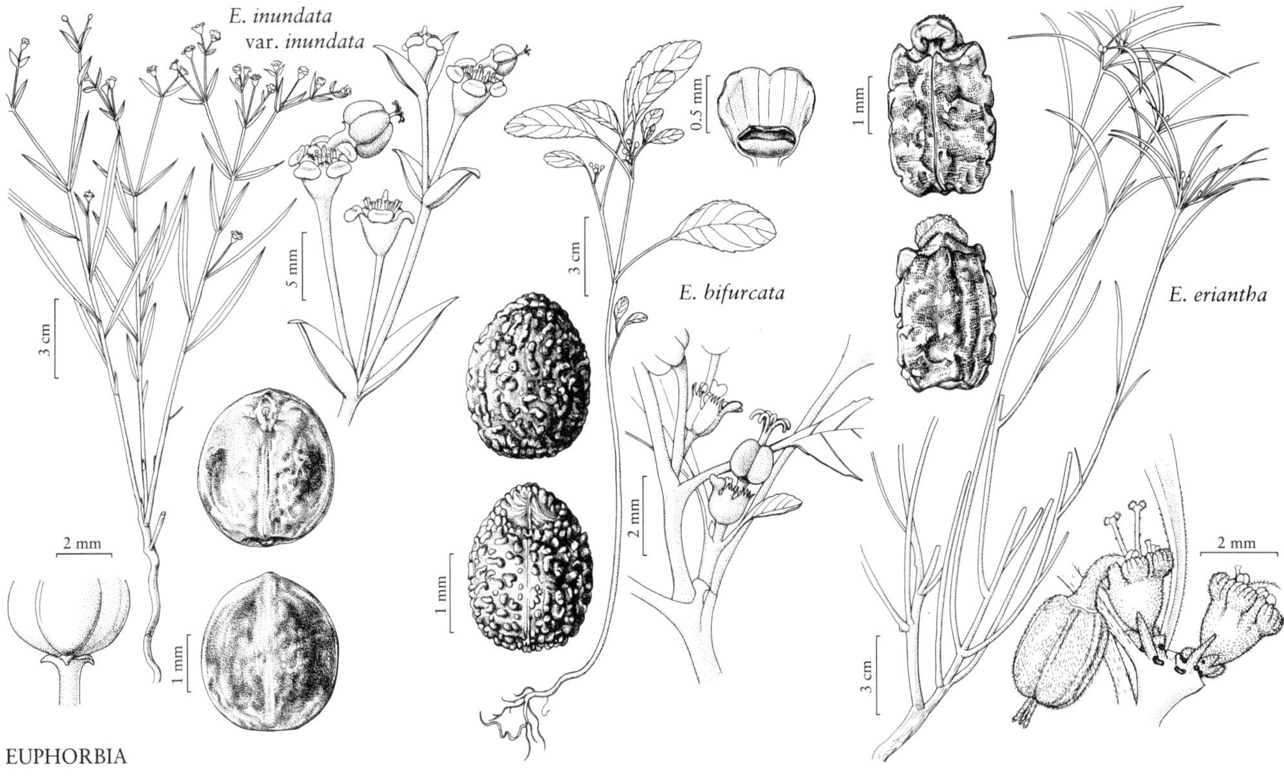

EUPHORBIA

lobes ovate, oblong, or nearly rectangular, 0.6–1 mm, laciniate-ciliate; glands red to greenish, oblong or nearly circular, 0.8–1 × 1.1–2 mm, distal margins crenulate-erose, undulate, or entire. **Staminate flowers** 20–25. **Pistillate flowers:** gynophore exserted 2.6–6.1 mm, calyxlike lobes triangular to subulate, 0.6–1.4 mm; styles connate ⅛ length, 1.3–2.2 mm. **Capsules** oblate-ovoid, 5.1–6.1 × 6.9–8.6 mm, 3-lobed; columella 3.9–4.7 mm. **Seeds** brown to blackish, ovoid-globose, circular or faintly 3- or 4-angled in cross section, 2.9–3.4 × 2.7–3 mm, smooth, base flattened or rounded, apex rounded or with inconspicuous blunt point.

Varieties 2 (2 in the flora): se United States.

Euphorbia inundata has a disjunct distribution in the southeastern United States. Allopatric, narrow-leaved populations from the west-central Florida peninsula are segregated as var. *garrettii* (E. L. Bridges and S. L. Orzell 2002).

1. Leaf blades usually linear to narrowly elliptic or narrowly lanceolate, rarely oblanceolate, (3–)4–14(–15) mm wide, length usually 5–10(–25) times width, apices obtuse to short-acute; se Mississippi to ne Florida . . . 127a. *Euphorbia inundata* var. *inundata*
1. Leaf blades narrowly linear, linear, or narrowly linear-lanceolate, 1.5–3.5(–4.5) mm wide, length 15–20(–50) times width, apices short-acute to acuminate; wc Florida. . . . . 127b. *Euphorbia inundata* var. *garrettii*

127a. Euphorbia inundata Torrey ex Chapman var. **inundata** [E] [F]

Leaf blades usually linear to narrowly elliptic or narrowly lanceolate, rarely oblanceolate, (30–)40–60(–115) × (3–)4–14 (–15) mm, length usually 5–10 (–25) times width, apex obtuse to short-acute.

Flowering and fruiting spring–fall. Grassy meadows, prairies, seepage-bogs, pine savannas, xeric oak woods, pine-oak scrub, often in seasonally inundated soils; 0–30 m; Ala., Fla., Ga., Miss.

Variety *inundata* is found in a variety of habitats from northeastern Florida and southern Georgia west to southeastern Mississippi (Jackson County).

127b. Euphorbia inundata Torrey ex Chapman var. **garrettii** E. L. Bridges & Orzell, Lundellia 5: 65, fig. 2. 2002 [E]

Leaves narrowly linear, linear, or narrowly linear-lanceolate, (25–)30–60(–100) × 1.5–3.5 (–4.5) mm, length 15–20(–50) times width, apex short-acute to acuminate.

Flowering and fruiting spring–fall. Pine savannas, pine-oak scrub, disturbed sandy soils, often in seasonally inundated soils; 10–30 m; Fla.

Variety *garrettii* is endemic to pineland habitats in the west-central Florida peninsula.

128. Euphorbia rosescens E. L. Bridges & Orzell, Lundellia 5: 71, fig. 4. 2002 • Rosy-pink or scrub spurge [C] [E]

Herbs, perennial, with thickened rootstock. **Stems** erect or ascending, 15–45 cm. **Leaves:** petiole indistinct, blade narrowly elliptic, elliptic, narrowly oblong, or obovate, 25–55 × 5–21 mm, thick and fleshy, base attenuate, apex acute or mucronulate; only midvein evident. **Cyathial arrangement:** terminal pleiochasial branches 3, 7–16 cm, 4–7 times 2-branched; pleiochasial bracts ovate to oblong, 17–31 × 13–14 mm, margins entire, apex acute, often also mucronulate; dichasial bracts ovate or lanceolate, 6–18 × 8–12 mm, margins entire, apex acute, often also mucronate; axillary cymose branches 1–2. **Cyathia:** peduncle 3.9–6.8 mm. **Involucre** obconic to campanulate, 2.4–3.6 × 2.5–3.6 mm, lobes ovate to oblong, 0.7–1 mm, ciliate; glands green to yellow, oblong or trapezoidal, 0.9–1.7 × 1.9–2.5 mm, distal margins crenulate-erose. **Staminate flowers** 25. **Pistillate flowers:** mature gynophore not seen; styles connate ½ length, 3.3–3.6 mm. **Mature capsules** not seen. **Seeds** not seen.

Flowering and fruiting spring–summer, rarely fall. Xeric oak and pine scrub, mostly on white sands, disturbed habitats; of conservation concern; 20–50 m; Fla.

Euphorbia rosescens is a narrow-endemic, gap-specialist known only from the southern portion of the Lake Wales Ridge in Highlands County. Based on leaf characteristics, it appears to be most similar to *E. telephioides*.

129. Euphorbia telephioides Chapman, Fl. South. U.S., 402. 1860 • Telephus spurge [C] [E]

Galarhoeus telephioides (Chapman) Small

Herbs, perennial, with thickened rootstock. **Stems** erect or ascending, 20–30 cm. **Leaves:** petiole usually absent or indistinct, occasionally to 5.5 mm, blade elliptic or obovate, 31–60 × 7–32 mm, thick and fleshy, base attenuate or cuneate, apex acute, obtuse, or mucronate; venation pinnate with 8–13 lateral veins, these sometimes obscure and only midvein evident. **Cyathial arrangement:** terminal pleiochasial branches (2–)3, 5–13 cm, 3–7 times 2-branched; pleiochasial bracts ovate to oblong, 17–31 × 8–15 mm, margins entire, apex acute or obtuse; dichasial bracts ovate, 5–19 × 3–9 mm, margins entire, apex acute or obtuse; axillary cymose branches 1–5. **Cyathia:** peduncle 3–9.2 mm (often exceeding subcyathial bracts). **Involucre** campanulate, 1.3–2.3 × 1.6–2.5 mm, lobes ovate to oblong, 0.5–0.7 mm, ciliate; glands purple-red, oblong to trapezoidal, 0.5–0.7 × 0.8–1.2 mm, distal margins entire or crenulate. **Staminate flowers** 25–30. **Pistillate flowers:** gynophore exserted 1.9–3.2 mm, calyxlike lobes triangular, 0.2–0.7 mm; styles connate ¼ length, 1.3–1.5 mm. **Capsules** ovoid-oblate, 5.2–5.6 × 6.6–8.3 mm, 3-lobed; columella 3.9–4.1 mm. **Seeds** blackish to dark brown, ovoid to globose-ovoid, circular in cross section, 3.2–3.5 × 2.6–3.1 mm, smooth, base rounded to flattened, apex rounded, occasionally with low point.

Flowering and fruiting spring–fall. Scrubby pine flatwoods, grasslands, disturbed areas, often in sandy soils; of conservation concern; 0–20 m; Fla.

Euphorbia telephioides, federally listed as threatened, is known only from Bay, Franklin, and Gulf counties in the Apalachicola region of the east-central Florida panhandle.

24f. EUPHORBIA Linnaeus sect. POINSETTIA (Graham) Baillon, Étude Euphorb., 284. 1858

Jess A. Peirson

Paul E. Berry

Victor W. Steinmann

Poinsettia Graham, Edinburgh New Philos. J. 20: 412. 1836

Herbs, annual or perennial [rarely shrubs or small trees], with taproot or tuberous rootstock. **Stems** erect or ascending, branched, terete, glabrous or hairy. **Leaves** opposite or alternate; stipules usually present, occasionally absent, at base of petiole; petiole present, glabrous or hairy; blade monomorphic (occasionally polymorphic in *E. cyathophora* and *E. heterophylla*), base symmetric, margins entire or toothed, flat to revolute, surfaces glabrous or variously hairy; venation pinnate, midvein often prominent. **Cyathial arrangement:** terminal monochasia, dichasia, or condensed pleiochasia with 1–3 primary branches; individual pleiochasial branches unbranched or few-branched at 1 or more successive nodes; bracts subtending pleiochasia (pleiochasial bracts) opposite or whorled, usually wholly green or with paler green, white, pink, or red at base, sometimes wholly white, pink, or red, similar in shape and size to distal leaves or distinctly different, those on branches and subtending cyathia (dichasial and subcyathial bracts) opposite, distinct; additional cymose branches occasionally present in axils of distal leaves, but alternately arranged and without whorled bracts. **Involucre** ± actinomorphic, not spurred; glands 1–3 (sometimes 4–5 in *E. eriantha*, *E. exstipulata*, and *E. radians*), sessile or stipitate, shallowly cupped to deeply concave; appendages absent or petaloid (*E. bifurcata*, *E. eriantha*, and *E. exstipulata*). **Staminate flowers** 3–25. **Pistillate flowers:** ovary glabrous or hairy; styles distinct, occasionally appearing connate at base, unbranched or 2-fid. **Seeds:** caruncle present or absent. $x = 7$.

Species ca. 30 (10 in the flora): North America, Mexico, West Indies, Central America, South America, Eurasia, Africa, Indian Ocean Islands, Pacific Islands, Australia.

Section *Poinsettia* belongs to *Euphorbia* subg. *Chamaesyce* (Gray) Reichenbach. An expanded sect. *Poinsettia* is recognized here to include three species that have previously often been included in sect. *Alectoroctonum* (Schlechtendal) Baillon (*E. bifurcata*, *E. eriantha*, and *E. exstipulata*). These three species differ from the so-called core *Poinsettia* by the presence of involucral gland appendages, but they possess the shallowly to deeply concave involucral glands and toothed leaves that are generally diagnostic for the broader section. Molecular phylogenetic analyses clearly unite these three species with the other members of sect. *Poinsettia* and not with sect. *Alectoroctonum* (Y. Yang et al. 2012).

SELECTED REFERENCES Dressler, R. L. 1961. A synopsis of *Poinsettia* (Euphorbiaceae). Ann. Missouri. Bot. Gard. 48: 329–341. Mayfield, M. H. 1997. A Systematic Treatment of *Euphorbia* Subg. *Poinsettia* (Euphorbiaceae). Ph.D. dissertation. University of Texas.

1. Leaf blades linear to linear-elliptic, margins entire or with 2–4 inconspicuous teeth near apex; involucral glands densely canescent, appendages divided into subulate segments, incurved and covering glands, densely canescent; styles unbranched . . . . 135. *Euphorbia eriantha*
1. Leaf blades linear, lanceolate, ovate, oblong, elliptic, or pandurate, margins conspicuously toothed (sometimes subentire in *E. cyathophora*, entire or few toothed in *E. pinetorum*); involucral glands glabrous, appendages absent or entire, undulate, slightly lobed, or divided into triangular segments, not incurved and covering glands, glabrous; styles 2-fid.
    2. Branches often arcuate; involucral gland appendages usually present, rarely absent.
        3. Leaf blades usually ovate, rarely oblong or elliptic, margins finely serrulate; petioles 15–49 mm; involucral glands 1(–3); ovaries glabrous; caruncles absent or rudimentary . . . . 130. *Euphorbia bifurcata*
        3. Leaf blades linear to narrowly elliptic or ovate, margins coarsely serrate; petioles 1–3 mm; involucral glands 4(–5); ovaries puberulent on keels; caruncles 0.1 × 0.2 mm . . . . 136. *Euphorbia exstipulata*
    2. Branches ± straight (except occasionally proximal branches arcuate in *E. davidii* and *E. dentata*); involucral gland appendages absent.
        4. Leaves usually opposite, occasionally alternate distally, blade margins coarsely crenate-dentate or doubly crenate; seeds with caruncles; pleiochasial bracts wholly green or with paler green, white, or mauve near base; annual herbs.
            5. Ovaries densely pilose; capsules pilose (often sparsely); involucral glands taller than wide, stipitate. . . . . 131. *Euphorbia cuphosperma*
            5. Ovaries and capsules glabrous or sparsely strigose; involucral glands shorter than wide, sessile.
                6. Hairs of abaxial leaf blade surface stiff, strongly tapered; capsules 4–4.8 mm wide; seeds angular in cross section, unevenly tuberculate . . . . . 133. *Euphorbia davidii*
                6. Hairs of abaxial leaf blade surface weak, filiform; capsules 3.5–4 mm wide; seeds rounded in cross section, evenly tuberculate . . . . 134. *Euphorbia dentata*
        4. Leaves usually alternate, occasionally opposite proximally, blade margins entire, subentire, or glandular-serrulate; seeds usually without caruncles, occasionally caruncles rudimentary; pleiochasial bracts green (purpurescent in *E. pinetorum*), often paler green, white, pink, or red at base, occasionally wholly white, pink, or red; annual or perennial herbs.
            7. Stems 5–20(–30) cm, from moniliform tuberous rootstocks; seeds 4–4.6 mm; perennial herbs. . . . . 139. *Euphorbia radians*
            7. Stems 20–100 cm, with taproots or woody rootstocks; seeds 2.1–3.1 mm; annual or perennial herbs.
                8. Pleiochasial bracts wholly green or paler green at base; involucral glands stipitate, opening round (occasionally flattened from pressing), with annular rim . . . . 137. *Euphorbia heterophylla*
                8. Pleiochasial bracts green or purpurescent, often white, pink, or red at base, occasionally wholly white, pink, or red; involucral glands sessile or substipitate, opening oblong (flattened without pressing), without annular rim.
                    9. Annual herbs with spreading taproots; leaf blades 4–40 mm wide, linear, lanceolate, elliptic, or broadly pandurate; bracts usually green with white, pink, or red at base, occasionally distal bracts wholly white, pink, or red, rarely all bracts wholly green; capsules green; involucral gland 1; widespread, including s Florida. . . . . . . 132. *Euphorbia cyathophora*
                    9. Perennial herbs with thickened, woody taproots; leaf blades 2.5–5 mm wide, lanceolate to linear; bracts wholly purpurescent green or with pink at base; capsules purpurescent; involucral glands 3(–5); Miami-Dade and Monroe counties, s Florida. . . . . 138. *Euphorbia pinetorum*

130. Euphorbia bifurcata Engelmann in W. H. Emory, Rep. U.S. Mex. Bound. 2(1): 190. 1859 • Forked spurge [F]

Herbs, annual, with slender, fibrous taproot. **Stems** erect, 20–70 cm, glabrous or with few scattered spreading hairs; branches arcuate, branching appearing dichotomous. **Leaves** usually alternate, occasionally opposite at proximalmost node; petiole 15–49 mm, glabrous; blade usually ovate, rarely oblong or elliptic, 13–54 × 7–38 mm, base usually rounded to broadly cuneate, rarely truncate, margins finely serrulate, apex obtuse, surfaces glabrous or with few scattered hairs; venation pinnate, midvein prominent. **Cyathial arrangement:** terminal dichasial branches 2, few-branched (weakly defined); pleiochasial bracts 2–3, opposite or whorled, wholly green, similar in shape and size to distal leaves; dichasial bracts smaller than distal leaves, often white at base. **Cyathia:** peduncle 0.9–3.5(–6.2) mm. **Involucre** tubular or obconic, 1–1.7 × 0.7–1.4 mm, glabrous except for few hairs on lobes; involucral lobes divided into several linear, smooth lobes; glands 1(–3), greenish, sessile and broadly attached, 0.3–0.4 × 0.4–0.8 mm, opening oblong to subcircular, glabrous; appendages petaloid, white, elliptic, oblong, transversely oblong, or forming thin, lunate rim on gland margin, not incurved and covering glands, 0.3–0.9 × 0.6–1.3 mm, entire, undulate or slightly lobed, glabrous. **Staminate flowers** 20–30. **Pistillate flowers:** ovary glabrous; styles 0.6–1 mm, 2-fid ½ length. **Capsules** oblate, 2.8–3.1 × 3.6–4.5 mm, glabrous; columella 1.9–2.4 mm. **Seeds** brown to blackish, ovoid, rounded in cross section, 1.9–2.4 × 1.5–1.8 mm, irregularly and coarsely tuberculate; caruncle absent or rudimentary.

Flowering and fruiting summer–fall. Riparian areas with cottonwoods and willows, pinyon pine woodlands, pine-oak forests, Douglas fir forests with pines; 1900–2300 m; N.Mex., Tex.; Mexico; Central America (Guatemala).

Euphorbia bifurcata is found in the mountains of southern New Mexico (Doña Ana, Grant, Lincoln, Otero, and Sierra counties) and trans-Pecos Texas (Brewster, Jeff Davis, and Presidio counties).

131. Euphorbia cuphosperma (Engelmann) Boissier in A. P. de Candolle and A. L. P. P. de Candolle, Prodr. 15(2): 73. 1862 • Hairy-fruit spurge [W]

Euphorbia dentata Michaux var. *cuphosperma* Engelmann in W. H. Emory, Rep. U.S. Mex. Bound. 2(1): 190. 1859; *Poinsettia cuphosperma* (Engelmann) Small

Herbs, annual, with taproot. **Stems** erect, 13–20 cm, both pilose to villous and densely strigillose; branches ± straight. **Leaves** usually opposite, occasionally alternate at distal nodes; petiole 3–15 mm, pilose; blade narrowly to broadly elliptic, or lanceolate to ovate, 30–80 × 10–15 mm, base cuneate to rounded, margins coarsely crenate-dentate, strigose, revolute to nearly flat, apex broadly acute to acuminate, or obtuse, abaxial surface pilose, adaxial surface sparsely strigose-hirsute; venation pinnate, midvein prominent. **Cyathial arrangement:** terminal pleiochasial branches usually 3, occasionally reduced to congested cyme, 1–2-branched; pleiochasial bracts 2–4, often whorled, wholly green or paler green at base, similar in shape and size to distal leaves or slightly narrower; dichasial bracts highly reduced. **Cyathia:** peduncle 0.5–0.8 mm. **Involucre** campanulate to slightly cylindric, 2.3 × 1.2 mm, glabrous; involucral lobes divided into several linear, smooth lobes; gland 1, yellow-green, stipitate, clavate, 1–1.2 × 0.8–0.9 mm, opening bilabiate and oblong, glabrous; appendages absent. **Staminate flowers** 3–5. **Pistillate flowers:** ovary pilose; styles 1.2 mm, 2-fid nearly entire length. **Capsules** broadly ovoid, 2.2–3 × 1.9–2.7 mm, 3-lobed, pilose; columella 2–2.4 mm. **Seeds** gray-brown to pale gray, pyramidally ovoid, angular in cross section, 2.3–2.6 × 2.4–2.6 mm, coarsely tuberculate, tubercles in 2 transverse rows; caruncle 0.2–0.4 mm. 2*n* = 56.

Flowering and fruiting summer–fall. Open montane and canyon forests, pinyon-juniper forests, montane grasslands, stream beds, disturbed habitats; 800–2000 m; Ariz., N.Mex.; Mexico; Central America (Guatemala).

132. Euphorbia cyathophora Murray, Commentat. Soc. Regiae Sci. Gott. 7: 81, plate 1. 1786 • Fire on the mountain, painted leaf [W]

Poinsettia cyathophora (Murray) Bartling

Herbs, annual, with spreading taproots. **Stems** erect or ascending, 20–100 cm, glabrous, sparsely pilose, or puberulent; branches ± straight. **Leaves** usually alternate, occasionally opposite proximally; petiole 2–20 mm, glabrous or pilose, or often hispid abaxially near blade junction; blade linear, lanceolate, elliptic, or wider leaves pandurate and unequally 4-lobed, occasionally polymorphic on single plants, 15–250 × 4–40 mm, base acute to cuneate, margins subulately glandular-serrulate distally, or sparsely glandular and subentire, hirtellous to glabrate, flat to revolute, apex acute to cuneate, abaxial surface sparsely pilose or glabrate, adaxial surface glabrous or sparsely puberulent; venation pinnate, midvein prominent. **Cyathial arrangement:** terminal pleiochasial branches (1–)3, 1–2-branched; pleiochasial bracts 2–3(–4), often as tight, involucrate whorl, usually green with white, pink, or red at base, occasionally distal bracts wholly white, pink, or red, rarely all bracts wholly green, similar in shape and size to distal leaves; dichasial bracts often colored, similar in shape and size to distal stem leaves or highly reduced. **Cyathia:** peduncle 1.6–2.8 mm. **Involucre** campanulate, occasionally broadly so, 1.8–2.8 × 2.2–2.8 mm, glabrous; involucral lobes triangularly 3–5 lobed; gland 1, yellow-green, sessile to substipitate and narrowly to broadly attached, 1–1.4 × 0.9–1.6 mm, opening oblong (flattened without pressing), without annular rim, glabrous; appendages absent. **Staminate flowers** 7–20. **Pistillate flowers:** ovary glabrous; styles 1.6 mm, 2-fid nearly entire length. **Capsules** green, depressed-globose to ellipsoid, 2.8–3.2 × 4–4.5 mm, 3-lobed, glabrous; columella 2–2.7 mm. **Seeds** black to ashy gray or light brown, cylindric to ovoid, rounded in cross section, 2.3–3.1 × 1.9–2.5 mm, uniformly tuberculate or tubercles arranged in median, transverse ridge in cylindric seeds; caruncle absent. $2n = 28, 56$.

Flowering and fruiting late spring–fall. Bottomland forests, stream and river banks, bases of bluffs, fallow fields, roadsides, open disturbed areas; 0–1800 m; Ala., Ark., Calif., Fla., Ga., Ill., Ind., Iowa, Kans., Ky., La., Md., Minn., Miss., Mo., Nebr., N.Mex., N.C., Ohio, Okla., S.C., S.Dak., Tenn., Tex., Utah, Va., Wis.; Mexico; West Indies; Central America; South America; introduced in Eurasia, Africa, Indian Ocean Islands, Pacific Islands, Australia.

Euphorbia cyathophora is native to the midwestern and southeastern United States, Mexico, the West Indies, and northern South America. Leaf shape can be polymorphic on individuals of this species, but not to the extent as in *E. heterophylla*.

133. Euphorbia davidii Subils, Kurtziana 17: 125, figs. 1, 2H–J. 1984 • Toothed spurge

Herbs, annual, with taproot. **Stems** erect or ascending, 20–70 cm, both coarsely and sparsely hirsute and closely strigillose; branches usually ± straight, occasionally proximal branches arcuate. **Leaves** usually opposite, occasionally alternate at distal nodes; petiole 7–25 mm, strigose; blade usually narrowly to broadly elliptic, occasionally lance-elliptic, 10–100 × 5–35 mm, base cuneate to attenuate, margins coarsely crenate-dentate, strigose, revolute to nearly flat, apex broadly acute to acuminate, or obtuse, abaxial surface strigose with stiff, strongly tapered hairs, adaxial surface sparsely strigose-hirsute; venation pinnate, midvein prominent. **Cyathial arrangement:** terminal pleiochasial branches usually 3, occasionally reduced to congested cyme, 1–2-branched; pleiochasial bracts 2–4, often whorled, green with diffuse greenish white to mauve near base, similar in shape and size to distal leaves or slightly narrower; dichasial bracts similar in shape to distal leaves but smaller, often highly reduced. **Cyathia:** peduncle 0.5–1 mm. **Involucre** cylindric, 2.5–3 × 1.3–1.8 mm, glabrous; involucral lobes divided into 5–7 linear, papillate lobes; gland 1, yellow-green, sessile and broadly attached, 0.9 × 1.3 mm, opening oblong, glabrous; appendages absent. **Staminate flowers** 5–8. **Pistillate flowers:** ovary glabrous or sparsely strigose; styles 1 mm, 2-fid ½ to nearly entire length. **Capsules** broadly ovoid, 2.9–3.3 × 4–4.8 mm, 3-lobed, glabrous; columella 2.2–2.7 mm. **Seeds** black to brown or pale gray, ovoid to triangular-ovoid, angular in cross section, 2.4–2.9 × 2.2–2.9 mm, low-tuberculate, tubercles irregularly arranged or in faint, transverse row; caruncle 0.9–1.1 mm. $2n = 56$.

Flowering and fruiting summer–fall. Forests, stream and riverbanks, prairies, roadsides and open disturbed areas; 200–1500 m; Ont., Que.; Ariz., Ark., Calif., Colo., Fla., Idaho, Ill., Ind., Iowa, Kans., Ky., La., Mass., Mich., Minn., Mo., Nebr., N.J., N.Mex., N.Y., N.C., Ohio, Okla., S.Dak., Tenn., Tex., Utah, Vt., Va., W.Va., Wis., Wyo.; Mexico (Chihuahua, Coahuila, Sonora); introduced in South America, Eurasia (China, Russia), Australia.

Euphorbia davidii is native from the southwestern United States and northern Mexico north through the southern Great Plains; it apparently is adventive elsewhere. The species is the weediest member of the *E. dentata* species group (following M. H. Mayfield 1997) and has become an agricultural weed in North America, South America (for example, Argentina), and in the Old World (particularly Australia and Russia). *Euphorbia davidii* can be distinguished from the closely similar *E. dentata* by its larger capsules and seeds, often more elliptic leaves, and shorter, stiffer hairs.

134. Euphorbia dentata Michaux, Fl. Bor.-Amer. 2: 211. 1803 • Toothed spurge

Poinsettia dentata (Michaux) Klotzsch & Garcke

Herbs, annual, with taproot. **Stems** erect or ascending, 15–60 cm, both pilose and inconspicuously strigillose; branches usually ± straight, occasionally proximal branches arcuate. **Leaves** usually opposite, occasionally alternate at distal nodes; petiole 5–20 mm, pilose; blade 30–70 × 4–35 mm, narrowly lanceolate to suborbiculate, usually broadest below middle, base usually acute to subobtuse, rarely subtruncate, margins coarsely crenate-dentate or doubly crenate, strigillose, flat to slightly revolute, apex broadly acute, abaxial surface long pilose with weak, filiform hairs, adaxial surface sparsely pilose to glabrate; venation pinnate, midvein prominent. **Cyathial arrangement:** terminal pleiochasial branches usually 3, occasionally reduced to congested cyme, 1–2-branched; pleiochasial bracts 2–4, often whorled, wholly green or paler green, white, or mauve at base, similar in shape and size to distal leaves or slightly narrower; dichasial bracts similar in shape to distal leaves but smaller (often highly reduced). **Cyathia:** peduncle 0.7–1 mm. **Involucre** campanulate, 3.8 × 1.8 mm, glabrous; involucral lobes divided into several linear, smooth lobes; glands (1–)2, green, sessile and broadly attached, 0.7–0.9 × 0.9–1.2 mm, opening oblong, glabrous; appendages absent. **Staminate flowers** 8–10. **Pistillate flowers:** ovary glabrous; styles 1.2 mm, 2-fid ½ to nearly entire length. **Capsules** depressed-globose, 2.5–2.8 × 3.5–4 mm, 3-lobed, glabrous; columella 1.8–2.1 mm. **Seeds** pale gray to black, ovoid, rounded in cross section, 2.1–2.7 × 1.7–2.1 mm, evenly minute-tuberculate; caruncle 0.4–0.6 mm. $2n = 28$.

Flowering and fruiting spring–fall. Bottomland forests, stream and river banks, bluffs, prairies, glades, fallow fields, roadsides, railroad cinders, open disturbed areas; 0–1000 m; Ont.; Ala., Ark., Ga., Ill., Ind., Iowa, Kans., Ky., La., Md., Mich., Miss., Mo., Nebr., N.C., Ohio, Okla., Pa., S.C., Tenn., Tex., Va., W.Va.; Mexico (Chihuahua, Coahuila, Nuevo León, Tamaulipas).

Euphorbia dentata is native from northern Mexico and the south-central United States north and east through the Ohio River Valley. Scattered occurrences in the southeastern United States likely represent adventive populations. Reports of *E. dentata* as a noxious weed (from the United States and the Old World) should most likely be attributed to introductions of *E. davidii.*

135. Euphorbia eriantha Bentham, Bot. Voy. Sulphur, 51. 1844 • Beetle spurge [F]

Herbs, annual or perennial, with slender to thick, woody taproot. **Stems** erect to ascending, 10–75 cm, glabrous or with scattered appressed hairs (especially near nodes); branches arcuate. **Leaves** alternate; petiole 0.1–0.4 mm, often indistinct, glabrous or sparsely pilose to shortly sericeous; blade linear to linear-elliptic, 20–55 × 1–3 mm, base attenuate, margins entire or with 2–4 inconspicuous teeth near apex, apex acute, abaxial surface pilose to shortly sericeous, adaxial surface usually glabrous, rarely pilose to shortly sericeous; only midvein conspicuous. **Cyathial arrangement:** terminal cymose branches 1, 1–2-branched; pleiochasial bracts 2–3, opposite or whorled, wholly green, similar in shape and size to distal leaves; dichasial bracts similar in shape to distal leaves or often highly reduced. **Cyathia:** peduncle 1–1.9 mm. **Involucre** obconic, 2.1–2.6 × 1.3–2 mm, canescent; involucral lobes triangular, obscured by hairs; glands (2–)4–5, green to maroon, color often obscured by hairs, sessile and broadly attached, 0.5–0.6 × 0.5–0.6 mm, opening oblong to nearly circular, densely canescent; appendages petaloid, whitish, hoodlike, incurved and covering glands, 0.5–1 × 0.5–0.8 mm, divided into 5–12 fringed, subulate segments, densely canescent. **Staminate flowers** 20–25. **Pistillate flowers:** ovary canescent; styles 0.9–1.5 mm, unbranched. **Capsules** oblong to ovoid, 4.4–4.9 × 3.5–4.1 mm, canescent (often with interspersed glabrescent patches); columella 3.5–3.9 mm. **Seeds** mottled black and gray or light brown, oblong or slightly ovoid, dorsiventrally compressed in cross section, 2.8–4.1 × 2–2.4 mm, irregularly pitted and tuberculate; caruncle 0.4–0.8 × 0.6–1.1 mm.

Flowering and fruiting year-round in response to sufficient rainfall. Desert scrub on rocky slopes and along washes; 60–800 m; Ariz., Calif., N.Mex., Tex.; Mexico (Baja California, Baja California Sur, Chihuahua, Coahuila, Durango, San Luis Potosí, Sonora).

EUPHORBIA

136. **Euphorbia exstipulata** Engelmann in W. H. Emory, Rep. U.S. Mex. Bound. 2(1): 189. 1859 • Squareseed or Clark Mountain spurge [F]

Euphorbia exstipulata var. *lata* Warnock & M. C. Johnston

Herbs, annual, with slender taproot. **Stems** erect, 5–26 cm, uniformly puberulent to hispidulous or glabrous; branches arcuate. **Leaves** opposite; petiole 1–3 mm, often indistinct, glabrous or puberulent; blade linear to narrowly elliptic or ovate, 14–42 × 3–28 mm, base attenuate, margins coarsely serrate, occasionally revolute, apex acute to obtuse, abaxial surface sparsely hispidulous to strigillose, adaxial surface glabrous; midvein conspicuous. **Cyathial arrangement:** terminal cymose or dichasial branches usually 1–2, occasionally reduced to monochasia, 1–2-branched; pleiochasial bracts 2–4, often whorled, wholly green or paler green at base, similar in shape and size to distal leaves or slightly narrower; dichasial bracts similar in shape to distal leaves but smaller or highly reduced. **Cyathia:** peduncle 1–1.9 mm. **Involucre** turbinate to campanulate, 1.1–1.5 × 1–1.3 mm, glabrous, pilose, or puberulent; involucral lobes divided into several linear lobes; glands 4(–5), yellow to pink, sessile and broadly attached, 0.2 × 0.3–0.4 mm, opening oblong to nearly circular, glabrous; appendages usually petaloid, white to pink, ovate to trapezoidal, occasionally absent, not incurved and covering glands, 0.2–0.4 × 0.3–0.8 mm, entire, undulate, or conspicuously divided into triangular segments, glabrous. **Staminate flowers** 10–12. **Pistillate flowers:** ovary puberulent on keels, styles 0.8–1.1 mm, 2-fid ½ to nearly entire length. **Capsules** broadly depressed-oblong to ovoid, 2.7–3.3 × 3.1–3.9 mm, puberulent (with appressed hairs usually concentrated on keels); columella 1.9–2.5 mm. **Seeds** white to gray or light brown, ovoid, bluntly 4-angled in cross section, 1.9–2.5 × 1.4–1.7 mm, tuberculate, often with 2 transverse ridges; caruncle 0.1 × 0.2 mm.

Flowering and fruiting summer–fall. Desert scrub, grasslands, mesquite savannas, oak and oak-juniper woodlands; 800–2000 m; Ariz., Calif., N.Mex., Tex., Utah, Wyo.; Mexico (Chihuahua, Coahuila, Durango, Guanajuato, San Luis Potosí, Sonora, Zacatecas).

Euphorbia exstipulata is native from Texas to California and northern Mexico. The species was found once in the late nineteenth century in Wyoming but has not been re-collected there. Broad-leaved plants have been segregated as var. *lata*, but the variation in leaf shape is continuous, and no varieties are formally recognized here.

137. Euphorbia heterophylla Linnaeus, Sp. Pl. 1: 453. 1753 • Mexican fireplant [F] [I] [W]

Euphorbia geniculata Ortega; *Poinsettia geniculata* (Ortega) Klotzsch & Garcke; *P. heterophylla* (Linnaeus) Klotzsch & Garcke

Herbs, annual, with taproot. **Stems** erect-ascending, 20–100 cm, sparsely pilose to villous; branches ± straight. **Leaves** usually alternate, occasionally opposite proximally; petiole 10–50 mm, pilose; blade narrowly lanceolate to elliptic or broadly obovate (then usually pandurate and 4-lobed), often polymorphic on single plants, 30–200 × 20–140 mm, base acute, margins sparsely glandular-serrulate, hirtellous, flat, apex acute to obtuse, abaxial surface sparsely appressed-pilose, adaxial surface sparsely pilosulous to glabrate; venation pinnate, midvein prominent. **Cyathial arrangement:** terminal dichasial branches usually 2, occasionally reduced to congested cyme, 1–2-branched (often congested and difficult to discern); pleiochasial bracts 2–4, often whorled, wholly green or paler green at base, similar in shape and size to distal leaves; dichasial bracts highly reduced, rarely absent in highly congested clusters. **Cyathia:** peduncle 0.9–1.5 mm. **Involucre** usually campanulate, occasionally nearly hemispheric, 1.5–1.9 × 1.2–1.8 mm, glabrous; involucral lobes divided into several linear, smooth lobes; gland 1, yellow-green, stipitate, clavate, 1–1.4 × 1–1.2 mm, opening circular (occasionally flattened from pressing), with annular rim, glabrous; appendages absent. **Staminate flowers** 8–15. **Pistillate flowers:** ovary glabrous or puberulent; styles 0.8–1.3 mm, 2-fid ½ to nearly entire length. **Capsules** broadly ovoid, 2.8–3.8 × 4–5.3 mm, 3-lobed, usually glabrous, rarely sparsely puberulent; columella 2.1–2.8 mm. **Seeds** brown-gray to ashy gray, broadly deltoid, 2.4–2.8 × 1.9–2.4 mm, angular in cross section, abaxial side strongly acute-carinate, tuberculate, with broad rounded tubercles in 2 rows; caruncle 0.1 mm. $2n = 28$.

Flowering and fruiting nearly year-round. Disturbed areas, roadsides; 0–500 m; introduced; Ala., Ariz., Calif., Fla., Ga., La., Miss., N.Mex., Tex.; Mexico; Central America; South America; introduced also in Eurasia, Africa.

Euphorbia heterophylla occurs from the southern United States, where it is likely naturalized, south through Mexico and Central America to South America. Because of its weediness, the precise native range in tropical and subtropical parts of the New World is not well understood. It has become widely established also in warm areas of the Old World. Leaf shape in this species is highly polymorphic within both populations and individuals. *Euphorbia heterophylla* can appear superficially similar to *E. cyathophora* but differs in its stipitate, circular involucral glands and its floral bracts that are at most very pale at the base (never colored as is typical in *E. cyathophora*).

138. Euphorbia pinetorum (Small) G. L. Webster, J. Arnold Arbor. 48: 403. 1967 • Pineland spurge [C] [E]

Poinsettia pinetorum Small, Fl. Miami, 111, 200. 1913

Herbs, perennial, with thickened, woody taproot. **Stems** erect, 30–100 cm, glabrous; branches ± straight. **Leaves** usually alternate, occasionally opposite proximally; petiole 0–1.5 mm, glabrous; blade narrowly lanceolate to linear, base long-attenuate, 30–120 × 2.5–5 mm, margins usually entire, occasionally with few inconspicuous teeth, revolute, apex narrowly acute, surfaces glabrous; venation obscurely pinnate, midvein prominent. **Cyathial arrangement:** terminal monochasial or dichasial branches 1–2, unbranched; pleiochasial bracts 2–3, often whorled, wholly purpurescent green or with pink at base, similar in shape and size to distal leaves or slightly narrower; dichasial bracts highly reduced. **Cyathia:** peduncle 1.5–3 mm. **Involucre** campanulate, 1.5–1.9 × 1.4–1.7 mm, glabrous; involucral lobes divided into broad, triangular segments; glands 3(–5), red to purple, sessile and broadly attached, 0.8–1.1 × 1.2–1.6 mm, opening oblong (flattened without pressing), without annular rim, glabrous; appendages absent. **Staminate flowers** 8–12. **Pistillate flowers:** ovary glabrous; styles 1.5 mm, 2-fid nearly entire length. **Capsules** purpurescent, depressed-globose, 2.8–3.2 × 3.6–4 mm, 3-lobed, glabrous; columella 2.6–2.9 mm. **Seeds** dark brown, cylindric-ovoid to ovoid, rounded in cross section, 2.1–2.4 × 2 mm, uniformly low-tuberculate, tubercles in median, transverse ridge; caruncle absent.

Flowering and fruiting year-round. Sandy soils in pinelands; of conservation concern; 0–10 m; Fla.

Euphorbia pinetorum has a restricted distribution in southern peninsular Florida, primarily in Miami-Dade and Monroe counties. The species is very similar to narrow-leaved forms of the closely related *E. cyathophora* but differs in its perennial habit, consistently unlobed leaves, and purpurescent involucral glands and cyathia.

139. Euphorbia radians Bentham, Pl. Hartw., 8. 1839 • Sun spurge [F]

Poinsettia radians (Bentham) Klotzsch & Garcke

Herbs, perennial, with moniliform tuberous rootstock. **Stems** erect, 5–20(–30) cm, usually glabrous, occasionally puberulent; branches ± straight. **Leaves** alternate; petiole 0–2 mm, glabrous or strigose; blade linear-lanceolate to ovate or broadly elliptic, 25–50 × 3–20 mm, unlobed, base rounded (tapered to petiole), margins with few glandular teeth, strigillose, flat to revolute, apex acute, abaxial surface coarsely strigose, adaxial surface strigose-hirsute; venation pinnate, midvein prominent. **Cyathial arrangement:** terminal pleiochasial branches usually 3, occasionally reduced to congested cyme, 1–2-branched (often highly condensed); pleiochasial bracts 6–8(–10), as tight involucrate whorl, wholly white to pale pink or red, usually narrower than distal leaves; dichasial bracts linear and highly reduced. **Cyathia:** peduncle 2–5.5 mm. **Involucre** broadly globose-cupulate, 1.7–2.1 × 2.2–2.5 mm, glabrous or puberulent; involucral lobes divided into triangular segments; glands 1–4(–5), white, sessile and broadly attached, 1.1 × 1.4 mm, opening oblong, glabrous; appendages absent. **Staminate flowers** 20–25. **Pistillate flowers:** ovary glabrous or puberulent; styles 3–4 mm, 2-fid ½ to nearly entire length. **Capsules** depressed-globose, 3.8–5 × 4–5 mm, 3-lobed, glabrous or puberulent; columella 3.6–4.5 mm. **Seeds** white, mottled brown to gray, ellipsoid, rounded in cross section, 4–4.6 × 2.4–3.2 mm, smoothly and broadly pitted or grooved; caruncle 0.1 mm.

Flowering and fruiting spring–summer. Pinyon-juniper woodlands, oak savannas, desert grasslands and scrub; 700–2500 m; Ariz., Tex.; Mexico.

Euphorbia radians is widely distributed but scattered from the Sonoran and Chihuahuan deserts south to Oaxaca in Mexico. The species is distinct among species in sect. *Poinsettia* in the flora area in its precocious habit, often flowering before the leaves emerge.

PICRODENDRACEAE Small
• Jamaica-walnut Family

Geoffrey A. Levin

Shrubs **[perennial herbs or trees]**, evergreen, monoecious or dioecious. **Leaves** alternate, opposite, or whorled, simple [palmately compound]; stipules absent [present]; petiole present [absent]; blade margins entire, dentate, or serrate; venation pinnate. **Inflorescences** unisexual [bisexual], axillary, thyrses, fascicles [cymes], or flowers solitary. **Flowers** unisexual; perianth hypogynous; hypanthium absent; sepals [3–]4–13, distinct; petals 0; nectary present [absent]; stamens [2–]4–10[–55], distinct [rarely connate], free; anthers dehiscing by longitudinal slits; pistil 1, (2–)3–4(–5)-carpellate, ovary superior, (2–)3–4(–5)-locular; placentation axile; ovules [1–]2 per locule, anatropous; styles (2–)3–4(–5)[–8], distinct, unbranched [rarely 2-fid]; stigmas (2–)3–4(–5)[–8]. **Fruits** capsules [drupes], schizocarpic with cocci separating from persistent columella, coccus dehiscence septicidal [loculicidal]. **Seeds** 1–2 per locule.

Genera 24, species ca. 90 (1 genus, 3 species in the flora): sw United States, Mexico, West Indies, Central America, South America, Asia, Africa, Pacific Islands, Australia; tropical and warm temperate regions.

The genera that make up Picrodendraceae traditionally have been treated as Euphorbiaceae subfam. Oldfieldioideae Eg. Köhler & G. L. Webster. This group has long been recognized as monophyletic because of its distinctive spiny pollen (G. A. Levin and M. G. Simpson 1994). Molecular data (Angiosperm Phylogeny Group 2003; K. Wurdack et al. 2004; C. C. Davis et al. 2005; Wurdack and Davis 2009) strongly support its monophyly and show that it is sister to Phyllanthaceae (formerly treated as Euphorbiaceae subfam. Phyllanthoideae) and more distant from Euphorbiaceae in the narrow sense.

1. TETRACOCCUS Engelmann ex Parry, W. Amer. Sci. 1: 13. 1885 • Four-pod spurge [Greek *tetra*, four, and *kokkos*, kernel or berry, alluding to 4-lobed capsule in *T. dioicus*]

W. John Hayden

Halliophytum I. M. Johnston

Shrubs, usually dioecious, sparingly hairy to glabrescent, hairs simple. **Leaves** often clustered on axillary short shoots; blade glabrescent. **Inflorescences:** staminate paniclelike or racemelike thyrses, or fascicles on short shoots; pistillate fascicles on short shoots or flowers solitary. **Pedicels** present or absent. **Staminate flowers:** sepals 4–10, sometimes weakly differentiated in 2 whorls; nectary intrastaminal, lobed; stamens 4–10. **Pistillate flowers:** sepals 5–13, sometimes differentiated in 2 whorls; nectary lobed. **Fruits** capsules. **Seeds** oblong, ovoid, or pyriform, with one flattened radial surface if 2 seeds develop per locule, otherwise both radial surfaces rounded, seed coat dry, smooth or wrinkled near hilum. $x = 12$.

Species 5 (3 in the flora): sw United States, n Mexico.

In addition to the three species treated for the flora area, *Tetracoccus* includes two Mexican species, *T. capensis* (I. M. Johnston) Croizat from the extreme tip of Baja California Sur and *T. fasciculatus* (S. Watson) Croizat from the Chihuahuan Desert.

SELECTED REFERENCE Dressler, R. L. 1954. The genus *Tetracoccus* (Euphorbiaceae). Rhodora 56: 45–61.

1. Leaves alternate and clustered on short shoots, blades 2–12 mm; staminate flowers in fascicles on short shoots; filaments glabrous; pistillate pedicels 1–3 mm; pistils usually 3-carpellate; seeds 1(–2) per locule, both radial surfaces usually rounded . . . . 3. *Tetracoccus hallii*
1. Leaves whorled, opposite, or subopposite, not clustered on short shoots, blades 15–30 mm; staminate flowers in paniclelike or racemelike thyrses; filaments villous basally; pistillate pedicels 6–15 mm; pistils usually 4-carpellate; seeds (1–)2 per locule, usually with one flattened radial surface.
    2. Leaf blades ovate to broadly elliptic, margins prominently serrate-dentate; staminate inflorescences congested paniclelike thyrses; pistillate sepals 8, in 2 series, greenish . . . . 1. *Tetracoccus ilicifolius*
    2. Leaf blades oblong or oblanceolate to linear, margins entire or remotely serrulate; staminate inflorescences sparingly branched racemelike thyrses; pistillate sepals 7–13, in 1 series, red . . . . 2. *Tetracoccus dioicus*

1. Tetracoccus ilicifolius Coville & Gilman, J. Wash. Acad. Sci. 26: 531. 1936 • Hollyleaf four-pod spurge [C] [E]

Shrubs to 1.5 m. **Leaves** opposite, not clustered on short shoots; blade ovate to broadly elliptic, 15–30 × 7–20 mm, base obtuse, margins prominently serrate-dentate, apex acute to obtuse. **Inflorescences:** staminate congested paniclelike thyrses, 15–35 mm; pistillate flowers solitary. **Pedicels:** staminate essentially absent; pistillate 8–15 mm, tomentose. **Staminate flowers:** sepals 7–9; stamens 7–9, filaments villous basally. **Pistillate flowers:** sepals 8, in 2 series, greenish, 2.5–4 mm, sparsely tomentose abaxially, densely tomentose adaxially; pistil usually 4-carpellate, 3 mm, tomentose. **Seeds** (1–)2 per locule, glossy brownish red, elliptic-oblong, 4–5 mm, usually with one flattened radial surface, smooth; caruncle present.

Flowering May–Jun. Desert scrub on limestone outcrops; of conservation concern; 600–1900 m; Calif.

Tetracoccus ilicifolius is restricted to the mountains flanking Death Valley.

TETRACOCCUS ◦ BISCHOFIA

2. **Tetracoccus dioicus** Parry, W. Amer. Sci. 1: 13. 1885 • Parry's tetracoccus [F]

Shrubs to 2 m. **Leaves** subopposite, opposite, or whorled, not clustered on short shoots; blade oblong or oblanceolate to linear, 20–30 × 1–4 mm, base obtuse to cuneate, margins entire or remotely serrulate, apex obtuse to acute. **Inflorescences:** staminate sparingly branched racemelike thyrses, to 20 mm; pistillate flowers solitary. **Pedicels:** staminate present; pistillate 6–15 mm, glabrous. **Staminate flowers:** sepals 6–10; stamens 5–10, filaments villous basally. **Pistillate flowers:** sepals 7–13, in 1 series (sometimes subequal sepals alternate), red, 3–5 mm, glabrate abaxially, minutely sericeous-tomentose adaxially; pistil usually 4-carpellate, 3 mm, crisp-tomentose. **Seeds** (1–)2 per locule, glossy brownish red, oblong, 4–5 mm, usually with one flattened radial surface, smooth; caruncle present.

Flowering Mar–Jun and sporadically year-round. Coastal chaparral and coastal scrub; 100–700 m; Calif.; Mexico (Baja California).

Tetracoccus dioicus is found in the flora area in Orange and San Diego counties.

3. **Tetracoccus hallii** Brandegee, Zoë 5: 229. 1906 • Hall's four-pod spurge [F]

Tetracoccus fasciculatus (S. Watson) Croizat var. *hallii* (Brandegee) Dressler

Shrubs to 2 m. **Leaves** alternate and clustered on short shoots; blade obovate to oblanceolate, 2–12 × 1–4 mm, base cuneate, margins entire, apex obtuse. **Inflorescences:** staminate flowers in fascicles of 2–20 on short shoots; pistillate flowers solitary or in fascicles of 2–6 on short shoots. **Pedicels:** staminate present; pistillate 1–3 mm, strigose. **Staminate flowers:** sepals 4–8; stamens 4–8, filaments glabrous. **Pistillate flowers:** sepals 5, in 1 series, greenish with red margins, 2–3 mm, sparsely strigose on both surfaces; pistil usually 3-carpellate, 1.5–2 mm, strigose. **Seeds** 1(–2) per locule, brown, ovoid to pyriform, 3–4 mm, both radial surfaces usually rounded, wrinkled near hilum; caruncle absent.

Flowering Mar–May and sporadically year-round. Desert slopes and dry washes on igneous substrates; 30–1200 m; Ariz., Calif., Nev.; Mexico (Baja California).

Tetracoccus hallii is very similar to the wholly Mexican calciphile *T. fasciculatus*.

PHYLLANTHACEAE Martinov • Leafflower Family

Geoffrey A. Levin

Herbs, shrubs, or trees, perennial or annual, deciduous or evergreen, monoecious or dioecious. **Leaves** alternate [rarely opposite], simple (pinnately compound in *Bischofia*); stipules present [rarely absent]; petiole usually present, sometimes absent; blade margins entire or crenate-serrate; venation pinnate. **Inflorescences** unisexual or bisexual, axillary [rarely supra-axillary or terminal], racemelike or paniclelike [spikelike] thyrses, cymes, fascicles, or glomes, or flowers solitary. **Flowers** unisexual; perianth hypogynous; hypanthium absent; sepals 4–6, distinct or connate basally to most of length; petals 0 or [4–]5[–6], distinct; nectary present or absent; stamens 2–5[–50], distinct or connate, free; anthers dehiscing by longitudinal slits; pistil 1, [2–]3–10[–15]-carpellate, ovary superior, [2–]3–10[–15]-locular, placentation axile; ovules 2 per locule, anatropous or hemitropous; styles [2–]3–10[–15], distinct or connate, unbranched or 2-fid; stigmas [2–]3–10[–15] (as many as style divisions). **Fruits** usually capsules, dehiscence septicidal, (usually schizocarpic with cocci separating from persistent columella, coccus usually dehiscent loculicidally), sometimes berries or drupes. **Seeds** 1–2 per locule.

Genera ca. 60, species ca. 2000 (7 genera, 23 species in the flora): North America, Mexico, West Indies, Central America, South America, Eurasia, Africa, Indian Ocean Islands, Pacific Islands, Australia; introduced in Bermuda, Atlantic Islands (Macaronesia); primarily tropical and warm temperate regions.

The genera that make up Phyllanthaceae traditionally have been treated as Euphorbiaceae subfam. Phyllanthoideae Beilschmied. Molecular data (Angiosperm Phylogeny Group 2003; K. Wurdack et al. 2004; C. C. Davis et al. 2005; Wurdack and Davis 2009; Z. Xi et al. 2012) strongly support its monophyly and show that it is sister to Picrodendraceae (formerly treated as Euphorbiaceae subfam. Oldfieldioideae Eg. Köhler & G. L. Webster), both of which are more distant from Euphorbiaceae in the narrow sense. *Drypetes*, the other genus in the flora area often included in Phyllanthoideae, belongs in Putranjivaceae (for example, see Wurdack and Davis).

Breynia, Glochidion, and most *Phyllanthus* species exhibit phyllanthoid branching (G. L. Webster 1956–1958), in which the main stems are orthotropic, indeterminate, persistent, and (beyond the first few seedling nodes) bear only scalelike leaves, while the ultimate branches are plagiotropic, of limited growth, deciduous (in woody species), and usually bear well-developed leaves. Most species with phyllanthoid branching produce flowers only on the ultimate branchlets. Because these branchlets often fall as a unit and resemble pinnate leaves, the flowers superficially appear to be borne on the leaves. In some species of *Phyllanthus*, the ultimate branches are essentially leafless and flattened into strikingly leaflike cladodes; their homology with branches is clearly demonstrated by the cymules of flowers along the margins.

Phylogenetic studies using DNA sequence data show that *Breynia*, *Glochidion*, and *Sauropus* Blume (also with phyllanthoid branching; not present in the flora area) are each monophyletic and that all three are derived from within *Phyllanthus* as usually treated, making that genus paraphyletic (K. Wurdack et al. 2004; H. Kathriarachchi et al. 2005, 2006; P. Hoffmann et al. 2006). The number of sequenced species currently is too few, and the taxonomic and nomenclatural problems too numerous, to produce a robust phylogenetic classification of this alliance of about 1200 species (Kathriarachchi et al. 2006). Thus in this treatment, *Breynia*, *Glochidion*, and *Phyllanthus* are maintained in their traditional senses. *Reverchonia*, which lacks phyllanthoid branching, is also derived from within *Phyllanthus* (Kathriarachchi et al. 2006), and here its single species is treated as *P. warnockii*.

Andrachne telephioides Linnaeus, native from southern Europe and northern Africa east to India, was collected in 1880 on a ballast dump in New Jersey. In the key below, this prostrate to procumbent perennial herb would key with the *Phyllanthus* species that lack phyllanthoid branching. *Andrachne telephioides* differs from *Phyllanthus* by having petals (minute in the pistillate flowers), pistillodes in the staminate flowers, and styles 2-fid to the base (versus 2-fid only distally).

SELECTED REFERENCES Hoffmann, P., H. Kathriarachchi, and K. Wurdack. 2006. A phylogenetic classification of Phyllanthaceae (Malpighiales; Euphorbiaceae sensu lato). Kew Bull. 61: 37–53. Kathriarachchi, H. et al. 2005. Molecular phylogenetics of Phyllanthaceae inferred from five genes (plastid *atp*B, *mat*K, 3'*ndh*F, *rbc*L, and nuclear PHYC). Molec. Phylogen. Evol. 36: 112–134. Levin, G. A. 1986b. Systematic foliar morphology of Phyllanthoideae (Euphorbiaceae). III. Cladistic analysis. Syst. Bot. 11: 515–530. Wurdack, K. et al. 2004. Molecular phylogenetic analysis of Phyllanthaceae (Phyllanthoideae pro parte, Euphorbiaceae sensu lato) using plastid *rbc*L sequences. Amer. J. Bot. 91: 1882–1900.

1. Leaves pinnately compound, blade margins crenate-serrate; inflorescences racemose to paniculate thyrses . . . . 1. *Bischofia*, p. 330
1. Leaves simple, blade margins entire; inflorescences glomes, fascicles, or cymules, or flowers solitary.
    2. Branching phyllanthoid; leaves on main stems scalelike, those on ultimate branchlets usually well developed, rarely scalelike.
        3. Pistils 5–10-carpellate; styles unbranched; sepals distinct; anther connectives extending beyond anthers as deltoid appendage . . . . 7. *Glochidion*, p. 347
        3. Pistils 3(–4)-carpellate; styles 2-fid; sepals connate basally or most of length; anther connectives not extending beyond anthers.
            4. Staminate sepals connate basally; nectaries present; seed coats dry. . . . . 5. *Phyllanthus* (in part), p. 335
            4. Staminate sepals connate most of length; nectaries absent; seed coats fleshy. . . . 6. *Breynia*, p. 345
    2. Branching not phyllanthoid; leaves all well developed.
        5. Herbs . . . . 5. *Phyllanthus* (in part), p. 335
        5. Shrubs or trees.
            6. Shrubs, 0.5–2 dm; leaf blades 1.5–3.5(–5) × 0.7–1.5 mm; sepals connate basally; pistillodes absent in staminate flowers . . . . 5. *Phyllanthus* (in part), p. 335
            6. Shrubs or trees, (2–)4–60(–80) dm; leaf blades 4–50(–80) × 3–30(–45) mm; sepals distinct; pistillodes present in staminate flowers.
                7. Leaves deciduous; fruits berries; petals absent; staminate nectary 5 glands . . . . 4. *Flueggea*, p. 334
                7. Leaves persistent; fruits capsules; petals present; staminate nectary annular.

[8. Shifted to left margin.—Ed.]

8. Staminate pedicels present; staminate flowers 1–2(–4) per fascicle; pistillate petals 0.4–0.8 mm; shrubs, (2–)4–10 dm . . . . 2. *Phyllanthopsis*, p. 331
8. Staminate pedicels rudimentary; staminate flowers 6–25 per glome; pistillate petals 1–2 mm; shrubs or trees, (5–)20–40(–80) dm . . . . 3. *Heterosavia*, p. 333

1. BISCHOFIA Blume, Bijdr. Fl. Ned. Ind. 17: 1168. 1827 • Bishopwood, javawood, toog [For Gottleib Wilhelm T. G. Bischoff, 1797–1854, German botanist] [I]

W. John Hayden

Microelus Wight & Arnott; *Stylodiscus* Bennett

Trees, usually dioecious, rarely monoecious, hairy, becoming glabrate, hairs simple; branching not phyllanthoid. **Leaves** deciduous, alternate, pinnately compound, leaflets 3(or 5), all well developed; stipules fugacious; blade margins crenate-serrate. **Inflorescences** unisexual, racemose or paniculate thyrses. **Pedicels** present. **Staminate flowers:** sepals 5, distinct; petals 0; nectary absent; stamens 5; filaments distinct, adnate to pistillode base; connectives not extending beyond anthers; pistillode short-stipitate, peltate, 5-angled. **Pistillate flowers:** sepals deciduous, 5, distinct; petals 0; nectary absent; pistil 3(–4)-carpellate; styles 3(–4), connate at base, unbranched. **Fruits** drupes. **Seeds** 1–2 per locule, rounded-trigonous; seed coat dry, smooth; caruncle absent. $x = 14$.

Species 2 (1 in the flora): introduced, Florida; s, e Asia, Pacific Islands, Australia; introduced also in s, e Africa.

Bischofia is distinctive in Phyllanthaceae by virtue of its pinnately compound leaves. H. K. Airy Shaw (1967b) placed it in his monogeneric Bischofiaceae, but embryological and anatomical data (A. K. Bhatnagar and R. N. Kapil 1973; G. A. Levin 1986b; A. M. W. Mennega 1987; G. L. Webster 1994) argue against its separation from other Phyllanthaceae. Molecular data reveal *Bischofia* to be a somewhat isolated early-divergent lineage of Phyllanthaceae subfam. Antidesmatoideae Hurusawa (H. Kathriarachchi et al. 2005; P. Hoffmann et al. 2006). *Bischofia javanica* is sometimes cultivated as a street tree in tropical regions. The second known species, *B. polycarpa* (H. Léveillé) Airy Shaw, is endemic to subtropical China.

SELECTED REFERENCES Airy Shaw, H. K. 1967b. Notes on the genus *Bischofia* Bl. (Bischofiaceae). Kew Bull. 21: 327–329. Bhatnagar, A. K. and R. N. Kapil. 1973. *Bischofia javanica*—its relationship with Euphorbiaceae. Phytomorphology 23: 264–267. Morton, J. F. 1984. Nobody loves the *Bischofia* anymore. Proc. Florida State Hort. Soc. 97: 241–244.

1. Bischofia javanica Blume, Bijdr. Fl. Ned. Ind. 17: 1168. 1827 [F] [I]

Trees to 15[–40] m; buttresses often present; heartwood, bark, and sap red. **Leaves:** stipules subulate, 5–10 mm; petiole 3–22 cm; petiolules present, that of terminal leaflet to 6.5 cm; leaflets usually elliptic, rarely obovate, 4–15 × 2–10 cm, base cuneate, margins with basal teeth sometimes glandular, surfaces glabrous. **Inflorescences** erect in flower, pendent, to 32 cm in fruit. **Pedicels:** staminate to 2.6 mm; pistillate to 11 mm in fruit, abscission zone near midpoint. **Staminate flowers** red in bud, yellow at anthesis, 2.5 mm diam.; sepals becoming reflexed, ovate, concave, 1.2 mm, nearly as wide; filaments 0.5 mm; anthers yellow, 1 mm. **Pistillate flowers** 2–3 × 1.5–2 mm; sepals ovate, 2–4 × 1 mm; staminodes to 0.5 mm; ovary green; styles 0.7 mm; stigmas erect in flower, spreading or reflexed in fruit, whitish, linear, to 5 mm. **Drupes** dark red to brown, 8–10 × 7–10 mm. **Seeds** oblong or curved, 4 × 3 × 3 mm. $2n = 196$.

Flowering Feb–Apr; fruiting Jul–Nov. Flatwoods, dry prairies, hardwood hammocks, strand swamps, disturbed areas; 0–50 m; introduced; Fla.; s, e Asia; Pacific Islands; Australia; introduced also in s, e Africa.

Bischofia javanica is invasive from southern Florida to Pinellas County; it is cultivated sporadically north to Alachua County.

Once promoted as a fast-growing ornamental and street tree, *Bischofia javanica* frequently colonizes cultivated ground, sprouting among shrubs and hedges; it also invades natural areas. In addition to its potential to displace native vegetation in Florida, *B. javanica* serves as host to a variety of destructive leaf spot diseases and insect pests, especially scales (J. F. Morton 1984). Pouchlike domatia inhabited by mites can sometimes be found in the angles of major veins on abaxial leaf surfaces (S. A. Rozario 1995). *Bischofia javanica* is also cultivated in southern California, where it has shown no invasive tendencies.

2. PHYLLANTHOPSIS (Scheele) Vorontsova & Petra Hoffmann, Kew Bull. 63: 47. 2008 • Maidenbush [Genus *Phyllanthus* and Greek *-opsis*, resembling]

Maria S. Vorontsova

Phyllanthus Linnaeus subgen. *Phyllanthopsis* Scheele, Linnaea 25: 584. 1853; *Andrachne* Linnaeus sect. *Phyllanthopsis* (Scheele) Müller Arg.

Shrubs, monoecious or dioecious, glabrous or hairy, hairs simple; branching not phyllanthoid. **Leaves** persistent, alternate, simple, all well developed; stipules persistent; blade margins entire. **Inflorescences** unisexual, fascicles or flowers solitary. **Pedicels** present. **Staminate flowers:** sepals 5, distinct; petals 5; nectary extrastaminal, annular, crenate; stamens 5; filaments distinct; connectives not extending beyond anthers; pistillode of 3 distinct segments. **Pistillate flowers:** sepals persistent, 5, distinct; petals 5; nectary annular, crenate; pistil 3(–4)-carpellate; styles 3(–4), distinct, usually 2-fid to ½ length, very rarely 2-fid to base. **Fruits** capsules. **Seeds** 2 per locule, rounded-trigonous; seed coat dry, smooth; caruncle usually absent, sometimes present. $x = 13$.

Species 2 (2 in the flora): s United States, n Mexico.

Members of *Phyllanthopsis* resemble *Andrachne* and *Leptopus* Decaisne due to their habits, petals, and discs, and generally have been included in one or the other of these genera, neither of which (in their currently accepted senses) occurs in the flora area. *Phyllanthopsis* was given generic status in 2008 after molecular phylogenetic studies revealed it to be an independent North American lineage closely related to the Asian *Leptopus* and *Actephila* Blume (M. S. Vorontsova et al. 2007; Vorontsova and P. Hoffmann 2008).

SELECTED REFERENCE Hoffmann, P. 1994. A contribution to the systematics of *Andrachne* sect. *Phyllanthopsis* and sect. *Pseudophyllanthus* compared with *Savia* s.l. (Euphorbiaceae) with special reference to floral morphology. Bot. Jahrb. Syst. 116: 321–331.

1. Leaf blades 4–10 × 3–6 mm, secondary vein pairs 1–3, obscure; stems highly branched, branches spreading, longest leafy branchlets to 10 cm .... 1. *Phyllanthopsis arida*
1. Leaf blades 7–25 × 6–20 mm, secondary vein pairs 4–7, clearly visible; stems sparsely branched, branches ascending, longest leafy branchlets to 30(–55) cm ..... 2. *Phyllanthopsis phyllanthoides*

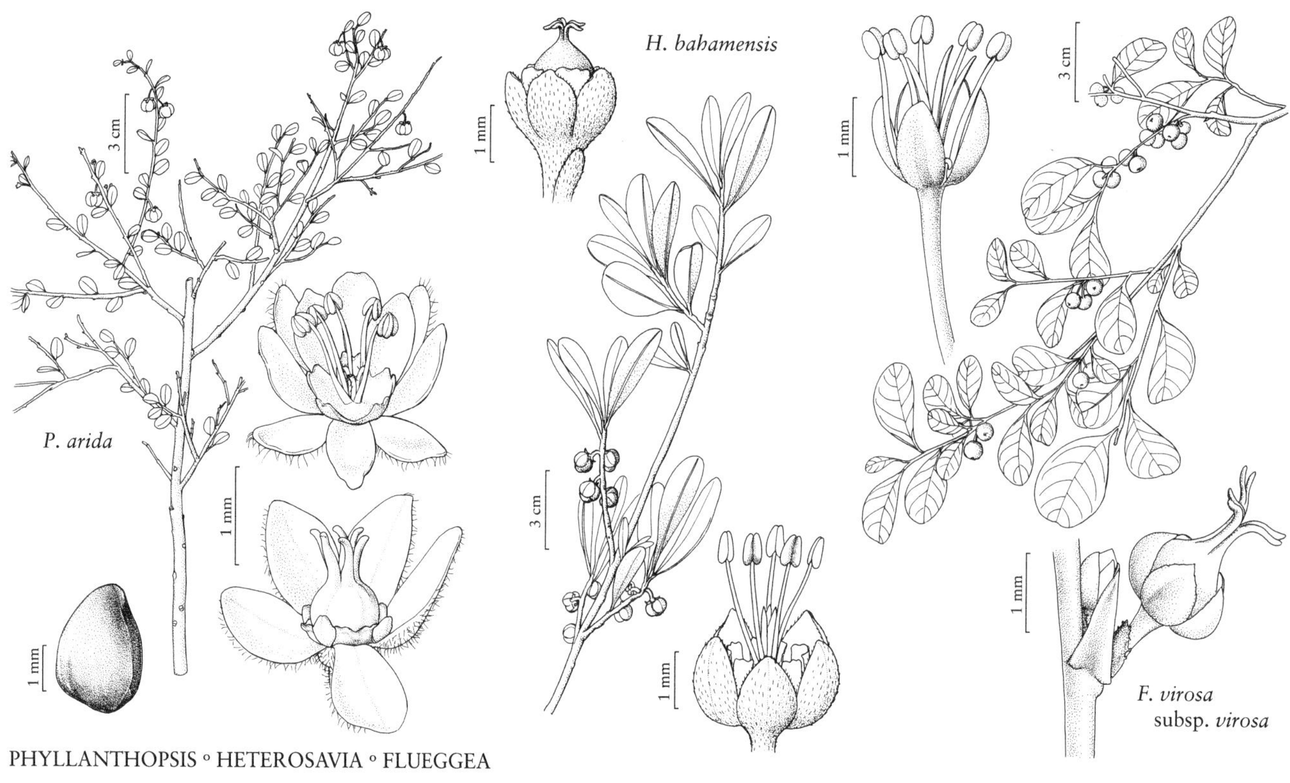

PHYLLANTHOPSIS ◦ HETEROSAVIA ◦ FLUEGGEA

1. **Phyllanthopsis arida** (Warnock & M. C. Johnston) Vorontsova & Petra Hoffmann, Kew Bull. 63: 47. 2008 C F

Savia arida Warnock & M. C. Johnston, SouthW. Naturalist 5: 3. 1960; *Andrachne arida* (Warnock & M. C. Johnston) G. L. Webster

Shrubs dioecious, (2–)4–6(–10) dm. **Stems** highly branched, branches spreading, longest leafy branchlets to 10 cm. **Leaves:** petiole 0.5–3 mm; blade elliptic, 4–10 × 3–6 mm, base cuneate to rounded, apex acute to rounded or mucronate, both surfaces pilose; secondary vein pairs 1–3, obscure. **Inflorescences:** staminate flowers usually solitary, sometimes 2–3 per fascicle; pistillate flowers usually solitary, rarely 2 per fascicle. **Staminate flowers:** sepals oblong to obovate, 1–1.5 mm; petals 0.8–1 mm; nectary 10- or irregularly crenulate; pistillode pilose. **Pistillate flowers:** sepals elliptic to oblong, 1–1.5 mm; petals 0.4–0.6 mm; nectary 5- to 10-crenate; pistil 3(–4)-carpellate. **Capsules** subglobose, 5–6 mm diam., shallowly 3(–4)-lobed; columella 2 mm. **Seeds** dark brown, 2.5–3 mm; caruncle absent.

Flowering and fruiting Jul–Sep. Scrub and rock crevices, limestone; of conservation concern; 1000–2500 m; Tex.; Mexico (Chihuahua, Coahuila).

Phyllanthopsis arida is rare; populations are highly localized, known only from El Solitario and the Dead Horse Mountains in the Big Bend region of Texas and two localities in nearby Mexico. Its distribution may be relictual (B. H. Warnock and M. C. Johnston 1960).

2. **Phyllanthopsis phyllanthoides** (Nuttall) Vorontsova & Petra Hoffmann, Kew Bull. 63: 47. 2008

Lepidanthus phyllanthoides Nuttall, Trans. Amer. Philos. Soc., n. s. 5: 175. 1835; *Andrachne phyllanthoides* (Nuttall) Müller Arg.; *A. reverchonii* J. M. Coulter; *A. roemeriana* (Scheele) Müller Arg.; *Leptopus phyllanthoides* (Nuttall) G. L. Webster; *Phyllanthus roemerianus* Scheele; *Savia phyllanthoides* (Nuttall) Pax & K. Hoffmann; *S. phyllanthoides* var. *reverchonii* (J. M. Coulter) Pax & K. Hoffmann; *S. phyllanthoides* var. *roemeriana* (Scheele) Pax & K. Hoffmann

Shrubs monoecious (often apparently dioecious), 5–10 dm. **Stems** sparsely branched, branches ascending, longest leafy branchlets to 30(–55) cm. **Leaves:** petiole 0–1 mm; blade elliptic or obovate to orbiculate, 7–25 × 6–20 mm, base cuneate to cordate, apex acute to rounded or mucronate, abaxial surface glabrous or pilose, adaxial surface glabrous or midvein pilose; secondary vein pairs 4–7, clearly visible. **Inflorescences:** staminate flowers solitary or 2(–4) per fascicle; pistillate flowers usually solitary, rarely 2 per fascicle. **Staminate flowers:**

sepals elliptic or oblong to obovate, 1.2–2.5 mm; petals 1–2 mm; nectary 5-, 10-, or irregularly crenate; pistillode glabrous. **Pistillate flowers:** sepals elliptic to obovate, 1.5–3 mm; petals 0.4–0.8 mm; nectary 10-crenate; pistil 3-carpellate. **Capsules** subglobose, 3.6–6.5 mm diam., shallowly 3-lobed; columella 2.5–3 mm. **Seeds** orange to brown, sometimes mottled, 1.8–3.5 mm; caruncle usually absent, sometimes present, white to pale orange, 0.5 mm diam. $2n = 26$.

Flowering and fruiting May–Sep. Exposed areas, riverbanks, open woodlands, limestone and dolomite; 0–200 m; Ala., Ark., Mo., Okla., Tenn., Tex.; Mexico (Nuevo León).

Phyllanthopsis phyllanthoides is morphologically variable, common in parts of its range, and sometimes cultivated as an ornamental.

3. HETEROSAVIA (Urban) Petra Hoffmann, Brittonia 60: 152. 2008 • Maidenbush [Greek *hetero*-, other or different from, and genus *Savia*]

Petra Hoffmann

Geoffrey A. Levin

Savia Willdenow sect. *Heterosavia* Urban, Symb. Antill. 3: 284. 1902

Shrubs or trees, dioecious, hairy, glabrescent, hairs simple; branching not phyllanthoid. **Leaves** persistent, alternate, simple, all well developed; stipules persistent; blade margins entire. **Inflorescences** unisexual, staminate glomes, pistillate flowers solitary. **Pedicels:** staminate rudimentary, pistillate present. **Staminate flowers:** sepals 5, distinct; petals 5; nectary extrastaminal, annular, crenate [entire]; stamens 5; filaments distinct or connate basally [to ½ length]; connectives not extending beyond anthers; pistillode 3-divided to base or nearly so. **Pistillate flowers:** sepals persistent, 5, distinct; petals 5; nectary annular, crenate [entire]; pistil 3(–4)-carpellate; styles (3–)4, distinct, 2-fid ½ length. **Fruits** capsules. **Seeds** 2 per locule, rounded-trigonous; seed coat dry, smooth; caruncle absent.

Species 4 (1 in the flora): Florida, West Indies (Greater Antilles).

Molecular phylogenetic analyses have shown that *Savia* as traditionally recognized is polyphyletic and should be treated as three genera, each assigned to a different tribe (H. Kathriarachchi et al. 2005; P. Hoffmann et al. 2006; Hoffmann 2008). *Heterosavia*, which includes the species formerly included in the Caribbean *Savia* sect. *Heterosavia*, belongs to Phyllantheae Dumortier and is sister to *Flueggea*.

SELECTED REFERENCE Hoffmann, P. 2008. Revision of *Heterosavia*, stat. nov., with notes on *Gonatogyne* and *Savia* (Phyllanthaceae). Brittonia 60: 136–166.

1. Heterosavia bahamensis (Britton) Petra Hoffmann, Brittonia 60: 153. 2008 • Bahama maidenbush [F]

Savia bahamensis Britton, Torreya 4: 104. 1904

Shrubs or trees (5–)20–40(–80) dm. **Leaves:** petiole 2–5(–8) mm; blade usually obovate or elliptic, sometimes oblong, (1–)2–4.5(–8) × (0.7–)1.2–2.5 (–4.5) cm, base acute, obtuse, or rounded, apex acute, obtuse, rounded, truncate, or retuse, surfaces usually glabrous, sometimes with scattered hairs. **Inflorescences:** staminate flowers (6–)10–25 per glome. **Pedicels:** pistillate 1–6 mm. **Staminate flowers:** sepals ovate or oblong, 1.5–2 mm, hairy abaxially, sometimes ciliate; petals spatulate to obdeltate, 1–1.5 mm, margins erose to lacerate, apex truncate; stamens 2–4 mm. **Pistillate flowers:** sepals orbiculate, ovate, or deltate, 1–2 mm, hairy abaxially, sometimes ciliate; petals spatulate to orbiculate or oblong, 1–2 mm, entire to slightly erose, sometimes ciliate. **Capsules** 6–7(–9) mm diam. **Seeds** brown to reddish, 3–5.5 mm.

Flowering and fruiting year-round. Hammocks, limestone; 0–10 m; Fla.; West Indies (Greater Antilles).

Heterosavia bahamensis is restricted in the flora area to the lower Florida Keys, where it may be locally abundant, forming monospecific thickets.

4. FLUEGGEA Willdenow, Sp. Pl. 4: 637, 757. 1806 (as Flüggea) • Bushweed [For Johannes Flüggé, 1775–1816, German botanist] [I]

Geoffrey A. Levin

Shrubs [trees], usually dioecious, rarely monoecious, glabrous [pubescent, hairs simple]; branching not phyllanthoid. **Leaves** deciduous [persistent], alternate, simple, all well developed; stipules deciduous [persistent]; blade margins entire. **Inflorescences** unisexual, fascicles [racemes or panicles, or pistillate flowers solitary]. **Pedicels** present. **Staminate flowers:** sepals [4–]5[–7], distinct; petals 0; nectary extrastaminal, [4–]5[–7] glands [annular, lobed]; stamens [4–]5[–7]; filaments distinct; connectives not extending beyond anthers; pistillode [2–]3-divided nearly to base [absent]. **Pistillate flowers:** sepals persistent, [4–]5[–7], distinct; petals 0; nectary annular, entire [angled or lobed]; pistil [2–]3[–4]-carpellate; styles [2–]3[–4], connate proximally, 2-fid. **Fruits** berries [capsules]. **Seeds** 2 per locule, rounded-trigonous; seed coat dry, reticulate or finely verrucose [smooth]; caruncle absent. x = 13.

Species 16 (1 in the flora): introduced, Florida; West Indies, South America, Europe, Asia, Africa, Indian Ocean Islands, Pacific Islands, Australia; primarily tropical and subtropical regions.

Flueggea is primarily an Old World genus, with only three species found in the Neotropics. Phylogenetic studies using DNA sequence data have shown it is monophyletic (with the inclusion of the monospecific *Richeriella* Pax & K. Hoffmann) and is sister to *Heterosavia*, although relatively few species have been sampled (H. Kathriarachchi et al. 2006). The clade of *Flueggea* plus *Heterosavia* is in turn sister to *Phyllanthus* in the broad sense.

SELECTED REFERENCE Webster, G. L. 1984. A revision of *Flueggea* (Euphorbiaceae). Allertonia 3: 259–312.

1. Flueggea virosa (Roxburgh ex Willdenow) Royle, Ill. Bot. Himal. Mts., 328. 1836 (as Fluggea) • Common or simpleleaf bushweed, Chinese waterberry [F] [I]

Phyllanthus virosus Roxburgh ex Willdenow, Sp. Pl. 4: 578. 1805

Subspecies 3 (1 in the flora): introduced, Florida; Asia, Africa, Indian Ocean Islands, Pacific Islands (Malesia), Australia.

1a. Flueggea virosa (Roxburgh ex Willdenow) Royle subsp. virosa [F] [I]

Shrubs 10–60 dm. **Stems** sharply angled when young. **Leaves:** petiole 2–9 mm; blade elliptic to obovate, 2–5 × 1–3 cm, base acute-cuneate, apex rounded to acute. **Inflorescences:** staminate 20–40-flowered, pistillate 3–10-flowered. **Pedicels:** staminate 3–6 mm, pistillate 1.5–5(–12) mm. **Staminate flowers:** sepals ovate to elliptic, 0.8–1.5 mm; pistillode 1.2–2.2 mm, nearly as long as stamens. **Pistillate flowers:** sepals ovate to elliptic, 0.7–1 mm. **Berries** white, 3–5 mm diam. **Seeds** brown, 1.7–2.4(–3) mm. **2*n*** = 26 (Africa, Asia).

Flowering and fruiting summer. Disturbed rocky pinelands; 0–10 m; introduced; Fla.; Asia; Africa; Pacific Islands (Malesia except New Guinea).

Subspecies *virosa* is sometimes cultivated as an ornamental and has become naturalized in Miami-Dade County, where it was first reported escaping in the 1970s. It has weedy tendencies in its native range. The roots, and to a lesser extent the twigs and leaves, are used medicinally in Africa to treat a wide variety of maladies, and the fruits are eaten fresh or made into an alcoholic beverage (J. R. S. Tabuti 2008). The other subspecies grow from northeast India to Myanmar (subsp. *himalaica* D. G. Long) and New Guinea and Australia [subsp. *melanthesoides* (F. Mueller) G. L. Webster].

5. PHYLLANTHUS Linnaeus, Sp. Pl. 2: 981. 1753; Gen. Pl. ed. 5, 422. 1754 • Leafflower [Greek *phyllon*, leaf, and *anthos*, flower, alluding to apparent production of flowers on leaves (actually plagiotropic branchlets) of some species]

Geoffrey A. Levin

Reverchonia A. Gray

Herbs, shrubs, or trees, annual or perennial, terrestrial (*P. fluitans* floating aquatic), usually monoecious, sometimes dioecious, glabrous or hairy, hairs simple [branched]; branching phyllanthoid or not. **Stems** erect to prostrate. **Leaves** persistent or deciduous, alternate, simple, all well developed, scalelike on main stems and well developed on ultimate branchlets, or rarely all scalelike; stipules persistent; blade margins entire. **Inflorescences** unisexual or bisexual, cymules or flowers solitary. **Pedicels** present, pistillate sometimes elongating in fruit. **Staminate flowers:** sepals 4–6, connate basally; petals 0; nectary extrastaminal, 4–6 glands (intrastaminal, annular, 4-lobed in *P. warnockii*); stamens 2–5[–15]; filaments distinct or partially to completely connate; connectives not extending beyond anthers; pistillode absent. **Pistillate flowers:** sepals persistent, (4–)5–6, connate basally; petals 0; nectary annular to cupular, entire or lobed, or distinct glands [absent]; pistil 3(–4)-carpellate; styles 3(–4), distinct or connate to ½ length, 2-fid [rarely unbranched]. **Fruits** capsules or drupes. **Seeds** 2 per locule, rounded-trigonous; seed coat dry, verrucose, papillate, ribbed, or smooth; caruncle absent. x = 8, 9, 13.

Species 800–850 (16 in the flora): North America, Mexico, West Indies, Central America, South America, Asia, Africa, Indian Ocean Islands, Pacific Islands, Australia; introduced in Bermuda, Atlantic Islands (Macaronesia); primarily tropical and subtropical regions.

Phyllanthus is by far the largest genus in Phyllanthaceae and shows tremendous diversity in habit, from trees to small annual herbs, including a floating aquatic herb. Most species exhibit phyllanthoid branching (G. L. Webster 1956–1958), with well-developed leaves and flowers produced only on the ultimate branchlets, which in woody species are deciduous, and scalelike leaves on all other stems (referred to as main stems in this treatment; see family discussion for more details). Phylogenetic studies using DNA sequence data suggest that phyllanthoid branching evolved once and has been lost repeatedly, including within the clades containing *P. caroliniensis* and *P. warnockii* (H. Kathriarachchi et al. 2006). These studies also indicated that *Phyllanthus* is paraphyletic and that few of the subgenera and sections used by Webster (1956–1958, 1967) are monophyletic (K. Wurdack et al. 2004; Kathriarachchi et al. 2005, 2006; P. Hoffmann et al. 2006). However, it is premature to revise the classification of the genus (Kathriarachchi et al. 2006) and the sequence of species used here generally follows the classification by Webster (1956–1958, 1967, 1970). Exceptions are *P. warnockii*, which he treated as the sole member of *Reverchonia* but molecular phylogenetic studies show to be embedded in *Phyllanthus* (Kathriarachchi et al. 2006), and *P. fluitans*, which he did not treat; the latter species appears to be closely related to *P. caroliniensis* (Kathriarachchi et al. 2006).

A number of *Phyllanthus* species are of economic importance. Some cladode-producing species, especially *P. angustifolius* and *P. epiphyllanthus* Linnaeus, are grown as ornamental shrubs in tropical and subtropical areas (and in hothouses elsewhere); the former species has become sparingly naturalized in south Florida. Otaheite (or Tahitian) gooseberry tree, *P. acidus*, is grown throughout the tropics for its tart drupes; it has naturalized in south Florida. The floating aquatic herb *P. fluitans* (floating spurge or red root floater) is increasingly popular for use in tropical fish aquaria. Both *P. amarus* and *P. urinaria* have become weeds throughout tropical and subtropical areas, including the southeastern United States, and the American

P. caroliniensis has become a weed in southeast Asia. Several annual species, notably *P. niruri* and *P. urinaria*, are widely used in folk medicine to treat a variety of ailments, especially urinary problems, and are now the subject of intense pharmacological research.

In *Phyllanthus*, taxa that are not consistently distinct morphologically but are geographically disjunct are recognized as subspecies (for example, *P. caroliniensis* subspp. *caroliniensis* and *saxicola*). Those that intergrade morphologically and geographically, as in *P. abnormis*, are treated as varieties.

SELECTED REFERENCES Webster, G. L. 1970. A revision of *Phyllanthus* (Euphorbiaceae) in the continental United States. Brittonia 22: 44–76. Webster, G. L. and K. I. Miller. 1963. The genus *Reverchonia* (Euphorbiaceae). Rhodora 65: 193–207.

1. Floating aquatic herbs . . . . 6. *Phyllanthus fluitans*
1. Terrestrial herbs, shrubs, or trees.
  2. Branching not phyllanthoid; leaves all well developed.
    3. Leaves distichous; staminate sepals 0.5–0.7 mm.
      4. Stems terete, not winged; capsules 1.6–2 mm diam.; seeds 0.7–1.1 mm . . . . 4. *Phyllanthus caroliniensis*
      4. Stems proximally terete, distally compressed, winged; capsules 2.8–3.2 mm diam.; seeds 1.3–1.5 mm . . . . 5. *Phyllanthus evanescens*
    3. Leaves spiral; staminate sepals 0.7–2.5 mm.
      5. Sepals dark reddish purple, medially incurved and distally spreading; stamens 2, filaments distinct; staminate nectary intrastaminal, annular, 4-lobed; capsules 7–9.8 mm diam.; seeds mottled, (4.4–)4.7–6.2(–6.6) mm. . . . . 14. *Phyllanthus warnockii*
      5. Sepals green, greenish yellow, or pale brown, sometimes suffused with red, flat; stamens 3, filaments connate ⅔+ length; staminate nectary extrastaminal, 6 glands; capsules 2–4 mm diam.; seeds uniformly colored, 0.9–1.8 mm.
        6. Shrubs; leaf blades 1.5–3.5(–5) × 0.7–1.5 mm, apices pungent; staminate pedicels 0.6–0.8 mm, pistillate 0.9–1.2 mm; capsules 2 mm diam.; seeds 0.9–1 mm . . . . 3. *Phyllanthus ericoides*
        6. Perennial herbs; leaf blades 5–24 × 1.5–10 mm, apices rounded, mucronulate, or apiculate; staminate pedicels 1.5–3.5 mm, pistillate 2.5–8 mm; capsules 2.7–4 mm diam.; seeds (1.1–)1.2–1.8 mm.
          7. Herbs with woody caudices; leaf blades 5–10 × 1.5–5 mm; pistillate sepals 1.5–2.5 mm; capsules 2.7–3.2 mm diam.; seeds (1.1–)1.2–1.4 (–1.5) mm . . . . 1. *Phyllanthus polygonoides*
          7. Herbs with rhizomes; leaf blades 10–24 × 5–10 mm; pistillate sepals 2.8–3.5 mm; capsules 4 mm diam.; seeds 1.7–1.8 mm. . . . . 2. *Phyllanthus liebmannianus*
  2. Branching phyllanthoid; leaves on main stems scalelike, on ultimate branchlets well developed or scalelike.
    8. Shrubs or trees.
      9. Ultimate branchlets terete; leaves on ultimate branchlets well developed, deciduous with branchlets; fruits drupes . . . . 8. *Phyllanthus acidus*
      9. Ultimate branchlets flat (cladodes); leaves on ultimate branchlets scalelike, caducous; fruits capsules . . . . 16. *Phyllanthus angustifolius*
    8. Herbs.
      10. Stamens 5, filaments distinct; pistillate pedicels flexuous and pendent in fruit . . . . 7. *Phyllanthus tenellus*
      10. Stamens 2–3, filaments connate; pistillate pedicels spreading in fruit.
        11. Leaf blade abaxial surfaces hispidulous near margins; staminate pedicels 0.1–0.2 mm, pistillate 0.3–0.5 mm; capsules ± tuberculate; seeds transversely ribbed . . . . 9. *Phyllanthus urinaria*
        11. Leaf blade abaxial surfaces glabrous or scabridulous; staminate pedicels 0.2–1.8 mm, pistillate (1–)1.2–7 mm; capsules smooth; seeds verrucose or longitudinally ribbed.

[12. Shifted to left margin.—Ed.]

12. Pistillate sepals 3–3.5 mm, pinnately veined; staminate sepals 1.5–3 mm; seeds verrucose, 1.5–1.8 mm; capsules 3.5 mm diam.; pistillate nectary annular, unlobed............ 10. *Phyllanthus niruri*
12. Pistillate sepals (0.5–)0.7–1.5 mm, 1-veined or obscurely veined; staminate sepals 0.3–1 mm; seeds longitudinally ribbed, 0.8–1.5 mm; capsules 1.7–2.7 mm diam.; pistillate nectary 3 glands or annular and 5–9-lobed.
    13. Perennial herbs; stipules of main stems dark brown, auriculate; capsules 1.7–1.9 mm diam.; seeds 0.8–0.9 mm.. 15. *Phyllanthus pentaphyllus*
    13. Annual herbs; stipules of main stems pale green to pale brown, not auriculate; capsules 1.9–2.7 mm diam.; seeds 0.9–1.5 mm.
        14. Ultimate branchlets narrowly winged, main stems angled; distal inflorescences of solitary pistillate flowers.. 11. *Phyllanthus fraternus*
        14. Ultimate branchlets not winged, main stems terete; distal inflorescences of 1 pistillate flower and 1–3 staminate flowers.
            15. Pistillate nectaries annular, 5–7-lobed; staminate sepals 5(–6); capsules 1.9–2.1 mm diam.; seeds 0.9–1 mm .... 12. *Phyllanthus amarus*
            15. Pistillate nectaries 3 glands; staminate sepals 5–6 in flowers of basal cymules, 4 in flowers of distal cymules; capsules 2.3–2.7 mm diam.; seeds 1.1–1.5 mm ....13. *Phyllanthus abnormis*

1. Phyllanthus polygonoides Nuttall ex Sprengel, Syst. Veg. 3: 23. 1826 • Smartweed or knotweed leafflower [F]

Herbs, perennial, with woody caudex, usually monoecious, rarely dioecious, 1–5 dm; branching not phyllanthoid. **Stems** terete, not winged, glabrous. **Leaves** spiral, all well developed; stipules auriculate, pink or red to medium brown, with hyaline margins; blade narrowly oblong to obovate, 5–10 × 1.5–5 mm, base obtuse, apex acute to mucronulate, both surfaces glabrous or scabridulous. **Inflorescences** cymules or flowers solitary, unisexual or bisexual, with 1(–2) pistillate flowers and/or 1–3 staminate flowers. **Pedicels:** staminate 1.5–3.5 mm, pistillate spreading in fruit, 2.5–7 mm. **Staminate flowers:** sepals (5–)6, greenish yellow, sometimes suffused with red, with white margins, flat, 0.7–1.3 mm; nectary extrastaminal, 6 glands; stamens 3, filaments connate ⅔ length. **Pistillate flowers:** sepals (5–)6, green with white margins, flat, 1.5–2.5 mm, pinnately veined; nectary annular, 6-lobed. **Capsules** 2.7–3.2 mm diam., smooth. **Seeds** uniformly brown, (1.1–)1.2–1.4(–1.5) mm, irregularly verrucose. **2*n*** = 16.

Flowering and fruiting spring–fall. Grasslands, grass-shrublands, glades, especially calcareous soils; 700–2000 m; Ariz., Ark., La., Mo., N.Mex., Okla., Tex.; n, c Mexico.

Phyllanthus polygonoides is closely related to *P. liebmannianus*. Although in the flora area they are allopatric and easily distinguished by the characters used in the key, the differences other than habit are all quantitative, and where the species overlap in parts of northeastern Mexico they can be difficult to separate.

2. Phyllanthus liebmannianus Müller Arg. in A. P. de Candolle and A. L. P. P. de Candolle, Prodr. 15(2): 366. 1866 [C]

Subspecies 2 (1 in the flora): Florida, e Mexico, Central America (Belize, Guatemala).

2a. Phyllanthus liebmannianus Müller Arg. subsp. **platylepis** (Small) G. L. Webster, Brittonia 22: 57. 1970 • Florida leafflower, pinewood dainties [C] [E]

Phyllanthus platylepis Small, Fl. S.E. U.S. ed. 2, 1347. 1913

Herbs, perennial, with rhizomes, monoecious (sometimes staminate), 3–5 dm, branching not phyllanthoid. **Stems** terete, not winged, glabrous. **Leaves** spiral, all well developed; stipules auriculate, pink to brown with pale margins; blade obovate, 10–24 × 5–10 mm, base acute, apex rounded, sometimes apiculate, both surfaces glabrous. **Inflorescences** cymules, bisexual or staminate, with 1 pistillate flower and 3–6 staminate flowers or with 7–9 staminate flowers. **Pedicels:** staminate 2–3.5 mm, pistillate spreading in fruit, 4.5–8 mm. **Staminate flowers:** sepals 6, greenish yellow, flat, 1.2–1.8 mm; nectary extrastaminal, 6 glands; stamens 3, filaments connate 85–90% length. **Pistillate flowers:** sepals 6, green with narrow pale margins, flat, 2.8–3.5 mm, pinnately veined; nectary 6-lobed. **Capsules** 4 mm diam., smooth. **Seeds** uniformly brown, 1.7–1.8 mm, irregularly verrucose.

Flowering and fruiting spring. Hammocks, flatwoods, limestone soils; of conservation concern; 0–20 m; Fla.

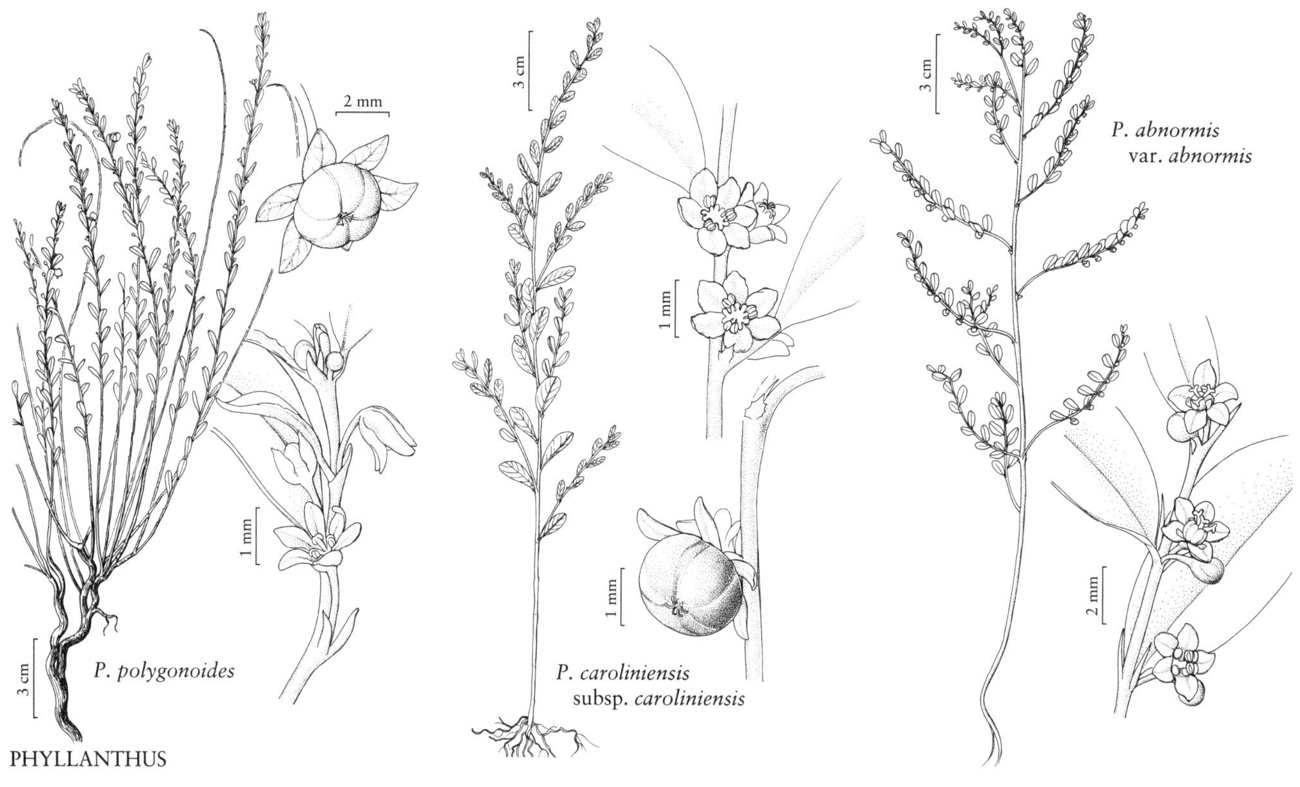

PHYLLANTHUS

Subspecies *platylepis* is restricted to northwestern peninsular Florida (Dixie, Lafayette, Levy, and Taylor counties), where its habitat is threatened by logging. Subspecies *liebmannianus* grows in eastern Mexico and northern Central America; its conservation status is secure. The differences between the subspecies are small, mainly that subsp. *liebmannianus* tends to have taller stems (to 1 m) that arise singly from the rhizome versus in clusters in subsp. *platylepis*, and the subspecies perhaps do not warrant recognition. See 1. *Phyllanthus polygonoides* for a discussion of the relationships between the two species.

3. Phyllanthus ericoides Torrey in W. H. Emory, Rep. U.S. Mex. Bound. 2(1): 193. 1859 • Heather leafflower [C]

Shrubs, monoecious, 0.5–2 dm; branching not phyllanthoid. **Stems** terete, not winged, glabrous. **Leaves** persistent, spiral, all well developed; stipules auriculate, pink with narrow white margins; blade narrowly oblanceolate, 1.5–3.5 (–5) × 0.7–1.5 mm, base obtuse to rounded, apex pungent, both surfaces glabrous. **Inflorescences** solitary flowers, staminate proximal, pistillate distal. **Pedicels:** staminate 0.6–0.8 mm, pistillate spreading in fruit, 0.9–1.2 mm. **Staminate flowers:** sepals 6, pale brownish green, flat, 0.9–1.2 mm; nectary extrastaminal, 6 glands; stamens 3, filaments connate throughout. **Pistillate flowers:** sepals 6, pale brownish green, flat, 1.1–1.4 mm, obscurely veined; nectary annular, 6-lobed. **Capsules** 2 mm diam., smooth. **Seeds** uniformly brown, 0.9–1 mm, verrucose.

Flowering and fruiting year-round, following rains. Desert scrublands on limestone rocks; of conservation concern; 600–700 m; Tex.; Mexico (Chihuahua, Coahuila).

Phyllanthus ericoides is known from only a few populations in Brewster and Terrell counties, Texas, and nearby Chihuahua and Coahuila, Mexico. Although stunted plants of the related *P. polygonoides* may resemble *P. ericoides*, the latter may be distinguished by its woody habit, pungent leaves, shorter pedicels, completely connate filaments, and smaller seeds. In addition, the woody base of *P. ericoides* is reported to smell like coffee with chicory (M. C. Johnston and B. H. Warnock 1963).

4. Phyllanthus caroliniensis Walter, Fl. Carol., 228. 1788 • Carolina leafflower [F]

Herbs, annual or perennial, without caudex or rhizomes, monoecious, 1–4.5 dm; branching not phyllanthoid. **Stems** terete, not winged [narrowly winged], glabrous or scabridulous. **Leaves** distichous; all well developed; stipules auriculate, pale brown or reddish brown; blade elliptic or oblong to obovate, 5–20(–30) × 2–10(–15) mm, base acute, apex obtuse to rounded and apiculate, both surfaces glabrous or scabridulous. **Inflorescences** cymules, bisexual, with 1–3(–5) pistillate flowers and 1–2 staminate flowers. **Pedicels:** staminate 0.5–1 mm, pistillate sharply reflexed in fruit, 0.5–1(–1.5) mm. **Staminate flowers:** sepals (5–)6, pale yellowish green, flat, 0.5–0.7 mm; nectary extrastaminal, 6 glands; stamens 3, filaments distinct. **Pistillate flowers:** sepals (5–)6(–7), green, often suffused with red, with scarious margins, flat, 0.6–1.4 mm, 1-veined; nectary cupular or annular, lobed or unlobed. **Capsules** 1.6–2 mm diam., smooth. **Seeds** uniformly brown, 0.7–1.1 mm, verrucose. 2*n* = 36 (subsp. *guianensis*, West Indies).

Subspecies 4 (2 in the flora): c, e United States, Mexico, West Indies, Central America, South America; introduced in se Asia.

Phyllanthus caroliniensis is the most widespread *Phyllanthus* in the flora area and in the Americas. In addition to the two subspecies in the flora, there are two others. Subspecies *guianensis* (Klotzsch) G. L. Webster, found in the West Indies, Central America, and northern South America, is similar to subsp. *caroliniensis* but distinguished by longer stipules (1.5–2 versus 0.8–12 mm) and staminate nectary glands that are longer than wide (versus as wide or wider than long). Subspecies *stenopterus* (Müller Arg.) G. L. Webster, of southern Central America and northern South America, is recognizable by its narrowly winged stems.

1. Pistillate sepals 0.2–0.3 mm wide; leaf blade vein reticulum clearly visible abaxially; stems glabrous; cymules with 1 staminate and (1–)2–3(–5) pistillate flowers; pistillate nectary cupular, unlobed, enclosing ovary ⅓–½ length . . . . . . 4a. *Phyllanthus caroliniensis* subsp. *caroliniensis*
1. Pistillate sepals (0.2–)0.3–0.5(–0.7) mm wide; leaf blade vein reticulum obscure or invisible abaxially; stems usually sparsely to densely scabridulous; cymules with 1–2 staminate and 1–2 pistillate flowers; pistillate nectary annular, unlobed or 6-lobed, enclosing ovary basally . . . . 4b. *Phyllanthus caroliniensis* subsp. *saxicola*

4a. Phyllanthus caroliniensis Walter subsp. **caroliniensis** [F]

Stems usually with 5+ lateral branches, these often branched, glabrous. **Leaf blades:** both surfaces glabrous; vein reticulum clearly visible abaxially. **Cymules** with 1 staminate and (1–)2–3(–5) pistillate flowers. **Pistillate flowers:** sepals linear-lanceolate or narrowly spatulate, (0.7–)0.8–1(–1.4) × 0.2–0.3 mm, apex acute; nectary cupular, unlobed, enclosing ovary ⅓–½ length at anthesis.

Flowering and fruiting summer–fall. Open, moist areas such as stream banks, lake and pond margins, forest openings, depressions in grasslands, disturbed sites; 0–600 m; Ala., Ark., Del., D.C., Fla., Ga., Ill., Ind., Kans., Ky., La., Md., Miss., Mo., N.J., N.C., Ohio, Okla., Pa., S.C., Tenn., Tex., Va., W.Va.; Mexico; West Indies; Central America; South America; introduced in se Asia.

Subspecies *caroliniensis* is found almost throughout the range of *Phyllanthus caroliniensis*. In Florida, it reaches its native southern limit in Hillsborough County [it was collected on Key West once in the late nineteenth century (*Curtis 185*, GH), where it presumably was an introduced waif]. It has also been found as a garden weed in San Diego, California (*Rebman 7115*, SD), and may become established in that state.

4b. Phyllanthus caroliniensis Walter subsp. **saxicola** (Small) G. L. Webster, Contr. Gray Herb. 176: 46. 1955 • Rock Carolina leafflower

Phyllanthus saxicola Small, Bull. New York Bot. Gard. 3: 428. 1905

Stems usually with 0–5 lateral branches, these usually unbranched, usually sparsely to densely scabridulous, sometimes glabrous. **Leaf blades:** both surfaces glabrous or scabridulous; vein reticulum obscure or invisible abaxially. **Cymules** with 1–2 staminate and 1–2 pistillate flowers. **Pistillate flowers:** sepals oblong to spatulate, 0.7–0.9 × (0.2–)0.3–0.5(–0.7) mm, apex rounded to subacute; nectary annular, unlobed or 6-lobed, enclosing ovary basally.

Flowering and fruiting spring–early winter. Pinewoods, hammocks, flatwoods, prairies, disturbed areas, over limestone; 0–10 m; Fla.; West Indies (Bahamas, Greater Antilles).

Subspecies *saxicola* is a variable entity and some plants can be difficult to distinguish from subsp. *caroliniensis*. Most plants of subsp. *saxicola* look quite distinct with their limited branching and generally smaller leaves (although leaf sizes overlap). They usually also can be recognized by their scabridulous stems, but occasional populations are glabrous. The obscure leaf venation and wider pistillate sepals are diagnostic. In the flora area, subsp. *saxicola* is restricted to southern peninsular Florida from Charlotte County south and is allopatric with subsp. *caroliniensis*. J. K. Small (1933) called this taxon *Phyllanthus pruinosus* Poeppig ex A. Richard, a synonym of *P. discolor* Poeppig ex Sprengel, a species restricted to Cuba.

5. **Phyllanthus evanescens** Brandegee, Zoë 5: 207. 1905 • Birdseed leafflower

Phyllanthus pudens L. C. Wheeler

Herbs, annual, monoecious, 1–5 dm; branching not phyllanthoid. **Stems** proximally terete, distally compressed, winged, scabridulous. **Leaves** distichous, all well developed; stipules auriculate, pale green with pale brown margins; blade elliptic or oblong, 8–20 × 2.5–10 mm, base obtuse to rounded, apex acute to obtuse, both surfaces glabrous. **Inflorescences** cymules, bisexual, with 1–2(–3) pistillate flowers and 1–3 staminate flowers. **Pedicels:** staminate 0.5–0.8 mm, pistillate sharply reflexed in fruit, (1–)1.4–1.8(–2.2) mm. **Staminate flowers:** sepals 5–6, pale brownish green with narrow white margins, flat, 0.5–0.7 mm; nectary extrastaminal, 5–6 glands; stamens 3, filaments connate basally to most of length. **Pistillate flowers:** sepals (5–)6, green (sometimes tinged pink) with narrow white margins, flat, 0.7–1.2 mm, 1-veined; nectary annular, unlobed. **Capsules** 2.8–3.2 mm diam., smooth. **Seeds** uniformly brown, 1.3–1.5 mm, irregularly verrucose.

Flowering and fruiting spring–fall. Coastal prairies, mesquite brushlands; 0–200 m; Ala., La., Tex.; Mexico; Central America (Nicaragua).

Plants from the United States and northeastern Mexico generally have been called *Phyllanthus pudens* and from the rest of Mexico and Central America *P. evanescens*. Characters used to distinguish these species (fruiting pedicel length and seed size) overlap broadly and recent authors treat them as synonyms (G. L. Webster 2001; V. W. Steinmann 2007). A report of *P. evanescens* (as *P. pudens*) from Arkansas (E. Sundell et al. 1999) is based on introduced plants in a nursery; it does not appear to have become established there. The species is introduced in Alabama, first collected there in 2012.

6. **Phyllanthus fluitans** Bentham ex Müller. Arg., Linnaea 32: 36. 1863 • Floating spurge, red root floater [I]

Herbs, perennial, floating aquatic, without caudex or rhizomes, monoecious, 0.5–13 dm; branching not phyllanthoid. **Stems** terete, not winged, glabrous. **Leaves** distichous, all well developed; stipules auriculate, pale brown; blade ± orbiculate, 9–17 mm diam., base cordate, apex rounded to shallowly emarginate, both surfaces papillate. **Inflorescences** cymules, bisexual, with 1–2 staminate and 1–2 pistillate flowers, or flowers solitary. **Pedicels:** staminate 0.5–1 mm, pistillate spreading in fruit, 0.5–1 mm. **Staminate flowers:** sepals (5–)6, white or greenish white, flat, 1–1.4 mm; nectary extrastaminal, (5–)6 glands; stamens 3, filaments distinct. **Pistillate flowers:** sepals (5–)6, white or greenish white, flat, 0.8–1.2 mm, 1-veined; nectary annular, unlobed to lobed. **Capsules** 2.5–3 mm diam., smooth. **Seeds** uniformly brown, 1–1.4 mm, verrucose.

Flowering and fruiting summer–fall. Slow-moving rivers, ponds; 0–10 m; introduced; Fla.; South America; introduced also in Mexico.

Phyllanthus fluitans, the only floating species in the genus, appears to be closely related to *P. caroliniensis* (H. Kathriarachchi et al. 2006). This popular aquarium plant was first discovered in the flora area in 2010 in the Peace River drainage, DeSoto County (G. J. Wilder and M. P. Sowinski 2010); it appears to be naturalized there despite intensive eradication efforts (Sowinski, pers. comm.).

7. **Phyllanthus tenellus** Roxburgh, Fl. Ind. ed. 1832, 3: 668. 1832 • Mascarene Island leafflower [I] [W]

Herbs, annual, monoecious, 2–5 dm; branching phyllanthoid. **Stems:** main stems terete, not winged, glabrous or scabridulous; ultimate branchlets subterete, not winged, glabrous or scabridulous. **Leaves on main stems** spiral, scalelike; stipules not auriculate, reddish brown. **Leaves on ultimate branchlets** distichous, well developed; stipules not auriculate, pale green or pink with paler margins; blade elliptic to obovate, 6–25 × 4–11 mm, base acute to rounded, apex acute to obtuse, both surfaces glabrous. **Inflorescences** cymules or flowers solitary, proximal bisexual with 1–2 pistillate flowers and 2–3 staminate flowers, distal with 1 pistillate flower. **Pedicels:** staminate 0.5–1.5 mm, pistillate flexuous, capillary, and pendent in fruit, (2.5–)3–8 mm. **Staminate flowers:** sepals 5, white except green midrib, flat,

0.4–0.7 mm; nectary extrastaminal, 5 glands; stamens 5, filaments distinct. **Pistillate flowers:** sepals 5, white except green midrib, flat, 0.6–0.8 mm, 1-veined; nectary annular, unlobed. **Capsules** 1.7–1.9 mm diam., smooth. **Seeds** uniformly brown, 0.8–0.9 mm, evenly papillate. $2n = 26$.

Flowering and fruiting spring–fall (year-round in southern areas). Fields, gardens, roadsides, other disturbed areas, especially on sandy soils; 10–500 m; introduced; Ala., Fla., Ga., La., Miss., N.C., S.C., Tenn., Tex., Va.; Asia; Africa; Indian Ocean Islands; introduced also in Mexico, West Indies, South America, Atlantic Islands (Macaronesia), Pacific Islands, Australia.

Phyllanthus tenellus is easily recognized by its long, capillary pistillate pedicels that are flexuous and pendent in fruit; it is native to the Mascarene Islands and perhaps to eastern Africa, other western Indian Ocean Islands, and the Arabian Peninsula, and is widely naturalized in tropical and subtropical regions worldwide. It appears to have been introduced into Florida in the 1920s and is continuing to spread. *Phyllanthus tenellus* has been reported from Arkansas (E. Sundell et al. 1999) and California as a nursery weed (G. F. Hrusa, pers. comm.), and from Oklahoma in flower beds (B. W. Hoagland, pers. comm.), and may be expected to become naturalized in those states.

8. Phyllanthus acidus (Linnaeus) Skeels, U.S.D.A. Bur. Pl. Industr. Bull. 148: 17. 1909 (as acida) • Otaheite or Tahitian gooseberry tree [I]

Averrhoa acida Linnaeus, Sp. Pl. 1: 428. 1753; *Cicca acida* (Linnaeus) Merrill; *C. disticha* Linnaeus

Trees, monoecious, 20–100 dm; branching phyllanthoid. **Stems:** main stems and ultimate branchlets terete, not winged, glabrous. **Leaves on main stems** deciduous, spiral, scalelike; stipules not auriculate, dark brown. **Leaves on ultimate branchlets** deciduous with branchlets, distichous, well developed; stipules not auriculate, dark brown; blade broadly ovate to ovate-lanceolate, (40–)50–90 × (20–)25–45 mm, base obtuse or rounded, apex acute, both surfaces glabrous. **Inflorescences** cymules on leafless short shoots, on old wood bisexual with 1–9 pistillate flowers and 25–40 staminate flowers, on new growth bisexual on proximal shoots with 1–2 pistillate flowers and 8–12 staminate flowers, staminate on distal shoots with 8–12 flowers. **Pedicels:** staminate 1.5–3 mm, pistillate spreading in fruit, 2.3–5(–6) mm. **Staminate flowers:** sepals 4, reddish purple with pink to white margins, flat, 1.1–1.4(–1.5) mm; nectary extrastaminal, 4 glands; stamens (3–)4, filaments distinct. **Pistillate flowers:** sepals 4, green to reddish purple with pink to white margins, flat, (1–)1.2–1.4 mm, 1-veined; nectary annular, 4-lobed. **Drupes** greenish yellow to white, (12–)15–20(–25) mm diam., smooth. **Seeds** uniformly brown, 3.3–3.5 mm, smooth. $2n = 26$ (West Indies).

Flowering and fruiting year-round. Disturbed sites; 0–10 m; introduced; Fla.; South America (Brazil); introduced also in Mexico, West Indies, Central America, elsewhere in South America, Asia, Africa.

Phyllanthus acidus is widely cultivated in the tropics and subtropics for its edible drupes. In the flora area, it is known sparingly from Collier and Monroe counties.

9. Phyllanthus urinaria Linnaeus, Sp. Pl. 2: 982. 1753 • Chamber bitter [I] [W]

Subspecies 2 (1 in the flora): introduced; Asia; introduced also in Mexico, West Indies, Central America, South America, Africa, Indian Ocean Islands, Pacific Islands, Australia; introduced widely in tropics and subtropics.

9a. Phyllanthus urinaria Linnaeus subsp. **urinaria** [I] [W]

Herbs, annual, monoecious, 1–5 dm; branching phyllanthoid. **Stems:** main stems angled, not winged, glabrous; ultimate branchlets compressed, winged, hirsutulous. **Leaves on main stems** spiral, scalelike; stipules auriculate, pale green. **Leaves on ultimate branchlets** distichous, well developed; stipules auriculate, pale green; blade usually oblong or oblong-obovate, sometimes linear, 6–25 × 2–9 mm, base obtuse, apex acute to obtuse and mucronulate, abaxial surface hispidulous near margins, adaxial surface glabrous, margins hispidulous. **Inflorescences** cymules or flowers solitary, unisexual, proximal with 1 pistillate flower, distal with 5–7 staminate flowers. **Pedicels:** staminate 0.1–0.2 mm, pistillate spreading in fruit, 0.3–0.5 mm. **Staminate flowers:** sepals 6, white, flat, 0.3–0.5 mm; nectary extrastaminal, 6 glands; stamens 3, filaments connate throughout. **Pistillate flowers:** sepals 6, yellowish green, sometimes suffused with red, flat, 0.6–0.9 mm, 1-veined; nectary annular, unlobed. **Capsules** 2–2.2 mm diam., ± tuberculate. **Seeds** uniformly brown, 1.1–1.2 mm, transversely ribbed, often with 1–3 lateral pits. $2n = 52$ (West Indies).

Flowering and fruiting year-round. Fields, gardens, roadsides, other disturbed areas; 10–500 m; introduced; Ala., Ark., Fla., Ga., Ill., Kans., La., Miss., N.C., S.C., Tenn., Tex., Va.; Asia; introduced also in Mexico, West Indies, Central America, South America, Africa, Indian Ocean Islands, Pacific Islands, Australia.

Phyllanthus urinaria probably is native only to Asia; it is now found in tropical and subtropical regions worldwide, where it is aggressively weedy in disturbed sites. It was first reported in the United States in the 1940s and spread rapidly. It is unique among *Phyllanthus* species in the flora area in that the leaves are sensitive to the touch, folding like the leaflets of the sensitive plant, *Mimosa pudica*. *Phyllanthus urinaria* has a long history of use in folk medicine to treat a wide variety of ailments and is the subject of intense pharmacological research. Subspecies *nudicarpus* Rossignol & Haicour, is restricted to eastern Asia and is not weedy.

10. Phyllanthus niruri Linnaeus, Sp. Pl. 2: 981. 1753 • Gale of the wind

Phyllanthus lathyroides Kunth; *P. niruri* subsp. *lathyroides* (Kunth) G. L. Webster

Herbs, annual, monoecious, 1–5 dm; branching phyllanthoid. **Stems:** main stems terete, not winged, glabrous; ultimate branchlets subterete, not winged, glabrous. **Leaves on main stems** spiral, scalelike; stipules not auriculate, brown. **Leaves on ultimate branchlets** distichous, well developed; stipules not auriculate, brown; blade elliptic, 11–20 × 4.5–9 mm, base obtuse to rounded, apex obtuse, both surfaces glabrous. **Inflorescences** cymules or flowers solitary, unisexual, proximal with 3–7 staminate flowers, distal with 1 pistillate flower. **Pedicels:** staminate 1.2–1.8 mm, pistillate spreading in fruit, 4–7 mm. **Staminate flowers:** sepals 5(–6), pale green, flat, 1.5–3 mm; nectary extrastaminal, 5(–6) glands; stamens 3, filaments connate ½ length. **Pistillate flowers:** sepals 5, green, flat, 3–3.5 mm, pinnately veined; nectary annular, unlobed. **Capsules** 3.5 mm diam., smooth. **Seeds** uniformly brown, 1.5–1.8 mm, verrucose. $2n = 26$ (Costa Rica).

Flowering and fruiting late summer–fall. River and stream banks, sand; 60–120 m; Tex; Mexico; West Indies; Central America; South America.

Phyllanthus niruri is found in the flora area only in DeWitt, Fayette, and Lavaca counties (and historically from Gonzales County, where it appears to be extirpated; L. E. Brown and S. J. Marcus 1998); it is widespread in the American tropics. Like *P. urinaria*, it is widely used in folk medicine and is the subject of intense pharmacological research. Plants from outside the West Indies and Caribbean northern South America often have been segregated as subsp. *lathyroides*; the differences are trivial, and recent authors (G. L. Webster 2001; V. W. Steinmann 2007) did not subdivide the species.

11. Phyllanthus fraternus G. L. Webster, Contr. Gray Herb. 176: 53. 1955 • Gulf leafflower [I]

Phyllanthus niruri Linnaeus var. *scabrellus* Müller Arg., Linnaea 32: 43. 1863

Herbs, annual, monoecious, 1–4 dm; branching phyllanthoid. **Stems:** main stems angled, not winged, glabrous; ultimate branchlets subterete, narrowly winged, ± scabridulous. **Leaves on main stems** spiral, scalelike; stipules not auriculate, pale green to nearly white. **Leaves on ultimate branchlets** distichous, well developed; stipules not auriculate, pale green to nearly white; blade elliptic-oblong, 6–11 × 3–5 mm, base cuneate to obtuse, apex rounded, both surfaces glabrous. **Inflorescences** cymules or flowers solitary, unisexual, proximal with 2–3 staminate flowers, distal with 1 pistillate flower. **Pedicels:** staminate 0.2–0.5 mm, pistillate spreading in fruit, 1.3–2 mm. **Staminate flowers:** sepals 6, white to pale yellow, flat, 0.4–0.7 mm; nectary extrastaminal, 6 glands; stamens 3, filaments connate throughout. **Pistillate flowers:** sepals 6, green with broad white margins, flat, (1–)1.2–1.5 mm, 1-veined; nectary annular, 6–9-lobed. **Capsules** 2.1 mm diam., smooth. **Seeds** uniformly brown, 0.9–1.1 mm, longitudinally ribbed.

Flowering and fruiting year-round. Fields, roadsides, gardens, other disturbed areas; 0–30 m; introduced; Fla., Ga., La., Miss., S.C., Tex.; s Asia (India, Pakistan); introduced also in West Indies, Bermuda, w Asia, Africa, Indian Ocean Islands.

Phyllanthus fraternus, native to Pakistan and northwest India, apparently was introduced into the United States in the 1950s, first in Louisiana, and has spread mainly along the Gulf Coast.

12. Phyllanthus amarus Schumacher & Thonning in H. C. F. Schumacher, Beskr. Guin. Pl., 421. 1827 • Carry me seed, dixie leafflower, gale of wind [I]

Herbs, annual, monoecious, 1–5 dm; branching phyllanthoid. **Stems:** main stems terete, not winged, glabrous; ultimate branchlets subterete, not winged, glabrous or slightly scabridulous proximally. **Leaves on main stems** spiral, scalelike; stipules not auriculate, pale brown. **Leaves on ultimate branchlets** distichous, well developed; stipules not auriculate, pale green with pale brown margins; blade elliptic-oblong to obovate, 5–11 × 3–6 mm, base obtuse or rounded, apex obtuse to rounded, often apiculate, both surfaces glabrous. **Inflorescences** cymules or flowers solitary, proximal staminate with 1–2 flowers, distal bisexual with 1 pistillate flower and 1 staminate flower. **Pedicels:** staminate 0.6–1.3 mm, pistillate spreading in fruit, (1–)1.2–2 mm. **Staminate flowers:** sepals 5(–6), pale green, flat, 0.3–0.6 mm; nectary extrastaminal, 5 glands; stamens (2–)3, filaments connate throughout. **Pistillate flowers:** sepals (5–)6, green with broad pale green to white margins, flat, 0.8–1.1 mm, 1-veined; nectary annular, 5(–7)-lobed. **Capsules** 1.9–2.1 mm diam., smooth. **Seeds** uniformly brown, 0.9–1 mm, longitudinally ribbed. $2n$ = 52 (Jamaica).

Flowering and fruiting year-round. Fields, roadsides, gardens, other disturbed areas; 0–10 m; introduced; Fla.; Mexico; West Indies; Central America; South America; introduced also in Asia, Africa, Indian Ocean Islands, Pacific Islands, Australia.

Phyllanthus amarus, a pantropical weed, appears to be native to the American tropics. In the flora area, it is naturalized only in southern Florida, where it apparently was introduced sometime in the second half of the nineteenth century. It has been found in nursery stock in California (G. F. Hrusa, pers. comm.) and may become naturalized there.

13. Phyllanthus abnormis Baillon, Adansonia 1: 42. 1860 [F]

Herbs, annual, sometimes becoming woody and then appearing to be perennial, monoecious, 1–5 dm; branching phyllanthoid. **Stems:** main stems persistent, terete, not winged, glabrous or scabridulous; ultimate branchlets deciduous, subterete, not winged, glabrous or scabridulous. **Leaves on main stems** spiral, scalelike; stipules not auriculate, pale reddish brown. **Leaves on ultimate branchlets** distichous, well developed; stipules not auriculate, greenish white; blade elliptic to oblong, 3–10 × 1–4 mm, base cuneate to subcordate, apex obtuse to emarginate, both surfaces glabrous or scabridulous. **Inflorescences** cymules, proximal staminate with 2 flowers, distal bisexual with 1 pistillate flower and 1–3 staminate flowers. **Pedicels:** staminate 0.7–1.5 mm, pistillate spreading in fruit, (1–)1.5–3(–3.5) mm. **Staminate flowers:** sepals 5–6 in basal cymules, 4 in distal cymules, pale yellowish green, sometimes suffused with red, flat, 0.5–1 mm; nectary extrastaminal, 4(–6) glands; stamens 2(–3 in basal cymules), filaments connate throughout. **Pistillate flowers:** sepals 5–6, green with nearly white margins, flat, (0.5–)0.7–1.1 mm, 1-veined; nectary 3 glands (2-fid distally, thus resembling 6 glands). **Capsules** 2.3–2.7 mm diam., smooth. **Seeds** uniformly brown, 1.1–1.5 mm, longitudinally ribbed.

Varieties 2 (2 in the flora): s United States, ne Mexico.

1. Pistillate nectary glands strongly unequal, spatulate, as long as or longer than broad; leaf blades glabrous on both surfaces or sparsely to moderately scabridulous abaxially; bisexual cymules with 1 staminate flower . . . . 13a. *Phyllanthus abnormis* var. *abnormis*
1. Pistillate nectary glands subequal, reniform, broader than long; leaf blades densely scabridulous on both surfaces; bisexual cymules with 1–3 staminate flowers . . . . 13b. *Phyllanthus abnormis* var. *riograndensis*

13a. Phyllanthus abnormis Baillon var. **abnormis** • Drummond's leafflower [F]

Phyllanthus garberi Small

Stems glabrous to densely scabridulous. **Leaf blades** glabrous on both surfaces or sparsely to moderately scabridulous abaxially. **Bisexual cymules** with 1 staminate flower. **Pistillate nectary** glands strongly unequal, spatulate, as long as or longer than broad.

Flowering and fruiting spring–fall (year-round in Florida). Open oak woodlands, prairies, barrens, dunes, always on sand; 0–1200 m; Fla., N.Mex., Okla., Tex.; Mexico (Tamaulipas).

Variety *abnormis* is disjunct between peninsular Florida and Texas, southwestern Oklahoma, southeastern New Mexico, and northern Tamaulipas. The Florida plants generally have three stamens in one of the two staminate flowers in the proximal cymules; both flowers have only two stamens in the western plants. The stems are consistently smooth in Florida, southern Texas, and Tamaulipas, but usually moderately to densely scabridulous elsewhere. The hypothesis by G. L. Webster (1970) that plants of var. *abnormis* with scabridulous stems show introgression from var. *riograndensis* is unlikely because populations of var. *abnormis* growing closest to var. *riograndensis* have smooth stems.

13b. Phyllanthus abnormis Baillon var. **riograndensis** G. L. Webster, Ann. Missouri Bot. Gard. 54: 198. 1967 • Rio Grande leafflower [E]

Stems densely scabridulous. **Leaf blades** densely scabridulous on both surfaces. **Bisexual cymules** with 1–3 staminate flowers. **Pistillate nectary** glands subequal, reniform, broader than long.

Flowering and fruiting spring–fall. Thornscrub, mesquite woodlands, sand, sandy loam; 20–200 m; Tex.

Variety *riograndensis* is found only in the Rio Grande Valley from Dimmit County to Hidalgo County, generally close to the river itself. Although it should be expected from adjacent Mexico, it has not been reported from there, perhaps because of limited collecting. The only consistent differences between vars. *riograndensis* and *abnormis* are the shape of the nectary glands in the pistillate flowers and the indumentum of the adaxial leaf surface; where the two varieties grow close together in the lower Rio Grande Valley, they also consistently differ in stem indumentum.

14. Phyllanthus warnockii G. L. Webster, Contr. Univ. Michigan Herb. 25: 235. 2007 • Sand reverchonia

Reverchonia arenaria A. Gray, Proc. Amer. Acad. Arts 16: 107. 1880, not *Phyllanthus arenarius* Beille 1927

Herbs, annual, monoecious, 2–5 dm; branching not phyllanthoid. **Stems** terete, not winged, glabrous. **Leaves** spiral, all well developed; stipules not auriculate, reddish purple to pale brown; blade elliptic to narrowly oblong-elliptic or nearly linear, (15–)20–40(–45) × (1.8–)2.5–8(–9) mm, base cuneate-attenuate, apex acute to mucronate, both surfaces glabrous. **Inflorescences** cymules, borne on lateral branches only, bisexual, with 1 pistillate flower and 4–6 staminate flowers. **Pedicels:** staminate 1.5–2.5 mm, pistillate spreading to sharply recurved, (2.5–)3.2–6.5(–8.7) mm. **Staminate flowers:** sepals 4, dark reddish purple, central portion sometimes paler, medially incurved and distally spreading (calyx appearing urceolate), 1.5–2.5 mm; nectary intrastaminal, annular, 4-lobed; stamens 2, filaments distinct. **Pistillate flowers:** sepals (5–)6, dark reddish purple, central portion sometimes paler and greenish purple, proximally flat to incurved and distally spreading, (1.3–)1.5–2.5(–2.9) mm, 1-veined; nectary annular, entire or 6-angled. **Capsules** 7–9.8 mm diam., smooth. **Seeds** mottled light and dark brown, (4.4–)4.7–6.2(–6.6) mm, 2 surfaces minutely papillate, 1 surface smooth. **2*n*** = 16.

Flowering and fruiting summer-fall. Dunes; 300–1800 m; Ariz., Colo., Kans., N.Mex., Okla., Tex., Utah; Mexico (Chihuahua).

Phyllanthus warnockii is endemic to quartz sand dunes. Although always recognized as close to *Phyllanthus*, it generally has been segregated as the monospecific *Reverchonia* because of its unique habit (leafy main stems and flowers restricted to lateral branches), calyx shape and color, staminate nectary, and embryo with linear cotyledons. Phylogenetic analysis of DNA sequence data shows that *P. warnockii* not only is embedded within *Phyllanthus*, but also that it is most closely related to *P. abnormis*, which also is restricted to sandy soil (H. Kathriarachchi et al. 2006). These results suggest that the peculiar habit of *P. warnockii* reflects partial loss of the phyllanthoid branching syndrome.

15. Phyllanthus pentaphyllus C. Wright ex Grisebach, Nachr. Königl. Ges. Wiss. Georg-Augusts-Univ. 1865: 167. 1865 • Fivepetal leafflower

Phyllanthus pentaphyllus var. *floridanus* G. L. Webster

Herbs, perennial, with woody caudex, dioecious or monoecious, often staminate and pistillate flowers on separate branchlets, 0.5–3 dm; branching phyllanthoid. **Stems:** main stems and ultimate branchlets terete, not winged, usually glabrous, rarely scabridulous. **Leaves on main stems** spiral, scalelike; stipules auriculate, dark brown. **Leaves on ultimate branchlets** distichous, well developed; stipules not auriculate, pale brown to brown; blade elliptic or obovate to suborbiculate, 2–8 × 1–5 mm, base acute to rounded, apex obtuse to rounded and apiculate, both surfaces glabrous or abaxial scabridulous. **Inflorescences** cymules or flowers solitary, unisexual, staminate distributed along branchlet, with (10–)15–20 flowers, pistillate distributed along branchlet or distal, with 1 flower. **Pedicels:** staminate 0.3–0.8 mm, pistillate spreading in fruit, (1–)1.2–1.8(–2.1) mm. **Staminate flowers:** sepals 5, pale yellow to white, flat, 0.7–0.8 mm; nectary extrastaminal, 5 glands; stamens 2, filaments connate ⅔ length. **Pistillate flowers:** sepals 5, green with broad white margins, flat, (0.7–)0.9–1.2 mm, obscurely veined; nectary annular, 5-lobed. **Capsules** 1.7–1.9 mm diam., smooth. **Seeds** uniformly brown, 0.8–0.9 mm, longitudinally ribbed. **2*n*** = 52.

Flowering and fruiting year-round. Rocky pinelands on limestone; 0–10 m; Fla.; West Indies (Bahamas, Cuba, Hispaniola).

G. L. Webster (1955, 1970) treated the Florida plants of *Phyllanthus pentaphyllus* as var. *floridanus*, distinguishing them from those of the West Indies based on the former being dioecious or at least having branchlets that produce flowers of only one sex, and the latter being monoecious with each branchlet bearing both staminate and pistillate flowers. However, he acknowledged that dioecious individuals are found in Cuba and that the distinction between the varieties was weak (Webster 1955, 1956–58). Examination of additional specimens shows that plants with monoecious branchlets are not uncommon in Florida and that dioecious plants are found in both Cuba and the Bahamas; therefore, var. *floridanus* is not recognized here.

G. L. Webster (1955) treated *Phyllanthus polycladus* Urban of Puerto Rico and the Lesser Antilles as a subspecies of *P. pentaphyllus*. These taxa seem amply different in habit, leaf blade shape and texture, and pistillate nectary shape, and are here considered separate species.

Phyllanthus pentaphyllus in the flora area is restricted to Miami-Dade and Monroe counties, where its rocky pineland habitat is threatened by development. Variety *floridanus*, when recognized, is regarded as threatened.

16. Phyllanthus angustifolius (Swartz) Swartz, Fl. Ind. Occid. 2: 1111. 1800 (as angustifolia) • Foliage flower, swordbush [I]

Xylophylla angustifolia Swartz, Prodr., 28. 1788

Shrubs, monoecious, 10–30 dm; branching phyllanthoid. **Stems:** main stems terete, not winged, glabrous; ultimate branchlets deciduous, flat cladodes to 1 cm wide, not winged, glabrous. **Leaves on main stems** deciduous, spiral, scalelike; stipules not auriculate, dark brown. **Leaves on ultimate branchlets** caducous, distichous, scalelike; stipules absent. **Inflorescences** cymules, bisexual or staminate, with 1–2(–3) pistillate and/or 2–5 staminate flowers. **Pedicels:** staminate (1–)2–6 mm, pistillate spreading in fruit, (1–)2–4(–7) mm. **Staminate flowers:** sepals 6, purplish pink to cream, flat, (0.8–)1–1.5(–2.3) mm; nectary extrastaminal, 6 distinct glands; stamens 3(–5), filaments connate ½–¾ length. **Pistillate flowers:** sepals 6, purplish pink to cream, flat, 1–1.5(–2.2) mm, 1-veined; nectary annular, undulate or crenulate. **Capsules** 3–4 mm diam., finely wrinkled. **Seeds** uniformly brown, 1.4–2.6 mm, irregularly verrucose.

Flowering and fruiting year-round. Disturbed hammocks; 0–10 m; introduced; Fla.; West Indies (Cayman Islands, Jamaica, Swan Islands).

Phyllanthus angustifolius is cultivated as an ornamental throughout the West Indies and in Florida. It has become naturalized in southern Florida (Miami-Dade and Monroe counties). The flattened ultimate branchlets are strikingly leaflike; their correct homology is demonstrated by the cymules of flowers borne along the margins. J. K. Small (1933) referred the Florida plants to *Xylophylla angustifolia* var. *linearis* Swartz, but that name is a synonym of *P. arbuscula* (Swartz) J. F. Gmelin, a West Indian species that also is cultivated in Florida but does not appear to have become naturalized there.

6. BREYNIA J. R. Forster & G. Forster, Char. Gen. Pl. ed. 2, 145, plate 73. 1776, name conserved • Snowbush [For Jacob Breyne, 1637–1697, and his son Johann Philipp Breyne, 1680–1764, Polish botanists] [I]

Geoffrey A. Levin

Shrubs [trees], monoecious [dioecious], glabrous [hairy, hairs simple]; branching phyllanthoid. **Leaves** persistent, alternate, simple, scalelike on main stems, well developed on ultimate branchlets; stipules persistent [deciduous]; blade margins entire. **Inflorescences** unisexual, staminate proximal, few-flowered fascicles or flowers solitary, pistillate distal, flowers solitary. **Pedicels** present. **Staminate flowers:** sepals 6, connate throughout [connate basally to most of length], with scales at rim of calyx tube [near bases of lobes]; petals 0; nectary absent; stamens 3; filaments connate; connectives not extending beyond anthers; pistillode absent. **Pistillate flowers:** sepals persistent, 6, connate basally; petals 0; nectary absent; pistil 3-carpellate; styles 3, distinct, 2-fid. **Fruits** capsules. **Seeds** 2 per locule, rounded-trigonous; seed coat fleshy, smooth; caruncle absent. $x = 13$.

Species 10–30 (1 in the flora): introduced, Florida; Asia, Indian Ocean Islands, Pacific Islands, Australia; introduced also in West Indies, Africa; tropical and subtropical regions.

Breynia exhibits phyllanthoid branching (G. L. Webster 1956–1958), with well-developed leaves and flowers produced only on the deciduous ultimate branchlets and scalelike leaves on all other stems (referred to as main stems in this treatment). Like *Glochidion* and some *Phyllanthus* species, *Breynia* has a pollination mutualism with thc moth genus *Epicephala* (reviewed in

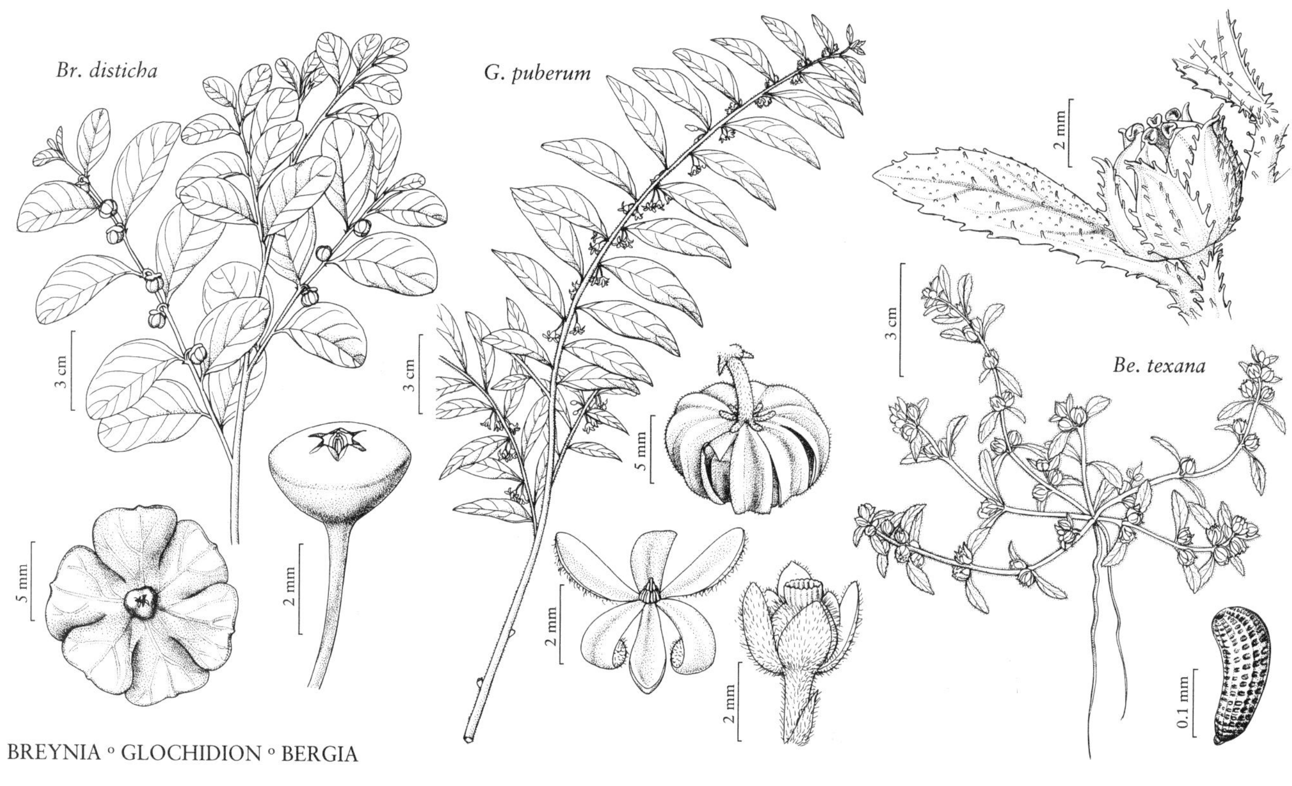

BREYNIA ◦ GLOCHIDION ◦ BERGIA

A. Kawakita and M. Kato 2009); see the discussion under 7. *Glochidion* for more information. *Breynia* is difficult taxonomically, hence the wide range in the number of species recognized within the genus.

Some *Breynia* species are grown as ornamental shrubs in tropical and subtropical areas; *B. disticha* is the most common and widespread.

1. **Breynia disticha** J. R. Forster & G. Forster, Char. Gen. Pl. ed. 2, 146, plate 73. 1776 [F] [I]

Breynia nivosa (W. Bull) Small

Shrubs, 0.5–3 m. **Leaves on main stems** spiral. **Leaves on ultimate branchlets** distichous; petiole 2–4 mm; blade elliptic to ovate or suborbiculate, 1.5–4 × 1–3 cm, base obtuse to rounded, apex rounded. **Pedicels:** staminate 6–9 mm, pistillate 3–9 mm. **Staminate flowers:** calyx top-shaped, truncate, 1.5–3 mm. **Pistillate flowers:** sepals obovate, 1–5 mm. **Capsules** 5 mm diam., smooth. **Seeds** brown, 2–3 mm. **2*n*** = 52 (Jamaica).

Flowering and fruiting summer. Disturbed sites; 0–10 m; introduced; Fla.; Pacific Islands; introduced also in West Indies, Indian Ocean Islands.

Breynia disticha, native to New Caledonia and Vanuatu, is widely grown in tropical and subtropical areas and occasionally becomes naturalized. In the flora area, it is limited to scattered counties in southern Florida. Most plants are cultivars with variegated leaves in various shades of green, white, red, and pink; naturalized populations may revert to uniformly green leaves. The calyx typically reproduces the colors of the leaves, sometimes even being variegated.

7. GLOCHIDION J. R. Forster & G. Forster, Char. Gen. Pl. ed. 2, 113, plate 57. 1776, name conserved • Needlebush, cheesetree [Greek *glochis*, projecting point or barb on an arrowhead, and *-idion*, smaller or little, alluding to pointed extension of anther connectives] [I]

Geoffrey A. Levin

Shrubs [trees], monoecious [dioecious], hairy [glabrous], hairs simple; branching phyllanthoid. **Leaves** deciduous [persistent], alternate, simple, scalelike on main stems, well developed on ultimate branchlets; stipules persistent; blade margins entire. **Inflorescences** unisexual or bisexual, fascicles [cymules], proximal mostly or entirely staminate, distal mostly or entirely pistillate. **Pedicels** present [absent]. **Staminate flowers:** sepals [5–]6, distinct; petals 0; nectary absent; stamens 3–4[–8]; filaments and anthers connate; connectives extending beyond anthers as deltoid appendages; pistillode absent. **Pistillate flowers:** sepals persistent, 6, distinct; petals 0; nectary absent; pistil [3–]5–10[–15]-carpellate; styles [3–]5–10[–15], connate nearly to apex, unbranched. **Fruits** capsules. **Seeds** 2 per locule, rounded-trigonous [ellipsoid]; seed coat ± fleshy, smooth; caruncle absent. x = 13, 16.

Species 200–300 (1 in the flora): introduced, Alabama; Asia, Pacific Islands, Australia; tropical and subtropical regions.

Glochidion exhibits phyllanthoid branching (G. L. Webster 1956–1958), with well-developed leaves and flowers produced only on the deciduous ultimate branchlets and scalelike leaves on all other stems (referred to as main stems in this treatment).

An obligate pollination mutualism similar to the well-known fig-fig wasp and yucca-yucca moth mutualisms has evolved between *Breynia*, *Glochidion*, and various *Phyllanthus* species [particularly subg. *Gomphidium* (Baillon) G. L. Webster] and the moth genus *Epicephala* (reviewed in A. Kawakita and M. Kato 2009). Species-specific moths pollinate the flowers nocturnally, with female moths collecting pollen on specialized proboscides and laying their eggs on the pistillate flowers. The developing larvae consume a portion of the seeds, leaving the remainder untouched. Phylogenetic analysis demonstrated that specialized behavior evolved once in *Epicephala* and that the mutualism has evolved repeatedly within Phyllantheae (Kawakita and Kato).

1. Glochidion puberum (Linnaeus) Hutchinson in C. S. Sargent, Pl. Wilson. 2: 518. 1916 [F] [I]

Agyneia pubera Linnaeus, Mant. Pl. 2: 296. 1771

Shrubs, 1–5 m. **Leaves on main stems** spiral. **Leaves on ultimate branchlets** distichous; petiole 1–4 mm; blade usually oblong, oblong-ovate, or obovate-oblong, rarely lanceolate, 3–9 × 1–3 cm, base cuneate to obtuse, apex acute or acuminate to obtuse or rounded. **Pedicels:** staminate 4–15 mm, pistillate 1–2 mm. **Staminate flowers:** sepals oblong to oblong-obovate, 2.5–3.5 mm. **Pistillate flowers:** sepals oblong, 2–3 mm. **Capsules** grooved, 8–15 mm diam., densely hairy. **Seeds** red, 4 mm. $2n$ = 64 (China).

Flowering and fruiting summer. Disturbed mixed forests; 10–20 m; introduced; Ala.; Asia (China, Japan, Taiwan).

Glochidion puberum is sometimes cultivated in tropical and subtropical areas and occasionally becomes naturalized. In the flora area, the species is limited to southern Alabama (Mobile County), where it was first found in the late twentieth century (M. L. Fearn and L. E. Urbatsch 2001). In its native range, roots, leaves, and fruit are used medicinally to treat a variety of ailments (Lai X. Z. et al. 2004), and the seed oil is used to make soap and as a lubricant (Li B. and M. G. Gilbert 2008).

ELATINACEAE Dumortier
• Waterwort Family

Gordon C. Tucker

Herbs [**subshrubs**], annual [short-lived perennial], synoecious [polygamous]. **Leaves** opposite [whorled], simple; stipules present; petiole present or absent; blade margins entire or serrulate; venation pinnate. **Inflorescences** axillary, usually cymes or flowers solitary, sometimes 2 [–3]-flowered clusters. **Flowers** bisexual [pistillate]; perianth and androecium hypogynous; hypanthium absent; sepals 2–5, distinct or connate basally; petals (0 or)2–5, distinct; nectary absent; stamens [0–]1–10, distinct, free; anthers dehiscing by longitudinal slits; pistil 1, 2–5-carpellate, ovary superior, 2–5-locular, placentation axile; ovules 2–33[–44] per locule, anatropous; styles 2–5, distinct; stigmas 2–5, capitate. **Fruits** capsules, dehiscence septicidal or irregular. **Seeds** 2–33[–44] per locule.

Genera 2, species ca. 50 (2 genera, 11 species in the flora): nearly worldwide; temperate and tropical regions.

The affinities of Elatinaceae have long been uncertain; relationships with Caryophyllaceae and Clusiaceae have been proposed (G. C. Tucker 1986). Recent molecular work, combined with a review of morphology, indicates a sister relationship with Malpighiaceae (C. C. Davis and M. W. Chase 2004; K. Wurdack and Davis 2009).

SELECTED REFERENCES Davis, C. C. and M. W. Chase. 2004. Elatinaceae are sister to Malpighiaceae; Peridiscaceae belong to Saxifragales. Amer. J. Bot. 91: 262–273. Tucker, G. C. 1986. The genera of Elatinaceae in the southeastern United States. J. Arnold Arbor. 67: 471–483.

1. Sepals 5, carinate; petals 5; inflorescences usually cymes, rarely solitary flowers, pedicels present; plants glandular-pubescent; stems solid or pithy . . . . 1. *Bergia*, p. 349
1. Sepals 2–4, not carinate; petals (0 or)2–4; inflorescences solitary flowers, pedicels present or absent, or if flowers 2 per node (*E. chilensis*) then pedicels absent; plants glabrous; stems with longitudinal air spaces . . . . 2. *Elatine*, p. 349

1. BERGIA Linnaeus, Mant. Pl. 2: 152, 241. 1771 • Water-fire [For Peter Jonas Bergius, 1730–1790, Swedish botanist and physician, student of Linnaeus]

Gordon C. Tucker

Herbs [subshrubs], terrestrial [aquatic], glandular-pubescent. **Stems** ascending to prostrate, solid or pithy, not rooting at nodes [rooting at proximal nodes]. **Leaves:** stipules nearly membranous; petiole present; blade margins serrulate. **Inflorescences** usually cymes, rarely flowers solitary. **Pedicels** present. **Flowers:** sepals 5, distinct, equal, carinate, apex acuminate or acute; petals 5, apex acute; stamens (5 or 7–)10; pistil 5-carpellate; ovary 5-locular, apex ± acute; styles 5; stigmas 5. **Capsules** cartilaginous. **Seeds** 3–6[–10] per locule, brown, slightly curved, surface obscurely reticulate with rectangular pits. $x = 6$.

Species ca. 25 (1 in the flora): w, c United States, n Mexico, Asia, Africa, Australia; introduced in South America; warm-temperate and tropical regions.

Bergia has its greatest diversity in Africa and Australia.

1. **Bergia texana** (Hooker) Seubert in W. G. Walpers, Repert. Bot. Syst. 1: 285. 1842 • Texas water-fire [F]

Merimea texana Hooker, Icon. Pl. 3: plate 278. 1840; *Elatine texana* (Hooker) Torrey & A. Gray

Herbs, (5–)10–30(–40) cm, often branched from base. **Leaves:** stipules 2-fid, 1–3 mm, lobes lanceolate, margins laciniate-denticulate; petiole 1–3 mm; blade oblong to oblong-lanceolate, 1–3 cm × 3–10 mm, base attenuate, apex acute or acuminate, surfaces glandular-puberulent or adaxial glabrous. **Inflorescences** (1–)3–5-flowered. **Pedicels** ± erect, (0.5–)3–8 mm, slender, glandular. **Flowers:** sepals greenish white, narrowly elliptic or lanceolate, 2–3.5 mm; petals white (sometimes reddish distally), obovate or elliptic, 2.5 mm. **Capsules** ovoid to nearly globose, 2–2.5(–3) mm wide. **Seeds** oblong, 0.4–0.6 mm; pits in 8–9 rows, 15 per row.

Flowering summer(–early fall). Muddy or sandy shores and flats, emergent shorelines of rivers, lakes, playas, pools, ditches, ponds; 0–1600 m; Ark., Calif., Colo., Idaho, Ill., Kans., La., Mo., Mont., Nebr., Nev., N.Mex., Okla., Oreg., S.Dak., Tex., Utah, Wash.; Mexico (Baja California, Baja California Sur, Nuevo León, Sinaloa, Sonora, Tamaulipas).

2. ELATINE Linnaeus, Sp. Pl. 1: 367. 1753; Gen. Pl. ed. 5, 172. 1754 • Waterwort, élatine [Greek name for a plant with firlike leaves]

Hamid Razifard

Gordon C. Tucker

Donald H. Les

Crypta Nuttall

Herbs, submersed or emergent aquatic, glabrous. **Stems** erect, ascending, decumbent, or prostrate, with longitudinal air spaces, rooting at nodes. **Leaves:** stipules membranous; petiole present or absent; blade margins entire, with hydathodes. **Inflorescences:** flowers usually solitary, sometimes 2[–3] per node. **Pedicels** present or absent. **Flowers:** sepals 2–4, connate basally, equal or 1 smaller, not carinate, apex obtuse; petals (0 or)2–4, apex obtuse; stamens [0–]1–8; pistil 2–4-carpellate, ovary 2–4-locular, apex truncate; styles 2–4; stigmas 2–4. **Capsules** membranous. **Seeds** 2–33[–44] per locule, brown to yellowish brown, straight or curved (nearly circular in *E. californica*), surface with hexagonal, rectangular, elliptic, or ± round pits (pits oriented with longer dimension at right angles to length of seed). $x = 9$.

Species ca. 25 (10 in the flora): North America, Mexico, South America, Europe, Asia, Africa, Pacific Islands, Australia.

SELECTED REFERENCES Duncan, W. H. 1964. New *Elatine* populations in the southeastern United States. Rhodora 66: 47–53. Fassett, N. C. 1939. Notes from the herbarium of the University of Wisconsin. No. 17. *Elatine* and other aquatics. Rhodora 41: 367–377. Fernald, M. L. 1917. The genus *Elatine* in eastern North America. Rhodora 19: 10–15. Fernald, M. L. 1941b. *Elatine americana* and *E. triandra*. Rhodora 43: 208–211.

1. Petals 2; capsules 2-locular . . . . 10. *Elatine minima*
1. Petals usually 3–4, sometimes 0; capsules 3–4-locular.
    2. Sepals 4; petals 4; capsules 4-locular.
        3. Seeds curved 90–180°; pedicels 1.5–2.5(–3.5) mm, recurved in fruit; w United States . . . . 1. *Elatine californica*
        3. Seeds straight or curved to 15°; pedicels 0.1–23 mm, erect; Quebec . . . . 2. *Elatine ojibwayensis*
    2. Sepals 2–3 (rarely 4 in *E. triandra*); petals 3 (sometimes 0 in *E. ambigua*); capsules 3-locular.
        4. Pedicels recurved in fruit, 0.5–2.5 mm . . . . 6. *Elatine ambigua*
        4. Pedicels erect, 0–0.5 mm.
            5. Stamens 1–6, number variable within plant . . . . 9. *Elatine heterandra*
            5. Stamens 3.
                6. Seed pits ± round, (9–)14–17 per row . . . . 8. *Elatine brachysperma*
                6. Seed pits angular-hexagonal, (13–)16–35 per row.
                    7. Seeds ellipsoid, pit length 3–5 times width; stipule margins entire . . . . 7. *Elatine chilensis*
                    7. Seeds oblong or slenderly cylindric, pit length 1–3 times width; stipule margins dentate.
                        8. Leaves reddish green; seed pit length 1–2 times width. . . . . 5. *Elatine rubella*
                        8. Leaves light green to green; seed pit length 2–3 times width.
                            9. Leaf blades linear, lanceolate or narrowly oblong, apices acute or obtuse. . . . . 3. *Elatine triandra*
                            9. Leaf blades obovate or broadly spatulate, apices rounded to shallowly emarginate . . . . 4. *Elatine americana*

1. Elatine californica A. Gray, Proc. Amer. Acad. Arts 13: 361, 364. 1878 • California waterwort

Elatine californica var. *williamsii* (Rydberg) Fassett; *E. williamsii* Rydberg

Herbs, submersed or growing on exposed but wet substrates, 1–5 cm. **Stems** decumbent to erect, branched. **Leaves** light green to green, sometimes becoming reddish in terrestrial plants; stipules lanceolate, 0.5–0.6 mm, margins dentate, apex acute; petiole 1–4 mm; blade obovate to oblanceolate, 4–12(–15) × 1.2–3 mm, base narrowly cuneate, apex acute to obtuse. **Pedicels** recurved in fruit, 1.5–2.5 (–3.5) mm. **Flowers:** sepals 4, usually equal, sometimes 1 reduced, oblong-ovate or widely lanceolate, 0.5–0.6 × 0.3–0.4 mm; petals 4, greenish white, slightly reddish, or pink, elliptic or ovate, 1–1.5 × 0.5–0.6 mm; stamens 8; styles 4. **Capsules** depressed-ovoid, 4-locular, 1.3–2.5 mm diam. **Seeds** 2–5 per locule, oblong to ellipsoid, curved 90–180°, 0.6–1 × 0.2 mm; pits elliptic, length 1–3 times width, in 6–10 rows, (16–)20–29 per row.

Flowering summer. Pools, pond shores, rice fields, stream banks; 0–1900(–2600) m; Ariz., Calif., Idaho, Mont., Nev., N.Mex., Oreg., Utah, Wash.; Mexico (Baja California).

Elatine californica is distinguished from other species of the genus by having long pedicels that are recurved in fruit and strongly curved seeds that can be nearly circular.

2. Elatine ojibwayensis Garneau, Canad. J. Bot. 84: 1040, fig. 1e,f. 2006 • Ojibway waterwort, élatine du lac Ojibway [C] [E]

Herbs, submersed, 1–2 cm. **Stems** decumbent to erect, highly branched. **Leaves** green; stipules lanceolate, 1 mm, margins dentate, apex acute; petiole 1–3 mm; blade oblong, 1.2–5 × 0.3–2 mm, base cuneate, apex acute to obtuse. **Pedicels** erect, 0.1–23 mm. **Flowers:** sepals 4, equal, elliptic, 0.9–1.3 × 0.5–0.6 mm; petals 4, pale purple, elliptic, 0.8–1 × 0.5 mm; stamens 8; styles 2. **Capsules** depressed-ovoid, 4-locular, 1–1.1 mm diam.

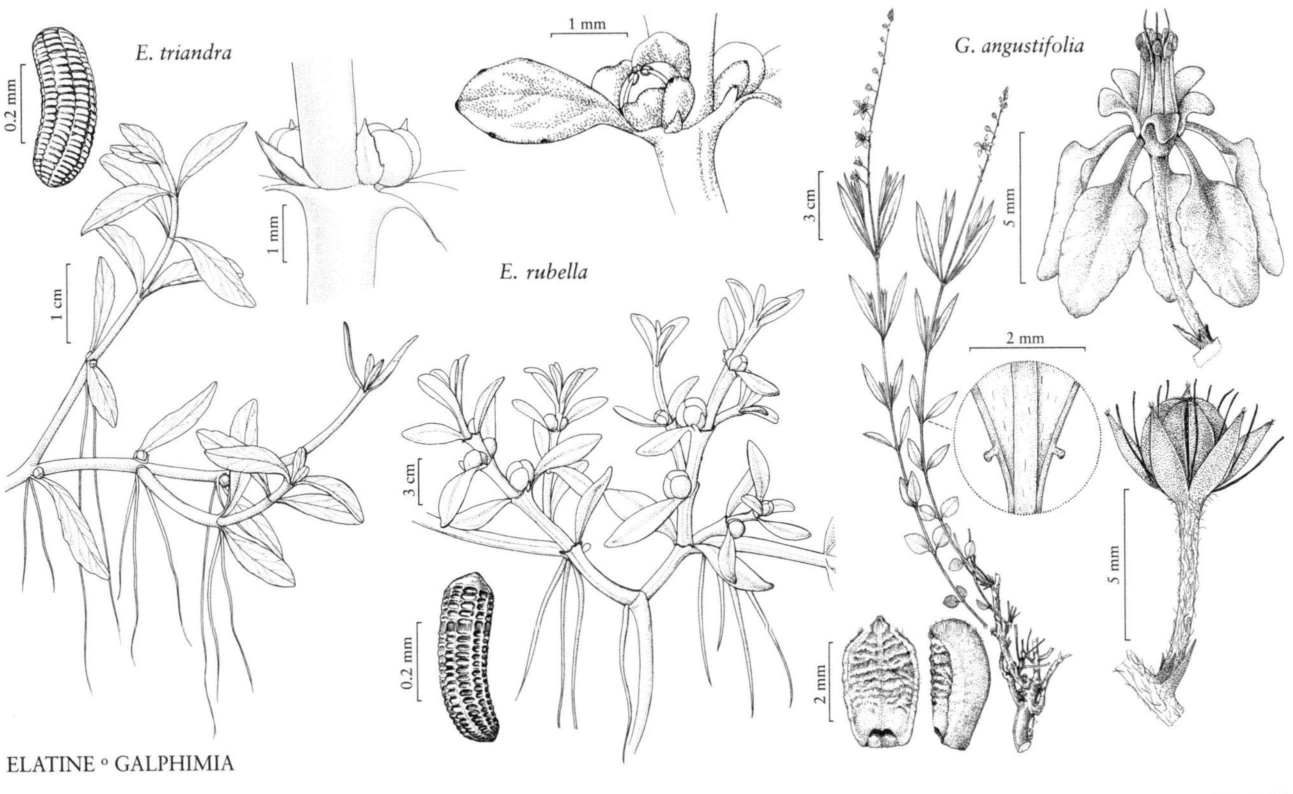

ELATINE ○ GALPHIMIA

Seeds 3 per locule, oblong to ellipsoid, straight or curved to 15°, 0.7–1 × 0.2–0.3 mm; pits rectangular, length 1–3 times width, in 12 rows, 27–37 per row.

Flowering summer. Riverbanks; 0–100 m; of conservation concern; Que.

Morphologically, *Elatine ojibwayensis* is very similar to *E. macropoda* Gussone of Europe and North Africa. Morphological and molecular examination of the holotype as well as the other specimens cited in the original description showed that three characters in M. Garneau's description need to be modified. Thus our description for this species differs from Garneau's observations of number of stamens '[4]', pedicel length '[0.5–0.7 mm]', and number of seed pits per row '[30–37]'. It is noteworthy that Garneau inadvertently included a specimen of *E. americana* (*J. Deshaye 91-1422*, QUE) in her new species description.

Elatine ojibwayensis is known from the Hudson Bay area near Rupert Bay and the Grande Rivière de la Baleine. It and *E. californica* are the only members of the genus with tetramerous flowers in the flora area.

3. **Elatine triandra** Schkuhr, Bot. Handb. 1: 345, plate 109b, fig. 2. 1789–1791 • Three-stamened waterwort [F] [I]

Elatine callitrichoides (Nylander) Kauffmann; *E. triandra* var. *callitrichoides* Nylander

Herbs, submersed, 2–10 cm. **Stems** prostrate, highly branched. **Leaves** light green to green; stipules triangular or ovate-lanceolate, 0.7–1 mm, margins dentate, apex acute or obtuse; petiole 0–3 mm; blade linear, lanceolate, or narrowly oblong, 3–10 × 1.5–3.5 mm, base attenuate, apex acute or obtuse. **Pedicels** erect, 0.3–0.4 mm. **Flowers:** sepals (2–)3(–4), usually 1 reduced, ovate, 0.5 (–0.7) × 0.2–0.4 mm; petals 3, white or reddish, widely ovate or elliptic, 0.6–1.2 × 0.5 mm; stamens 3; styles 3. **Capsules** compressed-globose, 3-locular, 1.2–1.7[–2] mm diam. **Seeds** 10–25 per locule, oblong, straight or slightly curved, 0.4–0.5 × 0.1–0.2 mm; pits angular-hexagonal, length 2–3 times width, in 6–8 rows, 16–25 per row. 2*n* = ca. 40.

Flowering summer–early fall. Shores, pools; 0–1100 (–2500) m; introduced; Alta., N.W.T., Ont., Sask.; Ala., Ariz., Calif., Colo., Conn., Ga., La., Maine, Mass., Minn., Nebr., Nev., N.J., N.Y., N.C., N.Dak., Okla., Pa., S.C., Tex., Utah, Va., Wis.; e, se Asia; introduced also in South America (Brazil), Europe, Africa, Australia.

Elatine triandra is a popular aquarium plant. Reports of *E. triandra* from Yukon were based on misidentification of *Callitriche hermaphroditica* (B. Bennett et al. 2010). *Elatine triandra* is a common weed in rice fields of south and southeast Asia (K. Moody 1989), thus rice farming or aquarium trade may have been responsible for its introduction.

4. **Elatine americana** (Pursh) Arnott, Edinburgh J. Nat. Geogr. Sci. 1: 431. 1830 • American waterwort, élatine américaine [E]

Peplis americana Pursh, Fl. Amer. Sept. 1: 238. 1813; *Elatine triandra* Schkuhr subsp. *americana* (Pursh) Á. Löve & D. Löve; *E. triandra* var. *americana* (Pursh) Fassett

Herbs, submersed, 1.5–5 cm. **Stems** ascending or prostrate, branched. **Leaves** green; stipules lanceolate, 0.5–0.6 mm, margins dentate, apex acute; petiole 0.5–3 mm; blade obovate to broadly spatulate, 3–8 × 0.5–3.5 mm, base attenuate or cuneate, apex rounded to shallowly emarginate. **Pedicels** erect, to 0.1 mm. **Flowers** cleistogamous when submersed; sepals (2–)3, equal or 1 reduced, oblong-ovate or widely lanceolate, 0.5 × 0.3–0.5 mm; petals 3, greenish white or pinkish white to reddish, elliptic or ovate, 0.6–1.2 × 0.5 mm; stamens 3; styles 3. **Capsules** oblong-ovoid, 3-locular, 1.5 mm diam. **Seeds** 9–12 per locule, slenderly cylindric, ± straight, 0.4–0.6 × (0.1–)0.2 mm; pits angular-hexagonal, length 2–3 times width, in 6(–8) rows, (13–)16–25(–30) per row. 2*n* = 36.

Flowering summer–early fall. Muddy shorelines including estuaries, in water to 50 cm deep; 0–300(–900) m; Man., N.B., N.W.T., Ont., Que.; Calif., Conn., Del., Ill., Maine, Mass., Mich., Minn., Mo., Mont., N.H., N.J., N.Y., N.C., Okla., Pa., R.I., S.Dak., Va.

5. **Elatine rubella** Rydberg, Mem. New York Bot. Gard. 1: 260. 1900 • Southwestern waterwort [E] [F]

Herbs, emergent on wet substrates, 1–6(–16) cm. **Stems** ascending or prostrate, branched. **Leaves** reddish green; stipules lanceolate, 0.5–0.6 mm, margins dentate, apex acute; petiole 0.5–1(–3) mm; blade oblong-lanceolate, 3–8 × (0.5–)1–2(–2.2) mm, base attenuate to rounded, apex obtuse to slightly emarginate. **Pedicels** erect, 0–0.5 mm. **Flowers:** sepals (2–)3, if 2, equal, if 3, 1 smaller, oblong-ovate or widely lanceolate, 1–1.5 × 1–1.5 mm; petals 3, pinkish white to reddish, elliptic or ovate, 1–1.5 × 0.5 mm; stamens 3; styles 3. **Capsules** oblong-ovoid, 3-locular, 1.5–2.5 mm diam. **Seeds** 10–20(–30) per locule, narrowly oblong, straight or curved to 15°, 0.5–0.7 × 0.1–0.3 mm; pits angular-hexagonal, length 1–2 times width, in 8–10 rows, 16–25(–30) per row.

Flowering summer. Muddy shores, shallow vernal pools, ditches, rice fields; 0–2900 m; Ariz., Calif., Colo., Idaho, Kans., Mont., Nebr., Nev., N.Mex., Oreg., S.Dak., Utah, Wyo.

Reports of *Elatine rubella* from southern British Columbia and Washington are based on misidentifications of *E. brachysperma* and *E. chilensis*, respectively.

6. **Elatine ambigua** Wight, Bot. Misc. 2: 103, suppl. plate 5. 1830 • Asian waterwort [I] [W]

Herbs, submersed, 1.5–4(–8) cm. **Stems** prostrate or erect, branched. **Leaves** green; stipules lanceolate, 1 mm, margins dentate, apex acute; petiole 0.1–1 mm; blade lanceolate or oblong-elliptic, 2–6 × 0.5–1.5 mm, base cuneate, apex obtuse to acute. **Pedicels** recurved in fruit, 0.5–2.5 mm. **Flowers:** sepals 3, equal, oblong-ovate or broadly lanceolate, 0.5 × 0.3 mm; petals 0 or 3, reddish, long-elliptic or ovate, 1–1.5 × 0.5 mm; stamens 3; styles 3. **Capsules** oblong-ovoid, 3-locular, 1 mm diam. **Seeds** 10 per locule, narrowly oblong, straight or curved 15–30°, 0.3 × 0.1 mm; pits obscure, round, length 1–4 times width, in 6 rows, 19–25(–30) per row.

Flowering late spring–summer. Pools, marshy places, rice fields; 0–100 m; introduced; Calif., Conn., Mass., S.C., Va.; s, se Asia; introduced also in Europe, s Pacific Islands, Australia.

Elatine ambigua is a popular aquarium plant usually misidentified as *E. trianda*. It is also a common weed in rice fields. This species probably was introduced to California, South Carolina, and Virginia through rice farming. Its introduction to Connecticut and Massachusetts probably was through aquarium disposal or fish stocking.

7. **Elatine chilensis** Gay, Fl. Chil. 1: 286. 1846 • Chilean waterwort [W]

Elatine gracilis H. Mason

Herbs, emergent on wet substrates, 0.5–10 cm. **Stems** ascending or prostrate, branched. **Leaves** usually green, sometimes reddish green; stipules triangular, 0.5–0.6 mm, margins entire, apex acute; petiole 1–3 mm; blade obovate to broadly spatulate, 3–4(–8) × 0.5–2.5 mm, base cuneate to rounded, apex obtuse to slightly emarginate.

Pedicels absent. **Flowers:** sepals 2–3, if 2, equal, if 3, 1 smaller, oblong, 0.5 × 0.3 mm; petals 3, pinkish white to reddish, elliptic to ovate, 1–1.5 × 0.5 mm; stamens 3; styles 3. **Capsules** depressed-ovoid, 3-locular, 1.5–2 mm diam. **Seeds** 23–33 per locule, ellipsoid, straight or curved to 20°, 0.4–0.5 × 0.1–0.2 mm; pits angular-hexagonal, length 3–5 times width, in 6 rows, 16–25 (–35) per row.

Flowering summer. Muddy shores (fresh or tidal), shallow vernal pools, ditches, rice fields; 0–2800 m; Ariz., Calif., Nev., N.Mex., Oreg., Wash.; South America.

The distribution of *Elatine chilensis* has an amphitropical disjunction between western North America and southern South America. The minute seeds of *E. chilensis* probably have been dispersed between the two regions by birds (S. Carlquist 1983).

8. Elatine brachysperma A. Gray, Proc. Amer. Acad. Arts 13: 361, 363. 1878 • Short-seed waterwort

Alsinastrum brachyspermum (A. Gray) Greene; *Elatine obovata* (Fassett) H. Mason; *E. triandra* Schkuhr var. *brachysperma* (A. Gray) Fassett; *E. triandra* var. *obovata* Fassett; *Potamopitys brachysperma* (A. Gray) Kuntze

Herbs, submersed or emersed on wet substrates, 0.5–3(–5) cm. **Stems** ascending or prostrate, branched. **Leaves** reddish green; stipules lanceolate, 0.5–0.6 mm, margins dentate, apex acute; petiole (0–)1–3 mm; blade narrowly oblong to ovate, 3–8 × 1–3 mm, base cuneate to rounded, apex rounded to obtuse. **Pedicels** erect, 0–0.4 mm. **Flowers:** sepals (2–)3, equal, oblong-ovate, 1.1–1.3 × 0.3 mm; petals 3, pinkish, elliptic or ovate, 1–1.5 × 0.5 mm; stamens 3; styles 3. **Capsules** globose, 3-locular, 1.1–1.7 mm diam. **Seeds** (6–)10–15 per locule, oblong, straight or curved to 30°, 0.3–0.5 × 0.1–0.3 mm; pits ± round, length 1–2 times width, in 6(–8) rows, (9–)14–17 per row.

Flowering summer–early fall. Muddy shores, often of ponds or reservoirs; 0–1500(–2200) m; B.C.; Ala., Ariz., Calif., Ga., Ill., La., Mont., Nebr., Nev., N.Mex., Ohio, Okla., Oreg., Tex., Wash., Wyo.; Mexico (Baja California); South America (Argentina).

9. Elatine heterandra H. Mason, Madroño 13: 240. 1956 • Mosquito waterwort [E]

Herbs, emersed on wet substrates, 0.5–3 cm. **Stems** ascending or prostrate, branched. **Leaves** usually green, sometimes reddish green; stipules ovate-lanceolate, 1 mm, margins irregularly toothed, apex acute; petiole 0.5–2 mm; blade obovate to broadly elliptic-oblong, 2–4 × 0.5–1.5 mm, base usually cuneate, sometimes rounded, apex acute to obtuse, notched. **Pedicels** absent. **Flowers:** sepals 2(–3), if 2, equal, if 3, 1 smaller, oblong, 1.1–1.3 × 0.3 mm; petals 3, pinkish, elliptic to ovate, 1–1.5 × 0.5 mm; stamens 1–6 (number variable within plant); styles 3. **Capsules** globose, 3-locular, 1.1–1.7 mm diam. **Seeds** 8–12 per locule, broadly ellipsoid to oblong, straight or curved 15–30°, 0.3–0.5 × 0.1–0.2 mm; pits rounded-hexagonal, length 1–2 times width, in 9–10 rows, 12–18 per row.

Flowering summer. Wet mud of montane ponds; (400–)1000–1500 m; Calif., N.Mex., Tex.

In California, *Elatine heterandra* is known from the North Coast Ranges and the Sierra Nevada; additionally, two disjunct localities are known for the species in McKinley County, New Mexico, and Kenedy County, Texas.

10. Elatine minima (Nuttall) Fischer & C. A. Meyer, Linnaea 10: 73. 1835 • Small waterwort, élatine naine [E]

Crypta minima Nuttall, J. Acad. Nat. Sci. Philadelphia 1: 117, plate 6, fig. 1. 1817

Herbs, submersed or emersed on wet substrates, 0.2–3(–10) cm. **Stems** erect or ascending, usually unbranched, sometimes 1–2 branched. **Leaves** green; stipules lanceolate, 0.2 mm, margins entire, apex rounded; petiole 0–1 mm; blade elliptic to oblanceolate or spatulate, 0.3–5(–11) × 0.7–1.8 (–4) mm, base rounded to truncate, apex obtuse-acute. **Pedicels** erect, 0–0.2 mm. **Flowers** cleistogamous when submersed; sepals 2, equal, ovate to oblong, 0.5 × 0.3 mm; petals 2, reddish, long-elliptic, elliptic, or ovate, 1–1.5 × 0.5 mm; stamens 2; styles 2. **Capsules** oblong-ovoid, 2-locular, 0.9–1.5 mm diam. **Seeds** 4–9 per locule, thickly cylindric to barrel-shaped, straight, 0.4–0.5 (–0.7) × 0.2(–0.3) mm; pits elliptic, length 1.2–1.6 times width, in 8–9 rows, (11–)15–20 per row.

Flowering summer. Shores (sometimes intertidal) or in water to 2 m deep, sandy substrates mixed with organic debris, abandoned sunfish nests; 0–400 m; St. Pierre and Miquelon; N.B., Nfld. and Labr., N.S., Ont., P.E.I., Que., Sask.; Conn., Del., Ill., Maine, Md., Mass., Mich., Minn., N.H., N.J., N.Y., N.C., Pa., R.I., S.C., Tenn., Vt., Va., Wis.

MALPIGHIACEAE Jussieu
• Malpighia Family

William R. Anderson†[1]

Herbs, subshrubs, shrubs, vines (twining, woody to herbaceous), **or trees,** perennial, evergreen or deciduous, synoecious [dioecious or functionally dioecious]; hairs unicellular, usually 2-armed and medifixed or submedifixed [basifixed or stellate]. **Leaves** opposite [whorled, subopposite, or alternate], simple; stipules present [absent]; petiole present [absent]; blade margins usually entire [lobed], sometimes pseudodentate [ciliate at location of marginal glands or with stout bristlelike hairs], often bearing multicellular glands on margin or abaxial [adaxial] surface; venation pinnate. **Inflorescences** terminal or axillary, racemes, panicles, umbels, corymbs, or thyrses, or flowers solitary. **Flowers** bisexual [rarely unisexual], radially or bilaterally symmetric, mostly all chasmogamous, sometimes both chasmogamous and cleistogamous; perianth and androecium hypogynous [perigynous]; hypanthium absent [present]; sepals 5, distinct or connate basally, usually glandular, sometimes eglandular; petals (in chasmogamous flowers) 5, posterior (flag) petal often different from lateral 4, distinct, mostly clawed; nectary absent; stamens (in chasmogamous flowers) (2–5)[6–]10[–20], distinct or connate proximally, free; anthers dehiscing by longitudinal slits [apical or subapical pores or very short slits]; pistil 1, (2–)3-carpellate, carpels nearly distinct to completely connate in ovary [connate throughout], ovary superior, (2–)3-locular, placentation apical; ovule 1 per locule, anatropous; styles (in chasmogamous flowers) (1–)3 (usually as many as carpels but sometimes fewer by reduction [or connation]), distinct [partially to completely connate]; stigmas 1–3 (1 per style). **Fruits** drupes or schizocarps splitting into mericarps, mericarps nutlets, thin-walled cocci, or bearing wings [or vascularized setae] [berries or dry and indehiscent]. **Seeds** 1 per locule or mericarp.

Genera ca. 75, species ca. 1300 (8 genera, 9 species in the flora): nearly worldwide; tropics and subtropics.

---

[1] Christiane Anderson completed revisions of this treatment in cooperation with the FNA editorial staff.

Malpighiaceae are far more numerous and diverse in the New World than in the Old World. Many are grown as ornamentals in warm areas of the world; they are intolerant of cold. *Malpighia emarginata* de Candolle produces a fruit that is rich in vitamin C, which has been exploited commercially as acerola. One of the most famous hallucinogens in the world, ayahuasca or caapi, is extracted from *Banisteriopsis caapi* (Grisebach) C. V. Morton, a vine native to South America but cultivated widely.

In Malpighiaceae, the ancestral inflorescence appears to be a raceme of cincinni, but in most genera the cincinni have been reduced to 1-flowered units. The pedicel is usually borne on a peduncle, the juncture marked by two bracteoles (W. R. Anderson 1981). This morphology technically makes many inflorescences in the family cymose (cymes, dichasia, or thyrses). For simplicity, however, inflorescence types are described here based on their gross morphological appearance, treating reduced cincinni as if they were single flowers.

Most New World taxa bear two (sometimes one) large multicellular glands abaxially on all five sepals or on three or four lateral sepals. Many Old World taxa and some New World taxa (including one in the flora) have the calyx glands much reduced in number and size or absent. Calyx glands always are absent from cleistogamous flowers.

SELECTED REFERENCES Anderson, W. R. 2004. Malpighiaceae. In: N. P. Smith et al., eds. 2004. Flowering Plants of the Neotropics. Princeton. Pp. 229–232. Anderson, W. R. 2013. Origins of Mexican Malpighiaceae. Acta Bot. Mex. 104: 107–156. Davis, C. C. and W. R. Anderson. 2010. A complete generic phylogeny of Malpighiaceae inferred from nucleotide sequence data and morphology. Amer. J. Bot. 97: 2031–2048. Niedenzu, F. 1928. Malpighiaceae. In: H. G. A. Engler, ed. 1900–1953. Das Pflanzenreich.... 107 vols. Berlin. Vols. 91, 93, 94[IV,141], pp. 1–870.

Key to plants with flowers

1. Flowers cleistogamous, to 1.5 mm diam., without visible petals, stamens, or styles .... 8. *Aspicarpa* (in part), p. 363
1. Flowers chasmogamous, 6+ mm diam., showy with visible petals, stamens, and styles.
    2. Styles 1(–2) per flower.
        3. Petals densely sericeous abaxially, glabrous adaxially, white or pink, except posterior petal proximally lemon yellow, distally white or pink; stamens 10, all fertile; s Florida .... 6. *Hiptage*, p. 361
        3. Petals glabrous, lemon or carrot yellow; stamens 2–5, 2 or 3 fertile, 0–3 staminodial; sw United States.
            4. Slender wiry twining vines .... 7. *Cottsia*, p. 362
            4. Subshrubs or non-twining perennial herbs .... 8. *Aspicarpa* (in part), p. 363
    2. Styles 3 per flower.
        5. Petals lemon yellow, sometimes becoming orange or red in age.
            6. Petals glabrous; calyx glands 0; styles slender, subulate; stigmas minute, terminal; subshrubs or shrubs .... 1. *Galphimia*, p. 356
            6. Petals abaxially white-sericeous or -tomentose; calyx glands 8 (anterior sepal eglandular, 4 lateral 2-glandular); styles stout, cylindric; stigmas large, on internal angle; woody vines or shrubs with scandent or trailing branches .... 4. *Callaeum*, p. 359
        5. Petals pink, pink and white, lavender, white, or red.
            7. Woody vines; s Florida .... 5. *Heteropterys*, p. 360
            7. Shrubs or small trees; s Florida, s Texas.
                8. Inflorescences terminal, racemes; pedicels sessile; stipules intrapetiolar; leaves eglandular; styles slender, subulate; stigmas minute; Florida .... 2. *Byrsonima*, p. 357
                8. Inflorescences axillary, umbels or corymbs; pedicels raised on peduncles; stipules interpetiolar; leaves bearing (0–)2–4 glands impressed in abaxial surface of blade; styles stout, cylindric; stigmas large; Texas .... 3. *Malpighia*, p. 358

Key to plants with fruits

1. Fruits drupes.
   2. Inflorescences terminal, racemes; pedicels sessile; stipules intrapetiolar; leaves eglandular; Florida . . . . 2. *Byrsonima*, p. 357
   2. Inflorescences axillary, umbels or corymbs; pedicels raised on peduncles; stipules interpetiolar; leaves bearing (0–)2–4 glands impressed in abaxial surface of blade; Texas. . . . . .3. *Malpighia*, p. 358
1. Fruits schizocarps, breaking apart into samaras, unwinged nutlets, or cocci.
   3. Mericarps cocci or nutlets, unwinged but bearing dorsal keel or crest.
      4. Mericarps bearing narrow dorsal keel, walls thin, brittle; calyx glands 0 . . . . . . 1. *Galphimia*, p. 356
      4. Mericarps with dorsal crest, walls thick, tough; calyx glands in chasmogamous flowers 8 or 10 (sepals all 2-glandular or anterior eglandular), in cleistogamous flowers 0. . . . . .8. *Aspicarpa*, p. 363
   3. Mericarps samaras.
      5. Samaras bearing 2 or 3 elongate or semicircular lateral wings, dorsal wing much smaller than lateral wings or absent.
         6. Lateral wings of samara 2, semicircular; dorsal wing well developed, like lateral wings but much smaller; Texas . . . . 4. *Callaeum*, p. 359
         6. Lateral wings of samara 3, elongate; dorsal wing mostly absent, occasionally present, much smaller than lateral wings; Florida. . . . . 6. *Hiptage*, p. 361
      5. Samaras bearing 1 elongate dorsal wing, short lateral winglets present or absent.
         7. Dorsal wing of samara thickened on abaxial edge, veins bending from it toward thinner adaxial edge; woody vines; Florida . . . . 5. *Heteropterys*, p. 360
         7. Dorsal wing of samara thickened on adaxial edge, veins bending toward thinner abaxial edge; slender wiry vines; sw United States . . . . 7. *Cottsia*, p. 362

1. GALPHIMIA Cavanilles, Icon. 5: 61, plate 489. 1799 • [Anagram of generic name *Malpighia*]

Thryallis Linnaeus 1762, name rejected [not *Thryallis* Martius 1829, name conserved]

Subshrubs or shrubs [occasionally small trees]. Leaves usually bearing glands proximally on blade margin [and/or on petiole]; stipules intrapetiolar, distinct. **Inflorescences** terminal, racemes [several grouped in panicle]. **Pedicels** sessile or raised on peduncles. **Flowers** all chasmogamous, 6+ mm diam., showy with visible petals, stamens, and styles; calyx glands 0 [1 very small gland at base of sinus between some or all adjacent sepals]; corollas nearly radial [moderately bilaterally symmetric], petals lemon yellow, becoming orange or red in age [suffused with red], glabrous [rarely hairy]; stamens 10, all fertile; anthers subequal; pistil 3-carpellate, carpels connate in ovary along broad adaxial faces; styles 3, subulate, slender; stigmas terminal, minute. **Fruits** schizocarps, breaking apart at maturity into 3 cocci; cocci unwinged, bearing narrow dorsal keel, walls thin, brittle. $x = 6$.

Species 26 (1 in the flora): Texas, Mexico, Central America, South America.

Galphimia is most diverse from Texas south to Nicaragua with 22 species in that region; only four species are known from South America south of the Amazon valley. *Galphimia gracilis* Bartling, native to eastern Mexico, is widely cultivated in the tropics and subtropics as an ornamental shrub.

SELECTED REFERENCE Anderson, C. E. 2007. Revision of *Galphimia* (Malpighiaceae). Contr. Univ. Michigan Herb. 25: 1–82.

1. Galphimia angustifolia Bentham, Bot. Voy. Sulphur, 9, plate 5. 1844 • Narrowleaf goldshower [F]

Galphimia linifolia A. Gray; *Thryallis angustifolia* (Bentham) Kuntze; *T. linifolia* (A. Gray) Kuntze

Subshrubs or small shrubs to 1 m. **Leaf blades** linear to narrowly elliptic or narrowly lanceolate or narrowly ovate to elliptic or ovate, larger blades 2.5–4 × 0.3–2 cm, surfaces glabrous or abaxial surface with scattered hairs proximally on midrib, usually bearing 2 glands proximally on margin above base. **Inflorescences** 5–18 cm, 4–13-flowered. **Pedicels** usually sessile, rarely raised on peduncles to 3.5 mm, much shorter than pedicels. **Flowers:** posterior petal sometimes a little longer than lateral petals; stamen filaments opposite sepals longer than opposite petals; anthers 0.7–0.9 mm. **Cocci** 3–3.5 mm. $2n = 24$.

Flowering and fruiting Apr–Nov. Open rocky shrublands, mostly on limestone; 60–700 m; Tex.; Mexico (Baja California Sur, Coahuila, Durango, Nuevo León, San Luis Potosí, Sinaloa, Sonora, Tamaulipas).

2. BYRSONIMA Richard ex Kunth in A. von Humboldt et al., Nov. Gen. Sp. 5(fol.): 114; 5(qto.): 147; plates 446–449. 1822 • [Greek *byrsa*, leather, alluding to use of bark of some species in tanning; meaning of suffix obscure]

Shrubs or trees [subshrubs]. Leaves eglandular; stipules intrapetiolar, distinct or partially [completely] connate. **Inflorescences** terminal, racemes [sometimes racemes of few-flowered cincinni]. **Pedicels** sessile [raised on short peduncles]. **Flowers** all chasmogamous, 6+ mm diam., showy with visible petals, stamens, and styles; calyx glands [0 or]10 (sepals all 2-glandular); corollas bilaterally symmetric, petals white or pink, becoming red in age [light or medium yellow or red], glabrous [rarely hairy]; stamens 10, all fertile; anthers subequal; pistil 3-carpellate, carpels completely connate in ovary; styles 3, subulate, slender; stigmas terminal [slightly internal], minute. **Fruits** drupes, yellow or brownish [orange, red, purple, blue, or blue-black], stone 1, 3-locular, wall hard, smooth or rugose. $x = 12$.

Species ca. 135 (1 in the flora): Florida, Mexico, West Indies, Central America, South America.

1. Byrsonima lucida (Miller) de Candolle in A. P. de Candolle and A. L. P. P. de Candolle, Prodr. 1: 580. 1824 • Locustberry [F]

Malpighia lucida Miller, Gard. Dict. ed. 8, Malpighia no. 9. 1768; *Byrsonima biflora* Grisebach; *B. cuneata* (Turczaninow) P. Wilson; *M. cuneata* Turczaninow

Shrubs or trees 1–6 m. **Leaves** often clustered at tips of shoots; blade obovate, larger blades 20–30 × 9–19(–24) mm, base cuneate or gradually narrowed, apex rounded or obtuse, surfaces very sparsely sericeous to soon glabrate. **Inflorescences** 2.5–5 cm, 6–10(–16)-flowered. **Pedicels** straight in bud, somewhat decurved in fruit. **Flowers:** anthers glabrous, locules rounded at apex, connectives equaling or exceeding locules to 0.3 mm; ovary glabrous. **Drupes** 8–12 mm diam. (dried), ovoid to spheroid with short apical beak when immature, glabrous.

Flowering most commonly Jan–Jun; fruiting Feb–Jul. Hammocks in dry rocky pinelands and sandy palm-pine woods; 0–10 m; Fla.; West Indies (Bahamas, Greater Antilles, Lesser Antilles, Virgin Islands).

Byrsonima lucida, native in the flora area only to Miami-Dade and Monroe counties, is widely cultivated in peninsular Florida as an ornamental shrub; it probably has little or no tolerance for frost or temperatures below freezing.

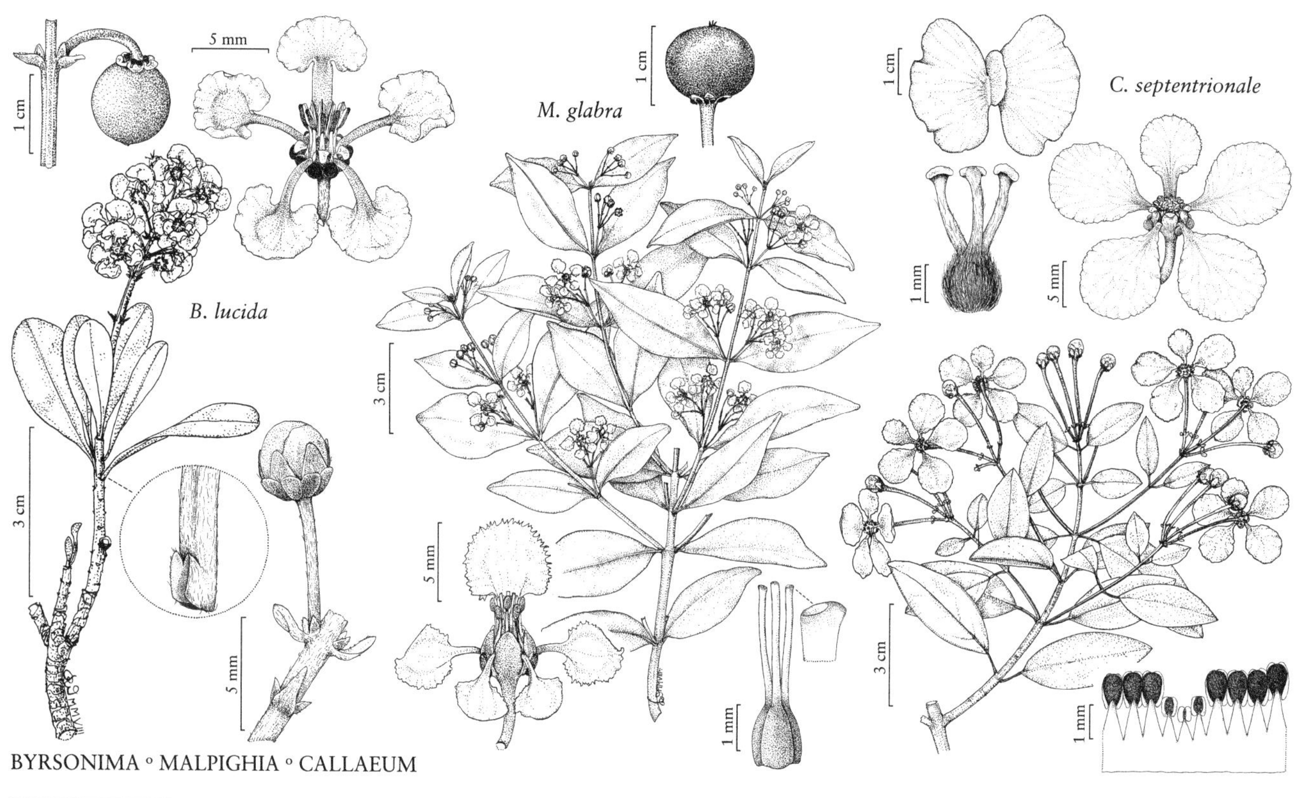

BYRSONIMA ◦ MALPIGHIA ◦ CALLAEUM

3. MALPIGHIA Linnaeus, Sp. Pl. 1: 425. 1753; Gen. Pl. ed. 5, 194. 1754 • [For Marcello Malpighi, 1628–1694, Italian anatomist]

Shrubs or small trees. Leaves usually bearing (0–)2–4[–10] glands impressed in abaxial surface of blade; stipules interpetiolar, mostly distinct. **Inflorescences** axillary, dense corymbs or umbels. **Pedicels** raised on peduncles. **Flowers** all chasmogamous, 6+ mm diam., showy with visible petals, stamens, and styles; calyx glands 6(–10) (3 sepals each bearing 2 large glands, others very rarely bearing 1–4 smaller glands); corollas bilaterally symmetric, petals pink, lavender, or white, glabrous [glabrate]; stamens 10, all fertile; anthers subequal or 2 opposite posterior-lateral petals larger; pistil 3-carpellate, carpels completely [rarely proximally] connate in ovary; styles 3, cylindric, stout; stigmas on internal angle or subterminal, large. **Fruits** drupes [berries or very rarely breaking into separate pyrenes], red [sometimes orange]; pyrenes 3, connate in center or distinct at maturity but then usually retained in common exocarp, walls hard, bearing rudimentary dorsal and lateral wings and sometimes rudimentary intermediate winglets or dissected outgrowths. $x = 10$.

Species ca. 50 (1 in the flora): Texas, Mexico, West Indies, Central America, South America.

Malpighia coccigera Linnaeus, dwarf- or Singapore-holly, native to the West Indies, is grown as an ornamental. *Malpighia emarginata*, acerola or Barbados cherry, native to Mexico and Central America, is widely cultivated for its fruits, which are rich in vitamin C.

1. Malpighia glabra Linnaeus, Sp. Pl. 1: 425. 1753 • Wild crapemyrtle 

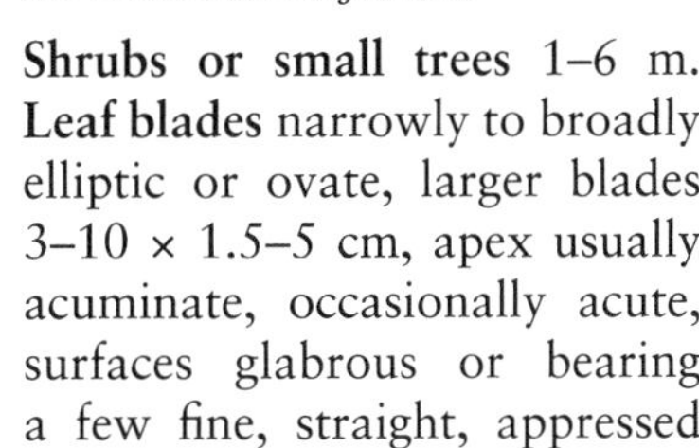

Malpighia punicifolia Linnaeus; *M. semeruco* A. Jussieu

Shrubs or small trees 1–6 m. **Leaf blades** narrowly to broadly elliptic or ovate, larger blades 3–10 × 1.5–5 cm, apex usually acuminate, occasionally acute, surfaces glabrous or bearing a few fine, straight, appressed hairs. **Inflorescences** 1.5–3(–3.5) cm, (3–)4–10(–12)-flowered. **Flowers:** petals pink or pink and white or lavender-pink; anthers glabrous; ovary glabrous; styles nearly straight, parallel or divergent distally, ± alike. **Drupes** 7–13 mm diam., spheroid. 2*n* = 20 (Costa Rica).

Flowering Sep–Apr; fruiting Oct–May. Roadside thickets, sandy plains; 0–100 m; Tex.; e, s Mexico; West Indies (Greater Antilles); Central America; South America.

Malpighia glabra, native in southernmost Texas, is rarely cultivated as an ornamental shrub in Texas, but many of the plants sold under that name are actually *M. emarginata*. *Malpighia emarginata* resembles *M. glabra*, but its leaves are usually rounded or obtuse at the apex and often emarginate or apiculate, and some pairs of leaves are crowded in dense shoots with very short internodes, while others are separated by much longer internodes (versus all more or less evenly spaced in *M. glabra*).

4. CALLAEUM Small in N. L. Britton et al., N. Amer. Fl. 25: 128. 1910 • [Greek *kallaion*, cockscomb, alluding to lobed or corrugated outgrowths on samara between lateral and dorsal wings in the type species, *C. nicaraguense*]

Shrubs or woody vines, branches scandent or trailing. **Leaves** bearing small glands on distal ½ of petiole and/or on blade margin near base; stipules borne on petiole at or just distal to base, distinct. **Inflorescences** axillary or terminal, few-flowered umbels, corymbs, or short pseudoracemes. **Pedicels** raised on peduncles. **Flowers** all chasmogamous, 6+ mm diam., showy with visible petals, stamens, and styles; calyx glands 8 (anterior sepal eglandular, 4 lateral 2-glandular); corollas bilaterally symmetric, petals lemon yellow, densely white-sericeous or -tomentose [glabrous] abaxially; stamens 10, all fertile [posterior 3 occasionally sterile]; anterior 7 anthers notably larger than posterior 3; pistil 3-carpellate, carpels completely connate in ovary; styles 3, cylindric, stout; stigmas on internal angle, oblate, large. **Fruits** schizocarps, breaking into 3 samaras; samaras bearing 2 semicircular lateral wings and 1 well-developed dorsal wing, like lateral wings but notably [somewhat] smaller; nut wall thick, tough. *x* = 10.

Species 11 (1 in the flora): Texas, Mexico, Central America (to Nicaragua), South America.

In his revision of *Callaeum*, D. M. Johnson (1986c) cited a specimen of *C. macropterum* (de Candolle) D. M. Johnson, *Palmer s.n.* in 1869 (NY), as coming from Arizona, "without definite locality." That species occurs in Sonora, Mexico, fairly close to the U.S. border, but no other collection is known from north of the border. R. McVaugh (1956b) said that in 1869 Palmer collected in Arizona but was in southernmost Arizona only a short time; from Tucson he took a coach south into Sonora, where he collected in several areas, including the Yaqui River. There are two sheets of *C. macropterum* at NY labeled as *Palmer s.n.* in 1869, one with a label giving the locality as "Arizona," the other with a label giving the locality as "Yaqui River Sonora." The specimens look as if they were parts of the same gathering. Given McVaugh's cautionary notes on the inaccuracies found in many of the labels placed on Palmer's collections after his return to Washington, I believe that the specimen cited by Johnson as coming from Arizona probably actually originated on the Yaqui River in Sonora, and am excluding *C. macropterum* from this treatment.

SELECTED REFERENCE Johnson, D. M. 1986c. Revision of the neotropical genus *Callaeum* (Malpighiaceae). Syst. Bot. 11: 335–353.

1. **Callaeum septentrionale** (A. Jussieu) D. M. Johnson, Syst. Bot. 11: 343. 1986 [F]

Hiraea septentrionalis A. Jussieu, Ann. Sci. Nat. Bot., sér. 2, 13: 259. 1840; *H. greggii* S. Watson; *Mascagnia septentrionalis* (A. Jussieu) Grisebach

Leaf blades lanceolate to elliptic-ovate, larger blades 20–70(–95) × 6–40 mm, base cuneate to rounded, apex acute or acuminate, abaxial surface tomentose or sparsely sericeous to glabrate, adaxial surface glabrous, bearing 1–3 small glands on each side near petiole apex or blade margin. **Pedicels** 6–24 mm, 1.8–5 times as long as peduncles. **Samaras** orbiculate or butterfly-shaped, lateral wings 10–42(–55) × 16–21 mm, dorsal wing 10–17 × 6 mm. **2*n*** = 20 (Mexico).

Flowering year-round, most commonly May–Sep; fruiting Jun–Oct. Brushy woodlands; 100–200 m; Tex.; e Mexico.

Callaeum septentrionale is distinctive in the flora area, being the only native species of Malpighiaceae with hairy petals and orbiculate or butterfly-shaped samaras with dominant lateral wings. The posterior three anthers are smaller than the anterior seven, and the stigma, borne on the internal angle of the style, is oblate (wider than high). The species has the ability to climb, but when growing without support it can be shrubby, with the branches then often trailing. *Callaeum septentrionale* is known in the flora area only from south of Laredo, but it is to be sought in southern Texas near the Mexican states of Nuevo León and Tamaulipas. It is sometimes cultivated as an ornamental in southern Arizona and southern California; *C. macropterum* (de Candolle) D. M. Johnson is also cultivated in southern Arizona. There is no evidence that either species has become naturalized in either state.

5. HETEROPTERYS Kunth in A. von Humboldt et al., Nov. Gen. Sp. 5(fol.): 126; 5(qto.): 163; plate 450. 1822 (as Heteropteris), name and orthography conserved • [Greek *heteros*, different, and *pteron*, wing, alluding to dorsal wing of samara being thickened on abaxial edge and bent upward, opposite of arrangement in other genera with dorsal-winged samaras] [I]

Banisteria Linnaeus, name rejected

Woody vines [shrubs or small trees]. **Leaves** bearing glands on blade [and/or petiole, rarely eglandular]; stipules borne on [beside] base of petiole [absent], distinct. **Inflorescences** axillary or terminal or both, umbels or corymbs [pseudoracemes], these single or grouped in racemes or panicles. **Pedicels** raised on peduncles [sessile]. **Flowers** all chasmogamous, 6+ mm diam., showy with visible petals, stamens, and styles; calyx glands 0 or 8 (sepals all eglandular or 4 lateral 2-glandular); corollas bilaterally symmetric, petals pink or pink and white [white, light yellow, bronze, or dark red], glabrous [rarely hairy]; stamens 10, all fertile; anthers ± alike; pistil 3-carpellate, carpels connate proximally in ovary; styles 3, cylindric; stigmas on internal angle [very rarely terminal], large. **Fruits** schizocarps, breaking into 3 samaras; samaras bearing 1 elongate dorsal wing [rarely rudimentary or absent] thickened on abaxial edge, veins bending toward thinner adaxial edge, and usually single or double crown of short, irregular lateral winglets [crests] on each side [winglets and crests absent]; nut wall thick, tough. $x = 10$.

Species 140+ (1 in the flora): introduced, Florida; Mexico, West Indies, Central America, South America, Africa.

Heteropterys is almost exclusively a New World genus, with a single species in West Africa.

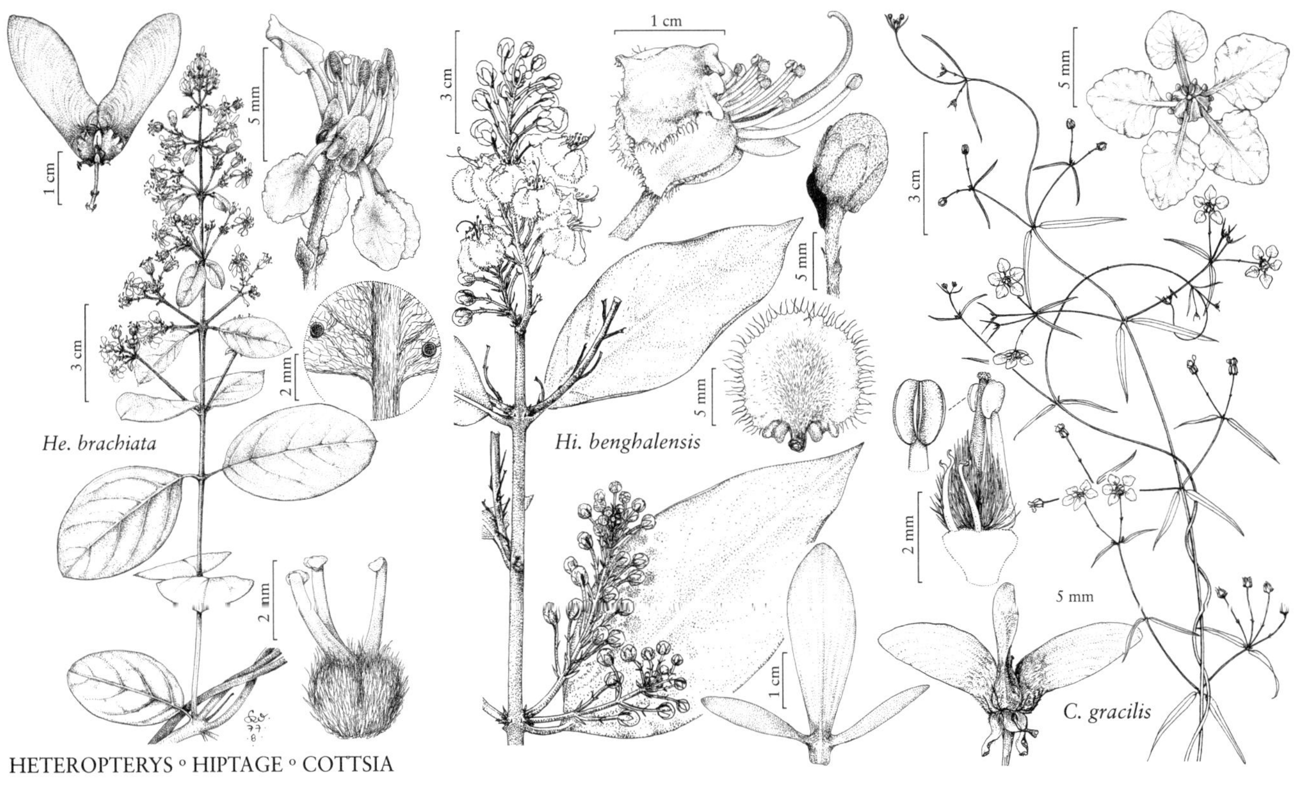

HETEROPTERYS ◦ HIPTAGE ◦ COTTSIA

1. Heteropterys brachiata (Linnaeus) de Candolle in A. P. de Candolle and A. L. P. P. de Candolle, Prodr. 1: 591. 1824 (as Heteropteris) F I

Banisteria brachiata Linnaeus, Sp. Pl. 1: 428. 1753; *Heteropterys beecheyana* A. Jussieu

Leaf blades elliptic, larger blades 4–8.5(–9.5) × (2–)3–6(–7) cm, base cuneate or rounded, apex rounded or slightly emarginate, obtuse, or acute and often apiculate, abaxial surface moderately to densely and usually persistently tomentose, occasionally tardily glabrescent, adaxial surface tomentose to glabrate, bearing 1–4 pairs of peltate glands abaxially near base and occasionally 1–2 additional pairs distally. **Inflorescences:** umbels or corymbs (2–)3–6-flowered. **Samaras** 18–32 mm; nut 3–4 mm diam., winglets 1–3.5 mm high.

Flowering and fruiting Nov–Dec. Hammocks; 0–10 m; introduced; Fla.; s Mexico; Central America; w South America.

Heteropterys brachiata is cultivated as an ornamental and locally naturalized in southernmost Florida.

6. HIPTAGE Gaertner, Fruct. Sem. Pl. 2: 169, plate 116, fig. 4. 1790, name conserved • [Greek *hiptamai*, to fly, alluding to wind-dispersed samaras] I

Shrubs or woody vines. Leaves usually bearing glands at apex of petiole or on blade at base or distally, on or near margin; stipules interpetiolar, occasionally minute or apparently lacking. **Inflorescences** axillary or terminal, usually elongate, occasionally condensed, racemes, these sometimes grouped in terminal panicles. **Pedicels** raised on peduncles. **Flowers** all chasmogamous, 6+ mm diam., showy with visible petals, stamens, and styles; calyx gland 1, decurrent, below posterior petal and between 2 adjacent sepals [2–10 (in 1–5 adjacent pairs), rarely 0]; corollas bilaterally symmetric, petals white or pink except posterior proximally yellow, distally white or pink [all white or pink], densely sericeous abaxially [in 1 species sparsely sericeous proximally], glabrous adaxially; stamens 10, all fertile; anther opposite anterior sepal largest, other 9 subequal; pistil 3-carpellate, carpels completely connate in ovary; style 1, on anterior carpel, tapered, curved toward posterior petal; stigma on internal angle or terminal but bent inward, small [large]. **Fruits** schizocarps, breaking into 3 samaras; samaras bearing 3 elongate lateral wings, 1 straddling plane of symmetry at apex of carpel, other 2 shorter, 1 on each side of plane of symmetry; dorsal wing usually absent, occasionally 1, much smaller than lateral wings; nut wall thick, tough.

Species 25+ (1 in the flora): introduced, Florida; se Asia (India & Pakistan to Taiwan), w Pacific Islands (Indonesia, Philippines).

1. Hiptage benghalensis (Linnaeus) Kurz, J. Asiat. Soc. Bengal, Pt. 2, Nat. Hist. 43: 136. 1874 • Hiptage F I

Banisteria benghalensis Linnaeus, Sp. Pl. 1: 427. 1753

Branches spreading. **Leaf blades** elliptic or ovate, larger blades 10–16 × 4–7(–9.5) cm, base cuneate, apex acuminate, abaxial surface glabrescent or sparsely sericeous, hairs short, straight, adaxial surface glabrous, bearing 2 larger glands at base and usually several small impressed glands in an inframarginal row. **Flowers** very fragrant; calyx gland 3–5 mm; petals long-fimbriate. **Samaras:** upper central wing 37–45(–52) mm, 2 lower lateral wings 17–27 mm.

Flowering and fruiting Apr–May. Hammocks; 0–10 m; introduced; Fla.; Asia.

Hiptage benghalensis is thought to be native from India and Sri Lanka to the Philippines, but it is difficult to know the true natural range because it has been cultivated as an ornamental for a long time and escapes readily, spreading aggressively and becoming a serious pest. The species is cultivated as an ornamental and locally naturalized in southernmost Florida.

7. COTTSIA Dubard & Dop, Rev. Gén. Bot. 20: [358–]359, fig. 1. 1908 • [Based on an anagram of Scott; for George Francis Scott Elliot, 1862–1934, Scottish botanist]

Vines, twining, wiry, slender, with woody base, sometimes seeming shrubby when grazed. **Leaves** usually bearing pair of stalked glands or eglandular processes on blade margin near base; stipules interpetiolar, distinct. **Inflorescences** terminal on lateral shoots, 2–4-flowered umbels. **Pedicels** raised on peduncles. **Flowers** all chasmogamous, 6+ mm diam., showy with visible petals, stamens, and styles; calyx glands usually 8 (anterior sepal usually eglandular, 4 lateral usually 2-glandular); corollas bilaterally symmetric, petals lemon yellow, glabrous; stamens [2] (4–)5, 2 fertile, opposite posterior-lateral sepals, [0–](2–)3 staminodial, opposite anterior and anterior-lateral sepals, (staminodial filaments rudimentary); fertile anthers subequal, staminodial anthers absent; pistil 3-carpellate, carpels completely connate in ovary; style 1, borne on anterior carpel, cylindric; stigma terminal, capitate [truncate], large. **Fruits** schizocarps, breaking into 3 samaras; samaras bearing 1 elongate dorsal wing thickened on adaxial edge, veins bending toward thinner abaxial edge, lateral wings absent; nut reticulate and often parallel-rugose on sides, wall thick, tough. $x = 10$.

Species 3 (1 in the flora): sw United States, n Mexico.

1. **Cottsia gracilis** (A. Gray) W. R. Anderson & C. Davis, Contr. Univ. Michigan Herb. 25: 163. 2007 [F]

Janusia gracilis A. Gray, Smithsonian Contr. Knowl. 3(5): 37. 1852

Leaf blades very narrowly lanceolate or elliptic, larger blades 12–40(–50) × (1.5–) 3–7(–9) mm, length 4–10 times width, base cuneate or rounded, apex acute, obtuse, or occasionally rounded, margins very often bearing few to many cilia or toothlike projections distally, surfaces persistently sericeous or sometimes glabrescent. **Pedicels** 4–8 mm, 0.7–1.8 times as long as peduncles. **Samaras** 9–15(–17) mm, nut 1.5–2.5 × 3–4.5 mm. $2n = 40$ (Mexico).

Flowering and fruiting Mar–Oct(–Jan). Open rocky slopes and deserts; 300–1600(–2100) m; Ariz., N.Mex., Tex.; Mexico (Baja California, Baja California Sur, Chihuahua, Durango, Nuevo León, Sonora, Zacatecas).

Cottsia gracilis is widespread in Arizona, absent only from the northeastern part of the state, but otherwise in the flora area is restricted to southern New Mexico and trans-Pecos Texas.

8. ASPICARPA Richard, Mém. Mus. Hist. Nat. 2: 396, plate 13. 1815 • [Greek *aspis*, shield, and *karpos*, fruit, alluding to shape of nutlet of *A. hirtella* in abaxial view]

Herbs or subshrubs, perennial, non-twining, base woody. **Leaves** eglandular; stipules interpetiolar, distinct. **Inflorescences** of chasmogamous flowers terminal, several-flowered umbels or corymbs, or axillary single flowers; of cleistogamous flowers axillary, clusters or single flowers. **Pedicels** of chasmogamous flowers raised on peduncles, of cleistogamous flowers ± sessile. **Flowers** chasmogamous and cleistogamous [all chasmogamous]. **Chasmogamous flowers** 6+ mm diam., showy with visible petals, stamens, and styles; calyx glands 8 or 10 (sepals all 2-glandular or anterior eglandular); corollas bilaterally symmetric, petals carrot yellow or lemon yellow, outermost sometimes with red blotch, glabrous; stamens 5, 3 fertile, opposite anterior and posterior-lateral sepals, 2 staminodial opposite anterior-lateral sepals, (staminodial filaments shorter to somewhat longer than those of fertile stamens), [all fertile]; fertile anthers subequal, staminodial anthers rudimentary; pistil 3-carpellate, carpels nearly distinct in ovary; styles 1(–2)[–3], 1 on anterior carpel, 2d style occasionally on 1 posterior carpel, cylindric; stigmas terminal, capitate or truncate, large. **Cleistogamous flowers** to 1.5 mm diam., without visible petals, stamens, or styles; calyx glands 0; petals 0 or 1–2, rudimentary; stamens represented by 1–2 minute sessile anthers; pistil 2-carpellate, carpels nearly distinct in ovary; style 0 or 1, rudimentary. **Fruits** schizocarps, breaking into nutlets (3 in chasmogamous flowers, 2 in cleistogamous flowers); nutlets unwinged, bearing dorsal crest and often narrower lateral crest around periphery; wall thick, tough. $x = 40$.

Species 9 (2 in the flora): sw United States, Mexico, South America.

Aspicarpa comprises three species in Mexico, with two of those in the flora area, and six species in South America south of the Amazon valley.

1. Chasmogamous flowers in (2–)4(–7)-flowered umbels, occasionally corymbs, terminating leafy shoots; petals carrot yellow; plants erect or decumbent with branches ascending; blades of larger leaves 6–23 mm wide, length 1.3–2.5 times width; staminodes equaling or longer than fertile stamens, surpassing sepals; nutlets with dorsal crest entire . . . . 1. *Aspicarpa hirtella*
1. Chasmogamous flowers borne singly in axils of full-sized leaves; petals lemon yellow, outermost sometimes with red blotch; plants erect or spreading-ascending; blades of larger leaves 2–8 mm wide, length (2.2–)2.7–7(–8.3) times width; staminodes shorter than fertile stamens, ± hidden by sepals; nutlets with dorsal crest coarsely toothed . . . . 2. *Aspicarpa hyssopifolia*

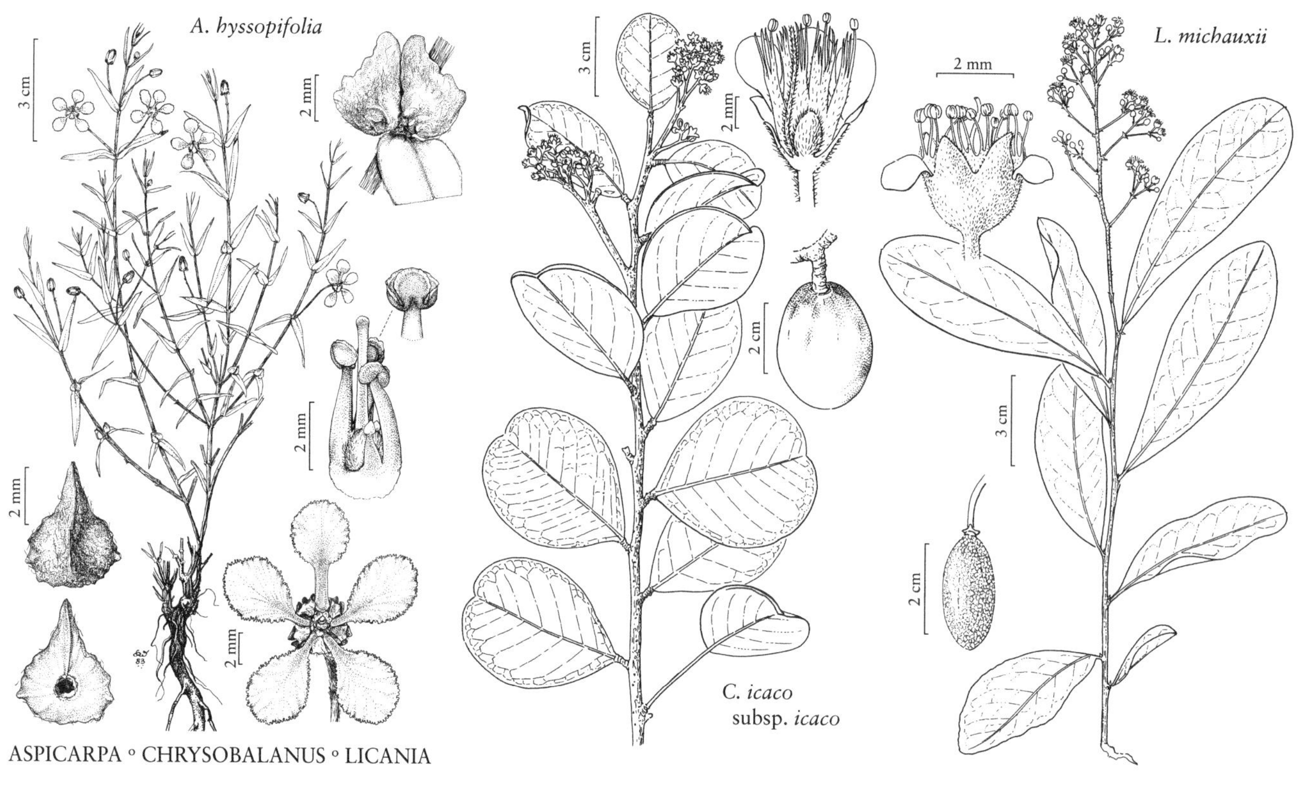

ASPICARPA ◦ CHRYSOBALANUS ◦ LICANIA

1. **Aspicarpa hirtella** Richard, Mém. Mus. Hist. Nat. 2: 396, plate 13. 1815

Aspicarpa longipes A. Gray; *A. urens* Lagasca

Herbs or subshrubs erect, 10–20 cm, or decumbent, branches 10–60(–100) cm. **Leaves:** petiole 1–3 mm; blade narrowly to broadly lanceolate or ovate, larger blades 15–45 × 6–23 mm, length 1.3–2.5 times width, base rounded or shallowly cordate, apex usually obtuse, occasionally acute, surfaces thinly sericeous or velutinous, hairs strongly appressed to V-shaped, more persistent abaxially than adaxially. **Chasmogamous flowers** in (2–)4(–7)-flowered umbels, occasionally corymbs, terminating leafy shoots; petals carrot yellow; staminodes equaling or longer than fertile stamens, surpassing sepals. **Cleistogamous flowers** often borne singly, or in clusters raised on slender axillary stalk, 5–55 mm, also sessile or subsessile in axils of full-sized leaves. **Nutlets** 3–3.5 mm diam., smooth or rugose, dorsal crest 0.5–0.8 mm, entire, extended forward 1–1.5 mm. **2*n*** = 80.

Flowering and fruiting (May–)Jul–Sep(–Oct). Dry rocky slopes among boulders, grassy slopes with oaks, pines, junipers; 1200–2000 m; Ariz., N.Mex., Tex.; Mexico.

In the flora area, *Aspicarpa hirtella* is restricted to southeastern Arizona, southwestern New Mexico, and trans-Pecos Texas.

2. **Aspicarpa hyssopifolia** A. Gray, Boston J. Nat. Hist. 6: 167. 1850 [F]

Herbs or subshrubs erect or spreading-ascending, 8–30 cm. **Leaves:** petiole 0.5 mm; blade narrowly lanceolate, larger blades 10–25 × 2–8 mm, length (2.2–)2.7–7(–8.3) times width, base rounded, apex acute or slightly obtuse, surfaces thinly sericeous to glabrescent, hairs persisting longest on margins and abaxial midrib. **Chasmogamous flowers** borne singly in axils of full-sized leaves; petals lemon yellow, outermost sometimes with red blotch; staminodes shorter than fertile stamens, ± hidden by sepals. **Cleistogamous flowers** sessile or subsessile in axils of full-sized leaves, sometimes absent. **Nutlets** 2.5–3 mm diam., rugose or bearing toothlike protuberances, dorsal crest 0.5–1 mm, coarsely toothed, extended forward 1.5–2 mm. **2*n*** = 80.

Flowering and fruiting Apr–Nov. Brushy, dry, calcareous slopes; 50–900 m; Tex.; Mexico (Coahuila, Nuevo León, Tamaulipas).

In the flora area, *Aspicarpa hyssopifolia* is found only in southern and western Texas.

CHRYSOBALANACEAE R. Brown

• Coco-plum Family

R. David Whetstone

Christopher F. Nixon†

Shrubs or trees, tardily deciduous to evergreen. **Leaves** alternate (2-ranked), simple; stipules present; petiole present, short; blade often coriaceous, margins entire or remotely toothed; venation pinnate. **Inflorescences** terminal or axillary, thyrses [cymes, racemes, or panicles]. **Flowers** bisexual; perianth and androecium perigynous; hypanthium free, well developed, densely hairy on both surfaces; sepals 5, distinct; petals [0 or 4–]5, distinct; nectary present, lining hypanthium; stamens [2–]14–22[–300], connate basally to proximally [distinct], free; anthers versatile, dehiscing by longitudinal slits; pistil 1, 3-carpellate with 1 [rarely 2–3] carpel developing, ovary superior, 1[–3]-locular, placentation basal; ovules 2 per locule, anatropous; style 1, basal; stigmas [1]3. **Fruits** drupes. **Seeds** 1 per fruit.

Genera 18, species ca. 530 (2 genera, 2 species in the flora): se United States, Mexico, West Indies, Central America, South America, s Asia, Africa, Pacific Islands, Australia.

Chrysobalanaceae was traditionally considered a subfamily of Rosaceae. Family status is supported by morphology (G. T. Prance 1972; Prance and C. A. Sothers 2003) and molecular data, which place it as a member of the Malpighiales and thus not closely related to Rosaceae (see, for example, M. W. Chase et al. 1993; N. Korotkova et al. 2009; K. Wurdack and C. C. Davis 2009). Only *Chrysobalanus icaco* is important commercially; it is planted as an ornamental and its fruits are eaten raw or bottled in syrup and sold (A. Cronquist 1981).

SELECTED REFERENCES Prance, G. T. 1970. The genera of Chrysobalanaceae in the southeastern United States. J. Arnold Arbor. 51: 521–528. Prance, G. T. 1972. Chrysobalanaceae. In: Organization for Flora Neotropica. 1968+. Flora Neotropica. 110+ nos. New York. No. 9. Prance, G. T. and C. A. Sothers. 2003. Chrysobalanaceae. In: Australian Biological Resources Study. 1999+. Species Plantarum: Flora of the World. 11+ parts. Canberra. Parts 9, 10.

1. Shrubs or trees, 1–5 m; leaf blades broadly elliptic, broadly ovate, or broadly obovate, length 1.2–1.5 times width; stamen filaments densely hairy proximally; endocarps longitudinally ribbed . . . . 1. *Chrysobalanus*, p. 366
1. Shrubs, to 0.3(–0.5) m; leaf blades oblanceolate, lanceolate, or narrowly oblong, length 2.5–3.5 times width; stamen filaments glabrous; endocarps not longitudinally ribbed . . . . . . 2. *Licania*, p. 366

1. CHRYSOBALANUS Linnaeus, Sp. Pl. 1: 513. 1753; Gen. Pl. ed. 5, 229. 1754 • [Greek *chrysos*, golden, and *balanos*, acorn or fruit, alluding to yellow fruits of some individuals of *C. icaco*]

Shrubs or trees. Leaves persistent; blade margins entire. **Thyrses** axillary or terminal. **Flowers:** petals 5; stamens [12–]14–22[–26], filaments connate basally to ⅓ length in groups, densely hairy proximally; ovary densely hairy; styles densely hairy. **Drupes** globose or broadly ellipsoid; endocarp longitudinally ribbed. $x = 11$.

Species 4 (1 in the flora): Florida, Mexico, West Indies, Central America, South America, w Africa; introduced in Pacific Islands; tropical and subtropical.

1. Chrysobalanus icaco Linnaeus, Sp. Pl. 1: 513. 1753 [F]

Subspecies 2 (1 in the flora): Florida, Mexico, West Indies, Central America, e South America, w Africa; introduced in Pacific Islands.

Subspecies *atacorensis* (A. Chevalier) F. White is found in western tropical Africa (G. T. Prance and C. A. Sothers 2003).

1a. Chrysobalanus icaco Linnaeus subsp. **icaco** • Coco-plum [F]

Chrysobalanus interior Small; *C. pellocarpus* G. Meyer

Shrubs or trees 1–5 m; stems usually arising singly. **Twigs** reddish, glabrate, lenticels elliptic, pith tan; bark striate. **Leaves:** stipules ovate, 0.8–2.5 mm; petiole 2–3 mm; blade broadly elliptic, broadly ovate, or broadly obovate, 3.5–6 × 3–5 cm, length 1.2–1.5 times width, base cuneate, obtuse, or rounded, margins revolute, apex emarginate, rounded or obtuse, surfaces glabrescent except for scattered hairs along midvein. **Thyrses:** rachis densely hairy; bracteoles caducous, sessile, ovate, 1.5 mm. **Flowers:** hypanthium 2.5–3 mm; sepals ovate to triangular, 1 mm, both surfaces densely strigose; petals white, spatulate to narrowly spatulate, 3.5–4 mm, glabrous, margins erose, apex obtuse. **Drupes** white, yellow-green, pink, red, dark purple, or black, (1.2–)1.5–2(–2.5) cm; endocarp 6-ribbed with secondary ribs. $2n = 22$.

Flowering year-round. Hammocks, beaches, frequently calcareous (shelly) sands; 0–10 m; Fla.; Mexico; West Indies; Central America; e South America; introduced in Pacific Islands.

Chrysobalanus icaco is known in the flora area only from southern Florida. Leaf morphology varies widely and has prompted a proliferation of names for this taxon. The fruits are eaten, and their flavor varies from a taste of marshmallow to apple; they have folk medicinal value in tropical areas. Although preserved as a foodstuff and used as an ornamental, the species has little current economic value in Florida; it is being investigated for cultivation there as a tropical fruit.

2. LICANIA Aublet, Hist. Pl. Guiane 1: 119, plate 45. 1775 • [Misspelled anagram of local French Guiana name *caligni*]

Geobalanus Small

Shrubs [trees]. Leaves persistent or deciduous; blade margins entire or irregularly crenate. **Thyrses** terminal [axillary]. **Flowers:** petals [0 or 4–]5; stamens [3–]14–16[–40], filaments connate basally to proximally [distinct], glabrous (rarely basally hairy); ovary glabrous or sparsely to densely hairy; styles glabrous [hairy]. **Drupes** obovoid, ellipsoid, or subglobose; endocarp not longitudinally ribbed. $x = 11$.

Species 192 (1 in the flora): se United States, Mexico, West Indies, Central America, South America, s Asia, w Africa.

1. Licania michauxii Prance, J. Arnold Arbor. 51: 526. 1970 • Gopher-apple, ground-oak, golden-apple [E] [F]

Chrysobalanus oblongifolius Michaux, Fl. Bor.-Amer. 1: 283. 1803, not *Licania oblongifolia* Standley 1937; *C. pallidus* (Small) L. B. Smith; *Geobalanus oblongifolius* (Michaux) Small; *G. pallidus* Small

Shrubs to 0.3(–0.5) m, forming colonies, strongly rhizomatous; aerial stems arising in clumps near growing tips of rhizomes. **Twigs** maroon to dark brown, densely hairy proximally, becoming sparsely hairy distally, lenticels round to elliptic, pith cream; bark not striate. **Leaves** persistent or deciduous; stipules cuneate, 1–1.5 mm, margins entire or remotely glandular-toothed; petiole 2–2.5 mm; blade oblanceolate, lanceolate, or narrowly oblong, 5–8 × 1.7–3 cm, length 2.5–3.5 times width, base cuneate, apex rounded (obscurely mucronate to mucronulate), surfaces glabrous except for scattered hairs along abaxial midvein. **Thyrses:** rachis densely white-pubescent; bracteoles caducous, sessile, ovate, 1.8 mm (base auriculate). **Flowers:** hypanthium 4 mm; sepals triangular, 1.5 mm, both surfaces densely hairy; petals white, narrowly spatulate, 1.5 mm, margins entire basally, erose to lobed apically, apex obtuse. **Drupes** yellow-green tinged with red, becoming pink to white when mature, 2–2.5 cm.

Flowering Feb–Sep; fruiting Mar–Oct. Pinelands, open scrub, primarily well-washed sandy soils in disturbed areas such as roadsides; 0–10 m; Ala., Fla., Ga., La., Miss., S.C.

Licania michauxii is a common species of roadsides and open woods on the outer coastal plain of the southeastern United States. Leaves are persistent in the southern part to deciduous in the northern part of its range. The common name ground-oak reflects superficial similarity to running oak (*Quercus pumila*), a sympatric species. Vegetative plants of the two species may be distinguished by hairs on leaves: stellate hairs are found on the oak; simple unbranched hairs occur on *L. michauxii*.

PUTRANJIVACEAE Endlicher
• Guiana-plum Family

Geoffrey A. Levin

Trees [shrubs], evergreen [deciduous], dioecious [rarely monoecious]. **Leaves** alternate [rarely opposite], simple; stipules present; petiole present; blade margins entire, dentate, or serrate; venation pinnate. **Inflorescences** unisexual, axillary [cauliflorous], fascicles or flowers solitary. **Flowers** unisexual; perianth hypogynous; hypanthium absent; sepals [3–]4–5[–7], distinct; petals 0; nectary present [absent]; stamens [2–]4–10[–50], distinct, free; anthers dehiscing by longitudinal slits; pistil 1, 1–2[–4]-carpellate; ovary superior, 1–2[–4]-locular, placentation axile; ovules 2 per locule, anatropous; styles [0]1–2[–4], distinct [connate basally]; stigmas 1–2[–4]; pistillode present [absent] in staminate flowers. **Fruits** drupes. **Seeds** 1 per locule or 1 per fruit by abortion.

Genera 2, species ca. 200 (1 genus, 2 species in the flora): Florida, Mexico, West Indies, Central America, South America, Asia, Africa, Indian Ocean Islands, Pacific Islands, Australia; tropical and subtropical regions.

Putranjiva Wallich, with three or four species in tropical and subtropical Asia differs from *Drypetes* in lacking a nectary, usually having only two or three stamens, and having conspicuously dilated, petaloid stigmas. These two genera were long considered to belong to Euphorbiaceae subfam. Phyllanthoideae Kosteletzky. DNA sequence data demonstrate that they form a clade in Malpighiales separate from the rest of Euphorbiaceae and should be treated as a distinct family (Angiosperm Phylogeny Group 2003, 2009; K. Wurdack et al. 2004; T. Tokuoka and H. Tobe 2006; Wurdack and C. C. Davis 2009); their distinctness is also supported by embryology and seed anatomy (W. Stuppy 1996; Tokuoka and Tobe 1999, 2001). Molecular data support a sister-group relationship between *Drypetes* and *Putranjiva*. *Sibangea* Oliver, a genus of three tropical African species, is sometimes distinguished from *Drypetes* by its pistillate sepals not touching in bud and persisting in fruit (versus imbricate and deciduous in *Drypetes*); molecular data show it to be embedded within *Drypetes* (Wurdack et al.).

Members of Putranjivaceae generally produce glucosinolates (mustard oils), often giving the plants a characteristic odor. They are the only plants outside the Capparales known to produce these chemicals (J. Rodman et al. 1998).

1. DRYPETES Vahl, Eclog. Amer. 3: 49. 1807 • [Probably from Greek *drypa*, dried olive or drupe, alluding to fruit]

Trees [shrubs]; trunks often fluted; indumentum of simple hairs. **Leaves** often subdistichous; stipules deciduous [persistent]; blade base oblique [rarely symmetrical]. **Pedicels** present. **Staminate flowers:** sepals 4–5[–7]; nectary intrastaminal, lobed [annular]; stamens 1–2 times number of sepals [or –50]; pistillode ± rudimentary. **Pistillate flowers:** sepals 4–5[–7]; nectary annular or lobed [absent]; styles 1 mm or less; stigmas dilated [2-fid, reniform, or subpeltate]. *x* = 10.

Species ca. 200 (2 in the flora): Florida, Mexico, West Indies, Central America, South America, Asia, Africa, Indian Ocean Islands, Pacific Islands, Australia; tropical and subtropical regions.

1. Ovaries 2-carpellate; stigmas 2; drupes red-orange at maturity, endocarps 0.5 mm thick, brittle; stamens 4(–5); leaves thick-papery, apices usually abruptly acute to acuminate, venation finely reticulate; buds not resinous . . . . 1. *Drypetes lateriflora*
1. Ovaries 1-carpellate; stigma 1; drupes white at maturity, endocarps 1–2 mm thick, bony; stamens 8(–10); leaves leathery, apices usually rounded to obtuse, if acute, not abruptly so, venation coarsely reticulate; buds resinous. . . . . 2. *Drypetes diversifolia*

1. Drypetes lateriflora (Swartz) Krug & Urban, Bot. Jahrb. Syst. 15: 357. 1892 • Guiana-plum [F]

Schaefferia lateriflora Swartz, Prodr., 38. 1788

Trees to 10 m. **Bark** light brown, rough, separating into scales. **Buds** hairy, not resinous. **Leaves:** petiole 0.3–1 cm; blade lanceolate to narrowly ovate-elliptic, 4–12 × 1.5–5 cm, thick-papery, base obtuse to acute, margins entire or minutely serrate, apex usually abruptly acute to acuminate; venation finely reticulate. **Staminate inflorescences:** flowers 8–25 per fascicle. **Pistillate inflorescences:** flowers 1–5 per fascicle. **Staminate flowers:** sepals 4(–5), ovate-elliptic, 1.5–2 mm, ciliate, otherwise sparsely hairy or glabrous; nectary deeply lobed; stamens 4(–5). **Pistillate flowers:** sepals like those of staminate flowers; nectary lobed; ovary 2-carpellate; styles 2; stigmas 2. **Drupes** red-orange at maturity, ovoid to subglobose, 10–13 × 8–11 mm; mesocarp 1–2 mm thick, fleshy; endocarp 0.5 mm thick, brittle. **Seeds** usually 1 per fruit, sometimes 1 per locule.

Flowering late winter–spring; fruiting spring–early summer. Tropical hammocks; 0–10 m; Fla.; e Mexico; West Indies; Central America.

Fruits of *Drypetes lateriflora* apparently are removed by animals soon after they ripen; specimens with mature fruits are rarely collected. *Drypetes lateriflora* is more common on the Florida mainland than in the Keys.

2. Drypetes diversifolia Krug & Urban, Bot. Jahrb. Syst. 15: 353. 1892 • Whitewood, milkbark

Trees to 7(–12) m. **Bark** white, often with irregular gray or light brown patches, smooth. **Buds** glabrous, resinous. **Leaves:** petiole 0.3–1.2 cm; blade ovate to elliptic, 5–9 × 2–4 cm, leathery, base rounded to obtuse, margins entire on adult leaves, spinose-dentate on leaves of seedlings and sprouts, apex usually rounded to obtuse, if acute, not abruptly so; venation coarsely reticulate. **Staminate inflorescences:** flowers (1–)4–7 per fascicle. **Pistillate inflorescences:** flowers 1–3 per fascicle. **Staminate flowers:** sepals (4–)5, oblong to ovate, 2.5–4 mm, densely hairy, not ciliate; nectary slightly lobed; stamens 8(–10). **Pistillate flowers:** sepals like those of staminate flowers; nectary annular; ovary 1-carpellate; style 1; stigma 1. **Drupes** white at maturity, ellipsoid to obovoid, 15–20 × 12–15 mm; mesocarp 2–3 mm thick, mealy; endocarp 1–2 mm thick, bony. **Seeds** 1 per fruit.

Flowering late spring; fruiting summer–late winter. Tropical hammocks; 0–10 m; Fla.; West Indies (Bahamas).

Because of its short blooming period and inconspicuous flowers, *Drypetes diversifolia* is rarely collected in flower. The fruits persist on the plants long after they ripen. *Drypetes diversifolia* is fairly common in the Florida Keys but rare on the mainland.

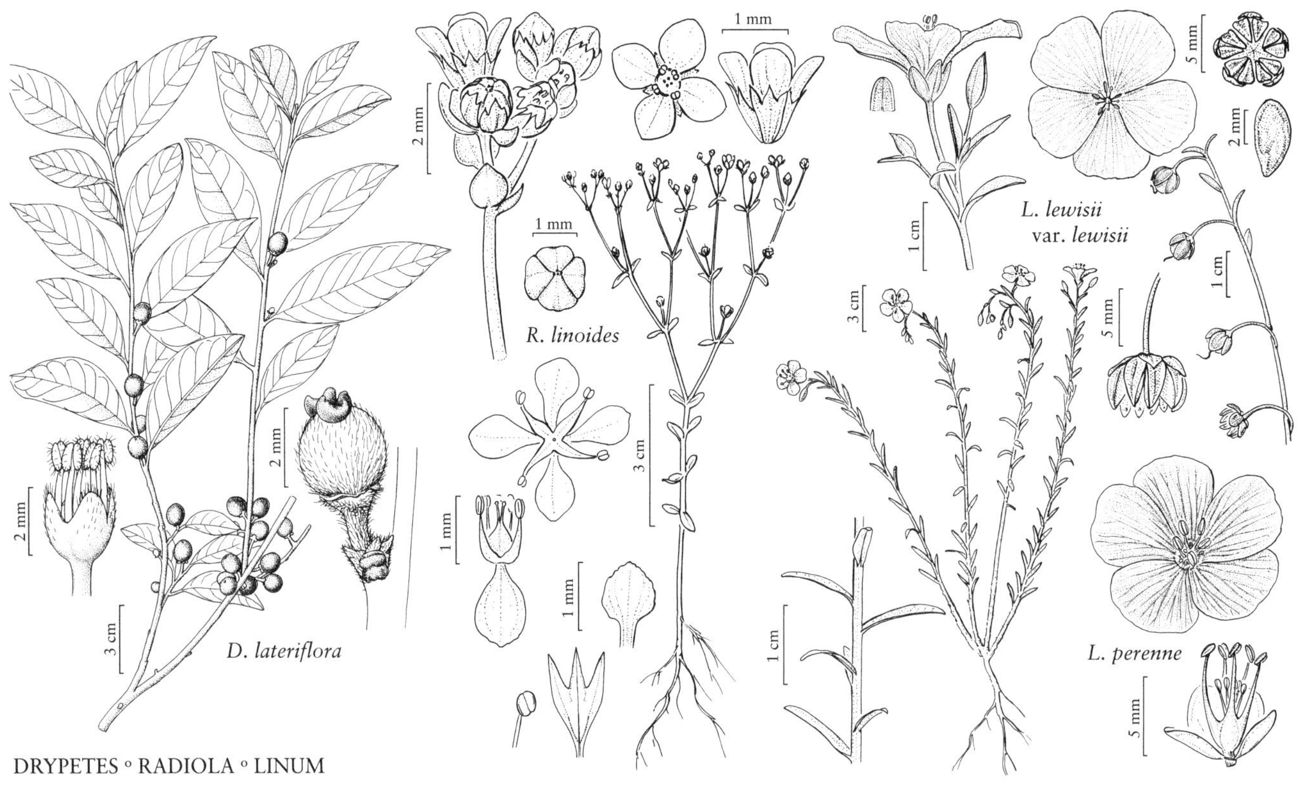

DRYPETES ◦ RADIOLA ◦ LINUM

LINACEAE de Candolle ex Perleb • Flax Family

Nancy R. Morin

Herbs or subshrubs [shrubs, trees, vines], annual, biennial, or perennial. **Leaves** alternate, opposite, or whorled, simple; stipules absent or present as small, dark, spheric glands; petiole usually absent, rarely present; blade margins entire, serrate, or denticulate; venation pinnate. **Inflorescences** terminal, racemes, panicles, or cymes (rarely thyrses or corymbs in *Linum*) [spikes]. **Flowers** bisexual; perianth and androecium hypogynous; hypanthium absent; sepals 4–5, connate basally [distinct]; petals 4–5, distinct or coherent basally, imbricate or convolute, bases sometimes with appendages; nectary extrastaminal; stamens 4–5[10], connate basally, filament tube and petal bases adherent or adnate [free]; anthers dehiscing by longitudinal slits; pistil 1, 2–5-carpellate, ovary superior, 4–5-locular, placentation axile or apical-axile; ovules 2 per locule, anatropous; styles 2–5, distinct or partly connate; stigmas 2–5. **Fruits** capsules, dehiscence septicidal, or indehiscent, or schizocarps breaking into 4 nutlets (*Sclerolinon*). **Seeds** 2 per locule, seed coat often mucilaginous.

Genera 10–14, species ca. 260 (4 genera, 52 species in the flora): North America, Mexico, West Indies, Bermuda, Central America, South America, Eurasia, Africa, Atlantic Islands, Pacific Islands, Australia.

Two subfamilies are generally recognized in Linaceae: the mostly herbaceous, temperate Linoideae Arnott (8 genera, ca. 240 species), in which all the genera in the flora area are placed, and the woody, mostly tropical Hugonoideae Reveal. Based on molecular phylogenetic analysis, J. R. McDill et al. (2009) concluded that Linaceae is a monophyletic group, as is Linoideae.

According to J. R. McDill (2009), *Cliococca* Babington, *Hesperolinon*, and *Sclerolinon* are nested within *Linum* sect. *Linopsis*, and collectively these are sister to *Radiola*; *Hesperolinon* and *Sclerolinon* are most closely related to Mexican and Central American species of *Linum*. McDill et al. (2009) noted that the relationships within this clade are not well-enough resolved or supported to warrant nomenclatural changes; McDill (2009) came to the same conclusion based on a much wider sample of species. The current generic circumscriptions are maintained here.

SELECTED REFERENCES McDill, J. R. 2009. Molecular Phylogenetic Studies in the Linaceae and *Linum*, with Implications for Their Systematics and Historical Biogeography. Ph.D. dissertation. University of Texas. McDill, J. R. et al. 2009. The phylogeny of *Linum* and Linaceae subfamily Linoideae, with implications for their systematics, biogeography, and evolution of heterostyly. Syst. Bot. 34: 386–405. McDill, J. R. and B. B. Simpson. 2011. Molecular phylogenetics of Linaceae with complete generic sampling and data from two plastid genes. Bot. J. Linn Soc. 165: 64–83. Rogers, C. M. 1975. Relationships of *Hesperolinon* and *Linum* (Linaceae). Madroño 23: 153–159. Rogers, C. M. 1984. Linaceae. In: N. L. Britton et al., eds. 1905+. North American Flora.... 47+ vols. New York. Ser. 2, part 12, pp. 1–54.

1. Sepals 4; petals 4 . . . . 1. *Radiola*, p. 372
1. Sepals 5; petals 5.
   2. Styles 5; fruits capsules, dehiscing into 5 or 10 segments. . . . . 2. *Linum*, p. 373
   2. Styles 2–3; fruits capsules dehiscing into 4 or 6 segments, schizocarps breaking into 4 nutlets, or indehiscent.
      3. Leaves: basal and proximal usually whorled, distal alternate or opposite; fruits capsules, dehiscing into 4 or 6 segments; styles 2–3, stigmas ± equal in width to styles; stipular glands present (exudate often red) or absent . . . . 3. *Hesperolinon*, p. 395
      3. Leaves: proximal opposite, distal sometimes alternate; fruits schizocarps, breaking into 4 nutlets, or indehiscent; styles 2, stigmas wider than styles; stipular glands absent. . . . . 4. *Sclerolinon*, p. 402

1. RADIOLA Hill, Brit. Herb., 227, plate 33 [upper center]. 1756 • Allseed [Latin *radiolus*, little ray, alluding to rayed capsules] [I]

Herbs, annual, glabrous. **Stems** erect, branched throughout. **Leaves** persistent, opposite; stipular glands absent; blade ovate to lanceolate, margins entire. **Inflorescences** dichasia. **Pedicels** not articulated. **Flowers:** sepals persistent, 4, connate at base, equal in size, margins entire, not glandular; petals 4, distinct, attached at rim of filament cup, white, appendages absent; stamens 4; staminodes 0; pistil 4-carpellate, ovary 4-locular; styles 4, distinct; stigmas capitate, wider than styles. **Fruits** capsules, dehiscing into 4 segments, each partially divided by an incomplete septum. **Seeds** 8, obovoid to ellipsoid, sometimes flattened on one side. ***x*** = 9.

Species 1: introduced; Europe, sw Asia, Africa, Atlantic Islands (Macaronesia).

1. Radiola linoides Roth, Tent. Fl. Germ. 1: 71. 1789 [F] [I]

Linum radiola Linnaeus, Sp. Pl. 1: 281. 1753; *Millegrana radiola* (Linnaeus) Druce

Herbs 3–10 cm; dichotomously branched. **Leaves** spreading; blade 2–6 mm, base clasping. **Inflorescences:** bracts usually leaflike, entire. **Pedicels** ascending. **Flowers:** sepals equal, oblong, apex deeply 3-fid; petals widely spreading, ovate, 1–1.5 mm, slightly narrowed at base; stamens 0.5–0.8 mm; anthers white. **2*n*** = 18.

Flowering Jul–Aug. Vernally damp, sandy, acidic, open ground, roadsides, logging trails, rocky outcrops; 0–100 m; introduced; N.B., N.S.; Maine; Europe; sw Asia; Africa; Atlantic Islands (Macaronesia).

Radiola linoides is abundant and widespread in Nova Scotia.

2. LINUM Linnaeus, Sp. Pl. 1: 277. 1753; Gen Pl. ed. 5, 135. 1754 • Flax [Latin *lin*, flax]

Herbs or subshrubs, annual, biennial, or perennial, glabrous or hairy. **Stems** usually erect or spreading to ascending, sometimes decumbent or ascending from decumbent base, unbranched or branched at base, throughout, or only in inflorescence. **Leaves** sometimes falling early, alternate or sometimes partially opposite or whorled; stipular glands present or absent; blade linear, linear-lanceolate, linear-oblanceolate, lanceolate, oblanceolate, elliptic, oblong, obovate, spatulate, or awl-shaped, margins glandular-toothed or entire, sometimes ciliate. **Inflorescences** usually panicles, racemes, or cymes, rarely thyrses or corymbs. **Pedicels** articulated or not. **Flowers:** sepals persistent or deciduous, 5, connate at base, equal or unequal in size, margins scarious, entire, or toothed, glandular or not, sometimes ciliate; petals 5, distinct or coherent at base, attached to filament cup at base, midway, or on or proximal to rim, blue, white, yellow, yellowish orange, orange, or salmon, rarely red or maroon, sometimes with darker bands near base, appendages absent or pouches formed on petal margins at base of claw; stamens 5; staminodes 0 or 5, as small deltate projections; pistil 5-carpellate, ovary 5-locular, or 10-locular by intrusion of false septa; styles 5, distinct or connate; stigmas capitate, linear, or clavate, wider than styles. **Fruits** capsules, usually 5-celled and dehiscing into 5 segments, sometimes each cell partially divided by incomplete or nearly complete false septum and dehiscing into 10 segments. **Seeds** 10, lenticular. x = 13, [15, 18].

Species ca. 180 (37 in the flora): nearly worldwide; temperate and subtropical regions.

C. M. Rogers (1963, 1964, 1968, 1982, 1984) published comprehensive studies of *Linum* in North America and Central America; he also studied *Linum* in South America (Rogers and R. Mildner 1976), southern Africa (Rogers 1981), and Madagascar (Rogers 1981b). This treatment draws largely on his work and follows his taxonomic arrangement, which is congruent, at least at the level of section, with the results in J. R. McDill et al. (2009). Species of *Linum* in the flora have been placed in three sections of the genus, out of a total of five sections worldwide.

SELECTED REFERENCES Harris, B. D. 1968. Chromosome numbers and evolution in North America species of *Linum*. Amer. J. Bot. 55: 1197–1204. Rogers, C. M. 1963. Yellow-flowered species of *Linum* in eastern North America. Brittonia 15: 97–122. Rogers, C. M. 1964. Yellow-flowered *Linum* (Linaceae) in Texas. Sida 1: 328–336. Rogers, C. M. 1968. Yellow-flowered species of *Linum* in Central America and western North America. Brittonia 20: 107–135.

1. Petals yellow, sometimes with maroon at base . . . . .2c. *Linum* sect. *Linopsis*, p. 378
1. Petals red, white, or blue.
   2. Sepal margins not glandular-toothed; petals usually blue or red to maroon, rarely white . . . . 2a. *Linum* sect. *Linum*, p. 373
   2. Sepal margins (at least inner) glandular-toothed; petals white . . . . .2b. *Linum* sect. *Cathartolinum*, p. 377

2a. Linum Linnaeus sect. Linum

Herbs, annual, biennial, or perennial. **Stems** usually terete (ridged in *L. grandiflorum*). **Leaves** alternate; stipular glands absent; blade margins entire, not glandular-toothed. **Flowers** homostylous or heterostylous; sepals persistent, margins scarious, entire, not glandular-toothed; petals usually blue or red to maroon, rarely white; staminodia present or absent; styles distinct, connate basally or to midlength; stigmas capitate, clavate, or linear. **Capsules** dehiscing into 10, 1-seeded segments; false septa incomplete. **Pollen** tricolpate.

Species ca. 50 (6 in the flora): North America, Mexico, Eurasia, n Africa, Pacific Islands (New Zealand), Australia; introduced in Central America, s South America.

SELECTED REFERENCE Mosquin, T. 1971. Biosystematic studies in the North American species of *Linum*, section *Adenolinum* (Linaceae). Canad. J. Bot. 49: 1379–1388.

1. Petals bright red to maroon, fading to purple . . . . 6. *Linum grandiflorum*
1. Petals usually blue, rarely white.
   2. Stigmas linear or clavate; margins of inner sepals minutely ciliate.
      3. Annuals; petals 10–15 mm; capsules 6–10 mm, apices rounded; seeds 4–6 mm . . . . 1. *Linum usitatissimum*
      3. Biennials or short-lived perennials; petals 6–10 mm; capsules 4–6 mm, apices very sharp-pointed; seeds 2.5–3 mm. . . . . 2. *Linum bienne*
   2. Stigmas capitate or ellipsoid-capitate; margins of inner sepals glabrous.
      4. Flowers heterostylous . . . . 5. *Linum perenne*
      4. Flowers homostylous.
         5. Perennials; styles 2–12 mm; capsule apices acute . . . . 3. *Linum lewisii*
         5. Annuals; styles 1–3 mm; capsule apices obtuse. . . . . 4. *Linum pratense*

1. Linum usitatissimum Linnaeus, Sp. Pl. 1: 277. 1753 • Common flax, lin cultivé [I] [W]

Herbs annual, 20–100 cm, glabrous or glabrate throughout. **Stems** erect, unbranched or few-branched at base (all flowering). **Leaves** divergent; blade linear to linear-lanceolate, 10–40 × 1.5–5 mm. **Inflorescences** open panicles. **Pedicels** erect in fruit, to 20–25 mm. **Flowers** homostylous; sepals ovate, 6–9 mm, margins of inner sepals minutely ciliate, outer ciliate, apex acuminate; petals usually blue, rarely white, obovate, 10–15 mm; stamens 5–7 mm; anthers 1–1.5 mm; staminodia present; styles distinct or connate at base, 3–6 mm; stigmas linear or clavate. **Capsules** ovoid to subglobose, 6–10 × 5–10 mm, apex rounded, dehiscing incompletely, segments falling freely, margins ciliate or not. **Seeds** 4–6 × 2.5–3 mm. **2*n*** = 30.

Flowering Apr–Sep. Disturbed areas, roadsides, abandoned homesteads, fields; 0–2400 m; introduced; Alta., B.C., Man., Nfld. and Labr. (Nfld.), N.W.T., N.S., Ont., Que., Sask.; Ala., Ariz., Ark., Calif., Colo., Conn., Del., D.C., Fla., Ga., Idaho, Ill., Ind., Iowa, Kans., Ky., La., Maine, Md., Mass., Mich., Minn., Miss., Mo., Mont., Nebr., N.H., N.J., N.Mex., N.Y., N.C., N.Dak., Ohio, Okla., Oreg., Pa., R.I., S.C., S.Dak., Tenn., Tex., Vt., Va., Wash., W.Va., Wis., Wyo.; Eurasia; introduced also in c Mexico, Central America, s South America, Pacific Islands (New Zealand).

Linum usitatissimum has been cultivated since antiquity, and it is this cultivated form that has naturalized in the wild. Flax fibers twisted to make rope or dyed for fabric dated 32,000–26,000 years before present were found in a cave in Dzudzuana, Georgia (E. Kvavadze et al. 2009). Stem fibers of *L. usitatissimum* are used to make linen; the seeds are pressed to produce linseed oil; the rest of the seeds are compacted into cakes and used as fodder. *Linum usitatissimum* is the only species in the flora area except *L. bienne* that has linear stigmas and minutely ciliate inner sepals. It can be distinguished from *L. bienne* by its larger, apically rounded capsules.

2. Linum bienne Miller, Gard. Dict. ed. 8, Linum no. 8. 1768 • Pale flax [I] [W]

Linum angustifolium Hudson

Herbs biennial or short-lived perennial (flowering 1st year), 6–60 cm, glabrous. **Stems** erect, usually branched from near base and in inflorescence. **Leaves:** blade linear to linear-lanceolate, 5–25 × 1–1.5 mm. **Inflorescences** open panicles. **Pedicels** 10–25 mm. **Flowers** homostylous; sepals ovate, 4–5.5 mm, margins of inner sepals minutely ciliate, outer glabrous, apex acute to acuminate; petals blue, obovate, 6–10 mm; stamens 4–5 mm; anthers 1–2.5 mm; staminodia present or absent; styles distinct, 2 mm; stigmas linear or clavate. **Capsules** broadly ovate to subglobose, 4–6 × 4–6 mm, apex very sharp-pointed, segments ± persistent on plant, margins ciliate. **Seeds** 2.5–3 × 1.5–2 mm. **2*n*** = 30.

Flowering Mar–Aug. Grasslands, woodlands, disturbed places; 0–1900 m; introduced; B.C.; Calif., Oreg., Pa.; Europe; n Africa; introduced also in South America (Argentina, Chile), Pacific Islands (New Zealand).

Linum bienne is thought to be the progenitor of *L. usitatissimum* (D. J. Ockendon 1971).

3. Linum lewisii Pursh, Fl. Amer. Sept. 1: 210. 1813 • Lewis's or wild blue flax [F]

Linum perenne Linnaeus subsp. *lewisii* (Pursh) Hultén; *L. perenne* var. *lewisii* (Pursh) Eaton & Wright

Herbs perennial, 5–80 cm, glabrous or glabrate throughout, ± glaucous. **Stems** erect to spreading or ascending, branched from near base and in inflorescence. **Leaves:** blade linear to linear-lanceolate or linear-oblanceolate, 5–30 × 0.5–3(–4.5) mm. **Inflorescences** open panicles or racemes. **Pedicels** 5–20 mm. **Flowers** homostylous; sepals elliptic or elliptic-ovate, 3.5–6 mm, margins glabrous, apex acute; petals usually blue, sometimes white, base whitish or yellowish, cuneate-obovate, 6–23 mm; stamens 3–10 mm; anthers 1–2.2 mm; staminodia present; styles distinct, 2–12 mm; stigmas thickened ellipsoid-capitate. **Capsules** ovoid to globose, 4–8 × 5–6 mm, apex acute, segments ± persistent on plant, margins arachnoid-ciliate. **Seeds** 2.5–5 × 1.5–3 mm. $2n = 18$.

Varieties 3 (3 in the flora): North America, n Mexico.

Linum lewisii is native to many habitats in western North America from northern Mexico to Alaska east to the Great Plains in the United States and to the west side of Hudson and James bays in Canada; it appears to be less common in the Great Basin. A component of wildflower seed mixes, the species may be expanding its range. Some authors have considered it conspecific with *L. perenne*, and many collections in herbaria are identified as *L. perenne* without an indication of variety; they are most likely *L. lewisii* var. *lewisii* (D. J. Ockendon 1971; C. M. Rogers 1984). Because of the prevalence of *L. bienne*, *L. perenne,* and *L. usitatissimum* in bird seed and wildflower mixes, it may be that these three non-natives are becoming more common than in the past. Capitate stigmas distinguish *L. lewisii* from *L. bienne* and *L. usitatissimum*, which have linear or clavate stigmas. Distinguishing *L. lewisii* from *L. perenne* is more difficult: the size of flower parts in the homostyled *L. lewisii* varies along elevational and latitudinal gradients, with smaller flowers and flower parts in higher elevations and higher latitudes; except in var. *lepagei*, the styles are always longer than the stamens. In the heterostyled *L. perenne*, populations usually include plants in which flowers have stamens much longer than the very short styles (short-styled form) and plants in which flowers have stamens much shorter than the very long styles, up to twice as long as the stamens (long-styled form).

C. A. Kearns and D. W. Inouye (1994) reported that *Linum lewisii* is facultatively autogamous but tends not to set seed in the absence of pollinators; small bees and flies are the most common pollinators. A. Cronquist et al. (1997b) reported unusual populations of *L. lewisii* on sandy soil in Nye County, Nevada, in the 40-Mile-Canyon drainage, that had persistent, ascending, pale blue petals with darker veins.

SELECTED REFERENCES Becker, J. D. T. 2010. Taxonomy of the *Linum lewisii* Complex in Canada Based on Macromorphology, Micromorphology, and Phytogeography. Honors thesis. University of Manitoba. Kearns, C. A. and D. W. Inouye. 1994. Fly pollination in *Linum lewisii* (Linaceae). Amer. J. Bot. 81: 1091–1095.

1. Petals mostly white; Hudson and James bay regions . . . . 3c. *Linum lewisii* var. *lepagei*
1. Petals usually blue; mostly c, w North America, not Hudson and James bay regions.
   2. Petals (8–)12–23 mm; styles 6–12 mm. . . . . 3a. *Linum lewisii* var. *lewisii*
   2. Petals 6–13 mm; styles 2–6 mm . . . . 3b. *Linum lewisii* var. *alpicola*

3a. Linum lewisii Pursh var. **lewisii** [F]

Herbs 15–80 cm. **Flowers:** petals usually blue, (8–)12–23 mm; styles 6–12 mm.

Flowering Apr–Aug. Semi-deserts, mesas, prairies, dry calcareous glades, barrens, alpine meadows, arctic tundra; 0–3600 m; Alta., B.C., Man., N.W.T., Nunavut, Ont., Que., Sask., Yukon; Alaska, Ariz., Ark., Calif., Colo., Idaho, Kans., La., Minn., Mo., Mont., Nebr., Nev., N.Mex., N.Dak., Okla., Oreg., S.Dak., Tex., Utah, Wash., W.Va., Wyo.; Mexico (Baja California, Chihuahua, Coahuila, Durango, Mexico, Nuevo León, Sonora).

T. Mosquin (1971) found that within populations of var. *lewisii* variation in height, branching, flower and fruit size, and the relative lengths of styles and stamens was low, whereas these characters differed considerably among populations. Populations of var. *lewisii* in the central and eastern United States (Louisiana, Missouri, West Virginia) are most likely recent introductions (C. M. Rogers 1984), as are populations in Ontario and Quebec (J. D. T. Becker 2010). Some populations in Minnesota appear to be hybrids between var. *lewisii* and *Linum perenne*.

3b. Linum lewisii Pursh var. **alpicola** Jepson, Fl. Calif. 2: 398. 1936 • Lewis's alpine flax, prairie flax [E]

Herbs 5–25 cm. **Flowers:** petals blue, 6–13 mm; styles 2–6 mm.

Flowering Jun–Aug. Alpine ridges; 2000–3700 m; Calif., Idaho, Nev., Utah.

3c. Linum lewisii Pursh var. **lepagei** (B. Boivin) C. M. Rogers, Phytologia 41: 448. 1979 • Lepage's flax, lin de Lepage [C] [E]

Linum lepagei B. Boivin, Naturaliste Canad. 75: 219. 1948

Herbs 10–30 cm. **Flowers:** petals mostly white, 8.4–10 (–14) mm; styles 2.6–5.1 mm.

Flowering Jul–Aug. Sandy or gravelly ridges, limestone outcrops on arctic shores; of conservation concern; 0–20 m; Man., Nunavut, Ont.

According to M. Oldham (pers. comm.), var. *lepagei* is one of the few Hudson Bay endemic vascular plants; it is widespread but local on the Ontario coast, where it always occurs on sandy or gravelly beach ridges. It grows in similar habitats on Akimiski Island in the James Bay region, Nunavut. J. D. T. Becker (2010) found that specimens of var. *lepagei* were smaller in general than those of var. *lewisii*. Stamens were longer than carpels in var. *lepagei* and shorter than carpels in var. *lewisii*. Specimens identified as var. *lewisii* from coastal and high-latitude localities in Yukon and Northwest Territories were similar to var. *lepagei*. Variety *lepagei* can be expected to occur in Quebec; specimens collected from Nunavut islands are only 7 km from the Quebec shoreline (Becker).

4. Linum pratense (Norton) Small in N. L. Britton et al., N. Amer. Fl. 25: 69. 1907 • Meadow or Norton's flax [E]

Linum lewisii Pursh var. *pratense* Norton, Trans. Acad. Sci. St. Louis 12: 38, plate 6. 1902

Herbs annual, 5–60 cm, glabrous. **Stems** ± spreading or ascending, or branches from base prostrate. **Leaves:** blade linear to linear-oblanceolate, 8–20 × 0.7–2.3 mm. **Inflorescences:** open panicles or racemes. **Pedicels** 8–25 mm. **Flowers** homostylous; sepals ovate, 3–5 mm, margins glabrous, apex acute; petals usually blue, rarely white, obovate, 5–14 mm; stamens 3–5 mm; anthers 0.4–1.3 mm; staminodia present; styles distinct, 1–3 mm; stigmas capitate. **Capsules** broadly ovate to subglobose, 4–6 mm diam., apex obtuse, segments persistent on plant, margins ciliate. **Seeds** 3–5 × 1.2–1.6 mm. $2n = 18$.

Flowering Mar–Jun. Sandy prairies, roadsides, disturbed areas, limestone; 1200–2000 m; Ariz., Colo., Kans., N.Mex., Okla., Tex.

In a study of pollination in *Linum pratense*, G. E. Uno (1984) observed that petals dropped soon after anthesis and the persistent sepals quickly moved inward, pressing the dehiscing anthers against the receptive stigmas. Small bees and flies were seen to visit flowers even after the petals fell. Uno noted sepals closing in both *L. lewisii* and *L. rigidum*, but in these species the stamens tend to be somewhat shorter than the styles, so self-pollination was less likely.

C. M. Rogers (1984) wrote that some plants of *Linum pratense* intergrade with *L. lewisii* in areas where their ranges overlap; however, in most of its range, *L. pratense* is the only blue-flowered *Linum* and can be distinguished from the occasional plant of *L. bienne* or *L. usitatissimum* by its lack of cilia on the inner sepals and its capitate stigmas.

SELECTED REFERENCE Uno, G. E. 1984. The role of persistent sepals in the reproductive biology in *Linum pratense* (Linaceae). SouthW. Naturalist 29: 429–434.

5. Linum perenne Linnaeus, Sp. Pl. 1: 277. 1753 • Blue flax [F] [I]

Herbs perennial, 20–100 cm, glabrous. **Stems** ascending or erect, usually unbranched. **Leaves:** blade linear or linear-lanceolate, 5–20 × 1–3 mm. **Inflorescences** much-branched panicles. **Pedicels** spreading, 5–25 mm. **Flowers** heterostylous; inner sepals ovate-lanceolate or ovate, 4.5–5.5 mm, margins glabrous, apex obtuse, outer ones lanceolate or ovate-lanceolate, 3.5–4.5 mm, narrower than inner ones, margins glabrous, apex acute or acuminate; petals blue, obovate or obovate-lanceolate, 10–25 mm; stamens 5 mm, anthers 2 mm (long-styled morph) or stamens 6.5 mm, anthers 1.8 mm (short-styled morph); styles distinct, 8 mm (long-styled morph) or 2.5 mm (short-styled morph); stigmas capitate. **Capsules** subglobose, 5–7 mm diam., apex acute to obtuse, segments persistent on plant, margins ciliate or not. **Seeds** 3–4.2 × 1.7–2 mm. $2n = 18$.

Flowering Mar–Aug. Disturbed areas; 100–1000 m; introduced; B.C., Ont., Yukon; Ariz., Colo., Idaho, Ill., Iowa, Maine, Mich., Mont., Nebr., Nev., N.Y., Ohio, Oreg., Pa., Utah, Va., W.Va., Wis.; Eurasia; introduced also in Mexico (Sonora).

Most collections in North America identified as *Linum perenne* are most likely *L. lewisii* var. *lewisii* (D. J. Ockendon 1971; C. M. Rogers 1984). According to Ockendon, *L. perenne* is often confused with *L. austriacum* Linnaeus in Europe; its exact native distribution is not known.

6. Linum grandiflorum Desfontaines, Fl. Atlant. 1: 277, plate 78. 1798 • Flowering or red or scarlet or crimson flax [I]

Herbs annual, 10–60 cm, glabrous, glaucous. **Stems** ascending or sometimes decumbent at base, usually freely branched. **Leaves:** blade linear to lanceolate or narrowly elliptic, 10–30 × 2–3(–7) mm. **Inflorescences** cymes, few-flowered. **Pedicels** 10–25 mm. **Flowers** heterostylous; sepals lanceolate, 7–11 mm, margins glabrous, apex acuminate; petals bright red to maroon, fading to purple, broadly obovate, 15–30 mm; stamens 8–10 mm; anthers 5 mm; staminodia not seen; styles connate proximal ½, 4.5 mm (short-styled) or 8–10 mm (long-styled); stigmas clavate. **Capsules** ovoid-globose, 6–7 mm diam., apex apiculate, segments persistent on plant, margins not seen. **Seeds** 2–3 × 0.5–1 mm. **2*n*** = 16.

Flowering Apr–Sep. Disturbed areas; 0–2700 m; introduced; Calif., Colo., Fla., Ky., Nebr., N.Y., Ohio, Pa., Tex., Utah; n Africa.

Linum grandiflorum occasionally escapes from gardens and persists along roadsides and trails. This showy garden plant has blue anthers.

2b. Linum Linnaeus sect. Cathartolinum (Reichenbach) Grisebach, Spic. Fl. Rumel. 1: 118. 1843 [I]

Cathartolinum Reichenbach, Handb. Nat. Pfl.-Syst., 306. 1837

Herbs, annual. **Stems** terete. **Leaves** opposite; stipular glands absent; blade margins entire, not glandular-toothed. **Flowers** homostylous; sepals persistent, margins scarious, at least inner glandular-toothed; petals white; staminodia present; styles distinct; stigmas capitate. **Capsules** dehiscing into 10, 1-seeded segments; false septa incomplete. **Pollen** tricolpate.

Species 1: introduced; Europe, w Asia, Atlantic Islands (Iceland); introduced also in South America, Pacific Islands, Australia.

7. Linum catharticum Linnaeus, Sp. Pl. 1: 281. 1753 • Fairy or purging flax, lin purgatif [I] [W]

Herbs 8–30 cm, glabrous. **Stems** erect, usually unbranched proximal to inflorescence, sometimes branched from decumbent base. **Leaves:** blade narrowly elliptic to oblanceolate or narrowly obovate or oblong, larger 5–18 × 1.4–3.1 mm, largest at midstem, reduced in size both proximally and distally, apex obtuse to acute. **Inflorescences** panicles. **Pedicels** 6–35 mm. **Flowers:** sepals broadly lanceolate to ovate, outer sepals 2–3 mm, inner sepals broader, shorter, margins of all or sometimes only of inner sparsely but conspicuously glandular-toothed, apex acute to acuminate; petals white or whitish, base yellowish, 2–5 mm, obovate; stamens 1 mm, anthers 0.2–0.3 mm; styles 0.5–1 mm. **Capsules** ovoid, 2–2.5 × 2 mm, fragile and subject to crushing when pressed; segments persistent on plant, septa margins ciliate. **Seeds** 1–1.5 × 0.6–0.8 mm. **2*n*** = 16.

Flowering Jun–Aug. Calcareous or sandy soils, fields, pastures, roadsides; 0–400 m; introduced; B.C., N.B., Nfld. and Labr. (Nfld.), N.S., Ont., P.E.I., Que.; Maine, Mass., Mich., N.H., N.Y., Pa., Vt.; Europe; w Asia; Atlantic Islands (Iceland); introduced also in South America (Argentina), Pacific Islands (New Zealand), Australia (Tasmania).

Linum catharticum has small, white, funnelform corollas, yellow anthers, and light green stigmas. It is the only *Linum* in the flora area with white petals and opposite leaves. The species is widespread in Europe and occurs only sporadically in the northern United States and most of its range in Canada. It may be native in Newfoundland and Nova Scotia, where it is well established.

2c. LINUM Linnaeus sect. LINOPSIS (Reichenbach) Engelmann, Smithsonian Contr. Knowl. 3(5): 25. 1852

Linopsis Reichenbach, Handb. Nat. Pfl.-Syst., 306. 1837; *Mesyniopsis* W. A. Weber

Herbs or subshrubs, annual or perennial, rarely with woody caudex. **Stems** terete, ridged, or sulcate. **Leaves** alternate, opposite, or whorled; stipular glands present or absent; blade margins entire or glandular-toothed, sometimes ciliate. **Flowers** homostylous; sepals persistent or deciduous, margins scarious or not, glandular-toothed or not; petals yellow, sometimes with maroon at base; staminodia present or absent; styles distinct or connate nearly to apex; stigmas usually capitate, rarely linear. **Capsules** dehiscing into 10, 1-seeded segments or 5, 2-seeded segments; false septa incomplete to complete. **Pollen** tricolpate or multiporate.

Species ca. 85 (30 in the flora): nearly worldwide.

SELECTED REFERENCE Rogers, C. M. 1982. The systematics of *Linum* sect. *Linopsis* (Linaceae). Pl. Syst. Evol. 140: 225–234.

1. Styles distinct.
   2. Stigmas linear [subsect. *Halolinum*] . . . . 37. *Linum trigynum*
   2. Stigmas capitate [subsect. *Linopsis*].
      3. Stipular glands present [ser. *Linopsis*].
         4. Proximal leaves in whorls of 4 . . . . 8. *Linum schiedeanum*
         4. Proximal leaves opposite or alternate.
            5. Styles 2–3 mm; petals 4–6.5 mm; Florida . . . . 9. *Linum arenicola*
            5. Styles 3–6.5 mm; petals 7–11 mm; New Mexico, Texas . . . . 10. *Linum rupestre*
      3. Stipular glands absent.
         6. False septa incomplete, proximal margins ciliate.
            7. Capsules turbinate, 2–3 mm; anthers 0.5–1 mm; c, e United States [ser. *Virginianum* (in part)] . . . . 13. *Linum intercursum*
            7. Capsules triangular-ovoid to broadly ovoid or ovoid-pyriform, 2.3–4 mm; anthers 1–2.5 mm; w United States [ser. *Neomexicana*].
               8. Annuals, 15–60 cm; inflorescences thyrses; styles 1.5–3 mm; capsule apices obtuse . . . . 11. *Linum neomexicanum*
               8. Perennials, 5–30 cm; inflorescences panicles or thyrses; styles 4–7 mm; capsule apices pointed . . . . 12. *Linum kingii*
         6. False septa nearly complete, proximal margins sparsely or not ciliate [ser. *Virginianum* (in part)].
            9. Capsules 2–3.9 mm, either pyriform to ovoid or subglobose, apices abruptly short-pointed, obtuse, or 5-apiculate.
               10. Leaves mostly opposite; capsules subglobose, apices abruptly short-pointed . . . . 16. *Linum westii*
               10. Leaves: proximal usually opposite, distal alternate, rarely all alternate; capsules pyriform to ovoid, apices obtuse or minutely 5-apiculate.
                   11. Margins of inner sepals entire; capsules 3.4–3.9 mm; seeds 2.8–3 mm . . . . 14. *Linum macrocarpum*
                   11. Margins of inner sepals glandular-toothed; capsules 2–3.4 mm; seeds 1.6–2.5 mm . . . . 15. *Linum floridanum*
            9. Capsules 1.3–2.3 mm, depressed-globose (broader than long) or globose, apices depressed.
               12. Margins of inner sepals usually glandular-toothed, rarely entire; mature capsule segments usually persistent on plant . . . . 17. *Linum medium*
               12. Margins of inner sepals eglandular or with a few small sessile glands; mature capsule segments falling freely.
                   13. Inflorescences corymbs; pedicels 1–10 mm; carpels flattened or ± concave abaxially . . . . 18. *Linum virginianum*
                   13. Inflorescences panicles; pedicels 0–4 mm; carpels convex abaxially . . . . 19. *Linum striatum*

1. Styles connate [subsect. *Rigida*].
   14. Capsules dehiscing into 10, 1-seeded segments; sepals persistent [ser. *Sulcata*].
      15. Sepals (3.1–)3.6–5(–7.3) mm, apices acuminate . . . . 20. *Linum sulcatum*
      15. Sepals 2.3–3.7 mm, apices acute . . . . 21. *Linum harperi*
   14. Capsules dehiscing into 5, 2-seeded segments; sepals deciduous or persistent [ser. *Rigida*].
      16. Sepal margins not glandular-toothed.
         17. Distal leaves and bracts not ciliate; capsule false septa entirely hyaline, or with cartilaginous portion very narrow, uniform, distal . . . . .22. *Linum hudsonioides*
         17. Distal leaves and bracts sparsely ciliate; capsule false septa with cartilaginous portion conspicuously broader near base . . . . 23. *Linum imbricatum*
      16. Sepal margins (some or all) glandular-toothed.
         18. Outer sepals ovate or obovate, margins undulate or crenate, with sessile gland near summit of each crenation . . . . 28. *Linum alatum*
         18. Outer sepals linear, linear-lanceolate, lanceolate, or narrowly ovate, margins glandular-toothed.
            19. False septa incomplete, proximal margins terminating in loose fringe; sepals persistent.
               20. Styles connate to within 0.8–3 mm of apex; pedicels (5–)20–30(–60) mm; stipular glands absent . . . . .24. *Linum subteres*
               20. Styles connate to within 0.2 mm of apex; pedicels 2–12 mm; stipular glands usually present . . . . 25. *Linum vernale*
            19. False septa complete, proximal margins not terminating in loose fringe; sepals usually deciduous.
               21. Plants gray-puberulent throughout or puberulent or glabrescent in proximal ⅓.
                  22. Herbs, annual or short-lived perennial, gray-puberulent throughout . . . . 26. *Linum puberulum*
                  22. Subshrubs, puberulent or glabrescent in proximal ⅓, otherwise glabrous . . . . 27. *Linum allredii*
               21. Plants glabrous, glabrate, scabrous, or puberulent or hirsutulous at base.
                  23. Sepals linear-lanceolate.
                     24. Petals obovate; stamens 5–7 mm; styles 4.5–7 mm. . . . . . . 29. *Linum aristatum*
                     24. Petals obcordate; stamens 4–5 mm; styles 3–4 mm. . . . . .31. *Linum lundellii* (in part)
                  23. Sepals lanceolate to narrowly ovate.
                     25. Stipular glands usually absent.
                        26. Styles 2.5–4 mm; petals 6–11 mm . . . . .30. *Linum compactum*
                        26. Styles 3–11 mm,; petals 10–18 mm.
                           27. Petals coppery yellow or orange, red-lined or with short pale to deep brown-red zone at base; styles 5–11 mm; sepal apices sharply acute to acuminate; w, c North America . . . . 34. *Linum rigidum* (in part)
                           27. Petals orange-yellow throughout; styles 4–6 mm; sepal apices short-awned; s Florida . . . . 36. *Linum carteri* (in part)
                     25. Stipular glands usually present at least at some nodes.
                        28. Styles 6–9.5 mm.
                           29. Petals with prominent wine-colored band proximal to middle. . . . . .33. *Linum elongatum*
                           29. Petals reddish below middle . . . . 35. *Linum berlandieri*
                        28. Styles 2–6 mm.
                           30. Sepal apices aristate; petals yellow to yellow-orange throughout . . . . .32. *Linum australe*
                           30. Sepal apices acute to acuminate or short-awned; petals yellow to orange, sometimes with reddish base.

[31. Shifted to left margin.—Ed.]

31. Sepal apices short-awned; s Florida . . . . 36. *Linum carteri* (in part)
31. Sepal apices acute to acuminate; c United States.
  32. Stamens 4–5 mm . . . . 31. *Linum lundellii* (in part)
  32. Stamens 6–8 mm . . . . 34. *Linum rigidum* (in part)

8. Linum schiedeanum Schlechtendal & Chamisso, Linnaea 5: 234. 1830 • Schiede's flax

Herbs, perennial, 20–70 cm, glabrous except for occasional hairs near nodes. **Stems** erect or spreading, branching at base and in inflorescence. **Leaves:** proximal in whorls of 4, distal alternate, or mostly whorled or mostly alternate, spreading to ascending; stipular glands present; blade lanceolate to oblanceolate, 10–20 × 2–6 mm, margins entire, of distal leaves ciliate, apex deltate-acute to obtuse. **Inflorescences** panicles; bracts ciliate. **Pedicels** 0–1 mm. **Flowers:** sepals persistent, lanceolate, 2–3.5 mm, margins of inner sepals scarious, glandular-toothed, apex acute; petals lemon yellow, oblanceolate to narrowly obcordate, 2.5–6 mm; stamens 2–5 mm; anthers 0.3–0.7 mm; staminodia usually present, sometimes absent; styles distinct, 1.6–3 mm; stigmas capitate. **Capsules** broadly ovoid, 1.5–2.5 × 2–2.5 mm, apex sharp-pointed (easily crushed), readily dehiscing into 10, 1-seeded segments, segments falling freely, false septa rudimentary, margins of true septa usually ciliate. **Seeds** 1–1.5 × 0.6–1 mm. **2*n*** = 36.

Flowering Jun–Aug. Open or semishaded areas, calcareous soils; 1200–2800 m; N.Mex., Tex.; e, ne, s Mexico.

Linum schiedeanum has yellow, broadly bowl-shaped corollas, yellow stamens, and yellow styles and stigmas. The styles are broadly incurved, following the line of the petals, and are held outside the ring of stamens. Staminodia in *L. schiedeanum* are low, deltoid, and usually two between adjacent stamens, sometimes one or none. J. R. McDill (2009) reported that *L. schiedeanum* formed a group (*L. schiedeanum* group) with four other species with whorled leaves occuring from the Guadalupe Mountains of western Texas south to Veracruz. C. M. Rogers (1984) noted that a compact form of *L. schiedeanum* from sunny areas might warrant more study.

9. Linum arenicola (Small) H. J. P. Winkler in H. G. A. Engler et al., Nat. Pflanzenfam. ed. 2, 19a: 116. 1931 • Sand flax [C] [E]

Cathartolinum arenicola Small in N. L. Britton et al., N. Amer. Fl. 25: 75. 1907

Herbs, perennial (flowering 1st year), 25–70 cm, glabrous. **Stems** erect, usually multiple from base, sometimes 1, unbranched or few-branched proximal to inflorescence, slender, wiry, prominently ribbed in inflorescence. **Leaves** early deciduous, alternate or basalmost opposite, appressed-ascending; stipular glands present, reddish, becoming dark; blade linear, 5–15 × 0.5–1.2 mm, margins entire or with scattered minute marginal glands, not ciliate, apex acute; 1-nerved. **Inflorescences** cymes. **Pedicels** 0–2 mm. **Flowers:** sepals persistent, lanceolate to ovate or inner ones sometimes obovate, outer sepals 2.5–3.6 mm, margins hyaline, not scarious, all glandular-toothed, apex acuminate; petals yellow, obovate, 4–6.5 mm; stamens 3 mm; anthers 0.3–0.7 mm; staminodia present or absent; styles distinct, 2–3 mm; stigmas capitate. **Capsules** pyriform, 2–2.5 mm diam., apex pointed, dehiscing readily into 10, 1-seeded segments, segments falling freely, false septa incomplete, margins of septa ciliate. **Seeds** 1–1.5 × 0.6–1 mm. **2*n*** = 36.

Flowering Feb–Jun(–Sep). Shallow soils of ephemeral pools, calcareous soils, slash pine woods over oölite, pine-palmetto rocklands, disturbed areas; of conservation concern; 0–10 m; Fla.

All parts of *Linum arenicola* flowers are yellow; the stamens are held close to the styles, with anthers at the same level as stigmas. The staminodia are low, deltoid, and less than 0.5 mm. *Linum arenicola* is only known from about nine sites in Miami-Dade County and the Florida Keys in Monroe County. Its habitat of pine rocklands has been almost completely destroyed by urban development and altered fire regimes. J. R. McDill (2009) reported that *L. arenicola* grouped with *L. rupestre* (southwestern United States), *L. flagellare* (Small) H. J. P. Winkler (south-central Mexico), and *L. bahamense* Northrop (Bahamas), all perennials with many branches arising from a woody taproot or caudex. *Linum arenicola* and *L. bahamense* both occur on calcareous soils, and C. M. Rogers (1984) considered them to be closely related.

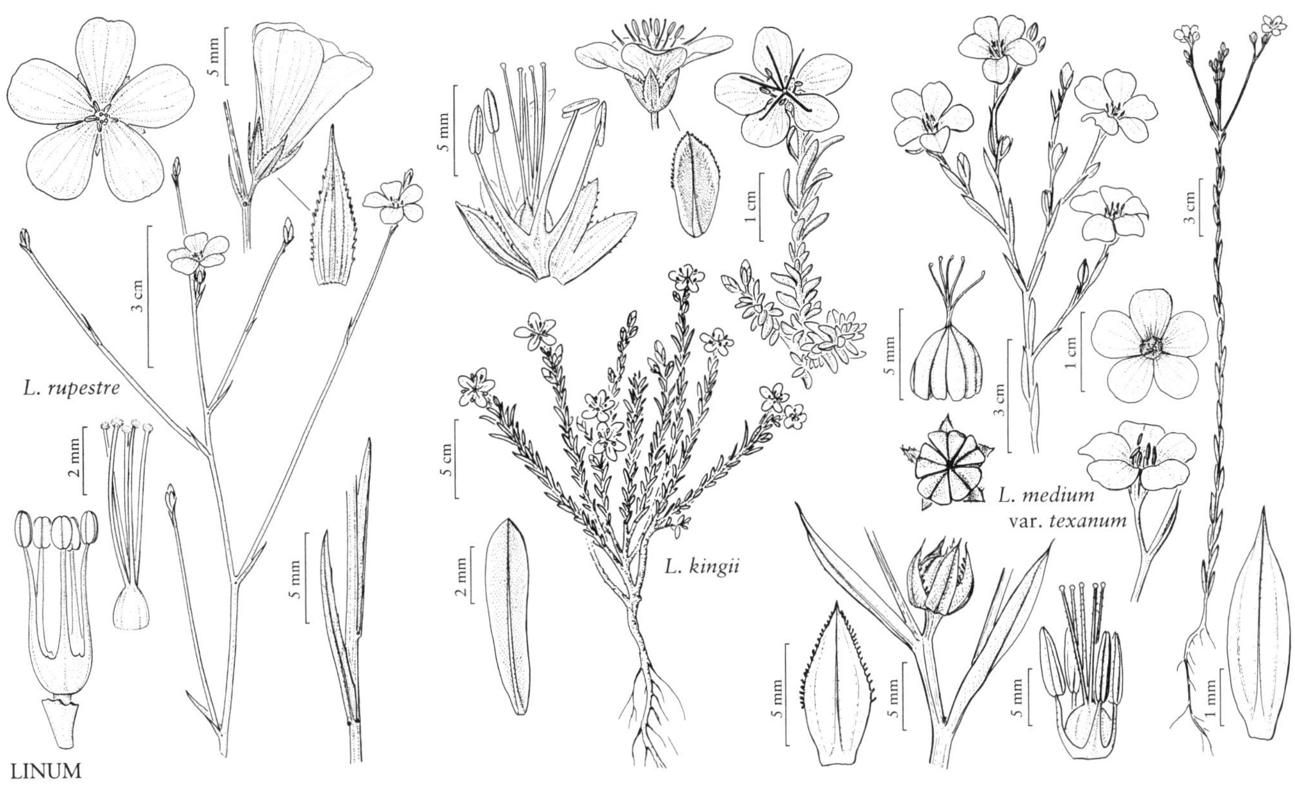

LINUM

10. Linum rupestre Engelmann ex A. Gray, Boston J. Nat. Hist. 6: 232. 1850 • Rock flax [F]

Herbs, perennial, 20–75 cm, glabrous or rarely sparsely hairy proximally. **Stems** erect, branched at base and in inflorescence. **Leaves** opposite near base or alternate throughout, appressed-ascending; stipular glands present; blade linear to linear-lanceolate, 8–20 × 0.5–2.1 mm, margins entire or with scattered minute marginal glands, not ciliate, apex acute; 1-nerved. **Inflorescences** panicles, with ascending to spreading branches. **Pedicels** 0–3 mm. **Flowers:** sepals persistent, lanceolate to ovate, 2.5–5 mm, margins of inner sepals narrowly scarious, conspicuously glandular-toothed, apex acute or acuminate; petals lemon yellow, oblanceolate or narrowly obcordate, 7–11 mm; stamens 2–8 mm; anthers 0.5–1 mm; staminodia present; styles distinct, 3–6.5 mm; stigmas capitate. **Capsules** ovoid, 2–3 × 2–2.5 mm, apex sharp-pointed, dehiscing readily into 10, 1-seeded segments, segments falling freely, false septa incomplete, false and true septa margins ciliate. **Seeds** 1.2–1.9 × 0.7–1.1 mm. **2*n*** = 36.

Flowering Apr–Aug. Sandy soils, rocky slopes and ledges, often on limestone; 150–1500 m; N.Mex., Tex.; Mexico (Chihuahua, Coahuila, Durango, Nuevo León); Central America (Guatemala).

Linum rupestre has narrowly funnelform corollas and yellow stamens and styles. The anthers and stigmas are held closely adjacent at the mouth of the corolla tube, below the broad, spreading limbs. The species occurs from southeastern New Mexico and central Texas to Guatemala. It often grows with *L. schiedeanum* in Texas and Mexico.

11. Linum neomexicanum Greene, Bot. Gaz. 6: 183. 1881 (as neo-mexicanum) • New Mexico yellow flax

Herbs, annual, 15–60 cm, glabrous, sometimes glaucous. **Stems** erect or shortly decumbent at base, becoming erect, branched from near base. **Leaves:** usually only proximalmost opposite, sometimes to midstem, distal alternate, erect to ± spreading; stipular glands absent; blade of proximal leaves narrowly to broadly elliptic-oblanceolate, 10–15 × 1.5–2.5 mm, margins entire, not ciliate, apex subacute to acuminate; 1-nerved. **Inflorescences** slender thyrses. **Pedicels** 1–4 mm. **Flowers:** sepals persistent, linear-lanceolate to lanceolate, 3–5.3 mm, margins not scarious, inner ones conspicuously glandular-toothed, outer entire or very sparsely glandular-toothed, apex acute to acuminate; petals yellow, oblanceolate, 4–7 mm; stamens 3–5 mm; anthers 1–1.5 mm; staminodia absent; styles distinct,

1.5–3 mm; stigmas capitate. **Capsules** triangular-ovoid to broadly ovoid, 2.5–3.5 × 2.7–3.5 mm, apex obtuse, somewhat tardily dehiscing into 10, 1-seeded segments, segments persistent on plant, false septa incomplete, proximal margins ciliate. **Seeds** 2–2.5 × 1.1–1.3 mm. **2*n*** = 26.

Flowering (Mar–)Jul–Sep(–Nov). Pine and oak woodlands; (600–)1300–2900 m; Ariz., N.Mex.; Mexico (Baja California, Chihuahua, Sonora).

Within its range, *Linum neomexicanum* is the only species of *Linum* with yellow flowers and distinct styles. Its inflorescence is more slender than other species. The corollas are nearly rotate and the styles may be at nearly right angles to the flower axis, to spreading, or to ultimately ascending. J. R. McDill (2009) found that *L. neomexicanum* is most closely related to *L. kingii.*

12. Linum kingii S. Watson, Botany (Fortieth Parallel), 49. 1871 • King's flax, perennial yellow flax [E] [F]

Cathartolinum kingii (S. Watson) Small; *Mesyniopsis kingii* (S. Watson) W. A. Weber

Herbs, perennial, caudex woody, 5–30 cm, glabrous and glaucous. **Stems** ascending to erect from decumbent base, branched from base. **Leaves:** alternate throughout or proximal opposite, divergent, erect or spreading; stipular glands absent; blade narrowly lanceolate, 5–25 × 1–3 mm, thick (basal leaves), margins entire, not ciliate, apex rounded to subacute; 1-nerved. **Inflorescences** panicles or thyrses. **Pedicels** 1–5 mm. **Flowers:** sepals persistent, lanceolate to ovate or broadly oblong, 2.5–4.5 mm, margins not scarious, inner glandular-toothed, outer entire or sparsely glandular-toothed near apex, apex acute to ± obtuse, not acuminate; petals bright yellow, oblanceolate to obovate, 5–12 mm; stamens 3–8 mm; anthers 1.5–2.5 mm; staminodia absent; styles distinct, 4–7 mm; stigmas capitate. **Capsules** ovoid-pyriform, 2.3–4 × 2.8–3.6 mm, apex pointed (easily crushed), freely dehiscing into 10, 1-seeded segments, segments persistent on plant, false septa incomplete, proximal margins ciliate. **Seeds** 2–2.7 × 1–1.4 mm. **2*n*** = 26.

Flowering May–Aug. Open slopes, often on barren alkaline clay or rocky calcareous substrates; 1400–3400 m; Colo., Idaho, Nev., Utah, Wyo.

Linum kingii is low, compact, and much branched. All parts of the flowers are yellow. The corolla is nearly rotate, the petals are abruptly narrowed to a claw, the styles are at right angles to the flower axis, and the anthers are relatively large. *Linum kingii* is extremely variable in habit and in size of floral and vegetative parts, even within a population or within a single plant (C. M. Rogers 1984).

13. Linum intercursum E. P. Bicknell, Bull. Torrey Bot. Club 39: 418. 1912 • Sandplain flax, Bicknell's yellow flax [E]

Cathartolinum intercursum (E. P. Bicknell) Small

Herbs or subshrubs, perennial, 20–92 cm, glabrous. **Stems** erect, unbranched proximal to inflorescence or few-branched at base. **Leaves:** proximalmost opposite, distalmost alternate, sometimes opposite nearly to inflorescence, erect to ascending; stipular glands absent; blade narrowly elliptic to oblanceolate, 8–27 × 1.2–5.6 mm, margins entire, not ciliate, apex acute; internal venation shown by transmitted light. **Inflorescences** panicles. **Pedicels** 0–5 mm. **Flowers:** sepals persistent, lanceolate, 2–3 mm, margins not scarious, entire, or inner and rarely outer sparsely glandular-toothed, apex sharp-pointed; petals yellow, obovate, 4–7 mm; stamens 3 mm; anthers 0.5–1 mm; staminodia absent; styles distinct, 1.5–2.5 mm; stigmas capitate. **Capsules** turbinate, 2–3 × 2–2.3 mm, apex acute or obtuse, dehiscing freely into 10, sharp-pointed, 1-seeded segments, segments persistent on plant, false septa incomplete, proximal margins sparsely but conspicuously ciliate. **Seeds** 1.3–1.8 × 0.6–0.9 mm. **2*n*** = 36.

Flowering Jun–Oct. Clay, sandy, siliceous, or peaty shores, dry, clay or sandy soils in clearings, coastal sandplains and shrubby grasslands, open forests, woodlands, roadsides, sometimes in alternately wet and dry, hardpan soils; 0–800 m; Ala., Conn., Del., D.C., Ga., Ind., Md., Mass., N.J., N.Y., N.C., Pa., R.I., S.C., Tenn., Va.

Linum intercursum is sometimes confused with *L. floridanum*, from which it differs by its pointed capsules and broader leaves. All parts of the flower of *L. intercursum* are yellow, and the corolla is nearly rotate.

14. Linum macrocarpum C. M. Rogers, Brittonia 15: 109, fig. 3(1–4). 1963 • Spring Hill flax [C] [E]

Herbs, perennial, 60–150 cm, glabrous. **Stems** erect, unbranched below inflorescence. **Leaves:** proximal opposite, distal alternate, ascending; stipular glands absent; blade of cauline leaves narrowly elliptic or linear-oblanceolate, 23 × 4 mm, margins entire, not ciliate, apex acute or apiculate. **Inflorescences** panicles. **Pedicels** 0–2.5 mm. **Flowers:** sepals persistent, lanceolate or oblanceolate to obovate, 2.8–3.8 mm, inner broader, somewhat shorter than outer, margins not scarious, entire, apex apiculate; petals yellow, obovate, 8–11 mm; stamens 6 mm; anthers 0.8 mm; staminodia absent; styles distinct, length unknown; stigmas capitate. **Capsules** ovoid, 3.4–3.9 × 3.2–3.5 mm, apex obtuse, dehiscing freely into 10, 1-seeded segments, segments persistent on plant, falling tardily, false septa nearly complete, proximal margins very sparsely and inconspicuously ciliate. **Seeds** 2.8–3 × 1.2 mm.

Flowering May–Jul. Pitcher-plant seepage bogs, wet longleaf and/or slash pine flatwoods and savannas; of conservation concern; 0–30 m; Ala., Fla., La., Miss.

Linum macrocarpum is known only from about 20 populations in Bay, Franklin, and Okaloosa counties in Florida; St. Tammany Parish in Louisiana; Hancock, Harrison, Jackson, and Stone counties in Mississippi; and Baldwin, Escambia, Mobile, and Washington counties in Alabama. Its range overlaps both varieties of *L. floridanum*, but *L. macrocarpum* may be distinguished by its larger capsules, larger seeds, and usually taller stems (B. A. Sorrie, pers. comm.).

15. Linum floridanum (Planchon) Trelease, Trans. Acad. Sci. St. Louis 5: 13. 1887

Linum virginianum Linnaeus var. *floridanum* Planchon, London J. Bot. 7: 480. 1848; *Cathartolinum floridanum* (Planchon) Small

Herbs, perennial, 20–110 cm, glabrous. **Stems** erect, usually unbranched, sometimes branched from base. **Leaves:** proximal usually opposite, distal alternate, rarely all alternate, appressed-ascending; stipular glands absent; blade linear-oblanceolate or oblanceolate, 10–20 × 1–3.2 mm, margins entire, not ciliate, apex sharply acute. **Inflorescences** corymbs. **Pedicels** 0.5–3.5 mm. **Flowers:** sepals persistent, narrowly lanceolate, outer sepals 2.5–4.5 mm, inner somewhat shorter, broader than outer, margins not scarious, inner conspicuously glandular-toothed, outer entire, apex acute; petals lemon yellow, obovate, 5.5–9.5 mm; stamens 2 mm; anthers 0.5–1.5 mm; staminodia absent; styles distinct, 2–4.5 mm; stigmas capitate. **Capsules** pyriform or ovoid, 2–3.4 × 2–3 mm, apex obtuse or minutely 5-apiculate, dehiscing freely into 10, 1-seeded segments, segments persistent on plant, false septa nearly complete, proximal margins not ciliate. **Seeds** 1.6–2.5 × 0.7–1.2 mm.

Varieties 2 (2 in the flora): c, se United States, West Indies (Jamaica).

Linum floridanum occurs only on the Atlantic and Gulf coastal plains. Leaves of the species are firm and opaque, and the veins are not shown in transmitted light. All parts of its flower are yellow except the anthers, which may have wine red coloring; the corolla is nearly rotate, and the styles are spreading.

1. Anthers 0.5–1.2 mm; capsules pyriform, 2–3 mm, walls relatively thin, apices obtuse . . . . 15a. *Linum floridanum* var. *floridanum*
1. Anthers 1–1.5 mm; capsules ovoid, 2.8–3.4 mm, walls relatively thick textured, apices minutely 5-apiculate . . . . 15b. *Linum floridanum* var. *chrysocarpum*

15a. Linum floridanum (Planchon) Trelease var. **floridanum** • Florida yellow flax

Cathartolinum macrosepalum Small

Flowers: anthers 0.5–1.2 mm. **Capsules** pyriform, 2–3 mm, walls relatively thin, apex obtuse. **Seeds** 1.6–2.1 mm. **2*n*** = 36.

Flowering Jun–Oct. Open pine and pine-palmetto woodlands, pine savannas, sandhill seeps; 0–150 m; Ala., D.C., Fla., Ga., La., Miss., N.C., S.C., Tex.; West Indies (Jamaica).

According to C. M. Rogers (1984), var. *floridanum* can be distinguished from *Linum intercursum* by its completely developed false septa that lack cilia and by its more numerous leaves. It can be distinguished from *L. medium* var. *texanum* (often misidentified as *L. floridanum*) by its pyriform capsule. Rogers noted that *L. floridanum* has subspheric, multiporate pollen, whereas *L. medium* var. *texanum* has subtriangular, tricolpate pollen. The records from the District of Columbia and Texas require confirmation; the species is otherwise not known from north of North Carolina or west of central Louisiana (B. A. Sorrie, pers. comm.).

15b. Linum floridanum (Planchon) Trelease var. **chrysocarpum** C. M. Rogers, Brittonia 15: 114, fig. 4(1, 2). 1963 • Yellow-fruited yellow flax [E]

Flowers: anthers 1–1.5 mm. **Capsules** ovoid, 2.8–3.3 mm, walls relatively thick textured, apex minutely 5-apiculate. **Seeds** 2.1–2.4 mm. **2*n*** = 36.

Flowering Jun–Oct. Wet pine and pine-palmetto savannas, pitcher plant seepage bogs; 0–50 m; Ala., Fla., Ga., La., Miss., N.C., S.C.

Variety *chrysocarpum* is much less abundant than var. *floridanum* (C. M. Rogers 1963) and may be distinguished by its larger capsules and seeds and usually yellow apex of the capsules versus red-purple apex in var. *floridanum*.

16. Linum westii C. M. Rogers, Brittonia 15: 114, figs. 3(8–11). 1963 • West's flax [C] [E]

Herbs, perennial, 43–50 cm, glabrous. **Stems** erect, unbranched proximal to inflorescence. **Leaves** mostly opposite, appressed-erect; stipular glands absent; blade elliptic to oblanceolate, cauline leaves 13–17 × 3–4 mm, margins entire, apex obtuse to acute. **Inflorescences** of few, few-flowered panicles, branches spreading-ascending, occupying ¼ or less of total height. **Pedicels** 0.5–2.9 mm, stout. **Flowers:** sepals persistent, inner ones broadly obovate, outer ovate, 3.1–3.6 mm, margins not scarious, inner glandular-toothed, outer entire, apex acute to acuminate; petals pale to bright yellow, ovate, 6–7 mm; stamen length unknown; anther length unknown; staminodia absent; styles distinct, 2–3.1 mm; stigmas capitate. **Capsules** subglobose, 2.6–3 × 2.8–3 mm, apex abruptly short-pointed, dehiscing into 10, 1-seeded segments, segments falling freely, false septa nearly complete, ± spongy, proximal margins not ciliate. **Seeds** not seen. $2n = 36$.

Flowering Jun–Jul. Wet depressions in pine palmetto flatwoods, cypress-gum ponds; of conservation concern; 0–20 m; Fla.

The broad, many-toothed inner sepals and subglobose, sharply pointed capsules set *Linum westii* apart from other species (C. M. Rogers 1984). In addition, its seeds are lunate, compared with narrowly elliptic seeds of *L. floridanum* and *L. macrocarpum*. *Linum westii* is known only from about 16 occurrences, all from the Florida panhandle and northeastern Florida (Clay, Franklin, Gulf, Jackson, Liberty, and Okaloosa counties, and possibly Bay County), with historical records from Baker and Calhoun counties. R. Kral (1973) considered a Chapman collection from Georgia to have been mislabelled. There are no vouchers for reports of *L. westii* from Mississippi.

17. Linum medium (Planchon) Britton in N. L. Britton and A. Brown, Ill. Fl. N. U.S. 2: 349. 1897 • Stiff yellow flax [F]

Linum virginianum Linnaeus var. *medium* Planchon, London J. Bot. 7: 480. 1848; *Cathartolinum medium* (Planchon) Small

Herbs, usually perennial, rarely annual, 10–80 cm, glabrous. **Stems** erect, usually multiple from base, unbranched proximal to inflorescence. **Leaves:** proximal 3–20 pairs opposite, distal alternate, rarely (in northern plants) opposite nearly to inflorescence, erect to appressed; stipular glands absent; blade narrowly lanceolate to oblanceolate, 10–25 × 1.5–5.5 mm, margins entire, not ciliate, apex obtuse or apiculate. **Inflorescences** corymbs. **Pedicels** 0–5 mm. **Flowers:** sepals persistent, lanceolate, inner somewhat shorter, broader than outer, outer sepals 2–5 mm, margins not scarious, inner usually glandular-toothed, rarely entire, outer entire, apex acute; petals lemon yellow, obovate, 4.5–8 mm; stamens 2.5 mm; anthers 0.5–1.3 mm; staminodia absent; styles distinct, 1–3 mm; stigmas capitate. **Capsules** depressed-globose, 1.6–2.3 × 2–2.5 mm, apex depressed, tardily (or readily in var. *medium*) dehiscing into 10, 1-seeded segments, segments usually persistent on plant, false septa nearly complete, proximal margins not ciliate. **Seeds** 1.3–1.7 × 0.6–0.8 mm.

Varieties 2 (2 in the flora): c, e North America, West Indies (Bahamas).

The corollas of *Linum medium* are broadly funnelform to nearly rotate, with all flower parts yellow except the brownish anthers.

1. Leaves opaque, apices obtuse; inner sepals usually sparsely glandular-toothed, sometimes entire . . . . 17a. *Linum medium* var. *medium*
1. Leaves ± translucent, apices minutely apiculate; inner sepals conspicuously glandular-toothed . . . . 17b. *Linum medium* var. *texanum*

17a. Linum medium (Planchon) Britton var. **medium** [E]

Leaves relatively thick, opaque, apex obtuse. **Flowers:** inner sepals usually sparsely glandular-toothed, sometimes entire. $2n = 72$.

Flowering Jul–Sep. Moist banks, dry fields; 0–300 m; Ont.; N.Y., Pa.

Variety *medium* occurs from around Georgian Bay to Lake Erie and western Lake Ontario. In New York, it is known from Chautauqua and Erie counties and in Pennsylvania from Erie County. Where it occurs, the variety is the only yellow-flowered representative of the genus. According to C. M. Rogers (1984), var. *medium* intergrades with var. *texanum* at the southern edge of its range.

17b. Linum medium (Planchon) Britton var. **texanum** (Planchon) Fernald, Rhodora 37: 428. 1935 • Texas or sucker flax [F]

Linum virginianum Linnaeus var. *texanum* Planchon, London J. Bot. 7: 481. 1848; *Cathartolinum curtissii* (Small) Small; *C. medium* (Planchon) Small var. *texanum* (Planchon) Moldenke; *L. striatum* Walter var. *texanum* (Planchon) B. Boivin

Leaves relatively thin, ± translucent, apex minutely apiculate. **Flowers:** inner sepals conspicuously glandular-toothed. **2*n*** = 36.

Flowering Mar–Aug. Damp, open, or somewhat shaded places, fields, roadsides; 0–1000 m; Ont.; Ala., Ark., Conn., Del., D.C., Fla., Ga., Ill., Ind., Iowa, Kans., Ky., La., Maine, Md., Mass., Mich., Miss., Mo., N.J., N.Y., N.C., Ohio, Okla., Pa., R.I., S.C., Tenn., Tex., Vt., Va., W.Va., Wis.; West Indies (Bahamas).

Variety *texanum* can be distinguished from other members of the genus in most of its range by its yellow flowers, distinct styles, and conspicuously glandular-toothed inner sepals. Its capsules also tend to persist longer than those of *Linum striatum* and *L. virginianum*, in which the capsules shatter early.

18. Linum virginianum Linnaeus, Sp. Pl. 1: 279. 1753 • Woodland flax, Virginia yellow flax [E]

Cathartolinum virginianum (Linnaeus) Reichenbach

Herbs, perennial, 15–80 cm, glabrous. **Stems** erect, branches 1–several from base, unbranched proximal to inflorescence. **Leaves:** proximal 4–10 pairs opposite, distal alternate, erect to spreading; stipular glands absent; blade of proximal leaves spatulate, central and distal elliptic, oblanceolate, or obovate, 15–25 × 3–7 mm, margins entire, not ciliate, apex acute to apiculate. **Inflorescences** corymbs. **Pedicels** 1–10 mm. **Flowers:** sepals persistent, lanceolate-ovate, inner shorter, broader, thinner than outer, outer sepals 2–4 mm, margins not scarious, inner sepals usually with a few small, sessile glands along margin distal to middle, rarely eglandular, outer ones entire, apex acute to acuminate; petals yellow, obovate (sometimes notched at apex), 3–5.5 mm; stamens 1.2–3 mm; anthers 0.5–1 mm; staminodia absent; styles distinct, 1–2 mm; stigmas capitate. **Capsules** globose, carpels flattened or ± concave abaxially, 1.3–1.8 × 2–2.5 mm, apex depressed, dehiscing freely into 10, 1-seeded segments, segments falling freely, false septa nearly complete, proximal margins usually sparsely and inconspicuously few-ciliate. **Seeds** 1–1.5 × 0.6–0.9 mm. **2*n*** = 36.

Flowering Jun–Oct. Open woods, fields, thickets, roadsides; 0–800 m; Ont.; Ala., Conn., Del., D.C., Ga., Ill., Ind., Iowa, Ky., Md., Mass., Mich., Mo., N.J., N.Y., N.C., Ohio, Pa., R.I., S.C., Tenn., Va., W.Va.

Linum virginianum lacks prominent marginal teeth on the inner sepals, thus distinguishing it from *L. striatum*. It has a less elongate inflorescence and lacks the ribbed branchlets found in *L. striatum* (C. M. Rogers 1984). The corollas of *L. virginianum* are nearly rotate; all parts of the flower are yellow except the brownish anthers. Its capsules shatter readily and often are absent on herbarium sheets.

19. Linum striatum Walter, Fl. Carol., 118. 1788 • Ridged yellow flax [E]

Cathartolinum striatum (Walter) Small; *Linum striatum* var. *multijugum* Fernald

Herbs, perennial, 25–100 cm, glabrous. **Stems** erect-ascending, unbranched or branched from base, unbranched proximal to inflorescence, conspicuously ribbed distally. **Leaves:** proximal 5–20 pairs opposite, distal opposite or alternate, erect to spreading; stipular glands absent; blade elliptic to oblanceolate or obovate, 15–35 × 4–10 mm, margins entire, not ciliate, apex obtuse or acute. **Inflorescences** elongate panicles. **Pedicels** 0–4 mm. **Flowers:** sepals persistent, lanceolate to ovate, 1.5–3.5 mm, margins not scarious, eglandular or inner with a few delicate, small, marginal glands, apex acute or apiculate; petals pale yellow, obovate, 2.7–4.6 mm; stamens 1.5–2 mm; anthers 0.3–0.7 mm; staminodia absent; styles distinct, 1.2–2 mm; stigmas capitate. **Capsules** globose, carpels convex abaxially, 1.3–1.9 × 1.8–2.3 mm, apex depressed, dehiscing freely into 10, 1-seeded segments, segments falling freely, false septa nearly complete, proximal margins not ciliate. **Seeds** 1–1.4 × 0.5–0.7 mm. **2*n*** = 36.

Flowering Jun–Oct. Open or semishaded areas, swamp forests and margins, seepage bogs; 0–500 m; Ont.; Ala., Ark., Conn., Del., D.C., Fla., Ga., Ill., Ind., Ky., La., Md., Mass., Mich., Miss., Mo., N.J., N.Y., N.C., Ohio, Okla., Pa., R.I., S.C., Tenn., Tex., Va., W.Va.

The branches of *Linum striatum* are conspicuously ribbed distally and the carpels are convex abaxially, compared with *L. virginianum*, which has smooth branches and carpels that are abaxially flattened. In *L. striatum*, the corolla is nearly rotate; all parts of the flower are yellow except the brownish anthers.

20. Linum sulcatum Riddell, W. J. Med. Phys. Sci., 10. 1836 • Grooved yellow flax, lin à rameaux sillonnés [E]

Cathartolinum sulcatum (Riddell) Small

Herbs, annual, 25–85 cm, glabrous. **Stems** erect to ascending, unbranched proximally, few- to many-branched above middle, conspicuously sulcate. **Leaves:** proximal 0–13 pairs opposite (often fallen at anthesis), distal alternate, appressed-ascending; stipular glands usually present, very rarely absent; blade linear to narrowly lanceolate, 7–30 × 1–3 mm, margins entire, distal leaves not ciliate, apex acute to subulate; midrib prominent, marginal nerves less conspicuous. **Inflorescences** open panicles; bracts glandular-toothed, not ciliate. **Pedicels** 1.3–4.7 mm. **Flowers:** sepals persistent, lanceolate, (3.1–)3.6–5(–7.3) mm, inner sepals more delicate than outer, shorter, margins not scarious, all very conspicuously glandular-toothed, apex acuminate, central and marginal veins conspicuous; petals pale yellow, obovate, 5–10 mm; stamens 3.3–5.7 mm; anthers 0.3–0.7 mm; staminodia absent; styles connate 0.2–1.8 mm at base, 2–4.5 mm; stigmas capitate. **Capsules** globose, 2.5–3.3 × 2.1–3 mm, apex rounded to acute, dehiscing freely into 10, sharp-pointed, 1-seeded segments; segments persistent on plant, false septa incomplete, margins prominently ciliate. **Seeds** 1.6–2.1 × 0.8–1.1 mm. $2n = 30$.

Flowering May–Sep. Sandy, gravelly fields, calcareous ledges and barrens, diabase barrens, cedar glades, prairies, alvars, sometimes in open woods, interdunal flats; 0–800 m; Man., Ont., Que.; Ala., Ark., Conn., Fla., Ga., Ill., Ind., Iowa, Kans., Ky., La., Md., Mass., Mich., Minn., Miss., Mo., Nebr., N.H., N.J., N.Y., N.C., N.Dak., Ohio, Okla., Pa., R.I., S.Dak., Tenn., Tex., Vt., Va., W.Va., Wis.

Linum sulcatum and *L. harperi* are the only species of the genus in eastern North America with styles united from the base to the middle, and all five sepals persistent and glandular-toothed margins. In *L. sulcatum*, all parts of the flower are yellow and the corolla is funnelform. Dried plants of *L. sulcatum* are pale green.

21. Linum harperi Small, Fl. S.E. U.S. 663, 1332. 1903 • Harper's flax [E]

Cathartolinum harperi (Small) Small; *Linum sulcatum* Riddell var. *harperi* (Small) C. M. Rogers

Herbs, annual, 25–85 cm, glabrous. **Stems** erect to ascending, unbranched proximally, few- to many-branched distal to middle. **Leaves:** proximal 0–13 pairs opposite (often fallen at anthesis), distal alternate, appressed-ascending; stipular glands usually present, very rarely absent; blade of proximal leaves oblanceolate or spatulate, of distal ones linear, 7–30 × 1–3 mm, margins entire, not ciliate, apex acute to subulate; midrib prominent, marginal nerves less conspicuous. **Inflorescences** racemelike; bracts glandular-toothed, not ciliate. **Pedicels** 1.3–4.7 mm. **Flowers:** sepals persistent, outer sepals oblong, 2.3–3.7 mm, margins not scarious, all very coarsely, irregularly glandular-toothed (inner sepals more closely and finely toothed than outer), apex acute; petals pale yellow, obovate, 5–10 mm; stamen length unknown; anthers 0.3–0.7 mm; staminodia absent; styles connate 0.2–1.8 mm at base, 2–4.5 mm; stigmas capitate. **Capsules** ovoid, 2.5–3.3 × 2.1–3 mm, apex rounded to acute, dehiscing freely into 10, sharp-pointed, 1-seeded segments, segments persistent on plant, false septa incomplete, margins prominently ciliate. **Seeds** 1.6–2.1 × 0.8–1.1 mm. $2n = 30$.

Flowering Jul–Aug. Dry pine barrens, clearings in pine flatwoods, calcareous soils or limestone outcrops; 0–100 m; Ala., Fla., Ga.

Dried plants of *Linum harperi* are dark purple-dotted distally. The species occurs in the center of the Florida panhandle, southwestern Georgia, and central Alabama.

22. Linum hudsonioides Planchon, London J. Bot. 7: 186. 1848 • Texas flax [E]

Herbs, annual, 5–30 cm, hirsutulous on angles distally, otherwise glabrous. **Stems** ascending to erect, branched from base. **Leaves** proximalmost opposite, distal alternate, imbricate throughout, proximal leaves spreading to ascending, distal closely appressed; stipular glands absent; blade awl-shaped, 5–10 × 0.5–1 mm, margins entire, distalmost narrowly scarious, not ciliate, apex of proximal leaves sharp-pointed, distal with short, slender terminal awn. **Inflorescences** panicles; bracts not ciliate. **Pedicels** 3–15 mm. **Flowers:** sepals persistent, lanceolate to ovate, 4.5–7 mm, margins broadly scarious, entire, or sparsely delicately toothed or, in age, ± lacerate, not glandular-toothed, apex conspicuously awn-tipped; petals yellow, with or without dark red base, obovate, 8–12 mm; stamens 5 mm; anthers 1–1.6 mm; staminodia absent; styles connate to 0.3–1.1 mm of apex, 2.7–6.3 mm; stigmas capitate. **Capsules** broadly ovoid, 2.7–3.5 × 2.8–3.6 mm, apex rounded, dehiscing into 5, 2-seeded segments, segments persistent on plant, false septa entirely hyaline, or with very narrow, uniform, distal cartilaginous portion, margins tomentose near apex. **Seeds** 2–2.7 × 1–1.2 mm. $2n = 30$.

Flowering Mar–Sep. Sandy or gravelly prairies; 100–1400 m; Kans., N.Mex., Okla., Tex.

The corollas of *Linum hudsonioides* are very broadly bowl-shaped to nearly rotate and yellow, sometimes with a broad wine red band near the base. The filaments and styles are yellow or dark pinkish, anthers are yellow, and stigmas are bright green to yellowish. Its stems are nearly smooth proximally, ribbed distally. *Linum hudsonioides* occurs mainly in west-central Texas, the trans-Pecos region of western Texas, Harding and San Miguel counties of northeastern New Mexico, and southwestern Oklahoma; it is known in Kansas from a single historic record from Sedgwick County.

23. Linum imbricatum (Rafinesque) Shinners, Field & Lab. 25: 32. 1957 • Tufted flax [E]

Nezera imbricata Rafinesque, New Fl. 4: 66. 1838

Herbs, annual, 3–30 cm, glabrous proximally, conspicuously short hirsute distally with stout-based hairs. **Stems** spreading to ascending or erect, branched from base. **Leaves:** proximalmost opposite, middle and distal alternate, closely imbricate, proximal leaves spreading-ascending, distal strongly appressed; stipular glands absent; blade linear-lanceolate, 5–9 × 0.5–1.2 mm, margins entire, distal leaves sparsely ciliate, apex short-awned; midrib cartilaginous. **Inflorescences** panicles; bracts ciliate. **Pedicels** 2–11 mm. **Flowers:** sepals persistent, ovate, 4.2–6.1 mm, margins broad, purplish, scarious, prominently toothed distally, not glandular-toothed, apex conspicuously awn-shaped; petals yellow, with or without dark red base, obovate, 6.5–8 mm; stamens 5 mm; anthers 0.6–1.2 mm; staminodia absent; styles connate to within 0.3–0.8 mm of apex, 2–4.3 mm; stigmas capitate. **Capsules** broadly ovoid, 2.6–3 × 2.9–3.3 mm, apex rounded, dehiscing into 5, 2-seeded segments, segments persistent on plant, false septa hyaline, with cartilaginous portion conspicuously broader near base, proximal margin appressed-pilose, otherwise glabrous. **Seeds** 2–2.6 × 1.1–1.5 mm. $2n = 30$.

Flowering Apr–May. Sandy or rocky open ground; 0–400 m; La., Okla., Tex.

Linum imbricatum stamens have red filaments and yellow pollen and anthers, a very showy combination against the broadly bowl-shaped, butter yellow corollas often with a broad, wine red base. The stems are terete proximally and strongly ribbed distally; the sepals persist even in fruit. *Linum imbricatum* occurs in southern Oklahoma, is scattered in Texas, and is known from one report from Acadia Parish, Louisiana.

24. Linum subteres (Trelease) H. J. P. Winkler in H. G. A. Engler et al., Nat. Pflanzenfam. ed. 2, 19a: 116. 1931 • Sprucemont flax, Utah yellow flax [E]

Linum aristatum Engelmann var. *subteres* Trelease in A. Gray et al., Syn. Fl. N. Amer. 1(1,2): 347. 1897; *L. leptopoda* A. Nelson

Herbs, annual or perennial, 15–50 cm, glabrous and glaucous. **Stems** stiffly spreading-ascending, branched at base and distal to middle. **Leaves** alternate or proximalmost opposite, crowded at base, appressed-ascending; stipular glands absent; blade oblanceolate to lanceolate or linear-lanceolate, 8–17 × 1.2–2.3 mm, margins entire, not ciliate, apex apiculate. **Inflorescences** few-flowered racemes. **Pedicels** (5–)20–30 (–60) mm. **Flowers:** sepals persistent, lanceolate to lance-ovate, 4.5–7 mm, margins narrowly scarious, inner sepals conspicuously toothed, outer ones very coarsely glandular-toothed, sometimes sparsely so, apex acuminate or narrowly acute; petals lemon yellow, obovate, 9–15 mm; stamens 5–7 mm; anthers 1–2 mm; staminodia absent; styles connate to within 0.8–3 mm of apex, 5.7–9 mm; stigmas capitate. **Capsules** ovoid (distinctly longer than broad), 3.5–4.6 × 2.5–3.1 mm, apex sharp-pointed, dehiscing completely into 5, 2-seeded segments (very easily crushed), segments persistent on plant, false septa incomplete, proximal margins terminating in loose fringe, cartilaginous plates at base of segments poorly developed. **Seeds** 2.5–3 × 0.9–1.2 mm. $2n = 30$.

Flowering May–Aug. Sandy soils, clay, sagebrush and pinyon-juniper zones; 1300–2200 m; Ariz., Nev., N.Mex., Utah.

Linum subteres is most closely related to *L. vernale*; it has lemon yellow petals, rather than orange to salmon-colored with a maroon base, and relatively thick, crowded, broad basal leaves (C. M. Rogers 1984). Leaves on the proximal half of each stem are closely spaced and imbricate; distal branches and inflorescences are widely spaced and subtended by closely appressed, relatively long, narrow leaves or bracts, giving the upper part of the plant a leafless look.

25. Linum vernale Wooton, Bull. Torrey Bot. Club 25: 452. 1898 (as vernall) • Chihuahuan or red-eye flax

Herbs, annual, 10–50 cm, glabrous. **Stems** ascending to erect, branched at base and in inflorescence. **Leaves** alternate or proximal leaves opposite, divergent to ascending; stipular glands usually present, sometimes absent; blade linear, 8–17 × 0.5–1.3 mm, margins entire, with widely spaced glandular hairs, apex acute. **Inflorescences** open panicles. **Pedicels** 2–12 mm. **Flowers:** sepals persistent, lanceolate to narrowly lanceolate, 4–7.5 mm, margins narrowly scarious or not, inner sepals abundantly glandular-toothed, outer sparsely toothed, apex narrowly acute; petals yellow-orange to salmon with maroon base, broadly obovate, 10–17 mm; stamens 4–8 mm; anthers 1–1.8 mm; staminodia absent; styles connate to within 0.2 mm of apex, 4–8 mm; stigmas capitate. **Capsules** ovoid, 3–4 × 2.5–3.2 mm, apex depressed, dehiscing completely into 5, 2-seeded segments (very easily crushed), segments persistent on plant, false septa incomplete, proximal margins terminating in loose fringe. **Seeds** 2–2.8 × 0.9–1.3 mm. **2*n*** = 30.

Flowering Mar–Sep. Limestone soils, bajadas, openings in scrublands and woodlands; 1200–2400 m; N.Mex., Tex.; Mexico (Chihuahua, Coahuila).

Corollas of *Linum vernale* are broadly bowl-shaped and yellow-orange to salmon with a maroon base. The filaments and styles are pale pink, and the stigmas are dark maroon. The pollen is bright yellow; on herbarium specimens, the anthers appear to be maroon.

26. Linum puberulum (Engelmann) A. Heller, Pl. World 1: 22. 1897 • Plains flax

Linum rigidum Pursh var. *puberulum* Engelmann, Smithsonian Contr. Knowl. 3(5): 25. 1852

Herbs, annual or short-lived perennial, 4–25 cm, densely and finely gray-puberulent throughout. **Stems** ascending, branched at base, herbaceous throughout. **Leaves** alternate or sometimes proximal leaves opposite, appressed-ascending; stipular glands present (conspicuous); blade linear, 7–20 × 0.6–1.5 mm, margins entire or distal leaves sparsely glandular-toothed, ciliate, apex acute; 1-nerved. **Inflorescences** open panicles. **Pedicels** 5–10 mm. **Flowers:** sepals falling tardily, lanceolate, 4–7 mm, margins of inner sepals scarious, glandular-toothed, apex acute to acuminate, puberulent at least on midrib; outer 3-nerved; petals yellowish orange to salmon, with maroon or reddish base, obcordate or broadly obovate, 9–15 mm; stamens 4–7 mm; anthers 0.6–1.4 mm; staminodia absent; styles connate nearly to apex, 3–7 mm; stigmas dark, capitate. **Capsules** ovoid-ellipsoid, 3.5–4 × 2.5–5 mm, apex obtuse, dehiscing into 5, 2-seeded segments, segments persistent on plant, false septa complete, proximal margin not terminating in loose fringe, distal part cartilaginous, margins ciliate. **Seeds** 1.5–3 × 0.9–1.3 mm. **2*n*** = 30.

Flowering May–Oct. Dry, open areas, rocky, sandy, limestone, gypsum, or sometimes clay soils; 300–2500 m; Ariz., Calif., Colo., Nebr., Nev., N.Mex., Tex., Utah, Wyo.; Mexico (Chihuahua, Coahuila, Sonora).

Corollas of *Linum puberulum* are broadly bowl-shaped. The filaments and styles are pale pink; the stigmas are dark maroon. The pollen is bright yellow; on herbarium specimens, the anthers are golden yellow to orangish yellow, drying darker. In some flowers of *L. puberulum*, the styles seem to be eccentric. C. M. Rogers (1968) noted that *L. puberulum* is the only hairy species of *Linum* in western North America with united styles; its gray indument and complete false septa differentiate it from *L. vernale*, which is glabrous and has incomplete false septa. *Linum puberulum* is fairly common in the Rocky Mountain foothills and high plains; it occurs in the mountains in the eastern Mojave Desert.

27. Linum allredii Sivinski & M. O. Howard, Phytoneuron 2011-33: 1, figs. 1, 3. 2011 • Allred's flax [C] [E]

Subshrubs, to 25 cm, puberulent or glabrescent in proximal ⅓, otherwise glabrous; roots relatively thick, lateral. **Stems** stiffly ascending, suffrutescent from woody branching base. **Leaves** alternate, tightly appressed or ascending to spreading; stipular glands present throughout, dark; blade linear to linear-lanceolate, proximal and midstem leaves 3–10 (–12) mm, distal leaves 3–7 × 0.6–1 mm, margins of proximal and midstem leaves entire, distal serrulate, teeth usually gland-tipped, not ciliate, apex of proximal and midstem leaves mucronate, distal acuminate-aristate. **Inflorescences** few-flowered panicles; bracts with irregular scarious margins. **Pedicels** 2–3 mm, conspicuously articulated. **Flowers:** sepals deciduous, lanceolate, 4.5–7 mm, margins not scarious, glandular-toothed, apex acute-aristate, glabrous; prominently 1-nerved; petals pumpkin yellow with a wide, pale, red band distal to a deeply wine red band at base, broadly obovate, 10–13 mm; stamens 5.5–7 mm; anthers 1.2–1.6 mm; staminodia absent; styles connate nearly to apex, 7–9 mm; stigmas dark, capitate. **Capsules** ovoid, 3.7–4 × 3 mm, apex obtuse, dehiscing into 5, 2-seeded segments, segments persistent on plant, false septa complete, translucent, proximal part membranaceous, not terminating in loose fringe, distal part cartilaginous, margins ciliate. **Seeds** 2.4–2.7 × 0.9–1.1 mm.

Flowering Apr. Gypsum soils; of conservation concern; 1100–1200 m; N.Mex., Tex.

Linum allredii is a rare endemic known from seven to 12 occurrences in the Yeso Hills border region of New Mexico and Texas, apparently restricted to gypsum soils. When Sivinski and Howard described this species, they noted that it occurs only on pale, sandy, biologically crusted gypsum distinct from adjacent, darker gypsum. The corollas are deeply bowl-shaped. The filaments and styles are the same pumpkin color as the petals, the stigmas are dark maroon, and the pollen is bright yellow. *Linum allredii* and *L. kingii* are the only species in the flora area growing from a woody base.

28. Linum alatum (Small) H. J. P. Winkler in H. G. A. Engler et al., Nat. Pflanzenfam. ed. 2, 19a: 116. 1931 • Winged flax

Cathartolinum alatum Small in N. L. Britton et al., N. Amer. Fl. 25: 81. 1907

Herbs, annual or short-lived perennial, 10–40 cm, scabrous or puberulent at base, otherwise glabrous. **Stems** spreading to suberect, branched at base. **Leaves** opposite near base or alternate throughout, divergent to widely ascending; stipular glands present; blade linear to narrowly linear-lanceolate, 10–30 × 1–3 mm, margins entire, ciliate, apex apiculate. **Inflorescences** panicles; bracts with irregular, scarious margins. **Pedicels** 3–8 mm, stout. **Flowers:** sepals deciduous, inner sepals somewhat shorter than outer, regularly and delicately glandular-toothed, outer sepals ovate or obovate, 6–8 mm, margins widely scarious, undulate or crenate, with sessile gland near apex of each crenation, apex conspicuously aristate; petals yellow, grading to reddish near base, obovate, 9–18 mm; stamens 5–8 mm; anthers 1–2 mm; with or without staminodia; styles connate nearly to apex, 5–10 mm; stigmas capitate. **Capsules** ovoid, 3.5–4.5 × 3–3.8 mm, apex obtuse, dehiscing into 5, 2-seeded segments, segments persistent on plant, false septa incomplete, united more than halfway, proximal part membranaceous with basal, 5-sided cartilaginous plates, distal part cartilaginous, constituting more than ½ of false septum, margins not ciliate. **Seeds** 2.3–2.8 × 1–1.3 mm. $2n = 30$

Flowering Mar–Sep. Open sandy areas, beaches; 0–300 m; La., Tex.; Mexico (Tamaulipas).

Linum alatum has broadly funnelform corollas that are deep yellow distally, grading through a diffuse pale wine red band of color, the red color extending along the petal veins. The filaments, anthers, styles, and stigmas are yellow. The distinct portions of the styles spread at nearly right angles to the style axis, and the styles are sometimes eccentric. The unique gland-tipped crenations of the sepal margins set *L. alatum* apart from other species. Its thick pedicels and thickened cartilaginous areas on the capsule also are distinctive. The stems of *L. alatum* are smooth proximally, strongly ribbed distally. The species occurs in Texas in the east-central, Gulf, and southern mesquite plains regions (and adjacent Tamaulipas) with one historical record from southwestern Louisiana.

29. Linum aristatum Engelmann in F. A. Wislizenus, Mem. Tour N. Mexico, 101. 1848 • Bristle or broom flax

Herbs, annual, 10–45 cm, glabrous. **Stems** stiffly spreading-ascending, slender, broomlike, branched throughout. **Leaves** alternate or proximalmost opposite, proximal leaves spreading, distal leaves appressed-ascending; stipular glands usually present; blade linear, 5–20 × 0.3–1.1 mm, margins entire, not ciliate, apex acute. **Inflorescences** diffuse panicles, branches relatively long, stiffly spreading-ascending. **Pedicels** 6–30 mm, slender. **Flowers:** sepals deciduous, outer linear-lanceolate, 5.5–9 mm, apex attenuate, inner somewhat broader, shorter, margins narrowly scarious, glandular-toothed, apex awn-tipped; petals faintly maroon at base, otherwise yellow to yellow-orange throughout, obovate, 8–12 mm; stamens 5–7 mm; anthers 0.7–1.1 mm; staminodia absent; styles connate nearly to apex, 4.5–7 mm; stigmas capitate. **Capsules** narrowly ellipsoid, 3.5–4.5 × 2.5–3 mm, thin-walled, apex obtuse, dehiscing into 5, 2-seeded segments, segments persistent on plant, false septa complete, proximal part membranaceous, not terminating in loose fringe, distal part cartilaginous, margins not, or only sparsely, ciliate. **Seeds** 2.5–3 × 0.8–1 mm. $2n = 30$.

Flowering May–Sep(–Oct). Open places, dry, sandy soils, sagebrush or pinyon-juniper zones; (300–)1100–3100 m; Ariz., Colo., N.Mex., Tex., Utah; Mexico (Chihuahua).

The corollas of *Linum aristatum* are broadly funnelform and almost entirely rich yellow to yellow-orange with a faint blush of maroon toward the base. The stamens and styles are yellow; the stigmas are greenish yellow. *Linum aristatum* is highly branched, giving it a bushy look. It can be recognized by narrowly ellipsoid capsules and long-attenuate sepals.

30. Linum compactum A. Nelson, Bull. Torrey Bot. Club 31: 241. 1904 • Wyoming flax [E]

Linum rigidum Pursh var. *compactum* (A. Nelson) C. M. Rogers

Herbs, annual, 5–30 cm, glabrous throughout or puberulent at base. **Stems** erect, branched from base and throughout, bushy. **Leaves** alternate, spreading to ascending; stipular glands absent; blade linear, 10–28 × 1–1.5 mm, margins entire or sparsely toothed on distal leaves, not ciliate, apex acute. **Inflorescences** dense panicles. **Pedicels** 2–7 mm. **Flowers:** sepals tardily deciduous, lanceolate, 5–9 mm, margins narrowly scarious, conspicuously glandular-toothed, apex acute to acuminate; petals yellow, obovate, 6–11 mm; stamens 4–6 mm; anthers 0.4–0.8 mm; staminodia absent; styles connate nearly to apex, 2.5–4 mm; stigmas capitate. **Capsules** ovoid, 3.5–4.4 × 2.7–3.5 mm, apex obtuse, dehiscing into 5, 2-seeded segments, segments persistent on plant, false septa complete, proximal part membranaceous, not terminating in loose fringe, distal part cartilaginous, margins not or only minutely ciliate. **Seeds** 2.6–3.1 × 1–1.3 mm. **2*n*** = 30.

Flowering Jun–Aug(–Sep). Sagebrush grasslands, ponderosa pine woodlands, meadows, prairies, rocky outcrops; 600–1800 m; Alta., Sask.; Colo., Ill., Kans., Mo., Mont., Nebr., N.Mex., N.Dak., Okla., S.Dak., Tex., Wyo.

Linum compactum is generally low-growing, much branched, and bushy; its flowers are smaller than those of *L. rigidum*. *Linum compactum* is found on the high plains in open areas. C. M. Rogers (1984) suggested that it might be closely related to *L. australe*, which differs in being more slender with more open habit and of pine forest habitats.

31. Linum lundellii C. M. Rogers, Sida 8: 184, fig. 4c. 1979 • Sullivan City flax

Herbs, annual, 10–40 cm, glabrate. **Stems** spreading to erect, few-branched. **Leaves** alternate, spreading; stipular glands moderately developed at proximal nodes, absent on distal nodes; blade linear, 5–30 × 0.5–1.5 mm, margins entire, not ciliate, apex acute. **Inflorescences** panicles. **Pedicels** 5–13 mm. **Flowers:** sepals deciduous, linear-lanceolate to lanceolate, 4–12 mm, margins narrowly scarious, glandular-toothed, apex acute to acuminate; petals yellow to orange salmon, faintly maroon banded near base, obcordate, 7–12 mm; stamens 4–5 mm; anthers 1–1.5 mm; staminodia absent; styles connate nearly to apex, 3–4 mm; stigmas dark, capitate. **Capsules** ovoid, 3.3–4 × 2.6–3.1 mm, apex obtuse, dehiscing into 5, 2-seeded segments, segments persistent on plant, false septa complete, proximal part membranaceous, not terminating in loose fringe, distal part cartilaginous, margins not ciliate. **Seeds** 2.5–2.7 × 1.1 mm. **2*n*** = 30.

Flowering Feb–Apr. Sandy loam in arroyos, gravelly hillsides, mesquite scrub woodlands; 0–100 m; Tex.; Mexico (Nuevo León, Tamaulipas).

Linum lundellii occurs in southern Texas and adjacent Tamaulipas (the collection from Nuevo León, *Mueller 470*, TEX, made at 2400 m, may be misidentified); it can be distinguished from other species by its relatively very short styles. C. M. Rogers (1968) identified a variable population of yellow-flowered plants that he included in *L. berlandieri* var. *filifloium* (then treated as *L. rigidum* var. *filifolium*). As a result of subsequent study of these plants, he concluded that *L. lundellii* and *L. elongatum* should be recognized as separate species. Rogers (1979) compared garden-grown plants of these three taxa and observed that the anthers of *L. lundellii* are at the same level as the stigmas at anthesis and that pollen had already been deposited on stigmas when the flowers opened, whereas styles of *L. berlandieri* var. *filifolium* and *L. elongatum* are much longer than the stamens, and pollen is not shed before anthesis.

32. Linum australe A. Heller, Bull. Torrey Bot. Club 25: 627. 1898 • Southern flax

Herbs, annual, 10–50 cm, puberulent near base, otherwise glabrous. **Stems** stiffly ascending-spreading, few- to many-branched. **Leaves** alternate, appressed; stipular glands present at basal nodes or throughout; blade linear, 7–20 × 0.5–1.9 mm, margins entire, not ciliate, apex aristate. **Inflorescences** racemes. **Pedicels** 3–15 mm. **Flowers:** sepals deciduous, lanceolate to narrowly ovate, 4–7 mm, margins scarious, delicately glandular-toothed, apex aristate; petals yellow to yellow-orange throughout, oblanceolate to narrowly obovate, 5–10 mm; stamens (3–)4–7 mm; anthers 0.4–1 mm; staminodia present or absent; styles connate nearly to apex, 2–5.7 mm; stigmas green, capitate. **Capsules** ovoid, 3.2–4.5 × 2.5–3.4 mm, relatively thick-walled and with characteristic thickened areas at apex in region of true septa, apex obtuse, dehiscing into 5, 2-seeded segments, segments persistent on plant, false septa complete, proximal part membranaceous, not terminating in loose fringe, distal part cartilaginous, margins ciliate. **Seeds** 2–3 × 0.8–1.3 mm.

Varieties 2 (2 in the flora): w North America, Mexico.

The stems of *Linum australe* are strongly ridged-sulcate to ribbed, especially distally. The corollas are broadly funnelform; petals are yellow to yellow-orange; stamens and styles are yellow; stigmas are bright to olive green. Staminodia in *L. australe* are short, deltoid, usually two between each pair of stamens, sometimes one or absent. *Linum australe* is the only species in its range that is glabrous beyond the base and has connate styles. It differs from *L. aristatum*, which it overlaps in the southern part of the range, in being much more highly branched and having more slender capsules. C. M. Rogers (1984) noted a compact form found in sunny areas from Wyoming northward that warrants more study.

1. Stipular glands present only at proximal nodes; stamens (3–)4–5 mm; styles (2–)2.7–3.3(–4) mm . . . . 32a. *Linum australe* var. *australe*
1. Stipular glands present at nodes throughout plant; stamens 5–7 mm; styles 3.6–5.7 mm . . . . 32b. *Linum australe* var. *glandulosum*

32a. Linum australe A. Heller var. **australe**

Leaves: stipular glands present only at proximal nodes. **Flowers:** stamens (3–)4–5 mm; styles (2–)2.7–3.3(–4) mm.

Flowering Jun–Sep. Sandy, cindery, or loamy soils; 1500–2800 m; Alta.; Ariz., Colo., Mont., Nev., N.Mex., Tex., Utah, Wyo.; Mexico (Baja California, Sonora).

Variety *australe* occurs in the Rocky Mountains from southern Alberta to southern Arizona and western Texas and Sonora. It is disjunct in the high elevations of the Sierra de San Pedro Mártir, Baja California.

32b. Linum australe A. Heller var. **glandulosum** C. M. Rogers, Sida 1: 336. 1964

Leaves: stipular glands present and prominent at nodes throughout plant. **Flowers:** stamens 5–7 mm; styles 3.6–5.7 mm.

Flowering May–Sep. Open woods and hillsides, often on red cinder soils; 1500–2300 m; Ariz., Tex.; Mexico (Chihuahua, Distrito Federal, Durango, Guanajuato, México, Puebla, Veracruz).

Variety *glandulosum* occurs on mountain slopes from southern Arizona and western Texas south to Puebla in Mexico (C. M. Rogers 1984).

33. Linum elongatum (Small) H. J. P. Winkler in H. G. A. Engler et al., Nat. Pflanzenfam. ed. 2, 19a: 116. 1931 • Laredo flax [F]

Cathartolinum elongatum Small in N. L. Britton et al., N. Amer. Fl. 25: 82. 1907

Herbs, perennial (sometimes flowering in 1st year), 15–30 cm, glabrous. **Stems** erect to ascending, branches diffuse. **Leaves** alternate, ascending; stipular glands present throughout or at distal nodes only; blade linear, 5–25 × 0.5–1 mm, margins entire, sometimes glandular, not ciliate, apex acute. **Inflorescences** panicles. **Pedicels** 10 mm. **Flowers:** sepals deciduous, mostly lanceolate, 6–11 mm, margins not scarious, all conspicuously glandular-toothed, apex attenuate; petals yellow-orange to salmon or brownish red, with prominent wine-colored band proximal to middle, broadly obovate, 14–18 mm; stamens 5–6 mm; anthers 1.5–2.5 mm; staminodia absent; styles connate nearly to apex, 7–9.5 mm; stigmas capitate. **Capsules** ovoid, 4 × 3 mm, apex obtuse, dehiscing into 5, 2-seeded segments, segment persistence unknown, apex blunt or subacute, minutely apiculate, false septa complete, proximal part membranaceous, not terminating in loose fringe, distal part cartilaginous, margins not ciliate. **Seeds** 3 × 1.2–1.3 mm. **2*n*** = 30.

Flowering Feb–May. Often on hard-packed, often red, sandy soils; 0–300 m; Tex.; Mexico (Nuevo León, Tamaulipas).

The stems of *Linum elongatum* are ribbed. In the flora area, it occurs in southern Texas, especially along the Rio Grande, and in south central Texas. The yellow-orange to salmon or brownish red petals with the distinctive wine-colored band proximal to the middle, brick red anthers, and wine-colored stigmas make the flowers of *L. elongatum* striking.

34. Linum rigidum Pursh, Fl. Amer. Sept. 1: 210. 1813 • Stiffstem flax

Herbs, annual, 15–50 cm, glabrous throughout or puberulent near base of stem. **Stems** erect, branches few, fastigiate or spreading-ascending. **Leaves** alternate, erect; stipular glands present or absent; blade linear, 11–30 × 0.7–1.6 mm, margins entire or distally sparsely toothed, not ciliate, apex acute. **Inflorescences** panicles or cymes. **Pedicels** 4.5–9 mm. **Flowers:** sepals deciduous, lanceolate to broadly lanceolate, 5.5–9.5 mm, margins of inner sepals conspicuously scarious, all conspicuously glandular-

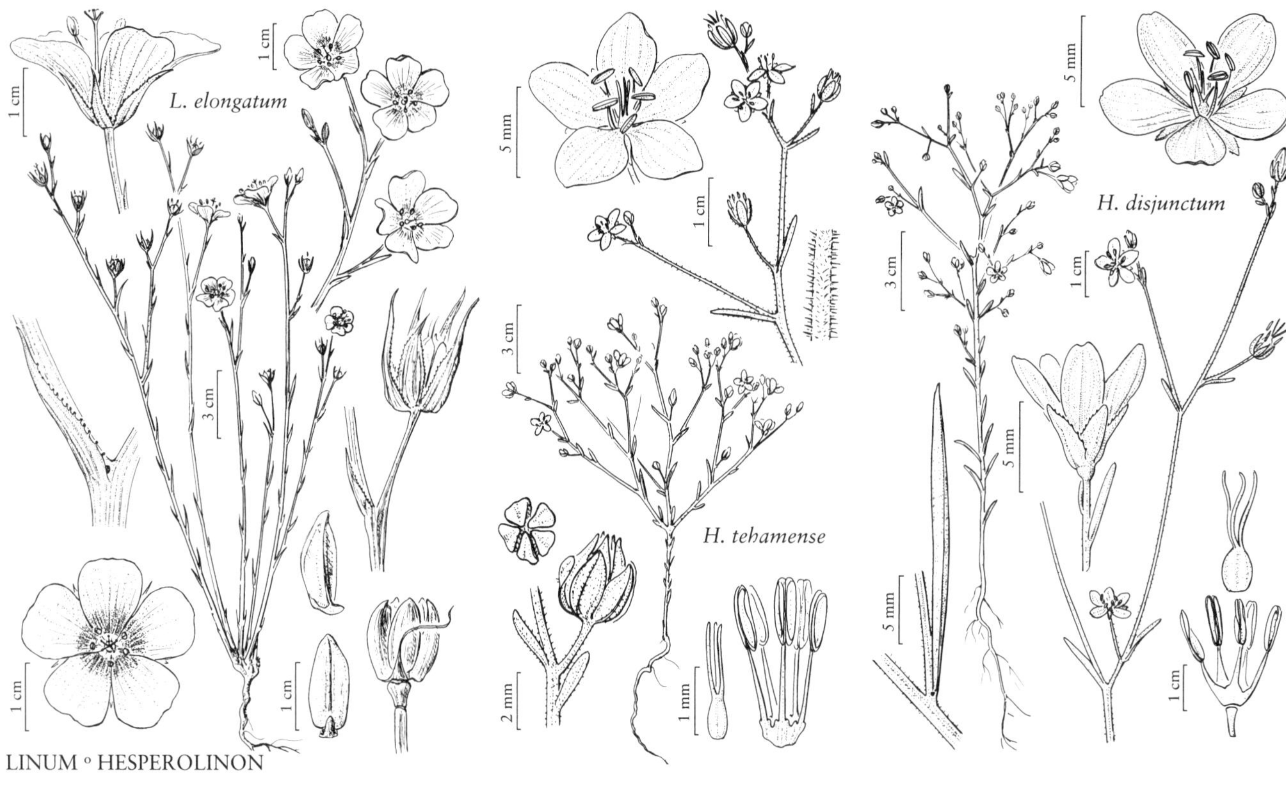

LINUM ◦ HESPEROLINON

toothed, apex sharply acute to acuminate; petals coppery yellow or orange, red-lined or with short pale to deep brown-red zone at base, obovate with short claw, 6–18 mm; stamens 6–8 mm; anthers 1–1.8(–2.3) mm; staminodia absent; styles connate nearly to apex, 3–11 mm; stigmas pale, capitate. **Capsules** ellipsoid, 3.5–4.5 × 2.6–3.4 mm, apex obtuse, dehiscing into 5, 2-seeded segments, segments blunt or subacute, persistent on plant, false septa complete, proximal part membranaceous, not terminating in loose fringe, distal part cartilaginous, margins not ciliate. **Seeds** 2.6–3.6 × 0.9–1.2 mm. $2n = 30$.

Varieties 2 (2 in the flora): w, c North America, n Mexico.

The corollas of *Linum rigidum* are funnelform at the base, opening out into a broader bowl. The styles and stamens are yellow, although the distinct portion of the style may be slightly maroon; the stigmas are grayish or light green. The stems of *L. rigidum* are angled. Its capsule walls are so thin that the dark seeds can be seen through them. C. M. Rogers (1984) noted that *L. australe*, *L. berlandieri*, *L. compactum*, and both varieties of *L. rigidum* are closely related, and that there is some overlap in character expressions, not easily resolved in a dichotomous key.

1. Stipular glands absent; styles 5–11 mm; stigmas light gray. . . . . 34a. *Linum rigidum* var. *rigidum*
1. Stipular glands nearly always present; styles 3–5 mm; stigmas light green . . . . .34b. *Linum rigidum* var. *simulans*

34a. Linum rigidum Pursh var. **rigidum**

Leaves: stipular glands absent. **Flowers:** stamens spreading; styles 5–11 mm; stigmas light gray. $2n = 30$.

Flowering May–Sep. Short grass prairie, fields, roadsides, on clay, rarely calcareous soils; (300–)1000–2300 m; Alta., Man., Sask.; Ark., Colo., Ill., Iowa, Kans., Minn., Mo., Mont., Nebr., N.Mex., N.Dak., Okla., S.Dak., Tex., Wis., Wyo.; Mexico (Coahuila, Durango).

34b. Linum rigidum Pursh var. **simulans** C. M. Rogers in N. L. Britton et al., N. Amer. Fl., ser. 2, 12: 38. 1984 [E]

Leaves: stipular glands nearly always present. **Flowers:** stamens erect; styles 3–5 mm; stigmas light green.

Flowering May–Sep. Sandy soils; 400–1900 m; Colo., Kans., Minn., Nebr., N.Dak.

35. Linum berlandieri Hooker, Bot. Mag. 63: plate 3480. 1836 (as berendieri) • Berlandier's yellow flax

Linum rigidum Pursh var. *berlandieri* (Hooker) Torrey & A. Gray

Herbs, usually annual, rarely perennial, 5–40 cm, glabrous except hirsutulous near base and sometimes on stem angles. **Stems** spreading-ascending, branching basally in inflorescence, sometimes throughout. **Leaves:** basal leaves opposite, or alternate throughout, spreading; stipular glands usually present; blade linear to linear-lanceolate, 10–25 × 1–4 mm, margins entire or few small teeth on distal leaves, not ciliate, apex acute; mostly 3-nerved. **Inflorescences** dense, ± flat-topped panicles. **Pedicels** 4–20 mm. **Flowers:** sepals deciduous, lanceolate, inner somewhat broader, 6–12 mm, margins of inner sepals densely and delicately glandular-toothed, outer ones scarious, conspicuously and more coarsely but less densely glandular-toothed than inner, apex acute or attenuate; petals yellow to orange, sometimes reddish below middle, broadly obovate, 11–19 mm; stamens 4–9 mm; anthers 1–2 mm; staminodia absent; styles connate nearly to apex, 6–9 mm; stigmas capitate. **Capsules** broadly ovoid to triangular-ovoid, tapering abruptly to flattened base, 3.6–4.7 × 3–4 mm, thick- or thin-walled, apex obtuse, dehiscing into 5, 2-seeded segments, segments blunt or subacute, persistent on plant, false septa complete, proximal part membranaceous, not terminating in loose fringe, distal part cartilaginous, margins not ciliate. **Seeds** 2.6–3.4 × 1–1.6 mm. $2n = 30$.

Varieties 2 (2 in the flora): w, c United States, n Mexico.

The stems of *Linum berlandieri* are ribbed. The corollas range from pale yellow to deep orange, with darker veins and little color banding to deeply maroon at base. Styles and stamens may be yellow or the distinct portions of the styles somewhat maroon. The capsule walls are opaque or translucent, thicker than those in *L. rigidum*.

C. M. Rogers (1984) noted that some populations of *Linum berlandieri* in western Texas are morphologically intermediate between the two varieties.

1. Sepals green; stigmas green; capsules relatively thick-walled, opaque . . . . 35a. *Linum berlandieri* var. *berlandieri*
1. Sepals grayish or purplish; stigmas dark gray; capsules relatively thin-walled, translucent . . . . 35b. *Linum berlandieri* var. *filifolium*

35a. Linum berlandieri Hooker var. **berlandieri** [E]

Flowers: sepals green; stigmas green. **Capsules** relatively thick-walled, opaque. $2n = 30$.

Flowering Apr–Aug. Sandy, gravelly soils; 100–2300 m; Ark., Colo., Kans., La., Nebr., N.Mex., Okla., Tex.

C. M. Rogers (1984) found that there were small, compact, much-branched, and very leafy plants, especially in the northern part of the range, and suggested that these needed further study. In southern Texas, Rogers found a form with comparatively few, large, spheric, marginal glands on the sepals.

35b. Linum berlandieri Hooker var. **filifolium** (Shinners) C. M. Rogers in N. L. Britton et al., N. Amer. Fl., ser. 2, 12: 39. 1984

Linum rigidum Pursh var. *filifolium* Shinners, Field & Lab. 17: 136. 1949

Flowers: sepals grayish or purplish; stigmas dark gray. **Capsules** relatively thin-walled, translucent. $2n = 30$.

Flowering Mar–Jun. Sandy, rocky, sometimes calcareous soils; 200–1600 m; N.Mex., Tex.; Mexico (Coahuila).

Variety *filifolium* occurs in southeastern New Mexico, central and western Texas, and adjacent Mexico.

36. Linum carteri Small, Bull. New York Bot. Gard. 3: 424. 1905 • Carter's flax [C] [E]

Cathartolinum carteri (Small) Small

Herbs, annual or short-lived perennial, 23–60 cm, puberulent or ± scabrous on angles near base or throughout. **Stems** erect, usually unbranched below inflorescence, sometimes branched at base. **Leaves** alternate, spreading; stipular glands present or absent; blade linear, 15–30 × 0.6–1.4 mm, margins entire or distalmost glandular-toothed, not ciliate, apex long-attenuate. **Inflorescences** ascending or spreading cymes. **Pedicels** 4.5–9 mm. **Flowers:** sepals deciduous, lanceolate, inner sepals slightly shorter than outer, outer ones 4.5–7 mm, margins of inner sepals narrowly scarious, all glandular-toothed, apex short-awned; petals orange-yellow, broadly obovate, 10–17 mm; stamens 4.5–7 mm; anthers 0.8–2.3 mm; staminodia absent; styles connate nearly to apex, 4–6 mm; stigmas capitate.

Capsules ovoid, 4–5 × 3.4–3.7 mm, thick textured, apex obtuse, dehiscing into 5, 2-seeded segments, segments persistent on plant, false septa complete, proximal part membranaceous, not terminating in loose fringe, distal part cartilaginous, margins ciliate. **Seeds** 2.3–2.8 × 1–1.3 mm. **2*n*** = 60.

Varieties 2 (2 in the flora): Florida.

The stems of *Linum carteri* are ribbed. The corollas are broadly bowl-shaped and all parts of the flower are yellow except the bright green stigmas.

1. Plants 23–30 cm; stems puberulent or scabrous on angles throughout; stipular glands usually present . . . . 36a. *Linum carteri* var. *carteri*
1. Plants 30–60 cm; stems sparsely puberulent or scabrous on angles only at base of plant; stipular glands usually absent . . . 36b. *Linum carteri* var. *smallii*

36a. Linum carteri Small var. **carteri** C E

Herbs 23–30 cm. **Stems** puberulent or scabrous on angles throughout. **Leaves:** stipular glands usually present, dark. **Flowers:** petals 10 mm.

Flowering year-round. Disturbed areas in pine rocklands on pockets in limestone rock surfaces; of conservation concern; 0–10 m; Fla.

Variety *carteri* is endemic to Miami-Dade County in southern Florida. It is known from nine sites, although it may have been extirpated at two of those. The variety is intolerant of shading and litter accumulation; fire suppression probably has reduced areas suitable for it.

J. Maschinski and D. Walters (2008) determined that var. *carteri* flowers throughout the year, more abundantly following rain. From studying plants in the wild at two sites, they found that var. *carteri* sometimes persists more than one year. Maschinski and Walters (2007) found higher densities of plants at a site that was routinely mowed, where competition from other plants was reduced. They found higher mortality in mowed sites, but Maschinski (2006) found greater capsule production in mowed sites.

36b. Linum carteri Small var. **smallii** C. M. Rogers, Sida 3: 210. 1968 C E

Herbs 30–60 cm. **Stems** sparsely puberulent or scabrous on angles only at base of plant. **Leaves:** stipular glands usually absent. **Flowers:** petals 11–17 mm.

Flowering Mar–Apr. Roadsides, prairies, pinelands, pine rocklands; of conservation concern; 0–10 m; Fla.

Variety *smallii* is known from 11 populations, most with 50 or fewer plants. The pinelands that are its native habitat are disappearing due to development, but some populations are persisting along roadsides and in Everglades National Park.

37. Linum trigynum Linnaeus, Sp. Pl. 1: 279. 1753 • French flax I

Linum gallicum Linnaeus

Herbs, annual, 10–50 cm, glabrous. **Stems** erect or spreading, few-branched. **Leaves** alternate, spreading to ascending; stipular glands absent; blade linear-lanceolate to narrowly elliptic, 5–10 × 1–1.5 mm, margins entire, not ciliate, apex acuminate. **Inflorescences** panicles. **Pedicels** 1–5 mm. **Flowers:** sepals persistent, lanceolate to ovate, 3–4 mm, margins of inner sepals broadly scarious, densely glandular-ciliate, glandular-toothed, apex acuminate to setaceous; petals lemon yellow, oblong to obovate, 4–6 mm; stamens 1.5 mm; anthers 0.3 mm; staminodia present or absent; styles distinct, 1 mm; stigmas linear. **Capsules** subglobose, 2 mm diam., apex sharp-pointed (easily crushed), readily dehiscing into 5, 2-seeded segments, segments persistent on plant, false septa incomplete, margins of true septa ciliate. **Seeds** 1.1 × 0.9–1 mm. **2*n*** = 20.

Flowering May–Jul. Grasslands; 100–200 m; introduced; Calif.; s Europe; w Asia; n Africa; introduced also in Pacific Islands (Hawaii, New Zealand), Australia.

Linum trigynum is one of three species in sect. *Linopsis* subsect. *Halolinum* (Planchon) C. M. Rogers. This section is characterized as having separate styles, linear stigmas, and incomplete false septa. *Linum trigynum* is homostylous; the other two species, *L. maritimum* Linnaeus and *L. tenue* Desfontaines, are heterostylous. Two populations of *L. trigynum* have been reported in Sonoma County on the Jenner and Fort Ross State Historic Park headlands, both with hundreds of individuals and apparently persisting. Where native, the species sometimes occurs on serpentine soils, and it is reported as a weed in western Australia.

3. HESPEROLINON (A. Gray) Small in N. L. Britton et al., N. Amer. Fl. 25: 84. 1907 • Western flax [Greek *hesperos*, western, and *linon*, flax]

Linum Linnaeus [unranked] *Hesperolinon* A. Gray, Proc. Amer. Acad. Arts 6: 521. 1865

Herbs, annual, glabrous, puberulent, or hoary. **Stems** usually erect, rarely decumbent, usually unbranched, sometimes branched, proximally (branches opposite or whorled), branched distally. **Leaves** falling early (persistent in *H. drymarioides*), in whorls of 4 at basal nodes, usually becoming irregularly whorled, opposite, or alternate proximally (distal to basal nodes) on main stem, alternate or opposite distally; stipular glands usually present (exudate often red), sometimes absent; blade threadlike to linear, oblong-lanceolate, ovate, or orbiculate, margins glandular-toothed, stipitate-glandular, or entire. **Inflorescences** open or dense, monochasial or dichasial cymes, sometimes helicoid or scorpioid. **Pedicels** articulated. **Flowers:** sepals persistent, 5, connate at base, equal or unequal in size, margins entire, stipitate-glandular or eglandular, surfaces glabrous or hairy; petals 5, distinct, attached near rim of cup between filament bases, yellow, white, or pink, base with 0 or 2 auricles flanking central ligule; stamens 5; staminodes 0; pistil 2–3-carpellate, ovary 4- or 6-locular by intrusion of incomplete false septa; styles 2–3, distinct; stigmas minute, linear, ± equal in width to styles. **Fruits** capsules, dehiscing into 4 or 6 segments. **Seeds** 4 or 6, oblong to clavate, triangular in cross section. $x = 18$.

Species 13 (13 in the flora): w United States, nw Mexico.

Hesperolinon was monographed by H. K. Sharsmith (1961); the treatment here is based primarily on that work.

Hesperolinon taxa are winter annuals; the plants flower in late spring, after most other associated species are in fruit. H. K. Sharsmith (1961) was first to describe the complex flower morphology. Appendages are formed at the base of the petal by expansion into a pair of triangular lobes bearing pockets or pouches in their margins. The pouches are usually swollen on their adaxial surfaces with glandular-papillate cells that form horizontal crests (the lateral appendages), sometimes meeting near the center of the claw and connecting with the base of the ligule, a central, erect, greenish yellow, fleshy, often hairy or glandular-papillate, tonguelike protuberance. The ligule arises over the petal midvein, usually just proximal to the point where the two lateral veins arise, or the ligule may be a longitudinal, glandular, sometimes hairy, thickening or fold in the lamina.

Flowers in *Hesperolinon* often are pseudocleistogamous with the anthers dehiscing and depositing pollen on the stigma in bud, which may account for the morphological uniformity within populations, and the large differences between populations of the same species (H. K. Sharsmith 1961). In areas where two or more species grow sympatrically, there usually is a difference in phenology, with one flowering before another, or the species occupying different microhabitats.

C. M. Rogers (1975) noted that whorled leaves, common in *Hesperolinon,* were found in *Linum* only in his *L. schiedeanum* complex (the most primitive group) in sect. *Linopsis*. *Linum schiedeanum* and *Hesperolinon* also share other character states, including stipular glands and distinct styles. Rogers noted that features shared with the southwestern *L. neomexicanum* include annual habit, distinct styles, auricles at the base of petals, some floral pigments (D. E. Giannasi and Rogers 1970), and petals attached at the top of the filament cup. J. R. McDill (2009) suggested that *Hesperolinon* is related to a clade including the Mexican endemic *L. mexicanum* Kunth and *L. guatemalense* Bentham of Guatemala and San Salvador, in what he termed the *L. mexicanum* group in *Linum* subsect. *Linopsis*, ser. *Linopsis*.

All species of *Hesperolinon* except *H. micranthum* are endemic to California, mostly restricted to the North and South Coast ranges, usually in chaparral of the inner Coast Ranges. H. K. Sharsmith (1961) considered Lake and Napa counties to be the center of diversity for the genus, and, except for *H. californicum* and *H. micranthum*, all species are found on the Jurassic Franciscan formation, most often on exposed serpentine components. *Hesperolinon micranthum* occurs on serpentine soil in the Coast Ranges and on volcanic flows in the northern Sierra Nevada and Modoc Plateau. At least eight of the species Sharsmith studied were obligate or near-obligate serpentine endemics and the remaining four species were facultative serpentine species.

Y. P. Springer (2009) compared occurrences of *Hesperolinon* on serpentine and nonserpentine soil types with a phylogeny based on gene sequences of four chloroplast loci from multiple populations of each species. For five of the species, sequences from different populations were assigned to different clades. Springer suggested this was due to either chloroplast capture or recent divergence of the species in *Hesperolinon*.

A. C. Schneider et al. (2016) sequenced plastid DNA and ITS using broader geographical sampling than Y. P. Springer (2009). They found that *Hesperolinon* had diversified mostly in the last one to two million years and that all species except *H. drymarioides*, which diverged first and is morphologically very distinctive, were placed in more than one clade. More study is needed to determine how genetic relationships relate to distribution and morphology in *Hesperolinon*.

SELECTED REFERENCES O'Donnell, R. 2010. The genus *Hesperolinon* (Linaceae): An introduction. Four Seasons 13(4): 1–61. Sharsmith, H. K. 1961. The genus *Hesperolinon*. Univ. Calif. Publ. Bot. 32: 235–314.

1. Petals usually yellow, sometimes fading to white, often veined or tinged orange or reddish; anthers yellow.
   2. Leaf bases keeled, clasping, margins with stalked glands on teeth in 1–2 rows . . . . 9. *Hesperolinon adenophyllum*
   2. Leaf bases flat, not clasping, margins without stalked glands.
      3. Styles 2(–3) (plants sometimes with both 2- and 3-carpellate flowers). . . . . 4. *Hesperolinon bicarpellatum* (in part)
      3. Styles (2–)3 (sometimes some 2-carpellate flowers on otherwise 3-carpellate plants).
         4. Styles 0.5–1(–1.8) mm, included; petals not or slightly spreading. . . . 2. *Hesperolinon clevelandii*
         4. Styles 2–6(–8) mm, exserted; petals widely spreading.
            5. Flowers clustered at inflorescence tips; stipular glands usually present at all nodes; petals (4–)6–7(–10) mm; filaments (4–)6–8 mm . . . . 11. *Hesperolinon breweri*
            5. Flowers widely scattered throughout inflorescence; stipular glands absent or present only at proximal nodes; petals (2–)3–5.5 mm; filaments 1.5–4 mm.
               6. Plants usually hoary throughout, sometimes glabrous except on stems; petals 3.5–5.5 mm; styles 3.5–4.5(–5) mm . . . . .6. *Hesperolinon tehamense*
               6. Plants glabrous, or stems puberulent only immediately distal to nodes; petals (2–)3–4.5 mm; styles 2–3.5 mm.
                  7. Pedicels (2–)10–12 mm (to 40 mm in fruit); sepals 1.5–2(–3) mm . . . . 4. *Hesperolinon bicarpellatum* (in part)
                  7. Pedicels 0.5–5 mm; sepals (1.5–)3 mm . . . . 5. *Hesperolinon sharsmithiae*
1. Petals white to pink, often veined or tinged darker pink or lavender to purple; anthers white to pink, rose, or purple.
   8. Proximal leaves whorled (usually in 4s) on main stem, blades ovate or orbiculate, 3–6 mm wide, margins minutely stipitate-glandular . . . . 10. *Hesperolinon drymarioides*
   8. Proximal leaves alternate on main stem, blades threadlike to linear or narrowly oblong, 0.5–2.5(–3) mm wide, margins eglandular (except sometimes in *H. congestum* and in *H. disjunctum*).

[9. Shifted to left margin.—Ed.]

9. Styles 2 . . . . 8. *Hesperolinon didymocarpum*
9. Styles (2–)3 (sometimes some 2-carpellate flowers on otherwise 3-carpellate plants).
    10. Inflorescences dense; pedicels 0.5–2(–5) mm at anthesis, to 10 mm in fruit; stipular glands present.
        11. Sepals glabrous; petals white or partly pink to white, irregularly veined or flushed with pink or rose pink . . . . 12. *Hesperolinon californicum*
        11. Sepals hairy; petals pink to rose . . . . 13. *Hesperolinon congestum*
    10. Inflorescences open; pedicels 1–15(–25) mm at anthesis, to 45 mm in fruit; stipular glands absent or minute, or present only at proximal nodes.
        12. Styles included; petals not or only slightly spreading, 1.5–3.5 mm (shorter than to equaling sepals). . . . . 1. *Hesperolinon micranthum*
        12. Styles exserted; petals widely spreading to reflexed, (3–)4–7 mm (longer than sepals).
            13. Pedicels bent, pendent in bud, 5–15(–25) mm in flower . . . . 3. *Hesperolinon spergulinum*
            13. Pedicels straight, not pendent in bud, 1–5(–8) mm in flower . . . . 7. *Hesperolinon disjunctum*

1. **Hesperolinon micranthum** (A. Gray) Small in N. L. Britton et al., N. Amer. Fl. 25: 85. 1907 • Small-flowered western or dwarf flax

Linum micranthum A. Gray, Proc. Amer. Acad. Arts 7: 333. 1868

Herbs 5–20(–50) cm, glabrous or stems puberulent just distal to nodes; branched throughout, branches from proximal nodes in whorls, distal nodes alternate, widely spreading. **Leaves** alternate; stipular glands present at proximal nodes, usually absent distally; blade linear or narrowly oblong, 10–20(–30) × 1.5–2.5(–3) mm, base flat, not clasping, margins without stalked glands, surfaces sometimes microscopically glandular. **Inflorescences:** cymes monochasial (scorpioid or helicoid), open, internodes long, flowers widely scattered; bract margins eglandular or glands inconspicuous. **Pedicels** (2–)5–8 mm (–25 mm in proximal axils), (5–)10–15(–45) mm in fruit, spreading at 45–60(–90)° angle, not or only slightly bent at apex. **Flowers:** sepals erect or reflexed at tip, lanceolate, 1–3(–4) mm, sometimes unequal, marginal glands absent or minute, surfaces glabrous; petals not or only slightly spreading, white to pink, sometimes streaked with deeper pink or rose purple, usually oblanceolate, sometimes obovate, 1.5–3.5 mm, apex slightly notched; cup white, rim usually glabrous, sometimes hairy, petal attachment often a prominent protuberance in sinuses; stamens included; filaments 1.5–2.5 mm; anthers white to deep purple, dehisced anthers (0.3–)0.5–0.8(–1) mm; ovary chambers 6; styles (2–)3, white, 0.5–1(–2) mm, included. 2*n* = 36.

Flowering Mar–Aug. Open areas, woodland margins, serpentine and nonserpentine soils, volcanic soils; 50–2000 m; Calif., Oreg.; Mexico (Baja California).

Hesperolinon micranthum is the most widespread species of the genus. It occurs in Oregon east of the Cascade Mountains and southern Blue Mountains to southern Oregon. In California, it occurs in the Klamath and Cascade ranges, Warner Mountains, Modoc Plateau, on the western slope of the Sierra Nevada, and in the North Coast and South Coast ranges, western Transverse Range, and Peninsular Ranges. Its southernmost population is near Cerro Matomi in Baja California. In much of its range, it is the only *Hesperolinon* species. It can be distinguished from other white-petaled species by its smaller flowers, included styles, and erect or only slightly spreading petals. In the southern part of its range, plants are generally smaller-flowered, have shorter pedicels, and more leaflike sepals; on the Modoc Plateau it grows almost exclusively on volcanic rock and plants are shorter, stockier, have broader leaves, shorter pedicels, and larger flowers. H. K. Sharsmith (1961) noted that any population may have some flowers in which appendages, auricles, or ligules are reduced or completely absent.

2. **Hesperolinon clevelandii** (Greene) Small in N. L. Britton et al., N. Amer. Fl. 25: 85. 1907 • Allen Springs dwarf flax [E]

Linum clevelandii Greene, Bull. Torrey Bot. Club 9: 121. 1882 (as clevelandi)

Herbs 5–20(–30) cm, glabrous or glabrate; unbranched proximally or proximal branches whorled, branches from distal nodes dichotomous, widely spreading. **Leaves** alternate; stipular glands very inconspicuous, present at proximal nodes, absent distally; blade linear or narrowly oblong, 10–13(–20) × 2–2.5 mm, base flat, not clasping, margins without stalked glands. **Inflorescences:** cymes monochasial (scorpioid or helicoid), open, branches unequal (main axis obvious), internodes long, flowers widely scattered; bract margins without prominent glands. **Pedicels** 5–25 mm, scarcely longer in fruit, spreading at 70–80(–90)° angles, scarcely bent at apex. **Flowers:** sepals erect or reflexed at tip, lanceolate, 1.5–2.5 mm, usually equal, sometimes one larger, marginal glands absent or minute, surfaces glabrous; petals not or slightly spreading at anthesis, yellow, often with reddish or orange streak on midvein, oblanceolate, sometimes obovate, 0.5–2.5(–4) mm, apex notched or erose; cup yellow, rim with petal attachment protruding prominently in sinus or strongly indented; stamens included; filaments 1–2 mm; anthers yellow, dehisced anthers 0.5–0.8(–1.2) mm; ovary chambers 6; styles 3, yellow, 0.5–1(–1.8) mm, included. **2*n*** = 36.

Flowering May–Jul. Chaparral margins, oak woodlands, ponderosa pine woodlands, serpentine or volcanic soils; 150–1400 m; Calif.

Hesperolinon clevelandii occurs in the inner North Coast Ranges from Mendocino to Napa counties and on the Mount Hamilton Range in Santa Clara and Stanislaus counties. It can be distinguished from *H. micranthum* by its yellow stamens and petals. The flowers in Mount Hamilton populations may be twice as large as those of other populations and might warrant recognition as a subspecies (H. K. Sharsmith 1961).

3. Hesperolinon spergulinum (A. Gray) Small in N. L. Britton et al., N. Amer. Fl. 25: 86. 1907 • Slender western or dwarf flax [E]

Linum spergulinum A. Gray, Proc. Amer. Acad. Arts 7: 333. 1868

Herbs 10–30(–50) cm, glabrous or glabrate; branches from distal nodes, alternate, widely spreading. **Leaves** alternate; stipular glands absent or minute; blade linear or narrowly oblong, 10–35 × 0.5–2(–2.5) mm, base flat, not clasping, margins without stalked glands. **Inflorescences:** cymes monochasial (helicoid), open, internodes long, flowers widely scattered; bract margins without prominent glands. **Pedicels** 5–15(–25) mm, 5–25 mm in fruit, pendent in bud, deflexed at 90° angle, slightly bent at apex. **Flowers:** sepals erect, not reflexed at tip, ovate, 1.5–2.5(–3.5) mm, equal, margins minutely gland-toothed, surfaces glabrous; petals widely spreading to reflexed, white or pale pink, usually darker-veined, obovate, 4–7 mm, apex obtuse; cup white, rim petal attachments in indentations; stamens exserted; filaments (3–)4–5(–7) mm; anthers pink to red-purple, white-margined, dehisced anthers 1.2–2 mm; ovary chambers 6; styles 3, white, 3.5–7 mm, exserted. **2*n*** = 36.

Flowering May–Aug. Chaparral or woodland margins, serpentine soils; 100–1000 m; Calif.

Hesperolinon spergulinum occurs in the central and southern North Coast Ranges; there are historical reports from Santa Clara County. The pendent buds, a result of the deflexed and sometimes downward-curved pedicel, and the relatively long styles and stamens are distinctive. The petal appendages are relatively well developed in *H. spergulinum*; the ligule may be as large as 1 mm and hairy.

4. Hesperolinon bicarpellatum (H. Sharsmith) H. Sharsmith, Univ. Calif. Publ. Bot. 32: 297. 1961 • Two-carpellate western or dwarf flax [C] [E]

Linum bicarpellatum H. Sharsmith, Madroño 8: 143. 1945

Herbs 10–30(–70) cm, puberulent on stems immediately distal to nodes, otherwise glabrous; branches usually alternate from distal nodes on short main axis, sometimes whorled from basal nodes, widely spreading. **Leaves** alternate; stipular glands present at proximal nodes, usually absent distally; blade threadlike to linear, 15–20(–30) × 1–1.5 mm, base flat, not clasping, margins without stalked glands. **Inflorescences:** cymes dichasial, open, internodes long, flowers widely scattered; bract margins without prominent glands. **Pedicels** (2–)10–12 mm, to 40 mm in fruit, spreading at 45–90° angle, not or only slightly bent at apex. **Flowers:** sepals all or outer 2 spreading, not reflexed at tip, lanceolate, 1.5–2(–3) mm, ± equal, marginal glands minute or absent, surfaces glabrous; petals horizontally widely spreading, bright yellow fading white, tinged with orange or red along veins, oblanceolate to obovate, (2–)3–4 mm, apex acute or apiculate; cup yellow, rim lobed between filaments and petal attachments; stamens exserted; filaments 2–3(–3.5) mm; anthers yellow, dehisced anthers 1.2–1.5 mm; ovary chambers 4; styles 2(–3), yellow, 2–3.5 mm, exserted. **2*n*** = 34.

Flowering May–Jul. Rocky slopes, chaparral in *Pinus sabiniana* belt, serpentine soils; of conservation concern; 60–1000 m; Calif.

Hesperolinon bicarpellatum grows in the southern part of the Inner North Coast Ranges of Lake and Napa counties. Plants in the northern part of its range consistently have two carpels; those in the southern part occasionally may also have three carpels, and sometimes both conditions occur on one plant.

5. Hesperolinon sharsmithiae R. O'Donnell, Madroño 53: 404, figs. 2, 3A. 2007 • Sharsmith's western or dwarf flax [C] [E]

Herbs 5–50 cm, puberulent on stems just distal to nodes, otherwise glabrous or glabrate; unbranched proximally, branches from distal nodes dichotomous, widely spreading. **Leaves** alternate; stipular glands present only at proximal nodes; blade linear, 15–20 × 1–1.5 mm, base flat, not clasping, margins without stalked glands. **Inflorescences:** cymes dichasial, open, internodes long, flowers scattered; bract margins sparsely glandular. **Pedicels** 0.5–2 mm near tips of branches, 4–5 mm in fruit, ascending, not bent at apex. **Flowers:** sepals erect, not reflexed at tip, oblanceolate, (1.5–)3 mm, equal, margins sparsely glandular, surfaces glabrous; petals widely spreading, yellow, veins sometimes red-streaked, oblanceolate to obovate, or almost oval, 3–4.5 mm, apex erose; cup yellow, rim lobed between filaments and petal attachments; stamens exserted; filaments 1.5–2.5 mm; anthers yellow, dehisced anthers 1–1.5 mm; ovary chambers 6; styles 3, color not known, 2–3.5 mm, exserted.

Flowering May–Jul. On serpentine soils, in chaparral, Sargent cypress forests; of conservation concern; 200–300 m; Calif.

Hesperolinon sharsmithiae is known from Lake and Napa counties in the Hunting Creek drainage, Cedar Roughs, and Butts and Pope Creek canyons (R. O'Donnell 2010). It grows in the vicinity of at least six other species of *Hesperolinon*: *H. bicarpellatum*, *H. californicum*, *H. clevelandii*, *H. didymocarpum*, *H. disjunctum*, and *H. spergulinum*. The species is distinguished from *H. bicarpellatum* by having three rather than two carpels and by its larger flowers. *Hesperolinon sharsmithiae* differs from *H. tehamense* in lacking the hoary indument and in having narrower petals, and it differs from *H. clevelandii* in having shorter styles and stamens. In his original description, O'Donnell noted that most of the collections annotated as *H. serpentinum* (a name never validly published) can be assigned to *H. sharsmithiae*. O'Donnell reported an area on the ridge between Pope Valley and Butts Canyon in which plants were intermediate between *H. bicarpellatum* and *H. sharsmithiae*; he noted that over a period of four years individuals of *H. sharsmithiae* increased in number and those of *H. bicarpellatum* decreased.

6. Hesperolinon tehamense H. Sharsmith, Univ. Calif. Publ. Bot. 32: 298, figs. 4f, 5w, 10g, 14f, 20. 1961 • Tehama County western flax, Paskenta Grade dwarf flax [C] [E] [F]

Herbs (2–)20–35(–50) cm, usually hoary throughout, hairs microscopic, straight, stiff, white, erect, sometimes glabrous except on stems immediately distal to nodes; branches from distal ⅓–⅔ of plant, alternate, widely spreading. **Leaves** alternate; stipular glands only at proximal nodes, inconspicuous, or absent; blade ± linear, 10–20(–30) × 1–2(–3) mm, base flat, not clasping, margins without stalked glands. **Inflorescences:** cymes mostly monochasial (scorpioid or helicoid), open, internodes long, flowers widely scattered; bract margins without prominent glands. **Pedicels** 0.5–2(–3) mm, to 6 mm in fruit, spreading at 45° angle, not bent at apex. **Flowers:** sepals erect, ± spreading at tip, lanceolate, 2–3 mm, equal, marginal glands tiny, surfaces glabrous or with hoary microscopic puberulence; petals widely spreading, distal ½ ± recurved, light or bright yellow, veins sometimes ± red, obovate, 3.5–5.5 mm, apex often deeply notched; cup yellow, rim hairy, rim with petal attachment in shallow sinus; stamens exserted; filaments 3–4 mm; anthers yellow, dehisced anthers 1.2–1.4 mm; ovary chambers 6; styles 3, yellow, 3.5–4.5(–5) mm, exserted. $2n = 34$.

Flowering May–Jul. Dry rocky hillsides in chaparral in *Pinus sabiniana* and *Quercus douglasii* woodlands, serpentine soils; of conservation concern; 100–1000 m; Calif.

Hesperolinon tehamense is found in the north and central Inner North Coast Ranges in Glenn and Tehama counties. Its range overlaps that of *H. disjunctum*, from which it differs in having petals bright yellow, sometimes tinged with orange or red, rather than petals white, pink, or lavender.

7. Hesperolinon disjunctum H. Sharsmith, Univ. Calif. Publ. Bot. 32: 300, figs. 1c, 3, 5i–l, 10e, f, 15c–e, 20. 1961 • Coast Range western flax [E] [F]

Herbs (3–)20–25(–30) cm, stout, usually hoary, hairs minute, straight, stiff, white, sometimes glabrous except on stems distal to nodes; branches usually from distal ⅓–⅔ of main axis, sometimes from base, alternate, widely spreading. **Leaves** alternate; stipular glands minute or absent; blade linear, 10–20 × 1–2 mm, base flat, not clasping, margins eglandular or minutely gland-toothed. **Inflorescences:** cymes usually monochasial (scorpioid), sometimes dichasial, open, internodes long, flowers widely scattered; bract margins eglandular or minutely gland-toothed. **Pedicels** 1–5(–8) mm, 5–10(–25) mm in fruit, straight in bud, spreading at 75–90° angle, not reflexed or bent at apex. **Flowers:** sepals erect, usually spreading at tip, lanceolate, 2–3 mm, equal, marginal glands small, surfaces glabrous or with hoary microscopic puberulence; petals widely spreading, white or pink to lavender-pink or rose pink, veins usually pink, usually obovate, sometimes oblanceolate, (3–)4–5(–6) mm, apex notched, sometimes deeply; cup white, rim with petal attachment in shallow sinus; stamens exserted; filaments (2.5–)3–3.5(–4) mm; anthers usually pink, sometimes deep rose, white-margined, dehisced anthers (0.8–)1.2–1.8(–2) mm; ovary chambers 6; styles 3, white, (2.5–)3–4(–5) mm, exserted. **2*n*** = 34.

Flowering Apr–Jul. Dry, rocky hillsides in chaparral in *Pinus sabiniana* belt, serpentine soils; 100–1000 m; Calif.

Hesperolinon disjunctum is found in the Inner North Coast Ranges and eastern San Francisco Bay area. Dwarf plants with short internodes and crowded branches sometimes occur intermixed with normal-sized plants. Restricted to serpentine soils, populations of *H. disjunctum* are often distant from each other; Sharsmith noted that morphological disjunction accompanies the geographical disjunction.

8. **Hesperolinon didymocarpum** H. Sharsmith, Univ. Calif. Publ. Bot. 32: 302, figs. 11h, 15a, 17a, 18b, 20. 1961 • Lake County western flax [C] [E]

Herbs 10–30 cm, glabrous except stems microscopically puberulent distal to nodes; branches from short main axis, alternate, spreading. **Leaves** alternate; stipular glands present on proximal nodes; blade threadlike to linear, 10–20(–30) × 1–1.5 mm, base flat, not clasping, margins eglandular. **Inflorescences:** cymes dichasial, open, internodes long, flowers widely scattered; bract margins eglandular. **Pedicels** (2–)5–8(–15) mm, 20+ mm in fruit, spreading at 45–90° angle, not reflexed or bent at apex. **Flowers:** sepals spreading, lanceolate, 2–3 mm, ± equal, marginal glands minute or absent, surfaces glabrous; petals horizontally spreading, white or light pink or pink to deep purplish pink and streaked with light or deeper pink, veins white, oblanceolate to obovate or nearly oval, 2.5–3(–4) mm, apex notched; cup white, rim lobed between petal attachment and filaments; stamens exserted; filaments 2.5–3 mm; anthers white to deep purplish pink bordered with white, dehisced anthers 1.2–2 mm; ovary chambers 4; styles 2, white, 2.5–4 mm, exserted.

Flowering May–Jun. Chaparral, grasslands, under scattered *Pinus sabiniana*, serpentine soils; of conservation concern; 100–200 m; Calif.

Hesperolinon didymocarpum is known only from fewer than ten populations in the Inner North Coast Ranges in Big Canyon Creek, Lake County.

9. **Hesperolinon adenophyllum** (A. Gray) Small in N. L. Britton et al., N. Amer. Fl. 25: 85. 1907 • Glandular western flax [C] [E]

Linum adenophyllum A. Gray, Proc. Amer. Acad. Arts 8: 624. 1873

Herbs (10–)15–25(–50) cm, glabrous except puberulent on stems just distal to nodes; branches from well-developed main axis, alternate, widely spreading. **Leaves** opposite or proximal whorled; stipular glands absent; blade linear to lanceolate, 5–15(–20) × 1.5–2.5 mm, base keeled, clasping, margins with stalked glands on teeth in 1–2 rows. **Inflorescences:** cymes monochasial to dichasial (scorpioid or helicoid), open, internodes long, flowers widely scattered along thin branchlets; bract margins with stalked glands. **Pedicels** (3–)5–10(–15) mm, to 25 mm in fruit, spreading at 45–90° angle, not bent at apex. **Flowers:** sepals erect, tips spreading, lanceolate, 2–3 mm, subequal, marginal glands few or absent, surfaces glabrous; petals horizontally spreading, yellow, often veined or tinged orange, fading white, oblanceolate, 3–4(–5) mm, apex notched; cup yellow, rim lobed between filaments and petal attachments; stamens exserted; filaments 2.5–3.5(–4) mm; anthers yellow, dehisced anthers 1.2–1.5 mm; ovary chambers 6; styles 3, yellow, 2.5–3(–4) mm, exserted. **2*n*** = 36.

Flowering May–Aug. Chaparral and brushy slopes on serpentine soils; of conservation concern; 150–1000 m; Calif.

Hesperolinon adenophyllum is found in the north and central North Coast Ranges, especially in Lake and Mendocino counties. It can be distinguished from all other species in the genus except *H. drymarioides* in having glandular-serrate leaves. The leaves of *H. adenophyllum* are lanceolate with large glands in one or two rows on the margin, compared with *H. drymarioides*, which has ovate to orbiculate leaves with several rows of relatively small, delicate glands on the margins.

10. Hesperolinon drymarioides (Curran) Small in N. L. Britton et al., N. Amer. Fl. 25: 84. 1907 • Drymaria-like western flax, drymary dwarf flax [C] [E]

Linum drymarioides Curran, Bull. Calif. Acad. Sci. 1: 152. 1885

Herbs (5–)15–20(–30) cm, puberulent, sometimes appearing hoary, stems less hairy distally; branches from distal nodes, alternate, virgate, 1 or 2 prostrate. **Leaves** persistent, whorled (usually in 4s), distal sometimes ± opposite; stipular glands absent; blade ovate or orbiculate, 4–8 × 3–6 mm, base keeled, clasping, margins minutely stipitate-glandular. **Inflorescences:** cymes dichasial and open proximally, monochasial and dense near tips, few-branched, internodes long proximally, short near apices, flowers congested; bract margins stipitate-glandular. **Pedicels** 0.5–2(–10) mm, scarcely longer in fruit, little spreading, not bent at apex. **Flowers:** sepals erect, tips spreading (even in bud), ovate to broadly lanceolate, 2.5–4 mm, usually equal, marginal glands inconspicuous, surfaces glabrous or with a few stiff, white hairs; petals widely spreading, white or lavender to pink, sometimes streaked with deeper pink or rose purple, obovate, 3–5(–6) mm, apex notched; cup white, rim lobed between filaments and petal attachment; stamens exserted; filaments 2.5–3 (–3.5) mm; anthers usually white, sometimes purple, white-margined, dehisced anthers (0.8–)1–1.5(–2) mm; ovary chambers 6; styles 3, white, (2–)2.5–3(–3.5) mm, exserted.

Flowering May–Jul. Dry chaparral ridges or woodlands, serpentine soils; of conservation concern; 100–1000 m; Calif.

Hesperolinon drymarioides is known from only 20 populations scattered along the eastern slopes of the Inner North Coast Ranges of Colusa, Lake, and Napa counties. It is the most distinctive species in the genus. The persistent, ovate to orbiculate leaves arranged in well-spaced whorls, separated by long internodes, to the top of the main axis, with margins outlined by rows of tiny, stipitate glands, set it apart from all other species. The species is also unusual in often producing one or two long, decumbent, cyme-producing branches at the base of the main stem (H. K. Sharsmith 1961).

11. Hesperolinon breweri (A. Gray) Small in N. L. Britton et al., N. Amer. Fl. 25: 85. 1907 • Brewer's western flax [C] [E]

Linum breweri A. Gray, Proc. Calif. Acad. Sci. 3: 102. 1864

Herbs (5–)20–40(–50) cm, glabrous or glabrate; branches from distal nodes, alternate, virgate (proximal unbranched main axis long in comparison to distal portion). **Leaves** alternate; stipular glands usually present at all nodes; blade linear, 10–25(–30) × (0.5–) 1–1.5 mm, base flat, not clasping, margins eglandular. **Inflorescences:** cymes monochasial, dense, sparingly branched, internodes sometimes all shortened, or long proximally, flowers clustered at inflorescence tips; bract margins eglandular. **Pedicels** 0.5–3 mm, scarcely longer in fruit, ascending, not bent at apex. **Flowers:** sepals erect, narrowly ovate, 3–4 mm, subequal, margins glandular-toothed, surfaces glabrous; petals widely spreading, yellow, often tinged with orange-red, obovate, (4–)6–7(–10) mm, apex obtuse, slightly erose; cup yellow, rim with petal attachments set in notches; stamens exserted; filaments (4–)6–8 mm; anthers yellow, dehisced anthers 1.5–2 mm; ovary chambers 6; styles 3, yellow, (3–)4–6(–8) mm, exserted. $2n = 36$.

Flowering May–Jun. Chaparral or grasslands, usually on serpentine soils, sometimes on nonserpentine soils; of conservation concern; 30–700 m; Calif.

Hesperolinon breweri is found in the Vaca Mountains and vicinity in the southern Inner North Coast Ranges in Napa and Solano counties, and in the northeastern San Francisco Bay area on Mount Diablo in Contra Costa County. It is similar to *H. californicum* and can be distinguished by its bright yellow petals tinged with orange and yellow dehisced anthers, filaments, and cup, rather than petals white tinged with pink, and pink or white dehisced anthers. *Hesperolinon breweri* occurs within or at the edge of chaparral; *H. californicum* is found in open grassy areas adjacent to chaparral.

12. Hesperolinon californicum (Bentham) Small in N. L. Britton et al., N. Amer. Fl. 25: 86. 1907 [E]

Linum californicum Bentham, Pl. Hartw., 299. 1849; *L. californicum* var. *confertum* A. Gray ex Trelease; *L. congestum* A. Gray var. *confertum* (A. Gray ex Trelease) Abrams

Herbs (10–)20–40(–50) cm, glabrous or glabrate; branches from distal stem nodes, alternate, virgate (proximal unbranched main axis long in comparison to distal portion). **Leaves** alternate; stipular glands present; blade threadlike to linear, 10–25 × 1–1.5(–2) mm, base flat, not clasping, margins eglandular. **Inflorescences:** cymes monochasial, dense, sparingly branched, internodes short and flowers condensed or internodes long near base and flowers condensed at apices; bract margins eglandular. **Pedicels** 0.5–2(–5) mm, to 10 mm in fruit, ascending, not bent at apex. **Flowers:** sepals erect, lanceolate to narrowly ovate, 3.5–4 mm, subequal, margins minutely glandular-toothed, surfaces glabrous; petals widely spreading, white or partly pink, irregularly veined or flushed with pink or rose pink, obovate, (4–)6–8(–12) mm, apex obtuse, slightly erose; cup white, rim with petal attachments set in deep notches; stamens exserted; filaments (4–)5–7(–8) mm; anthers white to rose, dehisced anthers 2(–3) mm; ovary chambers 6; styles 3, white, (4–)5–7(–10) mm, exserted. $2n = 36$.

Flowering Apr–Jul. Rocky areas, chaparral, grasslands, usually on serpentine soils; 30–1300 m; Calif.

Hesperolinon californicum usually occurs on serpentine soil in the central Inner Coast Ranges and on nonserpentine soil in the foothills east and west of the Sacramento Valley and west of the San Joaquin Valley.

13. Hesperolinon congestum (A. Gray) Small in N. L. Britton et al., N. Amer. Fl. 25: 86. 1907 • Marin dwarf flax [C] [E]

Linum congestum A. Gray, Proc. Amer. Acad. Arts 6: 521. 1865

Herbs (5–)15–30(–45) cm, glabrous or glabrate; branches from distal nodes, alternate, strongly virgate (proximal unbranched main axis usually long in comparison to branched distal portion). **Leaves** alternate; stipular glands present; blade linear, (5–)15–25 × 1–1.5(–2) mm, base flat, not clasping, margins eglandular or glands minute. **Inflorescences:** cymes monochasial, dense, sparingly branched, internodes sometimes all condensed or long proximally and condensed distally, flowers condensed at apices; bract margins eglandular. **Pedicels** 0.5–2(–5) mm, scarcely longer in fruit, ascending, not bent at apex. **Flowers:** sepals erect, ± reflexed at tip, lanceolate, 3–4 mm, equal, margins minutely glandular, surfaces hairy, hairs intertwined or matted; petals ± reflexed, pink to rose, oblanceolate, (3–)6–7(–8) mm, apex obtuse, sometimes slightly notched; cup white, rim with petal attachments set in deep notches between filaments; stamens exserted; filaments 4–5 mm; anthers rose to purple, dehisced anthers 1.5–2 mm; ovary chambers 6; styles 3, white, (3–)4–4.5(–5) mm, exserted. $2n = 36$.

Flowering Apr–Aug. Grasslands on serpentine soils; of conservation concern; 0–200 m; Calif.

Hesperolinon congestum is known from a narrow band within the Outer Coast Ranges from Marin County south to San Mateo County. It can be distinguished from all other species in the genus by its hairy sepals.

4. SCLEROLINON C. M. Rogers, Madroño 18: 182, figs. 2–6. 1966 • Hard flax [Greek *skleros*, hard, and *linon*, flax, alluding to fruit] [E]

Herbs, annual, glabrous, sometimes glaucous. **Stems** erect, usually unbranched, sometimes branched proximally and corymbosely branched distally. **Leaves** persistent, opposite, or distal sometimes alternate; stipular glands absent; blade oblong to elliptic, margins entire, or distal leaves sometimes serrate. **Inflorescences** cymes. **Pedicels** articulated. **Flowers:** sepals persistent, 5, connate at base, unequal in size, margins glandular-toothed, glabrous; petals 5, distinct, attached proximal to rim of cup between filament bases, yellow, appendages absent; stamens 5; staminodes 0; pistil 2-carpellate, ovary 4-locular, false septa complete and similar to true septa; styles 2, distinct, or connate at base; stigmas capitate, wider than styles. **Fruits** schizocarps, breaking into 4 nutlets, or indehiscent. **Seeds** 4, narrowly ovate. $x = 6$.

Species 1: w United States.

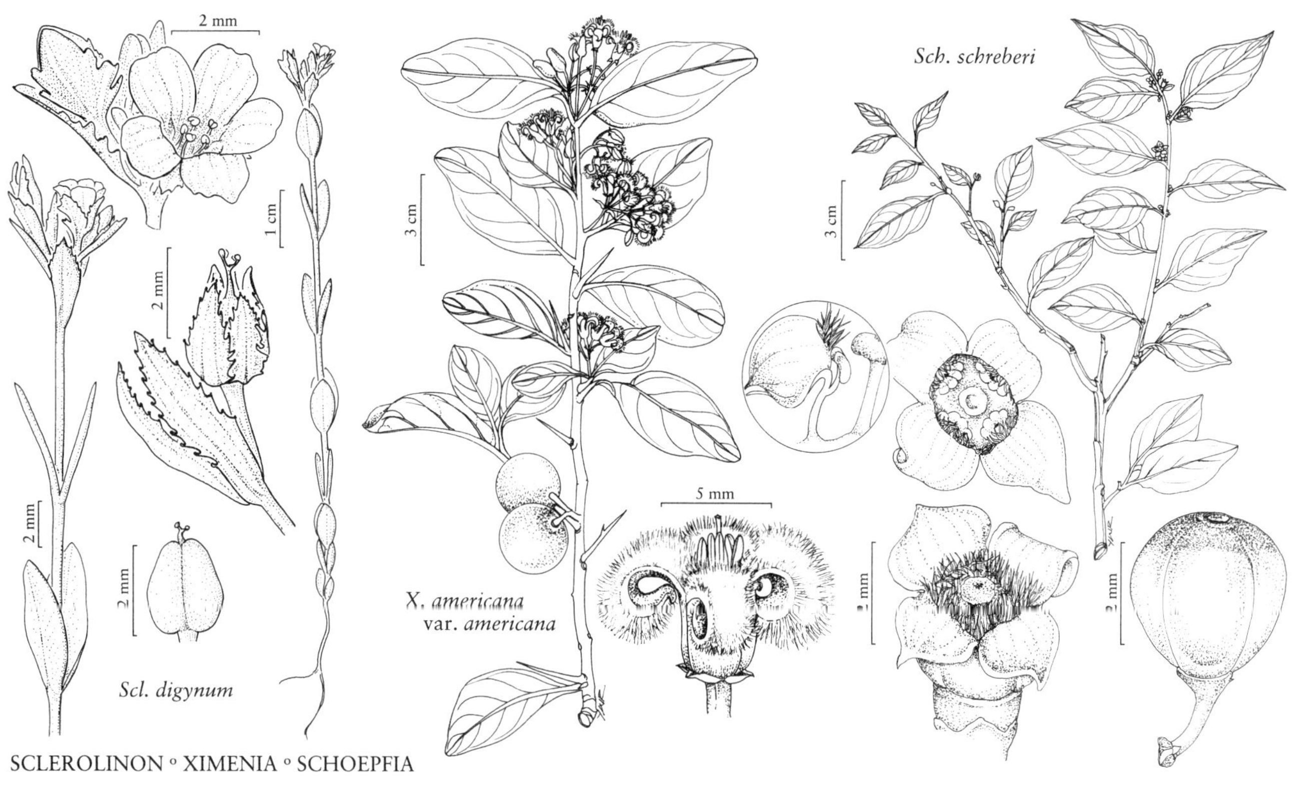

SCLEROLINON ◦ XIMENIA ◦ SCHOEPFIA

1. **Sclerolinon digynum** (A. Gray) C. M. Rogers, Madroño 18: 182. 1966 [E] [F]

Linum digynum A. Gray, Proc. Amer. Acad. Arts 7: 334. 1868; *Cathartolinum digynum* (A. Gray) Small

Herbs 6–20(–42) cm. **Leaves** erect or ascending; blade 5–17(–21) mm, base flat, nearly clasping, margins sessile-glandular, sometimes densely so. **Inflorescences:** bracts usually leaflike, margins glandular-serrate. **Pedicels** ascending, 1–2 mm. **Flowers:** sepals erect, (1.7–)2.3–2.9(–4.5) mm, inner shorter, lance-ovate, apex obtuse to acuminate, outer larger, triangular-ovate, apex obtuse; petals erect, obovate or obcordate, 3–4 mm, apex erose to apiculate, spreading; stamens 1–2 mm, included; anthers yellow, 0.1–0.3 mm; styles yellow, 0.3–0.7 mm, included. **Capsules** pyriform, broadly 4-angled, 1.6–2 × 1.4–1.9 mm, apex truncate, mucronate. **Seeds** dark brown to black, triangular in cross section. $2n$ = 8, 12.

Flowering Jun–Aug. Vernally moist meadows; 300–1900 m; Calif., Idaho, Oreg., Wash.

Sclerolinon digynum occurs along the Washington and Idaho border, the Cascade and Klamath ranges in Oregon and California, and the northern Sierra Nevada; it is disjunct in the Yosemite Valley region of the Sierra Nevada of California. The species has not been collected in Idaho for nearly 100 years; it may be extirpated from that part of its historic range. In his original description of *Sclerolinon*, Rogers considered *S. digynum* to be the most highly specialized species of Linaceae in North America. Chromosome numbers reported for *S. digynum* are $2n$ = 8 (P. H. Raven 1959, as *Cathartolinum digynum*) and $2n$ = 12 (C. M. Rogers et al. 1972); either would be the lowest count in the family (Rogers et al.).

XIMENIACEAE Horaninow • Hog Plum Family

Daniel L. Nickrent

Shrubs or trees, root parasites, evergreen, synoecious. **Leaves** alternate, simple; stipules absent; petiole present; blade margins entire; venation pinnate. **Inflorescences** axillary or at ends of short shoots (brachyblasts), umbels, subumbellate cymes, or fascicles [flowers solitary]. **Flowers** bisexual [functionally unisexual]; perianth and androecium hypogynous; hypanthium absent; sepals 4(–5), distinct, valvate; petals 4(–5), distinct, adaxial surface hairy [not hairy]; nectary present [absent]; stamens 4(–5) or 8 in 2 whorls, distinct, free; anthers dehiscing by longitudinal slits; pistil 1, [2–]4-carpellate, ovary superior, [2–]4-locular proximally, 1-locular distally, placentation free-central, pendulous; ovule 1 per locule, anatropous; style 1; stigma 1. **Fruits** drupes. **Seeds** 1 per locule.

Genera 4, species 13 (1 in the flora): nearly worldwide.

The genera in Ximeniaceae other than *Ximenia* are monospecific. The association of these four genera was first recognized by L. van den Oever (1984) following wood anatomical studies. This clade was also recovered in a cladistic analysis of 80 micro- and macromorphologic characters (V. Malécot et al. 2004). Members of Ximeniaceae share several anatomical features as well as umbellate inflorescences, 4-merous flowers, stamens in two whorls, and lipid-rich fruits. A clade containing three of the four genera was strongly supported in the molecular phylogenetic analysis by Malécot and D. L. Nickrent (2008), and this group was recognized as a family by Nickrent et al. (2010).

1. XIMENIA Linnaeus, Sp. Pl. 2: 1193. 1753; Gen. Pl. ed. 5, 500. 1754 • Hog plum [For Francisco Ximenes de Luna, 17th century Franciscan monk and botanist]

Shrubs or small trees, long shoots vegetative, short shoots fertile, arising from leaf axils of long shoots, each paired with a thorn. **Stems** glabrous. **Leaves** densely fascicled on short shoots, subcoriaceous, surfaces glabrous or puberulent. **Inflorescences:** bracts 0 or 2–4 at pedicel bases.

Pedicels present. **Flowers:** sepals minute, not accrescent in fruit; petals glabrous or puberulent abaxially, densely hairy adaxially; ovary elongate-conic or lanceoloid. **Drupes** yellow, orange, pink, or red, ellipsoid, oblong-ovoid, or globose. $x = 12$.

Species 10 (1 in the flora): Florida, Mexico, West Indies, Central America, South America, Asia, Africa, Indian Ocean Islands, Pacific Islands, Australia; subtropical and tropical regions.

Fruits of *Ximenia americana* and *X. caffra* Sonder are eaten either raw or cooked. In India, oil from the seeds of *X. americana* is used as a ghee substitute and the wood is used in place of sandalwood (see R. A. DeFilipps 1968 for other economic applications). Anticancer compounds known as ribosome-inactivating proteins have been found in *X. americana* (C. Voss et al. 2006). Long chain acetylenic acids in that species showed potential pesticidal activity (M. O. Fatope et al. 2000).

SELECTED REFERENCE DeFilipps, R. A. 1968. A Revision of *Ximenia* [Plum.] L. (Olacaceae). Ph.D. dissertation. Southern Illinois University, Carbondale.

1. Ximenia americana Linnaeus, Sp. Pl. 2: 1193. 1753 • Hog or Spanish or spiny plum, tallow wood [F]

Varieties 3 (1 in the flora): Florida, Mexico, West Indies, Central America, South America, Asia, Africa, Indian Ocean Islands, Pacific Islands, Australia.

1a. Ximenia americana Linnaeus var. **americana** [F]

Shrubs or small trees to 12 m. **Leaves:** blade elliptic, lanceolate, ovate, obovate, or orbiculate, 1.3–10 cm, apex retuse, obtuse, or acute, with or without 0.5–1 mm mucro; venation eucamptodromous. **Inflorescences** 2–10-flowered; peduncles 1–15 mm. **Pedicels** 4–12 mm. **Flowers:** sepals 0.5–4 mm, ciliate; petals yellow, pale yellow, yellowish green, or white, 4.5–12 mm, recurved at maturity; stamen filaments 2.5–6 mm; anthers 1.5–4.5 mm; style 2.5–5.5 mm. **Drupes** 1–3.5 × 1.1–3 cm. **Seeds** 1.5–2.5 × 1.1–2 cm. $2n = 24$.

Flowering Apr–May(–Nov); fruiting year-round. Pinelands, hammock margins, coastal scrub, coastal sand dunes; 0–30 m; Fla.; Mexico; West Indies; Central America; South America; Asia; Africa; Indian Ocean Islands; Pacific Islands; Australia.

R. A. DeFilipps (1968, 1969) recognized three varieties of *Ximenia americana*, with the pantropical var. *americana* occurring in peninsular Florida. The other varieties occur in Argentina (var. *argentinensis* DeFilipps) and Africa (var. *microphylla* Welwitsch ex Oliver). From greenhouse pot studies, DeFilipps (1969b) determined that *X. americana* is able to exist without a host, thus it should be considered a facultative hemiparasite. The flowers are fragrant and presumably insect pollinated. The anthers dehisce before the flower bud opens and the adaxial hairs on the petals may serve to present the pollen (P. B. Tomlinson 1980). Variety *americana* shows different growth forms on different substrates; plants on sandy coastal areas are sprawling shrubs with orbiculate, fleshy leaves, whereas plants in forests and scrublands are trees with oblong to oblanceolate, thin leaves.

Ximenia inermis Linnaeus, an illegitimate and superfluous name, pertains here.

SCHOEPFIACEAE Blume

• Whitewood Family

Daniel L. Nickrent

Shrubs or trees [perennial herbs], root parasites, evergreen, synoecious; glabrous [hairy]. **Leaves** alternate [2 in fascicle], simple; stipules absent; petiole present; blade margins entire; venation pinnate. **Inflorescences** axillary [terminal], cymes [thyrses, spikes, umbels] or flowers solitary. **Flowers** bisexual, heterostylous; perianth and androecium epigynous; epicalyx present; hypanthium completely adnate to ovary; sepals unknown number [0], connate into cuplike rim; petals [3–]4–5[–6], connate, post-staminal hairs present [absent]; nectary present; stamens [3–]4–5 [–6], opposite petals, distinct, adnate to corolla tube; anthers dehiscing by longitudinal slits; pistil 1, 2–3-carpellate, ovary inferior, 2–3-locular proximally, 1-locular distally, placentation free-central, ovules 3 per ovary, anatropous; style 1; stigma 1. **Fruits** drupes [nutlike achenes]. **Seeds** 1 per fruit.

Genera 3, species 50 (1 in the flora): Florida, Mexico, West Indies, Central America, South America, Asia.

Although *Schoepfia* is often classified in Olacaceae R. Brown, its placement has been dubious for over two centuries. Early molecular phylogenetic studies showed that this genus is more closely related to Loranthaceae Jussieu and Misodendraceae J. Agardh than to Olacaceae (D. L. Nickrent and V. Malécot 2001). Later work (J. P. Der and Nickrent 2008) confirmed this result and also showed the association with *Arjona* Commerson ex Cavanilles and *Quinchamalium* Molina from Andean South America, in agreement with P. van Tieghem (1896). The cupular bracts (epicalyx) subtending the flowers may be a synapomorphy between *Schoepfia* and *Quinchamalium.* These three genera form Schoepfiaceae in the classification of Santalales by Nickrent et al. (2010).

The homology of the reduced calyx in Schoepfiaceae and other Santalales, which often is referred to as a calyculus, has been controversial. D. L. Nickrent et al. (2010) and J. Kuijt (2013) argued that this structure is sepalar in origin, and that interpretation is followed here. The corolla in Schoepfiaceae and many other Santalales bears post-staminal hairs, which arise as a tuft opposite each anther and become attached to the anther by secretions.

1. SCHOEPFIA Schreber, Gen. Pl. 1: 129. 1789 • Whitewood [For Johann David Schoepf, 1752–1800, German physician and botanist]

Shrubs or trees, branching sympodial on distal shoots, glabrous [hairy]. **Leaves:** blade subcoriaceous, surfaces glabrous. **Inflorescences:** peduncle base with persistent imbricate bracts. **Pedicels** absent [present]. **Flowers** sweetly fragrant; epicalyx 3-lobed; nectary annular, fleshy, covering ovary distally; corolla cylindric, subcampanulate, or urceolate, lobes reflexed; filaments short, arising proximal to post-staminal hairs. **Drupes** subtended by persistent epicalyx, with remains of nectary and corolla at apex. $x = 12$.

Species 25 (1 in the flora): Florida, Mexico, West Indies, Central America, South America, Asia.

Twenty species of *Schoepfia* are Neotropical (H. Sleumer 1984) and are classified in sect. *Schoepfia*; five species are in sects. *Schoepfiopsis* (Miers) Engler and *Alloschoepfia* Sleumer of Asia (Sleumer 1935; K. R. Robertson 1982).

1. Schoepfia schreberi J. F. Gmelin, Syst. Nat. 1: 376. 1791 [F]

Schoepfia chrysophylloides (A. Richard) Planchon

Shrubs or trees, 1.3–7(–9) m; bark whitish, corky, fissured; branches striate, olive green to whitish, slender. **Leaf blades** lanceolate, ovate, or elliptic, 4–8 × 2–4 cm, brittle, base cuneate-attenuate to obtuse, apex acuminate, both surfaces shiny, ± tuberculate; venation brochidodromous, midrib sunken on both surfaces, lateral veins 4–6 pairs. **Inflorescences** 1–2 per axil, each a 2–3-flowered cyme or solitary flower; peduncle 1–4 mm. **Flowers** of two forms, pin and thrum; calyx rim entire or slightly lobed, 1 mm; corolla tube yellow or orange and lobes pink to red, urceolate to cylindric, 4.5 mm (pin flowers), cylindric to campanulate, (3.5–)4(–5) mm (thrum flowers); ovary 2-locular in 4-merous flowers, 3-locular in 5-merous flowers; stigma 2–3-lobed, at or distal to included anthers (pin flowers) or proximal to exserted anthers (thrum flowers). **Drupes** pink, orange, or red, subovoid to ellipsoid, (7–)10–13 × (6–)7–8 mm, with persistent rim of corolla at apex.

Flowering Oct–Mar; fruiting Oct–Mar, soon after flowering. Pinelands, coppices, hammocks, limestone and sand substrates; 0–10 m; Fla.; West Indies; n South America.

The species concept used here follows the broad view of H. Sleumer (1984), who considered many of the named variants of *Schoepfia schreberi* to be synonyms. Although they display variable vegetative morphology, flower and fruit morphology appears relatively constant. The floral biology of *S. schreberi* requires investigation. Sleumer indicated that anthers in pin flowers lack pollen; however, P. B. Tomlinson (1980) stated that the heterostylous condition is apparently efficient, given that trees of both forms set fruit. Root parasitism of ten host species was documented by C. R. Werth et al. (1979).

COMANDRACEAE Nickrent & Der • Bastard Toadflax Family

Daniel L. Nickrent

Herbs or subshrubs, perennial, root parasites, deciduous, synoecious or andromonoecious; glabrous. **Stems** horizontal rhizomes bearing erect fertile shoots. **Leaves** alternate, simple; stipules absent; petiole present or absent; blade margins entire; venation pinnate. **Inflorescences** unisexual or bisexual, axillary cymules or terminal thyrses. **Flowers** bisexual or unisexual; perianth and androecium epigynous; hypanthium completely adnate to ovary or adnate to ovary proximally, free distally; sepals 0; petals (4–)5(–7), usually connate basally, sometimes distinct, valvate, post-staminal hairs present; nectary present; stamens (4–)5(–7), opposite petals, distinct, free; anthers dehiscing by longitudinal slits; pistil 1, 1-carpellate, ovary inferior, 1-locular, placentation free-central, pendulous; ovules 2–4 per locule, anatropous; style 1; stigma 1. **Fruits** pseudodrupes (mesocarp hard, exocarp leathery or fleshy). **Seeds** 1 per fruit.

Genera 2, species 2 (2 in the flora): North America, Mexico, Eurasia.

Comandraceae have not been recognized at the family rank in past classifications, although *Comandra* was placed in its own invalidly published tribe Comandreae by P. van Tieghem (1896). *Comandra* and *Geocaulon* have a suite of generalized morphological features that can be found in many other genera of Santalaceae in the broad sense. Despite this, *Comandra* has a number of distinctive embryological features (M. Ram 1957; B. M. Johri and S. P. Bhatnagar 1960). A molecular phylogenetic analysis recovered a strongly supported *Comandra* and *Geocaulon* clade, which was recognized as a family by D. L. Nickrent et al. (2010). That clade was placed in a polytomy with other Santalaceae in the broad sense in that study. *Comandra* and *Geocaulon* are clearly closely related as revealed by molecular studies; however, M. L. Fernald (1928e) made a good case for maintaining them as separate genera. They differ in a number of features involving the rhizome, plant sex, hypanthium, disc shape, style length, and fruit type.

1. Inflorescences terminal thyrses; flowers bisexual; hypanthia free distally, funnel-shaped; nectaries lining hypanthium; styles filiform; pseudodrupe exocarps leathery . . . . 1. *Comandra*, p. 409
1. Inflorescences axillary cymules; flowers bisexual and staminate (plants andromonoecious); hypanthia completely adnate to ovaries; nectaries nearly flat; styles short-conic; pseudodrupe exocarps fleshy . . . . 2. *Geocaulon*, p. 411

1. COMANDRA Nuttall, Gen. N. Amer. Pl. 1: 157. 1818 • Bastard toadflax [Greek *kome*, hair, and *andros*, male, alluding to petal hairs that attach to anthers]

Herbs or subshrubs, perennial, synoecious. **Rhizomes** somewhat woody, white to beige or blue (then drying blackish), cortex corky or papery, loose exfoliating. **Leaves:** petiole short or absent. **Inflorescences** terminal, paniclelike or corymblike thyrses; cymules 3–5-flowered; prophyllar bracteole subtending each flower persistent. **Pedicels** present. **Flowers** bisexual, campanulate; hypanthium adnate to ovary proximally, free distally, funnel-shaped; petals (4–)5(–7), white, yellowing with age, ovate or oblong to lanceolate, reflexed upon maturation; nectary lining hypanthium, lobes small, alternating with filaments; styles filiform; stigmas capitate. **Pseudodrupes** usually multiple, petals persistent, forming neck at apex; exocarp leathery.

Species 1: North America, n Mexico, s Europe (Balkan peninsula); temperate regions.

Circumscription of species within *Comandra* has varied, as have opinions about whether *Geocaulon* is distinct. C. L. Hitchcock and A. Cronquist (1973) considered *Geocaulon* as a species of *Comandra*, whereas M. L. Fernald (1950) recognized separate genera. The treatment here follows the most comprehensive study of *Comandra* to date (M. A. Piehl 1965), which recognized a single variable species with four subspecies.

SELECTED REFERENCE Piehl, M. A. 1965. The natural history and taxonomy of *Comandra* (Santalaceae). Mem. Torrey Bot. Club 22(1): 1–97.

1. Comandra umbellata (Linnaeus) Nuttall, Gen. N. Amer. Pl. 1: 157. 1818 [F]

Thesium umbellatum Linnaeus, Sp. Pl. 1: 208. 1753

Leaf blades light green to grayish or bluish green, lanceolate, elliptic, or ovate, 0.7–5.3 cm, apex obtuse, acute, or acuminate. **Flowers:** hypanthium base not dilated.

Subspecies 4 (3 in the flora): North America, n Mexico, s Europe (Balkan peninsula).

Comandra umbellata is likely the most widespread Santalales species, occurring throughout the United States, southern Canada, and northern Mexico as well as in the Balkan peninsula, where subsp. *elegans* (Rochel ex Reichenbach) Piehl occurs. *Comandra umbellata* is the alternate host for comandra blister rust (*Cronartium comandrae*), which damages pines in North America.

1. Leaf blades thin, green, not glaucous; pseudodrupes 4–6 mm; rhizome cortices white to beige; Canada, c, e United States. . . . . 1a. *Comandra umbellata* subsp. *umbellata*
1. Leaf blades thin or thick, becoming ± succulent, green to grayish or bluish green, glaucous; pseudodrupes 5–9 mm; rhizome cortices blue, drying blackish; w North America.

[2. Shifted to left margin.—Ed.]

2. Leaf blade lateral veins obscure on abaxial surface; proximal part of aerial stems not overwintering; herbs 5–33 cm. . . . . 1b. *Comandra umbellata* subsp. *pallida*
2. Leaf blade lateral veins apparent on abaxial surface; proximal part of aerial stems overwintering; subshrubs 15–40 cm . . . . 1c. *Comandra umbellata* subsp. *californica*

1a. Comandra umbellata (Linnaeus) Nuttall subsp. **umbellata** [E] [F]

Comandra richardsiana Fernald

Herbs or subshrubs 7–40 cm. **Rhizomes:** cortex white to beige. **Aerial stems** usually branched, sometimes much-branched at base; proximal portions overwintering or not. **Leaves:** blade green, paler abaxially, not glaucous, lanceolate, oblanceolate, elliptic, or ovate, 0.7–5(–7.6) cm, thin, soft, base attenuate to acute, margins often slightly revolute, apex obtuse, sometimes apiculate; midrib and lateral veins conspicuous, protruding on abaxial surface. **Pedicels** 0–1.4 mm. **Flowers** funnel-shaped to almost rotate; petals lanceolate, lanceolate-oblong, or ovate, 2–3 mm; anthers 0.5 mm. **Pseudodrupes** dark to light brown, sometimes red-tinged, not glaucous, subglobose to globose, 4–6 mm, smooth. $2n = 28$.

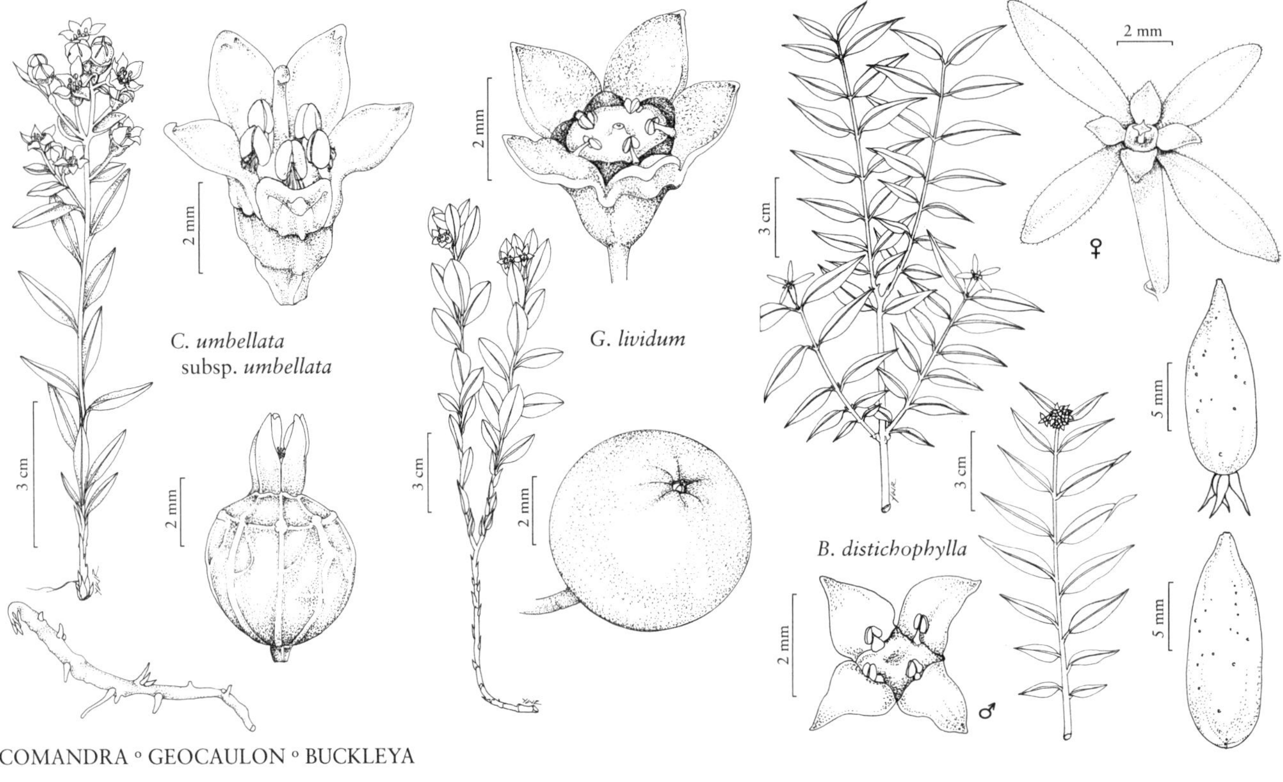

COMANDRA ◦ GEOCAULON ◦ BUCKLEYA

Flowering Mar–Aug. Swamps and bogs, rich mesic sites, dry, sandy or rocky soils, savannas, early successional forests; 0–700 m; Alta., B.C., Man., N.B., Nfld. and Labr., N.S., Ont., P.E.I., Que., Sask.; Ala., Ark., Conn., Del., D.C., Ga., Ill., Ind., Iowa, Kans., Ky., Maine, Md., Mass., Mich., Minn., Miss., Mo., Nebr., N.H., N.J., N.Y., N.C., N.Dak., Ohio, Okla., Pa., R.I., S.C., S.Dak., Tenn., Vt., Va., W.Va., Wis.

As discussed by M. A. Piehl (1965), subsp. *umbellata* exhibits a wide range of morphological variation, which has prompted various authors to name numerous species and subspecies. A common example is *Comandra richardsiana*, which according to M. L. Fernald (1950) represents a western and northern extreme form with a corymbose inflorescence with cymule branches ascending versus a paniculate inflorescence with cymule branches divergent. Intermediate forms between subsp. *umbellata* and subsp. *pallida* are reported from Alberta, British Columbia, Manitoba, Saskatchewan, Kansas, Minnesota, Missouri, Nebraska, North Dakota, and South Dakota.

1b. Comandra umbellata (Linnaeus) Nuttall subsp. **pallida** (A. de Candolle) Piehl, Mem. Torrey Bot. Club 22(1): 70. 1965

Comandra pallida A. de Candolle in A. P. de Candolle and A. L. P. P. de Candolle, Prodr. 14: 636. 1857; *C. umbellata* var. *pallida* (A. de Candolle) M. E. Jones

Herbs, 5–33 cm. **Rhizomes:** cortex blue, drying blackish. **Aerial stems** often much branched; proximal portions not overwintering. **Leaves:** blade green to grayish green, glaucous, linear to lanceolate, elliptic, or ovate, 0.9–4.2 cm, thick, firm, becoming ± succulent, base attenuate to cuneate, margins not revolute, apex attenuate to acute, often mucronulate or apiculate; midrib obscure or sometimes conspicuous or protruding on abaxial surface, lateral veins obscure. **Pedicels** 0–1.6 mm. **Flowers** funnel-shaped to rotate; petals oblong to lanceolate-oblong, 2.5–5 mm; anthers 0.7 mm. **Pseudodrupes** purplish brown to red, glaucous, subglobose to ovate, 5.5–9 mm, slightly roughened. $2n = 52$.

Flowering Mar–Jul. Open areas, sandy and rocky slopes, sagebrush communities, coniferous and deciduous forests, deserts; 150–2800 m; Alta., B.C., Man., N.W.T., Sask., Yukon; Ariz., Colo., Idaho, Kans., Mont., Nebr., Nev., N.Mex., N.Dak., Okla., Oreg., S.Dak., Tex., Utah, Wash., Wyo.; Mexico (Chihuahua, Coahuila).

As with the other subspecies, subsp. *pallida* is morphologically variable, this likely related to its diverse ecological habitats. Clinal variation and intergradation with the other subspecies that flank its eastern and western boundaries may also play a role (M. A. Piehl 1965). Intermediates between subsp. *pallida* and subsp. *californica* occur in Arizona, New Mexico, Nevada, Oregon, and Washington. Ecotypic variation apparently has a genetic component, as demonstrated by common garden experiments with seeds and vegetative propagules (Piehl). One extreme but widespread form is highly branched basally and has dimorphic leaves, with the proximal blades linear and the distal blades lanceolate or elliptic.

1c. Comandra umbellata (Linnaeus) Nuttall subsp. **californica** (Eastwood ex Rydberg) Piehl, Mem. Torrey Bot. Club 22(1): 65. 1965 E

Comandra californica Eastwood ex Rydberg, Fl. Rocky Mts. ed. 2, 1138. 1923; *C. umbellata* var. *californica* (Eastwood ex Rydberg) C. L. Hitchcock

Subshrubs, 15–40 cm. **Rhizomes:** cortex blue, drying blackish. **Aerial stems** often much branched; proximal portions overwintering. **Leaves:** blade light green to bluish or grayish green, slightly paler abaxially, glaucous, broadly elliptic, ovate, lanceolate, or linear, 1.7–5.3 cm, thin, becoming ± succulent, base acute to attenuate, margins rarely revolute, apex acute, often apiculate; midrib and lateral veins apparent and somewhat protruding on abaxial surface. **Pedicels** 0–2.5 mm. **Flowers** funnel-shaped to rotate; petals lanceolate, lanceolate-oblong, or ovate, 2–3.5 mm; anthers 0.6 mm. **Pseudodrupes** brown, not glaucous, subglobose, 5–7.5 mm, smooth.

Flowering Mar–Jul. Dry places, mountains, foothills, open conifer forests, oak woodlands, chaparral margins; 300–3000 m; B.C.; Ariz., Calif., Nev., N.Mex., Oreg., Wash.

Subspecies *californica* is frequent in the Sierra Nevada of California and east of the Cascade range in Oregon and Washington; it is found also on Vancouver Island in British Columbia. In Arizona, this subspecies is restricted to high elevations such as the Santa Catalina Mountains. Intergradation with subsp. *pallida* occurs over wide areas of Arizona, Nevada, New Mexico, Oregon, and Washington.

2. GEOCAULON Fernald, Rhodora 30: 23. 1928 • Northern comandra, false toadflax [Greek *ge*, earth, and *kaulos*, stalk, alluding to slightly subterranean and stemlike rhizome] E

Herbs, perennial, andromonoecious. **Rhizomes** not woody, reddish to dark brown, cortex smooth, not exfoliating. **Leaves:** petiole short. **Inflorescences** axillary (appearing terminal in early developmental stages but occupying middle axils with continued growth of main axis), cymules; cymules mostly 3-flowered; prophyllar bracteole subtending each flower caducous. **Pedicels** present. **Flowers** bisexual and staminate (central flower, or rarely 2 flowers, of dichasium usually bisexual, sometimes staminate, laterals staminate), campanulate to turbinate; hypanthium completely adnate to ovary; petals (4–)5, greenish to bronze (bisexual and unisexual flowers often differing in color), triangular or ovate; nectary nearly flat, lobes prominent, alternating with filaments; styles short-conic; stigmas slightly lobed. **Pseudodrupes** usually solitary; petal remains vestigial at apex; exocarp fleshy.

Species 1: North America.

Geocaulon is one of only two genera named by M. L. Fernald (the other being *Alcoceria*, Euphorbiaceae, now treated as a synonym of *Dalembertia* Baillon). Fernald stated that many features of *Geocaulon* are similar to *Nestronia* (Santalaceae); however, molecular data clearly indicate that *Geocaulon* is sister to *Comandra*.

1. Geocaulon lividum (Richardson) Fernald, Rhodora 30: 23. 1928 [E] [F]

Comandra livida Richardson in J. Franklin, Narr. Journey Polar Sea, 734. 1823

Stems: rhizomes 1.5–3 mm; aerial shoots 0.7–3 dm. **Leaf blades** elliptic, oblong, or obovate, 1.5–5 × 0.5–1 cm, apex obtuse to rounded, thin, flaccid, surfaces green, grayish green, or purplish. **Inflorescences:** peduncles 5 mm, expanding to 1.5 cm in fruit. **Pedicels** 1 mm. **Flowers** 4 mm diam.; filaments 0.5 mm; styles 0.3 mm. **Pseudodrupes** yellowish orange to scarlet, 6–10 mm. **Seeds** oily, fleshy. **2*n*** = 52.

Flowering late May–early Aug. Damp humus, *Sphagnum* bogs, wet coniferous forests; 70–2100 m; St. Pierre and Miquelon; Alta., B.C., Man., N.B., Nfld. and Labr., N.W.T., N.S., Nunavut, Ont., Que., Sask., Yukon; Alaska, Idaho, Maine, Mich., Minn., Mont., N.H., N.Y., Vt., Wash., Wis.

Although *Geocaulon lividum* is considered secure across its full range, in parts of the eastern United States it is of special concern (Maine), threatened (New Hampshire), or endangered (New York, Wisconsin).

Fernald described the sexual condition as androdioecious, but F. H. Smith and E. C. Smith (1943) stated that the central flowers of each cymule are pistillate and the laterals staminate, thus the species would be monoecious. Here it is considered andromonoecious, with the central flower (rarely two flowers) bisexual and the lateral staminate (and dropping after anthesis), or sometimes all the flowers staminate.

THESIACEAE Vest
• Thesium Family

Daniel L. Nickrent

Herbs or shrubs, perennial [annual], root parasites, deciduous, synoecious or dioecious. **Leaves** opposite, subopposite, or alternate, simple; stipules absent; petiole present or absent; blade margins entire or minutely serrulate; venation pinnate. **Inflorescences** unisexual or bisexual, terminal or axillary, cymes (sometimes umbel-like) or thyrses, or flowers solitary. **Flowers** bisexual or unisexual; perianth and androecium epigynous; hypanthium completely adnate to ovary, adnate to ovary proximally but free distally, or absent; sepals 0 or 4, distinct, valvate; petals 4–5, distinct [connate], valvate, post-staminal hairs present or absent; nectary present [absent]; stamens 4–5, opposite petals, distinct, free [adnate to petal bases]; anthers dehiscing by longitudinal slits; pistil 1, 3–4-carpellate, ovary inferior, 1-locular, placentation free-central, pendulous; ovules 2–4 per locule, anatropous; style 1; stigma 1. **Fruits** pseudodrupes (mesocarp hard, exocarp leathery or fleshy). **Seeds** 1 per fruit.

Genera 4, species ca. 370 (2 genera, 2 species in the flora): nearly worldwide.

The assemblage of genera comprising Thesiaceae received strong support from molecular phylogenetic analyses (J. P. Der and D. L. Nickrent 2008; Nickrent and M. A. García 2015). *Buckleya* is sister to the remaining genera, which are either endemic to or have their center of diversity in Africa. Because *Buckleya* occurs in eastern North America and Asia, the ancestor to the remaining genera apparently migrated to Africa prior to diversification.

1. Shrubs; leaves opposite or subopposite; flowers unisexual . . . . 1. *Buckleya,* p. 413
1. Perennial herbs; leaves alternate; flowers bisexual . . . . 2. *Thesium,* p. 415

1. BUCKLEYA Torrey, Amer. J. Sci. Arts 45: 170. 1843, name conserved • [For Samuel Botsford Buckley, 1809–1884, American botanist]

Shrubs, dioecious. **Stems** hairy. **Leaves** opposite or subopposite, distichous to decussate. **Inflorescences:** staminate 1–3[–5] dichasia, pistillate flowers solitary, subtended by 2 pairs of decussate, deciduous bracts. **Pedicels:** staminate present, pistillate absent. **Flowers** unisexual.

Staminate flowers: hypanthium absent; sepals 0; petals 4, post-staminal hairs absent; nectary squarish; stamens 4. **Pistillate flowers:** hypanthium completely adnate to ovary; sepals 4; petals 4; nectary squarish; ovules 3–4 per locule; stigma 4-lobed. **Pseudodrupes:** exocarp fleshy; pedicels not enlarging or becoming fleshy; sepals accrescent, deciduous [persistent], petals deciduous. x = 15.

Species 5 (1 in the flora): e United States, e Asia (China, Japan).

The morphologic nature of the two floral whorls appearing at the fruit apex has received various interpretations. R. K. F. Pilger (1935) referred to the outer whorl as bracts and suggested that they are prophylls from two aborted side flowers of a dichasium. A simpler explanation, followed here, is that the outer whorl represents sepals and the inner whorl petals (F. H. Smith and E. C. Smith 1943; J. P. Der and D. L. Nickrent 2008).

Buckleya is disjunct between eastern North America and Asia. Based on morphologic features such as the deciduous sepals, W. N. Carvell and W. H. Eshbaugh (1982) proposed that *B. distichophylla* is most closely related to *B. graebneriana* Diels of China, a hypothesis supported by the molecular phylogenetic analysis of Li J. H. et al. (2001).

SELECTED REFERENCES Carvell, W. N. and W. H. Eshbaugh. 1982. A systematic study of the genus *Buckleya* (Santalaceae). Castanea 47: 17–37. Li, J. H., D. E. Boufford, and M. J. Donoghue. 2001. Phylogenetics of *Buckleya* (Santalaceae) based on ITS sequences of nuclear ribosomal DNA. Rhodora 103: 137–150.

1. **Buckleya distichophylla** (Nuttall) Torrey, Amer. J. Sci. Arts 45: 170. 1843 • Buckleya, pirate-bush [E] [F]

Borya distichophylla Nuttall, Gen. N. Amer. Pl. 2: 232. 1818; *Nestronia distichophylla* (Nuttall) Kuntze

Shrubs rhizomatous, to 3–4 m. **Bark** light brown, smooth, with white lenticels. **Stems** green, terete to slightly 4-angled when young, puberulent; growth sympodial, terminal bud sometimes aborting giving false appearance of dichotomous branching. **Leaves:** petiole short or absent; blade lanceolate or ovate-lanceolate, to elliptic near base of branchlet, 1.2–9 × 0.3–2.9 cm distally, progressively smaller proximally, thin, base cuneate, margins entire, apex attenuate, surfaces puberulent especially on margins and midvein. **Inflorescences** terminal. **Pedicels:** staminate 1–4.2 mm. **Staminate flowers:** petals greenish, 1.2–2.5 × 1.1–2 mm, puberulent; nectary 1–1.8 mm diam. **Pistillate flowers:** sepals greenish, narrowly elliptic to lanceolate, 3–17 × 1–3 mm, apex acuminate, puberulent, venation conspicuously pinnate-reticulate; petals greenish, triangular, 1.3–3 × 1–22 mm, puberulent; ovary narrowly conic, 3.4–15 mm, puberulent; styles 0.7–1.6 mm. **Pseudodrupes** ellipsoid to obovoid, 0.9–3 × 0.4–1.7 cm, puberulent, with small whitish lenticels. $2n$ = 30.

Flowering Apr–May, Aug. Dry rocky or shaly outcrops and river bluffs; 500–1000 m; N.C., Tenn., Va.

Buckleya distichophylla is an uncommon shrub in the southern Appalachian Blue Ridge Mountains and adjacent Ridge and Valley region; the reason for its rarity is not understood. One hypothesis is that the plants require both direct sunlight and hemlock (*Tsuga canadensis*) hosts, a rare combination of conditions (A. E. Radford et al. 1968). But other explanations are called for, because *Buckleya* is found in stands lacking hemlock (W. N. Carvell and W. H. Eshbaugh 1982). Moreover, pot-grown plants showed seedling parasitism of 19 different tree species (L. J. Musselman and W. F. J. Mann 1979b); these authors suggested that *Buckleya* requires a distinctive habitat where hemlock happens to be dominant.

Buckleya distichophylla is in the Center for Plant Conservation's National Collection of Endangered Plants. It is not federally listed, but three states (North Carolina, Tennessee, Virginia) list it as threatened or endangered.

2. THESIUM Linnaeus, Sp. Pl. 1: 207. 1753; Gen. Pl. ed. 5, 97. 1754 • [Greek *thes*, laboring servant, alluding to simple appearance] [I]

Herbs [**shrubs**], perennial [annual], synoecious [dioecious]. **Stems** erect, hairy [glabrous]. **Leaves** alternate [opposite]. **Inflorescences** racemelike [spikelike] thyrses; flowers subtended by bract and 2 bracteoles [bract absent]. **Pedicels** absent [present]. **Flowers** bisexual [unisexual]; hypanthium adnate to ovary proximally, free distally; sepals 0 [present as small lobes or glands]; petals (4–)5, post-staminal hairs present; nectary lining hypanthium; stamens (4–)5; ovules 2–3 per locule; stigma capitate. **Pseudodrupes:** exocarp leathery; pedicel enlarging and becoming fleshy; hypanthium and petals persistent. x = 6–9.

Species ca. 365 (1 in the flora): introduced; nearly worldwide.

Thesium is closely related to the South African endemic *Osyridocarpos* A. de Candolle (J. P. Der and D. L. Nickrent 2008). For this reason and because Africa contains the highest species diversity, this region can be assumed to be the center of origin for *Thesium*.

1. **Thesium ramosum** Hayne, J. Bot. (Schrader) 1800(1): 30, plate 7 [left]. 1800 [F] [I] [W]

Herbs 15–40(–70) cm. **Stems:** caudex relatively unbranched, to 15 mm diam.; aerial stems 20–100+, to 2 mm diam. **Leaves:** blade scalelike proximally, linear to falcate distally, 30–45 (–91) × 1–3 mm; margins minutely serrulate; midvein prominent, lateral veins less prominent or absent. **Inflorescences** 10–20 cm; bract lanceolate, 1.2–3 times flower length; bracteoles linear, 0.5–1 times flower length. **Flowers** bell- or funnel-shaped, 3–4 mm; hypanthium 1 mm; petals white with green central adaxial stripe, 1–1.3 mm, margins with small lobes; stamens 1.5 mm, exserted; style plus stigma 1.5 mm. **Pseudodrupes** brown, 2–3 mm, longitudinally striate, glabrous; swollen pedicel reddish brown or amber, 1.5–2 mm; persistent perianth remnants 1–2 mm.

Flowering summer. Riparian areas, dry, rocky slopes, roadsides, pastureland, open fields; 400–1800 m; introduced; Alta.; Idaho, Mont., N.Dak.; Eurasia.

The native range of *Thesium ramosum* is in temperate Eurasia from the Balkans to Siberia (R. Hendrych 1964; A. G. Miller 1982). The species was first reported in North America in 1943 from Towner County, North Dakota. An herbarium specimen collected in 1993 (at RM) documents its occurrence in Teton County, Idaho. Later it was recorded also from Madison and Teton counties, Montana (L. J. Musselman and S. C. Haynes 1996), and in 2001 discovered near Calgary, Alberta, where it is well established in and around Fish Creek Provincial Park (I. D. MacDonald, pers. comm.). The historical data suggest that this species has potential for significant range expansion within North America. The diversity of hosts recorded for the Montana population indicates that it is a host generalist.

The North American plants were originally reported to be *Thesium linophyllon* (T. Van Bruggen 1986), but L. J. Musselman and S. C. Haynes (1996) determined that they are either *T. arvense* Horvátovszky or *T. longifolium* Turczaninow. A DNA sequence from the Montana population was identical to that of *T. ramosum* from Eurasia (M. García, pers. comm.). The North American populations differ from the Eurasian ones in bract size, persistent perianth remnant length relative to the pseudodrupe length, and fleshy pedicel color, but the significance of these features in this polymorphic taxon is not clear.

Thesium ramosum is called *T. arvense* in many European and North American publications, all following the recommendation of R. Hendrych (1961). That name, as pointed out by W. Gutermann (2009), is a superfluous new one for a Linnaean species and is therefore illegitimate.

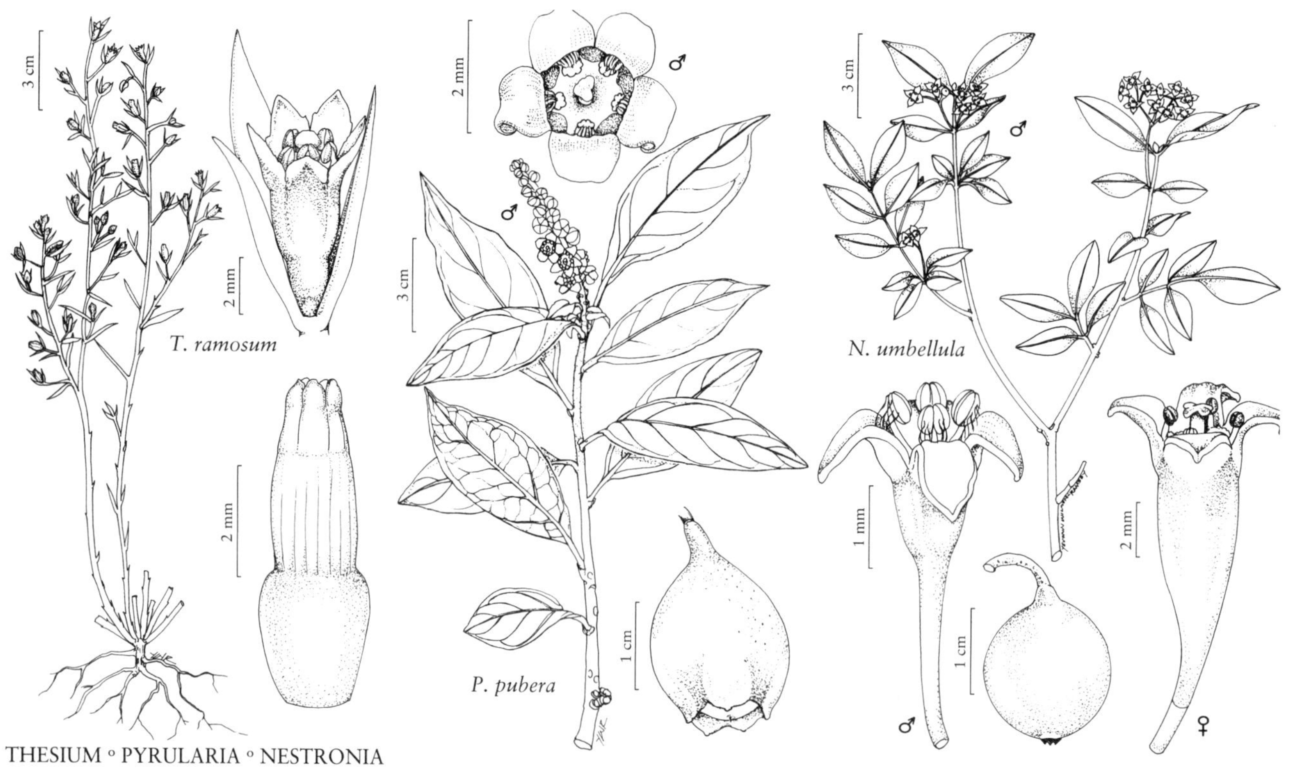

THESIUM ◦ PYRULARIA ◦ NESTRONIA

CERVANTESIACEAE Nickrent & Der • Buffalo Nut Family

Daniel L. Nickrent

Shrubs [**trees**], root parasites, deciduous, dioecious [synoecious, polygamous, trioecious]. **Leaves** alternate, simple; stipules absent; petiole present; blade margins entire [spinose]; venation pinnate. **Inflorescences** unisexual, terminal or axillary, racemes [spikes, cymes, panicles]. **Flowers** unisexual [bisexual]; perianth and androecium perigynous or epigynous; hypanthium completely adnate to ovary; sepals 0; petals (4–)5(–6), distinct, post-staminal hairs present; nectary present; stamens (4–)5(–6), opposite petals, distinct, adnate to petal base; anthers dehiscing by longitudinal slits; staminodes present in pistillate flowers; pistil 1, carpel number unknown; ovary ½ inferior (staminate flowers) or almost superior (pistillate flowers), becoming inferior during fruit development, 1-locular, placentation free-central, pendulous; ovules 2–3 per locule, anatropous; style 1; stigma 1; pistillode present in staminate flowers. **Fruits** pseudodrupes (mesocarp hard). **Seeds** 1 per fruit.

Genera 8, species 20 (1 in the flora): e, se United States, South America, Asia, Africa.

Cervantesiaceae show a trend from bisexual to fully unisexual flowers, with intermediate stages seen by the presence of staminodes and pistillodes, as well as trioecy in *Scleropyrum* Arnott. Carpel number in Cervantesiaceae is unclear, and the necessary anatomical studies have not been done. The fruits in this family (the largest in Santalales) have a hard or crustaceous mesocarp. Because the mesocarp is hard, not the endocarp, these are considered pseudodrupes, not true drupes. In *Jodina* Hooker & Arnott ex Meisner, the endocarp is apparently consumed during endosperm development (S. P. Bhatnagar and G. Sabharwal 1969); this interpretation is extended to other members of the family.

The affinity of the eight genera of Cervantesiaceae, first noted by H. U. Stauffer (1957, 1961), was confirmed using molecular phylogenetic methods (J. P. Der and D. L. Nickrent 2008; Z. S. Rogers et al. 2008). The segregation of Cervantesiaceae from Santalaceae in the broad sense follows the classification of Nickrent et al. (2010). *Pyrularia* is the only genus in the family with species occurring in both the Old and New Worlds.

1. PYRULARIA Michaux, Fl. Bor.-Amer. 2. 231. 1803 • Buffalo or oil nut [Genus *Pyrus* and Latin *-aria*, connecting, alluding to pear-shaped fruit]

Shrubs [trees], dioecious [polygamous]. **Leaf blades** herbaceous; venation brochidodromous, conspicuous. **Pedicels** present. **Flowers:** petals recurved [spreading]; nectary prolonged into scales between filament bases; styles long and cylindric (staminate flowers) or short and conic (pistillate flowers); stigma capitate, 2–3-lobed. **Pseudodrupes** crowned with expanded disc surrounded by swollen petal bases.

Species 2 (1 in the flora): e, se United States, Asia.

The sexual condition in *Pyrularia* is not clear, which is reflected in varied descriptions. The genus has been called dioecious (M. L. Fernald 1950), subdioecious (H. A. Gleason and A. Cronquist 1991), and polygamous (Xia N. H. and M. G. Gilbert 2003). The anthers and filaments in *P. pubera* are smaller on pistillate flowers, do not contain pollen, and are considered staminodes. Thus, at least this species is functionally dioecious. The polygamous condition reported for the Chinese species *P. edulis* (Wallich) A. de Candolle requires further investigation.

1. Pyrularia pubera Michaux, Fl. Bor.-Amer. 2: 233. 1803

E F

Shrubs rhizomatous, much branched, to 4 m; young growth minutely pilosulous. **Petioles** (5–)10(–19) mm. **Leaf blades** ovate-oblong, obovate, or elliptic, (4.2–)10(–21) × (2–)4(–8) cm, base acute to rounded, apex acute to acuminate, surfaces puberulent when young. **Inflorescences:** staminate terminal on axillary branches, erect, 3–8 cm, 15+-flowered; pistillate terminal or axillary, to 9-flowered; bracts caducous, pilose. **Staminate flowers** green, turbinate, 4 mm diam.; pistillode stigma above anthers. **Pistillate flowers** green, turbinate, 5–6 mm diam.; stigma at same height as staminode anthers. **Pseudodrupes** yellowish, pyriform or subglobose, 2–3 × 1–2 cm; exocarp splitting irregularly when mature, releasing mesocarp/seed. $2n = 38$.

Flowering May–Jul; fruiting Sep–Oct. Rich forests; 200–1400 m; Ala., Ga., Ky., N.Y., N.C., Pa., S.C., Tenn., Va., W.Va.

Pyrularia pubera can be locally abundant in the Blue Ridge Mountains and Appalachian Plateau, often forming dense stands in second-growth forests. The species is apparently a host generalist (D. J. Leopold and R. N. Muller 1983) and has been reported to parasitize planted fir trees (*Abies fraseri*) in Virginia (L. J. Musselman and S. C. Haynes 1996). The seeds are very high in oil. Cytotoxic and antimicrobial peptides called thionins are present in *P. pubera* (L. P. Vernon et al. 1985).

SANTALACEAE R. Brown
• Sandalwood Family

Daniel L. Nickrent

Shrubs or trees, root parasites [stem parasites], evergreen or deciduous, synoecious or dioecious [monoecious, andromonoecious]. **Leaves** opposite [alternate, whorled], simple; stipules absent; petiole present; blade margins entire; venation pinnate. **Inflorescences** unisexual or bisexual, axillary or terminal, thyrses or umbels [spikes, racemes, cymes, panicles, fascicles], or flowers solitary. **Flowers** bisexual or unisexual; perianth and androecium perigynous or epigynous; hypanthium adnate to ovary proximally, free distally [completely adnate to ovary]; sepals 0; petals 3–4(–5), distinct, valvate, post-staminal hairs present or absent; nectary present [absent]; stamens 3–4(–5), opposite petals, distinct, free; anthers dehiscing by longitudinal slits; staminodes present in pistillate flowers; pistil 1, 1–3-carpellate, ovary ½ inferior or inferior, 1–3-locular proximally, 1-locular distally, placentation free-central, pendulous; ovules 2–4 per ovary, anatropous; style 1; stigma 1. **Fruits** pseudodrupes (mesocarp hard). **Seeds** 1 per fruit.

Genera 11, species 67 (2 genera, 2 species in the flora): nearly worldwide.

This circumscription of Santalaceae in the strict sense derives from the molecular phylogenetic results of J. P. Der and D. L. Nickrent (2008). That study yielded a strongly supported clade of 11 genera that contains a diverse array of life forms from root parasitic trees and shrubs to aerially parasitic mistletoes, including three genera that previously have been segregated as Eremolepidaceae. *Nestronia*, the only genus of Santalaceae native to North America, is sister to *Colpoon* P. J. Bergius and *Rhoiacarpos* A. de Candolle, both South African endemics. These three genera, as well as *Osyris* Linnaeus, share a number of morphologic features, including habit, epigynous flowers, anther morphology, presence of nectaries, and fruit type; they previously have been classified in Santaleae A. de Candolle (= Osyrideae Reichenbach). Using phylogenetic trees calibrated for time with fossil evidence, it appears that *Nestronia* diverged from its Old World relatives in the Eocene.

SELECTED REFERENCE Der, J. P. and D. L. Nickrent. 2008. A molecular phylogeny of Santalaceae (Santalales). Syst. Bot. 33: 107–116.

1. Rhizomatous shrubs, to 1 m; inflorescences axillary, umbels or flowers solitary; petals white to greenish; se United States north of Florida . . . . 1. *Nestronia*, p. 420
1. Trees, to 9 m; inflorescences axillary and terminal, thyrses; petals rose to crimson red or maroon; Florida . . . . 2. *Santalum*, p. 420

1. NESTRONIA Rafinesque, New Fl. 3: 12. 1838 • Leachbrush [Greek *knestron*, name for *Daphne*] [E]

Darbya A. Gray

Shrubs, deciduous, probably dioecious. **Stems** glabrous. **Leaves:** petiole short. **Inflorescences** axillary, pedunculate umbels (staminate) or flowers solitary (pistillate). **Pedicels** present. **Flowers** unisexual; perianth and androecium epigynous; hypanthium in staminate flowers turbinate or campanulate, in pistillate flowers a short tube; petals 3–4(–5), post-staminal hairs present (staminate flowers) or absent (pistillate flowers); nectary lining hypanthium, wavy or slightly lobed distally; stamens 3–4(–5), in staminate flowers exserted, in pistillate flowers staminodial, ± included; ovary inferior, 3-locular proximally, 1-locular distally; ovules 2–3; styles in pistillate flowers conic; stigmas 3–4-lobed. **Pseudodrupes** spheric, without prominent perianth remnants at apex.

Species 1: se United States.

The sexual condition of *Nestronia* has been reported to be dioecious (A. E. Radford et al. 1968) or polygamodioecious (J. K. Small 1933; R. K. F. Pilger 1935; H. A. Gleason and A. Cronquist 1991). Flowers referred to as pistillate or bisexual possess staminal structures. Flowers from Virginia that have gynoecia were examined and had small anthers that lacked pollen. Therefore, these structures are staminodes, and the species is dioecious. Whether truly bisexual flowers exist in *Nestronia* remains to be determined. The pistillate flowers of *Nestronia* have been cited as an example of a receptacular inferior ovary by F. H. Smith and E. C. Smith (1942), but R. H. Eyde (1975) questioned this interpretation, favoring the more common appendicular inferior ovary.

1. Nestronia umbellula Rafinesque, New Fl. 3: 13. 1838 • Conjurer's nut, Indian olive [E] [F]

Darbya umbellulata A. Gray; *Nestronia quadriala* (Bentham & Hooker f.) Kuntze; *N. undulata* Rafinesque

Shrubs to 1 m, rhizomatous, forming large colonies. **Stems** forming shoots from crown. **Leaves:** petiole 2–5 mm; blade ovate-lanceolate, 2–7 × 0.7–3.2 cm (mean 4.2 × 1.8 cm), decreasing in size toward base of shoot, base acute, apex acute, abaxial surface glaucous, adaxial surface bright green. **Staminate inflorescences** 3–11-flowered. **Flowers** fragrant, staminate 2–3 mm, pistillate 6–10 mm, petals white to greenish, margins puberulent. **Pseudodrupes** yellowish green, spheric, 1–2 × 0.8–1.8 cm.

Flowering May–Jun. Moist and dry woodlands and stream banks; 20–500 m; Ala., Ga., Ky., N.C., S.C., Tenn., Va.

Although sometimes reported as specific to pines, *Nestronia* parasitizes a wide variety of host plants (D. D. Horn and R. Kral 1984). The species is apparently adapted to early successional habitats, thus natural disturbance may favor its spread and establishment (L. J. Musselman 1982; G. Libby and C. Bloom 1998).

Nestronia umbellula is listed as threatened, endangered, or of conservation concern by each state in which it occurs except Alabama, which does not provide regulatory protection to plants.

2. SANTALUM Linnaeus, Sp. Pl. 2: 349. 1753; Gen. Pl. ed. 5, 165. 1754 • Sandalwood [Greek *santalon*, derived from Sanskrit *chandana*, fragrant] [I]

Trees [shrubs], evergreen, synoecious. **Stems** glabrous. **Leaves:** petiole well developed. **Inflorescences** axillary and terminal, thyrses [cymes, racemes, umbels]. **Pedicels** present. **Flowers** bisexual; perianth and androecium perigynous [epigynous]; hypanthium campanulate [very short, funnelform, cylindric]; petals 4, post-staminal hairs present; nectary lining hypanthium, lobed distally; stamens 4, with hairs at filament base; ovary ½ inferior to almost superior [inferior], 1-locular; ovules 2–4; styles long, cylindric [short]; stigmas 3-lobed. **Pseudodrupes** rimmed by circular collar (scar or perianth remnants) at apex. $x = 10$.

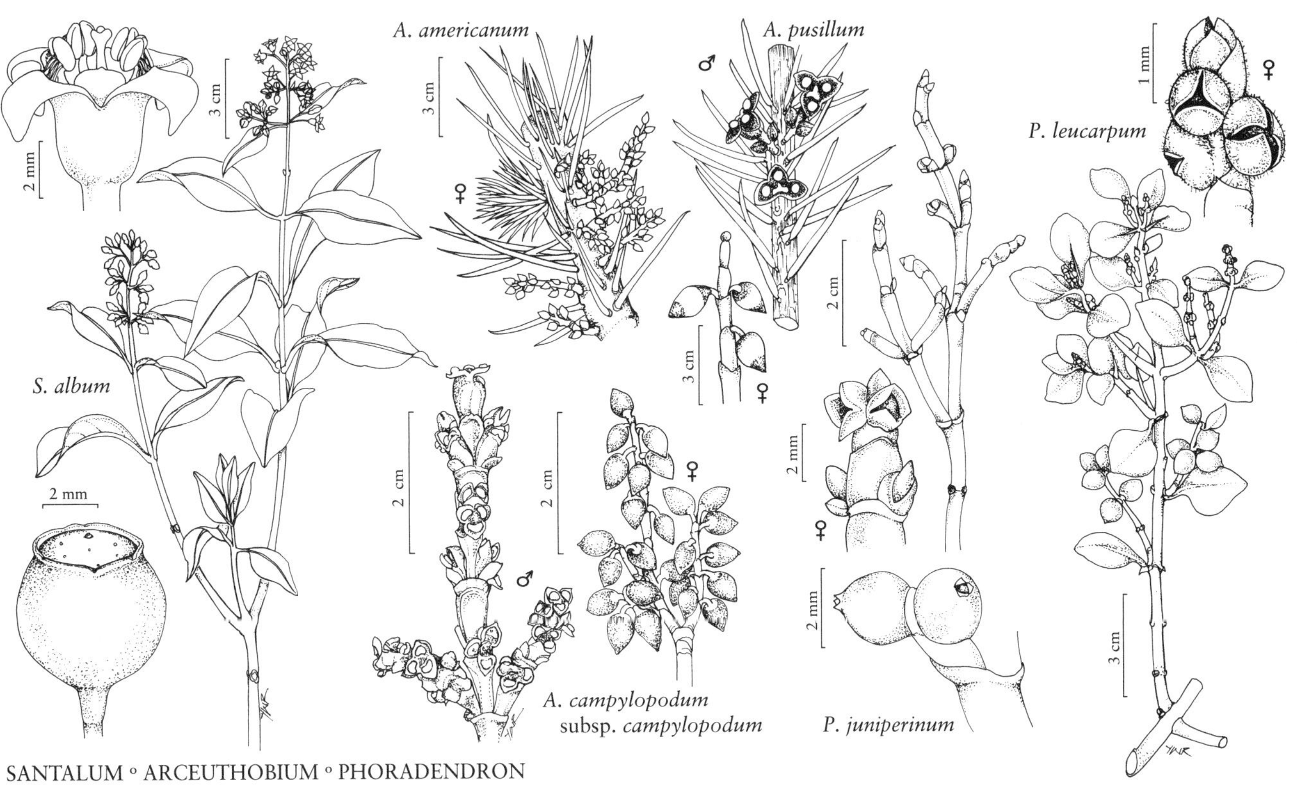

SANTALUM ◦ ARCEUTHOBIUM ◦ PHORADENDRON

Species 15 (1 in the flora): introduced, Florida; Asia (Indonesia), Pacific Islands (Papua New Guinea), Australia; introduced also elsewhere in Asia (India).

1. Santalum album Linnaeus, Sp. Pl. 2: 349. 1753 [F] [I]

Trees 4–9 m, upright or clambering, sprawling among other vegetation; branches slightly angular-striate. **Leaves:** petiole 2-ribbed, 1–1.8 cm; blade ovate to lanceolate-elliptic, 3.5–6 × 1.1–2.3 cm, apex obtuse or acute, adaxial surface shiny, darker green than abaxial surface; venation brochidodromous. **Inflorescences:** peduncles 4–11 mm. **Pedicels** to 1 mm. **Flowers** 3–4 mm; petals turning rose to crimson red or maroon through maturation, reflexed, triangular, 2 mm, post-staminal hairs white, long, coarse; nectary lobes reddish, ovate, prominent; stamens situated between nectary lobes; filaments longer than anther, positioning anthers above nectary lobes at height of stigma; styles 2 mm. **Pseudodrupes** nearly spheric, 7 × 8 mm, borne on short peduncle, often 1 per dichasium; epicarp dark red to black, mesocarp smooth. **Seeds** spheric. **2*n*** = 20.

Flowering Jul–Dec; fruiting Jul–Dec. Overgrown pine rocklands; 0 m; introduced; Fla.; Asia (Indonesia); introduced also elsewhere in Asia (India), Australia.

Santalum album has long been cultivated for its fragrant wood, and its oil is used in incense, perfume, cosmetics, and medicines (L. Hamilton and C. E. Conrad 1990).

Santalum album was introduced into cultivation after 1920 at the USDA Chapman Field Subtropical Horticulture Research Station in Miami-Dade County, Florida. The first collections documenting its escape from cultivation were made in 1989 and 1991 at that location. Later collections indicate that the species is common along the edges of a pine rockland remnant, thus indicating that a small colony had become established. An additional plant was also seen at the Deering Estate at Cutler (a county preserve) near Chapman Field (R. P. Wunderlin, pers. comm.). Given that flowers and fruits are present on vouchered specimens, *S. album* appears to be reproducing and spreading via its bird-dispersed fruits and seeds. This introduction demonstrates that sandalwood is not host specific, as none of its North American hosts in Florida are found within its natural range.

Santalum album is exceptional in the genus by having the ovary often appearing nearly superior. The ovaries in the remaining species of *Santalum* are half inferior or more.

VISCACEAE Batsch

• Christmas Mistletoe Family

Daniel L. Nickrent

Herbs, subshrubs, or shrubs, perennial, aerially parasitic on branches of angiosperms and gymnosperms, evergreen, dioecious or monoecious; roots present as haustorial endophytes. **Stems** erect or pendulous, brittle, nodes articulated. **Leaves** opposite [whorled], simple; stipules absent; petiole present or absent; blade brittle, fleshy, or scalelike, margins entire; venation pinnate or parallel [palmate]. **Inflorescences** unisexual or bisexual, axillary or terminal, spikelike thyrses or cymes [fascicles]. **Flowers** unisexual, radially symmetric or ± asymmetric; perianth epigynous; hypanthium absent (staminate flowers) or completely adnate to ovary (pistillate flowers); sepals 0; petals (2–)3–4(–6), usually distinct, sometimes connate basally, valvate, post-staminal hairs absent; nectary present or absent (staminate flowers), absent or not well defined (pistillate flowers); stamens (2–)3–4(–6), opposite petals, distinct, adnate to petals; anthers dehiscing by transverse slits or pores; pistil 1, 2[–3]-carpellate, ovary inferior, 0–1-locular, embryo sacs arising from placental nucellar complex; no true integumented ovules formed; style 0 or 1, very short; stigma 1, undifferentiated or 2[–3]-lobed [capitate]. **Fruits** berries (explosively dehiscent in *Arceuthobium*). **Seeds** 1 per fruit, viscin sticky or mucilaginous.

Genera 7, species ca. 575 (3 genera, 15 species in the flora): nearly worldwide.

To people in the northern hemisphere, the word mistletoe most often calls to mind a branch parasite in Viscaceae. In Europe, the Christmas mistletoe is typically *Viscum album*, whereas in North America, it is more often a species of *Phoradendron*. Mistletoe is associated historically with ancient Teutonic myths and Druidic rituals, but vestiges of these associations survive today as Yuletide kissing, a custom that dates to sixteenth-century England. Viscaceae represent just one of five clades that contain mistletoes, the others being Misodendraceae J. Agardh, Loranthaceae Jussieu, Santalaceae, and Amphorogynaceae Nickrent & Der (R. L. Mathiasen et al. 2008; D. L. Nickrent et al. 2010). Thus, the term mistletoe describes both a habit (an aerial parasite) and a taxonomic association (a member of Santalales). Early twentieth-century works (for example, A. Engler and K. Krause 1935) often treated Viscaceae as a subfamily of Loranthaceae, and indeed this practice persists today in some floras, herbaria, and popular treatments. But molecular phylogenetic work has greatly clarified relationships among the five mistletoe clades, showing that they originated independently at different time periods

(J. P. Der and Nickrent 2008; R. Vidal-Russell and Nickrent 2008). Thus, the mainly tropical family Loranthaceae, whose members often have large, bird-pollinated flowers, is only distantly related to Viscaceae with small, insect-pollinated flowers.

In North America, *Arceuthobium* and *Phoradendron* are native, and *Viscum* was introduced purposely and is now naturalized. By far, *Arceuthobium* is the most economically significant viscaceous mistletoe. In 1982, timber losses caused by species of *Arceuthobium* for western North America were estimated as 130 million cubic meters (D. B. Drummond 1982; F. G. Hawksworth and D. Wiens 1996). Although these losses are severe, particularly in areas managed for timber production, species of *Arceuthobium* also play important ecological roles in natural forest ecosystems, for example its witches' brooms provide nesting habitat for the northern spotted owl (R. Everett et al. 1997). Native trees as well as cultivated fruit and nut trees can be negatively impacted by *Phoradendron* in North America, Mexico, and Central and South America, but the extent of damage has not been quantified. In addition to its use as a Christmas holiday novelty, *Viscum* is also being explored for a variety of medicinal uses including treatments for some forms of cancer (A. Bussing 2000).

Viscaceae have been shown by molecular phylogenetic studies to be sister to Amphorogynaceae Der & Nickrent (previously a tribe of Santalaceae), which includes root and stem parasites (J. P. Der and D. L. Nickrent 2008; Su H. J. et al. 2015). Resolution of some relationships among the genera of Viscaceae had been seen from molecular studies conducted over a decade ago, but only recently has a clear picture emerged. *Viscum* and *Notothixos* Oliver are sister to the remaining genera, which include a clade consisting of *Korthalsella* Tieghem and *Ginalloa* Korthals and another of *Phoradendron/Dendrophthora* and *Arceuthobium*. From a biogeographic perspective, *Arceuthobium* is interesting because it is the only mistletoe to occur naturally in both the Old and New Worlds. A hypothesis by F. G. Hawksworth and D. Wiens (1972) is that *Arceuthobium* evolved in Asia and, following a trans-Beringian migration, underwent a massive radiation in the New World. Chloroplast spacer sequences show, however, that the New World species are most similar to *Phoradendron*, whereas the Old World species have accumulated numerous deletions (Nickrent and M. A. García 2009). The most parsimonious hypothesis is that both *Arceuthobium* and *Phoradendron* first evolved in the New World, possibly in the Mexican highlands, where both genera are today quite diverse, and *Arceuthobium* then colonized the Old World.

SELECTED REFERENCES Mathiasen, R. L. et al. 2008. Mistletoes: Pathology, systematics, ecology, and management. Pl. Dis. 92: 988–1006. Vidal-Russell, R. and D. L. Nickrent. 2008. The first mistletoes: Origins of aerial parasitism in Santalales. Molec. Phylogen. Evol. 47: 523–527.

1. Leaves scalelike; anthers 1-locular; berries on recurved pedicels, 2-colored, explosively dehiscent; seeds sticky when fruit dehisces; endosperm globose to pyriform .... 1. *Arceuthobium*, p. 424
1. Leaves well developed or scalelike; anthers 2-locular or multilocular; berries sessile, 1-colored, not explosively dehiscent; seeds mucilaginous when removed from fruit; endosperm ± flattened.
    2. Inflorescences spikelike thyrses, with intercalary meristems; flowers borne in cavities or grooves; anthers 2-locular; ovaries 1-locular; embryos oriented longitudinally... 2. *Phoradendron*, p. 434
    2. Inflorescences dichasial cymes, without intercalary meristems; flowers not borne in cavities or grooves; anthers multilocular; ovaries 0-locular; embryos oriented transversely .... 3. *Viscum*, p. 439

1. ARCEUTHOBIUM M. Bieberstein, Fl. Taur.-Caucas. 3: 629. 1819, name conserved • Dwarf mistletoe [Greek *arceuthos*, juniper, and *bios*, life, alluding to *A. oxycedri*, which parasitizes that host]

Herbs or subshrubs, dioecious; parasitic on branches of Pinaceae [Cupressaceae], infections localized, nonsystemic, or systemic, sometimes inducing witches' brooms. **Stems** multiple; primary branching fanlike, secondary branching fanlike or whorled. **Leaves** scalelike, connate. **Inflorescences** axillary or terminal, spikelike thyrses; flowers borne singly or in cymes in axils of subtending scales, not in cavities or grooves. **Staminate flowers:** petals 3–4(–6), triangular, distinct, radially symmetric or slightly asymmetric; stamens 3–4(–6); anthers 1-locular, dehiscing by transverse slits; nectary (= central cushion) present. **Pistillate flowers:** petals 2(–3), deltate, connate basally; ovary 1-locular; style very short to absent; stigma 2-lobed, secreting pollination droplet. **Berries** borne on recurved pedicels (when mature), explosively dehiscent from pedicellar end, 2-colored, smooth, glaucous, petal remnants not persisting. **Seeds** sticky when fruit dehisces, mucilaginous later after imbibing water; endosperm globose to pyriform; embryo oriented longitudinally. $x = 14$.

Species 29 (7 in the flora): North America, Mexico, West Indies, Central America, Eurasia, Africa, Atlantic Islands.

Species of *Arceuthobium* have received much attention because of their negative impact on commercially important conifers and also because of their fascinating life history. Most mistletoes rely on birds to disperse their seeds, but in all but one species of *Arceuthobium* (*A. verticilliflorum* Engelmann from Durango, Mexico), seeds are dispersed by explosive dehiscence. Another feature considered an evolutionary advancement is reduced photosynthesis accompanying greater dependence on host carbon. *Arceuthobium* can be considered on the brink of becoming a holoparasite (D. L. Nickrent and M. A. García 2009). It can form three basic types of infections on the host. The first, called a localized infection, is present as the mistletoe shoots emerge from an often swollen host branch at the original site of seedling penetration. In some species, the infections remain localized, but in many species they later develop into witches' brooms, with the parasite inducing prolific branching of the host distal to the point of infection. With nonsystemic witches' brooms, the mistletoe endophyte remains localized, whereas with systemic witches' brooms, the parasite endophyte occurs within the broom, sometimes existing adjacent to and dividing in synchrony with the host apical meristem (J. Kuijt 1960).

Arceuthobium has long been considered a taxonomically difficult group, mainly owing to losses and reductions in morphologic features used for classification. L. S. Gill (1935) was the first to propose a comprehensive treatment of the genus in the United States. In their classification, F. G. Hawksworth and D. Wiens (1972) utilized many of the same features as Gill, such as branching pattern, but did not agree with his host-form concept. Host specificity in *Arceuthobium* ranges from specialists (only one host known) to generalists that parasitize over a dozen species. Hawksworth and Wiens and D. L. Nickrent (1996) strongly advocated that mistletoe morphologic integrity was maintained irrespective of host being parasitized and recognized 28 and 42 species, respectively.

A classification of *Arceuthobium*, revised from D. L. Nickrent (1996), was given by Nickrent et al. (2004), who recognized 26 species in two subgenera and 11 sections. That system is followed here, considering only subg. *Vaginata* Hawksworth & Wiens because all North American *Arceuthobium* reside there. This classification differs most from F. G. Hawksworth and D. Wiens (1972) with regard to sect. *Campylopoda* Hawksworth & Wiens, whereas the latter authors recognized 13 species in this section, Nickrent et al. recognized only two.

SELECTED REFERENCES Hawksworth, F. G. and D. Wiens. 1972. Biology and Classification of Dwarf Mistletoes *(Arceuthobium)*. Washington. [Agric. Handb. 401.] Hawksworth, F. G. and D. Wiens, eds. 1996. Dwarf Mistletoes: Biology, Pathology, and Systematics. Washington. [Agric. Handb. 709.] Kuijt, J. 1955. Dwarf mistletoes. Bot. Rev. (Lancaster) 21: 569–628. Kuijt, J. 1960. Morphological aspects of parasitism in the dwarf mistletoes (*Arceuthobium*). Univ. Calif. Publ. Bot. 30: 337–436. Nickrent, D. L. 1996. Molecular systematics. In: F. G. Hawksworth and D. Wiens, eds. 1996. Dwarf Mistletoes: Biology, Pathology, and Systematics. Washington. Pp. 155–170. Nickrent, D. L. et al. 2004. A phylogeny of all species of *Arceuthobium* (Viscaceae) using nuclear and chloroplast DNA sequences. Amer. J. Bot. 91: 125–138. Nickrent, D. L. and M. A. García. 2009. On the brink of holoparasitism: Plastome evolution in dwarf mistletoes (*Arceuthobium*, Viscaceae). J. Molec. Evol. 68: 603–615. Nickrent, D. L., K. P. Schuette, and E. M. Starr. 1994. A molecular phylogeny of *Arceuthobium* based upon rDNA internal transcribed spacer sequences. Amer. J. Bot. 81: 1149–1160.

1. Secondary branching whorled; staminate pedicels present; lateral staminate flowers subglobose in bud; flowering (Mar–)Apr–Jun; principal hosts *Pinus banksiana* and *P. contorta* .... 1. *Arceuthobium americanum*
1. Secondary branching fanlike; staminate pedicels absent (present in *A. douglasii*); lateral staminate flowers lenticular or subglobose in bud; flowering Feb–Dec; hosts various Pinaceae.
    2. Flowering Jul–Dec.
        3. Parasites of pinyon pines (principally *Pinus edulis* and *P. monophylla*) .... 6. *Arceuthobium divaricatum*
        3. Parasites of other Pinaceae (*Abies*, *Picea*, *Pinus*, *Larix*, and *Tsuga*) .... 7. *Arceuthobium campylopodum*
    2. Flowering Feb–Jun(–Jul).
        4. Lateral staminate flowers subglobose in bud; plants forming systemic witches' brooms; principal host *Pseudotsuga menziesii* .... 2. *Arceuthobium douglasii*
        4. Lateral staminate flowers lenticular in bud; plants forming localized infections or systemic or nonsystemic witches' brooms; principle hosts *Picea* and *Pinus*.
            5. Secondary branches (1–)3 cm, rarely seen; plants forming systemic witches' brooms; principal host *Picea*; e North America .... 3. *Arceuthobium pusillum*
            5. Secondary branches 8–20(–27) cm; plants forming localized infections or nonsystemic witches' brooms; principal host *Pinus*; w North America.
                6. Staminate and pistillate plants dimorphic; principal host *Pinus leiophylla* var. *chihuahua*; Arizona, New Mexico .... 4. *Arceuthobium gillii*
                6. Staminate and pistillate plants not dimorphic; principal hosts primarily *Pinus ponderosa*, sometimes *P. durangensis*, and *P. engelmannii*; w United States .... 5. *Arceuthobium vaginatum*

1. Arceuthobium americanum Nuttall ex Engelmann, Boston J. Nat. Hist. 6. 214. 1850 • Lodgepole pine dwarf mistletoe [E] [F]

Razoumofskya americana (Nuttall ex Engelmann) Kuntze

Plants usually forming systemic witches' brooms, sometimes nonsystemic witches' brooms in secondary hosts. **Stems** yellowish to olive green; secondary branching whorled, branches 5–9(–30) cm, third internode 6–23 × 1–2 mm, dominant shoot 1–3 mm diam. at base. **Staminate pedicels** present. **Staminate flowers** radially symmetric, subglobose in bud, 2.2 mm diam.; petals 3(–4), same color as stems. **Berries** proximally olive green, distally yellowish to reddish brown, 3.5–4.5 × 1.5–2.5 mm. **Seeds** ellipsoid, 2.4 × 1.1 mm, endosperm green. $2n = 28$.

Flowering (Mar–)Apr–Jun; fruiting Aug–Sep. Coniferous forests, especially jack or lodgepole pine; 200–3400 m; Alta., B.C., Man., Ont., Sask.; Calif., Colo., Idaho, Mont., Nev., Oreg., Utah, Wash., Wyo.

Meiosis occurs in August, with fruits maturing 16 months after pollination; seeds germinate in May.

The principal hosts of *Arceuthobium americanum* are *Pinus contorta* var. *latifolia* in western North America, *P. contorta* var. *murrayana* in the Sierra Nevada and Cascade ranges of the western United States, and *P. banksiana* in western Canada. A study utilizing AFLPs (C. A. Jerome and B. A. Ford 2002) documented that the parasite exists as three genetic races that correspond to these host species. *Arceuthobium americanum* has the most extensive geographic range of any species of the genus and can utilize other species as secondary hosts, including *P. albicaulis*, *P. flexilis*, *P. jeffreyi*, and *P. ponderosa*, as well as a number of rare hosts. Although young infections may be localized, *A. americanum* eventually forms massive systemic witches' brooms. Interestingly, when parasitizing some secondary hosts, the brooms may become nonsystemic, possibly indicating partial breakdown of coordinated developmental pathways.

2. Arceuthobium douglasii Engelmann in J. T. Rothrock, Rep. U.S. Geogr. Surv., Wheeler, 253. 1879 • Douglas-fir dwarf mistletoe

Razoumofskya douglasii (Engelmann) Kuntze

Plants forming systemic witches' brooms. **Stems** yellowish green, olive green, orange-brown, or maroon; secondary branching fanlike, branches 2(–8) cm, third internode 2–6 × 1 mm, dominant shoot 1–1.5 mm diam. at base. **Staminate pedicels** present. **Staminate flowers** radially symmetric, subglobose in bud, 2.3 mm diam.; petals (2–)3(–4), reddish or purple. **Berries** proximally olive green to purplish, distally yellow, brownish orange, or maroon, 3.5–4.5 × 1.5–2 mm. **Seeds** ellipsoid, 2.4 × 1.1 mm, endosperm bright green. $2n = 28$.

Flowering (Mar–)Apr–May(–Jun); fruiting Aug–Sep (–Oct). Coniferous forests with Douglas fir; 300–3300 m; B.C.; Ariz., Calif., Colo., Idaho, Mont., Nev., N.Mex., Oreg., Tex., Utah, Wash., Wyo.; Mexico (Chihuahua, Coahuila, Durango, Nuevo León).

Staminate meiosis occurs in September, pistillate in April, with fruits maturing 17 to 18 months after pollination; seeds germinate in March.

Arceuthobium douglasii has the widest latitudinal distribution of any North American dwarf mistletoe. Its principal host is *Pseudotsuga menziesii*, but it is occasionally found also on *Abies amabilis*, *A. concolor*, *A. grandis*, *A. lasiocarpa*, *Picea engelmannii*, and *P. pungens*. Plants of *Arceuthobium douglasii* induce massive systemic witches' brooms that severely affect host growth and may eventually result in host mortality. Section *Minuta* Hawksworth & Wiens was erected to accommodate the two diminutive North American dwarf mistletoes, *A. douglasii* and *A. pusillum*. Isozymes and later DNA analyses showed that these two species are not closely related, thus their small shoot size and spring flowering evolved in parallel.

3. Arceuthobium pusillum Peck, Trans. Albany Inst. 7: 191. 1872 • Eastern dwarf mistletoe [E] [F]

Plants forming systemic witches' brooms. **Stems** green, orange, red, maroon, or brown; secondary branching (rarely seen) fanlike, branches 1(–3) cm, third internode 1–1.9 × 0.5–1.5 mm; dominant shoot 1 mm diam. at base; pistillate stems usually longer than staminate. **Staminate pedicels** absent. **Staminate flowers** radially symmetric, 1 terminating stem or 3 in dichasium, terminal subglobose in bud, lateral lenticular in bud, 1.7–2.2 mm diam.; petals (2–)3(–4), red to maroon. **Berries** proximally green, distally yellowish to reddish brown, 3 × 1.3–1.8 mm. **Seeds** pyriform to elliptic, 1.4–1.8 × 1–1.4 mm, endosperm bright green. $2n = 28$.

Flowering (Mar–)Apr–May(–Jun); fruiting Sep–Oct. Spruce forests within 500 m of coast and inland bogs, often within 2 km of lakes and rivers; 0–800 m; Man., N.B., Nfld. and Labr. (Nfld.), N.S., Ont., P.E.I., Que., Sask.; Conn., Maine, Mass., Mich., Minn., N.H., N.J., N.Y., Pa., R.I., Vt., Wis.

Staminate meiosis occurs in September, pistillate meiosis in May, with fruits maturing five months after pollination; seeds germinate in May to June.

In 1858, Thoreau wrote about the witches' brooms of Eastern dwarf mistletoe, over a decade before the species was actually described. Lucy B. Millington of Warrensburg, New York, recognized the mistletoe in 1871 and related her discovery via correspondence with C. H. Peck, who named the species in the following year (B. S. Smith 1992). The species was later found to be widespread in spruce forests throughout the Great Lakes states.

Arceuthobium minutum Engelmann, which pertains here, was published in 1871 but without a description, hence it is invalid.

Arceuthobium pusillum forms massive systemic witches' brooms that severely affect the vigor of its principal host, *Picea mariana* (black spruce). White spruce (*P. glauca*) and red spruce (*P. rubens*) are less commonly infected. Occasional to rare hosts include *Abies balsamea*, *Larix laricina*, *Pinus banksiana*, *Pinus resinosa*, and *Pinus strobus*. Molecular phylogenetic work revealed that *Arceuthobium pusillum* is most closely related to *A. bicarinatum* Urban of Hispaniola (D. L. Nickrent et al. 2004). Given that these two species differ greatly in size, this result demonstrates the dramatic morphologic changes that ancestors of *A. pusillum* underwent, possibly as adaptations to cold climates.

4. Arceuthobium gillii Hawksworth & Wiens, Brittonia 16: 55, figs. 1A, 2A. 1964 • Chihuahua pine dwarf mistletoe

Plants forming nonsystemic witches' brooms; staminate and pistillate plants dimorphic: staminate taller with open divaricate branching, pistillate shorter with dense branching. **Stems** olive green, greenish brown, greenish yellow, or orange; secondary branching fanlike, branches 8–15(–25) cm, third internode 5–18 × 2–4.5 mm, dominant shoot 2.5–8 mm diam. at base. **Staminate pedicels** absent. **Staminate flowers** slightly asymmetric, (proximal petal deflexed at anthesis), lenticular in bud, 2.5–4 mm diam.; petals 3, same color as stem abaxially, tawny reddish brown adaxially. **Berries** proximally olive green and conspicuously glaucous with bluish hue, distally greenish or yellowish brown, 4–5 × 2–3 mm. **Seeds** pyriform to ellipsoid, 3–4 × 2–3 mm, endosperm dark olive green. $2n = 28$.

Flowering Feb–Apr; fruiting Oct. Coniferous and mixed forests with Chihuahua pine; 1700–2700 m; Ariz., N.Mex.; Mexico (Chihuahua, Durango, Sinaloa, Sonora).

Meiosis occurs in September, with fruits maturing 19 to 20 months after pollination; seeds germinate in April.

In the flora area, *Arceuthobium gillii* is found in the Chiricahua, Huachuca, Santa Catalina, and Santa Rita mountains of Arizona, and the Animas Mountains of New Mexico; its principal host is *Pinus leiophylla* var. *chihuahua*. Subspecies *nigrum* Hawksworth & Wiens, which occurs in the Sierra Madre Occidental and Oriental of Mexico, was subsequently elevated to species rank as *A. nigrum* (Hawksworth & Wiens) Hawksworth & Wiens. The two species are apparently allopatric but occur in the same mountain range in northern Durango, Mexico. Both form nonsystemic witches' brooms on members of *Pinus* subsect. *Leiophylla* Loudon and are closely related as shown by molecular analyses (D. L. Nickrent et al. 2004).

5. Arceuthobium vaginatum (Humboldt & Bonpland ex Willdenow) J. Presl in F. Berchtold and J. S. Presl, Prir. Rostlin 2: 28. 1825

Viscum vaginatum Humboldt & Bonpland ex Willdenow, Sp. Pl. 4: 740. 1806

Subspecies 2 (1 in the flora): sw, wc United States, n Mexico.

5a. Arceuthobium vaginatum (Humboldt & Bonpland ex Willdenow) J. Presl subsp. **cryptopodum** (Engelmann) Hawksworth & Wiens, Brittonia 17: 230. 1965 • Southwestern dwarf mistletoe

Arceuthobium cryptopodum Engelmann, Boston J. Nat. Hist. 6: 214. 1850

Plants forming nonsystemic witches' brooms; staminate and pistillate plants not dimorphic. **Stems** green, yellow, orange, reddish brown, or dark purple; secondary branching fanlike, branches 10–20(–27) cm, third internode 4–16 × 2–4.5 mm, dominant shoot 2–10 mm diam. at base. **Staminate pedicels** absent. **Staminate flowers** radially symmetric, lenticular in bud, 2.5–3 mm diam.; petals 3–4(–6), green, greenish yellow, or light pink, sometimes differing abaxially and adaxially. **Berries** proximally olive green, distally brown, 4.5–5.5 × 2–3 mm. **Seeds** pyriform to ellipsoid, 2.8 × 1.8 mm, endosperm green or green and maroon. $2n = 28$.

Flowering (Apr–)May–Jun(–Jul); fruiting Jul–Aug (–Sep). Coniferous forests with ponderosa pine; 1700–3000 m; Ariz., Colo., N.Mex., Tex., Utah; Mexico (Chihuahua, Coahuila, Sonora).

Meiosis occurs in March to April, with fruits maturing 14 to 15 months after pollination; seeds germinate in August to September immediately after dispersal.

Subspecies *cryptopodum* occurs in the southwestern United States and northern Mexico, whereas subsp. *vaginatum* occurs in the Sierra Madre Occidental and Sierra Madre Oriental of Mexico. The two subspecies are sympatric in central Chihuahua, where some morphologically intermediate populations occur. The Mexican subspecies is often deeply pigmented (dark brown to black), has larger shoots and staminate flowers, and flowers in May and June.

Subspecies *cryptopodum* can be found parasitizing its most common principal host, *Pinus ponderosa* var. *scopulorum*, in nearly every location where this tree occurs. It forms nonsystemic witches' brooms that can be massive, increasing the longevity of the supporting branch, which would normally self-prune. This subspecies also parasitizes other principal hosts such as *P. durangensis* Roezl ex Gordon, *P. engelmannii* and *P. ponderosa* var. *arizonica*, as well as secondary and occasional hosts such as *P. aristata*, *P. contorta*, and *P. cooperi* C. E. Blanco. The parasite can be damaging to its host in some portions of its range, such as the Front Range of the Rocky Mountains in Colorado and south-central New Mexico.

6. Arceuthobium divaricatum Engelmann in J. T. Rothrock, Rep. U.S. Geogr. Surv., Wheeler, 253. 1879 • Pinyon dwarf mistletoe

Razoumofskya divaricata (Engelmann) Coville

Plants usually forming localized infections only, sometimes forming small nonsystemic witches' brooms. **Stems** olive green, orange, or reddish brown, glaucous; secondary branching fanlike, branches 8(–13) cm, slender, third internode 6–15 × 1–2 mm, dominant shoot 1.5–4 mm diam. at base. **Staminate pedicels** absent. **Staminate flowers** radially symmetric, lenticular in bud, 2.5 mm diam.; petals 3, light yellowish green. **Berries** proximally olive or grayish to bluish green, distally olive green to brown, glaucous, 3.5 × 2 mm. **Seeds** ellipsoid, 2 × 0.9 mm, endosperm bright green. $2n$ = 28.

Flowering Aug–Sep; fruiting Sep–Oct. Pinyon-juniper woodlands; 1200–3000 m; Ariz., Calif., Colo., Nev., N.Mex., Tex., Utah; Mexico (Baja California).

Meiosis occurs in July, with fruits maturing 13 months after pollination.

Arceuthobium divaricatum is parasitic only on pinyon pines. Its principal hosts are *Pinus edulis* and *P. monophylla*, but it can also parasitize *P. cembroides*, and *P. quadrifolia*. Most infections are localized; however, over time small witches' brooms can form that are cryptic given the shrubby habit of the host tree. Pinyon dwarf mistletoe was classified in sect. *Campylopoda* (F. G. Hawksworth and D. Wiens 1972), as was another parasite of pinyons from Mexico, *A. pendens* (Hawksworth and Wiens 1980). Isozyme data first showed that these species are not part of sect. *Campylopoda* (D. L. Nickrent 1996), and this was later confirmed with DNA evidence (Nickrent et al. 2004). Moreover, the two pinyon dwarf mistletoes are not closely related to each other; *A. divaricatum* is close to *A. douglasii*, and *A. pendens* is sister to *A. guatemalense* Hawksworth & Wiens of Mexico and Guatemala.

7. Arceuthobium campylopodum Engelmann, Boston J. Nat. Hist. 6: 214. 1850 • Western dwarf mistletoe [F]

Razoumofskya campylopoda (Engelmann) Kuntze

Plants forming localized infections only or nonsystemic witches' brooms. **Stems** yellow, yellowish green, green, olive green, brown, light tan, orange, red, maroon, or purple; secondary branching fanlike, branches 3–10(–22) cm, third internode 2–15(–23) × 1–3.5(–5) mm, dominant shoot 1–6 mm diam. at base. **Staminate pedicels** absent. **Staminate flowers** slightly asymmetric (distal lateral petals keeled and hooded, proximal petal not keeled, deflexed at anthesis), lenticular in bud, 2.3–3.3 mm diam.; petals 3–4(–6), yellowish green. **Berries** proximally yellowish green or olive green, distally yellow, orange, or brown, 3–5 × 2–3 mm. **Seeds** pyriform to ellipsoid, 3.3–5.5 × 2.3–4 mm, endosperm bright green. $2n$ = 28.

Subspecies 13 (13 in the flora): w North America, n Mexico.

Arceuthobium campylopodum is here considered a wide ranging and polymorphic species that parasitizes a number of hosts in Pinaceae throughout western North America and northern Mexico. Earlier taxonomic concepts, such as those of L. S. Gill (1935) and J. Kuijt (1955), recognized several forms of *A. campylopodum* that usually corresponded with principal host species. F. G. Hawksworth and D. Wiens (1972, 1996) elevated those forms to species, stating that morphologic integrity is maintained even when parasitizing nonprincipal hosts. Despite this claim, most of the characters they used to differentiate species were quantitative with continuous variation. The absence of hybridization was also given as evidence for separate species status; however, all species have the same chromosome number and very similar morphology, thus it is not clear how a hybrid would be recognized should it occur. Examination of dwarf mistletoe specimens without information on location and host can often result in an ambiguous identification.

During the 1980s and 1990s, biosystematic studies were conducted using isozyme electrophoresis, particularly within sect. *Campylopoda*. Results of these studies, summarized by D. L. Nickrent (1996), in some cases provided evidence for populational genetic divergence but in other cases did not. Following those studies, phylogenetic analyses of chloroplast *trn*L-F and nuclear rDNA ITS sequences were applied to species level questions (Nickrent et al. 1994, 2004). All members of sect. *Campylopoda* had essentially identical DNA sequences in these regions, in contrast with species in other sections, which showed greater genetic distances.

All taxa within sect. *Campylopoda* were examined with respect to their morphology, host associations, levels of sympatry, and genetic relationships (D. L. Nickrent 2012). That study concluded that all taxa of sect. *Campylopoda*, recognized as species by F. G. Hawksworth and D. Wiens (1996), are best viewed as ecotypes of a single variable species, *Arceuthobium campylopodum*. Two factors were used to justify describing the variants at the rank of subspecies: less than 20% of the time are the 13 subspecies sympatric with one another (they are generally geographically and ecologically isolated), and subspecies have already been described in *A. vaginatum*. For a complete list of synonyms for each subspecies, see Nickrent.

A morphometric study using discriminant function analysis was conducted by R. L. Mathiasen and S. C. Kenaley (2015) on four taxa within sect. *Campylopoda*: *Arceuthobium campylopodum* subspp. *campylopodum*, *littorum*, *occidentale*, and *siskiyouense*. With the exception of male plants of the last subspecies, this analysis showed that the 95% confidence intervals overlapped, indicating potential misclassification of the taxa assigned *a priori* to species. Differences in multivariate means indicate that some differentiation has occurred among these four taxa. An additional three taxa of sect. *Campylopdoda* that parasitize white pines were examined by B. P. Reif et al. (2015) using amplified fragment length polymorphisms: *A. campylopodum* subspp. *apachecum*, *blumeri*, and *cyanocarpum*. Support was low for genetic differentiation between the first two subspecies, whereas greater differentiation was seen between these and the third, in agreement with D. L. Nickrent et al. (2004). These two studies examined seven of the 13 taxa in sect. *Campylododa*, hence more work is needed simultaneously treating the entire complex. Although these authors argued that their results support classification at the species level, their results could also be viewed as evidence for early stages of genetic differentiation among widespread populations, thus equally supporting classification at the subspecific rank.

SELECTED REFERENCES Mathiasen, R. L. and S. C. Kenaley. 2015. A morphometric analysis of dwarf mistletoes in the *Arceuthobium campylopodum-occidentale* complex (Viscaceae). Madroño 62: 1–20. Nickrent, D. L. 2012. Justification for subspecies in *Arceuthobium campylopodum* (Viscaceae). Phytoneuron 2012-51: 1–11. Reif, B. P. et al. 2015. Genetic structure and morphological differentiation of three western North American dwarf mistletoes (*Arceuthobium*: Viscaceae). Syst. Bot. 40: 191–207.

1. Stems 3–5(–11) cm; principal hosts *Larix*, *Picea*, *Pinus albicaulis*, *P. aristata*, *P. flexilis*, *P. longaeva*, *P. strobiformis*, and *Tsuga*; plants forming witches' brooms; usually occurring at 1000–3000 m.
    2. Stems 3(–9) cm; principal host *Pinus*.
        3. Third internodes 5–7.2(–10) mm; staminate flowers 2.7 mm diam.; principal host *Pinus strobiformis*; Arizona, New Mexico . . . . 7c. *Arceuthobium campylopodum* subsp. *apachecum*
        3. Third internodes 2–5.2(–14) mm; staminate flowers 3 mm diam.; principal hosts *Pinus albicaulis*, *P. aristata*, *P. flexilis*, and *P. longaeva*; widely distributed in w United States but not Arizona . . . . .7f. *Arceuthobium campylopodum* subsp. *cyanocarpum*
    2. Stems 4–5(–11) cm; principal hosts *Larix*, *Picea*, and *Tsuga*.
        4. Staminate flowers 2.7 mm diam.; principal hosts *Larix* and *Tsuga*; British Columbia, nw United States. . . . . 7g. *Arceuthobium campylopodum* subsp. *laricis*
        4. Staminate flowers 2.3 mm diam.; principal host *Picea*; Arizona, New Mexico. .7i. *Arceuthobium campylopodum* subsp. *microcarpum*
1. Stems 5–12(–22) cm; principal hosts *Abies*, *Pinus attenuata*, *P. contorta*, *P. jeffreyi*, *P. lambertiana*, *P. monticola*, *P. muricata*, *P. ponderosa*, *P. radiata*, *P. sabiniana*, *P. strobiformis*, and *Tsuga*; plants forming localized infections only or forming witches' brooms; usually occurring at 0–2500 m.
    5. Principal hosts *Abies*, *Pinus contorta*, and *Tsuga*.
        6. Stems 8(–22) cm; third internodes 4–14 (–23) cm; principal hosts *Abies concolor* and *A. magnifica*; Arizona, California, Nevada, Oregon, Utah, Washington . . . . 7b. *Arceuthobium campylopodum* subsp. *abietinum*
        6. Stems 5–7(–13) cm; third internodes 4–9 (–16) cm; principal hosts *Abies*, *Pinus*, and *Tsuga*; British Columbia to California . . . . .7m. *Arceuthobium campylopodum* subsp. *tsugense*
    5. Principal host *Pinus* (subsp. *siskiyouense* rarely on *P. contorta*).
        7. Principal host *Pinus strobiformis*; plants usually forming localized infections only; Arizona. . . . . . 7d. *Arceuthobium campylopodum* subsp. *blumeri*
        7. Principal hosts other than *Pinus strobiformis*; plants forming localized infections only or forming witches' brooms; California, Idaho, Oregon, Washington.
            8. Principal hosts *Pinus* subg. *Strobus* (white and soft pines).
                9. Shoots yellow or green; plants forming witches' brooms; principal host *Pinus lambertiana*; California . . . . 7e. *Arceuthobium campylopodum* subsp. *californicum*
                9. Shoots olive green or brown; plants forming localized infections only; principal host *Pinus monticola*; nw California, sw Oregon . . . . .7j. *Arceuthobium campylopodum* subsp. *monticola*
            8. Principal hosts *Pinus* subg. *Pinus*.

[10. Shifted to left margin.—Ed.]

10. Plants usually forming witches' brooms.
    11. Third internodes 7–11(–22) × 1.5–2(–2.5) mm; staminate flowers: petals 3(–4); principal hosts *Pinus jeffreyi* and *P. ponderosa*; California, Idaho, Oregon, Washington............................ 7a. *Arceuthobium campylopodum* subsp. *campylopodum*
    11. Third internodes 10–15(–20) × 2–3.5(–5) mm; staminate flowers: petals 4; principal hosts *Pinus muricata* and *P. radiata*; coastal California ...... 7h. *Arceuthobium campylopodum* subsp. *littorum*
10. Plants usually forming localized infections only.
    12. Stems yellow or orange; principal host *Pinus sabiniana*; flowering Sep–Nov(–Dec); California (surrounding central valley and in Coast Ranges) . . . 7k. *Arceuthobium campylopodum* subsp. *occidentale*
    12. Stems brown; principal host *Pinus attenuata*; flowering Aug–Sep; nw California, sw Oregon.......7l. *Arceuthobium campylopodum* subsp. *siskiyouense*

7a. Arceuthobium campylopodum Engelmann subsp. **campylopodum** [F]

Plants usually forming witches' brooms. **Stems** yellow, green, olive green, orange, or brown, 8(–13) cm; third internode 7–11(–22) × 1.5–2(–2.5) mm, dominant shoot 1.5–5 mm diam. at base. **Staminate flowers** 3 mm diam.; petals 3(–4). **Berries** 5 × 3 mm.

Flowering Aug–Oct; fruiting Sep–Nov. Coniferous forests with ponderosa or Jeffrey pine; 30–2500 m; Calif., Idaho, Oreg., Wash.; Mexico (Baja California).

Meiosis occurs in July, with fruits maturing 13 months after pollination.

The distribution of subsp. *campylopodum* overlays a subset of the range of its principal host, *Pinus ponderosa*. The treatment of the host species by R. Kral (1993) recognized three varieties, two of which (*P. ponderosa* var. *ponderosa* and var. *scopulorum*) intergrade in Idaho, Montana, and Washington. With this taxonomic concept, the hosts of subsp. *campylopodum* would include *P. ponderosa* subsp. *scopulorum*. In contrast, The Gymnosperm Database (http://www.conifers.org/pi/Pinus_ponderosa.php) uses a modification of the classification by F. Lauria (1991), which suggests that *P. ponderosa* is composed of four subspecies: subspp. *ponderosa*, *benthamiana* (Hartweg) Silba, *brachyptera* (Hartweg) Silba, and *scopulorum*. Based on the geographical limits of these four subspecies, it appears that *Arceuthobium campylopodum* subsp. *campylopodum* is restricted to the *P. ponderosa* subspp. *ponderosa* and *benthamiana*. In addition to *P. ponderosa*, *P. jeffreyi* is another principal host. Both are heavily parasitized, thus this mistletoe is considered a major pathogen of these commercially important trees (F. G. Hawksworth and D. Wiens 1996). Secondary to rare hosts include *P. attenuata*, *P. contorta*, *P. coulteri*, *P. lambertiana*, and *P. sabiniana*.

7b. Arceuthobium campylopodum Engelmann subsp. **abietinum** (Engelmann) Nickrent, Phytoneuron 2012-51: 9. 2012 • Fir dwarf mistletoe

Arceuthobium douglasii Engelmann var. *abietinum* Engelmann in W. H. Brewer et al., Bot. California 2: 106. 1880; *Razoumofskya abietina* (Engelmann) Abrams

Plants forming witches' brooms. **Stems** yellow, green, olive green, orange, brown, or red, 8(–22) cm; third internode 4–14(–23) × 1.5–2(–4) mm, dominant shoot 1.5–6 mm diam. at base. **Staminate flowers** 2.5 mm diam.; petals 3(–4). **Fruits** 4 × 2 mm.

Flowering Jul–Aug(–Sep); fruiting Sep–Oct. Coniferous forests generally with fir; 0–2700 m; Ariz., Calif., Nev., Oreg., Utah, Wash.; Mexico (Chihuahua).

Meiosis occurs in July, with fruits maturing 13 to 14 months after pollination.

Subspecies *abietinum* includes forma speciales *concoloris* Hawksworth & Wiens, which parasitizes *Abies concolor* (white fir), and forma speciales *magnificae* Hawksworth & Wiens, which parasitizes *A. magnifica* (red fir). These forms were based upon inoculation studies showing that seeds of one form apparently will not infect the other's host species and vice versa (J. R. Parmeter and R. F. Scharpf 1963). Morphologically the two forms are extremely similar, with the former having a greater mean shoot height (10 versus 6 cm). The white fir dwarf mistletoe occurs throughout the above geographical range, whereas red fir dwarf mistletoe is restricted to California and southwestern Oregon. In addition to the above two species of fir, *Abies durangensis* Martínez and *A. grandis* are principal hosts; secondary to rare hosts include *A. lasiocarpa*, *Picea breweriana*, *Pinus ayacahuite* C. Ehrenberg ex Schlechtendal, *Pinus contorta*, *Pinus lambertiana*, and *Pinus monticola*.

7c. Arceuthobium campylopodum Engelmann subsp. **apachecum** (Hawksworth & Wiens) Nickrent, Phytoneuron 2012-51: 10. 2012 • Apache dwarf mistletoe

Arceuthobium apachecum Hawksworth & Wiens, Brittonia 22: 266. 1970

Plants forming witches' brooms. **Stems** yellow, green, or red, 3.5(–9) cm; third internode 5–7.2(–10) × 1–1.5(–2) mm, dominant shoot 1–2 mm diam. at base. **Staminate flowers** 2.7 mm diam.; petals 3–4(–5). **Fruits** 4 × 2.5 mm.

Flowering Jul–Sep; fruiting Aug–Oct. Coniferous forests with southwestern white pine; 2000–3000 m; Ariz., N.Mex.; Mexico (Chihuahua).

Meiosis occurs in July, with fruits maturing 13 months after pollination.

Subspecies *apachecum* is known only from a portion of the range of its only host tree, *Pinus strobiformis*. Subspecies *blumeri* also utilizes this host, but the two taxa are not sympatric.

7d. Arceuthobium campylopodum Engelmann subsp. **blumeri** (A. Nelson) Nickrent, Phytoneuron 2012-51: 9. 2012 • Blumer's dwarf mistletoe

Arceuthobium blumeri A. Nelson, Bot. Gaz. 56: 65. 1913

Plants usually forming localized infections only, rarely forming witches' brooms. **Stems** yellow, light green, or light tan, 6.5 (–18) cm; third internode 5–9.1(–14) × 1–1.6(–2) mm, dominant shoot 1–3 mm diam. at base. **Staminate** flowers 2.5–3 mm diam.; petals 3–4(–6). **Fruits** 4 × 2.5 mm.

Flowering Jul–Aug; fruiting Aug–Oct. Coniferous forests with southwestern white pine; 2100–3300 m; Ariz.; Mexico (Chihuahua, Durango, Nuevo León).

Meiosis occurs in July, with fruits maturing 13 to 14 months after pollination.

In addition to *Pinus strobiformis*, *P. ayacahuite* in Mexico is a principal host for subsp. *blumeri*. In the flora area, this subspecies occurs only in the Huachuca Mountains, but it is more widely distributed in the Sierra Madre Occidental of Chihuahua and Durango and is disjunct in the Sierra Madre Oriental on Cerro Potosí in Nuevo León. The molecular phylogenetic study by D. L. Nickrent et al. (2004) showed subsp. *blumeri* to be sister to the other members of sect. *Campylopoda*. Given that Mexico is likely the center of origin for *Arceuthobium* (Nickrent and M. A. García 2009), it is likely that subsp. *blumeri* represents an early diverging member of this species.

7e. Arceuthobium campylopodum Engelmann subsp. **californicum** (Hawksworth & Wiens) Nickrent, Phytoneuron 2012-51: 10. 2012 • Sugar pine dwarf mistletoe [E]

Arceuthobium californicum Hawksworth & Wiens, Brittonia 22: 266. 1970

Plants forming witches' brooms. **Stems** bright yellow or green, 6–8(–14) cm; third internode 6–10.5(–16) × 1–1.5(–2) mm, dominant shoot 1.5–4 mm diam. at base. **Staminate flowers** 3.3 mm diam.; petals 3–4. **Fruits** 4 × 2.5 mm.

Flowering Jul–Aug; fruiting Sep–Oct. Coniferous forests with sugar pine or western white pine; 600–2000 m; Calif.

Meiosis occurs in July, with fruits maturing 13 to14 months after pollination.

As the common name implies, subsp. *californicum* is parasitic primarily on *Pinus lambertiana*, secondarily on *P. monticola*. It is found from the Peninsular Ranges of San Diego County through the Sierra Nevada to the Cascade Range of Siskiyou County, as well as some locations in the Klamath Mountains. In some locations it is sympatric with subsp. *campylopodum*, and rarely both taxa can be found on the same host. It induces large witches' brooms on sugar pine and is considered a serious pathogen of that species.

7f. Arceuthobium campylopodum Engelmann subsp. **cyanocarpum** (A. Nelson ex Rydberg) Nickrent, Phytoneuron 2012-51: 9. 2012 • Limber pine dwarf mistletoe [E]

Razoumofskya cyanocarpa A. Nelson ex Rydberg, Fl. Colorado, 100. 1906; *Arceuthobium cyanocarpum* (A. Nelson ex Rydberg) J. M. Coulter & A. Nelson

Plants forming witches' brooms. **Stems** yellow, green, olive green, or brown, 3(–7) cm; third internode 2–5.2(–14) × 1–1.1(–1.5) mm, dominant shoot 1–2 mm diam. at base. **Staminate flowers** 3 mm diam.; petals 3(–4). **Fruits** 3.5 × 2.5 mm.

Flowering Jul–Sep; fruiting Aug–Sep. Coniferous forests; 1600–3100 m; Calif., Colo., Idaho, Mont., Nev., Oreg., Utah, Wyo.

Meiosis occurs in July, with fruits maturing 12 months after pollination.

Subspecies *cyanocarpum* is widely distributed at high elevations in the western United States from the Rocky Mountains to the Sierra Nevada of California. Its most common host is *Pinus flexilis*; however, *P. albicaulis*,

P. aristata, and *P. longaeva* are also listed as principal hosts, owing to their high incidence of infection. Additional secondary to rare hosts include *Picea engelmannii*, *Pinus balfouriana*, *Pinus contorta*, *Pinus monticola*, *Pinus ponderosa*, and *Tsuga mertensiana*. This mistletoe is a significant pathogen in many locations, sometimes resulting in massive host tree mortality.

Meiosis occurs in July, with fruits maturing 14 months after pollination.

Subspecies *littorum* parasitizes *Pinus muricata* and *P. radiata*, and occasionally *P. contorta* in Alameda, Mendocino, and Monterey counties. It has the largest shoots among the 13 subspecies of *Arceuthobium campylopodum*.

7g. Arceuthobium campylopodum Engelmann subsp. **laricis** (M. E. Jones) Nickrent, Phytoneuron 2012-51: 9. 2012 • Larch dwarf mistletoe [E]

Arceuthobium douglasii Engelmann var. *laricis* M. E. Jones, Biol. Ser. Bull. State Univ. Montana. 15: 25. 1910; *A. laricis* (M. E. Jones) H. St. John; *Razoumofskya laricis* (M. E. Jones) Piper

Plants forming witches' brooms. **Stems** green, olive green, maroon, or purple, 4(–6) cm; third internode 5–8(–14) × 1–1.3(–2.5) mm, dominant shoot 1.5–3 mm diam. at base. **Staminate flowers** 2.7 mm diam.; petals 3(–4). **Fruits** 3.5 × 2.5 mm.

Flowering Jul–Aug; fruiting (Aug–)Sep(–Oct). Coniferous forests, especially with western larch or mountain hemlock; 600–2300 m; B.C.; Idaho, Mont., Oreg., Wash.

Meiosis occurs in June, with fruits maturing 13 to 14 months after pollination.

Larix occidentalis and *Tsuga mertensiana* are the principal hosts for subsp. *laricis*; secondary to rare hosts include *Abies grandis*, *A. lasiocarpa*, *Picea engelmannii*, *Pinus albicaulis*, *Pinus contorta*, *Pinus monticola*, and *Pinus ponderosa*. This dwarf mistletoe is a major pathogen on larch in Idaho and Montana.

7h. Arceuthobium campylopodum Engelmann subsp. **littorum** (Hawksworth, Wiens & Nickrent) Nickrent, Phytoneuron 2012-51: 10. 2012 • Coastal dwarf mistletoe [E]

Arceuthobium littorum Hawksworth, Wiens & Nickrent, Novon 2: 206. 1992

Plants usually forming witches' brooms. **Stems** olive green or brown, 8–12(–20) cm; third internode 10–15(–20) × 2–3.5(–5) mm, dominant shoot 2–5 mm diam. at base. **Staminate flowers** 3 mm diam.; petals 4. **Fruits** 4–5 × 3 mm.

Flowering Aug–Sep; fruiting Sep–Oct. Closed-cone pine forests; 0–300 m; Calif.

7i. Arceuthobium campylopodum Engelmann subsp. **microcarpum** (Engelmann) Nickrent, Phytoneuron 2012-51: 10. 2012 • Western spruce dwarf mistletoe [E]

Arceuthobium douglasii Engelmann var. *microcarpum* Engelmann in J. T. Rothrock, Rep. U.S. Geogr. Surv., Wheeler, 253. 1879; *A. microcarpum* (Engelmann) Hawksworth & Wiens

Plants forming witches' brooms. **Stems** yellowish green, green, orange, red, maroon, or purple, 5(–11) cm; third internode 5–9.3(–16) × 1–1.5(–2) mm, dominant shoot 1.5–3 mm diam. at base. **Staminate flowers** 2.3 mm diam.; petals 3(–4). **Fruits** 3.5 × 2 mm.

Flowering (Jul–)Aug–Sep; fruiting (Aug–)Sep(–Oct). Coniferous forests, especially with blue or Engelmann spruce; 2400–3200 m; Ariz., N.Mex.

Meiosis occurs in July, with fruits maturing 12 to 13 months after pollination.

Subspecies *microcarpum* is a serious pathogen on its principal hosts, *Picea engelmannii* and *P. pungens*, as well as on *Pinus aristata* in northern Arizona. Rare hosts include *Abies lasiocarpa* and *Pinus strobiformis*. Interestingly, this mistletoe is not found in the central Rocky Mountains where its principal hosts are most abundant.

7j. Arceuthobium campylopodum Engelmann subsp. **monticola** (Hawksworth, Wiens & Nickrent) Nickrent, Phytoneuron 2012-51: 10. 2012 • Western white pine dwarf mistletoe [E]

Arceuthobium monticola Hawksworth, Wiens & Nickrent, Novon 2: 205. 1992

Plants forming localized infections only. **Stems** olive green or brown, 5–7(–10) cm; third internode 8–12(–15) × 1.5–1.7(–2) mm, dominant shoot 2–4 mm diam. at base. **Staminate flowers** 3 mm diam.; petals 3. **Fruits** 4–4.5 × 2–2.5 mm.

Flowering Jul–Aug; fruiting Oct–Nov. Coniferous forests, especially western white pine; 700–1900 m; Calif., Oreg.

Meiosis likely occurs in July, with fruits maturing 15 months after pollination.

The principal host of subsp. *monticola* is *Pinus monticola*; secondary to rare hosts include *Picea breweriana*, *Pinus jeffreyi*, and *Pinus lambertiana*. It is endemic to the Klamath and Siskiyou Mountains.

7k. Arceuthobium campylopodum Engelmann subsp. **occidentale** (Engelmann) Nickrent, Phytoneuron 2012-51: 10. 2012 • Digger pine dwarf mistletoe [E]

Arceuthobium occidentale Engelmann in J. T. Rothrock, Rep. U.S. Geogr. Surv., Wheeler, 254, 375. 1879

Plants usually forming localized infections only. **Stems** yellow or orange, 8(–17) cm; third internode 7–12.7(–18) × 1.5–1.8 (–3.5) mm, dominant shoot 1.5–5 mm diam. at base. **Staminate flowers** 3 mm diam.; petals 3–4. **Fruits** 4.5 × 3 mm.

Flowering Sep–Nov(–Dec); fruiting (Sep–)Oct–Jan (–Feb). Coniferous and mixed forests, especially digger pine; 30–1200 m; Calif.

Meiosis occurs in August, with fruits maturing 13 months after pollination.

Pinus sabiniana is the principal host of subsp. *occidentale*; secondary hosts include *P. attenuata*, *P. coulteri*, *P. jeffreyi*, and *P. ponderosa*, as well as some exotic species of pines. Subspecies *occidentale* occurs in the foothills surrounding the Central Valley and in the Coast Ranges.

7l. Arceuthobium campylopodum Engelmann subsp. **siskiyouense** (Hawksworth, Wiens & Nickrent) Nickrent, Phytoneuron 2012-51: 10. 2012 • Knobcone pine dwarf mistletoe [E]

Arceuthobium siskiyouense Hawksworth, Wiens & Nickrent, Novon 2: 204. 1992

Plants forming witches' brooms. **Stems** brown, 6–8(–10) cm; third internode 8–9(–15) × 2 mm, dominant shoot 2–2.5 mm diam. at base. **Staminate flowers** 3 mm diam.; petals 3–4. **Fruits** 4 × 2.5 mm.

Flowering Aug–Sep; fruiting Sep–Oct. Coniferous forests, especially closed-cone pine forests with knobcone pine; 400–1200 m; Calif., Oreg.

Meiosis likely occurs in July, with fruits maturing 13 months after pollination.

The principal host of subsp. *siskiyouense* is *Pinus attenuata*; rare hosts include *P. contorta*, *P. jeffreyi*, and *P. ponderosa*. Subspecies *siskiyouense* is endemic to the Klamath and Siskiyou mountains.

7m. Arceuthobium campylopodum Engelmann subsp. **tsugense** (Rosendahl) Nickrent, Phytoneuron 2012-51: 10. 2012 • Hemlock dwarf mistletoe [E]

Razoumofskya tsugensis Rosendahl, Minnesota Bot. Stud. 3: 272, plates 27, 28. 1903; *Arceuthobium tsugense* (Rosendahl) G. N. Jones; *A. tsugense* subsp. *amabilae* Mathiasen & C. M. Daugherty; *A. tsugense* subsp. *contortae* Wass & Mathiasen; *A. tsugense* subsp. *mertensianae* Hawksworth & Nickrent

Plants forming witches' brooms. **Stems** yellow, green, olive green, or purple, 5–7(–13) cm; third internode 4–9.2(–16) × 1–1.5(–2) mm, dominant shoot 1.5–4 mm diam. at base. **Staminate flowers** 2.8 mm diam.; petals 3–4. **Fruits** 3 × 3 mm.

Flowering Jul–Sep(–Oct); fruiting (Aug–)Sep–Nov. Coniferous forests; 0–2500 m; B.C.; Alaska, Calif., Idaho, Oreg., Wash.

Meiosis occurs in July, with fruits maturing 12 to 13 months after pollination.

Subspecies *tsugense* has the broadest host range among all subspecies of *Arceuthobium campylopodum*. F. G. Hawksworth and D. Wiens (1996) considered subsp. *tsugense* to be a distinct species with two subspecies, subspp. *tsugense* and *mertensianae*. Two additional subspecies have been named in *A. tsugense*: subsp. *amabilae* and subsp. *contortae*. Although the authors of these subspecies presented evidence of quantitative character variation and differences in phenology and host preferences, all variants are here considered host races of subsp. *tsugense* (and could be treated taxonomically as forms if so desired). The principal hosts of subsp. *tsugense* are *Abies amabilis*, *A. lasiocarpa*, *A. procera*, *Pinus contorta*, *Tsuga heterophylla*, and *T. mertensiana*. Secondary to rare hosts include *A. grandis*, *Picea breweriana*, *Picea engelmannii*, *Picea sitchensis*, *Pinus albicaulis*, *Pinus monticola*, and *Pseudotsuga menziesii*.

2. PHORADENDRON Nuttall, J. Acad. Nat. Sci. Philadelphia, n.s. 1: 185. 1848 • Mistletoe [Greek *phor*, thief, and *dendron*, tree, alluding to parasitism]

Subshrubs, evergreen, monoecious or dioecious; hemiparasitic on branches of woody angiosperms and gymnosperms, infections localized [systemic]. **Stems** single or multiple; branching percurrent (branches with single main axis) [pseudodichotomous]. **Leaves** scalelike or well developed. **Inflorescences** axillary or terminal, unisexual (bisexual in *P. rubrum*), spikelike thyrses with intercalary meristems; flowers borne in cavities or grooves. **Staminate flowers:** petals (2–)3(–4), triangular, distinct; stamens (2–)3(–4); anthers 2-locular, dehiscing by transverse slits; nectary absent. **Pistillate flowers:** petals (2–)3(–4), triangular, distinct; ovary 1-locular; style short; stigma undifferentiated [2-lobed]. **Berries** sessile, not explosively dehiscent, 1-colored, smooth or puberulent, petal remnants persisting at apex. **Seeds** mucilaginous when removed from fruit, endosperm flattened, ovate to elliptical in broadest outline; embryo oriented longitudinally. $x = 14$.

Species ca. 244 (7 in the flora): United States, Mexico, West Indies, Central America, South America.

Although not particularly diverse in the United States, *Phoradendron* underwent a massive radiation in Mexico, Central America, and South America. The first modern taxonomic treatment of the genus was by W. Trelease (1916), who named 240 species. More recently, J. Kuijt (2003) produced a monograph that included 234 species, but the similarity in number belies the fact that only 18 of the names accepted by Trelease were retained.

Phoradendron is closely related to *Dendrophthora* Eichler (a genus of 125 species), separated from it only by the presence of 2-locular versus 1-locular anthers. Whether these two genera are monophyletic remains to be tested using molecular methods (V. E. T. M. Ashworth 2000). Together *Phoradendron* and *Dendrophthora* are important components of mesic and arid environments in the New World, particularly because their fruits provide food for various bird species that disperse their seeds.

In the key and descriptions that follow, basal phyllotaxy refers to the orientation of the basal pair of leaves or cataphylls on a lateral branch. When those leaves or cataphylls are in the same plane as the main and lateral branch, basal phyllotaxy is median; when they are at right angles to that plane, it is transverse. Flower seriation refers to the arrangement of flowers on fertile internodes (J. Kuijt 2003). Different seriation may occur in staminate and pistillate inflorescences of the same species (for example in *Phoradendron californicum*). Flowers typically are arranged in one or more columns above each bract, and each set of columns is topped by a single median flower. Each fertile internode has two sets of flower columns, one above each of the opposite bracts. When two columns of flowers occur above the bract, the condition is called biseriate. When three columns of flowers occur, the condition is triseriate. Uniseriate and multiseriate conditions exist, but only outside the flora area. When only three flowers occur above each bract, seriation cannot always be determined (for example in *P. juniperinum*).

SELECTED REFERENCES Ashworth, V. E. T. M. 2000. Phylogenetic relationships in Phoradendreae (Viscaceae) inferred from three regions of the nuclear ribosomal cistron. I. Major lineages and paraphyly of *Phoradendron*. Syst. Bot. 25: 349–370. Ashworth, V. E. T. M. 2000b. Phylogenetic relationships in Phoradendreae (Viscaceae) inferred from three regions of the nuclear ribosomal cistron. II. The North American species of *Phoradendron*. Aliso 19: 41–53. Kuijt, J. 2003. Monograph of *Phoradendron* (Viscaceae). Syst. Bot. Monogr. 66. Trelease, W. 1916. The Genus *Phoradendron*: A Monographic Revision. Urbana. Wiens, D. 1964. Revision of the acataphyllous species of *Phoradendron*. Brittonia 16: 11–54.

1. Leaves scalelike.
   2. Stems green to olive green, glabrous; inflorescence fertile internodes usually 1; parasitic on gymnosperms, frequently *Calocedrus* and *Juniperus* . . . . 1. *Phoradendron juniperinum*
   2. Stems grayish green to reddish green (in full sun), densely hairy (hairs silvery white, closely appressed), becoming glabrate; inflorescence fertile internodes (1–)2–4(–6); parasitic on angiosperms, frequently *Prosopis*, *Senegalia*, or *Vachellia* . . . . 2. *Phoradendron californicum*
1. Leaves well developed.
   3. Basal phyllotaxy median.
      4. Stems densely stellate-hairy; plants dioecious; Arizona, New Mexico, Texas . . . . 3. *Phoradendron capitellatum*
      4. Stems glabrous; plants monoecious; Florida . . . . 4. *Phoradendron rubrum*
   3. Basal phyllotaxy transverse.
      5. Staminate and pistillate inflorescences 3–6 mm; pistillate inflorescence fertile internodes each 2-flowered; parasitic on *Abies*, *Cupressus*, and *Juniperus* . . . . 5. *Phoradendron bolleanum*
      5. Staminate and pistillate inflorescences 10–80 mm; pistillate inflorescence fertile internodes each (4–)6–11(–24)-flowered; parasitic on angiosperms.
         6. Flowering Oct–Mar; stem internodes 8–59 mm; berries glabrous. . . . 6. *Phoradendron leucarpum*
         6. Flowering Jul–Sep; stem internodes 1.5–3.8 mm; berries puberulent below petals . . . . 7. *Phoradendron villosum*

1. Phoradendron juniperinum A. Gray, Mem. Amer. Acad. Arts, n. s. 4: 58. 1849 • Juniper or incense cedar mistletoe [F]

Phoradendron juniperinum subsp. *libocedri* (Engelmann) Wiens; *P. juniperinum* var. *libocedri* Engelmann; *P. juniperinum* var. *ligatum* (Trelease) Fosberg; *P. libocedri* (Engelmann) Howell; *P. ligatum* Trelease

Subshrubs erect, 1–2(–2.5) dm, dioecious. **Stems** green to olive green, glabrous; internodes terete, 5–20 × 1.5–2.5 mm. **Leaves** green to olive green, scalelike; blade triangular, 2 mm, apex acute; basal phyllotaxy transverse. **Staminate inflorescences** 3–5 mm; peduncle with 1 internode, 3 mm; fertile internode usually 1, 6-flowered, seriation unknown, flowers 3 (2 proximal, 1 distal) per bract. **Pistillate inflorescences** 3–5 mm; peduncle with 1 internode, 2 mm; fertile internode 1, 2-flowered, flowers 1 per bract. **Flowers:** petals 3–4, 0.5–1 mm. **Berries** white or pinkish, globose to ellipsoid-globose, 4–5 × 3 mm, glabrous. **2*n*** = 28.

Flowering summer–early fall. Forests or woodlands with juniper or incense cedar; 800–2900 m; Ariz., Calif., Colo., Idaho, Nev., N.Mex., Oreg., Tex., Utah; Mexico (Baja California, Chihuahua, Coahuila, Sonora).

Phoradendron juniperinum is often classified as having two subspecies, subspp. *juniperinum* and *libocedri*. Subspecies *juniperinum* is found throughout the species' range as globose infections on various species of *Juniperus*. The larger, pendent parasites of *Calocedrus* from California have been recognized as subsp. *libocedri*. J. Kuijt (2003) argued that this habit could be a host response because intermediate morphologies are known; the two taxa are not recognized here.

2. Phoradendron californicum Nuttall, J. Acad. Nat. Sci. Philadelphia, n. s. 1: 185. 1848 • Mesquite mistletoe

Phoradendron californicum var. *distans* Trelease; *P. californicum* var. *leucocarpum* (Trelease ex Munz & I. M. Johnston) Jepson

Subshrubs pendent, 1–5(–20) dm, dioecious. **Stems** grayish green to reddish green (in full sun), densely hairy, hairs silvery white, closely appressed, becoming glabrate; internodes terete, 10–20(–30) × 1–1.7(–2.5) mm. **Leaves** grayish green, scalelike; blade triangular, 1.5–3 mm, apex acute; basal phyllotaxy transverse or median. **Staminate inflorescence** 5–25 mm, peduncle with 1(–2) internodes, each 0.5–3 mm; fertile internodes (1–)2–3(–5), each 6–14-flowered, biseriate, flowers 1–3 per column. **Pistillate inflorescences** 5–10 mm, elongating in fruit; peduncle with 1(–2) internodes, each 0.5–3 mm; fertile internodes (1–)2–4(–6), each 2-flowered, flowers 1 per bract. **Flowers:** petals 3(–4), 1–2 mm. **Berries** white, translucent yellowish, pinkish, orange-red, or maroon, globose, 3–6 × 3–6 mm, glabrous. **2*n*** = 28.

Flowering late fall–winter. Desert scrub or washes with mesquite or acacia; 0–1800 m; Ariz., Calif., Nev., Utah; Mexico (Baja California, Baja California Sur, Sinaloa, Sonora).

Phoradendron californicum, like *P. juniperinum*, bears only scalelike leaves, but this character has evolved independently in the two species (V. E. T. M. Ashworth 2000). *Phoradendron californicum* differs by its different hosts (legumes versus conifers) and inflorescences with more than one fertile internode. Molecular data indicate that *P. californicum* is not part of the acataphyllous Boreales group in the sense of W. Trelease (1916) but allied with cataphyllous tropical species (Ashworth). Varieties and host races have been proposed, but these are not recognized in the most recent monograph of the genus (J. Kuijt 2003). In addition to its primary hosts, *Prosopis*, *Senegalia*, and *Vachellia*, *Phoradendron californicum* has also been recorded from a number of other hosts including *Condalia*, *Dalea*, *Ebenopsis*, *Havardia*, *Larrea*, *Olneya*, *Parkinsonia*, and sometimes is hyperparasitic on *Psittacanthus*.

3. **Phoradendron capitellatum** Torrey ex Trelease, Phoradendron, 25, plate 17. 1916 • Downy mistletoe

Phoradendron bolleanum (Seemann) Eichler var. *capitellatum* (Torrey ex Trelease) Kearney & Peebles

Subshrubs erect but pendulous with age, 3–6 dm, dioecious. **Stems** green, densely hairy, hairs stellate, fine, white; internodes terete, 5–15 × 1–2 mm. **Leaves** green, well developed; petiole indistinct; blade narrowly elliptic to spatulate or oblanceolate, 8–15 × 1–3 mm, fleshy, base cuneate, apex rounded or acute; basal phyllotaxy median. **Staminate inflorescences** 3–5 mm; peduncle with 1 internode, 0.5–2.5 mm; fertile internodes 1–2, each 6-flowered, seriation unknown, flowers 3 (2 proximal, 1 distal) per bract. **Pistillate inflorescences** 3–5 mm; peduncle with 1 internode, 0.5–2.5 mm; fertile internodes 1–2, each 2-flowered, flowers 1 per bract, deeply embedded in axis. **Flowers:** petals 3, 1 mm. **Berries** pinkish white, globose, 3.5 × 3.5 mm, glabrous. **2*n*** = 28.

Flowering winter. Juniper-pinyon woodlands; 1000–2000 m; Ariz., N.Mex., Tex.; Mexico (Chihuahua, Sonora).

Phoradendron capitellatum is parasitic exclusively on *Juniperus* (for example, *J. monosperma*, *J. osteosperma*, and *J. pinchotii*). This species has sometimes been classified as a subspecies of *P. bolleanum*; however, it differs in flowering time, basal phyllotaxy, stellate pubescence, and inflorescences sometimes with more than one fertile internode. These two species, along with *P. juniperinum*, were supported as monophyletic using molecular data, although relationships among them were not fully resolved (V. E. T. M. Ashworth 2000). These parasites of conifers were all classified in sect. *Pauciflorae* Engler in the sense of D. Wiens (1964).

4. **Phoradendron rubrum** (Linnaeus) Grisebach, Fl. Brit. W.I., 314. 1860 • Narrow-leaved mistletoe

Viscum rubrum Linnaeus, Sp. Pl. 2: 1023. 1753

Subshrubs erect, 3.5–5 dm, monoecious. **Stems** green, glabrous; internodes quadrangular proximally, flattened distally, keeled proximally to nodes, 20–30 × 3(–7) mm. **Leaves** dull green, well developed; petiole (3–)5(–8) mm; blade obovate, elliptic, oblanceolate, or oblong-lanceolate, (40–)50.5(–90) × 20–40 mm, thin, base cuneate, apex rounded; basal phyllotaxy median. **Inflorescences** bisexual, staminate flowers few, irregularly placed among pistillate, to 25 mm; peduncle with 1(–2) internodes, each 3 mm; fertile internodes 3, each 6–18-flowered, biseriate, flowers 1–4 per column. **Flowers:** petals 3, 1 mm. **Berries** lemon yellow or orange [pink, red], ovoid to globose, 4 × 3 mm, glabrous.

Flowering year-round. Hammocks with West Indian mahogany; 0–500 m; Fla.; West Indies.

Phoradendron rubrum is a mainly Caribbean species that has been recorded in the flora area only from Key Largo, Monroe County. Its primary host is *Swietenia mahagoni*, but it has been found also on *Byrsonima*, *Guapira*, *Mangifera*, and *Pisonia*.

5. **Phoradendron bolleanum** (Seemann) Eichler in C. F. P. von Martius et al., Fl. Bras. 5(2): 134m. 1868 • Bollean mistletoe

Viscum bolleanum Seemann, Bot. Voy. Herald, 295, plate 63. 1856; *Phoradendron bolleanum* subsp. *densum* (Torrey ex Trelease) Wiens; *P. bolleanum* var. *densum* (Torrey ex Trelease) Fosberg; *P. bolleanum* subsp. *pauciflorum* (Torrey) Wiens; *P. densum* Torrey ex Trelease; *P. hawksworthii* Wiens; *P. pauciflorum* Torrey

Subshrubs erect, forming globose clumps, to 10 dm, dioecious. **Stems** green, brown, reddish brown, or orange, glabrous or slightly puberulent, hairs simple; internodes terete, to 2 cm. **Leaves** green, well developed, glabrous or slightly puberulent, hairs simple; petiole very short or absent; blade terete to narrowly oblong or oblanceolate, 7–35 × 1–10 mm, thin, base slightly tapered, apex acute-apiculate to rounded; basal phyllotaxy transverse. **Staminate inflorescences** 3–6 mm, glabrous or slightly puberulent, hairs simple; peduncle with 1 internode, 1–2 mm; fertile internode usually 1, 6–20-flowered, triseriate, flowers 1–3 per column or not in columns. **Pistillate inflorescences** 3–6 mm, elongating in fruit, glabrous or slightly puberulent, hairs simple; peduncle with 1 internode, 1.5–3 mm; fertile internodes 1(–2), each 2-flowered, flowers 1 per bract. **Flowers:** petals 3–4, 1–2 mm. **Berries** white to pink, ovoid, 3.5–5 × 3.5–5 mm, glabrous. **2*n*** = 28.

Flowering May–Aug. Coniferous forests; 300–3000 m; Ariz., Calif., N.Mex., Oreg., Tex.; Mexico; Central America (Guatemala).

This treatment follows J. Kuijt (2003), who considered *Phoradendron bolleanum* to be a widespread and variable species complex. The three main taxa that often have been recognized as species are *P. bolleanum* in the narrow sense, *P. densum*, and *P. pauciflorum*. *Phoradendron bolleanum* in the narrow sense has small, narrow leaves and frequently parasitizes *Juniperus* as well as *Arbutus*. The *P. pauciflorum* variant has broad leaves and mostly parasitizes *Abies concolor*. The *P. densum* variant tends to have leaves that are intermediate between those of the other two variants; it ranges from Oregon to Mexico and parasitizes *Cupressus* and *Juniperus*. Molecular analyses indicate that *P. bolleanum* is not monophyletic unless *P. minutifolium* Urban is synonymized with the other variants. Hybrids between *P. bolleanum* and *P. juniperinum* produce plants closely resembling *P. minutifolium* (D. Wiens and M. DeDecker 1972). As pointed out by Kuijt, molecular studies will be required to determine species boundaries within this complex.

6. Phoradendron leucarpum (Rafinesque) Reveal & M. C. Johnston, Taxon 38: 107. 1989 • Oak or American mistletoe [F]

Viscum leucarpum Rafinesque, Fl. Ludov., 79. 1817; *Phoradendron coloradense* Trelease; *P. eatonii* Trelease; *P. flavens* Grisebach subsp. *macrophyllum* (Engelmann) A. E. Murray; *P. flavens* var. *macrophyllum* Engelmann; *P. flavens* var. *tomentosum* (de Candolle) Engelmann; *P. flavescens* (Pursh) Nuttall ex A. Gray; *P. leucarpum* subsp. *angustifolium* (Kuijt) J. R. Abbott & R. L. Thompson; *P. leucarpum* subsp. *macrophyllum* (Engelmann) J. R. Abbott & R. L. Thompson; *P. leucarpum* subsp. *tomentosum* (de Candolle) J. R. Abbott & R. L. Thompson; *P. longispicum* Trelease; *P. macrotomum* Trelease; *P. serotinum* (Rafinesque) M. C. Johnston; *P. serotinum* subsp. *macrophyllum* (Engelmann) Kuijt; *P. serotinum* var. *macrophyllum* (Engelmann) M. C. Johnston; *P. serotinum* var. *macrotomum* (Trelease) M. C. Johnston; *P. serotinum* subsp. *tomentosum* (de Candolle) Kuijt; *P. tomentosum* (de Candolle) Engelmann ex A. Gray; *P. tomentosum* subsp. *macrophyllum* (Engelmann) Wiens; *P. tomentosum* var. *macrophyllum* (Engelmann) L. D. Benson

Subshrubs erect, 4–10 dm, dioecious. **Stems** green, grayish green, or yellowish green, hairy, hairs simple or stellate, white or yellow, becoming glabrate; internodes terete, 8–59 × 1–3 mm. **Leaves** bright green, yellowish green, or grayish green, well developed, hairy, hairs simple or stellate; petiole 3–8 mm; blade obovate, spatulate, ovate, ovate-elliptic, or nearly orbiculate, 14–48 × 8–30 mm, thin to thick and rigid, base cuneate to obtuse, apex rounded; basal phyllotaxy transverse. **Staminate inflorescences** 10–80 mm, hairy, hairs simple or stellate; peduncle with 1 internode, 2–4 mm; fertile internodes 2–7, each (15–)29–39(–62)-flowered, triseriate, becoming irregular, flowers 1–10 per column. **Pistillate inflorescences** 10–80 mm, hairy, hairs simple or stellate; peduncle with 1 internode, 2–4 mm; fertile internodes 2–6, each (4–)6–11(–20)-flowered, triseriate, flowers 1–3 per column. **Flowers:** petals 3, 1 mm. **Berries** white, oblong to globose, 3–6 × 2–5 mm, glabrous.

Flowering Oct–Mar. Hardwood forests and woodlands; 0–1800 m; Ala., Ariz., Ark., Calif., Del., D.C., Fla., Ga., Ill., Ind., Kans., Ky., La., Md., Miss., Mo., N.J., N.Mex., N.Y., N.C., Ohio, Okla., Oreg., Pa., S.C., Tenn., Tex., Va., W.Va.; Mexico (Chihuahua, Coahuila, Durango, Nuevo León, San Luis Potosí, Sonora, Tamaulipas).

J. Kuijt (2003) used the name *Phoradendron serotinum*, based on the name *Viscum serotinum* Rafinesque (1820), not *P. leucarpum*, which is based on the earlier name by the same author, *V. leucarpum* (1817). A proposal to conserve the later name (D. L. Nickrent et al. 2010b) was not accepted, thus the name *P. leucarpum* has priority.

Phoradendron leucarpum has a convoluted taxonomic history, reflecting not only various species concepts but also complex evolutionary and ecological processes. Among the 234 species of *Phoradendron*, J. Kuijt (2003) recognized subspecies only in *P. leucarpum* (as *P. serotinum*). In addition to the typical subspecies from eastern Texas eastward, they are subsp. *augustifolium* from Mexico, subsp. *macrophyllum* from eastern Texas through New Mexico and Arizona to California and Oregon, and subsp. *tomentosum*, with about the same distribution as subsp. *macrophyllum* but also extending into Mexico. Kuijt noted that in some geographic areas, such as east-central Texas, the putative subspecies show a continuum of morphological intergradation.

A population genetic and morphometric study of this complex was undertaken by A. K. Hawkins (2010). Principal component analyses using the characters that J. Kuijt (2003) considered to be diagnostic of the subspecies, such as leaf size, color, and venation, as well as the type and density of hairs present on young vegetative and reproductive tissues, in addition to host species, did not result in clusters corresponding to the four described subspecies. Moreover, F_{ST} analyses of microsatellites showed significant interpopulational differentiation that did not match the subspecies that

Kuijt recognized. Because morphological and molecular analyses show that subspecies, at least as defined by Kuijt, cannot be differentiated in *Phoradendron leucarpum*, no subspecies are accepted here.

Phoradendron leucarpum is the only species of the genus found east of Texas. It parasitizes over 60 species of native and introduced trees, especially *Acer*, *Fraxinus*, *Juglans*, *Nyssa*, *Platanus*, *Populus*, *Quercus*, *Salix*, and *Ulmus*.

SELECTED REFERENCE Hawkins, A. K. 2010. Subspecific Classification within *Phoradendron serotinum* (Viscaceae): Development of Microsatellite Markers for Assessment of Population Genetic Structure. M.S. thesis. Sam Houston State University.

7. Phoradendron villosum (Nuttall) Nuttall ex Engelmann, Boston J. Nat. Hist. 6: 212. 1850 • Pacific or oak mistletoe

Viscum villosum Nuttall in J. Torrey and A. Gray, Fl. N. Amer. 1: 654. 1840; *Phoradendron flavens* Grisebach var. *villosum* (Nuttall) Engelmann

Subshrubs erect, to 1 dm, dioecious. **Stems** green or olive green, grayish from pubescence, hairy, hairs stellate, becoming hirtellous to glabrate; internodes terete, 1.5–3.8 × 1–3 mm. **Leaves** green or olive green, grayish from pubescence, stellate-hairy; blade obovate-elliptic to orbiculate, 13–45 × 9–22 mm, thin to thick and fleshy, base subtruncate to acute, apex rounded to acute; basal phyllotaxy transverse. **Staminate inflorescences** 10–80 mm, stellate-hairy; peduncle with 1 internode, 2–4 mm; fertile internodes 2–5, each (14–)26(–44)-flowered, triseriate, becoming irregular, flowers 2–7 per column. **Pistillate inflorescences** 10–80 mm, stellate-hairy; peduncle with 1 internode, 2–4 mm; fertile internodes 2–3, each (6–)11(–24)-flowered, triseriate, flowers 1–4 per column. **Flowers:** petals 3–4, 1 mm. **Berries** white to pink, oblong to globose, 3 × 3 mm, puberulent below petals.

Subspecies 3 (2 in the flora): w, sw, sc United States, n Mexico.

Phoradendron villosum was treated by J. Kuijt (2003) as a synonym of *P. serotinum* subsp. *tomentosum* (= *P. leucarpum* in this treatment), but *P. villosum* has a different, non-overlapping flowering time, shorter stem internodes, and hairy berries. In addition, *P. villosum* averages fewer pistillate inflorescence fertile internodes than *P. leucarpum*. Molecular studies by V. E. T. M. Ashworth (2000, 2000b) showed that *P. villosum* did not form part of the strongly supported *P. leucarpum* clade. For these reasons, they are here considered distinct species, as was done by D. Wiens (1964).

Subspecies *flavum* (I. M. Johnston) Wiens, distinguished by its yellow leaf hairs, is found in Coahuila and Durango, Mexico.

1. Hairs on young stems, leaf blades, and inflorescences of one type, uniformly distributed and of similar lengths; pistillate inflorescence internodes each (7–)12(–24)-flowered; California, Oregon . . . . 7a. *Phoradendron villosum* subsp. *villosum*
1. Hairs on young stems, leaf blades, and inflorescences of two types, some in dense clusters, relatively longer, others uniformly distributed, relatively shorter; pistillate inflorescence internodes each (6–)8(–12)-flowered; Arizona, New Mexico, Texas . . . . .7b. *Phoradendron villosum* subsp. *coryae*

7a. Phoradendron villosum (Nuttall) Nuttall ex Engelmann subsp. **villosum**

Phoradendron villosum var. *rotundifolium* Trelease

Hairs on young stems, leaf blades, and inflorescences of one type, uniformly distributed and of similar lengths. **Staminate inflorescences:** internodes each (15–)29(–45)-flowered. **Pistillate inflorescences:** internodes each (7–)12(–24)-flowered. $2n = 28$.

Flowering Jul–Sep. Oak woodlands; 20–2200 m; Calif., Oreg.; Mexico (Baja California).

Subspecies *villosum* is parasitic mainly on oaks, including *Quercus agrifolia*, *Q. douglasii*, and *Q. lobata* (V. E. Thompson and B. E. Mahall 1983), but can be found on *Aesculus*, *Arctostaphylos*, *Betula*, *Populus*, *Robinia*, and *Salix*. The subspecies is distributed from the Coast Ranges and Willamette Valley of Oregon south through the Sierra Nevada and Coast and Transverse ranges of California into northern Baja California.

7b. Phoradendron villosum (Nuttall) Nuttall ex Engelmann subsp. **coryae** (Trelease) Wiens, Brittonia 16: 45. 1964

Phoradendron coryae Trelease, Phoradendron, 43, plate 44. 1916; *P. villosum* var. *coryae* (Trelease) B. L. Turner

Hairs on young stems, leaf blades, and inflorescences of two types, some in dense clusters, relatively longer, others uniformly distributed, relatively shorter. **Staminate inflorescences:** internodes each (14–)25(–44)-flowered. **Pistillate inflorescences:** internodes each (6–)8(–12)-flowered.

Flowering Jul–Sep. Oak woodlands; 400–2400 m; Ariz., N.Mex., Tex.; Mexico (Chihuahua, Coahuila, Sonora).

In the flora area, subsp. *coryae* is found in Arizona south of the Mogollon Rim, through central New Mexico to the Chisos and Davis mountains of western Texas. Its principal hosts are *Quercus* species.

The molecular study by V. E. T. M. Ashworth (2000b) showed that *Phoradendron villosum* is paraphyletic because *P. scaberrimum* Trelease forms a clade with subsp. *coryae*. All three of these taxa parasitize *Quercus*, and a shared indel supported their phylogenetic relationship.

3. VISCUM Linnaeus, Sp. Pl. 2: 1023. 1753; Gen. Pl. ed. 5, 448. 1754 • Christmas mistletoe [Latin name for mistletoe, alluding to viscid fruits] [I]

Shrubs [herbs], dioecious [monoecious]; parasitic on branches of woody angiosperms and gymnosperms, infections localized. **Stems** single or multiple; branching pseudodichotomous [percurrent]. **Leaves** well developed [scalelike]. **Inflorescences** axillary or terminal, dichasial cymes. **Staminate flowers:** petals (3–)4(–6), triangular, distinct; stamens (3–)4(–6); anthers multilocular, dehiscing by numerous pores; nectary absent. **Pistillate flowers:** petals (3–)4(–6), triangular, distinct; ovary 0-locular; style absent [short-conic]; stigma poorly differentiated [capitate]. **Berries** sessile in bracteal cup [pedicel present, not recurved], not explosively dehiscent, 1-colored, smooth [warty], scars of petal remnants at apex. **Seeds** mucilaginous when removed from fruit, endosperm slightly flattened, ovate to elliptic in broadest outline; embryo oriented transversely. $x = 14$.

Species ca. 130 (1 in the flora): introduced; Eurasia, Africa, Indian Ocean Islands, Australia.

Viscum is widely distributed in the Old World and is present in North America via purposeful introduction. The genus is most diverse in tropical and southern Africa, where various species form a decreasing aneuploid series (from $x = 14$) (D. Wiens 1975). Higher gametic chromosome numbers are the result of polyploidy, which is relatively uncommon in Viscaceae. Several species of dioecious *Viscum* show translocation heterozygosity that determines plant sexuality and sex ratios in populations (Wiens and B. A. Barlow 1979; A. Aparicio 1993).

SELECTED REFERENCES Bussing, A., ed. 2000. Mistletoe: The genus *Viscum*. Amsterdam. Gill, L. S. 1935. *Arceuthobium* in the United States. Trans. Connecticut Acad. Arts 32: 111–245. Wiens, D. and B. A. Barlow. 1979. Translocation heterozygosity and the origin of dioecy in *Viscum*. Heredity 42: 201–222.

1. Viscum album Linnaeus, Sp. Pl. 2: 1023. 1753 [F] [I] [W]

Subspecies 3 (1 in the flora): introduced; Europe, Asia.

Viscum album is found throughout Europe and western Asia and has been divided into subsp. *album* on woody angiosperms, subsp. *abietis* (Wiesbaur) Abromeit on *Abies*, and subsp. *austriacum* (Wiesbaur) Vollmann on *Picea* and *Pinus* (D. Zuber 2004).

1a. Viscum album Linnaeus subsp. album • European mistletoe [F] [I] [W]

Shrubs forming globose infections, to 2 m diam. **Leaves** green to yellowish green; blade oblanceolate or obovate-oblong, 3–8 cm, leathery, base attenuate, apex rounded, surfaces glabrous; venation parallel, veins 3–7. **Staminate inflorescences** simple dichasia, consisting of 3 flowers subtended by 2 bracts fused into cupule. **Pistillate inflorescences** compound dichasia, usually 3-flowered; primary peduncle 4–6 mm; lateral flowers each subtended by 1 bract; terminal flower subtended by 2 bracts at apex of 2 mm secondary peduncle. **Staminate flowers** sessile, 5 mm; petals yellow; anthers white, covering most of petal adaxial surface. **Pistillate flowers** sessile,

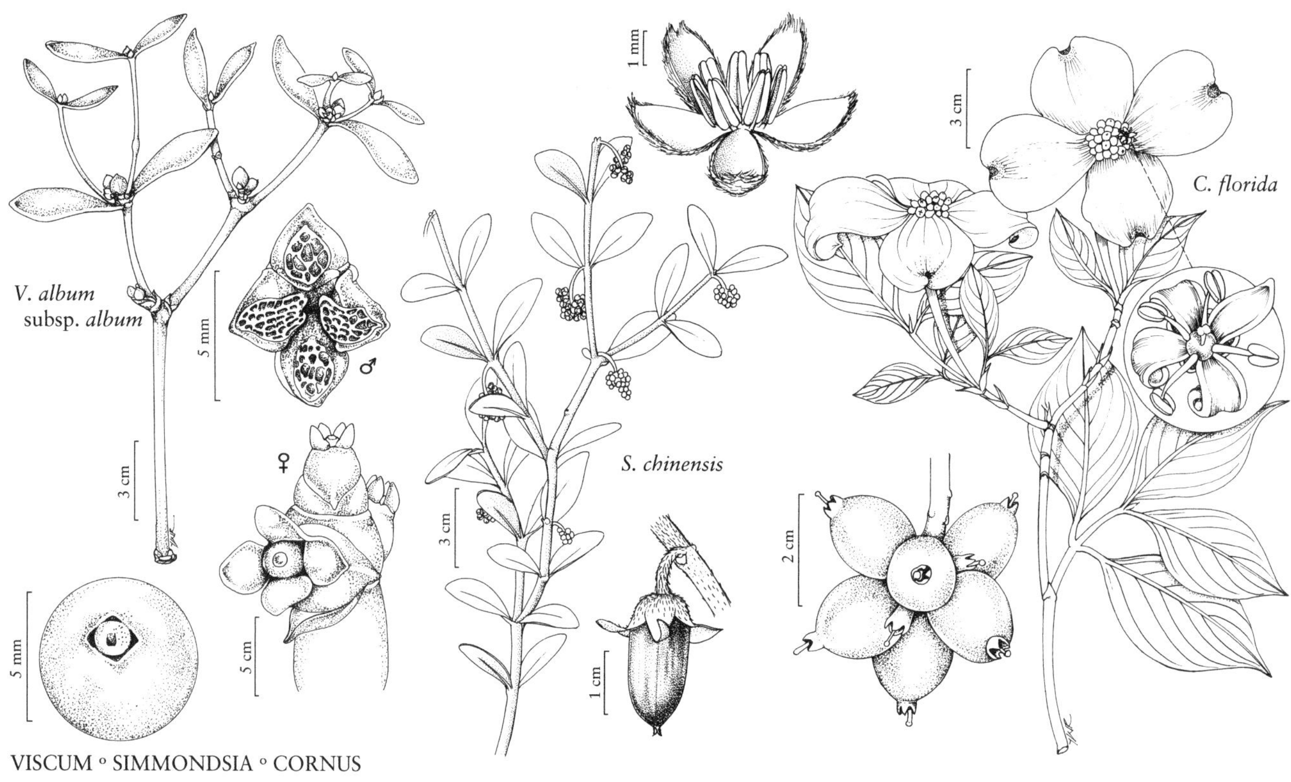

VISCUM ◦ SIMMONDSIA ◦ CORNUS

somewhat flattened, 2 mm; stigma cushion-shaped. **Berries** white, globose, 6–10 mm, apex with dark markings representing stigma surrounded by petal scars. **2*n*** = 20.

Flowering Feb–Apr; fruiting Nov–Dec. Orchards, urban forests; 60–100 m; introduced; B.C.; Calif.; Europe; w Asia.

Subspecies *album* parasitizes many different tree species, but is most often found on *Crataegus*, *Malus*, *Populus*, *Robinia*, and *Tilia*. It was inoculated onto trees in the Sebastopol area, Sonoma County, California, around 1900 by the horticulturalist Luther Burbank, who wished to cultivate it for Christmas decorations (R. F. Scharpf and W. McCartney 1975). Although this source remained localized for several decades, by the 1980s the population had been spread to 114 km^2 by birds such as cedar waxwings (*Bombycilla cedrorum*) and American robins (*Turdus migratorius*) (F. G. Hawksworth and Scharpf 1986). A survey in 1996 showed the mistletoe infestation occupied 220 km^2 but remained localized to Sonoma County (Scharpf 2002). In 1988, European mistletoe was found in Victoria, British Columbia, parasitizing *Malus*, but the population had apparently been purposely introduced more than 37 years before (C. E. Dorworth 1989). This naturalized mistletoe is categorized by the California Department of Food and Agriculture as a Noxious Weed List B because it has potential to negatively impact orchard crops. *Viscum album* is being studied extensively because it contains lectins that appear to have anticancer efficacy (A. Bussing 2000).

SIMMONDSIACEAE Tieghem

• Jojoba Family

Lynn J. Gillespie

Shrubs, evergreen, dioecious. **Leaves** opposite, simple; stipules absent; petiole present or absent; blade margins entire; venation pinnate. **Inflorescences** unisexual, axillary or terminal, glomerate cymes (staminate) or flowers usually solitary, rarely in 2–3-flowered clusters (pistillate). **Flowers** unisexual; perianth hypogynous; hypanthium absent; sepals (4–)5(–6), distinct; petals 0; nectary absent; stamens (8–)10(–12), distinct, free; anthers dehiscing by longitudinal slits; pistil 1, 3-carpellate; ovary superior, 3-locular, placentation apical-axile; ovule 1 per locule, apotropous; styles 3, distinct; stigmas 3. **Fruits** capsules, dehiscence loculicidal. **Seeds** 1(–3) per fruit.

Species 1: sw United States, nw Mexico.

The status of Simmondsiaceae has long been controversial; it has previously been included in the Euphorbiales, either submerged in Buxaceae or as a separate family near Buxaceae (A. Cronquist 1981), or in the Hamamelidae as its own order (A. L. Takhtajan 1997). Recent molecular phylogenetic evidence supports recognition as a monogeneric family with a near-basal position in the Caryophyllales (P. Cuénoud et al. 2002; Angiosperm Phylogeny Group 2003; K. W. Hilu et al. 2003; S. F. Brockington et al. 2009).

SELECTED REFERENCE Gentry, H. S. 1958. The natural history of jojoba (*Simmondsia chinensis*) and its cultural aspects. Econ. Bot. 12: 261–291.

1. SIMMONDSIA Nuttall, London J. Bot. 3: 400, plate 16. 1844 • Jojoba, goatnut, pignut [For Thomas Williams Simmonds, d. 1804, English naturalist]

Shrubs, indumentum of slender, mostly appressed, simple hairs. **Leaves** jointed near base; blade secondary veins steeply ascending from near base, usually basally distinct, distally obscure. **Inflorescences:** staminate terminal, often pseudoaxillary (terminal but overtopped by lateral branch and appearing axillary), pedunculate, 10–20-flowered; pistillate axillary, pedunculate, 1(–3)-flowered; bracts present. **Pedicels** present or absent. **Staminate flowers:** sepals oblanceolate. **Pistillate flowers:** sepals persistent, spreading, triangular-ovate, enlarged in fruit. **Capsules** nutlike, ovoid or ellipsoid, usually ± obtusely 3-angled; pericarp tough-leathery. **Seeds** ovoid. $x = 13$.

Species 1: sw United States, nw Mexico; desert and semidesert areas.

1. Simmondsia chinensis (Link) C. K. Schneider, Ill. Handb. Laubholzk. 2: 141. 1907 F

Buxus chinensis Link, Enum. Hort. Berol. Alt. 2: 386. 1822; *Simmondsia californica* Nuttall

Shrubs to 2(–3) m, mostly densely branched; young branches hairy. **Leaves** dull pale green or gray-green, sometimes glaucous; petiole 0–2 mm; blade elliptic or oblong-ovate, (1.5–)2–5.5 × (0.5–)1–3 cm, thick-leathery (often becoming wrinkled when dry), base acute to obtuse, apex rounded, obtuse, or acute, surfaces sparsely hairy to glabrate. **Staminate inflorescences:** peduncles 0.2–2 cm, recurved; cymes 1–2 cm diam. **Pistillate inflorescences:** peduncle (pedicel in solitary flowers) 0.2–2 cm, recurved; bracts 3–4, linear-lanceolate or lanceolate, 2–10 mm. **Pedicels:** staminate 0–2 mm, pistillate usually absent. **Staminate flowers** greenish yellow; sepals 2–4 mm, densely hairy abaxially; stamen filaments 0.1–0.3 mm; anthers 1.5–2 mm. **Pistillate flowers** pale green, pendent; sepals (in fruit) triangular-ovate, basal ½ appressed, distal ½ abruptly narrowed and spreading, 1–2 cm, surfaces densely hairy; styles exserted. **Capsules** shiny greenish brown, 15–25 × 9–11 mm, apex rounded, usually apiculate. **Seeds** reddish brown or brown, 13–17 mm. $2n = 26$.

Flowering late winter–spring; fruiting summer–fall. Desert scrub, coastal sage scrub, alluvial fans, arroyo margins, rocky, gravelly, or sandy slopes; 0–1500 m; Ariz., Calif., N.Mex.; Mexico (Baja California, Baja California Sur, Sonora).

In the flora area, *Simmondsia chinensis* is native to arid areas from southern California east through Arizona to extreme southwestern New Mexico (Hildago County); despite its specific epithet, it is not native to China. It is very drought tolerant and able to withstand extreme daily temperature fluctuations. Its thick, leathery leaves are usually held erect, maximizing interception of solar radiation during the cooler part of the day and minimizing it during the heat of midday. Flowering and fruiting times are dependent on timing, duration, and intensity of the winter and spring rains, but generally flowers are at anthesis March to April and fruits mature about six months later. Flower buds are initiated during development of new leafy shoots in winter and generally remain dormant until early spring (or longer under drought conditions). Capsules may reach full size several months before the seeds mature and disperse. The evergreen foliage is important as year-round forage for livestock and wild animals, and the fruits are eaten by wild mammals and birds.

Simmondsia chinensis is cultivated as a commercial crop in desert and semidesert areas (Australia, North and South America, and the Middle East) for its oil-rich seeds. Jojoba oil is an unusual liquid wax ester, more similar to whale oil than it is to most vegetable oils, and is used primarily in cosmetics. The species does not appear to have escaped or become naturalized outside its native range.

CORNACEAE Berchtold & J. Presl
• Dogwood Family

Zack E. Murrell
Derick B. Poindexter

Herbs, shrubs, or trees, perennial, deciduous, synoecious [dioecious]; hairs unbranched or 2-armed (occasionally 1 arm absent). **Leaves** usually opposite, sometimes alternate, simple; stipules absent; petiole usually present, sometimes absent; blade margins entire; venation pinnate (eucamptodromous). **Inflorescences** axillary or terminal, cymes, umbels, or capitula. **Flowers** bisexual [unisexual]; perianth and androecium epigynous; hypanthium completely adnate to ovary; sepals 4(–5), distinct or slightly connate; petals 4(–5), distinct, valvate; nectary present, intrastaminal; stamens 4(–5), distinct, free; anthers dehiscing by longitudinal slits; pistil 1, [1–]2[–4]-carpellate, ovary inferior, [1–]2[–4]-locular, placentation apical; ovules 1 per locule, apotropous to epitropous; style 1; stigmas 2. **Fruits** drupes, rarely fused into a syncarp. **Seeds** 1(–2) per fruit.

Genus 1, species ca. 60 (20 species in the flora): North America, Mexico, Central America, n, w South America, Eurasia, Africa; predominately northern boreal and temperate regions, also high elevations in subtropical and tropical regions.

Delimitation of Cornaceae has varied over the years, with many treatments circumscribing the family broadly to include *Nyssa* and several small Asian genera along with *Cornus* (for example, Xiang Q. Y. et al. 1998). DNA sequence data from intensive taxon sampling support narrower circumscription, with *Cornus* as the sole member of the family and sister to the Asian Alangiaceae de Candolle (Xiang et al. 2011); that treatment is followed here. *Aucuba*, sometimes included in Cornaceae, is here placed in Garryaceae; for *Aucuba*, see page 553 in this volume.

1. CORNUS Linnaeus, Sp. Pl. 1: 117. 1753; Gen. Pl. ed. 5, 54. 1754 • Dogwood [Latin *cornu*, horn, alluding to the hard wood]

Herbs, shrubs, or trees, clonal from rhizomes, rooting from decumbent branches, or aclonal; hairs 1-celled, arms either short and ornamented with micropapillae and calcium carbonate crystals, or long, erect, curling, and twisted. **Leaves:** blade lanceolate to broadly ovate; abaxial surface often papillate. **Inflorescences:** bracts adnate to inflorescence branches, distal portions either minute and caducous or expanding into showy, nonchlorophyllous involucres. **Pedicels** present or absent. **Flowers:** hypanthium turbinate or urceolate; petals spreading or recurved, usually cream, rarely purple; stamens exserted; anthers dorsifixed, versatile. **Drupes** globose, subglobose, or ellipsoid, slightly fleshy. $x = 11$.

Species ca. 60 (20 in the flora): North America, Mexico, Central America, n, w South America, Eurasia, Africa; predominately northern boreal and temperate regions, also high elevations in subtropical and tropical regions.

Cornus as treated here is a monophyletic genus (Z. E. Murrell 1993; Xiang Q. Y. et al. 2006) that has at various times been more narrowly circumscribed by other authors who have chosen to recognize morphological variation in this diverse group as worthy of generic segregation [for example, *Arctocrania* (Endlicher) Nakai, *Benthamia* Lindley (not A. Richard), *Benthamidia* Spach, *Chamaepericlymenum* Hill, *Cynoxylon* (Rafinesque) Small, *Eukrania* Rafinesque, *Macrocarpium* (Spach) Nakai, *Swida* Opiz, and *Thelycrania* (Dumortier) Fourreau]. *Cornus* is retained here as a coherent group, maintaining subgenera as more appropriate biological units for recognition of this variation.

Some North American members of *Cornus* are susceptible to fungal pathogens that may cause severe species decline, such as Dogwood Anthracnose (*Discula destructiva*) in association with *C. florida* and *C. nuttallii*, or the less virulent but still destructive Cryptodiaporthe Canker (*Cryptodiaporthe corni*), which is restricted to *C. alternifolia.*

SELECTED REFERENCES Eyde, R. H. 1988. Comprehending *Cornus*: Puzzles and progress in the systematics of dogwoods. Bot. Rev. (Lancaster) 54: 233–351. Ferguson, I. K. 1966. Notes on the nomenclature of *Cornus*. J. Arnold Arbor. 47: 100–105. Ferguson, I. K. 1966b. The Cornaceae of the southeastern United States. J. Arnold Arbor. 47: 106–116. Murrell, Z. E. 1992. Systematics of the Genus *Cornus* (Cornaceae). Ph.D. dissertation. Duke University. Murrell, Z. E. 1993. Phylogenetic relationships in *Cornus* (Cornaceae). Syst. Bot. 18: 469–495.

1. Bracts petaloid, subtending inflorescences.
   2. Perennial herbs; inflorescences congested cymes; pedicels present . . . 1c. *Cornus* subg. *Arctocrania*, p. 446
   2. Trees; inflorescences capitula; pedicels absent.
      3. Drupes within inflorescence fused into a syncarp . . . . 1a. *Cornus* subg. *Syncarpea*, p. 444
      3. Drupes distinct . . . . 1b. *Cornus* subg. *Cynoxylon*, p. 445
1. Bracts not petaloid, subtending inflorescences or not.
   4. Inflorescences umbels, bracts well developed, subtending inflorescence and enclosing it over winter. . . . . 1d. *Cornus* subg. *Cornus*, p. 449
   4. Inflorescences cymes, bracts minute, subtending primary and secondary inflorescence branches.
      5. Branches and leaves alternate; stone apex with cavity. . . . . .1e. *Cornus* subg. *Mesomora*, p. 450
      5. Branches and leaves usually opposite, rarely whorled, subopposite, or alternate at some nodes; stone apex rounded, pointed, or with slight dimple. . . 1f. *Cornus* subg. *Thelycrania*, p. 450

1a. CORNUS Linnaeus subg. SYNCARPEA (Nakai) Q. Y. Xiang, Acta Phytotax. Sin. 25: 128. 1987 [I]

Benthamia Lindley subg. *Syncarpea* Nakai, Bot. Mag. (Tokyo) 23: 41. 1909; *Dendrobenthamia* Hutchinson

Trees [shrubs]; rhizomes absent. **Branches and leaves** opposite. **Inflorescences** capitula; bracts 4, well developed, petaloid, subtending inflorescence but not enclosing it over winter, 2 pairs of scalelike bracts enclosing inflorescence over winter. **Pedicels** absent. **Drupes** within inflorescence fused into a syncarp; stone apex pointed.

Species 5 (1 in the flora): introduced; Asia.

1. Cornus kousa Bürger ex Hance, J. Linn. Soc., Bot. 13: 105. 1872 • Kousa dogwood [I]

Benthamia japonica Siebold & Zuccarini, Fl. Jap. 1: 38, plate 16. 1836, not *Cornus japonica* Thunberg 1784; *Benthamidia japonica* (Siebold & Zuccarini) H. Hara; *Dendrobenthamia japonica* (Siebold & Zuccarini) Hutchinson

Trees to 10 m, flowering at 2 m. **Stems** solitary or clustered, bark mottled gray and tan, of thin exfoliating plates; branches gray-maroon with longitudinal fissures, bark flaky; branchlets gray, densely appressed-hairy; lenticels pale, round or lenticular spots. **Leaves:** petiole 5–12 mm; blade elliptic, ovate, or widely ovate, 5–8 × 2–5 cm, base cuneate, apex long acuminate, abaxial surface pale green, appressed-hairy, tufts of erect hairs in axils of secondary veins, adaxial surface dark green, appressed-hairy; secondary veins 4–6 per side, evenly spaced, vein also present on leaf margin. **Inflorescences** subglobose, 0.9–1.4 cm diam., 40–75-flowered; peduncle 40–60 mm; petaloid bracts white to slightly yellow, narrowly ovate, rhombic, or trullate, 4–7 × 3–6 cm, base cuneate, apex acuminate. **Flowers:** hypanthium appressed-hairy; sepals 1–1.5 mm; petals cream or yellow-green, 3–4 mm. **Syncarps** red, subglobose, 20–27 mm diam.; stones ellipsoid, 7–9 × 4–5 mm, smooth.

Flowering Apr–Jul; fruiting Aug–Oct. Disturbed areas, lawns; 0–1000 m; introduced; N.Y., N.C.; Asia (China, Japan, Korea).

Cornus kousa is frequently planted in the flora area as an ornamental and for its edible fruit. Though not well established, it has been noted to spread from cultivation.

1b. Cornus Linnaeus subg. Cynoxylon (Rafinesque) Rafinesque, Alsogr. Amer., 59. 1838

Cornus sect. *Cynoxylon* Rafinesque, Med. Fl. 1: 132. 1828; *Benthamidia* Spach; *Cynoxylon* (Rafinesque) Small

Shrubs or trees; rhizomes absent. **Branches and leaves** opposite. **Inflorescences** capitula; bracts 4–6, well developed, petaloid, subtending inflorescence and surrounding or enclosing it over winter. **Pedicels** absent. **Drupes** distinct; stone apex dimpled to rounded.

Species 2 (2 in the flora): North America, e Mexico.

1. Drupes angular in cross section, closely appressed to each other; petaloid bracts 4–6 per inflorescence; w North America . . . . 2. *Cornus nuttallii*
1. Drupes round in cross section, spreading from each other; petaloid bracts 4 per inflorescence; e North America . . . . 3. *Cornus florida*

2. Cornus nuttallii Audubon, Ornithol. Biogr. 4: 482. 1838 (as nuttalli) • Pacific or mountain dogwood [E] [W]

Benthamidia nuttallii (Audubon) Moldenke; *Cynoxylon nuttallii* (Audubon) Shafer

Shrubs or trees, to 20 m, flowering at 2 m. **Stems** clustered, bark forming square to rectangular plates 0.5–1 cm wide; branchlets green, maroon, or dark red, appressed-hairy; lenticels pale round or lenticular spots, with dark central pore. **Leaves:** petiole 5–17 mm; blade elliptic, ovate, or obovate, 6–15 × 3–9.5 cm, base broadly cuneate, apex short acuminate, abaxial surface pale green, appressed-hairy, tufts of erect hairs present in axils of secondary veins, adaxial surface yellow green, appressed-hairy; secondary veins 4–6 per side, most arising from proximal ½. **Inflorescences** convex, 1.5–3 cm diam., 40–75-flowered, subtended by pair of reduced leaves and pair of cataphylls; peduncle 40–60 mm; petaloid bracts 4–6, surrounding but not enclosing inflorescence through winter, white or tinged with pink, rhombic to obovate, 2.5–7 × 1.7–6 cm, apex short acuminate. **Flowers:** hypanthium appressed-hairy; sepals 1–1.5 mm; petals cream or yellow-green, rarely purple distally, 3–4 mm. **Drupes** red, closely appressed to each other, angular in cross section, 10–17 × 5–7.5 mm; stone ellipsoid, 7–9 × 4–5 mm, smooth. $2n = 22$.

Flowering Apr–Jul, sometimes Sep–Oct; fruiting Aug–Oct. Dry to moist coastal coniferous and montane forests; 0–2400(–3000) m; B.C.; Calif., Idaho, Oreg., Wash.

3. **Cornus florida** Linnaeus, Sp. Pl. 1: 117. 1753
• Flowering dogwood F W

Benthamidia florida (Linnaeus) Spach; *Cynoxylon floridum* (Linnaeus) Small

Trees, to 20 m, flowering at 2 m. **Stems** clustered, occasionally decumbent and rooting at nodes, bark corky, forming rectangular plates 0.5–1 cm wide; branchlets green, maroon, or red, appressed-hairy; lenticels maroon swellings. **Leaves:** petiole 3–20 mm; blade ovate, elliptic, or obovate, 5–12 × 2–7 cm, base cuneate to rounded, apex abruptly acuminate, abaxial surface whitish, appressed-hairy, tufts of erect hairs present in axils of secondary veins, adaxial surface dark green, appressed-hairy; secondary veins 5–7 per side, most arising from proximal ½. **Inflorescences** flat-topped, 1–2 cm diam., 15–30-flowered, subtended by 2 pairs of cataphylls; peduncle 10–20 mm; petaloid bracts 4, surrounding and enclosing inflorescence through winter, white or tinged with red and with brown or white callous at apex, obovate to obcordate, 2–6 × 1–4.5 cm, apex rounded or emarginate. **Flowers:** hypanthium appressed-hairy; sepals 0.5–0.8 mm; petals cream or yellow-green, 3–3.5 mm. **Drupes** usually red, rarely yellow, drying black, spreading from each other, round in cross section, 13–18 × 6–9 mm; stone ellipsoid, 10–12 × 4–7 mm, smooth. $2n$ = 44.

Flowering Mar–Jun; fruiting Aug–Oct. Deciduous, mixed, and pine forests; 0–2000 m; Ont.; Ala., Ark., Conn., Del., D.C., Fla., Ga., Ill., Ind., Kans., Ky., La., Maine, Md., Mass., Mich., Miss., Mo., N.H., N.J., N.Y., N.C., Ohio, Okla., Pa., R.I., S.C., Tenn., Tex., Vt., Va., W.Va.; e Mexico.

Mexican populations of *Cornus florida* sometimes have been treated as *C. urbinia* Rose or *C. florida* subsp. *urbinia* (Rose) Rickett, distinguished primarily by bract size, shape, and apical cohesion following expansion. North American and Mexican populations overlap in these characters and are treated here as a single taxon.

Cornus florida, the state tree of Missouri and Virginia and the state flower of North Carolina, is an understory tree that can form spectacular displays when flowering. Cultivars with pink to red bracts are often planted as ornamentals. Dogwood anthracnose is causing serious declines in *C. florida* throughout its range.

1c. Cornus Linnaeus subg. Arctocrania (Endlicher) Reichenbach, Deut. Bot. Herb.-Buch., 143. 1841

Cornus [unranked] *Arctocrania* Endlicher, Gen. Pl. 10: 798. 1839; *Arctocrania* (Endlicher) Nakai; *Chamaepericlymenum* Hill

Herbs, perennial; rhizomes present. **Branches and leaves** opposite, leaves at distalmost node appearing whorled in some species. **Inflorescences** congested cymes; bracts 4, well developed, petaloid, subtending inflorescence. **Pedicels** present. **Drupes** distinct; stone apex rounded.

Species 3 (3 in the flora): North America, Eurasia.

Species definitions within subg. *Arctocrania* have been controversial, reflecting an apparently complicated history of post-glacial range shifts, hybridization, and polyploidization in these long-lived clonal plants (Z. E. Murrell 1994). Many treatments (for example, E. Hultén 1937; J. A. Calder and R. L. Taylor 1965) had broad concepts of *Cornus suecica* and especially *C. canadensis*, and considered the intermediates to be hybrids, sometimes calling them *C.* ×*intermedia* (Farr) Calder & Roy L. Taylor. J. F. Bain and K. E. Denford (1979) recognized *C. unalaschkensis* as a tetraploid species derived from hybridization between *C. canadensis* and *C. suecica* and the remaining intermediates as *C. canadensis* × *C. suecica*. C. Gervais and M. Blondeau (2003) applied the name *C.* ×*lapagei* Gervais & Blondeau to these hybrids. A morphometric analysis by Murrell showed five morphological groups: the morphological extremes, *C. canadensis* and *C. suecica*; the tetraploid intermediate species, *C. unalaschkensis*; and two groups of intermediates that were considered introgressive hybrids between *C. canadensis* and *C. suecica* (or perhaps rarely between one of these species and *C. unalaschkensis*); that treatment is followed here.

Identification of many specimens may be difficult because of the relatively high frequency of introgressants (approximately half of specimens examined by Z. E. Murrell 1994). Some intermediates are found in places where *Cornus canadensis* and *C. suecica* currently occur somewhat close together in Alaska, British Columbia, Greenland, Newfoundland and Labrador, Northwest Territories, Nova Scotia, Quebec, and St. Pierre and Miquelon. However, plants that mostly resemble *C. canadensis* but have petals that are purple on their distal third or chlorophyllous leaves at the second node from the apex (and non-chlorophyllous scale leaves at the third node), and sometimes only opposite leaves, are found at scattered locations throughout the range of *C. canadensis* but far from the current range of *C. suecica*, apparently reflecting the presence of the latter in the past (Murrell).

1. Petals cream; leaves at 2d node from apex non-chlorophyllous, scalelike; sepals cream, membranous, apices rounded . . . . 4. *Cornus canadensis*
1. Petals purple or cream proximally, purple distally; leaves at 2d node from apex chlorophyllous or non-chlorophyllous proximally and chlorophyllous distally, well developed; sepals purple or mottled purple and cream, thick, apices rounded or acute.
    2. Petals purple; hypanthia very sparsely hairy; distalmost leaves similar size to those at 2 more proximal nodes . . . . 5. *Cornus suecica*
    2. Petals cream proximally, purple distally; hypanthia densely hairy; distalmost leaves much larger than those at 2 more proximal nodes. . . . . 6. *Cornus unalaschkensis*

4. Cornus canadensis Linnaeus, Sp. Pl. 1: 118. 1753
• Bunchberry, dwarf cornel, quatre-temps [F] [W]

Arctocrania canadensis (Linnaeus) Nakai; *Chamaepericlymenum canadense* (Linnaeus) Ascherson & Graebner; *Cornella canadensis* (Linnaeus) Rydberg; *Cornus canadensis* subsp. *pristina* Gervais & Blondeau; *C. cyananthus* Rafinesque; *Cynoxylon canadense* (Linnaeus) J. H. Schaffner; *Mesomora canadensis* (Linnaeus) Lunell

Stems erect, green, 5–25 cm, appressed-hairy; nodes 4–6, internodes progressively longer distally; branches only at distalmost node, much shorter than distal internodes so stems appear unbranched. **Leaves** at all but distalmost node non-chlorophyllous, opposite, scalelike, caducous, at distalmost node chlorophyllous, appearing to be in whorl of 6 (sometimes 4 on sterile stems), well developed, persistent; distalmost leaves: petiole 0.5–2.8 mm; blade obovate, ovate, elliptic, or rhombic, 2–7 × 1–4.5 cm, apex acute or short acuminate, abaxial surface pale green, sparsely appressed-hairy to glabrate, adaxial surface green, appressed-hairy; secondary veins 2–3 per side, all arising from proximal ½. **Inflorescences** 12–40-flowered; peduncle 10–30 mm; primary branches 0.5–2.5 mm; bracts greenish white to white, occasionally red-tipped or red-tinged, ± equal, ovate, 5–15 × 5–15 mm, apex acute to acuminate. **Pedicels** 0.5–3 mm, sparsely appressed-hairy. **Flowers:** hypanthium cream, 1–2 mm, densely appressed-hairy; sepals cream, turning purple as fruit matures, 0.1–0.3 mm, apex rounded, membranous, glabrous, eglandular; petals cream, 1–2 mm, apical awn 0.3–1.2 mm; nectary cream or purplish black. **Drupes** 5–15 per infructescence, red, globose, 6–9 mm; stone ovoid, 2.3–3.3 × 1.7–2.3 mm, smooth, apex rounded. $2n$ = 22, 44.

Flowering May–Jul; fruiting Aug–Oct. Dry to moist broadleaf or coniferous forests, roadbanks, marshes, bogs; 0–3400 m; Greenland; St. Pierre and Miquelon; Alta., B.C., Man., N.B., Nfld. and Labr., N.W.T., N.S., Nunavut, Ont., P.E.I., Que., Sask., Yukon; Alaska, Colo., Conn., Idaho, Ill., Ind., Iowa, Maine, Md., Mass., Mich., Minn., Mont., N.H., N.J., N.Mex., N.Y., N.Dak., Ohio, Pa., R.I., S.Dak., Vt., Va., Wash., W.Va., Wis., Wyo.; Asia.

C. Gervais and M. Blondeau (2003) showed that *Cornus canadensis* includes both diploid and tetraploid populations, which can be distinguished solely by pollen size and chromosome number. The diploids, which they called subsp. *pristina*, are found in Alaska and western and northern Canada (including the northern parts of Labrador and Quebec), and the tetraploids, which they called subsp. *canadensis*, are found in eastern Canada and the northeastern United States. Unfortunately, Gervais and Blondeau did not sample populations from most of the species' distribution in the United States.

Reports of *Cornus canadensis* from California and Oregon (for example, C. L. Hitchcock and A. Cronquist 1973; J. R. Shevock 2012) represent *C. unalaschkensis*.

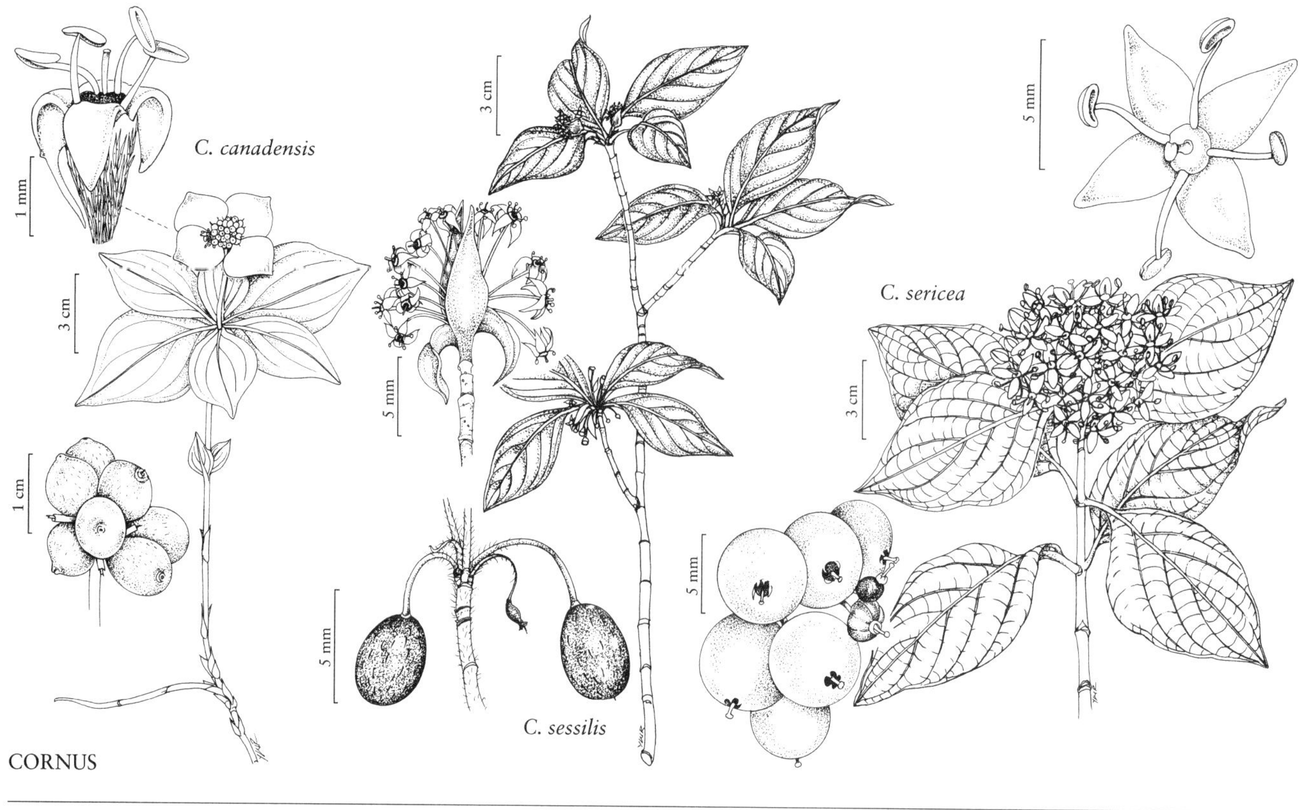

CORNUS

5. **Cornus suecica** Linnaeus, Sp. Pl. 1: 118. 1753 • Lapland cornel, cornouiller de Suède

Arctocrania suecica (Linnaeus) Nakai; *Chamaepericlymenum suecicum* (Linnaeus) Ascherson & Graebner; *Cornella suecica* (Linnaeus) Rydberg; *Cornus biramis* Stokes; *C. herbacea* Linnaeus

Stems erect, green, 5–22 cm, appressed-hairy; nodes 4–7, internodes only slightly longer distally; branches at distal 1–3 nodes, ca. as long as distal internodes. **Leaves** at basal node nonchlorophyllous, opposite, scalelike, caducous, at remaining nodes chlorophyllous, opposite, well developed, persistent; distalmost leaves similar in size to those at 2 more proximal nodes; petiole 0–7.5 mm; blade ovate or oval, 1.3–4 × 0.7–2.8 cm, apex broadly acute or obtuse, abaxial surface pale green, moderately appressed-hairy, adaxial surface bluish green, appressed-hairy; secondary veins 3 per side, all arising from near base, occasionally distalmost vein bifurcating in basal ⅕. **Inflorescences** 5–17-flowered; peduncle 8–20 mm; primary branches 0–0.4 mm; bracts white, often red- or purple-tinged, ± equal, ovate, 8–12 × 4–7.5 mm, apex slightly acute. **Pedicels** 0.7–1.5 mm, moderately appressed-hairy proximally, densely at base of hypanthium. **Flowers:** hypanthium deep purple, 1–2 mm, very sparsely appressed-hairy; sepals deep purple, 0.2–0.5 mm, apex acute, thick, densely hairy on margins, densely glandular; petals purple, 1.2–1.5 mm, apical awn 0.5 mm; nectary purple or black. **Drupes** 3–8 per infructescence, red, globose, 5–8 mm; stone usually globose, 2.1–3.4 mm diam., longitudinally grooved, apex rounded. 2*n* = 22.

Flowering May–Jul; fruiting Aug–Oct. Maritime heath, tundra, or copse, along streams, peat banks, bogs, marshes; 0–500 m; Greenland; St. Pierre and Miquelon; B.C., Man., Nfld. and Labr., N.W.T., N.S., Nunavut, Que.; Alaska; Eurasia.

6. **Cornus unalaschkensis** Ledebour, Fl. Ross. 2: 378. 1844 • Western cordilleran bunchberry [E]

Arctocrania unalaschkensis (Ledebour) Nakai; *Chamaepericlymenum unalaschkense* (Ledebour) Rydberg; *Cornella unalaschkensis* (Ledebour) Rydberg; *Swida unalaschkensis* (Ledebour) A. Heller

Stems erect, green, 6–22 cm, appressed-hairy; nodes 4–6, internodes progressively longer distally; branches only at distalmost node, much shorter than distal internodes so stems appear unbranched. **Leaves** at proximal 2–4 nodes nonchlorophyllous, opposite, ± scalelike, caducous (rarely chlorophyllous at 3d node from apex but much smaller than more distal leaves), at 2d node from apex nonchlorophyllous proximally, chlorophyllous distally, opposite, well developed, persistent, at distalmost node chlorophyllous, appearing to be in whorl of 6, well developed, persistent; distalmost leaves much bigger than those at 2 more proximal nodes; petiole 0–3.4 mm; blade ovate to elliptic, 3.5–8 × 0.9–4 cm, apex acute or short acuminate, abaxial surface pale green, hairs sparsely appressed-hairy, adaxial surface green, appressed-hairy; secondary veins 3 per side, all arising from proximal ½. **Inflorescences** 20–40-flowered; peduncle 13–30 mm; primary branches 0–2 mm; bracts greenish white or white, often red-tipped, unequal, 2 ovate, 21–30 × 12–13 mm, 2 suborbiculate, 17–1.9 × 13–16 mm, apex acuminate. **Pedicels** 0.4–1.6 mm, sparsely appressed-hairy or glabrous. **Flowers:** hypanthium cream to mottled purple, 1.2–2 mm, densely appressed-hairy; sepals mottled purple and cream, 0.1–0.4 mm, apex rounded or acute, thick, sparsely hairy on margin, densely glandular; petals cream proximally, purple distally, 1.5–1.8 mm, apical awn 0.4–0.6 mm; nectary dark purple or black. **Drupes** 10–20 per inflorescence, red, globose, 6–8 mm; stone globose or subglobose, 2.7–3.4 × 2.1–3.4 mm, longitudinally grooved, apex slightly pointed. $2n$ = 44.

Flowering May–Aug; fruiting Aug–Oct. Maritime copse or heath, maritime coniferous forests and bog woodlands, moist broadleaf or coniferous forests; 0–3000 m; Alta., B.C., Yukon; Alaska, Calif., Idaho, Oreg., Wash.

1d. Cornus Linnaeus subg. Cornus

Shrubs or trees; rhizomes absent. **Branches and leaves** opposite. **Inflorescences** umbels; bracts 4, well developed, not petaloid, subtending inflorescence and enclosing it over winter. **Pedicels** present. **Drupes** distinct; stone apex pointed to slightly rounded.

Species 4 (2 in the flora): w United States, Eurasia, introduced in c, e United States.

1. Petals greenish yellow; bracts lanceolate, apices acute; drupes purple-black when mature . . . 7. *Cornus sessilis*
1. Petals bright yellow; bracts ovate, apices obtuse with apiculate tip; drupes red when mature . . . 8. *Cornus mas*

7. Cornus sessilis Torrey in E. M. Durand, Pl. Pratten. Calif., 89. 1855 • Blackfruit dogwood [E] [F]

Shrubs or trees to 5 m, flowering at 2 m. **Stems** solitary; bark corky; branches splotched with maroon, brown, or red, eventually splitting along longitudinal swellings; branchlets green, densely appressed-hairy; lenticels maroon swellings, often erupting with corky surface. **Leaves:** petiole 5–10 mm, base encircling twig; blade elliptic, 4–9 × 2–4 cm, base cuneate, apex acute or short acuminate, abaxial surface yellow-green, appressed-hairy, tufts of erect hairs in axils of secondary veins, adaxial surface dark green, sparsely appressed-hairy; secondary veins 4–5 per side, most usually arising from proximal ½. **Inflorescences** 10–15-flowered; peduncle 0–1 mm; bracts tan or brown, lanceolate, 0.5–1 cm, apex acute. **Pedicels** lax, apex flared. **Flowers:** hypanthium narrowly conic, appressed-hairy; sepals 0.1–0.5 mm; petals greenish yellow, lanceolate, 3–4 mm. **Drupes** maturing from green to yellow, red, then purple-black, ellipsoid, 10–15 × 5–7.5 mm; stone widely fusiform, 8–12 × 4–6 mm, with 2 lateral grooves on distal ⅔. $2n$ = 20.

Flowering Mar–Jun; fruiting Aug–Sep. Moist ravines and stream banks; 60–2000 m; Calif., Oreg.

Cornus sessilis is restricted to northern California, generally in the Klamath Range, Cascade Range, and northern Sierra Nevada, and southern Oregon in the Siskiyou Mountains.

8. Cornus mas Linnaeus, Sp. Pl. 1: 117. 1753 • European cornel, cornelian cherry [I]

Cornus mascula Linnaeus

Shrubs or trees to 5 m, flowering at 2 m. **Stems** solitary, branching profusely from lower trunk, bark of thin broad plates that shed sporadically, leaving a mottled gray-tan to red color; branches splotched with maroon, brown, or red, eventually splitting along longitudinal swellings; branchlets green, densely appressed-hairy; lenticels maroon swellings, often erupting with corky surface. **Leaves:** petiole 5–10 mm, base encircling twig; blade elliptic, 4–9 × 2–4 cm, base cuneate, apex acute or short acuminate, abaxial surface yellow-green, appressed-hairy, tufts of erect hairs in axils of secondary veins, adaxial surface dark green, sparsely appressed-hairy; secondary veins

4–5 per side, most usually arising from proximal ½. **Inflorescences** 10–15-flowered; peduncle 5–10 mm; bracts tan or brown, ovate, 0.5–1 cm, apex obtuse with apiculate tip. **Pedicels** lax, apex flared. **Flowers:** hypanthium narrowly conic, appressed-hairy; sepals 0.1–0.5 mm; petals bright yellow, lanceolate, 3–4 mm. **Drupes** maturing from green to yellow, then red, ellipsoid, 10–15 × 5–7.5 mm; stone widely fusiform, 8–12 × 4–6 mm, with 2 lateral grooves on distal ⅔. **2*n*** = 18.

Flowering Apr–Jun; fruiting Aug–Sep. Disturbed areas, woodland margins; 0–1000 m; introduced; Ill., N.Y., Pa.; Europe; Asia.

Cornus mas is regularly utilized in horticulture throughout North America and is a rare local escape.

1e. Cornus Linnaeus subg. Mesomora Rafinesque, Alsogr. Amer., 58 [as Mesomera], 62, 76. 1838

Swida Opiz

Shrubs or trees; rhizomes absent. **Branches and leaves** alternate. **Inflorescences** cymes; bracts numerous, minute, not petaloid, subtending primary and secondary inflorescence branches, proximal portion adnate to branch, free portion caducous. **Pedicels** present. **Drupes** distinct; stone apex with cavity.

Species 2 (1 in the flora): e North America, e Asia.

Cornus controversa Hemsley is widespread in eastern Asia.

9. Cornus alternifolia Linnaeus f., Suppl. Pl., 125. 1782 • Pagoda or alternate-leaf dogwood, cornouiller à feuilles alternes [E]

Bothrocaryum alternifolium (Linnaeus f.) Pojarkova; *Cornus alterna* Marshall; *Swida alternifolia* (Linnaeus f.) Small

Shrubs or trees to 12 m, flowering at 2 m. **Stems** clustered; bark thick, corky, plates rectangular, 0.5–1 cm wide; branchlets usually green to yellowish green, sometimes reddish brown, glabrous; lenticels lenticular, then splitting periderm longitudinally. **Leaves:** petiole 20–50 mm; blade narrowly to broadly ovate or obovate, 4–12 × 2.3–7 cm, base usually cuneate, rarely rounded, apex abruptly acuminate or cuspidate, abaxial surface yellow-green, papillose, appressed-hairy, adaxial surface dark green, glabrate; secondary veins 5–6 per side, most arising from proximal ½. **Inflorescences** flat to hemispheric, 3–15 cm diam., 50–100-flowered; peduncle 30–60 mm; branches and pedicels yellow-green, turning red in fruit, alternate on central axis, proximal 2–3 orders with minute bracts. **Flowers:** hypanthium appressed-hairy; sepals 0–0.2 mm; petals cream, 2.5–4 mm. **Drupes** blue, globose, 5–8 mm diam.; stone subglobose, laterally compressed, 5–6 × 5–6 × 4 mm, slightly ribbed. **2*n*** = 20.

Flowering Apr–Jun; fruiting Jun–Aug. Deciduous hardwood forests, usually mesic or dry-mesic, loamy soils, rocky slopes; 10–2000 m; St. Pierre and Miquelon; Man., N.B., Nfld. and Labr. (Nfld.), N.S., Ont., P.E.I., Que.; Ala., Ark., Conn., Del., D.C., Fla., Ga., Ill., Ind., Iowa, Ky., Maine, Md., Mass., Mich., Minn., Miss., Mo., N.H., N.J., N.Y., N.C., Ohio, Pa., R.I., S.C., Tenn., Vt., Va., W.Va., Wis.

Cornus alternifolia persists around abandoned homesteads in Nebraska and occasionally escapes, but does not appear to be naturalized there. A putative hybrid in Nova Scotia between *C. alternifolia* and *C. sericea* was named *C. ×acadiensis* by M. L. Fernald (1941c).

1f. Cornus Linnaeus subg. Thelycrania (Dumortier) C. K. Schneider, Ill. Handb. Laubholzk. 2: 437. 1909

Cornus [unranked] *Thelycrania* Dumortier, Fl. Belg., 83. 1827

Shrubs or trees; rhizomes present or absent. **Branches and leaves** usually opposite, rarely whorled, subopposite, or alternate at some nodes. **Inflorescences** cymes; bracts numerous, minute, not petaloid, subtending primary and secondary inflorescence branches, proximal portion adnate to branch, free portion caducous. **Pedicels** present. **Drupes** distinct; stone apex rounded, pointed, or with slight dimple.

Species ca. 30 (11 in the flora): North America, Mexico, Central America, n, w South America, Eurasia.

What is treated here as subg. *Thelycrania* has almost invariably been called subg. *Kraniopsis* Rafinesque. That name is invalid, however, because C. S. Rafinesque (1838) apparently treated his new taxon as a genus, not a subgenus, and no publication validating the name at the subgeneric rank has been located.

1. Leaf secondary veins usually (4–)5–9 per side.
    2. Bark splitting longitudinally, appearing braided; lenticels not protruding on 2d year branches; leaf blade abaxial surfaces without tufts of erect hairs in axils of secondary veins.
        3. Leaf blade bases cuneate, abaxial surface hairs all appressed and rigid, secondary veins evenly spaced . . . . . 10. *Cornus obliqua*
        3. Leaf blade bases rounded or truncate, abaxial surface hairs both appressed and rigid, and erect and curling, most secondary veins arising from proximal ½ of blade . . . . . 11. *Cornus amomum*
    2. Bark loosely verrucose; lenticels protruding on 2d year branches; leaf blade abaxial surfaces with tufts of erect hairs in axils of secondary veins.
        4. Areas surrounding lenticels suffused with purple on older branches; leaf blades suborbiculate or broadly ovate, secondary veins 7–9 per side, evenly spaced, tertiary veins usually prominent . . . . . 12. *Cornus rugosa*
        4. Areas surrounding lenticels not suffused with purple on older branches; leaf blades lanceolate, elliptic, or ovate, secondary veins 5–7 per side, most arising from proximal ½ of blade, tertiary veins not prominent.
            5. Leaf blade abaxial surface hairs appressed except in and near axils of secondary veins . . . . . 13. *Cornus sericea*
            5. Leaf blade abaxial surface hairs (except those on veins) all erect . . . . . 14. *Cornus occidentalis*
1. Leaf secondary veins usually 3–4(–5) per side.
    6. Leaf blade abaxial surface hairs curved upward, erect, or both erect and appressed.
        7. Drupes purple-black; rhizomes absent; stems clustered . . . . . 15. *Cornus sanguinea*
        7. Drupes blue to whitish blue or white; rhizomes present; stems solitary.
            8. Petioles 2–7 mm; leaf secondary veins evenly spaced; drupes blue to whitish blue . . . . . 16. *Cornus asperifolia*
            8. Petioles 8–25 mm; most leaf secondary veins arising from proximal ½ of blade; drupes white . . . . . 17. *Cornus drummondii*
    6. Leaf blade abaxial surface hairs appressed.
        9. Rhizomes present; stems solitary; lenticels protruding and extruding tissue on 2d year branches; bark appearing verrucose; drupes usually white, occasionally pale blue . . . . . 18. *Cornus racemosa*
        9. Rhizomes absent; stems clustered; lenticels not or seldom protruding or extruding tissue on 2d year branches; bark appearing braided; drupes blue, pale blue, violet plumbeous, or blue violet, often turning whitish blue to white.
            10. Basal secondary leaf veins arising 1–2 mm from blade bases; e, c United States . . . . . 19. *Cornus foemina*
            10. Basal secondary leaf veins arising 5–10 mm from blade bases; California, Oregon . . . . . 20. *Cornus glabrata*

10. Cornus obliqua Rafinesque, W. Rev. & Misc. Mag. 1: 229. 1819 • Pale dogwood, cornouiller oblique [E]

Cornus amomum Miller subsp. *obliqua* (Rafinesque) J. S. Wilson; *C. amomum* var. *schuetzeana* (C. A. Meyer) Rickett; *C. purpusii* Koehne; *Swida purpusii* (Koehne) A. Heller

Shrubs, to 5 m, flowering at 1.5 m; rhizomes absent. **Stems** clustered, branches occasionally arching to ground and rooting at nodes; bark green-tan or maroon-tan, not corky, appearing braided, splitting longitudinally; branchlets green abaxially, maroon to green adaxially, turning red-maroon in fall, densely erect-hairy when young; lenticels not protruding on 2d year branches, area surrounding them not suffused with purple on older branches; pith tan or brown. **Leaves:** petiole 6–20 mm; blade lanceolate to narrowly elliptic, 4–12 × 1–5 cm, base cuneate, apex acuminate, abaxial surface pale whitish yellow, hairs white, all appressed and rigid, tufts of erect hairs absent in axils of secondary veins, midvein and secondary veins densely tomentose, adaxial surface dark green, hairs appressed; secondary veins (4–)5–6 per side, evenly spaced, tertiary veins not prominent. **Inflorescences** flat-topped or convex, 2–7 cm diam., peduncle 20–70 mm; branches and pedicels green or greenish yellow, turning maroon in fruit. **Flowers:** hypanthium densely appressed-hairy, especially at base; sepals 1–2.3 mm; petals cream, 3.8–5 mm. **Drupes** blue, portion in direct sunlight bleached white, globose, 5–9 mm diam.; stone globose, 4–6 mm diam., irregularly longitudinally ridged, apex pointed. $2n = 22$.

Flowering May–Aug; fruiting Aug–Oct. Alluvial woods, river and stream banks, wet meadows, marshes, ditches; 0–1500 m; N.B., Ont., Que.; Ark., Conn., D.C., Ill., Ind., Iowa, Kans., Ky., Maine, Md., Mass., Mich., Minn., Mo., Nebr., N.H., N.J., N.Y., N.Dak., Ohio, Okla., Pa., R.I., S.Dak. Tenn., Vt., Va., W.Va., Wis.

H. W. Rickett (1934) argued that the description of *Cornus obliqua* by Rafinesque is inadequate to associate that name with a species; that assessment is not accepted here. The description by Rafinesque of the plants having reddish brown, slightly rugose bark, narrowly elliptic to lanceolate discs, and whitish yellow abaxial leaf surfaces, along with the cited locality of the Kentucky River, clearly delineates this species. Rafinesque was the first to divide the blue-fruited dogwood of L. Plukenet (1691–1705, part 4) into two species.

Cornus obliqua and *C. amomum* can be distinguished not only by the differences included in the key above, but also by their abaxial leaf cuticle, which is coronulate in *C. obliqua* but not in *C. amomum*, but seeing this character requires high magnification. However, in much of the area where they are sympatric (Connecticut, District of Columbia, Maine, Maryland, Massachusetts, New Hampshire, New York, Tennessee, and Vermont), many individuals show intermediate leaf blade abaxial surface and hair morphology. They are detected by having both the whitish leaf blade abaxial surface and appressed hairs of *C. obliqua* with occasional scattered erect, often tan to brown hairs, similar to those typical for *C. amomum*. J. S. Wilson (1964) concluded that differences in leaf surface and hair morphology are environmentally based (sun versus shade), whereas work by Z. E. Murrell (1992) documented a geographical basis for the differences. The geographical zone of intermediacy needs greater scrutiny to determine its full extent and whether the intermediacy of many plants represent hybridization or incomplete speciation.

A putative hybrid between *Cornus obliqua* and *C. racemosa*, reported from Massachusetts, Ohio, and Pennsylvania, has been called *C.* ×*arnoldiana* Rehder [= *Swida arnoldiana* (Rehder) Soják].

11. Cornus amomum Miller, Gard. Dict. ed. 8, Cornus no. 5. 1768 • Kinnikinnik [E]

Swida amomum (Miller) Small

Shrubs, to 5 m, flowering at 1.5 m; rhizomes absent. **Stems** clustered, branches occasionally arching to ground and rooting at nodes; bark green-tan or maroon-tan, not corky, appearing braided, splitting longitudinally; branchlets green abaxially, maroon to green adaxially, turning red-maroon in fall, densely erect-hairy when young; lenticels not protruding on 2d year branches, area surrounding them not suffused with purple on older branches; pith tan or brown. **Leaves:** petiole 8–25 mm; blade broadly ovate, 8–15 × 4–9 cm, base rounded or truncate, apex abruptly acuminate, abaxial surface yellow-green, hairs brown, tan, or white, both appressed and rigid and others erect and curling on same leaf, tufts of erect hairs absent in axils of secondary veins, midvein and secondary veins densely tomentose, adaxial surface light to dark green, hairs appressed; secondary veins (4–)5–6 per side, most arising from proximal ½, tertiary veins perpendicular to secondary veins, ladderlike. **Inflorescences** flat-topped or convex, 2–8 cm diam., peduncle 15–80 mm; branches and pedicels green or greenish yellow, turning maroon in fruit. **Flowers:** hypanthium densely appressed-hairy, especially at base; sepals 1.3–2 mm; petals cream, 3–5 mm. **Drupes** blue, portion in direct sunlight bleached white, globose, 5–9 mm diam.; stone globose, 4–6 mm diam., irregularly longitudinally ridged, apex pointed. $2n = 22$.

Flowering May–Aug; fruiting Aug–Oct. Alluvial woods, river and stream banks, wet meadows, marshes, ditches; 0–1500 m; Ala., Conn., Del., D.C., Fla., Ga., Ill., Ind., Iowa, Ky., Maine, Md., Mass., Mich., Miss., Mo., N.H., N.J., N.Y., N.C., Ohio, Pa., R.I., S.C., Tenn., Vt., Va., W.Va.

The confusion regarding the name of this taxon dates to the description and plate by L. Plukenet (1691–1705, parts 3, 4) of the "Amomum Nova Angliae quorundum." In his protologue, Miller cited Plukenet and was the first to recognize the ovate-leaved, blue-fruited dogwood of eastern North America. O. A. Farwell (1931) and H. W. Rickett (1934) emphasized the essay by Miller following his description, which indicated red shoots and a whitish undersurface to the leaves; Farwell concluded that *Cornus amomum* is the correct name for the red-osier dogwood, treated here as *C. sericea*, whereas Rickett decided that the red shoots and whitish leaf undersurface comments by Miller were a misprint meant instead for *C. candidissima*, treated here as a synonym of *C. racemosa*. Because the majority of the description by Miller fits *C. amomum* better than *C. sericea*, the interpretation by Rickett is followed here.

Intermediates between *Cornus amomum* and 10. *C. obliqua* are common where their ranges overlap; see the latter species for further discussion.

Putative hybrids between *Cornus amomum* and *C. racemosa* have been called *C.* ×*arnoldiana* Rehder; these have been reported from Massachusetts, Missouri, Ohio, and Pennsylvania.

12. Cornus rugosa Lamarck in J. Lamarck et al., Encycl. 2: 115. 1786 • Roundleaf dogwood, cornouiller rugueux [E]

Cornus circinata L'Héritier; *C. tomentulosa* Michaux; *Swida circinata* (L'Héritier) Small; *S. rugosa* (Lamarck) Rydberg

Shrubs, to 5 m, flowering at 1 m; rhizomes present. **Stems** solitary, 5–10 dm apart; bark pink, light maroon, or green, not corky, loosely verrucose; branchlets yellow-green, with scattered hairs; lenticels protruding on 2d year branches, area surrounding them suffused with purple on older branches; pith white or tan. **Leaves:** petiole 10–23 mm; blade suborbiculate or broadly ovate, 7–15 × 5–14 cm, base usually subcordate to broadly cuneate, rarely nearly truncate, apex abruptly acuminate, abaxial surface pale green, hairs erect, dense, tufts of erect hairs present in axils of secondary veins, adaxial surface dark green, hairs appressed or erect; secondary veins 7–9 per side, evenly spaced, tertiary veins usually prominent giving leaf a wrinkled appearance. **Inflorescences** flat-topped, 5–7 cm diam., peduncle 18–35 mm; branches and pedicels pink, turning red in fruit. **Flowers:** hypanthium constricted below sepals, appressed-hairy; sepals 0.2–0.4 mm; petals white, 2.6–3.8 mm. **Drupes** pale blue, globose, 5–8 mm diam.; stone globose, 4 mm diam., slightly ribbed, apex dimpled. **2*n*** = 22.

Flowering May–Jul; fruiting late Aug–Oct. Wooded slopes, forests, stream banks, lake shores; 0–2000 m; Man., N.B., N.S., Ont., P.E.I., Que.; Conn., Del., Ill., Ind., Iowa, Maine, Md., Mass., Mich., Minn., N.H., N.J., N.Y., Ohio, Pa., R.I., Tenn., Vt., Va., W.Va., Wis.

Cornus rugosa appears to be extirpated from Tennessee (B. E. Wofford, pers. comm.). Putative hybrids in Michigan between *C. rugosa* and *C. racemosa* have been called *C.* ×*friedlanderi* W. H. Wagner. Putative hybrids between *C. rugosa* and *C. sericea* have been called *C.* ×*slavinii* Rehder, and reported from Maine, New York, Ontario, Prince Edward Island, and Wisconsin.

13. Cornus sericea Linnaeus, Mant. Pl. 2: 199. 1771 • Red-osier dogwood, cornouiller stolonifère, hart rouge [F]

Cornus alba Linnaeus subsp. *baileyi* (J. M. Coulter & W. H. Evans) Wangerin; *C. alba* subsp. *stolonifera* (Michaux) Wangerin; *C. baileyi* J. M. Coulter & W. H. Evans; *C. instolonea* A. Nelson; *C. interior* (Rydberg) N. Petersen; *C. nelsonii* Rose; *C. stolonifera* Michaux; *Swida interior* Rydberg; *S. stolonifera* (Michaux) Rydberg; *S. stolonifera* var. *riparia* Rydberg

Shrubs, to 4 m, flowering at 1 m; rhizomes absent. **Stems** clustered, branches occasionally arching to the ground and rooting at nodes; bark yellow to red, not corky, loosely verrucose; branchlets bright red, reddish brown, maroon, or green, occasionally green in winter and maroon in summer, appressed-hairy when young; lenticels protruding on 2d year branches, area surrounding them not suffused with purple on older branches; pith white. **Leaves:** petiole 5–38 mm; blade lanceolate, elliptic, or ovate, 3.5–20 × 1.5–12 cm, base cuneate, apex acute to acuminate, abaxial surface white, hairs appressed except near secondary vein axils, tufts of erect hairs present in axils of secondary veins, adaxial surface green, hairs appressed, sparse; secondary veins 5–7 per side, most arising from proximal ½, tertiary veins not prominent. **Inflorescences** flat-topped, 3–6 cm diam., peduncle 20–40 mm; branches and pedicels green to yellow-green, turning maroon in fruit. **Flowers:** hypanthium densely appressed-hairy; sepals 0.2–0.6 mm; petals white to cream, 2.5–4 mm. **Drupes** white, globose or subglobose, 6–10 mm diam.; stone subglobose, laterally compressed, 4–6 × 4–6 × 1.5–3 mm, furrowed laterally, apex rounded. **2*n*** = 22.

Flowering May–Jun and Sep–Oct; fruiting Aug–Oct. Wet meadows, thickets, edges of mesic upland forests, fens, marshes, swamps, stream banks, lake shores, river banks; 0–2500 m; St. Pierre and Miquelon; Alta., B.C., Man., N.B., Nfld. and Labr., N.W.T., N.S., Ont., P.E.I., Que., Sask., Yukon; Alaska, Ariz., Calif., Colo., Conn., Del., Idaho, Ill., Ind., Iowa, Kans., Ky., Maine, Md., Mass., Mich., Minn., Mont., Nebr., Nev., N.H., N.J., N.Mex., N.Y., N.Dak., Ohio, Oreg., Pa., R.I., S.Dak., Utah, Vt., Va., Wash., W.Va., Wis., Wyo.; Mexico (Chihuahua, Durango, Nuevo León); introduced in w Europe.

As the synonymy implies, *Cornus sericea* has received considerable attention from taxonomists wishing to subdivide the species, presumably in order to make it more comprehensible. Most of the divisions have been based upon indumentum and stone differences, although habit has also been used. Although one of the synonyms and one of the common names imply a stoloniferous habit, the species is not stoloniferous; evidently, branch tips infrequently arching to the ground and rooting at the nodes led to confusion regarding the growth habit. H. W. Rickett (1944b) examined the morphology of the various forms, varieties, and subspecies, and found extensive overlap using fruit shape and indumentum differences. It is not known whether the variation is due to primary differentiation or secondary intergradation, and the complex is treated here as a single species. There is little doubt that the European species *C. alba* Linnaeus is closely related to *C. sericea* and should be included in any future studies of this species complex.

The name *Cornus stolonifera* has sometimes been applied to *C. sericea* (for example, H. W. Rickett 1944b) because the description by Linnaeus of the latter could apply to several currently recognized species. F. R. Fosberg (1942) lectotypified *C. sericea*, establishing that it applies to this species.

Cornus sericea is commonly planted as an ornamental and occasionally escapes; plants in suburban areas and in highly acidic soils are suspected as non-natural occurrences. Putative hybrids between *C. sericea* and *C. rugosa* have been called *C.* ×*slavinii* Rehder, and are reported from Maine, New York, Ontario, Prince Edward Island, and Wisconsin.

14. Cornus occidentalis (Torrey & A. Gray) Coville, Contr. U.S. Natl. Herb. 4: 117. 1893 • Creek dogwood [E]

Cornus sericea Linnaeus var. *occidentalis* Torrey & A. Gray, Fl. N. Amer. 1: 652. 1840; *C. alba* Linnaeus var. *occidentalis* (Torrey & A. Gray) B. Boivin; *C. californica* C. A. Meyer; *C. californica* var. *pubescens* J. F. Macbride; *C. sericea* subsp. *occidentalis* (Torrey & A. Gray) Fosberg

Shrubs, to 3 m, flowering at 1 m; rhizomes absent. **Stems** clustered; bark green to maroon or reddish brown, not corky, loosely verrucose; branchlets red, reddish brown, maroon, or green, mostly erect-hairy when young; lenticels protruding on 2d year branches, area surrounding them not suffused with purple on older branches; pith white. **Leaves:** petiole 10–20 mm; blade ovate, 5–12 × 2–6 cm, base cuneate to rounded, apex slightly acuminate, abaxial surface white, hairs erect, especially dense near midvein, tufts of erect hairs present in axils of secondary veins, vein hairs brown, appressed, adaxial surface green, hairs appressed; secondary veins 6–7 per side, most arising from proximal ½, tertiary veins not prominent. **Inflorescences** flat-topped, 3–6 cm diam., peduncle 20–40 mm; branches and pedicels maroon. **Flowers:** hypanthium densely erect-hairy; sepals 0.3–0.7 mm; petals white, 3.5–4 mm. **Drupes** white, globose, 8 mm diam.; stone oblate-ellipsoid, 4–5 × 5–7 mm, with 3 ridges on each face, furrowed laterally, apex pointed.

Flowering May–Jun and Sep–Oct; fruiting Aug–Oct. Wet meadows, bogs, marshes, stream banks, lake shores, river banks; 0–2000 m; B.C.; Alaska, Calif., Idaho, Mont., Nev., Oreg., Wash.

F. R. Fosberg (1942) and H. W. Rickett (1944b) examined variation within the *Cornus sericea* complex in North America. Fosberg regarded the whole complex as a single species, stating that the differences in indumentum were not sufficient to distinguish two taxa. In contrast, Rickett concluded that there were two species, *C. occidentalis* and *C. sericea*, and that treatment is followed here.

The illegitimate name *Cornus pubescens* Nuttall has sometimes been used for this species.

15. Cornus sanguinea Linnaeus, Sp. Pl. 1: 117. 1753 • Bloodtwig dogwood [I]

Swida sanguinea (Linnaeus) Opiz

Shrubs, to 3 m, flowering at 1 m; rhizomes absent. **Stems** clustered; bark maroon to reddish brown, not corky, loosely verrucose; branchlets reddish brown to yellow to maroon, appressed-hairy when young; lenticels protruding on 2d year branches; pith white. **Leaves:** petiole 5–38 mm; blade elliptic to narrowly ovate, 4–17 × 1.5–12 cm, base attenuate, apex acuminate, abaxial surface white, hairs erect and appressed, tufts of erect hairs present in axils of secondary veins, adaxial surface green, hairs erect; secondary veins 3–5 per side, most arising from proximal ½. **Inflorescences** flat-topped, 3–6 cm diam., peduncle 20–40 mm; branches and pedicels green, turning maroon in fruit. **Flowers:** hypanthium densely appressed-hairy; sepals 0.2–0.6 mm; petals white, 2.5–4 mm. **Drupes** purple-black, globose or subglobose, 6–10 mm diam.; stone globose, 4–6 mm diam., smooth or slightly grooved, apex rounded. $2n = 22$.

Flowering Apr–Jun; fruiting Aug–Sep. Disturbed areas; 0–1000 m; introduced; Mass., Pa., Wash.; Europe; w Asia.

Cornus sanguinea is frequently planted across North America and occasionally has become naturalized.

16. Cornus asperifolia Michaux, Fl. Bor. Amer. 1: 93. 1803 • Eastern roughleaf dogwood [E]

Cornus excelsa Kunth var. *beyrichiana* C. A. Meyer; *C. foemina* Miller subsp. *microcarpa* (Nash) J. S. Wilson; *C. microcarpa* Nash; *C. sericea* Linnaeus var. *asperifolia* (Michaux) de Candolle; *C. stricta* Lamarck var. *asperifolia* (Michaux) Feay; *Swida asperifolia* (Michaux) Small; *S. microcarpa* (Nash) Small

Shrubs, to 4 m, flowering at 1.5 m; rhizomes present. **Stems** solitary, 1–5 dm apart; bark gray, splitting into small plates; branchlets green to bronze, often tinged with maroon, densely pubescent; lenticels inconspicuous on new growth, periderm around them swelling to form broad raised areas on 2d year branches; pith white. **Leaves:** petiole 2–7 mm; blade elliptic to ovate, 3–8.5 × 2–4 cm, base usually rounded, sometimes cuneate, apex acute, abaxial surface pale green, hairs erect, curling, white, adaxial surface dark green, hairs spreading to erect, occasionally 1 arm appressed; secondary veins 3–4 per side, evenly spaced. **Inflorescences** pyramidal, 2–5 cm diam., peduncle 15–45 mm; branches and pedicels yellow-green, turning maroon in fruit. **Flowers:** hypanthium densely appressed-hairy; sepals 0.2–0.8 mm; petals white, 1.7–2.4 mm. **Drupes** blue to whitish blue, globose, 4–7 mm diam.; stone globose, 3–5 mm diam., smooth or slightly grooved, apex rounded.

Flowering Apr–Jun; fruiting Aug–Sep. Marl or limestone outcrops, hammocks, swamp margins; 0–100 m; Ala., Fla., Ga., Miss., N.C., S.C.

G. V. Nash (1896b) collected *Cornus asperifolia* at River Junction, Florida; based on the conflicting reports of fruit colors given by A. W. Chapman (1860) and J. M. Coulter and W. H. Evans (1890) for the two rough-leaved dogwoods (*C. asperifolia* and *C. drummondii*), Nash decided to name the rough-leaved dogwood of Florida with blue fruit as *C. microcarpa*. However, the description by Michaux, even without a reference to fruit color, cannot apply to *C. drummondii*, because the locality is given as "Carolinae inferioris," and *C. drummondii* does not occur in South Carolina.

17. Cornus drummondii C. A. Meyer, Cornus-Arten, 20. 1845 (as drummondi) • Rough-leaved dogwood [E]

Cornus asperifolia Michaux var. *drummondii* (C. A. Meyer) J. M. Coulter & W. H. Evans; *C. priceae* Small; *Swida drummondii* (C. A. Meyer) Soják; *S. priceae* (Small) Small

Shrubs or trees, to 6 m, flowering at 1.5 m; rhizomes present. **Stems** solitary, 1–6 dm apart; bark pinkish gray, appearing braided, becoming corky, checkered with square or rectangular plates 2–5 mm wide; branchlets yellow-green abaxially, pink-maroon adaxially, densely erect-hairy when young; lenticels inconspicuous on new growth, protruding and splitting longitudinally on 2d year branches; pith brown, white, or tan. **Leaves:** petiole 8–25 mm; blade lanceolate to ovate, 2–12 × 1.2–7.7 cm, base cuneate, truncate, or cordate, apex abruptly acuminate, abaxial surface pale green, hairs curved upward, dense, adaxial surface gray-green, hairs curved upward or appressed; secondary veins 3–4(–5) per side, most arising from proximal ½. **Inflorescences** flat-topped or convex, 3–8 cm diam., peduncle 20–40 mm; branches and pedicels pink or red, fading yellow when dried, central part of pedicels white after fruit falls. **Flowers:** hypanthium densely appressed-hairy; sepals 0.4–1.3 mm; petals white, 2.3–3.2 mm. **Drupes** white, globose or subglobose, 4–7.5 mm diam.; stone subglobose, 3–6 mm diam., smooth or slightly grooved, apex rounded. $2n = 22$.

Flowering Apr–Jul; fruiting Aug–Oct. Limestone barrens, limestone outcrops, dry woodlands, rocky stream banks, prairies, old fields, roadsides, meadows, swamp margins; 0–1500 m; Ont.; Ala., Ark., Ga., Ill.,

Ind., Iowa, Kans., Ky., La., Mich., Miss., Mo., Nebr., N.Y., Ohio, Okla., Pa., S.Dak., Tenn., Tex., Wis.

Several workers considered the erect hairs on the adaxial leaf surface of *Cornus drummondii* and *C. asperifolia* to indicate relatedness, and treated these two species as one (J. Torrey and A. Gray 1838–1843; Gray 1856; J. M. Coulter and W. H. Evans 1890; W. Wangerin 1910). However, A. Wood (1861), following W. T. Feay, linked *C. asperifolia* with *C. foemina*, and J. S. Wilson (1964) also segregated the southern coastal plain rough-leaved dogwood from the continental rough-leaved dogwood as *C. foemina* subsp. *microcarpa* and *C. drummondii*, respectively.

J. K. Small described *Cornus priceae* from Bowling Green, Kentucky, with small drupes and stones. However, extensive collections from the type locality and examination of the type material revealed that the drupe and stone sizes are within the normal range of variation of *C. drummondii.*

18. Cornus racemosa Lamarck in J. Lamarck et al., Encycl. 2: 116. 1786 • Gray dogwood, cornouiller à grappes [E]

Cornus albida Ehrhart; *C. foemina* Miller subsp. *racemosa* (Lamarck) J. S. Wilson; *C. gracilis* Koehne; *C. paniculata* L'Heritier; *C. paniculata* var. *albida* (Ehrhart) Pursh; *C. paniculata* var. *radiata* Pursh; *Swida candidissima* Small; *S. racemosa* (Lamarck) Moldenke

Shrubs, to 5 m, flowering at 0.7 m; rhizomes present. **Stems** solitary, 2–10 dm apart; bark gray, brittle, verrucose, frequently forming small plates; branchlets pinkish brown, turning green-maroon, and later gray-maroon, 2 proximal internodes densely pubescent, distal internodes sparsely appressed-hairy; lenticels pale lenticular spots on new growth, uniformly scattered, especially on the distal internodes, protruding and extruding corky tissue on 2d year branches; pith white in 1st year branches, brown in older branches. **Leaves:** petiole 2–8 mm; blade lanceolate to ovate, 1.5–8 × 0.5–4 cm, base cuneate, apex acuminate to obtuse, abaxial surface pale green, adaxial surface green, turning maroon in full sun, both surfaces with hairs appressed, white to ferruginous; secondary veins 3–4(–5 on some sucker shoots) per side, evenly spaced. **Inflorescences** usually elongate, sometimes convex or pyramidal, 2.5–5 cm diam., peduncle 20–40 mm; branches and pedicels yellow-green, turning bright red. **Flowers:** hypanthium appressed-hairy; sepals 0.2–0.5 mm; petals white, 2.3–3 mm. **Drupes** usually white, occasionally pale blue, oblate-ellipsoid, laterally compressed, 4–8 mm diam.; stone oblate-ellipsoid, 3–6.5 mm diam., smooth, apex rounded to slightly pointed. $2n = 22$.

Flowering May–Jul; fruiting Aug–Oct. Fields, meadows, roadsides, fencerows, swamp margins; 0–1500 m; Man., Ont., Que.; Conn., D.C., Ill., Ind., Iowa, Ky., Maine, Md., Mass., Mich., Minn., Mo., Nebr., N.H., N.J., N.Y., N.C., Ohio, Pa., R.I., S.Dak., Tex., Vt., Va., W.Va., Wis.

Cornus racemosa, much like *C. foemina* and *C. obliqua*, has a coronulate or minutely papillate abaxial leaf surface, visible under high magnification. *Cornus gracilis* may be a hybrid between *C. racemosa* and *C. foemina*. Hybrids between *C. racemosa* and 10. *C. obliqua* and 12. *C. rugosa* are discussed under those species.

19. Cornus foemina Miller, Gard. Dict. ed. 8, Cornus no. 4. 1768 • Stiff dogwood [E]

Cornus stricta Lamarck; *Swida foemina* (Miller) Small; *S. stricta* (Lamarck) Small

Shrubs, to 8 m, flowering at 1.5 m; rhizomes absent. **Stems** clustered; bark gray-brown, becoming gray-black, corky, appearing braided, splitting longitudinally, checkered; branchlets deep red, often pale green abaxially, or completely green or bronze if shaded, 2 proximal internodes densely pubescent, distal internodes sparsely appressed-hairy; lenticels pale circular spots on new growth, usually more common on distal portion of internodes, often overlapping, localized to form longitudinal bands, splitting periderm but not protruding or extruding tissue on 2d year branches, periderm swelling around them and usually over large contiguous areas; pith white in 1st year branches, tan in older branches. **Leaves:** petiole 5–16 mm; blade lanceolate, elliptic, or oblanceolate, 3.5–11 × 1–6 cm, base cuneate to rounded, apex abruptly acuminate to an obtuse tip, abaxial surface pale green, adaxial surface dark green, midvein turning red or maroon, both surfaces with hairs appressed, sparse, glabrate by late summer; secondary veins 3–4 per side, evenly spaced, basal vein arising 1–2 mm from blade base. **Inflorescences** flat-topped, convex, or pyramidal, 2–8 cm diam., peduncle 15–45 mm; branches and pedicels greenish yellow, turning red in fruit. **Flowers:** hypanthium densely appressed-hairy; sepals 0.4–1 mm; petals cream, 2.6–3.8 mm. **Drupes** pale blue, violet plumbeous, or blue violet, often turning whitish blue to white, globose or oblate-ellipsoid, 5–6 × 6–10 mm; stone globose to oblong, 3–3.7 × 3.7–5 mm, slightly ribbed, apex rounded.

Flowering Mar–Jun; fruiting Aug–Oct. Marshes, swamps, river and stream banks, pocosin margins, interdune swales, wet ditches; 0–1500 m; Ala., Ark., Del., Fla., Ga., Ill., Ind., Ky., La., Md., Miss., Mo., N.J., N.C., Okla., S.C., Tenn., Tex., Va.

Cornus foemina has a coronulate or minutely papillate abaxial leaf surface, visible under high magnification, like that of *C. obliqua* and *C. racemosa*.

Cornus cyanocarpus J. F. Gmelin 1791, a parahomonym (thus illegitimate) of *C. cyanocarpa* Moench 1785, pertains here.

20. Cornus glabrata Bentham, Bot. Voy. Sulphur, 18. 1844 • Smooth Dogwood [E]

Cornus costulata Jepson; *Swida catalinensis* Millspaugh

Shrubs, to 3 m, flowering at 1 m; rhizomes absent. **Stems** clustered, branches at colony margin decumbent or trailing and rooting at the nodes, forming large thickets; bark gray, flaky, not corky, appearing braided, splitting longitudinally; branchlets brown, pink, or maroon, appressed-hairy; lenticels inconspicuous, swelling but seldom protruding or extruding tissue on 2d year branches, periderm swelling around them and often over large contiguous areas; pith brown. **Leaves:** petiole 3–7 mm; blade lanceolate, elliptic, or obovate, 3–5 × 1.5–2.5 cm, base cuneate, apex abruptly acuminate or rounded, abaxial surface pale green, adaxial surface gray-green, both surfaces with hairs appressed, sparse; secondary veins 3–4(–5) per side, evenly spaced, basal vein arising 5–10 mm from blade base. **Inflorescences** flat-topped or slightly convex, 2.5–4.5 cm diam., peduncle 10–20 mm; branches and pedicels green or maroon. **Flowers:** hypanthium densely appressed-hairy; sepals 0.4–0.8 mm; petals cream, 3–5 mm. **Drupes** blue, bleaching white in direct sun, globose or subglobose, 6–9 mm diam.; stone globose to subglobose, 4–6 mm diam., smooth, apex rounded. 2*n* = 22.

Flowering May–Jun; fruiting Aug–Oct. Stream banks, roadsides, fields, meadows; 50–1500 m; Calif., Oreg.

Cornus glabrata, which ranges from southern Oregon through most of transmontane California, grows in two habits. Along streams and in moist areas, the branches are trailing, often vinelike, and rooting at the nodes. H. McMinn (1939) reported this growth habit for cultivated plants. A second form is found in drier roadsides and fields, where the stems are erect or grow horizontally, rooting at the nodes and forming dense thickets. This growth pattern is accompanied by smaller, thicker leaves and erect branches. There is no doubt this is a single species, because the authors have observed a single clone on a stream bank, with half the clone growing erect, with small leaves, and half trailing into the stream bed, with large leaves.

NYSSACEAE Jussieu ex Dumortier • Tupelo Family

Gordon C. Tucker

Shrubs or trees, deciduous, usually monoecious or polygamous, rarely dioecious. **Leaves** alternate, simple; stipules absent; petiole present; blade margins entire or sparsely, irregularly dentate; venation pinnate. **Inflorescences** unisexual, axillary [terminal], umbel-like or glomerate, or flowers solitary. **Flowers** bisexual or unisexual; perianth and androecium epigynous; hypanthium absent (staminate flowers), completely adnate to ovary (bisexual and pistillate flowers); sepals 5, connate; petals 5(–10), distinct; nectary present, intrastaminal; stamens (4–)6–12(–15) in 1–2 whorls, distinct, free; anthers dehiscing by longitudinal slits; pistil 1, 1(–2)[–3]-carpellate; ovary inferior, 1(–2)[–3]-locular, placentation apical; ovule 1 per locule, anatropous; styles 1(–2)[–3], connate basally; stigmas 1(–2)[–3]. **Fruits** drupes. **Seeds** 1 per locule.

Genera 2, species ca. 15 (1 genus, 5 species in the flora): c, e North America, Mexico, Central America (Costa Rica, Panama), e Asia.

Taxonomic opinions about the status and circumscription of Nyssaceae have varied. *Nyssa* and the small eastern Asian genera *Camptotheca* Decaisne and *Davidia* Baillon sometimes have been included in an expanded Cornaceae, a view supported by earlier phylogenetic studies (Xiang Q. Y. et al. 1998). The narrower circumscription used here, with the family comprising only *Camptotheca* and *Nyssa* but still close to Cornaceae, follows R. H. Eyde (1966) and is supported by phylogenetic studies by Xiang et al. (2011).

SELECTED REFERENCES Eyde, R. H. 1963. Morphological and paleobotanical studies of the Nyssaceae. I. A survey of the modern species and their fruits. J. Arnold Arbor. 44: 1–59. Eyde, R. H. 1966. The Nyssaceae in the southeastern United States. J. Arnold Arbor. 47: 117–125.

1. NYSSA Linnaeus, Sp. Pl. 2: 1058. 1753; Gen. Pl. ed. 5, 478. 1754 • Tupelo [Classical Greek name for a water nymph, alluding to habitat]

Gordon C. Tucker

Tracy J. Park

Shrubs or trees; bark gray brown, rough, ridged; twigs with transverse diaphragms; winter buds scaly. **Leaves:** petiole terete or winged; blade usually elliptic to oblanceolate or obovate, rarely ovate, base cuneate to rounded. **Pedicels:** staminate present or absent, bisexual and pistillate absent [present]. **Flowers:** sepals forming a low rim; petals greenish to greenish white; style subulate or conic. **Drupes** usually blue-black (sometimes reddish purple in *N. aquatica*; yellow, orange, or red in *N. ogeche*), topped by persistent remnants of sepals; mesocarp juicy, acidic.

Species ca. 13 (5 in the flora): c, e North America, Mexico, Central America (Costa Rica, Panama), e Asia.

Plant sexuality in *Nyssa* is complicated and often difficult to determine. R. E. Burckhalter (1992) described all North American species as dioecious, without elaboration. The most widely distributed species in the flora area, *N. sylvatica*, has both staminate and morphologically bisexual flowers and appears to be androdioecious, but is functionally dioecious, as anthers of bisexual flowers do not dehisce (R. H. Eyde 1963; M. L. Cipollini and E. W. Stiles 1991). Similarly, the Chinese *N. yunnanensis* W. Q. Yin ex H. N. Qin & Phengklai appears to be androdioecious but is functionally dioecious; in this case pollen from the morphologically bisexual flowers is inaperturate and inviable, rendering the trees that bear these flowers functionally pistillate (Sun B.-L. et al. 2009). *Nyssa ogeche* appears to be polygamodioecious, and other North American species appear to be monoecious or polygamomonoecious, but their reproductive biology has not been examined carefully.

Measurements of leaves refer to fully expanded ones from mid shoot (earlier leaves typically are smaller).

SELECTED REFERENCES Burckhalter, R. E. 1992. The genus *Nyssa* (Cornaceae) in North America: A revision. Sida 15: 323–342. Wen, J. and T. F. Stuessy. 1993. The phylogeny and biogeography of *Nyssa* (Cornaceae). Syst. Bot. 18: 68–79.

1. Drupes 22–28 mm; ovaries hairy; staminate pedicels absent.
   2. Petioles 30–60 mm; drupes punctate, black to blue or reddish purple . . . . 1. *Nyssa aquatica*
   2. Petioles 6–20(–35) mm; drupes smooth, yellow, orange, or red . . . . 2. *Nyssa ogeche*
1. Drupes 7–14 mm; ovaries glabrous; staminate pedicels present.
   3. Shrubs or trees, 2–5 m; Florida panhandle region . . . . .5. *Nyssa ursina*
   3. Trees, 5–30 m; widespread in c, e North America, including Florida.
      4. Leaf blades herbaceous, obovate to elliptic, apices acute to acuminate, margins entire proximally and 0–1(–3)-toothed distally; bark ± regularly longitudinally and transversely fissured; pistillate and bisexual inflorescences (2–)3–5(–8)-flowered . . . . .3. *Nyssa sylvatica*
      4. Leaf blades subcoriaceous, usually oblanceolate to narrowly elliptic, rarely ovate, apices obtuse to acute, margins usually entire, rarely coarsely dentate distally; bark irregularly fissured; pistillate and bisexual inflorescences 1–3-flowered . . . . 4. *Nyssa biflora*

1. Nyssa aquatica Linnaeus, Sp. Pl. 2: 1058. 1753
• Water tupelo, cotton gum [E]

Trees, (6–)10–30 m, base buttressed, proximal limbs spreading to slightly drooping, crown irregular; bark irregularly fissured; twigs puberulent; pith white. **Leaves:** petiole 30–60 mm; blade elliptic to ovate, 8–11(–18) × (2.5–)3.5–5.5(–7.3) cm, herbaceous, base cuneate to rounded, margins entire proximally and usually coarsely dentate distally, apex usually acute, rarely acuminate, often mucronulate, abaxial surface glabrous or puberulent especially along veins, adaxial surface glabrous or sparsely hairy. **Inflorescences:** peduncle 1–1.5 cm (elongating to 3 cm in fruit), glabrous; staminate (1–)2–5-flowered, pistillate and bisexual 1–2-flowered. **Staminate pedicels** absent. **Flowers:** ovary densely hairy. **Drupes** black to blue or reddish purple, glaucous, oblong, 24–27 mm, punctate; stone 15–20 mm, roughened, often longitudinally ridged.

Flowering spring. Swamps, floodplain forests, moist woods; 0–100 m; Ala., Ark., D.C., Fla., Ga., Ill., Ky., La., Miss., Mo., N.C., S.C., Tenn., Tex., Va.

SELECTED REFERENCE Shea, M. M., P. M. Dixon, and R. R. Sharitz. 1993. Size differences, sex ratio, and spatial distribution of male and female water tupelo, *Nyssa aquatica* (Nyssaceae). Amer. J. Bot. 80: 26–30.

2. Nyssa ogeche W. Bartram ex Marshall, Arbust. Amer., 97. 1785 • Ogeechee tupelo or lime [E]

Nyssa acuminata Small; *N. ogeche* var. *acuminata* (Small) Eyles

Trees, 10–30 m, trunks of old individuals usually grotesquely formed, abruptly branching into multiple large stems 1–6 m above ground, proximal limbs spreading to slightly drooping, crown irregular; bark irregularly fissured; twigs puberulent. **Leaves:** petiole 6–20(–35) mm; blade elliptic to narrowly obovate, 5–12 × 3–5.5 cm, herbaceous, base cuneate to rounded, margins entire, rarely coarsely dentate distally, apex acute to obtuse, rarely acuminate, abaxial surface glabrous or puberulent, adaxial surface glabrous. **Inflorescences:** peduncle 1.5–2 cm, densely hairy; staminate (1–)2–5-flowered, bisexual and pistillate 1–2-flowered. **Staminate pedicels** absent. **Flowers:** ovary hairy. **Drupes** yellow, orange, or red, not glaucous, oblong-ellipsoid, 22–28 mm, smooth; stone 18–20 mm, with several roughened, longitudinal papery ridges.

Flowering spring. Blackwater stream banks, depression swamps, moist woods; 0–100 m; Ala., Fla., Ga., S.C.

Thousands of hectares of *Nyssa ogeche* have been planted in bee farms along the lower Apalachicola River in Florida and around swamps where it grows naturally (R. M. Burns and B. H. Honkala 1990, vol. 2). Bees collect nectar from the trees to make tupelo honey. The mature fruit, known as Ogeche lime, has a subacid flavor and is made into preserves.

3. Nyssa sylvatica Marshall, Arbust. Amer., 97. 1785
• Black or sour gum, pepperidge [F]

Trees, 5–30 m, proximal limbs spreading to slightly drooping, crown irregular; bark ± regularly longitudinally and transversely fissured; twigs puberulent. **Leaves:** petiole 5–15(–30) mm; blade obovate to elliptic, 5.5–12 (–16) × (2.2–)3.5–6.5(–9.1) cm, herbaceous, base cuneate to rounded, margins entire proximally and 0–1(–3)-toothed distally, apex acute to acuminate, abaxial surface glabrous or puberulent, adaxial surface glabrous. **Inflorescences:** peduncle 3.6–4.3 cm, sparsely hairy or glabrous; staminate (1–)2–5(–7)-flowered, pistillate and bisexual (2–)3–5(–8)-flowered. **Staminate pedicels** present. **Flowers:** ovary glabrous. **Drupes** black to blue, glaucous, ovoid to ellipsoid, 8–12 mm, rough or smooth; stone 7–9 mm, with several low, rounded longitudinal ridges.

Flowering spring. Well-drained sites, swamps (especially in northern part of range), saturated longleaf pine savannas, moist to mesic (or dry-mesic) woods; 0–1100(–1600) m; Ont.; Ala., Ark., Conn., Del., D.C., Fla., Ga., Ill., Ind., Ky., La., Maine, Md., Mass., Mich., Miss., Mo., N.H., N.J., N.Y., N.C., Ohio, Okla., Pa., R.I., S.C., Tenn., Tex., Vt., Va., W.Va., Wis.; c Mexico.

Nyssa sylvatica is widely planted as an ornamental, appreciated especially for its fall color.

SELECTED REFERENCES Batra, S. W. T. 1999. Native bees (Hymenoptera: Apoidea) in native trees: *Nyssa sylvatica* Marsh. (Cornaceae). Proc. Entomol. Soc. Washington 101: 449–457. Cipollini, M. L. and E. W. Stiles. 1991. Costs of reproduction in *Nyssa sylvatica*: Sexual dimorphism in reproductive frequency and nutrient flux. Oecologia 86: 585–593.

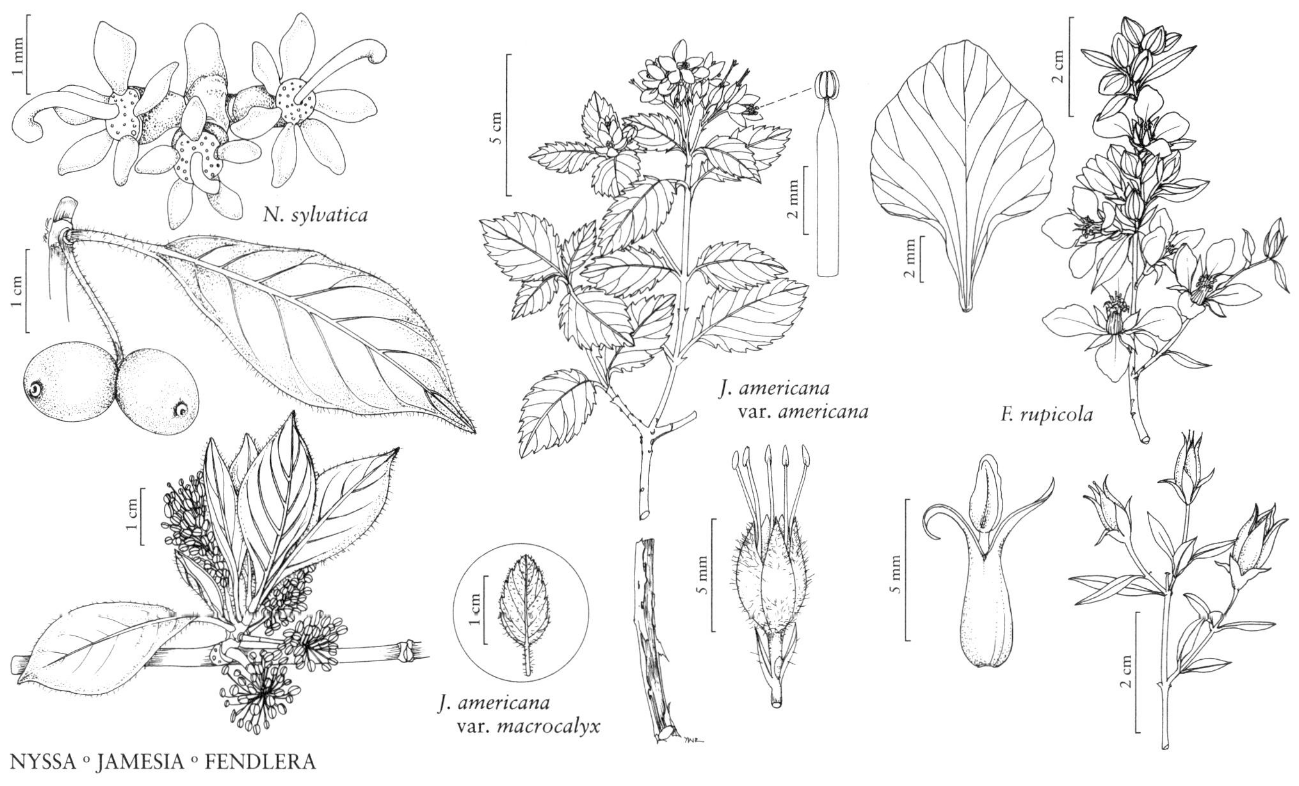

NYSSA ◦ JAMESIA ◦ FENDLERA

4. Nyssa biflora Walter, Fl. Carol., 253. 1788 • Swamp black or sour gum [E]

Nyssa sylvatica Marshall subsp. *biflora* (Walter) A. E. Murray; *N. sylvatica* var. *biflora* (Walter) Sargent

Trees, 10–30 m, base often buttressed in larger individuals, proximal limbs spreading to slightly drooping, crown irregular; bark irregularly fissured; twigs glabrous or puberulent. **Leaves:** petiole 7–10 mm; blade oblanceolate to narrowly elliptic, rarely ovate, 3.7–7.2 × 1.6–3.5 cm, subcoriaceous, base cuneate to rounded, margins usually entire, rarely coarsely dentate distally, apex obtuse to acute, abaxial surface glabrous or puberulent (primarily along veins), adaxial surface glabrous. **Inflorescences:** peduncle 3.2–5.5 cm, sparsely hairy; staminate (1–)2–8-flowered, pistillate and bisexual 1–3-flowered. **Staminate pedicels** present. **Flowers:** ovary glabrous. **Drupes** usually black, sometimes to blue, glaucous, ovoid, 7–14 mm, smooth; stone 7–9 mm, with several low, rounded longitudinal ridges.

Flowering spring. Swamps, flatwood depressions and ponds, bogs, wet streamheads, seepage slopes, often in sites with standing water during part of the year or with organic soils saturated year-round, less often in mesic forests; 0–100(–200) m; Ala., Ark., Del., Fla., Ga., Ill., Ky., La., Md., Miss., Mo., N.J., N.C., S.C., Tenn., Tex., Va.

5. Nyssa ursina Small, Torreya 27: 92. 1927 • Bear or Apalachicola tupelo [C] [E]

Nyssa biflora Walter var. *ursina* (Small) D. B. Ward; *N. sylvatica* Marshall var. *ursina* (Small) J. Wen & Stuessy

Shrubs or trees, 2–5 m, crown typically intricately branched; bark irregularly fissured; twigs usually glabrous, rarely puberulent. **Leaves:** petiole (4–)5–9 mm; blade narrowly elliptic to oblanceolate, rarely to ovate, 3–7 × 1–2 cm, coriaceous, base cuneate to rounded, margins entire, apex obtuse, abaxial surface glabrous or puberulent (primarily along veins), adaxial surface glabrous. **Inflorescences:** peduncle 3.2–5.5 cm, sparsely hairy or glabrous; staminate (1–)2–5-flowered, pistillate and bisexual 1–2-flowered. **Staminate pedicels** present. **Flowers:** ovary glabrous. **Drupes** usually black, rarely blue, glaucous, globose, 7–11 mm, smooth; stone 6–8 mm, with several low, rounded longitudinal ridges.

Flowering spring. Open savannas, depressions in flatwoods; of conservation concern; 0–70 m; Fla.

Nyssa ursina is limited to six counties in the panhandle region of Florida. It occurs together with *N. biflora* throughout its limited range, which supports recognizing it as a distinct species.

HYDRANGEACEAE Dumortier • Hydrangea Family

Craig C. Freeman

Subshrubs, shrubs, trees, or vines [herbs], evergreen or deciduous. **Leaves** usually opposite, sometimes whorled [alternate], simple; stipules absent; petiole present or absent; blade sometimes palmately lobed, margins entire, serrate, serrulate, dentate, denticulate, or crenate; venation pinnate or acrodromous (*Fendlera*, *Fendlerella*, *Philadelphus*, *Whipplea*). **Inflorescences** terminal or axillary, cymes, panicles, racemes, or corymbs, or flowers solitary. **Flowers** bisexual [unisexual], or sometimes marginal ones sterile, radially symmetric (bisexual ones) or bilaterally symmetric with enlarged petaloid sepals (sterile ones); perianth and androecium nearly hypogynous, perigynous, or epigynous; hypanthium completely adnate to ovary or adnate to ovary proximally, free distally; sepals 4–12, distinct or connate basally; petals 4–12, connate basally [entirely, then calyptrate]; nectary usually present, rarely absent; stamens 8–200, usually distinct, sometimes connate proximally, free; anthers dehiscing by longitudinal slits; pistil 1, 2–12-carpellate, ovary less than ½ inferior, ½ inferior, or completely inferior, 1–12-locular, placentation usually axile proximally, parietal distally, rarely strictly axile or parietal; ovules 1–50 per locule, anatropous; styles 1–12, distinct or connate proximally to most of length; stigmas (1–)2–12. **Fruits** capsules [berries], dehiscence septicidal, loculicidal, interstylar, or intercostal. **Seeds** 1–50 per locule, funicular appendage present (*Fendlerella*, *Whipplea*) or absent.

Genera 17, species ca. 240 (9 genera, 25 species in the flora): North America, Mexico, Central America, South America, Eurasia, Pacific Islands.

A. Cronquist (1981) placed Hydrangeaceae among a group of woody families traditionally allied with Saxifragaceae. Phylogenetic studies consistently place Hydrangeaceae in the Cornales and sister to Loasaceae (A. L. Hempel et al. 1995; D. E. Soltis et al. 1995; L. Hufford et al. 2001; Hufford 2004). Within Hydrangeaceae, the western North American genera *Fendlera* and *Jamesia* form a clade (subfam. Jamesioideae L. Hufford) that is sister to the rest of the family (subfam. Hydrangeoideae Burnett) (Hufford et al.; Hufford). Subfamily Hydrangeoideae comprises two tribes: Philadelpheae de Candolle ex Duby and Hydrangeeae de Candolle. North American genera in the former are *Carpenteria*, *Deutzia*, *Fendlerella*, *Philadelphus*, and *Whipplea*. A molecular phylogenetic study by Y. De Smet et al. (2015) clarified relationships within Hydrangeeae, found *Hydrangea* to be polyphyletic, and promoted adoption of a broader concept of *Hydrangea* that includes the eight other genera in the tribe. The two North American genera in the tribe, *Decumaria* and *Hydrangea*, are circumscribed here in their traditional senses.

The Hydrangeaceae are well represented in the paleobotanical record dating back to the Upper Cretaceous but best represented in the Tertiary (L. Hufford 2004). Some genera are sources of popular introduced or native ornamentals, including *Carpenteria*, *Deutzia*, *Hydrangea*, and *Philadelphus*. Some ornamentals have become established outside of cultivation in the flora area. A few North American Hydrangeaceae have reputed medicinal (D. E. Moerman 1998) or toxicologic (G. E. Burrows and R. J. Tyrl 2001) properties.

Trichomes in most Hydrangeaceae consist of a long, unicellular portion, often borne on a multicellular base. The unicellular portion often bears tubercles on its surface. Sometimes instead of tubercles, it bears long extensions, making the trichome appear branched or dendritic. Such trichomes are here referred to as branched.

SELECTED REFERENCES De Smet, Y. et al. 2015. Molecular phylogenetics and new (infra)generic classification to alleviate polyphyly in tribe Hydrangeeae (Cornales: Hydrangeaceae). Taxon 64: 741–752. Hufford, L. 1995. Seed morphology of Hydrangeaceae and its phylogenetic implications. Int. J. Plant Sci. 156: 555–580. Hufford, L. 1997. A phylogenetic analysis of Hydrangeaceae based on morphological data. Int. J. Plant Sci. 158: 652–672. Hufford, L. 1998. Early development of androecia in polystemonous Hydrangeaceae. Amer. J. Bot. 85: 1057–1067. Hufford, L. 2004. Hydrangeaceae. In: K. Kubitzki et al., eds. 1990+. The Families and Genera of Vascular Plants. 10+ vols. Berlin etc. Vol. 6, pp. 202–215. Hufford, L., M. L. Moody, and D. E. Soltis. 2001. A phylogenetic analysis of Hydrangeaceae based on sequences of the plastid gene *matK* and their combination with *rbcL* and morphological data. Int. J. Plant Sci. 162: 835–846. Small, J. K. and P. A. Rydberg. 1905. Hydrangeaceae. In: N. L. Britton et al., eds. North American Flora.... 1905+. New York. 47+ vols. Vol. 22, pp. 159–178. Soltis, D. E., Xiang Q. Y., and L. Hufford. 1995. Relationships and evolution of Hydrangeaceae based on *rbcL* sequence data. Amer. J. Bot. 82: 504–514. Spongberg, S. A. 1972. The genera of Saxifragaceae in the southeastern United States. J. Arnold Arbor. 53: 409–498.

1. Woody vines . . . . 9. *Decumaria*, p. 489
1. Subshrubs, shrubs, or trees.
    2. Twigs with stellate and simple trichomes. . . . . 5. *Deutzia*, p. 471
    2. Twigs glabrous or with simple or, sometimes, branched trichomes, never with stellate trichomes.
        3. Flowers both sterile and bisexual . . . . 8. *Hydrangea* (in part), p. 486
        3. Flowers all bisexual.
            4. Stamens (11–)13–90 or 150–200.
                5. Sepals 4; petals 4 (or 8+ in some horticultural forms); ovaries inferior to ½ inferior, 4-locular; styles 1 or 4; leaves deciduous; stamens (11–)13–90 . . . . 6. *Philadelphus*, p. 473
                5. Sepals 5–7; petals 5–7(–8); ovaries nearly superior, 5–7-locular; styles 1; leaves persistent; stamens 150–200 . . . . 7. *Carpenteria*, p. 485
            4. Stamens 8–12.
                6. Stems prostrate to decumbent . . . . 4. *Whipplea*, p. 469
                6. Stems erect, ascending, or spreading.
                    7. Filament apices 2-lobed, lobes prolonged beyond anthers; seeds (1–)2–4(–6) per locule . . . . 2. *Fendlera*, p. 466
                    7. Filament apices not 2-lobed; seeds 1 or 10–50 per locule.
                        8. Inflorescences 100–1000-flowered; capsule dehiscence interstylar, creating pore at base of styles . . . . 8. *Hydrangea* (in part), p. 486
                        8. Inflorescences 1–35-flowered; capsule dehiscence septicidal.
                            9. Leaf blade margins usually crenate to dentate, rarely entire; blades ovate or broadly ovate to obovate, rhombic, or suborbiculate, venation pinnate; seeds 25–50 per locule. . . . . . 1. *Jamesia*, p. 464
                            9. Leaf blade margins entire; blades elliptic to lanceolate, oblanceolate, obovate, or linear-oblong, venation acrodromous; seeds 1 per locule. . . . . 3. *Fendlerella*, p. 468

1. JAMESIA Torrey & A. Gray, Fl. N. Amer. 1: 593. 1840, name conserved • Cliffbush, waxflower [For Edwin P. James, 1797–1861, American physician and naturalist on the Stephen Harriman Long expeditions of 1819 & 1820]

Craig C. Freeman

Shrubs. Stems ascending or spreading. **Bark** exfoliating in sheets, strips, or strings. **Branches** ascending, spreading, or descending; twigs with simple trichomes. **Leaves** deciduous, opposite; petiole present; blade ovate or broadly ovate to obovate, rhombic, or suborbiculate, herbaceous, margins usually crenate to dentate, rarely entire, plane; venation pinnate. **Inflorescences** terminal, cymes, 2–35-flowered, sometimes flowers solitary; peduncle absent or present. **Pedicels** present. **Flowers** bisexual; perianth and androecium perigynous; hypanthium completely adnate to ovary, hemispheric, not ribbed in fruit; sepals persistent, 4–5, erect, lanceolate to narrowly ovate, sparsely to densely strigose, canescent, or sericeous; petals 4–5, imbricate, spreading, white or pink, obovate or oblanceolate, base clawed, sometimes obscurely, surfaces hairy; stamens 8 or 10; filaments distinct, dorsiventrally flattened, gradually or abruptly tapered from base to narrow apex, apex not 2-lobed; anthers depressed-ovate; pistil 3–5-carpellate, ovary to ½ inferior, partially 3–5-locular initially, becoming 1-locular; placentation axile proximally, parietal distally; styles persistent, (2–)3–5, distinct. **Capsules** ovoid or conic, indurate, dehiscence basipetally septicidal to middle of fruit. **Seeds** 25–50 per locule, orangish brown or tan, ellipsoid. $x = 16$.

Species 2 (2 in the flora): w United States, n Mexico.

This treatment of *Jamesia* is essentially an adaptation of that by N. H. Holmgren and P. K. Holmgren (1989).

Fossil leaves referred to *Jamesia* have been identified in Oligocene sediments from Colorado and Montana; the identity of material from the latter site is ambiguous (N. H. Holmgren and P. K. Holmgren 1989).

In the key and descriptions that follow, tooth number is per leaf.

SELECTED REFERENCE Holmgren, N. H. and P. K. Holmgren. 1989. A taxonomic study of *Jamesia* (Hydrangeaceae). Brittonia 41: 335–350.

1. Sepals 5; petals 5; stamens 10; inflorescences usually 3–35-flowered, cymose, rarely 1–2-flowered on lateral branches; leaf blade margins (5–)9–51(–69)-toothed . . . . 1. *Jamesia americana*
1. Sepals 4; petals 4; stamens 8; inflorescences (1–)3-flowered; leaf blade margins usually 3–13(–16)-toothed, rarely entire . . . . 2. *Jamesia tetrapetala*

1. Jamesia americana Torrey & A. Gray, Fl. N. Amer. 1: 593. 1840 • American cliffbush [F]

Edwinia americana (Torrey & A. Gray) A. Heller

Stems 5–20(–40) dm. **Bark** exfoliating readily or tardily in reddish brown or blackish sheets, or gray to brown strips or strings. **Branches** spreading or ascending, often stunted and straggly; twigs ascending-strigose. **Leaves:** petiole (1–)2–18(–54) mm, ascending-strigose to canescent or sericeous; blade ovate or broadly ovate to obovate, rhombic, or suborbiculate, (0.7–)1.3–8(–10) × (0.5–)1–6.3(–8.5) cm, base cuneate to obtuse or rounded, usually asymmetric, margins crenate to dentate, (5–)9–51(–69)-toothed, apex obtuse to rounded, abaxial surface moderately to densely canescent or sericeous, adaxial sparsely strigose to glabrescent. **Inflorescences** cymose, usually 3–35-flowered, rarely 1–2-flowered on lateral branches; peduncle 0–30 mm, sparsely to densely strigose. **Pedicels** 1.5–8 mm, sparsely to densely strigose. **Flowers:** hypanthium 1.3–2 × 2.5–4.5 mm, sparsely to densely strigose; sepals 5, lanceolate to deltate-ovate, 1.5–7(–8) × 1.1–2 mm, margins usually entire, sometimes 2–3-lobed apically, apex acute to obtuse, abaxial surface sparsely to densely strigose; petals 5, white or pink, (4–)5.5–11(–11.5) × 3–4.5 mm, sparsely to densely strigose or canescent, especially distally; stamens 10; filaments (2–)2.7–10 × (0.2–)0.5–1 mm; anthers 0.7–1.1 mm; styles (2–)3–5, 3–8 mm. **Capsules** 3.5–5.5(–7) × 2–3.8 mm. **Seeds** 0.6–1 mm. **2*n*** = 32.

Varieties 4 (4 in the flora): w United States, n Mexico.

N. H. Holmgren and P. K. Holmgren (1989) recognized subsp. *americana*, which is widespread in the eastern part of the species' range and included only var. *americana*, and subsp. *californica* (Small) A. E. Murray, which occurs in the western part of the range and included three allopatric varieties: *macrocalyx*, *rosea*, and *zionis*.

The Chiricahua Apache and Mescalero Apache of the southwestern United States and northern Mexico used the seeds as food (D. E. Moerman 1998). The species has been cultivated as an ornamental since 1862 (R. A. Vines 1960).

1. Bark exfoliating readily in reddish brown or blackish sheets; leaf blade margins crenate to dentate, teeth (13–)19–51(–69); inflorescences (7–)15–35-flowered; se Arizona, c Colorado, New Mexico, se Wyoming . . . . 1a. *Jamesia americana* var. *americana*
1. Bark exfoliating tardily in gray or brown strips or strings; leaf blade margins dentate, teeth (5–)9–27 (–33); inflorescences (1–)3–12(–19)-flowered; s California, s Nevada, Utah.
   2. Petals usually pink, rarely white; sepals (1.5–) 1.9–4(–6) mm; s California, s Nevada . . . . 1c. *Jamesia americana* var. *rosea*
   2. Petals white or pinkish white; sepals (2.5–) 3–7(–8) mm; Utah.
      3. Leaf blades (0.7–)1.2–3(–3.9) × (0.5–) 0.8–1.8(–3.2) cm; n Utah . . . . 1b. *Jamesia americana* var. *macrocalyx*
      3. Leaf blades (2.5–)3–5(–5.5) × (1.3–)2–4 (–4.5) cm; sw Utah . . . . 1d. *Jamesia americana* var. *zionis*

1a. Jamesia americana Torrey & A. Gray var. **americana** [F]

Bark exfoliating readily in reddish brown or blackish sheets. **Leaves:** petiole (1–)3–18 (–54) mm; blade ovate, (1–)2–8 (–10) × (1–)1.3–6.3(–8.5) cm, margins crenate to dentate, teeth (13–)19–51(–69). **Inflorescences** (7–)15–35-flowered. **Flowers:** sepals triangular-ovate, 1.5–3 (–3.8) mm; petals white, 6–11 mm; styles 3.5–5(–7) mm. **2*n*** = 32.

Flowering May–Aug(–Oct). Stream banks, forested canyon bottoms, rocky slopes, cliffs; 1500–3000(–3800) m; Ariz., Colo., N.Mex., Wyo.; Mexico (Chihuahua, Nuevo León).

1b. Jamesia americana Torrey & A. Gray var. **macrocalyx** (Small) Engler in H. G. A. Engler and K. Prantl, Nat. Pflanzenfam. ed. 2, 18a: 196. 1930 [C] [E] [F]

Edwinia macrocalyx Small in N. L. Britton et al., N. Amer. Fl. 22: 176. 1905

Bark exfoliating tardily in gray strips or strings. **Leaves:** petiole 2–8(–12) mm; blade ovate to obovate or rhombic, (0.7–)1.2–3 (–3.9) × (0.5–)0.8–1.8(–3.2) cm, margins dentate, teeth (5–)11–21 (–27). **Inflorescences** usually (2–)3–9(–13)-flowered, sometimes flowers solitary. **Flowers:** sepals lanceolate to triangular-ovate, (2.5–)3–7 mm; petals white or pinkish white, (6.5–)7.5–11.5 mm; styles 4–6.7 mm.

Flowering Jun–Aug. Cliffs, crevices, rocky slopes; of conservation concern; (1700–)2100–3300(–3700) m; Utah.

Variety *macrocalyx* is known from Juab, Salt Lake, Tooele, Utah, and Wasatch counties.

1c. Jamesia americana Torrey & A. Gray var. **rosea** Purpus ex C. K. Schneider, Ill. Handb. Laubholzk. 1: 375. 1905 [E]

Bark exfoliating tardily in brown or gray strips or strings. **Leaves:** petiole 1–6(–14) mm; blade ovate to elliptic or suborbiculate, (0.8–)1.3–3(–5) × (0.5–)1–2.2(–2.9) cm, margins dentate, teeth (9–)11–23(–29). **Inflorescences** (4–)6–10(–19)-flowered. **Flowers:** sepals lanceolate, (1.5–)1.9–4(–6) mm; petals usually pink, rarely white, (4–)5.5–9.5 mm; styles 3–6.5(–7.8) mm.

Flowering Jun–Aug. Cliffs, crevices, rocky slopes, talus; 2000–3700 m; Calif., Nev.

1d. Jamesia americana Torrey & A. Gray var. **zionis** N. H. Holmgren & P. K. Holmgren, Brittonia 41: 344, fig. 2G–I. 1989 [C] [E]

Bark exfoliating tardily in brown or gray strips or strings. **Leaves:** petiole 4–18 mm; blade ovate to broadly ovate or suborbiculate, (2.5–)3–5(–5.5) × (1.3–)2–4(–4.5) cm, margins dentate, teeth 13–27(–33). **Inflorescences** 4–12-flowered. **Flowers:** sepals lanceolate, (3.6–)4–7(–8) mm; petals white or pinkish white, 5.5–7 (–11) mm; styles (4.5–)6–8 mm.

Flowering Jun–Aug. Moist cliffs, hanging gardens; of conservation concern; 1200–2000(–2500) m; Utah.

Variety *zionis* is known from Iron, Kane, and Washington counties.

2. **Jamesia tetrapetala** N. H. Holmgren & P. K. Holmgren, Brittonia 41: 348, fig. 3J–K. 1989 • Four-petal cliffbush [C] [E]

Stems 5–20(–40) dm. **Bark** exfoliating in gray, orangish gray, or brownish strips or strings. **Branches** spreading or descending, often stunted and straggly; twigs densely spreading- or retrorse-pilose. **Leaves:** petiole 1–4(–7) mm, ascending-strigose to canescent or sericeous; blade ovate or obovate, (0.5–)1–2(–2.7) × (0.2–)0.8–1.2(–1.6) cm, base obtuse to usually rounded, sometimes cuneate, usually asymmetric, margins usually dentate, rarely entire, teeth 3–13(–16), apex obtuse to rounded, abaxial surface moderately to densely canescent or sericeous, adaxial sparsely strigose to glabrescent. **Inflorescences** 1(–3)-flowered; peduncle 2–10 mm, canescent or subsericeous. **Pedicels** 1–5 mm, canescent or subsericeous. **Flowers:** hypanthium 2.5–3.5 × 1–1.5 mm, canescent or sericeous; sepals 4, lanceolate to deltate-ovate, (3–)5–9 × (1.5–)2–2.8 mm, margins usually entire, rarely 2-lobed apically, apex acute to obtuse, abaxial surface canescent or sericeous; petals 4, white with pinkish margin or entirely pink, 5–13(–15) × 3–4.9 mm, sparsely to densely strigose or canescent, especially distally; stamens 8; filaments (4–)5–8 × 0.8–1.4 mm; anthers 0.8–1.1 mm; styles 3–5, 4.8–7.7 mm. **Capsules** 3–5.5(–7) × 2.5–3.5 mm. **Seeds** 0.6–0.8 mm.

Flowering Jun–Aug. Limestone cliffs, crevices, talus; of conservation concern; 2000–3300 m; Nev., Utah.

Jamesia tetrapetala is known from the Grant, Highland, and Snake ranges in eastern Nevada and the House Range in western Utah.

2. FENDLERA Engelmann & A. Gray, Smithsonian Contr. Knowl. 3(5): 77, plate 5. 1852 • Fendler-bush [For August Fendler, 1813–1883, German-born plant collector in North and South America, early botanical explorer of southwestern United States]

Ronald L. McGregor†

James Henrickson

Shrubs. Stems ascending or spreading. **Bark** exfoliating in grayish or reddish strings or strips, or sometimes not exfoliating. **Branches** erect or arching; twigs with simple and minutely branched trichomes. **Leaves** deciduous, opposite, sometimes clustered on short shoots; petiole present, relatively short; blade linear, elliptic, lanceolate, oblong, ovate, or falcate, herbaceous or coriaceous, margins entire, plane or strongly revolute; venation acrodromous, lateral veins sometimes obscure. **Inflorescences** terminal, flowers solitary; peduncle absent. **Pedicels** present. **Flowers** bisexual; perianth and androecium nearly hypogynous; hypanthium completely adnate to ovary, turbinate, broadly campanulate, or hemispheric, weakly or strongly 4- or 8-ribbed in fruit; sepals persistent, 4, eventually erect or strongly recurved, usually triangular, sometimes ovate, glabrous or hairy; petals 4, imbricate, spreading, white, sometimes tinged pink or red, broadly spatulate, base clawed, surfaces finely pubescent; stamens 8; filaments distinct, dorsiventrally flattened, gradually tapered toward apex, apex 2-lobed, lobes prolonged beyond anthers; anthers oblong; pistil 4(–5)-carpellate, ovary to ½ inferior, 4(–5)-locular; placentation axile; styles persistent, 4(–5), distinct. **Capsules** ovoid-ellipsoid, ± cartilaginous, dehiscence basipetally septicidal to middle of fruit. **Seeds** (1–)2–4(–6) per locule, reddish brown, ellipsoid. $x = 11$.

Species 2 (2 in the flora): sw United States, n Mexico.

A. J. Rehder (1920) found that the species of *Fendlera* show great uniformity in floral characters and may be distinguished easily by their leaves. N. H. Holmgren and P. K. Holmgren (1997) also reported that floral characters are remarkably uniform.

B. L. Turner (2001) recognized five species of *Fendlera*, including four in the flora. In addition to the species recognized here, B. L. Turner also recognized *F. falcata* and *F. wrightii* in our area; as noted later, these are best included within *F. rupicola*. The fifth species of Turner is *F. tamaulipana*, a taxon from Tamaulipas, Mexico; this species is here included in *F. linearis*.

Inflorescences of *Fendlera* have been described as solitary flowers or as clusters of two, three, or five flowers, racemes, or dichasial cymes. The inflorescence consists of a single flower terminating a leafy branch; these flower-bearing branches are sometimes aggregated. Frequently, the lateral branches are much-reduced and consist of a relatively short axis bearing two relatively small leaves and terminated by a flower. More rarely, the flowers appear nearly sessile in leaf axils, subtended by scales; this phenomenon is most frequent in *F. linearis*.

As in some species of *Philadelphus*, the buds on the long shoots of *Fendlera* are hidden in pouches of tissue at the bases of the petioles.

SELECTED REFERENCE Turner, B. L. 2001. Taxonomic revision of the genus *Fendlera* (Hydrangeaceae). Lundellia 4: 1–11.

1. Leaf blades coriaceous, linear, 1.3–2 mm wide, margins strongly revolute, touching midvein, abaxial surfaces hidden, adaxial surfaces with scattered, appressed trichomes; midveins 0.5–1 mm wide, flat . . . . . 1. *Fendlera linearis*
1. Leaf blades herbaceous to coriaceous, usually elliptic, lanceolate, oblong, ovate, or falcate, rarely linear, 2–10 mm wide, margins plane to revolute, not touching midvein, abaxial surfaces visible, strigose, sometimes with understory of minute, finely branched trichomes, adaxial surfaces with appressed trichomes and/or short, erect trichomes; midveins 0.2–0.3 mm wide, raised . . . . . 2. *Fendlera rupicola*

1. Fendlera linearis Rehder, J. Arnold Arbor. 1: 205. 1920 • Narrow-leaf Fendler-bush

Fendlera rigida I. M. Johnston; *F. tamaulipana* B. L. Turner

Stems 5–30 dm. **Branches** ± thorn-tipped; twigs densely strigose and with minute, branched trichomes. **Leaves:** blade linear, 7–25(–35) × 1.3–2 mm, coriaceous, base cuneate, margins strongly revolute, touching midvein, apex acute, mucronulate, abaxial surface hidden, adaxial surface usually sparsely sericeous, sometimes glabrate, trichomes appressed, to 0.4 mm; midvein 0.5–1 mm wide, flat. **Pedicels** 2–6 mm, strigose and with minute, branched trichomes. **Flowers:** hypanthium and calyx tube 1.5–3.5 mm; sepals becoming reflexed, triangular, 3–6 × 1.5–3 mm, sparsely sericeous and with minute, branched trichomes abaxially; petals white, 5–11 × 4–7 mm, claw 3–4.5 mm, blade ovate, round-obovate, or deltate-ovate, margins ± erose; filaments 3.5–8 × 2–2.5 mm; anthers 2–4 mm; styles 1–1.2 mm. **Capsules** ovoid, 9–13 × 5–6 mm. **Seeds** 3–5 mm.

Flowering May–Oct. Limestone walls, slopes, rocky woodlands, chaparral, gypsum hills; 1100–1900 m; Tex.; Mexico (Chihuahua, Coahuila, Nuevo León, Tamaulipas).

Fendlera linearis is well marked, although variable in habit and in flower and fruit sizes. In Texas, it is known only from Brewster and Presidio counties. It reportedly is browsed heavily by deer and other animals.

2. Fendlera rupicola Engelmann & A. Gray, Smithsonian Contr. Knowl. 3(5): 77, plate 5. 1852 • Cliff Fendler-bush [F]

Fendlera falcata Thornber ex Wooton & Standley; *F. rupicola* var. *falcata* (Thornber ex Wooton & Standley) Rehder; *F. rupicola* var. *wrightii* Engelmann & A. Gray; *F. tomentella* Thornber ex Wooton & Standley; *F. wrightii* (Engelmann & A. Gray) A. Heller

Stems 5–30 dm. **Branches** never becoming thorn-tipped; twigs sericeous-strigose, trichomes appressed, loosely appressed, and/or minute and branched, glabrescent. **Leaves:** blade usually elliptic, lanceolate, oblong, ovate, or falcate, rarely linear, 10–38 × 2–10 mm, herbaceous to coriaceous, base cuneate to rounded, margins plane or slightly to strongly revolute, not touching midvein, apex obtuse, acute, or slightly attenuate, abaxial surface visible, sericeous-strigose, trichomes often coarse, 0.2–0.6 mm, sometimes also with understory of minute, branched trichomes, adaxial surface glabrous or sparsely to moderately sericeous-strigose, trichomes ± appressed, 0.2–0.6 mm, and/or hispidulous, trichomes erect, 0.1–0.2 mm; midvein 0.2–0.3 mm wide, raised. **Pedicels** 1–25 mm, glabrous or sparsely sericeous-strigose, trichomes ± appressed, and/or with minute, branched hairs. **Flowers:** hypanthium and calyx tube 2–3 mm; sepals erect or reflexed, narrowly to broadly triangular, 4–8.5 × 1.5–4 mm, sparsely to moderately sericeous-strigose and/or with minute, branched trichomes abaxially; petals white or tinged pink or red, 8–15 × 5–14 mm, claw 3–4 mm, blade usually ovate or ovate-rhombic, rarely oblanceolate, obovate, or oval, margins erose; filaments 3–8 × 1–2.5 mm; anthers 2–3.5 mm; styles 4.5–6 mm. **Capsules** ovoid to ellipsoid, 7–15 × 5–8 mm. **Seeds** 5–8 mm.

Flowering Apr–Jun (Aug–Oct). Limestone bluffs and ledges, sandstone hillsides, granitic soils, rocky and brushy slopes, among boulders, desert grasslands, chaparral, oak and juniper-oak woodlands; 700–2800 m; Ariz., Colo., Nev., N.Mex., Tex., Utah; Mexico (Chihuahua, Coahuila, Sonora).

B. L. Turner (2001) divided *Fendlera rupicola* into three species. He restricted *F. rupicola* to plants from central Texas with large, thin leaf blades with flat margins and sparse vestiture, and contrasted those with plants from more arid areas farther west with leaf blades that are smaller, more often coriaceous, and revolute margined, with the surfaces sparsely to densely strigose, sometimes also hirtellous adaxially. Turner assigned these more western plants to two species. The first, *F. wrightii*, has abaxial leaf blade surfaces with an overstory of coarse, appressed to ascending trichomes and an understory of minute, branched trichomes; the adaxial surface is strigose and hispidulous, especially near the margins. According to Turner, this species occurs from trans-Pecos Texas west to Arizona and southeastern Nevada, north to extreme southern Colorado, and south to Chihuahua and Sonora, Mexico. The second western species, *F. falcata*, has similar leaves, but the abaxial surfaces lack the understory of minute, branched trichomes, and sometimes they are glabrous or only strigose adaxially. Maps provided by Turner show the distribution of *F. falcata* broadly overlapping that of *F. wrightii* but extending farther north in Colorado and into southeastern Utah. Despite this extensive sympatry, Turner reported that the two species rarely occur together, stating that in trans-Pecos Texas and southeastern New Mexico, for example, *F. falcata* occurred at higher elevations and on igneous substrates, whereas *F. wrightii* occurred at lower elevations on limestone substrates. Turner did not examine populations from elsewhere and considered many of the Sonoran specimens he examined, in addition to some from northern Arizona, to be intermediate between *F. falcata* and *F. wrightii*.

In contrast to the treatment by B. L. Turner (2001), N. H. Holmgren and P. K. Holmgren (1997) considered size, shape, and vestiture of leaves to be too variable to warrant recognition of infraspecific taxa in *Fendlera rupicola*. Their conclusion is supported by examination of numerous specimens and study of some populations (J. Henrickson, pers. obs.). The density of branched trichomes on the abaxial leaf blade surfaces varies from fairly dense to sparse, and even leaves that appear to have no branched trichomes are sometimes found to have them next to the midvein when examined with SEM. Despite some correlation between the presence or absence of the branched understory trichomes and blade thickness, the development of revolute margins, and the density of adaxial vestiture, variation in all these characters is essentially continuous. Furthermore, populations containing both plants with and without branched understory trichomes are more common than Turner reported, and the alleged difference between *F. falcata* and *F. wrightii* in substrate preference does not hold, even in Texas. Plants agreeing with Turner's concept of *F. rupicola*, which he considered to be restricted to Texas, occur in New Mexico west to the Grand Canyon area of Arizona, and some specimens from shaded habitats in Arizona are nearly identical to the type of *F. rupicola*. Consequently, a broad concept of *F. rupicola* is adopted here.

Fendlera rupicola is reportedly a favorite browse of deer, goats, sheep, and cattle. D. E. Moerman (1998) reported that early Native Americans used plants for drugs and as ceremonial items and the hard wood for making arrow shafts. Vegetatively, some plants of *F. rupicola* are nearly impossible to distinguish from similar forms of *Philadelphus microphyllus*.

3. FENDLERELLA (Greene) A. Heller, Bull. Torrey Bot. Club 25: 626. 1898 • Yerba desierto [For August Fendler, 1813–1883, botanical collector, and Latin *-ella*, honor]

Ronald L. McGregor†

Fendlera Engelmann & A. Gray [unranked] *Fendlerella* Greene, Bull. Torrey Bot. Club 8: 26. 1881

Shrubs. **Stems** ascending or spreading. **Bark** exfoliating in grayish strips or flakes. **Branches** spreading; twigs with simple trichomes. **Leaves** deciduous, opposite, sometimes clustered on short shoots; petiole present, relatively short, those of opposite leaves connate-perfoliate; blade elliptic to lanceolate, oblanceolate, obovate, or linear-oblong, ± coriaceous, margins entire, plane or revolute; venation acrodromous, lateral veins sometimes obscure. **Inflorescences** terminal, compound cymes, (3–)5–26(–34)-flowered; peduncle present. **Pedicels** present. **Flowers** bisexual; perianth and androecium perigynous; hypanthium completely adnate to ovary, turbinate-campanulate, not ribbed in fruit; sepals persistent, (4–)5, erect, lanceolate, strigose; petals (4–)5, valvate, spreading, white to yellowish, oblanceolate, base clawed, surfaces usually glabrous, sometimes sericeous-pilose abaxially; stamens (8–)10; filaments distinct, dorsiventrally flattened, gradually or abruptly tapered from base to tip, apex not 2-lobed; anthers globose; pistil (2–)3(–5)-carpellate, ovary slightly inferior, (2–)3(–5)-locular; placentation axile; styles persistent, (2–)3(–5), distinct. **Capsules** lanceoloid, cartilaginous, dehiscence basipetally septicidal to near base of fruit. **Seeds** 1 per locule, reddish brown, with a white cap, ellipsoid.

Species 4 (1 in the flora): sw, sc United States, Mexico.

Three other species have been named from Mexico. These differ from *Fendlerella utahensis* in having leaf blades that have denser pubescence and more revolute margins.

1. Fendlerella utahensis (S. Watson) A. Heller, Bull. Torrey Bot. Club 25: 626. 1898 • Utah Fendler-bush [F]

Whipplea utahensis S. Watson, Amer. Naturalist 7: 300. 1873; *Fendlerella utahensis* var. *cymosa* (Greene ex Wooton & Standley) Kearney & Peebles

Stems to 10 dm. **Twigs** reddish to orangish, strigose. **Leaves:** petiole 0.1–0.3 mm, pilose in axils; blade 5–20(–25) × 1–5(–6.5) mm, base cuneate to rounded, apex obtuse to acute, abaxial surface glabrate or strigose to tomentose, adaxial glabrate or strigose. **Inflorescences** congested to open, bracteate; peduncle 2–10 mm, strigose. **Pedicels** 0.8–3(–4.5) mm, strigose. **Flowers:** hypanthium 0.5–2 × 0.9–1.2 mm; sepals 0.8–2.2 × 0.2–0.6 mm; petals 2–4 × 0.8–1.2 mm, margins erose; filaments 1.2–3 × 0.2–0.4 mm, those opposite sepals longer than those opposite petals; anthers 0.2–0.5 mm. **Capsules** 4–6.1 × 1.5–1.9 mm. **Seeds** 2–3 mm.

Flowering Apr–Jun (Aug–Sep). Cracks and crevices, usually in limestone, sometimes sandstone, outcrops, sandy soils of mixed desert scrub, pine communities; 1200–2800 m; Ariz., Calif., Colo., Nev., N.Mex., Tex., Utah; Mexico (Chihuahua, Nuevo León).

Some plants in the southern range of *Fendlerella utahensis* (southern Arizona, southern New Mexico, trans-Pecos Texas, and northern Mexico) have leaves that average narrower, longer, and more acute than leaves on plants to the north. These have been recognized as var. *cymosa*. Intergradation is common; the variety is not recognized here.

4. WHIPPLEA Torrey in War Department [U.S.], Pacif. Railr. Rep. 4(5): 90, plate 7. 1857 • Modesty, yerba de silva [For Lieutenant Amiel Weeks Whipple, 1816–1863, commander of Pacific Railroad Expedition 1853 & 1854] [E]

Ronald L. McGregor†

Subshrubs. Stems prostrate to decumbent. **Bark** exfoliating in grayish or grayish brownish strips. **Branches** erect to decumbent; twigs with simple trichomes. **Leaves** marcescent, opposite; petiole absent or present; blade elliptic, ovate, or ovate-elliptic, herbaceous, margins entire or shallowly crenate-serrate, plane; venation acrodromous. **Inflorescences** terminal, racemes or racemose cymes, 4–12-flowered; peduncle present. **Pedicels** present. **Flowers** bisexual; perianth and androecium nearly hypogynous; hypanthium adnate to ovary proximally, free distally, hemispheric, weakly 9–11-ribbed in fruit; sepals eventually deciduous, (4–)5–6, erect, narrowly oblong or oblong-lanceolate, appressed-pubescent abaxially; petals 4–6, valvate, reflexed, white, obovate or oblong, base clawed, surfaces glabrous; stamens 8–12; filaments distinct, dorsiventrally flattened, abruptly tapered medially, apex not 2-lobed; anthers elliptic; pistil (2–)4–5-carpellate, ovary slightly inferior, (3–)4–5-locular; placentation axile; styles caducous,

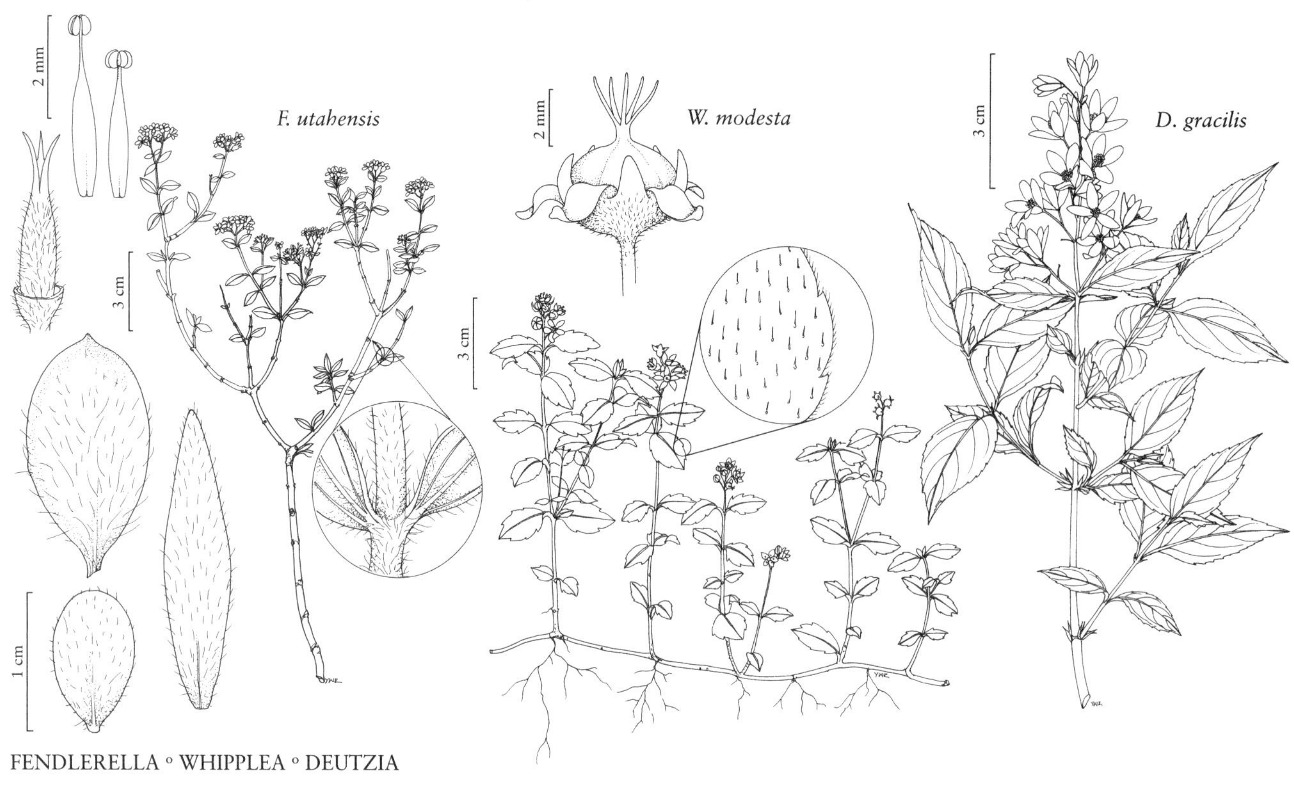

FENDLERELLA ◦ WHIPPLEA ◦ DEUTZIA

(3–)4–5, distinct or connate basally. **Capsules** depressed-spheric, cartilaginous, dehiscence basipetally septicidal, carpels separating entirely. **Seeds** 1 per locule, brown or reddish brown, ellipsoid, trigonous.

Species 1: w United States.

1. Whipplea modesta Torrey in War Department [U.S.], Pacif. Railr. Rep. 4(5): 90, plate 7. 1857 [E] [F]

Stems to 10 dm; adventitious roots at proximal nodes. **Flowering branches** weak, (0.4–)0.6–1.5 dm, appressed-pubescent. **Leaves:** petiole 0–3 mm, pilose; blade 10–40 × 10–30 mm, base rounded or tapered, apex obtuse, abaxial surface strigose, adaxial with white, pustule-based trichomes. **Inflorescences** congested to open; peduncle 20–50 mm, appressed-pubescent. **Pedicels** 0.5–2.5 mm, appressed-pubescent. **Flowers** odorless; hypanthium 1.4–2 × 2.3–2.6 mm; sepals 1.5–2 × 0.2–0.5 mm; petals 2.5–4 × 1–1.5 mm; filaments 1.2 × 0.5 mm; anthers 0.4–0.6 mm; styles 1–1.2 mm. **Capsules** 1.5–2 × 2–2.5 mm. **Seeds** 1–1.5 mm.

Flowering Mar–Jun. Dry, rocky sites, open to sparsely forested areas; (30–)400–1300(–1700) m; Calif., Oreg., Wash.

Whipplea modesta grows on the west side of the Cascade and Coast ranges. In Washington, it is known only from Clallam County.

5. DEUTZIA Thunberg, Nov. Gen. Pl., 19. 1781 • [For Johann van der Deutz, ca. 1743–1784, Dutch merchant and patron of Carl Peter Thunberg] [I]

Ronald L. McGregor†

Shrubs. Stems erect to spreading. **Bark** exfoliating in grayish, brownish, or reddish sheets or flakes. **Branches** erect, ascending, spreading, or arching; twigs with stellate and simple trichomes. **Leaves** deciduous or semideciduous, opposite; petiole present; blade lanceolate, elliptic, or ovate, herbaceous, margins serrulate to crenate-denticulate, plane; venation pinnate. **Inflorescences** terminal, corymbose cymes, panicles, or racemes, usually (2–)5–60-flowered, rarely flowers solitary; peduncle present or absent. **Pedicels** present. **Flowers** bisexual; perianth and androecium epigynous; hypanthium completely adnate to ovary, campanulate or ovoid, not ribbed in fruit; sepals deciduous, 5, erect, triangular to broadly ovate, glabrous or abaxially short-hairy or stellate-pubescent; petals 5 [10 in some cultivars], valvate or imbricate, spreading, white, pink, or purplish, elliptic to oblong, base clawed, surfaces glabrous or hairy; stamens 10, in 2 whorls; filaments distinct, dorsiventrally flattened or terete, gradually or abruptly tapered medially or filiform, apex 2-lobed (lobes much shorter than [exceeding] anthers) or not lobed; anthers globose; pistil 3–5-carpellate, ovary completely inferior, 3–5-locular; placentation usually axile proximally and parietal distally, sometimes strictly parietal; styles persistent, 3(–4), distinct. **Capsules** hemispheric, cartilaginous, dehiscence acropetally septicidal along septum at base of fruit and also apically. **Seeds** 10–20 per locule, dark brown, ellipsoid. $x = 13$.

Species ca. 60 (3 in the flora): introduced; Mexico, Asia (China, Himalaya, Japan, Korea), Pacific Islands (Philippines).

A. J. Rehder (1940) recognized 23 species (including eight named hybrids), 27 varieties, and about 27 cultivars of *Deutzia* growing mostly at the Arnold Arboretum. L. H. Bailey et al. (1976) listed taxa only in the horticultural trade and cited 34 species and 36 cultivars that had been listed in the previous two decades. M. A. Dirr (1998) reported that *Deutzia* has lost favor as an ornamental over the years. He noted that identification of species, particularly cultivars, borders on the impossible, and nothing is clear-cut in the world of *Deutzia* identification. He listed seven species and 22 cultivars, including nine cultivars of *D. scabra*.

In cultivation, deutzias may spread by suckers; several species have escaped and become marginally naturalized. M. L. Fernald (1950) reported that *Deutzia gracilis* and *D. scabra* were beginning to spread to thickets and roadsides; there is no evidence that either species has become widely naturalized. H. A. Gleason and A. Cronquist (1991) noted that *D. scabra* rarely escapes from cultivation.

1. Inflorescences corymbose cymes; petals broadly ovate or suborbiculate, imbricate; filaments terete, filiform . . . . . 3. *Deutzia parviflora*
1. Inflorescences usually racemes or panicles, sometimes flowers solitary (*D. gracilis*); petals narrowly elliptic to oblong or oblong-lanceolate, valvate; filaments dorsiventrally flattened, narrowly oblong or subulate.
    2. Leaf blade abaxial surfaces glabrous or glabrate; inflorescences glabrous . . . . . 1. *Deutzia gracilis*
    2. Leaf blade abaxial surfaces densely stellate-pubescent; inflorescences stellate-pubescent . . . . . 2. *Deutzia scabra*

1. Deutzia gracilis Siebold & Zuccarini, Fl. Jap. 1: 22, plate 8. 1835 [F] [I]

Shrubs 5–20 dm. **Branches** spreading or arching. **Leaves:** petiole 2–8 mm, glabrous; blade lanceolate, elliptic-lanceolate, oblong-lanceolate, or broadly ovate-lanceolate, 30–35 × 15–30 mm, base cuneate or rounded, margins unevenly serrate, apex acuminate, abaxial surface light green, glabrous or glabrate, adaxial dark green, sparsely stellate-pubescent. **Inflorescences** usually racemes or narrow panicles, open, (2–)10–50-flowered, sometimes flowers solitary, 8–12 × 3–6 cm, glabrous; peduncle 5–10 mm, glabrous. **Pedicels** 5–10 mm, glabrous. **Flowers** faintly fragrant, 15–20 mm; hypanthium campanulate, 2.5–3 × 2.5–3 mm, glabrous or short-hairy; sepals triangular, 1.5–1.7 × 1–1.5 mm, apex obtuse, surfaces glabrous or glabrate; petals valvate, white, oblong or oblong-lanceolate, 10–12(–15) × 4–6 mm, glabrous or sparsely stellate-pubescent abaxially, glabrous adaxially; filaments dorsiventrally flattened, narrowly oblong or subulate, outer 5–6 mm, apex 2-lobed or not lobed, inner 3–4 mm, apex 2-lobed; styles 3, 4–7 mm. **Capsules** hemispheric, 6–7 × 5–6 mm. **Seeds** 1.2–1.5 mm. **2*n*** = 26 (Asia).

Flowering Apr–Jun. Waste areas, roadsides; 10–900 m; introduced; Ga., Md.; Asia (China, Japan).

2. Deutzia scabra Thunberg, Nov. Gen. Pl., 20. 1781 [I]

Deutzia crenata Siebold & Zuccarini

Shrubs 10–30 dm. **Branches** erect to ascending. **Leaves:** petiole 1–3 mm, sparsely to densely stellate-pubescent; blade ovate-lanceolate to ovate, 30–80 × 15–50 mm, base rounded or broadly cuneate, margins crenate-denticulate, apex acute to acuminate, abaxial surface light green, densely stellate-pubescent, trichomes 10–20-rayed, adaxial dark green, stellate-pubescent, trichomes 4–6-rayed. **Inflorescences** racemes or panicles, loose, 5–50-flowered, 4–10 × 2–3 cm, stellate-pubescent; peduncle absent. **Pedicels** 2–10 mm, sparsely stellate-pubescent. **Flowers** faintly fragrant, 8–20 mm; hypanthium campanulate, 2.5–5 × 3–4 mm, densely stellate-pubescent; sepals triangular to ovate, 1.5–2.5 × 1–2.2 mm, apex obtuse, surfaces stellate-pubescent; petals valvate, white or pinkish, narrowly elliptic to oblong, 7–15 × 2.5–3 mm, stellate-pubescent abaxially, glabrous or sparsely stellate-pubescent adaxially; filaments dorsiventrally flattened, narrowly oblong, outer 7–9 mm, apex 2-lobed, inner 5–6 mm, apex 2-lobed; styles 3(–4), 5–11 mm. **Capsules** 3.8–5 × 4–5 mm. **Seeds** 1.5–2 mm. **2*n*** = 26, 130 (Asia).

Flowering Apr–Jul. Roadsides, waste areas, homesteads, parks; 10–200 m; introduced; Ark., Conn., Fla., Ga., Ill., Ky., Md., Mass., Mich., N.J., N.Y., N.C., Pa., R.I., Utah, Vt., Va., W.Va.; e Asia (Japan).

Deutzia crenata and *D. scabra* are highly variable and have long been variously interpreted. A. J. Rehder (1920), assisted by H. O. Juel of Uppsala University, studied the three specimens at UPS that Thunberg had named *D. scabra*. He selected one as the type but said it was the same as the later-described *D. crenata*. J. Ohwi (1965) listed eight synonyms for *D. scabra* and interpreted *D. crenata* as consisting of three varieties, which included six synonyms (including "*D. scabra* Thunberg in part"). In view of the above, and the many cultivars that have been developed, *D. crenata* is treated here as a synonym of *D. scabra*.

3. Deutzia parviflora Bunge, Enum. Pl. China Bor., 31. 1833 [I]

Shrubs 10–20 dm. **Branches** ascending to erect. **Leaves:** petiole 2–8 mm, glabrous; blade ovate or elliptic to ovate-lanceolate, 30–90 × 20–40(–50) mm, base cuneate or rounded, margins serrulate, apex acute to acuminate, abaxial surface green, glabrous or stellate-pubescent, trichomes 6–9-rayed, adaxial green, glabrous or stellate-pubescent, trichomes 3–5-rayed. **Inflorescences** corymbose cymes, open, 10–60-flowered, 5–10 × 4–7 cm, glabrous; peduncle 20–80 mm, villous or stellate-pubescent. **Pedicels** 10–15 mm, stellate-pubescent. **Flowers** faintly fragrant, 9–10 mm; hypanthium ovoid, 2.2–3 × 1–1.3 mm, densely stellate-pubescent; sepals ovate-deltate or broadly ovate, 0.9–1.2 × 0.3–0.4 mm, apex obtuse, surfaces densely stellate-pubescent; petals imbricate, white to pink, broadly ovate or suborbiculate, 2–7 × 3–5 mm, stellate-pubescent; filaments terete, filiform, outer 4–5 mm, apex not lobed, inner 3–4 mm, apex 2-lobed; styles 3, 2–3 mm. **Capsules** hemispheric, 2–3 × 2–3 mm. **Seeds** 0.9–1.3 mm. **2*n*** = 26, 78 (Asia).

Flowering Apr–Jun. Disturbed areas, moist ravines; 100–200 m; introduced; Ga.; Asia (n China, Korea).

Deutzia ×*lemoinei* Lemoine, the hybrid between *D. parviflora* and *D. gracilis*, often has been grown in the flora area and is considered one of the hardiest cultivars (A. J. Rehder 1940; M. A. Dirr 1998). Three varieties of *D. parviflora* are recognized in China; plants in the flora area seem closest to var. *parviflora*.

6\. PHILADELPHUS Linnaeus, Sp. Pl. 1: 470. 1753; Gen. Pl. ed. 5, 211. 1754 • Mock orange, syringa [Greek *phil-*, loving, and *adelphos*, brother, traditionally (but on uncertain grounds) considered to be an honorific for Ptolemy Philadelphus, 309–246 BCE, King of Ptolemaic Egypt]

Alan S. Weakley

James Henrickson

Shrubs. **Stems** erect, ascending, arching, or spreading, decussately branched. **Bark** tight or exfoliating in grayish, brown, or reddish brown sheets. **Branches** erect, ascending, or spreading, often arching; twigs glabrous or with simple trichomes. **Leaves** winter- or drought-deciduous, opposite; petiole present; blade ovate, elliptic-ovate, elliptic, suborbiculate, lanceolate, or linear-lanceolate, herbaceous, subcoriaceous, or coriaceous, margins entire or serrulate to serrate, often irregularly and variably so, plane or revolute; venation acrodromous, secondarily and distally pinnate. **Inflorescences** terminal, sometimes appearing axillary when 1-flowered, cymes, cymose racemes, or cymose panicles, or flowers solitary, 1–49-flowered; peduncle present. **Pedicels** present. **Flowers** bisexual; perianth and androecium perigynous to epigynous; hypanthium completely adnate to ovary, turbinate, obconic, or hemispheric, weakly or strongly 4- or 8-ribbed in fruit; sepals usually persistent, 4, spreading or reflexed, deltate to triangular-acuminate, villous, strigose, or glabrous; petals 4 (or 8+ in some horticultural forms), imbricate, spreading to ascending, white to cream colored, rarely purple-maculate, drying yellowish, oblong-obovate, obovate, or orbiculate, base sessile and tapered, or minutely clawed, surfaces glabrous [rarely hairy]; stamens (11–)13–90; filaments distinct or irregularly connate into groups proximally, dorsiventrally flattened proximally, gradually or abruptly tapered from base to apex, apex not 2-lobed, although sometimes slightly notched; anthers depressed-ovate or transversely oblong; pistil 4-carpellate, ovary inferior to ½ inferior, 4-locular; placentation axile proximally, parietal distally; styles persistent, 1 or 4, connate proximally to completely; stigmas 4. **Capsules** turbinate, obconic to obovoid, hemispheric, subglobose, or oblong-ovoid, coriaceous, persistent and gradually deteriorating, dehiscence loculicidal. **Seeds** 10+ per locule, rusty brown, fusiform, sometimes caudate. $x = 13$.

Species ca. 25 (9 in the flora): w, se North America, Mexico, Central America, Europe, Asia.

Philadelphus has a relictual distribution in western and southeastern North America, Mexico, and Central America (from southwestern Canada south in the western cordillera to Panama); southern Europe (perhaps only by human introduction); the Caucasus; and eastern Asia. It is naturalized elsewhere, including most temperate areas of the western and eastern hemispheres, and in Hawaii, where *P. karwinskianus* Koehne is invasive.

Hu S. Y. (1954–1956) relied on vestiture of the leaves, twigs, pedicels, flowers, and fruits to distinguish species in *Philadelphus*. She extended and expanded the traditional use of such characters by C. D. Beadle (1902), P. A. Rydberg (1905), and C. L. Hitchcock (1943). However, with more specimens now available, it is clear that these characters are variable and poorly correlated with one another. Basing taxa on permutations of such characters often has resulted in sympatric taxa that lack geographic and/or ecologic coherence. Herein, the recognition of biologically meaningful taxonomic units is attempted; this means that many previously named taxa are being combined in general agreement with recent, more local or regional, assessments in North America, which generally recognize fewer taxa (C. L. Hitchcock et al. 1955–1969, vol. 3; W. C. Martin and C. R. Hutchins 1980, vol. 1; C. F. Quibell 1993; N. H. Holmgren and P. K. Holmgren 1997; C. K. Frazier 1999; B. L. Turner 2006).

Hu S. Y. (1954–1956) recognized four subgenera in *Philadelphus*: *Deutzioides* S. Y. Hu, *Gemmatus* (Koehne) S. Y. Hu, *Macrothyrsus* S. Y. Hu, and *Philadelphus* (as *Euphiladelphus*). A preliminary phylogenetic study of *Philadelphus* based on ITS sequences provided cladistic support for three of the four subgenera (with the reassignment of *P. microphyllus* from subg. *Philadelphus* to subg. *Gemmatus*) and no support for recognition of subg. *Macrothyrsus* (A. E. Weakley 2002). Guo Y. L. (2013) conducted a more detailed phylogenetic study of *Carpenteria* and *Philadelphus* utilizing both nuclear and chloroplast markers and found three lineages: two species sampled from subg. *Deutzioides* (*P. mearnsii* and *P. texensis* var. *texensis*); *Carpenteria*; and the remainder of *Philadelphus* (32 accessions, including *P. hirsutus*, which had been previously considered to be a component of subg. *Deutzioides*). These results suggest the potential inclusion of *Carpenteria* in *Philadelphus* and also that characters (such as buds exposed versus in nodal pouches) that have been used to distinguish subg. *Deutzioides* and subg. *Macrothyrsus* from subg. *Philadelphus* may be plesiomorphic. In this treatment, species one through four belong to subg. *Deutzioides*, the basal clade in *Philadelphus*, which is restricted to southeastern North America and southwestern North America into northern Mexico; species five (*P. microphyllus*) belongs to subg. *Gemmatus*, which occurs from southwestern United States south to Panama; and species six through nine belong to subg. *Philadelphus*, of eastern North America, northwestern North America, southern Europe, the Caucasus, and eastern Asia.

Within each of these three subgeneric clades, the morphology of *Philadelphus* is very conservative; relatively few morphological characters are useful in distinguishing taxa in each subgenus. Additionally, W. Bangham (1929) studied the chromosomes of about 40 taxa and found no variation in number and complete compatibility in all hybrids that he studied; E. K. Janaki Ammal (1951) reported some pairing irregularities in hybrids.

A number of horticultural forms have been developed, are planted in temperate areas, and may be found as local escapes. Given the morphological variability of *Philadelphus* species and the uncertain origins of some of these plants, the horticultural forms are difficult to deal with by conventional taxonomic means. An example is *P.* ×*virginalis* Rehder, which is alleged to be a hybrid of *P.* ×*lemoinei* hort. (itself alleged to be a hybrid of the Old World *P. coronarius* and southwestern North American *P. microphyllus*) and *P.* ×*nivalis* Jacques (itself possibly a hybrid of *P. coronarius* and eastern North American *P. pubescens*); C. A. Stace (2010b) considered this to be the most widely cultivated, persistent, and presumably established *Philadelphus* taxon in the British Isles. *Philadelphus* ×*virginalis* was reported as escaping locally in Lenawee County, Michigan, by E. G. Voss and A. A. Reznicek (2012). Neither reliable identification of cultivars of this kind nor determination of their genetic origins are currently possible. Some idea of the cultivated entities can be gained from A. J. Rehder (1940), Hu S. Y. (1954–1956), D. Wright (1980), G. Krüssmann (1984–1986, vol. 2), and M. H. A. Hoffman (1996).

Philadelphus tomentosus Royle, a native of the Himalayan region, has been reported as naturalized in the flora area (M. A. Vincent and A. W. Cusick 1998) based on specimens from Allen and Paulding counties, Ohio. Because of the close similarity of all *Philadelphus* taxa, those specimens cannot be definitely identified as *P. tomentosus*. They appear to be *P. pubescens*, a native North American species widely cultivated and naturalized in eastern North America.

In some species of *Philadelphus*, the axillary buds are enclosed in pouches of cortical and epidermal tissue at the petiole bases. The petiole abscises distal to the pouch so that the bud is hidden and proximal to the leaf scar, only becoming visible as it expands. In other species, the buds are exposed, and the petiole abscises proximal to the bud; these buds are distal to the leaf scars, which is the condition normally seen in most plants. Whether the axillary buds are exposed (species one to four) or hidden in pouches (species five to nine) is best observed at nodes of mature leaves on vigorous long shoots and may not be apparent at nodes of young leaves or on short shoots.

SELECTED REFERENCES Beadle, C. D. 1902. Studies in *Philadelphus*. Biltmore Bot. Stud. 1: 159–161. Frazier, C. K. 1999. A taxonomic study of *Philadelphus* (Hydrangeaceae) as it occurs in New Mexico. New Mexico Bot. Newslett. 13: 1–6. Hitchcock, C. L. 1943. The xerophyllous species of *Philadelphus* in southwestern North America. Madroño 7: 36–56. Hoffman, M. H. A. 1996. Cultivar classification of *Philadelphus* L. (Hydrangeaceae). Acta Bot. Neerl. 45: 199–209. Hu, S. Y. 1954–1956. A monograph of the genus *Philadelphus*. J. Arnold Arbor. 35: 275–333; 36: 52–109, 325–368; 37: 15–90. Rydberg, P. A. 1905. *Philadelphus*. In: N. L. Britton et al., eds. 1905+. North American Flora.... 47+ vols. New York. Vol. 22, pp. 162–175. Turner, B. L. 2006. Species of *Philadelphus* (Hydrangeaceae) from trans-Pecos Texas. Lundellia 9: 34–40. Weakley, A. E. 2002. Evolutionary Relationships within the Genus *Philadelphus* (Hydrangeaceae): A Molecular Phylogenetic and Biogeographic Analysis. M.A. thesis. University of North Carolina. Wright, D. 1980. *Philadelphus*. Plantsman 2: 104–116.

1. Axillary buds hidden in pouches; styles 4, connate proximally, cylindric, lobes 0.5–8 mm.
    2. Styles 2.5–5.5(–7) mm, lobes 0.5–2.5 mm; leaf blades (0.5–)0.8–3(–5.5) × (0.2–)0.3–1.3(–3.3) cm; filaments often connivent-connate in irregular clusters; w Texas westward. . . . . 5. *Philadelphus microphyllus*
    2. Styles 4–16 mm, lobes 1–8 mm; leaf blades 1.5–12(–16) × 1–7(–11) cm; filaments distinct; widespread in temperate North America.
        3. Inflorescences cymes or racemes, or flowers solitary, 1–3(–9)-flowered; stamens 60–90; styles 10–16 mm, lobes 0.8–1 mm wide . . . . 6. *Philadelphus inodorus*
        3. Inflorescences usually cymose racemes or panicles, sometimes flowers solitary, (1–)5–49-flowered; stamens 20–50; styles 4–10 mm, lobes 0.3–0.9 mm wide.
            4. Leaf blade abaxial surfaces moderately to densely strigose, or tomentose to villous, hairs twisted, main vein axils and main veins often more densely strigose-tomentose; hypanthia and sepal abaxial surfaces usually sparsely to densely strigose or villous; bark usually gray, tight; inflorescences (1–)5–9-flowered . . . . 7. *Philadelphus pubescens*
            4. Leaf blade abaxial surfaces glabrous or sparsely strigose, hairs usually appressed-ascending, not twisted, main vein axils often moderately to densely strigose-tomentose; hypanthia and sepal abaxial surfaces glabrous or sparsely villous or strigose; bark reddish, soon exfoliating in flakes or strips; inflorescences (1–)5–49-flowered.
                5. Inflorescences (1–)7–49-flowered; style lobes 1–4 mm; larger leaf blades usually less than 6 × 2.5 cm; w North America . . . . 8. *Philadelphus lewisii*
                5. Inflorescences 5–7(–9)-flowered; style lobes 3–8 mm; larger leaf blades usually greater than 6 × 2.5 cm; e North America . . . . 9. *Philadelphus coronarius*
1. Axillary buds exposed; styles 1, clavate.
    6. Styles 4–6 mm; leaf blades 2–8 cm; e United States (Arkansas, eastward) . . . . 4. *Philadelphus hirsutus*
    6. Styles 1.9–3.2(–3.5) mm; leaf blades 0.5–3.3(–4.7) cm; sc, sw United States (c Texas, westward).
        7. Leaf blade adaxial surfaces strigose, abaxial surfaces with coarse, appressed hairs only, without understory of coiled-crisped hairs.
            8. Hairs on abaxial leaf surface longer and more dense than those of adaxial surface; leaf blades (1–)1.9–3(–4.1) × (0.4–)0.5–1.1(–1.8) cm; c Texas. . . . . 2. *Philadelphus texensis* (in part)
            8. Hairs on adaxial and abaxial leaf surfaces ± equal in length and density; leaf blades 0.5–1.7(–3) × 0.1–0.6(–1.1) cm; New Mexico, w Texas . . . . 3. *Philadelphus mearnsii*
        7. Leaf blade adaxial surfaces strigose to sericeous-strigose, abaxial surfaces with both coarse, appressed hairs and understory of coiled-crisped hairs.
            9. Leaf blade adaxial surfaces with both scattered coarse, appressed hairs and many shorter, erect, often slender hairs . . . . 1. *Philadelphus serpyllifolius* (in part)
            9. Leaf blade adaxial surfaces with appressed hairs only.
                10. Leaf blade adaxial surfaces with (4–)5–9 appressed hairs per mm of surface width; w Texas . . . . 1. *Philadelphus serpyllifolius* (in part)
                10. Leaf blade adaxial surfaces with 1–3 appressed hairs per mm of surface width; c Texas . . . . 2. *Philadelphus texensis* (in part)

1. Philadelphus serpyllifolius A. Gray, Smithsonian Contr. Knowl. 3(5): 77. 1852 • Thyme-leaf mock orange [F]

Philadelphus serpyllifolius var. *intermedius* B. L. Turner

Shrubs, 5–10(–20) dm. **Stems** light reddish brown, weathering gray and striate, stiffly divaricately to loosely branched, moderately strigose and appressed villous-sericeous; internodes (0.5–)1.3–3(–4.5) cm; short-shoot spurs sometimes present; axillary buds exposed. **Leaves:** petiole (1–)2–4(–5.5) mm; blade gray-white abaxially, green adaxially, lance-ovate or oblong-ovate to broadly ovate, (0.5–)1.2–2.5(–3) × (0.2–)0.4–0.8(–1.4) cm, herbaceous to subcoriaceous, margins entire, plane, often drying revolute, abaxial surface sparsely to moderately sericeous-strigose, hairs ± appressed, coarse to slender, 0.6–1.1 mm, with dense understory of white, slender, curled-crisped hairs, adaxial surface sparsely to moderately sericeous-strigose, hairs scattered, appressed, coarse, 0.5–1.2 mm, (these (4–)5–9 per mm of width in populations that have been called var. *intermedius*), also hirsute, hairs slender, usually erect, rarely appressed, 0.1–0.3(–0.4) mm. **Inflorescences:** flowers solitary, produced from previous year's long shoots. **Pedicels** 1–2 mm, strigose-sericeous to villous. **Flowers:** hypanthium strigose-sericeous or villous; sepals ovate, 3–5.5 × 2–3.2 mm, apex obtuse, acuminate to ± caudate, abaxial surface strigose-sericeous or villous, adaxial surface sparsely sericeous but villous along margins; petals white, oblong-ovate to ovate, 4–9(–11) × 3–6 mm; stamens 14–22; filaments sometimes proximally connivent, (1.5–)2–4.5 mm; style 1, clavate, 2.4–3.2 mm, slender base 0.5–0.7 mm; stigmatic portion 1.3–1.7 × 0.9–1.1 mm, apex slightly lobed. **Capsules** turbinate-globose, 4–5.2 × 4–5 mm, sepals persistent on distal ⅓, capsule distal surface usually impressed in 4(–8) vertical lines. **Seeds** not caudate, to 1.1 mm.

Flowering Apr–Jun; fruiting May–Nov. Rocky igneous and limestone slopes, bluffs, pinyon-oak-juniper zones; 1100–1900(–2300) m; Tex.; Mexico (Coahuila).

In two populations of *Philadelphus serpyllifolius* from Brewster County (Glass Mountains, Sierra Madera), the adaxial leaf vestiture has both long and short appressed hairs (the shorter hairs not erect as in typical *P. serpyllifolius*). These populations also show variation in development of the understory of tightly coiled hairs on the abaxial leaf surface, with some collections lacking the understory of coiled hairs. These have been recognized as var. *intermedius*.

2. Philadelphus texensis S. Y. Hu, J. Arnold Arbor. 37: 54. 1956 [C]

Shrubs, 6–15 dm. **Stems** spreading, green to yellowish, weathering grayish and striate, divaricately branched, moderately to sparsely strigose-sericeous; internodes 1.5–5.5 cm (long shoots), 0.2–2.2 cm (lateral branches); axillary buds exposed. **Leaves:** petiole 2–4.5(–8) mm; blade grayish or whitish abaxially, green adaxially, usually narrowly lanceolate to lance-ovate, sometimes ovate, (1–)1.6–3.3(–4.7) × (0.4–)0.5–1.1 (–2.3) cm, ± herbaceous, margins usually entire, rarely sparsely serrulate, plane, sometimes drying revolute, abaxial surface densely and loosely strigose, hairs arched-twisted, 0.3–1.2 mm, with or without understory of coiled-crisped, slender hairs between raised veins, adaxial surface sparsely to moderately strigose, hairs antrorsely appressed, 0.4–0.9 mm, 1–3 hairs per mm of width. **Inflorescences:** flowers solitary. **Pedicels** 1–2 (–4) mm, loosely strigose or glabrous. **Flowers:** hypanthium loosely strigose or glabrous; sepals ovate to ovate-lanceolate, 2.5–5.5 × 1–2.5 mm, apex obtuse to long acuminate-caudate, abaxial surface loosely strigose or glabrous, adaxial surface glabrous except ciliate and villous along distal margins; petals white, oblong-ovate to ovate, 8–10(–14) × 3.5–5(–10.5) mm, margins entire to undulate, apex obtuse to notched; stamens (11–) 14–16(–24); filaments distinct, 1–4.7 mm; style 1, clavate, 1.9–3(–3.5) mm, slender base 0.5–2 mm; stigmatic portion obovoid, 1.2–1.7 × 1.1 mm. **Capsules** turbinate-spheroid, 3.5–5.5 × 3.5–5 mm, sepals at distal ⅓, tardily falling, capsule distal surface smooth or impressed in 4 vertical lines. **Seeds** short-caudate distally, 1–1.2 mm.

Varieties 2 (2 in the flora): Texas, n Mexico.

A. E. Weakley (2002) found that the ITS sequences of *Philadelphus texensis* and *P. mearnsii* do not differ, suggesting that they are very closely related.

1. Leaf blade abaxial surfaces strigose and with understory of slender, white, tightly coiled-crisped hairs . . . . .2a. *Philadelphus texensis* var. *texensis*
1. Leaf blade abaxial surfaces strigose, without understory of white, tightly coiled-crisped hairs . . . . 2b. *Philadelphus texensis* var. *ernestii*

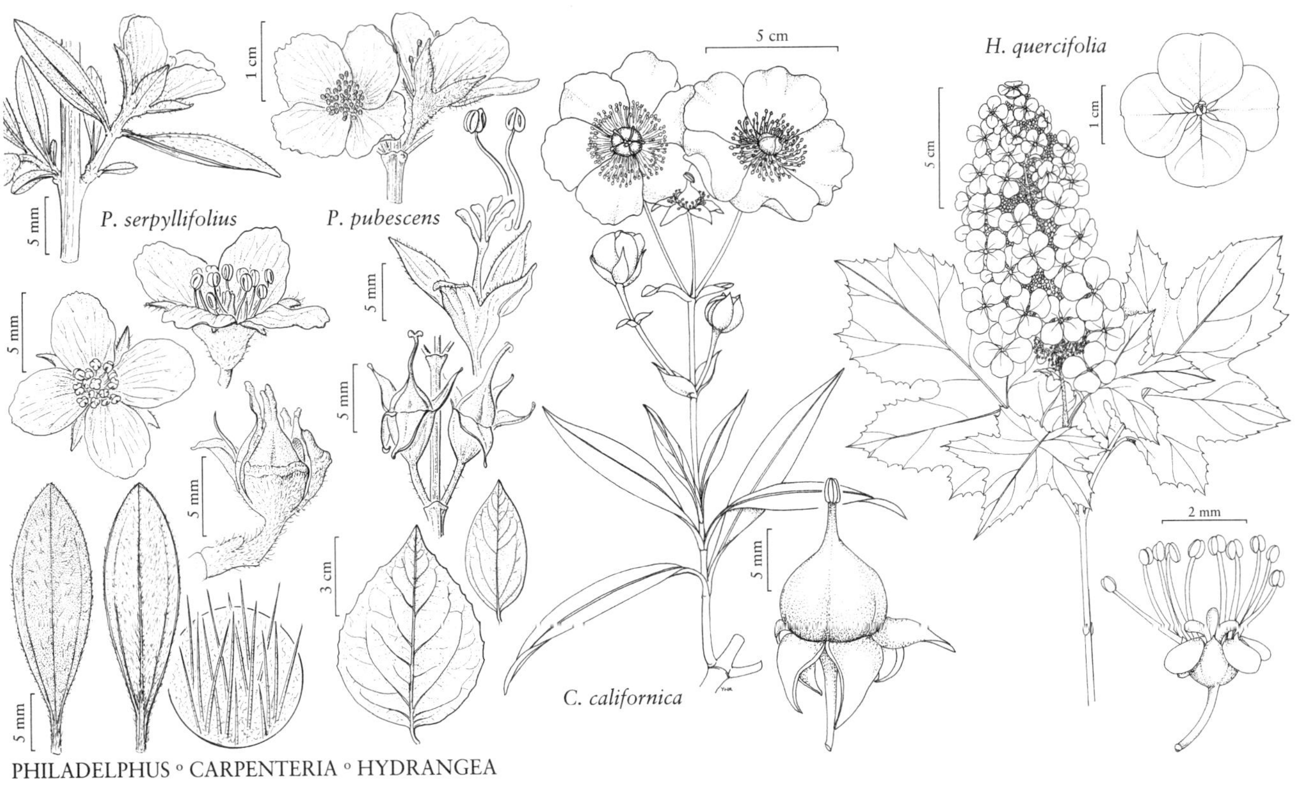

PHILADELPHUS ◦ CARPENTERIA ◦ HYDRANGEA

2a. Philadelphus texensis S. Y. Hu var. **texensis** • Texas mock orange [C]

Philadelphus texensis var. *coryanus* S. Y. Hu

Leaf blades (1–)1.6–3.3(–4.7) × (0.5–)0.6–1(–2.3) cm, abaxial surface loosely strigose, hairs 0.4–0.8 mm, with understory of white, more slender, tightly coiled-crisped hairs.

Flowering Apr–May; fruiting Jun–Oct. Limestone slopes and ravines, slopes in oak-juniper woodlands; of conservation concern; 200–400 m; Tex.; Mexico (Coahuila).

Variety *texensis* has a more westward range than var. *ernestii*; it is known from Bandera, Bexar, Edwards, Kendall, Medina, Real, and Uvalde counties in central Texas and is rare in the mountains of central Coahuila, Mexico. The two varieties are identical in all important characteristics, differing primarily in the presence or absence of the highly coiled-crisped understory hairs on the abaxial leaf surfaces. There are several collections of var. *texansis* in which most leaves lack understory vestiture, as in var. *ernestii*, but the understory vestiture occurs on some leaves or branches indicating the expression or non-expression of a simple genetic trait. In both varieties, sepals can be glabrous or sparsely sericeous, even within a population.

2b. Philadelphus texensis S. Y. Hu var. **ernestii** (S. Y. Hu) Henrickson, Phytoneuron 2016-8: 5. 2016 • Canyon mock orange [C]

Philadelphus ernestii S. Y. Hu, J. Arnold Arbor. 37: 50. 1956

Leaf blades (1–)1.9–3(–4.1) × (0.4–)0.5–1.1(–1.8) cm, abaxial surface loosely strigose, hairs 0.2–0.5(–0.7) mm, without underlying layer of white finely coiled-crisped hairs.

Flowering Apr–Jun; fruiting May–Oct. Limestone cliffs, ravines, slopes in oak-juniper woodlands; of conservation concern; 300–600 m; Tex.; Mexico (Coahuila).

Variety *ernestii* occurs in Bandera, Blanco, Hays, Kendall, and Travis counties in central Texas and in northern Coahuila, Mexico.

3. Philadelphus mearnsii W. H. Evans ex Rydberg in N. L. Britton et al., N. Amer. Fl. 22: 174. 1905 • Mearns's mock orange

Philadelphus hitchcockianus S. Y. Hu; *P. mearnsii* subsp. *bifidus* C. L. Hitchcock

Shrubs, 5–15(–40) dm. **Stems** tan-brown, weathering gray, striate, stiffly divaricately branched, strigose-sericeous; internodes 0.1–3.5 cm; short shoots often present; axillary buds exposed. **Leaves:** petiole 0.5–4 mm; blade green or gray-green abaxially and adaxially, oblong-lanceolate, linear-lanceolate, or elliptic to ovate, 0.5–1.7(–3) × 0.1–0.6(–1.1) cm, coriaceous, margins entire, ± revolute, surfaces ± equally moderately strigose, hairs appressed, coarse, 0.2–0.6(–0.9) mm, without understory of coiled-crisped hairs; veins inconspicuous. **Inflorescences:** flowers solitary, produced from previous year's long shoots, often appearing axillary. **Pedicels** 0.5–2 mm. **Flowers:** hypanthium sparsely to moderately strigose-sericeous or glabrous; sepals ovate-lanceolate, 2–3.5 (–4.2) × 1.3–2.3 mm, apex acute to acuminate-caudate, abaxial surface sparsely to moderately strigose-sericeous or glabrous, adaxial surface glabrous except villous along distal margins; petals white, oblong-lanceolate to broadly oblong-ovate, (5–)7–8.5(–10) × 2.5–4.5(–5.8) mm; stamens 13–18(–24); filaments distinct, 1.3–4 mm; style 1, clavate, 2–3.2 mm, narrow base 0.5–1.5 mm; stigmatic portion 1.3–2 mm, distally lobed to 0.5 mm. **Capsules** ovoid-turbinate, 3–5.5 × 3.2–5.5 mm, sepals ± persistent at distal ⅓ or more distally, capsule distal surface impressed in 4(–8) radial lines. **Seeds** not caudate, 1–1.2 mm.

Flowering Apr–Jun; fruiting May–Nov. Limestone mountains, oak-pinyon zones; 1300–1800(–2200) m; N.Mex., Tex.; Mexico (Coahuila).

Philadelphus hitchcockianus, with its hypanthium and abaxial sepal surfaces glabrous, is here synonymized with *P. mearnsii*, which has its hypanthium and abaxial sepal surfaces moderately sericeous-strigose; substantial variation in this character is found in several populations. Hu S. Y. (1954–1956) alleged that *P. hitchcockianus* and *P. mearnsii* also differ in fruit size; however, that character is variable and overlapping.

4. Philadelphus hirsutus Nuttall, Gen. N. Amer. Pl. 1: 301. 1818 • Cliff mock orange [E]

Philadelphus hirsutus var. *intermedius* S. Y. Hu; *P. hirsutus* var. *nanus* S. Y. Hu; *P. sharpianus* S. Y. Hu; *P. sharpianus* var. *parviflorus* S. Y. Hu

Shrubs, 5–20 dm. **Stems** brown, gray, or stramineous, branched, glabrous to densely strigose; bark deciduous, exfoliating or flaking, reddish; branches sprawling; axillary buds exposed. **Leaves:** petiole 1–9 mm; blade white or gray abaxially, green adaxially, broadly lanceolate to broadly ovate or narrowly to broadly elliptic, 2–8 × 1–5.5 cm, herbaceous, base narrowly cuneate to rounded, margins entire or irregularly to regularly serrate, crenate, or dentate, plane, abaxial surface slightly to densely loosely strigose, hairs often gently curved or arched-twisted, 0.7–1.5 mm, adaxial surface glabrate to moderately strigose, hairs usually evenly distributed, when sparse, sometimes distributed mainly along major veins, 0.2–0.8 mm, either thick-based and tightly antrorsely appressed or less thick-based, longer and looser, similar to abaxial hairs. **Inflorescences** cymes or cymose racemes, or flowers solitary, 1–3(–5)-flowered, proximal 2(–4) flowers often in axils of leaves or bracts. **Pedicels** 2–11 mm, moderately to densely strigose. **Flowers:** hypanthium glabrous or densely strigose, hairs tightly or loosely appressed; sepals ovate-lanceolate or triangular, 3–6 × 2–4 mm, apex acuminate, abaxial surface sparsely to densely strigose, adaxial surface glabrous except densely villous distally; petals white, oblong, obovate, or orbiculate, 5–13 × 4–13 mm; stamens 14–35; filaments distinct, 3–9 mm; anthers 1 × 0.7 mm; style 1, clavate, 4–6 mm; stigmatic portion 2–3 mm. **Capsules** obconic to obovoid, (3–)4–7 × 3–6 mm. **Seeds** not caudate, 0.9–1.2 mm. 2*n* = 26.

Flowering Apr–May; fruiting Jun–Aug. Bluffs, rock outcrops, seepage areas over rock, stream banks, particularly over calcareous sedimentary or mafic metamorphic or igneous rocks; 100–800 m; Ala., Ark., Ga., Ky., Md., Miss., N.C., S.C., Tenn., Va.

The occurrence of *Philadelphus hirsutus* in Maryland is believed to be due to an introduction. *Philadelphus hirsutus* is the most xerophytic of the native southeastern species of *Philadelphus*, often occurring on rock outcrops with only seasonal moisture. It is sometimes confused with *P. pubescens*; the undivided style and exposed buds are diagnostic.

5. Philadelphus microphyllus A. Gray, Mem. Amer. Acad. Arts, n. s. 4: 54. 1849 • Small-leaf mock orange

Shrubs, 5–12(–20) dm. **Stems** copper to reddish brown, stiffly to loosely branched, appressed villous-sericeous, ± strigose, hairs often red-gland based, or glabrous; epidermis soon or tardily deciduous exposing cortex and striate bundle caps; bark grayish; internodes (0.1–)1–2.5(–6) cm; short-shoot spurs not present; axillary buds hidden in pouches. **Leaves:** petiole 1–2(–4) mm; blade greenish or whitish abaxially, green adaxially, linear-lanceolate, narrowly ovate to ovate, (0.5–)0.8–3(–5.5) × (0.2–)0.3–1.3(–3.3) cm, herbaceous to coriaceous, margins usually entire, rarely sparsely serrulate, plane or revolute upon drying, abaxial surface short sericeous-strigose, or sericeous-villous with longer hairs, or with ascending to erect hairs, sometimes with dense to moderate understory of slender curled hairs, adaxial surface glabrous, glabrate, ± sericeous-strigose, villous, or with erect hairs. **Inflorescences** usually solitary flowers, sometimes 3–5-flowered cymes. **Pedicels** 0.5–3 mm. **Flowers:** hypanthium glabrous, sericeous-strigose basally or throughout, or weakly to densely villous to densely lanate with mixed strigose and villous vestiture, with understory of slender curled hairs; sepals ovate to lanceolate, (2.5–)4–8.5(–10) × (2.5–)3–4.3(–5) mm, apex acute to acuminate-caudate, abaxial surface glabrous, sericeous-strigose, or weakly to densely villous to densely lanate with mixed strigose and villous vestiture, with understory of slender curled hairs, adaxial surface glabrous except villous along distal margins; petals white [marked purple near base], oblong-obovate to broadly ovate, (5.8–)7–16(–21) × (5.3–)6–11(–15) mm, margins entire or erose-undulate, apex ± acute, rounded, or notched; stamens 26–64; filaments often connivent-connate in irregular clusters in proximal 0.5–4 mm, 1.8–8 mm, of unequal length, glabrous; anthers yellowish, 0.7–1.2 mm; styles 4, connate proximally, cylindric, 2.5–5.5(–7) mm, lobes sometimes connate proximally in pairs, 0.5–2.5 mm; stigmatic surfaces extending from adaxial lobes onto abaxial lobes and down to cylindric style. **Capsules** oblong-globose or globose-turbinate, (3.6–)5–8(–9.5) × (3.5–)4–7(–9.5) mm, sepals persisting at equator or more distally, capsule distal surface often impressed in 4(–8) radiating lines. **Seeds** short caudate distally, 1.5–2.5 mm.

Varieties 5 (4 in the flora): sw United States, Mexico.

Within *Philadelphus microphyllus* as treated here, P. A. Rydberg (1905) recognized nine species, C. L. Hitchcock (1943) one species with eight subspecies, and Hu S. Y. (1954–1956) 11 species and four varieties, based on vestiture, leaf size and shape, and floral differences. Four varieties are recognized here within the flora area, with a fifth restricted to Mexico and without the needed varietal combination.

Two characters are particularly important in distinguishing the varieties of *Philadelphus microphyllus*: adaxial leaf blade cuticle thickness and vestiture. Adaxial leaf blade cuticles can be thin and papillate, closely reflecting the adaxial epidermis cells (as seen at 30–40×) or can be thick and smooth. Leaves with thin cuticles dry brown due to brownish granules developing in the epidermis; those with thick cuticles dry gray-green, olive green, or yellowish green without granules in the epidermis cells. Sometimes both types occur in a leaf in either a tight or bold mosaic pattern or the leaf blade may be papillate and brown only along its margins.

Vestiture is mostly sericeous-strigose on leaves, stems, hypanthia, and sepals. The appressed hairs can be slender, short or long (0.2–1.5 mm), appressed, loosely appressed, ascending, or erect. The larger hairs have slender bases that allow them to be strictly appressed, but in some taxa, the base (upon drying) lifts and twists the hair upward, leaving the hairs oriented in many different directions; we refer to this condition as chaotic vestiture. In more densely vestitured plants, very slender, elongate, wavy-curved hairs form an understory beneath the more or less dense sericeous-strigose vestiture.

1. Hypanthia and sepal abaxial surfaces glabrous or sparsely to moderately sericeous, hairs not obscuring epidermis.
   2. Leaf blade adaxial surfaces sparsely sericeous-strigose with appressed or slightly ascending slender hairs; leaf blade margins entire; inflorescences 1(–3)-flowered; capsules 4.4–6 mm; w United States, including se Arizona and sw New Mexico. . . . . 5a. *Philadelphus microphyllus* var. *microphyllus* (in part)
   2. Leaf blade adaxial surfaces sparsely to moderately strigose-sericeous with appressed or loosely appressed thick hairs mixed with erect hairs, or all hairs erect; leaf blade margins usually entire, on larger leaf blades sometimes sparsely serrulate; inflorescences 1–3(–5)-flowered; capsules 5–8 mm; mountains of se Arizona, sw New Mexico . . . . 5c. *Philadelphus microphyllus* var. *madrensis*
1. Hypanthia and sepal abaxial surfaces moderately to densely sericeous-strigose or villous-lanate, often with understory of thinner ± curled hairs usually completely obscuring epidermis except sometimes in fruit.
   3. Leaf blade abaxial surfaces with hairs usually erect and chaotically oriented, sometimes appressed; mountains of sc New Mexico . . . . 5b. *Philadelphus microphyllus* var. *argyrocalyx*
   3. Leaf blade abaxial surfaces with hairs appressed or loosely appressed; w United States, but mostly not sc New Mexico.

[4. Shifted to left margin.—Ed.]

4. Leaf blade adaxial cuticles forming mosaic of thin, papillate areas and thick, smooth areas, adaxial surfaces drying mosaic of brown and yellowish gray-green, or cuticles uniformly thin, papillate, adaxial surfaces drying dark brown .... 5a. *Philadelphus microphyllus* var. *microphyllus* (in part)
4. Leaf blade adaxial cuticles thick, smooth, or papillate near margins, adaxial surfaces drying olive green or yellowish gray-green.
   5. Leaf blade abaxial surfaces with appressed or loosely appressed hairs 0.5–1.2 mm, adaxial surfaces with only appressed or slightly ascending hairs 0.3–0.7 mm; w Texas to sw Arizona .... 5a. *Philadelphus microphyllus* var. *microphyllus* (in part)
   5. Leaf blade abaxial surfaces with appressed or loosely appressed hairs 0.3–0.7 mm, adaxial surfaces with appressed hairs 0.1–0.6 mm and often with shorter erect hairs 0.1–0.3 mm; California, adjacent w Nevada............ .... 5d. *Philadelphus microphyllus* var. *pumilus*

5a. Philadelphus microphyllus A. Gray var. **microphyllus**

Philadelphus argenteus Rydberg; *P. argyrocalyx* Wooton var. *argenteus* (Rydberg) Engler; *P. crinitus* (C. L. Hitchcock) S. Y. Hu; *P. microphyllus* subsp. *argenteus* (Rydberg) C. L. Hitchcock; *P. microphyllus* var. *argenteus* (Rydberg) Kearney & Peebles; *P. microphyllus* subsp. *crinitus* C. L. Hitchcock; *P. microphyllus* var. *crinitus* (C. L. Hitchcock) B. L. Turner; *P. microphyllus* var. *linearis* S. Y. Hu; *P. microphyllus* subsp. *occidentalis* (A. Nelson) C. L. Hitchcock; *P. microphyllus* var. *occidentalis* (A. Nelson) Dorn; *P. microphyllus* var. *ovatus* S. Y. Hu; *P. minutus* Rydberg; *P. nitidus* A. Nelson; *P. occidentalis* A. Nelson; *P. occidentalis* var. *minutus* (Rydberg) S. Y. Hu

Leaf blades (0.8–)1–2.2(–4) × (0.3–)0.4–1(–1.8) cm, herbaceous to coriaceous, margins entire, abaxial surface gray-white, often mottled with brown, sparsely to moderately sericeous-strigose, hairs appressed or loosely appressed, 0.2–0.6 mm, or coarser, to 0.5–1.2 mm, or when vestiture dense, with understory of slender, wavy-curved hairs, or chaotically hirsute-pilose (through introgression with var. *argyrocalyx*), marginal hairs often more erect, adaxial surface green, drying dark brown (cuticle thin, finely papillate), mosaic of dark brown and yellowish gray-green (cuticle forming mosaic of thin, papillate areas and thick, smooth areas), or rarely uniformly yellowish gray-green (cuticle thick, smooth), usually sparsely sericeous-strigose, hairs appressed or slightly ascending, slender, 0.3–0.7 mm, sometimes glabrate or glabrous, rarely weakly hirtellous or with some erect hairs near base. **Inflorescences** 1(–3)-flowered. **Flowers** 14–32 mm diam.; hypanthium and sepal abaxial surfaces glabrous, sparsely sericeous proximally or in 4 lines extending from pedicels to alternate sepals, or moderately sericeous throughout, hairs appressed, coarse to fine, not obscuring epidermis, or densely sericeous-strigose with understory of thinner curved-wavy hairs completely obscuring epidermis; sepals (3.5–)6.5–8 mm; petals (5.8–)8–10(–15) × (5.3–)6–8(–10) mm; stamens (26–)34–46. **Capsules** turbinate, 4.5–6 × 4.5–6 mm.

Flowering Jun–Aug; fruiting Jul–Nov. Arid montane scrub, hardwood or pine-oak woodlands, yellow pine-fir forests, limestone and rhyolitic substrates, canyons, open slopes, bluffs, canyons; 1600–2700 m; Ariz., Calif., Colo., Nev., N.Mex., Tex., Utah, Wyo.; Mexico (Baja California, Coahuila).

Four species that were recognized by P. A. Rydberg (1905) and Hu S. Y. (1954–1956) on the basis of hypanthium and sepal vestiture are here combined: *Philadelphus argenteus* and *P. crinitus* (densely lanate-sericeous-strigose, with underlying hairs more slender and coiled), *P. microphyllus* (glabrous or sparsely sericeous in four vertical lines), and *P. occidentalis* (uniformly sericeous-strigose, vestiture not completely obscuring the epidermis). The vestiture patterns are often mixed in populations, although regional patterns exist. Most plants in the northern range (northern Arizona, Colorado, southern Nevada, northern New Mexico, Utah, Wyoming) have sparsely sericeous-strigose adaxial leaf surfaces and a slightly denser vestiture on the abaxial surfaces; the adaxial leaf surfaces have a thin cuticle, and the leaves dry a brown color. In more arid zones (southeast Arizona, southwestern New Mexico, trans-Pecos Texas, and Baja California, Mexico), the adaxial epidermis often has a thicker, smoother cuticle or a mosaic of areas of thin and thick cuticle, and the leaves dry a gray- or olive green color.

The taxon *crinitus*, first named as a subspecies of *Philadelphus microphyllus*, is based on a population from the rhyolitic Davis Mountains in trans-Pecos Texas. Plants assigned to this taxon have densely vestitured hypanthia and sepals; longer hairs on the abaxial leaf surface (to 1.2 mm), with an underlying layer of thinner, wavy hairs; and thick adaxial cuticles. They are usually distinct from adjacent var. *microphyllus*. However, similar plants occur in the nearby limestone Guadalupe Mountains and again in southeastern Arizona (Huachuca and Santa Rita mountains) and in the intervening areas. Many from the latter region have been considered *P. argenteus* (with densely vestitured hypanthium and sepals, often similar leaf vestiture, and thick cuticles), but plants referable to *P. argenteus* occur scattered among plants of var. *microphyllus* and do not form large uniform populations; the *P. argenteus* form is here considered a more strongly vestitured phase within an expanded var. *microphyllus*, which then also includes var. *crinitus*.

5b. Philadelphus microphyllus A. Gray var. **argyrocalyx** (Wooton) Henrickson, J. Bot. Res. Inst. Texas 1: 901. 2007 • New Mexican mock orange C E

Philadelphus argyrocalyx Wooton, Bull. Torrey Bot. Club 25: 452. 1898; *P. wootonii* S. Y. Hu

Leaf blades (1.2–)1.5–3(–4.5) × (0.5–)0.7–1.3(–1.9) cm, herbaceous, margins entire, sometimes ± revolute, abaxial surface whitish gray, sericeous initially, but hairs usually becoming erect and chaotic, sometimes remaining appressed, straight or slightly curved, 0.3–0.7 mm, hairs longer (0.7–1.1 mm) and more common along main veins, adaxial surface brown (cuticle thin, papillate) or gray-green (cuticle thick, smooth), glabrous or sparsely to moderately strigose, hairs usually loosely appressed to spreading, rarely erect, 0.3–0.6 mm. **Inflorescences** 1(–3)-flowered. **Flowers** 23–35(–40) mm diam.; hypanthium and sepal abaxial surfaces ± densely white sericeous-strigose and villous-lanate, hairs straight and curved to loosely coiled, and with understory of slender curved-crisped hairs, hairs usually completely obscuring epidermis except sometimes in fruit; sepals 4.5–7 mm; petals 10–16(–21) × 7–11(–15) mm; stamens (35–)49–64. **Capsules** globose to oblong-globose, 6–9.5 × 6–9.5 mm.

Flowering Jun–Aug; fruiting Jul–Nov. Limestone and igneous substrates, rocky areas, canyons, drainages, pinyon chaparral, juniper-oak and ponderosa-oak woodlands, limber pine and fir forests; of conservation concern; 2000–2700 m; N.Mex.

The distribution of var. *argyrocalyx* is centered in Lincoln and Otero counties. Plants show considerable variation in characters formerly used to separate *Philadelphus argyrocalyx* and *P. wootonii*, with both glabrous and sericeous (occasionally to pilose-villous) adaxial leaf surfaces and both narrowly and broadly obovate petals, sometimes even on one plant; this supports including *P. wootonii* within *P. argyrocalyx*. The distinctive, erect, rather chaotically oriented hairs on the abaxial leaf surface occur in 93% of the specimens seen; the remainder have a sericeous-strigose abaxial leaf-surface vestiture of the same type of hair (J. Henrickson, pers. obs.). This chaotic vestiture occurs also in similar plants scattered throughout central and northern New Mexico, but in combination with glabrous or sparsely sericeous hypanthia and sepals as in var. *microphyllus*; these are placed in an expanded var. *microphyllus*.

5c. Philadelphus microphyllus A. Gray var. **madrensis** (Hemsley) Henrickson, J. Bot. Res. Inst. Texas 1: 901. 2007 • Montane mock orange

Philadelphus madrensis Hemsley, Bull. Misc. Inform. Kew 1908: 251. 1908

Leaf blades (0.7–)1.6–3(–5.5) × (0.4–)0.7–1.3(–3.3) cm, herbaceous to coriaceous, margins usually entire, on larger leaves sometimes sparsely serrulate, abaxial surface whitish gray, sparsely to moderately sericeous, hairs appressed, loosely appressed, ascending, or chaotic, ± coarse, straight or wavy-curved, 0.5–1.5 mm, infrequently more densely lanate with understory of more slender, wavy-curved hairs, adaxial surface green, drying gray-green or brown (cuticle thick, smooth or mottled smooth and papillate), sparsely to moderately strigose-sericeous, hairs both appressed or loosely appressed, thick, 0.2–1 mm, and erect, 0.2–0.7 mm, or all erect. **Inflorescences** 1–3(–5)-flowered. **Flowers** 25–37 mm diam.; hypanthium and sepal abaxial surfaces moderately, often loosely, sericeous, hairs wavy, weakly appressed or erect to spreading, 0.4–1.2 mm, not obscuring epidermis; sepals (3.5–)5–8.5(–10) mm; petals 9–12(–16) × 6–11 mm; stamens 32–46+. **Capsules** globose-turbinate, 5–8 × 3.8–6.5 mm.

Flowering Jun–Aug; fruiting Jul–Nov. Oak, pine, and fir forests, open, rocky rhyolitic ridges; 1800–2700 m; Ariz., N.Mex.; Mexico (Chihuahua, Durango, Sonora).

Variety *madrensis* is primarily Mexican, centered in the Sierra Madre Occidental of Chihuahua, Durango, and Sonora, entering the United States in southeastern Arizona (Chiricahua Mountains, Cochise County; Pinaleño Mountains, Graham County; Catalina Mountains, Pima County); and southwestern New Mexico (Burro Mountains). In populations within the flora area, over 90% have erect, sometimes chaotically oriented, hairs on the adaxial leaf surface, and in the Chiricahua Mountains, nearly half have erect hairs on the abaxial leaf surface, and some have chaotic hairs throughout leaves, hypanthia, and young stems. Some plants have a dense vestiture of loosely appressed elongate hairs (to 1.5 mm) on the abaxial leaf surface as in the type from Durango, Mexico. Plants in the higher mountains have relatively large, often weakly serrulate leaves, and typically flowers are in threes at stem tips (but only solitary in the Pinaleño Mountains). Some plants around Rucker Canyon and Fly Peak in the Chiricahua Mountains are vegetatively similar but have sericeous vestiture and may be considered introgressants with var. *microphyllus*. The characteristics of var. *madrensis* and the southern phase of var. *microphyllus* are mixed in plants found in the Rincon and Santa Rita Mountains in southeastern Arizona.

5d. Philadelphus microphyllus A. Gray var. **pumilus** (Rydberg) Henrickson, Phytoneuron 2016-8: 6. 2016 • Straw-stemmed mock orange [E]

Philadelphus pumilus Rydberg in N. L. Britton et al., N. Amer. Fl. 22: 173. 1905; *P. microphyllus* subsp. *pumilus* (Rydberg) C. L. Hitchcock; *P. microphyllus* subsp. *stramineus* (Rydberg) C. L. Hitchcock; *P. pumilus* var. *ovatus* S. Y. Hu; *P. stramineus* Rydberg

Leaf blades (0.5–)0.8–1.7(–2.7) × (0.2–)0.3–0.6(–1.1) cm, ± coriaceous, margins entire, drying revolute, abaxial surface mottled grayish, moderately to densely sericeous-strigose, hairs appressed or loosely appressed, 0.3–0.7 mm, understory hairs slender, wavy-curved, adaxial surface green, drying grayish olive green, sometimes fading to yellow-tan (cuticle thick, smooth or papillate near margins), uniformly sericeous, hairs appressed, 0.1–0.6 mm, often also some hairs erect, shorter (0.1–0.3 mm), marginal hairs often spreading. **Inflorescences** 1(–3)-flowered. **Flowers** 15–30 mm diam.; hypanthium and sepal abaxial surfaces moderately to densely sericeous-strigose, hairs 0.4–0.5 mm, often with slender, wavy-coiled understory hairs usually completely obscuring epidermis, rarely nearly glabrous; sepals (2.5–)3.5–6 mm; petals 7–11(–13.5) × 6–8(–11) mm; stamens 26–42. **Capsules** globose, 3.6–6 × 3.5–6 mm.

Flowering May–Jul; fruiting Jun–Nov. Limestone and granite substrates, pinyon, oak, and juniper woodlands, limber pine forests; 1600–3000 m; Calif., Nev.

A taxon of desert mountains of eastern California and portions of adjacent Nevada, var. *pumilus* is characterized by its stout, short branching and for developing a thorny aspect, with sun-bleached stramineous stems (in first-year stems, the epidermis falls away and the cortex dries exposing the initial bundle caps) and dry leaves that are grayish olive green to yellow-tan, and associated with a nearly continuous, thickened cuticle. On the eastern slopes of the Spring (Charleston) Mountains of southwestern Nevada, plants of var. *pumilus* intergrade with var. *microphyllus*. The two varieties are distinguished by their dried-leaf colors: a greenish color in var. *pumilus* versus a dull brown color in var. *microphyllus*.

Plants in dry valleys in southwestern Utah and northeastern Arizona (Apache County) with thick cuticles and gray-green dried leaves, and with subglabrous hypanthia and sepals are similar to var. *pumilus*; they are placed in the expanded var. *microphyllus*.

6. Philadelphus inodorus Linnaeus, Sp. Pl. 1: 470. 1753 • Scentless mock orange [E]

Philadelphus floridus Beadle; *P. floridus* var. *faxonii* Rehder; *P. gloriosus* Beadle; *P. grandiflorus* Willdenow; *P. inodorus* var. *carolinus* S. Y. Hu; *P. inodorus* var. *grandiflorus* (Willdenow) A. Gray; *P. inodorus* var. *laxus* (Schrader ex de Candolle) S. Y. Hu; *P. inodorus* var. *strigosus* Beadle; *P. strigosus* (Beadle) Rydberg

Shrubs, 20–40 dm. **Stems** brown, gray, or stramineous, branched, 20–40 dm, glabrous or very sparsely strigose, especially at nodes; bark reddish, exfoliating or flaking; branches erect to arching; axillary buds hidden in pouches. **Leaves:** petiole 1–8 mm; blade broadly lanceolate to broadly ovate, or narrowly to broadly elliptic, (3.5–)5–12(–14) × (1.4–)2–5.3(–7) cm, herbaceous, base narrowly cuneate to rounded, margins entire or irregularly to regularly serrate, crenate, or dentate, plane, abaxial surface usually glabrous or moderately strigose, rarely moderately to densely strigose-tomentose in main vein axils, sometimes sparsely strigose on main veins, rarely sparsely strigose on secondary and tertiary veins as well, adaxial surface glabrous or very sparsely strigose, especially near base and margins. **Inflorescences** cymes or racemes, or flowers solitary, 1–3(–9)-flowered, proximal 2 flowers sometimes in axils of nearly normal to much reduced (bracteal) leaves, if 1-flowered, with articulation between peduncle and pedicel revealing that it is a 1-flowered cyme by reduction. **Pedicels** 3–8 mm, glabrous or slightly strigose. **Flowers:** hypanthium usually glabrous, rarely moderately strigose; sepals ovate or ovate-lanceolate, 7–14 × 5–8 mm, apex acuminate to acute, abaxial surface usually glabrous, rarely moderately strigose, adaxial surface glabrous except densely villosulous distally; petals white, oblong, obovate, or orbiculate, 15–25(–30) × 10–22 mm; stamens 60–90; filaments distinct, 5–11 mm; anthers 1–1.5 × 1 mm; styles 4, connate proximally, cylindric, 10–16 mm, lobes 4–8 × 0.8–1 mm; stigmatic surfaces 3–4.5 mm. **Capsules** obconic to obovoid, 10–13 × 7–10 mm. **Seeds** caudate, 2–3 mm. $2n = 26$.

Flowering Apr–May; fruiting Jun–Aug. Stream banks, bluffs, cliffs, rock outcrops; 0–1000 m; Ont.; Ala., Ark., Conn., Fla., Ga., Ill., Ind., Kans., Ky., La., Md., Mass., Mich., Miss., Mo., N.J., N.Y., N.C., Ohio, Pa., R.I., S.C., Tenn., Va., W.Va., Wis.

The native distribution of *Philadelphus inodorus* was originally narrower than the current range. It is native in Alabama, Florida, Georgia, Kentucky, Mississippi, North Carolina, South Carolina, Tennessee, Virginia, and West Virginia, whereas it is considered introduced in the remaining places listed.

7. **Philadelphus pubescens** Loiseleur, Herb. Gén. Amat. 4: 268. 1820 • Downy mock orange [E] [F]

Philadelphus gattingeri S. Y. Hu; *P. intectus* Beadle; *P. intectus* var. *pubigerus* S. Y. Hu; *P. latifolius* Schrader; *P. pubescens* var. *intectus* (Beadle) A. H. Moore; *P. pubescens* var. *verrucosus* (Schrader) S. Y. Hu; *P. verrucosus* Schrader

Shrubs, 10–65 dm. **Stems** erect to ascending, green, older stems gray, glabrous or sparsely strigose or villous, especially at nodes; bark tight, not exfoliating or flaking, gray; branches erect; axillary buds hidden in pouches. **Leaves:** petiole 1–12(–20) mm; blade broadly lanceolate to broadly ovate, or narrowly to broadly elliptic, (3–)5–10(–16) × 1.6–7(–11) cm, base narrowly cuneate to rounded, margins entire or irregularly to regularly serrate, crenate, or dentate, plane, abaxial surface moderately to densely strigose or tomentose to villous, hairs twisted, main vein axils and main veins often more densely strigose-tomentose, adaxial surface glabrous or very sparsely strigose, especially near base and margins. **Inflorescences** usually cymose racemes or cymose panicles, sometimes flowers solitary, (1–)5–9-flowered, proximal 2 or 4 flowers often in axils of nearly normal to much reduced (bracteal) leaves. **Pedicels** 3–8 mm, glabrous or moderately strigose. **Flowers:** hypanthium sparsely to densely strigose or villous; sepals ovate, ovate-lanceolate, or triangular, 5–8 × 3–5 mm, apex acute to acuminate, abaxial surface moderately to densely strigose or villous, adaxial surface densely villous distally; petals white, oblong, obovate, or orbiculate, 12–21 × 8–12 mm; stamens 25–50; filaments distinct, 5–11 mm; anthers 1.5 × 1 mm; styles 4, connate proximally, cylindric, 6–10 mm, lobes 1.5–5 × 0.4–0.7 mm; stigmatic surfaces 1.5–4 mm. **Capsules** obconic to obovoid, 6–11 × 4–7 mm. **Seeds** caudate, 3–4 mm. $2n = 26$.

Flowering May–Jul; fruiting Jun–Oct. Cliffs, rock outcrops, bluffs, rocky slopes, old homesites, suburban woodlands, stream banks; 0–1000 m; Ont.; Ala., Ark., Conn., Ga., Ill., Iowa, Kans., Ky., La., Md., Mass., Mich., Miss., Mo., N.Y., Ohio, Okla., Tenn., Tex., Va., Wyo.

Philadelphus pubescens is cultivated beyond its native range, which is believed to have been Alabama, Arkansas, Georgia, Illinois, Kentucky, Louisiana, Mississippi, Missouri, Oklahoma, Tennessee, and Texas; it is considered introduced in the remaining places listed.

8. **Philadelphus lewisii** Pursh, Fl. Amer. Sept. 1: 329. 1813 • Lewis's mock orange [E]

Philadelphus californicus Bentham; *P. confusus* Piper; *P. cordifolius* Lange; *P. gordonianus* Lindley; *P. gordonianus* var. *columbianus* (Koehne) Rehder; *P. helleri* Rydberg; *P. insignis* Carriére; *P. lewisii* var. *angustifolius* (Rydberg) S. Y. Hu; *P. lewisii* subsp. *californicus* (Bentham) Munz; *P. lewisii* var. *ellipticus* S. Y. Hu; *P. lewisii* subsp. *gordonianus* (Lindley) Munz; *P. lewisii* var. *gordonianus* (Lindley) Jepson; *P. lewisii* var. *helleri* (Rydberg) S. Y. Hu; *P. lewisii* var. *intermedius* (A. Nelson) S. Y. Hu; *P. lewisii* var. *oblongifolius* S. Y. Hu; *P. lewisii* var. *parvifolius* Torrey; *P. lewisii* var. *platyphyllus* (Rydberg) A. H. Moore; *P. oreganus* Nuttall ex S. Y. Hu; *P. trichothecus* S. Y. Hu; *P. zelleri* S. Y. Hu

Shrubs, 15–40 dm. **Stems** erect to ascending, green, weathering brown, gray, or stramineous, glabrous or sparsely strigose (especially at nodes); bark deciduous, exfoliating or flaking, reddish; axillary buds hidden in pouches, sometimes apex exposed, especially on vigorous sprout-shoots. **Leaves:** petiole 1–6 mm; blade broadly lanceolate to broadly ovate, or narrowly to broadly elliptic, 1.5–10 × 1–5 cm, larger blades usually less than 6 × 2.5 cm, base narrowly cuneate to rounded or cordate, margins entire or irregularly to regularly serrate, crenate, or dentate, plane or slightly revolute, abaxial surface glabrous or sparsely strigose, hairs usually appressed-ascending, not twisted, main vein axils often densely strigose-tomentose, main veins sometimes sparsely strigose, secondary and tertiary veins rarely sparsely strigose, adaxial surface glabrous or sparsely to moderately strigose, especially near base and margins. **Inflorescences** usually cymose racemes or cymose panicles, sometimes flowers solitary, (1–)7–49-flowered, proximal 2, 4, or 6 flowers often in axils of nearly normal to much reduced (bracteal) leaves. **Pedicels** 3–8 mm, glabrous or moderately strigose. **Flowers:** hypanthium glabrous or sparsely strigose, hairs scattered or concentrated on veins; sepals ovate, ovate-lanceolate, or triangular, 5–8 × 3–5 mm, apex acute to acuminate, abaxial surface glabrous or sparsely strigose, adaxial surface glabrous except densely villous distally; petals white, oblong, obovate, or orbiculate, 5–20(–25) × 4–15 mm; stamens 25–40(–50); filaments distinct, 5–11 mm; anthers 2 × 1.5 mm; styles 4, cylindric, 4–8 mm, connate proximally, lobes 1–4 × 0.4–0.5 mm; stigmatic surfaces 1–3.5 mm. **Capsules** obconic to obovoid, 7–11 × 5–7 mm. **Seeds** caudate, 3 mm. $2n = 26$.

Flowering May–Jul; fruiting Jun–Sep. Cliffs, rock outcrops, slopes in pine woodlands and forests, stream banks, talus, seasonally dry ravines; 0–2500 m; Alta., B.C.; Calif., Idaho, Mont., Oreg., Wash.

Philadelphus lewisii, the state flower of Idaho, is rarely cultivated beyond its native range; there is no evidence that it is naturalized in other parts of North America.

A broadly defined *Philadelphus lewisii* is here recognized with some reluctance; it includes all previously named northwestern entities of subg. *Philadelphus*. Hu S. Y. (1954–1956) recognized 16 entities at varietal and specific rank, arrayed in two subgenera. C. L. Hitchcock et al. (1955–1969, vol. 3) described *P. lewisii* as being extremely variable in both vegetative and floral characters and further stated that most ecological or geographic races do not merit taxonomic recognition. Most recent floristic treatments have followed Hitchcock et al. or in some cases recognized two or three entities. The treatment of Hitchcock et al. is followed here, and there does not appear to be a case for the recognition of the two possible exceptions they mentioned (vars. *gordonianus* and *parvifolius*). *Philadelphus californicus*, *P. cordifolius*, and *P. insignis* were separated by Hu from *P. lewisii* at a subgeneric level based primarily on exposed buds and many-flowered inflorescences. The exposed buds are seen sporadically in *P. lewisii*, including outside the distribution accorded *P. californicus*, and are not evolutionarily cognate with the exposed buds of subg. *Deutzioides*; based on examination of herbarium material, they seem to be an expression of extra-vigorous growth, but additional study of plants in the field is needed. Some plants from the Sierra Nevada in California, considered by Hu to be occupied solely by *P. californicus*, are indistinguishable from *P. lewisii* in flower number, bud exposure, or any other character. The elaborated inflorescences of some plants in this part of California are striking; for now *P. californicus* and relatives are treated as part of a broadly defined *P. lewisii*. In her phylogenetic analysis based on ITS sequence, A. E. Weakley (2002) also found that *P. californicus* was not distinguished from *P. lewisii*.

9. Philadelphus coronarius Linnaeus, Sp. Pl. 1: 470. 1753 • Sweet mock orange, seringa commun [I]

Philadelphus caucasicus Koehne

Shrubs, 10–40 dm. **Stems** erect to ascending, green, older stems brown, branched, glabrous or sparsely strigose (especially at nodes); bark deciduous, exfoliating or flaking, reddish; branches erect to arching; axillary buds hidden in pouches, sometimes apex exposed, especially on vigorous sprout-shoots. **Leaves:** petiole 1–6 mm; blade usually broadly lanceolate to broadly ovate, or narrowly to broadly elliptic, rarely narrowly lanceolate in horticultural forms, 3–10 × 2–6 cm, larger blades usually greater than 6 × 2.5 cm, base narrowly cuneate to rounded, margins entire to irregularly or regularly serrate, crenate, or dentate, plane, abaxial surface glabrous or sparsely strigose, hairs usually appressed-ascending, not twisted, main vein axils often moderately to densely strigose-tomentose, main veins sometimes sparsely strigose, secondary and tertiary veins rarely sparsely strigose, adaxial surface glabrous or sparsely strigose, especially near base and margins. **Inflorescences** cymose racemes, 5–7(–9)-flowered, proximal 2 flowers often in axils of nearly normal to much reduced (bracteal) leaves. **Pedicels** 3–20 mm, glabrous or sparsely strigose. **Flowers:** hypanthium glabrous or sparsely strigose to villous; sepals ovate, ovate-lanceolate, or triangular, 4–8 × 2.5–5 mm, apex acute to acuminate, abaxial surface glabrous or sparsely strigose, adaxial surface glabrous except densely villosulous distally; petals white to cream, oblong, obovate, or orbiculate, 5–25 × 5–22 mm; stamens 20–50; filaments distinct, 4–9 mm; anthers 1–1.5 × 0.7–1 mm; styles 4, connate proximally, cylindric, 7–10 mm, lobes 3–8 × 0.3–0.9 mm; stigmatic surfaces 1–4 mm. **Capsules** obconic to obovoid, 7–11 × 4–7 mm. **Seeds** caudate, 3 mm. **2*n*** = 26.

Flowering May–Jul; fruiting Jul–Sep. Old home sites, suburban woodlands, stream banks; 0–1000 m; introduced; N.B., Ont., Que.; Conn., Ga., Ill., Ind., Kans., Maine, Md., Mass., Mich., Minn., Mo., N.H., N.J., N.Y., N.C., Ohio, Pa., R.I., S.C., Vt., Va., Wis.; Eurasia.

Philadelphus coronarius is likely to occur in states and provinces other than those listed because it is widely cultivated and may escape.

Philadelphus caucasicus is here provisionally considered to be a synonym of *P. coronarius*, though further study in their native area is needed. Both are native in the Caucasus region, according to Hu S. Y. (1954–1956), and the only character by which she separated them (vestiture on the disc and style) is variable in other taxa in the genus and seems of doubtful taxonomic meaning in the case of these two sympatric, putative taxa. The native distribution of *P. coronarius* is unclear. Hu considered it native in southern Europe and the Caucasus Mountains of southern Russia, Armenia, Georgia, and Azerbaijan. D. A. Webb (1993) emphasized the uncertainty of its native range, mentioning that stations in Europe where it is undoubtedly native are very few; it may be that this species is native only in the Caucasus and was brought early to Europe for ornament.

7. CARPENTERIA Torrey, Proc. Amer. Assoc. Advancem. Sci. 4: 192. 1851 • California tree-anemone [For William Marbury Carpenter, 1811–1848, Louisiana physician and botanist] [C] [E]

Craig C. Freeman

Shrubs. Stems erect, ascending, or spreading. **Bark** exfoliating in grayish sheets or strips. **Branches** ascending or spreading; twigs with simple trichomes. **Leaves** persistent, opposite; petiole present; blade lanceolate to narrowly elliptic or oblong, herbaceous to coriaceous, margins entire or obscurely denticulate, usually revolute, sometimes plane; venation pinnate. **Inflorescences** terminal or axillary, cymes, 3–9(–13)-flowered; peduncle present. **Pedicels** present. **Flowers** bisexual; perianth and androecium nearly hypogynous; hypanthium adnate to ovary proximally, free distally, patelliform, not ribbed in fruit; sepals persistent, 5–7, spreading, ovate to ovate-lanceolate or triangular, sparsely to densely appressed-pubescent abaxially; petals 5–7(–8), imbricate, spreading to reflexed, white, ovate, round, or depressed-elliptic, base sessile or obscurely clawed, surfaces glabrous; stamens 150–200; filaments distinct, essentially terete, not noticeably tapered from base to apex, apex not 2-lobed; anthers depressed-ovate; pistil 5–7-carpellate, ovary nearly superior, 5–7-locular; placentation axile proximally, parietal distally; style persistent, 1. **Capsules** conic to depressed-spheric, corticate to coriaceous, dehiscence basipetally septicidal to near base of fruit, delayed apically by persistent style. **Seeds** 50 per locule, brown or reddish brown, ellipsoid. x = 10.

Species 1: California.

Phylogenetic analyses place *Carpenteria* as sister to *Philadelphus* (D. E. Soltis et al. 1995; L. Hufford 1997) or as a clade nested within a paraphyletic *Philadelphus* (Guo Y. L. et al. 2013). Androecium development in the two genera is unique in Hydrangeaceae (Hufford 1998).

SELECTED REFERENCE Cheatham, N. H. 1974. *Carpenteria*, the mystery shrub. Fremontia 1: 3–8.

1. Carpenteria californica Torrey, Smithsonian Contr. Knowl. 6(2): 12, plate 7. 1853 [C] [E] [F]

Stems 10–30(–40) dm. **Twigs** glabrous or sparsely to moderately appressed pubescent. **Leaves:** petiole (2–)5–10 mm, appressed pubescent abaxially, glabrous or sparsely pubescent adaxially; blade 2.7–8.5(–10) × 0.8–2.5 (–3.6) cm, base attenuate, apex acute to rounded, abaxial surface whitish, densely appressed-tomentulose, adaxial green, glabrous. **Inflorescences** open; peduncle 1.5–5 cm, glabrous or sparingly appressed-pubescent; bracts prominent, sessile. **Pedicels** 1.5–5.2 mm, sparsely to densely strigose. **Flowers** fragrant; hypanthium 2–4 × 8–13 mm; sepals (6–)8–14 × (3–)4–6 mm, margins entire, apex acute to acuminate, abaxial surface glabrous or sparingly pubescent except for tomentulose tip; petals 20–40 × 20–36(–40) mm; filaments 4–9 × 0.2–0.6 mm; anthers 0.9–1.1 mm; style 3.5–6 mm. **Capsules** 10–12 × 8–12 mm. **Seeds** 0.7–1 mm. **2*n*** = 20.

Flowering Apr–Jun. Dry, rocky slopes in chaparral and lower ponderosa pine woodlands; of conservation concern; 300–1500 m; Calif.

Carpenteria californica is a fire-adapted shrub endemic to the central and southern Sierra Nevada Foothills between the Kings and San Joaquin rivers, Fresno County (R. Casamajor 1950). Rare in the wild, it has a long history of cultivation in both Europe and the United States, including elsewhere in California (N. H. Cheatham 1974), where it is prized for its lightly scented, showy flowers and attractive, persistent foliage. The plant does well in sunny or partially shaded, sheltered sites with well-drained soils.

8. HYDRANGEA Linnaeus, Sp. Pl. 1: 397. 1753; Gen. Pl. ed. 5, 189. 1754

• Graybeard, sevenbark, hortensia [Greek *hydor*, water, and *angeion*, diminutive of *angos*, vessel or container, alluding to shape of mature, dehisced capsule]

Craig C. Freeman

Shrubs or trees. Stems erect, ascending, or spreading. **Bark** exfoliating in grayish, brown, or reddish brown sheets. **Branches** erect, ascending, or spreading, sometimes arching; twigs with simple or branched trichomes. **Leaves** deciduous, opposite or 3-whorled; petiole present; blade ovate, elliptic-ovate, elliptic, or suborbiculate, sometimes lobed, herbaceous, margins serrate to serrulate, plane; venation pinnate. **Inflorescences** terminal, cymose panicles, 100–1000-flowered; peduncle present. **Pedicels** present. **Flowers** bisexual or marginal ones often sterile (these with a petaloid, salverform calyx); perianth and androecium epigynous or perigynous; hypanthium campanulate or hemispheric, completely adnate to ovary, weakly or strongly 7–10(–11)-ribbed in fruit; sepals persistent, 5, spreading or reflexed, deltate to shallowly triangular, usually glabrous, rarely abaxially sparsely hairy; petals 5, valvate, spreading or reflexed, white to yellowish white, ovate-lanceolate, elliptic, oblong, spatulate, or narrowly ovate to ovate, base sessile, surfaces glabrous; stamens 10; filaments distinct, terete or flattened proximally, gradually or abruptly tapered from base to apex, apex not 2-lobed; anthers depressed-ovoid or transversely oblong; pistil 2–4-carpellate, ovary completely inferior or nearly so, or ½ inferior, 2–4-locular; placentation axile proximally, parietal distally; styles persistent, 2–4, distinct or connate to middle or distally. **Capsules** hemispheric, suburceolate, or oblong-ovoid, coriaceous, dehiscence interstylar, creating elliptic to circular pore at base of styles. **Seeds** 10–40 per locule, light brown to dark brown, fusiform or ellipsoid. $x = 18$.

Species ca. 29 (5 in the flora): United States, Mexico, Central America, w South America, e Asia.

Hydrangea enjoys considerable esteem as an ornamental shrub, especially for its prominent, sterile flowers. North American species have been cultivated in Europe since before the mid 1700s (W. L. Stern 1978). Besides the species treated here, popular ornamentals in North America are *H. anomala* D. Don, *H. aspera* D. Don, *H. heteromalla* D. Don, *H. involucrata* Siebold, *H. macrophylla* (Thunberg) Seringe, and *H. serrata* (Thunberg) Seringe. Among these, *H. macrophylla* may be the most widely grown; M. A. Dirr (2004) listed nearly 170 cultivars of this species. Surprisingly, it has not escaped from cultivation.

Tubercles, comprising crystals of calcium carbonate, often are visible on leaf trichomes (G. W. Burkett 1932). They are observed most easily at magnifications greater than 30×, and some taxonomic utility has been ascribed to their presence and abundance.

Toxic and medicinal properties are attributed to some native and cultivated species of *Hydrangea* (J. M. Kingsbury 1964; W. L. Stern 1978; D. E. Moerman 1998; G. E. Burrows and R. J. Tyrl 2001). These possibly are related to various alkaloids present in roots and leaves of some species.

Y. De Smet et al. (2015) found *Hydrangea* to be polyphyletic and promoted adoption of a broader, monophyletic concept of *Hydrangea* that includes all eight genera in tribe Hydrangeeae. Both North American genera in the tribe, *Decumaria* and *Hydrangea*, are circumscribed here in their traditional senses.

SELECTED REFERENCES Dirr, M. A. 2004. Hydrangeas for American Gardens. Portland. McClintock, E. 1957. A monograph of the genus *Hydrangea*. Proc. Calif. Acad. Sci., ser. 4, 29: 147–256. Pilatowski, R. E. 1980. A Taxonomic Study of the *Hydrangea arborescens* Complex. M.S. thesis. North Carolina State University. Pilatowski, R. E. 1982. A taxonomic study of the *Hydrangea arborescens* complex. Castanea 47: 84–98. St. John, H. 1921. A critical revision of *Hydrangea arborescens*. Rhodora 23: 203–208. Stern, W. L. 1978. Comparative anatomy and systematics of woody Saxifragaceae. *Hydrangea*. Bot. J. Linn. Soc. 76: 83–113.

1. Leaf blades (3–)5(–7)-lobed, margins coarsely serrate; inflorescences ovoid to conic . . . . 4. *Hydrangea quercifolia*
1. Leaf blades unlobed, margins dentate, serrate, or serrulate; inflorescences dome-shaped to hemispheric or conic to ovoid.
   2. Ovaries ½ inferior; inflorescences usually conic, sometimes ovoid. . . . . 5. *Hydrangea paniculata*
   2. Ovaries completely inferior or nearly so; inflorescences dome-shaped to hemispheric.
      3. Leaf blade abaxial surfaces white or grayish, densely tomentose, trichomes at 40× either smooth, 1–3 mm, or sparsely tuberculate, 0.3–1 mm . . . . 3. *Hydrangea radiata*
      3. Leaf blade abaxial surfaces green, glabrous or glabrate, or grayish, uniformly velutinous, pilose, or tomentose, trichomes at 40× conspicuously tuberculate, 0.3–1 mm.
         4. Leaf blade abaxial surfaces green, glabrous or glabrate, or sparsely hirsute along midveins and sometimes along lateral veins . . . . 1. *Hydrangea arborescens*
         4. Leaf blade abaxial surfaces grayish, uniformly velutinous, pilose, or tomentose . . . . 2. *Hydrangea cinerea*

1. Hydrangea arborescens Linnaeus, Sp. Pl. 1: 397. 1753 • Wild or smooth hydrangea [E]

Hydrangea arborescens var. *oblonga* Torrey & A. Gray

Shrubs, 10–30 dm. **Twigs** strigose to hirsute, trichomes white. **Leaves** opposite; petiole 1.4–8.5(–11.5) cm, glabrous or glabrous abaxially and sparsely tomentose adaxially; blade ovate, elliptic-ovate, or broadly ovate, (2.7–)6–17.8 × (1.4–)2.5–12(–15.5) cm, unlobed, base cordate, truncate, or cuneate, margins dentate to serrate, apex acute to acuminate, abaxial surface green, glabrous or glabrate, or sparsely hirsute along midvein and sometimes along lateral veins, trichomes at 40× conspicuously tuberculate, 0.3–1 mm, adaxial surface green, glabrous or sparsely hirsute. **Inflorescences** compact, 100–500-flowered, dome-shaped to hemispheric, (3.3–)4–14 × 3.6–12 cm; peduncle 1.5–7.8 cm, sparsely tomentose. **Pedicels** 1–2.5 mm, glabrous or sparsely hirsute. **Sterile flowers** absent or present, white, greenish white, or yellowish white, tube 6–16 mm, lobes 3–4(–5), obovate to broadly ovate, round, or elliptic, 3.6–15 × 2.2–14 mm. **Bisexual flowers:** hypanthium adnate to ovary to near its apex, 0.7–1 × 0.8–1.2 mm, strongly 8–10(–11)-ribbed in fruit, glabrous; sepals deltate to triangular, 0.2–0.5 × 0.2–0.5 mm, margins entire, apex acute to acuminate, abaxial surface glabrous; petals caducous, white to yellowish white, elliptic to narrowly ovate, 1–1.5 × 0.6–1.1 mm; filaments 2–4.5 × 0.1–0.2 mm; anthers 0.3–0.5 mm; pistils 2(–3)-carpellate, ovary completely inferior or nearly so; styles 2(–3), distinct, 0.9–1.2 mm. **Capsules** hemispheric, 1.2–2.1 × 1.7–2.5 mm. **Seeds** 0.3–0.6(–0.8) mm. **2*n*** = 36.

Flowering (May–)Jun–Jul(–Aug). Moist to dry deciduous forests and woods, moist slopes, shaded bluffs, ledges, stream banks; 70–2000 m; N.B., N.S.; Ala., Ark., Conn., Del., D.C., Fla., Ga., Ill., Ind., Ky., La., Md., Mass., Miss., Mo., N.J., N.Y., N.C., Ohio, Okla., Pa., S.C., Tenn., Va., W.Va.

Hydrangea arborescens has escaped from cultivation in Connecticut, Massachusetts, New Brunswick, and Nova Scotia; it is not native in those states or provinces.

E. McClintock (1957) circumscribed *Hydrangea arborescens* as comprising three partly sympatric subspecies: *arborescens*, *discolor*, and *radiata*. R. E. Pilatowski (1980, 1982) concluded that these were best treated as three species (*H. arborescens*, *H. cinerea*, and *H. radiata*), citing chemical, morphological, reproductive, and geographic discontinuities among the taxa. Most herbarium specimens are easily referred to one of these species; occasional specimens appear intermediate between *H. arborescens* and *H. cinerea*.

The Cherokee and Delaware tribes used bark and occasionally leaves from *Hydrangea arborescens* to prepare infusions or poultices to treat various internal and external ailments. The Cherokee used peeled twigs and branches to make tea or cooked twigs and branches as a vegetable (D. E. Moerman 1998).

2. Hydrangea cinerea Small, Bull. Torrey Bot. Club 25: 148. 1898 • Ashy hydrangea [E]

Hydrangea arborescens Linnaeus var. *deamii* H. St. John; *H. arborescens* subsp. *discolor* (Seringe) E. M. McClintock; *H. arborescens* var. *discolor* Seringe

Shrubs, 10–30 dm. **Twigs** sparsely hirsute to pilose, trichomes white. **Leaves** opposite; petiole 0.8–8.6 cm, sparsely to densely hirsute abaxially, densely hirsute adaxially; blade ovate, elliptic-ovate, or broadly ovate, (2.5–)4.2–14.9 × (1.7–)2.1–12 cm, unlobed, base cordate, truncate, or cuneate, margins dentate to serrate, apex acute to acuminate, abaxial surface grayish, uniformly velutinous, pilose, or tomentose, trichomes at 40× conspicuously tuberculate, 0.3–1 mm, adaxial surface green, sparsely hirsute. **Inflorescences** compact,

150–500-flowered, dome-shaped to hemispheric, 3.5–14.5 × 3–16.5 cm; peduncle 0.5–6 cm, sparsely to densely tomentose. **Pedicels** 0.8–2.8 mm, glabrous or sparsely to densely hirsute. **Sterile flowers** absent or present, white or greenish white, tube 4–21 mm, lobes 3–4(–5), obovate to broadly ovate, round, or elliptic, 6.5–15 × 2.5–14 mm. **Bisexual flowers:** hypanthium usually adnate to ovary to near its apex, rarely ca. ⅔ up ovary, 0.7–1 × 0.8–1.2 mm, strongly 8–10-ribbed in fruit, glabrous; sepals deltate to triangular, 0.3–0.5 × 0.3–0.6 mm, margins entire, apex acute to acuminate, abaxial surface glabrous; petals caducous, white to yellowish white, elliptic to narrowly ovate, 1.2–1.5 × 0.7–1.1 mm; filaments 1.8–4.5 × 0.1–0.2 mm; anthers 0.3–0.5 mm; pistils 2(–3)-carpellate, ovary completely inferior or nearly so; styles 2(–3), distinct, 0.8–1.1 mm. **Capsules** hemispheric, (1.2–)1.5–2.3 × (1.2–)2–2.8 mm. **Seeds** 0.3–0.6(–0.8) mm. **2*n*** = 36.

Flowering May–Jul(–Aug). Deciduous upland and bottomland forests, shaded cliffs, ravines, streambeds; 100–700 m; Ala., Ark., Ga., Ill., Ind., Kans., Ky., Mass., Mo., N.C., Okla., S.C., Tenn.

Hydrangea cinerea has escaped from cultivation in Massachusetts.

The Cherokee used infusions of bark or roots as antiemetics, emetics, and cathartics, and as gynecological and liver aids (D. E. Moerman 1998).

3. Hydrangea radiata Walter, Fl. Carol., 251. 1788 • Silverleaf or snowy hydrangea E

Hydrangea arborescens Linnaeus subsp. *radiata* (Walter) E. M. McClintock

Shrubs, 10–30 dm. **Twigs** glabrous or sparsely to densely tomentose, trichomes usually white, sometimes brown or orangish brown. **Leaves** opposite; petiole (0.7–)1.1–8 cm, glabrous or glabrous abaxially and sparsely to densely tomentose adaxially; blade narrowly to broadly ovate or elliptic, 4–15.2 × 1.3–11.2 cm, unlobed, base truncate to cuneate, margins dentate to serrate, apex acute to acuminate, abaxial surface white or grayish, densely tomentose, trichomes at 40× either smooth, 1–3 mm, or sparsely tuberculate, 0.3–1 mm, adaxial surface green, glabrous or sparsely hirsute along major veins. **Inflorescences** compact, (100–)200–500-flowered, dome-shaped to hemispheric, 4–13 × 5–14 cm; peduncle 1.6–8.3 cm, sparsely to densely tomentose. **Pedicels** 0.8–3.2 mm, glabrous or sparsely hirsute. **Sterile flowers** usually present, rarely absent, white or greenish white, tube 5–14 mm, lobes 3–4(–5), obovate to broadly ovate or round, 5–18 × 3–15 mm. **Bisexual flowers:** hypanthium adnate to ovary to near its apex, 0.9–1.2 × 1–1.3 mm, strongly 8–10-ribbed in fruit, glabrous; sepals deltate to triangular, 0.3–0.8 × 0.3–0.5 mm, margins entire, apex acute to acuminate, abaxial surface glabrous; petals caducous, white to yellowish white, elliptic to narrowly ovate, 1.2–1.6 × 0.7–1.1 mm; filaments 1.5–4.5 × 0.1–0.2 mm; anthers 0.3–0.5 mm; pistil 2(–3)-carpellate, ovary completely inferior or nearly so; styles 2(–3), distinct, (0.6–)0.9–1.5 mm. **Capsules** hemispheric, (1.3–)1.7–2.8 × 1.8–3 mm. **Seeds** 0.3–0.6(–0.8) mm. **2*n*** = 36.

Flowering May–Jul. Moist deciduous forests, ravines, banks, rocky slopes, cliffs; 200–1200 m; Ga., N.C., S.C., Tenn.

4. Hydrangea quercifolia W. Bartram, Travels Carolina, 382, plate 7. 1791 • Oak-leaf hydrangea E F

Shrubs, 10–20 dm. **Twigs** densely tomentose, trichomes brown or orangish brown, sometimes white. **Leaves** opposite; petiole (1–)1.5–8 (–12.6) cm, densely tomentose; blade suborbiculate to ovate, (2.8–)7–26.4 × (2.6–)6.5–26.5 cm, (3–)5(–7)-lobed, base truncate to cuneate, margins coarsely serrate, apex acute to acuminate, abaxial surface grayish, pilose to tomentose, trichomes at 40× either smooth, 1–4 mm, or tuberculate, 0.4–1 mm, adaxial surface green, glabrous or sparsely hirsute. **Inflorescences** open or compact, 500–1000-flowered, ovoid to conic, 9–32 × (6–)8–14 cm; peduncle 4.3–7.3 cm, sparsely tomentose to hirsute. **Pedicels** 1–3 mm, usually glabrous, rarely sparsely hirsute. **Sterile flowers** present, white, greenish white, pink, or reddish, tube 11–31 mm, lobes 4(–5), obovate to broadly ovate or round, 6–20 × 5–20 mm. **Bisexual flowers:** hypanthium adnate to ovary to near its apex, 0.9–1.3 × 1.2–2.5 mm, strongly (7–)8–10-ribbed in fruit, glabrous; sepals deltate to shallowly triangular, (0.3–)0.5–0.9 × 0.5–1 mm, margins entire, apex acute to acuminate, abaxial surface glabrous; petals caducous, white or yellowish white, elliptic, oblong, or spatulate, 1.5–2.6 × 0.7–1.4 mm; filaments 2.5–5(–6) × 0.1–0.2 mm; anthers 0.3–0.6 mm; pistil 2–4-carpellate, ovary completely inferior; styles 2–4, distinct, 1.5–2.5 mm. **Capsules** hemispheric to sub-urceolate, 1.5–2.5 × 2–2.5 mm. **Seeds** 0.6–0.8 mm. **2*n*** = 36.

Flowering (Apr–)May–Jul. Deciduous forests, pine-oak forests, ravines, ledges; 20–400 m; Ala., Conn., Fla., Ga., La., Miss., N.C., S.C., Tenn.

Hydrangea quercifolia is a popular ornamental in the eastern United States, prized for its distinctive leaves, showy inflorescences, and fall color. It is introduced and naturalized in Connecticut and North Carolina, and in South Carolina is native only in the extreme western part and naturalized elsewhere. The brownish color often seen in the trichomes is imparted by tanninlike substances (W. L. Stern 1978). M. A. Dirr (2004) listed nearly 40 cultivars derived from the species.

5. **Hydrangea paniculata** Siebold, Nov. Actorum Acad. Caes. Leop.-Carol. Nat. Cur. 14(suppl.): 691. 1829 • Panicled hydrangea [I]

Shrubs or trees, 5–70 dm. **Twigs** appressed-pubescent, trichomes white. **Leaves** opposite or 3-whorled; petiole 0.6–2.4 cm, glabrous or sparsely hirsute; blade elliptic to ovate, (5–)6.4–15 × 3–6(–10) cm, unlobed, base rounded to truncate or cuneate, margins serrate to serrulate, apex usually acute, sometimes acuminate, abaxial surface green, sparsely hirsute, often densely hirsute along veins, trichomes at 40× smooth, 0.5–1 mm, adaxial surface green, glabrate or sparsely hirsute, sometimes moderately hirsute along veins. **Inflorescences** open or compact, 250–1000-flowered, usually conic, sometimes ovoid, 7–25 × 6–11.5 cm; peduncle 2–5.5 cm, sparsely to densely hirsute. **Pedicels** 1–4 mm, glabrous or sparsely hirsute proximally. **Sterile flowers** present, white or greenish white, sometimes tinged pink or blue, tube 9–16 mm, lobes 4(–5), elliptic to round, (6–)10–21 × (4–)8–15 mm. **Bisexual flowers:** hypanthium adnate to ovary to near its middle, 1–1.8 × 1.2–2.2 mm, weakly 7–10-ribbed in fruit, usually glabrous, rarely sparsely hairy; sepals shallowly triangular, 0.3–0.7 × 0.5–1.2 mm, margins entire, apex acute to acuminate, abaxial surface glabrous; petals tardily deciduous, spreading or reflexed, white or yellowish white, elliptic, ovate, or ovate-lanceolate, 3–4 × 1.2–1.5 mm; filaments 1.8–5 × 0.1–0.3 mm; anthers 0.3–0.6 mm; pistil 2–4-carpellate, ovary ½ inferior; styles 2–4, connate to middle or distally, 2–3.5 mm. **Capsules** oblong-ovoid, 3.5–5 × 2–3 mm. **Seeds** caudate at both ends, 1.7–3 mm. $2n$ = 72, 108 (Japan).

Flowering Jul–Sep. Disturbed woods, roadsides, wooded swamps; 30–500 m; introduced; Ont.; Conn., Maine, Mass., N.H., N.Y., N.C., Pa., R.I., Va., W.Va.; e Asia.

M. A. Dirr (2004) listed 26 primary cultivars derived from *Hydrangea paniculata*.

9. DECUMARIA Linnaeus, Sp. Pl. ed. 2, 2: 1663. 1763 • Climbing-hydrangea, wood-vamp • [Latin *decumae*, tenths, and *-aria*, possessing, alluding to sometimes 10-merous flowers]

Ronald L. McGregor†

Woody vines. Stems climbing, sometimes trailing and forming loose nonflowering mats. **Bark** exfoliating in grayish or reddish brown sheets, strips, or strings. **Branches** spreading laterally or declining; twigs glabrous or with simple trichomes. **Leaves** deciduous or semideciduous, opposite; petiole present; blade ovate, elliptic, obovate, oblanceolate, subround, or round, herbaceous, margins usually entire, rarely dentate or lobed, plane; venation pinnate. **Inflorescences** terminal, on shoots of the season, corymbs, (20–)50–100-flowered; peduncle present. **Pedicels** present. **Flowers** bisexual; perianth and androecium epigynous; hypanthium completely adnate to ovary, turbinate, strongly 7–12-ribbed in fruit; sepals persistent, 7–12, erect, triangular, glabrous; petals 7–12, valvate, spreading, white, lanceolate, elliptic, or oblong, base essentially sessile, surfaces glabrous; stamens 20–30; filaments distinct, dorsiventrally flattened, linear, tapering abruptly just proximal to apex, apex not 2-lobed; anthers suborbiculate to ovate; pistil 6–12-carpellate, ovary completely inferior, 6–12-locular; placentation axile proximally, parietal distally; style persistent, 1. **Capsules** turbinate, cartilaginous, dehiscence intercostal, lateral walls separating from ribs, eventually leaving cagelike remnants. **Seeds** 10–20 per locule, yellow, fusiform. x = 14.

Species 2 (1 in the flora): se United States, Asia (China).

A molecular phylogenetic study of tribe Hydrangeeae (Y. De Smet et al. 2015) found *Hydrangea* to be polyphyletic. The authors promoted adoption of a broader, monophyletic concept of *Hydrangea* that includes all eight genera in the tribe, including *Decumaria*. *Decumaria* is treated here in its traditional sense.

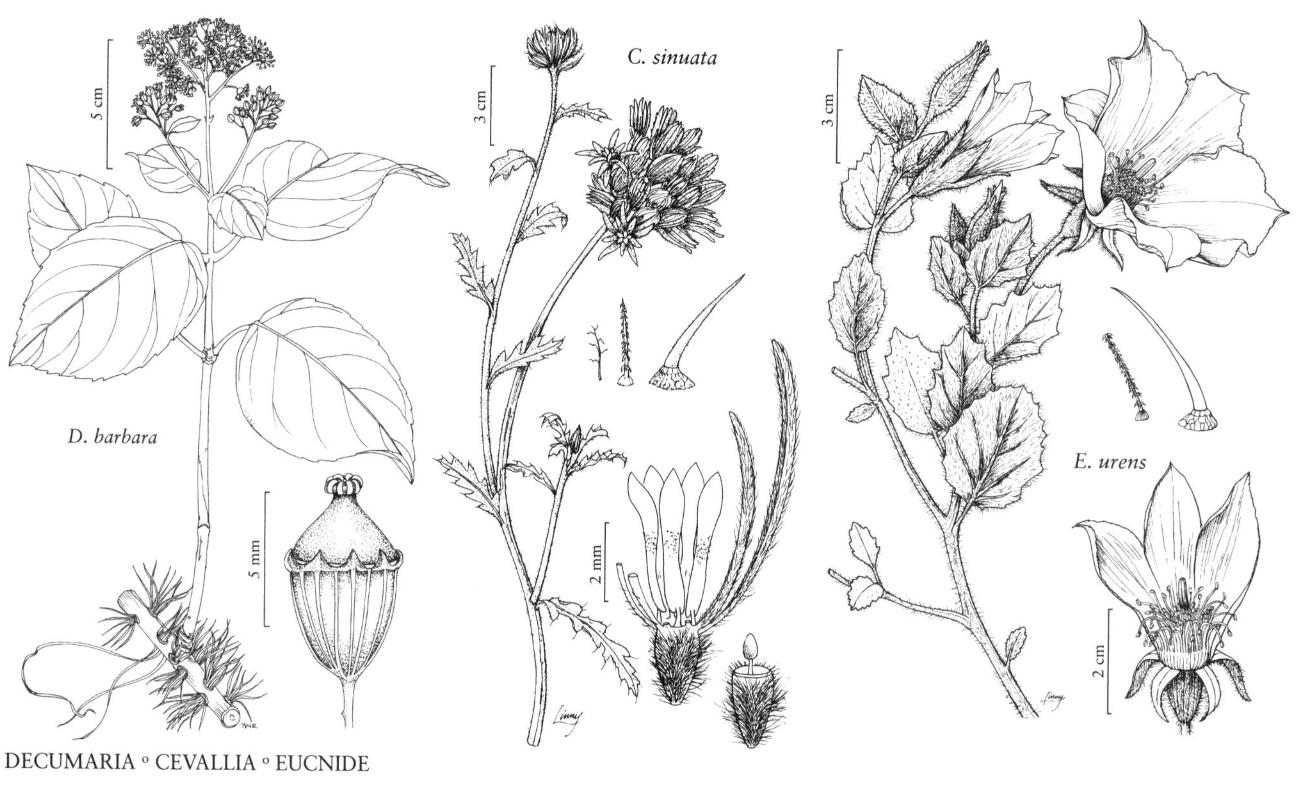

DECUMARIA ◦ CEVALLIA ◦ EUCNIDE

1. Decumaria barbara Linnaeus, Sp. Pl. ed. 2, 2: 1663. 1763 [E] [F]

Stems to 100 dm; adventitious roots usually present. **Twigs** glabrous. **Leaves:** petiole slightly winged proximally, 10–30 mm, glabrous or short-pubescent; blade 10–12 × 6–8 cm, base cuneate, truncate, cordate, or rounded, apex acute, obtuse, rounded, or mucronate, abaxial surface light green, finely pubescent along veins, adaxial dark green, glabrous. **Inflorescences** compact, congested to open, 3–8 × 4–10 cm; peduncle 2–6 cm, glabrous. **Pedicels** 1–6 mm, glabrous. **Flowers** faintly fragrant; hypanthium 1.5–2.2 × 0.5–2 mm; sepals 0.2–1.3 × 0.2–0.5 mm, margins entire, apex acute to acuminate, surfaces glabrous abaxially; petals 2.8–3.2 × 1.2–1.7 mm; filaments 3–5 × 0.2–0.3 mm; anthers 0.7–1 mm; style stout, broad at base, 0–2 mm; stigmatic lines 7–12, radiating. **Capsules** 3.2–5.5 × 3–5 mm. **Seeds** 1.5–3 mm. **2*n*** = 28.

Flowering Apr–Jun; fruiting Jul–Oct. Very moist to wet woodlands, swamps, rich woods; 0–400 m; Ala., Ark., Del., Fla., Ga., La., Miss., N.Y., N.C., S.C., Tenn., Va.

Decumaria barbara is a rare escape from cultivation in New York. In the southeastern United States, *D. barbara* is found mostly on the Costal Plain.

Decumaria sinensis Oliver, of central China, differs from *D. barbara* in being a low-climbing vine with semipersistent or persistent leaves, stigmas almost completely sessile, and in plants growing at elevations of 600–1300 m.

LOASACEAE Jussieu
• Stickleaf Family

Larry D. Hufford

Herbs, subshrubs, shrubs, [**trees**], annual, biennial, or perennial, evergreen (leaves withering in age), scabrid, trichomes (1) unbranched and smooth, knobby, notched, or antrorsely or retrorsely barbed, (2) dendritic, and/or (3) stinging. **Leaves** alternate [opposite], simple; stipules absent; petiole present or absent; blade sometimes lobed, margins entire, serrate, dentate, or crenate; venation pinnate, basal secondary veins commonly present. **Inflorescences** terminal, cymes, thyrses, racemes, panicles, or flowers solitary. **Flowers** bisexual, usually radially symmetric, rarely bilaterally symmetric (by bilateral symmetry of corolla, androecium, or gynoecium); perianth and androecium epigynous; hypanthium adnate to ovary proximally and free distally, or completely adnate to ovary; sepals 5, radially symmetric, distinct or connate basally; petals 5, radially or bilaterally symmetric, distinct or connate, sometimes postgenitally coherent; nectary absent or present, distal on ovary; stamens 5–150+, distinct, free or adnate to petal bases, bilaterally or radially symmetric; anthers dehiscing by longitudinal slits; staminodes absent or present, sometimes petaloid and petals appearing to be 6+; pistil 1, 3–7-carpellate, ovary inferior, 1-locular, placentation parietal, subapical, or apical; ovules 1–60+ per locule, anatropous (epitropous); style 1; stigma 1, 2–5(–7)-lobed (lobes usually as many as carpels, except in pseudomonomerous Gronovioideae). **Fruits** capsules, dehiscence by apical valves [splitting longitudinally], or cypselae, sepals or sepals and petals persisting. **Seeds** 1–60+ per fruit.

Genera 20, species ca. 350 (4 genera, 94 species in the flora): North America, Mexico, West Indies, Central America, South America, sw Asia (w Arabian Peninsula), Africa, Pacific Islands (Galápagos Islands, Marquesas Islands); introduced in Atlantic Islands (Cape Verde Islands).

Loasaceae are in Cornales and are the sister family of Hydrangeaceae. The two families likely diverged in western North America or Mexico in the Late Cretaceous to Paleocene, 92–58 million years before present (J. J. Schenk and L. Hufford 2010). Most extant species of Loasaceae in western North America date to the Oligocene or more recent times, with the Miocene and later periods being especially rich phases for speciation (Schenk and Hufford; J. Grissom and Hufford, unpubl.). In Loasaceae, *Eucnide* and *Schismocarpus* S. F. Blake, a genus restricted to southern Mexico, are the earliest diverging lineages, and the cliff-dwelling habits of these two genera may be plesiomorphic for the family (Hufford et al. 2003). The species-rich genus *Mentzelia* is well supported as the sister of a clade that consists of the taxonomically depauperate genera *Cevallia, Fuertesia* Urban, *Gronovia* Linnaeus, and *Petalonyx*. Phylogenetic studies support the monophyly of subfamilies Gronovioideae Fenzl and Loasoideae Gilg as conceived by I. Urban and E. Gilg (1900), but their early concept of Mentzelioideae as including both *Eucnide* and *Mentzelia* (and also *Schismocarpus* according to S. F. Blake 1918b; Gilg 1925b) is paraphyletic (Hufford et al.). To enable classification based on monophyly, circumscription of Mentzelioideae should be restricted to *Mentzelia*. The *Mentzelia* plus Gronovioideae clade is sister to the more taxon rich Loasoideae, which is especially prevalent in Andean South America (Hufford et al.).

SELECTED REFERENCES Brown, D. K. and R. B. Kaul. 1981. Floral structure and mechanism in Loasaceae. Amer. J. Bot. 68: 361–372. Carlquist, S. 1984. Wood anatomy of Loasaceae with relation to systematics, habit and ecology. Aliso 10: 583–602. Ernst, W. R. and H. J. Thompson. 1963. The Loasaceae in the southeastern United States. J. Arnold Arbor. 44: 138–142. Hufford, L. 1990. Androecial ontogeny and the problem of monophyly of Loasaceae. Canad. J. Bot. 68: 402–419. Moody, M. L. and L. Hufford. 2000. Floral ontogeny and morphology of *Cevallia*, *Fuertesia*, and *Gronovia* (Loasaceae subfamily Gronovioideae). Int. J. Pl. Sci. 161: 869–883. Poston, M. E. and J. W. Nowicke. 1993. Pollen morphology, trichome types, and relationships of the Gronovioideae (Loasaceae). Amer. J. Bot. 80: 689–704. Weigend, M. 2004. Loasaceae. In: K. Kubitzki et al., eds. 1990+. The Families and Genera of Vascular Plants. 9+ vols. Berlin etc. Vol. 6, pp. 239–254. Weigend, M., J. Kufer, and A. A. Mueller. 2000. Phytochemistry and the systematics and ecology of Loasaceae and Gronoviaceae (Loasales). Amer. J. Bot. 87: 1202–1210.

1. Stamens 5; seeds 1 per fruit; fruits cypselae.
    2. Inflorescences headlike thyrses, peduncles to 1 dm; perianth whorls similar, sepals and petals white to yellowish abaxially, yellow adaxially, linear-lanceolate; petals densely hairy on both surfaces; stamen filaments dorsiventrally flattened, linear, shorter than anthers; anthers with distal connective extension; stigmas densely hairy .... 1. *Cevallia*, p. 492
    2. Inflorescences racemes or panicles, peduncles inconspicuous; perianth whorls differentiated, sepals green, lanceolate, petals white, spatulate; petals glabrous except hairy abaxially on midribs; stamen filaments filiform, longer than anthers; anthers without distal connective extension; stigmas papillate .... 4. *Petalonyx*, p. 543
1. Stamens 8–50+; seeds (1–)2–60+ per fruit; fruits capsules.
    3. Pistils 3-carpellate; stigmas 3-lobed .... 3. *Mentzelia* (in part), p. 496
    3. Pistils 5–7-carpellate; stigmas 5–7-lobed.
        4. Hypanthia completely adnate to ovaries; petals connate proximally to ½+ length; stamen filaments filiform; pedicels elongating in fruit; seeds cylindric to ovoid, to 1 mm, not winged .... 2. *Eucnide*, p. 493
        4. Hypanthia adnate to ovaries proximally, free distally; petals distinct or connate basally; 5 outermost stamen filaments dorsiventrally flattened, spatulate; pedicels not elongating in fruit; seeds dorsiventrally flattened, 2.3–4 mm, winged .... 3. *Mentzelia* (in part), p. 496

1. CEVALLIA Lagasca, Varied. Ci. 2(4): 35. 1805 • Stingleaf or stinging serpent [For Pedro Cevallos, 1760–1840, Spanish statesman and diplomat]

Larry D. Hufford

Herbs or subshrubs, perennial; trichomes (1) pointed with surfaces ± smooth, knobby, or notched, (2) retrorsely barbed along shaft and at apex, (3) dendritic, and (4) stinging. **Stems** erect, sometimes becoming decumbent. **Leaves** cauline; petiole present or absent; blade ovate-elliptic, lobed, margins entire. **Inflorescences** headlike thyrses; peduncle to 1 dm. **Pedicels** not elongating in fruit. **Flowers:** hypanthium adnate to ovary proximally, free distally; perianth whorls similar; sepals white to yellowish abaxially, yellow adaxially, distinct, linear-lanceolate, longer than petals; petals white to yellowish abaxially, yellow adaxially, distinct, linear-lanceolate, spreading, both surfaces densely hairy; nectary absent; stamens 5, included;

filaments monomorphic, dorsiventrally flattened, linear, shorter than anthers; anthers with distal connective extension; staminodes absent; pistil pseudomonomerous, placenta subapical; stigma ovoid, 3–5-lobed, densely hairy. **Fruits** cypselae, urceolate-ovoid, straight; sepals and petals persistent. **Seeds** 1, ovoid. $x = 7$.

Species 1: sw, sc United States, n Mexico.

In Gronovioideae, *Cevallia* is sister to a clade consisting of *Fuertesia* and *Gronovia*.

The distinctive stamens, which have inflated, tonguelike extensions beyond the pollen sacs, are unique in the family.

1. **Cevallia sinuata** Lagasca, Varied. Ci. 2(4): 36. 1805 [F]

Herbs or subshrubs to 6 dm. **Leaves:** petiole 0–3[–8] mm; blade to 6 × 3 cm, pinnately lobed ± ½ to midrib, base usually oblique, acute to obtuse, apex acute with rounded tip; midvein and secondary veins prominent. **Flowers:** perianth densely covered with long, pointed hairs; sepals 8–9 mm; petals 6–8 mm; stamen filaments 1 mm. $2n = 14, 26$.

Flowering Apr–Oct. Gypsum and limestone hills, gravelly flats, open desert scrub, Tamaulipan thorn scrub, grasslands; 130–2000 m; Ariz., N.Mex., Okla., Tex.; Mexico (Chihuahua, Coahuila, Durango, Zacatecas).

The small flowers of *Cevallia sinuata* have an architecture that promotes deposition of self-pollen on the densely hairy stigma. The absence of a nectary in the flowers may reflect selection for self-pollination or for a pollen-flower pollination syndrome (a so-called pollen flower mimic in the sense of S. Vogel 1978) that possibly involves secondary pollen presentation on the stigma. Various insects, especially bees but also butterflies, have been observed to visit the flowers (W. S. Davis and H. J. Thompson 1967; A. M. Powell et al. 1977). The flowers undoubtedly self-pollinate, but it is unclear whether they self-fertilize or may be self-incompatible.

A. M. Powell et al. (1977) noted that populations of $n = 7$ and $n = 13$ cytotypes can be found, and that the latter have a larger geographic distribution and wider ecological tolerance compared to the former. They hypothesized that $n = 13$ populations were derived via polyploidization followed by aneuploidy.

2. EUCNIDE Zuccarini, Index Seminum (München) 1844: [4]. 1844, name conserved • Rock or pretty nettle, stingbush [Greek *eu*-, good or pretty, and *knide*, nettle, alluding to stinging trichomes and showy flowers]

Larry D. Hufford

Sympetaleia A. Gray

Herbs or subshrubs, annual or perennial; trichomes (1) pointed with surfaces ± smooth, (2) retrorsely barbed along shaft and at apex or only at apex, and (3) stinging. **Stems** usually erect or spreading, rarely prostrate or pendent on cliffs. **Leaves** cauline; petiole present; blade ovate, lobed or unlobed, margins crenate or dentate. **Inflorescences** dichasia and monochasia [solitary flowers]; peduncle inconspicuous [conspicuous]. **Pedicels** elongating in fruit. **Flowers:** hypanthium completely adnate to ovary; perianth whorls differentiated; sepals green, distinct, lanceolate, straplike, or narrowly ovate, shorter than petals; petals white, green, or yellow [reddish orange], connate proximally to ½+ length, spatulate or ovate, spreading or erect (then corolla essentially tubular) [erect proximally, divaricate distally (corolla salverform)], glabrous except apices sparsely hairy; nectary distal on ovary; stamens 15–150+, exserted or included; filaments monomorphic, filiform, longer or shorter than anthers; anthers without distal connective extension; staminodes absent; pistil 5-carpellate, placentae parietal; stigma lingulate, 5-lobed, papillate. **Fruits** capsules, dehiscing by apical valves [splitting longitudinally], cup-shaped, straight; sepals persistent. **Seeds** many, cylindric to ovoid, not dorsiventrally flattened, to 1 mm, not winged. $x = 21$.

Species 14 (4 in the flora): sw, sc United States, Mexico, Central America (Guatemala).

Eucnide was placed in Mentzelioideae by I. Urban and E. Gilg (1900), Gilg (1925b), and H. J. Thompson and W. R. Ernst (1967); however, this subfamily is paraphyletic. *Eucnide* has been placed in molecular phylogenetic studies as sister to the rest of Loasaceae (L. Hufford et al. 2003).

Three sections of *Eucnide* were recognized by H. J. Thompson and W. R. Ernst (1967). Section *Mentzeliopsis* H. J. Thompson & W. R. Ernst, consisting only of *E. urens*, was distinguished on the basis of its floral architecture in which all stamens are shorter than the style and not exserted beyond the corolla. The numerous stamens of *E. urens* are also tightly positioned around the style. Section *Sympetaleia* (A. Gray) H. J. Thompson & W. R. Ernst consists of three species that are endemic to the Baja California Peninsula and surrounding islands, except for *E. rupestris*, which has a distribution that extends into extreme southern California and southwestern Arizona and to Sinaloa and Sonora, Mexico. Species of sect. *Sympetaleia* have stamens with monothecate, bisporangiate anthers in contrast to other members of the genus, which have more conventional bithecate, tetrasporangiate anthers. All other species of *Eucnide* were placed in sect. *Eucnide* by Thompson and Ernst.

Among the North American species of *Eucnide*, *E. urens* is the only species found in the Mojave Desert, where it is centered (H. J. Thompson and W. R. Ernst 1967). The other North American species are found in the Chihuahuan and Sonoran deserts and in adjacent areas. All species are found in similar cliff or rocky slope habitats (uncommonly in arroyos and washes). Some species, such as *E. bartonioides*, have fruit pedicels that are negatively phototropic and elongate extensively, which appear to be adaptations for dispersal on cliffs (Thompson and Ernst).

Most species of *Eucnide* are self-pollinating (H. J. Thompson and W. R. Ernst 1967), although they generally have some spatial separation between the stigmas and anthers soon after the flowers open that allows for cross-pollination (L. Hufford 1988). Only taxa with the largest flowers, such as *E. bartonioides* var. *bartonioides* (possibly pollinated by hawk moths) and *E. urens* (pollinated by the melittid bee, *Hesperaster laticeps*), appear to be strictly outcrossing (Thompson and Ernst).

SELECTED REFERENCES Hufford, L. 1987. Inflorescence architecture of *Eucnide* (Loasaceae). Madroño 34: 18–28. Hufford, L. 1988. Potential roles of scaling and post-anthesis developmental changes in the evolution of floral forms of *Eucnide* (Loasaceae). Nordic J. Bot. 8: 147–157. Hufford, L. 1988b. Seed coat morphology of *Eucnide* and other Loasaceae. Syst. Bot. 13: 154–167. Waterfall, U. T. 1959. A revision of *Eucnide*. Rhodora 61: 231–243.

1. Corollas essentially tubular, petals to 15 mm, connate to 9 mm, 2-colored, yellow to brown basally, green distally; stamen filaments less than 5 mm, inserted in upper portion of corolla tube . . . . 3. *Eucnide rupestris*
1. Corollas funnelform or rotate, petals 10–55 mm, connate to 5 mm, 1-colored, white, cream, or yellow; stamen filaments 10–60 mm, inserted at base of corolla.
   2. Petals white to cream; stamens included, most aggregated around style, only longer, outermost stamens spreading away from style. . . . . 4. *Eucnide urens*
   2. Petals yellow; stamens exserted, spreading away from style.
      3. Corollas funnelform; filaments 13–60 mm . . . . 1. *Eucnide bartonioides*
      3. Corollas rotate; filaments 7–16 mm . . . . 2. *Eucnide lobata*

1. Eucnide bartonioides Zuccarini, Index Seminum (München) 1844: [4]. 1844 • Yellow stingbush

Herbs or subshrubs, annual or perennial, moundlike to spindle-shaped (wider than tall). **Leaves:** blade shallowly to prominently lobed. **Pedicels** (fruiting) 3–35+ cm, frequently recurved. **Flowers** radially symmetric or slightly bilaterally symmetric through upward curvature of stamen filaments; corolla funnelform, petals connate to 3 mm, 1-colored, yellow, spatulate, 10–55[–58] mm; stamens 15–150+, inserted at base of corolla, exserted, spreading away from style; filaments 13–60 mm, longer than anthers.

Varieties 2 (2 in the flora): Texas, n Mexico.

SELECTED REFERENCE Turner, B. L. 2012b. Taxonomy of *Eucnide bartonioides* (Loasaceae) complex in Texas. Phytologia 94: 305–309.

1. Petals 30–55 mm; stamens (55–)70–150+, filaments 30–60 mm . . . . 1a. *Eucnide bartonioides* var. *bartonioides*
1. Petals 10–25 mm; stamens 15–70, filaments 13–30 (–35) mm . . . 1b. *Eucnide bartonioides* var. *edwardsiana*

1a. Eucnide bartonioides Zuccarini var. **bartonioides**

Herbs or subshrubs, annual or perennial, moundlike to spindle-shaped (wider than tall). **Flowers:** petals connate 0.5–3 mm, [20–]30–55[–58] mm; stamens (55–)70–150+, filaments 30–60 mm. $2n = 42$.

Flowering Jan–Dec. Rocky slopes and cliffs, arroyos, limestone and volcanic substrates; 500–1500 m; Tex.; Mexico (Chihuahua, Coahuila, Nuevo León, San Luis Potosí, Tamaulipas).

Variety *bartonioides* is known in the flora area from Brewster, Edwards, Jeff Davis, Presidio, Terrell, Uvalde, Val Verde, and Webb counties.

1b. Eucnide bartonioides Zuccarini var. **edwardsiana** B. L. Turner, Phytologia 94: 306, fig. 1. 2012 [E]

Herbs, annual (or perennial?), spindle-shaped (wider than tall), sparsely branched or unbranched. **Flowers:** petals connate to 0.5 mm, 10–25 mm; stamens 15–70, filaments 13–30 (–35) mm.

Flowering Mar–Nov. Limestone cliffs; 200–300 m; Tex.

H. J. Thompson and W. R. Ernst (1967) called attention to two different floral size morphs among populations of *Eucnide bartonioides*, and Turner recognized them formally as different varieties. Thompson, in a personal communication to Turner, suggested that the small-flowered var. *edwardsiana* may be a highly autogamous derivative of the larger flowered form, and observations for this treatment showed the stigma of this form to be closely surrounded by dehisced anthers. Turner indicated that var. *edwardsiana*, which is found on the Edwards Plateau (Comal, Edwards, Hays, Kerr, Llano, and Travis counties), grades into var. *bartonioides* in Edwards County. Flowers of var. *edwardsiana* from Edwards County are at the large end of their size range, and a survey of specimens showed that both relatively large-flowered plants and more typical forms of var. *edwardsiana* occur near each other in some localities. Turner characterized var. *edwardsiana* as perennial, but most of the examined specimens appear to be annuals. Across its range in Mexico, *E. bartonioides* displays variation in flower size, with small-flowered plants that appear to be annuals found in different regions. These are likely to be separate evolutionary origins of small-flowered annuals that converge on the characteristics of var. *edwardsiana*.

2. Eucnide lobata (Hooker) A. Gray, Boston J. Nat. Hist. 6: 192. 1857 • Lobed-leaf stingbush

Microsperma lobatum Hooker, Icon. Pl. 3: plate 234. 1839 (as lobata); *M. grandiflorum* Groenland

Subshrubs, moundlike to spindle-shaped (wider than tall), branches decumbent. **Leaves:** blade lobed. **Pedicels** (fruiting) to 10 cm, recurved. **Flowers** radially symmetric or slightly bilaterally symmetric through orientation of style; corolla rotate, petals connate 3 mm, 1-colored, yellow, spatulate, to 40 mm; stamens 50–120, inserted at base of corolla, exserted, spreading away from style; filaments 7–16 mm, longer than anthers. $2n = 42$.

Flowering Mar–Dec. Steep rocky hills of gypsum; 50 m; Tex.; ne, c Mexico.

Eucnide lobata is known in the flora area from Starr County.

3. **Eucnide rupestris** (Baillon) H. J. Thompson & W. R. Ernst, J. Arnold Arbor. 48: 86. 1967

Loasella rupestris Baillon, Bull. Mens. Soc. Linn. Paris 1: 650. 1887; *Sympetaleia rupestris* (Baillon) A. Gray ex S. Watson

Herbs, annual, spindle-shaped (often nearly as wide as tall). **Leaves:** blade slightly lobed. **Pedicels** (fruiting) to 2 cm, usually nodding. **Flowers** radially symmetric; corolla essentially tubular, petals connate 9 mm, 2-colored, yellow to brown basally, green distally, ovate, to 15 mm; stamens 15, inserted in upper portion of corolla tube, included, projecting toward style; filaments to 5 mm, shorter than anthers. $2n = 42$.

Flowering Mar–May. Rocky slopes, washes; 10–500 m; Ariz., Calif.; Mexico (Baja California, Sinaloa, Sonora).

Eucnide rupestris is endemic to the Sonoran Desert.

4. **Eucnide urens** Parry, Amer. Naturalist 9: 144. 1875 • Desert stingbush [F]

Mentzelia urens Parry ex A. Gray, Proc. Amer. Acad. Arts 10: 71. 1874, not Vellozo 1831

Subshrubs, moundlike (wider than tall). **Leaves:** blade usually unlobed, sometimes inconspicuously lobed. **Pedicels** (fruiting) less than 3 cm, usually curved to nodding. **Flowers** radially symmetric; corolla funnelform, petals connate 5 mm, 1-colored, white to cream, spatulate, to 45 mm; stamens 50+, inserted at base of corolla, included, most aggregated around style, only longer, outermost stamens spreading away from style; filaments 10–20 mm, longer than anthers. $2n = 42$.

Flowering Mar–Jul. Clefts in cliffs, rocky slopes, wash margins, limestone, desert scrub; -50–2000 m; Ariz., Calif., Nev., Utah; Mexico (Baja California).

Eucnide urens is found primarily in the Mojave Desert but extends into surrounding areas. *Asydates inyoensis*, soft-wing flower beetles of the family Melyridae, have been found in flowers of *E. urens*; they collect pollen on dorsal setae and likely serve as pollinators.

3. MENTZELIA Linnaeus, Sp. Pl. 1: 516. 1753; Gen. Pl. ed. 5, 233. 1754 • Blazingstar, stickleaf [For Christian Mentzel, 1622–1701, German botanist]

Larry D. Hufford

John J. Schenk

Joshua M. Brokaw

Herbs or subshrubs [shrubs or trees], annual, biennial, or perennial; trichomes (1) pointed with surfaces ± smooth or antrorsely barbed and (2) retrorsely barbed along shaft and at apex or only at apex. **Stems** erect, clambering, or decumbent. **Leaves** basal and cauline or cauline; petiole present or absent; blade hastate, deltate, cordate, ovate, elliptic, lanceolate, linear, spatulate, oblanceolate, obovate, or orbiculate, lobed or unlobed, margins dentate, serrate, crenate, or entire. **Inflorescences** dichasia or flowers solitary; peduncle inconspicuous. **Pedicels** not elongating in fruit. **Flowers:** hypanthium adnate to ovary proximally, free distally; perianth whorls differentiated; sepals green, connate basally, lanceolate to narrowly ovate, shorter than petals; petals white, yellow, or orange, sometimes red proximally, distinct or connate basally, spatulate, ovate, elliptic, oblanceolate, or obovate, spreading to erect, glabrous or hairy abaxially, on margins, or on apices; nectary distal on ovary; stamens 8–45+, exserted or included; filaments monomorphic, filiform or dorsiventrally flattened and linear, or heteromorphic, outer dorsiventrally flattened and linear, elliptic, spatulate, or oblanceolate [lanceolate], inner filiform or dorsiventrally flattened and linear, longer than anthers; anthers without distal connective extension; staminodes present or absent; pistil 3-carpellate (5–7-carpellate in *M. decapetala*), placentae parietal; stigma lingulate, 3-lobed (5–7-lobed in *M. decapetala*), papillate. **Fruits** capsules, dehiscing by apical valves [splitting longitudinally], cup-shaped, lingulate, subcylindric, cylindric, ovoid, urceolate, clavate, or funnelform, straight or curved, sometimes S-shaped; sepals persistent. **Seeds** (1–)2–60+, ovoid, oblong, bottle-shaped, pyriform, irregularly polygonal, or trigonal prisms, dorsiventrally flattened or not, 0.5–4.5 mm, winged or not winged. $x = 9$.

Species ca. 95 (85 in the flora): w, c, se North America, Mexico, West Indies, Central America, South America; introduced in Atlantic Islands (Cape Verde Islands).

Mentzelia is monophyletic and sister to Gronovioideae (M. L. Moody et al. 2001; L. Hufford et al. 2003). These studies recovered clades that correspond to the six described sections of *Mentzelia*, including the five treated here and the monospecific sect. *Dendromentzelia* Urban & Gilg, which consists only of *M. arborescens* Urban & Gilg of Michoacán and Oaxaca, Mexico.

Mentzelias in North America often are ruderal or found in low-productivity, disturbance-prone environments, such as arroyos, sand dunes, cliffs, or talus slopes. They frequently are associated with distinctive substrates, including gypsum, limestone, serpentine, or volcanic tuff, and some might be edaphically restricted (J. B. Glad 1976; H. J. Thompson and A. M. Powell 1981; C. M. Christy 1997; N. H. Holmgren and P. K. Holmgren 2002; J. M. Brokaw et al. 2015).

Changes in chromosome number, through both aneuploidy and polyploidy, have been important in the evolution of *Mentzelia*. Aneuploidy has been especially important in reproductive isolation in sections *Bartonia* (haploid chromosome numbers include 9, 10, 11, and 18) and *Bicuspidaria* (haploid chromosome numbers include 9, 10). In contrast, aneuploidy has not been found in sections *Mentzelia* or *Trachyphytum*, although polyploidy has been found in both. In sect. *Mentzelia*, polyploidy has been reported only in *M. asperula*. Polyploidy has played a more important role in reproductive isolation and diversification in sect. *Trachyphytum*, in which diploids, tetraploids, hexaploids, and octoploids have been reported (H. J. Thompson and H. Lewis 1953).

Floral forms differ among the sections of *Mentzelia*, which may be indicative of reproductive shifts that were important in the differentiation of the sectional lineages. Reproductive ecology has been investigated best in sect. *Bicuspidaria*, in which *Xeralictus bicuspidariae*, a robust oligolectic bee, is the main pollinator of *M. hirsutissima*, *M. tricuspis*, *M. tridentata*, and is one of three main pollinators of *M. involucrata* (G. S. Daniels 1970). Only *M. reflexa*, which has markedly different flowers compared to other species of sect. *Bicuspidaria*, is not pollinated by *X. bicuspidariae*. Instead, another oligolectic bee, *Perdita koebelei*, is its primary pollinator (Daniels). Bees have been suggested to be the primary pollinators of other sections of *Mentzelia*, except for the few large-flowered species of sect. *Bartonia* in which hawk moth pollination has been observed (H. J. Thompson 1960; A. R. Moldenke 1976; R. J. Hill 1977). Many polyploid species of sect. *Trachyphytum* have small flowers and reportedly self-fertilize (J. E. Zavortink 1966).

Seed surface features are important in *Mentzelia* taxonomy. In sections *Dendromentzelia*, *Mentzelia*, and *Micromentzelia*, which compose a basal grade in *Mentzelia*, seed coat cells are longer than wide. In contrast, the derived clade consisting of sections *Bartonia*, *Bicuspidaria*, and *Trachyphytum* has seed coat cells that are polygonal and largely isodiametric. The shapes of anticlinal and outer periclinal walls of seed coat cells are similar in sections *Bicuspidaria* and *Trachyphytum* but differ among the other sections of North American species. Seed coat cell anticlinal wall shape, which is particularly useful in distinguishing species in sect. *Bartonia*, may be straight, wavy (curves along walls forming less than 45° angles), or sinuous (curves along walls forming at least 45° angles, often forming U-shaped undulations).

Leaf and bract margins in *Mentzelia* usually are toothed or lobed to some degree, ranging from serrate or dentate (divided relatively shallowly) through pinnate (defined here as divided partway to the midvein) to pinnatisect (divided nearly or to the midvein).

1. Stamen filaments all or mostly distally 2-lobed . . . . 3b. *Mentzelia* sect. *Bicuspidaria* (in part), p. 524
1. Stamen filaments unlobed, or 5 outermost distally 2-lobed and 5–15 inner filaments unlobed (*M. micrantha*, sect. *Trachyphytum*).
   2. Stamen filaments dorsiventrally flattened, 1–3(–5) outermost petaloid; seeds not winged . . . . 3b. *Mentzelia* sect. *Bicuspidaria* (in part), p. 524
   2. Stamen filaments filiform or outer dorsiventrally flattened, inner filiform, 0 or 5+ outermost petaloid, if 5+, seeds winged.
      3. Ovules and seeds oriented perpendicular to long axis of ovary; seeds winged; 5 outermost stamen filaments usually petaloid, sometimes not . . . 3a. *Mentzelia* sect. *Bartonia*, p. 498
      3. Ovules and seeds oriented parallel to long axis of ovary; seeds not winged (occasionally winged in *M. lindleyi*); 5 outermost stamen filaments not petaloid.
         4. Plants perennial.
            5. Leaf blades, at least some, with broad basal lobes, margins flat; capsules usually subcylindric, clavate, lingulate, or funnelform, sometimes ovoid; seeds oblong, ovoid, or pyriform, coat cells oblong . . . . 3c. *Mentzelia* sect. *Mentzelia* (in part), p. 527
            5. Leaf blades without broad basal lobes, margins revolute; capsules ovoid to urceolate; seeds ovoid to bottle-shaped, coat cells polygonal . . . . 3d. *Mentzelia* sect. *Micromentzelia*, p. 530
         4. Plants annual.
            6. Leaf blades, at least some, with broad basal lobes; seeds oblong, ovoid, or pyriform, dorsiventrally flattened to trigonal, coat cells oblong, much longer than wide, anticlinal walls sinuous; basal rosette of leaves absent, proximalmost internodes 10+ mm . . . . 3c. *Mentzelia* sect. *Mentzelia* (in part), p. 527
            6. Leaf blades without broad basal lobes; seeds triangular prisms or irregularly polygonal, coat cells polygonal, ± isodiametric, anticlinal walls straight; basal rosette of leaves present, persistent or not, proximalmost internodes to 5 mm . . . . 3e. *Mentzelia* sect. *Trachyphytum*, p. 531

3a. MENTZELIA Linnaeus sect. BARTONIA (Torrey & A. Gray) Bentham & Hooker f., Gen. Pl. 1: 804. 1867 • Western star

John J. Schenk

Larry D. Hufford

Mentzelia [unranked] *Bartonia* Torrey & A. Gray, Fl. N. Amer. 1: 534. 1840

Herbs, winter annual, biennial, or short-lived perennial. **Leaves:** basal rosette present, not persistent, proximalmost internodes to 10+ mm; blade without broad basal lobes, margins entire, toothed, pinnate, or pinnatisect, flat. **Flowers:** stamens all fertile or 5+ outer sterile; filaments heteromorphic, outer dorsiventrally flattened, linear, elliptic, spatulate, or oblanceolate, usually petaloid, sometimes not, inner filiform, unlobed; ovules (and seeds) oriented perpendicular to long axis of ovary. **Capsules** cup-shaped to lingulate or cylindric, straight. **Seeds** ovoid, flattened dorsiventrally, winged; seed coat cells polygonal, ± isodiametric, anticlinal walls straight, wavy, or sinuous.

Species 51 (51 in the flora): w, c North America, n Mexico, s South America; arid to mesic regions.

In the following keys and descriptions, three trichome types are defined as follows. Simple grappling-hook trichomes have a cylindric stalk that is unbarbed except at the apex, which terminates with four (or sometimes fewer) retrorse barbs. Complex grappling-hook trichomes have a tapered stalk that is retrorsely barbed along its length and capped usually by four retrorse barbs. Needlelike trichomes have swollen bases resting on a pedestal formed by a ring of epidermal cells, are relatively long and pointed, and bear antrorse barbs or knobs along their length.

Leaf blade intersinus distance is measured across the leaf blade between opposite sinuses; for leaves with entire margins, it is the width of the leaf blade. Anther papillae and seed coat anticlinal wall shape can be observed at 10× magnification; greater magnification (30–40×) is needed to see seed coat cell papillae in species with numerous small papillae.

SELECTED REFERENCE Schenk, J. J. and L. Hufford. 2011. Phylogeny and taxonomy of *Mentzelia* section *Bartonia* (Loasaceae). Syst. Bot. 36: 711–720.

1. Petals white.
   2. Anther epidermis papillate.
      3. Leaf blade widest intersinus distances 9.5–29 mm .... 1. *Mentzelia hualapaiensis* (in part)
      3. Leaf blade widest intersinus distances 0.8–9.1 mm.
         4. 5 outermost stamens without anthers; New Mexico, Texas ....2. *Mentzelia humilis*
         4. 5 outermost stamens with anthers; Arizona .... 3. *Mentzelia canyonensis*
   2. Anther epidermis smooth.
      5. Petals 13–22.7 mm wide; stamens white to yellow .... 4. *Mentzelia decapetala*
      5. Petals 1.9–10.3 mm wide; stamens white.
         6. Petals 22.6–49 × 3.6–10.3 mm; bracts pinnate..........................5. *Mentzelia nuda*
         6. Petals 14.7–22(–24.4) × 1.9–4.4 mm; bracts entire or dentate ....6. *Mentzelia strictissima*
1. Petals light to golden yellow.
   7. Petals hairy abaxially.
      8. Branches perpendicular to stem, especially proximal and mid-stem, or slightly antrorse, especially distal, ± equal; Montrose County, Colorado .... 7. *Mentzelia paradoxensis*
      8. Branches antrorse, proximal longer than distal, all usually extending to near the distal end of plant; Arizona, Colorado (including Montrose County), New Mexico, Utah.
         9. 5 outermost stamens with anthers; margins of proximal leaf blades with teeth 0.5–4 mm; Colorado ....8. *Mentzelia marginata*
         9. 5 outermost stamens usually without anthers; margins of proximal leaf blades with teeth or lobes 1.2–6(–9) mm, always some more than 4 mm; Arizona, Colorado, New Mexico, Utah .... 9. *Mentzelia cronquistii*
   7. Petals glabrous abaxially.
      10. Anther epidermis papillate.
         11. 5 outermost stamens with anthers.
            12. Leaf blades 4–12.2 mm wide, always some more than 10 mm wide, margins usually entire, rarely dentate, teeth 0(–4), 0.1–1.4 mm; seed coat cell papillae 3–5 per cell; Utah .... 10. *Mentzelia argillosa*
            12. Leaf blades 1.5–8.8 mm wide, margins entire, dentate, or pinnate, teeth or lobes 0–8 (always some leaves with 5+ teeth or lobes), 0.1–2 mm (always some more than 1.4 mm); seed coat cell papillae 5–8 per cell; Nevada ....11. *Mentzelia argillicola*
         11. 5 outermost stamens without anthers.
            13. Leaf blade widest intersinus distances 9.5–29 mm...... 1. *Mentzelia hualapaiensis* (in part)
            13. Leaf blade widest intersinus distances 1–4.2 mm.
               14. Second whorl of stamens without anthers .... 12. *Mentzelia perennis*
               14. Second whorl of stamens with anthers.

15. Seed coat anticlinal cell walls straight, papillae 38–45 per cell; Texas . . . . 13. *Mentzelia saxicola* (in part)
15. Seed coat anticlinal cell walls sinuous, papillae 6–12 per cell; New Mexico . . . . 14. *Mentzelia todiltoensis*

[10. Shifted to left margin.—Ed.]

10. Anther epidermis smooth.
    16. Capsules longitudinally ridged . . . . 15. *Mentzelia rusbyi*
    16. Capsules not longitudinally ridged.
        17. Stems multiple, from subterranean branching caudices.
            18. 5 outermost stamens without anthers . . . . 16. *Mentzelia springeri*
            18. 5 outermost stamens with anthers.
                19. Bases of distal leaf blades clasping.
                    20. Leaf blade surfaces densely hairy, whitish, abaxial surfaces with complex grappling-hook and needlelike trichomes; Ash Meadows, Nye County, Nevada . . . . 17. *Mentzelia leucophylla*
                    20. Leaf blade surfaces moderately hairy, green, abaxial surfaces with complex grappling-hook trichomes only; se California and s Nevada (including Nye County).
                        21. Leaf blades 5.8–9.1 mm wide, widest intersinus distances 4.6–7.1 mm; margins of distal leaf blades with 0–6 teeth; seed coat cell papillae 15–46 per cell . . . . 18. *Mentzelia tiehmii*
                        21. Leaf blades 7.6–41.2 mm wide, widest intersinus distances 5.1–35.3 mm; margins of distal leaves with 6–16 teeth; seed coat cell papillae 6–17 per cell . . . . 19. *Mentzelia oreophila*
                19. Bases of distal leaf blades not clasping.
                    22. Margins of proximal leaf blades 3-fid or pinnate to pinnatisect (rarely entire in *M. librina*).
                        23. Filaments of 5 outermost stamens broadly spatulate, strongly clawed.
                            24. Margins of proximal leaf blades pinnate to pinnatisect, lobes 4–12 . . . . 20. *Mentzelia uintahensis*
                            24. Margins of proximal leaf blades usually 3-fid, rarely entire or pinnate, teeth or lobes usually 2, rarely 0 or 4 . . . . 21. *Mentzelia librina*
                        23. Filaments of 5 outermost stamens narrowly spatulate or oblanceolate, slightly clawed.
                            25. Petals 2.9–3.8 mm wide; filaments of 5 outermost stamens 1.3–1.9 mm wide; Utah . . . . 22. *Mentzelia flumensevera*
                            25. Petals 5.1–9 mm wide; filaments of 5 outermost stamens 2–4.5 mm wide; Colorado . . . . 23. *Mentzelia multicaulis*
                    22. Margins of proximal leaf blades entire or dentate to serrate.
                        26. Stems glabrescent, smooth to the touch . . . . 24. *Mentzelia polita*
                        26. Stems hairy.
                            27. Margins of proximal leaves entire.
                                28. Petals 7.4–11 × 2.6–3.4 mm; margins of distal leaf blades entire; filaments of 5 outermost stamens slightly clawed; Arizona . . . . 25. *Mentzelia memorabilis*
                                28. Petals 12.5–17.2 × 6.1–11.2 mm; margins of distal leaf blades entire or dentate; filaments of 5 outermost stamens strongly clawed; Colorado . . . . 26. *Mentzelia rhizomata*
                            27. Margins of proximal leaf blades dentate or serrate, with 4–12 teeth.
                                29. Petals 6.7–9.2 mm wide; seed coat cell papillae 11–15 per cell . . . . 27. *Mentzelia goodrichii*
                                29. Petals 2.3–5.9 mm wide; seed coat cell papillae 3–6 per cell.

30. Stems zigzag; margins of proximal leaf blades with 4–6 teeth, each 0.5–0.6 × 1.2–1.5 mm . . . . . . 28. *Mentzelia shultziorum*
30. Stems straight; margins of proximal leaf blades with 8–12 teeth, each 1–4 × 1.7–6 mm . . . . 29. *Mentzelia puberula*

[17. Shifted to left margin.—Ed.]

17. Stems solitary or, if multiple, from ground-level caudices.
    31. 5 outermost stamens not petaloid, with anthers, filaments linear to narrowly elliptic or narrowly spatulate, 0.5–2(–2.5) mm wide, if narrowly spatulate, to 1.5 mm wide.
        32. Petals 23.5–70 mm; styles 20.4–57 mm; 5 outermost stamens 17.7–55 mm . . . . 30. *Mentzelia laevicaulis*
        32. Petals (6.1–)7.3–15.8(–18) mm; styles 3.7–13 mm; 5 outermost stamens 4–13.2 mm.
            33. Petals 11.7–15.8(–18) mm; 5 outermost stamens 10.2–13.2 mm . . . . 31. *Mentzelia inyoensis*
            33. Petals (6.1–)7.3–10 mm; 5 outermost stamens 4–8 mm . . . . 32. *Mentzelia candelariae*
    31. 5 outermost stamens petaloid, with or without anthers, filaments narrowly spatulate to elliptic or oblanceolate, 1–9.7 mm wide, if narrowly spatulate and less than 1.5 mm wide, then without anthers.
        34. Margins of distal leaf blades pinnatisect (sometimes pinnate in *M. laciniata*).
            35. Petals 30–42.2 mm; 5 outermost stamens 27–37.4 mm; styles 24–32.4 mm . . . . 33. *Mentzelia conspicua*
            35. Petals 8.3–23.8(–26) mm; 5 outermost stamens 6.5–20.4 mm; styles 6.1–12.5 (–14) mm.
                36. Petals 8.3–13 mm; 5 outermost stamens 6.5–10.7 mm; Colorado, Nevada, Utah. . . . . 34. *Mentzelia lagarosa*
                36. Petals 13–23.8(–26) mm; 5 outermost stamens 10.2–20.4 mm; Arizona, Colorado, New Mexico.
                    37. Lobes of distal leaf blades strongly antrorse; branches upcurved; Arizona, New Mexico . . . . 35. *Mentzelia holmgreniorum*
                    37. Lobes of distal leaf blades slightly antrorse or perpendicular to leaf axis; branches straight; Arizona, Colorado, New Mexico.
                        38. Seed coat cell papillae 42–48 per cell; Chuska Mountains of Arizona and New Mexico. . . . . 36. *Mentzelia filifolia*
                        38. Seed coat cell papillae 5–14 per cell; Colorado, New Mexico (not Chuska Mountains) . . . . 37. *Mentzelia laciniata*
        34. Margins of distal leaf blades entire, dentate, serrate, or pinnate.
            39. Bases of distal leaf blades clasping (rarely a few not clasping in *M. pterosperma*).
                40. Capsules cup-shaped; Arizona, California, Colorado, Nevada, Utah . . . . 38. *Mentzelia pterosperma*
                40. Capsules cylindric; Kansas, Missouri, Oklahoma, Texas . . . . 39. *Mentzelia albescens*
            39. Bases of distal leaf blades not clasping (sometimes a few clasping in *M. mexicana*).
                41. Filaments of 5 outermost stamens strongly clawed; on black (to sometimes reddish) volcanic cinder cone soils in San Francisco Volcanic Field, Coconino County, Arizona . . . . 40. *Mentzelia collomiae*
                41. Filaments of 5 outermost stamens slightly clawed; not on black (to sometimes reddish) volcanic cinder cone soils in San Francisco Volcanic Field, Coconino County, Arizona.
                    42. Seed coat cell anticlinal walls straight.
                        43. Capsule length more than 2 times diam. (occasionally a few slightly less in *M. reverchonii*).
                            44. Styles 8.7–11.4(–13) mm; Colorado, New Mexico, Oklahoma, Texas . . . . 41. *Mentzelia reverchonii*
                            44. Styles 5.6–7.8 mm; Montana, Utah, Wyoming . . . . 42. *Mentzelia pumila*
                        43. Capsule length to 2 times diam. (sometimes a few more in *M. longiloba*).

45. Capsule bases tapering; seed coat cell papillae 4–6 per cell . . . . 43. *Mentzelia longiloba* (in part)
45. Capsule bases rounded; seed coat cell papillae 8–45 per cell.
46. Seed coat cell papillae 38–45 per cell; leaf blade widest intersinus distances 1.6–3.4 mm . . . . 13. *Mentzelia saxicola* (in part)
46. Seed coat cell papillae 8–12 per cell; leaf blade widest intersinus distances 3.4–19 mm . . . . 44. *Mentzelia mexicana*

[42. Shifted to left margin.—Ed.]

42. Seed coat cell anticlinal walls wavy to sinuous.
47. Margins of proximal leaves with 4–12 teeth or lobes.
48. Plants candelabra-form; proximal branches not decumbent; largest leaf trichomes with pearly white bases; leaf blade margin teeth with proximal side antrorse, distal side perpendicular to leaf axis; Arizona, Nevada, Utah . . . . 45. *Mentzelia integra*
48. Plants bushlike; proximal branches decumbent; largest leaf trichomes without pearly white bases; leaf blade margin teeth or lobes perpendicular to leaf axis; c Colorado . . . . 46. *Mentzelia densa*
47. Margins of proximal leaves with 14–38 teeth or lobes (10–38 in *M. speciosa*).
49. Seed coat cell anticlinal walls wavy; petals golden yellow; leaf blades 43.6–233 mm, always some 146+ mm . . . . 47. *Mentzelia speciosa*
49. Seed coat cell anticlinal walls sinuous; petals light to golden yellow; leaf blades 32.8–125(–146) mm.
50. 5 outermost stamens with anthers.
51. Leaf blade widest intersinus distances 2.3–7.1 mm, always some leaves 4+ mm; branches upcurved . . . . 43. *Mentzelia longiloba* (in part)
51. Leaf blade widest intersinus distances 1–2.9 mm; branches straight. . . 48. *Mentzelia sivinskii*
50. 5 outermost stamens without anthers.
52. Leaf blade widest intersinus distances 1–3.9 mm; petals light yellow . . .49. *Mentzelia procera*
52. Leaf blade widest intersinus distances 2.1–13.7(–14) mm, always some leaves 4+ mm; petals light to golden yellow.
53. Seed coat cell papillae 10–21 or 67–106 per cell; Arizona, California, Colorado, New Mexico, Utah . . . . 43. *Mentzelia longiloba* (in part)
53. Seed coat cell papillae 29–48 per cell; Colorado, Nebraska, New Mexico, Wyoming.
54. Branches distal or along entire stem, straight . . . . 50. *Mentzelia chrysantha*
54. Branches distal, upcurved . . . . .51. *Mentzelia multiflora*

1. Mentzelia hualapaiensis J. J. Schenk, W. C. Hodgson & L. Hufford, Brittonia 62: 1, figs. 1, 2A,C–E. 2010 • Hualapai blazingstar [E]

Herbs perennial, bushlike, with subterranean caudices. **Stems** multiple, erect, zigzag or straight; branches distal or along entire stem, distal longest or all ± equal, antrorse, upcurved; hairy. **Leaves:** blade 17–94 × 12–32 mm, widest intersinus distance 9.5–29 mm; proximal broadly spatulate to elliptic, margins serrate, teeth 6–12, slightly antrorse or perpendicular to leaf axis, 0.4–1.5 mm; distal broadly spatulate, elliptic, or obovate, base not clasping, margins serrate, teeth 8–10, slightly antrorse or perpendicular to leaf axis, 1–3.6 mm; abaxial surface with complex grappling-hook and occasionally with simple grappling-hook and needlelike trichomes, adaxial surface with needlelike trichomes. **Bracts:** margins entire. **Flowers:** petals white to light yellow, 9.4–17.4(–20) × 1.7–4.3 mm, apex acute to rounded, glabrous abaxially; stamens white to light yellow, 5 outermost petaloid, filaments narrowly spatulate, slightly clawed, 8.5–18 × 1–1.3(–2.9) mm, without anthers, second whorl with anthers; anthers straight after dehiscence, epidermis papillate; styles 5.8–9 mm. **Capsules** cup-shaped, 4.9–10.3 × 4.6–8.6 mm, base rounded, not longitudinally ridged. **Seeds:** coat anticlinal cell walls straight to slightly wavy, papillae 8–14 per cell.

Flowering Mar–Jun(–Sep). Rocky desert scrub, gravelly canyon bottoms, steep travertine, shaley slopes associated with Muav limestone formations; 400–1400 m; Ariz.

Mentzelia hualapaiensis is distributed in Coconino and Mohave counties along the Grand Canyon and its side canyons between the Upper and Lower Granite Gorge.

2. Mentzelia humilis (Urban & Gilg) J. Darlington, Ann. Missouri Bot. Gard. 21: 155. 1934 [E]

Mentzelia pumila Torrey & A. Gray var. *humilis* Urban & Gilg, Nova Acta Acad. Caes. Leop.-Carol. German. Nat. Cur. 76: 93. 1900; *Nuttallia humilis* (Urban & Gilg) Rydberg; *Touterea humilis* (Urban & Gilg) Rydberg

Herbs perennial, bushlike, with ground-level caudices. **Stems** solitary or multiple, erect, straight; branches distal or along entire stem, distal longest or all ± equal, antrorse; straight to upcurved; hairy. **Leaves:** blade 25–95 × 5.5–28 (–36.2) mm, widest intersinus distance 0.8–9.1 mm; proximal spatulate to oblanceolate or elliptic, margins pinnate to pinnatisect, lobes 4–16(–20), slightly antrorse or perpendicular to leaf axis, 2.3–11.8(–16.4) mm; distal elliptic, lanceolate, spatulate, or linear, base not clasping, margins entire or dentate to pinnatisect, teeth or lobes (0–)4–16, slightly antrorse or perpendicular to leaf axis, 2.3–13.8(–16.9) mm; abaxial surface with simple grappling-hook, needlelike, and sometimes complex grappling-hook trichomes, adaxial surface with simple grappling-hook and needlelike trichomes. **Bracts:** margins usually entire, rarely pinnate. **Flowers:** petals white, 10.3–13(–28.6) × 1.4–4 mm, apex acute, glabrous abaxially; stamens white, 5 outermost petaloid, filaments linear to narrowly spatulate, slightly clawed, 8.8–19(–22.3) × 0.7–3.3 mm, without anthers, second whorl without anthers; anthers straight after dehiscence, epidermis papillate; styles 6.5–11.3 mm. **Capsules** cup-shaped, 5.3–10.2 × (4.2–)5.2–8.6 mm, base rounded, not longitudinally ridged. **Seeds:** coat anticlinal cell walls wavy, papillae 6–12 per cell.

Varieties 2 (2 in the flora): sc United States.

The basionym of *Mentzelia humilis* is often cited as *M. multiflora* var. *humilis* A. Gray (1852). However, in that publication Gray indicated accepted names in Roman capitals (see ipni.org), whereas the varietal name is in lower case, indicating that he was using it as a descriptive term rather than a scientific name. Valid publication of the basionym thus must be attributed to Urban and Gilg.

1. Leaf blade widest intersinus distances 0.8–2.6 (–3.3) mm; lobes of proximal leaves 1–2.9 mm wide .... 2a. *Mentzelia humilis* var. *humilis*
1. Leaf blade widest intersinus distances 1.6–9.1 mm (always on some leaves greater than 2.6 mm); lobes of proximal leaves (1.1–)2.9–5.1 mm wide .... 2b. *Mentzelia humilis* var. *guadalupensis*

2a. Mentzelia humilis (Urban & Gilg) J. Darlington var. **humilis** • Gypsum blazingstar [E]

Nuttallia gypsea Wooton & Standley

Leaf blades: widest intersinus distance 0.8–2.6(–3.3) mm; proximal oblanceolate to elliptic, margins pinnatisect, lobes 4–16(–20), 2.3–11.8 (–16.4) × 1–2.9 mm; distal elliptic, lanceolate, or linear, margins entire, dentate, or pinnatisect, teeth or lobes (0–)4–16, 2.3–13.8(–16.9) × 0.9–1.8 mm. **Flowers:** petals 11.8–21.3(–24.6) × 1.4–3.7 mm; 5 outermost stamens linear to narrowly spatulate, 10.6–19(–22.3) × 0.7–3.3 mm. **Seed coat:** papillae 7–12 per cell. $2n = 20$.

Flowering Apr–Oct. Sparsely vegetated areas in dry grasslands, roadsides, level areas or gentle slopes, gravelly, clayey, and sandy gypsum substrates; 900–1600 m; N.Mex., Tex.

Variety *humilis* occurs in western Texas and south-central New Mexico. Its range approaches var. *guadalupensis* in Eddy County, New Mexico, but the two varieties are not known to co-occur.

2b. Mentzelia humilis (Urban & Gilg) J. Darlington var. **guadalupensis** Spellenberg, Sida 18: 995, fig. 2. 1999 • Guadalupe Mountain blazingstar [E]

Leaf blades: widest intersinus distance 1.6–9.1 mm (always on some leaves greater than 2.6 mm); proximal spatulate to elliptic, margins pinnate, lobes 4–14, 2.5–6(–8.4) × (1.1–) 2.9–5.1 mm; distal elliptic, lanceolate, or spatulate, margins pinnate, lobes 4–10, 2.8–6 × 1.1–4 mm. **Flowers:** petals 10.3–13(–28.6) × 1.6–2.3(–4) mm, 5 outermost stamens narrowly spatulate, 8.8–11 (–23.3) × 1–1.4(–1.9) mm. **Seed coat:** papillae 6–9 per cell.

Flowering Jul–Sep. Sparsely vegetated knolls, gypsum clay substrates; 1400–1600 m; N.Mex.

Variety *guadalupensis* is distributed along the west side of the Guadalupe Mountains in Eddy and Otero counties.

3. Mentzelia canyonensis J. J. Schenk, W. C. Hodgson & L. Hufford, Brittonia 65: 410, figs. 1,2. 2013 • Grand Canyon blazingstar [E]

Herbs perennial, bushlike, with subterranean caudices or rhizomes. **Stems** multiple, erect or decumbent, zigzag; branches along entire stem, all ± equal, antrorse, upcurved; hairy. **Leaves:** blade 16–31 × 3.9–9.2 mm, widest intersinus distance 3–8.3 mm; proximal oblanceolate to elliptic, margins entire or dentate, teeth 0–4(–6), perpendicular to leaf axis, 0.2–1.3 mm; distal oblanceolate to elliptic, base not clasping, margins entire or dentate, teeth 0–4, perpendicular to leaf axis, 0.4–2 mm; abaxial surface with simple grappling-hook (occasionally absent) and complex grappling-hook trichomes, adaxial surface with complex grappling-hook and needlelike trichomes. **Bracts:** margins entire. **Flowers:** petals white, 9–15 × 1.8–2.4 mm, apex rounded to acute, glabrous abaxially; stamens white, 5 outermost petaloid, filaments narrowly spatulate, slightly clawed, 7.1–13 × 0.5–1.4 mm, with anthers, second whorl with anthers; anthers twisted or straight after dehiscence, epidermis papillate; styles 4.8–7.3 mm. **Capsules** cup-shaped, 4.5–8(–13) × 4.5–7 mm, base rounded, not longitudinally ridged. **Seeds:** coat anticlinal cell walls straight, papillae usually 6–9 per cell.

Flowering May–Nov. Loose, rocky soils primarily on steep, degraded slopes, less frequently on stream bottoms, Muav limestones and Bright Angel shales; 800–1600 m; Ariz.

Mentzelia canyonensis is known only from the Grand Canyon and Little Colorado River canyon in Coconino County.

4. Mentzelia decapetala (Pursh) Urban, Ber. Deutsch. Bot. Ges. 10: 263. 1892 • Gumbo-lily, Ten-petalled western star [E] [W]

Bartonia decapetala Pursh, Bot. Mag. 36: plate 1487. 1812

Herbs biennial or perennial, bushlike, perennials with ground-level caudices. **Stems** solitary, erect, straight; branches distal or along entire stem, proximal or distal longest, antrorse, straight to upcurved; hairy. **Leaves:** blade 72–295 × 14–45 mm, widest intersinus distance 10.1–23.3 mm; proximal oblanceolate or elliptic, margins serrate to pinnate, teeth or lobes 16–26, slightly antrorse, 1–16.5 mm; distal elliptic to lanceolate, base clasping or not, margins serrate to pinnate, teeth or lobes 9–20, slightly antrorse, 5.3–13.7 mm; surfaces with needlelike trichomes. **Bracts:** margins pinnate. **Flowers:** petals white, 47–75 × 13–22.7 mm, apex acute to attenuate, glabrous abaxially; stamens white to yellow, 5 outermost petaloid, filaments spatulate, strongly clawed, 48–75 × 12–23 mm, without anthers, second whorl with anthers; anthers straight after dehiscence, epidermis smooth; styles 36–53 mm. **Capsules** cylindric, 30–43 × 12–17 mm, base tapering, not longitudinally ridged. **Seeds** 2.3–4 mm; coat anticlinal cell walls straight, papillae 4–10 per cell. $2n = 22$.

Flowering Jun–Aug(–Oct). Rock outcrops, slopes, dry short-grass prairies, riverbanks, roadsides, loam, limestone, sandy, silty, clayey, and gravelly soils; 300–2400 m; Alta., Man., Sask.; Colo., Ill., Iowa, Kans., Minn., Mont., Nebr., N.Mex., Okla., S.Dak., Tex., Utah, Wyo.

Mentzelia decapetala is introduced in Grundy County, Illinois. It appears to be native throughout the rest of its distribution.

5. Mentzelia nuda (Pursh) Torrey & A. Gray, Fl. N. Amer. 1: 535. 1840 • Goodmother, naked blazingstar, or naked western star [E] [F] [W]

Bartonia nuda Pursh, Fl. Amer. Sept. 1: 328; 2: 749. 1813; *Mentzelia stricta* (Osterhout) G. W. Stevens

Herbs winter annual, biennial, or perennial, candelabra-form, perennials with ground-level caudices. **Stems** solitary (or multiple as wound response), erect, straight; branches distal, distal longest, antrorse, straight; hairy. **Leaves:** blade 37–120 × 5–24.4 mm, widest intersinus distance 4.8–19.6 mm; proximal oblanceolate to elliptic, margins serrate, teeth 14–30, slightly antrorse, 0.6–4.9 mm; distal elliptic to lanceolate, base not clasping, margins serrate, teeth 12–30, slightly antrorse, 0.7–4.7 mm; abaxial surface with simple grappling-hook, complex grappling-hook, and needlelike trichomes, adaxial surface with simple grappling-hook and needlelike trichomes. **Bracts:** margins pinnate. **Flowers:** petals white, 22.6–49 × 3.6–10.3 mm, apex acute, glabrous abaxially, stamens white, 5 outermost petaloid, filaments narrowly spatulate, slightly clawed, 20–47 × 3–10.2 mm, without anthers, second whorl without anthers; anthers straight after dehiscence, epidermis smooth; styles 11.5–18.5 mm. **Capsules** cylindric, 14.5–29 × 6.9–12.3 mm, base tapering or rounded, not longitudinally ridged. **Seeds:** coat anticlinal cell walls wavy, papillae 4–8 per cell. $2n = 20$.

Flowering Jun–Nov. Disturbed roadsides, hillsides, stream banks, sandy and rocky soils; 100–2300 m; Colo., Kans., Minn., Mo., Nebr., N.Mex., Okla., S.Dak., Tex., Wyo.

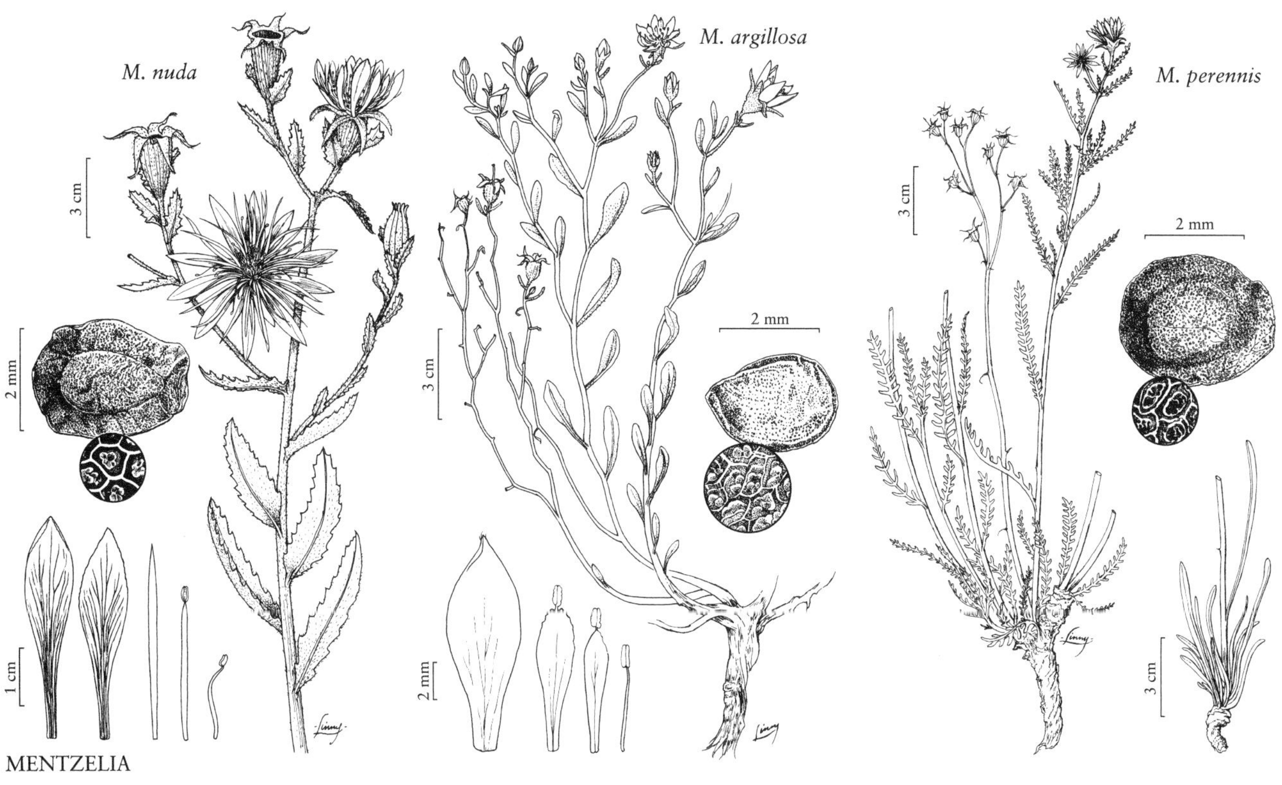

MENTZELIA

Mentzelia nuda is morphologically similar to, and phylogenetically near, *M. strictissima* (J. J. Schenk 2009), but it can be distinguished from the latter by its much larger flowers and capsules. The two species have adjacent ranges in New Mexico and Texas but do not overlap despite apparently similar habitat requirements. *Mentzelia strictissima* occurs west and south of *M. nuda* and abuts against the southernmost portion of the Rocky Mountains. In the northern portion of its range, *M. nuda* too approaches the foothills of the southern Rocky Mountains and extends eastward into the plains. *Mentzelia nuda* was collected once in northeastern Illinois in 1901 but apparently did not become established there. Reports of *M. nuda* from Arizona are based on misidentified material of *M. laevicaulis* and *M. rusbyi*.

6. **Mentzelia strictissima** (Wooton & Standley) J. Darlington, Ann. Missouri Bot. Gard. 21: 163. 1934 • Grassland blazingstar [E]

Nuttallia strictissima Wooton & Standley, Contr. U.S. Natl. Herb. 16: 150. 1913

Herbs biennial, candelabraform. **Stems** solitary (or multiple as wound response), erect, straight; branches distal, distal longest, antrorse, straight; hairy. **Leaves:** blade 28–70 × 5.8–17.2 mm, widest intersinus distance 3.4–11.4 mm; proximal oblanceolate, lanceolate, or elliptic, margins dentate to pinnate, teeth or lobes 10–20, perpendicular to leaf axis, 0.9–3.1 mm; distal lanceolate, base usually not clasping, rarely clasping, margins usually dentate, occasionally serrate, teeth 8–18, usually perpendicular to leaf axis, occasionally slightly antrorse, 0.6–5.4 mm; abaxial surface with simple grappling-hook, complex grappling-hook, and needlelike trichomes, adaxial surface with simple grappling-hook and needlelike trichomes. **Bracts:** margins entire or toothed. **Flowers:** petals white, 14.7–22(–24.4) × 1.9–4.4 mm, apex acute, glabrous abaxially; stamens white, 5 outermost petaloid, filaments narrowly spatulate, slightly clawed, 13.3–23.1 × 1.3–3.1(–4.4) mm, without anthers, second whorl without anthers; anthers straight after dehiscence, epidermis smooth; styles 7.6–12.9 mm. **Capsules** cup-shaped to cylindric, 10.2–20.1 × 6.7–9.6 mm, base rounded to occasionally tapering, not longitudinally ridged. **Seeds:** coat anticlinal cell walls wavy, papillae 3–8 per cell. **2*n*** = 20.

Flowering May–Oct. Arid grasslands; 800–1800 m; N.Mex., Tex.

See discussion under 5. *Mentzelia nuda*.

7. **Mentzelia paradoxensis** J. J. Schenk & L. Hufford, Madroño 57: 249, fig. 2A. 2010 • Paradox Valley blazingstar [E]

Herbs biennial, usually cylindric, rarely candelabra-form. **Stems** solitary, erect, straight; branches distal or along entire stem, ± equal, perpendicular to stem, especially proximal and mid-stem, or slightly antrorse, especially distal, straight; hairy. **Leaves:** blade 38.1–95 × 6–17 (–26) mm, widest intersinus distance 1.6–4.7 mm; proximal oblanceolate, lanceolate, or elliptic, margins dentate to serrate or pinnate, teeth or lobes 10–22, slightly antrorse or perpendicular to leaf axis, 2–6.8 mm; distal elliptic to lanceolate, base not clasping, margins dentate to serrate or pinnate, teeth or lobes 8–16, slightly antrorse or perpendicular to leaf axis, 2.4–8.5 mm; abaxial surface with simple grappling-hook, complex grappling-hook, and needlelike trichomes, adaxial surface with needlelike and occasionally simple grappling-hook trichomes. **Bracts:** margins entire. **Flowers:** petals golden yellow, 8.3–14.5 (–17.2) × 1.7–5.3 mm, apex acute to rounded, hairy abaxially; stamens golden yellow, 5 outermost petaloid, filaments narrowly spatulate, slightly clawed, 4.9–12.5 × 1.2–3.8 mm, with or without anthers, second whorl with anthers; anthers straight or occasionally twisted after dehiscence, epidermis papillate or not; styles 5.4–10.4 mm. **Capsules** cup-shaped, 5–9 × 3.7–6.5 mm, base rounded, not longitudinally ridged. **Seeds:** coat anticlinal cell walls straight to slightly wavy, papillae 6–11 per cell. **2*n*** = 20.

Flowering Jun–Sep. Sparsely vegetated gypsum knolls; 1500–2000 m; Colo.

Mentzelia paradoxensis is known only from white gypsum hills in Paradox Valley, Montrose County.

8. **Mentzelia marginata** (Osterhout) H. J. Thompson & Prigge, Great Basin Naturalist, 46: 549. 1986 • Colorado blazingstar [E]

Nuttallia marginata Osterhout, Bull. Torrey Bot. Club 49: 183. 1922

Herbs winter annual or biennial, candelabra-form. **Stems** solitary, erect, straight; branches mostly distal, proximal longer than distal, all usually extending to near the distal end of plant, antrorse, straight; hairy. **Leaves:** blade 22–115 × 2.8–10.9 (–17.2) mm, widest intersinus distance 2.2–7(–9) mm; proximal oblanceolate, elliptic, or lanceolate, margins dentate, teeth 10–24(–38), perpendicular to leaf axis, 0.5–4 mm; distal elliptic to lanceolate, base not clasping, margins dentate, teeth 8–20, perpendicular to leaf axis, 0.3–5.6 mm; abaxial surface with simple grappling-hook, complex grappling-hook, and usually with needlelike trichomes, adaxial surface with simple grappling-hook and needlelike trichomes. **Bracts:** margins entire. **Flowers:** petals golden yellow, 8–14.4 × 2.1–3.9 mm, apex acute, hairy abaxially; stamens golden yellow, 5 outermost petaloid, filaments narrowly spatulate, slightly clawed, 5.5–11.1 × 1.2–2.9 mm, with anthers, second whorl with anthers; anthers straight after dehiscence, epidermis papillate; styles 5.3–10 mm. **Capsules** cylindric, 7–14.6 × 3.5–6.7 mm, base tapering, not longitudinally ridged. **Seeds:** coat anticlinal cell walls straight to slightly wavy, papillae 4–10 per cell. **2*n*** = 20.

Flowering May–Aug. Steep roadside banks, steep cliffs; 1500–2000 m; Colo.

Mentzelia marginata occurs on the western edge of Colorado in Delta, Garfield, Mesa, and Montrose counties. Reports of this species from Utah are based on specimens treated here as *M. cronquistii*.

9. **Mentzelia cronquistii** H. J. Thompson & Prigge, Great Basin Naturalist 46: 550, figs. 2, 3A,B, 4A. 1986 • Cronquist's blazingstar [E]

Mentzelia marginata (Osterhout) H. J. Thompson & Prigge var. *cronquistii* (H. J. Thompson & Prigge) N. H. Holmgren & P. K. Holmgren; *Nuttallia cronquistii* (H. J. Thompson & Prigge) W. A. Weber

Herbs biennial, candelabra-form. **Stems** solitary, erect, straight; branches distal or along entire stem, proximal longer than distal, all usually extending to near the distal end of plant, antrorse, straight; hairy. **Leaves:** blade 21–101 × 4–10.7(–19.3) mm, widest intersinus distance

1.8–7.3(–12.6) mm; proximal oblanceolate, lanceolate, or elliptic, margins dentate to pinnate, teeth or lobes 8–28, perpendicular to leaf axis, 1.2–6(–9) mm, always some more than 4 mm; distal oblanceolate, elliptic, or lanceolate, base not clasping, margins dentate to pinnate, teeth or lobes 6–18, perpendicular to leaf axis, 1.1–5 (–6.9) mm; abaxial surface with simple grappling-hook, complex grappling-hook, and needlelike trichomes, adaxial surface with simple grappling-hook, needlelike, and occasionally complex grappling-hook trichomes. **Bracts:** margins entire or toothed to pinnate. **Flowers:** petals golden yellow, 9–16.6 × 2.5–5.1 mm, apex usually rounded, rarely acute, hairy abaxially; stamens golden yellow, 5 outermost petaloid, filaments narrowly spatulate, slightly clawed, 7.3–13.2 × 1.8–4.3 mm, usually without, rarely with anthers, second whorl with anthers; anthers straight after dehiscence, epidermis papillate or not; styles 6.5–10 mm. **Capsules** cup-shaped, 5.9–10.6(–11.4) × 5–7.6 mm, base tapering to rounded, not longitudinally ridged. **Seeds:** coat anticlinal cell walls wavy, papillae 8–13 per cell. $2n = 20$.

Flowering May–Nov. Sandy and rocky soils, washes, roadside banks, steep slopes; 800–2300 m; Ariz., Colo., N.Mex., Utah.

10. Mentzelia argillosa J. Darlington, Ann. Missouri Bot. Gard. 21: 153. 1934 • Arapien blazingstar [C] [E] [F]

Nuttallia argillosa (J. Darlington) W. A. Weber

Herbs perennial, bushlike, with subterranean caudices. **Stems** multiple, erect, zigzag or straight; branches distal or along entire stem, distal longest or all ± equal, antrorse, straight; hairy. **Leaves:** blade 15.4–40 (–86) × 4–12.2 mm, always some more than 10 mm wide, widest intersinus distance 4–12.2 mm; proximal oblanceolate to spatulate, margins usually entire, rarely dentate, teeth 0(–4), perpendicular to leaf axis, 0.1–1 mm; distal oblanceolate, spatulate, or elliptic, base not clasping, margins usually entire, rarely dentate, teeth 0(–4), perpendicular to leaf axis, 0.3–1.4 mm; abaxial surface with simple grappling-hook, complex grappling-hook, and needlelike trichomes, adaxial surface with simple grappling-hook and needlelike trichomes. **Bracts:** margins entire. **Flowers:** petals golden yellow, 8.2–12.2(–15.2) × 2.4–5 mm, apex acute, glabrous abaxially; stamens golden yellow, 5 outermost petaloid, filaments narrowly spatulate, slightly clawed, 5.6–11 (–13.5) × 1.4–3 mm, with anthers, second whorl with anthers; anthers usually twisted after dehiscence, epidermis papillate; styles 5.5–9.4 mm. **Capsules** cup-shaped, 5–8.4 × 3.1–6 mm, base tapering to rounded, not longitudinally ridged. **Seeds:** coat anticlinal cell walls straight, papillae 3–5 per cell. $2n = 22$.

Flowering Jun–Aug. Sparsely vegetated steep cliffs or slopes composed of gypsum-rich clayey and gravelly soils; of conservation concern; 1600–1900 m; Utah.

Mentzelia argillosa is endemic to the Arapien Shale formation in Sevier and Sanpete counties.

11. Mentzelia argillicola N. H. Holmgren & P. K. Holmgren, Syst. Bot. 27: 751, fig. 3. 2002 • Pioche blazingstar [C] [E]

Herbs perennial, bushlike, with subterranean caudices. **Stems** multiple, erect or decumbent, straight; branches along entire stem, distal longest, antrorse, straight; hairy. **Leaves:** blade (13.5–)16.4–33.3 × 1.5–8.8 mm, widest intersinus distance 1.5–6.1 mm; proximal oblanceolate to spatulate, margins entire or pinnate, lobes 0–8 (always some leaves with 5+ lobes), perpendicular to leaf axis, 0.5–2 mm, always some more than 1.4 mm; distal oblanceolate, elliptic, lanceolate, or linear, base not clasping, margins entire, dentate, or pinnate, teeth or lobes 0–6, perpendicular to leaf axis, 0.1–2 mm; abaxial surface with complex grappling-hook, needlelike, and occasionally simple grappling-hook trichomes, adaxial surface with needlelike and usually simple grappling-hook trichomes. **Bracts:** margins entire. **Flowers:** petals golden yellow, 7.2–11.9 × 2.7–4.9 mm, apex acute to rounded, glabrous abaxially; stamens golden yellow, 5 outermost petaloid, filaments narrowly spatulate, slightly clawed, 5.5–10.3 × 1.4–2.7 mm, with anthers, second whorl with anthers; anthers usually twisted after dehiscence, epidermis papillate; styles 5–9 mm. **Capsules** cup-shaped, 4.9–7.6 × 3.4–5.5 mm, base rounded to tapering, not longitudinally ridged. **Seeds:** coat anticlinal cell walls straight, papillae 5–8 per cell. $2n = 22$.

Flowering Jun–Aug. Sparsely vegetated cliffs and knolls, gypsum-rich clayey soils; of conservation concern; 1400–1900 m; Nev.

Mentzelia argillicola is known only from Lincoln County.

12. Mentzelia perennis Wooton, Bull. Torrey Bot. Club 25: 260. 1898 • Perennial blazingstar [E] [F]

Herbs perennial, bushlike, with ground-level caudices. **Stems** multiple, erect, straight; branches distal, distal longest, antrorse, straight; hairy. **Leaves:** blade 25–100 × 1–18.1 mm, widest intersinus distance 1–4.2 mm; proximal spatulate to elliptic, margins entire or dentate to pinnatisect, teeth or lobes 0–26, slightly antrorse, 0.6–9.5 mm; distal oblanceolate, elliptic, or lanceolate, base not clasping, margins entire, serrate, or pinnatisect, teeth or lobes 0–24, slightly antrorse, 0.4–7.1 mm; abaxial surface with needlelike and occasionally simple and complex grappling-hook trichomes, adaxial surface with simple grappling-hook and needlelike trichomes. **Bracts:** margins usually entire, occasionally pinnate. **Flowers:** petals light yellow, 11.4–19(–22.7) × 1.8–3.6 (–5) mm, apex acute, glabrous abaxially; stamens light yellow, 5 outermost petaloid, filaments narrowly spatulate, slightly clawed, 11–18.1(–22.7) × (0.5–) 1–3.7 mm, without anthers, second whorl without anthers; anthers straight after dehiscence, epidermis papillate; styles 7.5–11.4 mm. **Capsules** cup-shaped, 6.8–12 × (4–)4.8–9.1 mm, base tapering or rounded, not longitudinally ridged. **Seeds:** coat anticlinal cell walls wavy, papillae 6–14 per cell. **2*n*** = 18.

Flowering Jul–Aug. Roadsides, hillside slopes, gypsum-rich soils; 1200–2200 m; N.Mex.

Mentzelia perennis is known from central New Mexico.

13. Mentzelia saxicola H. J. Thompson & Zavortink, Wrightia 4: 22. 1968 • El Paso blazingstar

Herbs usually biennial, rarely perennial, bushlike or candelabra-form, perennials with ground-level caudices. **Stems** solitary or multiple, erect, straight; branches distal or along entire stem, distal longest or all ± equal, antrorse, upcurved; hairy. **Leaves:** blade 21–44.4 × 5.9–11.2 mm, widest intersinus distance 1.6–3.4 mm; proximal oblanceolate to elliptic, margins pinnate, lobes 6–14, slightly antrorse, 2.4–3.9 mm; distal oblanceolate, elliptic, or lanceolate, base not clasping, margins pinnate, lobes 4–12, slightly antrorse, 1.8–3.9 mm; abaxial surface with simple grappling-hook and complex grappling-hook trichomes, adaxial surface with simple grappling-hook and needlelike trichomes. **Bracts:** margins entire. **Flowers:** petals light to golden yellow, 11.7–16.6 × 3.8–4.8 mm, apex rounded, glabrous abaxially; stamens light to golden yellow, 5 outermost petaloid, filaments narrowly spatulate, slightly clawed, 10.4–14.6 × 2–3.8 mm, usually without, rarely with, anthers, second whorl with anthers; anthers twisted or straight after dehiscence, epidermis papillate or not; styles 6.8–9 mm. **Capsules** cup-shaped, 7–11.9 × 5.7–7.3 mm, length to 2 times diam., base rounded, not longitudinally ridged. **Seeds:** coat anticlinal cell walls straight, papillae 38–45 per cell. **2*n*** = 20.

Flowering Mar–Oct. Dry roadsides, slopes; 500–1500 m; Tex.; Mexico (Chihuahua, Coahuila, Zacatecas).

Mentzelia saxicola occurs in the flora area in El Paso, Hudspeth, and Presidio counties.

14. Mentzelia todiltoensis N. D. Atwood & S. L. Welsh, W. N. Amer. Naturalist 65: 365, fig. 1. 2005 • Jemez Mountains blazingstar C E

Herbs biennial or perennial, bushlike, with ground-level caudices. **Stems** solitary or multiple, erect, straight; branches distal, distal longest, antrorse, straight; hairy. **Leaves:** blade 41–121 × 1.5–40 mm, widest intersinus distance 1.1–3.3 mm; proximal oblanceolate, elliptic, or linear, margins entire or serrate to pinnatisect, teeth or lobes 0–26, slightly antrorse, (0.9–) 2.4–18.6 mm; distal oblanceolate, elliptic, lanceolate, or linear, base not clasping, margins entire, serrate, or pinnatisect, teeth or lobes 0–22, slightly antrorse, 0.8–18 mm; abaxial surface with needlelike and occasionally simple grappling-hook and/or complex grappling-hook trichomes, adaxial surface with needlelike trichomes. **Bracts:** margins entire or pinnate. **Flowers:** petals light to golden yellow, (10.4–)11.7–24.6 × 1.8–5.1 mm, apex acute to rounded, glabrous abaxially; stamens light to golden yellow, 5 outermost petaloid, filaments narrowly spatulate, slightly clawed, 10–21 × 1.4–4 mm, without anthers, second whorl with anthers; anthers twisted or occasionally straight after dehiscence, epidermis papillate; styles 5.5–12.7 mm. **Capsules** cup-shaped to cylindric, 6.7–20.2 × 4.5–8.5 mm, base tapering or rounded, not longitudinally ridged. **Seeds:** coat anticlinal cell walls sinuous, papillae 6–12 per cell. **2*n*** = 20.

Flowering Jul–Oct. Hillside slopes, hilltops, hard gypsum-rich clayey soils; of conservation concern; 1600–2200 m; N.Mex.

Mentzelia todiltoensis occurs in north-central New Mexico.

15. Mentzelia rusbyi Wooton, Bull. Torrey Bot. Club 25: 261. 1898 • Rusby's blazingstar E

Nuttallia rusbyi (Wooton) Rydberg

Herbs biennial or perennial, candelabra-form, perennials with ground-level caudices. **Stems** solitary, erect, straight; branches distal, distal longest, antrorse, straight; hairy. **Leaves:** blade 41–123 × 6.4–26.4 mm, widest intersinus distance 4.8–18 mm; proximal spatulate, elliptic, or lanceolate, margins dentate, teeth 10–24, perpendicular to leaf axis, 0.6–3.6 mm; distal lanceolate, base not clasping, margins dentate, teeth 10–22, perpendicular to leaf axis, 1.2–6 mm; abaxial surface with simple grappling-hook, complex grappling-hook, and needlelike trichomes, adaxial surface with simple grappling-hook and needlelike trichomes. **Bracts:** margins pinnate. **Flowers:** petals light yellow, 11.8–23.8 × 3–6.9 mm, apex acute, glabrous abaxially; stamens light yellow, 5 outermost petaloid, filaments narrowly spatulate, slightly clawed, 9.2–19.5 × 1.5–4.5 mm, usually without, rarely with, anthers, second whorl with anthers; anthers straight after dehiscence, epidermis smooth; styles 9–14 mm. **Capsules** cylindric, (13–)18.8–29 × 7–10.5 mm, base tapering to rounded, longitudinally ridged. **Seeds:** coat anticlinal cell walls sinuous, papillae 23–54 per cell. $2n = 20$.

Flowering Jul–Aug(–Sep). Mesic habitats, moist washes, roadsides, roadcuts, steep to gentle slopes, disturbed sites, rocky soils composed of sand and loam; 1800–3100 m; Ariz., Colo., N.Mex., Utah, Wyo.

16. Mentzelia springeri (Standley) Tidestrom in I. Tidestrom and M. T. Kittell, Fl. Ariz. New Mex., 288. 1941 • Springer's or Santa Fe blazingstar [E]

Nuttallia springeri Standley, Proc. Biol. Soc. Wash. 26: 115. 1913

Herbs perennial, bushlike, with subterranean caudices. **Stems** multiple, erect or decumbent, straight; branches along entire stem, distal longest or all ± equal, antrorse, upcurved; hairy. **Leaves:** blade 18–56 × 1.9–5.9 mm, widest intersinus distance 0.7–2.1 mm; proximal oblanceolate to elliptic, margins dentate to pinnate, teeth or lobes 6–12, perpendicular to leaf axis, 0.8–2.1 mm; distal elliptic, lanceolate, or linear, base not clasping, margins usually dentate, rarely entire, teeth (0–)4–6, perpendicular to leaf axis, 0.4–2.1 mm; abaxial surface with simple grappling-hook, complex grappling-hook, and needlelike trichomes, adaxial surface with simple grappling-hook, needlelike, and rarely complex grappling-hook trichomes. **Bracts:** margins entire. **Flowers:** petals golden yellow, 8.7–14.2 × 3.1–4.7 mm, apex rounded, glabrous abaxially; stamens golden yellow, 5 outermost petaloid, filaments narrowly spatulate, slightly clawed, 7–13.1 × 2–3.3 mm, without anthers, second whorl with anthers; anthers straight after dehiscence, epidermis smooth; styles 6.2–8.3 mm. **Capsules** cup-shaped to cylindric, 5.9–10.3 × 3.8–4.8 mm, base rounded to occasionally tapering, not longitudinally ridged. **Seeds:** coat anticlinal cell walls wavy, papillae 24–27 per cell. $2n = 22$.

Flowering Mar–Aug. Sparsely vegetated, steep talus and pumice slopes; 1600–2200 m; N.Mex.

Mentzelia springeri is known only from Los Alamos and Sandoval counties.

17. Mentzelia leucophylla Brandegee, Bot. Gaz. 27: 448. 1899 • Ash Meadows blazingstar [C] [E]

Herbs biennial or perennial, bushlike, perennials with subterranean caudices. **Stems** multiple, erect, straight; branches distal or along entire stem, distal longest or all ± equal, antrorse, straight; hairy. **Leaves:** blade 15–61 × 8–28.5 mm, widest intersinus distance 6–23.8 mm; proximal obovate, ovate, or broadly elliptic, margins usually dentate to serrate, rarely entire, teeth (0–)6–20, slightly antrorse or perpendicular to leaf axis, 0.4–4 mm; distal deltate to cordate, base clasping, margins usually dentate, rarely entire, teeth (0–)6–12, perpendicular to leaf axis, 0.5–2 mm; abaxial surface with complex grappling-hook and needlelike trichomes, adaxial surface with needlelike trichomes, both surfaces whitish, densely hairy. **Bracts:** margins entire. **Flowers:** petals golden yellow, 9.2–13.2 × 2.8–5 mm, apex rounded, glabrous abaxially; stamens golden yellow, 5 outermost petaloid, filaments narrowly spatulate, slightly clawed, 6–10.6 × 1.3–2.6 mm, with anthers, second whorl with anthers; anthers twisted or straight after dehiscence, epidermis smooth; styles 6–8.1 mm. **Capsules** cup-shaped, 5–9.5 × 6–8.6 mm, base rounded, not longitudinally ridged. **Seeds:** coat anticlinal cell walls straight, papillae 11–13 per cell. $2n = 36$.

Flowering Apr–Sep. Barren washes, rock ledges, gypsum with alkaline outcrops; of conservation concern; 600–700 m; Nev.

Mentzelia leucophylla is known only from the Ash Meadows of Nye County (N. H. Holmgren et al. 2005). It is morphologically similar to *M. oreophila*, which occurs in Nye County and elsewhere, but the two taxa are not known to co-occur. Trichome density, despite being difficult to quantify, provides perhaps the easiest method to differentiate the two taxa, with the densely hairy leaves of *M. leucophylla* appearing whitish. In addition to characters provided in the key, *M. leucophylla*, which is up to 7 dm tall, is often more robust than *M. oreophila*, which is no more than 6 dm tall.

18. Mentzelia tiehmii N. H. Holmgren & P. K. Holmgren, Syst. Bot. 27: 748, fig. 1. 2002 • Jerry's blazingstar [C] [E]

Herbs perennial, bushlike, with subterranean caudices. **Stems** multiple, erect, straight; branches along entire stem, all ± equal, antrorse, upcurved; hairy. **Leaves:** blade 19–47.5 × 5.8–9.1 mm, widest intersinus distance 4.6–7.1 mm; proximal oblanceolate to elliptic, margins dentate, teeth 6–20, perpendicular to leaf axis, 0.5–1.9 mm; distal lanceolate, deltate, or cordate, base clasping, margins entire or dentate, teeth 0–6, perpendicular to leaf axis, 0.4–1.4 mm; abaxial surface with complex grappling-hook trichomes, adaxial surface with complex grappling-hook and needlelike trichomes, both surfaces green, moderately hairy. **Bracts:** margins entire. **Flowers:** petals golden yellow, 7.7–10.8 × 1.8–4.8 mm, apex rounded, glabrous abaxially; stamens golden yellow, 5 outermost petaloid, filaments narrowly spatulate, slightly clawed, 7–7.3 × 1.2–2 mm, with anthers, second whorl with anthers; anthers twisted after dehiscence, epidermis smooth; styles 5.8–8 mm. **Capsules** cup-shaped, 4–6.6 × 4–7.4 mm, base rounded, not longitudinally ridged. **Seeds:** coat anticlinal cell walls straight, papillae 15–46 per cell.

Flowering Jun–Aug. White, sparsely vegetated clay knolls rich in gypsum; of conservation concern; 1500–1600 m; Nev.

Mentzelia tiehmii occurs in Lincoln and Nye counties.

19. Mentzelia oreophila J. Darlington, Ann. Missouri Bot. Gard. 21: 175. 1934 • Argus blazingstar [E]

Herbs perennial, bushlike, with subterranean caudices. **Stems** multiple, erect, straight; branches distal or along entire stem, distal longest or all ± equal, antrorse, upcurved; hairy. **Leaves:** blade 17–103 × 7.6–41.2 mm, widest intersinus distance 5.1–35.3 mm; proximal oblanceolate to elliptic, margins serrate, teeth 6–22, slightly antrorse, 0.4–5.3 mm; distal elliptic, lanceolate, or deltate, base clasping, margins serrate, teeth 6–16, slightly antrorse, 0.5–4.2 mm; abaxial surface with complex grappling-hook trichomes, adaxial surface with complex grappling-hook and needlelike trichomes, both surfaces green, moderately hairy. **Bracts:** margins entire. **Flowers:** petals golden yellow, 7–14.5(–16.2) × 1.7–5.2 mm, apex rounded, glabrous abaxially; stamens golden yellow, 5 outermost petaloid, filaments narrowly spatulate, slightly clawed, 5.3–11.1(–15.4) × 0.6–4.4 mm, with anthers, second whorl with anthers; anthers twisted after dehiscence, epidermis smooth; styles 3.8–8.1 mm. **Capsules** cup-shaped, 5.8–9 × 5.2–8.8 mm, base rounded, not longitudinally ridged. **Seeds:** coat anticlinal cell walls straight, papillae 6–17 per cell. **2*n*** = 22.

Flowering Feb–Oct. Sparsely vegetated slopes, roadcuts, loose, rocky and sandy limestone soils; 400–1600 m; Calif., Nev.

Mentzelia oreophila is found in Inyo, Riverside, and San Bernardino counties, California, and Clark, Esmeralda, Lincoln, and Nye counties, Nevada. The California populations have a smaller stature than those in Nevada, and phylogenetic analysis indicated that the species is potentially polyphyletic.

20. Mentzelia uintahensis (N. H. Holmgren & P. K. Holmgren) J. J. Schenk & L. Hufford, Novon 19: 120. 2009 • Uintah blazingstar [E]

Mentzelia multicaulis (Osterhout) J. Darlington var. *uintahensis* N. H. Holmgren & P. K. Holmgren, Syst. Bot. 27: 758, fig. 5. 2002

Herbs perennial, bushlike, with subterranean caudices or rhizomes. **Stems** multiple, erect, zigzag or straight; branches distal or along entire stem, distal longest or all ± equal, antrorse, upcurved; hairy. **Leaves:** blade 17–56 × 5.8–28 mm, widest intersinus distance 1–4.9(–6) mm; proximal oblanceolate to elliptic, margins pinnate to pinnatisect, lobes 4–12, antrorse, 1.9–7.7 mm; distal elliptic to lanceolate, base not clasping, margins pinnate to pinnatisect, lobes 4–12, antrorse, 2.6–13.3 mm; abaxial surface with simple grappling-hook and occasionally complex grappling-hook and needlelike trichomes, adaxial surface with simple grappling-hook and needlelike trichomes. **Bracts:** margins entire. **Flowers:** petals golden yellow, 8.5–15.2 × 3.8–7.9 mm, apex acute to rounded, glabrous abaxially; stamens golden yellow, 5 outermost petaloid, filaments broadly spatulate, strongly clawed, 5–10.4 × 2.8–6 mm, with anthers, second whorl with anthers; anthers usually twisted after dehiscence, epidermis smooth; styles 5.8–8.5 mm. **Capsules** cup-shaped, 4.2–8.8 × 3.6–5.8 mm, base tapering to rounded, not longitudinally ridged. **Seeds:** coat anticlinal cell walls straight, papillae 4–7 per cell. **2*n*** = 22.

Flowering May–Sep. Sparsely vegetated steep talus slopes and roadcuts; 1500–2800 m; Colo., Utah.

Mentzelia uintahensis is known from northwestern Colorado (Mesa, Moffat, and Rio Blanco counties) and northeastern Utah (Carbon, Duchesne, and Uintah counties).

21. Mentzelia librina (K. H. Thorne & F. J. Smith) J. J. Schenk & L. Hufford, Novon 19: 119. 2009 • Book Cliffs stickleaf [C] [E]

Mentzelia multicaulis (Osterhout) J. Darlington var. *librina* K. H. Thorne & F. J. Smith, Great Basin Naturalist 46: 556, fig. 1. 1986

Herbs perennial, bushlike, with subterranean caudices or rhizomes. **Stems** multiple, erect, zigzag; branches distal or along entire stem, distal longest or all ± equal, antrorse, straight to upcurved; hairy. **Leaves:** blade 14–26.3 × 5.1–10.8 mm, widest intersinus distance 2.3–5.2 mm; proximal oblanceolate to elliptic, margins usually 3-fid, rarely entire or pinnate, teeth or lobes usually 2, rarely 0 or 4, perpendicular to leaf axis, 0.8–2.7 mm; distal elliptic, base not clasping, margins 3-fid to occasionally pinnate, teeth or lobes usually 2, occasionally 4, perpendicular to leaf axis, 1.5–3.5 mm; abaxial surface with simple grappling-hook and complex grappling-hook trichomes, adaxial surface with simple grappling-hook and needlelike trichomes. **Bracts:** margins usually entire, rarely pinnate. **Flowers:** petals golden yellow, 8.3–13.6 × 2.6–4.8(–6.2) mm, apex acute, glabrous abaxially; stamens golden yellow, 5 outermost petaloid, filaments broadly spatulate, strongly clawed, 6–8(–9.1) × 2.3–4.4 mm, with anthers, second whorl with anthers; anthers straight after dehiscence, epidermis smooth; styles 5–7.9 mm. **Capsules** cup-shaped, 4–7.3 × 3.3–5.8 mm, base rounded, not longitudinally ridged. **Seeds:** coat anticlinal cell walls straight, papillae 4–5 per cell.

Flowering Jun–Sep. Barren gray soils, steep shale slopes; of conservation concern; 1700–2100 m; Utah.

Mentzelia librina in narrowly distributed along the border between Carbon and Emery counties.

22. Mentzelia flumensevera (N. H. Holmgren & P. K. Holmgren) J. J. Schenk & L. Hufford, Novon 19: 118. 2009 • Sevier canyon stickleaf [C] [E]

Mentzelia multicaulis (Osterhout) J. Darlington var. *flumensevera* N. H. Holmgren & P. K. Holmgren, Syst. Bot. 27: 761, fig. 6E–G. 2002

Herbs perennial, bushlike, with subterranean caudices or rhizomes. **Stems** multiple, erect or decumbent, zigzag; branches along entire stem, all ± equal, antrorse, upcurved; hairy. **Leaves:** blade 11.2–22 × 5.2–10.2 mm, widest intersinus distance 1.2–2.4 mm; proximal oblanceolate to elliptic, margins pinnate, lobes 4–6, perpendicular to leaf axis, 2.3–4.1 mm; distal elliptic, base not clasping, margins pinnate, lobes 2–4, perpendicular to leaf axis, 1.4–3 mm; abaxial surface with complex grappling-hook and needlelike trichomes, adaxial surface with complex grappling-hook and needlelike trichomes. **Bracts:** margins entire. **Flowers:** petals golden yellow, 10–11.2 × 2.9–3.8 mm, apex rounded, glabrous abaxially; stamens golden yellow, 5 outermost petaloid, filaments narrowly spatulate, slightly clawed, 7.3–8.4 × 1.3–1.9 mm, with anthers, second whorl with anthers; anthers straight after dehiscence, epidermis smooth; styles 5.8–8.7 mm. **Capsules** cup-shaped, 5.6–7.1 × 4–6 mm, base rounded, not longitudinally ridged. **Seeds:** coat anticlinal cell walls straight, papillae 3–7 per cell.

Flowering Jul–Aug(–Oct). Sparsely vegetated, steep talus slopes composed of gravel and clay, white soils; of conservation concern; 1700–1900 m; Utah.

Mentzelia flumensevera is narrowly distributed across the Piute and Sevier county line.

23. Mentzelia multicaulis (Osterhout) J. Darlington, Ann. Missouri Bot. Gard. 21: 156. 1934 • Multiple-branched or manystem blazingstar [E] [F]

Touterea multicaulis Osterhout, Bull. Torrey Bot. Club 30: 236. 1903

Herbs perennial, bushlike, with subterranean caudices or sometimes rhizomes. **Stems** multiple, erect or decumbent, zigzag or straight; branches distal or along entire stem, distal longest or all ± equal, antrorse, straight to upcurved; hairy. **Leaves:** blade 20–49(–57) × 4.2–19 mm, widest intersinus distance 1.1–5.3 mm; proximal oblanceolate to elliptic, margins pinnate to pinnatisect, lobes 4–10, slightly antrorse, 0.7–7.6 mm; distal linear to lanceolate, base not clasping, margins entire or pinnate, lobes 0–10, slightly antrorse, 1.9–7.8 mm; abaxial surface with simple grappling-hook trichomes, adaxial surface with simple grappling-hook and needlelike trichomes. **Bracts:** margins entire. **Flowers:** petals golden yellow, 10.6–17.9 × 5.1–9 mm, apex obtuse, glabrous abaxially; stamens golden yellow, 5 outermost petaloid, filaments oblanceolate, slightly clawed, 6.5–10.5 × 2–4.5 mm, with anthers, second whorl with anthers; anthers twisted or straight after dehiscence, epidermis smooth; styles 6–11.5 mm. **Capsules** cup-shaped to cylindric, 6–13.6 × 3.4–6 mm, base tapering to rounded, not longitudinally ridged. **Seeds:** coat anticlinal cell walls straight, papillae 4–9 per cell. $2n = 22$.

Flowering Jun–Aug. Sparsely vegetated steep slopes, drainage gullies, roadcuts; 2000–2500 m; Colo.

Mentzelia multicaulis is known to occur in Eagle, Grand, and Summit counties.

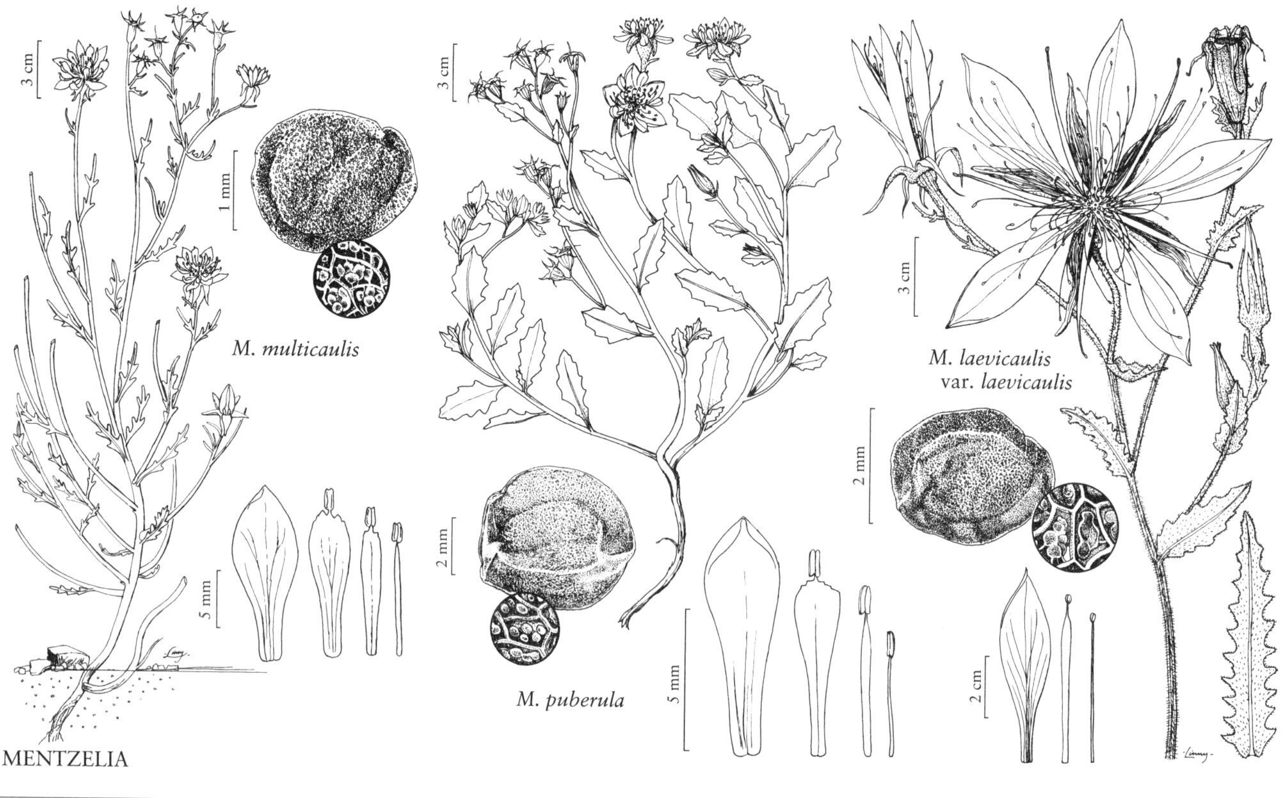

MENTZELIA

24. Mentzelia polita A. Nelson, Bot. Gaz. 47: 427. 1909 • Polished blazingstar [C] [E]

Herbs perennial, bushlike, with subterranean caudices. **Stems** multiple, erect, zigzag or straight; branches along entire stem, distal longest, antrorse, straight to upcurved; glabrescent, smooth to touch. **Leaves:** blade 16–84 × 2.2–10.6 mm, widest intersinus distance 2.2–7 mm; proximal oblanceolate, margins usually entire, occasionally dentate, teeth 0–6(–14), perpendicular to leaf axis, 1–2 mm; distal elliptic, lanceolate, or linear, base not clasping, margins usually entire, rarely dentate, teeth 0(–6), perpendicular to leaf axis, 0.6–2 mm; abaxial surface with complex grappling-hook and infrequently needlelike trichomes, adaxial surface with needlelike trichomes. **Bracts:** margins entire. **Flowers:** petals light to golden yellow, 7.8–11(–14.2) × 1.9–2.8 (–4.2) mm, apex rounded, glabrous abaxially; stamens white to light yellow, 5 outermost petaloid, filaments narrowly spatulate, slightly clawed, 5.9–9.2(–10.1) × 1.2–2.7 mm, with anthers, second whorl with anthers; anthers twisted after dehiscence, epidermis smooth; styles 5.2–7.6 mm. **Capsules** cup-shaped, 4.5–8.8 × 5.8–8.8 mm, base rounded, not longitudinally ridged. **Seeds:** coat anticlinal cell walls straight, papillae 15–24 per cell. $2n = 22$.

Flowering Apr–Aug. Dry washes, arroyos, steep slopes; 500–1500 m; of conservation concern; Calif., Nev.

Mentzelia polita is known only from the Clark Mountains of San Bernardino County, California, and the Spring Mountains of Clark County, Nevada.

25. Mentzelia memorabilis N. H. Holmgren & P. K. Holmgren, Syst. Bot. 27: 753, fig. 4. 2002 (as memorabalis) • Nine-eleven blazingstar [C] [E]

Herbs perennial, bushlike, with subterranean caudices. **Stems** multiple, erect, straight; branches distal, distal longest, antrorse, straight; hairy. **Leaves:** blade 17–36.6 × 1.1–4.8 mm, widest intersinus distance 1.1–4.8 mm; proximal linear to oblanceolate, margins entire; distal linear, base not clasping, margins entire; abaxial surface with simple grappling-hook, complex grappling-hook, and needlelike trichomes, adaxial surface with simple grappling-hook and needlelike trichomes. **Bracts:** margins entire. **Flowers:** petals golden yellow, 7.4–11 × 2.6–3.4 mm, apex rounded, glabrous abaxially; stamens golden yellow, 5 outermost petaloid, filaments narrowly spatulate, slightly clawed, 5.5–9.1 × 1.4–2.7 mm, with anthers, second whorl with anthers; anthers twisted or straight after dehiscence, epidermis smooth; styles

4.8–8 mm. **Capsules** cup-shaped, 3.6–5.3 × 3.1–4.2 mm, base rounded, not longitudinally ridged. **Seeds:** coat anticlinal cell walls straight, papillae 6–10 per cell.

Flowering Jun–Sep. Barren gypsum-clay outcrops and hilltops; of conservation concern; 1400–1700 m; Ariz.

Mentzelia memorabilis occurs on the Uinkaret Plateau in Mohave County.

26. Mentzelia rhizomata Reveal, Syst. Bot. 27: 763, figs. 1, 2. 2002 • Roan Cliffs blazingstar C E

Nuttallia rhizomata (Reveal) W. A. Weber & R. C. Wittmann

Herbs perennial, bushlike, with rhizomes. **Stems** multiple, erect or decumbent, zigzag; branches along entire stem, distal longest, antrorse, upcurved; hairy. **Leaves:** blade 8.1–29 × 4.1–12.7 mm, widest intersinus distance 4.1–7.4 mm; proximal oblanceolate, margins entire; distal elliptic, base not clasping, margins entire or dentate, teeth 0–8, perpendicular to leaf axis, 0.6–2.7 mm; abaxial surface with simple grappling-hook and needlelike trichomes, adaxial surface with needlelike trichomes. **Bracts:** margins entire. **Flowers:** petals golden yellow, 12.5–17.2 × 6.1–11.2 mm, apex rounded to obtuse, glabrous abaxially; stamens golden yellow, 5 outermost petaloid, filaments narrowly spatulate, strongly clawed, 7–9.5 × 2.7–6 mm, with anthers, second whorl with anthers; anthers straight after dehiscence, epidermis smooth; styles 7–10 mm. **Capsules** cup-shaped, 7–10 × 6.5–8 mm, base rounded, not longitudinally ridged. **Seeds:** coat anticlinal cell walls straight, papillae 9–13 per cell.

Flowering Jul–Aug. Sparsely vegetated, steep talus slopes with loose gravelly soils; of conservation concern; 1700–2100 m; Colo.

Mentzelia rhizomata is known only from Garfield County.

27. Mentzelia goodrichii K. H. Thorne & S. L. Welsh, Rhodora 95: 407, fig. 14. 1993 • Goodrich's blazingstar C E

Herbs perennial, bushlike, with subterranean caudices or rhizomes. **Stems** multiple, erect or decumbent, zigzag or straight; branches along entire stem, distal longest, antrorse, straight; hairy. **Leaves:** blade 23–35.3 × 6.5–13(–16) mm, widest intersinus distance 3.7–11.3 mm; proximal oblanceolate, margins dentate, teeth 4–10, perpendicular to leaf axis, 0.5–1.8 mm; distal elliptic to lanceolate, base not clasping, margins dentate to pinnate, teeth or lobes 4–8, perpendicular to leaf axis, 1.3–3.6 mm; abaxial surface with simple grappling-hook, complex grappling-hook, and needlelike trichomes, adaxial surface with simple grappling-hook and needlelike trichomes. **Bracts:** margins entire. **Flowers:** petals golden yellow, 16.1–18.5 × 6.7–9.2 mm, apex rounded, glabrous abaxially; stamens golden yellow, 5 outermost petaloid, filaments narrowly spatulate, slightly clawed, 7.8–9.3 × 2.3–3.3 mm, with anthers, second whorl with anthers; anthers straight after dehiscence, epidermis smooth; styles 6.3–8.6 mm. **Capsules** cup-shaped, 8–12 × 6–7.2 mm, base rounded, not longitudinally ridged. **Seeds:** coat anticlinal cell walls straight, papillae 11–15 per cell.

Flowering Jul–Aug. Steep slopes composed of Green River Formation shale; of conservation concern; 1900–2700 m; Utah.

Mentzelia goodrichii is known only from Duchesne County.

28. Mentzelia shultziorum Prigge, Great Basin Naturalist 46: 361, figs. 1, 3, 4. 1986 • Shultz's blazingstar C E

Herbs perennial, bushlike, with subterranean caudices. **Stems** multiple, erect or decumbent, zigzag; branches along entire stem, distal longest, antrorse, straight to upcurved; hairy. **Leaves:** blade 13.6–33.3 × 7.2–18.5 mm, widest intersinus distance 6.2–14.8 mm; proximal elliptic, margins dentate, teeth 4–6, perpendicular to leaf axis, 0.5–0.6 x 1.2–1.5 mm; distal oblanceolate to elliptic, base usually not clasping, rarely clasping, margins dentate to serrate, teeth 4–8, slightly antrorse or perpendicular to leaf axis, 0.6–2.5 mm; abaxial surface with simple grappling-hook and complex grappling-hook trichomes, adaxial surface with simple grappling-hook and needlelike trichomes. **Bracts:** margins entire or slightly toothed. **Flowers:** petals golden yellow, 7.7–14.8 × 2.7–5.9 mm, apex acute, glabrous abaxially; stamens golden yellow, 5 outermost petaloid, filaments narrowly spatulate, strongly clawed, 6–11 × 1.9–5.2 mm, with anthers, second whorl with anthers; anthers straight after dehiscence, epidermis smooth; styles 5.4–9.7 mm. **Capsules** cup-shaped, 4.2–8 × 4.9–7.6 mm, base rounded, not longitudinally ridged. **Seeds:** coat anticlinal cell walls straight, papillae 4–6 per cell.

Flowering Aug–Sep. Steep, barren slopes of white, green, and red clayey soils; of conservation concern; 1200–1600 m; Utah.

Mentzelia shultziorum occurs only on outcrops above Onion Creek and at the head of Castle Valley in Grand County.

29. Mentzelia puberula J. Darlington, Ann. Missouri Bot. Gard. 21: 177. 1934 • Pubescent blazingstar [F]

Herbs perennial, bushlike, with subterranean caudices. **Stems** multiple, erect or decumbent, straight; branches along entire stem, distal longest, antrorse, upcurved; hairy. **Leaves:** blade 18.8–62.7 × 11–36 mm, widest intersinus distance 8.9–28.8 mm; proximal broadly elliptic to obovate, margins dentate to serrate, teeth 8–12, slightly antrorse or perpendicular to leaf axis, 1–4 × 1.7–6 mm; distal ovate or obovate, base not clasping, margins dentate to serrate, teeth 6–12, slightly antrorse or perpendicular to leaf axis, 1–3.5 mm; abaxial surface with complex grappling-hook trichomes, adaxial surface with needlelike trichomes. **Bracts:** margins entire. **Flowers:** petals golden yellow, 6.7–9.5(–12.6) × 2.3–5.8 mm, apex rounded, glabrous abaxially; stamens golden yellow, 5 outermost petaloid, filaments narrowly spatulate, slightly clawed, 5.4–9.2(–11) × 1.3–2.7 mm, with anthers, second whorl with anthers; anthers twisted after dehiscence, epidermis smooth; styles 4.1–6.8(–9) mm. **Capsules** cup-shaped, 5–9.8 × 5–8 mm, base rounded, not longitudinally ridged. **Seeds:** coat anticlinal cell walls straight, papillae 3–5 per cell. **2*n*** = 20, 22.

Flowering Feb–Oct. Bases of steep cliffs in crevices composed of basalt, granite, and limestone, steep gravelly and sandy slopes, sandy washes; 90–2200 m; Ariz., Calif., Nev.; Mexico (Baja California, Sonora).

Populations of *Mentzelia puberula* are associated with the Colorado River in Arizona, California, Nevada, and northern Mexico. Disjunct populations also occur in the Gila Mountains of Arizona.

30. Mentzelia laevicaulis (Douglas) Torrey & A. Gray, Fl. N. Amer. 1: 535. 1840 • Giant blazingstar [E] [F] [W]

Bartonia laevicaulis Douglas in W. J. Hooker, Fl. Bor.-Amer. 1: 221, plate 69. 1832

Herbs biennial or perennial (in rosette stage), bushlike. **Stems** solitary, erect, straight; branches distal or along entire stem, proximal or distal longest, antrorse, upcurved; hairy or glabrescent. **Leaves:** blade 17.4–196 × 6.7–40.4 mm, widest intersinus distance 2.4–24.6 mm; proximal oblanceolate, lanceolate, or elliptic, margins pinnate, lobes 14–44, slightly antrorse, 2.5–8.9 mm; distal lanceolate, base clasping or not, margins pinnate, lobes 6–42, slightly antrorse, 2.1–13.8 mm; abaxial surface with simple grappling-hook, complex grappling-hook, and needlelike trichomes, adaxial surface with needlelike trichomes. **Bracts:** margins entire or pinnate. **Flowers:** petals golden yellow, 23.5–70 × 3–17.3 mm, apex acute, glabrous abaxially; stamens golden yellow, 5 outermost not petaloid, filaments linear to narrowly elliptic, not clawed, 17.7–55 × 0.5–2(–2.5) mm, with anthers, second whorl with anthers; anthers straight after dehiscence, epidermis smooth; styles 20.4–57 mm. **Capsules** cup-shaped to cylindric, 10.6–43 × 6.3–11.1 mm, base tapering or rounded, not longitudinally ridged. **Seeds:** coat anticlinal cell walls straight, papillae 4–11 per cell. **2*n*** = 22.

Varieties 2 (2 in the flora): w North America.

1. Petals 43–70 mm; 5 outermost stamens 39–55 mm. . . . 30a. *Mentzelia laevicaulis* var. *laevicaulis*
1. Petals 23.5–40 mm; 5 outermost stamens 17.7–29 mm. . . 30b. *Mentzelia laevicaulis* var. *parviflora*

30a. Mentzelia laevicaulis (Douglas) Torrey & A. Gray var. **laevicaulis** [E] [F]

Mentzelia acuminata (Rydberg) Tidestrom

Leaf blades 39.8–154 × 8.4–40.4 mm, widest intersinus distance 4.9–24.6 mm; proximal oblanceolate to elliptic, margin lobes 18–44, 2.5–8.9 × 3.6–9.6 mm; distal with base clasping or non-clasping, margin lobes 14–42, 2.1–6.1 × 3.2–6.3 mm; adaxial surfaces with needlelike trichomes. **Bracts** 13–29 × 1–5 mm, margins entire or pinnate. **Flowers:** petals 43–70 × 8–17.3 mm; 5 outermost stamens 39–55 × 1–2(–2.5) mm. **Capsules** cylindric, 22–43 × 8–11.1 mm. **Seed coat:** papillae 4–11 per cell.

Flowering May–Oct. Dry streambed bottoms and banks, gravel bars, hillside slopes, roadsides, roadcuts, sand dunes; 0–3000 m; Ariz., Calif., Idaho, Mont., Nev., Oreg., Utah, Wash., Wyo.

30b. Mentzelia laevicaulis (Douglas) Torrey & A. Gray var. **parviflora** (Douglas) C. L. Hitchcock in C. L. Hitchcock et al., Vasc. Pl. Pac. N.W. 3: 455. 1961 • Small-flowered blazingstar [E]

Bartonia parviflora Douglas in W. J. Hooker, Fl. Bor.-Amer. 1: 221. 1832; *Mentzelia brandegeei* S. Watson; *M. douglasii* H. St. John

Leaf blades 17.4–196 × 6.7–24.7 mm, widest intersinus distance 2.4–7.2 mm; proximal oblanceolate, lanceolate, or elliptic, margin lobes 14–24, 3.5–6.3 × 3.1–6.6 mm; distal with base clasping, margin lobes 6–14, 2.2–13.8

× 0.8–6.2 mm; adaxial surfaces with simple grappling-hook and needlelike trichomes. **Bracts** 8.4–25 × 1–1.3 mm, margins entire or toothed. **Flowers:** petals 23.5–40 × 3–9 mm; 5 outermost stamens 17.7–29 × 0.5–1.4 mm. **Capsules** cup-shaped to cylindric, 19–28 × 6.3–9.6 mm. **Seed coat:** papillae 4–6 per cell.

Flowering Jun–Aug. Roadsides, slopes, riverbanks, dry sandy and gravelly soils; 200–600 m; B.C.; Idaho, Wash.

Variety *parviflora* was first described at the specific level by Douglas as *Bartonia parviflora*. The combination *Mentzelia parviflora* (Douglas) J. F. McBride is a later homonym of *M. parviflora* A. Heller, which is a synonym of *M. albicaulis* in sect. *Trachyphytum*.

31. Mentzelia inyoensis H. J. Thompson & Prigge, Madroño 51: 379, figs. 1, 2. 2004 • White Mountain or Inyo blazingstar [C] [E]

Herbs biennial, candelabra-form. **Stems** solitary, erect, straight; branches usually distal, occasionally along entire stem, distal usually longest, antrorse, straight; hairy. **Leaves:** blade 16.7–89 × 6.3–14.2(–20) mm, widest intersinus distance 3.1–9.7 mm; proximal oblanceolate to elliptic, margins serrate to pinnate, teeth or lobes 10–24, slightly antrorse, 1.4–3.4 mm; distal lanceolate, base usually not clasping, occasionally clasping, margins serrate to pinnate, teeth or lobes 8–14, slightly antrorse, 0.7–7 mm; abaxial surface with complex grappling-hook, and occasionally simple grappling-hook and needlelike, trichomes, adaxial surface with needlelike and occasionally simple grappling-hook trichomes. **Bracts:** margins usually entire, sometimes toothed. **Flowers:** petals golden yellow, 11.7–15.8(–18) × 2–4.8 mm, apex acute, glabrous abaxially; stamens golden yellow, 5 outermost not petaloid, filaments linear, not clawed, 10.2–13.2 × 0.6–1.6 mm, with anthers, second whorl with anthers; anthers straight after dehiscence, epidermis smooth; styles 6.7–13 mm. **Capsules** usually cylindric, rarely cup-shaped, 10.4–25 × 5.5–8.3 mm, base tapering or rounded, not longitudinally ridged. **Seeds:** coat anticlinal cell walls straight, papillae 3–6 per cell. $2n = 22$.

Flowering Jun–Aug. Sparsely vegetated, gravelly slopes, gypsum or ash soils; of conservation concern; 1400–2000 m; Calif., Nev.

Mentzelia inyoensis is known only from Inyo and Mono counties, California, and Churchill and Esmeralda counties, Nevada.

32. Mentzelia candelariae H. J. Thompson & Prigge, Phytologia 55: 281, figs. 1, 3. 1984 • Candelaria blazingstar [E]

Herbs biennial, candelabra-form. **Stems** solitary, erect, straight; branches distal, distal longest, antrorse, upcurved; hairy. **Leaves:** blade 26–74 × 6–19.2 mm, widest intersinus distance 4.5–11.4 mm; proximal obovate, oblanceolate, or lanceolate, margins serrate, teeth 8–14, slightly antrorse, 0.6–4.5 mm; distal elliptic to lanceolate, base not clasping, margins serrate, teeth 6–12, slightly antrorse, 1.3–4.1 mm; abaxial surface with simple grappling-hook, complex grappling-hook, and occasionally needlelike trichomes, adaxial surface with simple grappling-hook and needlelike trichomes. **Bracts:** margins entire. **Flowers:** petals golden yellow, (6.1–)7.3–10 × 1.6–3 mm, apex acute, glabrous abaxially; stamens golden yellow, 5 outermost not petaloid, filaments linear to narrowly spatulate, not clawed, 4–8 × 0.7–1.5 mm, with anthers, second whorl with anthers; anthers straight after dehiscence, epidermis smooth; styles 3.7–7 mm. **Capsules** cup-shaped to cylindric, 10.4–16 × 6.6–9 mm, base tapering or rounded, not longitudinally ridged. **Seeds:** coat anticlinal cell walls straight, papillae 6–16 per cell. $2n = 22$.

Flowering May–Jun. Sparsely vegetated washes, steep slopes, hilltops, gravelly, clayey, and sandy soils composed of volcanic ash; 1100–2000 m; Nev.

Mentzelia candelariae is known from Churchill, Esmeralda, Mineral, Nye, and Pershing counties.

33. Mentzelia conspicua Todsen, Sida 18: 819, fig. 1. 1999 • Remarkable blazingstar [C] [E]

Herbs biennial, candelabra-form. **Stems** solitary, erect, straight; branches distal, distal longest, antrorse, straight; hairy. **Leaves:** blade 69–195 × 11.6–55.7 mm, widest intersinus distance 1.3–2.5(–3.5) mm; proximal oblanceolate or elliptic, margins pinnatisect, lobes 16–26, slightly antrorse or perpendicular to leaf axis, 5.1–15.1 mm; distal elliptic, base not clasping, margins pinnatisect, lobes 10–20, slightly antrorse or perpendicular to leaf axis, 6.2–12.3(–22.5) mm; abaxial surface with needlelike and occasionally simple grappling-hook trichomes, adaxial surface with needlelike trichomes. **Bracts:** margins entire. **Flowers:** petals golden yellow, 30–42.2 × 5.7–10.8 mm, apex acute, glabrous abaxially; stamens golden yellow, 5 outermost petaloid, filaments narrowly spatulate, slightly clawed, 27–37.4 × 4.8–9.7 mm, without anthers, second whorl with anthers; anthers straight after dehiscence, epidermis smooth; styles 24–32.4 mm. **Capsules** cylindric, 15–26 × 5–7.2 mm, base tapering to rounded, not longitudinally ridged. **Seeds:** coat anticlinal cell walls sinuous, papillae 8–12 per cell. **2*n*** = 20.

Flowering Jun–Sep. Slopes, pinyon pine and juniper woodlands, grasslands, sparsely vegetated soils composed of red and brown loam; of conservation concern; 1800–2400 m; N.Mex.

Mentzelia conspicua is known from Rio Arriba, Sandoval, and Torrance counties.

34. Mentzelia lagarosa (K. H. Thorne) J. J. Schenk & L. Hufford, Madroño 57: 247. 2010 • Slender-lobed blazingstar [E]

Mentzelia pumila (Nuttall) Torrey & A. Gray var. *lagarosa* K. H. Thorne, Great Basin Naturalist 46: 558, fig. 1. 1986

Herbs biennial, candelabra-form. **Stems** solitary, erect, straight; branches distal, distal longest, antrorse, straight; hairy. **Leaves:** blade 11.3–103 × 4.8–20.1 mm, widest intersinus distance 1.2–5.7 mm; proximal oblanceolate to elliptic, margins pinnate to pinnatisect, lobes 8–20, slightly antrorse or perpendicular to leaf axis, 1.4–8.2 mm; distal elliptic to lanceolate, base not clasping, margins pinnatisect, lobes 6–16, slightly antrorse or perpendicular to leaf axis, 1.6–7.5 mm; abaxial surface with simple grappling-hook, complex grappling-hook, and occasionally needlelike trichomes, adaxial surface with simple grappling-hook and needlelike trichomes. **Bracts:** margins usually entire, sometimes toothed or pinnate. **Flowers:** petals golden yellow, 8.3–13 × 2.2–5.4 mm, apex acute or rounded, glabrous abaxially; stamens golden yellow, 5 outermost petaloid, filaments narrowly spatulate, slightly clawed, 6.5–10.7 × 1.7–4.3 mm, without anthers, second whorl with anthers; anthers straight after dehiscence, epidermis smooth; styles 6.1–10.2 mm. **Capsules** cylindric, 12.1–21.2 × 4.9–7.6 mm, base tapering or rounded, not longitudinally ridged. **Seeds:** coat anticlinal cell walls wavy, papillae 29–31 per cell. **2*n*** = 22.

Flowering Jun–Aug(–Oct). Sparsely vegetated hills, slopes, knolls, white ash and limestone soils; 1500–2500 m; Colo., Nev., Utah.

Mentzelia lagarosa is allopatric with two of the three species most similar to it, namely *M. filifolia* and *M. holmgreniorum*, and nearly allopatric with the third, *M. laciniata*. Where the ranges of *M. lagarosa* and *M. laciniata* overlap in western Colorado, they can be distinguished by petal length [8.3–13 mm in *M. lagarosa* versus 14–23.8(–26) mm in *M. laciniata*], outermost stamen length (6.5–10.7 mm in *M. lagarosa* versus 12–20 mm in *M. laciniata*), and number of seed coat cell papillae (29–31 per cell in *M. lagarosa* versus 5–14 per cell in *M. laciniata*); in addition, *M. lagarosa* bears both simple grappling-hook and needlelike trichomes on its adaxial leaf blade surfaces, whereas leaf blades of *M. laciniata* bear only needlelike trichomes adaxially. In the *Intermountain Flora*, N. H. Holmgren et al. (2005) treated *M. lagarosa* as a synonym of *M. multiflora*, but J. J. Schenk and L. Hufford (2011) showed not only that *M. lagarosa* is distinct from *M. multiflora*, but also that the latter does not occur in the intermountain region.

35. Mentzelia holmgreniorum J. J. Schenk & L. Hufford, Madroño 57: 252, fig. 2C. 2010 • Holmgrens' blazingstar [C] [E]

Herbs biennial, candelabra-form. **Stems** solitary, erect, straight; branches distal or along entire stem, distal or proximal longest, antrorse, upcurved; hairy. **Leaves:** blade 42–89 × 11–31.9 mm, widest intersinus distance 2.3–3.6 mm; proximal oblanceolate to elliptic, margins pinnatisect, lobes 14–20, strongly antrorse, 4.9–14.4 mm; distal lanceolate, base not clasping, margins pinnatisect, lobes 12–18, strongly antrorse, 4.2–12.4 mm; abaxial surface with simple grappling-hook, complex grappling-hook, and needlelike trichomes, adaxial surface with simple grappling-hook and needlelike trichomes. **Bracts:** margins pinnate. **Flowers:** petals golden yellow, 13.5–18.8 × 5.2–6.6 mm, apex rounded, glabrous abaxially; stamens golden yellow, 5 outermost petaloid, filaments narrowly spatulate, slightly clawed, 11.1–16 × 2.7–5 mm, without anthers, second whorl with anthers; anthers straight after dehiscence, epidermis smooth; styles 8.4–10.6 mm. **Capsules** cylindric, 13.1–14.6 × 5.8–6.9 mm, base tapering, not longitudinally ridged. **Seeds:** coat anticlinal cell walls sinuous, papillae 26–51 per cell. **2*n*** = 20.

Flowering Jun–Aug. Dry sandy washes, roadsides, disturbed areas; of conservation concern; 1400–2300 m; Ariz., N.Mex.

Mentzelia holmgreniorum is known from Apache, Coconino, and Navajo counties, Arizona, and Catron County, New Mexico. It is allopatric from the species most similar to it (*M. filifolia*, *M. laciniata*, and *M. lagarosa*), occurring south and west of all three; it differs from all these species in having upcurved rather than straight branches, and from *M. filifolia* and *M. laciniata* in having both simple grappling-hook and needlelike trichomes (versus only needlelike trichomes) on its adaxial leaf surfaces. In addition, *M. holmgreniorum* differs from *M. filifolia* in having leaf blades with greater intersinus distances (2.3–3.6 versus 1–2.4 mm) and wider lobes (1.6–2.5 versus 0.8–1.4 mm), from *M. laciniata* in having pinnate bracts (versus usually entire, rarely pinnate), and from *M. lagarosa* in having leaf blade lobes that are strongly antrorsely oriented (versus slightly antrorsely or perpendicular to the leaf axis) and flowers with larger petals (13.5–18.8 × 5.2–6.6 versus 8.3–13 × 2.2–5.4 mm) and longer outermost stamens (11.1–16 versus 6.5–10.7 mm).

36. Mentzelia filifolia J. J. Schenk & L. Hufford, Madroño 57: 251, figs. 1D, 2B. 2010 • Narrow-leaved blazingstar [C] [E]

Herbs biennial, candelabra-form. **Stems** solitary, erect, straight; branches distal, distal longest, antrorse, straight; hairy. **Leaves:** blade 43–94(–115) × (4.3–)7.5–36 mm, widest intersinus distance 1–2.4 mm; proximal oblanceolate to elliptic, margins pinnatisect, lobes 8–20, perpendicular to leaf axis, 3.2–12(–15.7) mm; distal oblanceolate to elliptic, base not clasping, margins pinnatisect, lobes 8–20, perpendicular to leaf axis, 5.6–17.1 mm; abaxial surface with simple grappling-hook, complex grappling-hook, and occasionally needlelike trichomes, adaxial surface with needlelike trichomes. **Bracts:** margins entire or pinnate. **Flowers:** petals golden yellow, 13–18.5 × 3.7–6.1 mm, apex acute, glabrous abaxially; stamens golden yellow, 5 outermost petaloid, filaments narrowly spatulate, slightly clawed, 10.2–14.7(–18) × (1.3–)2.5–4.4 mm, without anthers, second whorl with anthers; anthers straight after dehiscence, epidermis smooth; styles 10–12.5(–14) mm. **Capsules** cylindric, 11–19 × 5–7.5 mm, base tapering, not or diminutively longitudinally ridged. **Seeds:** coat anticlinal cell walls sinuous, papillae 42–48 per cell. $2n = 20$.

Flowering Jul–Aug. Roadcuts, slopes, dark loam, rocky soils; of conservation concern; 2100–2200 m; Ariz., N.Mex.

Mentzelia filifolia is known from the Chuska Mountains of Apache County, Arizona, and McKinley and San Juan counties, New Mexico.

37. Mentzelia laciniata (Rydberg) J. Darlington, Ann. Missouri Bot. Gard. 21: 173. 1934 • Cut-leaf blazingstar [E]

Touterea laciniata Rydberg, Bull. Torrey Bot. Club 31: 565. 1904

Herbs biennial, bushlike or candelabra-form. **Stems** solitary, erect, straight; branches distal or along entire stem, distal or proximal longest, antrorse, straight; hairy. **Leaves:** blade 52–112 × (5.4–)8.3–25 mm, widest intersinus distance 1.4–4 mm; proximal oblanceolate or elliptic, margins pinnatisect, lobes 8–20, slightly antrorse, 4.2–7.4(–10.7) mm; distal oblanceolate, elliptic, or lanceolate, base not clasping, margins usually pinnatisect, sometimes pinnate, especially near apex, lobes 8–18, slightly antrorse, 3–10.8 mm; abaxial surface with simple grappling-hook, complex grappling-hook, and needlelike trichomes, adaxial surface with needlelike trichomes. **Bracts:** margins usually entire, rarely pinnate. **Flowers:** petals golden yellow, 14–23.8(–26) × 3.8–7.4 mm, apex acute to rounded, glabrous abaxially; stamens golden yellow, 5 outermost petaloid, filaments narrowly spatulate to elliptic, slightly clawed, 12–20.4 × 2.5–4.9 mm, usually without, rarely with, anthers, second whorl with anthers; anthers straight after dehiscence, epidermis smooth; styles 9.2–17.7 mm. **Capsules** cylindric, 12–20.2 × 4.5–8.1 mm, base tapering, not longitudinally ridged. **Seeds:** coat anticlinal cell walls sinuous, papillae 5–14 per cell. $2n = 20$.

Flowering Jun–Sep. Dry hillsides, roadcuts, roadsides, sandy or clayey soils; 1400–2300 m; Colo., N.Mex.

Mentzelia laciniata is found in southwestern Colorado and northwestern New Mexico, where it does not extend as far west as the Chuska Mountains.

38. Mentzelia pterosperma Eastwood, Proc. Calif. Acad. Sci., ser. 2, 6: 290. 1896 • Wing-seed blazingstar [E]

Herbs winter annual or biennial, candelabra-form. **Stems** solitary, erect, straight; branches distal, distal longest, antrorse, straight, hairy. **Leaves:** blade 13.8–78 × 3.8–20 mm, widest intersinus distance 2.5–15.8 mm; proximal oblanceolate to elliptic, margins entire or serrate to pinnate, teeth or lobes (0–)8–14(–22), slightly antrorse, 0.3–5.3 mm; distal elliptic to lanceolate, base usually clasping, rarely a few not clasping, margins usually serrate to pinnate, occasionally entire, teeth or lobes (0–)6–18, slightly antrorse, 2.8–4.3(–6.1) mm; abaxial surface with simple grappling-hook, occasionally complex grappling-hook, and rarely needlelike trichomes, adaxial surface occasionally with simple grappling-hook and needlelike trichomes. **Bracts:** margins usually entire, rarely pinnate. **Flowers:** petals golden yellow, 6.4–17(–20) × 2–5.3 mm, apex usually acute, occasionally rounded, glabrous abaxially; stamens golden yellow, 5 outermost petaloid, filaments linear to oblanceolate, slightly clawed, 6.2–14.2(–17.5) × 1–3.9 mm, usually without, rarely with, anthers, second whorl with anthers; anthers straight after dehiscence, epidermis smooth; styles 6–11 mm. **Capsules** cup-shaped, 7.3–13.7 × 5.5–9.8 mm, base rounded, not longitudinally ridged. **Seeds:** coat anticlinal cell walls straight, papillae 5–18 per cell. $2n = 22$.

Flowering Apr–Jun(–Jul). Disturbed soils, washes, sand dunes, roadcuts, badland knolls, clayey soils, gravelly soils with sandy or gypsum-rich clay; 300–1900 m; Ariz., Calif., Colo., Nev., Utah.

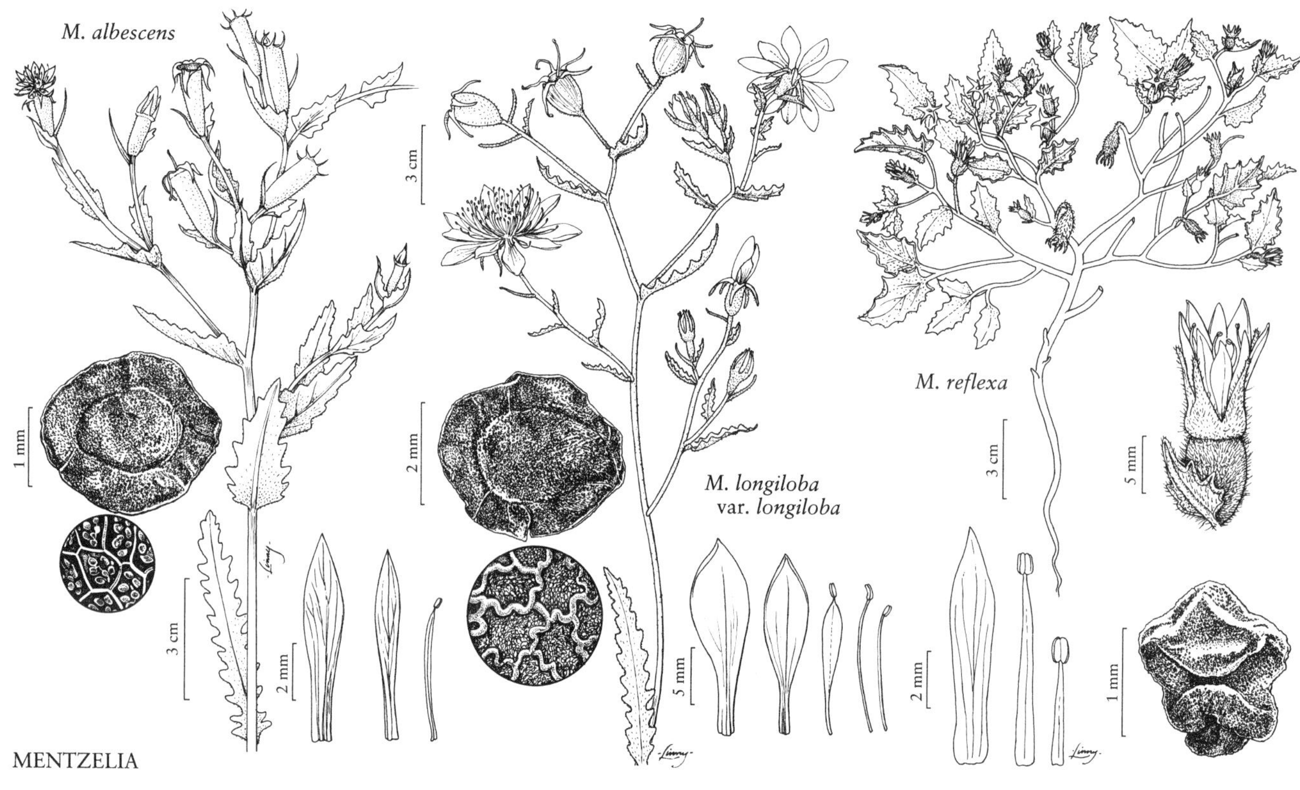

Mentzelia pterosperma occurs in the Colorado Plateau and Mojave Desert. Mojave Desert populations occur from eastern San Bernardino and southeastern Inyo counties, California, through Clark County, Nevada. Colorado Plateau populations are common in northern Arizona and extend northward to Utah and western Colorado. Populations in Utah occur in habitats composed of gravelly soils that are sandy or gypsum-rich clay, whereas Arizona and Nevada populations occur primarily on clayey soils, which sometimes contain gypsum.

39. Mentzelia albescens (Gillies ex Arnott) Bentham & Hooker f. ex Grisebach, Abh. Königl. Ges. Wiss. Göttingen 19: 150. 1874 • Wavyleaf blazingstar [F]

Bartonia albescens Gillies ex Arnott, Edinburgh J. Nat. Geogr. Sci. 3: 273. 1831; *B. wrightii* (A. Gray) Walpers

Herbs biennial, candelabraform. **Stems** solitary, erect, straight; branches distal, distal longest, antrorse, straight; hairy. **Leaves:** blade 31–92(–157) × 10.5–27.6(–41) mm, widest intersinus distance 5.2–23.3 (–29) mm; proximal lanceolate or elliptic, margins serrate to pinnate, teeth or lobes 8–22, slightly antrorse, 1.8–6 mm; distal lanceolate, base clasping, margins usually serrate to pinnate, occasionally entire, teeth or lobes (0–)10–20, slightly antrorse, 1.4–7.6 mm; abaxial surface with simple grappling-hook, complex grappling-hook, and needlelike trichomes, adaxial surface with simple grappling-hook and needlelike trichomes. **Bracts:** margins entire. **Flowers:** petals golden yellow, 5.7–9.2 × 1.3–3 mm, apex usually acute, occasionally rounded, glabrous abaxially; stamens golden yellow, 5 outermost petaloid, filaments narrowly spatulate, slightly clawed, 4.7–8.4 × 1–2.4 mm, without anthers, second whorl with anthers; anthers straight after dehiscence, epidermis smooth; styles 3.5–5.4 mm. **Capsules** cylindric, 13.6–23.5 × 5.1–7.8 mm, base tapering, not or slightly longitudinally ridged. **Seeds:** coat anticlinal cell walls straight, papillae 4–17 per cell. **2*n*** = 22.

Flowering May–Nov. Dry grasslands, xeric habitats of arroyos, roadsides, roadcuts, washes, chat piles, slopes; 200–1600 m; Kans., Mo., Okla., Tex.; South America (Argentina, Chile).

Phylogenetic analyses (J. J. Schenk and L. Hufford 2011) recovered representative populations of *Mentzelia albescens* from Texas and South America in a monophyletic group. Phylogenetic placement of these populations among lineages in sect. *Bartonia*, as well as a lack of morphological differentiation among North American and South American populations, suggests a recent dispersal to South America. In the flora area, this species is native to central and western Texas and introduced in Kansas, Missouri, and Oklahoma.

40. Mentzelia collomiae Christy, Novon 7: 25. 1997 • Sunset Crater blazingstar [C] [E]

Herbs biennial, candelabra-form. **Stems** solitary, erect, straight; branches distal, distal longest, antrorse, straight; hairy. **Leaves:** blade 17.8–55.2 × 3–16.6(–26.3) mm, widest intersinus distance 1.4–4.9 mm; proximal oblanceolate, lanceolate, or elliptic, margins dentate to pinnate, teeth or lobes 8–18, perpendicular to leaf axis, 0.6–5.3 mm; distal elliptic to lanceolate, base not clasping, margins dentate to pinnate, teeth or lobes 4–12, perpendicular to leaf axis, 0.7–6.8 mm; abaxial surface with simple grappling-hook, complex grappling-hook, and needlelike trichomes, adaxial surface with simple grappling-hook and needlelike trichomes. **Bracts:** margins usually entire, occasionally toothed. **Flowers:** petals golden yellow, 8.5–16.2 × 2.6–6.4 mm, apex acute to rounded, glabrous abaxially; stamens golden yellow, 5 outermost petaloid, filaments narrowly spatulate, strongly clawed, (6–)9.9–13.8 × (1–)2.4–4.3 mm, without anthers, second whorl with anthers; anthers straight after dehiscence, epidermis smooth; styles (5–)7.8–11 mm. **Capsules** cylindric, 9.3–16.1 × 3.6–5.9 mm, base tapering, not longitudinally ridged. **Seeds:** coat anticlinal cell walls sinuous, papillae 51–59 per cell. **2*n*** = 22.

Flowering Jun–Aug. Sparsely vegetated level ground, slopes, black (to sometimes reddish) volcanic cinder cone soils; of conservation concern; 1300–2500 m; Ariz.

Mentzelia collomiae is known only from the San Francisco Volcanic Field of Coconino County.

41. Mentzelia reverchonii (Urban & Gilg) H. J. Thompson & Zavortink, Wrightia 4: 24. 1968 • Reverchon's blazingstar

Mentzelia pumila Torrey & A. Gray var. *reverchonii* Urban & Gilg, Nova Acta Acad. Caes. Leop.-Carol. German. Nat. Cur. 76: 94. 1900; *M. hintoniorum* B. L. Turner & A. L. Hempel; *Nuttallia reverchonii* (Urban & Gilg) W. A. Weber

Herbs biennial or perennial, candelabra-form, perennials with ground-level caudices. **Stems** solitary, erect, straight; branches distal, distal longest, antrorse, straight to upcurved; hairy. **Leaves:** blade 24.2–54 × 6.6–29 mm, widest intersinus distance 5.4–13.5(–22.4) mm; proximal oblanceolate to elliptic, margins dentate to pinnate, teeth or lobes 10–20, perpendicular to leaf axis, 1.1–5 mm; distal lanceolate, base not clasping, margins dentate, teeth 8–14, perpendicular to leaf axis, 0.6–3.8 mm; abaxial surface with simple grappling-hook, complex grappling-hook, and occasionally needlelike trichomes, adaxial surface with simple grappling-hook and needlelike trichomes. **Bracts:** margins usually toothed, occasionally entire. **Flowers:** petals golden yellow, (9.6–)11.1–14.1(–22) × 2.5–4.2(–5.5) mm, apex acute to rounded, glabrous abaxially; stamens golden yellow, 5 outermost petaloid, filaments narrowly spatulate, slightly clawed, 9.4–13.4(–20) × 2.2–3.7(–5) mm, without anthers, second whorl with anthers; anthers straight after dehiscence, epidermis smooth; styles 8.7–11.4(–13) mm. **Capsules** usually cylindric, rarely cup-shaped, 9.4–22 × 6–9.1 mm, length usually more than, occasionally a few slightly less than, 2 times diam., base tapering, not longitudinally ridged. **Seeds:** coat anticlinal cell walls straight, papillae 5–7 per cell. **2*n*** = 18.

Flowering Apr–Oct. Grasslands on eroded riverbanks, roadsides, roadcuts, sparsely vegetated hillside slopes, sandy, gravelly, clayey, occasionally gypsum soils; 50–600 m; Colo., N.Mex., Okla., Tex.; Mexico (Coahuila, Nuevo León).

42. Mentzelia pumila Torrey & A. Gray, Fl. N. Amer. 1: 535. 1840 • Dwarf blazingstar [E] [W]

Herbs biennial, candelabra-form. **Stems** solitary, erect, straight; branches distal, distal longest, antrorse, straight; hairy. **Leaves:** blade 19–82 × 4.8–22.1 mm, widest intersinus distance 2.7–6.7 mm; proximal oblanceolate to elliptic, margins serrate to pinnate, teeth or lobes 10–22, slightly antrorse or perpendicular to leaf axis, 1–5.5 mm; distal elliptic to lanceolate, base not clasping, margins pinnate, lobes 8–14, usually perpendicular to leaf axis, 0.8–7 mm; abaxial surface with simple grappling-hook and complex grappling-hook trichomes, adaxial surface with simple grappling-hook and needlelike trichomes. **Bracts:** margins entire or toothed. **Flowers:** petals golden yellow, 7.6–12.3 × 2–3.8 mm, apex usually acute, occasionally rounded, glabrous abaxially; stamens golden yellow, 5 outermost petaloid, filaments narrowly spatulate, slightly clawed, 6.4–11.8 × 1.1–2.7 mm, without anthers, second whorl with anthers; anthers straight after dehiscence, epidermis smooth; styles 5.6–7.8 mm. **Capsules** cylindric, 10.8–20 × 5.3–7.6 mm, length more than 2 times diam., base tapering, not longitudinally ridged. **Seeds:** coat anticlinal cell walls straight, papillae 4–5 per cell. **2*n*** = 22.

Flowering May–Aug. Hillside slopes, sandy and clayey soils; 1100–2500 m; Colo., Mont., Wyo.

The name *Mentzelia pumila* has been applied variously in regional floras and in other taxonomic treatments of *Mentzelia*; however, we follow R. J. Hill

(1975) and N. H. Holmgren et al. (2005), which are consistent with our phylogenetic results (J. J. Schenk and L. Hufford 2011), in treating *M. pumila* as a species found only in Wyoming and adjacent areas of Colorado and Montana.

43. **Mentzelia longiloba** J. Darlington, Ann. Missouri Bot. Gard. 21: 176. 1934 [F]

Mentzelia multiflora (Nuttall) A. Gray subsp. *longiloba* (J. Darlington) Felger

Herbs biennial or perennial, bushlike or candelabra-form, perennials with ground-level caudices. **Stems** solitary, erect, straight; branches distal or along entire stem, distal or proximal longest, antrorse, upcurved; hairy. **Leaves:** blade 35–112 × 4.3–27.1 mm, widest intersinus distance 2.3–15.3 mm, always on some leaves 4+ mm; proximal oblanceolate to elliptic, margins dentate to pinnate, teeth or lobes 12–30(–50), perpendicular to leaf axis or antrorse, 1–9 mm; distal lanceolate, base not clasping, margins serrate to pinnate, teeth or lobes 8–28, perpendicular to slightly antrorse, 0.8–7.5 mm; abaxial surface with simple grappling-hook, complex grappling-hook, and generally needlelike trichomes, adaxial surface with simple grappling-hook and needlelike trichomes. **Bracts:** margins entire or toothed to pinnate. **Flowers:** petals golden yellow, 11.3–20.4 × 2.9–7.2[–8.9] mm, apex acute to rounded, glabrous abaxially; stamens golden yellow, 5 outermost petaloid, filaments narrowly spatulate, slightly clawed, 10.4–17.8 × 1.9–5.2[–5.6] mm, with or without anthers, second whorl with anthers; anthers straight after dehiscence, epidermis smooth; styles (5.4–)8.5–10.9(–13) mm. **Capsules** cup-shaped to cylindric, [7.6–]9.1–16.4 × 5.7–9.2 mm, length usually to 2 times diam., sometimes a few more, base tapering to rounded, not longitudinally ridged. **Seeds:** coat anticlinal cell walls straight or sinuous, papillae 4–106 per cell.

Varieties 4 (3 in the flora): sw, sc United States, n Mexico.

Variety *pinacatensis* J. J. Schenk & L. Hufford is known from the Mexican state of Sonora.

1. Seed coat anticlinal cell walls straight, papillae 4–6 per cell; margins of proximal leaf blades dentate . . . 43b. *Mentzelia longiloba* var. *chihuahuensis*
1. Seed coat anticlinal cell walls sinuous, papillae 10–106 per cell; margins of proximal leaf blades pinnate.
    2. 5 outermost stamens without anthers; leaf blade margin teeth or lobes perpendicular to leaf axis; seed coat cell papillae 67–106 per cell. . . . . .43a. *Mentzelia longiloba* var. *longiloba*
    2. 5 outermost stamens with anthers; leaf blade margin teeth or lobes slightly antrorse; seed coat cell papillae 10–21 per cell. . . . . 43c. *Mentzelia longiloba* var. *yavapaiensis*

43a. **Mentzelia longiloba** J. Darlington var. **longiloba** • Dune blazingstar [F]

Leaves: blade 38–112 × 10–24.6 mm, widest intersinus distance 3.3–13.7 mm; proximal with margins pinnate, lobes 12–30, perpendicular to leaf axis, 1.4–6.6 × 1.9–6.1 mm; distal with margins pinnate, lobes 10–28, perpendicular to leaf axis, 2.3–8.6 × 1.8–5.8 mm. **Bracts:** margins entire or toothed. **Flowers:** petals 12.2–20.4 × 3.7–7.2 mm; 5 outermost stamens without anthers. **Seeds** 3.3–4 mm; seed coat anticlinal cell walls sinuous, papillae 67–106 per cell. $2n = 18$.

Flowering Feb–Oct. Stable sand dunes, hills, washes; 40–1400 m; Ariz., Calif., Colo., N.Mex., Utah; Mexico (Baja California, Sonora).

43b. **Mentzelia longiloba** J. Darlington var. **chihuahuaensis** J. J. Schenk & L. Hufford, Madroño 57: 256. 2010 • Chihuahuan blazingstar

Leaves: blade 35–110 × 7.4–27.1 mm, widest intersinus distance 2.3–15.3 mm; proximal with margins dentate, teeth 12–20, perpendicular to leaf axis, 2.3–9 × 4–5 mm; distal with margins serrate to pinnate, teeth or lobes 10–14, slightly antrorse, 2.9–7.5 × 3.7–4.8 mm. **Bracts:** margins entire or pinnate. **Flowers:** petals 11.3–16.3 × 3.1–5.1 mm; 5 outermost stamens without anthers. **Seeds** 2.9–3.2 mm; seed coat anticlinal cell walls straight, papillae 4–6 per cell.

Flowering Aug–Nov. Roadsides, sand dunes, dry clayey or sandy soils; 500–1600 m; N.Mex., Tex.; Mexico (Chihuahua, Coahuila).

43c. Mentzelia longiloba J. Darlington var. **yavapaiensis** J. J. Schenk & L. Hufford, Madroño 57: 258, figs. 4C, 5B. 2010 • Yavapai blazingstar [E]

Leaves: blade 39–72.4 × 4.3–19.1 mm, widest intersinus distance 2.3–7.1 mm; proximal with margins pinnate, lobes 14–24, slightly antrorse, 1–6.5 × 0.8–3.2 mm; distal with margins pinnate, lobes 8–20, slightly antrorse, 0.8–5.8 × 0.7–2.8 mm. **Bracts:** margins entire. **Flowers:** petals 11.6–18.4 × 2.9–6.8 mm; 5 outermost stamens with anthers. **Seeds** 3–3.4 mm; seed coat anticlinal cell walls sinuous, papillae 10–21 per cell. **2*n*** = 18.

Flowering Mar–Oct. Sandy washes, roadsides in grasslands; 400–1700 m; Ariz.

Variety *yavapaiensis* is found in all counties in Arizona.

44. Mentzelia mexicana H. J. Thompson & Zavortink, Wrightia 4: 21. 1968 • Mexico blazingstar

Herbs biennial, bushlike or candelabra-form. **Stems** solitary, erect, straight; branches distal or along entire stem, distal or proximal longest, antrorse, upcurved; hairy. **Leaves:** blade 24–82 × 11.7–29.1 mm, widest intersinus distance 3.4–19 mm; proximal oblanceolate or elliptic, margins dentate, serrate, or pinnate, teeth or lobes 6–16, slightly antrorse or perpendicular to leaf axis, 1.4–5.1 mm; distal elliptic to lanceolate, base usually not clasping, sometimes a few clasping, margins dentate, serrate, or pinnate, teeth or lobes 6–12, slightly antrorse or perpendicular to leaf axis, 2.5–6.7 mm; abaxial surface with simple grappling-hook, complex grappling-hook, and usually needlelike trichomes, adaxial surface with simple grappling-hook and needlelike trichomes. **Bracts:** margins entire. **Flowers:** petals golden yellow, 10.2–15.3(–17) × 2.5–6.2 mm, apex rounded, glabrous abaxially; stamens golden yellow, 5 outermost petaloid, filaments narrowly spatulate, slightly clawed, 9.9–14.4(–16.7) × 2–4.8 mm, without anthers, second whorl with anthers; anthers straight after dehiscence, epidermis smooth; styles 4.5–10.9 mm. **Capsules** cup-shaped, 7.1–12.8 × 5.6–8.4 mm, length to 2 times diam., base rounded, not longitudinally ridged. **Seeds:** coat anticlinal cell walls straight, papillae 8–12 per cell. **2*n*** = 18.

Flowering Mar–Oct. Arroyos, knolls, steep slopes, gypsum and limestone clay and shale; 700–1500 m; Tex.; Mexico (Chihuahua, Coahuila, Nuevo León, Zacatecas).

In the flora area, *Mentzelia mexicana* occurs in Brewster, Hudspeth, and Presidio counties.

45. Mentzelia integra (M. E. Jones) Tidestrom, Contr. U.S. Natl. Herb. 25: 363. 1925 • Virgin blazingstar [E]

Mentzelia multiflora (Nuttall) A. Gray var. *integra* M. E. Jones, Proc. Calif. Acad. Sci., ser. 2, 5: 689. 1895; *Nuttallia lobata* Rydberg 1913, not *M. lobata* (Hooker) Hemsley 1880

Herbs winter annual or biennial, candelabra-form. **Stems** solitary, erect, straight, branches distal or along entire stem, distal or proximal longest, antrorse, straight, proximal not decumbent; hairy or glabrescent. **Leaves:** blade (23–)29–78.4 × 3–14.6 mm, widest intersinus distance 1.6–8.5 mm; proximal oblanceolate or elliptic, margins dentate to serrate, teeth 4–12, proximal sides slightly antrorse, distal sides perpendicular to leaf axis, 0.8–5.3 mm; distal oblanceolate, elliptic, lanceolate, or linear, base not clasping, tapered, margins usually dentate to serrate, rarely entire, teeth (0–)2–10, proximal sides antrorse, distal sides perpendicular to leaf axis, 0.4–5.3 mm; abaxial surface with simple grappling-hook, complex grappling-hook, and needlelike trichomes, largest trichomes with pearly white bases, adaxial surface with needlelike trichomes. **Bracts:** margins entire. **Flowers:** petals golden yellow, 8.6–13.9(–17.4) × 2.9–4.7(–6.1) mm, apex rounded, glabrous abaxially; stamens golden yellow, 5 outermost petaloid, filaments narrowly spatulate, slightly clawed, 6.7–13.2(–15.4) × 1.7–4.3 mm, usually without, rarely with, anthers, second whorl with anthers; anthers straight after dehiscence, epidermis smooth; styles 4.8–11.3 mm. **Capsules** cup-shaped, 6.2–11.4(–13) × 6–8.4 mm, base rounded, not longitudinally ridged. **Seeds:** coat anticlinal cell walls wavy, papillae 5–13 per cell. **2*n*** = 20.

Flowering May–Sep. Roadsides, outcrops, hillsides, washes, dunes, sandy, gravelly, or volcanic soils; 800–1800 m; Ariz., Nev., Utah.

Mentzelia integra is found in northwestern Arizona, southeastern Nevada, and southwestern Utah.

46. Mentzelia densa Greene, Pittonia 3: 99. 1896 • Royal Gorge blazingstar, Arkansas Canyon stickleaf [C] [E]

Nuttallia densa (Greene) Greene

Herbs biennial, bushlike. **Stems** solitary, decumbent to erect, straight; branches along entire stem, distal or proximal longest, antrorse, straight or upcurved, proximal decumbent; hairy. **Leaves:** blade 32–86 × 5.8–15 mm, widest intersinus distance 1.7–5.2(–7.1) mm; proximal oblanceolate to elliptic, margins dentate to pinnate, teeth or lobes 6–12, perpendicular to leaf axis, 1.2–6.3 mm; distal elliptic to lanceolate, base not clasping, margins dentate to pinnate, teeth or lobes 4–10, perpendicular to leaf axis, 1.8–5.2 mm; abaxial surface with simple grappling-hook, complex grappling-hook, and occasionally needlelike trichomes, largest trichomes without pearly white bases, adaxial surface with simple grappling-hook and needlelike trichomes. **Bracts:** margins entire or toothed. **Flowers:** petals golden yellow, 14.2–19.8 × 3.8–6.5 mm, apex acute to rounded, glabrous abaxially; stamens golden yellow, 5 outermost petaloid, filaments narrowly spatulate, slightly clawed, 12–18.1 × 2.7–4.8 mm, with or without anthers, second whorl with anthers; anthers straight after dehiscence, epidermis smooth; styles 8.3–11.1 mm. **Capsules** cylindric, 12.2–18 × 4.3–8.3 mm, base tapering, not longitudinally ridged. **Seeds:** coat anticlinal cell walls wavy, papillae 7–17 per cell. $2n = 20$.

Flowering Jun–Aug. Moist canyon walls, lower talus slopes, gravelly and sandy soils; of conservation concern; 1600–2400 m; Colo.

Mentzelia densa is known only from Chaffee and Fremont counties. It is in the Center for Plant Conservation's National Collection of Endangered Plants.

47. Mentzelia speciosa Osterhout, Bull. Torrey Bot. Club 28: 689. 1901 • Jeweled blazingstar, showy western star [E]

Mentzelia aurea Osterhout, Bull. Torrey Bot. Club 28: 644. 1901, not Nuttall 1818; *M. sinuata* (Rydberg) R. J. Hill; *Nuttallia sinuata* (Rydberg) Daniels; *N. speciosa* (Osterhout) Greene; *Touterea sinuata* Rydberg

Herbs biennial, bushlike or candelabra-form. **Stems** solitary, erect, straight; branches distal or along entire stem, distal or proximal longest, antrorse, upcurved; hairy. **Leaves:** blade 43.6–233 × 9.2–34.7 mm, always some longer than 146 mm, widest intersinus distance 5.2–17 mm; proximal oblanceolate to elliptic, margins serrate to pinnate, teeth or lobes 10–38, slightly antrorse, 1.5–9.4 mm; distal lanceolate, base not clasping, margins serrate to pinnate, teeth or lobes 8–24, slightly antrorse, 2.9–10.3 mm; abaxial surface with simple grappling-hook, complex grappling-hook, and needlelike trichomes, adaxial surface with simple grappling-hook and needlelike trichomes. **Bracts:** margins entire. **Flowers:** petals golden yellow, 14.4–28.6 × 3.8–8 mm, apex acute to rounded, glabrous abaxially; stamens golden yellow, 5 outermost petaloid, filaments narrowly spatulate, slightly clawed, 13–21.3 × 2.9–6.8 mm, without anthers, second whorl with anthers; anthers straight after dehiscence, epidermis smooth; styles 8.9–15.2 mm. **Capsules** cylindric, 13.1–31.1 × 5.7–9.4 mm, base tapering to rounded, not longitudinally ridged. **Seeds:** coat anticlinal cell walls wavy, papillae 4–13 per cell. $2n = 18, 20$.

Flowering Jun–Sep. Dry hillside slopes, roadcuts, roadsides, reddish, rocky soils; 1500–3000 m; Colo., Wyo.

Mentzelia sinuata is treated here as a synonym of *M. speciosa*. The morphological characters previously used to differentiate these taxa are much more variable and overlapping than previous authors have identified, including within material from type populations. Although typical morphological forms of each taxon can be found, intermediate forms are found throughout the putative species' highly overlapping ranges, often in a single population. The main character used to differentiate the two species has been chromosome number, with *M. sinuata* having $2n = 18$ and *M. speciosa* having $2n = 20$. Until a cytological study is conducted on populations throughout the range, *M. speciosa* is regarded as a single species that contains two chromosomal lineages.

48. Mentzelia sivinskii J. J. Schenk & L. Hufford, Madroño 57: 253, fig. 4A. 2010 • Sivinski's blazingstar [E]

Mentzelia linearifolia S. L. Welsh & N. D. Atwood

Herbs biennial, bushlike or candelabra-form. **Stems** solitary, erect, straight; branches distal or along entire stem, distal or proximal longest, antrorse, straight; hairy. **Leaves:** blade 32.8–112.2 × 2.9–11.4 mm, widest intersinus distance 1–2.9 mm; proximal oblanceolate to elliptic, margins pinnate, lobes 18–24, perpendicular to leaf axis, 0.8–4 mm; distal elliptic to lanceolate, base not clasping, margins pinnate, lobes 6–16, perpendicular to leaf axis, 1–5.1 mm; abaxial surface with simple grappling-hook, complex grappling-hook, and needlelike trichomes, adaxial surface with needlelike and occasionally simple grappling-hook trichomes. **Bracts:** margins entire. **Flowers:** petals light to golden yellow, 9–14.7 × 3.1–6.4 mm, apex rounded, glabrous abaxially; stamens light to golden yellow, 5 outermost petaloid, filaments narrowly spatulate, slightly clawed, 6.3–11.5 × 2.4–4.9 mm, with anthers, second whorl with anthers; anthers straight after dehiscence, epidermis smooth; styles 4.6–9.9 mm. **Capsules** cup-shaped, 8.2–12.7 × 5.1–7.7 mm, base tapering to rounded, not longitudinally ridged. **Seeds:** coat anticlinal cell walls sinuous, papillae 12–21 per cell. **2*n*** = 18.

Flowering Jun–Aug. Knolls, slopes, and grassland roadsides, gypsum or brown clayey soils; 1500–1900 m; Colo., N.Mex.

Mentzelia sivinskii is known from La Plata County, Colorado, and Rio Arriba and San Juan counties, New Mexico.

49. Mentzelia procera (Wooton & Standley) J. J. Schenk & L. Hufford, Madroño 57: 247. 2010 • Upright blazingstar

Nuttallia procera Wooton & Standley, Contr. U.S. Natl. Herb. 16: 150. 1913; *Mentzelia pumila* Torrey & A. Gray var. *procera* (Wooton & Standley) J. Darlington

Herbs biennial, usually candelabra-form, rarely bushlike. **Stems** solitary, erect, straight; branches distal or occasionally along entire stem, distal or proximal longest, antrorse, straight or upcurved; hairy. **Leaves:** blade 38.3–108 × 5.3–14.9 mm, widest intersinus distance 1.7–3.9 mm; proximal oblanceolate to elliptic, margins pinnate, lobes 14–22(–26), perpendicular to leaf axis, 1.8–4.5 mm; distal elliptic to lanceolate, base not clasping, margins pinnate, lobes 6–18, perpendicular to leaf axis, 1.3–4.5 mm; abaxial surface with simple grappling-hook, complex grappling-hook, and occasionally needlelike trichomes, adaxial surface with simple grappling-hook and needlelike trichomes. **Bracts:** margins entire or toothed to pinnate. **Flowers:** petals light yellow, 9.5–16.8 × 3.4–5.6 mm, apex usually rounded, rarely acute, glabrous abaxially; stamens light yellow, 5 outermost petaloid, filaments narrowly spatulate, slightly clawed, 8.2–14.6 × 2.3–3.4 mm, without anthers, second whorl with anthers; anthers straight after dehiscence, epidermis smooth; styles 7.1–12.6 mm. **Capsules** cup-shaped to cylindric, 9.8–18.8 × 5.2–7.3 mm, base tapering, not longitudinally ridged. **Seeds:** coat anticlinal cell walls sinuous, papillae 28–68 per cell.

Flowering Jul–Aug. Dry hillsides, roadcuts, roadsides, sandy, clayey, or silty soils; 1400–2500 m; Ariz., Colo., N.Mex.; n Mexico.

50. Mentzelia chrysantha Engelmann, Bull. U.S. Geol. Geogr. Surv. Territ. 2: 237. 1876 • Golden blazingstar [C] [E]

Nuttallia chrysantha (Engelmann) Greene

Herbs biennial, bushlike or candelabra-form. **Stems** solitary, erect, straight; branches distal or along entire stem, distal or proximal longest, antrorse, straight; hairy. **Leaves:** blade 38–113 × 8–20.5(–25.2) mm, widest intersinus distance 5.4–12.8 mm; proximal oblanceolate, lanceolate, or elliptic, margins serrate to pinnate, teeth or lobes 14–22, slightly antrorse, 0.6–4.6 mm; distal elliptic or lanceolate, base not clasping, margins serrate to pinnate, teeth or lobes 6–22, slightly antrorse, 0.6–4 mm; abaxial surface with simple grappling-hook, complex grappling-hook, and needlelike trichomes, adaxial surface with simple grappling-hook and needlelike trichomes. **Bracts:** margins toothed. **Flowers:** petals light to golden yellow, 12.5–21.4 × 3.8–6.7 mm, apex acute to rounded, glabrous abaxially; stamens light to golden yellow, 5 outermost petaloid, filaments narrowly spatulate, slightly clawed, 11.5–18.6 × 2–5.8 mm, without anthers, second whorl with anthers; anthers straight after dehiscence, epidermis smooth; styles 8.1–13.1 mm. **Capsules** cup-shaped to cylindric, 9.4–16 × 5.1–7.7 mm, base tapering, not longitudinally ridged. **Seeds:** coat anticlinal cell walls sinuous, papillae 29–35 per cell. **2*n*** = 20.

Flowering Jul–Sep. Steep hillsides, washes, clayey soils, sometimes rich in gypsum; of conservation concern; 1600–1900 m; Colo.

Mentzelia chrysantha occurs in Custer and Fremont counties.

51. Mentzelia multiflora (Nuttall) A. Gray, Mem. Amer. Acad. Arts, n. s. 4: 48. 1849 • Adonis blazingstar, many-flowered western star [E] [W]

Bartonia multiflora Nuttall, Proc. Acad. Nat. Sci. Philadelphia 4: 23. 1848; *Mentzelia lutea* Greene; *Nuttallia multiflora* (Nuttall) Greene

Herbs biennial, candelabra-form. **Stems** solitary, erect, straight; branches distal, distal longest, antrorse, upcurved; hairy. **Leaves:** blade 35.9–125(–146) × 2.2–26.2 mm, widest intersinus distance 2.1–11.9(–14) mm, always on some leaves 4+ mm; proximal oblanceolate, lanceolate, or elliptic, margins pinnate, lobes 14–30, slightly antrorse or perpendicular to leaf axis, 1.5–4.6 (–8.9) mm; distal lanceolate, base not clasping, margins pinnate, lobes 10–26, slightly antrorse or perpendicular to leaf axis, 1.3–9.5 mm; abaxial surface with simple grappling-hook, complex grappling-hook, and usually needlelike trichomes, adaxial surface with simple grappling-hook or needlelike trichomes. **Bracts:** margins entire or toothed to pinnate. **Flowers:** petals light to golden yellow, (11.4–)13.8–24.4(–26.9) × 4–7.5 mm, apex rounded, glabrous abaxially; stamens light to golden yellow, 5 outermost petaloid, filaments narrowly spatulate, slightly clawed, (10.6–)12–21.7 × 2.1–5.1 mm, without anthers, second whorl with anthers; anthers straight after dehiscence, epidermis smooth; styles 7.7–15.8 mm. **Capsules** cylindric, 11.2–24.7 × 5.6–8.7 mm, base tapering to rounded, not longitudinally ridged. **Seeds:** coat anticlinal cell walls sinuous, papillae 34–48 per cell. $2n$ = 18.

Flowering Jul–Sep. Dry roadsides, hillsides, washes, clayey, rocky, or sandy soils; 1200–2100 m; Colo., Nebr., N.Mex., Wyo.

Mentzelia multiflora has been considered one of the most widespread species in sect. *Bartonia*, a result of treating the species as a "garbage bin" for populations that lack features characteristic of more specialized species. The phylogenetic study by J. J. Schenk and L. Hufford (2011) showed that populations consistent with the type of *M. multiflora* are centered in the southern Rocky Mountains, especially along their eastern front, and the species notably does not occur in Arizona, California, Nevada, or Utah, in which it regularly has been described in regional floras. In the intermountain region, many specimens previously determined as *M. multiflora* are *M. longiloba*. In southeastern New Mexico and Texas, many specimens previously determined as *M. multiflora* are likely to be *M. procera* or *M. longiloba* var. *chihuahuaensis*.

3b. Mentzelia Linnaeus sect. Bicuspidaria S. Watson, Proc. Amer. Acad. Arts 20: 367. 1885

Joshua M. Brokaw

Herbs, annual. **Leaves:** basal rosette present, not persistent, proximalmost internodes to 10+ mm; blade without broad basal lobes, margins toothed to pinnate, flat. **Flowers:** stamens all fertile or 1–3(–5) outermost sterile, filaments dorsiventrally flattened, ± monomorphic, linear, or heteromorphic, 1–3(–5) outermost oblanceolate, petaloid, inner linear, distally 2-lobed or unlobed; ovules (and seeds) oriented perpendicular or ± parallel to long axis of ovary. **Capsules** cup-shaped, cylindric, or ovoid, straight. **Seeds** pyriform to ovoid and flattened dorsiventrally, not winged; seed coat cells polygonal, ± isodiametric, anticlinal walls straight or wavy.

Species 5 (5 in the flora): sw United States, nw Mexico.

SELECTED REFERENCE Daniels, G. S. 1970. The Floral Biology and Taxonomy of *Mentzelia* Section *Bicuspidaria* (Loasaceae). Ph.D. dissertation. University of California, Los Angeles.

1. 1–3(–5) outermost stamens staminodial and petaloid (flowers appearing to have 6–8(–10) petals); stamen filaments unlobed; petals lanceolate to oblanceolate, 6–12 mm, apices acute; styles 5–6.5 mm . . . . 54. *Mentzelia reflexa*
1. Stamens all fertile, none petaloid (flowers appearing to have 5 petals); stamen filaments all or most distally 2-lobed; petals obovate, 10–65 mm, apices mucronate; styles 6–30 mm.
    2. Bracts conspicuous, concealing pedicels, ovaries, and capsules, white with green margins . . . . 53. *Mentzelia involucrata*
    2. Bracts inconspicuous, not concealing pedicels, ovaries, or capsules, green.
        3. Stamen filament lobes 0.6–2.5 mm, anther stalks usually shorter; seeds constricted and grooved at middle . . . . 55. *Mentzelia tricuspis*
        3. Stamen filament lobes 0.1–0.7 mm, anther stalks equal or longer; seeds widest at middle, not grooved.
            4. Petioles absent in cauline leaves, blade bases often cordate-clasping; capsules erect; far sw Sonoran Desert . . . . 52. *Mentzelia hirsutissima*
            4. Petioles present in distalmost cauline leaves, blade bases not cordate-clasping; proximal capsules nodding, distal erect; c Mojave Desert . . . . 56. *Mentzelia tridentata*

52. Mentzelia hirsutissima S. Watson, Proc. Amer. Acad. Arts 12: 252. 1877 • Hairy blazingstar

Mentzelia hirsutissima var. *stenophylla* (Urban & Gilg) I. M. Johnston; *M. stenophylla* Urban & Gilg

Herbs (5–)15–30(–40) cm. **Basal leaves:** petiole present or absent; blade oblanceolate, to 11 cm; margins shallowly lobed, lobes rounded. **Cauline leaves:** petiole absent; blade ovate to lanceolate, to 11 cm, base often cordate-clasping, margins deeply to shallowly lobed, lobes acute. **Bracts** green, inconspicuous, not concealing pedicel, ovary, or capsule. **Flowers:** petals yellow to orange, obovate, 12–31 mm, apex mucronate; stamens 4–12 mm, less than ½ petal length, all fertile, none petaloid (flowers appearing to have 5 petals); filaments ± monomorphic, linear, most or all distally 2-lobed, lobes to 0.3 mm; anther stalk longer than filament lobes; style 6–15 mm. **Capsules** cylindric to cup-shaped, 13–25 × 5–8 mm, erect. **Seeds** widest at middle, not grooved; seed coat anticlinal cell walls deeply wavy.

Flowering Apr–May. Washes, fans, slopes, desert scrub; 0–800 m; Calif.; Mexico (Baja California, Baja California Sur, Sonora).

Mentzelia hirsutissima is superficially similar to *M. tricuspis* and *M. tridentata*, but these taxa can be distinguished geographically. *Mentzelia hirsutissima* is distributed primarily on the Baja California Peninsula and occurs in California only in San Diego and far western Imperial counties, whereas *M. tricuspis* and *M. tridentata* occur north and east of this distribution. California populations previously have been called *M. hirsutissima* var. *stenophylla*. However, G. S. Daniels (1970) found that characters used to distinguish varieties of *M. hirsutissima* were not stable within populations.

53. Mentzelia involucrata S. Watson, Proc. Amer. Acad. Arts 20: 367. 1885 • Whitebract blazingstar

Mentzelia involucrata var. *megalantha* I. M. Johnston

Herbs 7–35(–45) cm. **Basal leaves:** petiole present or absent; blade lanceolate, to 10(–16) cm; margins shallowly to deeply lobed, lobes acute. **Cauline leaves:** petiole absent; blade ovate to lanceolate, to 10(–16) cm, base often cordate-clasping, margins deeply to shallowly lobed, lobes acute. **Bracts** white with green margins, conspicuous, concealing pedicel, ovary, and capsule. **Flowers:** petals white to pale yellow, usually with orange veins, obovate, 13–65 mm, apex mucronate; stamens 4–26 mm, less than ½ petal length, all fertile, none petaloid (flowers appearing to have 5 petals); filaments ± monomorphic, linear, distally 2-lobed, lobes 0.5–2 mm; anther stalk shorter or longer than filament lobes; style 8–30 mm. **Capsules** cylindric to cup-shaped, 14–25(–30) × 5–10 mm, erect. **Seeds** usually constricted proximal to middle; seed coat anticlinal cell walls wavy. $2n = 18$.

Flowering Jan–May. Washes, fans, slopes, desert scrub; 50–900 m; Ariz., Calif.; Mexico (Baja California, Sonora).

Mentzelia involucrata is easily distinguished from other species in sect. *Bicuspidaria* by its large white bracts, which are strikingly similar to those of *M. congesta* in sect. *Trachyphytum*. Phylogenetic studies show that these species are not closely related, suggesting a homoplasious origin of the characteristic (J. M. Brokaw and L. Hufford 2010). Populations exhibiting petals longer than 3 cm have previously been called var. *megalantha*. However, a study by G. S. Daniels (1970) suggested that insufficient geographic and morphological discontinuities exist for recognition of varieties.

P. A. Munz (1974) mistakenly cited var. *megalantha* as *Mentzelia involucrata* subsp. *megalantha* I. M. Johnston, a name never validly published.

54. Mentzelia reflexa Coville, Proc. Biol. Soc. Wash. 7: 74. 1892 • Reflexed blazingstar [E] [F]

Herbs 2–20 cm. **Basal leaves:** petiole present; blade oblanceolate to lanceolate, to 10 cm; margins shallowly lobed, lobes rounded. **Cauline leaves:** petiole usually present, rarely absent; blade broadly ovate to lanceolate, to 10 cm, base not cordate-clasping, margins dentate to serrate. **Bracts** green, inconspicuous, not concealing pedicel, ovary, or capsule. **Flowers:** petals pale yellow to white, lanceolate to oblanceolate, 6–12 mm, apex acute; stamens 3–12 mm (fertile 3–8 mm), ½ to ± equal petal length, 1–3(–5) outermost staminodial and petaloid (flowers appearing to have 6–8(–10) petals); filaments heteromorphic, 1–3(–5) outermost oblanceolate, inner linear, all unlobed; style 5–6.5 mm. **Capsules** cylindric to ovoid, 9–13 × 5–7 mm, proximal nodding, distal erect. **Seeds** constricted and grooved at middle; seed coat anticlinal cell walls straight, obscure. $2n = 20$.

Flowering Mar–May. Washes, rocky flats, slopes, roadsides, desert scrub; 0–1600 m; Calif., Nev.

Mentzelia reflexa has been placed in other sections, including *Octopetaleia* Urban & Gilg (I. Urban and E. Gilg 1900) and *Bartonia* (J. Darlington 1934), due to the presence of three petaloid staminodes, giving an appearance of eight nearly equal petals (specimens

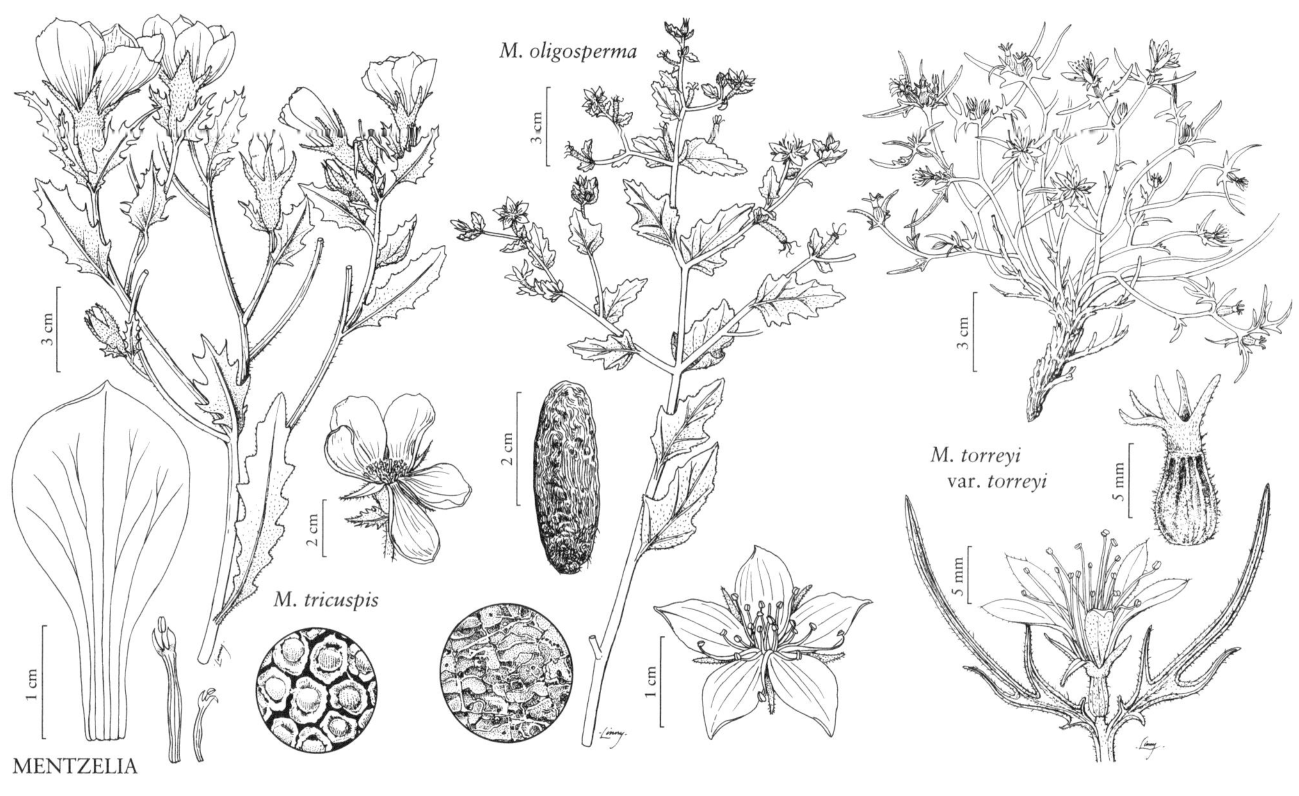

subsequently have been found with as few as one and rarely as many as five staminodes). However, G. S. Daniels (1970) placed it in sect. *Bicuspidaria* based on characteristics including leaf, capsule, and seed shapes, which are very similar to those of *M. tricuspis* and *M. tridentata*. Phylogenetic studies subsequently have supported inclusion of *M. reflexa* in sect. *Bicuspidaria* (L. Hufford et al. 2003).

Mentzelia reflexa is found in the eastern Mojave Desert.

55. **Mentzelia tricuspis** A. Gray, Amer. Naturalist 9: 271. 1875 • Spinyhair or three-pointed blazingstar E F

Herbs 5–30 cm. **Basal leaves:** petiole present; blade lanceolate, to 12 cm; margins shallowly lobed, lobes rounded. **Cauline leaves:** petiole usually present, rarely absent; blade broadly ovate to lanceolate, to 12 cm, base not cordate-clasping, margins dentate to serrate. **Bracts** green, inconspicuous, not concealing pedicel, ovary, or capsule. **Flowers:** petals white to pale yellow, obovate, 10–30(–50) mm, apex mucronate; stamens 7–17 mm, less than ½ petal length, all fertile, none petaloid (flowers appearing to have 5 petals); filaments ± monomorphic, linear, distally 2-lobed, lobes 0.6–2.5 mm; anther stalk usually shorter than filament lobes; style 10–15 mm. **Capsules** cylindric to ovoid, 9–18(–23) × 5–8 mm, proximal nodding, distal erect. **Seeds** constricted and grooved at middle; seed coat anticlinal cell walls straight to slightly wavy, conspicuous. $2n = 20$.

Flowering Mar–May. Sandy or gravelly slopes, washes, desert scrub; 150–1300 m; Ariz., Calif., Nev.

Mentzelia tricuspis is morphologically most similar to *M. tridentata* with differences largely limited to the lengths of the stamen filament lobes (and thus relative lengths of the anther stalks) and shapes of the seeds. However, the species are also distinct geographically, with *M. tricuspis* found in the southeastern Mojave Desert and northwestern Sonoran Desert and *M. tridentata* in the central to western Mojave Desert.

56. **Mentzelia tridentata** (Davidson) H. J. Thompson & J. E. Roberts, Phytologia 21: 287. 1971 • Threetooth blazingstar C E

Acrolasia tridentata Davidson, Bull. S. Calif. Acad. Sci. 9: 71. 1910; *Mentzelia tricuspis* A. Gray var. *brevicornuta* I. M. Johnston

Herbs 5–25 cm. **Basal leaves:** petiole present; blade lanceolate, to 9 cm; margins shallowly lobed, lobes rounded. **Cauline leaves:** petiole present, rarely absent, present in distalmost leaves; blade broadly ovate to lanceolate, to 9 cm, base not cordate-clasping, margins dentate to serrate. **Bracts** green, inconspicuous, not concealing pedicel, ovary, or capsule. **Flowers:** petals white to pale yellow, obovate, 10–40 mm, apex mucronate; stamens 6–15 mm, less than ½ petal length, all fertile, none petaloid (flowers appearing to have 5 petals); filaments ± monomorphic, linear, most or all distally 2-lobed, lobes 0.1–0.7 mm; anther stalk longer than or equal to filament lobes; style 9–13 mm. **Capsules** cylindric to ovoid, 9–18 × 5–8 mm, basal nodding, distal erect. **Seeds** widest at middle, not grooved; seed coat anticlinal cell walls straight or wavy. $2n = 20$.

Flowering Apr–May. Sandy or gravelly slopes, washes, desert scrub; of conservation concern; 600–1300 m; Calif.

Mentzelia tridentata is known from the central and western Mojave Desert in Inyo, Kern, and San Bernardino counties. See 55. *M. tricuspis* for comparison with *M. tridentata.*

3c. Mentzelia Linnaeus sect. Mentzelia

Larry D. Hufford

Herbs [subshrubs or shrubs], annual or perennial. **Leaves:** basal rosette absent, proximalmost internodes 10+ mm; blade, at least some, with broad basal lobes, margins entire, toothed, or crenate, flat. **Flowers:** stamens all fertile; filaments monomorphic, filiform, or heteromorphic, outer dorsiventrally flattened, spatulate [linear or lanceolate], not petaloid, inner filiform, unlobed; ovules (and seeds) oriented parallel to long axis of ovary. **Capsules** usually subcylindric, clavate, lingulate, or funnelform, sometimes ovoid, straight or curved. **Seeds** oblong, ovoid, or pyriform, dorsiventrally flattened to trigonal, not winged; seed coat cells oblong, much longer than wide, anticlinal walls sinuous.

Species ca. 23 (7 in the flora): United States, Mexico, West Indies, Central America, South America; introduced in Atlantic Islands (Cape Verde Islands).

Reports of *Mentzelia hispida* Willdenow from Texas (for example, by B. L. Turner et al. 2003) appear to be misidentifications; *M. hispida* is regarded here as limited to Mexico (see R. McVaugh 2001c), although its distribution approaches the southwestern border of the United States.

In the descriptions that follow, stem length is measured to the first terminal flower.

1. Capsule bases rounded or cuneate, capsules and pedicels well differentiated; fruiting pedicels less than 1 mm diam.
    2. Stamen filaments heteromorphic, 5 outermost narrowly spatulate, inner filiform . . . . 59. *Mentzelia floridana*
    2. Stamen filaments monomorphic, filiform. . . . . 61. *Mentzelia lindheimeri*
1. Capsule bases tapering gradually, capsules and pedicels not well differentiated; fruiting pedicels absent or more than 1 mm diam.
    3. Plants perennial, with caudices or tubers; seeds (1–)2–3(–4) per capsule; petals hairy abaxially on distal ½.
        4. Petioles to 3 mm in proximal leaves, absent in distal leaves; capsule walls thick, woody; seeds without transverse folds; plants with caudices. . . . . 62. *Mentzelia oligosperma*
        4. Petioles mostly 3–20 mm, less than 3 mm only on smallest, distalmost leaves; capsule walls thin, brittle; seeds with transverse folds; plants with tubers . . . 63. *Mentzelia pachyrhiza*
    3. Plants annual, without caudices or tubers; seeds 5–12 per capsule; petals hairy on apex or abaxially near or at apex.
        5. Stamen filaments monomorphic, filiform; leaf blades to 4.5 × 3.5 cm. . . . . 58. *Mentzelia asperula*
        5. Stamen filaments heteromorphic, 5 outermost slightly to narrowly spatulate, inner filiform; leaf blades to 14–18 × 4.5–10 cm.
            6. Stamens 20–30; fruiting pedicels 1–3 mm . . . . 57. *Mentzelia aspera*
            6. Stamens 8–12; fruiting pedicels 0.5–0.8 mm . . . . 60. *Mentzelia isolata*

57. Mentzelia aspera Linnaeus, Sp. Pl. 1: 516. 1753 • Rough stickleaf, dal-pega

Herbs annual, without caudices or tubers. **Stems** erect to decumbent, to 30 cm. **Leaves:** petiole 10–65 mm; blade hastate to ovate, usually basally lobed, sometimes unlobed, to 18 × 10 cm, base cuneate to truncate, margins serrate to crenate, apex acute. **Pedicels** (fruiting) 1–3 × 2 mm (often appearing absent because thick and continuous with capsule). **Flowers:** petals orange or yellow, 5–15 × 3–7 mm, apex cuspidate, hairy abaxially at apex; stamens 20–30, 5 mm, filaments heteromorphic, 5 outermost narrowly spatulate, inner filiform; style 5 mm. **Capsules** subcylindric to clavate, 9–30 × 3–3.5 mm, base tapering gradually, capsule and pedicel not well differentiated. **Seeds** 5–9 per capsule, pyriform to oblong, without transverse folds. **2*n*** = 20.

Flowering Aug–Oct. Arroyo and canyon bottoms, grasslands, desert scrub, riparian cottonwood and willow vegetation; 100–2000 m; Ariz.; Mexico; West Indies; Central America; South America; introduced in Atlantic Islands (Cape Verde Islands).

Mentzelia aspera is uncommon in southern Arizona. It is the most widespread species of the genus and is regarded as weedy by some authors (H. J. Thompson and A. M. Powell 1981).

58. Mentzelia asperula Wooton & Standley, Contr. U.S. Natl. Herb. 16: 148. 1913 • Mountain stickleaf, Organ Mountain blazingstar

Herbs annual, without caudices or tubers. **Stems** erect, to 25 cm. **Leaves:** petiole to 12 mm (proximal leaves), absent (distal leaves); blade usually ovate to hastate, or smallest distal elliptic or lanceolate, basally lobed or unlobed, to 4.5 × 3.5 cm, base cuneate or obtuse to truncate, margins serrate, apex acute. **Pedicels** (fruiting) 1–3 × 2 mm. **Flowers:** petals orange, 5–8 × 3–5 mm, apex cuspidate, hairy abaxially at apex; stamens 10–20, 5–8 mm, filaments monomorphic, filiform; style 3–5 mm. **Capsules** subcylindric to clavate, 12–25 × 3–5 mm, base tapering gradually, capsule and pedicel not well differentiated. **Seeds** (7–)8–10(–12) per capsule, pyriform, without transverse folds. **2*n*** = 20, 40.

Flowering Aug–Oct. Rocky limestone or igneous slopes, arroyo bottoms, grasslands, oak woodlands; 0–1800 m; Ariz., N.Mex., Tex.; n, c Mexico.

Populations consistent with the form and geographic region of the type specimen of *Mentzelia asperula* (collected in southwestern New Mexico) have been recovered in two clades of sect. *Mentzelia*. In one clade, *M. asperula* is closely related to *M. isolata*; in the second clade, it is closely related to *M. gypsophila* B. L. Turner of northern Mexico (J. Grissom 2014). The polyphyly of *M. asperula* likely represents convergence on rapid developmental times and small, self-fertilizing flowers. In southeastern Arizona, *M. asperula* can be difficult to distinguish from 60. *M. isolata*; see the discussion of the latter for more information.

59. Mentzelia floridana Nuttall ex Torrey & A. Gray, Fl. N. Amer. 1: 533. 1840 • Florida stickleaf, poorman's patches

Herbs perennial, with caudices. **Stems** erect, becoming decumbent, to 60 cm. **Leaves:** petiole to 25 mm; blade usually hastate or ovate, distal sometimes elliptic, basally lobed or unlobed, to 8.4 × 5.5 cm, base usually truncate, sometimes obtusely cuneate, margins usually serrate or dentate to crenate, sometimes entire, apex acute. **Pedicels** (fruiting) 0.6–4 × less than 1 mm. **Flowers:** petals creamy yellow to orange, 6.5–13 × 3.5–7 mm, apex cuspidate, hairy on apex; stamens 20–35, 6–11 mm, filaments heteromorphic, 5 outermost narrowly spatulate, inner filiform; style 8–10 mm. **Capsules** usually lingulate to funnelform, rarely ovoid, 10–18 × 4–5.8 mm, base rounded or cuneate, capsule and pedicel well differentiated. **Seeds** (4–)6–8 per capsule, pyriform, without transverse folds. **2*n*** = 20.

Flowering Sep–May. Beaches, dunes, sand flats along ocean and rivers, coastal hammocks, disturbed areas, roadsides, shell mounds; 0–10 m; Fla.; West Indies (Bahamas).

Mentzelia floridana, which is widespread in peninsular Florida, belongs to a clade restricted to the Gulf coastal and Caribbean region; it is most closely related to *M. gracilis* Urban & Gilg of Mexico and to *M. lindheimeri* of Texas.

60. Mentzelia isolata Gentry, Brittonia 6: 322. 1948 • Isolated blazingstar

Herbs annual, without caudices or tubers. **Stems** erect, to 70 cm. **Leaves:** petiole to 25 mm (proximal leaves), absent (distal leaves); blade usually lanceolate to hastate, rarely elliptic, basally lobed, sometime with 2 pairs of lobes, or unlobed, to 14 × 4.5 cm, base obliquely obtuse to acute, margins shallowly serrate, apex acute. **Pedicels** (fruiting) 0.5–0.8 × 1–2 mm (often appearing absent because thick and continuous with capsule). **Flowers:** petals orange, 5–7 × 2.3–4 mm, apex cuspidate, hairy on apex and abaxially near apex; stamens 8–12,

4–5.3 mm, filaments heteromorphic, 5 outermost narrowly spatulate, inner filiform; styles 3.8–5.5 mm. **Capsules** clavate or funnelform, 12–27 × 3–5 mm, base tapering gradually, capsule and pedicel not well differentiated. **Seeds** 8–12 per capsule, pyriform, without transverse folds. **2*n*** = 20.

Flowering Aug–Oct. Arroyo and canyon bottoms, rocky slopes; 1000–1900 m; Ariz.; Mexico (Sinaloa, Sonora).

Mentzelia isolata intergrades with both *M. aspera* and *M. asperula* in southern Arizona. Typically, *M. isolata* can be distinguished from both *M. aspera* and *M. asperula* because its leaf blades are more than two times as long as wide, whereas those of the latter two species are less than two times as long as wide.

61. Mentzelia lindheimeri Urban & Gilg, Nova Acta. Acad. Caes. Leop.-Carol. German. Nat. Cur. 76: 54. 1900 • Lindheimer's or Texas stickleaf [E]

Mentzelia texana Urban & Gilg

Herbs perennial, with caudices. **Stems** erect to decumbent or clambering, to 50 cm. **Leaves:** petiole to 35 mm; blade usually hastate to ovate, sometimes smallest distal elliptic, basally lobed or unlobed, to 12 × 7.5 cm, base truncate to obtusely cuneate, margins serrate, dentate, or crenulate, apex acute. **Pedicels** (fruiting) 1–5 × less than 1 mm. **Flowers:** petals yellow to orange, 6.5–17 × 4–9.5 mm, apex cuspidate, hairy on apex and abaxially near apex; stamens (10–)20–45, 5–12(–20) mm, filaments monomorphic, filiform; style 3.5–13 mm. **Capsules** usually clavate to funnelform, sometimes slightly ovoid, 8–18 × 3.3–4.5 mm, base rounded or cuneate, capsule and pedicel well-differentiated. **Seeds** (4–)5–10 per capsule, pyriform, without transverse folds. **2*n*** = 20.

Flowering Feb–Nov. Sand flats, dunes, coastal mud flats, limestone gravels or faces; 10–130(–2000) m; Tex.

As with most species of sect. *Mentzelia* first described by Urban and Gilg, the delimitation of *Mentzelia lindheimeri* has not been well understood. Phylogenetic results indicate that populations consistent with the types of *M. lindheimeri* and *M. texana* are part of a clade restricted to Gulf coastal areas from Florida to northeastern Mexico that also includes *M. floridana* and *M. gracilis*. Populations consistent with the *M. lindheimeri* and *M. texana* types overlap, and we treat *M. texana* as a synonym of the former. *Mentzelia lindheimeri* as treated here is restricted mostly to subtropical southeastern Texas. It is rare in trans-Pecos Texas, although we identified populations of *M. lindheimeri* in the Davis Mountains. Texas collections annotated as *M. incisa* Urban & Gilg by Thompson and Zavortink, which served as the basis for Texas reports of that species (for example, D. S. Correll and M. C. Johnston 1970; B. L. Turner et al. 2003) are treated here as *M. lindheimeri*.

62. Mentzelia oligosperma Nuttall ex Sims, Bot. Mag. 42: plate 1760. 1815 • Chickenthief [E] [F]

Mentzelia aurea Nuttall

Herbs perennial, with caudices. **Stems** erect, to 50 cm. **Leaves:** petiole to 3 mm (proximal leaves), absent (distal leaves); blade ovate to hastate, rarely elliptic, basally lobed or unlobed, to 10 × 5 cm, base usually cuneate to truncate, sometimes acute, margins usually serrate, sometimes crenate, rarely entire, apex acute. **Pedicels** (fruiting) 0–2.5 × 2 mm. **Flowers:** petals orange, (6–)8–18.5 × (3–)4–10.5 mm, apex cuspidate, hairy abaxially on distal ½; stamens 15–45, 5–9.5 mm, filaments monomorphic, filiform; styles 5–10 mm. **Capsules** subcylindric to clavate, (5–)7–17 × 2–3.5 mm, base tapering gradually, capsule and pedicel not well differentiated, walls thick, woody. **Seeds** (1–)2–3 (–4) per capsule, oblong, without transverse folds. **2*n*** = 20, 22.

Flowering Mar–Oct. Limestone, gypsum, or sandstone rock outcrops or cliffs, clay or loam flats, grasslands, savannas; 0–1800 m; Ariz., Ark., Colo., Ill., Kans., Mo., Nebr., N.Mex., Okla., S.Dak., Tex., Wyo.

After *Mentzelia aspera*, *M. oligosperma* may be the most widespread member of sect. *Mentzelia*. Occurring widely across the Great Plains, it also extends eastward into Missouri and western Illinois and across southwestern New Mexico into southeastern Arizona.

63. Mentzelia pachyrhiza I. M. Johnston, J. Arnold Arbor. 21: 71. 1940 • Big-rooted stickleaf

Herbs perennial, with tubers. **Stems** erect, to 50 cm. **Leaves:** petiole mostly 3–20 mm, less than 3 mm only on smallest, distalmost leaves; blade ovate to hastate, basally lobed or unlobed, to 4 × 3 cm, base usually acute to obtuse, sometimes attenuate or hastate, margins usually serrate, sometimes crenate, apex acute. **Pedicels** (fruiting) (0–)2 × 2 mm. **Flowers:** petals orange, 5–11.5 × 3–5 mm, apex acute to rounded, hairy abaxially on distal ½; stamens 15–45, 4–8 mm, filaments usually monomorphic, filiform, rarely heteromorphic, outermost slightly spatulate, inner filiform; style 3–7.5 mm. **Capsules** subcylindric to clavate, 6–10 × 1.5–2.3 mm, base tapering gradually, capsule and pedicel not well differentiated, walls thin, brittle. **Seeds** 2–3 per capsule, oblong, with transverse folds. **2*n*** = 22.

Flowering May–Nov. Steep limestone cliffs, gravelly slopes of gypseous clayey soils, desert scrub and *Larrea* communities; 900–2000 m; Tex.; Mexico (Chihuahua, Coahuila, Durango, Nuevo León).

Mentzelia pachyrhiza, native to the Chihuahuan Desert, reaches the flora area only in southern Brewster and Presidio counties. In southwestern Texas and northeastern Chihuahua, where *M. pachyrhiza* and *M. oligosperma* have overlapping ranges, H. J. Thompson and A. M. Powell (1981) reported that the former was found at elevations usually below 1100 m in the *Larrea* zone, below the elevation of junipers and *M. oligosperma*. Thompson and Powell allied *M. pachyrhiza* with *M. oligosperma* and the South American *M. grisebachii* Urban & Gilg, now treated as a synonym of *M. parvifolia* Urban & Gilg ex Kurtz (M. Weigend 2007b). This placement is consistent with phylogenetic studies (J. Grissom and L. Hufford, unpubl.), which show that *M. pachyrhiza* is the sister species of *M. parvifolia*; they together are sister to *M. oligosperma* and the Mexican *M. pattersonii* B. L. Turner. I. M. Johnston's (1940) assertion that the large tuber of *M. pachyrhiza* is unique in sect. *Mentzelia* is incorrect. Tubers, often carrot-shaped, are common among the perennial species of the section.

3d. Mentzelia Linnaeus sect. Micromentzelia Urban & Gilg in H. G. A. Engler and K. Prantl, Nat. Pflanzenfam. 100[III,6a]: 110. 1894 [E]

Larry D. Hufford

Herbs or subshrubs, perennial. **Leaves:** basal rosette absent, proximalmost internodes 5 mm; blade without broad basal lobes, margins entire, toothed, or pinnatisect, revolute. **Flowers:** stamens all fertile, filaments monomorphic, filiform, unlobed; ovules (and seeds) oriented parallel to long axis of ovary. **Capsules** ovoid to urceolate, straight. **Seeds** ovoid to bottle-shaped, mostly trigonal, not winged; seed coat cells polygonal, longer than wide, anticlinal walls sinuous.

Species 1: w United States.

64. Mentzelia torreyi A. Gray, Proc. Amer. Acad. Arts 10: 72. 1874 • Torrey's blazingstar [E] [F]

Herbs or subshrubs with ground-level or subterranean caudices. **Stems** 5–25 cm. **Leaves:** petiole absent; blade ovate or obovate, to 45 × 35 mm, intersinus distance 2–4 mm. **Pedicels** 0(–1.5) mm. **Flowers:** petals yellow to orange, narrowly spatulate to obovate, 4.5–17 × 1.2–5 mm, apex rounded; stamens 25–45, 7–21 mm; styles 7–18 mm. **Capsules** 3–7 × 3–4.5 mm. **Seeds** 3–9 per capsule, edges prominently to inconspicuously ridged. **2*n*** = 28.

Varieties 2 (2 in the flora): w United States.

In the phylogenetic study by L. Hufford et al. (2003), *Mentzelia torreyi* was recovered as the sister to the rest of the genus.

1. Petals 8–17 mm, yellow; longest stamens to 1 mm longer than petals . . . . 64a. *Mentzelia torreyi* var. *torreyi*
1. Petals 4.5–8 mm, orangish yellow, orange, or burnt orange; longest stamens usually much more than 1 mm longer than petals . . . . 64b. *Mentzelia torreyi* var. *acerosa*

64a. Mentzelia torreyi A. Gray var. **torreyi** [E] [F]

Flowers: petals 8–17 mm, yellow; stamens 7–18 mm, longest to 1 mm longer than petals.

Flowering May–Jul. Sagebrush and desert scrub; 1000–2200 m; Calif., Nev.

64b. Mentzelia torreyi A. Gray var. **acerosa** (M. E. Jones) Barneby, Leafl. W. Bot. 5: 63. 1947 [E]

Mentzelia acerosa M. E. Jones, Contr. W. Bot. 17: 30 1930

Flowers: petals 4.5–8 mm, orangish yellow, orange, or burnt orange; stamens 7–14 mm, longest usually much more than 1 mm longer than petals.

Flowering May–Jul. Sagebrush scrub; 700–1000 m; Idaho.

Variety *acerosa* is known from southwestern Idaho (Ada, Elmore, Gooding, Owyhee, and Twin Falls counties).

3e. Mentzelia Linnaeus sect. Trachyphytum (Torrey & A. Gray) Bentham & Hooker f., Gen Pl. 1: 804. 1867

Joshua M. Brokaw

Mentzelia [unranked] *Trachyphytum* Torrey & A. Gray, Fl. N. Amer. 1: 534. 1840

Herbs, annual. **Leaves:** basal rosette present, persistent or not, proximalmost internodes to 5 mm; blade without broad basal lobes, margins entire, toothed, pinnate, or pinnatisect, flat. **Flowers:** stamens all fertile, filaments monomorphic, filiform, or heteromorphic, 5 outermost dorsiventrally flattened, linear or elliptic, not petaloid, inner filiform, usually unlobed, rarely distally 2-lobed; ovules (and seeds) oriented parallel to long axis of ovary. **Capsules** cylindric or clavate, axillary straight to strongly curved, sometimes S-shaped, terminal straight to slightly curved. **Seeds** triangular prisms with grooves along longitudinal edges or irregularly polygonal (angular or rounded), not winged (occasionally winged in *M. lindleyi*); seed coat cells polygonal, ± isodiametric, anticlinal walls straight.

Species ca. 22 (21 in the flora): w North America, nw Mexico, South America (Argentina, Chile).

Both homoploid hybridization and allopolyploidy have played important roles in the diversification of sect. *Trachyphytum* (J. M. Brokaw 2009; Brokaw and L. Hufford 2010, 2010b). H. J. Thompson and H. Lewis (1953) and J. E. Zavortink (1966) reported diploid, tetraploid, hexaploid, and octoploid species in the section. The small desert and grassland annuals of this section are often difficult to distinguish due to the simplicity of their shoot systems, inter-population variability, and morphological intermediacy of allopolyploids. In contrast, homoploid hybridization appears to be rare among existing populations. Evidence of homoploid hybridization has been found primarily in molecular studies (Brokaw and Hufford 2010) and has been observed in natural populations only among hexaploid species (Zavortink).

Seed characteristics have been some of the most reliable characters for discrimination of the common, widespread species but often require an advanced developmental stage for observation. However, the two major clades in sect. *Trachyphytum*, Affines [containing *Mentzelia affinis*, *M. dispersa*, and *M. micrantha* (J. M. Brokaw and L. Hufford 2010)] and Trachyphyta [containing all other diploids and most polyploids (Brokaw and Hufford 2010, 2010b)], can be distinguished relatively early in seed development. Even when immature, plants of the Affines clade have a single triangular ovule in the cross-sectional view of the top of the ovary. With the exception of the narrowly distributed *M. packardiae*, immature plants of the Trachyphyta clade have two or more ovules at the top of the ovary. Further, these two groups differ in the size of their seed coat cells, which can be distinguished early in seed development: cells of the Affines clade are small and barely discernible at 10× magnification, whereas those of the Trachyphyta clade are large and clearly visible at 10× magnification, resulting in a tessellate to tuberculate appearance of the seed coat.

The only species in sect. *Trachyphytum* not found in the flora area is the South American *Mentzelia bartonioides* (C. Presl) Urban & Gilg.

SELECTED REFERENCES Brokaw, J. M. 2009. Phylogeny of *Mentzelia* Section *Trachyphytum*: Origins and Evolutionary Ecology of Polyploidy. Ph.D. dissertation. Washington State University. Brokaw, J. M. and L. Hufford. 2010. Phylogeny, introgression, and character evolution of diploid species in *Mentzelia* section *Trachyphytum* (Loasaceae). Syst. Bot. 35: 601–617. Brokaw, J. M. and L. Hufford. 2010b. Origins and introgression of polyploid species in *Mentzelia* section *Trachyphytum* (Loasaceae). Amer. J. Bot. 97: 1457–1473.

1. Seeds in 1 row distal to mid fruit, triangular prisms.
    2. Stamen filaments heteromorphic, 5 outermost elliptic, distally 2-lobed, inner filiform, unlobed . . . . 75. *Mentzelia micrantha*
    2. Stamen filaments monomorphic, filiform, unlobed.
        3. Petals 10–20 mm; styles 6–14 mm; Malheur County, Oregon . . . . 81. *Mentzelia packardiae*
        3. Petals 2–12 mm; styles 2–6.5 mm; w North America, including Malheur County, Oregon.
            4. Blades of basal leaves usually deeply to moderately lobed, sinuses extending more than ¼ to midvein, rarely entire; styles 3–6.5 mm; capsules often prominently longitudinally ribbed; Arizona, s California, below 1200 m . . . . 65. *Mentzelia affinis*
            4. Blades of basal leaves usually dentate, sinuses extending less than ¼ to midvein, or entire, rarely deeply lobed; styles 2–3.5(–5); capsules usually inconspicuously longitudinally ribbed; w North America, including Arizona and s California, where above 1200 m . . . . 70. *Mentzelia dispersa*
1. Seeds in 2+ rows distal to mid fruit, irregularly polygonal or occasionally triangular prisms proximal to mid fruit.
    5. Basal leaves not persisting; blade margins of proximalmost remaining leaves (proximal cauline) dentate or entire; usually on barren, alkaline or saline soils.
        6. Petals 8–12 mm; sepals 3–5.5 mm; Idaho, Nevada, Oregon . . . . 76. *Mentzelia mollis*
        6. Petals 2–4 mm; sepals 1–3 mm; Colorado, New Mexico, Utah . . . . 84. *Mentzelia thompsonii*
    5. Basal leaves persisting; blade margins of proximalmost leaves (basal) usually deeply to shallowly lobed, rarely entire; on wide variety of substrates.
        7. Bract margins entire, bracts green.
            8. Petals (6–)8–25 mm.
                9. Sepals (7–)9–16 mm; styles 7–15 mm; e Kern County and nw San Bernardino County, California . . . . 71. *Mentzelia eremophila* (in part)
                9. Sepals (2–)3–8(–10) mm; styles 4–10 mm; Arizona, e California, Nevada.
                    10. Seeds without recurved flap over hilum; seed coat cell outer periclinal wall domes on seed edges more than ½ as tall as wide at maturity . . . . 73. *Mentzelia jonesii*
                    10. Seeds usually with recurved flap over hilum; seed coat cell outer periclinal wall domes on seed edges less than ½ as tall as wide at maturity . . . . 79. *Mentzelia nitens*
            8. Petals 2–8 mm.
                11. Seeds dark brown or tan and moderately to densely dark-mottled; seed coat cell outer periclinal walls domed, domes on seed edges more than ½ as tall as wide at maturity . . . . 66. *Mentzelia albicaulis* (in part)
                11. Seeds tan, not or occasionally sparsely dark-mottled; seed coat cell outer periclinal walls flat to slightly convex, or if domed, domes on seed edges less than ½ as tall as wide.
                    12. Capsules 6–15 mm, axillary curved to 20°; 2000–2500 m; Mono County, California . . . . 77. *Mentzelia monoensis* (in part)
                    12. Capsules 11–31 mm, axillary curved to 250°; 30–1700 m; sw United States, but not Mono County, California.
                        13. Blade margins of basal leaves usually shallowly lobed, lobes rounded; seed coat cell outer periclinal walls flat or slightly convex . . . . 69. *Mentzelia desertorum*
                        13. Blade margins of basal leaves usually deeply lobed, lobes pointed; seed coat cell outer periclinal walls domed . . . . 80. *Mentzelia obscura*
        7. Bract margins toothed or lobed, or if entire, bracts green with white base.
            14. Bracts green.
                15. Styles 15–35 mm; stamens 11–40 mm, filaments heteromorphic, 5 outermost linear, inner filiform.
                    16. Petals usually elliptic to ovate, rarely obovate, 8–17(–21) mm wide; west slope of Sierra Nevada, California . . . . 68. *Mentzelia crocea*
                    16. Petals obovate, (12–)16–33 mm wide; Coast Ranges, California . . . 74. *Mentzelia lindleyi*

15. Styles 2–15 mm; stamens 3–11 mm, filaments monomorphic, filiform.
    17. Petals 3–7(–10) mm; styles 2–6 mm.
        18. Petals red to orange proximally, orange to orange-yellow distally; styles (3–)3.5–6 mm . . . . 85. *Mentzelia veatchiana* (in part)
        18. Petals orange proximally, yellow distally; styles usually less than 3.5 mm.
            19. Bract margins 3-lobed or entire, lateral lobes never prominent; capsules 8–28(–35) mm (longest capsules usually more than 15 mm), axillary curved to 180°; 0–2300 m. . . 66. *Mentzelia albicaulis* (in part)
            19. Bract margins usually 3–7-lobed, rarely entire, lateral lobes usually prominent; capsules 6–17(–20) mm, axillary curved to 45°; 600–3400 m . . . . .78. *Mentzelia montana* (in part)
    17. Petals 8–25 mm; styles 5–15 mm.
        20. Petals yellow; bracts usually entire, rarely 2-lobed; w Mojave Desert, e Kern County, nw San Bernardino County, California . . . . 71. *Mentzelia eremophila* (in part)
        20. Petals red to orange proximally, orange to yellow distally; bracts 3–7-lobed; s San Joaquin Valley, Inner Coast Ranges, s Sierra Nevada foothills, California.
            21. Petals orange proximally, yellow distally; Fresno, Monterey, and San Benito counties, California, 200–1400 m, and s San Luis Obispo, Santa Barbara, and Ventura counties, California, 1500–1700 m . . . . 72. *Mentzelia gracilenta* (in part)
            21. Petals red to orange proximally, orange to yellow distally; Kern, San Luis Obispo, Santa Barbara, and Tulare counties, California, 200–1400 m . . . . 82. *Mentzelia pectinata*

[14. Shifted to left margin.—Ed.]

14. Bracts green with white base or mostly white with green margins.
    22. Bracts concealing capsules, mostly white with green margins . . . . .67. *Mentzelia congesta*
    22. Bracts not concealing capsules, green with white base.
        23. Petals usually 8+ mm, usually orange, rarely yellow, proximally, yellow distally; seed coat cell outer periclinal wall domes on seed edges ½ as tall as wide at maturity.
            24. Plants of grasslands, pine-oak woodlands; Coast and Transverse ranges, California. . . . . 72. *Mentzelia gracilenta* (in part)
            24. Plants of desert scrub, Joshua-tree woodlands; ne Los Angeles and w Riverside counties, California . . . . 83. *Mentzelia ravenii* (in part)
        23. Petals usually less than 5 mm, red to orange proximally, orange to yellow distally; if petals 4+ mm, then seed coat cell outer periclinal wall domes on seed edges more than ½ as tall as wide at maturity.
            25. Styles usually less than 3.5 mm.
                26. Bracts 3–4.1 × 1.1–1.7 mm, margins entire; seeds not mottled, seed coat cell outer periclinal wall domes on seed edges less than ½ as tall as wide at maturity. . . . . 77. *Mentzelia monoensis* (in part)
                26. Bracts 5.9–9.2 × 1.7–5 mm, margins usually 3–7-lobed, rarely entire; seeds moderately to densely dark-mottled, seed coat cell outer periclinal wall domes on seed edges more than ½ as tall as wide at maturity . . . . .78. *Mentzelia montana* (in part)
            25. Styles usually 3.5+ mm.
                27. Petals red to orange proximally, orange to orange-yellow distally; bracts usually 3–7-lobed, rarely entire . . . . 85. *Mentzelia veatchiana* (in part)
                27. Petals orange proximally, yellow distally; bracts 3–7-lobed or entire.
                    28. Plants of desert scrub, Joshua-tree woodlands; ne Los Angeles and w Riverside counties, California . . . . 83. *Mentzelia ravenii* (in part)
                    28. Plants of oak-pine woodlands, grasslands; Arizona, California, Nevada, Oregon . . . . 85. *Mentzelia veatchiana* (in part)

65. Mentzelia affinis Greene, Pittonia 2: 103. 1890 • Yellowcomet

Plants candelabra-form, 5–40(–50) cm. **Basal leaves** persisting; petiole present or absent; blade linear-lanceolate to linear, margins usually deeply to moderately lobed, sinuses extending ¼+ to midvein, rarely entire. **Cauline leaves:** petiole absent; blade ovate-lanceolate to lanceolate, to 17 cm, margins usually dentate or entire, rarely deeply lobed. **Bracts** green, ovate to lanceolate, 2.7–6.6 × 0.9–2.1 mm, width ⅕–⅓ length, not concealing capsule, margins 3-lobed or entire. **Flowers:** sepals 1–7 mm; petals yellow to orange proximally, yellow distally, 4–12 mm, apex acute; stamens 20+, 3–6.5 mm, filaments monomorphic, filiform, unlobed; styles 3–6.5 mm. **Capsules** narrowly cylindric, 7–32 × 1–3 mm, axillary curved to 90° at maturity, often prominently longitudinally ribbed. **Seeds** 10–20, in 1 row distal to mid fruit, tan, dark-mottled or not, triangular prisms, surface ± smooth under 10× magnification; recurved flap over hilum absent; seed coat cell outer periclinal wall flat. $2n = 18$.

Flowering Mar–May. Sandy, rocky, or gray-white silty soils, grasslands, creosote-bush scrub, Joshua-tree or saguaro woodlands; 0–1200 m; Ariz., Calif.; Mexico (Sonora).

Herbarium specimens of *Mentzelia affinis* are often difficult to distinguish from those of *M. dispersa* despite distinct evolutionary histories (J. M. Brokaw and L. Hufford 2010). Several characters, including flower size, leaf margins, and capsule surfaces, differ substantially between these species, but habitat is the most dependable diagnostic character. Verified populations of *M. affinis* have not been found above 1200 m in desert habitats, and grassland populations are usually restricted to much lower elevations. Sympatric populations of *M. affinis* and *M. dispersa* have not been found, and, in areas of range overlap in southern California, *M. dispersa* has not been found below 1200 m or in desert vegetation.

66. Mentzelia albicaulis (Douglas) Douglas ex Torrey & A. Gray, Fl. N. Amer. 1: 534. 1840 • Whitestem blazingstar [W]

Bartonia albicaulis Douglas in W. J. Hooker, Fl. Bor.-Amer. 1: 222. 1832; *Mentzelia mojavensis* H. J. Thompson & J. E. Roberts

Plants wandlike or candelabra-form, (2–)10–40(–50) cm. **Basal leaves** persisting; petiole present or absent; blade linear-lanceolate to linear, margins deeply to shallowly lobed. **Cauline leaves:** petiole absent; blade ovate-lanceolate to linear, to 15 cm, margins deeply to shallowly lobed or entire. **Bracts** green, ovate to linear, 3.7–8.6 × 0.8–3.9 mm, width ⅙–⅔ length, not concealing capsule, margins 3-lobed or entire, lateral lobes never prominent. **Flowers:** sepals 1–5 mm; petals orange proximally, yellow distally, 3–7(–8) mm, apex usually acute, rarely retuse; stamens 20+, 3–5 mm, filaments monomorphic, filiform, unlobed; styles 2–5 mm. **Capsules** clavate, 8–28(–35) × 1.5–3.5 mm, longest capsules usually 15+ mm, axillary curved to 180° at maturity, usually inconspicuously longitudinally ribbed. **Seeds** 10–30, in 2+ rows distal to mid fruit, dark brown or tan, moderately to densely dark-mottled, usually irregularly polygonal, occasionally triangular prisms proximal to mid fruit, surface tuberculate under 10× magnification; recurved flap over hilum absent; seed coat cell outer periclinal wall domed, domes on seed edges more than ½ as tall as wide at maturity. $2n = 54, 72$.

Flowering Mar–Aug. Sand dunes, gravel fans, washes, desert scrub, sagebrush or antelope bitterbrush scrub, open ponderosa pine woodlands, pinyon/juniper woodlands; 0–2300 m; B.C., Sask.; Ariz., Calif., Colo., Idaho, Mont., Nebr., Nev., N.Mex., Oreg., S.Dak., Tex., Utah, Wash., Wyo.; Mexico (Baja California, Chihuahua, Sonora).

Mentzelia albicaulis is the most widespread species in sect. *Trachyphytum* and exhibits extensive morphological variation. Most populations of *M. albicaulis* are octoploid; however, hexaploids from southern California that have been called *M. mojavensis* and occasionally *M. californica* are also treated here as *M. albicaulis*. Two tetraploids in sect. *Trachyphytum*, *M. montana* and *M. obscura*, also have been treated previously as *M. albicaulis* (N. H. Holmgren et al. 2005). Both exhibit morphological forms and distributions overlapping with *M. albicaulis*. However, in most cases these species can be distinguished without chromosome counts, and their distinctiveness has been supported by phylogenetic analyses (J. M. Brokaw and L. Hufford 2010b).

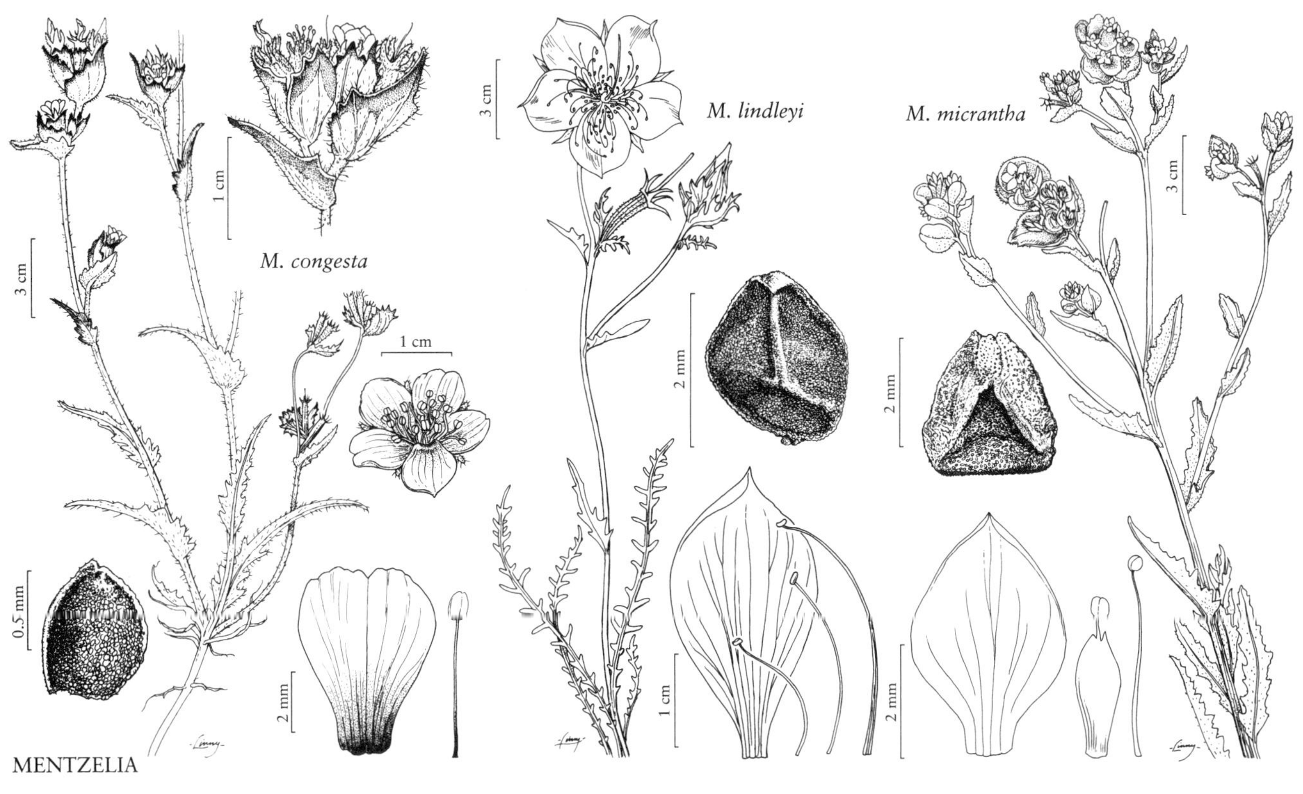

67. Mentzelia congesta Torrey & A. Gray, Fl. N. Amer. 1: 534. 1840 • United or congested blazingstar, flower baskets [E] [F]

Acrolasia davidsoniana Abrams; *Mentzelia congesta* var. *davidsoniana* (Abrams) J. F. Macbride

Plants wandlike or candelabraform, 7–40(–50) cm. **Basal leaves** persisting; petiole present or absent; blade linear-lanceolate to linear, margins usually deeply to shallowly lobed, rarely entire. **Cauline leaves:** petiole present or absent (proximal leaves), absent (distal leaves); blade linear-lanceolate to linear (proximal leaves), ovate-lanceolate (distal leaves), to 9 cm, margins deeply to shallowly lobed or entire. **Bracts** mostly white with green margins, conspicuous and showy, outer broadly ovate, inner obovate, outer 7–14.7 × 4.2–10.4 mm, inner 4.1–5.8 × 2.7–3.5 mm, width ½–⅞ length, concealing capsule, margins usually deeply 3–9-lobed to toothed, rarely entire. **Flowers:** sepals 1–4 mm; petals orange proximally, pale yellow to yellow distally, 2–8(–10) mm, apex retuse; stamens 20–40, 1.5–5 mm, filaments monomorphic, filiform, unlobed; styles 1.5–5 mm. **Capsules** cylindric or clavate, 5–12 × 2–3 mm, axillary curved to 10° at maturity, usually inconspicuously longitudinally ribbed. **Seeds** 10–25, in 2+ rows distal to mid fruit, tan, dark-mottled or not, irregularly polygonal, surface tessellate under 10× magnification; recurved flap over hilum absent; seed coat cell outer periclinal wall domed, domes on seed edges less than ½ as tall as wide at maturity. **2*n*** = 18.

Flowering May–Jul. Disturbed slopes, sagebrush scrub, pinyon/juniper woodlands, pine forests; 1200–2700 m; Calif., Idaho, Nev., Oreg.

Mentzelia congesta is easily distinguished from other species in sect. *Trachyphytum* by its large, mostly white bracts. However, the bracts of *M. congesta* are strikingly similar to those of *M. involucrata* in sect. *Bicuspidaria* (see discussion under 53. *M. involucrata*). Phylogenetic studies suggest that the genome of *M. congesta* is present in more than half of the polyploid taxa in sect. *Trachyphytum*, including all species that exhibit bracts with partially white bases (J. M. Brokaw and L. Hufford 2010b).

68. Mentzelia crocea Kellogg, Proc. Calif. Acad. Sci. 7: 110. 1877 • Sierra blazingstar [E]

Mentzelia lindleyi Torrey & A. Gray subsp. *crocea* (Kellogg) C. B. Wolf

Plants candelabra-form, 30–70 (–100) cm. **Basal leaves** persisting; petiole present or absent; blade lanceolate to linear, margins usually shallowly, rarely deeply, lobed. **Cauline leaves:** petiole absent; blade ovate to lanceolate, to 40 cm, margins deeply to shallowly lobed. **Bracts** green, broadly ovate to lanceolate, 6–12.7 × 3.5–7.7 mm, width 1/2–4/5 length, not concealing capsule, margins 3–10-lobed. **Flowers:** sepals 7–20 mm; petals usually orange, rarely yellow, proximally, yellow distally, usually elliptic to ovate, rarely obovate, 21–42 × 8–17(–21) mm, apex acute; stamens 20+, 11–40 mm, filaments heteromorphic, 5 outermost linear, inner filiform, unlobed; styles 20–35 mm. **Capsules** clavate, 20–35 × 3–5 mm, axillary curved to 45° at maturity, usually inconspicuously longitudinally ribbed. **Seeds** 30–40, in 2+ rows distal to mid fruit, tan, dark-mottled, irregularly polygonal, surface minutely tessellate under 10× magnification; recurved flap over hilum absent; seed coat cell outer periclinal wall domed, domes on seed edges less than 1/2 as tall as wide at maturity. 2*n* = 36.

Flowering May–Jun. Rocky slopes, roadsides, grasslands, oak-pine woodlands; 150–1700 m; Calif.

Mentzelia crocea and *M. lindleyi* are very similar morphologically and are most reliably distinguished geographically, with *M. crocea* occurring on the west slope of the Sierra Nevada and *M. lindleyi* occurring in the Coast Ranges. Low fertility of interspecific crosses supports recognition as separate species (H. J. Thompson 1960), which has been substantiated by genetic differences and non-sister relationships of the two species (J. M. Brokaw and L. Hufford 2010b).

69. Mentzelia desertorum (Davidson) H. J. Thompson & J. E. Roberts, Phytologia 21: 280. 1971 • Desert blazingstar [C]

Acrolasia desertorum Davidson, Bull. S. Calif. Acad. Sci. 5: 16. 1906

Plants candelabra-form, 5–40 cm. **Basal leaves** persisting; petiole present or absent; blade linear, margins usually shallowly lobed, lobes rounded. **Cauline leaves:** petiole absent; blade ovate-lanceolate to linear, to 12 cm, margins shallowly lobed or entire. **Bracts** green, ovate to ovate-lanceolate, 3.6–4.7 × 1.6–2.3 mm, width 1/3–1/2 length, not concealing capsule, margins entire. **Flowers:** sepals 2–4 mm; petals yellow to orange proximally, yellow distally, 2–6 mm, apex acute or rounded; stamens 10–30, 2–4 mm, filaments monomorphic, filiform, unlobed; styles 2–4 mm. **Capsules** clavate, 12–27 × 1–2.5 mm, axillary curved to 180° at maturity, usually inconspicuously longitudinally ribbed. **Seeds** 10–50, in 2+ rows distal to mid fruit, tan, usually not, occasionally sparsely, dark-mottled, usually irregularly polygonal, occasionally triangular prisms proximal to mid fruit, surface tessellate under 10× magnification; recurved flap over hilum absent; seed coat cell outer periclinal wall flat to slightly convex. 2*n* = 18.

Flowering Feb–Mar. Sandy flats, washes, creosote-bush scrub; of conservation concern; 30–1000 m; Ariz., Calif.; Mexico (Baja California, Sonora).

Mentzelia desertorum, a diploid, is most similar morphologically to the tetraploid *M. obscura*, and the two species may be difficult to distinguish prior to seed maturation. However, many populations of *M. desertorum* have narrow basal leaves with short, widely spaced lobes that are unique among species within sect. *Trachyphytum*.

70. Mentzelia dispersa S. Watson, Proc. Amer. Acad. Arts 11: 137. 1876 • Nada stickleaf, bushy or Nevada blazingstar [E]

Mentzelia albicaulis (Douglas) Douglas ex Torrey & A. Gray var. *integrifolia* S. Watson, Botany (Fortieth Parallel), 114. 1871; *M. dispersa* var. *compacta* (A. Nelson) J. F. Macbride; *M. dispersa* var. *latifolia* (Rydberg) J. F. Macbride; *M. dispersa* var. *obtusa* Jepson; *M. pinetorum* A. Heller

Plants wandlike or candelabra-form, 10–40(–50) cm. **Basal leaves** persisting; petiole present or absent; blade elliptic to linear, margins usually dentate, sinuses extending less than 1/4 to midvein, or entire, rarely deeply lobed. **Cauline leaves:** petiole present or absent (proximal leaves), absent (distal leaves); blade elliptic to linear (proximal leaves), orbiculate to linear (distal leaves), to 10 cm, margins usually dentate, sinuses extending less than 1/4 to midvein, or entire, rarely deeply lobed. **Bracts** green, orbiculate to ovate, 2.1–6.5 × 1.1–3 mm, width 1/3–7/8 length, not concealing capsule, margins 3-lobed or entire. **Flowers:** sepals 1–3.5 mm; petals usually yellow, rarely orange, proximally, yellow distally, 2–6(–8) mm, apex rounded; stamens 20–40, 2–4.5 mm, filaments monomorphic, filiform, unlobed; styles 2–3.5(–5) mm. **Capsules** narrow-cylindric, 7–30 × 1–2.5 mm, axillary curved to 30° at maturity, usually inconspicuously longitudinally ribbed. **Seeds** 10–20, in 1 row distal to mid fruit, tan, dark-mottled or not, triangular prisms, surface ± smooth under 10× magnification; recurved flap over hilum absent; seed coat cell outer periclinal wall flat. $2n$ = 18, 36, 72.

Flowering (Apr–)May–Aug(–Sep). Loamy to sandy or rocky slopes, grasslands, scrub, dry forests, roadsides; 400–3100 m; B.C.; Ariz., Calif., Colo., Idaho, Mont., Nev., N.Dak., Oreg., S.Dak., Utah, Wash., Wyo.

Mentzelia dispersa is the only polyploid species solely derived from the Affines clade and may be an autopolyploid complex (J. M. Brokaw and L. Hufford 2010, 2010b). Morphological characters that consistently distinguish cytotypes within *M. dispersa* have not been found. *Mentzelia dispersa* is most easily confused with *M. affinis* (see discussion under 65. *M. affinis*) but is phylogenetically distinct (Brokaw and Hufford 2010, 2010b).

71. Mentzelia eremophila (Jepson) H. J. Thompson & J. E. Roberts, Phytologia 21: 281. 1971 • Solitary or pinyon blazingstar [E]

Mentzelia lindleyi Torrey & A. Gray var. *eremophila* Jepson, Man. Fl. Pl. Calif., 650. 1925

Plants candelabra-form, (7–)30–50(–60) cm. **Basal leaves** persisting; petiole present or absent; blade linear-lanceolate to linear, margins very deeply lobed, lobes slender. **Cauline leaves:** petiole absent; blade ovate-lanceolate to linear, to 15 cm, margins deeply to shallowly lobed or entire. **Bracts** green, ovate to lanceolate, 4.8–12.4 × 0.9–3.5 mm, width 1/8–1/2 length, not concealing capsule, margins usually entire, rarely 2-lobed. **Flowers:** sepals (7–)9–16 mm; petals yellow, 12–25 mm, apex acute or mucronate; stamens 20+, 3–10 mm, filaments monomorphic, filiform, unlobed; styles 7–15 mm. **Capsules** clavate, 19–40 × 2–3.5 mm, axillary curved to 270° at maturity, usually inconspicuously longitudinally ribbed. **Seeds** 30–60, in 2+ rows distal to mid fruit, tan, usually dark-mottled, usually irregularly polygonal, occasionally triangular prisms proximal to mid fruit, surface colliculate under 10× magnification; recurved flap over hilum usually present; seed coat cell outer periclinal wall domed, domes on seed edges less than 1/2 as tall as wide at maturity. $2n$ = 18.

Flowering Mar–May. Rocky slopes, washes, canyons, creosote-bush scrub; 600–1300 m; Calif.

Mentzelia eremophila is narrowly distributed in eastern Kern and northwestern San Bernardino counties. It is morphologically similar and closely related to *M. nitens* (J. M. Brokaw and L. Hufford 2010). However, *M. eremophila* generally has longer sepals, petals, and styles, and populations of *M. nitens* have not been found south of Inyo County in California.

72. Mentzelia gracilenta Torrey & A. Gray, Fl. N. Amer. 1: 534. 1840 • Slender or grass blazingstar [E]

Plants wandlike or candelabraform, (3–)20–60(–70) cm. **Basal leaves** persisting; petiole present or absent; blade linear-lanceolate to linear, margins deeply to shallowly lobed. **Cauline leaves:** petiole absent; blade ovate-lanceolate to linear, to 13 cm, margins deeply to shallowly lobed or entire. **Bracts** usually green with white base, occasionally green, obovate to ovate, 7.1–11.3 × 3–6.2 mm, width 2/5–3/5 length, not concealing capsule, margins 3–12-lobed. **Flowers:** sepals 3–8 mm; petals usually orange, rarely yellow, proximally, yellow distally, (7–)8–18 mm, apex retuse; stamens 20+, 5–11 mm, filaments monomorphic, filiform, unlobed; styles 5–11 mm. **Capsules** cylindric or clavate, 9–15(–23) × 3–5 mm, axillary curved to 20° at maturity, usually inconspicuously longitudinally ribbed. **Seeds** 15–25, in 2+ rows distal to mid fruit, tan, dark-mottled or not, usually irregularly polygonal, occasionally triangular prisms proximal to mid fruit, surface tuberculate under 10× magnification; recurved flap over hilum absent; seed coat cell outer periclinal wall domed, domes on seed edges 1/2 as tall as wide at maturity. $2n$ = 36.

Flowering Apr–May. Serpentine talus, gray-white calcium-rich soils, grasslands, pine-oak woodlands; 200–1700 m; Calif.

Mentzelia gracilenta occurs in the southern Coast Ranges and western Transverse Ranges (Fresno, Monterey, and San Benito counties below 1400 m, and southern San Luis Obispo, Santa Barbara, and Ventura counties above 1500 m). Populations in the northern portion of its range are commonly associated with serpentine or other stressful substrates. *Mentzelia gracilenta* is very similar morphologically to *M. ravenii*, but it does not occur in desert communities, as does *M. ravenii*. Furthermore, these taxa appear to be phylogenetically distinct (J. M. Brokaw and L. Hufford 2010b).

73. Mentzelia jonesii (Urban & Gilg) H. J. Thompson & J. E. Roberts, Phytologia 21: 282. 1971 • Jones's blazingstar [E]

Mentzelia albicaulis (Douglas) Douglas ex Torrey & A. Gray var. *jonesii* Urban & Gilg, Nova Acta Acad. Caes. Leop.-Carol. German. Nat. Cur. 76: 29. 1900; *M. californica* H. J. Thompson & J. E. Roberts; *M. nitens* Greene var. *jonesii* (Urban & Gilg) J. Darlington

Plants candelabra-form, 20–40(–50) cm. **Basal leaves** persisting; petiole present or absent; blade linear-lanceolate to linear, margins deeply lobed to dentate. **Cauline leaves:** petiole absent; blade ovate-lanceolate to linear, to 15 cm, margins deeply to shallowly lobed or entire. **Bracts** green, ovate to lanceolate, 3.7–5.9 × 0.8–3.9 mm, width 1/8–2/3 length, not concealing capsule, margins entire. **Flowers:** sepals (2–)4–8(–10) mm; petals yellow to orange proximally, yellow distally, (6–)8–22 mm, apex acute to rounded; stamens 20+, 3–10 mm, filaments monomorphic, filiform, unlobed; styles 4–10 mm. **Capsules** clavate, 15–38 × 2–4 mm, axillary curved to 180° or S-shaped at maturity, usually inconspicuously longitudinally ribbed. **Seeds** 15–30, in 2+ rows distal to mid fruit, tan, dark-mottled, usually irregularly polygonal, occasionally triangular prisms proximal to mid fruit, surface tuberculate under 10× magnification; recurved flap over hilum absent; seed coat cell outer periclinal wall domed, domes on seed edges more than 1/2 as tall as wide at maturity. **2*n*** = 36, 54.

Flowering Mar–May. Sandy to rocky washes, fans, or flats, creosote-bush scrub, Joshua-tree or saguaro woodlands; 200–1500 m; Ariz., Calif., Nev., Utah.

Mentzelia jonesii is widespread throughout southeastern California and western and southern Arizona. However, tetraploid populations have been found only in California and far northwestern Arizona. Hexaploids in California have previously been called *M. californica*. The ranges of *M. jonesii* and the diploid, *M. nitens*, overlap in southern Inyo County, California, and these species are difficult to distinguish morphologically without mature seeds; sepal and petal lengths, which have sometimes been used to distinguish them, overlap completely.

74. Mentzelia lindleyi Torrey & A. Gray, Fl. N. Amer. 1: 533. 1840 • Lindley's blazingstar [E] [F]

Bartonia aurea Lindley, Edwards's Bot. Reg. 22: plate 1831. 1836, not *Mentzelia aurea* Nuttall 1818

Plants candelabra-form, 30–70 cm. **Basal leaves** persisting; petiole present or absent; blade lanceolate to linear, margins usually deeply lobed, lobes rounded. **Cauline leaves:** petiole absent; blade ovate to lanceolate, to 17 cm, margins deeply to shallowly lobed. **Bracts** green, ovate to lanceolate, 6.1–14.2 × 3.6–5.1 mm, width 1/3–2/3 length, not concealing capsule, margins 3–7-lobed. **Flowers:** sepals 9–19 mm; petals orange proximally, yellow distally, obovate, 20–40 × (12–)16–33 mm, apex mucronate; stamens 20+, 12–30 mm, filaments heteromorphic, 5 outermost linear, inner filiform, unlobed; styles 15–24 mm. **Capsules** clavate, 25–40 × 4–5 mm, axillary curved to 90° at maturity, usually inconspicuously longitudinally ribbed. **Seeds** 30–40, in 2+ rows distal to mid fruit, tan, dark-mottled or not, irregularly polygonal, surface minutely tessellate under 10× magnification; recurved flap over hilum absent; seed coat cell outer periclinal wall domed, domes on seed edges less than 1/2 as tall as wide at maturity. **2*n*** = 36.

Flowering May–Jun. Rocky, open slopes, coastal sage scrub, oak-pine woodlands; 90–1400 m; Calif.

Naturally occurring populations of *Mentzelia lindleyi* are limited primarily to the Coast Ranges of the San Francisco Bay area. However, commercial distribution of seeds has resulted in casual alien populations of *M. lindleyi* throughout the southwestern United States. See 68. *M. crocea* for discussion of similarities between these species.

75. Mentzelia micrantha (Hooker & Arnott) Torrey & A. Gray, Fl. N. Amer. 1: 535. 1840 • San Luis or chaparral blazingstar [F]

Bartonia micrantha Hooker & Arnott, Bot. Beechey Voy., 343, plate 85. 1839

Plants wandlike or candelabra-form, 10–80 cm. **Basal leaves** persisting; petiole present or absent; blade lanceolate to linear, margins irregularly deeply lobed to dentate. **Cauline leaves:** petiole present or absent (proximal leaves), absent (distal leaves); blade lanceolate to linear (proximal leaves), orbiculate to lanceolate (distal leaves), to 18 cm, margins irregularly deeply lobed to dentate proximally, dentate or entire distally. **Bracts** green, orbiculate to ovate, 3.4–6.6 × 2.5–5.9 mm, width 3/4 to ± equal length, often concealing capsule, margins sinuate or entire. **Flowers:** sepals 1–3 mm; petals yellow, 2–5 mm, apex acute; stamens 10–20, 1.5–4 mm, filaments heteromorphic, 5 outermost elliptic, distally 2-lobed, inner filiform, unlobed; styles 2–3(–5) mm. **Capsules** cylindric, 6–13 × 1.5–2.5 mm, axillary curved to 20° at maturity, usually inconspicuously, occasionally prominently, longitudinally ribbed. **Seeds** 4–10, in 1 row distal to mid fruit, dark brown or tan, dark-mottled, triangular prisms, surface ± smooth to minutely tessellate under 10× magnification; recurved flap over hilum absent; seed coat cell outer periclinal wall flat. **2*n*** = 18.

Flowering Apr–Jun. Open, often recently-burned or disturbed chaparral or oak woodlands; 0–2300 m; Calif.; Mexico (Baja California).

Mentzelia micrantha is easily distinguished from other species in sect. *Trachyphytum* by the presence of two lateral lobes on the filaments of the five outermost stamens. This characteristic is distinct from the filament lobes of some species in sect. *Bicuspidaria*, which occur on all or most stamens. Phylogenetic studies have found that *M. micrantha* is not closely related to species in sect. *Bicuspidaria* (L. Hufford et al. 2003; J. M. Brokaw and Hufford 2010).

76. Mentzelia mollis M. Peck, Leafl. W. Bot. 4: 183. 1945 • Smooth or soft blazingstar C E

Plants candelabra-form, 3–15 (–20) cm. **Basal leaves** not persisting. **Cauline leaves:** petiole present (proximal leaves), absent (distal leaves); blade lanceolate to linear (proximal leaves), ovate to lanceolate (distal leaves), to 6 cm, margins dentate or entire (proximal leaves), entire (distal leaves). **Bracts** green, ovate to elliptic, 5–8.5 × 2–5 mm, width ⅖–⅗ length, not concealing capsule, margins entire. **Flowers:** sepals 3–5.5 mm; petals yellow to orange proximally, yellow distally, 8–12 mm, apex rounded; stamens 20+, 3–8 mm, filaments monomorphic, filiform, unlobed; styles 7–9 mm. **Capsules** cylindric or clavate, 5–22 × 2–4 mm, axillary curved to 45° at maturity, often prominently longitudinally ribbed. **Seeds** 15–25, in 2+ rows distal to mid fruit, tan, dark-mottled, irregularly polygonal, surface smooth to minutely tessellate under 10× magnification; recurved flap over hilum absent; seed coat cell outer periclinal wall flat to slightly convex. $2n = 36$.

Flowering Apr–Jul. Barren, sodic or calcic clay slopes and bluffs derived from volcanic ash; of conservation concern; 800–1500 m; Idaho, Nev., Oreg.

Mentzelia mollis is narrowly distributed in eastern Malheur County, Oregon, and western Owyhee County, Idaho, and disjunctly in the Black Rock Range of Humboldt County, Nevada. Recent phylogenetic studies support treatment of these disjunct populations as a single species (J. M. Brokaw and L. Hufford 2010b). In both ranges, *M. mollis* is predominantly limited to barren soils with high salinity. *Mentzelia mollis* is listed as endangered by the Oregon Department of Agriculture and is in the Center for Plant Conservation's National Collection of Endangered Plants.

77. Mentzelia monoensis J. M. Brokaw & L. Hufford, Madroño 58: 57, figs. 1,2A,3. 2011 • Mono Craters blazingstar E

Plants candelabra-form, 10–30 cm. **Basal leaves** persisting; petiole present or absent; blade linear-lanceolate to linear, margins usually moderately to shallowly lobed, rarely entire. **Cauline leaves:** petiole present or absent (proximal leaves), absent (distal leaves); blade linear-lanceolate to linear (proximal leaves), ovate to linear (distal leaves), to 13 cm, margins usually moderately to shallowly lobed, rarely entire. **Bracts** green, sometimes with white base, ovate, 3–4.1 × 1.1–1.7 mm, width ¼–½ length, not concealing capsule, margins entire. **Flowers:** sepals 2–3 mm; petals orange proximally, yellow distally, 2–4 mm, apex retuse or rounded; stamens 10–30, 2–3 mm, filaments monomorphic, filiform, unlobed; styles 2–3 mm. **Capsules** cylindric or clavate, 6–15 × 2–3 mm, axillary curved to 20° at maturity, usually inconspicuously longitudinally ribbed. **Seeds** 15–30, in 2+ rows distal to mid fruit, tan, not dark-mottled, usually irregularly polygonal, occasionally triangular prisms proximal to mid fruit, surface colliculate under 10× magnification; recurved flap over hilum absent; seed coat cell outer periclinal wall domed, domes on seed edges less than ½ as tall as wide at maturity. $2n = 54$.

Flowering May–Aug. Coarse pumice soils on open slopes, sagebrush or bitterbrush scrub, pine forests; 2000–2500 m; Calif.

Mentzelia monoensis is narrowly distributed predominantly south of Mono Lake and north of Lake Crowley in Mono County, California, and is most commonly found in soils derived from the eruptions of the Mono Craters (J. M. Brokaw et al. 2015). Phylogenetic studies suggest that this hexaploid is the only allopolyploid derived from representatives of both the "Affines" and "Trachyphyta" clades (Brokaw and L. Hufford 2010b). *Mentzelia monoensis* is morphologically similar to sympatric populations of *M. montana*. However, the bracts of *M. monoensis* are more often unlobed and green throughout. Furthermore, seeds of *M. monoensis* have tan, unmottled coats that are always composed of cells that are rounded, appearing as shallow domes. In contrast, seeds of *M. montana* have mottled coats with cells that stand out as rough, pointed knobs along the seed edges.

78. Mentzelia montana (Davidson) Davidson in A. Davidson and G. L. Moxley, Fl. S. Calif., 240. 1923 • Mountain blazingstar

Acrolasia montana Davidson, Bull. S. Calif. Acad. Sci. 5: 18. 1906

Plants wandlike or candelabra-form, (5–)20–40(–50) cm. **Basal leaves** persisting; petiole present or absent; blade lanceolate to linear, margins usually deeply to shallowly lobed, rarely entire. **Cauline leaves:** petiole present or absent (proximal leaves), absent (distal leaves); blade lanceolate to linear (proximal leaves), ovate-lanceolate to linear (distal leaves), to 13 cm, margins deeply to shallowly lobed or entire. **Bracts** usually green with prominent white base conspicuously extending outwards from midvein, rarely green, usually obovate, rarely lanceolate, 5.9–9.2 × 1.7–5 mm, width ⅕–⅔ length, not concealing capsule, margins usually 3–7-lobed, lateral lobes usually prominent, rarely entire. **Flowers:** sepals 1–4; petals orange proximally, yellow distally, 2–6(–8) mm, apex retuse or rounded; stamens 20–40, 2–7 mm, filaments monomorphic, filiform, unlobed; styles 1.5–3.5(–6) mm. **Capsules** cylindric or clavate, 6–17(–20) × 2–3 mm, axillary curved to 45° at maturity, usually inconspicuously longitudinally ribbed. **Seeds** 15–35, in 2+ rows distal to mid fruit, tan, moderately to densely dark-mottled, usually irregularly polygonal, occasionally triangular prisms proximal to mid fruit, surface tuberculate under 10× magnification; recurved flap over hilum absent; seed coat cell outer periclinal wall domed, domes on seed edges more than ½ as tall as wide at maturity. **2*n*** = 36.

Flowering Apr–Aug. Open, disturbed slopes or flats, grasslands, sagebrush scrub, coniferous forests; 600–3400 m; Ariz., Calif., Colo., Idaho, Nev., N.Mex., Oreg., Tex., Utah, Wash., Wyo.; Mexico (Baja California).

Mentzelia montana is widely distributed and, in portions of its range, difficult to distinguish from *M. albicaulis*. *Mentzelia montana* is morphologically intermediate to *M. albicaulis* and *M. congesta*, but ecologically more similar to *M. congesta* (J. M. Brokaw 2009). *Mentzelia montana* generally occurs at higher elevations than *M. albicaulis* and is best distinguished morphologically from *M. albicaulis* by capsule and bract characteristics. Capsules of *M. montana* are usually not longer than 17 mm or curved more than 45°, whereas those of *M. albicaulis* are often longer and more curved. Both species may have bracts with some lobes and a whitish base, but only *M. montana* has populations in which these features are prominent. Sepal and petal lengths, which have sometimes been used to distinguish these species, overlap completely.

79. Mentzelia nitens Greene, Fl. Francisc., 234. 1891 • Shining blazingstar [E]

Plants candelabra-form, 5–20(–35) cm. **Basal leaves** persisting; petiole present or absent; blade lanceolate to linear, margins usually deeply to shallowly lobed, rarely entire. **Cauline leaves:** petiole absent; blade ovate-lanceolate to linear, to 15 cm, margins deeply to shallowly lobed or entire. **Bracts** green, lanceolate, 4.5–8.3 × 1.2–3.6 mm, width ⅕–½ length, not concealing capsule, margins entire. **Flowers:** sepals 3–8 mm; petals yellow to orange proximally, yellow distally, (7–)8–18 mm, apex rounded or acute; stamens 20+, 3–8 mm, filaments monomorphic, filiform, unlobed; styles 4–8 mm. **Capsules** clavate, 13–26 × 2–3.5 mm, axillary curved to 180° at maturity, usually inconspicuously longitudinally ribbed. **Seeds** 15–40, in 2+ rows distal to mid fruit, tan, usually dark-mottled, usually irregularly polygonal, occasionally triangular prisms proximal to mid fruit, surface colliculate under 10× magnification; recurved flap over hilum usually present; seed coat cell outer periclinal wall domed, domes on seed edges less than ½ as tall as wide at maturity. **2*n*** = 18.

Flowering Apr–Jun. Sandy washes, rocky slopes, desert scrub; 400–2000 m; Calif., Nev.

Mentzelia nitens is similar to both *M. eremophila* and *M. jonesii* but exhibits little distributional overlap with either species. See 71. *M. eremophila* and 73. *M. jonesii* for discussion of similarities. Reports of *M. nitens* from Arizona are based on specimens treated here as *M. jonesii*.

80. Mentzelia obscura H. J. Thompson & J. E. Roberts, Phytologia 21: 284. 1971 • Pacific blazingstar

Plants candelabra-form, 8–45 cm. **Basal leaves** persisting; petiole present or absent; blade linear-lanceolate to linear, margins usually irregularly deeply lobed, lobes pointed. **Cauline leaves:** petiole absent; blade ovate-lanceolate to linear, to 15(–22) cm, margins few-lobed or entire. **Bracts** green, ovate to ovate-lanceolate, 2.9–8.2 × 1.1–1.9 mm, width ⅛–½ length, not concealing capsule, margins entire. **Flowers:** sepals 2–6 mm; petals yellow to orange proximally, yellow distally, 3–8 mm, apex rounded or acute; stamens 20–40, 2–7 mm, filaments monomorphic, filiform, unlobed; styles 2–6 mm. **Capsules** clavate, 11–31 × 1.5–3 mm, axillary curved to 250° at maturity, usually inconspicuously longitudinally ribbed. **Seeds** 15–50, in 2+ rows distal to mid fruit, tan, usually not, occasionally sparsely, dark-mottled, usually irregularly polygonal, occasionally triangular prisms proximal to mid fruit, surface colliculate under 10× magnification; recurved flap over hilum absent; seed coat cell outer periclinal wall domed, domes on seed edges less than ½ as tall as wide at maturity. **2*n*** = 36.

Flowering Feb–May. Sandy to rocky washes or slopes, desert scrub, Joshua-tree woodlands, roadsides; 200–1700 m; Ariz., Calif., Nev., Utah; Mexico (Baja California, Sonora).

Mentzelia obscura is morphologically intermediate to *M. desertorum* and *M. albicaulis* and is known to occur in mixed populations with both species. Reliable discrimination among these species usually requires mature seeds.

81. Mentzelia packardiae Glad, Madroño 23: 289, figs. 2C,D. 1976 • Packard's blazingstar [C] [E]

Plants candelabra-form, (10–) 20–45 cm. **Basal leaves** persisting or not. **Cauline leaves:** petiole present or absent (proximal leaves), absent (distal leaves); blade lanceolate to linear (proximal leaves), ovate to linear (distal leaves), to 10 (–14) cm, margins dentate or entire (proximal leaves), entire (distal leaves). **Bracts** green, ovate to elliptic, 4.3–8.3 × 1.9–3.2 mm, width ⅓–⅔ length, not concealing capsule, margins entire. **Flowers:** sepals 4–10 mm; petals orange proximally, yellow distally, 10–20 mm, apex rounded to retuse; stamens 20+, 4–13 mm, filaments monomorphic, filiform, unlobed; styles 6–14 mm. **Capsules** narrow-cylindric, 8–35 × 2.5–4 mm, axillary curved to 45° at maturity, often prominently longitudinally ribbed. **Seeds** 10–20, in 1 row distal to mid fruit, tan, dark-mottled or not, triangular prisms, surface ± smooth to minutely tessellate under 10× magnification; recurved flap over hilum absent; seed coat cell outer periclinal wall flat to slightly convex. **2*n*** = 72.

Flowering May–Jul. Yellow to whitish green ash-tuff soils, coarse gravels, steep, open to grassy slopes; of conservation concern; 800–1300(–2000) m; Oreg.

Mentzelia packardiae is the most narrowly distributed species in sect. *Trachyphytum*, known only from the Leslie Gulch area in eastern Malheur County. During most years, *M. packardiae* is predominantly limited to barren, ash-derived gravel slopes. A reported collection from Elko County, Nevada, is most likely from a large-flowered population of *M. dispersa* (N. H. Holmgren et al. 2005). *Mentzelia packardiae* can be distinguished from *M. dispersa* by its larger flowers with longer petals [10–20 versus 2–6(–8) mm] and styles [6–14 versus 2–3.5(–5) mm], and its wider capsules (2.5–4 versus 1–2.5 mm).

Mentzelia packardiae is listed as threatened by the Oregon Department of Agriculture and is in the Center for Plant Conservation's National Collection of Endangered Plants.

82. Mentzelia pectinata Kellogg, Proc. Calif. Acad. Sci. 3: 40, fig. 9. 1863 • San Joaquin blazingstar [E]

Plants candelabra-form, (8–) 20–50(–60) cm. **Basal leaves** persisting; petiole present or absent; blade lanceolate to linear, margins deeply to shallowly lobed. **Cauline leaves:** petiole absent; blade ovate to linear, to 12 cm, margins deeply lobed to dentate. **Bracts** green, ovate to lanceolate, 6.6–12.8 × 1.9–6.8 mm, width ⅕–⅔ length, not concealing capsule, margins 3–7-lobed. **Flowers:** sepals 3–13 mm; petals red to orange proximally, orange to yellow distally, 8–22 mm, apex mucronate, rounded, or retuse; stamens 20+, 4–11 mm, filaments monomorphic, filiform, unlobed; styles 5–13 mm. **Capsules** clavate, 12–35 × 2–4 mm, axillary curved to 90° at maturity, usually inconspicuously longitudinally ribbed. **Seeds** 20–40, in 2+ rows distal to mid fruit, tan, dark-mottled or not, usually irregularly polygonal, occasionally triangular prisms proximal to mid fruit, surface tuberculate under 10× magnification; recurved flap over hilum absent; seed coat cell outer periclinal wall domed, domes on seed edges more than ½ as tall as wide at maturity. **2*n*** = 18.

Flowering Mar–May. Slopes of sandy or gray-white silty soils, grasslands, oak savannas, uncommonly juniper woodlands; 200–1400 m; Calif.

Mentzelia pectinata occurs in Kern, San Luis Obispo, Santa Barbara, and Tulare counties, around the southern rim of the San Joaquin Valley, extending into the Inner Coast Ranges and the southern foothills of the Sierra Nevada. Petal color varies from orange to yellow, and fully fertile artificial hybrids have been obtained between populations representing the extreme phenotypes (J. E. Zavortink 1966). Phylogenetic studies suggest that *M. pectinata* and *M. congesta* have hybridized to form several allopolyploid species (J. M. Brokaw and L. Hufford 2010b) despite their current allopatric distributions.

83. Mentzelia ravenii H. J. Thompson & J. E. Roberts, Phytologia 21: 285. 1971 • Raven's blazingstar [E]

Plants candelabra-form, (5–) 20–45 cm. **Basal leaves** persisting; petiole present or absent; blade linear-lanceolate, margins deeply to shallowly lobed. **Cauline leaves:** petiole absent; blade ovate-lanceolate to lanceolate, to 18 cm, margins deeply lobed to dentate. **Bracts** green with prominent white base usually conspicuously extending outwards from midvein, ovate, 5.6–8.4 × 1.8–3.9 mm, width ⅕–⅔ length, not concealing capsule,

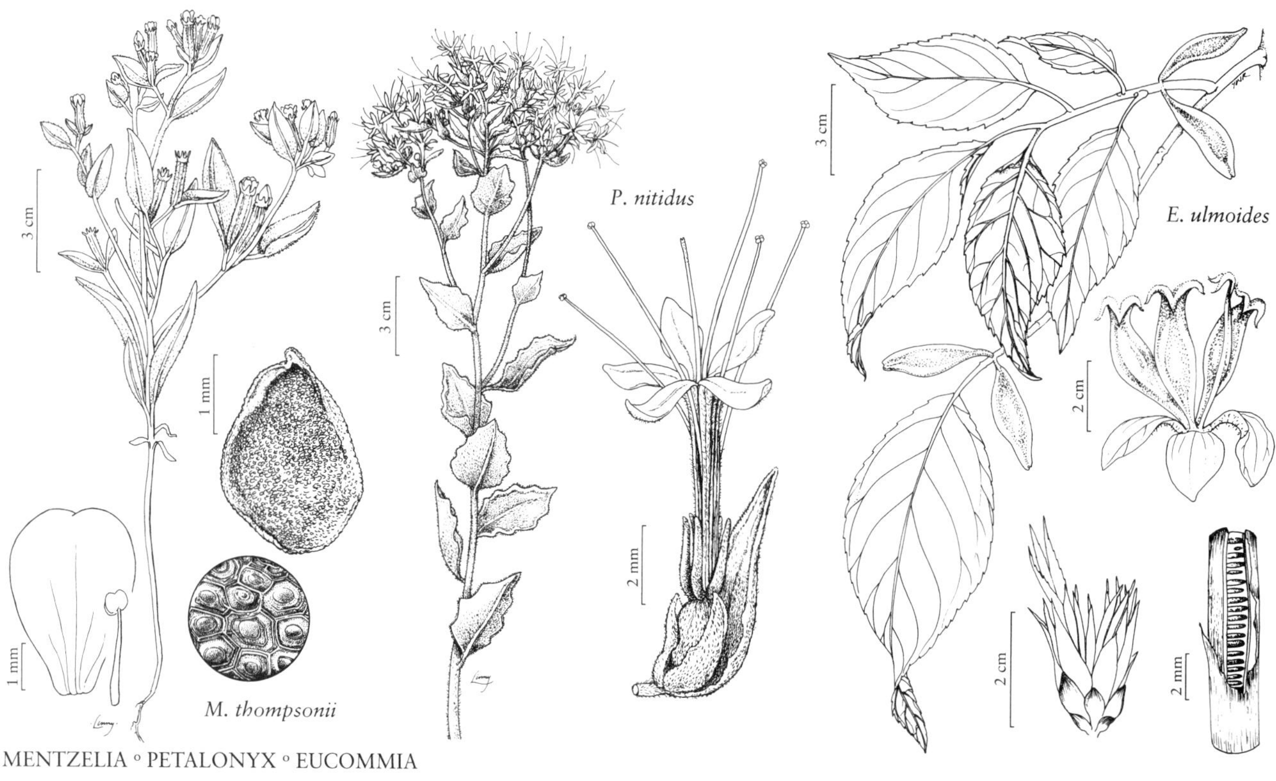

MENTZELIA ◦ PETALONYX ◦ EUCOMMIA

margins 3–5-lobed. **Flowers:** sepals 2–6 mm; petals orange proximally, yellow distally, 5–11(–13) mm, apex retuse; stamens 20+, 3–7 mm, filaments monomorphic, filiform, unlobed; styles 3.5–8 mm. **Capsules** clavate, 8–23 × 2–3 mm, axillary curved to 45° at maturity, usually inconspicuously longitudinally ribbed. **Seeds** 15–30, in 2+ rows distal to mid fruit, tan, dark-mottled, usually irregularly polygonal, occasionally triangular prisms proximal to mid fruit, surface tuberculate under 10× magnification; recurved flap over hilum absent; seed coat cell outer periclinal wall domed, domes on seed edges more than or equal to ½ as tall as wide at maturity. $2n = 36$.

Flowering Mar–May. Sandy desert foothills, roadsides, desert scrub, Joshua-tree woodlands; 300–1200 m; Calif.

Mentzelia ravenii is narrowly distributed, with most populations limited to desert foothills on the northern edge of the San Gabriel Mountains in northeastern Los Angeles County. Populations of *M. ravenii* with relatively large flowers with yellow petals are often found growing under desert shrubs in mixed populations with *M. veatchiana*, which has relatively small flowers with orange petals in this area. A few tetraploid populations from western Riverside County have also been called *M. ravenii* (J. E. Zavortink 1966); further work is needed to confirm that these represent the same species. See 72. *M. gracilenta* for discussion of similarities it shares with *M. ravenii*.

84. Mentzelia thompsonii Glad, Madroño 23: 289, figs. 2A,B. 1976 • Thompson's stickleaf [E] [F]

Acrolasia humilis Osterhout, Bull. Torrey Bot. Club 49: 183. 1922, not *Mentzelia humilis* (Urban & Gilg) J. Darlington 1934; *A. thompsonii* (Glad) W. A. Weber

Plants wandlike or candelabraform, 5–20 cm. **Basal leaves** not persisting. **Cauline leaves:** petiole present or absent (proximal leaves), absent (distal leaves); blade ovate to linear (proximal leaves), ovate to lanceolate (distal leaves), to 6 cm, margins entire. **Bracts** green, elliptic to oblanceolate, 3.3–6.2 × 1.5–3.2 mm, width ¼–⅞ length, not concealing capsule, margins entire. **Flowers:** sepals 1–3 mm; petals yellow, 2–4 mm, apex retuse to rounded; stamens 10–30, 2–3.5 mm, filaments monomorphic, filiform, unlobed; styles 1.5–3 mm. **Capsules** cylindric or clavate, 5–16(–20) × 2–4 mm, axillary curved to 45° at maturity, often prominently longitudinally ribbed. **Seeds** 10–25, in 2+ rows distal to mid fruit, tan, dark-mottled, irregularly polygonal, surface smooth to minutely tessellate under 10× magnification; recurved flap over hilum absent; seed coat cell outer periclinal wall domed, domes on seed edges less than ½ as tall as wide at maturity. $2n = 18$.

Flowering Apr–Jun. Barren clay to silt slopes, usually on Mancos Shale barrens; 1300–2000 m; Colo., N.Mex., Utah.

Mentzelia thompsonii was originally described as *Acrolasia humilis*, which has caused some confusion with 2. *M. humilis* of sect. *Bartonia* (J. J. Schenk 2009). Subsequently, *M. thompsonii* was incorrectly treated as a synonym of *M. humilis* by J. T. Kartesz and C. A. Meacham (1999), leading to lack of recognition of this easily distinguished species in several plant databases and herbaria. *Mentzelia thompsonii* is distributed predominantly along the Colorado-Utah border south to the Four Corners region.

85. **Mentzelia veatchiana** Kellogg, Proc. Calif. Acad. Sci. 2: 99, fig. 28. 1863 • Veatch's blazingstar [E]

Mentzelia albicaulis (Douglas) Douglas ex Torrey & A. Gray var. *veatchiana* (Kellogg) Urban & Gilg

Plants candelabra-form, (5–)20–50 cm. **Basal leaves** persisting; petiole present or absent; blade linear-lanceolate, margins deeply to shallowly lobed. **Cauline leaves:** petiole absent; blade ovate-lanceolate to lanceolate, to 17 cm, margins usually deeply lobed to dentate, rarely entire. **Bracts** usually green with prominent white base usually conspicuously extending outwards from midvein, rarely green, usually ovate, rarely lanceolate, 3.3–6.2 × 1.5–3.2 mm, width ¼–⅞ length, not concealing capsule, margins usually 3–7-lobed, rarely entire. **Flowers:** sepals 2–5 mm; petals red to orange proximally, orange to yellow distally, 4–7 (–10) mm, apex retuse; stamens 20+, 3–7 mm, filaments monomorphic, filiform, unlobed; styles (3–)3.5–6 mm. **Capsules** clavate, 8–28 × 2–4 mm, axillary curved to 70° at maturity, usually inconspicuously longitudinally ribbed. **Seeds** 15–35, in 2+ rows distal to mid fruit, tan, dark-mottled, usually irregularly polygonal, occasionally triangular prisms proximal to mid fruit, surface tuberculate under 10× magnification; recurved flap over hilum absent; seed coat cell outer periclinal wall domed, domes on seed edges more than or equal to ½ as tall as wide at maturity. $2n = 54$.

Flowering Mar–Jun. Loamy to sandy soils, grasslands, desert scrub, oak-pine woodlands; 200–2500 m; Ariz., Calif., Nev., Oreg.

Mentzelia veatchiana is the most common and widely distributed hexaploid species in sect. *Trachyphytum*. It exhibits considerable morphological variation and can be difficult to distinguish from *M. montana* in northern California. Like the larger-flowered *M. pectinata*, *M. veatchiana* has interfertile populations with petal colors ranging from orange to yellow (J. E. Zavortink 1966). When bearing orange petals, *M. veatchiana* is easily distinguished from other species. Reports of *M. veatchiana* from Utah are based on specimens treated here as *M. montana*.

4. PETALONYX A. Gray, Pl. Nov. Thurb., 319. 1854 • Sandpaper plant [Greek *petalon*, petal, and *onyx*, claw, alluding to distinctive petal morphology]

Larry D. Hufford

Subshrubs or shrubs; trichomes (1) pointed with surfaces ± smooth, notched, or antrorsely barbed, and (2) retrorsely barbed along shaft and at apex. **Stems** erect or spreading. **Leaves** cauline; petiole present or absent; blade ovate, elliptic, lanceolate, or falcate, unlobed, margins dentate, serrate, crenate, or entire. **Inflorescences** racemes or panicles; peduncle inconspicuous. **Pedicels** not elongating in fruit. **Flowers:** hypanthium adnate to ovary proximally, free distally; perianth whorls differentiated; sepals green, connate basally, lanceolate, shorter than petals; petals white, distinct or connate, spatulate or obovate, erect proximally, spreading to divaricate distally (corolla salverform or appearing so), glabrous except hairy abaxially on midribs; nectary distal on ovary; stamens 5, exserted, filaments monomorphic, filiform, longer than anthers; anthers without distal connective extension; staminodes absent [present]; pistil pseudomonomerous, placenta subapical; stigma lingulate, 2-lobed, papillate. **Fruits** cypselae, ± clavate, straight; sepals persistent. **Seeds** 1, ovoid. $x = 23$.

Species 5 (4 in the flora): w United States, n Mexico.

Petalonyx belongs to Gronovioideae, which is characterized by relatively small flowers that have five stamens, one ovule, and indehiscent fruits. *Petalonyx* is most closely related to a clade that consists of *Cevallia*, *Fuertesia* Urban, and *Gronovia* Linnaeus (L. Hufford et al. 2003). *Petalonyx crenatus* A. Gray ex S. Watson, from Coahuila, Mexico, is the only species of the genus known only from outside the flora area; it can be distinguished from the species treated here by having two anther-bearing stamens and three shorter staminodes, rather than having five anther-bearing stamens.

SELECTED REFERENCE Davis, W. S. and H. J. Thompson. 1967. A revision of *Petalonyx* (Loasaceae) with a consideration of affinities in the subfamily Gronovioideae. Madroño 19: 1–18.

1. Leaf blades usually elliptic, sometimes falcate, to 5 mm wide, margins entire; petal claws distinct; stamens exserted distally (not laterally between petal claws) .... 1. *Petalonyx linearis*
1. Leaf blades ovate, or if elliptic then to 15–30 mm wide, margins usually serrate, dentate, or crenate, sometimes entire; petal claws postgenitally distally coherent; stamens exserted laterally through slits between petal claws.
    2. Petioles absent; leaf blades with marked size dimorphism, much larger on main stems than on fertile branches; petals to 6.5 mm. .... 4. *Petalonyx thurberi*
    2. Petioles 0.5–4 mm; leaf blades without marked size dimorphism; petals 6–15 mm.
        3. Inflorescences 10–30-flowered; petals 6–11 mm; branches of current season 11–37 cm .... 2. *Petalonyx nitidus*
        3. Inflorescences 35–65-flowered; petals 10–15 mm; branches of current season to 13 cm .... 3. *Petalonyx parryi*

1. Petalonyx linearis Greene, Bull. Calif. Acad. Sci. 1: 188. 1885 • Narrow-leaf sandpaper plant

Subshrubs, bushy, to 16 dm; branches of current season 10–38 cm. **Leaves:** petiole 0–1 mm; blade usually elliptic, sometimes falcate, without marked size dimorphism, to 42 × 5 mm, base acute, margins entire, apex rounded to acute. **Inflorescences** 60-flowered. **Flowers** ± radially symmetric; petals spatulate, 5 mm, claws distinct; stamens exserted distally (not laterally between petal claws). $2n = 46$.

Flowering Mar–Jun. Sandy and gravelly canyon and arroyo bottoms, creosote bush scrub; 20–1000 m; Ariz., Calif.; Mexico (Baja California, Sonora).

Petalonyx linearis is found in the Sonoran Desert.

2. Petalonyx nitidus S. Watson, Amer. Naturalist 7: 300. 1873 • Shiny-leaf sandpaper plant [E] [F]

Subshrubs or shrubs, bushy to moundlike, to 6 dm; branches of current season 11–37 cm. **Leaves:** petiole 1–4 mm; blade ovate, without marked size dimorphism, to 35 × 28 mm, base acute to obtuse, margins serrate to dentate, apex acute. **Inflorescences** 10–30-flowered. **Flowers** slightly bilaterally symmetric through curvature along length of flower; petals spatulate, 6–11 mm, claws postgenitally distally coherent, forming slitted corolla tube; stamens exserted laterally through slits between petal claws. $2n = 46$.

Flowering Mar–Aug. Sandy, gravelly, or rocky canyon slopes, arroyo bottoms, scrub; 400–2200 m; Ariz., Calif., Nev., Utah.

SELECTED REFERENCE Hufford, L. 1989. Structure of the inflorescence and flower of *Petalonyx linearis*. Pl. Syst. Evol. 163: 211–226.

3. Petalonyx parryi A. Gray, Proc. Amer. Acad. Arts 10: 72. 1874 • Parry's sandpaper plant [C] [E]

Shrubs, bushy to moundlike, to 15 dm; branches of current season to 13 cm. **Leaves:** petiole 0.5–3.5 mm; blade ovate to elliptic, without marked size dimorphism, to 40 × 30 mm, base acute to rounded, margins usually crenate to serrate, sometimes small leaves entire, apex acute. **Inflorescences** 35–65-flowered. **Flowers** strongly bilaterally symmetric; petals spatulate, 10–15 mm, claws postgenitally distally coherent, forming slitted corolla tube; stamens exserted laterally through slits between petal claws. $2n = 46$.

Flowering Apr–Jul. Wash bottoms, desert plains, usually white to gray, clayey soils; of conservation concern; 400–1300 m; Ariz., Nev., Utah.

4. Petalonyx thurberi A. Gray, Pl. Nov. Thurb., 319. 1854 • Thurber's sandpaper plant

Shrubs, bushy to moundlike, to 10 dm; branches of current season 12–45 cm. **Leaves:** petiole absent; blade ovate to elliptic, with marked size dimorphism, to 45 × 15 mm, much larger on main stems than on fertile branches, base acute to rounded, margins usually serrate or crenate, distal often entire, apex acute. **Inflorescences** to 40-flowered. **Flowers** conspicuously bilaterally symmetric; petals spatulate, to 6.5 mm, claws postgenitally distally coherent, forming slitted corolla tube; stamens exserted laterally through slits between petal claws. $2n = 46$.

Subspecies 2 (2 in the flora): w United States, n Mexico.

W. S. Davis and H. J. Thompson (1967) called attention to geographical variation in *Petalonyx thurberi* and distinguished subsp. *gilmanii*, which is restricted to washes in Inyo County, California, from the widespread subsp. *thurberi.* Subspecies *gilmanii* has flowers on the small side of those found among other populations of *P. thurberi*, although floral attributes do not readily distinguish between the two named subspecies. Davis and Thompson also noted another form that has relatively small leaves on and closely appressed to the inflorescence-bearing stems, although they did not formally distinguish this variant with a name. Morphometric and phylogeographic studies are warranted in *P. thurberi* to test whether morphological variation is significant, geographically partitioned, and associated with genetically isolated lineages.

1. Hairs on inflorescence-bearing stems ± retrorse . . . . 4a. *Petalonyx thurberi* subsp. *thurberi*
1. Hairs on inflorescence-bearing stems ± erect . . . . 4b. *Petalonyx thurberi* subsp. *gilmanii*

4a. Petalonyx thurberi A. Gray subsp. **thurberi**

Plants to 8 dm; branches of current season to 45 cm; hairs on inflorescence-bearing stems stiff, ± retrorse. **Leaves** to 45 × 15 mm. **Flowers:** petals to 6.5 mm; stamens 6–10 mm; style 5–11 mm. $2n = 46$.

Flowering Mar–Nov. Sandy washes, sand dunes; 10–1600 m; Ariz., Calif., Nev.; Mexico (Baja California, Sonora).

4b. Petalonyx thurberi A. Gray subsp. **gilmanii** (Munz) W. S. Davis & H. J. Thompson, Madroño 19: 15. 1967 [E]

Petalonyx gilmanii Munz, Leafl. W. Bot. 2: 69, plate 1, figs. 4–6. 1938

Plants to 10 dm; branches of current season to 21 cm; hairs on inflorescence-bearing stems soft, ± erect. **Leaves** to 20 × 13 mm. **Flowers:** petals to 4.1 mm; stamens 5–8 mm; style 4–6 mm. $2n = 46$.

Flowering May–Sep. Sandy washes, sand dunes; 400–1000 m; Calif.

Subspecies *gilmanii* is found only in Inyo County.

EUCOMMIACEAE Engler • Hardy Rubber-tree Family

Michael A. Vincent

Trees, deciduous, dioecious; latex present. **Leaves** alternate, simple; stipules absent; petiole present; blade margins serrate; venation pinnate. **Inflorescences** unisexual, axillary, fascicles (staminate) or flowers solitary (pistillate). **Flowers** unisexual; hypanthium absent; sepals 0; petals 0; nectary absent; stamens 5–12, distinct; anthers dehiscing by longitudinal slits; pistil 1, 2-carpellate; ovary superior, 1-locular; placentation apical; ovules 2 per locule, anatropous; style 0; stigmas 2. **Fruits** samaras. **Seeds** 1 per fruit.

Genus 1, species 1: introduced; Asia (c, w China); introduced and cultivated widely.

Placement of *Eucommia* has been controversial, with various authors allying it with Cornales, Euphorbiaceae, Hamamelidaceae, Trochodendraceae Eichler, and Ulmaceae (reviewed in M. A. Vincent 2002). Most recent DNA analyses support recognizing Eucommiaceae as a distinct family in Garryales (D. E. Soltis et al. 2000; K. Bremer et al. 2001). Fossil remains of *Eucommia* have been found in North America, Europe, and Asia, evidence that the genus was widespread in the Cenozoic and that its present-day distribution is relictual (V. B. Call and D. L. Dilcher 1997).

SELECTED REFERENCES Call, V. B. and D. L. Dilcher. 1997. The fossil record of *Eucommia* (Eucommiaceae) in North America. Amer. J. Bot. 84: 798–814. Vincent, M. A. 2002. *Eucommia ulmoides* (Hardy rubber-tree, Eucommiaceae) as an escape in North America. Michigan Bot. 41: 141–145. Zhang, Y. L., Wang F. H., and Chien N. F. 1988. A study on pollen morphology of *Eucommia ulmoides* Oliver. Acta Phytotax. Sin. 26: 367–370.

1. EUCOMMIA Oliver, Hooker's Icon. Pl. 20: plate 1950. 1890 • [Greek *eu*, good or well, and *kommi*, gum, alluding to abundant latex in younger tissues of plant] I

Trees, indumentum of simple hairs. **Pedicels** present. **Staminate flowers:** stamen filaments 0.9–1.2 mm. **Pistillate flowers:** stigmas decurrent, reflexed-spreading. **Seeds** elongate-oblong, seed coat dry. $x = 17$.

Species 1: introduced; Asia (wc, w China); introduced and cultivated widely.

1. **Eucommia ulmoides** Oliver, Hooker's Icon. Pl. 20: plate 1950. 1890 • Hardy rubber-tree [F] [I]

Trees to 20 m; bark gray-brown, ridged and furrowed; branchlets brown, hairy when young, becoming glabrate, pith septate. **Leaves:** petiole 1–2.5 cm; blade elliptical to ovate, 5–20 × 2.5–8 cm, base usually truncate to broadly cordate or cuneate, sometimes oblique, apex acuminate, surfaces hairy when young, soon glabrate; latex strands visible in carefully torn leaf. **Staminate flowers:** anthers 10–12 mm. **Pistillate flowers:** pistil 10–12 mm. **Samaras** brown, 25–32 mm, apex emarginate. **Seeds** 12–15 × 3 mm. $2n$ = 34.

Flowering (Mar–)Apr. Disturbed woods, fence rows; 200–300 m; introduced; Ind., N.Y., Ohio; Asia (China); introduced widely.

Eucommia ulmoides, first reported as an escape in North America by M. A. Vincent (2002), has been put to many uses, including lumber, firewood, and a medicinal tonic (duzhong or tu-chung) made from the bark (C. S. Sargent 1913–1917; T. Forrest 1995; D. J. Mabberley 2008). Leaf and stem extracts have been shown to have potential medicinal value (K. Metori et al. 1997, 1998; T. Nakamura et al. 1997; Y. Nakazawa et al. 1997; H. J. Jeon et al. 1998; Y. Li et al. 1998). *Eucommia ulmoides* contains a latex (gutta-percha) that has been used in China for lining oil pipelines, insulating electrical lines, and filling teeth (Mabberley).

Eucommia ulmoides is sometimes used as a street or lawn tree. It is drought resistant, disease-free, and easily propagated by seed or cuttings; it is hardy in USDA Zones 4–7 (A. J. Rehder 1940; M. A. Dirr 1990).

GARRYACEAE Lindley

• Silktassel Family

Guy L. Nesom

Shrubs or trees, evergreen, dioecious. **Leaves** opposite (decussate), simple; stipules absent; petiole present, basally connate with that of opposite leaf; blade margins entire, serrate, or dentate; venation pinnate. **Inflorescences** unisexual, axillary aments or terminal panicles. **Flowers** unisexual; perianth epigynous; hypanthium absent (staminate flowers), completely adnate to ovary (pistillate flowers); sepals 2, 4, or rudimentary, distinct or connate proximally; petals 0 or 4, distinct; nectary absent or present, intrastaminal, sometimes rudimentary; stamens 4, distinct, free; anthers dehiscing by longitudinal slits; pistil 1, 1-carpellate, ovary inferior, 1-locular, placentation apical; ovules 1–2 per locule, anatropous; styles 1–2(–3), distinct; stigmas 1–2(–3). **Fruits** drupes or berries. **Seeds** 1–2 per fruit.

Genera 2, species 27 (2 genera, 9 species in the flora): w United States, Mexico, West Indies (Cuba, Hispaniola, Jamaica), Central America, e Asia; some species of both genera are cultivated and introduced widely.

Aucuba and *Garrya* are strikingly different, especially in their inflorescences, and *Aucuba* usually has been placed in Cornaceae or in Aucubaceae. However, plants from the two genera have many chemical similarities (A. Liston 2003) and can be readily intergrafted. B. Bremer et al. (2002) and Angiosperm Phylogeny Group (2003) united them in Garryaceae on the basis of molecular, chemical, and morphological evidence indicating that the two have a sister relationship. Aucubaceae was maintained as distinct from Garryaceae in the *Flora of China* (Xiang Q. Y. and D. E. Boufford 2005).

1. Inflorescences axillary aments; styles 2(–3), linear-lanceolate . . . . 1. *Garrya*, p. 549
1. Inflorescences terminal panicles; style 1, cylindric . . . . 2. *Aucuba*, p. 553

1. GARRYA Douglas ex Lindley, Edwards's Bot. Reg. 20: plate 1686. 1834 • [For Nicholas Garry, 1782–1856, deputy-governor of the Hudson's Bay Company from 1822–1835, diarist of his 1821 travels in the Northwest Territories, friend and benefactor of David Douglas]

Shrubs or trees. Leaves: blade flat to concave-convex, coriaceous, margins entire, flat, revolute, or strongly undulate. **Inflorescences** axillary, aments; bracts opposite, distinct or connate basally; bracteoles 0. **Pedicels:** staminate present, pistillate absent. **Staminate flowers:** sepals 4, valvate, apically connivent by intertwined hairs, linear to lanceolate-oblong; petals 0; nectary absent or vestigial; stamens alternate with sepals, anthers basifixed. **Pistillate flowers:** sepals 2, sometimes rudimentary; petals 0; nectary absent; ovules 2; styles 2(–3), erect or recurved, linear-lanceolate; stigmas decurrent on adaxial surfaces of styles. **Fruits** berries, dark blue to black, drying whitish-gray, subglobose to ovoid, fleshy, becoming brittle. **Seeds** (1–)2 per fruit. $x = 11$.

Species 17 (8 in the flora): w, sc United States, Mexico, West Indies (Cuba, Hispaniola, Jamaica), Central America; some cultivated.

Two sections of *Garrya* have been recognized (see key below). Species of sect. *Garrya* range from Washington to Baja California and New Mexico, with one outlier in Guatemala, *G. corvorum* Standley & Steyermark. Section *Fadyenia* de Candolle ranges from trans-Pecos Texas to Arizona and from Mexico (including Baja California) to Central America plus the West Indies (Cuba, Hispaniola, and Jamaica). In addition to the features noted in the key, in sect. *Garrya* pistillate flowers bear a pair of small appendages at the ovary apex near the style base, whereas in sect. *Fadyenia* pistillate flowers occasionally produce two foliaceous bracts partially adnate to the ovary. The phylogenetic distinction between the two sections has been confirmed by D. O. Burge (2011).

The staminate flowers in pendulous aments of *Garrya* are similar to those in other genera specialized for wind-pollination. The staminate aments are more flexible than the pistillate and more responsive to wind currents. P. A. Munz (1959) noted that in California many of the species intergrade extensively where their distributions overlap, and hybridization and intraspecific variation underlie continuing uncertainty about delimitations of taxa. Natural hybrids, however, have not been reported elsewhere. The artificial hybrids *G.* ×*issaquahensis* Talbot de Malahide ex E. C. Nelson (*G. elliptica* × *G. fremontii*) and *G.* ×*thuretii* Carrière (*G. elliptica* × *G. fadyenii* Hooker) have been bred for garden planting.

The common name silktassel alludes to the sericeous vestiture of the long, showy aments in sect. *Garrya*. Plants are cultivated mainly for the foliage and showy staminate aments, and cultivars are mostly staminate plants propagated from cuttings.

Stem extracts of *Garrya laurifolia* Bentham are toxic but are used as an antidiarrhetic throughout rural Mexico, and bark extracts were reportedly used by Native Americans to treat fever (G. V. Dahling 1978).

SELECTED REFERENCES Burge, D. O. 2011. Molecular phylogenetics of *Garrya* (Garryaceae). Madroño 58: 249–255. Dahling, G. V. 1978. Systematics and evolution of *Garrya*. Contr. Gray Herb. 209: 1–104. Eyde, R. H. 1964. Inferior ovary and generic affinities of *Garrya*. Amer. J. Bot. 51: 1083–1092. Liston, A. 2003. A new interpretation of floral morphology in *Garrya* (Garryaceae). Taxon 52: 271–276.

1. Staminate aments 1–3 cm; pistillate aments loose, internodes 4+ mm, erect, sometimes branched; pistillate bracts distinct or connate basally, each usually subtending 1 flower, at least proximal similar in size and shape to distal leaves; Arizona, New Mexico, Texas [*Garrya* sect. *Fadyenia*].

2. Leaf blade abaxial surfaces glabrous or sparsely strigose . . . . 1. *Garrya wrightii*
2. Leaf blade abaxial surfaces persistently sparsely to densely puberulent-tomentulose to tomentulose with coiling to recurved hairs.
   3. Leaf blades 1.6–4(–5.5) × 0.7–2.5 cm, margins undulate, ± muricate-roughened, especially distal to middle, adaxial surfaces usually persistently tomentose, sometimes ± glabrescent . . . . 2. *Garrya goldmanii*
   3. Leaf blades 4.5–8 × 2.5–5 cm, margins flat, smooth, adaxial surfaces glabrate or glabrous . . . . 3. *Garrya lindheimeri*

1. Staminate aments 3–20 cm; pistillate aments compact, internodes to 1 mm, pendulous, unbranched; pistillate bracts connate proximally into deep cup, at least at proximal nodes each subtending 3 flowers, differing in size and shape from leaves; Arizona, California, Nevada, New Mexico, Oregon, Utah, Washington [*Garrya* sect. *Garrya*].
   4. Leaf blade abaxial surfaces usually densely, sometimes becoming sparsely, closely tomentose, hairs curled or crisped, interwoven.
      5. Leaf blade apices rounded to obtuse; pistillate and staminate aments 8–15 cm. . . . 4. *Garrya elliptica*
      5. Leaf blade apices acuminate; pistillate and staminate aments 2.5–7 cm . . . . 5. *Garrya veatchii*
   4. Leaf blade abaxial surfaces glabrous, glabrate, or sparsely to densely strigose-sericeous or strigose, hairs antrorsely appressed.
      6. Leaf blades abaxially whitish, adaxially yellow-green to gray-green, dull; berries densely strigose-sericeous, sometimes glabrate toward base . . . . 6. *Garrya flavescens*
      6. Leaf blades abaxially green, adaxially bright to olive green, glossy; berries glabrous, glabrate, or sparsely strigose near apex.
         7. Staminate aments 7–20 cm; leaf blade abaxial surfaces glabrous or sparsely strigose . . . . 7. *Garrya fremontii*
         7. Staminate aments 5–7 cm; leaf blade abaxial surfaces densely strigose-sericeous . . . . 8. *Garrya buxifolia*

1. Garrya wrightii Torrey in War Department [U.S.], Pacif. Railr. Rep. 4(5): 136. 1857 • Wright's silktassel [F]

Shrubs or trees, 1–3(–4) m, branchlets sparsely strigose, glabrescent. **Leaves:** blade yellowish green, flat, elliptic to broadly elliptic, oblong-elliptic, or elliptic-ovate, (2.5–)3–5.5 × (1–)1.5–3 cm, length 1.2–2.5 times width, margins flat, roughened or muriculate, apex obtuse to acute and mucronulate, abaxial surface glabrous, glabrate, or sparsely strigose, adaxial surface dull, glabrous or glabrate. **Aments:** staminate 1–2 cm; pistillate loose, internodes 4+ mm, sometimes branched, erect, 2–4 cm; pistillate bracts distinct or connate basally, each usually subtending 1 flower, narrowly lanceolate-ovate to linear-oblanceolate, at least proximal similar in size and shape to distal leaves, strigose. **Berries** 5–7 mm diam., sparsely strigose, glabrate, not glaucous.

Flowering (Mar–)Apr–Aug. Cliff crevices, bluffs, canyons, among boulders, open slopes, pinyon-pine, pine-oak, oak, oak-*Nolina-Agave*, and evergreen oak-mountain mahogany woodlands; (500–)1300–2500 m; Ariz., N.Mex., Tex.; Mexico (Chihuahua, Coahuila, Sonora).

2. Garrya goldmanii Wooton & Standley, Contr. U.S. Natl. Herb. 16: 157. 1913 • Eggleaf silktassel

Garrya ovata Bentham subsp. *goldmanii* (Wooton & Standley) Dahling; *G. ovata* var. *goldmanii* (Wooton & Standley) B. L. Turner

Shrubs, (0.5–)1–2.5 m, branchlets puberulent, glabrescent. **Leaves:** blade gray-green, flat to concave-convex, elliptic to oblong-elliptic, ovate-elliptic, or broadly lanceolate, 1.6–4(–5.5) × 0.7–2.5 cm, length 2 times width, margins undulate, ± muricate-roughened and with callose rim, especially distally, apex obtuse to subacute and mucronulate, both surfaces usually persistently densely tomentulose, hairs coiling to recurved, or adaxial sometimes barely glossy, ± glabrescent. **Aments:** staminate 2–3 cm; pistillate loose, internodes 4+ mm, sometimes branched, erect, 1.5–3 cm; pistillate bracts distinct or connate basally, each usually subtending 1 flower, elliptic to ovate, at least proximal similar in size and shape to distal leaves, closely puberulent with curly or crisped hairs. **Berries** 4–8 mm diam., glabrous, not glaucous.

Flowering Mar–Apr. Ledges, bluffs, slopes, talus, canyons, limestone substrates, ash woods, oak scrub, oak-pine-juniper woodlands; 1400–2400(–2700) m; N.Mex., Tex.; Mexico (Chihuahua, Coahuila).

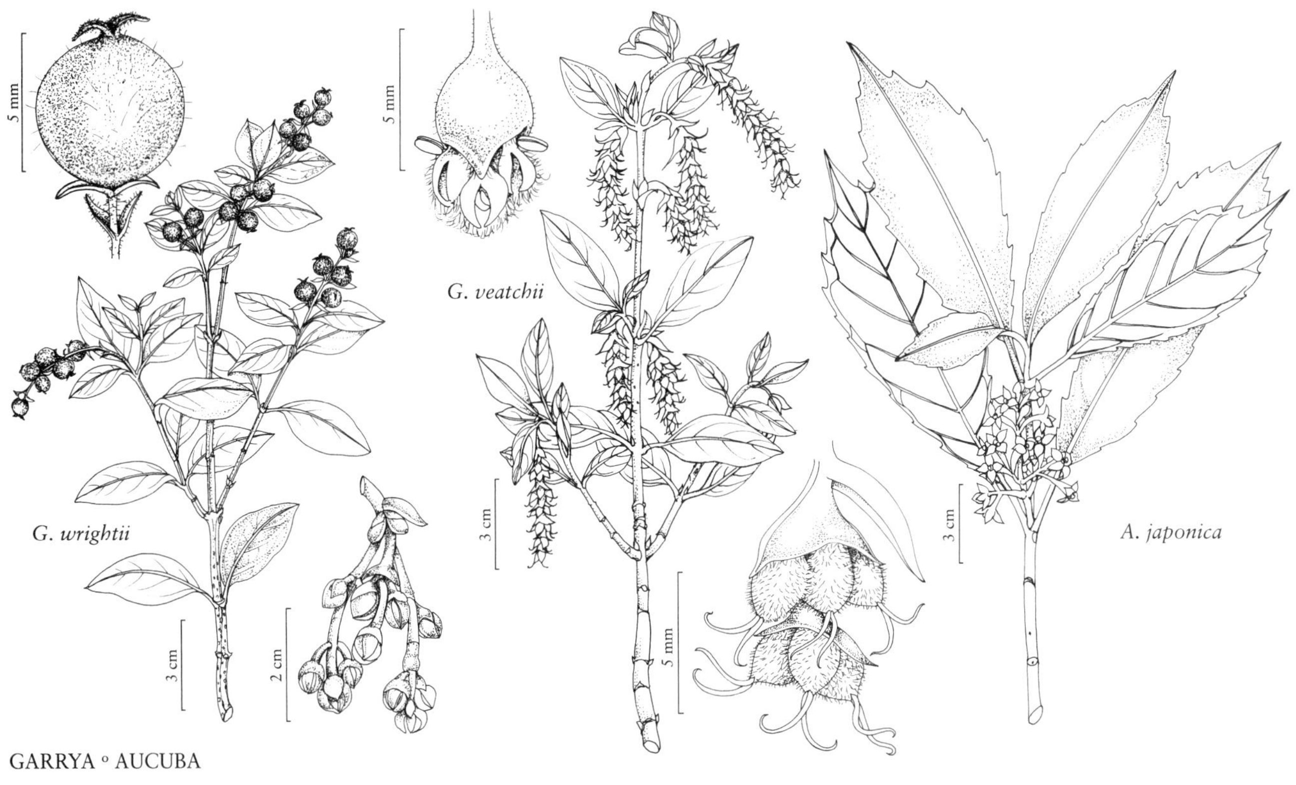

GARRYA ◦ AUCUBA

Garrya goldmanii, which in the flora area is found in trans-Pecos Texas and southern New Mexico, has been treated as a subspecies of *G. ovata* (G. V. Dahling 1978), along with *G. ovata* subsp. *mexicana* Dahling, but all three taxa are distinct and justifiably treated at specific rank (G. L. Nesom 2012e). Both *G. ovata* and *G. mexicana* (Dahling) G. L. Nesom (*G. ovata* subsp. *mexicana*) are restricted to Mexico.

SELECTED REFERENCE Nesom, G. L. 2012e. Notes on the *Garrya ovata* complex (Garryaceae). Phytoneuron 2012-97: 1–6.

3. **Garrya lindheimeri** Torrey in War Department [U.S.], Pacif. Railr. Rep. 4(5): 136. 1857 • Lindheimer's silktassel

Garrya ovata Bentham subsp. *lindheimeri* (Torrey) Dahling; *G. ovata* var. *lindheimeri* (Torrey) J. M. Coulter & W. H. Evans

Shrubs or trees, 1–3.5(–5) m, branchlets puberulent, glabrescent. **Leaves:** blade green, flat, oblong-elliptic to broadly elliptic or obovate, 4.5–8 × 2.5–5 cm, length 2 times width, margins flat, smooth, without callose rim, apex rounded and mucronulate, abaxial surface persistently sparsely to densely puberulent-tomentulose, hairs coiling to recurved, adaxial surface glossy, glabrous or glabrate. **Aments:** staminate 2–3 cm; pistillate loose, internodes 4+ mm, sometimes branched, erect, 2–8 cm; pistillate bracts distinct or connate basally, each usually subtending 1 flower, elliptic to ovate, at least proximal similar in size and shape to distal leaves, minutely puberulent, hairs tightly coiling to strongly recurved. **Berries** 5–10 mm diam., glabrous, usually glaucous. **2*n*** = 22.

Flowering Mar–May. Rocky hills, ledges, cliffs, bluffs, canyons, ravines, along streams, limestone substrates, usually in oak-juniper woodlands; 200–400 m; Tex.; Mexico (Nuevo León).

Judgments have varied regarding the rank of *Garrya lindheimeri*, as illustrated by the synonymy above. Evidence counter to uniting it with *G. ovata* includes their ecological distinction and allopatry over most of their ranges. Morphological differences between *G. lindheimeri* and *G. ovata* are at least comparable to those among some of the California species, which intergrade to a greater extent than these do.

Garrya lindheimeri and *G. goldmanii* are allopatric in Texas, with the former restricted to the Edwards Plateau and adjacent Lampasas Cut Plain and the latter found only in trans-Pecos Texas. They become sympatric in Coahuila, Mexico, and remain distinct although hybrids and perhaps introgressants may be formed. *Garrya goldmanii* occurs in more xeric habitats, as indicated by the differences in distribution and ecology in Texas, and the ecological distinction also apparently exists in Coahuila. Their distinctness, even where sympatric, implies a degree of reproductive isolation and provides rationale for maintaining both at specific rank.

4. Garrya elliptica Douglas ex Lindley, Edwards's Bot. Reg. 20: plate 1686. 1834 • Wavyleaf silktassel [E]

Shrubs or trees, 2–7 m, branchlets short-villous. **Leaves:** blade grayish to silvery white abaxially, green adaxially, flat to concave-convex, usually elliptic, sometimes oval to ovate-lanceolate, (2–)6–10(–12) × 1.4–7.2 cm, length 1.2–2.6 times width, margins ± undulate, often revolute and appearing crenate or dentate, smooth, apex rounded to obtuse, abaxial surface usually densely, sometimes becoming sparsely, closely tomentose, hairs curled or crisped, interwoven, adaxial surface glossy, glabrous. **Aments:** staminate 8–15 cm; pistillate compact, internodes to 1 mm, unbranched, pendulous, 8–15 cm; pistillate bracts connate proximally into deep cup, at least at proximal nodes each subtending 3 flowers, ovate-triangular, differing in size and shape from leaves, densely silky-villous. **Berries** 7–11 mm diam., densely tomentose, glabrescent, not glaucous. **2*n*** = 22.

Flowering (Dec–)Jan–Mar. Sea cliffs, sand dunes and hills, sandy riverbanks, gravelly sand, chaparral, closed-cone pine forests; 0–900(–1600) m; Calif., Oreg.

As noted by T. F. Daniel (1993), the fruiting aments of *Garrya elliptica* are wider (18–28 mm) than those of *G. veatchii* (13–18 mm); apparent intermediates between the two have been noted in closed-cone pine forests in San Luis Obispo County (F. Hrusa, pers. comm.). In their area of sympatry they differ in habitat.

5. Garrya veatchii Kellogg, Proc. Calif. Acad. Sci. 5: 40. 1873 • Canyon silktassel [F]

Shrubs or small trees, 0.8–3.5 m, branchlets white-tomentose, glabrescent. **Leaves:** blade green, flat to concave-convex, lanceolate to ovate, ovate-elliptic, or obovate-elliptic, 2.5–7(–9) × 0.8–3(–5) cm, length 1.3–3.9 times width, margins flat to slightly revolute, sometimes slightly undulate, smooth, apex acuminate, abaxial surface densely, closely tomentose, hairs curled or crisped, interwoven, adaxial surface glossy, glabrous. **Aments:** staminate 3–7 cm; pistillate compact, internodes to 1 mm, unbranched, pendulous, 2.5–6 cm; pistillate bracts connate proximally into deep cup, at least at proximal nodes each subtending 3 flowers, deltate, differing in size and shape from leaves, floccose. **Berries** 7–8 mm diam., tomentose, glabrescent, not glaucous.

Flowering Feb–Apr. Rocky slopes, gravelly alluvium, sandy soils, dry stream beds, chaparral; 200–2300 m; Calif.; Mexico (Baja California).

6. Garrya flavescens S. Watson, Amer. Naturalist 7: 301. 1873 • Ashy silktassel

Garrya congdonii Eastwood; *G. flavescens* subsp. *congdonii* (Eastwood) Dahling; *G. flavescens* subsp. *pallida* (Eastwood) Dahling; *G. flavescens* var. *pallida* (Eastwood) Bacigalupi ex Ewan; *G. mollis* Greene; *G. veatchii* Kellogg var. *flavescens* (S. Watson) J. M. Coulter & W. H. Evans

Shrubs, 1.5–3(–6) m, branchlets strigose-sericeous, glabrescent. **Leaves:** blade whitish abaxially (because of white epidermis), yellow-green to gray-green adaxially, flat to concave-convex, elliptic to obovate-elliptic to oval, (2–)3–8 × 1–4.5 cm, length 1.3–3.3 times width, margins flat to undulate, smooth, apex obtuse to rounded, abaxial surface sparsely to densely strigose-sericeous, hairs antrorsely appressed, adaxial surface dull, sparsely strigose to glabrate. **Aments:** staminate 3–8 cm; pistillate compact, internodes to 1 mm, unbranched, pendulous, 2–5 cm; pistillate bracts connate proximally into deep cup, at least at proximal nodes each subtending 3 flowers, broadly ovate-deltate with acuminate-recurved apices, differing in size and shape from leaves, densely strigose-sericeous. **Berries** 5–8 mm diam., densely strigose-sericeous, sometimes glabrate toward base, not glaucous.

Flowering Feb–Apr. Desert slopes, chaparral, pinyon-juniper woodlands, pine-oak woodlands; 400–2800 m; Ariz., Calif., Nev., N.Mex., Utah; Mexico (Baja California).

The white abaxial leaf epidermis is distinctive of *Garrya flavescens*. Hairs on the abaxial leaf surfaces vary in length and density, but all are relatively straight and parallel. Plants with slightly wavy hairs, said to be diagnostic for *G. congdonii*, occur in the western portion of the range (most often on serpentine but not restricted to it); intergrades are numerous, and recognition of *G. congdonii* even at varietal rank appears to be subjective. Subspecies *congdonii* and subsp. *pallida* (the types of both taxa from California) together as a single evolutionary unit might be treated at subspecific rank and distinct from subsp. *flavescens* (the type from Arizona), but intergrades make even this difficult.

7. **Garrya fremontii** Torrey in War Department [U.S.], Pacif. Railr. Rep. 4(5): 136. 1857 • Bearbrush [E]

Garrya fremontii var. *laxa* Eastwood; *G. rigida* Eastwood

Shrubs, 1–3(–5) m, branchlets strigose-sericeous, glabrate. **Leaves:** blade green abaxially, bright green adaxially, generally flat, ovate-elliptic to broadly lanceolate, obovate-elliptic, oblong, or suborbiculate, 2.5–12 × 1.1–7 cm, length 1.3–2.4 times width, margins flat to narrowly revolute, not undulate, smooth, apex rounded to obtuse, abaxial surface glabrous or sparsely strigose, hairs antrorsely appressed, adaxial surface glossy, glabrous; strongly reticulate-veined adaxially. **Aments:** staminate 7–20 cm; pistillate compact, internodes to 1 mm, unbranched, pendulous, 4–6 cm; pistillate bracts connate proximally into deep cup, at least at proximal nodes each subtending 3 flowers, triangular with acuminate-recurved apices, differing in size and shape from leaves, densely pilose to glabrate, ciliate. **Berries** 5–6 mm diam., glabrous or glabrate, not glaucous.

Flowering Jan–Apr(–May). Rocky slopes, volcanic slopes, serpentine, chaparral, oak-pine, yellow pine, and mixed conifer-*Arbutus* woodlands; 100–2000(–2800) m; Calif., Nev., Oreg., Wash.

8. **Garrya buxifolia** A. Gray, Proc. Amer. Acad. Arts 7: 349. 1868 • Dwarf silktassel [E]

Shrubs 0.5–2(–3) m, branchlets moderately strigose-sericeous, glabrate. **Leaves:** blade green abaxially, bright to olive green adaxially, flat to concave-convex, ovate-elliptic or obovate-elliptic to suborbiculate, 1–5(–6.5) × 0.9–3.3 cm, length 1.3–2.3 times width, margins flat, smooth, apex rounded to obtuse, abaxial surface densely strigose-sericeous, hairs antrorsely appressed, adaxial surface glossy, glabrous. **Aments:** staminate 5–7 cm; pistillate compact, internodes to 1 mm, unbranched, pendulous, 3–9 cm; pistillate bracts connate proximally into deep cup, at least at proximal nodes each subtending 3 flowers, triangular to oblong-acuminate, differing in size and shape from leaves, strigose-sericeous. **Berries** 4–6 mm diam., glabrous or sparsely strigose near apex, not glaucous.

Flowering Mar–Apr. Serpentine, chaparral, yellow-pine forests; 50–2200 m; Calif., Oreg.

Garrya buxifolia occurs in northwestern California and southwestern Oregon. Its relatively narrow geographical range is essentially parapatric with that of *G. flavescens* where the two meet in northwestern California. Without observation of the diagnostic fruit vestiture, the green-glossy and completely glabrous adaxial leaf surfaces of *G. buxifolia* usually distinguish it.

2. AUCUBA Thunberg, Nov. Gen. Pl., 61. 1783 • [Latinized Japanese name *aokiba*] [I]

Shrubs [trees]. Leaves: blade flat, coriaceous [subcoriaceous], margins dentate or entire [serrate or glandular-serrate], flat. **Inflorescences** terminal, panicles; bracts opposite, distinct; bracteoles 1–2 per flower. **Pedicels** present. **Staminate flowers:** sepals 4, triangular to suborbiculate, minute; petals 4, apex acuminate [caudate]; nectary present, annular; stamens alternate with petals, anthers dorsifixed. **Pistillate flowers:** sepals 4, triangular to suborbiculate, minute; petals 4; nectary present, annular; ovule 1; style 1, cylindric; stigma capitate, sometimes slightly 2(–4)-lobed. **Fruits** drupes, red, drying black, cylindric to ovoid, fleshy. **Seeds** 1 per fruit. $x = 8$.

Species 10 (1 in the flora): introduced; e Asia; introduced also in Europe, Pacific Islands (New Zealand), Australia.

Species of *Aucuba* are commonly planted in temperate gardens because of their evergreen habit, glossy leaves, and brightly colored drupes. Inflorescences may sometimes appear axillary but actually are terminal in association with branches arising from the axils of the subtending pair of leaves.

SELECTED REFERENCES Abe, T. 2001. Flowering phenology, display size, and fruit set in an understory dioecious shrub, *Aucuba japonica* (Cornaceae). Amer. J. Bot. 88: 455–461. Iwashina, T., K. Kamenosono, and H. Hatta. 1997. Flavonoid glycosides from leaves of *Aucuba japonica* and *Helwingia japonica* (Cornaceae): Phytochemical relationship with the genus *Cornus*. J. Jap. Bot. 72: 337–346.

1. Aucuba japonica Thunberg, Nov. Gen. Pl., 62. 1783
• Aucuba, Japanese-laurel, spotted-laurel [F] [I]

Shrubs 1–3 m. **Leaves:** blade usually narrowly elliptic to ovate-elliptic, seldom widely lanceolate, 8–20 × 5–12 cm, base subrounded or broadly cuneate, margins dentate on distal ½ or entire, apex acuminate, abaxial surface light green, adaxial surface green, sometimes yellow-variegated, glossy. **Inflorescences:** staminate 7–10 cm, pistillate (1–)2–3 cm. **Flowers:** petals valvate, purplish red to dark red, ovate to ovate-lanceolate or elliptic-lanceolate, 3.5–4.5 × 2–2.5 mm, apex acuminate; nectary slightly 4-lobed, fleshy. **Drupes** ovoid, 2 cm. **2*n*** = 16, 32.

Flowering Mar–Apr. Thickets, brushy hillsides, near botanical gardens, suburban woodlands; 0–200 m; introduced; N.C., Wash.; e Asia (se China, Japan, Korea, Taiwan); introduced also in Europe, Pacific Islands (New Zealand), Australia.

Aucuba japonica has been cultivated in the United States for nearly a century, but naturalized plants appear to be recent escapes from cultivation in Orange County, North Carolina (A. S. Weakley 2012), and King County, Washington (P. F. Zika and A. L. Jacobson 2005).

Frequently planted cultivars of *Aucuba japonica* with dark green and yellow variegated leaves are var. *variegata* Dombrain.

Literature Cited

Robert W. Kiger, Editor

This is a consolidated list of all works cited in volume 12, whether as selected references, in text, or in nomenclatural contexts. In citations of articles, both here and in the taxonomic treatments, and also in nomenclatural citations, the titles of serials are rendered in the forms recommended in G. D. R. Bridson and E. R. Smith (1991), Bridson (2004), and Bridson and D. W. Brown (http://fmhibd.library.cmu.edu/HIBD-DB/bpho/findrecords.php). When those forms are abbreviated, as most are, cross references to the corresponding full serial titles are interpolated here alphabetically by abbreviated form. In nomenclatural citations (only), book titles are rendered in the abbreviated forms recommended in F. A. Stafleu and R. S. Cowan (1976–1988) and Stafleu et al. (1992–2009). Here, those abbreviated forms are indicated parenthetically following the full citations of the corresponding works, and cross references to the full citations are interpolated in the list alphabetically by abbreviated form. Two or more works published in the same year by the same author or group of coauthors will be distinguished uniquely and consistently throughout all volumes of *Flora of North America* by lower-case letters (b, c, d, ...) suffixed to the date for the second and subsequent works in the set. The suffixes are assigned in order of editorial encounter and do not reflect chronological sequence of publication. The first work by any particular author or group from any given year carries the implicit date suffix "a"; thus, the sequence of explicit suffixes begins with "b". Works missing from any suffixed sequence here are ones cited elsewhere in the *Flora* that are not pertinent in this volume.

Aagesen, D. et al. 2005. Phylogeny of the tribe Colletieae (Rhamnaceae)—A sensitivity analysis of the plastid region *trn*L-*trn*F combined with morphology. Pl. Syst. Evol. 250: 197–214.

Abe, T. 2001. Flowering phenology, display size, and fruit set in an understory dioecious shrub, *Aucuba japonica* (Cornaceae). Amer. J. Bot. 88: 455–461.

Abh. Königl. Akad. Wiss. Berlin = Abhandlungen der Königlichen Akademie der Wissenschaften in Berlin.

Abh. Königl. Ges. Wiss. Göttingen = Abhandlungen der Königlichen Gesellschaft der Wissenschaften zu Göttingen.

Abh. Math.-Phys. Cl. Königl. Bayer. Akad. Wiss. = Abhandlungen der Mathematisch-physikalischen Classe der Königlich bayerischen Akademie der Wissenschaften.

Abrams, L. and R. S. Ferris. 1923–1960. Illustrated Flora of the Pacific States: Washington, Oregon, and California. 4 vols. Stanford.

Acta Bot. Mex. = Acta Botanica Mexicana.

Acta Bot. Neerl. = Acta Botanica Neerlandica.

Acta Bot. Yunnan. = Acta Botanica Yunnanica. [Yunnan Zhiwu Yanjiu.]

Acta Phytotax. Sin. = Acta Phytotaxonomica Sinica. [Chih Wu Fen Lei Hsüeh Pao.]

Acta Univ. Upsal., Symb. Bot. Upsal. = Acta Universitatis Upsaliensis. Symbolae Botanicae Upsalienses.

Adansonia = Adansonia; Recueil Périodique d'Observations Botaniques.

Airy Shaw, H. K. 1967. Notes on Malaysian and other Asiatic Euphorbiaceae LXXII. Generic segregation in the affinity of *Aleurites* J. R. et G. Forster. Kew Bull. 20: 393–395.

Airy Shaw, H. K. 1967b. Notes on the genus *Bischofia* Bl. (Bischofiaceae). Kew Bull. 21: 327–329.

Aiton, W. 1789. Hortus Kewensis; or, a Catalogue of the Plants Cultivated in the Royal Botanic Garden at Kew. 3 vols. London. (Hort. Kew.)

Allem, A. C. 1994. The origin of *Manihot esculenta* Crantz (Euphorbiaceae). Genet. Resources Crop Evol. 41: 133–150.

Allertonia = Allertonia; a Series of Occasional Papers.

Allred, K. W. 2002b. *Bernardia* (Euphorbiaceae) in New Mexico. New Mexico Bot. Newslett. 24: 5.

Alsogr. Amer.—See: C. S. Rafinesque 1838

Amer. J. Bot. = American Journal of Botany.

Amer. J. Sci. Arts = American Journal of Science, and Arts.

Amer. Midl. Naturalist = American Midland Naturalist; Devoted to Natural History, Primarily That of the Prairie States.

Amer. Monthly Mag. & Crit. Rev. = American Monthly Magazine and Critical Review.

Amer. Naturalist = American Naturalist....

Anales Hist. Nat. = Anales de Historia Natural.

Anderson, C. E. 2007. Revision of *Galphimia* (Malpighiaceae). Contr. Univ. Michigan Herb. 25: 1–82.

Anderson, W. R. 1981. The botany of the Guayana Highland. Part XI. Malpighiaceae. Mem. New York Bot. Gard. 32: 21–305.

Anderson, W. R. 2004. Malpighiaceae. In: N. P. Smith et al., eds. 2004. Flowering Plants of the Neotropics. Princeton. Pp. 229–232.

Anderson, W. R. 2013. Origins of Mexican Malpighiaceae. Acta Bot. Mex. 104: 107–156.

Angiosperm Phylogeny Group. 2003. An update of the Angiosperm Phylogeny Group classification for the orders and families of flowering plants: APG II. Bot. J. Linn. Soc. 141: 399–436.

Angiosperm Phylogeny Group. 2009. An update of the Angiosperm Phylogeny Group classification for the orders and families of flowering plants: APG III. Bot. J. Linn. Soc. 161: 105–121.

Angwin, P., ed. 2002. Proceedings of the Forty-eighth Western International Forest Disease Work Conference. 2000. August 14–18. Waikoloa, Hawaii. Redding and Kona.

Ann. Bot. Fenn. = Annales Botanici Fennici.

Ann. Bot. (König & Sims) = Annals of Botany. [Edited by König & Sims.]

Ann. Bot. (Oxford) = Annals of Botany. (Oxford.)

Ann. Lyceum Nat. Hist. New York = Annals of the Lyceum of Natural History of New York.

Ann. Missouri Bot. Gard. = Annals of the Missouri Botanical Garden.

Aparicio, A. 1993. Sex-determining and floating translocation complexes in *Viscum cruciatum* Sieber ex Boiss. (Viscaceae) in southern Spain. Some evolutionary and ecological comments. Bot. J. Linn. Soc. 111: 359–369.

Aquatic Bot. = Aquatic Botany; International Scientific Journal Dealing with Applied and Fundamental Research on Submerged, Floating and Emergent Plants in Marine and Freshwater Ecosystems.

Arbust. Amer.—See: H. Marshall 1785

Arch. Biochem. Biophys. = Archives of Biochemistry and Biophysics.

Arch. Naturgesch. (Berlin) = Archiv für Naturgeschichte.

Armbruster, W. S. 1985. Patterns of character divergence and the evolution of reproductive ecotypes of *Dalechampia scandens* (Euphorbiaceae). Evolution 39: 733–752.

Arnoldia (Jamaica Plain) = Arnoldia; a Continuation of the Bulletin of Popular Information.

Ashworth, V. E. T. M. 2000. Phylogenetic relationships in Phoradendreae (Viscaceae) inferred from three regions of the nuclear ribosomal cistron. I. Major lineages and paraphyly of *Phoradendron*. Syst. Bot. 25: 349–370.

Ashworth, V. E. T. M. 2000b. Phylogenetic relationships in Phoradendreae (Viscaceae) inferred from three regions of the nuclear ribosomal cistron. II. The North American species of *Phoradendron*. Aliso 19: 41–53.

Atlantic J. = Atlantic Journal, and Friend of Knowledge.

Aublet, J. B. 1775. Histoire des Plantes de la Guiane Françoise.... 4 vols. Paris. [Vols. 1 and 2: text, paged consecutively; vols. 3 and 4: plates.] (Hist. Pl. Guiane)

Audubon, J. J. 1831–1839. Ornithological Biography.... 5 vols. Philadelphia and Edinburgh. (Ornithol. Biogr.)

Australian Biological Resources Study. 1999+. Species Plantarum: Flora of the World. 11+ parts. Canberra.

Autik. Bot.—See: C. S. Rafinesque 1840

B. M. C. Evol. Biol. = B M C Evolutionary Biology.

Bader, N. E. 1999. Pollen Analysis of Death Valley Core DV93-1: A Closeup of Marine Oxygen Isotope Stage 6 and Glacial Termination II. M.S. thesis. University of Arizona.

Bailey, L. H. 1934. The species of grapes peculiar to North America. Gentes Herbarum 3: 154–244.

Bailey, L. H., E. Z. Bailey, and Bailey Hortorium Staff. 1976. Hortus Third. A Concise Dictionary of Plants Cultivated in the United States and Canada. New York.

Baillon, H. E. 1858. Étude Générale du Groupe des Ephorbiacées.... 1 vol. + atlas. Paris. (Étude Euphorb.)

Bain, J. F. and K. E. Denford. 1979. The herbaceous members of the genus *Cornus* in NW North America. Bot. Not. 132: 121–129.

Baldwin, B. G. et al., eds. 2012. The Jepson Manual: Vascular Plants of California, ed. 2. Berkeley.

Bangham, W. 1929. The chromosomes of some species of the genus *Philadelphus*. J. Arnold Arbor. 10: 167–169.

Bartram, W. 1791. Travels through North and South Carolina, Georgia, East and West Florida, the Cherokee Country, the Extensive Territories of the Muscogulges, or Creek Confederacy, and the Country of the Chactaws.... Philadelphia. (Travels Carolina)

Beadle, C. D. 1902. Studies in *Philadelphus*. Biltmore Bot. Stud. 1: 159–161.

Becker, J. D. T. 2010. Taxonomy of the *Linum lewisii* Complex in Canada Based on Macromorphology, Micromorphology, and Phytogeography. Honors thesis. University of Manitoba.

Beier, B.-A. 2005. A revision of *Fagonia* (Zygophyllaceae). Syst. Biodivers. 3: 221–263.

Beier, B.-A. et al. 2004. Phylogeny and taxonomy of the subfamily Zygophylloideae (Zygophyllaceae) based on molecular and morphological data. Pl. Syst. Evol. 240: 11–40.

Beier, B.-A. et al. 2004b. Phylogenetic relationships and biogeography of the desert plant genus *Fagonia* (Zygophyllaceae), inferred by parsimony and Bayesian model averaging. Molec. Phylogen. Evol. 33: 91–108.

Beitr. Biol. Pflanzen = Beiträge zur Biologie der Pflanzen.

Bennett, B. A. et al. 2010. New records of vascular plants in the Yukon Territory VIII. Canad. Field-Naturalist 124: 1–27.

Benson, G. T. 1930. The trees and shrubs of western Oregon. Contr. Dudley Herb. 2: 1–170.

Benson, L. D. and R. A. Darrow. 1954. The Trees and Shrubs of the Southwestern Deserts, ed. 2. Tucson. [Second edition of L. D. Benson and R. A. Darrow 1945 with title change.] (Trees Shrubs Southw. Deserts)

Bentham, G. 1839[–1857]. Plantas Hartwegianas Imprimis Mexicanas.... London. [Issued by gatherings with consecutive signatures and pagination.] (Pl. Hartw.)

Bentham, G. 1844[–1846]. The Botany of the Voyage of H.M.S. Sulphur, under the Command of Captain Sir Edward Belcher...during the Years 1836–1842. 6 parts. London. [Parts paged consecutively.] (Bot. Voy. Sulphur)

Bentham, G. and J. D. Hooker. 1862–1883. Genera Plantarum ad Exemplaria Imprimis in Herbariis Kewensibus Servata Definita. 3 vols. London. (Gen. Pl.)

Ber. Deutsch. Bot. Ges. = Berichte der Deutschen botanischen Gesellschaft.

Ber. Schweiz. Bot. Ges. = Bericht der Schweizerischen botanischen Gesellschaft.

Berchtold, F. and J. S. Presl. 1823–1835. O Přirozenosti Rostlin aneb Rostlinář.... 3 vols. Prague. [Vols. 2 and 3 by Presl only.] (Prir. Rostlin)

Berlandier, J. L. [1832.] Memorias de la Comision de Limites.... [Matamoros.] (Mem. Comis. Limites)

Berry, P. E. et al. 2005. Molecular phylogenetics of the giant genus *Croton* (Euphorbiaceae sensu stricto) using ITS and *trn*L-F DNA sequence data. Amer. J. Bot. 92: 1520–1534.

Berry, P. E., V. W. Steinmann, and Y. Yang. 2011. Proposal to conserve the name *Euphorbia acuta* Engelm. against *E. acuta* Bellardi ex Colla. Taxon 60: 603–604.

Beskr. Guin. Pl.—See: H. C. F. Schumacher 1827

Betancourt, J. L., T. R. Van Devender, and P. S. Martin, eds. 1990. Packrat Middens: The Last 40,000 Years of Biotic Change. Tucson.

Bhatnagar, A. K. and R. N. Kapil. 1973. *Bischofia javanica*—its relationship with Euphorbiaceae. Phytomorphology 23: 264–267.

Bhatnagar, S. P. and G. Sabharwal. 1969. Morphology and embryology of *Iodina rhombifolia* Hook. & Arn. Beitr. Biol. Pflanzen 45: 465–479.

Bijdr. Fl. Ned. Ind.—See: C. L. Blume 1825–1826[–1827]

Biltmore Bot. Stud. = Biltmore Botanical Studies; a Journal of Botany Embracing Papers by the Director and Associates of the Biltmore Herbarium.

Biol. Pharm. Bull. = Biological and Pharmaceutical Bulletin.

Biol. Ser. Bull. State Univ. Montana = Biological Series of the Bulletin of the State University of Montana.

Blakelock, R. A. 1951. A synopsis of the genus *Euonymus* L. Kew Bull. 6: 210–290.

Blume, C. L. 1825–1826[–1827]. Bijdragen tot de Flora van Nederlandsch Indië. 17 parts. Batavia. [Parts paged consecutively.] (Bijdr. Fl. Ned. Ind.)

Blumea = Blumea; Tidjschrift voor die Systematiek en die Geografie der Planten (A Journal of Plant Taxonomy and Plant Geography).

Boissier, P. E. 1860. Centuria Euphorbiarum.... Leipzig and Paris. (Cent. Euphorb.)

Bol. Inst. Bot. Univ. Guadalajara = Boletín del Instituto de Botánica, Universidad de Guadalajara.

Bol. Soc. Brot. = Boletim da Sociedade Broteriana.

Bolmgren, K. and B. Oxelman. 2004. Generic limits in *Rhamnus* s.l. L. (Rhamnaceae) inferred from nuclear and chloroplast DNA sequence phylogenies. Taxon 53: 383–390.

Boston J. Nat. Hist. = Boston Journal of Natural History.

Bot. Beechey Voy.—See: W. J. Hooker and G. A. W. Arnott [1830–]1841

Bot. California—See: W. H. Brewer et al. 1876–1880

Bot. Gall.—See: J. É. Duby [1828–1830]

Bot. Gaz. = Botanical Gazette; Paper of Botanical Notes.

Bot. Handb.—See: C. Schkuhr [1787–]1791–1803

Bot. J. Linn. Soc. = Botanical Journal of the Linnean Society.

Bot. Jahrb. Syst. = Botanische Jahrbücher für Systematik, Pflanzengeschichte und Pflanzengeographie.

Bot. Mag. = Botanical Magazine; or, Flower-garden Displayed.... [Edited by Wm. Curtis.] [With vol. 15, 1801, title became Curtis's Botanical Magazine; or....]

Bot. Mag. (Tokyo) = Botanical Magazine. [Shokubutsu-gaku Zasshi.] (Tokyo.)

Bot. Misc. = Botanical Miscellany.

Bot. Not. = Botaniska Notiser.

Bot. Rev. (Lancaster) = Botanical Review, Interpreting Botanical Progress.

Bot. Voy. Herald—See: B. Seemann 1852–1857

Bot. Voy. Sulphur—See: G. Bentham 1844[–1846]

Botany (Fortieth Parallel)—See: S. Watson 1871

Bothalia = Bothalia; a Record of Contributions from the National Herbarium, Union of South Africa.

Boyd, R. S. 2007. Response to fire of *Ceanothus roderickii* (Rhamnaceae), a federally endangered California endemic shrub. Madroño 54: 13–21.

Boyd, S., T. S. Ross, and L. Arnseth. 1991. *Ceanothus ophiochilus* (Rhamnaceae): A distinctive, narrowly endemic species from Riverside County, California. Phytologia 70: 28–41.

Bradea = Bradea; Boletim do Herbarium Bradeanum.

Bradley, C. E. 1956. Yerba de la fleche: Arrow and fish poison of the American Southwest. Econ. Bot. 10: 363–366.

Bremer, K. et al. 2001. A phylogenetic analysis of 100+ genera and 50+ families of euasterids based on morphological and molecular data with notes on possible higher morphological synapomorphies. Pl. Syst. Evol. 229: 137–169.

Brewer, W. H. et al. 1876–1880. Geological Survey of California.... Botany.... 2 vols. Cambridge, Mass. (Bot. California)

Bridges, E. L. and S. L. Orzell. 2002. *Euphorbia* (Euphorbiaceae) section *Tithymalus* subsection *Inundatae* in the southeastern United States. Lundellia 5: 59–78.

Bridson, G. D. R. 2004. BPH-2: Periodicals with Botanical Content. 2 vols. Pittsburgh.

Bridson, G. D. R. and E. R. Smith. 1991. B-P-H/S. Botanico-Periodicum-Huntianum/Supplementum. Pittsburgh.

Brit. Herb.—See: J. Hill 1756[–1757]

Britton, N. L. et al., eds. 1905+. North American Flora.... 47+ vols. New York. [Vols. 1–34, 1905–1957; ser. 2, parts 1–13+, 1954+.] (N. Amer. Fl.)

Britton, N. L. and A. Brown. 1896–1898. An Illustrated Flora of the Northern United States, Canada and the British Possessions from Newfoundland to the Parallel of the Southern Boundary of Virginia, and from the Atlantic Ocean Westward to the 102d Meridian.... 3 vols. New York. (Ill. Fl. N. U.S.)

Britton, N. L. and A. Brown. 1913. An Illustrated Flora of the Northern United States, Canada and the British Possessions from Newfoundland to the Parallel of the Southern Boundary of Virginia, and from the Atlantic Ocean Westward to the 102d Meridian..., ed. 2. 3 vols. New York.

Brittonia = Brittonia; a Journal of Systematic Botany....

Brizicky, G. K. 1964. The genera of Celastraceae in the southeastern United States. J. Arnold Arbor. 45: 206–234.

Brizicky, G. K. 1964b. The genera of Rhamnaceae in the southeastern United States. J. Arnold Arbor. 45: 439–463.

Brizicky, G. K. 1964c. A further note on *Ceanothus herbaceus* versus *C. ovatus*. J. Arnold Arbor. 45: 471–473.

Brizicky, G. K. 1965. The genera of Vitaceae in the southeastern United States. J. Arnold Arbor. 46: 48–67.

Brockington, S. F. et al. 2009. Phylogeny of the Caryophyllales sensu lato: Revisiting hypotheses on pollination biology and perianth differentiation in the core Caryophyllales. Int. J. Pl. Sci. 170: 627–643.

Brokaw, J. M. 2009. Phylogeny of *Mentzelia* Section *Trachyphytum:* Origins and Evolutionary Ecology of Polyploidy. Ph.D. dissertation. Washington State University.

Brokaw, J. M. and L. Hufford. 2010. Phylogeny, introgression, and character evolution of diploid species of *Mentzelia* section *Trachyphytum* (Loasaceae). Syst. Bot. 35: 601–617.

Brokaw, J. M. and L. Hufford. 2010b. Origins and introgression of polyploid species in *Mentzelia* section *Trachyphytum* (Loasaceae). Amer. J. Bot. 97: 1457–1473.

Brokaw, J. M., T. A. Johnson, and C. H. Hofsommer. 2015. Edaphic specialization in the cryptic species *Mentzelia monoensis* (Loasaceae). Madroño 62: 88–100.

Brongniart, A. T. 1826. Mémoire sur la Famille des Rhamnées.... Paris. (Mém. Fam. Rhamnées)

Brown, D. K. and R. B. Kaul. 1981. Floral structure and mechanism in Loasaceae. Amer. J. Bot. 68: 361–372.

Brown, L. E. and K. N. Gandhi. 1989. Notes on the flora of Texas with additions, range extensions, and one correction. Phytologia 67: 394–399.

Brown, L. E. and S. J. Marcus. 1998. Notes on the flora of Texas with additions and other significant records. Sida 18: 315–324.

Browne, P. 1756. The Civil and Natural History of Jamaica.... London. (Civ. Nat. Hist. Jamaica)

Brummitt, R. K. and C. E. Powell, eds. 1992. Authors of Plant Names. A List of Authors of Scientific Names of Plants, with Recommended Standard Forms of Their Names, Including Abbreviations. Kew.

Bryson, C. T. and D. A. Skojac. 2011. An annotated checklist of the vascular flora of Washington County, Mississippi. J. Bot. Res. Inst. Texas 5: 855–866.

Bull. Calif. Acad. Sci. = Bulletin of the California Academy of Sciences.

Bull. Canad. Bot. Assoc. = Bulletin, Canadian Botanical Association.

Bull. Mens. Soc. Linn. Paris = Bulletin Mensuel de la Société Linnéenne de Paris.

Bull. Misc. Inform. Kew = Bulletin of Miscellaneous Information, Royal Gardens, Kew.

Bull. New York Bot. Gard. = Bulletin of the New York Botanical Garden.

Bull. S. Calif. Acad. Sci. = Bulletin of the Southern California Academy of Sciences.

Bull. Soc. Bot. France = Bulletin de la Société Botanique de France.

Bull. Soc. Imp. Naturalistes Moscou = Bulletin de la Société Impériale des Naturalistes de Moscou.

Bull. Soc. Neuchâtel. Sci. Nat. = Bulletin de la Société Neuchâteloise de Sciences Naturelles.

Bull. Torrey Bot. Club = Bulletin of the Torrey Botanical Club.

Bull. Univ. Mus. Univ. Tokyo = Bulletin, University Museum, University of Tokyo. [Sogo Kenkyu Shiryokan.]

Bull. U.S. Geol. Geogr. Surv. Territ. = Bulletin of the United States Geological and Geographical Survey of the Territories.

Bunge, A. A. [1833.] Enumeratio Plantarum, Quas in China Boreali Collegit.... St. Petersburg. [Preprinted from Mém. Acad. Imp. Sci. St.-Pétersbourg Divers Savans 2: 75–148. 1835.] (Enum. Pl. China Bor.)

Burch, D. G. 1965. A Taxonomic Revision of the Genus *Chamaesyce* in the Caribbean. Ph.D. dissertation. University of Florida.

Burch, D. G. 1966. The application of the Linnaean names of some New World species of *Euphorbia* subgenus *Chamaesyce*. Rhodora 67: 155–166.

Burckhalter, R. E. 1992. The genus *Nyssa* (Cornaceae) in North America: A revision. Sida 15: 323–342.

Burge, D. O. 2011. Molecular phylogenetics of *Garrya* (Garryaceae). Madroño 58: 249–255.

Burge, D. O. and S. R. Manchester. 2008. Fruit morphology, fossil history, and biogeography of *Paliurus* (Rhamnaceae). Int. J. Pl. Sci. 169: 1066–1085.

Burge, D. O. and K. Zhukovsky. 2013. Taxonomy of the *Ceanothus vestitus* complex (Rhamnaceae). Syst. Bot. 38: 406–417.

Burger, W. C. and M. J. Huft. 1995. Flora Costaricensis: Family #113, Euphorbiaceae. Fieldiana, Bot., n. s. 36.

Burkett, G. W. 1932. Anatomical studies within the genus *Hydrangea*. Proc. Indiana Acad. Sci. 41: 83–95.

Burns, R. M. and B. H. Honkala. 1990. Silvics of North America. 2 vols. Washington. [Agric. Handb. 654.]

Burrows, G. E. and R. J. Tyrl. 2001. Toxic Plants of North America. Ames.

Bussing, A., ed. 2000. Mistletoe: The Genus *Viscum.* Amsterdam.

Cacho, N. I. et al. 2010. Are spurred cyathia a key innovation? Molecular systematics and trait evolution in the slipper-spurges (Pedilanthus clade - *Euphorbia,* Euphorbiaceae). Amer. J. Bot. 97: 493–510.

Calder, J. A. and R. L. Taylor. 1965. New taxa and nomenclatural changes with respect to the flora of the Queen Charlotte Islands, British Columbia. Canad. J. Bot. 43: 1387–1400.

Call, V. B. and D. L. Dilcher. 1997. The fossil record of *Eucommia* (Eucommiaceae) in North America. Amer. J. Bot. 84: 798–814.

Callihan, R. H., S. L. Carson, and R. T. Dobbins. 1995. NAWEEDS, Computer-aided Weed Identification for North America. Illustrated User's Guide plus Computer Floppy Disk. Moscow, Idaho.

Campos López, E., T. J. Mabry, and S. Fernández Tavizon. 1979. *Larrea.* Mexico City.

Canad. Field-Naturalist = Canadian Field-Naturalist.

Canad. J. Bot. = Canadian Journal of Botany.

Canad. J. Pl. Sci. = Canadian Journal of Plant Science.

Cancer Res. = Cancer Research.

Candolle, A. L. P. P. de and C. de Candolle, eds. 1878–1896. Monographiae Phanerogamarum.... 9 vols. Paris. (Monogr. Phan.)

Candolle, A. P. de and A. L. P. P. de Candolle, eds. 1823–1873. Prodromus Systematis Naturalis Regni Vegetabilis.... 17 vols. Paris etc. [Vols. 1–7 edited by A. P. de Candolle, vols. 8–17 by A. L. P. P. de Candolle.] (Prodr.)

Cannon, W. A. 1910. The root habits and parasitism of *Krameria canescens* A. Gray. Publ. Carnegie Inst. Wash. 129: 5–24.

Canotia = Canotia; a New Journal of Arizona Botany.

Caribbean J. Sci. = Caribbean Journal of Science.

Carlquist, S. 1983. Intercontinental dispersal. Sonderb. Naturwiss. Vereins Hamburg 7: 37–47.

Carlquist, S. 1984. Wood anatomy of Loasaceae with relation to systematics, habit and ecology. Aliso 10: 583–602.

Carr, W. R. and M. H. Mayfield. 1993. *Chamaesyce velleriflora* (Euphorbiaceae) new to Texas. Sida 15: 550–551.

Carvell, W. N. and W. H. Eshbaugh. 1982. A systematic study of the genus *Buckleya* (Santalaceae). Castanea 47: 17–37.

Casamajor, R. 1950. One of California's rare endemics. Aliso 2: 115–118.

Castanea = Castanea; Journal of the Southern Appalachian Botanical Club.

Cat. Pl. Cub.—See: A. H. R. Grisebach 1866

Catling, P. M. and G. Mitrow. 2005. The dune race of *Vitis riparia* in Ontario: Taxonomy, conservation and biogeography. Canad. J. Pl. Sci. 85: 407–415.

Catling, P. M. and G. Mitrow. 2012. Major invasive alien plants of natural habitats in Canada. 3. Leafy spurge, wolf's-milk, euphorbe ésule. *Euphorbia esula* L. Bull. Canad. Bot. Assoc. 45: 24–32.

Cavanaugh, K. C. et al. 2014. Poleward expansion of mangroves is a threshold response to decreased frequency of extreme cold events. Proc. Natl. Acad. Sci. U.S.A. 111: 723–727.

Cavanilles, A. J. 1791–1801. Icones et Descriptiones Plantarum, Quae aut Sponte in Hispania Crescunt, aut in Hortis Hospitantur. 6 vols. Madrid. (Icon.)

Ceanothus—See: M. Van Rensselaer and H. McMinn 1942

Cely, J. E. 1979. The ecology and distribution of banana waterlily and its utilization by canvasback ducks. Proc. Annual Conf. SouthE. Assoc. Fish Wildlife Agencies 33: 43–47.

Cent. Euphorb.—See: P. E. Boissier 1860

Chapman, A. W. 1860. Flora of the Southern United States.... New York. (Fl. South. U.S.)

Chapman, A. W. 1883. Flora of the Southern United States..., ed. 2. New York. (Fl. South. U.S. ed. 2)

Chapman, A. W. 1892. Flora of the Southern United States..., ed. 2 reprint 2. New York, Cincinnati, and Chicago. (Fl. South. U.S. ed. 2 repr. 2)

Char. Gen. Pl. ed. 2—See: J. R. Forster and G. Forster 1776

Chase, M. W. et al. 1993. Phylogenetics of seed plants: An analysis of nucleotide sequences from the plastid gene *rbc*L. Ann. Missouri Bot. Gard. 80: 528–580.

Cheatham, N. H. 1974. *Carpenteria,* the mystery shrub. Fremontia 1: 3–8.

Chen, Y. L. and C. Schirarend. 2007. Rhamnaceae. In: Wu Z. and P. H. Raven, eds. 1994+. Flora of China. 20+ vols. Beijing and St. Louis. Vol. 12, pp. 115–168.

Chen, Z. D., Ren H., and Wen J. 2007. Vitaceae. In: Wu Z. and P. H. Raven, eds. 1994+. Flora of China. 20+ vols. Beijing and St. Louis. Vol. 12, pp. 173–222.

Christie, K. et al. 2006. Vascular plants of Arizona: Rhamnaceae (buckthorn family). Canotia 2: 23–46.

Christy, C. M. 1997. A new species of *Mentzelia* section *Bartonia* (Loasaceae) from Arizona. Novon 7: 25–26.

Cipollini, M. L. and E. W. Stiles. 1991. Costs of reproduction in *Nyssa sylvatica:* Sexual dimorphism in reproductive frequency and nutrient flux. Oecologia 86: 585–593.

Civ. Nat. Hist. Jamaica—See: P. Browne 1756

Class-book Bot. ed. s.n.(b)—See: A. Wood 1861

Coile, N. C. 1988. Taxonomic Studies on the Deciduous Species of *Ceanothus* L. (Rhamnaceae). Ph.D. dissertation. University of Georgia.

Coker, W. C. 1912. The Plant Life of Hartsville, S. C. Columbia. (Pl. Life Hartsville)

Collectanea—See: N. J. Jacquin 1786[1787]–1796[1797]

Comeaux, B. L. 1987. Studies on *Vitis champinii.* Proc. Texas Grape Growers Assoc. 11: 158–162.

Comeaux, B. L. and P. R. Fantz. 1987. Nomenclatural clarification of the name *Vitis simpsonii* Munson (Vitaceae). Sida 12: 279–286.

Comeaux, B. L. and Lu J. 2000. Distinction between *Vitis blancois* and *V. cinerea* var. *tomentosa* (Vitaceae). Sida 19: 123–131.

Comeaux, B. L., W. B. Nesbitt, and P. R. Fantz. 1987. Taxonomy of the native grapes of North Carolina. Castanea 52: 197–215.

Comley, E. 2008. Noteworthy collections: Kentucky. Castanea 73: 151.

Commentat. Soc. Regiae Sci. Gott. = Commentationes Societatis Regiae Scientiarum Gottingensis.

Compt. Rend. Hebd. Séances Acad. Sci. = Comptes Rendus Hebdomadaires des Séances de l'Académie des Sciences.

Conard, S. G. et al. 1985. The Role of the Genus *Ceanothus* in Western Forest Ecosystems. Portland [U.S.D.A. Forest Serv., Gen. Techn. Rep. PNW-182.]

Contr. Dudley Herb. = Contributions from the Dudley Herbarium of Stanford University.

Contr. Gray Herb. = Contributions from the Gray Herbarium of Harvard University. [Some numbers reprinted from (or in?) other periodicals, e.g. Rhodora.]

Contr. Univ. Michigan Herb. = Contributions from the University of Michigan Herbarium.

Contr. U.S. Natl. Herb. = Contributions from the United States National Herbarium.

Contr. W. Bot. = Contributions to Western Botany.

Cook, A. D., P. R. Atsatt, and C. A. Simon. 1971. Doves and dove weed: Multiple plant defenses against avian predation. BioScience 21: 277–281.

Cooperrider, T. S. 1995. The Dicotyledoneae of Ohio: Linaceae through Campanulaceae. Columbus. [The Vascular Flora of Ohio, vol. 2(2).]

Cornus-Arten—See: C. A. von Meyer 1845

Correll, D. S. and M. C. Johnston. 1970. Manual of the Vascular Plants of Texas. Renner, Tex.

Coulter, J. M. and W. H. Evans. 1890. A revision of North American Cornaceae. Bot. Gaz. 15: 30–31, 34–36, 88–89.

Crantz, H. J. N. von. 1766. Institutiones Rei Herbariae.... 2 vols. Vienna. (Inst. Rei Herb.)

Croizat, L. 1942. A study of *Manihot* in North America. J. Arnold Arbor. 23: 216–225.

Cronquist, A. 1981. An Integrated System of Classification of Flowering Plants. New York.

Cronquist, A. et al. 1972–2012. Intermountain Flora. Vascular Plants of the Intermountain West, U.S.A. 6 vols. in 8. New York and London. (Intermount. Fl.)

Cronquist, A., N. H. Holmgren, and P. K. Holmgren. 1997b. Linaceae. In: A. Cronquist et al. 1972–2012. Intermountain Flora. Vascular Plants of the Intermountain West, U.S.A. 6 vols. in 8. New York and London. Vol. 3, part A, pp. 293–300.

Cuénoud, P. et al. 2002. Molecular phylogenetics of Caryophyllales based on nuclear 18S rDNA and plastid *rbc*L, *atp*B, and *mat*K DNA sequences. Amer. J. Bot. 89: 132–144.

Cytologia = Cytologia; International Journal of Cytology.

Dagb. Ostind. Resa—See: P. Osbeck 1757

Dahling, G. V. 1978. Systematics and evolution of *Garrya*. Contr. Gray Herb. 209: 1–104.

Daniel, T. F. 1993. *Garrya*. In: J. C. Hickman, ed. 1993. The Jepson Manual. Higher Plants of California. Berkeley, Los Angeles, and London. Pp. 664–665.

Daniels, G. S. 1970. The Floral Biology and Taxonomy of *Mentzelia* Section *Bicuspidata* (Loasaceae). Ph.D. dissertation. University of California, Los Angeles.

Darlington, J. 1934. A monograph of the genus *Mentzelia*. Ann. Missouri Bot. Gard. 21: 103–227.

Darwiniana = Darwiniana; Carpeta del "Darwinion."

Davidson, A. and G. L. Moxley. 1923. Flora of Southern California.... Los Angeles. (Fl. S. Calif.)

Davis, C. C. et al. 2005. Explosive radiation of Malpighiales supports a mid-Cretaceous origin of modern tropical rain forests. Amer. Naturalist 165: E36–E65.

Davis, C. C. and W. R. Anderson. 2010. A complete generic phylogeny of Malpighiaceae inferred from nucleotide sequence data and morphology. Amer. J. Bot. 97: 2031–2048.

Davis, C. C. and M. W. Chase. 2004. Elatinaceae are sister to Malphigiaceae; Peridiscaceae belong to Saxifragales. Amer. J. Bot. 91: 262–273.

Davis, P. H., ed. 1965–2000. Flora of Turkey and the East Aegean Islands. 11 vols. Edinburgh.

Davis, W. S. and H. J. Thompson. 1967. A revision of *Petalonyx* (Loasaceae) with a consideration of affinities in the subfamily Gronovioideae. Madroño 19: 1–18.

de Carvalho, R., M. Guerra, and P. C. L. de Carvalho. 1999. Occurrence of spontaneous triploidy in *Manihot esculenta* Crantz. Cytologia 64: 137–140.

de Lange, P. J. et al. 2005. Vascular flora of Norfolk Island: Some additions and taxonomic notes. New Zealand J. Bot. 43: 563–596.

De-Nova, J. A. and V. Sosa. 2007. Phylogenetic relationships and generic delimitation in *Adelia* (Euphorbiaceae s.s.) inferred from nuclear, chloroplast, and morphological data. Taxon 56: 1027–1036.

De-Nova, J. A., V. Sosa, and V. W. Steinmann. 2007. A synopsis of *Adelia* (Euphorbiaceae s.s.). Syst. Bot. 32: 583–595.

De Smet, Y. et al. 2015. Molecular phylogenetics and new (infra)generic classification to alleviate polyphyly in tribe Hydrangeeae (Cornales: Hydrangeaceae). Taxon 64: 741–752.

Deam, C. C. 1940. Flora of Indiana. Indianapolis.

DeFilipps, R. A. 1968. A Revision of *Ximenia* [Plum.] L. (Olacaceae). Ph.D. dissertation. Southern Illinois University, Carbondale.

DeFilipps, R. A. 1969. *Ximenia americana* (Olacaceae) in Angola and South West Africa. Bol. Soc. Brot., ser. 2, 43: 193–200.

DeFilipps, R. A. 1969b. Parasitism in *Ximenia* (Olacaceae). Rhodora 71: 439–443.

Dehgan, B. 2012. *Jatropha*. In: Organization for Flora Neotropica. 1968+. Flora Neotropica. 110+ nos. New York. No. 110.

Dehgan, B. and B. Schutzman. 1994. Contributions toward a monograph of neotropical *Jatropha:* Phenetic and phylogenetic analysis. Ann. Missouri Bot. Gard. 81: 349–367.

Dehgan, B. and G. L. Webster. 1979. Morphology and infrageneric relationships of the genus *Jatropha* (Euphorbiaceae). Univ. Calif. Publ. Bot. 74: 1–73.

Delong, M. K. et al. 2005. Floristic survey of a highly disturbed wetland within Shaker Median Park, Beachwood (Cuyahoga County), Ohio. Ohio J. Sci. 105: 102–115.

Dendrologie—See: K. H. E. Koch 1869–1873

Denkschr. Königl. Akad. Wiss. München = Denkschriften der Königlichen Akademie der Wissenschaften zu München.

Denton, M. F. 1973. A monograph of *Oxalis*, section *Ionoxalis* (Oxalidaceae) in North America. Publ. Mus. Michigan State Univ., Biol. Ser. 4: 455–615.

Der, J. P. and D. L. Nickrent. 2008. A molecular phylogeny of Santalaceae (Santalales). Syst. Bot. 33: 107–116.

Dertien, J. R. and M. R. Duvall. 2009. Biogeography and divergence in *Guaiacum sanctum* (Zygophyllaceae) revealed in chloroplast DNA: Implications for conservation in the Florida Keys. Biotropica 41: 120–127.

Descr. Icon. Pl. Hung.—See: F. Waldstein and P. Kitaibel [1799–]1802–1812

Desfontaines, R. L. [1798–1799.] Flora Atlantica sive Historia Plantarum, Quae in Atlante, Agro Tunetano et Algeriensi Crescunt. 2 vols. in 9 parts. Paris. (Fl. Atlant.)

Deut. Bot. Herb.-Buch—See: H. G. L. Reichenbach 1841

DeWalt, S. J., E. Siemann, and W. E. Rogers. 2011. Geographic distribution of genetic variation among native and introduced populations of Chinese tallow tree, *Triadica sebifera* (Euphorbiaceae). Amer. J. Bot. 98: 1128–1138.

Diggs, G. M., B. L. Lipscomb, and R. J. O'Kennon. 1999. Shinners' and Mahler's Illustrated Flora of North Central Texas. Fort Worth. [Sida Bot. Misc. 16.]

Dirr, M. A. 1990. Manual of Woody Landscape Plants..., ed. 4. Champaign.

Dirr, M. A. 1998. Manual of Woody Landscape Plants..., ed. 5. Champaign.

Dirr, M. A. 2004. Hydrangeas for American Gardens. Portland.

Don, G. 1831–1838. A General History of the Dichlamydeous Plants.... 4 vols. London. (Gen. Hist.)

Dorsey, B. L. et al. 2013. Phylogenetics, morphological evolution, and classification of *Euphorbia* subgenus *Euphorbia*. Taxon 62: 291–315.

Dorworth, C. E. 1989. European mistletoe (*Viscum album* subsp. *album*) in Canada. Pl. Dis. 73: 444.

Drapiez, P. A. J., ed. 1833–1838. Encyclographie du Règne Végétal. 6 vols. Brussels. (Encyclogr. Règne Vég.)

Dressler, R. L. 1954. The genus *Tetracoccus* (Euphorbiaceae). Rhodora 56: 45–61.

Dressler, R. L. 1957. The genus *Pedilanthus* (Euphorbiaceae). Contr. Gray Herb. 182: 1–188.

Dressler, R. L. 1961. A synopsis of *Poinsettia* (Euphorbiaceae). Ann. Missouri Bot. Gard. 48: 329–341.

Drummond, D. B. 1982. Timber Loss Estimates for the Coniferous Forests of the United States Due to Dwarf Mistletoes. Fort Collins.

Duby, J. É. [1828–1830]. Aug. Pyrami de Candolle Botanicon Gallicum.... Editio Secunda. 2 vols. Paris. (Bot. Gall.)

Dugal, A. W. 1989. Unusual forms of the black buckthorn *(Rhamnus frangula)* in Ottawa-Carleton. Trail & Landscape 23: 119–121.

Dugal, A. W. 1992. Leitrim Albion Road Wetlands: Part 2. Trail & Landscape 26: 64–94.

Dumortier, B. C. J. 1827. Florula Belgica, Operis Majoris Prodromus.... Tournay. (Fl. Belg.)

Duncan, W. H. 1964. New *Elatine* populations in the southeastern United States. Rhodora 66: 47–53.

Duncan, W. H. and J. T. Kartesz. 1981. Vascular Flora of Georgia: An Annotated Checklist. Athens, Ga.

Durand, E. M. 1855. Plantae Prattenianae Californicae.... Philadelphia. [Preprinted from J. Acad. Nat. Sci. Philadelphia, n. s. 3: 79–104. 1855.] (Pl. Pratten. Calif.)

Durand, E. M. and T. C. Hilgard. 1854. Plantae Heermannianae.... Philadelphia. [Preprinted from J. Acad. Nat. Sci. Philadelphia, n. s. 3: 37–46. 1855.] (Pl. Heermann.)

Eclog. Amer.—See: M. Vahl 1796[1797]–1807

Econ. Bot. = Economic Botany; Devoted to Applied Botany and Plant Utilization.

Edinburgh J. Bot. = Edinburgh Journal of Botany.

Edinburgh J. Nat. Geogr. Sci. = The Edinburgh Journal of Natural and Geographical Science.

Edinburgh New Philos. J. = Edinburgh New Philosophical Journal.

Edwards's Bot. Reg. = Edwards's Botanical Register....

Eiten, G. 1955. Typification of the names "*Oxalis corniculata* L." and "*Oxalis stricta* L." Taxon 4: 99–105.

Eiten, G. 1963. Taxonomy and regional variation of *Oxalis* section *Corniculatae*. I. Introduction, keys and synopsis of the species. Amer. Midl. Naturalist 69: 257–309.

Elliott, S. [1816–]1821–1824. A Sketch of the Botany of South-Carolina and Georgia. 2 vols. in 13 parts. Charleston. (Sketch Bot. S. Carolina)

Emory, W. H. 1848. Notes of a Military Reconnoissance, from Fort Leavenworth, in Missouri, to San Diego, in California, Including Part of the Arkansas, Del Norte, and Gila Rivers.... Made in 1846–7, with the Advanced Guard of the "Army of the West." Washington. (Not. Milit. Reconn.)

Emory, W. H. 1857–1859. Report on the United States and Mexican Boundary Survey, Made under the Direction of the Secretary of the Interior. 2 vols. in parts. Washington. (Rep. U.S. Mex. Bound.)

Encycl.—See: J. Lamarck et al. 1783–1817

Encyclogr. Règne Vég.—See: P. A. J. Drapiez 1833–1838

Endlicher, S. L. 1836–1840[–1850]. Genera Plantarum Secundum Ordines Naturales Disposita. 18 parts + 5 suppls. in 6 parts. Vienna. [Paged consecutively through suppl. 1(2); suppls. 2–5 paged independently.] (Gen. Pl.)

Endress, P. K. 2010. Flower structure and trends of evolution in eudicots and their major subclades. Ann. Missouri Bot. Gard. 97: 541–583.

Engler, H. G. A., ed. 1900–1953. Das Pflanzenreich.... 107 vols. Berlin. [Sequence of vol. (Heft) numbers (order of publication) is independent of the sequence of series and family (Roman and Arabic) numbers (taxonomic order).] (Pflanzenr.)

Engler, H. G. A. et al., eds. 1924+. Die natürlichen Pflanzenfamilien..., ed. 2. 26+ vols. Leipzig and Berlin. (Nat. Pflanzenfam. ed. 2)

Engler, H. G. A. and K. Krause. 1935. Loranthaceae. In: H. G. A. Engler et al., eds. 1924+. Die natürlichen Pflanzenfamilien..., ed. 2. 26+ vols. Leipzig and Berlin. Vol. 16b, pp. 98–203.

Engler, H. G. A. and K. Prantl, eds. 1887–1915. Die natürlichen Pflanzenfamilien.... 254 fascs. Leipzig. [Sequence of fasc. (Lieferung) numbers (order of publication) is independent of the sequence of division (Teil) and subdivision (Abteilung) numbers (taxonomic order).] (Nat. Pflanzenfam.)

Enum. Hort. Berol. Alt.—See: J. H. F. Link 1821–1822

Enum. Pl. China Bor.—See: A. A. Bunge [1833]

Enum. Syst. Pl.—See: N. J. Jacquin 1760

Ernst, W. R. and H. J. Thompson. 1963. The Loasaceae in the southeastern United States. J. Arnold Arbor. 44: 138–142.
Erythea = Erythea; a Journal of Botany, West American and General.
Esser, H.-J. 1998. New combinations in *Microstachys* (Euphorbiaceae). Kew Bull. 53: 955–959.
Esser, H.-J. 2002. A revision of *Triadica* Lour. (Euphorbiaceae). Harvard Pap. Bot. 7: 17–21.
Étude Euphorb.—See: H. E. Baillon 1858
Euphorb. Gen.—See: A. H. L. de Jussieu 1824
Eur. J. Forest Pathol. = European Journal of Forest Pathology.
Everett, R. et al. 1997. Structure of northern spotted owl nest stands and their historical conditions on the eastern slope of the Pacific Northwest Cascades. Forest Ecol. Managem. 94: 1–14.
Evolution = Evolution; International Journal of Organic Evolution.
Eyde, R. H. 1963. Morphological and paleobotanical studies of the Nyssaceae. I. A survey of the modern species and their fruits. J. Arnold Arbor. 44: 1–59.
Eyde, R. H. 1964. Inferior ovary and generic affinities of *Garrya*. Amer. J. Bot. 51: 1083–1092.
Eyde, R. H. 1966. The Nyssaceae in the southeastern United States. J. Arnold Arbor. 47: 117–125.
Eyde, R. H. 1975. The bases of angiosperm phylogeny: Floral anatomy. Ann. Missouri Bot. Gard. 62: 521–537.
Eyde, R. H. 1988. Comprehending *Cornus:* Puzzles and progress in the systematics of dogwoods. Bot. Rev. (Lancaster) 54: 233–351.
F. A. S. E. B. J. = F A S E B Journal; Official Publication of the Federation of American Societies for Experimental Biology.
Farwell, O. A. 1931. Concerning some species of *Cornus* of Philip Miller. Rhodora 33: 72.
Fassett, N. C. 1939. Notes from the herbarium of the University of Wisconsin. No. 17. *Elatine* and other aquatics. Rhodora 41: 367–377.
Fatope, M. O., O. A. Adoum, and Y. Takeda. 2000. C-18 acetylenic fatty acids of *Ximenia americana* with potential pesticidal activity. J. Agric. Food Chem 48: 1872–1874.
Fearn, M. L. and L. E. Urbatsch. 2001. *Glochidion puberum* (Euphorbiaceae) naturalized in southern Alabama. Sida 19: 711–714.
Felger, R. S. 2000. Flora of the Gran Desierto and Río Colorado of Northwestern Mexico. Tucson.
Felger, R. S. 2007. Living resources at the center of the Sonoran Desert: Regional uses of plants and animals by Native Americans. In: R. S. Felger and B. Broyles, eds. 2007. Dry Borders: Great Natural Reserves of the Sonoran Desert. Salt Lake City. Pp. 147–192.
Felger, R. S. et al. 2012. Flora of Tinajas Altas, Arizona—A century of botanical forays and forty thousand years of *Neotoma* chronicles. J. Bot. Res. Inst. Texas 6: 157–257.
Felger, R. S. and B. Broyles, eds. 2007. Dry Borders: Great Natural Reserves of the Sonoran Desert. Salt Lake City.
Fer, A. et al., eds. 2001. Proceedings of the 7th International Parasitic Weed Symposium. Nantes.
Ferguson, A. M. 1901. Crotons of the United States. Rep. (Annual) Missouri Bot. Gard. 1901: 33–73.
Ferguson, I. K. 1966. Notes on the nomenclature of *Cornus*. J. Arnold Arbor. 47: 100–105.
Ferguson, I. K. 1966b. The Cornaceae of the southeastern United States. J. Arnold Arbor. 47: 106–116.
Fernald, M. L. 1917. The genus *Elatine* in eastern North America. Rhodora 19: 10–15.
Fernald, M. L. 1928e. Contributions from the Gray Herbarium of Harvard University, LXXIX. I. *Geocaulon*, a new genus of the Santalaceae. Rhodora 30: 21–24.
Fernald, M. L. 1941b. *Elatine americana* and *E. triandra*. Rhodora 43: 208–211.
Fernald, M. L. 1941c. A hybrid *Cornus* from Cape Breton. Rhodora 43: 411–412.
Fernald, M. L. 1950. Gray's Manual of Botany, ed. 8. New York.
Fernández, R. 1996. Rhamnaceae. In: J. Rzedowski and G. C. de Rzedowski, eds. 1991+. Flora del Bajío y de Regiones Adyacentes. 99+ fascs. Pátzcuaro. Fasc. 43, pp. 1–68.
Field & Lab. = Field & Laboratory.
Fieldiana, Bot. = Fieldiana: Botany.
Fl. Amer. Sept.—See: F. Pursh [1813]1814
Fl. Ariz. New Mex.—See: I. Tidestrom and T. Kittell 1941
Fl. Atlant.—See: R. L. Desfontaines [1798–1799]
Fl. Belg.—See: B. C. J. Dumortier 1827
Fl. Bor.-Amer.—See: A. Michaux 1803; W. J. Hooker [1829–] 1833–1840
Fl. Bras.—See: C. F. P. von Martius et al. 1840–1906
Fl. Bras. Merid.—See: A. St.-Hilaire et al. 1824[–1833]
Fl. Brit. W. I.—See: A. H. R. Grisebach [1859–]1864
Fl. Calif.—See: W. L. Jepson 1909–1943
Fl. Carol.—See: T. Walter 1788
Fl. Chil.—See: C. Gay 1845–1854
Fl. Cochinch.—See: J. de Loureiro 1790
Fl. Colorado—See: P. A. Rydberg 1906
Fl. Florida Keys—See: J. K. Small 1913c
Fl. Francisc.—See: E. L. Greene 1891–1897
Fl. Ind.—See: W. Roxburgh 1820–1824
Fl. Ind. ed. 1832—See: W. Roxburgh 1832
Fl. Ind. Occid.—See: O. P. Swartz 1797–1806
Fl. Jap.—See: P. F. von Siebold and J. G. Zuccarini 1835 [–1870]
Fl. Ludov.—See: C. S. Rafinesque 1817
Fl. Miami—See: J. K. Small 1913b
Fl. N. Amer.—See: J. Torrey and A. Gray 1838–1843
Fl. Neotrop.—See: Organization for Flora Neotropica 1968+
Fl. Rocky Mts. ed. 2—See: P. A. Rydberg 1923
Fl. Ross.—See: C. F. von Ledebour [1841]1842–1853
Fl. S. Calif.—See: A. Davidson and G. L. Moxley 1923
Fl. S.E. U.S.—See: J. K. Small 1903
Fl. S.E. U.S. ed. 2—See: J. K. Small 1913
Fl. South. U.S.—See: A. W. Chapman 1860
Fl. South. U.S. ed. 2—See: A. W. Chapman 1883
Fl. South. U.S. ed. 2 repr. 2—See: A. W. Chapman 1892
Fl. Taur.-Caucas.—See: F. A. Marschall von Bieberstein 1808–1819
Fl. W. Calif.—See: W. L. Jepson 1901
Flexner, S. B. and L. C. Hauck, eds. 1987. The Random House Dictionary of the English Language, ed. 2 unabridged. New York.

Flora = Flora; oder (allgemeine) botanische Zeitung. [Vols. 1–16, 1818–1833, include "Beilage" and "Ergänzungsblätter"; vols. 17–25, 1834–1842, include "Beiblatt" and "Intelligenzblatt."]
Flora of North America Editorial Committee, eds. 1993+. Flora of North America North of Mexico. 19+ vols. New York and Oxford.
Flora, Morphol. Geobot. Ecophysiol. = Flora. Morphology, Geobotany, Ecophysiology.
Florida Entomol. = Florida Entomologist.
Florida Trees—See: J. K. Small 1913d
Forest Ecol. Managem. = Forest Ecology and Management.
Forrest, T. 1995. Two thousand years of eating bark: *Magnolia officinalis* var. *biloba* and *Eucommia ulmoides* in traditional Chinese medicine. Arnoldia (Jamaica Plain) 55: 13–18.
Forster, J. R. and G. Forster. 1776. Characteres Generum Plantarum, Quas in Itinere ad Insulas Maris Australis..., ed. 2. London. (Char. Gen. Pl. ed. 2)
Fosberg, F. R. 1942. *Cornus sericea* L. (*C. stolonifera* Michx.). Bull. Torrey Bot. Club 69: 583–589.
Franklin, J. et al. 1823. Narrative of a Journey to the Shores of the Polar Sea, in the Years 1819, 20, 21 and 22. London. [Richardson: Appendix VII. Botanical appendix, pp. [729]–768, incl. bryophytes by Schwägrichen, algae and lichens by Hooker.] (Narr. Journey Polar Sea)
Frazier, C. K. 1999. A taxonomic study of *Philadelphus* (Hydrangeaceae) as it occurs in New Mexico. New Mexico Bot. Newslett. 13: 1–6.
Fremontia = Fremontia; Journal of the California Native Plant Society.
Fross, D. and D. H. Wilken. 2006. *Ceanothus*. Portland.
Fruct. Sem. Pl.—See: J. Gaertner 1788–1791[–1792]
Gadek, P. A. et al. 1996. Sapindales: Molecular delimitation and infraordinal groups. Amer. J. Bot. 83: 802–811.
Gaertner, J. 1788–1791[–1792]. De Fructibus et Seminibus Plantarum.... 2 vols. Stuttgart and Tübingen. [Vol. 1 in 1 part only, 1788. Vol. 2 in 4 parts paged consecutively: pp. 1–184, 1790; pp. 185–352, 353–504, 1791; pp. 505–520, 1792.] (Fruct. Sem. Pl.)
Galil, J. 1968. Vegetative dispersal in *Oxalis cernua*. Amer. J. Bot. 55: 787–792.
Gard. & Forest = Garden and Forest; a Journal of Horticulture, Landscape Art and Forestry.
Gard. Chron. = Gardener's Chronicle.
Gard. Dict. ed. 8—See: P. Miller 1768
Gard. Dict. Abr. ed. 4—See: P. Miller 1754
Gardner, A. G. et al. 2012. Diversification of the American bulb-bearing *Oxalis* (Oxalidaceae): Dispersal to North America and modification of the tristylous breeding system. Amer. J. Bot. 99: 152–164.
Gay, C. 1845–1854. Historia Física y Política de Chile.... Botánica [Flora Chilena]. 8 vols., atlas. Paris and Santiago. (Fl. Chil.)
Gaz. Lit. México = Gazeta de Literatura de México.
Geltman, D. V. 1998. Taxonomic notes on *Euphorbia esula* (Euphorbiaceae) with special reference to its occurrence in the east part of the Baltic region. Ann. Bot. Fenn. 35: 113–117.
Geltman, D. V. et al. 2011. Typification and synonymy of the species of *Euphorbia* subgenus *Esula* (Euphorbiaceae) native to the United States and Canada. J. Bot. Res. Inst. Texas 5: 143–151.
Gen. Amer. Bor.—See: A. Gray 1848–1849
Gen. Euphorb.—See: A. Radcliffe-Smith 2001
Gen. Hist.—See: G. Don 1831–1838
Gen. N. Amer. Pl.—See: T. Nuttall 1818
Gen. Pl.—See: G. Bentham and J. D. Hooker 1862–1883; S. L. Endlicher 1836–1840[–1850]; J. C. Schreber 1789–1791
Gen. Pl. ed. 5—See: C. Linnaeus 1754
Gen. Sp. Pl.—See: M. Lagasca y Segura 1816b
Genet. Resources Crop Evol. = Genetic Resources and Crop Evolution; an International Journal.
Gentes Herbarum = Gentes Herbarum; Occasional Papers on the Kinds of Plants.
Gentry, H. S. 1958. The natural history of jojoba *(Simmondsia chinensis)* and its cultural aspects. Econ. Bot. 12: 261–291.
Gerrath, J., U. Posluszny, and L. Melville. 2015. Taming the Wild Grape: Botany and Horticulture in the Vitaceae. Cham.
Gervais, C. and M. Blondeau. 2003. Cytogèographies des *Cornus* herbacés (Cornaceae) du Nord de L'Amerique: Deux nouveaux taxons. Bull. Soc. Neuchâtel. Sci. Nat. 126: 33–44.
Giannasi, D. E. and C. M. Rogers. 1970. Taxonomic significance of floral pigments in *Linum* (Linaceae). Brittonia 22: 163–174.
Gil-Ad, N. L. and A. A. Reznicek. 1997. Evidence for hybridization of two Old World *Rhamnus* species—*R. cathartica* and *R. utilis* (Rhamnaceae)—in the New World. Rhodora 99: 1–22.
Gilg, E. 1925b. Loasaceae. In: H. G. A. Engler et al., eds. 1924+. Die natürlichen Pflanzenfamilien..., ed. 2. 26+ vols. Leipzig and Berlin. Vol. 21, pp. 522–543.
Gill, L. S. 1935. *Arceuthobium* in the United States. Trans. Connecticut Acad. Arts 32: 111–245.
Gillespie, L. J. 1994. Pollen morphology and phylogeny of the Plukenetieae (Euphorbiaceae). Ann. Missouri Bot. Gard. 81: 317–348.
Glad, J. B. 1976. Taxonomy of *Mentzelia mollis* and allied species (Loasaceae). Madroño 23: 283–292.
Gleason, H. A. and A. Cronquist. 1991. Manual of Vascular Plants of Northeastern United States and Adjacent Canada, ed. 2. Bronx.
Global Change Biol. = Global Change Biology.
Gmelin, J. F. 1791[–1792]. Caroli à Linné...Systema Naturae per Regna Tria Naturae.... Tomus II. Editio Decima Tertia, Aucta, Reformata. 2 parts. Leipzig. (Syst. Nat.)
Godara, R. K., B. J. Williams, and E. P. Webster. 2011. Texasweed *(Caperonia palustris)* can survive and reproduce in 30-cm flood. Weed Technol. 25: 667–673.
Godfrey, R. K. 1988. Trees, Shrubs, and Woody Vines of Northern Florida and Adjacent Georgia and Alabama. Athens, Ga.
Godfrey, R. K. and J. W. Wooten. 1981. Aquatic and Wetland Plants of Southeastern United States: Dicotyledons. Athens, Ga.
Gorham, A. L. 1953. The question of fertilization in *Smilacina racemosa* (L.) Desf. Phytomorphology 3: 44–50.

Gornall, R. J. and J. E. Wentworth. 1993. Chromosome numbers and locations of populations of *Parnassia palustris* from the British Isles. New Phytol. 123: 383–388.

Govaerts, R., D. G. Frodin, and A. Radcliffe-Smith. 2000. World Checklist and Bibliography of Euphorbiaceae (and Pandaceae). Kew. (World Checklist Bibliogr. Euphorbiaceae)

Graham, S. A. 1964. The genera of Rhizophoraceae and Combretaceae in the southeastern United States. J. Arnold Arbor. 45: 285–301.

Gray, A. 1848. A Manual of the Botany of the Northern United States.... Boston, Cambridge, and London. (Manual)

Gray, A. 1848–1849. Genera Florae Americae Boreali-orientalis Illustrata. The Genera of the Plants of the United States.... 2 vols. Boston, New York, and London. (Gen. Amer. Bor.)

Gray, A. 1852. Plantae Wrightianae Texano-Neo-Mexicanae. Part 1. Smithsonian Contr. Knowl. 3(5): 1–146.

Gray, A. 1854. Plantae Novae Thurberianae.... Cambridge, Mass. [Preprinted from Mem. Amer. Acad. Arts, n. s. 5: 297–328. 1855.] (Pl. Nov. Thurb.)

Gray, A. 1856. A Manual of the Botany of the Northern United States..., ed. 2. New York. (Manual ed. 2)

Gray, A. 1867. A Manual of the Botany of the Northern United States..., ed. 5. New York and Chicago. [Pteridophytes by D. C. Eaton.] (Manual ed. 5)

Gray, A., S. Watson, B. L. Robinson, et al. 1878–1897. Synoptical Flora of North America. 2 vols. in parts and fascs. New York etc. [Vol. 1(1,1), 1895; vol. 1(1,2), 1897; vol. 1(2), 1884; vol. 2(1), 1878.] (Syn. Fl. N. Amer.)

Great Plains Flora Association. 1986. Flora of the Great Plains. Lawrence, Kans.

Greene, E. L. 1891–1897. Flora Franciscana. An Attempt to Classify and Describe the Vascular Plants of Middle California. 4 parts. San Francisco. [Parts paged consecutively.] (Fl. Francisc.)

Greene, E. L. 1894. Manual of the Botany of the Region of San Francisco Bay.... San Francisco. (Man. Bot. San Francisco)

Grisebach, A. H. R. 1843–1845[–1846]. Spicilegium Florae Rumelicae et Bithynicae.... 2 vols. in 6 parts. Braunschweig. [Vols. paged independently but parts numbered consecutively.] (Spic. Fl. Rumel.)

Grisebach, A. H. R. [1859–]1864. Flora of the British West Indian Islands. 7 parts. London. [Parts paged consecutively.] (Fl. Brit. W. I.)

Grisebach, A. H. R. 1866. Catalogus Plantarum Cubensium Exhibens Collectionem Wrightianam Aliasque Minores ex Insula Cuba Missas. Leipzig. (Cat. Pl. Cub.)

Grissom, J. 2014. Phylogenetic and Phylogeographic Studies in *Mentzelia*. M.S. thesis. Washington State University.

Gu, C. Z. and U.-M. Hultgård. 2001. *Parnassia*. In: Wu Z. and P. H. Raven, eds. 1994+. Flora of China. 20+ vols. Beijing and St. Louis. Vol. 8, pp. 358–379.

Guo, Y. L. et al. 2013. Molecular phylogenetic analysis suggests paraphyly and early diversification of *Philadelphus* (Hydrangeaceae) in western North America: New insights into affinity with *Carpenteria*. J. Syst. Evol. 51: 545–563.

Gutermann, W. 2009. Notulae nomenclaturales 29–40 (zur Nomenklatur von Gefäßpflanzen Österreichs). Phyton (Horn) 49: 77–92.

Hamilton, L. and C. E. Conrad. 1990. Proceedings of the Symposium on Sandalwood in the Pacific: April 9–11, 1990, Honolulu, Hawaii. Berkeley. [U.S.D.A. Forest Serv., Gen. Techn. Rep. PSW-122.]

Handb. Nat. Pfl.-Syst.—See: H. G. L. Reichenbach 1837

Handel-Mazzetti, H. 1929–1937. Symbolae Sinicae. Botanische Ergebnisse der Expedition der Akademie der Eissenschaften in Wien nach Südwest-China. 1914/1918. 7 parts. Vienna. (Symb. Sin.)

Hardig, T. M. et al. 2000b. Diversification of the North American shrub genus *Ceanothus* (Rhamnaceae). Conflicting phylogenies from nuclear ribosomal DNA and chloroplast DNA. Amer. J. Bot. 87: 108–123.

Harling, G., B. Sparre, and L. Andersson, eds. 1973+. Flora of Ecuador. 89+ nos. Göteborg. [Nos. 1–4 published as Opera Bot., B, 1–4.]

Harris, B. D. 1968. Chromosome numbers and evolution in North American species of *Linum*. Amer. J. Bot. 55: 1197–1204.

Harvard Pap. Bot. = Harvard Papers in Botany.

Hastings, J. R., R. M. Turner, and D. K. Warren. 1972. An Atlas of Some Plant Distributions in the Sonoran Desert. Tucson.

Haston, E. M. et al. 2007. A linear sequence of Angiosperm Phylogeny Group II families. Taxon 56: 7–12.

Hawkins, A. K. 2010. Subspecific Classification within *Phoradendron serotinum* (Viscaceae): Development of Microsatellite Markers for Assessment of Population Genetic Structure. M.S. thesis. Sam Houston State University.

Hawksworth, F. G. and D. Wiens. 1970. New taxa and nomenclatural changes in *Arceuthobium* (Viscaceae). Brittonia 22: 265–269.

Hawksworth, F. G. and D. Wiens. 1972. Biology and Classification of Dwarf Mistletoes *(Arceuthobium)*. Washington. [Agric. Handb. 401.]

Hawksworth, F. G. and D. Wiens. 1980. A new species of *Arceuthobium* (Viscaceae) from central Mexico. Brittonia 32: 348–352.

Hawksworth, F. G. and D. Wiens, eds. 1996. Dwarf Mistletoes: Biology, Pathology, and Systematics. Washington. [Agric. Handb. 709.]

Hawksworth, F. G. and R. F. Scharpf. 1986. Spread of European mistletoe *(Viscum album)* in California, U.S.A. Eur. J. Forest Pathol. 16: 1–5.

Haworth, A. H. 1812. Synopsis Plantarum Succulentarum.... London. (Syn. Pl. Succ.)

Heikens, A. L. 2003. Conservation Assessment for Illinois Wood-sorrel (*Oxalis illinoensis* Schwegm.) Milwaukee.

Hempel, A. L. et al. 1995. Implications of *rbc*L sequence data for higher order relationships of the Loasaceae and the anomalous aquatic plant *Hydrostachys* (Hydrostachyaceae). Pl. Syst. Evol. 194: 25–37.

Hendrych, R. 1961. Nomenclatural remarks on *Thesium ramosum*. Taxon 10: 20–23.

Hendrych, R. 1964. Santalaceae. In: T. G. Tutin et al., eds. 1964–1980. Flora Europaea. 5 vols. Cambridge. Vol. 1, pp. 70–72.

Herb. Gén. Amat.—See: J. L. A. Loiseleur-Deslongchamps [1814–]1816–1827

Heredity = Heredity; an International Journal of Genetics.

Herndon, A. 1993. A revision of the *Chamaesyce deltoidea* (Euphorbiaceae) complex of southern Florida. Rhodora 95: 38–51.

Herndon, A. 1993b. Notes on *Chamaesyce* (Euphorbiaceae) in Florida. Rhodora 95: 352–368.

Hickman, J. C., ed. 1993. The Jepson Manual. Higher Plants of California. Berkeley, Los Angeles, and London.

Hill, J. 1756[–1757]. The British Herbal: An History of Plants and Trees.... 52 fascs. London. [Fascicles paged and plates numbered consecutively.] (Brit. Herb.)

Hill, J. 1768. Hortus Kewensis. London. (Hort. Kew.)

Hill, R. J. 1975. A Biosystematic Study of the Genus *Mentzelia* (Loasaceae) in Wyoming and Adjacent States. M.S. thesis. University of Wyoming.

Hill, R. J. 1977. Variability of soluble seed proteins in populations of *Mentzelia* L. (Loasaceae) from Wyoming and adjacent states. Bull. Torrey Bot. Club 104: 93–101.

Hilu, K. W. 2003. Angiosperm phylogeny based on *mat*K sequence information. Amer. J. Bot. 90: 1758–1776.

Hist. Fis. Cuba—See: R. de la Sagra 1840–1855

Hist. Phys. Cuba, Pl. Vasc.—See: A. Richard 1841–1851

Hist. Pl. Guiane—See: J. B. Aublet 1775

Hist. Pl. Remarq. Brésil—See: A. St.-Hilaire 1824[–1826]

Hitchcock, A. S. 1894. A Key to the Spring Flora of Manhattan.... Manhattan, Kans. (Key Spring Fl. Manhattan)

Hitchcock, C. L. 1943. The xerophyllous species of *Philadelphus* in southwestern North America. Madroño 7: 36–56.

Hitchcock, C. L. et al. 1955–1969. Vascular Plants of the Pacific Northwest. 5 vols. Seattle. [Univ. Wash. Publ. Biol. 17.] (Vasc. Pl. Pacif. N.W.)

Hitchcock, C. L. and A. Cronquist. 1973. Flora of the Pacific Northwest: An Illustrated Manual. Seattle.

Hoffman, M. H. A. 1996. Cultivar classification of *Philadelphus* L. (Hydrangeaceae). Acta Bot. Neerl. 45: 199–209.

Hoffmann, P. 1994. A contribution to the systematics of *Andrachne* sect. *Phyllanthopsis* and sect. *Pseudophyllanthus* compared with *Savia* s.l. (Euphorbiaceae) with special reference to floral morphology. Bot. Jahrb. Syst. 116: 321–331.

Hoffmann, P. 2008. Revision of *Heterosavia*, stat. nov., with notes on *Gonatogyne* and *Savia* (Phyllanthaceae). Brittonia 60: 136–166.

Hoffmann, P., H. Kathriarachchi, and K. Wurdack. 2006. A phylogenetic classification of Phyllanthaceae (Malpighiales; Euphorbiaceae sensu lato). Kew Bull. 61: 37–53.

Holmgren, N. H. and P. K. Holmgren. 1989. A taxonomic study of *Jamesia* (Hydrangeaceae). Brittonia 41: 335–350.

Holmgren, N. H. and P. K. Holmgren. 1997. Hydrangeaceae. In: A. Cronquist et al. 1972–2012. Intermountain Flora. Vascular Plants of the Intermountain West, U.S.A. 6 vols. in 8. New York and London. Vol. 3, part A, pp. 5–12.

Holmgren, N. H. and P. K. Holmgren. 2002. New mentzelias (Loasaceae) from the intermountain region of western United States. Syst. Bot. 27: 747–762.

Holmgren, N. H., P. K. Holmgren, and A. Cronquist. 2005. Loasaceae. In: A. Cronquist et al. 1972–2012. Intermountain Flora. Vascular Plants of the Intermountain West, U.S.A. 6 vols. in 8. New York and London. Vol. 2, part B, pp. 81–118.

Hooker, J. D. 1865b. Saxifrageae. In: G. Bentham and J. D. Hooker. 1862–1883. Genera Plantarum ad Exemplaria Imprimis in Herbariis Kewensibus Servata Definita. 3 vols. London. Vol. 1, pp. 629–655.

Hooker, W. J. [1829–]1833–1840. Flora Boreali-Americana; or, the Botany of the Northern Parts of British America.... 2 vols. in 12 parts. London, Paris, and Strasbourg. (Fl. Bor.-Amer.)

Hooker, W. J. and G. A. W. Arnott. [1830–]1841. The Botany of Captain Beechey's Voyage; Comprising an Account of the Plants Collected by Messrs Lay and Collie, and Other Officers of the Expedition, during the Voyage to the Pacific and Bering's Strait, Performed in His Majesty's Ship Blossom, under the Command of Captain F. W. Beechey...in the Years 1825, 26, 27, and 28. 10 parts. London. [Parts paged and plates numbered consecutively.] (Bot. Beechey Voy.)

Hooker's Icon. Pl. = Hooker's Icones Plantarum....

Hoover, R. F. 1970. The Vascular Plants of San Luis Obispo County, California. Berkeley.

Horn, D. D. and R. Kral. 1984. *Nestronia umbellula* Raf. (Santalaceae), a new state record for Tennessee. Castanea 49: 69–73.

Horn, J. W. et al. 2012. Phylogenetics and the evolution of major structural characters in the giant genus *Euphorbia* L. (Euphorbiaceae). Molec. Phylogen. Evol. 63: 305–326.

Horne, H. E., T. W. Barger, and G. L. Nesom. 2013. Two South American species of *Oxalis* (Oxalidaceae) naturalized in Alabama and the USA, first report. Phytoneuron 2013-54: 1–7.

Hort. Bot. Vindob.—See: N. J. Jacquin 1770–1776

Hort. Kew.—See: W. Aiton 1789; J. Hill 1768

Hou, D. 1955. A revision of the genus *Celastrus*. Ann. Missouri Bot. Gard. 42: 215–302.

Howard, R. A. 1981. Three experiences with the manchineel (*Hippomane* spp., Euphorbiaceae). Biotropica 13: 224–227.

Howell, J. T. 1940. Studies in *Ceanothus* 3–5. Leafl. W. Bot. 2: 228–240, 259–262, 285–289.

Hu, S. Y. 1954–1956. A monograph of the genus *Philadelphus*. J. Arnold Arbor. 35: 275–333; 36: 52–109, 325–368; 37: 15–90.

Hufford, L. 1987. Inflorescence architecture of *Eucnide* (Loasaceae). Madroño 34: 18–28.

Hufford, L. 1988. Potential roles of scaling and post-anthesis developmental changes in the evolution of floral forms of *Eucnide* (Loasaceae). Nordic J. Bot. 8: 147–157.

Hufford, L. 1988b. Seed coat morphology of *Eucnide* and other Loasaceae. Syst. Bot. 13: 154–167.

Hufford, L. 1989. Structure of the inflorescence and flower of *Petalonyx linearis*. Pl. Syst. Evol. 163: 211–226.

Hufford, L. 1990. Androecial ontogeny and the problem of monophyly of Loasaceae. Canad. J. Bot. 68: 402–419.

Hufford, L. 1995. Seed morphology of Hydrangeaceae and its phylogenetic implications. Int. J. Pl. Sci. 156: 555–580.

Hufford, L. 1997. A phylogenetic analysis of Hydrangeaceae based on morphological data. Int. J. Pl. Sci. 158: 652–672.

Hufford, L. 1998. Early development of androecia in polystemonous Hydrangeaceae. Amer. J. Bot. 85: 1057–1067.

Hufford, L. 2004. Hydrangeaceae. In: K. Kubitzki et al., eds. 1990+. The Families and Genera of Vascular Plants. 10+ vols. Berlin etc. Vol. 6, pp. 202–215.

Hufford, L. et al. 2003. The major clades of Loasaceae: Phylogenetic analysis using the plastid *mat*K and *trn*L-*trns*F regions. Amer. J. Bot. 90: 1215–1228.

Hufford, L., M. L. Moody, and D. E. Soltis. 2001. A phylogenetic analysis of Hydrangeaceae based on sequences of the plastid gene *mat*K and their combination with *rbc*L and morphological data. Int. J. Pl. Sci. 162: 835–846.

Huft, M. J. 1979. A Monograph of *Euphorbia* Section *Tithymalopsis*. Ph.D. dissertation. University of Michigan.

Hultén, E. 1937. Flora of the Aleutian Islands and Westernmost Alaska Peninsula with Notes on the Flora of Commander Islands.... Stockholm.

Hultgård, U.-M. 1987. *Parnassia palustris* L. in Scandinavia. Acta Univ. Upsal., Symb. Bot. Upsal. 28: 1–128.

Humboldt, A. von, A. J. Bonpland, and C. S. Kunth. 1815[1816]–1825. Nova Genera et Species Plantarum Quas in Peregrinatione Orbis Novi Collegerunt, Descripserunt.... 7 vols. in 36 parts. Paris. (Nov. Gen. Sp.)

Huntia = Huntia; a Yearbook of Botanical and Horticultural Bibliography [later: a Journal of Botanical History].

Hylander, N. 1957. On cut-leaved and small-leaved forms of *Alnus glutinosa* and *A. incana*. Svensk Bot. Tidskr. 51: 437–453.

Icon.—See: A. J. Cavanilles 1791–1801

Icon. Pl. = Icones Plantarum....

Ill. Bot. Himal. Mts.—See: J. F. Royle [1833–]1839[–1840]

Ill. Fl. N. U.S.—See: N. L. Britton and A. Brown 1896–1898

Ill. Handb. Laubholzk.—See: C. K. Schneider [1904–]1906–1912

Index Seminum (München) = Delectus Seminum Hortus Monacensis.

Ingram, J. W. 1980. The generic limits of *Argythamnia* (Euphorbiaceae) defined. Gentes Herbarum 11: 427–436.

Ingrouille, M. J. et al. 2002. Systematics of Vitaceae from the viewpoint of plastid *rbc*L DNA sequence data. Bot. J. Linn. Soc. 138: 421–432.

Inst. Rei Herb.—See: H. J. N. von Crantz 1766

Int. J. Pl. Sci. = International Journal of Plant Sciences.

Intermount. Fl.—See: A. Cronquist et al. 1972–2012

Intr. Hist. Nat.—See: J. A. Scopoli 1777

Islam, M. B. and R. P. Guralnick. 2015. Generic placement of the former *Condaliopsis* (Rhamnaceae) species. Phytotaxa 236: 25–39.

Islam, M. B. and M. P. Simmons. 2006. A thorny dilemma: Testing alternative intrageneric classifications within *Ziziphus* (Rhamnaceae). Syst. Bot. 31: 826–842.

Iter Hispan.—See: P. Loefling 1758

Ives, J. C. 1861. Report upon the Colorado River of the West, Explored in 1857 and 1858 by Lieutenant Joseph C. Ives.... 5 parts, appendices. Washington. (Rep. Colorado R.)

Iwashina, T., K. Kamenosono, and H. Hatta. 1997. Flavonoid glycosides from leaves of *Aucuba japonica* and *Helwingia japonica* (Cornaceae): Phytochemical relationship with the genus *Cornus*. J. Jap. Bot. 72: 337–346.

J. Acad. Nat. Sci. Philadelphia = Journal of the Academy of Natural Sciences of Philadelphia.

J. Agric. Food Chem = Journal of Agricultural and Food Chemistry.

J. Arizona Acad. Sci. = Journal of the Arizona Academy of Science.

J. Arnold Arbor. = Journal of the Arnold Arboretum.

J. Asiat. Soc. Bengal, Pt. 2, Nat. Hist. = Journal of the Asiatic Society of Bengal. Part 2, Natural History.

J. Bot. Res. Inst. Texas = Journal of the Botanical Research Institute of Texas.

J. Bot. (Schrader) = Journal für die Botanik. [Edited by H. A. Schrader.] [Volumation indicated by nominal year date and vol. number for that year (1 or 2); e.g. 1800(2).]

J. Chem. Ecol. = Journal of Chemical Ecology.

J. Elisha Mitchell Sci. Soc. = Journal of the Elisha Mitchell Scientific Society.

J. Exp. Bot. = Journal of Experimental Botany; an Official Organ of the Society for Experimental Biology.

J. Forest. (Washington) = Journal of Forestry. (Washington.)

J. Jap. Bot. = Journal of Japanese Botany.

J. Kansas Entomol. Soc. = Journal of the Kansas Entomological Society.

J. Korean Soc. Agric. Chem. Biotechnol. = Journal of the Korean Society of Agricultural Chemistry and Biotechnology. [Han'guk Nongwha Hakhoe Chi.]

J. Linn. Soc., Bot. = Journal of the Linnean Society. Botany.

J. Molec. Evol. = Journal of Molecular Evolution.

J. Pl. Res. = Journal of Plant Research. [Shokubutsu-gaku zasshi.]

J. Roy. Hort. Soc. = Journal of the Royal Horticultural Society.

J. Syst. Evol. = Journal of Systematics and Evolution.

J. Torrey Bot. Soc. = Journal of the Torrey Botanical Society.

J. Trop. Ecol. = Journal of Tropical Ecology.

J. Wash. Acad. Sci. = Journal of the Washington Academy of Sciences.

Jacquin, N. J. 1760. Enumeratio Systematica Plantarum, Quas in Insulis Caribaeis Vicinaque Americes Continente Detexit Novas.... Leiden. (Enum. Syst. Pl.)

Jacquin, N. J. 1763. Selectarum Stirpium Americanarum Historia.... Vienna. (Select. Stirp. Amer. Hist.)

Jacquin, N. J. 1770–1776. Hortus Botanicus Vindobonensis.... 3 vols. Vienna. (Hort. Bot. Vindob.)

Jacquin, N. J. 1786[1787]–1796[1797]. Collectanea ad Botanicam, Chemiam, et Historiam Naturalem Spectantia.... 5 vols. Vienna. (Collectanea)

Jacquin, N. J. 1794. *Oxalis*. Vienna, London, and Leiden. (Oxalis)

Janaki Ammal, E. K. 1951. Chromosomes and the evolution of garden *Philadelphus*. J. Roy. Hort. Soc. 76: 269–275.

Jard. Malmaison—See: É. P. Ventenat 1803–1804[–1805]

Jarvis, C. E. 2007. Order out of Chaos: Linnaean Plant Names and Their Types. London.

Jasieniuk, M. and M. J. Lechowicz. 1987. Spatial and temporal variation in chasmogamy and cleistogamy in *Oxalis montana* (Oxalidaceae). Amer. J. Bot. 74: 1672–1680.

Jeon, H. J. et al. 1998. Growth-inhibiting effects of various traditional drinks of plant origin on human intestinal bacteria. J. Korean Soc. Agric. Chem. Biotechnol. 41: 605–607.

Jepson, W. L. 1901. A Flora of Western Middle California.... Berkeley. (Fl. W. Calif.)

Jepson, W. L. 1909–1943. A Flora of California.... 3 vols. in 12 parts. San Francisco etc. [Pagination consecutive within each vol.; vol. 1 page sequence independent of part number sequence (chronological); part 8 of vol. 1 (pp. 1–32, 579–index) never published.] (Fl. Calif.)

Jepson, W. L. [1923–1925.] A Manual of the Flowering Plants of California.... Berkeley. (Man. Fl. Pl. Calif.)

Jerome, C. A. and B. A. Ford. 2002. The discovery of three genetic races of the dwarf mistletoe *Arceuthobium americanum* (Viscaceae) provides insight into the evolution of parasitic angiosperms. Molec. Ecol. 11: 387–405.

Johnson, D. M. 1986c. Revision of the neotropical genus *Callaeum* (Malpighiaceae). Syst. Bot. 11: 335–353.

Johnston, I. M. 1940. New phanerogams from Mexico. II. J. Arnold Arbor. 21: 67–75.

Johnston, L. A. 1975. Revision of the *Rhamnus serrata* complex. Sida 6: 67–79.

Johnston, M. C. 1962. Revision of *Condalia* including *Microrhamnus* (Rhamnaceae). Brittonia 14: 332–368.

Johnston, M. C. 1963. The species of *Ziziphus* indigenous to United States and Mexico. Amer. J. Bot. 50: 1020–1027.

Johnston, M. C. 1968. Botanical name of the tea sageretia of China. J. Arnold Arbor. 49: 377–379.

Johnston, M. C. 1969. *Colubrina stricta* Engelmann ex M. C. Johnston (Rhamnaceae), new species from Texas, Nuevo León and Coahuila. SouthW. Naturalist 14: 257.

Johnston, M. C. 1971. Revision of *Colubrina* (Rhamnaceae). Brittonia 23: 2–53.

Johnston, M. C. 1975. Synopsis of *Canotia* (Celastraceae) including a new species from the Chihuahuan Desert. Brittonia 27: 119–122.

Johnston, M. C. 1990. Vascular Plants of Texas: A List, Up-dating the Manual of the Vascular Plants of Texas, ed. 2. Austin.

Johnston, M. C. and L. A. Johnston. 1978. *Rhamnus*. In: Organization for Flora Neotropica. 1968+. Flora Neotropica. 110+ nos. New York. No. 20, pp. 1–96.

Johnston, M. C. and B. H. Warnock. 1963. *Phyllanthus* and *Reverchonia* (Euphorbiaceae) in far western Texas. SouthW. Naturalist 8: 15–22.

Johri, B. M. and S. P. Bhatnagar. 1960. Embryology and taxonomy of the Santalales I. Proc. Natl. Inst. Sci. India, B 26: 199–220.

Jones, G. N. and G. D. Fuller. 1955. Vascular Plants of Illinois. Urbana. (Vasc. Pl. Illinois)

Jussieu, A. H. L. de. 1824. De Euphorbiacearum Generibus.... Paris. (Euphorb. Gen.)

Just's Bot. Jahresber. = Just's botanischer Jahresbericht; systematisch geordnetes Repertorium der botanischen Literatur aller Länder.

Kartesz, J. T. and C. A. Meacham. 1999. Synthesis of the North American Flora, ver. 1.0. Chapel Hill. [CD-ROM.]

Kathriarachchi, H. et al. 2005. Molecular phylogenetics of Phyllanthaceae inferred from five genes (plastid, *atp*B, *mat*K, 3' *ndh*F, *rbc*L, and nuclear PHYC). Molec. Phylogen. Evol. 36: 112–134.

Kathriarachchi, H. et al. 2006. Phylogenetics of tribe Phyllantheae (Phyllanthaceae; Euphorbiaceae sensu lato) based on nrITS and plastid *mat*K DNA sequence data. Amer. J. Bot. 93: 637–655.

Kawakita, A. and M. Kato. 2009. Repeated independent evolution of obligate pollination mutualism in the Phyllantheae-*Epicephala* association. Proc. Roy. Soc. Biol. Sci. Ser. B 276: 417–426.

Kearney, T. H. and R. H. Peebles. 1942. Flowering Plants and Ferns of Arizona.... Washington.

Kearney, T. H. and R. H. Peebles. 1960. Arizona Flora, ed. 2. Berkeley.

Kearns, C. A. and D. W. Inouye. 1994. Fly pollination in *Linum lewisii* (Linaceae). Amer. J. Bot. 81: 1091–1095.

Kew Bull. = Kew Bulletin.

Key Spring Fl. Manhattan—See: A. S. Hitchcock 1894

Kiger, R. W. and D. M. Porter. 2001. Categorical Glossary for the Flora of North America Project. Pittsburgh.

Kingsbury, J. M. 1964. Poisonous Plants of the United States and Canada. Englewood Cliffs.

Klein, F. K. 1970. Chemotaxonomic evidence for a new form of *Ceanothus incanus* Nutt. Four Seasons 3(3): 19–22.

Koch, K. H. E. 1869–1873. Dendrologie. Bäume, Sträucher und Halbsträucher, welche in Mittel- und Nord-Europa im Freien kultivirt werden. 2 vols. in 3. Erlangen. (Dendrologie)

Koller, G. L. and J. H. Alexander. 1979. The raisin tree—its use, hardiness and size. Arnoldia (Jamaica Plain) 39: 6–15.

Korotkova, N. et al. 2009. Phylogeny of the eudicot order Malpighiales: Analysis of a recalcitrant clade with sequences of the *pet*D group II intron. Pl. Syst. Evol. 282: 201–228.

Krähenbühl, M., Yuan Y. M., and P. Küpfer. 2002. Chromosome and breeding system evolution of the genus *Mercurialis* (Euphorbiaceae): Implications of ITS molecular phylogeny. Pl. Syst. Evol. 234: 155–169.

Kral, R. 1973. Some notes on the flora of the southern states, particularly Alabama and middle Tennessee. Rhodora 75: 366–410.

Kral, R. 1993. *Pinus*. In: Flora of North America Editorial Committee, eds. 1993+. Flora of North America North of Mexico. 19+ vols. New York and Oxford. Vol. 2, pp. 373–398.

Kruckeberg, A. R. 1996. Gardening with Native Plants of the Pacific Northwest, ed. 2. Seattle.

Krüssmann, G. 1984–1986. Manual of Cultivated Broad-leaved Trees and Shrubs. 3 vols. Beaverton.

Kubitzki, K. et al., eds. 1990+. The Families and Genera of Vascular Plants. 10+ vols. Berlin etc.

Kuijt, J. 1955. Dwarf mistletoes. Bot. Rev. (Lancaster) 21: 569–628.

Kuijt, J. 1960. Morphological aspects of parasitism in the dwarf mistletoes *(Arceuthobium)*. Univ. Calif. Publ. Bot. 30: 337–436.

Kuijt, J. 1969. The Biology of Parasitic Flowering Plants. Berkeley.

Kuijt, J. 2003. Monograph of *Phoradendron* (Viscaceae). Syst. Bot. Monogr. 66.

Kuijt, J. 2013. Prophyll, calyculus, and perianth in Santalales. Blumea 57: 248–252.

Kuntze, O. 1891–1898. Revisio Generum Plantarum Vascularium Omnium atque Cellularium Multarum.... 3 vols. Leipzig etc. [Vol. 3 in 3 parts paged independently; parts 1 and 3 unnumbered.] (Revis. Gen. Pl.)

Kurylo, J. S. 2007. *Rhamnus cathartica:* Native and naturalized distribution and habitat preferences. J. Torrey Bot. Soc. 134: 420–430.

Kvavadze, E. et al. 2009. 30,000-year-old wild flax fibers. Science, ser. 2, 325: 1359.

L'Héritier de Brutelle, C.-L. 1788[1789–1792]. Sertum Anglicum.... 4 fascs. Paris. [All text in fasc. 1; plates numbered consecutively.] (Sert. Angl.)

Lagasca y Segura, M. 1816b. Genera et Species Plantarum.... Madrid. (Gen. Sp. Pl.)

Lai, X. Z., Yang Y. B., and Shan X. L. 2004. The investigation of euphorbiaceous medicinal plants in southern China. Econ. Bot. 58: S307–S320.

Lamarck, J. et al. 1783–1817. Encyclopédie Méthodique. Botanique.... 13 vols. Paris and Liège. [Vols. 1–8, suppls. 1–5.] (Encycl.)

Laport, R. G., R. L. Minckley, and J. Ramsey. 2012. Phylogeny and cytogeography of the North American creosote bush (*Larrea tridentata,* Zygophyllaceae). Syst. Bot. 37: 153–164.

Latiff, A. 1981. Studies in Malaysian Vitaceae V. The genus *Cayratia* in the Malay Peninsula. Sains Malaysiana 10: 129–139.

Lauria, F. 1991. Taxonomy, systematics, and phylogeny of *Pinus* subsection *Ponderosae* Loudon (Pinaceae). Alternative concepts. Linzer Biol. Beitr. 23: 129–202.

Leafl. Bot. Observ. Crit. = Leaflets of Botanical Observation and Criticism.

Leafl. W. Bot. = Leaflets of Western Botany.

Ledebour, C. F. von. [1841]1842–1853. Flora Rossica sive Enumeratio Plantarum in Totius Imperii Rossici Provinciis Europaeis, Asiaticis, et Americanis Hucusque Observatarum.... 4 vols. Stuttgart. (Fl. Ross.)

Lefèvre, I., E. Corréal, and S. Lutts. 2005. Cadmium tolerance and accumulation in the noxious weed *Zygophyllum fabago*. Canad. J. Bot. 83: 1655–1662.

Leicht-Young, S. A. et al. 2007. Distinguishing native (*Celastrus scandens* L.) and invasive (*C. orbiculatus* Thunb.) bittersweet species using morphological characteristics. J. Torrey Bot. Soc. 134: 441–450.

Leopold, D. J. and R. N. Muller. 1983. Hosts of *Pyrularia pubera* Michaux (Santalaceae) in the field and in culture. Castanea 48: 138–145.

Levin, G. A. 1986b. Systematic foliar morphology of Phyllanthoideae (Euphorbiaceae). III. Cladistic analysis. Syst. Bot. 11: 515–530.

Levin, G. A. 1995. Systematics of the *Acalypha californica* complex (Euphorbiaceae). Madroño 41: 254–265.

Levin, G. A. 1999. Evolution in the *Acalypha gracilens/monococca* complex (Euphorbiaceae): Morphological analysis. Syst. Bot. 23: 269–287.

Levin, G. A. 1999b. Notes on *Acalypha* (Euphorbiaceae) in North America. Rhodora 101: 217–233.

Levin, G. A. and M. G. Simpson. 1994. Phylogenetic implications of pollen ultrastructure in the Oldfieldioideae (Euphorbiaceae). Ann. Missouri Bot. Gard. 81: 203–238.

Li, B. and M. G. Gilbert. 2008. *Glochidion*. In: Wu Z. and P. H. Raven, eds. 1994+. Flora of China. 20+ vols. Beijing and St. Louis. Vol. 11, pp. 193–202.

Li, J. H., D. E. Boufford, and M. J. Donoghue. 2001. Phylogenetics of *Buckleya* (Santalaceae) based on ITS sequences of nuclear ribosomal DNA. Rhodora 103: 137–150.

Li, Y. et al. 1998. The promoting effects of geniposidic acid and aucubin in *Eucommia ulmoides* Oliver leaves on collagen synthesis. Biol. Pharm. Bull. 21: 1306–1310.

Lia, V. V. et al. 2001. Molecular phylogeny of *Larrea* and its allies (Zygophyllaceae): Reticulate evolution and the probable time of creosote bush arrival to North America. Molec. Phylogen. Evol. 21: 309–320.

Libby, G. and C. Bloom. 1998. *Nestronia umbellula* Rafinesque (Santalaceae) from the Highland Rim of Kentucky. Castanea 63: 161–164.

Link, J. H. F. 1821–1822. Enumeratio Plantarum Horti Regii Berolinensis Altera.... 2 parts. Berlin. (Enum. Hort. Berol. Alt.)

Linnaea = Linnaea; ein Journal für die Botanik in ihrem ganzen Umfange.

Linnaeus, C. 1753. Species Plantarum.... 2 vols. Stockholm. (Sp. Pl.)

Linnaeus, C. 1754. Genera Plantarum, ed. 5. Stockholm. (Gen. Pl. ed. 5)

Linnaeus, C. 1758[–1759]. Systema Naturae per Regna Tria Naturae..., ed. 10. 2 vols. Stockholm. (Syst. Nat. ed. 10)

Linnaeus, C. 1762–1763. Species Plantarum..., ed. 2. 2 vols. Stockholm. (Sp. Pl. ed. 2)

Linnaeus, C. 1766–1768. Systema Naturae per Regna Tria Naturae..., ed. 12. 3 vols. Stockholm. (Syst. Nat. ed. 12)

Linnaeus, C. 1767[–1771]. Mantissa Plantarum. 2 parts. Stockholm. [Mantissa [1] and Mantissa [2] Altera paged consecutively.] (Mant. Pl.)

Linnaeus, C. f. 1781[1782]. Supplementum Plantarum Systematis Vegetabilium Editionis Decimae Tertiae, Generum Plantarum Editionis Sextae, et Specierum Plantarum Editionis Secundae. Braunschweig. (Suppl. Pl.)

Linzer Biol. Beitr. = Linzer Biologische Beiträge.

Liston, A. 2003. A new interpretation of floral morphology in *Garrya* (Garryaceae). Taxon 52: 271–276.

Liu, Q. R. and M. F. Watson. 2008. *Oxalis*. In: Wu Z. and P. H. Raven, eds. 1994+. Flora of China. 20+ vols. Beijing and St. Louis. Vol. 11, pp. 2–6.

Liu, X. Q. et al. 2015. Phylogeny of the *Ampelocissus-Vitis* clade in Vitaceae supports the New World origin of the grape genus. Molec. Phylogen. Evol., doi: 10.1016/j.ympev.2015.10.013.

Loefling, P. 1758. Iter Hispanicum, Eller Resa til Spanska Ländern uti Europa och America, Förrättad Iffrån År 1751 til År 1756 ... Utgifven Efter Dess Frånfälle af Carl Linnaeus. Stockholm. (Iter Hispan.)

Loiseleur-Deslongchamps, J. L. A. [1814–]1816–1827. Herbier Géneral de l'Amateur.... 8 vols. Paris. (Herb. Gén. Amat.)

Lombardi, J. A. 2000. Vitaceae: Géneros *Ampelocissus, Ampelopsis* e *Cissus*. In: Organization for Flora Neotropica. 1968+. Flora Neotropica. 110+ nos. New York. No. 80.

London J. Bot. = London Journal of Botany.

Long, R. W. and O. Lakela. 1971. A Flora of Tropical Florida: A Manual of the Seed Plants and Ferns of Southern Peninsular Florida. Coral Gables. [Reprinted 1976, Miami.]

Lookadoo, S. E. and A. J. Pollard. 1991. Chemical contents of stinging trichomes of *Cnidoscolus texanus*. J. Chem. Ecol. 17: 1909–1916.

Loureiro, J. de. 1790. Flora Cochinchinensis.... 2 vols. Lisbon. [Vols. paged consecutively.] (Fl. Cochinch.)

Lourteig, A. 1975. Oxalidaceae extra-Austroamericanae: 1. *Oxalis* L. sectio *Thamnoxys* Planchon. Phytologia 29: 449–471.

Lourteig, A. 1979. Oxalidaceae extra-Austroamericanae: 2. *Oxalis* L. sectio *Corniculatae* DC. Phytologia 42: 57–198.

Lourteig, A. 1980. Oxalidaceae. In: R. E. Woodson Jr. et al., eds. 1943–1981. Flora of Panama. 41 fascs. St. Louis. [Ann. Missouri Bot. Gard. 67: 823–850.]

Lourteig, A. 1982. Oxalidaceae extra-Austroamaricanae: 4. *Oxalis* L. sectio *Articulatae* Knuth. Phytologia 50: 130–142.

Lourteig, A. 1994. *Oxalis* L. subgénero *Thamnoxys* (Endl.) Reiche emend. Lourt. Bradea 7: 1–199.

Lourteig, A. 2000. *Oxalis* subgeneros *Monoxalis* (Small) Lourt., *Oxalis* y *Trifidus* Lourt. Bradea 7: 201–629.

Lu, L. M. et al. 2013. Phylogeny of the non-monophyletic *Cayratia* Juss. (Vitaceae) and implications for character evolution and biogeography. Molec. Phylogen. Evol. 68: 502–515.

Lundell, C. L. 1942–1969. Flora of Texas. 3 vols. in parts. Dallas and Renner, Tex.

Lundell, C. L. 1971. Studies of American plants. III. Celastraceae. Wrightia 4: 157–159.

Lundellia = Lundellia; Journal of the Plant Resources Center of the University of Texas at Austin.

Ma, J. S. 2001. A revision of *Euonymus* (Celastraceae). Thaiszia 11: 1–264.

Ma, J. S. 2010. A review of leafy spurge, *Euphorbia esula* (Euphorbiaceae), the most aggressive invasive alien in North America. Acta Bot. Yunnan. 32(suppl. 17): 19–45.

Mabberley, D. J. 2008. The Plant-book: A Portable Dictionary of the Vascular Plants..., ed. 3. Cambridge.

Mabry, T. J., J. H. Hunziker, and D. R. DiFeo, eds. 1977. Creosote Bush: Biology and Chemistry of *Larrea* in New World Deserts. Stroudsburg, Pa.

Macbride, J. F. 1918. A revision of *Mirabilis*, subgenus *Hesperonia*. Contr. Gray Herb. 56: 20–24.

Madroño = Madroño; Journal of the California Botanical Society [from vol. 3: a West American Journal of Botany].

Magee, D. W. and H. E. Ahles. 1999. Flora of the Northeast: A Manual of the Vascular Flora of New England and Adjacent New York. Amherst.

Malécot, V. et al. 2004. A morphological cladistic analysis of Olacaceae. Syst. Bot. 29: 569–586.

Malécot, V. and D. L. Nickrent. 2008. Molecular phylogenetic relationships of Olacaceae and related Santalales. Syst. Bot. 33: 97–106.

Man. Bot. San Francisco—See: E. L. Greene 1894

Man. Fl. Pl. Calif.—See: W. L. Jepson [1923–1925]

Man. S.E. Fl.—See: J. K. Small 1933

Mant.—See: J. A. Schultes and J. H. Schultes 1822–1827

Mant. Pl.—See: C. Linnaeus 1767[–1771]

Mantese, A. and D. Medan. 1993. Anatomía y arquitectura foliares de *Colletia* y *Adolphia* (Rhamnaceae). Darwiniana 32: 91–97.

Manual—See: A. Gray 1848

Manual ed. 2—See: A. Gray 1856

Manual ed. 5—See: A. Gray 1867

Marschall von Bieberstein, F. A. 1808–1819. Flora Taurico-Caucasica.... 3 vols. Charkow. (Fl. Taur.-Caucas.)

Marshall, H. 1785. Arbustrum Americanum. The American Grove.... Philadelphia. (Arbust. Amer.)

Martin, W. C. and C. R. Hutchins. 1980. A Flora of New Mexico. 2 vols. Vaduz.

Martius, C. F. P. von, A. W. Eichler, and I. Urban, eds. 1840–1906. Flora Brasiliensis. 15 vols. in 40 parts, 130 fascs. Munich, Vienna, and Leipzig. [Vols. and parts numbered in systematic sequence, fascs. numbered independently in chronological sequence.] (Fl. Bras.)

Maschinski, J. 2006. Demography of *Linum carteri* var. *carteri* growing in disturbed and undisturbed sites. In: J. Maschinski et al. 2006. Conservation of South Florida Endangered and Threatened Flora: 2005 Program at Fairchild Tropical Garden. Gainesville. Pp. 82–86.

Maschinski, J. et al. 2006. Conservation of South Florida Endangered and Threatened Flora: 2005 Program at Fairchild Tropical Garden. Gainesville.

Maschinski, J. et al. 2007. Conservation of South Florida Endangered and Threatened Flora: 2006–2007 Program at Fairchild Tropical Garden. Gainesville.

Maschinski, J. et al. 2008. Conservation of South Florida Endangered and Threatened Flora: 2007–2008 Program at Fairchild Tropical Garden. Gainesville.

Maschinski, J. and D. Walters. 2007. Demography of *Linum carteri* var. *carteri* growing in disturbed and undisturbed sites. In: J. Maschinski et al. 2007. Conservation of South Florida Endangered and Threatened Flora: 2006–2007 Program at Fairchild Tropical Garden. Gainesville. Pp. 55–61.

Maschinski, J. and D. Walters. 2008. Growth to maturity of *Linum carteri* var. *carteri* at two sites. In: J. Maschinski et al. 2008. Conservation of South Florida Endangered and Threatened Flora: 2007–2008 Program at Fairchild Tropical Garden. Gainesville. Pp. 27–29.

Mathiasen, R. L. et al. 2008. Mistletoes: Pathology, systematics, ecology, and management. Pl. Dis. 92: 988–1006.

Mathiasen, R. L. and S. C. Kenaley. 2015. A morphometric analysis of dwarf mistletoes in the *Arceuthobium campylopodum-occidentale* complex (Viscaceae). Madroño 62: 1–20.

Mayfield, M. H. 1991. *Euphorbia johnstonii* (Euphorbiaceae), a new species from Tamaulipas, Mexico, with notes on *Euphorbia* subsection *Acutae*. Sida 14: 573–579.

Mayfield, M. H. 1997. A Systematic Treatment of *Euphorbia* subg. *Poinsettia* (Euphorbiaceae). Ph.D. dissertation. University of Texas.

Mayfield, M. H. 2013. Four new annual species of *Euphorbia* section *Tithymalus* from North America. J. Bot. Res. Inst. Texas 7: 633–647.

McCartney, R. D., K. Wurdack, and J. Moore. 1989. The genus *Lindera* in Florida. Palmetto 9: 3–8.

McClintock, E. 1957. A monograph of the genus *Hydrangea*. Proc. Calif. Acad. Sci. 29: 147–256.

McDill, J. R. 2009. Molecular Phylogenetic Studies in the Linaceae and *Linum*, with Implications for Their Systematics and Historical Biogeography. Ph.D. dissertation. University of Texas.

McDill, J. R. et al. 2009. The phylogeny of *Linum* and Linaceae subfamily Linoideae, with implications for their systematics, biogeography, and evolution of heterostyly. Syst. Bot. 34: 386–405.

McDill, J. R. and B. B. Simpson. 2011. Molecular phylogenetics of Linaceae with complete generic sampling and data from two plastid genes. Bot. J. Linn. Soc. 165: 64–83.

McKenna, M. J. et al. 2011. Delimitation of the segregate genera of *Maytenus* s.l. (Celastraceae) based on morphological and molecular characters. Syst. Bot. 36: 922–932.

McMinn, H. 1939. An Illustrated Manual of California Shrubs.... San Francisco.

McMinn, H. 1944. The importance of field hybrids in determining the species in the genus *Ceanothus*. Proc. Calif. Acad. Sci., ser. 4, 25: 323–356.

McVaugh, R. 1943b. Edward Palmer's collections in Arizona in 1869, 1876, and 1877. Part I. General discussion: Itinerary and sources. Amer. Midl. Naturalist 29: 768–775.

McVaugh, R. 1945. The genus *Jatropha* in America: Principal intergeneric groups. Bull. Torrey Bot. Club 72: 271–294.

McVaugh, R. 1945b. The jatrophas of Cervantes and of the Sessé & Mociño Herbarium. Bull. Torrey Bot. Club 72: 31–41.

McVaugh, R. 1956b. Edward Palmer: Plant Explorer of the American West. Norman.

McVaugh, R. 1995. Euphorbiacearum sertum Novo-galiciarum revisarum. Contr. Univ. Michigan Herb. 20: 173–215.

McVaugh, R. 1998. Botanical results of the Sessé and Moçiño expedition (1797–1803). 6. Reports and records from western Mexico, 1790–1792. Bol. Inst. Bot. Univ. Guadalajara 6: 1–178.

McVaugh, R. 2001c. Loasaceae. In: R. McVaugh and W. R. Anderson, eds. 1974+. Flora Novo-Galiciana: A Descriptive Account of the Vascular Plants of Western Mexico. 8+ vols. Ann Arbor. Vol. 3, pp. 696–714.

McVaugh, R. and W. R. Anderson, eds. 1974+. Flora Novo-Galiciana: A Descriptive Account of the Vascular Plants of Western Mexico. 8+ vols. Ann Arbor.

McVaugh, R. and T. H. Kearney. 1943. Edward Palmer's collections in Arizona in 1869, 1876, and 1877. Part II. A consideration of some Palmer collections cited in the "Flowering Plants and Ferns of Arizona." Amer. Midl. Naturalist 29: 775–778.

Med. Fl.—See: C. S. Rafinesque 1828[–1830]

Med. Repos. = Medical Repository.

Medan, D. and C. Schirarend. 2004. Rhamnaceae. In: K. Kubitzki et al., eds. 1990+. The Families and Genera of Vascular Plants. 10+ vols. Berlin etc. Vol. 6, pp. 320–338.

Medley, M. E. 1993. An Annotated Catalog of the Known or Reported Vascular Flora of Kentucky. Ph.D. dissertation. University of Louisville.

Meisner, C. F. 1837–1843. Plantarum Vascularium Genera.... 2 parts in 14 fascs. Leipzig. [Parts (1: "Tab. Diagn."; 2: "Commentarius") issued together incrementally within fascs. and paged independently, fascs. paged consecutively for each part.] (Pl. Vasc. Gen.)

Mém. Acad. Imp. Sci. Saint Pétersbourg, Sér. 7 = Mémoires de l'Académie Impériale des Sciences de Saint Pétersbourg, Septième Série.

Mém. Acad. Imp. Sci. St. Pétersbourg Hist. Acad. = Mémoires de l'Académie Impériale des Sciences de St. Pétersbourg. Avec l'Histoire de l'Académie.

Mem. Amer. Acad. Arts = Memoirs of the American Academy of Arts and Science.

Mem. Boston Soc. Nat. Hist. = Memoirs Read before the Boston Society of Natural History; Being a New Series of the Boston Journal of Natural History.

Mem. Comis. Limites—See: J. L. Berlandier [1832]

Mém. Fam. Rhamnées—See: A. T. Brongniart 1826

Mém. Mus. Hist. Nat. = Mémoires du Muséum d'Histoire Naturelle.

Mem. New York Bot. Gard. = Memoirs of the New York Botanical Garden.

Mém. Soc. Sci. Phys. Nat. Bordeaux = Mémoires de la Société des Sciences Physiques et Naturelles de Bordeaux.

Mem. Torrey Bot. Club = Memoirs of the Torrey Botanical Club.

Mem. Tour N. Mexico—See: F. A. Wislizenus 1848

Méndez-Robles, M. D. et al. 2004. Chemical composition and current distribution of "Azafrán de Bolita" (*Ditaxis heterantha* Zucc., Euphorbiacee): A food pigment producing plant. Econ. Bot. 58: 530–535.

Mennega, A. M. W. 1987. Wood anatomy of the Euphorbiaceae, in particular of the subfamily Phyllanthoideae. Bot. J. Linn. Soc. 94: 111–126.

Merriam-Webster. 1988. Webster's New Geographical Dictionary. Springfield, Mass.

Metori, K., M. Furutsu, and S. Takahashi. 1997. The preventive effect of ginseng with du-zhong leaf on protein metabolism in aging. Biol. Pharm. Bull. 20: 237–242.

Metori, K., S. Y. Tanimoto, and S. Takahashi. 1998. Promotive effect of *Eucommia* leaf extract on collagen synthesis in rats. Nat. Med. 52: 465–469.

Meyer, C. A. von. 1845. Über einige *Cornus*-Arten.... St. Petersburg. (Cornus-Arten)

Michael, P. 1964. The identity and origin of varieties of *Oxalis pes-caprae* L. naturalized in Australia. Trans. Roy. Soc. South Australia 88: 167–173.

Michaux, A. 1803. Flora Boreali-Americana.... 2 vols. Paris and Strasbourg. (Fl. Bor.-Amer.)

Michigan Bot. = Michigan Botanist.

Milby, T. H. 1971. Floral anatomy of *Krameria lanceolata.* Amer. J. Bot. 58: 569–576.

Miller, A. G. 1982. *Thesium.* In: P. H. Davis, ed. 1965–2000. Flora of Turkey and the East Aegean Islands. 11 vols. Edinburgh. Vol. 7, pp. 536–544.

Miller, K. I. 1964. A Taxonomic Study of the Species of *Tragia* in the United States. Ph.D. thesis. Purdue University.

Miller, K. I. and G. L. Webster. 1967. A preliminary revision of *Tragia* (Euphorbiaceae) in the United States. Rhodora 69: 241–305.

Miller, L. W. 1964. A Taxonomic Study of the Species of *Acalypha* in the United States. Ph.D. dissertation. Purdue University.

Miller, P. 1754. The Gardeners Dictionary.... Abridged..., ed. 4. 3 vols. London. (Gard. Dict. Abr. ed. 4)

Miller, P. 1768. The Gardeners Dictionary..., ed. 8. London. (Gard. Dict. ed. 8)

Millspaugh, C. F. 1892. Medicinal Plants. 2 vols. Philadelphia.

Minnesota Bot. Stud. = Minnesota Botanical Studies.

Mission Vitic. Amér.—See: P. Viala 1889

Moerman, D. E. 1998. Native American Ethnobotany. Portland.

Mohlenbrock, R. H. 2014. Vascular Flora of Illinois: A Field Guide, ed. 4. Carbondale.

Moldenke, A. R. 1976. California pollination ecology and vegetation types. Phytologia 34: 305–361.

Molec. Ecol. = Molecular Ecology.

Molec. Phylogen. Evol. = Molecular Phylogenetics and Evolution.

Molina, G. I. 1782. Saggio sulla Storia Naturale del Chili.... Bologna. (Sag. Stor. Nat. Chili)

Monogr. Phan.—See: A. L. P. P. de Candolle and C. de Candolle 1878–1896

Montagna, P. A. et al. 2011. Coastal impacts. In: J. Schmandt et al. 2011. The Impact of Global Warming on Texas, ed. 2. Austin. Pp. 96–123.

Moody, K. 1989. Weeds Reported in Rice in South and Southeast Asia. Los Baños.

Moody, M. L. et al. 2001. Phylogenetic relationships of Loasaceae subfamily Gronovioideae inferred from *mat*K and ITS sequence data. Amer. J. Bot. 88: 326–336.

Moody, M. L. and L. Hufford. 2000. Floral ontogeny and morphology of *Cevallia, Fuertesia,* and *Gronovia* (Loasaceae subfamily Gronovioideae). Int. J. Pl. Sci. 161: 869–883.

Moore, M. O. 1991. Classification and systematics of eastern North American *Vitis* L. (Vitaceae) north of Mexico. Sida 14: 339–367.

Moreno, M. T. et al., eds. 1996. Advances in Parasitic Plant Research. Córdoba.

Morgan, D. R. and D. E. Soltis. 1993. Phylogenetic relationships among members of Saxifragaceae sensu lato based on *rbc*L sequence data. Ann. Missouri Bot. Gard. 80: 631–660.

Morton, J. F. 1968. Plants associated with esophageal cancer in Curaçao. Cancer Res. 28: 2268–2271.

Morton, J. F. 1984. Nobody loves the *Bischofia* anymore. Proc. Florida State Hort. Soc. 97: 241–244.

Mosquin, T. 1971. Biosystematic studies in the North American species of *Linum,* section *Adenolinum* (Linaceae). Canad. J. Bot. 49: 1379–1388.

Mühlberg, H. 1982. The Complete Guide to Water Plants: A Reference Book. Translated from the German by Ilse Lindsay; Revised by Colin D. Roe. New York.

Munson, T. V. 1893. *Vitis baileyana.* Denison. [2 unnumbered pp.] (Vitis baileyana)

Munson, T. V. 1909 Foundations of American Grape Culture. New York.

Munz, P. A. 1959. A California Flora. Berkeley and Los Angeles.

Munz, P. A. 1974. A Flora of Southern California. Berkeley.

Murray, J. A. 1784. Caroli à Linné Equitis Systema Vegetabilium.... Editio Decima Quarta.... Göttingen. (Syst. Veg. ed. 14)

Murrell, Z. E. 1992. Systematics of the Genus *Cornus* (Cornaceae). Ph.D. dissertation. Duke University.

Murrell, Z. E. 1993. Phylogenetic relationships in *Cornus* (Cornaceae). Syst. Bot. 18: 469–495

Murrell, Z. E. 1994. Dwarf dogwoods: Intermediacy and the morphological landscape. Syst. Bot. 19: 539–556.

Musselman, L. J. 1982. The Santalaceae of Virginia. Castanea 47: 276–283.

Musselman, L. J. and S. C. Haynes. 1996. Santalaceae with weed potential in the United States. In: M. T. Moreno et al., eds. 1996. Advances in Parasitic Plant Research. Córdoba. Pp. 521–527.

Musselman, L. J. and W. F. Mann. 1979b. Notes on seed germination and parasitism of seedlings of *Buckleya distichophylla* (Santalaceae). Castanea 44: 108–113.

Mutat. Res. = Mutation Research.

N. Amer. Euphorbia—See: J. B. S. Norton 1899

N. Amer. Fl.—See: N. L. Britton et al. 1905+

N. Amer. Sylv.—See: T. Nuttall 1842–1849

Nachr. Königl. Ges. Wiss. Georg-Augusts-Univ. = Nachrichten von der Königlichen Gesellschaft der Wissenschaften und der Georg-Augusts-Universität.

Nair, B. R. and P. Kuriachan. 2004. Cytogenetic evidence of the evolution of *Oxalis corniculata* var. *atropurpurea* Planch. Cytologia 69: 149–153.

Nair, N. C. and K. V. Mani. 1960. Organography and floral anatomy of species of Vitaceae. Phytomorphology 10: 138–144.

Nakamura, T. et al. 1997. Antimutagenicity of Tochu tea (an aqueous extract of *Eucommia ulmoides* leaves): 1. The clastogen-suppressing effects of Tochu tea in CHO cells and mice. Mutat. Res. 388: 7–20.

Nakazawa, Y. et al. 1997. Effect of eucommia leaf (*Eucommia ulmoides* Oliver leaf; du-zhong yge) extract on blood pressure: 1. Effect on blood pressure in spontaneous rats (SHR). Nat. Med. 51: 392–398.

Narr. Journey Polar Sea—See: J. Franklin et al. 1823

Nash, G. V. 1896b. Notes on some Florida plants.—II. Bull. Torrey Bot. Club 23: 95–108.

Nasir, E. and S. I. Ali. 1970+. Flora of West Pakistan. 219+ nos. Karachi and Islamabad. [Title changed to Flora of Pakistan from No. 132, 1980.]

Nat. Hist. = Natural History; the Magazine of the American Museum of Natural History.

Nat. Med. = Natural Medicines. [Shoyakugaku Zasshi.]
Nat. Pflanzenfam.—See: H. G. A. Engler and K. Prantl 1887–1915
Nat. Pflanzenfam. ed. 2—See: H. G. A. Engler et al. 1924+
Naturaliste Canad. = Naturaliste Canadien. Bulletin de Recherches, Observations et Découvertes se Rapportant à l'Histoire Naturelle du Canada.
Nature = Nature; a Weekly Illustrated Journal of Science.
Navaro, A. M. and W. H. Blackwell. 1990. A revision of *Paxistima* (Celastraceae). Sida 14: 231–249.
Neff, J. L. and B. B. Simpson. 1981. Oil-collecting structures in the Anthophoridae (Hymenoptera): Morphology, function, and use in systematics. J. Kansas Entomol. Soc. 54: 95–123.
Neogenyton—See: C. S. Rafinesque 1825
Nesom, G. L. 1993h. *Sageretia mexicana* (Rhamnaceae): A new species from southwestern Mexico. Phytologia 75: 369–376.
Nesom, G. L. 2009b. Again: Taxonomy of yellow-flowered caulescent *Oxalis* (Oxalidaceae) in eastern North America. J. Bot. Res. Inst. Texas 3: 727–738.
Nesom, G. L. 2009c. Notes on *Oxalis* sect. *Corniculatae* (Oxalidaceae) in the southwestern United States. Phytologia 91: 527–533.
Nesom, G. L. 2009d. Taxonomic notes on acaulescent *Oxalis* (Oxalidaceae) in the United States. Phytologia 91: 501–526.
Nesom, G. L. 2009e. Notes on the taxonomy of *Maytenus phyllanthoides* (Celastraceae). Phytologia 91: 64–68.
Nesom, G. L. 2013. Taxonomic note on *Colubrina* (Rhamnaceae). Phytoneuron 2013-4: 1–21.
Nesom, G. L., D. D. Spaulding, and H. E. Horne. 2014. Further observations on the *Oxalis dillenii* group (Oxalidaceae). Phytoneuron 2014-12: 1–10.
New Fl.—See: C. S. Rafinesque 1836[–1838]
New Mexico Bot. Newslett. = New Mexico Botanist Newsletter.
New Phytol. = New Phytologist; a British Botanical Journal.
New Zealand J. Bot. = New Zealand Journal of Botany.
Nickrent, D. L. 1996. Molecular systematics. In: F. G. Hawksworth and D. Wiens, eds. 1996. Dwarf Mistletoes: Biology, Pathology, and Systematics. Washington. Pp. 155–170.
Nickrent, D. L. 2012. Justification for subspecies in *Arceuthobium campylopodum* (Viscaceae). Phytoneuron 2012-51: 1–11.
Nickrent, D. L. et al. 2004. A phylogeny of all species of *Arceuthobium* (Viscaceae) using nuclear and chloroplast DNA sequences. Amer. J. Bot. 91: 125–138.
Nickrent, D. L. et al. 2010. A revised classification of Santalales. Taxon 59: 538–558.
Nickrent, D. L. et al. 2010b. Proposal to conserve the name *Viscum serotinum (Phoradendron serotinum)* against *Viscum leucarpum* (Viscaceae). Taxon 59: 1903–1904.
Nickrent, D. L. and M. A. Garcia. 2009. On the brink of holoparasitism: Plastome evolution in dwarf mistletoes (*Arceuthobium*, Viscaceae). J. Molec. Evol. 68: 603–615.
Nickrent, D. L. and M. A. Garcia. 2015. *Lacomucinaea,* a new monotypic genus in Thesiaceae (Santalales). Phytotaxa 224: 173–184.
Nickrent, D. L. and V. Malécot. 2001. A molecular phylogeny of Santalales. In: A. Fer et al., eds. 2001. Proceedings of the 7th International Parasitic Weed Symposium. Nantes. Pp. 69–74.
Nickrent, D. L., K. P. Schuette, and E. M. Starr. 1994. A molecular phylogeny of *Arceuthobium* based upon rDNA internal transcribed spacer sequences. Amer. J. Bot. 81: 1149–1160.
Nie, Z. L. et al. 2012. Evolution of the intercontinental disjunctions in six continents in the *Ampelopsis* clade of the grape family (Vitaceae). B. M. C. Evol. Biol. 12: 1–17.
Niedenzu, F. 1928. Malpighiaceae. In: H. G. A. Engler, ed. 1900–1953. Das Pflanzenreich.... 107 vols. Berlin. Vols. 91, 93, 94[IV,141], pp. 1–870.
Nobs, M. A. 1963. Experimental studies on species relationships in *Ceanothus.* Publ. Carnegie Inst. Wash. 623.
Nordic J. Bot. = Nordic Journal of Botany.
Norton, J. B. S. 1899. North American Species of *Euphorbia,* section *Tithymalus.* St. Louis. [Preprinted from Rep. (Annual) Missouri Bot. Gard. 11: 85–144, plates 11–52. 1900.] (N. Amer. Euphorbia)
Norton, J. B. S. 1900. A revision of the American species of *Euphorbia* of the section *Tithymalus* occurring north of Mexico. Rep. (Annual) Missouri Bot. Gard. 11: 85–144.
Not. Milit. Reconn.—See: W. H. Emory 1848
Notizbl. Bot. Gart. Berlin-Dahlem = Notizblatt des Botanischen Gartens und Museums zu Berlin-Dahlem.
Nov. Actorum Acad. Caes. Leop.-Carol. Nat. Cur. = Novorum Actorum Academiae Caesareae Leopoldinae-Carolinae Naturae Curiosorum.
Nov. Gen. Pl.—See: C. P. Thunberg 1781–1801
Nov. Gen. Sp.—See: A. von Humboldt et al. 1815[1816]–1825
Nov. Gen. Sp. Pl.—See: E. F. Poeppig and S. L. Endlicher 1835–1845
Nov. Pl. Descr. Dec.—See: C. G. Ortega 1797–1800
Nova Acta Acad. Caes. Leop.-Carol. German. Nat. Cur. = Nova Acta Academiae Caesareae Leopoldino-Carolinae Germanicae Naturae Curiosorum.
Novon = Novon; a Journal for Botanical Nomenclature.
Nuttall, T. 1818. The Genera of North American Plants, and Catalogue of the Species, to the Year 1817.... 2 vols. Philadelphia. (Gen. N. Amer. Pl.)
Nuttall, T. 1842–1849. The North American Sylva.... 3 vols. Philadelphia. (N. Amer. Sylv.)
O'Donnell, R. 2010. The genus *Hesperolinon* (Linaceae): An introduction. Four Seasons 13(4): 1–61.
O'Kennon, R. J. 1991. *Paliurus spina-christi* (Rhamnaceae) new for North America in Texas. Sida 14: 606–609.
Ockendon, D. J. 1971. Taxonomy of the *Linum perenne* group in Europe. Watsonia 8: 205–235.
Ohio J. Sci. = Ohio Journal of Science.
Ohwi, J. 1965. Flora of Japan (in English).... A Combined, Much Revised, and Extended Translation by the Author of His...Flora of Japan (1953) and...Flora of Japan—Pteridophyta (1957). Edited by Frederick G. Meyer... and Egbert H. Walker.... Washington. [Reprinted 1984, Washington.]

Olsen, K. and B. A. Schaal. 1999. Evidence on the origin of cassava: Phylogeography of *Manihot esculenta*. Proc. Natl. Acad. Sci. U.S.A. 96: 5586–5591.

Olsen, K. and B. A. Schaal. 2001. Microsatellite variation in cassava (*Manihot esculenta,* Euphorbiaceae) and its wild relatives: Further evidence for a southern Amazonian origin of domestication. Amer. J. Bot. 88: 131–142.

Oregon State Monogr., Stud. Bot. = Oregon State Monographs. Studies in Botany.

Organization for Flora Neotropica. 1968+. Flora Neotropica. 110+ nos. New York. (Fl. Neotrop.)

Ornduff, R. 1972. The breakdown of trimorphic incompatibility in *Oxalis* section *Corniculatae.* Evolution 26: 52–65.

Ornduff, R. 1986. The origin of weediness in *Oxalis pes-caprae*. Amer. J. Bot. 73: 779–780.

Ornduff, R. 1987. Reproductive systems and chromosome races of *Oxalis pes-caprae* L. and their bearing on the genesis of a noxious weed. Ann. Missouri Bot. Gard. 74: 79–84.

Ornithol. Biogr.—See: J. J. Audubon 1831–1839

Ortega, C. G. 1797–1800. Novarum, aut Rariorum Plantarum Horti Reg. Botan. Matrit. Descriptionum Decades.... 10 decades in 4 parts. Madrid. [Parts paged consecutively.] (Nov. Pl. Descr. Dec.)

Osbeck, P. 1757. Dagbok öfwer en Ostindisk Resa Åren 1750, 1751, 1752. Stockholm. (Dagb. Ostind. Resa)

Oswald, V. H. and L. Ahart. 1994. Manual of the Vascular Plants of Butte County, California. Sacramento.

Oudejans, R. C. H. M. 1993. World Catalogue of Species Names Published in the Tribe Euphorbieae (Euphorbiaceae) with Their Geographical Distribution: Cumulative Supplement I. Scherpenzeel. (World Cat. Euphorb. Cum. Suppl. I)

Oxalis—See: N. J. Jacquin 1794

Pacif. Railr. Rep.—See: War Department 1855–1860

Pacific Regional Wood Anatomy Conference. 1984. Proceedings of Pacific Regional Wood Anatomy Conference: October 1–7, 1984, Tsukuba, Ibaraki, Japan. Ibaraki.

Pallas, P. S. 1771–1776. Reise durch verschiedene Provinzen des russischen Reichs.... 3 vols. St. Petersburg. (Reise Russ. Reich.)

Pareek, O. P. 2001. Fruits for the Future 2—Ber. Southampton.

Park, K. R. 1998. Monograph of *Euphorbia* sect. *Tithymalopsis* (Euphorbiaceae). Edinburgh J. Bot. 55: 161–208.

Parmeter, J. R. and R. F. Scharpf. 1963. Dwarf mistletoe on red fir and white fir in California. J. Forest. (Washington) 61: 371–374.

Patterson, D. T. et al. 1989. Composite List of Weeds. Champaign.

Pattison, R. P. and R. N. Mack. 2007. Potential distribution of the invasive tree *Triadica sebifera* (Euphorbiaceae) in the United States: Evaluating CLIMEX predictions with field trials. Global Change Biol. 14: 813–826.

Peck, J. H. 2003. Arkansas flora: Additions, reinstatements, exclusions, and re-exclusions. Sida 20: 1737–1757.

Peirson, J. A. et al. 2014. Phylogenetics and taxonomy of the New World leafy spurges, *Euphorbia* section *Tithymalus* (Euphorbiaceae). Bot. J. Linn. Soc. 175: 191–228.

Pélabon, C. et al. 2005. Effects of crossing distance on offspring fitness and developmental stability in *Dalechampia scandens* (Euphorbiaceae). Amer. J. Bot. 92: 842–851.

Pemberton, R. W. and H. Liu. 2008. Naturalized orchid bee pollinates resin-reward flowers in Florida: Novel and known mutualisms. Biotropica 40: 714–718.

Pemberton, R. W. and H. Liu. 2008b. Naturalization of *Dalechampia scandens* in southern Florida. Caribbean J. Sci. 44: 417–419.

Persoon, C. H. 1805–1807. Synopsis Plantarum.... 2 vols. Paris and Tubingen. (Syn. Pl.)

Pflanzenr.—See: H. G. A. Engler 1900–1953

Philipp. J. Sci. = Philippine Journal of Science.

Phillips, R. B. 1980. Systematics of *Parnassia* L. (Parnassiaceae): Generic Overview and Revision of North American Taxa. Ph.D. dissertation. University of California, Berkeley.

Phoradendron—See: W. Trelease 1916

Phytologia = Phytologia; Designed to Expedite Botanical Publication.

Phytomorphology = Phytomorphology; an International Journal of Plant Morphology.

Phyton (Horn) = Phyton; Annales Rei Botanica.

Piehl, M. A. 1965. The natural history and taxonomy of *Comandra* (Santalaceae). Mem. Torrey Bot. Club 22: 1–97.

Pilatowski, R. E. 1980. A Taxonomic Study of the *Hydrangea arborescens* Complex. M.S. thesis. North Carolina State University.

Pilatowski, R. E. 1982. A taxonomic study of the *Hydrangea arborescens* complex. Castanea 47: 84–98.

Pilger, R. K. F. 1935. Santalaceae. In: H. G. A. Engler et al., eds. 1924+. Die natürlichen Pflanzenfamilien..., ed. 2. 26+ vols. Leipzig and Berlin. Vol. 16b, pp. 52–91.

Pl. Bras. Icon. Descr.—See: J. E. Pohl [1826–]1827–1831 [–1833]

Pl. Dis. = Plant Disease; International Journal of Applied Plant Pathology.

Pl. Dis. Reporter = Plant Disease Reporter.

Pl. Hartw.—See: G. Bentham 1839[–1857]

Pl. Heermann.—See: E. M. Durand and T. C. Hilgard 1854

Pl. Life Hartsville—See: W. C. Coker 1912

Pl. Nov. Thurb.—See: A. Gray 1854

Pl. Pratten. Calif.—See: E. M. Durand 1855

Pl. Syst. Evol. = Plant Systematics and Evolution.

Pl. Vasc. Gen.—See: C. F. Meisner 1837–1843

Pl. Wilson.—See: C. S. Sargent 1913–1917

Pl. World = Plant World.

Planchon, J. É. 1887. Ampelidae. In: A. L. P. P. de Candolle and C. de Candolle, eds. 1878–1896. Monographiae Phanerogamarum.... 9 vols. Paris. Vol. 5(2), pp. 305–654.

Plantsman = The Plantsman.

Plukenet, L. 1691–1705. Phytographia sive Illustriorum & Miniis Cognitarum Icones Tabulis Aeneis Summâ Diligentiâ Elaboratae.... 7 parts. London. [Pars Prior, Pars Altera, 1691; Pars Tertia, 1692; [Pars Quarta], 1694; Almagestum, 1696; Almagesti...Mantissa, 1700; Almatheum, 1705.]

Poeppig, E. F. and S. L. Endlicher. 1835–1845. Nova Genera ac Species Plantarum Quas in Regno Chilensi Peruviano et in Terra Amazonica.... 3 vols. Leipzig. (Nov. Gen. Sp. Pl.)

Pohl, J. E. [1826–]1827–1831[–1833]. Plantarum Brasiliae Icones et Descriptiones.... 2 vols. in parts. Vienna. [Vols. paged independently, plates numbered consecutively.] (Pl. Bras. Icon. Descr.)

Pool, A. 2014. Taxonomic revision of *Gouania* (Rhamnaceae) for North America. Ann. Missouri Bot. Gard. 99: 490–552.

Pooler, M. R., R. L. Dix, and J. Feely. 2002. Interspecific hybridizations between the native bittersweet, *Celastrus scandens,* and the introduced invasive species, *C. orbiculatus*. SouthE. Naturalist 1: 69–76.

Porcher, R. D. and D. A. Rayner. 2002. A Guide to the Wildflowers of South Carolina. Columbia.

Porter, D. M. 1963. The taxonomy and distribution of the Zygophyllaceae of Baja California, Mexico. Contr. Gray Herb. 192: 99–135.

Porter, D. M. 1968. The basic chromosome number in *Tribulus* (Zygophyllaceae). Wasmann J. Biol. 26: 5–6.

Porter, D. M. 1969. The genus *Kallstroemia* (Zygophyllaceae). Contr. Gray Herb. 198: 41–163.

Porter, D. M. 1971. Notes on the floral glands in *Tribulus* (Zygophyllaceae). Ann. Missouri Bot. Gard. 58: 1–5.

Porter, D. M. 1972. The genera of Zygophyllaceae in the southeastern United States. J. Arnold Arbor. 53: 531–552.

Porter, D. M. 1974b. Disjunct distributions in the New World Zygophyllaceae. Taxon 23: 339–346.

Porter, D. M. 2005. Zygophyllaceae. In: G. Harling et al., eds. 1973+. Flora of Ecuador. 89+ nos. Göteborg. No. 75, pp. 17–29.

Poston, M. E. and J. W. Nowicke. 1993. Pollen morphology, trichome types, and relationships of the Gronovioideae (Loasaceae). Amer. J. Bot. 80: 689–704.

Powell, A. M., S. A. Powell, and A. S. Tomb. 1977. Cytotypes in *Cevallia sinuata*. SouthW. Naturalist 21: 433–441.

Prance, G. T. 1970. The genera of Chrysobalanaceae in the southeastern United States. J. Arnold Arbor. 51: 521–528.

Prance, G. T. 1972. Chrysobalanaceae. In: Organization for Flora Neotropica. 1968+. Flora Neotropica. 110+ nos. New York. No. 9.

Prance, G. T. and C. A. Sothers. 2003. Chrysobalanaceae. In: Australian Biological Resources Study. 1999+. Species Plantarum: Flora of the World. 11+ parts. Canberra. Parts 9, 10.

Prenner, G. and P. J. Rudall. 2007. Comparative ontogeny of the cyathium in *Euphorbia* (Euphorbiaceae) and its allies: Exploring the organ-flower-inflorescence boundary. Amer. J. Bot. 94: 1612–1629.

Pringle, J. S. 2010. Nomenclature of the thicket creeper, *Parthenocissus inserta* (Vitaceae). Michigan Bot. 49: 73–78.

Prir. Rostlin—See: F. Berchtold and J. S. Presl 1823–1835

Proc. Acad. Nat. Sci. Philadelphia = Proceedings of the Academy of Natural Sciences of Philadelphia.

Proc. Amer. Acad. Arts = Proceedings of the American Academy of Arts and Sciences.

Proc. Amer. Assoc. Advancem. Sci. = Proceedings of the American Association for the Advancement of Science.

Proc. Annual Conf. SouthE. Assoc. Fish Wildlife Agencies = Proceedings of the Annual Conference Southeastern Association of Fish and Wildlife Agencies.

Proc. Annual Meetings Soc. Promot. Agric. Sci. = Proceedings of the Annual Meeting, Society for the Promotion of Agricultural Science.

Proc. Biol. Soc. Wash. = Proceedings of the Biological Society of Washington.

Proc. Calif. Acad. Sci. = Proceedings of the California Academy of Sciences.

Proc. Davenport Acad. Nat. Sci. = Proceedings of the Davenport Academy of Natural Sciences.

Proc. Florida State Hort. Soc. = Proceedings of the (Annual Meeting), Florida State Horticultural Society.

Proc. Indiana Acad. Sci. = Proceedings of the Indiana Academy of Science.

Proc. Natl. Acad. Sci. U.S.A. = Proceedings of the National Academy of Sciences of the United States of America.

Proc. Natl. Inst. Sci. India, B = Proceedings of the National Institute of Sciences of India. Part B, Biological Sciences.

Proc. Roy. Soc. Biol. Sci. Ser. B = Proceedings of the Royal Society. Biological Sciences Series B.

Proc. Texas Grape Growers Assoc. = Proceedings of the Texas Grape Growers Association.

Prodr.—See: A. P. de Candolle and A. L. P. P. de Candolle 1823–1873; O. P. Swartz 1788

Prodr. Stirp. Chap. Allerton—See: R. A. Salisbury 1796

Publ. Carnegie Inst. Wash. = Publications of the Carnegie Institution of Washington.

Publ. Field Mus. Nat. Hist., Bot. Ser. = Publications of the Field Museum of Natural History. Botanical Series.

Publ. Mus. Michigan State Univ., Biol. Ser. = Publications of the Museum. Michigan State University. Biological Series.

Pugnaire, F. I., F. S. Chapin, and T. M. Hardig. 2006. Evolutionary changes and correlations among functional traits in *Ceanothus* in response to Mediterranean condition. Web Ecol. 6: 17–26.

Punt, W. 1962. Pollen morphology of the Euphorbiaceae with special reference to taxonomy. Wentia 7: 1–116.

Puri, V. 1952. Placentation in angiosperms. Bot. Rev. (Lancaster) 18: 603–621.

Pursh, F. [1813]1814. Flora Americae Septentrionalis; or, a Systematic Arrangement and Description of the Plants of North America. 2 vols. London. (Fl. Amer. Sept.)

Putz, N. 1994. Vegetative spreading of *Oxalis pes-caprae* (Oxalidaceae). Pl. Syst. Evol. 191: 57–67.

Qaiser, M. and S. Nazimuddin. 1981. *Sageretia* (Rhamnaceae). In: E. Nasir and S. I. Ali, eds. 1970+. Flora of West Pakistan. 219+ nos. Karachi and Islamabad. No. 140.

Quibell, C. F. 1993. Philadelphaceae. In: J. C. Hickman, ed. 1993. The Jepson Manual. Higher Plants of California. Berkeley, Los Angeles, and London. Pp. 816–818.

Radcliffe-Smith, A. 1973. Allomorphic female flowers in the genus *Acalypha* (Euphorbiaceae). Kew Bull. 28: 525–529.

Radcliffe-Smith, A. 1985. Taxonomy of North American leafy spurge. In: A. K. Watson, ed. 1985. Leafy Spurge. Champaign. Pp. 14–25.

Radcliffe-Smith, A. 2001. Genera Euphorbiacearum. Kew. (Gen. Euphorb.)

Radford, A. E., H. E. Ahles, and C. R. Bell. 1968. Manual of the Vascular Flora of the Carolinas. Chapel Hill.

Rafinesque, C. S. 1817. Florula Ludoviciana; or, a Flora of the State of Louisiana. Translated, Revised, and Improved, from the French of C. C. Robin.... New York. (Fl. Ludov.)

Rafinesque, C. S. 1825. Neogenyton, or Indication of Sixty-six New Genera of Plants of North America. [Lexington, Ky.] (Neogenyton)

Rafinesque, C. S. 1828[–1830]. Medical Flora; or, Manual of the Medical Botany of the United States of North America. 2 vols. Philadelphia. (Med. Fl.)

Rafinesque, C. S. 1836[–1838]. New Flora and Botany of North America.... 4 parts. Philadelphia. [Parts paged independently.] (New Fl.)

Rafinesque, C. S. 1838. Alsographia Americana, or an American Grove.... Philadelphia. (Alsogr. Amer.)

Rafinesque, C. S. 1838b. Sylva Telluriana. Mantis. Synopt. ...Being a Supplement to the Flora Telluriana. Philadelphia. (Sylva Tellur.)

Rafinesque, C. S. 1840. Autikon Botanikon. 3 parts. Philadelphia. [Parts paged consecutively.] (Autik. Bot.)

Ram, M. 1957. Morphological and embryological studies in the family Santalaceae. I. *Comandra umbellata* (Linnaeus) Nutt. Phytomorphology 7: 24–35.

Ramírez-Amezcua, Y. 2011. Relaciones Filogenéticas en *Argythamnia* (Euphorbiaceae) Sensu Lato. Tesis de Maestría. Universidad Michoacana de San Nicolás de Hidalgo.

Rancho Santa Ana Bot. Gard. Monogr., Bot. Ser. = Rancho Santa Ana Botanic Garden. Monographs. Botanical Series.

Raven, P. H. 1959. Documented chromosome numbers of plants. Madroño 15: 49–52.

Reed, C. F. 1964. A flora of the chrome and manganese ore piles at Canton, in the port of Baltimore, Maryland, and at Newport News, Virginia, with descriptions of genera and species new to the flora of eastern United States. Phytologia 10: 321–406.

Rehder, A. J. 1920. New species, varieties and combinations from the herbarium and the collections of the Arnold Arboretum. J. Arnold Arbor. 1: 191–210.

Rehder, A. J. 1940. Manual of Cultivated Trees and Shrubs Hardy in North America..., ed. 2. New York.

Reichenbach, H. G. L. 1837. Handbuch des natürlichen Pflanzensystems.... Dresden and Leipzig. (Handb. Nat. Pfl.-Syst.)

Reichenbach, H. G. L. 1841. Der deutsche Botaniker.... Erster Band. Das Herbarienbuch. Dresden and Leipzig. [Alt. title: Repertorium Herbarii....] (Deut. Bot. Herb.-Buch)

Reif, B. P. et al. 2015. Genetic structure and morphological differentiation of three western North American dwarf mistletoes (*Arceuthobium:* Viscaceae). Syst. Bot. 40: 191–207.

Reise Russ. Reich.—See: P. S. Pallas 1771–1776

Ren, H. et al. 2011. Phylogenetic analysis of the grape family (Vitaceae) based on the noncoding plastid *trn*C-*pet*N, *trn*H-*psb*A, and *trn*L-F sequences. Taxon 60: 629–637.

Rep. (Annual) Board Supervisors Louisiana State Seminary Learning Military Acad. = Annual Report of the Board of Supervisors of the Louisiana State Seminary of Learning and Military Academy....

Rep. (Annual) Missouri Bot. Gard. = Report (Annual) of the Missouri Botanical Garden.

Rep. Colorado R.—See: J. C. Ives 1861

Rep. U.S. Geogr. Surv., Wheeler—See: J. T. Rothrock 1878[1879]

Rep. U.S. Mex. Bound.—See: W. H. Emory 1857–1859

Repert. Bot. Syst.—See: W. G. Walpers 1842–1847

Repert. Spec. Nov. Regni Veg. = Repertorium Specierum Novarum Regni Vegetabilis.

Rev. Gén. Bot. = Revue Générale de Botanique.

Rev. Vitic. = Revue de Viticulture.

Revis. Gen. Pl.—See: O. Kuntze 1891–1898

Rhodora = Rhodora; Journal of the New England Botanical Club.

Richard, A. 1841–1851. Histoire Physique, Politique et Naturelle de l'Ile de Cuba...Botanique. Plantes Vasculaires. 1 vol. + atlas. Paris. (Hist. Phys. Cuba, Pl. Vasc.)

Richards, A. J., ed. 1978. The Pollination of Flowers by Insects. London.

Richardson, J. E. et al. 2000. A phylogenetic analysis of Rhamnaceae using *rbc*L and *trn*L-F plastid DNA sequences. Amer. J. Bot. 87: 1309–1324.

Richardson, J. E. et al. 2000b. A revision of the tribal classification of Rhamnaceae. Kew Bull. 55: 311–340.

Rickett, H. W. 1934. *Cornus amomum* and *Cornus candidissima*. Rhodora 36: 269–274.

Rickett, H. W. 1944b. *Cornus stolonifera* and *Cornus occidentalis*. Brittonia 5: 149–159.

Riina, R. et al. 2013. A worldwide molecular phylogeny and classification of the leafy spurges, *Euphorbia* subgenus *Esula* (Euphorbiaceae). Taxon 62: 316–342.

Robertson, K. R. 1975. The Oxalidaceae in the southeastern United States. J. Arnold Arbor. 56: 223–239.

Robertson, K. R. 1982. The genera of Olacaceae in the southeastern United States. J. Arnold Arbor. 63: 387–399.

Rodman, J. E. et al. 1998. Parallel evolution of glucosinolate biosynthesis inferred from congruent nuclear and plastid gene phylogenies. Amer. J. Bot. 85: 997–1006.

Roemer, J. J., J. A. Schultes, and J. H. Schultes. 1817[–1830]. Caroli a Linné...Systema Vegetabilium...Editione XV.... 7 vols. Stuttgart. (Syst. Veg.)

Rogers, C. M. 1963. Yellow-flowered species of *Linum* in eastern North America. Brittonia 15: 97–122.

Rogers, C. M. 1964. Yellow-flowered *Linum* (Linaceae) in Texas. Sida 1: 328–336.

Rogers, C. M. 1968. Yellow-flowered species of *Linum* in Central America and western North America. Brittonia 20: 107–135.

Rogers, C. M. 1975. Relationships of *Hesperolinon* and *Linum* (Linaceae). Madroño 23: 153–159.

Rogers, C. M. 1979. A new species of *Linum* from southern Texas and adjacent Mexico. Sida 8: 181–187.

Rogers, C. M. 1981. A revision of the genus *Linum* in southern Africa. Nordic J. Bot. 1: 711–722.

Rogers, C. M. 1981b. Notes on the genus *Linum* in Madagascar. Pl. Syst. Evol. 139: 155–157.

Rogers, C. M. 1982. The systematics of *Linum* sect. *Linopsis* (Linaceae). Pl. Syst. Evol. 140: 225–234.

Rogers, C. M. 1984. Linaceae. In: N. L. Britton et al., eds. 1905+. North American Flora.... 47+ vols. New York. Ser. 2, part 12, pp. 1–54.

Rogers, C. M. and R. Mildner. 1976. South American *Linum*, a summary. Rhodora 78: 761–766.

Rogers, C. M., R. Mildner, and B. D. Harris. 1972. Some additional chromosome numbers in the Linaceae. Brittonia 24: 313–316.

Rogers, D. J. 1951. A revision of *Stillingia* in the New World. Ann. Missouri Bot. Gard. 38: 207–259.

Rogers, D. J. 1963. Studies of *Manihot esculenta* Crantz and related species. Bull. Torrey Bot. Club 90: 43–54.

Rogers, D. J. and S. G. Appan. 1973. *Manihot, Manihotoides* (Euphorbiaceae). In: Organization for Flora Neotropica. 1968+. Flora Neotropica. 110+ nos. New York. No. 13.

Rogers, Z. S., D. L. Nickrent, and V. Malécot. 2008. *Staufferia* and *Pilgerina:* Two new arborescent genera of Santalaceae from Madagascar. Ann. Missouri Bot. Gard. 95: 391–404.

Rose, S. L. 1980. Mycorrhizal associations of some actinomycete nodulated nitrogen-fixing plants. Canad. J. Bot. 58: 1449–1454.

Rosenfeldt, S. and B. G. Galati. 2009. The structure of the stigma and the style of *Oxalis* spp. (Oxalidaceae). J. Torrey Bot. Soc. 136: 33–45.

Rossetto, M. et al. 2001. Intergeneric relationships in the Australian Vitaceae: New evidence from cpDNA analysis. Genet. Resources Crop Evol. 48: 307–314.

Rossetto, M. et al. 2002. Is the genus *Cissus* (Vitaceae) monophyletic? Evidence from plastid and nuclear ribosomal DNA. Syst. Bot. 27: 522–533.

Roth, A. W. 1788–1800. Tentamen Florae Germanicae.... 3 vols. in 5 parts. Leipzig. (Tent. Fl. Germ.)

Rothrock, J. T. 1878[1879]. Report upon United States Geographical Surveys West of the One Hundredth Meridian, in Charge of First Lieut. Geo. M. Wheeler.... Vol. 6—Botany. Washington. (Rep. U.S. Geogr. Surv., Wheeler)

Roxburgh, W. 1820–1824. Flora Indica; or Descriptions of Indian Plants.... 2 vols. Serampore. (Fl. Ind.)

Roxburgh, W. 1832. Flora Indica; or, Descriptions of Indian Plants. 3 vols. Serampore. (Fl. Ind. ed. 1832)

Royle, J. F. [1833–]1839[–1840]. Illustrations of the Botany and Other Branches of the Natural History of the Himalayan Mountains and of the Flora of Cashmere.... 1 vol. text and 1 vol. plates. London. (Ill. Bot. Himal. Mts.)

Rozario, S. A. 1995. Associations between mites and leaf domatia: Evidence from Bangladesh, south Asia. J. Trop. Ecol. 11: 99–108.

Ruiz López, H. 1799. Disertación Sobre la Raiz de la Ratánhia, Específico Singular Contra los Fluxos de Sangre. Madrid.

Rydberg, P. A. 1905. *Philadelphus*. In: N. L. Britton et al., eds. 1905+. North American Flora.... 47+ vols. New York. Vol. 22, pp. 162–175.

Rydberg, P. A. 1906. Flora of Colorado.... Fort Collins, Colo. (Fl. Colorado)

Rydberg, P. A. 1923. Flora of the Rocky Mountains and Adjacent Plains, ed. 2. New York. (Fl. Rocky Mts. ed. 2)

Rzedowski, J. and G. C. de Rzedowski, eds. 1991+. Flora del Bajío y de Regiones Adyacentes. 99+ fascs. Pátzcuaro.

Sag. Stor. Nat. Chili—See: G. I. Molina 1782

Sage, R. F., P.-A. Christin, and E. J. Edwards. 2011. The C4 plant lineages of planet Earth. J. Exp. Bot. 62: 3155–3169.

Sage, T. L. et al. 2011. The occurrence of C2 photosynthesis in *Euphorbia* subgenus *Chamaesyce* (Euphorbiaceae). J. Exp. Bot. 62: 3183–3195.

Sagra, R. de la. 1840–1855. Historia Fisica Politica y Natural de la Isla de Cuba.... 12 vols. Paris and Madrid. (Hist. Fis. Cuba)

Sagun, V. G., G. A. Levin, and P. C. van Welzen. 2010. Revision and phylogeny of *Acalypha* (Euphorbiaceae) in Malesia. Blumea 55: 21–60.

Sains Malaysiana = Sains Malaysiana; Malaysian Journal of Natural Sciences.

Salisbury, R. A. 1796. Prodromus Stirpium in Horto ad Chapel Allerton Vigentium.... London. (Prodr. Stirp. Chap. Allerton)

Salter, T. M. 1944. The Genus *Oxalis* in South Africa, a Taxonomic Revision. Cape Town.

Sargent, C. S. [1902–]1905–1913. Trees and Shrubs.... 2 vols. in parts. Boston and New York. (Trees & Shrubs)

Sargent, C. S., ed. 1913–1917. Plantae Wilsonianae: An Enumeration of the Woody Plants Collected in Western China for the Arnold Arboretum...During the Years 1907, 1908, and 1910 by E. H. Wilson. 3 vols. in parts. Cambridge, Mass. [Parts paged consecutively within vols.] (Pl. Wilson.)

Scharpf, R. F. 2002. *Viscum album* in California—An update. In: P. Angwin, ed. 2002. Proceedings of the Forty-eighth Western International Forest Disease Work Conference. 2000. August 14–18. Waikoloa, Hawaii. Redding and Kona. Pp. 46–47.

Scharpf, R. F. and W. McCartney. 1975. *Viscum album* in California—Its introduction, establishment and spread. Pl. Dis. Reporter 59: 257–262.

Schenk, J. J. 2009. A Systematic Monograph of *Mentzelia* Section *Bartonia* (Loasaceae): Phylogeny, Diversity, and Divergence Times. Ph.D. dissertation. Washington State University.

Schenk, J. J. and L. Hufford. 2010. Effects of substitution models on divergence time estimates: Simulations and an empirical study of model uncertainty using Cornales. Syst. Bot. 35: 578–592.

Schenk, J. J. and L. Hufford. 2011. Phylogeny and taxonomy of *Mentzelia* section *Bartonia* (Loasaceae). Syst. Bot. 36: 711–720.

Schirarend, C. and M. N. Olabi. 1994. Revision of the genus *Paliurus* Tourn. ex Mill. (Rhamnaceae). Bot. Jahrb. Syst. 116: 333–359.

Schirarend, C. and P. Hoffmann. 1993. Untersuchungen zur Blutenmorphologie der Gattung *Reynosia* Griseb. (Rhamnaceae). Flora, Morphol. Geobot. Ecophysiol. 188: 275–286.

Schkuhr, C. [1787–]1791–1803. Botanisches Handbuch.... 3 vols. Wittenberg. (Bot. Handb.)

Schmandt, J., G. R. North, and J. Clarkson. 2011. The Impact of Global Warming on Texas, ed. 2. Austin.

Schmelzer, G. H. and A. Gurib-Fakim, eds. 2008. Medicinal Plants. Wageningen. [Incl. CD-ROM database.]

Schneider, A. C. et al. 2016. Pleistocene radiation of the serpentine-adapted genus *Hesperolinon* and other divergence times in Linaceae (Malpighiales). Amer. J. Bot. 103: 221–232.

Schneider, C. K. [1904–]1906–1912. Illustriertes Handbuch der Laubholzkunde.... 2 vols. in 12 fascs. Jena. (Ill. Handb. Laubholzk.)

Schneider, E. L. 1982. Notes on the floral biology of *Nymphaea elegans* (Nymphaeaceae) in Texas. Aquatic Bot. 12: 197–200.

Schreber, J. C. 1789–1791. Caroli a Linné...Genera Plantarum.... 2 vols. Frankfurt am Main. (Gen. Pl.)

Schultes, J. A. and J. H. Schultes. 1822–1827. Mantissa.... Systematis Vegetabilium Caroli a Linné ex Editione Joan. Jac. Roemer...et Jos. Aug. Schultes.... 3 vols. Stuttgart. (Mant.)

Schumacher, H. C. F. 1827. Beskrivelse af Guinddiske Planter.... Copenhagen. (Beskr. Guin. Pl.)

Schweickerdt, H. G. 1937. An account of the South African species of *Tribulus* Tourn. ex Linn. Bothalia 3: 159–178.

Science = Science; an Illustrated Journal [later: a Weekly Journal Devoted to the Advancement of Science].

Scoggan, H. J. 1978–1979. The Flora of Canada. 4 parts. Ottawa. [Natl. Mus. Nat. Sci. Publ. Bot. 7.]

Scopoli, J. A. 1777. Introductio ad Historiam Naturalem.... Prague. (Intr. Hist. Nat.)

Seemann, B. 1852–1857. The Botany of the Voyage of H.M.S. Herald...during the Years 1845–51. 10 parts. London. [Parts paged consecutively.] (Bot. Voy. Herald)

Select. Stirp. Amer. Hist.—See: N. J. Jacquin 1763

Sert. Angl.—See: C.-L. L'Héritier de Brutelle 1788[1789–1792]

Sharsmith, H. K. 1961. The genus *Hesperolinon.* Univ. Calif. Publ. Bot. 32: 235–314.

Shea, M. M., P. M. Dixon, and R. R. Sharitz. 1993. Size differences, sex ratio, and spatial distribution of male and female water tupelo, *Nyssa aquatica* (Nyssaceae). Amer. J. Bot. 80: 26–30.

Sheahan, M. C. and M. W. Chase. 1996. A phylogenetic analysis of Zygophyllaceae R. Br. based on morphological, anatomical and *rbc*L DNA sequence data. Bot. J. Linn. Soc. 122: 279–300.

Sheahan, M. C. and M. W. Chase. 2000. Phylogenetic relationships within Zygophyllaceae based on DNA sequences of three plastid regions, with special emphasis on Zygophylloideae. Syst. Bot. 25: 371–384.

Shevock, J. R. 2012. Cornaceae. In: B. G. Baldwin et al., eds. 2012. The Jepson Manual: Vascular Plants of California, ed. 2. Berkeley. P. 664.

Shinners, L. H. 1964. *Cayratia japonica* (Vitaceae) in southeastern Louisiana: New to the United States. Sida 1: 384.

Shreve, F. and I. L. Wiggins. 1964. Vegetation and Flora of the Sonoran Desert. 2 vols. Stanford.

Sida = Sida; Contributions to Botany.

Siebold, P. F. von and J. G. Zuccarini. 1835[–1870]. Flora Japonica.... 2 vols. in 30 parts. Leiden. [Vols. paged and parted independently, plates numbered consecutively.] (Fl. Jap.)

Simmons, M. P. 2004b. Celastraceae. In: K. Kubitzki et al., eds. 1990+. The Families and Genera of Vascular Plants. 10+ vols. Berlin etc. Vol. 6, pp. 29–64.

Simmons, M. P. et al. 2001. Phylogeny of the Celastraceae inferred from phytochrome B gene sequence and morphology. Amer. J. Bot. 88: 313–325.

Simmons, M. P. et al. 2001b. Phylogeny of the Celastraceae inferred from 26S nuclear ribosomal DNA, phytochrome B, *rbc*L, *atp*B, and morphology. Molec. Phylogen. Evol. 19: 353–366.

Simmons, M. P. et al. 2008. Phylogeny of the Celastreae (Celastraceae) and the relationships of *Catha edulis* (qat) inferred from morphological characters and nuclear and plastid genes. Molec. Phylogen. Evol. 48: 745–757.

Simpson, B. B. 1989. Krameriaceae. In: Organization for Flora Neotropica. 1968+. Flora Neotropica. 110+ nos. New York. No. 49, pp. 1–108.

Simpson, B. B. 1991. The past and present uses of rhatany (*Krameria,* Krameriaceae). Econ. Bot. 45: 397–409.

Simpson, B. B. 2007. Krameriaceae. In: K. Kubitzki et al., eds. 1990+. The Families and Genera of Vascular Plants. 10+ vols. Berlin etc. Vol. 9, pp. 207–212.

Simpson, B. B. 2013. *Krameria bicolor,* the correct name for *Krameria grayi* (Krameriaceae). Phytoneuron 2013-62: 1.

Simpson, B. B. et al. 2004. Species relationships in *Krameria* (Krameriaceae) based on ITS sequences and morphology: Implications for character utility and biogeography. Syst. Bot. 29: 97–108.

Simpson, B. B., J. L. Neff, and D. S. Seigler. 1977. *Krameria,* free fatty acids and oil-collecting bees. Nature 267: 150–151.

Sketch Bot. S. Carolina—See: S. Elliott [1816–]1821–1824

Sleumer, H. 1935. Olacaceae. In: H. G. A. Engler et al., eds. 1924+. Die natürlichen Pflanzenfamilien..., ed. 2. 26+ vols. Leipzig and Berlin. Vol. 16b, pp. 5–32.

Sleumer, H. 1984. Olacaceae. In: Organization for Flora Neotropica. 1968+. Flora Neotropica. 110+ nos. New York. No. 38, pp. 1–159.

Small, J. K. 1903. Flora of the Southeastern United States.... New York. (Fl. S.E. U.S.)

Small, J. K. 1913. Flora of the Southeastern United States..., ed. 2. New York. (Fl. S.E. U.S. ed. 2)

Small, J. K. 1913b. Flora of Miami.... New York. (Fl. Miami)

Small, J. K. 1913c. Flora of the Florida Keys.... New York. (Fl. Florida Keys)

Small, J. K. 1913d. Florida Trees: A Handbook of the Native and Naturalized Trees of Florida. New York. (Florida Trees)

Small, J. K. 1933. Manual of the Southeastern Flora, Being Descriptions of the Seed Plants Growing Naturally in Florida, Alabama, Mississippi, Eastern Louisiana, Tennessee, North Carolina, South Carolina and Georgia. New York. (Man. S.E. Fl.)

Small, J. K. and P. A. Rydberg. 1905. Hydrangeaceae. In: N. L. Britton et al., eds. 1905+. North American Flora.... 47+ vols. New York. Vol. 22, pp. 159–178.

Smith, A. C. 1940. The American species of Hippocrateaceae. Brittonia 3: 356–367.

Smith, B. S. 1992. Lucy Bishop Millington, nineteenth-century botanist: Her life and letters to Charles Horton Peck, State Botanist of New York. Huntia 8: 111–153.

Smith, F. H. and E. C. Smith. 1942. Anatomy of the inferior ovary of *Darbya.* Amer. J. Bot. 29: 464–471.

Smith, F. H. and E. C. Smith. 1943. Floral anatomy of the Santalaceae and related forms. Oregon State Monogr., Stud. Bot. 5: 1–93.

Smith, N. P. et al., eds. 2004. Flowering Plants of the Neotropics. Princeton.

Smithsonian Contr. Knowl. = Smithsonian Contributions to Knowledge.

Soejima, A. and J. Wen. 2006. Phylogenetic analysis of the grape family (Vitaceae) based on three chloroplast markers. Amer. J. Bot. 93: 278–287.

Soltis, D. E. et al. 2000. Angiosperm phylogeny inferred from 18S rDNA, *rbc*L, and *atp*B sequences. Bot. J. Linn. Soc. 133: 381–461.

Soltis, D. E., Xiang Q. Y., and L. Hufford. 1995. Relationships and evolution of Hydrangeaceae based on *rbc*L sequence data. Amer. J. Bot. 82: 504–514.

Sonderb. Naturwiss. Vereins Hamburg = Sonderbände des Naturwissenschaftlichen Vereins in Hamburg.

Sosa, V. and M. W. Chase. 2003. Phylogenetics of Crossosomataceae based on *rbc*L sequence data. Syst. Bot. 28: 96–105.

SouthE. Naturalist = Southeastern Naturalist.

SouthW. Naturalist = Southwestern Naturalist.

Sp. Pl.—See: C. Linnaeus 1753; C. L. Willdenow et al. 1797–1830

Sp. Pl. ed. 2—See: C. Linnaeus 1762–1763

Spic. Fl. Rumel.—See: A. H. R. Grisebach 1843–1845[–1846]

Spongberg, S. A. 1972. The genera of Saxifragaceae in the southeastern United States. J. Arnold Arbor. 53: 409–498.

Sprengel, K. [1824–]1825–1828. Caroli Linnaei...Systema Vegetabilium. Editio Decima Sexta.... 5 vols. Göttingen. [Vol. 4 in 2 parts paged independently; vol. 5 by A. Sprengel.] (Syst. Veg.)

Springer, Y. P. 2009. Do extreme environments provide a refuge from pathogens? A phylogenetic test using serpentine flax. Amer. J. Bot. 96: 2010–2021.

St. John, H. 1921. A critical revision of *Hydrangea arborescens*. Rhodora 23: 203–208.

St.-Hilaire, A. 1824[–1826]. Histoire des Plantes les Plus Remarquables du Brésil.... 6 parts. Paris. [Parts paged and plates numbered consecutively.] (Hist. Pl. Remarq. Brésil)

St.-Hilaire, A., J. Cambessèdes, and A. H. L. de Jussieu. 1824[–1833]. Flora Brasiliae Meridionalis.... 3 vols. in 24 parts. Paris. [Vols. paged independently, parts and plates numbered consecutively.] (Fl. Bras. Merid.)

Stace, C. A. 2010b. New Flora of the British Isles, ed. 3. Cambridge.

Stafleu, F. A. et al. 1992–2009. Taxonomic Literature: A Selective Guide to Botanical Publications and Collections with Dates, Commentaries and Types. Supplement. 8 vols. Königstein.

Stafleu, F. A. and R. S. Cowan. 1976–1988. Taxonomic Literature: A Selective Guide to Botanical Publications and Collections with Dates, Commentaries and Types, ed. 2. 7 vols. Utrecht etc.

Stauffer, H. U. 1957. Santalales-Studien I. Zur Stellung der Gattung *Okoubaka* Pellegrin et Normand. Ber. Schweiz. Bot. Ges. 67: 422–427.

Stauffer, H. U. 1961. Santalales-Studien VII. Südamerikanische Santalaceae I. *Acanthosyris, Cervantesia* und *Jodina.* Vierteljahrsschr. Naturf. Ges. Zürich 106: 406–412.

Steenis, C. G. G. J. van, ed. 1948+. Flora Malesiana.... Series I. Spermatophyta. 18+ vols., some in parts. Djakarta and Leiden.

Steinmann, V. W. 2003. The submersion of *Pedilanthus* into *Euphorbia* (Euphorbiaceae). Acta Bot. Mex. 65: 44–50.

Steinmann, V. W. 2007. Phyllanthaceae. In: J. Rzedowski and G. C. de Rzedowski, eds. 1991+. Flora del Bajío y de Regiones Adyacentes. 99+ fascs. Pátzcuaro. Fasc. 152, pp. 1–35.

Steinmann, V. W. and R. S. Felger. 1997. The Euphorbiaceae of Sonora, Mexico. Aliso 16: 1–71.

Steinmann, V. W. and G. A. Levin. 2011. *Acalypha herzogiana* (Euphorbiaceae), the correct name for an intriguing and commonly cultivated species. Brittonia 63: 500–504.

Steinmann, V. W. and J. M. Porter. 2002. Phylogenetic relationships in Euphorbieae (Euphorbiaceae) based on ITS and *ndh*F sequence data. Ann. Missouri Bot. Gard. 89: 453–490.

Stern, W. L. 1978. Comparative anatomy and systematics of woody Saxifragaceae. *Hydrangea.* Bot. J. Linn. Soc. 76: 83–113.

Stuppy, W. 1996. Systematische Morphologie und Anatomie der Samen der biovulaten Euphorbiaceen. Doctoral thesis. Universität Kaiserslautern.

Stuppy, W. et al. 1999. A revision of the genera *Aleurites* J. R. Forst. & G. Forst., *Reutealis* Airy Shaw and *Vernicia* Lour. (Euphorbiaceae). Blumea 44: 73–98.

Su, H. J. et al. 2015. Phylogenetic relationships of Santalales with insights into the origins of holoparasitic Balanophoraceae. Taxon 64: 491–506.

Sun, B.-L. et al. 2009. Cryptic dioecy in *Nyssa yunnanensis* (Nyssaceae), a critically endangered species from tropical eastern Asia. Ann. Missouri Bot. Gard. 96: 672–684.

Sundell, E. et al. 1999. Noteworthy vascular plants from Arkansas. Sida 18: 877–887.

Suppl. Pl.—See: C. Linnaeus f. 1781[1782]

Sussenguth, K. 1953. Vitaceae. In: H. G. A. Engler et al., eds. 1924+. Die natürlichen Pflanzenfamilien..., ed. 2. 26+ vols. Leipzig and Berlin. Vol. 20d, pp. 174–333.

Svensk Bot. Tidskr. = Svensk Botanisk Tidskrift Utgifven af Svenska Botaniska Föreningen.

Swain, M. D. and T. Beer. 1977. Explosive seed dispersal in *Hura crepitans* L. (Euphorbiaceae). New Phytol. 78: 695–708.

Swartz, O. P. 1788. Nova Genera & Species Plantarum seu Prodromus.... Stockholm, Uppsala, and Åbo. (Prodr.)

Swartz, O. P. 1797–1806. Flora Indiae Occidentalis.... 3 vols. Erlangen and London. (Fl. Ind. Occid.)

Swink, F. and G. S. Wilhelm. 1994. Plants of the Chicago Region: An Annotated Checklist of the Vascular Flora..., ed. 4. Indianapolis.

Sylva Tellur.—See: C. S. Rafinesque 1838b

Symb. Antill.—See: I. Urban 1898–1928

Symb. Bot.—See: M. Vahl 1790–1794

Symb. Sin.—See: H. Handel-Mazzetti 1929–1937

Syn. Fl. N. Amer.—See: A. Gray et al. 1878–1897

Syn. Pl.—See: C. H. Persoon 1805–1807
Syn. Pl. Succ.—See: A. H. Haworth 1812
Syst. Biodivers. = Systematics and Biodiversity.
Syst. Bot. = Systematic Botany; Quarterly Journal of the American Society of Plant Taxonomists.
Syst. Bot. Monogr. = Systematic Botany Monographs; Monographic Series of the American Society of Plant Taxonomists.
Syst. Nat.—See: J. F. Gmelin 1791[–1792]
Syst. Nat. ed. 10—See: C. Linnaeus 1758[–1759]
Syst. Nat. ed. 12—See: C. Linnaeus 1766[–1768]
Syst. Veg.—See: J. J. Roemer et al. 1817[–1830]; K. Sprengel [1824–]1825–1828
Syst. Veg. ed. 14—See: J. A. Murray 1784
Tabuti, J. R. S. 2008. *Flueggea virosa.* In: G. H. Schmelzer and A. Gurib-Fakim, eds. 2008. Medicinal Plants. Wageningen. Pp. 305–308.
Takhtajan, A. L. 1997. Diversity and Classification of Flowering Plants. New York.
Taxon = Taxon; Journal of the International Association for Plant Taxonomy.
Tent. Fl. Germ.—See: A. W. Roth 1788–1800
Thaiszia = Thaiszia; Journal of Botany.
Thoday, D. 1926. The contractile roots of *Oxalis incarnata.* Ann. Bot. (Oxford) 40: 571–583.
Thomas, R. D. and C. M. Allen. 1993–1998. Atlas of the Vascular Flora of Louisiana. 3 vols. Baton Rouge.
Thompson, H. J. 1960. A genetic approach to the taxonomy of *Mentzelia lindleyi* and *M. crocea* (Loasaceae). Brittonia 12: 81–93.
Thompson, H. J. and W. R. Ernst. 1967. Floral biology and systematics of *Eucnide* (Loasaceae). J. Arnold Arbor. 48: 56–88.
Thompson, H. J. and H. Lewis. 1953. Chromosome numbers in *Mentzelia* (Loasaceae). Madroño 13: 102–107.
Thompson, H. J. and A. M. Powell. 1981. Loasaceae of the Chihuahuan Desert region. Phytologia 49: 16–32.
Thompson, V. E. and B. E. Mahall. 1983. Host specificity by a mistletoe, *Phoradendron villosum* (Nutt.) Nutt. subsp. *villosum,* on three oak species in California. Bot. Gaz. 144: 124–131.
Thorne, R. F. and R. Scoggin. 1978. *Forsellesia* Greene (*Glossopetalon* Gray), a third genus in the Crossosomataceae, Rosineae, Rosales. Aliso 9: 171–178.
Thunberg, C. P. 1781–1801. Nova Genera Plantarum.... 16 diss. Uppsala. [Paged consecutively.] (Nov. Gen. Pl.)
Tidestrom, I. and T. Kittell. 1941. A Flora of Arizona and New Mexico.... Washington. (Fl. Ariz. New Mex.)
Tieghem, P. van. 1896. Sur les phanérogames à ovule sans nucelle, formant le groupe des innucellées ou Santalinées. Bull. Soc. Bot. France 43: 543–577.
Tobe, H. and P. H. Raven. 1993. Embryology of *Acanthothamnus, Brexia* and *Canotia* (Celastraceae): A comparison. Bot. J. Linn. Soc. 112: 17–32.
Tokuoka, T. 2007. Molecular phylogenetic analysis of Euphorbiaceae sensu stricto based on plastid and nuclear DNA sequences and ovule and seed character evolution. J. Pl. Res. 120: 511–522.
Tokuoka, T. and H. Tobe. 1999. Embryology of tribe Drypeteae, an enigmatic taxon of Euphorbiaceae. Pl. Syst. Evol. 215: 189–208.
Tokuoka, T. and H. Tobe. 2001. Ovules and seeds in subfamily Phyllanthoideae (Euphorbiaceae): Structure and systematic implications. J. Pl. Res. 114: 75–92.
Tokuoka, T. and H. Tobe. 2006. Phylogenetic analyses of Malpighiales using plastid and nuclear DNA sequences, with particular reference to the embryology of Euphorbiaceae sens. str. J. Pl. Res. 119: 599–616.
Tomlinson, P. B. 1980. The Biology of Trees Native to Tropical Florida. Allston, Mass.
Tomlinson, P. B. 1986. The Botany of Mangroves. Cambridge.
Torrey, J. and A. Gray. 1838–1843. A Flora of North America.... 2 vols. in 7 parts. New York, London, and Paris. (Fl. N. Amer.)
Torreya = Torreya; a Monthly Journal of Botanical Notes and News.
Tortosa, R. D. 1993. Revisión del género *Adolphia* (Rhamnaceae: Colletieae). Darwiniana 32: 185–189.
Trail & Landscape = Trail and Landscape; Publication Concerned with Natural History and Conservation.
Trans. Acad. Sci. St. Louis = Transactions of the Academy of Science of St. Louis.
Trans. Albany Inst. = Transactions of the Albany Institute.
Trans. Amer. Philos. Soc. = Transactions of the American Philosophical Society Held at Philadelphia for Promoting Useful Knowledge.
Trans. Connecticut Acad. Arts = Transactions of the Connecticut Academy of Arts and Sciences.
Trans. Kansas Acad. Sci. = Transactions of the Kansas Academy of Science.
Trans. Roy. Soc. South Australia = Transactions of the Royal Society of South Australia.
Travels Carolina—See: W. Bartram 1791
Trees & Shrubs—See: C. S. Sargent [1902–]1905–1913
Trees Shrubs Southw. Deserts—See: L. D. Benson and R. A. Darrow 1954
Trelease, W. 1916. The Genus *Phoradendron:* A Monographic Revision. Urbana. (Phoradendron)
Trias-Blasi, A., J. A. N. Parnell, and T. R. Hodkinson. 2012. Multi-gene region phylogenetic analysis of the grape family (Vitaceae). Syst. Bot. 37: 941–950.
Tröndle, D. et al. 2010. Molecular phylogeny of the genus *Vitis* (Vitaceae) based on plastid markers. Amer. J. Bot. 97: 1168–1178.
Trop. Woods = Tropical Woods....
Trudy Bot. Inst. Akad. Nauk S.S.S.R, Ser. 1, Fl. Sist. Vyssh. Rast. = Trudy Botanicheskogo Instituta Akademii Nauk S S S R. Ser. 1, Flora i Sistematika Vysshikh Rastenii.
Tucker, G. C. 1986. The genera of Elatinaceae in the southeastern United States. J. Arnold Arbor. 67: 471–483.
Tucker, G. C. 1995. The Vascular Flora of Southeastern Connecticut: An Annotated Checklist. New Haven.
Turner, B. L. 1958. Chromosome numbers of the genus *Krameria:* Evidence for familial status. Rhodora 60: 101–106.
Turner, B. L. 1994d. Regional variation in the North American elements of *Oxalis corniculata* (Oxalidaceae). Phytologia 77: 1–7.
Turner, B. L. 2001. Taxonomic revision of the genus *Fendlera* (Hydrangeaceae). Lundellia 4: 1–11.

Turner, B. L. 2001b. Biological status of *Argythamnia laevis* (Euphorbiaceae). Sida 19: 621–622.

Turner, B. L. 2006. Species of *Philadelphus* (Hydrangeaceae) from trans-Pecos Texas. Lundellia 9: 34–40.

Turner, B. L. 2011. Persistence of the weed *Euphorbia exigua* in Texas. Phytoneuron 2011-20: 1–3.

Turner, B. L. 2012b. Taxonomy of *Eucnide bartonioides* (Loasaceae) complex in Texas. Phytologia 94: 305–309.

Turner, B. L. et al. 2003. Atlas of the Vascular Plants of Texas. 2 vols. Fort Worth. [Sida Bot. Misc. 24.]

Tutin, T. G. 1968. *Mercurialis.* In: T. G. Tutin et al., eds. 1964–1980. Flora Europaea. 5 vols. Cambridge. Vol. 2, p. 212.

Tutin, T. G. et al., eds. 1964–1980. Flora Europaea. 5 vols. Cambridge.

Tutin, T. G. et al., eds. 1993+. Flora Europaea, ed. 2. 1+ vol. Cambridge and New York.

Tyrl, R. J. et al. 2010. Keys and Descriptions of the Vascular Plants of Oklahoma. Noble.

U.S.D.A. Bur. Pl. Industr. Bull. = U S Department of Agriculture. Bureau of Plant Industry. Bulletin.

Univ. Calif. Publ. Bot. = University of California Publications in Botany.

University of Chicago Press. 1993. The Chicago Manual of Style, ed. 14. Chicago.

Uno, G. E. 1984. The role of persistent sepals in the reproductive biology in *Linum pratense* (Linaceae). SouthW. Naturalist 29: 429–434.

Urban, I., ed. 1898–1928. Symbolae Antillanae seu Fundamenta Florae Indiae Occidentalis.... 9 vols. Berlin etc. (Symb. Antill.)

Urban, I. and E. Gilg. 1900. Monographia Loasacearum. Nova Acta Acad. Caes. Leop.-Carol. German. Nat. Cur. 76.

Urtecho, R. J. 1996. A Taxonomic Study of the Mexican Species of *Tragia* (Euphorbiaceae). Ph.D. dissertation. University of California, Davis.

Vahl, M. 1790–1794. Symbolae Botanicae.... 3 vols. Copenhagen. (Symb. Bot.)

Vahl, M. 1796[1797]–1807. Eclogae Americanae seu Descriptiones Plantarum Praesertim Americae Meridionalis.... 3 parts. Copenhagen. (Eclog. Amer.)

Van Bruggen, T. 1986. Santalaceae. In: Great Plains Flora Association. 1986. Flora of the Great Plains. Lawrence, Kans. Pp. 531–532.

van den Oever, L. 1984. Comparative wood anatomy of the Olacaceae. In: Pacific Regional Wood Anatomy Conference. 1984. Proceedings of Pacific Regional Wood Anatomy Conference: October 1–7, 1984, Tsukuba, Ibaraki, Japan. Ibaraki. Pp. 177–178.

Van Devender, T. R. 1990. Late Quaternary vegetation and climate of the Sonoran Desert, United States and Mexico. In: J. L. Betancourt et al., eds. 1990. Packrat Middens: The Last 40,000 Years of Biotic Change. Tucson. Pp. 134–165.

van Ee, B. W. et al. 2006. Phylogeny and biogeography of *Croton alabamensis* (Euphorbiaceae), a rare shrub from Texas and Alabama, using DNA sequence and AFLP data. Molec. Ecol. 15: 2735–2751.

van Ee, B. W. and P. E. Berry. 2009. The circumscription of *Croton* section *Crotonopsis* (Euphorbiaceae), a North American endemic. Harvard Pap. Bot. 14: 61–70.

van Ee, B. W. and P. E. Berry. 2010. Taxonomy and phylogeny of *Croton* section *Heptalon* (Euphorbiaceae). Syst. Bot. 35: 151–167.

van Ee, B. W. and P. E. Berry. 2010b. Typification notes for *Croton* (Euphorbiaceae). Harvard Pap. Bot. 15: 73–84.

van Ee, B. W., P. E. Berry, and S. Ginzbarg. 2009. An assessment of the varieties of *Croton glandulosus* (Euphorbiaceae) in the United States. Harvard Pap. Bot. 14: 45–59.

van Ee, B. W., R. Riina, and P. E. Berry. 2011. A revised infrageneric classification and molecular phylogeny of New World *Croton* (Euphorbiaceae). Taxon 60: 791–823.

Van Rensselaer, M. and H. McMinn. 1942. *Ceanothus.* Santa Barbara. (Ceanothus)

Varied. Ci. = Variedades de Ciencias, Literatura y Artes.

Vasc. Pl. Illinois—See: G. N. Jones and G. D. Fuller 1955

Vasc. Pl. Pacif. N.W.—See: C. L. Hitchcock et al. 1955–1969

Vasek, F. C. 1980. Creosote bush: Long-lived clones in the Mojave Desert. Amer. J. Bot. 67: 246–255.

Veldkamp, J. F. 1971. Oxalidaceae. In: C. G. G. J. van Steenis, ed. 1948+. Flora Malesiana.... Series I. Spermatophyta. 18+ vols., some in parts. Djakarta and Leiden. Vol. 7, pp. 151–178.

Ventenat, É. P. 1803–1804[–1805]. Jardin de la Malmaison.... 2 vols. Paris. [Plates numbered consecutively.] (Jard. Malmaison)

Verh. Batav. Genootsch. Kunst. = Verhandelingen van het Bataviaasch Genootschap van Kunsten en Wetenschappen.

Vernon, L. P. et al. 1985. A toxic thionin from *Pyrularia pubera:* Purification, properties, and amino acid sequence. Arch. Biochem. Biophys. 238: 18–29.

Viala, P. 1889. Une Mission Viticole en Amérique.... Montpellier. (Mission Vitic. Amér.)

Vidal-Russell, R. and D. L. Nickrent. 2008. The first mistletoes: Origins of aerial parasitism in Santalales. Molec. Phylogen. Evol. 47: 523–527.

Vierteljahrsschr. Naturf. Ges. Zürich = Vierteljahrsschrift der Naturforschenden Gesellschaft in Zürich.

Vigne Amér. Vitic. Eur. = La Vigne Américaine (et la Viticulture en Europe); sa Culture, son Avenir en Europe.

Vincent, M. A. 2002. *Eucommia ulmoides* (hardy rubber-tree, Eucommiaceae) as an escape in North America. Michigan Bot. 41: 137–141.

Vincent, M. A. and A. W. Cusick. 1998. New records of alien species in the Ohio vascular flora. Ohio J. Sci. 98(2): 10–17.

Vines, R. A. 1960. Trees, Shrubs, and Woody Vines of the Southwest. Austin.

Vinson, S. B., G. W. Frankie, and H. J. Williams. 1996. Chemical ecology of bees of the genus *Centris.* Florida Entomol. 79: 102–129.

Vitis baileyana—See: T. V. Munson 1893

Vogel, S. 1978. Evolutionary shifts from reward to deception in pollen flowers. In: A. J. Richards, ed. 1978. The Pollination of Flowers by Insects. London. Pp. 89–96.

Vorontsova, M. S. et al. 2007. Molecular phylogenetics of tribe Porantheréae (Phyllanthaceae; Euphorbiaceae sensu lato). Amer. J. Bot. 94: 2026–2040.

Vorontsova, M. S. and P. Hoffmann. 2008. A phylogenetic classification of tribe Poranthereae (Phyllanthaceae; Euphorbiaceae sensu lato). Kew Bull. 63: 41–59.

Voss, C. et al. 2006. Identification and characterization of riproximin, a new type II ribosome-inactivating protein with antineoplastic activity from *Ximenia americana*. F. A. S. E. B. J. 20: 1194–1196.

Voss, E. G. 1972–1996. Michigan Flora.... 3 vols. Bloomfield Hills and Ann Arbor.

Voss, E. G. and A. A. Reznicek. 2012. Field Manual of Michigan Flora. Ann Arbor.

W. Amer. Sci. = West American Scientist.

W. J. Med. Phys. Sci. = The Western Journal of the Medical and Physical Sciences.

W. N. Amer. Naturalist = Western North American Naturalist.

W. Rev. & Misc. Mag. = Western Review and Miscellaneous Magazine.

Waldstein, F. and P. Kitaibel. [1799–]1802–1812. Descriptiones et Icones Plantarum Rariorum Hungariae. 3 vols. in parts. Vienna. (Descr. Icon. Pl. Hung.)

Walpers, W. G. 1842–1847. Repertorium Botanices Systematicae.... 6 vols. Leipzig. (Repert. Bot. Syst.)

Walter, T. 1788. Flora Caroliniana, Secundum Systema Vegetabilium Perillustris Linnaei Digesta.... London. (Fl. Carol.)

Wan, Y. et al. 2013. A phylogenetic analysis of the grape genus (*Vitis* L.) reveals broad reticulation and concurrent diversification during Neogene and Quaternary climate change. B. M. C. Evol. Biol. 13: 141.

Wangerin, W. 1910. *Cornus*. In: H. G. A. Engler, ed. 1900–1953. Das Pflanzenreich.... 107 vols. Berlin. Vol. 41[IV,229], pp. 43–70.

War Department [U.S.]. 1855–1860. Reports of Explorations and Surveys, to Ascertain the Most Practicable and Economical Route for a Railroad from the Mississippi River to the Pacific Ocean. Made under the Direction of the Secretary of War, in 1853[–1856].... 12 vols. in 13. Washington. (Pacif. Railr. Rep.)

Ward, D. B. 2004. Keys to the flora of Florida–9, *Oxalis* (Oxalidaceae). Phytologia 86: 32–41.

Ward, D. B. 2006b. Keys to the flora of Florida—13, *Vitis* (Vitaceae). Phytologia 88: 216–223.

Ward, D. B. and A. K. Gholson. 1987. The hidden abundance of *Lepuropetalon spathulatum* (Saxifragaceae) and its first reported occurrence in Florida. Castanea 52: 59–66.

Warnock, B. H. and M. C. Johnston. 1960. The genus *Savia* (Euphorbiaceae) in extreme western Texas. SouthW. Naturalist 5: 1–6.

Wasmann J. Biol. = Wasmann Journal of Biology.

Waterfall, U. T. 1959. A revision of *Eucnide*. Rhodora 61: 231–243.

Watson, A. K., ed. 1985. Leafy Spurge. Champaign.

Watson, S. 1871. United States Geological Expolration [sic] of the Fortieth Parallel. Clarence King, Geologist-in-charge. [Vol. 5] Botany. By Sereno Watson.... Washington. [Botanical portion of larger work by C. King.] [Botany (Fortieth Parallel)]

Watson, W. 1903. A hardy rubber-yielding tree. Gard. Chron. 842: 104.

Watsonia = Watsonia; Journal of the Botanical Society of the British Isles.

Weakley, A. E. 2002. Evolutionary Relationships within the Genus *Philadelphus* (Hydrangeaceae): A Molecular Phylogenetic and Biogeographic Analysis. M.A. thesis. University of North Carolina.

Weakley, A. S. 2010. Flora of the Southern and Mid-Atlantic States. Working Draft of 8 March 2010. Chapel Hill.

Weakley, A. S. 2012. Flora of the Southern and Mid-Atlantic States. Working Draft of 30 November 2012. Chapel Hill.

Weakley, A. S., J. C. Ludwig, and J. F. Townsend. 2012. Flora of Virginia. Fort Worth.

Web Ecol. = Web Ecology.

Webb, D. A. 1993. Hydrangeaceae. In: T. G. Tutin et al., eds. 1993+. Flora Europaea, ed. 2. 1+ vol. Cambridge and New York. Vol. 1, p. 460.

Webster, G. L. 1955. Studies of the Euphorbiaceae, Phyllanthoideae. I. Taxonomic notes on the West Indian species of *Phyllanthus*. Contr. Gray Herb. 176: 45–63.

Webster, G. L. 1956–1958. A monographic study of the West Indian species of *Phyllanthus*. J. Arnold Arbor. 37: 91–122, 217–268, 340–359; 38: 51–80, 170–198, 295–373; 39: 49–100, 111–212.

Webster, G. L. 1967. The genera of Euphorbiaceae in the southeastern United States. J. Arnold Arbor. 48: 303–361, 363–430.

Webster, G. L. 1970. A revision of *Phyllanthus* (Euphorbiaceae) in the continental United States. Brittonia 22: 44–76.

Webster, G. L. 1975. Systematics of photosynthetic carbon fixation pathways in *Euphorbia*. Taxon 24: 27–33.

Webster, G. L. 1983. A botanical Gordian knot: The case of *Ateramnus* and *Gymnanthes* (Euphorbiaceae). Taxon 32: 304–305.

Webster, G. L. 1984. A revision of *Flueggea* (Euphorbiaceae). Allertonia 3: 259–312.

Webster, G. L. 1993. A provisional synopsis of the sections of the genus *Croton* (Euphorbiaceae). Taxon 42: 793–823.

Webster, G. L. 1993b. Euphorbiaceae. In: J. C. Hickman, ed. 1993. The Jepson Manual. Higher Plants of California. Berkeley, Los Angeles, and London. Pp. 567–577.

Webster, G. L. 1994. Synopsis of the genera and suprageneric taxa of Euphorbiaceae. Ann. Missouri Bot. Gard. 81: 33–144.

Webster, G. L. 1994b. Classification of the Euphorbiaceae. Ann. Missouri Bot. Gard. 81: 3–32.

Webster, G. L. 2001. Synopsis of *Croton* and *Phyllanthus* (Euphorbiaceae) in western tropical Mexico. Contr. Univ. Michigan Herb. 23: 353–388.

Webster, G. L. 2014. Euphorbiaceae. In: K. Kubitzki et al., eds. 1990+. The Families and Genera of Vascular Plants. 10+ vols. Berlin etc. Vol. 11, pp. 51–216.

Webster, G. L. and W. S. Armbruster. 1991. A synopsis of the neotropical species of *Dalechampia*. Bot. J. Linn. Soc. 105: 137–177.

Webster, G. L. and K. I. Miller. 1963. The genus *Reverchonia* (Euphorbiaceae). Rhodora 65: 193–207.

Weed Technol. = Weed Technology; a Journal of the Weed Science Society of America.

Weigend, M. 2004. Loasaceae. In: K. Kubitzki et al., eds. 1990+. The Families and Genera of Vascular Plants. 10+ vols. Berlin etc. Vol. 6, pp. 239–254.

Weigend, M. 2007b. Systematics of the genus *Mentzelia* (Loasaceae) in South America. Ann. Missouri Bot. Gard. 94: 655–689.

Weigend, M., J. Kufer, and A. A. Mueller. 2000. Phytochemistry and the systematics and ecology of Loasaceae and Gronoviaceae. Amer. J. Bot. 87: 1202–1210.

Weller, S. G. and M. F. Denton. 1976. Cytogeographic evidence for the evolution of distyly from tristyly in the North American species of *Oxalis* section *Ionoxalis*. Amer. J. Bot. 63: 120–125.

Wells, P. V. and J. H. Hunziker. 1977. Origin of the creosote bush *(Larrea)* deserts of southwestern North America. Ann. Missouri Bot. Gard. 63: 843–861.

Welsh, S. L. et al., eds. 1993. A Utah Flora, ed. 2. Provo.

Welsh, S. L. et al., eds. 2003. A Utah Flora, ed. 3. Provo.

Wen, J. 2007. Vitaceae. In: K. Kubitzki et al., eds. 1990+. The Families and Genera of Vascular Plants. 10+ vols. Berlin etc. Vol. 9, pp. 467–479.

Wen, J. et al. 2007. Phylogeny of Vitaceae based on the nuclear *GAI1* gene sequences. Canad. J. Bot. 85: 731–745.

Wen, J. et al. 2013. Diversity and evolution of Vitaceae in the Philippines. Philipp. J. Sci. 142: 223–244.

Wen, J. et al. 2013b. Transcriptome sequences resolve deep relationships of the grape family. PLOS ONE, 17 Sept, DOI: 10.1371/journal.pone.0074394.

Wen, J., J. K. Boggan, and Z. L. Nie. 2014. Synopsis of *Nekemias* Raf., a segregate genus from *Ampelopsis* Michx. (Vitaceae) disjunct between eastern/southeastern Asia and eastern North America, with ten new combinations. PhytoKeys 42: 11–19.

Wen, J. and T. F. Stuessy. 1993. The phylogeny and biogeography of *Nyssa* (Cornaceae). Syst. Bot. 18: 68–79.

Werth, C. R., W. V. Baird, and L. J. Musselman. 1979. Root parasitism in *Schoepfia* Schreb. (Olacaceae). Biotropica 11: 140–143.

Wheeler, L. C. 1941. *Euphorbia* subgenus *Chamaesyce* in Canada and the United States exclusive of southern Florida. Rhodora 43: 97–154, 168–205, 223–286.

Wheeler, L. C. 1943. History and orthography of the celastraceous genus "*Paxistima*" Rafinesque. Amer. Midl. Naturalist 29: 792–795.

Wherry, E. T., J. M. Fogg, Jr., and H. A. Wahl. 1979. Atlas of the Flora of Pennsylvania. Philadelphia.

Wiegand, K. M. 1925. *Oxalis corniculata* and its relatives in North America. Rhodora 27: 113–124, 133–139.

Wiens, D. 1964. Revision of the acataphyllous species of *Phoradendron*. Brittonia 16: 11–54.

Wiens, D. 1975. Chromosome numbers in African and Madagascan Loranthaceae and Viscaceae. Bot. J. Linn. Soc. 71: 295–310.

Wiens, D. and B. A. Barlow. 1979. Translocation heterozygosity and the origin of dioecy in *Viscum*. Heredity 42: 201–222.

Wiens, D. and M. DeDecker. 1972. Rare natural hybridization in *Phoradendron* (Viscaceae). Madroño 21: 395–402.

Wiggins, I. L. 1980. Flora of Baja California. Stanford.

Wilder, G. J. and M. P. Sowinski. 2010. *Phyllanthus fluitans* Benth. (Euphorbiaceae) a newly reported invasive species in Florida. Wildland Weeds 13(4): 14–15.

Wildland Weeds = Wildland Weeds; a Quarterly Publication of the Florida Exotic Pest Plant Council.

Willdenow, C. L., C. F. Schwägrichen, and J. H. F. Link. 1797–1830. Caroli a Linné Species Plantarum.... Editio Quarta.... 6 vols. Berlin. [Vols. 1–5(1), 1797–1810, by Willdenow; vol. 5(2), 1830, by Schwägrichen; vol. 6, 1824–1825, by Link.] (Sp. Pl.)

Wilson, E. O. and T. Eisner. 1968. Lignumvitae: Relict island. Nat. Hist. 77: 52–57.

Wilson, J. S. 1964. Variation of three taxonomic complexes of the genus *Cornus* in the eastern United States. Trans. Kansas Acad. Sci. 67: 747–817.

Wislizenus, F. A. 1848. Memoir of a Tour to Northern Mexico, Connected with Col. Doniphan's Expedition, in 1846 and 1847.... Washington. (Mem. Tour N. Mexico)

Wolf, C. B. 1938. The North American species of *Rhamnus*. Rancho Santa Ana Bot. Gard. Monogr., Bot. Ser. 1.

Wood, A. 1861. A Class-book of Botany.... New York. (Class-book Bot. ed. s.n.(b))

Woodson, R. E. Jr., R. W. Schery, et al., eds. 1943–1981. Flora of Panama. 41 fascs. St. Louis. [Fascs. published as individual issues of Ann. Missouri Bot. Gard. and aggregating 8 nominal parts + introduction and indexes.]

Woolfendon, W. S. 1996. Late-Quaternary Vegetation History of the Southern Owens Valley Region, Inyo County, California. Ph.D. dissertation. University of Arizona.

World Cat. Euphorb. Cum. Suppl. I—See: R. C. H. M. Oudejans 1993

World Checklist Bibliogr. Euphorbiaceae—See: R. Govaerts et al. 2000

Wright, D. 1980. *Philadelphus*. Plantsman 2: 104–116.

Wrightia = Wrightia; a Botanical Journal.

Wu, Z. and P. H. Raven, eds. 1994+. Flora of China. 20+ vols. Beijing and St. Louis.

Wunderlin, R. P. 1982. Guide to the Vascular Plants of Central Florida. Tampa.

Wurdack, K. 2006. The lectotypification and 19th century history of *Croton alabamensis* (Euphorbiaceae s.s.). Sida 22: 469–483.

Wurdack, K. et al. 2004. Molecular phylogenetic analysis of Phyllanthaceae (Phyllanthoideae pro parte, Euphorbiacee sensu lato) using plastid *rbc*L sequences. Amer. J. Bot. 91: 1882–1900.

Wurdack, K. and C. C. Davis. 2009. Malpighiales phylogenetics: Gaining ground on one of the most recalcitrant clades in the angiosperm tree of life. Amer. J. Bot. 96: 1551–1570.

Wurdack, K., P. Hoffman, and M. W. Chase. W. 2005. Molecular phylogenetic analysis of uniovulate Euphorbiaceae (Euphorbiaceae sensu stricto) using plastid *rbc*L and *trn*L-F DNA sequences. Amer. J. Bot. 92: 1397–1420.

Xi, Z. et al. 2012. Phylogenetics and a posteriori data partitioning resolve the Cretaceous angiosperm radiation Malpighiales. Proc. Natl. Acad. Sci. U.S.A. 109: 17519–17524.

Xia, N. H. and M. G. Gilbert. 2003. Santalaceae. In: Wu Z. and P. H. Raven, eds. 1994+. Flora of China. 20+ vols. Beijing and St. Louis. Vol. 5, pp. 208–219.

Xiang, Q. Y. et al. 2006. Species level phylogeny of the genus *Cornus* (Cornaceae) based on molecular and morphological evidence—Implications for taxonomy and Tertiary intercontinental migration. Taxon 55: 9–30.

Xiang, Q. Y. and D. E. Boufford. 2005. Aucubaceae. In: Wu Z. and P. H. Raven, eds. 1994+. Flora of China. 20+ vols. Beijing and St. Louis. Vol. 14, pp. 222–226.

Xiang, Q. Y., D. E. Soltis, and P. S. Soltis. 1998. Phylogenetic relationships of Cornaceae and close relatives inferred from *mat*K and *rbc*L sequences. Amer. J. Bot. 85: 285–297.

Xiang, Q. Y., D. T. Thomas, and Xiang Q. P. 2011. Resolving and dating the phylogeny of Cornales—Effects of taxon sampling, data partitions, and fossil calibrations. Molec. Phylogen. Evol. 59: 123–138.

Yang, T. W. 1970. Major chromosome races of *Larrea divaricata* in North America. J. Arizona Acad. Sci. 6: 41–45.

Yang, Y. et al. 2012. Molecular phylogenetics and classification of *Euphorbia* subgenus *Chamaesyce* (Euphorbiaceae). Taxon 61: 764–789.

Yang, Y. and P. E. Berry. 2011. Phylogenetics of the *Chamaesyce* clade (*Euphorbia,* Euphorbiaceae): Reticulate evolution and long-distance dispersal in a prominent C4 lineage. Amer. J. Bot. 98: 1486–1503.

Yatskievych, G. 1999–2013. Steyermark's Flora of Missouri. 3 vols. Jefferson City.

Zavortink, J. E. 1966. A Revision of *Mentzelia* Section *Trachyphytum* (Loasaceae). Ph.D. dissertation. University of California, Los Angeles.

Zhang, L. B. and M. P. Simmons. 2006. Phylogeny and delimitation of the Celastrales inferred from nuclear and plastid genes. Syst. Bot. 31: 122–137.

Zhang, N., J. Wen, and E. A. Zimmer. 2015. Congruent deep relationships in the grape family (Vitaceae) based on sequences of chloroplast genomes and mitochondrial genes via genome skimming. PLOS ONE, 14 Dec, DOI: 10.1371/journal.pone.0144701.

Zhang, Y. L., Wang F. H., and Chien N. F. 1988. A study on pollen morphology of *Eucommia ulmoides* Oliver. Acta Phytotax. Sin. 26: 367–370.

Zika, P. F. and A. L. Jacobson. 2005. Noteworthy collections. Washington. Madroño 52: 209–212.

Zoë = Zoë; a Biological Journal.

Zuber, D. 2004. Biological flora of central Europe: *Viscum album* L. Flora 199: 181–203.

Index

Names in *italics* are synonyms, casually mentioned hybrids, or plants not established in the flora. Page numbers in **boldface** indicate the primary entry for a taxon. Page numbers in *italics* indicate an illustration. Roman type is used for all other entries, including author names, vernacular names, and accepted scientific names for plants treated as established members of the flora.

Flora of North America — Index to families/volumes of vascular plants, current as of April 2016.
Bolding denotes published volume: page number.